INDEX OF TABLES

Acceleration due to gravity, 106
Compressibility, 218
Contact angle, 258
Convection coefficient, 320
Critical point, 339
Density, 244, 292
Dielectric constant, 523
Double refraction, 812
Elastic moduli, 216
Electromagnetic spectrum, 716
Electron configurations, 884
Electron shells, 883
Expansion, coefficient of linear, 290
 coefficient of volume, 291
Friction, coefficient of, 31
Fundamental constants, 938
Fundamental particles, 924
Heat capacity, gases, 357
 metals, 300
Heat of fusion and vaporization, 305
International system of units, 930–931

Index of refraction, 719
Magnetic susceptibility, 671
Moment of inertia, 185
Musical scales, 441
Noise levels, 438
Periodic table, 936
Prefixes, unit, 3
Remanence and coercivity, 675
Resistivity, 537
 temperature coefficient of, 539
Solar system, 938
Spectrum colors, 716
Stable nuclei, 900, 902, 907
Surface tension, 255
Thermal conductivity, 315
Trigonometric functions, 935
Triple point, 338
Unit conversion factors, 937
Uranium decay series, 910
Vapor pressure of water, 341
Viscosity, 273

UNIVERSITY PHYSICS

SIXTH EDITION

Francis W. Sears
Late Professor Emeritus
Dartmouth College

Mark W. Zemansky
Professor Emeritus
City College of the
City University of New York

Hugh D. Young
Professor of Physics
Carnegie-Mellon University

ADDISON-WESLEY PUBLISHING COMPANY

Reading, Massachusetts · Menlo Park, California
London · Amsterdam · Don Mills, Ontario · Sydney

This book is in the
ADDISON-WESLEY SERIES IN PHYSICS

Sponsoring editor: Robert L. Rogers
Production editor: Mary Cafarella
                     Marion E. Howe
Designer: Robert Rose
            Marie E. McAdam
Illustrator: Oxford Illustrators, Ltd.
Cover design: Ann Scrimgeour Rose

Library of Congress Cataloging in Publication Data

Sears, Francis Weston, 1898–
   University physics.

   Includes index.
   1. Physics.   I. Zemansky, Mark Waldo, 1900–
II. Young, Hugh D.   III. Title.
QC21.2.S36   1981   530   81-17551
ISBN 0-201-07195-9   AACR2

Copyright © 1982, 1976, 1970, 1964, 1955, 1949 by Addison-Wesley Publishing Company, Inc. Philippines copyright 1982, 1976, 1970, 1964, 1955, 1949 by Addison-Wesley Publishing Company, Inc.

All rights reserved. No part of this publication may be reproduced, stored in a retrieval system, or transmitted, in any form or by any means, electronic, mechanical, photocopying, recording, or otherwise, without the prior written permission of the publisher. Printed in the United States of America. Published simultaneously in Canada.

BCDEFGHIJ-RN-898765432

PREFACE

University Physics is intended to provide a broad introduction to physics at the beginning college level for students of science and engineering who are taking an introductory calculus course concurrently. Primary emphasis is placed on physical principles and the development of problem-solving ability rather than on historical background or specialized applications. The complete text may be taught in an intensive two- or three-semester course, and the book is also adaptable to a wide variety of shorter courses. Numerous worked-out examples and an extensive collection of problems are included with each chapter. *University Physics* is available as a single volume or as two separate parts. Part I includes mechanics, heat, and sound, and Part II includes electricity and magnetism, optics, and atomic and nuclear physics.

In this new edition, the basic philosophy and outline and the balance between depth of treatment and breadth of subject-matter coverage are unchanged from previous editions. We have tried to preserve those features and characteristics that users of previous editions have found desirable, while incorporating a number of changes, some quite extensive, that should substantially enhance the book's pedagogical usefulness. Here are some of the most important changes:

1. A new single-column two-color format has been adopted, and all the figures have been redrawn for two-color treatment. The second color is used not for cosmetic effect but to add clarity to figures through color-coding of vectors, materials, and other significant features.

    Color is also used to identify important equations and to set off examples. Thus the second color should make a very substantial improvement in the book's usefulness as a learning tool for students.

2. A list of thought-provoking questions, many of them related to everyday experience, has been included with each chapter, about 700 questions in all. These should help students to attain a deeper understanding of principles and to relate these principles to their everyday lives and should also serve as effective springboards for class discussion.

3. The problem collections have been carefully reviewed and expanded, especially with the aims of filling gaps in problem coverage, providing additional straightforward "confidence-builder" problems, and furnishing applications of elementary calculus to physical problems. The book now includes about 1600 problems, an increase of 140 from the previous edition. Worked-out examples in the text have also been reviewed, and about 100 new examples have been included. The authors have resisted the temptation to key problems to specific sections of the text. Learning to select the principles appropriate for a specific problem is, after all, part of learning to solve problems. In addition, many problems require material from more than one section.

4. Unit vectors are introduced in the opening chapter and are used to a limited extent where appropriate. Their use in later chapters of the text is optional, however; instructors who wish to omit unit vectors completely can do so without loss of continuity.

5. There are several minor changes in the outline. The chapter on rigid-body equilibrium has been placed following dynamics of a particle and just before rotational motion. However, users who want to go into this material immediately following equilibrium of a point may still do so. The thermodynamics material has been split into two chapters, one each for the first and second laws. The introductory material on optics has been somewhat condensed and consolidated into a single chapter. The relativity chapter has been moved to the end of the book to accompany the other chapters on twentieth-century physics, which have been considerably expanded. The number of chapters is now 47, but the overall length of the book is about the same as that of the previous edition.

6. Many sections have been extensively rewritten to improve clarity and continuity. New introductory material has been prepared for nearly every chapter; this will provide the student with a sense of perspective and ease his or her entry into the chapter. The first chapter is completely new; it now includes discussion of units, unit conversions, and significant figures, as well as an introduction to vector addition in the context of displacement vectors, preceding the more abstract addition of forces. Products of vectors are also introduced here. All of the material in the areas of magnetic fields and forces has been rewritten with the aim of making it less formal (and less formidable) than in the previous edition.

7. The treatment of modern physics topics has been considerably expanded. There is a completely new chapter on quantum mechanics. The discussion of nuclear fission and fusion has been expanded, as has the treatment of fundamental particles and the associated conservation and symmetry principles. Several topics have been added, including lasers, integrated circuits, superheavy nuclei, quarks, and several others.

8. Some material that is outdated or of peripheral importance has been de-emphasized; a few examples are the Wheatstone bridge, the potentiometer, thermoelectricity, and some aspects of magnetic materials.

The text is adaptable to a wide variety of course outlines. The entire text can be used for an intensive course two or three semesters in length. For a less intensive course, many instructors will want to omit certain chapters or sections to tailor the book to their individual needs. The format of this edition facilitates this kind of flexibility. For example, any or all of the chapters on relativity, hydrostatics, hydrodynamics, acoustics, magnetic properties of matter, electromagnetic waves, optical instruments, and several others can be omitted without loss of continuity. In addition, some sections that are unusually challenging or out of the mainstream have been identified with an asterisk preceding the section title. These too may be omitted without loss of continuity.

Conversely, however, many topics that were regarded a few years ago as of peripheral importance and were purged from introductory courses have now come to the fore again in the life sciences, earth and space sciences, and environmental problems. An instructor who wishes to stress these kinds of applications will find this text a useful source for discussion of the appropriate principles.

In any case, it should be emphasized that instructors should not feel constrained to work straight through the book from cover to cover. Many chapters are, of course, inherently sequential in nature, but within this general limitation instructors are encouraged to select from among the contents those chapters that fit their needs, omitting material that is not relevant for the objectives of a particular course.

Again we wish to thank our many colleagues and students who have contributed suggestions for this new edition. In particular, Professors William M. Cloud (Eastern Illinois University), James R. Gaines (Ohio State University), and A. Lewis Ford (Texas A. and M. University) have read the entire manuscript, and their critical and constructive comments are greatly appreciated. In addition, Professors Malcolm D. Cole and Charles McFarland (University of Missouri at Rolla) have read portions of the manuscript and have provided many valuable suggestions. One of the authors (H.D.Y.) offers special thanks to his students and colleagues at Carnegie-Mellon University for their many helpful comments. He acknowledges a special debt of gratitude to Professors Robert Eisenstein, Robert Kraemer, and Frederick Messing of Carnegie-Mellon for many stimulating discussions about the book and about physics pedagogy generally, and to Professor Kraemer for major contributions to the sections on high-energy physics. The kindness and helpfulness of these people will not soon be forgotten.

As usual, we welcome communications from readers concerning our book, and especially concerning any errors or deficiencies that may remain in this edition.

New York M.Z.
Pittsburgh H.D.Y.
November 1981

AVAILABLE SUPPLEMENTS

The following supplementary materials are available for use by students:

Study Guide	James R. Gaines and William F. Palmer (Ohio State University, Columbus)
Solutions Guide	A. Lewis Ford (Texas A&M University, College Station)

The following supplement is available to instructors:

Answers to Even-Numbered Problems	(The answers to the odd-numbered problems are in the back of the book.)

CONTENTS

PART I Mechanics, Heat, and Sound

1
Units, Physical Quantities, and Vectors

1-1	Introduction	1
1-2	Standards and units	1
1-3	Unit consistency and conversions	4
1-4	Precision and significant figures	5
1-5	Vectors and vector addition	6
1-6	Components of vectors	9
* 1-7	Unit vectors	11
* 1-8	Products of vectors	12

2
Equilibrium of a Particle

2-1	Introduction	18
2-2	Force	18
2-3	Equilibrium. Newton's first law	20
2-4	Discussion of Newton's first law	22
2-5	Newton's third law of motion	23
2-6	Idealized models	24
2-7	Equilibrium of a particle	25
2-8	Friction	29

3
Motion on a Straight Line

3-1	Motion	39
3-2	Average velocity	39
3-3	Instantaneous velocity	40
3-4	Average and instantaneous acceleration	42, 45
3-5	Motion with constant acceleration	48
* 3-6	Velocity and coordinate by integration	50
3-7	Freely falling bodies	52
3-8	Relative velocity	

4
Newton's Second Law. Gravitation

4-1	Introduction	59
4-2	Newton's second law. Mass	59
4-3	Systems of units	62
4-4	Newton's law of gravitation	64
4-5	Mass and weight	66
4-6	Applications of Newton's second law	68

5
Motion in a Plane

5-1	Motion in a plane	83
5-2	Average and instantaneous velocity	83
5-3	Average and instantaneous acceleration	85
5-4	Components of acceleration	85
5-5	Motion of a projectile	88
5-6	Circular motion	95
5-7	Centripetal force	96
5-8	Motion in a vertical circle	99
5-9	Motion of a satellite	101
* 5-10	Effect of the earth's rotation on g	105
5-11	Relative velocity	107

6

Work and Energy

6-1	Work	115
6-2	Work done by a varying force	118
6-3	Work and kinetic energy	120
6-4	Gravitational potential energy	122
6-5	Elastic potential energy	128
6-6	Conservative and dissipative forces	131
* 6-7	Internal work	132
* 6-8	Internal potential energy	133
6-9	Power	134
6-10	Power and velocity	135
* 6-11	Mass and energy	135

7

Impulse and Momentum

7-1	Impulse and momentum	143
7-2	Conservation of momentum	147
7-3	Collisions	148
7-4	Inelastic collisions	149
7-5	Elastic collisions	151
7-6	Recoil	154
7-7	Center of mass	155
* 7-8	Rocket propulsion	157
* 7-9	Generalizations	159

8

Equilibrium. Moment of a Force

8-1	Moment or torque of a force	166
8-2	The second condition for equilibrium	167
8-3	Center of gravity	170
8-4	Couples	172
* 8-5	Vector torque	173

9

Rotation

9-1	Introduction	179
9-2	Angular velocity	179
9-3	Angular acceleration	181
9-4	Rotation with constant angular acceleration	181
9-5	Relation between angular and linear velocity and acceleration	182
9-6	Kinetic energy of rotation. Moment of inertia	183
* 9-7	Calculation of moments of inertia	186
9-8	Work and power in rotational motion	189
9-9	Torque and angular acceleration	190
* 9-10	Rotation above a moving axis	192
* 9-11	Parallel-axis theorem	194
9-12	Angular momentum and angular impulse	195
9-13	Conservation of angular momentum	196
* 9-14	Vector representation of angular quantities	200
9-15	Comparison of linear and angular motion	202

10

Elasticity

10-1	Stress	211
10-2	Strain	214
10-3	Elasticity and plasticity	214
10-4	Elastic modulus	216
10-5	The force constant	219

11

Periodic Motion

11-1	Introduction	223
11-2	Elastic restoring forces	224
11-3	Basic concepts	224
11-4	Equations of simple harmonic motion	225
11-5	Circle of reference	232
11-6	Motion of a body suspended from a coil spring	234
* 11-7	Angular harmonic motion	235
11-8	The simple pendulum	236
* 11-9	The physical pendulum	238

12

Fluid Statics

12-1	Density	244
12-2	Pressure in a fluid	245
12-3	Pressure gauges	247
12-4	Pumps	248
12-5	Archimedes' principle	249
* 12-6	Forces against a dam	252
12-7	Surface tension	253
12-8	Pressure difference across a surface film	255
* 12-9	Contact angle and capillarity	257

13
Fluid Dynamics

13-1	Introduction	264
13-2	The equation of continuity	265
13-3	Bernoulli's equation	266
13-4	Applications of Bernoulli's equation	267
13-5	Viscosity	271
13-6	Poiseuille's law	273
13-7	Stokes' law	276
13-8	Reynolds number	276

14
Temperature and Expansion

14-1	Concept of temperature	283
14-2	Thermometers	285
14-3	Defining a temperature scale	287
14-4	The Celsius, Rankine, and Fahrenheit scales	288
14-5	Thermal expansion	289
*14-6	Thermal stresses	292

15
Quantity of Heat

15-1	Heat transfer	297
15-2	Quantity of heat	298
15-3	Heat capacity	299
15-4	Experimental values of heat capacities	302
15-5	Change of phase	303
15-6	Examples	307

16
Heat Transfer

16-1	Conduction	313
*16-2	Radial heat flow	317
16-3	Convection	318
16-4	Radiation	321
16-5	Stefan-Boltzmann law	321
16-6	The ideal radiator	323

17
Thermal Properties of Matter

17-1	Equations of state	328
17-2	The ideal gas	329
17-3	pVT-surface for an ideal gas	333
*17-4	pVT-surface for a real substance	334
17-5	Phase diagrams	336
17-6	Triple point and critical point	338
17-7	Vapor pressure	340
*17-8	Humidity	340
17-9	The bubble chamber	342

18
The First Law of Thermodynamics

18-1	Energy and work in thermodynamics	346
18-2	Work in volume changes	347
18-3	Heat in volume changes	348
18-4	Internal energy	349
18-5	Adiabatic process	352
18-6	Isochoric process	352
18-7	Isothermal process	352
18-8	Isobaric process	353
18-9	Throttling process	353
18-10	Differential form of the first law	354
18-11	Internal energy of an ideal gas	355
18-12	Heat capacities of an ideal gas	355
18-13	Adiabatic process for an ideal gas	358

19
The Second Law of Thermodynamics

19-1	Heat engines	364
19-2	Internal-combustion engines	366
19-3	Steam engines	368
19-4	The refrigerator	368
19-5	The second law of thermodynamics	369
19-6	The Carnot cycle	371
*19-7	The Kelvin temperature scale	376
*19-8	Absolute zero	377
*19-9	Entropy	378
*19-10	Entropy and the second law	380
19-11	Energy conversion	382

20
Molecular Properties of Matter

20-1	Molecular theory of matter	386
20-2	Avogadro's number	388
20-3	Properties of matter	389
20-4	Kinetic theory of an ideal gas	389
20-5	Molar heat capacity of a gas	394
*20-6	Distribution of molecular speeds	395
20-7	Crystals	396
*20-8	Heat capacity of a crystal	398

21
Mechanical Waves

21-1	Introduction	403
21-2	Periodic waves	404
21-3	Mathematical description of a wave	406
21-4	Speed of a transverse wave	408
21-5	Speed of a longitudinal wave	411
21-6	Adiabatic character of a longitudinal wave	413
* 21-7	Water waves	415

22
Vibrating Bodies

22-1	Boundary conditions for a string	420
22-2	Superposition and standing waves	422
22-3	Vibration of a string fixed at both ends	424
22-4	Longitudinal standing waves	426
22-5	Vibrations of organ pipes	427
* 22-6	Vibrations of rods and plates	428
22-7	Interference of longitudinal waves	429
22-8	Resonance	430

23
Acoustic Phenomena

23-1	Sound waves	434
23-2	Intensity	436
23-3	Intensity level and loudness	437
23-4	Quality and pitch	439
* 23-5	Musical intervals and scales	440
23-6	Beats	442
* 23-7	The Doppler effect	443
* 23-8	Radiation from a piston	448
23-9	Applications of acoustic phenomena	451

PART II Electricity and Magnetism; Heat, Light, and Atomic Physics

24
Coulomb's Law

24-1	Electric charges	457
24-2	Atomic structure	458
24-3	Conductors and insulators	460
24-4	Charging by induction	460
24-5	Coulomb's law	461
24-6	Electrical interactions	466

25
The Electric Field; Gauss's Law

25-1	The electric field	469
25-2	Calculation of electric field	473
25-3	Field lines	479
25-4	Gauss's law	480
25-5	Applications of Gauss's law	484

26
Potential

26-1	Electrical potential energy	494
26-2	Potential	497
26-3	Calculation of potential differences	500
26-4	Equipotential surfaces	503
26-5	Potential gradient	504
26-6	The Millikan oil-drop experiment	506
26-7	The electronvolt	508
* 26-8	The cathode-ray oscilloscope	509

27
Capacitance. Properties of Dielectrics

27-1	Capacitors	516
27-2	The parallel-plate capacitor	517
27-3	Capacitors in series and parallel	519
27-4	Energy of a charged capacitor	520
* 27-5	Effect of a dielectric	522
* 27-6	Molecular theory of induced charges	526
* 27-7	Polarization and displacement	528

28
Current, Resistance, and Electromotive Force

28-1	Current	534
28-2	Resistivity	537
28-3	Resistance	539
28-4	Electromotive force	541
28-5	Current–voltage relations	549
28-6	Work and power in electrical circuits	550
28-7	Physiological effects of currents	554
* 28-8	The electric field of the earth	555
* 28-9	Theory of metallic conduction	556

29
Direct-Current Circuits and Instruments

29-1	Resistors in series and in parallel	562
29-2	Kirchhoff's rules	564

29-3	Ammeters and voltmeters	567
29-4	The ohmmeter	571
* 29-5	The R-C series circuit	571
* 29-6	Displacement current	574

30
The Magnetic Field

30-1	Magnetism	583
30-2	The magnetic field	584
30-3	Magnetic field lines; magnetic flux	587
30-4	Motion of charged particles in magnetic fields	588
30-5	Thomson's measurement of e/m	589
* 30-6	Isotopes and mass spectroscopy	591

31
Magnetic Forces on Current-Carrying Conductors

31-1	Force on a conductor	596
31-2	The Hall effect	597
31-3	Force and torque on a complete circuit	599
31-4	The pivoted-coil galvanometer	603
* 31-5	The direct-current motor	604
* 31-6	The electromagnetic pump	605

32
Magnetic Field of a Current

32-1	Sources of magnetic field	609
32-2	Magnetic field of a moving charge	609
32-3	Magnetic field of a current element. The Biot law	612
32-4	Magnetic field of a long straight conductor	613
32-5	Force between parallel conductors; the ampere	615
32-6	Magnetic field of a circular loop	616
32-7	Ampère's law	619
32-8	Applications of Ampère's law	621
* 32-9	Magnetic fields and displacement currents	623

33
Induced Electromotive Force

33-1	Motional electromotive force	630
33-2	Faraday's law	634
33-3	Induced electric fields	638
33-4	Lenz's law	640
33-5	Eddy currents	642

34
Inductance

34-1	Mutual inductance	649
34-2	Self-inductance	650
34-3	Energy in an inductor	652
34-4	The R-L circuit	653
34-5	The L-C circuit	656
34-6	The R-L-C circuit	658

35
Magnetic Properties of Matter

35-1	Magnetic materials	664
35-2	Magnetic permeability	665
35-3	Molecular theory of magnetism	666
* 35-4	Magnetization and magnetic intensity	668
* 35-5	Ferromagnetism	671
* 35-6	Magnetic domains	672
* 35-7	Hysteresis	673
* 35-8	The magnetic field of the earth	675

36
Alternating Currents

36-1	Introduction	678
36-2	Circuits containing resistance, inductance, or capacitance	679
36-3	The R-L-C series circuit	682
36-4	Average and root-mean-square values; ac instruments	684
36-5	Power in ac circuits	686
36-6	Series resonance	688
36-7	Parallel circuits	690
* 36-8	The transformer	690

37
Electromagnetic Waves

37-1	Introduction	696
* 37-2	Speed of an electromagnetic wave	698
37-3	Energy in electromagnetic waves	700
* 37-4	Electromagnetic waves in matter	703
37-5	Sinusoidal waves	703
37-6	Standing waves	705
* 37-7	Radiation from an antenna	707

38
The Nature and Propagation of Light

38-1	Nature of light	711
38-2	Sources of light	712
38-3	The speed of light	713
38-4	The electromagnetic spectrum	715
38-5	Waves, wave fronts, and rays	717
38-6	Reflection and refraction	717
38-7	Total internal reflection	721
38-8	Huygens' principle	722
38-9	Dispersion	725
* 38-10	Absorption	726
* 38-11	Illumination	727

39
Images Formed by a Single Surface

39-1	Introduction	733
39-2	Reflection at a plane surface	733
39-3	Reflection at a spherical mirror	735
39-4	Focal point and focal length	740
39-5	Graphical methods	741
39-6	Refraction at a plane surface	742
39-7	Refraction at a spherical surface	744
39-8	Summary	747

40
Lenses and Optical Instruments

40-1	Images as objects	750
40-2	The thin lens	751
40-3	Diverging lenses	754
40-4	Graphical methods	755
40-5	Images as objects for lenses	757
* 40-6	Lens aberrations	757
40-7	The eye	759
* 40-8	Defects of vision	761
40-9	The magnifier	762
40-10	The camera	763
40-11	The projector	764
40-12	The compound microscope	765
40-13	Telescopes	766

41
Interference and Diffraction

41-1	Interference and coherent sources	773
41-2	Young's experiment and Pohl's experiment	776
* 41-3	Intensity distribution in interference fringes	778
41-4	Interference in thin films. Newton's rings	780
41-5	Thin coatings on glass	783
* 41-6	The Michelson interferometer	784
41-7	Fresnel diffraction	787
41-8	Fraunhofer diffraction from a single slit	789
41-9	The diffraction grating	794
* 41-10	Diffraction of x-rays by a crystal	797
* 41-11	The resolving power of optical instruments	798
* 41-12	Holography	800

42
Polarization

42-1	Polarization	806
42-2	Malus' law and percent polarization	808
42-3	Polarization by reflection	810
* 42-4	Double refraction	811
42-5	Polarizers	812
* 42-6	The scattering of light	814
42-7	Circular and elliptical polarization	815
42-8	Optical stress analysis	818
42-9	Optical activity	818

43
Relativistic Mechanics

43-1	Invariance of physical laws	822
43-2	Relative nature of simultaneity	824
43-3	Relativity of time	825
43-4	Relativity of length	827
* 43-5	The Lorentz transformation	829
43-6	Momentum	831
43-7	Work and energy	832
43-8	Relativity and newtonian mechanics	835

44
Photons, Electrons, and Atoms

44-1	Emission and absorption of light	840
44-2	Thermionic emission	841
44-3	The photoelectric effect	843
44-4	Line spectra	846
44-5	Energy levels	848
44-6	Atomic spectra	850
44-7	The laser	852
44-8	X-ray production and scattering	854

45
Quantum Mechanics

45–1	The Bohr atom	862
45–2	Wave nature of particles	867
45–3	The electron microscope	869
* 45–4	Probability and uncertainty	870
* 45–5	Wave functions	874
45–6	Electron spin	876

46
Atoms, Molecules, and Solids

46–1	The exclusion principle	881
46–2	Atomic structure	883
46–3	Diatomic molecules	885
46–4	Molecular spectra	887
46–5	Structure of solids	888
46–6	Properties of solids	890
46–7	Semiconductors	892
46–8	Semiconductor devices	893

47
Nuclear Physics

47–1	The nuclear atom	897
47–2	Properties of nuclei	899
47–3	Natural radioactivity	903
47–4	Nuclear stability	906
47–5	Radioactive transformations	908
47–6	Nuclear reactions	911
47–7	Particle accelerators	912
47–8	Nuclear fission	916
47–9	Nuclear fusion	918
47–10	Fundamental particles	920
* 47–11	High-energy physics	922
47–12	Radiation and the life sciences	926

Appendixes

A	The international system of units	T–1
B	Useful mathematical relations	T–3
C	The Greek alphabet	T–5
D	Natural trigonometric functions	T–6
E	Periodic table of the elements	T–7
F	Unit conversion factors	T–8
G	Numerical constants	T–9

Answers to Odd-Numbered Problems   A–1

Index   I–1

Endpapers

(Front) Appendix F   Unit conversion factors
(Rear) Appendix G   Numerical constants

MECHANICS, HEAT, AND SOUND
PART I

UNITS, PHYSICAL QUANTITIES, AND VECTORS 1

1-1 Introduction

Physics is an *empirical* study. Everything we know about the physical world and about the principles that govern its behavior has been learned through *observations* of the phenomena of nature. The ultimate test of any physical theory is its agreement with observations and measurements of physical phenomena.

Thus physics is inherently a science of *measurement*. Lord Kelvin (1824–1907), one of the pioneers in investigating energy relations in heat and thermal phenomena, stated this principle eloquently:

> I often say that when you can measure what you are speaking about, and express it in numbers, you know something about it; but when you cannot express it in numbers, your knowledge is of a meagre and unsatisfactory kind; it may be the beginning of knowledge, but you have scarcely, in your thoughts, advanced to the stage of *science,* whatever the matter may be.

Any number or set of numbers used for a quantitative description of a physical phenomenon is called a *physical quantity*. To *define* a physical quantity we must either specify a procedure for *measuring* the quantity, or specify a way to *calculate* the quantity from other quantities that can be measured. For example, we can define *distance* and *time* by describing procedures for measuring them, and then define the *speed* of a moving body as the distance traveled divided by the time of travel.

A definition in terms of a procedure for measuring the defined quantity is called an *operational definition*. Certain fundamental quantities can be defined *only* with operational definitions. The fundamental quantities of mechanics are usually taken to be mass, length, and time. For other areas of physics other fundamental quantities, including temperature, electric charge, and luminous intensity, will be introduced later.

1-2 Standards and units

In measuring a quantity, we always compare it with some established reference standard. To say that a rope is 30 meters long is to say that it is 30 times as long as an object whose length has been *defined* to be one meter. Such a standard is called a *unit* of the quantity; thus the *meter* is a unit of distance, and the *second* is a unit of time.

For precision in making measurements, it is essential to have precise and reproduceable definitions of the units of measurement. When the metric system was established in 1791 by the Paris Academy of Sciences, the meter was originally defined as one ten-millionth of the distance from the equator to the North Pole, and the second as the time for a pendulum one meter long to swing from one side to the other. In later years these definitions have been modified and greatly refined.

Since 1889 the definitions of the basic units have been established by an international organization called the General Conference on Weights and Measures, to which all the major countries of the world send representatives. The system of units defined by this organization, based on the metric system, has been known officially since 1960 as the *International System,* abbreviated SI because of the French equivalent, *Système International.*

Two essential characteristics of a unit standard are *permanence* and *reproduceability*. Thus in 1889, a meter bar of platinum–iridium alloy was chosen as the *standard of length;* this alloy is particularly stable chemically. However, the use of such a bar as a world standard is cumbersome; replicas must be made and compared with the world standard periodically. On October 14, 1960, the General Conference changed the standard of length to an *atomic constant,* namely, *the wavelength of the orange-red light emitted by the individual atoms of krypton-86* in a tube filled with krypton gas in which an electrical discharge is maintained. One meter is defined to be 1,650,763.73 times the wavelength of this light. Such a standard is more precise and more readily reproduced in various locations than is one based on a specific material object.

The standard of *mass* is the mass of a cylinder of platinum–iridium alloy, designated as *one kilogram,* kept at the International Bureau of Weights and Measures at Sèvres, near Paris. An atomic standard of mass has not yet been adopted because it is not yet possible to measure masses on an atomic scale with as great precision as on a macroscopic scale.

Until 1960 the standard of *time* was based on the mean solar day, the time interval between successive arrivals of the sun at its highest point, averaged over a year. In 1967 an atomic standard was adopted. The two lowest energy states of the cesium atom have slightly different energies, depending on whether the spin of the outermost electron is parallel or antiparallel to the nuclear spin. Electromagnetic radiation (microwaves) of precisely the proper frequency causes transitions from one of these states to the other. One second is now defined as the time required for 9,192,631,770 cycles of this radiation. Figure 1–1 shows a cesium frequency standard, which may be used as a clock.

Once the fundamental units are defined, it is easy to introduce larger and smaller units for the same physical quantities. In the metric (SI) system these additional units are always related to the fundamental ones by multiples of 10 or $\frac{1}{10}$. Thus one kilometer (1 km) is 1000 meters, one centimeter (1 cm) is $\frac{1}{100}$ meter, and so on. The multiplicative factors are most conveniently expressed in exponential notation; thus $1000 = 10^3$, $\frac{1}{1000} = 10^{-3}$, and so on. The names of the additional units are always derived by adding a prefix to the name of the fundamental unit. For example, the prefix "kilo-", abbreviated k, always means a unit

1-1 NBS-6 is the latest of six generations of primary atomic frequency standards developed by the National Bureau of Standards (NBS). Consisting of a 6-meter cesium beam tube, NBS-6 achieves an accuracy of better than one part in 10^{13}, and when operated as a clock, can keep time to within 3 millionths of a second per year. (Courtesy National Bureau of Standards.)

larger by a factor of 1000; thus

$$1 \text{ kilometer} = 1 \text{ km} = 10^3 \text{ meters} = 10^3 \text{ m},$$
$$1 \text{ kilogram} = 1 \text{ kg} = 10^3 \text{ grams} = 10^3 \text{ g},$$
$$1 \text{ kilowatt} = 1 \text{ kW} = 10^3 \text{ watts} = 10^3 \text{ W}.$$

Table 1-1 lists the standard SI prefixes with their meanings and abbreviations. We note that most of these are multiples of 10^3.

Table 1-1 Prefixes for powers of ten

Power of ten	10^{-18}	10^{-15}	10^{-12}	10^{-9}	10^{-6}	10^{-3}	10^{-2}	10^{3}	10^{6}	10^{9}	10^{12}	10^{15}	10^{18}
Prefix	atto-	femto-	pico-	nano-	micro-	milli-	centi-	kilo-	mega-	giga-	tera-	peta-	exa-
Abbreviation	a	f	p	n	μ	m	c	k	M	G	T	P	E

In pronunciation of unit names with prefixes, there is *always* an accent on the *first* syllable; some examples are *kil*-o-gram, *kil*-o-meter, *cen*-ti-meter, and *mic*-ro-meter. Accenting "kilometer" on the *second* syllable is not correct!

Following are several examples of the use of multiples of 10 and their prefixes, and some additional time units.

$1 \text{ nanometer} = 1 \text{ nm} = 10^{-9} \text{ m}$ (used by optical designers),

$1 \text{ micrometer} = 1 \text{ }\mu\text{m} = 10^{-6} \text{ m}$ (used commonly in biology),

$1 \text{ millimeter} = 1 \text{ mm} = 10^{-3} \text{ m}$ and

$1 \text{ centimeter} = 1 \text{ cm} = 10^{-2} \text{ m}$ } (used most often),

$1 \text{ kilometer} = 1 \text{ km} = 10^3 \text{ m}$ (a common European unit of distance).

$1 \text{ microgram} = 1 \text{ }\mu\text{g} = 10^{-9} \text{ kg},$

$1 \text{ milligram} = 1 \text{ mg} = 10^{-6} \text{ kg},$

$1 \text{ gram} = 1 \text{ g} = 10^{-3} \text{ kg}.$

$1 \text{ nanosecond} = 1 \text{ ns} = 10^{-9} \text{ s},$

$1 \text{ microsecond} = 1 \text{ }\mu\text{s} = 10^{-6} \text{ s},$

$1 \text{ millisecond} = 1 \text{ ms} = 10^{-3} \text{ s}.$

$1 \text{ minute} = 1 \text{ min} = 60 \text{ s},$

$1 \text{ hour} = 1 \text{ hr} = 3600 \text{ s},$

$1 \text{ day} = 1 \text{ day} = 86,400 \text{ s}.$

Finally, we mention the British system of units, used only in the United States and the British Commonwealth and rapidly being replaced by SI in the latter. The British units are now officially *defined* in terms of SI units, as follows:

Length: 1 inch = 2.54 cm (exactly)

Mass: 1 pound-mass = 0.45359237 kg (exactly).

The fundamental British unit of time is the second, with the same definition as in the SI. In the British system, the *pound* is a unit of *force*, and is a force equal to the weight of one pound-mass, under specified conditions. In this book the pound-mass is not used, and the pound is *always* a unit of force.

In physics, British units are used only in mechanics and thermodynamics; there is *no* British system of electrical units. For illustrative purposes, a few problems in each chapter in the first half of this book use British units, but the last half of the book uses only SI units.

1-3 Unit consistency and conversions

Relations among physical quantities are often expressed by equations in which the quantities are represented by algebraic symbols. An algebraic symbol always denotes both a *number* and a *unit*. For example, s might represent a distance of 10 m, t a time of 5 s, and v a speed of 2 m/s or 2 m·s^{-1}. (In this book negative exponents will usually be used with units to avoid use of the fraction bar.)

An equation must always be *dimensionally consistent;* this means that two terms may be added or equated only if they have the same units. For example, if a body moving with constant speed v travels a distance s in a time t, these quantities are related by the equation.

$$s = vt. \tag{1-1}$$

If s is measured in meters, then the product vt must also have units of meters. Using the above numbers as an example, we may write

$$10 \text{ m} = (2 \text{ m·s}^{-1})(5 \text{ s}).$$

The unit (s^{-1}) or (1/s) cancels the unit (s) on the right side; thus the right side *does* have units of meters (m), as it must. This illustrates the fact that in calculations, units are treated just like algebraic symbols with respect to multiplication, division, and cancellation.

In all calculations needed for problems, units should *always* be written with all numbers and treated as in the above example. This provides a very useful check on calculations; if at some stage in a calculation an expression is found to have inconsistent units, an error has been made somewhere. In all worked-out numerical examples throughout this book, units will be carried through the calculations, and the student is strongly urged to follow this practice in solving problems.

The algebraic properties of units provide a convenient procedure for converting a quantity from one unit to another. Equality is sometimes used to represent the same physical quantity expressed in two different units. For example, to say that 1 min = 60 s does not mean that the number 1 is equal to the number 60, but rather that 1 min represents the same physical time interval as 60 s. Thus one may multiply by 1 min and then divide by 60 s, or multiply by the quantity (1 min/60 s), without changing the physical meaning. To find the number of seconds in 3 min, we write

$$3 \text{ min} = (3 \text{ min}) \left(\frac{60 \text{ s}}{1 \text{ min}} \right) = 180 \text{ s}.$$

Similarly, converting 50 km·h^{-1} (50 kilometers per hour) into meters per second,

$$50 \text{ km} \cdot \text{h}^{-1} = (50 \text{ km} \cdot \text{h}^{-1})\left(\frac{1000 \text{ m}}{1 \text{ km}}\right)\left(\frac{1 \text{ h}}{3600 \text{ s}}\right) = 13.89 \text{ m} \cdot \text{s}^{-1}.$$

1-4 Precision and significant figures

Measurements can never be made with absolute precision; physical quantities obtained from experimental observations always have some uncertainty. A distance measured with an ordinary ruler or meter stick is usually precise only to the nearest millimeter, while a precision micrometer caliper can measure distances dependably to 0.01 mm or even less. The precision of a number is often indicated by following it with the symbol $\pm$ and a second number indicating the maximum likely error. If the diameter of a steel rod is given as 56.47 ± 0.02 mm, this means that the true value is unlikely to be less than 56.45 mm or greater than 56.49 mm. The terms *likely* and *unlikely* can also be given more precise meaning by use of statistical concepts, but these are beyond the scope of this book.

Precision can also be expressed in terms of the maximum likely *fractional* error or *percent* error. A resistor labeled "47 ohms, 10%" probably has a true resistance differing from 47 ohms by no more than 10% of 47 ohms or about 5 ohms; that is, the true value is between about 42 and 52 ohms. In the steel rod example above, the fractional uncertainty is (0.02 mm)/(56.47 mm) or about 0.00035; the percent uncertainty is (0.00035)(100%) or about 0.035%.

When numbers having uncertainties or errors are used to compute other numbers, these too will be uncertain. It is especially important to understand this when a number obtained from measurements is to be compared with a value obtained from a theoretical prediction. Suppose a student wants to verify the value of π, the ratio of circumference to diameter of a circle. The correct value, to ten digits, is 3.141592654. He draws a circle and measures its diameter and circumference to the nearest millimeter, obtaining the values 135 mm and 424 mm, respectively. He punches these into his pocket calculator and obtains the quotient 3.140740741. Does this agree with the true value or not?

To answer this question we must first recognize that at least the last six digits in the student's result are meaningless because they imply a greater precision in the result than is possible with his measurements. The number of meaningful digits in a number is called the number of *significant figures;* in general, no numerical result can have more significant figures than do the numbers from which it is computed. Thus the student's value of π has only *three significant figures* and should be stated simply as 3.14, or conceivably as 3.141 (rounded to four figures). Within the limit of three significant figures, the student's value of π *does* agree with the true value.

In the examples and problems in this book, it is usually assumed that the numerical values given are precise to three or at most four significant figures, and thus the answers should be stated to at most four significant figures. The student will usually do the arithmetic with a pocket calculator having a display with five to ten digits. To write down a ten-digit result from a calculation with numbers having three significant figures is not only unnecessary; it is genuinely wrong because it misrepresents the precision of the results. Such results must *always* be rounded to keep only the correct number of significant figures or, in doubtful cases, at most one more.

In calculations with very large or very small numbers, significant-figure considerations are simplified greatly by use of powers-of-ten notation, sometimes called *scientific* notation. The distance from the earth to the sun is about 149,000,000,000 m, but to write the number in this form gives no indication of the number of significant figures. Certainly not all 12 are significant! Instead, we move the decimal point 11 places to the left (corresponding to dividing by 10^{11}) and multiply by 10^{11}. That is,

$$149{,}000{,}000{,}000 \text{ m} = 1.49 \times 10^{11} \text{ m}.$$

In this form it is clear that the number of significant figures is three.

Similar considerations are applicable when very large or very small numbers have to be multiplied or divided. For example, the energy E corresponding to the mass m of an electron is given by the equation

$$E = mc^2, \tag{1-2}$$

where c is the speed of light. The appropriate numbers are $m = 9.11 \times 10^{-31}$ kg and $c = 3.00 \times 10^8$ m·s^{-1}. We find

$$\begin{aligned} E &= (9.11 \times 10^{-31} \text{ kg})(3.00 \times 10^8 \text{ m·s}^{-1})^2 \\ &= (9.11)(3.00)^2 (10^{-31})(10^8)^2 \text{ kg·m}^2\text{·s}^{-2} \\ &= (82.0)(10^{[-31+(2\times 8)]}) \text{ kg·m}^2\text{·s}^{-2}. \end{aligned}$$

Many pocket calculators use scientific notation and thus save the trouble of adding exponents; but the student should be able to do such calculations by hand when necessary. Incidentally, it should be noted that the value of c has three significant figures even though two of them are zeros. To greater precision, $c = 2.997925 \times 10^8$ m·s^{-1}; thus it would *not* be correct to write $c = 3.000 \times 10^8$ m·s^{-1}.

1-5 Vectors and vector addition

Some physical quantities are described completely by a single number with a unit; examples are time, temperature, mass, density, and quantity of electric charge. Many other quantities, however, have a *directional* quality that cannot be described by a single number. A familiar example is *velocity;* to describe the motion of a body we must say not only *how fast* it is moving but also *in what direction*. Force is another example; when we push or pull on a body, we are said to exert a force on it. To describe a force, we need to describe the *direction* in which it acts, as well as its *magnitude,* which is a quantitative description of "how much" or "how hard" the force pushes or pulls, in terms of a standard unit of force. The SI unit of force is the *newton* (N), and the British unit of force is the pound-force (lbf);

$$1 \text{ lbf} = 4.448 \text{ N}.$$

A physical quantity described by a single number (possibly with an algebraic sign, as for electric charge) is called a *scalar* quantity, and a quantity having both magnitude (the "how much" aspect of the quantity) and direction is called a *vector* quantity. Calculations with scalar quantities involve the operations of ordinary arithmetic, but calculations with vector quantities are somewhat different. Because vector quantities play an essential role in all areas of physics, we turn now to a discussion of their nature and the operation of *vector addition*.

We begin with a vector quantity called *displacement*. When a particle, which we represent as a point, moves from one location in space to another, it is said to undergo a displacement. In Fig. 1–2a, the change of position from point P_1 to point P_2 is represented graphically by the directed line segment $\overrightarrow{P_1P_2}$, with an arrowhead to represent the sense of the motion. Displacement is a vector quantity because to describe it completely we must state not only how far the particle moves but also *in what direction;* a displacement of 3 km north is not the same as a displacement of 3 km southeast.

Usually it is most convenient to represent a displacement by a single letter, such as *A* in Fig. 1–2a. To distinguish symbols representing vector quantities from those for scalar quantities, vector symbols are always printed using boldface type, as shown. In handwriting, vector symbols are usually underlined or written with an arrow above, as shown in the figure, to indicate that they represent vector quantities.

The vector from point P_3 to point P_4 in Fig. 1–2b has the same length and direction as that from P_1 to P_2, so these two displacements are *equal*, even though they start at different points. By definition, two vector quantities are equal if they have the same magnitude (length) and direction, no matter where they are located in space. The vector **B**, however, is *not* equal to **A** because its direction is opposite to that of **A**. To represent the relationship of the two, we define the *negative* of a vector quantity as a vector quantity having the same magnitude but opposite direction to that of the original vector. The negative of vector quantity **A** is denoted as **−A**, and we use a boldface "minus" to emphasize the vector nature of the quantities. Thus the relation between **A** and **B** may be written as **A = −B** or **B = −A**. Note that a boldface "equals" sign is also used to emphasize that equality of two vector quantities is not the same relationship as equality of scalar quantities.

Displacement is always a straight line segment, directed from the starting point to the endpoint, even though the path taken by the particle may be curved. Thus in Fig. 1–3, when the point moves from P_1 to P_2 along the curved path shown, the displacement is the vector **A** shown. Also, when it continues on to P_3 and then returns to P_1, the displacement for the entire trip is zero.

The *magnitude* of a vector quantity (its length in the case of displacement vectors) is represented by the same letter used for the vector but in light italic type rather than boldface italic. An alternative notation is the vector symbol with vertical bars on both sides. Thus

$$\text{Magnitude of } \mathbf{A} = A = |\mathbf{A}|. \tag{1–3}$$

By definition, the magnitude of a vector quantity is a scalar (a single number) and is always positive.

When a particle undergoes a displacement, **A**, followed by a second displacement **B**, as shown in Fig. 1–4a, the total effect is the same as though it had started at the same initial point and undergone a single displacement **C**, as shown. We call displacement **C** the *vector sum of*

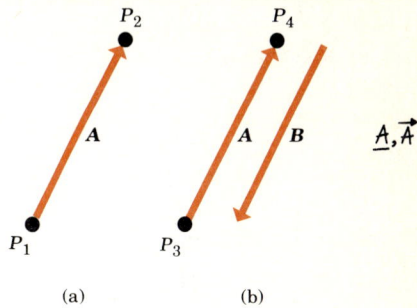

1–2 (a) Vector **A** is the displacement from point P_1 to point P_2. (b) The displacement from P_3 to P_4 is equal to that from P_1 to P_2, but displacement **B** is the negative of displacement **A**.

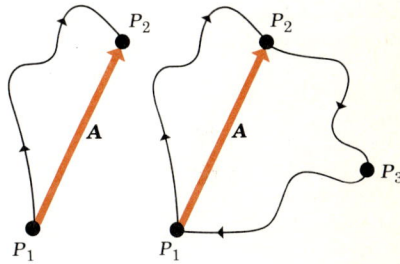

1–3 A displacement is always a straight line segment, directed from the starting point to the endpoint, even if the actual path is curved. When a point ends at the same place it started, the displacement is zero.

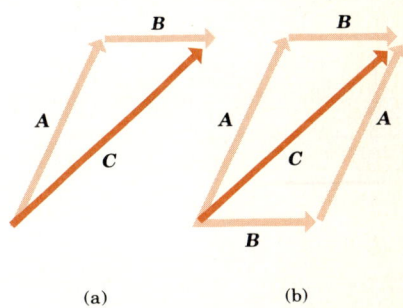

1–4 (a) Vector **C** is the vector sum of vectors **A** and **B**. (b) The order in vector addition is immaterial.

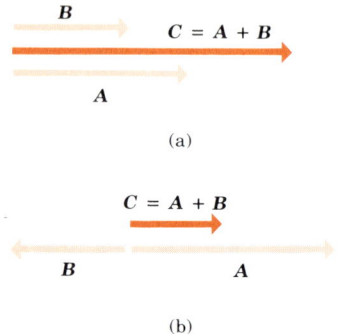

1-5 Vector sum of (a) two parallel vectors, and of (b) two antiparallel vectors.

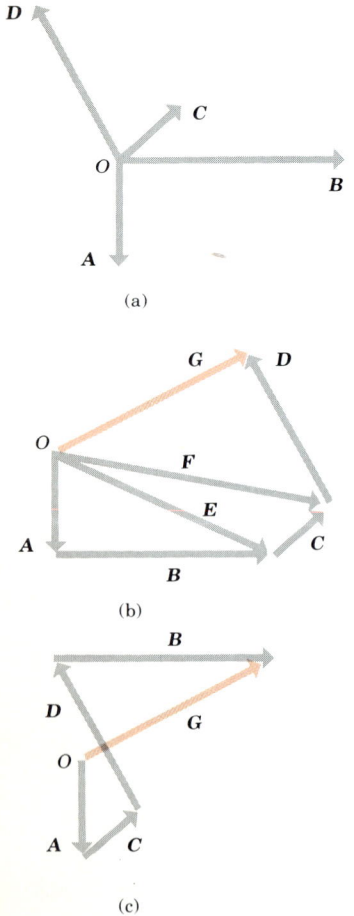

1-6 Polygon method of vector addition.

displacements A and B; the relationship is expressed symbolically as

$$C = A + B. \qquad (1\text{-}4)$$

The boldface "plus" sign emphasizes that obtaining the sum of (adding) two vector quantities is not the same operation as adding two scalar quantities.

If the displacements are made in the reverse order, as in Fig. 1-4b, with B first and A second, the result is the same, as the figure shows. Thus we may write

$$C = B + A \quad \text{and} \quad A + B = B + A. \qquad (1\text{-}5)$$

Thus vector addition is *commutative;* the order of the terms is immaterial. Formally, we have also used the *transitive* property of vector equality: Two vector quantities each equal to a third are also equal to each other.

Figure 1-4b also suggests an alternative graphical representation of the vector sum; when vectors A and B are both drawn from a common point, vector C is the diagonal of a parallelogram constructed with A and B as two adjacent sides.

The vector sum is often called the *resultant;* thus the vector sum C of vectors A and B can be called the resultant displacement.

Figure 1-5 illustrates a special case in which two vectors are parallel, as in (a), or antiparallel, as in (b). If they are parallel, the magnitude of the vector sum C equals the sum of the magnitudes of A and B. If they are antiparallel, the magnitude of the vector sum equals the *difference* of the magnitudes of A and B. The vectors in Fig. 1-5 have been displaced slightly sidewise to show them more clearly, but they actually lie along the same geometric line.

When more than two vectors are to be added, we may first find the vector sum of any two, add this vectorially to the third, and so on. This process is illustrated in Fig. 1-6, which shows in part (a) four vectors A, B, C, and D. In Fig. 1-6b, vectors A and B are first added, giving a vector sum E; vectors E and C are then added by the same process, to obtain the vector sum F; finally, F and D are added, to obtain the vector sum

$$G = A + B + C + D.$$

Evidently the vectors E and F need not have been drawn; we need only draw the given vectors in succession, with the tail of each at the head of the one preceding it, and complete the polygon by a vector G from the *tail* of the first to the *head* of the last vector. The order in which the vectors are drawn makes no difference, as shown in Fig. 1-6c; the reader may wish to try other possibilities.

Diagrams for addition of displacement vectors need not be drawn actual size; it is often convenient to use a scale in which the distance on the diagram is *proportional* to the actual distance, such as 1 cm for 5 km, just as on a map. Furthermore, such scales can also be used with other vector quantities whose units are not distance units. For example, in a diagram for force vectors (discussed in Sec. 2-2), one might use a scale in which a vector 1 cm long represents a force of magnitude 5 N. Then a 20-N force is represented by a vector 4 cm long with the appropriate direction. Thus sums of vector quantities can always be obtained by drawing scale diagrams and measuring the appropriate distances and angles. This is seldom done in practice because there are better methods available. Some of these are discussed in the next section.

A displacement or other vector quantity may be multiplied by a scalar quantity. The displacement $2\mathbf{A}$ is by definition a displacement in the same direction as that of $\mathbf{A}$ but twice the magnitude. In some cases the scalar quantity is a physical quantity having units. The reader may be familiar with the relationship $\mathbf{F} = m\mathbf{a}$; force (a vector quantity) is equal to the product of mass (a scalar quantity) and acceleration (a vector quantity). The magnitude of the force is equal to the mass multiplied by the magnitude of acceleration, and the unit of the magnitude of force is the product of the unit of mass and that of the magnitude of acceleration.

The special case of multiplication by -1 has already been mentioned; $(-1)\mathbf{A} = -\mathbf{A}$ is by definition a vector having the same magnitude as $\mathbf{A}$ but opposite direction. This provides the basis for defining vector *subtraction*. The difference $\mathbf{A} - \mathbf{B}$ of two vectors is defined to be the vector sum of $\mathbf{A}$ and $-\mathbf{B}$:

$$\mathbf{A} - \mathbf{B} = \mathbf{A} + (-\mathbf{B}). \tag{1-6}$$

The boldface $\mathbf{+}$, $\mathbf{-}$, and $\mathbf{=}$ signs remind us of the vector nature of these operations.

1-6 Components of vectors

Addition and subtraction of vectors are greatly facilitated by the use of *components*. To define components we use a rectangular (cartesian) coordinate axis system as in Fig. 1–7. Any vector lying in the xy-plane can be represented as the sum of a vector parallel to the x-axis and one parallel to the y-axis. These two vectors are labeled $\mathbf{A}_x$ and $\mathbf{A}_y$ in the figure; they are called the *rectangular component vectors* of vector $\mathbf{A}$. The relationship is expressed formally as

$$\mathbf{A} = \mathbf{A}_x + \mathbf{A}_y. \tag{1-7}$$

Each component vector lies along a coordinate-axis direction, by definition, so a single number is sufficient to describe each component vector. Thus the number A_x is, apart from a possible negative sign, the *magnitude* of the component vector $\mathbf{A}_x$. We further specify that A_x is positive when $\mathbf{A}_x$ points in the *positive* axis direction and negative when in the opposite direction. The two numbers A_x and A_y are called simply the *components* of $\mathbf{A}$.

If the magnitude A of the vector $\mathbf{A}$ and its direction, given by angle θ in Fig. 1–7, are known, the components can be calculated. From the definitions of the trigonometric functions,

$$\frac{A_x}{A} = \cos \theta \quad \text{and} \quad \frac{A_y}{A} = \sin \theta;$$

$$A_x = A \cos \theta \quad \text{and} \quad A_y = A \sin \theta.$$

The components of a vector are themselves numbers, and they may be positive or negative. In Fig. 1–8, the component B_x is negative, since its direction is opposite to that of the positive x-axis, and also since the

1-7 Vectors $\mathbf{A}_x$ and $\mathbf{A}_y$ are the rectangular components of $\mathbf{A}$ in the directions of the x- and y-axes.

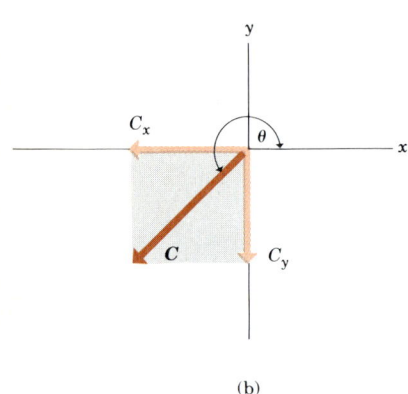

1-8 Components of a vector may be positive or negative numbers.

cosine of an angle in the second quadrant is negative; B_y is positive, but both C_x and C_y are negative.

Either the magnitude and direction or the components of a vector quantity can be used to describe it, and either provides a *complete* description. Equations (1-8) show how to obtain the components if the magnitude and direction are given; conversely, if the components are given, the magnitude and direction can be obtained. Applying the Pythagorean theorem to Fig. 1-7, we find

$$A = \sqrt{A_x^2 + A_y^2}. \qquad (1\text{-}9)$$

Also, from the definition of the tangent of an angle,

$$\tan\theta = \frac{A_y}{A_x} \quad \text{and} \quad \theta = \arctan\frac{A_y}{A_x}. \qquad (1\text{-}10)$$

There is a slight complication in applying Eq. (1-10). Suppose $A_x = 2$ m and $A_y = -2$ m. Then $\tan\theta = -1$. But there are two angles having a tangent of -1: 135° and 315° (or $-45°$). To decide which is correct we must look at the individual components; because A_x is positive and A_y negative, the angle must be in the fourth quadrant; thus $\theta = 315°$ (or $-45°$) is the correct value.

The use of components provides an efficient means of calculating the vector sum of several vectors. In principle such sums may always be found using diagrams such as those in Figs. 1-4 and 1-6; but this requires either making and measuring a scale diagram (which is difficult to do with precision) or the trigonometric solution of oblique triangles (which can become quite complicated). The component method, by contrast, requires only right triangles and simple computations, and can be carried out with great precision.

Figure 1-9 shows two vectors ***A*** and ***B*** and their vector sum (resultant) ***C***, along with the x- and y-components of all three vectors. It can be seen from the diagram that the x-component C_x of the vector sum is simply the sum $(A_x + B_x)$ of the x-components of the vectors being added, and similarly for the y-components. Thus

$$\begin{aligned} C_x &= A_x + B_x, \\ C_y &= A_y + B_y. \end{aligned} \qquad (1\text{-}11)$$

Thus once the components of ***A*** and ***B*** are known, perhaps by use of Eqs. (1-8), the components of the vector sum can be computed. Then, if needed, the magnitude and direction of ***C*** can be obtained from Eqs. (1-9) and (1-10).

This discussion of the sum of two vectors can easily be extended to any number. Let ***R*** be the vector sum of ***A***, ***B***, ***C***, ***D***, ***E***, ... Then

$$\begin{aligned} R_x &= A_x + B_x + C_x + D_x + E_x + \ldots, \\ R_y &= A_y + B_y + C_y + D_y + E_y + \ldots. \end{aligned} \qquad (1\text{-}12)$$

Example An airplane flies 20.0 km in a direction 60° north of east, then 30.0 km straight east, then 10.0 km straight north. How far and in what direction is the plane from the starting point?

Solution The situation is shown in Fig. 1-10. We have chosen the x-axis as east and the y-axis as north, the usual choice for maps. Let ***A*** be the first displacement, ***B*** the second, ***C*** the third, and ***R*** the vector sum

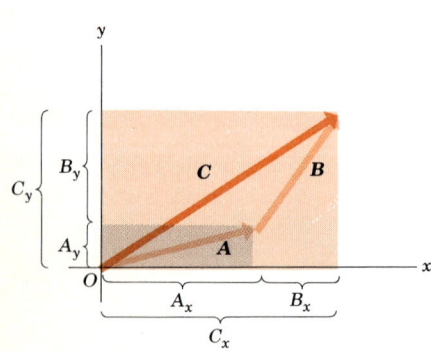

1-9 C_x is the x-component of the vector sum ***C*** of vectors ***A*** and ***B***, and is equal to the sum of the x-components of ***A*** and ***B***. The y-components are similarly related.

or resultant displacement. The components of A are:

$$A_x = (20.0 \text{ km})(\cos 60°) = 10.0 \text{ km},$$
$$A_y = (20.0 \text{ km})(\sin 60°) = 17.3 \text{ km}.$$

The components of all the displacements and the calculations can be arranged systematically, as in Table 1-2.

Table 1-2

Distance	Angle	x-component	y-component
$A = 20.0$ km	60°	10.0 km	17.3 km
$B = 30.0$ km	0°	30.0 km	0
$C = 10.0$ km	90°	0	10.0 km
		$R_x = 40.0$ km	$R_y = 27.3$ km

$$R = \sqrt{(40.0 \text{ km})^2 + (27.3 \text{ km})^2} = 48.4 \text{ km};$$
$$\theta = \arctan \frac{27.3 \text{ km}}{40.0 \text{ km}} = 34.3°.$$

1-10 Three successive displacements A, B, and C, and the resultant or vector sum displacement $R = A + B + C$.

The above discussion has been confined to vectors lying in the xy-plane, but it is easily generalized to vectors which do not all lie in a single plane. We introduce a z-axis perpendicular to the xy-plane; then in general a vector A has components $A_x, A_y,$ and A_z in the three coordinate directions. The magnitude A is given by

$$A = \sqrt{A_x^2 + A_y^2 + A_z^2}. \tag{1-13}$$

*1-7 Unit vectors

A unit vector is a vector having a magnitude of unity, with no units. Its only purpose is to describe a direction in space. Unit vectors provide a convenient notation in some expressions involving components of vectors.

In an xy-coordinate system, let us introduce a unit vector i in the direction of the positive x-axis, and a unit vector j in the direction of the positive y-axis. Then the relationships between the component vectors and the corresponding components, introduced in the beginning of Sec. 1-6, can be expressed as follows:

$$A_x = A_x i, \qquad A_y = A_y j. \tag{1-14}$$

Similarly, vector A may be written in terms of its components as

$$A = A_x i + A_y j. \tag{1-15}$$

It is important to understand that this and the preceding equations are *vector* equations.

The sum of two vectors A and B, each represented in terms of its components, can be expressed using unit vectors as follows:

$$A = A_x i + A_y j, \qquad B = B_x i + B_y j,$$
$$C = A + B$$
$$= (A_x i + A_y j) + (B_x i + B_y j)$$
$$= (A_x + B_x) i + (A_y + B_y) j. \tag{1-16}$$

In the last line we note that the coefficient of i is the x-component of C, and that the coefficient of j is its y-component. Equation (1–16) restates the content of Eqs. (1–11) in the form of a single vector equation rather than two ordinary equations. As such, it provides an alternative and sometimes more convenient form for expressing vector relationships.

If the vectors do not all lie in the xy-plane, then a third component is needed. We introduce a third unit vector k in the z-axis direction. The generalized forms of Eqs. (1–15) and (1–16) are

$$A = A_x i + A_y j + A_z k, \qquad (1\text{–}17)$$

$$C = (A_x + B_x)i + (A_y + B_y)j + (A_z + B_z)k. \qquad (1\text{–}18)$$

In later chapters, unit vectors will sometimes be used to describe particular directions in space, even without reference to a specific set of coordinate axes. For example, the direction of a line which is perpendicular to a particular plane can be described by defining a unit vector n (for *normal*) which is perpendicular to the plane. This situation arises often in the study of electric and magnetic fields in the second part of this book.

*1–8 Products of vectors

Many physical relationships are expressed most simply by use of *products* of vectors. Because vectors are not ordinary numbers, the concept of ordinary multiplication is not directly applicable to vectors. Just as addition of vectors is different from addition of ordinary numbers, so it is with multiplication. Indeed, two different kinds of vector products are commonly used. The first, called the *scalar product,* yields a result which is a scalar quantity, while the second, the *vector product,* yields another vector as its result.

The scalar product of two vectors A and B is defined as follows. The two vectors are drawn from a common point, as in Fig. 1–11a. The angle between their directions is θ, as shown. The scalar product, denoted by $A \cdot B$, is defined as

$$A \cdot B = AB \cos \theta = |A||B| \cos \theta. \qquad (1\text{–}19)$$

Because of this notation, the scalar product is also called the *dot product.* It is a number, not a vector, and it may be positive or negative. When θ is between zero and 90°, the scalar product is positive; when θ is between 90° and 180°, it is negative, and when $\theta = 90°$, $A \cdot B = 0$. *The scalar product of two perpendicular vectors is always zero.*

Vector B may be represented in terms of a component parallel to A and a component perpendicular to A, as shown in Fig. 1–11b; the component parallel to A is $(B \cos \theta)$. Thus, from Eq. (1–19), $A \cdot B$ is equal to the component of B parallel to A, multiplied by the magnitude of A. Alter-

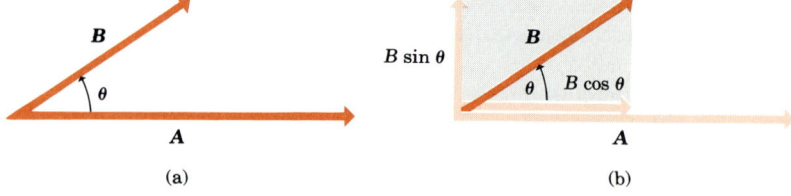

1–11 (a) Two vectors drawn from a common starting point to define their scalar product. (b) $B \cos \theta$ is the component of B in the direction of A, and $A \cdot B$ is the product of this component with the magnitude of A.

natively, it is also the component of **A** parallel to **B**, multiplied by the magnitude of **B**.

If the components of **A** and **B** are known, their scalar product can be calculated. This is done most easily by using the unit vector representation introduced in Sec. 1-7. First,

$$\mathbf{A}\cdot\mathbf{B} = (A_x\mathbf{i} + A_y\mathbf{j} + A_z\mathbf{k})\cdot(B_x\mathbf{i} + B_y\mathbf{j} + B_z\mathbf{k}). \quad (1\text{-}20)$$

We expand the product of the two parentheses on the right, obtaining nine terms in all, as follows:

$$\begin{aligned}\mathbf{A}\cdot\mathbf{B} = &A_x\mathbf{i}\cdot B_x\mathbf{i} + A_x\mathbf{i}\cdot B_y\mathbf{j} + A_x\mathbf{i}\cdot B_z\mathbf{k} \\ &+ A_y\mathbf{j}\cdot B_x\mathbf{i} + A_y\mathbf{j}\cdot B_y\mathbf{j} + A_y\mathbf{j}\cdot B_z\mathbf{k} \\ &+ A_z\mathbf{k}\cdot B_x\mathbf{i} + A_z\mathbf{k}\cdot B_y\mathbf{j} + A_z\mathbf{k}\cdot B_z\mathbf{k}.\end{aligned} \quad (1\text{-}21)$$

Each of these terms contains the scalar product of two vectors which are either parallel or perpendicular. For example, in $A_x\mathbf{i}\cdot B_x\mathbf{i}$, the two vectors are parallel, the angle between them is zero, its cosine is unity, and the scalar product is simply the product A_xB_x of the magnitudes. But in $A_x\mathbf{i}\cdot B_y\mathbf{j}$, the two vectors are perpendicular, the angle is 90°, the cosine is zero, and the scalar product is zero. Thus six of the nine terms are zero, and the three that survive give simply

$$\mathbf{A}\cdot\mathbf{B} = A_xB_x + A_yB_y + A_zB_z. \quad (1\text{-}22)$$

Example 1 Find the angle between the two vectors

$$\mathbf{A} = 2\mathbf{i} + 3\mathbf{j} + 4\mathbf{k}, \qquad \mathbf{B} = \mathbf{i} - 2\mathbf{j} + 3\mathbf{k}.$$

Solution We have

$$\begin{aligned}A_x &= 2 & B_x &= 1 \\ A_y &= 3 & B_y &= -2 \\ A_z &= 4 & B_z &= 3\end{aligned}$$

The scalar product is given by either of Eqs. (1-19) and (1-22). Equating these two and rearranging, we obtain

$$\cos\theta = \frac{A_xB_x + A_yB_y + A_zB_z}{AB}. \quad (1\text{-}23)$$

In our example,

$$A_xB_x + A_yB_y + A_zB_z = (2)(1) + (3)(-2) + (4)(3) = 8,$$
$$A = \sqrt{2^2 + 3^2 + 4^2} = \sqrt{29}, \qquad B = \sqrt{1^2 + (-2)^2 + 3^2} = \sqrt{14},$$
$$\cos\theta = \frac{8}{\sqrt{29}\sqrt{14}} = 0.397,$$

and $\theta = 66.6°$.

Our first physical application of the scalar product will come in Chapter 6 with the concept of work. When a constant force **F** is applied to a body which undergoes a displacement **d**, the work W (a scalar quantity) done by the force is given by

$$W = \mathbf{F}\cdot\mathbf{d}. \quad (1\text{-}24)$$

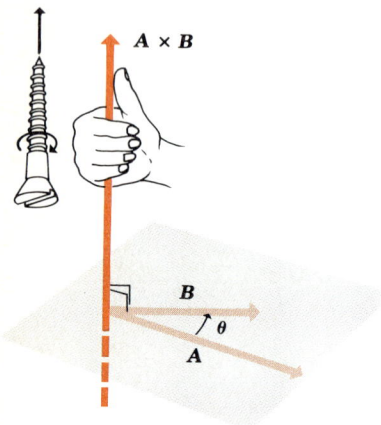

1-12 Vectors A and B lie in a plane; the vector product $A \times B$ is perpendicular to this plane, in a direction determined by the "righthand rule."

The *vector product* of two vectors A and B is denoted by $A \times B$. To define the vector product (also called the *cross product* because of this notation) we again draw A and B from a common point. The two vectors then lie in a plane. The vector product is defined as a vector quantity with a direction perpendicular to this plane (i.e., perpendicular to both A and B) and a magnitude given by $AB \sin \theta$. That is, if $C = A \times B$, then

$$C = AB \sin \theta. \tag{1-25}$$

There are always *two* directions perpendicular to a given plane. To distinguish between these, we imagine rotating vector A about the perpendicular line until it is aligned with B (choosing the smaller of the two possible angles). We then curl the fingers of the right hand around this perpendicular line so that the fingertips point in the direction of rotation; the thumb then gives the direction of the vector product. This rule is shown in Fig. 1-12. Alternatively, the direction of the vector product is the direction a righthand-thread screw advances if turned in the sense from A toward B, as shown in the figure.

Similarly, the direction of the vector product $B \times A$ is determined by rotating B into A in Fig. 1-12. This yields a result *opposite* to that for $A \times B$. The vector product is not commutative! In fact, for any two vectors A and B,

$$A \times B = -B \times A. \tag{1-26}$$

If the components of A and B are known, the vector product can be calculated using a procedure similar to that for the scalar product. We expand the expression

$$\begin{aligned} A \times B &= (A_x i + A_y j + A_z k) \times (B_x i + B_y j + B_z k) \\ &= A_x i \times B_x i + A_x i \times B_y j + A_x i \times B_z k \\ &\quad + A_y j \times B_x i + A_y j \times B_y j + A_y j \times B_z k \\ &\quad + A_z k \times B_x i + A_z k \times B_y j + A_z k \times B_z k. \end{aligned} \tag{1-27}$$

All terms in which the same unit vector appears twice, such as $A_x i \times B_x i$, are zero because they are products of two parallel vectors, the angle between them is zero, and its sine is zero. To evaluate the others, we refer to the axis system of Fig. 1-13a. We find, for example, $i \times j = k$ and $j \times i = -k$. Thus $A_x i \times B_y j = A_x B_y k$, and so on. We obtain finally

$$\begin{aligned} A \times B &= (A_y B_z - A_z B_y) i + (A_z B_x - A_x B_z) j \\ &\quad + (A_x B_y - A_y B_x) k. \end{aligned} \tag{1-28}$$

If $C = A \times B$, the components of C are given by

$$\begin{aligned} C_x &= A_y B_z - A_z B_y, \\ C_y &= A_z B_x - A_x B_z, \\ C_z &= A_x B_y - A_y B_x. \end{aligned} \tag{1-29}$$

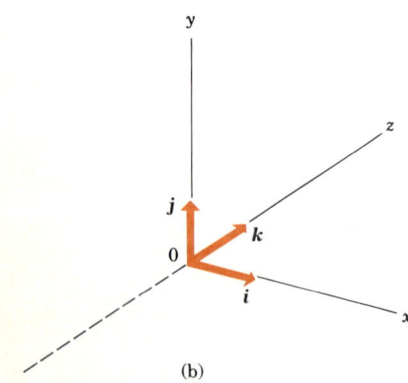

1-13 (a) A righthanded coordinate system, in which $i \times j = k, j \times k = i$, and $k \times i = j$. (b) A lefthanded coordinate system, in which $i \times j = -k$, and so on. Usually only righthanded systems are used.

Our first physical application of the vector product will come in Chapter 8 with the definition of torque or moment of a force. It will also be used extensively in the chapters on magnetic fields. From Eq. (1-25), we note in particular that *the vector product of two parallel vectors is always zero.*

An interesting problem arises in connection with the vector product. In Fig. 1–13a, suppose the direction of the z-axis is reversed, giving the axis system shown in Fig. 1–13b. Then, as the reader may verify, the definition of the vector product gives $i \times j = -k$ instead of $i \times j = k$. If *two* axis directions are reversed, we obtain again $i \times j = k$, and if all three are reversed, $i \times j = -k$. Thus there are two kinds of coordinate axis systems, differing in the signs of the products of unit vectors. In using the vector product, one must specify which kind of system is being used, to avoid ambiguity.

The axis systems in which $i \times j = k$ are called *righthanded systems;* the usual practice is to avoid the above ambiguity by agreeing to use *only* righthanded systems. This practice will be followed throughout this book.

Example 2 Vector A has magnitude 6 units and is in the direction of the $+x$-axis; vector B has magnitude 4 units and lies in the xy-plane, making an angle of 30° with the $+x$-axis and an angle of 60° with the $+y$-axis. Find the vector product $A \times B$.

Solution From Eq. (1–25), the magnitude of the vector product is

$$AB \sin \theta = (6)(4)(\sin 30°) = 12.$$

From the righthand rule, the direction of $A \times B$ is that of the $+z$-axis. Alternatively, we may write the components of A and B and use Eqs. (1–29):

$$A_x = 6, \quad A_y = 0, \quad A_z = 0,$$
$$B_x = 4 \cos 30° = 2\sqrt{3}, \quad B_y = 4 \cos 60° = 2, \quad B_z = 0.$$

If $C = A \times B$, then

$$C_x = (0)(0) - (0)(2) = 0,$$
$$C_y = (0)(2\sqrt{3}) - (6)(0) = 0,$$
$$C_z = (6)(2) - (0)(2\sqrt{3}) = 12.$$

Thus the vector has only a z-component, and hence lies along the z-axis. The magnitude agrees with the above result.

Questions

1-1 What are the units of the number π?

1-2 The rate of climb of a mountain trail was described as 150 meters per kilometer. How can this be expressed as a number with no units?

1-3 Suppose you are asked to compute the cosine of 3 meters. Is this possible?

1-4 Hydrologists describe the rate of volume flow of rivers in "second-feet." Is this unit technically correct? If not, what would be a correct unit?

1-5 Does a vector having zero length have a direction?

1-6 What is your weight in newtons?

1-7 What is your height in centimeters?

1-8 What physical phenomena (other than a pendulum or cesium clock) could be used to define a time standard?

1-9 Could some atomic quantity be used for a definition of a unit of mass? What advantages or disadvantages would this have compared to the one-kilogram platinum cylinder kept at Sèvres?

1-10 How could you measure the thickness of a sheet of paper with an ordinary ruler?

1-11 Can two vectors having different lengths have a vector sum of zero? What length restrictions are required for three vectors to have a vector sum of zero?

1-12 What is the displacement when a car travels from the north side of a circular race track of radius 500 m to the south side? When it makes one complete circle around the track?

1-13 What are the units of volume? If a student tells you a cylinder of radius r and height h has volume given by $\pi r^3 h$, explain why this cannot be right.

1-14 An angle (measured in radians) is a number with no units, since it is a ratio of two lengths. Think of other geometrical or physical quantities that are unitless.

1-15 Can a vector quantity ever have components different from zero but a magnitude of zero?

1-16 One sometimes speaks of the "direction of time," evolving from past to future. Does this mean that time is a vector quantity?

1-17 Is the scalar product of two vectors commutative? Explain.

1-18 What is the scalar product of a vector with itself? The vector product?

Problems

1-1 Starting with the definition 1 in. = 2.54 cm, compute the number of kilometers in one mile, to five significant figures.

1-2 The density of water is 1 g·cm^{-3}. What is this value in kilograms per cubic meter?

1-3 Convert the following speeds, as indicated.

a) 60 mi·hr^{-1} to feet per second;

b) 100 km·hr^{-1} to meters per second.

1-4 What is the mass in kilograms of a person weighing 170 lb?

1-5 Compute the number of seconds in a day, and in a year (365 da).

1-6 What is the percent error in each of the following approximations to π?

a) 22/7   b) 355/113

1-7 What is the percent error in the approximate statement 1 yr = $\pi \times 10^7$ s?

1-8 Estimate the percent error in measuring

a) a distance of about 50 cm with a meter stick;

b) a mass of about 1 g with a chemical balance;

c) a time interval of about 4 min with a stopwatch.

1-9 The mass of the earth is 5.98×10^{24} kg, and its radius is 6.38×10^6 m. Compute the density of the earth, using powers-of-ten notation and the correct number of significant figures.

1-10 The piston displacement of a certain automobile engine is given as 2.0 liters. Using only the facts that 1 liter = 1000 cm^3 and 1 in. = 2.54 cm, express this volume in cubic inches.

1-11 Using the definition 1 in. = 2.54 cm, compute the number of kilometers in one mile; comment on the precision of your result.

1-12 An angle is given, to one significant figure, as 5°, meaning that its value is between 4.5° and 5.5°. Find the corresponding range of possible values of the cosine of the angle. Is this a case where there are more significant figures in the result than in the input data?

1-13 Two points P_1 and P_2 are described by their x- and y-coordinates, (x_1, y_1) and (x_2, y_2), respectively. Show that the components of the displacement A from P_1 to P_2 are $A_x = x_2 - x_1$ and $A_y = y_2 - y_1$. Also derive expressions for the magnitude and direction of this displacement.

1-14 When two vectors A and B are drawn from a common point, the angle between them is θ. Show that the magnitude of their vector sum is given by

$$\sqrt{A^2 + B^2 + 2AB \cos \theta}.$$

1-15 Find the magnitude and direction of the vector represented by each of the following pairs of components:

a) $A_x = 3$ cm, $A_y = -4$ cm;

b) $A_x = -5$ m, $A_y = -12$ m;

c) $A_x = -2$ km, $A_y = 3$ km.

1-16 A delivery truck drives 1 mile north, then 2 miles east, then 3 miles northwest. Determine the resultant displacement

a) by drawing a scale diagram;

b) by using components.

1-17 A bug starts at the center of a 12-inch phonograph record and crawls along a straight radial line to the edge. While this is happening the record turns through an angle of 45°. Draw a sketch of the situation and describe the magnitude and direction of the bug's displacement.

1-18 A cave explorer is surveying a cave. He follows a passage 100 m straight east, then 50 m in a direction 30° west of north, then 150 m at 45° west of south. After a fourth unmeasured displacement he finds himself back where he started. Using a scale drawing, determine the fourth displacement (magnitude and direction).

1-19 Find graphically the magnitude and direction of the vector sum of the three forces in Fig. 1–14. Use the polygon

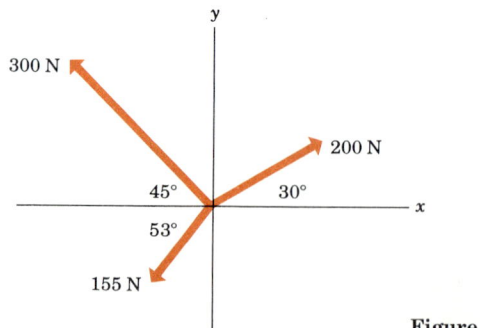

Figure 1–14

method. Check the precision of your result by using the component method.

1-20 Find graphically the vector sum $A + B$ and the vector difference $A - B$ in Fig. 1-15.

Figure 1-15

1-21 Find the vector sum $A + B$ and the vector difference $A - B$ in Fig. 1-15, using components.

1-22 Vector A is 2 in. long and is 60° above the x-axis in the first quadrant. Vector B is 2 in. long and is 60° below the x-axis in the fourth quadrant. Find graphically

a) the vector sum $A + B$, and

b) the vector differences $A - B$ and $B - A$.

1-23 Obtain the vector sum and differences requested in Problem 1-22, using the method of components.

1-24 Vector A has components $A_x = 2$ cm, $A_y = 3$ cm, and vector B has components $B_x = 4$ cm, $B_y = -2$ cm. Find

a) the components of the vector sum $A + B$;

b) the magnitude and direction of $A + B$;

c) the components of the vector difference $A - B$;

d) the magnitude and direction of $A - B$.

1-25 A car drives 5 mi east, then 4 mi south, then 2 mi west. Find the magnitude and direction of the resultant displacement.

1-26 A sailboat sails 2 km east, then 4 km southeast, then an additional distance in an unknown direction. Its final position is 5 km directly east of the starting point. Find the magnitude and direction of the third leg of the journey.

1-27 Vector M, of magnitude 5 cm, is at 36.9° counterclockwise from the $+x$-axis. It is added to vector N, and the resultant is a vector of magnitude 5 cm, at 53.1° counterclockwise from the $+x$-axis. Find

a) the components of N;

b) the magnitude and direction of N.

1-28 A vector A of length 10 units makes an angle of 30° with a vector B of length 6 units. Find the magnitude of the vector difference $A - B$ and the angle it makes with vector A:

a) by the triangle method;

b) by the method of rectangular resolution.

1-29 Two vectors A and B have the same magnitude. Under what circumstances does the vector sum $A + B$ have the same magnitude as A and B? When does the vector difference $A - B$ have this magnitude?

1-30 Given two vectors $A = 2i + 3j$ and $B = i - 2j$, obtain the following:

a) Find the magnitude of each vector.

b) Write an expression for the vector sum, using unit vectors.

c) Find the magnitude and direction of the vector sum.

d) Write an expression for the vector difference $A - B$, using unit vectors.

e) Find the magnitude and direction of the vector difference $A - B$.

1-31 Given two vectors $A = 2i + 3j + 4k$ and $B = i - 2j + 3k$, obtain the following:

a) Find the magnitude of each vector.

b) Write an expression for the vector sum, using unit vectors.

c) Find the magnitude of the vector sum.

d) Write an expression for the vector difference $A - B$, using unit vectors.

e) Find the magnitude of the vector difference $A - B$. Is this the same as the magnitude of $B - A$? Explain.

1-32 Write out a multiplication table for the scalar products of all possible pairs of unit vectors, such as $i \cdot i = ?$, $i \cdot j = ?$, and so on.

1-33 Write out a multiplication table for the vector products of all possible pairs of unit vectors, such as $i \times i = ?$, $i \times j = ?$, and so on, using a righthanded coordinate system.

1-34 Find the angle between the diagonal of a cube and an edge, using a method similar to that in the example of Sec. 1-8. Is this the same as the angle between the diagonal of a face and an edge?

1-35 Find the scalar product of the two vectors given in Problem 1-31.

1-36 Find the vector product of the two vectors given in Problem 1-31. What is the magnitude of this vector product?

1-37 Obtain a *unit vector* perpendicular to the two vectors given in Problem 1-31.

1-38 Find the angle between the two vectors $A = 3i + 4j + 5k$ and $B = 3i + 4j - 5k$.

1-39 Consider the two repeated vector products $i \times (i \times j)$ and $(i \times i) \times j$.

a) Are these two products equal?

b) Can you generalize this result for this type of repeated product?

1-40 Prove that for any three vectors A, B, and C, $A \cdot (B \times C) = (A \times B) \cdot C$.

2 EQUILIBRIUM OF A PARTICLE

2-1 Introduction

The science of mechanics is based on three natural laws relating force and motion. These were clearly stated for the first time by Sir Isaac Newton (1642–1727) and were published in 1686 in his *Philosophiae Naturalis Principia Mathematica* (The Mathematical Principles of Natural Science). Many men had preceded Newton in his field, however; the most outstanding was Galileo Galilei (1564–1642) who, in his studies of accelerated motion, had laid much of the groundwork for Newton's three laws.

2-2 Force

Force is a central concept in all of physics. When we push or pull on a body, we are said to exert a *force* on it. Forces can also be exerted by inanimate objects: a stretched spring exerts forces on the bodies to which its ends are attached; compressed air exerts a force on the walls of its container; a locomotive exerts a force on the train it is pulling or pushing. The force of gravitational attraction exerted on every physical body by the earth is called the *weight* of the body. Gravitational forces (and electrical and magnetic forces also) can act through empty space without contact. A force on an object resulting from direct contact with another object is called *a contact force;* viewed on an atomic scale, contact forces arise chiefly from electrical attraction and repulsion of the electrons and nuclei making up the atoms of material.

As mentioned in Sec. 1–5, force is a *vector* quantity. To describe a force, we need to describe the *direction* in which it acts, as well as its *magnitude,* which is a quantitative description of "how much" or "how hard" the force pushes or pulls, in terms of a standard unit of force. In Chapter 4 we shall see how a unit of force can be defined in terms of the units of mass, length, and time. As mentioned in Sec. 1–5, the SI unit of force is the *newton,* abbreviated N. In the British system the unit of force is the *pound-force,* equal to about 4.448 N. In the cgs system of units, the unit of force is the *dyne,* equal to 10^{-5} N.

2-1 Force may be exerted on the box by either (a) pulling it, or (b) pushing it.

An instrument commonly used to measure forces is the spring balance, which consists of a coil spring, enclosed in a case for protection, carrying at one end a pointer that moves over a scale. A force exerted on the balance changes the length of the spring, and the change can be read on the scale.

One way to calibrate such a balance is to make a number of bodies each having a weight of exactly 1 N. Then, when two, three, or more of these are suspended simultaneously from the balance, the total force stretching the spring is 2 N, 3 N, and so on, and the corresponding positions of the pointer can be labeled 2 N, 3 N, etc. The calibrated spring balance can then be used to measure the magnitude of an unknown force.

Suppose we slide a box along the floor by pulling it with a string or pushing it with a stick, as in Fig. 2-1. Our point of view is that the motion of the box is caused not by the *objects* that push or pull on it, but by the *forces* these objects exert. The forces in the two cases can be represented as in Fig. 2-2; the labels indicate the magnitude and direction of the force, and the length of the arrow, to some chosen scale, also shows the magnitude.

Some vector quantities, of which force is one, are not *completely* specified by their magnitude and direction alone. The effect of a force depends also on the position of the point at which the force is applied. For example, when one pushes horizontally against a door, the effectiveness of a given force in setting the door in motion depends on the distance of the point of application from the line of the hinges, about which the door rotates. We return to this consideration in Chapter 8.

Now consider the following physical problem. Two forces, represented by the vectors F_1 and F_2 in Fig. 2-3, are applied simultaneously at the same point A of a body. Is it possible to produce the same effect by applying a *single* force at A, and if so, what should be its magnitude and direction? The question can be answered only by experiment; investigation shows that a single force, represented in magnitude, direction, and line of action by the *vector sum* R of the original forces, is in all respects equivalent to them. This single force is called the resultant of the original forces. Hence the mathematical process of *vector addition* of two force vectors corresponds to the physical operation of finding the *resultant of two forces,* simultaneously applied at a given point.

The fact that forces can be combined by vector addition is of the utmost importance, as we shall see in the following chapters. Furthermore, this fact also permits a force to be represented by means of *components,* as we have done with displacements in Sec. 1-6. In Fig. 2-4a, force F is exerted on a body at point O. The rectangular components of F in the directions Ox and Oy are F_x and F_y; and it is found that simultaneous

2-2 Force diagram for the forces acting on the box in Fig. 2-1.

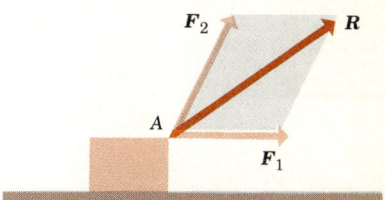

2-3 A force represented by the vector R, equal to the vector sum of F_1 and F_2, produces the same effect as the forces F_1 and F_2 acting simultaneously.

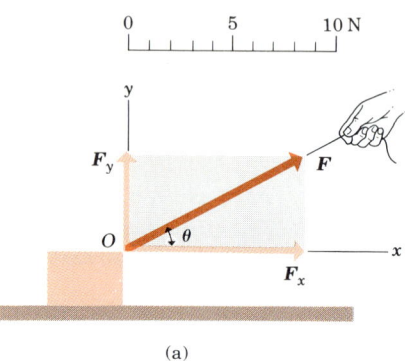

application of the forces F_x and F_y, as in Fig. 2–4b, is equivalent in all respects to the effect of the original force. *Any force can be replaced by its rectangular components, acting at the same point.*

As a numerical example, let

$$F = 10.0 \text{ N}, \qquad \theta = 30°.$$

Then,

$$F_x = F \cos \theta = (10.0 \text{ N})(0.866) = 8.66 \text{ N},$$
$$F_y = F \sin \theta = (10.0 \text{ N})(0.500) = 5.00 \text{ N},$$

and the effect of the original 10.0-N force is equivalent to the simultaneous application of a horizontal force, to the right, of 8.66 N, and a lifting force of 5.00 N.

The axes used to obtain rectangular components of a vector need not be vertical and horizontal. For example, Fig. 2–5 shows a block being pulled up an inclined plane by a force F, represented by its components F_x and F_y, parallel and perpendicular to the sloping surface of the plane.

We shall often have occasion to find the vector sum of several forces acting on a body, which again may be called their *resultant*. The Greek letter Σ is often used in a shorthand notation for this sum. Thus if the forces are labeled F_1, F_2, F_3, and so on, and their resultant is R, then the operation

$$R = F_1 + F_2 + F_3 + \cdots$$

is often abbreviated

$$R = \sum F, \tag{2–1}$$

where ΣF is read "sum of the forces." This means, of course, the vector sum. In terms of components, we may write

$$R_x = \sum F_x, \qquad R_y = \sum F_y, \tag{2–2}$$

in which ΣF_x is read as the sum of the *x*-components of the forces. We also note that a boldface Σ is used in Eq. (2–1) as a reminder that the sum is a vector sum, and a lightface Σ is used in Eqs. (2–2) because components are ordinary numbers.

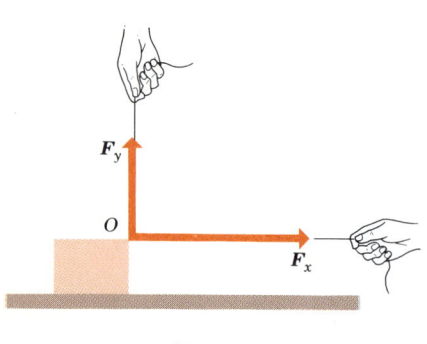

2–4 The inclined force F may be replaced by its rectangular components F_x and F_y. $F_x = F \cos \theta$, $F_y = F \sin \theta$.

2–3 Equilibrium. Newton's first law

One effect of a force is to alter the state of motion of the body. This motion can be considered as made up of its motion as a whole, or its *translational* motion, together with any *rotational* motion the body may have. In the most general case, a single force acting on a body produces a change in both its translational and rotational motion. However, when several forces act on a body simultaneously, their effects can compensate one another, with the result that there is no change in either the translational or rotational motion. When this is the case, the body is said to be in *equilibrium*. This means (1) that the body as a whole either remains at rest or moves in a straight line with constant speed, and (2) that the body is either not rotating at all or is rotating at a constant rate.

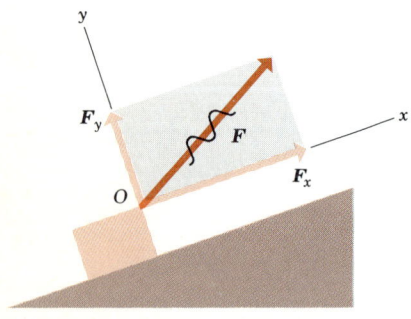

2–5 F_x and F_y are the rectangular components of F, parallel and perpendicular to the sloping surface of the inclined plane.

In Fig. 2–6 a flat, rigid body of arbitrary shape lies on a horizontal surface having negligible friction. If a single force F_1 acts on the body, as in Fig. 2–6a, and if the body is originally at rest, it at once starts to move and to rotate clockwise. If it is originally in motion, the effect of the force is to change the translational motion of the body in magnitude or direction (or both) and to increase or decrease its rate of rotation. In either case, the body does not remain in equilibrium.

Equilibrium can be maintained, however, by the application of a second force F_2 as in Fig. 2–6b, provided that F_2 is *equal in magnitude* to F_1, is opposite in direction to F_1, and has the same *line of action* as F_1. The resultant of F_1 and F_2 is then zero. If the lines of action of the two forces are *not* the same, as in Fig. 2–6c, the body will be in translational but not in rotational equilibrium. (The forces then form a *couple*; see Sec. 8–4.)

If the forces F_1 and F_2 are equal in magnitude and opposite in direction, then one equals the negative of the other. That is,

$$F_2 = -F_1.$$

Then, if R represents the resultant of F_1 and F_2,

$$R = F_1 + F_2 = F_1 - F_1 = 0.$$

For brevity, we shall often speak of two forces as simply being "equal and opposite," meaning that their magnitudes are equal and that one is the negative of the other.

This discussion can be generalized immediately to a body with any number of forces acting on it. We have noted that forces may be combined according to the rules of vector addition. When a body is in equilibrium, the vector sum, or resultant, of all the forces acting on it must be zero. Each component of the resultant must therefore be zero. Hence, for a body in equilibrium,

$$R = \sum F = 0,$$

or

$$\sum F_x = 0, \qquad \sum F_y = 0. \tag{2–3}$$

These equations represent *the first condition for equilibrium*.

The *second condition for equilibrium*, to be developed more fully in Chapter 8, is a mathematical statement of the fact that the forces must have no tendency to rotate the body. The first condition of equilibrium ensures that a body shall be in *translational* equilibrium; the second, that it be in *rotational* equilibrium. The statement that a body is in complete equilibrium when both conditions are satisfied is the essence of Newton's first law of motion. Newton did not state his first law in exactly these words. His original statement (translated from the Latin in which the *Principia* was written) reads:

Every body continues in its state of rest, or of uniform motion in a straight line, unless it is compelled to change that state by forces impressed on it.

Although rotational motion was not explicitly mentioned by Newton, it is clear from his work that he fully understood the conditions that the forces must satisfy when the rotation is zero or is constant.

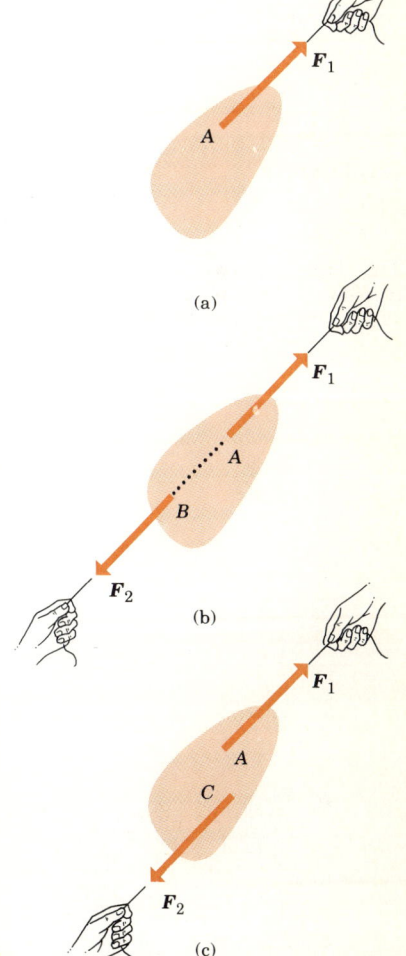

2–6 A rigid body acted on by two forces is in equilibrium if the forces are equal in magnitude, opposite in direction, and have the same line of action, as in part (b).

2–4 Discussion of Newton's first law

Newton's first law of motion is not as self-evident as it may seem. In the first place, this law asserts that in the absence of any applied force a body either remains at rest or moves uniformly in a straight line. It follows that *once a body has been set in motion it is no longer necessary to exert a force on it to maintain it in motion*.

This assertion appears to be contradicted by everyday experience. Suppose we exert a force with the hand to move a book along a horizontal tabletop. After the book has left our hand, and we are no longer exerting a force on it, it does *not* continue to move indefinitely, but slows down and eventually comes to rest. If we wish to keep it moving uniformly we must continue to exert some forward force on it. But this force is required only because a frictional force is exerted on the sliding book by the tabletop, in a direction opposite to the motion of the book. The smoother the surfaces of the book and table, the smaller the frictional force and the smaller the force we must exert to keep the book moving. The first law asserts that if the frictional force could be eliminated completely, no forward force at all would be required to keep the book moving, once it had been set in motion. More than that, however, the law implies that if the *resultant* force on the book is zero, as it is when the frictional force is balanced by an equal forward force, the book also continues to move uniformly. In other words, *zero resultant force is equivalent to no force at all*.

Furthermore, the first law defines by implication what is known as an *inertial reference system*. To understand what is meant by this term we must recognize that the motion of a given body can be specified only with respect to, or relative to, some other body. Its motion relative to one body may be very different from that relative to another. Thus a passenger in an aircraft that is making its take-off run may continue at rest relative to the aircraft but be moving faster and faster relative to the earth.

A *reference system* means a set of coordinate axes attached to (or moving with) some specified body or bodies. Suppose we consider a reference system attached to the aircraft referred to above. Everyone knows that during the take-off run, while the aircraft is going faster and faster, a passenger feels the back of his seat pushing him forward, although he remains at rest relative to a reference system attached to the aircraft. Newton's first law does not therefore correctly describe the situation; a forward force *does* act on the passenger, but nevertheless (relative to the aircraft) he remains at rest.

Suppose, on the other hand, that the passenger is standing in the aisle on roller skates. He then starts to move *backward,* relative to the aircraft, when the take-off run begins, even though no backward force acts on him. Newton's first law again does not correctly describe the facts.

We can now define an inertial reference system as one relative to which a body *does* remain at rest or move uniformly in a straight line when no force (or no resultant force) acts on it. That is, *an inertial reference system* is one in which Newton's first law correctly describes the motion of a body on which zero resultant force acts.

An aircraft gaining speed during take-off is evidently *not* an inertial system. For many purposes a reference system attached to the earth can be considered an inertial system, although it is not precisely so, due to effects related to the earth's rotation and other motions. But if a frame of reference *is* inertial, then a second frame moving uniformly (i.e., with constant speed in a straight line, without rotation) relative to it is *also* inertial, since any body in uniform motion as seen by an observer in the first frame will also appear to an observer in the second frame to be in uniform motion. The speed and direction appear different to the two observers, but both observe that Newton's first law is obeyed.

Thus there is no single, unique inertial frame of reference—there are infinitely many; but the motion of any one relative to any other is uniform in the above sense. It follows that the concepts of "absolute rest" and "absolute motion" have no physical meaning. Experiments show that a frame of reference at rest relative to the so-called *fixed stars* is inertial, within the limits of experimental error, and so every frame in uniform motion relative to this one is also inertial. An aircraft in steady flight over the earth is just as suitable an inertial frame of reference as the earth itself.

Finally, Newton's first law contains a qualitative definition of the concept of *force,* or at least of one aspect of the force concept, as "that which changes the state of motion of a body." (Of course, forces also produce other effects, such as changing the length of a coil spring.) When a body at rest relative to the earth is observed to start moving, or when a moving body speeds up, slows down, or changes its direction, we can conclude that a resultant force is acting on it. This effect of a force can be used to define the ratio of two forces and to define a *unit* of force, and we shall show in Chapter 4 how this is done.

2-5 Newton's third law of motion

Any given force is but one aspect of a mutual interaction between *two* bodies. It is found that *whenever one body exerts a force on another, the second always exerts on the first a force that is equal in magnitude, is opposite in direction, and has the same line of action.** A single isolated force is therefore an impossibility.

The two forces involved in every interaction between two bodies are often called an "action" and a "reaction," but this does not imply any difference in their nature, or that one force is the "cause" and the other its "effect." *Either* force may be considered the "action," and the other the "reaction" to it.

This property of forces was stated by Newton in his *third law of motion.* In his words,

To every action there is always opposed an equal reaction: or, the mutual actions of two bodies upon each other are always equal, and directed to contrary parts.

As an example, suppose that a man pulls on one end of a rope attached to a block, as in Fig. 2–7. The weight of the block and the force exerted on it by the surface are not shown. The block may or may not be in equilibrium. The resulting action–reaction pairs of forces are indicated in the figure. (The lines of action of all the forces lie along the rope; the force vectors have been offset from this line to show them more

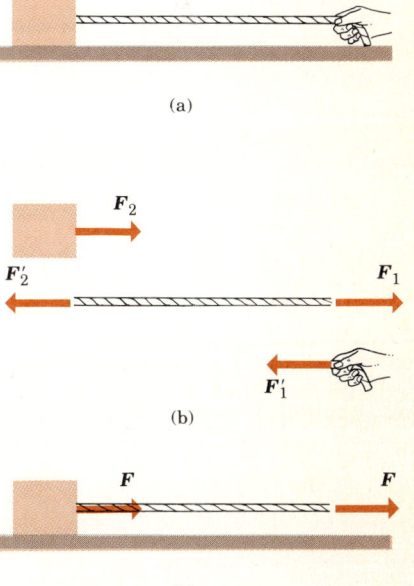

2-7 (a) A man pulls on a rope attached to a block. The forces that the rope exerts on the block and on the man are shown. (b) Separate diagrams showing the forces acting on the block, on the rope, and on the man. (c) If the rope is in equilibrium so that $F'_2 = F_1$, the rope can be considered to transmit a force from the man to the block, and vice versa.

---

* There are instances, such as the motion of electrically charged particles, where this law is not obeyed, but it is correct for all the macroscopic forces we shall encounter in mechanics.

clearly.) Vector F_1 represents the force exerted *on* the rope *by* the man. Its reaction is the equal and opposite force F'_1 exerted *on* the man *by* the rope. Vector F_2 represents the force exerted on the block by the rope. The reaction to it is the equal and opposite force F'_2, exerted on the rope by the block:

$$F'_1 = -F_1, \qquad F'_2 = -F_2. \tag{2-4}$$

It is very important to realize that the forces F_1 and F'_2, although they are opposite in direction and have the same line of action, do *not* constitute an action–reaction pair. For one thing, both of these forces act on the *same* body (the rope) while an action and its reaction necessarily act on *different* bodies. Furthermore, the forces F_1 and F'_2 are not necessarily equal in magnitude. If the block and rope are moving to the right with increasing speed, the rope is not in equilibrium and F_1 is greater in magnitude than F'_2. Only in the special case when the rope remains at rest or moves with constant speed are the forces F_1 and F'_2 equal in magnitude, but this is an example of Newton's *first* law, not his *third*. Even when the speed of the rope is changing, however, the action–reaction forces F_1 and F'_1 are equal in magnitude to *each other*, and the action–reaction forces F_2 and F'_2 are equal in magnitude to *each other*, although then F_1 is not equal to F'_2.

In the special case when the rope is in equilibrium, and when no forces act on it except those at its ends, F'_2 equals $-F_1$ by Newton's *first* law. Since F_2 *always* equals $-F'_2$ by Newton's *third* law, then in this special case F_2 also equals F_1 and the force exerted on the block by the rope is equal to the force exerted on the rope by the man. The rope can therefore be considered to "transmit" to the block, without change, the force exerted on it by the man, as in Fig. 2-7c. This point of view is often useful, but it is important to remember that it applies only under the restricted conditions as stated.

A body such as the rope in Fig. 2-7, which is subjected to pulls at its ends, is said to be in *tension*. The tension at any point equals the force exerted at that point. Thus in Fig. 2-7b the tension at the righthand end of the rope equals the magnitude of F_1 (or of F'_1) and the tension at the lefthand end equals the magnitude of F_2 (or of F'_2). If the rope is in equilibrium and if no forces act except at its ends, the tension is the same at both ends. If, for example, the magnitudes of F_1 and F_2 are each 50 N, the tension in the rope is 50 N (*not* 100 N).

2-6 Idealized models

The phenomena of nature are seldom simple; often a phenomenon involves several interrelated principles, and the relationships can be extremely complex. Often one must make simplifying assumptions and approximations in order to facilitate analysis of such a phenomenon and to focus attention on its most significant aspects. This process of simplifying and idealizing is called making a *model*, and models play an essential role in applications of principles of physics.

Here is an example. A baseball is thrown into the air, and we want to calculate where it lands and with what velocity. The ball has a somewhat irregular surface; it is spinning as well as moving through the air. It is affected by the force of gravity, which decreases slightly as the ball ascends. Additional effects are caused by air resistance (friction) and by buoyancy in the air. There may be variable air currents that complicate these effects further.

Clearly, the analysis is hopelessly complicated if we try to include all these effects; we need to make a simplified model. If we approximate the ball as a uniform sphere and assume it moves through still air, we can calculate the air resistance and buoyancy forces, although these are still complex. The buoyant force is usually much smaller than the force of gravity, and if the ball moves slowly enough the friction force is also small, so we may wish to omit these completely. We may feel that the shape and spinning motion are irrelevant for an overall description of motion; if so, we can pretend that the ball is a *particle* (a point mass without any size, shape, or spin). Finally, we can assume that the gravitational force is constant during the motion. Then at last the problem is simple enough to permit detailed analysis. Such problems are considered in Chapter 5.

Because of the approximations and simplifications, one should not expect this analysis to predict *precisely* how the ball will move; but of course an approximate description is a lot better than no description at all, which may be the alternative if the problem is so complex as to defy analysis.

The concept of an idealized model plays an important role in equilibrium problems. In general the forces acting on a body *do not* all have lines of action passing through a single point, and as a result the body undergoes both translational and rotational motion. However, if in a particular problem the rotational motion is irrelevant, then the body may be represented as a particle or point. In such a case, all the forces are considered to act at this point.

2-7 Equilibrium of a particle

In the remainder of this chapter, we shall consider only examples and problems involving the equilibrium of particles. It is surprising how many situations of interest and importance in engineering, in the life and earth sciences, and in everyday life involve the equilibrium of particles. In the analysis of such problems, the following systematic problem-solving procedure is strongly recommended.

1. Make a simple line sketch of the apparatus or structure, showing dimensions and angles.

2. Choose some object (a knot in a rope, for example) as the particle in equilibrium. Draw a separate diagram of this object and show by arrows (use a colored pencil) *all* of the forces exerted *on* it by other bodies. This is called the *force diagram* or *free-body diagram*. When a system is composed of several particles, it may be necessary to construct a separate free-body diagram for each one. Do *not* show, in the free-body diagram of a chosen particle, any of the forces exerted *by* it. These forces (which are the reactions to the forces acting *on* the chosen particle) all act *on other bodies* and appear in the free-body diagrams of those bodies.

3. Construct a set of rectangular axes and resolve into rectangular components all forces acting on the particle. Cross out lightly those forces that have been resolved.

4. Set the algebraic sum of all x-forces (or force components) equal to zero, and the algebraic sum of all y-forces (or components) equal to zero. This provides two independent equations, which can be solved simultaneously for two unknown quantities (which may be forces, angles, distances, etc.).

A remark about notation is in order. Thus far, vector quantities have been denoted by boldface italic symbols, and this usage will be continued throughout this book. However, it is sometimes convenient in diagrams to label a vector by using a lightface italic symbol to represent the *magnitude* of the quantity, and to describe its direction (if that is not obvious in the diagram) separately, perhaps by use of an angle. When both light and bold symbols appear on a diagram, confusion can be avoided by recalling that a bold symbol always represents the vector quantity itself, while a light symbol represents either the magnitude of a vector quantity or a component.

A force that will be encountered in many problems is the *weight* of a body, that is, the force of gravitational attraction exerted on the body by the earth. This is but one aspect of a mutual interaction between the earth and the body. At the same time, the body attracts the earth with a force equal in magnitude and opposite in direction to the force the earth exerts on the body. Thus if a body weighs 10 N, the earth pulls down on it with a force of 10 N, and the body pulls *up* on the earth with a force of 10 N, another example of an action–reaction pair.

Example 1 To begin with a simple example, consider the body in Fig. 2–8a, hanging at rest from the ceiling by a vertical cord. Part (b) of the figure is the free-body diagram for the body. The forces on it are its weight $\boldsymbol{w}_1$ and the upward force $\boldsymbol{T}_1$ exerted on it by the cord. If we take the x-axis horizontal and the y-axis vertical, there are no x-components of force. The y-components are those associated with the forces $\boldsymbol{T}_1$ and $\boldsymbol{w}_1$. Force $\boldsymbol{T}_1$ acts in the positive y-direction, and its y-component is just the magnitude T_1, a positive (scalar) quantity. But $\boldsymbol{w}_1$ acts in the negative y-direction, and its y-component is the *negative* of the magnitude w_1. Thus the algebraic sum of y-components is $T_1 - w_1$. Then from the equilibrium condition, $\Sigma F_y = 0$, we have

$$\Sigma F_y = T_1 - w_1 = 0,$$
$$T_1 = w_1. \quad \text{(First law)}$$

Let us emphasize again that the forces $\boldsymbol{w}_1$ and $\boldsymbol{T}_1$ are *not* an action–reaction pair, although they are equal in magnitude, opposite in direction, and have the same line of action. The weight $\boldsymbol{w}_1$ is a force of attraction exerted on the body by the earth. Its reaction is an equal and opposite force of attraction exerted on the earth by the body. This reaction is one of the set of forces acting *on the earth*, and therefore it does not appear in the free-body diagram of the suspended block.

The reaction to the force $\boldsymbol{T}_1$ is a downward force, $\boldsymbol{T}_1'$, of equal magnitude:

$$T_1 = T_1'. \quad \text{(Third law)}$$

The force $\boldsymbol{T}_1'$ is shown in part (c), which is the free-body diagram of the *cord*. The other forces on the cord are its own weight $\boldsymbol{w}_2$ and the upward

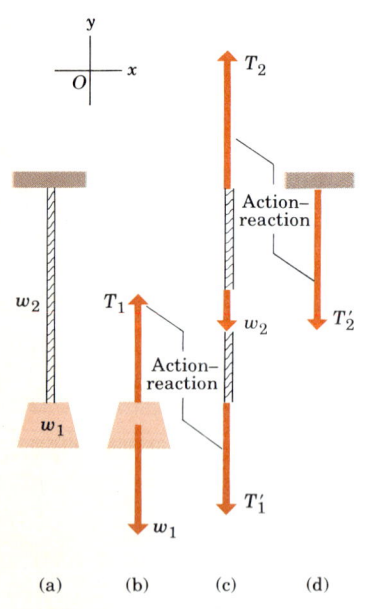

2–8 (a) Block hanging at rest from vertical cord. (b) The block is isolated and all forces acting on it are shown. (c) Forces on the cord. (d) Downward force on the ceiling. Lines connect action–reaction pairs.

force T_2 exerted on its upper end by the ceiling. The y-component of T_2 is positive, while those of w_2 and T_1' are negative. Since the cord is also in equilibrium,

$$\sum F_y = T_2 - w_2 - T_1' = 0$$
$$T_2 = w_2 + T_1'. \quad \text{(First law)}$$

The reaction to T_2 is the downward force T_2' in part (d), exerted on the ceiling by the cord:

$$T_2 = T_2'. \quad \text{(Third law)}$$

As a numerical example, let the body weigh 20 N and the cord weigh 1 N. Then

$$T_1 = w_1 = 20 \text{ N},$$
$$T_1' = T_1 = 20 \text{ N},$$
$$T_2 = w_2 + T_1' = 1 \text{ N} + 20 \text{ N} = 21 \text{ N},$$
$$T_2' = T_2 = 21 \text{ N}.$$

If the weight of the cord were so small as to be negligible, then in effect no forces would act on it except at its ends. The forces T_2 and T_2' would then each equal 20 N and, as explained earlier, the cord could be considered to transmit a 20-N force from one end to the other without change. We could then consider the upward pull of the cord on the block as an "action" and the downward pull on the ceiling as its "reaction." The tension in the cord would then be 20 N.

Example 2 In Fig. 2–9a a block of weight w hangs from a cord, which is knotted at point O to two other cords, one fastened to the ceiling, the other to the wall. We wish to find the tensions in these three cords, assuming the weights of the cords to be negligible.

Solution To use the conditions of equilibrium to determine an unknown force, we must consider some body that is in equilibrium and on which the desired force acts. The suspended block is one such body; as shown in the preceding example, the tension T_1 in the vertical cord supporting the block is equal in magnitude to the weight of the block. The other cords do not exert forces on the block (because they are not attached directly to it) but they do act on the knot at O. Hence we consider the *knot* as a particle in equilibrium, considering the weight of the knot itself as negligible.

Free-body diagrams for the block and the knot are shown in Fig. 2–9b, where T_1, T_2, and T_3 are the *magnitudes* of the forces shown; the directions of these forces are indicated by the vectors on the diagram. An x-y coordinate axis system is also shown, and the force of magnitude T_3 has been resolved into its x- and y-components.

Considering the block first, we note that there are no x-components of force. The y-component of force exerted by the cord is in the positive y-direction and is just T_1. The y-component of the weight is, however, in the negative y-direction and is $-w$. The equilibrium condition for the block, that the algebraic sum of y-components of force must be zero, is

$$T_1 + (-w) = 0.$$

2–9 (a) A block of weight w is suspended from a cord knotted at O to two other cords. (b) Free-body diagrams for the block and for the knot, showing the components of force acting on each.

28 EQUILIBRIUM OF A PARTICLE

Hence,
$$T_1 = w.$$

The tension in the vertical cord equals the weight of the block, as we have already noted.

We now consider the knot at O. There are both x- and y-components of force acting on it, so there are two separate equilibrium conditions. The algebraic sum of the x-components must be zero, and the algebraic sum of y-components must separately be zero. (Note that x- and y-components are *never* added together in a single equation.) We find:

$$\sum F_x = 0: \quad T_3 \cos 60° - T_2 = 0.$$
$$\sum F_y = 0: \quad T_3 \sin 60° - T_1 = 0.$$

Because $T_1 = w$, the second equation can be rewritten as:

$$T_3 = \frac{T_1}{\sin 60°} = \frac{w}{\sin 60°} = 1.155\, w.$$

This result can now be used in the first equation:

$$T_2 = T_3 \cos 60° = (1.155\, w) \cos 60° = 0.577\, w.$$

Thus all three tensions can be expressed as multiples of the weight of the block, which is assumed to be known. To summarize,

$$T_1 = w,$$
$$T_2 = 0.577\, w,$$
$$T_3 = 1.155\, w.$$

We note that T_3 is greater than the weight of the block; if this seems strange, one should note that it must be large enough so that its vertical component is equal to w in magnitude, so the force itself must have somewhat larger magnitude than this.

Example 3 In Fig. 2–10, block A of weight w_1 rests on a frictionless inclined plane of slope angle θ. A flexible cord is attached to the right face of the block, passes over a frictionless pulley, and is attached to a second block B of weight w_2. The weight of the cord and friction in the pulley are negligible. If w_1 and θ are given, find the weight w_2 for which the system is in equilibrium, that is, for which it remains at rest or moves in either direction at constant speed.

Solution Free-body diagrams for the two blocks are shown in parts (a) and (c) of the figure. Here, as in preceding examples, we label and refer to forces by using light italic symbols, which in equations represent the *magnitudes* of these forces. The forces on block B are its weight w_2 and the force T exerted on it by the cord. Because it is in equilibrium,

$$T = w_2.$$

Block A is acted on by its weight w_1, the force T exerted on it by the cord, and the force n exerted on it by the plane. We use the same symbol T for the force exerted on each block by the cord because, as discussed above, these forces are equivalent to an action–reaction pair and have

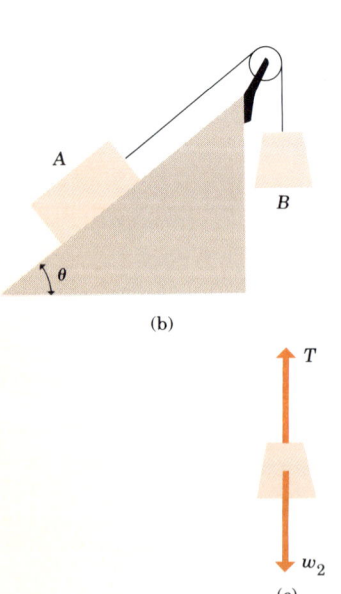

2–10 Forces on a block in equilibrium on a frictionless inclined plane. (a) Free-body diagram for block A; (b) the physical situation; (c) free-body diagram for block B.

the same magnitude. The pulley changes the directions of the forces exerted by the cord but not their magnitudes; and the two forces at the ends have the same magnitude even though their directions are different. If there is no friction, the force n can have no component along the plane that would resist the motion of the block on the plane; hence it must be perpendicular or *normal* to the surface of the plane.

It is simplest to choose x- and y-axes parallel and perpendicular to the surface of the plane, because then only the weight w_1 needs to be resolved into components. The conditions of equilibrium give

$$\sum F_x = T - w_1 \sin \theta = 0,$$

$$\sum F_y = n - w_1 \cos \theta = 0.$$

Thus if $w_1 = 100$ N and $\theta = 30°$, we find

$$w_2 = T = w_1 \sin \theta = (100 \text{ N})(0.500) = 50 \text{ N},$$

and

$$n = w_1 \cos \theta = (100 \text{ N})(0.866) = 86.6 \text{ N}.$$

Note carefully that, *in the absence of friction,* the same weight w_2 of 50 N is required whether the system remains at rest or moves with constant speed in *either* direction. This is *not* the case when friction is present.

2-8 Friction

Whenever the surface of one body slides over that of another, *each* body exerts a frictional force on the other, parallel to the surfaces. The force *on* each body is opposite to the direction of its motion relative to the other. Thus when a block slides from left to right along the surface of a table, a frictional force to the left acts on the block and an equal force toward the right acts on the table. Frictional forces may also act when there is no relative motion. A horizontal force on a heavy packing case resting on the floor may not be enough to set the case in motion, because of an equal and opposite frictional force exerted on the case by the floor.

In Fig. 2-11a a block is at rest on a horizontal surface, in equilibrium under the action of its weight w and the upward force P exerted on it by

2-11 The magnitude of the friction force $\mathcal{F}_s$ is less than or equal to $\mu_s n$ when there is no relative motion, and is equal to $\mu_k n$ when motion exists.

the surface. The lines of action of *w* and *P* have been displaced slightly to show these forces more clearly.

Suppose now that a cord is attached to the block as in Fig. 2-11b and the tension *T* in the cord is gradually increased. Provided the tension is not too great, the block remains at rest. The force *P* exerted on the block by the surface is inclined toward the left as shown, since its horizontal component opposes *T*. The component of *P* parallel to the surface is called the *force of static friction*, $\mathcal{F}_s$. The other component, n, is the *normal* force exerted on the block by the surface. From the conditions of equilibrium, the force of static friction $\mathcal{F}_s$ is equal in magnitude and opposite in direction to the force *T*, and the normal force n is equal and opposite to the weight *w*.

As the force *T* is increased further, a limiting value is reached at which the block breaks away from the surface and starts to move. In other words, there is a certain *maximum* value which the force of static friction $\mathcal{F}_s$ can have. Figure 2-11c is the force diagram when *T* is just below its critical value. If the force *T* exceeds this value, the block is no longer in equilibrium.

Experiment shows that, for a given pair of surfaces, the magnitude of the maximum value of $\mathcal{F}_s$ depends on that of the normal force n; when the normal force increases, a greater force is required to make the block slide. In general the details of this relationship can be quite complex, but in some cases it is approximately true that the maximum value of $\mathcal{F}_s$ is *directly proportional* to n, with proportionality factor μ_s, called the *coefficient of static friction*. The actual force of static friction can have any magnitude between zero (when there is no applied force parallel to the surface) and a maximum value given by $\mu_s n$. Thus,

$$\mathcal{F}_s \leq \mu_s n. \tag{2-5}$$

The equality sign holds only when the applied force *T*, parallel to the surface, has such a value that motion is about to start (Fig. 2-11c). When *T* is less than this value (Fig. 2-11b), the inequality sign holds and the magnitude of the friction force must be computed from the conditions of equilibruim.

As soon as sliding begins, it is found that the friction force *decreases*. This new friction force, for a given pair of surfaces, also depends on the magnitude of the normal force, and again it is convenient, though not always precise, to represent the relation as a proportionality. The proportionality factor, μ_k, is called the *coefficient of sliding friction* or *kinetic* friction. Thus, when the block is in motion, the force of sliding or kinetic friction is given by

$$\mathcal{F}_k = \mu_k n. \tag{2-6}$$

This is illustrated in Fig. 2-11d.

The force *P* exerted on the block by the surface is called the *contact force*. The frictional force $\mathcal{F}$ is always the component of the contact force *parallel* to the surface, and the normal force n is always the component *normal* (perpendicular) to the surface. We use special script symbols for these quantities to emphasize their special role in representing the contact force. In addition, use of the script letter n prevents our confusing this symbol for normal force with the abbreviation for newton, N. In many problems to be considered later, the normal force is *not* equal in magnitude to the weight of the body, because the surface is not horizontal or because there are additional components of force normal to the surface. Nevertheless, n always denotes the normal component of contact force exerted by the surface on the body. By definition, n and $\mathcal{F}$ for a given body are always perpendicular. Therefore Eqs. (2-5) and (2-6) are not *vector* equations, but relations between the *magnitudes* n and $\mathcal{F}$ of these forces.

The coefficients of static and kinetic friction depend primarily on the nature of the surfaces in contact, being relatively large if the surfaces are rough, and small if they are smooth. The coefficient of sliding friction varies somewhat with the relative velocity, but for simplicity we shall assume it to be independent of velocity. It is also nearly independent of the contact area. Equations (2-5) and (2-6) are useful empirical relations, but do not represent *fundamental* physical laws like Newton's laws. Typical numerical values are given in Table 2-1.

Table 2-1 lists coefficients of friction only for solids. Liquids and gases show frictional effects also, but the simple equation $\mathcal{F} = \mu n$ does not hold. It will be shown in Chapter 13 that there exists a property of liquids and gases called *viscosity*, which determines the friction force between two surfaces sliding over each other with a layer of liquid or gas between them. Gases have the lowest viscosities of all materials at normal temperatures, and therefore, to reduce friction to a value close to zero, it is convenient to have an object slide on a layer of gas.

This is the principle of the *hovercraft*, a type of vehicle containing powerful blowers that maintain a cushion of air between the vehicle and the ground so that it literally floats on a layer of air. A familiar laboratory example of the principle is the linear air track, conceived originally by H. V. Neher and R. B. Leighton at the California Institute of Technology, and developed further by John Stull of Alfred University. Elastic bumpers are provided for the inverted V-shaped riders. (See Fig. 2-12a.) When one rider only is placed on the air track and is given a push, it will hit the stationary bumper and then proceed to move back and forth many times before coming to rest. The frictional force is velocity-dependent, but at typical speeds the effective coefficient of friction is of the order of 0.001.

A similar device is the frictionless air table shown in Fig. 2-12b, developed by Harold A. Daw of New Mexico State University. The pucks, made of plastic, are supported by over 1600 tiny air jets about an inch apart. Two-dimensional collisions, both elastic and inelastic, may be demonstrated on this table. The same principle is used in "air-hockey" games.

Table 2-1 Coefficients of friction

Materials	Static, μ_s	Kinetic, μ_k
Steel on steel	0.74	0.57
Aluminum on steel	0.61	0.47
Copper on steel	0.53	0.36
Brass on steel	0.51	0.44
Zinc on cast iron	0.85	0.21
Copper on cast iron	1.05	0.29
Glass on glass	0.94	0.4
Copper on glass	0.68	0.53
Teflon on Teflon	0.04	0.04
Teflon on steel	0.04	0.04

Example 1 In Fig. 2-11, suppose that the block weighs 20 N, that the tension T can be increased to 8 N before the block starts to slide, and that a force of 4 N will keep the block moving at constant speed once it has been set in motion. Find the coefficients of static and kinetic friction.

Solution From Fig. 2-11c and the data above, we have

$$\sum F_y = n - w = n - 20 \text{ N} = 0,$$

$$\sum F_x = T - \mathcal{F}_s = 8 \text{ N} - \mathcal{F}_s = 0,$$

$$\mathcal{F}_s = \mu_s n \quad \text{(motion impends)}.$$

32 EQUILIBRIUM OF A PARTICLE

(a)

(b)

2-12 (a) The Ealing-Stull linear air track. Inverted Y-shaped sliders ride on a layer of air streaming through many fine holes in the inverted V-shaped surface. (b) The Ealing-Daw two-dimensional air table. Plastic pucks slide on a cushion of air issuing from more than a thousand minute holes in the tabletop. (Courtesy of the Ealing Corporation.)

Hence we have

$$\mu_s = \frac{\mathcal{F}_s}{n} = \frac{8\text{ N}}{20\text{ N}} = 0.40.$$

From Fig. 2-11d, we have

$$\sum F_y = n - w = n - 20\text{ N} = 0,$$
$$\sum F_x = T - \mathcal{F}_k = 4\text{ N} - \mathcal{F}_k = 0,$$
$$\mathcal{F}_k = \mu_k n \quad \text{(motion exists)}.$$

Hence

$$\mu_k = \frac{\mathcal{F}_k}{n} = \frac{4\text{ N}}{20\text{ N}} = 0.20.$$

Example 2 What is the friction force if the block in Fig. 2-11b is at rest on the surface and a horizontal force of 5 N is exerted on it?

Solution We have

$$\sum F_x = T - \mathcal{F}_s = 5\text{ N} - \mathcal{F}_s = 0, \quad \text{(First Law)}$$
$$\mathcal{F}_s = 5\text{ N}.$$

Note that in this case $\mathcal{F}_s < \mu_s n$.

Example 3 What force T, at an angle of 30° above the horizontal, is required to drag a block weighing 20 N to the right at constant speed, as in Fig. 2-13, if the coefficient of kinetic friction between block and surface is 0.20?

Solution The forces on the block are shown in the diagram. From the first condition of equilibrium,

$$\sum F_x = T \cos 30° - 0.2\, n = 0,$$
$$\sum F_y = T \sin 30° + n - 20\text{ N} = 0.$$

Simultaneous solution gives

$$T = 4.14\text{ N}, \quad n = 17.9\text{ N}.$$

Note that in this example the normal force n is *not* equal to the weight of the block, but is less than the weight by the vertical component of the force T.

Example 4 In Fig. 2-14, a block has been placed on an inclined plane and the slope angle θ of the plane has been adjusted until the block slides down the plane at constant speed, once it has been set in motion. Find the angle θ.

Solution The forces on the block are its weight w and the normal and frictional components of the force exerted by the plane. Take axes per-

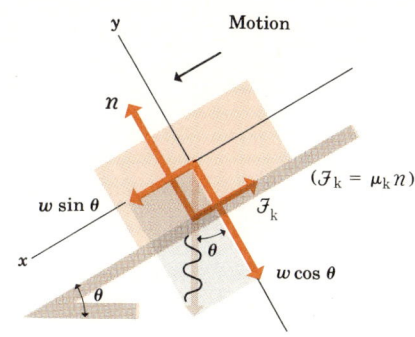

2-13 Forces on a block being dragged to the right on a level surface at constant speed.

2-14 Forces on a block sliding down an inclined plane (with friction) at constant speed.

pendicular and parallel to the surface of the plane. Then

$$\sum F_x = \mu_k n - w \sin \theta = 0,$$

$$\sum F_y = n - w \cos \theta = 0.$$

Hence

$$\mu_k n = w \sin \theta, \quad n = w \cos \theta.$$

Dividing the former by the latter, we get

$$\mu_k = \frac{\sin \theta}{\cos \theta} = \tan \theta.$$

It follows that a block, regardless of its weight, slides down an inclined plane with constant speed if the tangent of the slope angle of the plane equals the coefficient of kinetic friction. Measurement of this angle then provides a simple experimental method for determining the coefficient of kinetic friction.

Questions

2-1 Can a body be in equilibrium when only one force acts on it?

2-2 A helium balloon hovers in midair, neither ascending nor descending. Is it in equilibrium? What forces act on it?

2-3 If the two ends of a rope in equilibrium are pulled with forces of equal magnitude and opposite direction, why is the total tension in the rope not zero?

2-4 A horse is hitched to a wagon. Since the wagon pulls back on the horse just as hard as the horse pulls on the wagon, why doesn't the wagon remain in equilibrium, no matter how hard the horse pulls?

2-5 A clothesline is hung between two poles, and then a shirt is hung near the center. No matter how tightly the line is stretched, it always sags a little at the center. Explain why.

2-6 A man sits in a chair that is suspended from a rope. The rope passes over a pulley suspended from the ceiling and the man holds the other end of the rope in his hands. What is the tension in the rope, and what force does the chair exert on the man?

2-7 How can pushing *down* on a bicycle pedal make the bicycle move *forward*?

2-8 A car is driven up a steep hill at constant speed. Discuss all the forces acting on the car; in particular, what pushes it up the hill?

34 EQUILIBRIUM OF A PARTICLE

2-9 Can the coefficient of friction ever be greater than unity? If so, give an example; if not, explain why not.

2-10 A block rests on an inclined plane with enough friction to prevent sliding down. To start the block moving, is it easier to push it up the plane, down the plane, or sideways? Why?

2-11 In pushing a box up a ramp, is it better to push horizontally or to push parallel to the ramp?

2-12 In stopping a car on an icy road, is it better to push the brake pedal hard enough to "lock" the wheels and make them slide, or to push gently so the wheels continue to roll? Why?

2-13 When one stands with bare feet in a wet bathtub, the grip feels fairly secure, and yet a catastrophic slip is quite possible. Discuss this situation with respect to the two coefficients of friction.

2-14 The horrible squeak made by a piece of chalk held against a blackboard at the wrong angle results from alternate sticking and slipping of the chalk against the blackboard. Interpret this phenomenon in terms of the two coefficients of friction. Can you think of other examples of "slip-stick" behavior?

Problems

2-1

a) Find graphically the horizontal and vertical components of a 40-N force the direction of which is 50° above the horizontal to the right. Let 1 cm = 5 N.

b) Check your results by calculating the components.

2-2 A box is pushed along the floor as in Fig. 2-1b by a force of 40 N making an angle of 30° with the horizontal. Using a scale of 1 cm = 5 N, find the horizontal and vertical components of the force by the graphical method. Check your results by calculating the components.

2-3 A block is dragged up an inclined plane of slope angle 20° by a force F making an angle of 30° with the plane.

a) How large a force F is necessary in order that the component F_x parallel to the plane shall be 16 N?

b) How large will the component F_y then be? Solve graphically, letting 1 cm = 2 N.

2-4 The three forces shown in Fig. 1-14 act on a body located at the origin.

a) Find the x- and y-components of each of the three forces.

b) Use the method of rectangular resolution to find the resultant of the forces.

c) Find the magnitude and direction of a fourth force that must be added to make the resultant force zero. Indicate the fourth force by a diagram.

2-5 Use the method of rectangular resolution to find the resultant of the following set of forces and the angle it makes with the positive x-axis: 200 N, along the x-axis toward the right; 300 N, 60° above the x-axis to the right; 100 N, 45° above the x-axis to the left; 200 N, along the negative y-axis.

2-6 The resultant of four forces is 1000 N in the direction 30° west of north. Three of the forces are 400 N, 60° north of east; 200 N, south; and 400 N, 53° west of south. Find the rectangular components of the fourth force.

2-7 Two forces, F_1 and F_2, act at a point. The magnitude of F_1 is 8 N and its direction is 60° above the x-axis in the first quadrant. The magnitude of F_2 is 5 N and its direction is 53° below the x-axis in the fourth quadrant.

a) What are the horizontal and vertical components of the resultant force?

b) What is the magnitude of the resultant?

c) What is the magnitude of the vector difference $F_1 - F_2$?

2-8 Two forces, F_1 and F_2, act upon a body in such a manner that the resultant force R has a magnitude equal to that of F_1 and makes an angle of 90° with F_1. Let $F_1 = R = 10$ N. Find the magnitude of the second force, and its direction (relative to F_1).

2-9 Two men and a boy want to push a crate in the direction marked x in Fig. 2-15. The two men push with forces F_1 and F_2, whose magnitudes and directions are indicated in the figure. Find the magnitude and direction of the smallest force that the boy should exert.

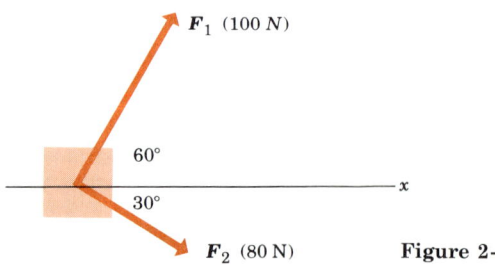

Figure 2-15

2-10 Two men pull horizontally on ropes attached to a post, the angle between the ropes being 45°. If man A exerts a force of 300 N and man B a force of 200 N, find the magnitude of the resultant force and the angle it makes with A's pull. Solve:

a) graphically, using a scale diagram, and

b) analytically, using components.

Let 1 cm = 50 N in (a).

2-11 Imagine that you are holding a book weighing 4 N at rest on the palm of your hand. Complete the following sentences.

a) A downward force of magnitude 4 N is exerted on the book by _____.

b) An upward force of magnitude _____ is exerted on _____ by the hand.

c) Is the upward force (b) the reaction to the downward force (a)?

d) The reaction to force (a) is a force of magnitude _____, exerted on _____ by _____. Its direction is _____.

e) The reaction to force (b) is a force of magnitude _____, exerted on _____ by _____. Its direction is _____.

f) That the forces (a) and (b) are equal and opposite is an example of Newton's _____ law.

g) That forces (b) and (e) are equal and opposite is an example of Newton's _____ law.

Suppose now that you exert an upward force of magnitude 5 N on the book.

h) Does the book remain in equilibrium?

i) Is the force exerted on the book by the hand equal and opposite to the force exerted on the book by the earth?

j) Is the force exerted on the book by the earth equal and opposite to the force exerted on the earth by the book?

k) Is the force exerted on the book by the hand equal and opposite to the force exerted on the hand by the book?

Finally, suppose that you snatch your hand away while the book is moving upward.

l) How many forces then act on the book?

m) Is the book in equilibrium?

n) What balances the downward force exerted on the book by the earth?

2-12 A block is given a push along a tabletop, and slides off the edge of the table.

a) What forces are exerted on it while it is falling from the table to the floor?

b) What is the reaction to each force, that is, on what body and by what body is the reaction exerted? Neglect air resistance.

2-13 Two 10-N weights are suspended at opposite ends of a rope that passes over a light frictionless pulley. The pulley is attached to a chain that goes to the ceiling.

a) What is the tension in the rope?

b) What is the tension in the chain?

2-14 In Fig. 2-16, a man lifts a weight w by pulling down on a rope with a force F. The upper pulley is attached to the ceiling by a chain, and the lower pulley is attached to the weight by another chain. If $w = 400$ N, find the tension in each chain and the force F if the weight is lifted at constant speed. Assume the weights of the rope, pulleys, and chains to be negligible.

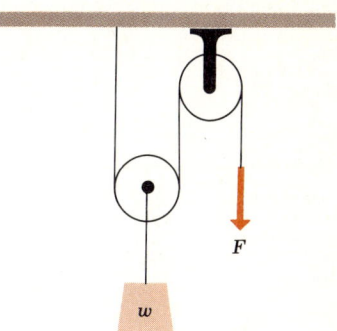

Figure 2-16

2-15 In Fig. 2-9a, let the weight of the hanging block be 50 N.

a) Find the tension in each cord.

b) If the 60° angle is changed to 45°, find the tension in each cord.

2-16 A picture frame is hung against a wall, suspended by two wires attached to its upper corners. If the two wires make the same angle with the vertical, what must this angle be for the tension in each wire to equal the weight of the frame?

2-17 Figure 2-17 shows a technique called rappelling, used by mountaineers for descending vertical rock faces. The climber sits in a rope seat, and the rappel rope passes through a friction device attached to this seat. Suppose the rock is perfectly smooth, and the climber's feet push horizontally into the rock; if the climber's weight is 800 N, find the tension in the rope and the force his feet exert on the rock face.

Figure 2-17

2-18 Figure 2-18 illustrates a mountaineering technique called a Tyrolean traverse. A rope is stretched tightly between two points, and the climber slides across the rope.

The climber's weight is 800 N, and the breaking strength of the rope (typically nylon, 11 mm diameter) is 20,000 N.

a) If the angle θ is 15°, find the tension in the rope.

b) What is the smallest value the angle θ can have if the rope is not to break?

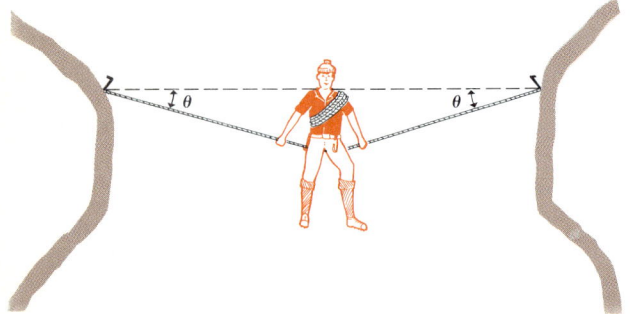

Figure 2-18

2-19 A man proposes to pull his car out of a mudhole on a country road by stretching a rope between the front of his car and a solid object directly in front of the car, and then pushing sideways at the midpoint of the rope. Suppose that the rope is 50 m long, and that a sideways force of 300 N is required to displace the midpoint sideways 5 m. Determine the tension in the rope and the forward component of force on the car.

2-20 Find the tension in each cord in Fig. 2-19 if the weight of the suspended body is 200 N.

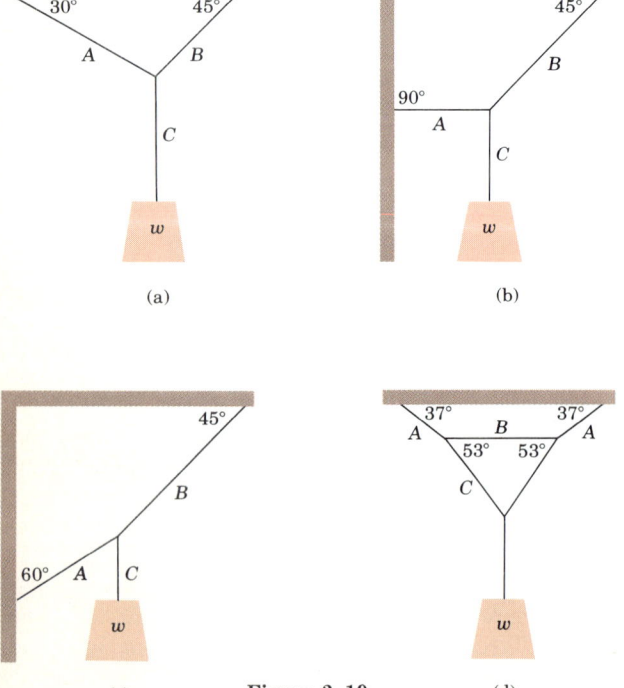

Figure 2-19

2-21 In Fig. 2-20, find the weight of the suspended body if the tension in the diagonal string is 20 N.

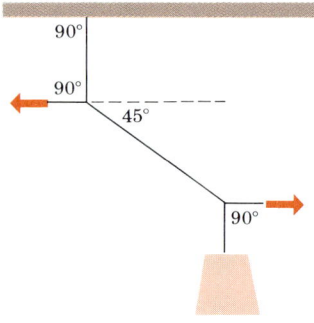

Figure 2-20

2-22 In each of the arrangements in Fig. 2-21, find each tension and the weight of the suspended body if the indicated tension T is 10 N.

Figure 2-21

2-23

a) Block A in Fig. 2-22 weighs 100 N. The coefficient of static friction between the block and the surface on which it rests is 0.30. The weight w is 20 N and the system is in equilibrium. Find the friction force exerted on block A.

b) Find the maximum weight w for which the system will remain in equilibrium.

Figure 2-22

2-24 A block hangs from a cord 10 m long. A second cord is tied to the midpoint of the first, and a horizontal pull equal to half the weight of the block is exerted on it, the second cord being always kept horizontal.

a) How far will the block be pulled to one side?

b) How far will it be lifted?

2-25 A flexible rope of weight w hangs between two hooks at the same height, as shown in Fig. 2-23. At each end the rope makes an angle θ with the horizontal.

a) What is the magnitude and direction of the force F exerted by the rope on the hook at the left?

b) What is the tension T in the rope at its lowest point?

Figure 2-23

2-26 A 30-N block is pulled at constant speed up a frictionless inclined plane by a weight of 10 N hanging from a cord attached to the block and passing over a frictionless pulley at the top of the plane. (See Fig. 2-10.) Find

a) the slope angle of the plane,
b) the tension in the cord, and
c) the normal force exerted on the block by the plane.

2-27

a) A block rests upon a rough horizontal surface. A horizontal force T is applied to the block and is slowly increased from zero. Draw a graph with T along the x-axis and the friction force f along the y-axis, starting at $T = 0$ and showing the region of no motion, the point where motion impends, and the region where motion exists.

b) A block of weight w rests on a rough horizontal plank. The slope angle of the plank θ is gradually increased until the block starts to slip. Draw two graphs, both with θ along the x-axis. In one graph show the ratio of the normal force to the weight $\mathcal{N}/w$ as a function of θ. In the second graph, show the ratio of the friction force to the weight $\mathcal{F}/w$. Indicate the region of no motion, the point where motion impends, and the region where motion exists.

2-28 A block weighing 20 N rests on a horizontal surface. The coefficient of static friction between block and surface is 0.40 and the coefficient of sliding friction is 0.20.

a) How large is the friction force exerted on the block?
b) How great will the friction force be if a horizontal force of 5 N is exerted on the block?
c) What is the minimum force that will start the block in motion?
d) What is the minimum force that will keep the block in motion once it has been started?
e) If the horizontal force is 10 N, how great is the friction force?

2-29 A block is pulled to the right at constant velocity by a 10-N force acting 30° above the horizontal. The coefficient of sliding friction between the block and the surface is 0.5. What is the weight of the block?

2-30 A block weighing 14 N is placed on an inclined plane and connected to a 10-N block by a cord passing over a small frictionless pulley, as in Fig. 2-10. The coefficient of sliding friction between the block and the plane is $(\frac{1}{7})$. For what two values of θ will the system move with constant velocity? [*Hint:* $\cos\theta = \sqrt{1 - \sin^2\theta}$.]

2-31 A block weighing 100 N is placed on an inclined plane of slope angle 30° and is connected to a second hanging block of weight w by a cord passing over a small frictionless pulley, as in Fig. 2-10. The coefficient of static friction is 0.40 and the coefficient of sliding friction is 0.30.

a) Find the weight w for which the 100-N block moves up the plane at constant speed.
b) Find the weight w for which it moves down the plane at constant speed.
c) For what range of values of w will the block remain at rest?

2-32 What force P at an angle ϕ above the horizontal is needed to drag a box of weight w at constant speed along a level floor if we are given that the coefficient of sliding friction between box and floor is μ?

2-33 A safe weighing 2000 N is to be lowered at constant speed down skids 4 m long, from a truck 2 m high.

a) If the coefficient of sliding friction between safe and skids is 0.30, will the safe need to be pulled down or held back?
b) How great a force parallel to the skids is needed?

2-34

a) If a force of 86 N parallel to the surface of a 20° inclined plane will push a 120-N block up the plane at constant speed, what force parallel to the plane will push it down at constant speed?
b) What is the coefficient of sliding friction?

2-35 Block A in Fig. 2-24 weighs 4 N and block B weighs 8 N. The coefficient of sliding friction between all surfaces is 0.25. Find the force P necessary to drag block B to the left at constant speed

a) if A rests on B and moves with it,
b) if A is held at rest, and
c) if A and B are connected by a light flexible cord passing around a fixed frictionless pulley.

Figure 2-24

2-36 Block A, of weight w, slides down an inclined plane S of slope angle 37° at constant velocity while the plank B, also of weight w, rests on top of A. The plank is attached by a cord to the top of the plane (Fig. 2–25).

a) Draw a diagram of all the forces acting on block A.

b) If the coefficient of kinetic friction is the same between the surfaces A and B and between S and A, determine its value.

and the coefficient of sliding friction between each block and the surface is 0.5. Block C descends with constant velocity.

a) Draw two separate force diagrams showing the forces acting on A and on B.

b) Find the tension in the rope connecting blocks A and B.

c) What is the weight of block C?

Figure 2–25

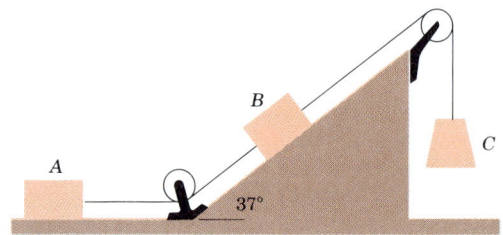

Figure 2–26

2-37 Two blocks, A and B, are placed as in Fig. 2–26 and connected by ropes to block C. Both A and B weigh 20 N

MOTION ON A STRAIGHT LINE 3

3-1 Motion

Mechanics deals with the relations of force, matter, and motion. The preceding chapters have been concerned with forces, and we are now ready to discuss mathematical methods for describing motion. This branch of mechanics is called *kinematics*.

Motion may be defined as a continuous change of position. In actual motion of an extended body, different points of the body move along different paths, but we shall consider first a description of motion in terms of motion of a single point. Such a model is adequate whenever rotation and similar complications are absent, or when the body is small enough to be considered a point.

The simplest motion to describe is motion of a point along a straight line, which we shall take to coincide with a coordinate axis. In later chapters we shall see that more general motions in space can always be represented by means of their projections onto the three coordinate axes. Thus, while displacement is in general a vector quantity, as discussed in Chapter 1, we consider first situations in which only one component of displacement is different from zero.

3-2 Average velocity

We consider a particle moving along the x-axis, as in Fig. 3–1a. Its distance from the origin O is described by the coordinate x, which varies with time. At a time t_1 the particle is at point P, where its coordinate is x_1, and at time t_2 it is at point Q, where its coordinate is x_2. The displacement during the time interval from t_1 to t_2 is the vector from P to Q; the x-component of this vector is $(x_2 - x_1)$, and all other components are zero.

It is convenient to represent the quantity $x_2 - x_1$, the *change* in x, by means of a notation using the Greek letter Δ (capital delta) to designate a change in any quantity. Thus we write

$$\Delta x = x_2 - x_1, \tag{3-1}$$

3–1 (a) Particle moving on the x-axis. (b) Coordinate–time graph of the motion. The average velocity between t_1 and t_2 equals the slope of the chord pq. The instantaneous velocity at p equals the slope of the tangent at p.

in which Δx is not a product but is to be interpreted as a single symbol representing the change in the quantity x. Similarly, we denote the time interval from t_1 to t_2 as $\Delta t = t_2 - t_1$.

The *average velocity* of the particle is defined as the ratio of the displacement Δx to the time interval Δt. We represent average velocity by the letter v with a bar to signify average value. Thus

$$\bar{v} = \frac{x_2 - x_1}{t_2 - t_1} = \frac{\Delta x}{\Delta t}. \tag{3-2}$$

Strictly speaking, this defines the x-component of the average velocity, which is really a vector quantity. However, in this chapter all vectors have *only* x-components, so it is not essential to distinguish between a vector and its x-component. In the following chapters we shall return to this distinction in the more general context of two-dimensional motion.

In Fig. 3–1b the coordinate x is graphed as a function of time. It is important to understand that the curve in this figure does *not* represent the path of the particle in space; as Fig. 3–1a shows, the path is a straight line. Rather, the graph provides a convenient representation of the change of position with time. The points on the coordinate–time graph corresponding to points P and Q are labeled p and q.

In Fig. 3–1b the average velocity is represented by the *slope* of the chord pq, that is, the ratio of vertical to horizontal intervals on the right triangle of which pq is the hypotenuse, evaluated using the scales and units with which x and t are plotted. The slope is the ratio of the "rise," $x_2 - x_1$ or Δx, to the "run," $t_2 - t_1$ or Δt.

Equation (3–2) can be written

$$x_2 - x_1 = \bar{v}(t_2 - t_1). \tag{3-3}$$

Since our time-measuring device can be started at any instant, we can let $t_1 = 0$ and let t_2 be any arbitrary later time t. Then if x_0 is the coordinate when $t = 0$ (called the *initial position*), and x is the coordinate at time t, Eq. (3–3) becomes

$$x - x_0 = \bar{v}t. \tag{3-4}$$

If the particle is at the origin when $t = 0$, then $x_0 = 0$ and Eq. (3–4) simplifies further to

$$x = \bar{v}t. \tag{3-5}$$

3–3 Instantaneous velocity

The velocity of a particle at some one instant of time, or at some one point of its path, is called its *instantaneous velocity*. This concept requires careful definition.

Suppose we wish to find the instantaneous velocity of the particle in Fig. 3–1 at the point P. The average velocity between points P and Q is associated with the entire displacement Δx, and with the entire time interval Δt. Imagine the second point Q to be taken closer and closer to the first point P, and let the average velocity be computed over these shorter and shorter displacements and time intervals. The instantaneous velocity at the first point can then be defined as the *limiting value* of the average velocity when the second point is taken closer and closer to the first. Although the displacement then becomes extremely small, the time interval by which it is divided becomes small also, and the quotient therefore is not necessarily a small quantity.

In the notation of calculus, the limiting value of $\Delta x/\Delta t$ as Δt approaches zero is written dx/dt, and is called the *derivative* of x with respect to t. Thus instantaneous velocity is defined as

$$v = \lim_{\Delta t \to 0} \frac{\Delta x}{\Delta t} = \frac{dx}{dt}. \tag{3-6}$$

Since Δt is necessarily positive, it follows that v has the same algebraic sign as Δx. Hence a positive velocity indicates motion toward the right along the x-axis, if we use the usual convention of signs.

Strictly speaking, Eq. (3-6) defines the x-component of the instantaneous velocity, which in more general motion must be treated as a vector quantity. However, in one-dimensional motion along a straight line, it is customary to call v the instantaneous velocity. When the term velocity is used without a qualifying adjective, one nearly always means instantaneous rather than average velocity.

As point Q approaches point P in Fig. 3-1a, point q approaches point p in Fig. 3-1b. In the limit, the slope of the chord pq equals the slope of the tangent to the curve at point p, due allowance being made for the scales to which x and t are plotted. *The instantaneous velocity at any point of a coordinate-time graph therefore equals the slope of the tangent to the graph at that point.* If the tangent slopes upward to the right, its slope is positive, the velocity is positive, and the motion is toward the right. If the tangent slopes downward to the right, the velocity is negative. At a point where the tangent is horizontal, its slope is zero and the velocity is zero.

If we express distance in meters and time in seconds, velocity is expressed in meters per second ($m \cdot s^{-1}$). Other common units of velocity are feet per second ($ft \cdot s^{-1}$), centimeters per second ($cm \cdot s^{-1}$), miles per hour ($mi \cdot hr^{-1}$), and knots (1 knot = 1 nautical mile or 6080 ft per hour).

Example Suppose the motion of the particle in Fig. 3-1 is described by the equation $x = 20 \text{ cm} + (4 \text{ cm} \cdot s^{-2})t^2$.

a) Find the displacement of the particle in the time interval between $t_1 = 2$ s and $t_2 = 5$ s.

At time $t_1 = 2$ s, the position is

$$x_1 = 20 \text{ cm} + (4 \text{ cm} \cdot s^{-2})(2 \text{ s})^2 = 36 \text{ cm}.$$

At time $t_2 = 5$ s,

$$x_2 = 20 \text{ cm} + (4 \text{ cm} \cdot s^{-2})(5 \text{ s})^2 = 120 \text{ cm}.$$

The displacement is therefore

$$x_2 - x_1 = 120 \text{ cm} - 36 \text{ cm} = 84 \text{ cm}.$$

b) Find the average velocity in this time interval:

$$\bar{v} = \frac{x_2 - x_1}{t_2 - t_1} = \frac{84 \text{ cm}}{3 \text{ s}} = 28 \text{ cm} \cdot s^{-1}.$$

c) Find the instantaneous velocity at time $t_1 = 2$ s.

The position at time $t = 2\text{ s} + \Delta t$ is

$$x = 20 \text{ cm} + (4 \text{ cm}\cdot\text{s}^{-2})(2\text{ s} + \Delta t)^2$$
$$= 36 \text{ cm} + (16 \text{ cm}\cdot\text{s}^{-1})\,\Delta t + (4 \text{ cm}\cdot\text{s}^{-2})(\Delta t)^2.$$

The displacement during the interval Δt is

$$\Delta x = 36 \text{ cm} + (16 \text{ cm}\cdot\text{s}^{-1})\,\Delta t + (4 \text{ cm}\cdot\text{s}^{-2})(\Delta t)^2 - 36 \text{ cm}$$
$$= (16 \text{ cm}\cdot\text{s}^{-1})\,\Delta t + (4 \text{ cm}\cdot\text{s}^{-2})(\Delta t)^2.$$

The average velocity during Δt is

$$\bar{v} = \frac{\Delta x}{\Delta t} = 16 \text{ cm}\cdot\text{s}^{-1} + (4 \text{ cm}\cdot\text{s}^{-1})\,\Delta t.$$

For the instantaneous velocity at $t = 2$ s, we let Δt approach zero in the expression for $\bar{v}$: $v = 16 \text{ cm}\cdot\text{s}^{-1}$. This corresponds to the slope of the tangent at point p in Fig. 3–1b.

The instantaneous velocity may also be obtained directly by calculating the derivative of x with respect to t. Starting again with

$$x = 20 \text{ cm} + (4 \text{ cm}\cdot\text{s}^{-2})t^2,$$

and using Eq. (3–6) along with the familiar formula for the derivative with respect to t of t^n, that is,

$$\frac{d}{dt}t^n = nt^{n-1},$$

we find

$$v = \frac{dx}{dt} = \frac{d}{dt}[20 \text{ cm} + (4 \text{ cm}\cdot\text{s}^{-2})t^2] = 2(4 \text{ cm}\cdot\text{s}^{-2})t.$$

The instantaneous velocity at time $t_1 = 2$ s is

$$v = (2)(4 \text{ cm}\cdot\text{s}^{-2})(2 \text{ s}) = 16 \text{ cm}\cdot\text{s}^{-1}.$$

The term *speed* has two different meanings. It may mean the *magnitude* of the instantaneous velocity; two automobiles traveling at $50 \text{ mi}\cdot\text{hr}^{-1}$, one north and the other south, both have speeds of $50 \text{ mi}\cdot\text{hr}^{-1}$. In another sense, the *average speed* of a body means the total length of path covered, divided by the elapsed time. Thus, if an automobile travels 90 mi in 3 hr, its average speed is $30 \text{ mi}\cdot\text{hr}^{-1}$, even if the trip starts and ends at the same point. The average *velocity,* in the latter case, would be zero, since the total displacement is zero.

3-4 Average and instantaneous acceleration

When the velocity of a moving body changes with time, the body is said to have an *acceleration*. Just as velocity is a quantitative description of rate of change of position with time, so acceleration is a quantitative description of rate of change of velocity with time.

Considering again the motion of a particle along the x-axis, we suppose that at time t_1 the particle is at point P and has velocity v_1, and that at a later time t_2 it is at point Q and has the velocity v_2.

The *average acceleration* $\bar{a}$ of the particle as it moves from P to Q is defined as the ratio of the change in velocity to the elapsed time:

$$\bar{a} = \frac{v_2 - v_1}{t_2 - t_1} = \frac{\Delta v}{\Delta t}, \quad \text{average} \quad (3\text{-}7)$$

where t_1 and t_2 are the times corresponding to the velocities v_1 and v_2. Again, strictly speaking v_1 and v_2 are values of the x-component of instantaneous velocity, and Eq. (3-7) defines the x-component of average acceleration.

Example 1 If $t_1 = 2\,\text{s}$ and $t_2 = 4\,\text{s}$, find the average acceleration for each of the following cases:

a) $v_1 = 8\,\text{m}\cdot\text{s}^{-1}$, $v_2 = 12\,\text{m}\cdot\text{s}^{-1}$;
b) $v_1 = 16\,\text{m}\cdot\text{s}^{-1}$, $v_2 = 12\,\text{m}\cdot\text{s}^{-1}$;
c) $v_1 = -4\,\text{m}\cdot\text{s}^{-1}$, $v_2 = -10\,\text{m}\cdot\text{s}^{-1}$;
d) $v_1 = -16\,\text{m}\cdot\text{s}^{-1}$, $v_2 = -8\,\text{m}\cdot\text{s}^{-1}$.

Solution

a) $\bar{a} = \dfrac{12\,\text{m}\cdot\text{s}^{-1} - 8\,\text{m}\cdot\text{s}^{-1}}{4\,\text{s} - 2\,\text{s}} = 2\,\text{m}\cdot\text{s}^{-2}$.

b) $\bar{a} = \dfrac{12\,\text{m}\cdot\text{s}^{-1} - 16\,\text{m}\cdot\text{s}^{-1}}{4\,\text{s} - 2\,\text{s}} = -2\,\text{m}\cdot\text{s}^{-2}$.

c) $\bar{a} = \dfrac{-10\,\text{m}\cdot\text{s}^{-1} - (-4\,\text{m}\cdot\text{s}^{-1})}{4\,\text{s} - 2\,\text{s}} = -3\,\text{m}\cdot\text{s}^{-2}$.

d) $\bar{a} = \dfrac{-8\,\text{m}\cdot\text{s}^{-1} - (-16\,\text{m}\cdot\text{s}^{-1})}{4\,\text{s} - 2\,\text{s}} = +4\,\text{m}\cdot\text{s}^{-2}$.

We note in particular that, when the point moves in the negative direction with increasing speed (c), the acceleration is negative and that, when it moves in the negative direction with decreasing speed (d), the acceleration is positive.

Figure 3-2a shows a particle moving along the x-axis, with velocity v_1 at point P (at time t_1) and velocity v_2 at point Q at a later time t_2. Figure 3-2b is a graph of instantaneous velocity v plotted as a function of time, points p and q corresponding to positions P and Q in Fig. 3-2a. The average acceleration is represented by the slope of the chord pq, computed using the appropriate scales and units of the graph.

The *instantaneous acceleration* of a body, that is, its acceleration at some one instant of time or at some one point of its path, is defined in the same way as instantaneous velocity. Let the second point Q in Fig. 3-2a be taken closer and closer to the first point P, and let the average acceleration be computed over shorter and shorter intervals of time. The instantaneous acceleration at the first point is defined as the limiting value of the average acceleration when the second point is taken closer and closer to the first:

$$a = \lim_{\Delta t \to 0} \frac{\Delta v}{\Delta t} = \frac{dv}{dt}. \quad \text{instantaneous} \quad (3\text{-}8)$$

3-2 (a) Particle moving on the x-axis. (b) Velocity-time graph of the motion. The average acceleration between t_1 and t_2 equals the slope of the chord pq. The instantaneous acceleration at P equals the slope of the tangent at p.

Instantaneous acceleration plays an important part in the laws of mechanics; average acceleration is less frequently used. From now on when the term "acceleration" is used we shall understand it to mean instantaneous acceleration.

When a particle moves in a curved path, the *direction* of its velocity changes, and this change in direction also gives rise to an acceleration, as will be explained in Chapter 5.

As point Q approaches point P in Fig. 3–2a, point q approaches point p in Fig. 3–2b, and the slope of the chord pq approaches the slope of the tangent to the velocity–time graph at point p. *The instantaneous acceleration at any point of the graph therefore equals the slope of the tangent to the graph at that point.*

The acceleration $a = dv/dt$ can be expressed in various ways. Since $v = dx/dt$, it follows that

$$a = \frac{dv}{dt} = \frac{d}{dt}\left(\frac{dx}{dt}\right) = \frac{d^2x}{dt^2}. \tag{3-9}$$

The acceleration is therefore the *second* derivative of the coordinate with respect to time.

We can also use the chain rule and write

$$a = \frac{dv}{dt} = \frac{dv}{dx}\frac{dx}{dt} = v\frac{dv}{dx}, \tag{3-10}$$

which expresses the acceleration in terms of the *spatial* rate of change in velocity, dv/dx.

If we express velocity in meters per second and time in seconds, acceleration is expressed in meters per second, per second ($\text{m} \cdot \text{s}^{-1} \cdot \text{s}^{-1}$). This is usually written as $\text{m} \cdot \text{s}^{-2}$, and is read "meters per second squared." Other common units of acceleration are feet per second squared ($\text{ft} \cdot \text{s}^{-2}$) and centimeters per second squared ($\text{cm} \cdot \text{s}^{-2}$).

Example 2 Suppose the velocity of the particle in Fig. 3–2 is given by the equation

$$v = 10 \text{ cm} \cdot \text{s}^{-1} + (2 \text{ cm} \cdot \text{s}^{-3})t^2.$$

a) Find the change in velocity of the particle in the time interval between $t_1 = 2$ s and $t_2 = 5$ s.

At time $t_1 = 2$ s,

$$v_1 = 10 \text{ cm} \cdot \text{s}^{-1} + (2 \text{ cm} \cdot \text{s}^{-3})(2 \text{ s})^2 = 18 \text{ cm} \cdot \text{s}^{-1}.$$

At time $t_2 = 5$ s,

$$v_2 = 10 \text{ cm} \cdot \text{s}^{-1} + (2 \text{ cm} \cdot \text{s}^{-3})(5 \text{ s})^2 = 60 \text{ cm} \cdot \text{s}^{-1}.$$

The change in velocity is therefore

$$v_2 - v_1 = 60 \text{ cm} \cdot \text{s}^{-1} - 18 \text{ cm} \cdot \text{s}^{-1} = 42 \text{ cm} \cdot \text{s}^{-1}.$$

b) Find the average acceleration in this time interval.

$$\bar{a} = \frac{v_2 - v_1}{t_2 - t_1} = \frac{42 \text{ cm} \cdot \text{s}^{-1}}{3 \text{ s}} = 14 \text{ cm} \cdot \text{s}^{-2}.$$

This corresponds to the slope of the chord pq in Fig. 3–2b.

c) Find the instantaneous acceleration at time $t_1 = 2$ s.

At time $t = 2\text{ s} + \Delta t$,

$$v = 10\text{ cm}\cdot\text{s}^{-1} + (2\text{ cm}\cdot\text{s}^{-3})(2\text{ s} + \Delta t)^2$$
$$= 18\text{ cm}\cdot\text{s}^{-1} + (8\text{ cm}\cdot\text{s}^{-2})\Delta t + (2\text{ cm}\cdot\text{s}^{-3})(\Delta t)^2.$$

The change in velocity during Δt is

$$\Delta v = 18\text{ cm}\cdot\text{s}^{-1} + (8\text{ cm}\cdot\text{s}^{-2})\Delta t + (2\text{ cm}\cdot\text{s}^{-3})(\Delta t)^2 - 18\text{ cm}\cdot\text{s}^{-1}$$
$$= (8\text{ cm}\cdot\text{s}^{-2})\Delta t + (2\text{ cm}\cdot\text{s}^{-3})(\Delta t)^2.$$

The average acceleration during Δt is

$$\bar{a} = \frac{\Delta v}{\Delta t} = 8\text{ cm}\cdot\text{s}^{-2} + (2\text{ cm}\cdot\text{s}^{-3})\Delta t.$$

The instantaneous acceleration at $t = 2$ s, obtained by letting Δt approach zero, is $a = 8\text{ cm}\cdot\text{s}^{-2}$. This corresponds to the slope of the tangent at point p in Fig. 3–2b.

The instantaneous acceleration can also be obtained directly by taking the time derivative of the instantaneous velocity: Starting with $v = 10\text{ cm}\cdot\text{s}^{-1} + (2\text{ cm}\cdot\text{s}^{-3})t^2$ and using Eq. (3–8), we find

$$a = \frac{dv}{dt} = \frac{d}{dt}[10\text{ cm}\cdot\text{s}^{-1} + (2\text{ cm}\cdot\text{s}^{-3})t^2] = 2(2\text{ cm}\cdot\text{s}^{-3})t.$$

The instantaneous acceleration when $t = 2$ s is

$$a = (2)(2\text{ cm}\cdot\text{s}^{-3})(2\text{ s}) = 8\text{ cm}\cdot\text{s}^{-2}.$$

3–5 Motion with constant acceleration

The simplest kind of accelerated motion is straight-line motion in which the acceleration is constant, that is, in which the velocity changes at the same rate throughout the motion. The velocity–time graph is then a straight line, as in Fig. 3–3, the velocity increasing by equal amounts in equal intervals of time. The slope of a chord between any two points on the line is the same as the slope of a tangent at any point, and the average and instantaneous accelerations are equal. Hence in Eq. (3–7) the average acceleration $\bar{a}$ can be replaced by the constant acceleration a, and we have

$$a = \frac{v_2 - v_1}{t_2 - t_1}. \tag{3-11}$$

Now let $t_1 = 0$ and let t_2 be any arbitrary later time t. Let v_0 represent the velocity when $t = 0$ (called the *initial* velocity), and let v be the velocity at the later time t. Then the preceding equation becomes

$$a = \frac{v - v_0}{t - 0},$$

or

$$v = v_0 + at. \tag{3-12}$$

This equation can be interpreted as follows: The acceleration a is the constant rate of change of velocity, or the change of velocity per unit

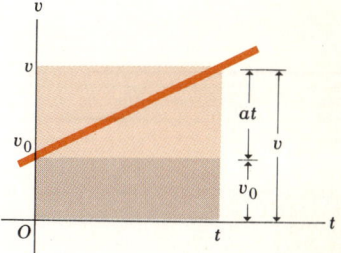

3–3 Velocity-time graph for rectilinear motion with constant acceleration.

time. The term at is the product of the change in velocity per unit time, a, and the duration of the time interval, t; therefore it equals the *total* change in velocity. The velocity v at the time t then equals the velocity v_0 at the time $t = 0$, plus the change in velocity at. Graphically, the ordinate v at time t, in Fig. 3-3, can be considered as the sum of two segments: one of length v_0 equal to the initial velocity, the other of length at equal to the change in velocity in time t.

An alternative route to Eq. (3-12) is as follows. We know that v is some function of t. What must that function be, in order for its derivative with respect to t to equal the constant a? (The function whose derivative is a given function is sometimes called the *antiderivative* of that function; the more usual term is *indefinite integral*.) One possibility is the function at; its derivative with respect to t is the constant a. But the derivative of any constant is zero, so the derivative of the function $at + C$ (where C is a constant) is also equal to a. Thus we conclude that the function for v has the form

$$v = at + C. \tag{3-13}$$

Now this function must also satisfy the additional requirement that at time $t = 0$ it yields the value v_0. Substituting $t = 0$ and equating the result to v_0, we find

$$v_0 = a(0) + C, \quad \text{or } C = v_0. \tag{3-14}$$

Putting all this together, we again obtain

$$v = v_0 + at.$$

A similar procedure may be used to find an expression for the position x of the particle as a function of time, if the position at time $t = 0$ is denoted by x_0. We know that $v = dx/dt$; what function of t must x be, in order for its derivative with respect to t to equal $v_0 + at$? That is, what is the antiderivative of $v_0 + at$? The function

$$x = v_0 t + \tfrac{1}{2}at^2 + D, \tag{3-15}$$

where D is any constant, satisfies this requirement; this may be checked easily by taking the derivative of Eq. (3-15).

We also know that at time $t = 0$, Eq. (3-15) must give the value x_0. Substituting the value $t = 0$ and equating the result to x_0, we find $x_0 = D$. Combining this with Eq. (3-15), we obtain

$$x = x_0 + v_0 t + \tfrac{1}{2}at^2. \tag{3-16}$$

We may also combine Eqs. (3-12) and (3-16) to obtain a relation for x, v, and a that does not contain t. We first solve Eq. (3-12) for t and then substitute the resulting expression into Eq. (3-16) and simplify:

$$t = \frac{v - v_0}{a},$$

$$x = x_0 + v_0\left(\frac{v - v_0}{a}\right) + \frac{1}{2}a\left(\frac{v - v_0}{a}\right)^2.$$

Transfer the term x_0 to the left side and multiply through by $2a$:

$$2a(x - x_0) = 2v_0 v - 2v_0^2 + v^2 - 2v_0 v + v_0^2,$$

or, finally,

$$v^2 = v_0^2 + 2a(x - x_0). \tag{3-17}$$

Equations (3–12), (3–16), and (3–17) are the *equations of motion with constant acceleration.*

The curve in Fig. 3–4 is the coordinate–time graph for motion with constant acceleration. That is, it is a graph of Eq. (3–16); the curve is a *parabola*. The slope of the tangent at $t = 0$ equals the initial velocity v_0, and the slope of the tangent at time t equals the velocity v at that time. It is evident that the slope continually increases, and measurements would show that the *rate* of increase is constant, that is, that the acceleration is constant.

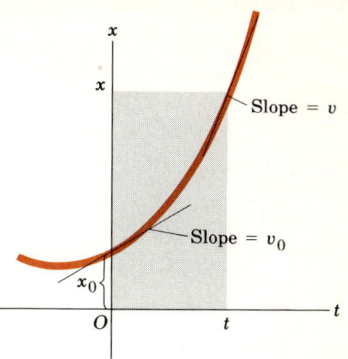

3–4 Coordinate-time graph for motion with constant acceleration.

Example A body moves along the x-axis with constant acceleration $a = 4 \text{ m} \cdot \text{s}^{-2}$. At time $t = 0$ it is at $x = 5 \text{ m}$ and has velocity $v = 3 \text{ m} \cdot \text{s}^{-1}$.

a) Find the position and velocity at time $t = 2 \text{ s}$.

b) Where is the body when its velocity is $5 \text{ m} \cdot \text{s}^{-1}$?

Solution With reference to Eqs. (3–12), (3–16), and (3–17), we have $x_0 = 5 \text{ m}$, $v_0 = 3 \text{ m} \cdot \text{s}^{-1}$, $a = 4 \text{ m} \cdot \text{s}^{-2}$.

a) From Eq. (3–16),

$$x = x_0 + v_0 t + \tfrac{1}{2}at^2$$
$$= 5 \text{ m} + (3 \text{ m} \cdot \text{s}^{-1})(2 \text{ s}) + \tfrac{1}{2}(4 \text{ m} \cdot \text{s}^{-2})(2 \text{ s})^2$$
$$= 19 \text{ m}.$$

From Eq. (3–12),

$$v = v_0 + at$$
$$= 3 \text{ m} \cdot \text{s}^{-1} + (4 \text{ m} \cdot \text{s}^{-2})(2 \text{ s})$$
$$= 11 \text{ m} \cdot \text{s}^{-1}$$

b) From Eq. (13–17),

$$v^2 = v_0^2 + 2a(x - x_0),$$
$$(5 \text{ m} \cdot \text{s}^{-1})^2 = (3 \text{ m} \cdot \text{s}^{-1})^2 + 2(4 \text{ m} \cdot \text{s}^{-2})(x - 5 \text{ m}),$$
$$x = 7 \text{ m}.$$

Alternatively, we may use Eq. (3–12) to find first the *time* when $v = 5 \text{ m} \cdot \text{s}^{-1}$:

$$5 \text{ m} \cdot \text{s}^{-1} = 3 \text{ m} \cdot \text{s}^{-1} + (4 \text{ m} \cdot \text{s}^{-2})(t),$$
$$t = \tfrac{1}{2} \text{ s}.$$

Then from Eq. (3–16),

$$x = 5 \text{ m} + (3 \text{ m} \cdot \text{s}^{-1})(\tfrac{1}{2} \text{ s}) + \tfrac{1}{2}(4 \text{ m} \cdot \text{s}^{-2})(\tfrac{1}{2} \text{ s})^2$$
$$= 7 \text{ m}.$$

A special case of motion with constant acceleration is that in which the acceleration is zero. The *velocity* is then constant and the equations of motion become simply

$$v = \text{constant},$$
$$x = x_0 + vt.$$

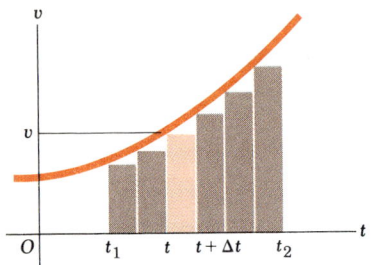

3-5 The area under a velocity-time graph equals the displacement.

*3-6 Velocity and coordinate by integration

When the acceleration is not constant, the preceding equations are not applicable. Even so, some of the same graphical considerations are useful; velocity is still the slope of the displacement-versus-time graph, and acceleration the slope of the velocity-versus-time graph. Figure 3-5 shows a motion in which the acceleration is not constant. We divide the total time into a large number of intervals, calling a typical one Δt. Let the velocity during that interval be v. Of course the velocity changes during Δt, but if the interval is very small, the change will also be very small. The displacement during that interval, neglecting the variation of v, is given by

$$\Delta x = v\, \Delta t.$$

This corresponds graphically to the area of the shaded strip with height v and width Δt, that is, the *area* under the curve corresponding to the interval Δt. Since the total displacement in any interval (say, t_1 to t_2) is the sum of the displacements in the small subintervals, this is given graphically by the *total* area under the curve between the vertical lines t_1 and t_2. In the limit, when all the Δt's become very small and their number very large, this is simply the *integral* of v (which is in general a function of t) from t_1 to t_2. Thus if x_1 is the position at time t_1 and x_2 that at t_2,

$$x_2 - x_1 = \int_{x_1}^{x_2} dx = \int_{t_1}^{t_2} v\, dt. \tag{3-18}$$

An exactly similar analysis with the acceleration-versus-time curve, where a is in general a function of t, shows that if v_1 is the velocity at time t_1 and v_2 the velocity at time t_2, the change in velocity Δv during a small time interval Δt is approximately equal to $a\, \Delta t$, and the total change in velocity $(v_2 - v_1)$ during the interval $t_2 - t_1$ is given by

$$v_2 - v_1 = \int_{v_1}^{v_2} dv = \int_{t_1}^{t_2} a\, dt. \tag{3-19}$$

Example An automobile travels along a straight highway. Its acceleration is given as a function of time by

$$a = 2.0\text{ m}\cdot\text{s}^{-2} - (0.1\text{ m}\cdot\text{s}^{-3})t.$$

The position at time $t = 0$ is given by $x_0 = 0$, and the velocity at time $t = 0$ is $v_0 = 10\text{ m}\cdot\text{s}^{-1}$.

a) Derive expressions for the velocity and position as functions of time.

b) At what time is the velocity of the car greatest?

c) What is the maximum velocity?

Solution a) We use the indefinite integral concept. First,

$$v = \int a\, dt = (2.0\text{ m}\cdot\text{s}^{-2})t - \tfrac{1}{2}(0.1\text{ m}\cdot\text{s}^{-3})t^2 + C.$$

But $v_0 = 10\text{ m}\cdot\text{s}^{-1}$; substituting this, with $t = 0$, we find $C = 10\text{ m}\cdot\text{s}^{-1}$. Thus

$$v = (2.0\text{ m}\cdot\text{s}^{-2})t - \tfrac{1}{2}(0.1\text{ m}\cdot\text{s}^{-3})t^2 + 10\text{ m}\cdot\text{s}^{-1}.$$

Next,

$$x = \int v\, dt = \tfrac{1}{2}(2.0\ \text{m}\cdot\text{s}^{-2})t^2 - \tfrac{1}{6}(0.1\ \text{m}\cdot\text{s}^{-3})t^3 + (10\ \text{m}\cdot\text{s}^{-1})t + D.$$

But $x_0 = 0$; substituting this, with $t = 0$, we find $D = 0$. Thus

$$x = \tfrac{1}{2}(2.0\ \text{m}\cdot\text{s}^{-2})t^2 - \tfrac{1}{6}(0.1\ \text{m}\cdot\text{s}^{-3})t^3 + (10\ \text{m}\cdot\text{s}^{-1})t.$$

b) The maximum value of v occurs at the time when $dv/dt = 0$. Taking the derivative of the above expression for v and setting it equal to zero, we obtain

$$0 = (2.0\ \text{m}\cdot\text{s}^{-2}) - (0.1\ \text{m}\cdot\text{s}^{-3})t,$$

$$t = \frac{2.0\ \text{m}\cdot\text{s}^{-2}}{0.1\ \text{m}\cdot\text{s}^{-3}} = 20\ \text{s}.$$

This result can also be obtained immediately from the expression for a by noting that a is positive between $t = 0$ and $t = 20$ s, and negative after that. It is zero at $t = 20$ s, corresponding to the fact that $dv/dt = 0$ at this time. Thus the automobile speeds up until $t = 20$ s and then begins to slow down again.

c) The maximum velocity is obtained by substituting $t = 20$ s (the time when the maximum velocity is attained) into the general velocity equation:

$$v_{\text{max}} = (2.0\ \text{m}\cdot\text{s}^{-2})(20\ \text{s}) - \tfrac{1}{2}(0.1\ \text{m}\cdot\text{s}^{-3})(20\ \text{s})^2 + 10\ \text{m}\cdot\text{s}^{-1}$$
$$= 40\ \text{m}\cdot\text{s}^{-1}.$$

As shown in Sec. 3-5, the concept of the indefinite integral can be used for a simple derivation of Eq. (3-16). If $a =$ constant, then

$$\frac{dv}{dt} = a = \text{constant}, \qquad v = \int dv = \int a\, dt = at + C. \qquad (3\text{-}20)$$

Also,

$$v = \frac{dx}{dt} = at + C, \qquad x = \int v\, dt = \tfrac{1}{2}at^2 + Ct + D, \qquad (3\text{-}21)$$

where C and D are integration constants. By insisting that the expression for v must yield the value v_0 and that for x the value x_0 at time $t = 0$, we find that $C = v_0$ and $D = x_0$.

Finally, the graphical considerations leading to Eq. (3-18) can be used for still another derivation of Eq. (3-16). We let $t_1 = 0$ and let t_2 be any general later time t; the corresponding values of x are x_0 and x. The graph of v as a function of t is given in Fig. 3-6. The total displacement $(x - x_0)$ from $t = 0$ to a later time t is given by the total area under the line, which can be obtained by computing the rectangular and triangular areas and adding. Carrying out this program, we find:

$$x - x_0 = v_0 t + \tfrac{1}{2}(t)(at) = v_0 t + \tfrac{1}{2}at^2,$$

in agreement with Eq. (3-16).

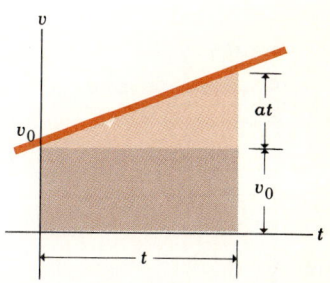

3-6 The area under a velocity-time graph equals the displacement $x - x_0$.

3-7 Freely falling bodies

The most common example of motion with (nearly) constant acceleration is that of a body falling toward the earth. The motion of a falling body has attracted the attention of philosophers and scientists since ancient times. Aristotle asserted in his writings that heavy objects fall faster than light objects, in proportion to their weight. Galileo argued that the motion of a falling body should be nearly independent of its weight. According to legend, he tested his reasoning experimentally by dropping cannonballs and bullets from the Leaning Tower of Pisa, although there is no reference to such experiments in his own writing.

In any event, the motion of falling of bodies has been studied with great care and precision by scientists in more recent times. In the absence of air resistance it is found that all bodies, regardless of their size or weight, fall with the same acceleration at the same point on the earth's surface; and if the distance covered is small compared to the radius of the earth, the acceleration remains constant throughout the fall. The effect of air resistance and the decrease in acceleration with altitude will be neglected. This idealized motion is spoken of as "free fall," although the term includes rising as well as falling motion.

The acceleration of a freely falling body is called *acceleration due to gravity,* or the *acceleration of gravity,* and is denoted by the letter g. At or near the earth's surface its magnitude is approximately $32 \text{ ft} \cdot \text{s}^{-2}$, $9.8 \text{ m} \cdot \text{s}^{-2}$, or $980 \text{ cm} \cdot \text{s}^{-2}$. More precise values, and small variations with latitude and elevation, will be considered later. On the surface of the moon, the acceleration of gravity is due to the attractive force exerted on a body by the moon rather than earth. On the moon, $g = 1.67 \text{ m} \cdot \text{s}^{-2} = 5.47 \text{ ft} \cdot \text{s}^{-2}$. Near the surface of the *sun*, $g = 274 \text{ m} \cdot \text{s}^{-2}$!

Example 1 A body is released from rest and falls freely. Compute its position and velocity after 1, 2, 3, and 4 s. Take the origin O at the elevation of the starting point, the y-axis vertical, and the upward direction as positive.

Solution The initial coordinate y_0 and the initial velocity v_0 are both *zero*. The acceleration is downward, in the negative y-direction, so $a = -g = -32 \text{ ft} \cdot \text{s}^{-2}$.

From Eqs. (3-16) and (3-12),
$$y = v_0 t + \tfrac{1}{2} a t^2 = 0 - \tfrac{1}{2} g t^2 = (-16 \text{ ft} \cdot \text{s}^{-2}) t^2,$$
$$v = v_0 + at = 0 - gt = (-32 \text{ ft} \cdot \text{s}^{-2}) t.$$

When $t = 1 \text{ s}$,
$$y = (-16 \text{ ft} \cdot \text{s}^{-2})(1 \text{ s})^2 = -16 \text{ ft},$$
and
$$v = (-32 \text{ ft} \cdot \text{s}^{-2})(1 \text{ s}) = -32 \text{ ft} \cdot \text{s}^{-1}.$$

The body is therefore 16 ft below the origin (y is negative) and has a downward velocity (v is negative) of magnitude $32 \text{ ft} \cdot \text{s}^{-1}$.

The position and velocity at 2, 3, and 4 s are found in the same way. The results are illustrated in Fig. 3-7.

Example 2 A ball is thrown (nearly) vertically upward from the cornice of a tall building, leaving the thrower's hand with a speed of

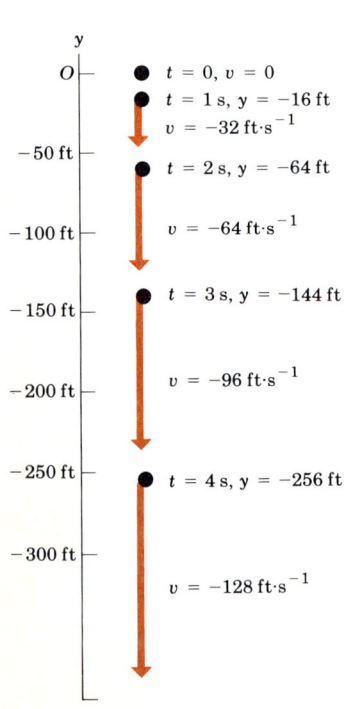

3-7 Position and velocity of a freely falling body.

48 ft·s⁻¹ and just missing the cornice on the way down, as in Fig. 3-8. The downward path is shown displaced to the right for clarity. Find (a) the position and velocity of the ball, 1 s and 4 s after leaving the thrower's hand; (b) the velocity when the ball is 20 ft above its starting point; (c) the maximum height reached and the time at which it is reached. Take the origin at the elevation at which the ball leaves the thrower's hand, the y-axis vertical and positive upward.

Solution The initial position y_0 is zero. The initial velocity v_0 is $+48\ \text{ft}\cdot\text{s}^{-1}$, and the acceleration is $-32\ \text{ft}\cdot\text{s}^{-2}$. The velocity at any time is

$$v = v_0 + at = 48\ \text{ft}\cdot\text{s}^{-1} + (-32\ \text{ft}\cdot\text{s}^{-2})t. \qquad (3\text{-}22)$$

The coordinate at any time is

$$y = v_0 t + \tfrac{1}{2}at^2 = (48\ \text{ft}\cdot\text{s}^{-1})t + \tfrac{1}{2}(-32\ \text{ft}\cdot\text{s}^{-2})t^2. \qquad (3\text{-}23)$$

The velocity at any coordinate is

$$v^2 = v_0^2 + 2ay = (48\ \text{ft}\cdot\text{s}^{-1})^2 + 2(-32\ \text{ft}\cdot\text{s}^{-2})y. \qquad (3\text{-}24)$$

a) When $t = 1$ s, Eqs. (3-22) and (3-23) give

$$y = +32\ \text{ft}, \qquad v = +16\ \text{ft}\cdot\text{s}^{-1}.$$

3-8 Position and velocity of a body thrown vertically upward.

The ball is 32 ft above the origin (y is positive) and it has an upward velocity (v is positive) of 16 ft·s⁻¹ (less than the initial velocity, as expected).

When $t = 4$ s, again from Eqs. (3-22) and (3-23),

$$y = -64\ \text{ft}, \qquad v = -80\ \text{ft}\cdot\text{s}^{-1}.$$

The ball has passed its highest point and is 64 ft *below* the origin (y is negative). It has a *downward* velocity (v is negative) of magnitude 80 ft·s⁻¹. Note that it is not necessary to find the highest point reached, or the time at which it was reached. The equations of motion give the position and velocity at *any* time, whether the ball is on the way up or the way down.

b) When the ball is 20 ft above the origin

$$y = +20\ \text{ft}$$

and, from Eq. (3-24),

$$v^2 = 1024\ \text{ft}^2\cdot\text{s}^{-2}, \qquad v = \pm 32\ \text{ft}\cdot\text{s}^{-1}.$$

The ball passes this point *twice*, once on the way up and again on the way down. The velocity on the way up is $+32\ \text{ft}\cdot\text{s}^{-1}$, and on the way down it is $-32\ \text{ft}\cdot\text{s}^{-1}$.

c) At the highest point, $v = 0$. Hence, from Eq. (3-24),

$$0 = (48\ \text{ft}\cdot\text{s}^{-1})^2 - (64\ \text{ft}\cdot\text{s}^{-2})y,$$

and

$$y = +36\ \text{ft}.$$

The time can now be found either from Eq. (3-22), setting $v = 0$, or from Eq. (3-23), setting $y = 36$ ft. From either equation, we get

$$t = 1.5\ \text{s}.$$

Although at the highest point the velocity is instantaneously zero, the *acceleration* at this point is still -32 ft·s^{-2}. The ball stops for an instant, but its velocity is continuously changing; the acceleration is constant throughout.

Figure 3-9 is a "multiflash" photograph of a freely falling golf ball. This photograph was taken with aid of a stroboscopic light source, developed originally by Dr. Harold E. Edgerton of the Massachusetts Institute of Technology. This source produces a series of intense flashes of light. The interval between successive flashes is controllable at will, and the duration of each flash is so short (a few millionths of a second) that there is no blur in the image of even a rapidly moving body. The camera shutter is left open during the entire motion, and as each flash occurs, the position of the ball at that instant is recorded on the photographic film.

The equally spaced light flashes subdivide the motion into equal time intervals Δt. Since the time intervals are all equal, the average velocity of the ball between any two flashes is directly proportional to the separation of its corresponding images in the photograph. If the velocity were constant, the images would be equally spaced. The increasing separation of the images during the fall shows that the velocity is continually increasing or the motion is accelerated. By comparing two successive displacements of the ball, we can find the *change* in velocity in the corresponding time interval. Careful measurement, preferably on an enlarged print, shows that this change in velocity is the same in each time interval. In other words, the motion is one of *constant* acceleration.

3-8 Relative velocity

The velocity of a body, like its position, must be described with reference to the origin of a coordinate system. Ordinarily this origin is taken to be fixed in some other body, but this second body may be in motion relative to a third, and so on. Thus when we speak of "the velocity of an automobile," we usually mean its velocity relative to the earth. But the earth is in motion relative to the sun, the sun is in motion relative to some other star, and so on.

Suppose a long train of flatcars is moving to the right along a straight level track, as in Fig. 3-10, and that a daring automobile driver is driving to the right along the flatcars. In Fig. 3-10, v_{FE} represents the velocity of the flatcars F relative to the earth E, and v_{AF} the velocity of the automobile A relative to the flatcars F. The velocity of the automobile relative to the earth, v_{AE}, is evidently equal to the *sum* of the relative velocities v_{AF} and v_{FE}:

$$v_{AE} = v_{AF} + v_{FE}. \qquad (3\text{-}25)$$

3-9 Multiflash photograph (retouched) of freely falling golf ball.

3-10 Vector v_{FE} is the velocity of the flatcars F relative to the earth E, and v_{AF} is the velocity of automobile A relative to the flatcars.

Thus if the flatcars are traveling relative to the earth at $15 \text{ m} \cdot \text{s}^{-1}$ ($= v_{\text{FE}}$) and the automobile is traveling relative to the flatcars at $20 \text{ m} \cdot \text{s}^{-1}$ ($= v_{\text{AF}}$), the velocity of the automobile relative to the earth (v_{AE}) is $35 \text{ m} \cdot \text{s}^{-1}$.

We note that in straight-line motion the velocity v_{AE} is the *algebraic* sum of v_{AF} and v_{FE}. Thus if the automobile were traveling to the *left* with a velocity of $20 \text{ m} \cdot \text{s}^{-1}$ relative to the flatcars, $v_{\text{AF}} = -20 \text{ m} \cdot \text{s}^{-1}$ and the velocity of the automobile relative to the earth would be $-5 \text{ m} \cdot \text{s}^{-1}$; that is, it would be traveling to the left, relative to the earth.

Example 1 An automobile driver A, traveling relative to the earth at $65 \text{ km} \cdot \text{hr}^{-1}$ on a straight level road, is ahead of motorcycle officer B traveling in the same direction at $80 \text{ km} \cdot \text{hr}^{-1}$. What is the velocity of B relative to A?

Solution We have given

$$v_{\text{AE}} = 65 \text{ km} \cdot \text{hr}^{-1}, \quad v_{\text{BE}} = 80 \text{ km} \cdot \text{hr}^{-1},$$

and we wish to find v_{BA}.

From the rule for combining velocities, we get:

$$v_{\text{BA}} = v_{\text{BE}} + v_{\text{EA}}.$$

But

$$v_{\text{EA}} = -v_{\text{AE}},$$

so

$$\begin{aligned} v_{\text{BA}} &= v_{\text{BE}} - v_{\text{AE}} \\ &= 80 \text{ km} \cdot \text{hr}^{-1} - 65 \text{ km} \cdot \text{hr}^{-1} \\ &= 15 \text{ km} \cdot \text{hr}^{-1}, \end{aligned}$$

and the officer is overtaking the driver at $15 \text{ km} \cdot \text{hr}^{-1}$.

Example 2 How would the relative velocity be altered if B were ahead of A?

Solution Not at all. The relative *positions* of the bodies do not matter. The velocity of B relative to A is still $+15 \text{ mi} \cdot \text{hr}^{-1}$, but he is now pulling ahead of A at this rate.

In more general motion, where the velocities do not all lie along a straight line, *vector* addition must be used to combine velocities. We shall return to problems of this sort at the end of Chapter 5.

Questions

3-1 When a particle moves in a circular path, how many coordinates are required to describe its position, assuming that the radius of the circle is given?

3-2 What is meant by the statement that space is three-dimensional?

3-3 Does the speedometer of a car measure speed or velocity?

3-4 Some European countries have highway speed limits of $100 \text{ km} \cdot \text{hr}^{-1}$. What is the equivalent number of miles per hour?

3-5 A student claims that a speed of $60 \text{ mi} \cdot \text{hr}^{-1}$ is the same as $88 \text{ ft} \cdot \text{s}^{-1}$. Is this relationship exact or only approximate?

3-6 In a given time interval, is the total displacement of a particle equal to the product of the average velocity and the time interval, even when the velocity is not constant?

3-7 Under what conditions is average velocity equal to instantaneous velocity? (constant velocity)

3-8 When one flies in an airplane at night in smooth air, there is no sensation of motion even though the plane may be moving at 500 mi·hr⁻¹. Why is this?

3-9 An automobile is traveling north. Can it have a velocity toward the north and at the same time have an acceleration toward the south?

3-10 A ball is thrown straight up in the air. What is its acceleration at the instant it reaches its highest point?

3-11 Is the acceleration of a car greater when the accelerator is pushed to the floor or when the brake pedal is pushed hard?

3-12 Under constant acceleration, the average velocity of a particle is half the sum of its initial and final velocities. Is this still true if the acceleration is *not* constant?

3-13 A baseball is thrown straight up in the air. Is the acceleration greater while it is being thrown, or after it is thrown?

3-14 How could you measure the acceleration of an automobile using only instruments located within the automobile?

3-15 If the initial position and initial velocity of a vehicle are known, and a record is kept of the acceleration at each instant, can its position after a certain time be computed from this data? Explain how this might be done.

Problems

3-1 Suppose that a runner on a straight track covers a distance of 1 mile in exactly 4 minutes. What was his average velocity in

  a) mi·hr⁻¹?   b) ft·s⁻¹?   c) cm·s⁻¹?

3-2 A body moves along a straight line; its distance from the origin at any instant is given by the equation $x = (8 \text{ cm} \cdot \text{s}^{-1})t - (3 \text{ cm} \cdot \text{s}^{-2})t^2$, where x is in centimeters and t is in seconds. Find the average velocity of the body in the interval from $t = 0$ to $t = 1$ s, and in the interval from $t = 0$ to $t = 4$ s.

3-3 The motion of a certain body along the x-axis is described by the equation $x = (10 \text{ cm} \cdot \text{s}^{-2})t^2$. Compute the instantaneous velocity of the body at time $t = 3$ s. Let Δt first equal 0.1 s, then 0.01 s, and finally 0.001 s. What limiting value do the results seem to be approaching?

3-4 An automobile is provided with a speedometer calibrated to read m·s⁻¹ rather than mi·hr⁻¹. The following series of speedometer readings was obtained during a start.

Time (s)	0	2	4	6	8	10	12	14	16
Velocity (m·s⁻¹)	0	0	2	5	10	15	20	22	22

  a) Compute the average acceleration during each 2-s interval. Is the acceleration constant? Is it constant during any part of the time?

  b) Make a velocity-time graph of the data above, using scales of 1 cm = 1 s horizontally, and 1 cm = 2 m·s⁻¹ vertically. Draw a smooth curve through the plotted points. What distance is represented by 1 cm²? What is the displacement in the first 8 s? What is the acceleration when $t = 8$ s? When $t = 13$ s? When $t = 15$ s?

3-5 The graph in Fig. 3-11 shows the velocity of a body plotted as a function of time.

  a) Find the instantaneous acceleration at $t = 3$ s, at $t = 7$ s, and at $t = 11$ s.

  b) How far does the body go in the first 5 s? The first 9 s? The first 13 s?

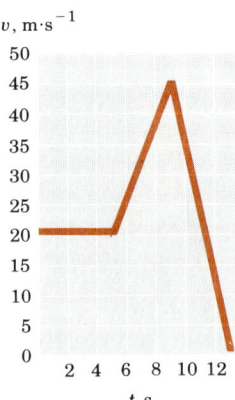

Figure 3-11

3-6 Figure 3-12 is a graph of the acceleration of a body moving on the x-axis. Sketch the graphs of its velocity and coordinate as functions of time, if $x = v = 0$ when $t = 0$.

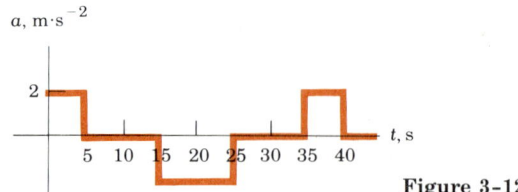

Figure 3-12

3-7 Figure 3-13 is a graph of the coordinate of a body moving on the x-axis. Sketch the graphs of its velocity and acceleration as functions of time.

Figure 3-13

3-8 Each of the following changes in velocity takes place in a 10-s interval. What are the magnitude, the algebraic sign, and the direction of the average acceleration in each interval?

a) At the beginning of the interval a body is moving toward the right along the x-axis at $5 \text{ m} \cdot \text{s}^{-1}$, and at the end of the interval it is moving toward the right at $20 \text{ m} \cdot \text{s}^{-1}$.

b) At the beginning it is moving toward the right at $20 \text{ m} \cdot \text{s}^{-1}$, and at the end it is moving toward the right at $5 \text{ m} \cdot \text{s}^{-1}$.

c) At the beginning it is moving toward the left at $5 \text{ m} \cdot \text{s}^{-1}$, and at the end it is moving toward the left at $20 \text{ m} \cdot \text{s}^{-1}$.

d) At the beginning it is moving toward the left at $20 \text{ m} \cdot \text{s}^{-1}$, and at the end it is moving toward the left at $5 \text{ m} \cdot \text{s}^{-1}$.

e) At the beginning it is moving toward the right at $20 \text{ m} \cdot \text{s}^{-1}$, and at the end it is moving toward the left at $20 \text{ m} \cdot \text{s}^{-1}$.

f) At the beginning it is moving toward the left at $20 \text{ m} \cdot \text{s}^{-1}$, and at the end it is moving toward the right at $20 \text{ m} \cdot \text{s}^{-1}$.

g) In which of these cases is the particle speeding up? In which cases is it slowing down?

3-9 The makers of a certain automobile advertise that it will accelerate from 15 to 50 $\text{mi} \cdot \text{hr}^{-1}$ in 13 s. Compute

a) the acceleration in $\text{ft} \cdot \text{s}^{-2}$, and

b) the distance the car travels in this time, assuming the acceleration to be constant.

3-10 An airplane taking off from a landing field has a run of 500 m. If it starts from rest, moves with constant acceleration, and makes the run in 30 s, with what velocity in $\text{m} \cdot \text{s}^{-1}$ did it take off?

3-11 An automobile starts from rest and acquires a velocity of $40 \text{ km} \cdot \text{hr}^{-1}$ in 15 s.

a) Compute the acceleration in kilometers per hour per second, assuming it to be constant.

b) If the automobile continues to gain velocity at the same rate, how many more seconds are needed for it to acquire a velocity of $60 \text{ km} \cdot \text{hr}^{-1}$?

c) Find the distances covered by the automobile in parts (a) and (b).

3-12 A body moving with constant acceleration covers the distance between two points 60 m apart in 6 s. Its velocity as it passes the second point is $15 \text{ m} \cdot \text{s}^{-1}$.

a) What is the acceleration?

b) What is its velocity at the first point?

3-13 A ball is released from rest and rolls down an inclined plane, requiring 4 s to cover a distance of 100 cm.

a) What is its acceleration in $\text{cm} \cdot \text{s}^{-2}$?

b) How many centimeters would it have fallen vertically in the same time?

3-14 The "reaction time" of the average automobile driver is about 0.7 s. (The reaction time is the interval between the perception of a signal to stop and the application of brakes.) If an automobile can accelerate at $-5 \text{ m} \cdot \text{s}^{-2}$, compute the total distance covered in coming to a stop after a signal is observed

a) from an initial velocity of $15 \text{ m} \cdot \text{s}^{-1}$,

b) from an initial velocity of $30 \text{ m} \cdot \text{s}^{-1}$.

3-15 At the instant the traffic lights turn green, an automobile that has been waiting at an intersection starts ahead with a constant acceleration of $2 \text{ m} \cdot \text{s}^{-2}$. At the same instant a truck, traveling with a constant velocity of $10 \text{ m} \cdot \text{s}^{-1}$, overtakes and passes the automobile.

a) How far beyond its starting point will the automobile overtake the truck?

b) How fast will it be traveling?

3-16 The engineer of a passenger train traveling at $30 \text{ m} \cdot \text{s}^{-1}$ sights a freight train whose caboose is 200 m ahead on the same track. The freight train is traveling in the same direction as the passenger train with a velocity of $10 \text{ m} \cdot \text{s}^{-1}$. The engineer of the passenger train immediately applies the brakes, causing a constant acceleration of $-1 \text{ m} \cdot \text{s}^{-2}$, while the freight train continues with constant speed.

a) Will there be a collision?

b) If so, where will it take place?

3-17 A sled starts from rest at the top of a hill and slides down with a constant acceleration. The sled is 140 ft from the top of the hill 2 s after passing a point which is 92 ft from the top. Four seconds after passing the 92-ft point it is 198 ft from the top, and 6 s after passing the point it is 266 ft from the top.

a) What is the average velocity of the sled during each of the 2-s intervals after passing the 92-ft point?

b) What is the acceleration of the sled?

c) What was the velocity of the sled when it passed the 92-ft point?

d) How long did it take to go from the top to the 92-ft point?

e) How far did the sled go during the first second after passing the 92-ft point?

f) How long does it take the sled to go from the 92-ft point to the midpoint between the 92-ft and the 140-ft mark?

g) What is the velocity of the sled as it passes the midpoint in part (f)?

3-18 A subway train starts from rest at a station and accelerates at a rate of $2 \text{ m} \cdot \text{s}^{-2}$ for 10 s. It then runs at constant speed for 30 s, and slows down at $-4 \text{ m} \cdot \text{s}^{-2}$ until it stops at the next station. Find the *total* distance covered.

3-19 A body starts from rest, moves in a straight line with constant acceleration, and covers a distance of 64 m in 4 s.

a) What is the final velocity?

b) How much time was required to cover half the total distance?

c) What is the distance covered in one-half the total time?

d) What is the velocity when half the total distance has been covered?

e) What is the velocity after one-half the total time?

3-20 The speed of an automobile going north is reduced from 30 to $20 \text{ m} \cdot \text{s}^{-1}$ in a distance of 125 m. Find

a) the magnitude and direction of the acceleration, assuming it to be constant,

b) the elapsed time, and

c) the distance in which the car can be brought to rest from $20 \text{ m} \cdot \text{s}^{-1}$, assuming the acceleration of part (a).

3-21 An automobile and a truck start from rest at the same instant, with the automobile initially at some distance behind the truck. The truck has a constant acceleration of $2 \text{ m} \cdot \text{s}^{-2}$ and the automobile an acceleration of $3 \text{ m} \cdot \text{s}^{-2}$. The automobile overtakes the truck after the truck has moved 75 m.

a) How long does it take the auto to overtake the truck?

b) How far was the auto behind the truck initially?

c) What is the velocity of each when they are abreast?

3-22

a) With what velocity must a ball be thrown vertically upward in order to rise to a height of 20 m?

b) How long will it be in the air?

3-23 A ball is thrown vertically downward from the top of a building, leaving the thrower's hand with a velocity of $10 \text{ m} \cdot \text{s}^{-1}$.

a) What will be its velocity after falling for 2 s?

b) How far will it fall in 2 s?

c) What will be its velocity after falling 10 m?

d) If it moved a distance of 1 m while in the thrower's hand, find its acceleration while in his hand (assuming it is constant).

e) If the ball was released at a point 40 m above the ground, in how many seconds will it strike the ground?

f) What will the velocity of the ball be when it strikes the ground?

3-24 A balloon, rising vertically with a velocity of $5 \text{ m} \cdot \text{s}^{-1}$, releases a sandbag at an instant when the balloon is 20 m above the ground.

a) Compute the position and velocity of the sandbag at the following times after its release: $\frac{1}{4}$ s, $\frac{1}{2}$ s, 1 s, 2 s.

b) How many seconds after its release will the bag strike the ground?

c) With what velocity will it strike?

3-25 A stone is dropped from the top of a tall cliff, and 1 s later a second stone is thrown vertically downward with a velocity of $20 \text{ m} \cdot \text{s}^{-1}$. How far below the top of the cliff will the second stone overtake the first?

3-26 A ball dropped from the cornice of a building takes 0.25 s to pass a window 3 m high. How far is the top of the window below the cornice?

3-27 A ball is thrown nearly vertically upward from a point near the cornice of a tall building. It just misses the cornice on the way down, and passes a point 160 ft below its starting point 5 s after it leaves the thrower's hand.

a) What was the initial velocity of the ball?

b) How high did it rise above its starting point?

c) What were the magnitude and direction of its velocity at the highest point?

d) What were the magnitude and direction of its acceleration at the highest point?

e) What was the magnitude of its velocity as it passed a point 64 ft below the starting point?

3-28 A juggler performs in a room whose ceiling is 3 m above the level of his hands. He throws a ball vertically upward so that it just reaches the ceiling.

a) With what initial velocity does he throw the ball?

b) What time is required for the ball to reach the ceiling?

He throws a second ball upward with the same initial velocity, at the instant that the first ball is at the ceiling.

c) How long after the second ball is thrown do the two balls pass each other?

d) When the balls pass each other, how far are they above the juggler's hands?

3-29 An object is thrown vertically upward. It has a speed of $10 \text{ m} \cdot \text{s}^{-1}$ when it has reached one-half its maximum height.

a) How high does it rise?

b) What are its velocity and acceleration 1 s after it is thrown?

c) 3 s after?

d) What is the average velocity during the first half second?

3–30 A student determined to test the law of gravity for himself walks off a skyscraper 300 m high, stopwatch in hand, and starts his free fall (zero initial velocity). Five seconds later, Superman arrives at the scene and dives off the roof to save the student.

a) What must Superman's initial velocity be in order that he catch the student just before the ground is reached?

b) What must be the height of the skyscraper so that even Superman can't save him? (Assume that Superman's acceleration is that of any free falling body.)

3–31 A ball is thrown vertically upward from the ground and a student gazing out of the window sees it moving upward past him at 5 m·s^{-1}. The window is 10 m above the ground.

a) How high does the ball go above the ground?

b) How long does it take to go from a height of 10 m to its highest point?

c) Find its velocity and acceleration $\frac{1}{2}$ s after it left the ground.

3–32 A ball is thrown vertically upward from the ground with a velocity of 30 m·s^{-1}.

a) How long will it take to rise to its highest point?

b) How high does the ball rise?

c) How long after projection will the ball have a velocity of 10 m·s^{-1} upward?

d) Of 10 m·s^{-1} downward?

e) When is the displacement of the ball zero?

f) When is the magnitude of the ball's velocity equal to half its velocity of projection?

g) When is the magnitude of the ball's displacement equal to half the greatest height to which it rises?

h) What are the magnitude and direction of the acceleration while the ball is moving upward?

i) While moving downward?

j) When at the highest point?

3–33 A ball rolling on an inclined plane moves with constant acceleration. One ball is released from rest at the top of an inclined plane 18 m long and reaches the bottom 3 s later. At the same instant that the first ball is released, a second ball is projected upward along the plane from its bottom with a certain initial velocity. The second ball is to travel part way up the plane, stop, and return to the bottom so that it arrives simultaneously with the first ball.

a) Find the acceleration.

b) What must be the initial velocity of the second ball?

c) How far up the plane will it travel?

3–34 The rocket-driven sled Sonic Wind No. 2, used for investigating the physiological effects of large accelerations, runs on a straight, level track 3500 ft long. Starting from rest, it can reach a speed of 1000 mi·hr^{-1} in 1.8 s.

a) Compute the acceleration, assuming it to be constant.

b) What is the ratio of this acceleration to that of a free falling body, g?

c) What is the distance covered?

d) A magazine article states that at the end of a certain run the speed of the sled was decreased from 632 mi·hr^{-1} to zero in 1.4 s, and that during this time its passenger was subjected to more than 40 times the pull of gravity (that is, the acceleration was greater than 40 g). Are these figures consistent?

3–35 The first stage of a rocket to launch an earth satellite will, if fired vertically upward, attain a speed of 4000 mi·hr^{-1} at a height of 26 mi above the earth's surface, at which point its fuel supply will be exhausted.

a) Assuming constant acceleration, find the time to reach a height of 26 mi.

b) How much higher would the rocket rise if it continued to "coast" vertically upward?

3–36 Suppose the acceleration of gravity were only 1.0 m·s^{-2}, instead of 10 m·s^{-2}.

a) Estimate the height to which you could jump vertically from a standing start.

b) How high could you throw a baseball?

c) Estimate the maximum height of a window from which you would care to jump to a concrete sidewalk below. (Each story of an average building is about 4 m high.)

d) With what speed, in miles per hour, would you strike the sidewalk?

e) How many seconds would be required?

3–37 A hypothetical spaceship takes a straight-line path from the earth to the moon, a distance of about 400,000 km. Suppose it accelerates at 10 m·s^{-2} for the first 10 min of the trip, then travels at constant speed until the last 10 min, when it accelerates at -10 m·s^{-2}, just coming to rest as it reaches the moon.

a) What is the maximum speed attained?

b) What total time is required for the trip?

c) What fraction of the total distance is traveled at constant speed?

3–38 The motion of a particle along a straight line is described by the function

$$x = (6 \text{ m}) + (4 \text{ m·s}^{-2})t^2 - (1 \text{ m·s}^{-4})t^4.$$

Assume that t is positive.

a) Find the position, velocity, and acceleration at time $t = 2$ s.

b) During what time interval is the velocity positive?

c) During what time interval is x positive?

d) What is the maximum positive velocity attained by the particle.

3-39 The acceleration of a particle is given by $a = (2 \text{ m·s}^{-4})t^2$. If it is at rest at the origin at time $t = 0$, find its position, velocity, and acceleration at time $t = 1$ s.

3-40 The acceleration of a particle is given by $a = (2 \text{ m·s}^{-3})t$. If its velocity at time $t = 1$ s is 5 m·s^{-1},

a) What is its velocity at time $t = 2$ s?

b) If the particle's position at time $t = 1$ s is 6 m, what is its position at time $t = 2$ s?

3-41 The position of a particle is given by $x = A \sin \omega t$, where A and ω are constants.

a) Find the velocity of the body as a function of time.

b) Find the acceleration as a function of time.

c) Find the velocity as a function of x. At what point is the velocity greatest? Least?

d) Find the acceleration as a function of x. At what point is the acceleration greatest? Least?

e) What is the maximum distance of the particle from the origin ($x = 0$)?

f) What is the maximum velocity?

g) What is the maximum acceleration?

h) Sketch graphs of x, v, and a as functions of time.

3-42 The acceleration of a body suspended from a spring and oscillating vertically is $a = -Ky$, where K is a constant and y is the coordinate measured from the equilibrium position. Suppose that a body moving in this way is given an initial velocity v_0 at the coordinate y_0. Find the expression for the velocity v of the body as a function of its coordinate y. [*Hint:* Use the expression $a = v \, dv/dy$.]

3-43 The motion of a body falling from rest in a resisting medium is described by the equation

$$\frac{dv}{dt} = A - Bv,$$

where A and B are constants. In terms of A and B, find

a) the initial acceleration, and

b) the velocity at which the acceleration becomes zero (the terminal velocity).

c) Show that the velocity at any time t is given by

$$v = \frac{A}{B}(1 - e^{-Bt}).$$

3-44 After the engine of a moving motorboat is cut off, the boat has an acceleration in the opposite direction to its velocity and directly proportional to the square of its velocity. That is, $dv/dt = -kv^2$, where k is a constant.

a) Show that the magnitude v of the velocity at a time t after the engine is cut off is given by

$$\frac{1}{v} = \frac{1}{v_0} + kt.$$

b) Show that the distance x traveled in a time t is

$$x = \frac{1}{k} \ln (v_0 k t + 1).$$

c) Show that the velocity after traveling a distance x is

$$v = v_0 e^{-kx}.$$

As a numerical example, suppose the engine is cut off when the velocity is 6 m·s^{-1}, and that the velocity decreases to 3 m·s^{-1} in a time of 15 s.

d) Find the numerical value of the constant k, and the unit in which it is expressed.

e) Find the acceleration at the instant the engine is cut off.

f) Construct graphs of x, v, and a for a time of 20 s.

3-45 A "moving sidewalk" in an airport terminal building moves 1 m·s^{-1} and is 150 m long. If a man steps on at one end and walks 2 m·s^{-1} relative to the moving sidewalk, how much time does he require to reach the opposite end if he walks

a) in the same direction the sidewalk is moving;

b) in the opposite direction?

3-46 Two piers A and B are located on a river, one mile apart. Two men must make round trips from pier A to pier B and return. One man is to row a boat at a velocity of 4 mi·hr^{-1} relative to the water, and the other man is to walk on the shore at a velocity of 4 mi·hr^{-1}. The velocity of the river is 2 mi·hr^{-1} in the direction from A to B. How long does it take each man to make the round trip?

3-47 The driver of a car wishes to pass a truck that is traveling at a constant speed of 20 m·s^{-1} (about 50 mi·hr^{-1}). The car's maximum acceleration at this speed is 0.5 m·s^{-2}. Initially the vehicles are separated by 25 m, and the car pulls back into the truck's lane after it is 25 m ahead of the truck. The car is 5 m long and the truck 20 m.

a) How much time is required for the car to pass the truck?

b) What distance does the car travel during this time?

c) What is the final speed of the car, assuming its acceleration while passing the truck is constant?

3-48 Two cars, A and B, travel in a straight line. The distance of A from the starting point is given as a function of time by $x_A = (4 \text{ m·s}^{-1})t + (1 \text{ m·s}^{-2})t^2$; the distance of B from the starting point is $x_B = (2 \text{ m·s}^{-2})t^2 + (2 \text{ m·s}^{-3})t^3$.

a) Which car is ahead just after they leave the starting point?

b) At what times are the cars at the same point?

c) At what times is the velocity of B relative to A zero?

d) At what times is the distance from A to B neither increasing nor decreasing?

NEWTON'S SECOND LAW. GRAVITATION 4

4-1 Introduction

In the preceding chapters we have discussed separately the concepts of force and acceleration. In equilibrium problems we have made use of Newton's first law, which states that when the resultant force on a body is zero, the acceleration of the body is also zero. The next logical step is to ask how a body behaves when the resultant force on it is *not* zero. The answer to this question is contained in Newton's *second* law, which states that when the resultant force is not zero the body moves with accelerated motion, and that with a given force, the acceleration depends on a property of the body known as its *mass*.

This part of mechanics, which includes the study both of motion and of the forces that bring about the motion, is called *dynamics*. In its broadest sense, dynamics includes nearly the whole of mechanics. Statics is limited to special cases in which the acceleration is zero, and kinematics deals with motion only.

It will be assumed in this chapter that all velocities are small compared with the velocity of light, so that relativistic considerations do not arise. We shall also assume that, unless otherwise stated, all velocities and accelerations are measured relative to an inertial reference system (Sec. 2-4), and shall consider only straight-line motion. Motion in a curved path will be discussed in the next chapter.

4-2 Newton's second law. Mass

We know from experience that an object at rest never starts to move by itself; a push or pull must be exerted on it by some other body. Similarly, a force is required to slow down or stop a body already in motion, and to make a moving body deviate from straight-line motion requires a sideways force. All these processes (speeding up, slowing down, or changing direction) involve a change in either the magnitude or direction of the velocity. Thus in each case the body *accelerates,* and an external force must act on it to produce the acceleration.

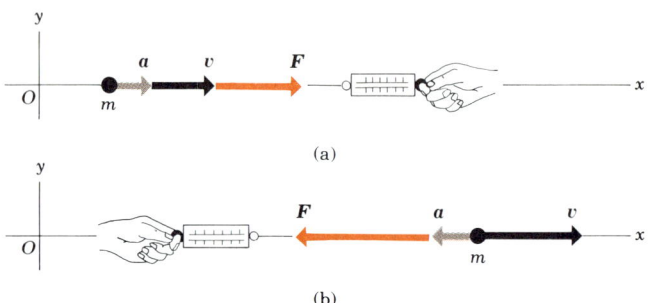

4-1 The acceleration a is proportional to the force F and is in the same direction as the force.

Let us consider several fundamental experiments. A small body, which we may model as a particle, rests on a level frictionless surface, and moves to the right along the x-axis of an inertial reference system, as in Fig. 4-1a. A constant horizontal force F is applied to the body. This force might be supplied by a spring balance, as described in Sec. 2-2, with the spring stretched a constant amount. We find that while the force is acting, the velocity of the body increases at a constant rate; that is, the body moves with *constant acceleration*, say a_1. When the force is removed, the acceleration becomes zero; the body then moves with constant velocity, as discussed in Sec. 2-4.

When the experiment is repeated with a different body but with the same force (as determined by stretching the spring the same amount as before), the acceleration is found to be again constant, but it has a different value, say a_2. We may assign to each body a numerical quantity that describes how much or how little it accelerates under a given force; we call this quantity the *mass* of the body, denoted by the symbol m. We may then define the *ratio* of masses of two bodies as the reciprocal of the ratio of their accelerations under the action of a given force. Thus

$$\frac{m_1}{m_2} = \frac{a_2}{a_1}.$$

If the first body has greater acceleration than the second, it has *smaller* mass. Thus the mass of a body is directly related to the property commonly called *inertia*.

We may define a standard unit of mass with reference to the mass of a particular body. For example, we might choose a chunk of platinum defined to have a mass of precisely one kilogram, as described in Sec. 1-2. Then the above equation may be used to *measure* the mass of any other body by comparison with the standard mass.

Another property of mass is demonstrated by measuring the masses of two bodies, using the procedure just described, and then fastening them together and measuring the mass of the composite body. If m_1 and m_2 are the individual masses, the mass of the composite body is always found to be $m_1 + m_2$. This very important result, which is reasonable but could not have been predicted in advance, shows that mass is an *additive* quantity, and that it is directly correlated with quantity of matter. Indeed, the concept of mass is one way to give the term "quantity of matter" a precise meaning.

Next, we may vary the force acting on a body, and measure the variation in acceleration. For example, when two identical spring balances, stretched the same amount, both pull simultaneously in the same direction on a body, the acceleration is found to be exactly twice as great as when only one spring, stretched the same amount, is applied. Thus the principle of superposition of force, which we have already used in Chapter 2 for equilibrium problems, is also valid in the present situation.

The results of the various experiments involving different values of force, mass, and acceleration can be summarized by saying that the acceleration of a body is directly proportional to the force acting on it and inversely proportional to its mass. That is,

$$a = C\frac{F}{m},$$

where C is some constant of proportionality. It is convenient to choose a unit of force such that $C = 1$; the above equation then becomes

$$F = ma. \tag{4-1}$$

This equation expresses the relationship of F, m, and a, and also defines the unit of force in any unit system as the force required to give unit acceleration (in whatever units of length and time are used) to unit mass.

In Fig. 4–1b, the velocity of the body is also toward the right, but the *force* is toward the left. Under these conditions the body moves more and more slowly to the right until it stops and then reverses its direction of motion and begins to move more and more rapidly toward the left, as long as the force continues to be applied. During this entire process the direction of the acceleration is toward the *left,* in the same direction as the force **F**. Hence we conclude that the *direction* of the acceleration is the *same* as that of the force, regardless of the direction of the velocity.

In the above discussion the particle moves along a straight line, the x-axis, and the force also lies along this direction. This is of course a special case; more generally, the force may also have a component in the y-direction, and the particle's motion need not be confined to a straight line. Furthermore, more than one force may act on the particle. Thus this formulation needs to be generalized to include motion in a plane or in space and the possibility of several forces acting simultaneously.

Experiment shows that when several forces act on a particle at the same time, the acceleration is the same as would be produced by a single force equal to the vector sum of these forces. This sum is usually most conveniently handled using the method of components. Thus when several forces act on a particle moving along the x-axis,

$$\sum F_x = ma. \tag{4-2}$$

Anticipating the more general discussion of velocity and acceleration to be given in Chapter 5, we remark briefly that when a particle moves in a plane, with position described by coordinates (x, y), the velocity is a vector quantity whose components v_x and v_y are equal to the time rates of change of x and y, respectively, and the acceleration is a vector quantity with components a_x and a_y equal to the time rates of change of v_x and v_y, respectively. Then a more general formulation of the relation of force to acceleration is

$$\sum F_x = ma_x, \quad \sum F_y = ma_y. \tag{4-3}$$

This pair of equations is equivalent to the single *vector* equation

$$\Sigma F = ma, \qquad (4\text{-}4)$$

where we write the lefthand side explicitly as ΣF to emphasize that the acceleration is determined by the vector sum of all the forces acting on the particle.

Equation (4-4), or the equivalent pair of equations (4-3), is the mathematical statement of Newton's *second law of motion*. The acceleration of a body (the rate of change of its velocity) is equal to the resultant of all forces exerted on the particle, divided by its mass, and has the same direction as the resultant force.

The acceleration must always be measured relative to an inertial frame of reference, as defined in Sec. 2-4. Also, it is necessary to state the law for a *particle,* because when a force acts on an extended body the body may also *rotate*. In that case, different particles in the body have different accelerations. We return to this more general problem in Chapter 9.

In this chapter only straight-line motion is considered, and the line of motion is taken to be the x-axis. In these problems v_y and a_y are always zero. Individual forces may have y-components, but the *sum* of the y-components must always be zero. In Chapter 5 we consider more general motion in which v_y, a_y, and ΣF_y need not be zero.

4-3 Systems of units

The preceding discussion and the equation $\Sigma F = ma$ show that the units used to measure force, mass, and acceleration must be such that unit force imparts unit acceleration to unit mass; whatever system of units is chosen must meet this requirement.

Several systems of units are in common use in the United States; one is the meter-kilogram-second (mks) system, which also forms the basis of the Système International (SI) introduced in Chapter 1. Other systems are the British or foot-pound-second system and the centimeter-gram-second (cgs) system. This book uses principally SI units, not only because this system is convenient and widely used in scientific work, but also because there seems little doubt that SI will eventually be adopted worldwide for commerce and industry as well as scientific work. The United States and Britain are the only two major countries that do not use the metric system exclusively, and Britain is now starting to convert.

In the mks (SI) system, the mass of the standard kilogram is the unit of mass, and the unit of acceleration is $1 \text{ m} \cdot \text{s}^{-2}$. The unit force in this system is then *that force that gives a standard kilogram an acceleration of* $1 \text{ m} \cdot \text{s}^{-2}$. This force is called *one newton* (1 N). One newton is approximately equal to one-quarter of a pound-force (more precisely, 1 N = 0.22481 lb.) Thus in the mks system,

$$F \text{ (N)} = m \text{ (kg)} \times a \text{ (m} \cdot \text{s}^{-2}).$$

The mass unit in the *centimeter-gram-second* (cgs) system is one gram, equal to $\frac{1}{1000}$ kg, and the unit of acceleration is $1 \text{ cm} \cdot \text{s}^{-2}$. The unit force in this system is then *that force that gives a body whose mass is* one gram, an acceleration of $1 \text{ cm} \cdot \text{s}^{-2}$. This force is called *one dyne* (1 dyn). Since $1 \text{ kg} = 10^3 \text{ g}$ and $1 \text{ m} \cdot \text{s}^{-2} = 10^2 \text{ cm} \cdot \text{s}^{-2}$, it follows that $1 \text{ N} = 10^5 \text{ dyn}$. In the cgs system,

$$F \text{ (dyn)} = m \text{ (g)} \times a \text{ (cm} \cdot \text{s}^{-2}).$$

In the mks and cgs systems, we first selected units of mass and acceleration, and defined the unit of force in terms of these. In the *British engineering system*, we first select a unit of force (1 lb) and a unit of acceleration ($1 \text{ ft} \cdot \text{s}^{-2}$), and then define the unit of mass as *the mass of a body whose acceleration is $1 \text{ ft} \cdot \text{s}^{-2}$ when the resultant force on the body is 1 lb*. This unit of mass is called *one slug*. (The origin of the name is obscure; it may have arisen from the concept of mass as inertia or *sluggishness*.) Then in the engineering system,

$$F \text{ (lb)} = m \text{ (slugs)} \times a \text{ (ft} \cdot \text{s}^{-2}).$$

The pound is used in everyday life as a unit of quantity of matter (e.g., a pound of butter) but properly speaking it is a unit of *force* or *weight*. Thus a pound of butter is that quantity which has a weight of 1 lb. A useful fact in converting between SI and British units is that an object with a *mass* of 1 kg has a *weight* of about 2.205 lb.

The units of force, mass, and acceleration in the three systems are summarized in Table 4–1.

Table 4–1

System of units	Force	Mass	Acceleration
mks	newton (N)	kilogram (kg)	$\text{m} \cdot \text{s}^{-2}$
cgs	dyne (dyn)	gram (g)	$\text{cm} \cdot \text{s}^{-2}$
engineering	pound (lb)	slug	$\text{ft} \cdot \text{s}^{-2}$

It should be emphasized that because of the above definitions there is an equivalence between force units and those of mass times acceleration. For example, $1 \text{ N} = 1 \text{ kg} \cdot \text{m} \cdot \text{s}^{-2}$; and in checking equations for consistency of units it is always appropriate to replace "N" wherever it appears by "$\text{kg} \cdot \text{m} \cdot \text{s}^{-2}$". Similarly, $1 \text{ lb} = 1 \text{ slug} \cdot \text{ft} \cdot \text{s}^{-2}$, and this conversion can be used in unit checks.

Example 1 A constant horizontal force of 2 N is applied to a body of mass 4 kg, resting on a level frictionless surface. The acceleration of the body is

$$a = \frac{F}{m} = \frac{2 \text{ N}}{4 \text{ kg}} = 0.5 \text{ N} \cdot \text{kg}^{-1} = 0.5 \text{ (kg} \cdot \text{m} \cdot \text{s}^{-2}) \cdot \text{kg}^{-1}$$
$$= 0.5 \text{ m} \cdot \text{s}^{-2}.$$

Since the force is constant, the acceleration is constant also. Hence if the initial position and velocity of the body are known, the velocity and position at any later time can be found from the equations of motion with constant acceleration.

Example 2 A body of mass 0.2 kg is given an initial velocity of $0.4 \text{ m} \cdot \text{s}^{-1}$ toward the right along a level laboratory tabletop. The body is observed to slide a distance of 1.0 m along the table before coming to rest. What was the magnitude and direction of the friction force $\mathcal{F}$ acting on it?

Solution In the absence of further information, let us assume that the friction force is constant. The acceleration is then constant also; from the equations of motion with constant acceleration, we have

$$v^2 = v_0^2 + 2ax, \qquad 0 = (0.4 \text{ m} \cdot \text{s}^{-1})^2 + (2a)(1.0 \text{ m}),$$
$$a = -0.08 \text{ m} \cdot \text{s}^{-2}.$$

The negative sign means that the acceleration is toward the *left* (although the velocity is toward the right). The friction force on the body is

$$\mathcal{F} = ma = (0.2 \text{ kg})(-0.08 \text{ m} \cdot \text{s}^{-2}) = -0.016 \text{ N},$$

and is toward the left also. (A force of equal magnitude, but directed toward the right, is exerted on the table by the sliding body.)

Example 3 The acceleration of a certain body is found to be 5 ft·s^{-2} when the resultant force on the body is 20 lb. The mass of the body is

$$m = \frac{F}{a} = \frac{20 \text{ lb}}{5 \text{ ft} \cdot \text{s}^{-2}} = 4 \text{ (slug} \cdot \text{ft} \cdot \text{s}^{-2}) \cdot \text{ft}^{-1} \cdot \text{s}^2$$
$$= 4 \text{ slugs}.$$

4-4 Newton's law of gravitation

We have already encountered the force of gravitational attraction between a body and the earth. We now wish to study gravitation in more detail.

The law of universal gravitation was discovered by Newton, and he published it in 1686. Newton's law of gravitation may be stated:

> Every particle of matter in the universe attracts every other particle with a force that is directly proportional to the product of the masses of the particles and inversely proportional to the square of the distance between them.

Thus

$$F_g = G \frac{m_1 m_2}{r^2}, \tag{4-5}$$

where F_g is the gravitational force on either particle, m_1 and m_2 are their masses, r is the distance between them, and G is a universal constant called the *gravitational constant*, whose numerical value depends on the units in which force, mass, and length are expressed.

The gravitational forces acting on the particles form an action-reaction pair. Although the masses of the particles may be different, forces of *equal* magnitude act on each, and the action line of both forces lies along the line joining the particles.

Newton's law of gravitation refers to the force between two *particles*. It can also be shown, however, that *the gravitational force exerted on or by a homogeneous sphere is the same as if the entire mass of the sphere were concentrated in a point at its center*. Thus if the earth were a homogeneous sphere, of mass m_E, the force exerted by it on a small body of mass m, at a distance r from its center, would be

$$F_g = G \frac{m m_E}{r^2},$$

provided the body lies outside the earth, i.e., that r is greater than the radius of the earth. A force of the same magnitude would be exerted *on* the earth *by* the body.

At points *inside* the earth these statements need to be modified. If one could drill a hole to the center of the earth and measure the force of gravity on a body at various distances from the center, the force would be found to *decrease* as the center is approached, rather than increasing as $1/r^2$. Qualitatively, it is easy to see why this should be so; as the body enters the interior of the earth (or other spherical body), some of the earth's mass is on the side of the body opposite from the center of the earth and pulls the body in the opposite direction. Exactly at the center of the earth, the gravitational force on the body is zero!

The magnitude of the gravitational constant G can be found experimentally by measuring the force of gravitational attraction between two bodies of known masses m_1 and m_2, at a known separation. For bodies of moderate size the force is extremely small, but it can be measured with an instrument called a *torsion balance*, invented by the Rev. John Michell and first used for this purpose by Sir Henry Cavendish in 1798. The same type of instrument was also used by Coulomb for studying forces of electrical and magnetic attraction and repulsion.

The Cavendish balance consists of a light, rigid T-shaped member (Fig. 4–2) supported by a fine vertical fiber such as a quartz thread or a thin metallic ribbon. Two small spheres, each of mass m_1, are mounted at the ends of the horizontal portion of the T, and a small mirror M, fastened to the vertical portion, reflects a beam of light onto a scale. To use the balance, two large spheres, each of mass m_2, are brought up to the positions shown. The forces of gravitational attraction between the large and small spheres twist the system through a small angle, thereby moving the reflected light beam along the scale.

By use of an extremely fine fiber, the deflection of the mirror may be made sufficiently large so that the gravitational force can be measured quite accurately. The gravitational constant, measured in this way, is found to be

$$G = 6.670 \times 10^{-11} \, \text{N} \cdot \text{m}^2 \cdot \text{kg}^{-2}.$$

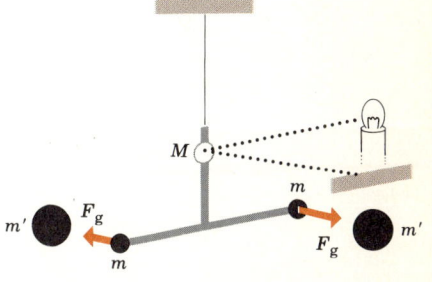

4–2 Principle of the Cavendish balance.

Example 1 The mass m_1 of one of the small spheres of a Cavendish balance is 0.001 kg, the mass m_2 of one of the large spheres is 0.5 kg, and the center-to-center distance between the spheres is 0.05 m. The gravitational force on each sphere is

$$F_g = 6.67 \times 10^{-11} \, \text{N} \cdot \text{m}^2 \cdot \text{kg}^{-2} \frac{(0.001 \, \text{kg})(0.5 \, \text{kg})}{(0.05 \, \text{m})^2}$$

$$= 1.33 \times 10^{-11} \, \text{N},$$

or about one hundred-billionth of a newton!

Example 2 Suppose the spheres in Example 1 are placed 0.05 m from each other at a point in space far removed from all other bodies. What is the acceleration of each, relative to an inertial system?

Solution The acceleration a_1 of the smaller sphere is

$$a_1 = \frac{F_g}{m_1} = \frac{1.33 \times 10^{-11} \text{ N}}{1.0 \times 10^{-3} \text{ kg}} = 1.33 \times 10^{-8} \text{ m} \cdot \text{s}^{-2}.$$

The acceleration a_2 of the larger sphere is

$$a_2 = \frac{F_g}{m_2} = \frac{1.33 \times 10^{-11} \text{ N}}{0.5 \text{ kg}} = 2.67 \times 10^{-11} \text{ m} \cdot \text{s}^{-2}.$$

In this case, the accelerations are *not* constant because the gravitational force increases as the spheres approach each other.

4-5 Mass and weight

The *weight* of a body can now be defined more generally than in the preceding chapters as *the resultant gravitational force exerted on the body by all other bodies in the universe.* At or near the surface of the earth, the force of the earth's attraction is much greater than that of any other body, and the weight can be considered as arising solely from the gravitational attraction of the earth. Similarly, at the surface of the moon, or of another planet, the weight of a body results almost entirely from the gravitational attraction of the moon or the planet. Thus if the earth were a homogeneous sphere of radius R and mass m_E, the weight w of a small body of mass m at or near its surface would be

$$w = F_g = \frac{Gmm_E}{R^2}. \tag{4-6}$$

The weight of a given body varies by a few tenths of a percent from point to point on the earth's surface, partly because of local deposits of ore, oil, or other substances whose density differs from the average, and partly because the earth is not a perfect sphere but is flattened somewhat at the poles. Also, the weight of a given body decreases inversely with the square of its distance from the earth's center. At a radial distance of two earth radii, for example, it has decreased to one-quarter of its value at the earth's surface.

The *apparent* weight of a body at the surface of the earth differs slightly in magnitude and direction from the earth's force of gravitational attraction because of the rotation of the earth about its axis. For the present, we shall ignore the small difference between the apparent weight of a body and the force of the earth's gravitational attraction, and shall assume that the earth is an inertial reference system. Then when a body is allowed to fall freely, the force accelerating it is its weight w, and the acceleration produced by this force is the acceleration due to gravity, g. The general relation

$$\sum F = ma$$

therefore becomes, for the special case of a freely falling body,

$$w = mg. \tag{4-7}$$

Since

$$w = mg = G\frac{mm_E}{R^2},$$

it follows that
$$g = \frac{Gm_E}{R^2}, \tag{4-8}$$
showing that the acceleration due to gravity is the same for *all* bodies (since m cancels out) and is very nearly constant (since G and m_E are constants and R varies only slightly from point to point on the earth).

The weight of a body is a force, and must be expressed in terms of the unit of force in the particular system of units one is using. Thus in the mks system the unit of weight is 1 N, in the cgs system it is 1 dyn, and in the engineering system it is 1 lb. Equation (4-7) indicates the relation between the mass and weight of a body in any consistent set of units.

For example, the weight of a standard kilogram, at a point where $g = 9.80 \text{ m} \cdot \text{s}^{-2}$, is
$$w = mg = (1 \text{ kg})(9.80 \text{ m} \cdot \text{s}^{-2}) = 9.80 \text{ N}.$$
At a second point where $g = 9.78 \text{ m} \cdot \text{s}^{-2}$, the weight is
$$w = 9.78 \text{ N}.$$

Thus, unlike the *mass* of a body, which is constant, the *weight* varies from one point to another. On the moon, where $g = 1.67 \text{ m} \cdot \text{s}^{-2}$, the weight is 1.67 N, but the mass is still 1 kg.

When Eq. (4-8) is solved for m_E, we find
$$m_E = \frac{R^2 g}{G},$$
where R is the earth's radius. All of the quantities on the right are known, so the mass of the earth, m_E, can be calculated. Taking $R = 6370 \text{ km} = 6.37 \times 10^6 \text{ m}$, and $g = 9.80 \text{ m} \cdot \text{s}^{-2}$, we find
$$m_E = 5.96 \times 10^{24} \text{ kg}.$$

The volume of the earth is
$$V = \tfrac{4}{3}\pi R^3 = 1.08 \times 10^{21} \text{ m}^3.$$

The mass of a body divided by its volume is its average *density*. (The density of water is $1 \text{ g} \cdot \text{cm}^{-3} = 1000 \text{ kg} \cdot \text{m}^{-3}$.) The average density of the earth is therefore
$$\frac{m_E}{V} = 5520 \text{ kg} \cdot \text{m}^{-3}.$$

The density of most rock near the earth's surface, such as granites and gneisses, is about $3 \text{ g} \cdot \text{cm}^{-3} = 3000 \text{ kg} \cdot \text{m}^{-3}$, so the interior of the earth must have much higher density. Some basaltic rock found on the surface has a density of about $5 \text{ g} \cdot \text{cm}^{-3}$.

The mass of a body can be measured in several different ways. One way is to use the relationship $m = F/a$; one may apply a measured force to the body, measure its acceleration, and obtain the mass as the ratio of force to acceleration. This method is used frequently in measuring masses of atomic and subatomic particles.

Another method uses a comparison technique, finding by trial some other body whose mass is equal to that of the given body but which is already known. Often it is easier to compare weights than masses. At a given point on the earth's surface, all bodies fall freely with the same acceleration g. Since the weight w of a body equals the product of its mass m and the acceleration g, it follows that if, at the same point, the weights of the two bodies are equal, their *masses* are equal also. The *equal-arm balance* is an instrument by means of which one can determine very precisely when the weights of two bodies are equal, and hence when their masses are equal.

Thus the property of matter called *mass* makes itself evident in two very different ways. The force of gravitational attraction between two particles is said to be proportional to the product of their masses; in this sense mass can be considered as *that property of matter by virtue of which every particle exerts a force of attraction on every other particle*. We may call this property *gravitational mass*. On the other hand, Newton's second law is concerned with the fact that a force (not necessarily gravitational) must be exerted on a particle in order to accelerate it. This property can be called *inertial mass*. It is not at all obvious that the gravitational mass of a particle should be the same as its inertial mass, but experiment shows that the two *are* in fact the same. That is, if we have to push twice as hard on body A as we do on body B to produce a given acceleration, then the force of gravitational attraction between body A and some third body C is twice as great as the gravitational attraction between body B and body C, at the same distance. Therefore the property represented by m in Newton's second law can be operationally defined as the result obtained by the prescribed methods of using an equal-arm balance.

Finally, we note that the SI units for mass and weight are frequently misused in everyday conversation. Expressions such as "This box weighs 6 kg" are nearly universal. What is meant, of course, is that the *mass* of the box (determined indirectly by weighing!) is 6 kg. This usage is so common that there is probably no hope of straightening things out, but it is important to recognize that the term *weight* is often used when *mass* is meant.

4-6 Applications of Newton's second law

We now give a number of applications of Newton's second law to specific problems. In all these examples, and in the problems at the end of the chapter, it will be assumed that the acceleration due to gravity is 9.80 m·s^{-2} or 32.0 ft·s^{-2}, unless otherwise specified.

These examples will also be used to illustrate a systematic problem-solving procedure which the student is urged to study carefully and apply in the solution of assigned problems. The procedure consists of the following steps:

1. Select a body for analysis. Newton's second law will be applied to this body.
2. Draw a free-body diagram just as with the problems of Chapter 2. Be sure to include all the forces acting *on* the chosen body, but be equally careful *not* to include any force exerted *by* the body on some other body. Some of the forces may be unknown; label them with algebraic symbols. Usually one of the forces will be the body's weight. Often it is useful to label this immediately as *mg* rather than

w. If a numerical value of mass is given, the corresponding numerical value of weight may be computed.

3. Show location of coordinate axes explicitly in the free-body diagram, and then determine components of forces with reference to these axes. When a force is represented in terms of its components, cross out the original force so as not to include it twice.
4. If more than one body is involved, the above steps must be carried out for each body, and in addition there may be geometrical relationships between the motions of two or more bodies. Express these in algebraic form.
5. Write the Newton's-second-law equations for each body, usually Eqs. (4–3), and solve to find the unknown quantities.
6. Check special cases or extreme values of quantities, where possible, and compare your intuitive expectations with the results for these particular cases. Ask, "Does this result make sense?"

Example 1 A block whose mass is 10 kg rests on a horizontal surface. What constant horizontal force T is required to give it a velocity of $4 \text{ m} \cdot \text{s}^{-1}$ in 2 s? Assume that the block starts from rest, that the friction force between the block and the surface is constant and equal to 5 N, and that all forces act at the center of the block. (See Fig. 4–3.)

Solution The mass of the block is given. The y-component of its acceleration is zero. The x-component of acceleration can be found from the data on the velocity acquired in a given time. Since the forces are constant, the x-acceleration is constant and, from the equations of motion with constant acceleration,

$$a_x = \frac{v - v_0}{t} = \frac{4 \text{ m} \cdot \text{s}^{-1} - 0}{2 \text{ s}} = 2 \text{ m} \cdot \text{s}^{-2}.$$

The resultant of the x-forces is

$$\sum F_x = T - \mathcal{F},$$

and that of the y-forces is

$$\sum F_y = n - w.$$

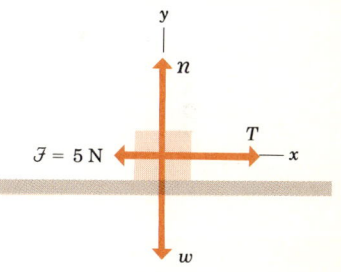

4–3 Force diagram for block in Example 1.

(Note that the *magnitude* of the friction force, $\mathcal{F} = 5$ N, is a positive quantity, but that the *component* of this force in the x-direction is negative, equal to $-\mathcal{F}$ or -5 N. Similarly, the magnitude of the weight is the positive quantity w, but the y-component of this force is $-w$.) Hence, from Newton's second law in component form,

$$T - \mathcal{F} = ma_x, \quad n - w = ma_y = 0.$$

From the second equation, we find that

$$n = w = mg = (10 \text{ kg})(9.80 \text{ m} \cdot \text{s}^{-2}) = 98.0 \text{ N},$$

and from the first,

$$T = \mathcal{F} + ma_x = 5 \text{ N} + (10 \text{ kg})(2 \text{ m} \cdot \text{s}^{-2}) = 25 \text{ N}.$$

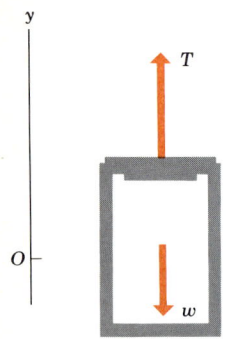

4-4 The resultant vertical force is $T - w$.

Example 2 An elevator and its load have a total mass of 800 kg. Find the tension T in the supporting cable when the elevator, originally moving downward at 10 m·s⁻¹, is brought to rest with constant acceleration in a distance of 25 m. (See Fig. 4-4.)

Solution The weight of the elevator is

$$w = mg = (800 \text{ kg})(9.8 \text{ m·s}^{-2}) = 7840 \text{ N}.$$

From the equations of motion with constant acceleration,

$$v^2 = v_0^2 + 2ay, \qquad a = \frac{v^2 - v_0^2}{2y}.$$

The initial velocity v_0 is -10 m·s⁻¹; the final velocity v is zero. If we take the origin at the point where the acceleration begins, then $y = -25$ m. Hence,

$$a = \frac{0 - (-10 \text{ m·s}^{-1})^2}{2(-25 \text{ m})} = 2 \text{ m·s}^{-2}.$$

The acceleration is therefore positive (upward). From the free-body diagram (Fig. 4-4) the resultant force is

$$\sum F = T - w = T - 7840 \text{ N}.$$

Since $\sum F = ma$,

$$T - 7840 \text{ N} = (800 \text{ kg})(2 \text{ m·s}^{-2}) = 1600 \text{ N},$$
$$T = 9440 \text{ N}.$$

The tension must be *greater* than the weight by 1600 N to cause the upward acceleration while the elevator is stopping.

Example 3 In the above example, with what force do the feet of a passenger press downward on the floor, if the passenger's mass is 80 kg?

Solution We first find the force the floor exerts *on* the passenger's feet, which is the reaction force to the force required. The forces on the passenger are his weight $w = mg = (80 \text{ kg})(9.8 \text{ m·s}^{-2}) = 784 \text{ N}$ and the upward force F caused by the floor. Thus Newton's second law applied to the man, whose acceleration is the same as that of the elevator, gives

$$F - 784 \text{ N} = (80 \text{ kg})(2 \text{ m·s}^{-2}) = 160 \text{ N},$$

and

$$F = 944 \text{ N}.$$

Thus the man pushes down on the floor with a force of 944 N while the elevator is stopping, and he *feels* a greater strain in his legs and feet than while standing at rest. If he stands on a bathroom scale calibrated in newtons, what will the scale read?

4-5 A block on a frictionless inclined plane.

Example 4 What is the acceleration of a block on a frictionless plane inclined at an angle θ with the horizontal?

Solution The only forces acting on the block are its weight $w = mg$ and the normal force n exerted by the plane (Fig. 4-5).

Take axes parallel and perpendicular to the surface of the plane and resolve the weight into x- and y-components. Then

$$\sum F_y = \mathcal{N} - mg \cos \theta,$$

$$\sum F_x = mg \sin \theta.$$

But we know that $a_y = 0$, so from the equation $\sum F_y = ma_y$, we find that $\mathcal{N} = mg \cos \theta$. From the equation $\sum F_x = ma_x$, we have

$$mg \sin \theta = ma_x,$$

and

$$a_x = g \sin \theta.$$

The mass does not appear in the final result, which means that any block, regardless of its mass, will slide down a frictionless inclined plane with an acceleration of $g \sin \theta$. In particular, when $\theta = 0$, $a_x = 0$, and when the plane is vertical, $\theta = 90°$ and $a_x = g$, as we should expect.

Example 5 Refer to Fig. 2–7 (Chapter 2). Let the mass of the block be 4 kg and that of the rope be 0.5 kg. If the force F_1 is 9 N, what are the forces F'_1, F_2, and F'_2? The surface on which the block moves is level and frictionless.

Solution We know from Newton's third law that $F_1 = F'_1$ and that $F_2 = F'_2$. Hence $F'_1 = 9$ N. The force F_2 could be computed by applying Newton's second law to the block, if its acceleration were known, or the force F'_2 could be computed by applying this law to the rope if its acceleration were known. The acceleration is not given, but it can be found by considering the block and rope together as a single system. The vertical forces on this system need not be considered. Since there is no friction, the resultant *external* force acting *on* the system is the force F_1. (The forces F_2 and F'_2 are *internal* forces when we consider block and rope as a single system, and the force F'_1 does not act on the system, but *on the man*.) Then, from Newton's second law,

$$\sum F = ma,$$
$$9 \text{ N} = (4 \text{ kg} + 0.5 \text{ kg})a,$$
$$a = 2 \text{ m} \cdot \text{s}^{-2}.$$

We can now apply Newton's second law to the block:

$$\sum F = ma,$$
$$F_2 = (4 \text{ kg})(2 \text{ m} \cdot \text{s}^{-2}) = 8 \text{ N}.$$

If we consider the rope alone, the resultant force on it is

$$\sum F = F_1 - F'_2 = 9 \text{ N} - F'_2,$$

and from the second law,

$$9 \text{ N} - F'_2 = (0.5 \text{ kg})(2 \text{ m} \cdot \text{s}^{-2}) = 1 \text{ N},$$
$$F'_2 = 8 \text{ N}.$$

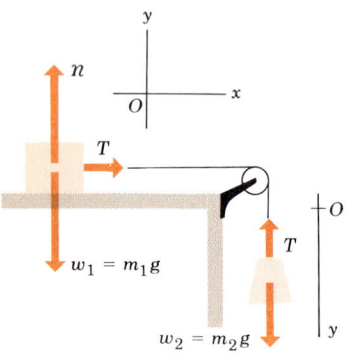

4–6 Force diagram for block on a frictionless horizontal surface and for the hanging block.

In agreement with Newton's third law, which was tacitly used when the forces F_2 and F'_2 were omitted in considering the system as a whole, we find that F_2 and F'_2 are equal in magnitude. Note, however, that the forces F_1 and F'_2 are *not* equal and opposite (the rope is not in equilibrium) and that these forces are *not* an action–reaction pair.

Example 6 In Fig. 4–6, a block of mass m_1 and weight $w_1 = m_1 g$ moves on a level frictionless surface. It is connected by a light flexible cord passing over a small frictionless pulley to a second hanging block of mass m_2 and weight $w_2 = m_2 g$. What is the acceleration of the system, and what is the tension in the cord connecting the two blocks?

Solution The diagram shows the forces acting on each block. The forces exerted on the blocks by the cord can be considered an action–reaction pair, so we have used the same symbol T for each. For the block on the surface,

$$\sum F_x = T = m_1 a_x,$$
$$\sum F_y = n - m_1 g = m_1 a_y = 0.$$

Applying Newton's second law to the hanging block, we obtain

$$\sum F_y = m_2 g - T = m_2 a_y.$$

We note that there is no necessity to use the same coordinate axes for the two bodies. It is convenient to take the $+y$ direction as downward for m_2, so both bodies move in positive axis directions.

Now assuming that the string does not stretch, the speeds of the two bodies at any instant are equal, and so their accelerations have the same magnitude. That is, a_x for m_1 equals a_y for m_2. We can thus denote this common magnitude as simply a. Then the two equations containing accelerations are

$$T = m_1 a,$$
$$m_2 g - T = m_2 a.$$

These are two simultaneous equations for the two unknowns T and a. Adding the two equations eliminates T, giving

$$m_2 g = m_1 a + m_2 a = (m_1 + m_2) a$$

and

$$a = \frac{m_2}{m_1 + m_2} g.$$

Substituting this back into the first equation yields

$$T = \frac{m_1 m_2}{m_1 + m_2} g.$$

We see that the tension T is *not* equal to the weight $m_2 g$ of mass m_2, but is *less* by a factor of $m_1/(m_1 + m_2)$. Further thought shows that T cannot be equal to $m_2 g$; if it were, m_2 would be in equilibrium, and it isn't!

It is interesting to check some special cases. If $m_1 = 0$, we would expect that m_2 would fall freely and there would be no tension in the string. The equations do give $T = 0$ and $a = g$ for this case. Also, if $m_2 = 0$, we expect no tension and no acceleration; for this case the equations give $T = 0$ and $a = 0$. Thus in these two cases the general analytical results agree with intuitive expectations.

Example 7 A body of mass m is suspended from a spring balance attached to the roof of an elevator, as shown in Fig. 4–7. What is the reading of the balance if the elevator has an acceleration a, relative to the earth? Consider the earth's surface to be an inertial reference system.

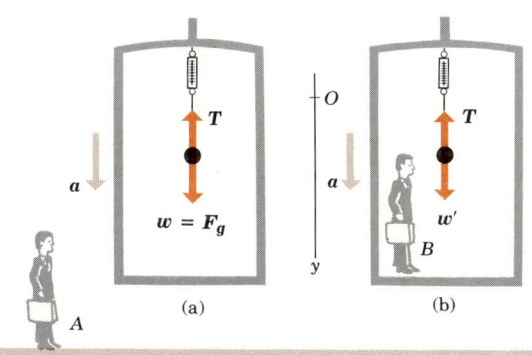

4-7 (a) To observer A, the body has a downward acceleration a, and he writes $w - T = ma$. (b) To observer B, the acceleration of the body is zero. He writes $w' = T$.

Solution The forces on the body are its weight $w = mg$ (the gravitational force F_g exerted on it by the earth) and the upward force T exerted on it by the balance. The body is at rest relative to the elevator, and hence has an acceleration a relative to the earth. (We take the downward direction as positive.) The resultant force on the body is $w - T$, so, from Newton's second law,

$$w - T = ma, \qquad T = w - ma = m(g - a).$$

By Newton's third law, the body pulls down on the balance with a force equal and opposite to T, or equal to $(w - ma)$, so the balance reading equals $(w - ma)$.

If the same body were suspended in equilibrium from a balance attached to the earth, the balance reading would equal the weight w. To an observer riding in the accelerating elevator, the body *appears* to be in equilibrium and hence *appears* to be acted on by a downward force w' equal in magnitude to the balance reading, as in Fig. 4–7b. This *apparent* force w' can be called the *apparent weight* of the body. Then

$$w' = w - ma = m(g - a). \qquad (4\text{–}9)$$

If the elevator is at rest, or moving vertically (either up or down) with constant velocity, $a = 0$ and the apparent weight equals the true weight. If the acceleration is downward, as in Fig. 4–7, so that a is positive, the apparent weight is less than the true weight; the body appears "lighter." If the acceleration is *upward*, a is negative, the apparent weight is greater than the true weight, and the body appears "heavier." These effects can be felt by taking a few steps in an elevator that is starting to move up or down after a stop. If the elevator falls freely,

$a = g$, and since the true weight w also equals mg, the apparent weight is zero and the body *appears* "weightless." It is in this sense that an astronaut orbiting the earth in a space capsule is said to be "weightless."

The effect of this condition is exactly the same as though the body were in outer space with no gravitational force at all. The physiological effect of prolonged weightlessness is an interesting medical problem that is just beginning to be explored, with the advent of manned spacecraft. Gravity plays a role in blood distribution in the body, and one reaction to weightlessness is decrease in volume of blood through increased excretion of water. In some cases astronauts returning to earth have experienced temporary impairment of sense of balance and a greater tendency toward motion sickness.

Example 8 Figure 4–8a represents a simple *accelerometer*. A small body is fastened at one end of a light rod pivoted freely at the point P. When the system has an acceleration a toward the right, the rod makes an angle θ with the vertical. (In a practical instrument, some form of damping must be provided to keep the rod from swinging when the acceleration changes. The rod might hang in a tank of mineral oil.) What is the relation of a to θ?

Solution As shown in the free-body diagram, Fig. 4–8b, two forces are exerted on the body: its weight $w = mg$ and the tension T in the rod. (We neglect the weight of the rod.) The resultant horizontal force is

$$\sum F_x = T \sin \theta,$$

and the resultant vertical force is

$$\sum F_y = T \cos \theta - mg.$$

The x-acceleration is the acceleration a of the system, and the y-acceleration is zero. Hence

$$T \sin \theta = ma, \quad T \cos \theta = mg.$$

When the first equation is divided by the second, we get

$$a = g \tan \theta,$$

and the acceleration a is proportional to the tangent of the angle θ.

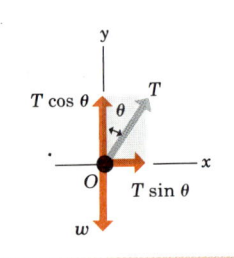

4–8 (a) A simple accelerometer. (b) The forces on the body are w and T.

*****Example 9** Sometimes a frictional force opposing the motion of a body is not constant but increases with the body's speed. Two familiar examples are air resistance on a falling body and the viscous resisting force on a body falling through a liquid. For small speeds this force f is sometimes approximately proportional to speed v:

$$f = kv,$$

where k is a proportionality constant that depends on the dimensions of the body and the viscosity of the fluid. Taking the positive direction to be downward and neglecting any force associated with buoyancy in the fluid, we find that the net vertical component of force is $mg - kv$; Newton's second law gives

$$mg - kv = ma.$$

When the body first starts to move, $v = 0$, the resisting force is zero, and the initial acceleration is $a = g$. As the speed increases, the resisting force also increases, until finally it equals the weight in magnitude. At this time $mg - kv = 0$, the acceleration becomes zero, and there is no further increase in speed. The final speed, also called the *terminal* speed, is given by this equation; solving for v, we find

$$v_t = \frac{mg}{k}.$$

Figure 4–9 shows how the acceleration, velocity, and position vary with time.

To find the relation between the speed and the time during the interval before the terminal speed is reached, we go back to Newton's second law, which we rewrite as

$$m\frac{dv}{dt} = mg - kv.$$

After rearranging terms and replacing mg/k by v_t, we find

$$\frac{dv}{v - v_t} = -\frac{k}{m}dt.$$

We now integrate both sides, noting that $v = 0$ when $t = 0$:

$$\int_0^v \frac{dv}{v - v_t} = -\frac{k}{m}\int_0^t dt,$$

whence

$$\ln\frac{v_t - v}{v_t} = -\frac{k}{m}t,$$

or

$$1 - \frac{v}{v_t} = e^{-(k/m)t},$$

and finally

$$v = v_t(1 - e^{-(k/m)t}).$$

From this result, expressions for the acceleration and position as functions of time can be derived. The derivations are left as an exercise; the results are

$$a = ge^{-(k/m)t},$$

$$y = v_t t + \frac{m}{k}(1 - e^{-(k/m)t}).$$

Now look again at Fig. 4–9, which shows graphs of these three relations.

4–9 Graphs of acceleration, velocity, and position versus time for a body falling in a viscous fluid.

Questions

4–1 When a heavy weight is lifted by a string that is barely strong enough, the weight can be lifted by a steady pull, but if the string is jerked it will break. Why?

4–2 When a car accelerates, starting from rest, where is the force applied to the car to cause its acceleration? By what other body is the force exerted?

4-3 When a car stops suddenly, the passengers are thrown forward, away from their seats. What force causes this motion?

4-4 For medical reasons it is important for astronauts in outer space to determine their body mass at regular intervals. Devise a scheme for measuring body mass in a zero-gravity environment.

4-5 In SI units, suppose the fundamental units had been chosen to be force, length, and time instead of mass, length, and time. What would be the units of mass, in terms of the fundamental units?

4-6 A car traveling on a level road with coefficient of friction μ can never have an acceleration greater in magnitude than μg. Explain.

4-7 When a bullet is fired from a gun, what is the origin of the force that accelerates the bullet?

4-8 Why can a person dive into water from a height of 10 m without injury, while a person who jumps off the roof of a 10-m building and lands on a concrete street is likely to be seriously injured?

4-9 A passenger in a bus notices that a ball in the aisle that has been at rest suddenly starts to move toward the rear of the bus. Think of two different possible explanations, and devise a way to decide which is correct.

4-10 Two bodies on the two sides of an equal-arm balance exactly balance each other. If the balance is placed in an elevator and given an upward acceleration, do they still balance?

4-11 If a man in an elevator drops his briefcase but it does not fall to the floor, what can he conclude about the elevator's motion?

4-12 Suppose you are sealed inside a box on the planet Vulcan. After some time you notice that objects do not fall to the floor when released. How can you determine whether you and the box are in free fall or whether the Vulcans have somehow found a way to turn off the force of gravity?

4-13 Scales for weighing objects can be classified as those that use springs and those using standard weights to balance unknown weights. Which group would have greater accuracy when used in an accelerating elevator? When used on the moon? Does it matter whether you are trying to determine mass or weight?

4-14 Because of air resistance, two objects of unequal mass do *not* fall at precisely the same rate. If two bodies of identical shape but unequal mass are dropped simultaneously from the same height, which one reaches the ground first?

4-15 In discussing the Cavendish balance, a student claimed that the reason the small pivoted masses move toward the larger stationary masses rather than the larger toward the smaller is that the larger masses exert a stronger gravitational pull than the smaller. Please comment.

4-16 Since the earth is constantly attracted toward the sun by the gravitational interaction, why does it not fall into the sun and burn up?

4-17 "It's not the fall that hurts you; it's the sudden stop at the bottom." Translate this saying into the language of Newton's laws of motion.

4-18 Cavendish described his measurement of the gravitational constant G as "weighing the earth." This is an apt description not of the experiment itself but of the significance of the result. Why?

4-19 Give arguments either for or against the statement that "the only reason an apple falls downward to meet the earth instead of the earth falling upward to meet the apple is that the earth, being so much more massive, exerts the greater pull."

4-20 Automotive engineers, in discussing the vertical motion of an automobile driving over a bump, call the time derivative of the acceleration the "jerk." Why is this a useful quantity in characterizing the riding qualities of an automobile?

Problems

In solving the problems of this chapter, the student is urged to use the systematic problem-solving procedures outlined in Sec. 4-6.

4-1
a) What is the mass of a body that weighs 1 N at a point where $g = 9.80$ m·s^{-2}?
b) What is the mass of a body that weighs 1 dyn at a point where $g = 980$ cm·s^{-2}?

4-2
a) At what distance from the earth's center would a standard kilogram weigh 1 N?
b) At what distance would a 1-g body weigh 1 dyn?

4-3 If action and reaction are always equal in magnitude and opposite in direction, why don't they always cancel each other and leave no net force for acceleration of the body?

4-4 The mass of a certain object is 10 kg.
a) What would its mass be if taken to the planet Mars?
b) Is the expression $F = ma$ valid on Mars?
c) Newton's second law is sometimes written in the form $F = wa/g$ instead of $F = ma$. Would this expression be valid on Mars?

4-5 A constant horizontal force of 40 N acts on a body on a *smooth horizontal* plane. The body starts from rest and is observed to move 100 m in 5 s.
  a) What is the mass of the body?
  b) If the force ceases to act at the end of 5 s, how far will the body move in the next 5 s?
  c) What are the answers for (a) and (b) if the force is 10 lb and the body moves 250 ft in 5 s?

4-6 A .22 rifle bullet, traveling at 360 m·s^{-1}, strikes a block of soft wood, which it penetrates to a depth of 0.1 m. The mass of the bullet is 1.8 g. Assume a constant retarding force.
  a) How much time was required for the bullet to stop?
  b) What was the accelerating force, in newtons? in pounds?

4-7 A body of mass 15 kg rests on a frictionless horizontal plane and is acted on by a horizontal force of 30 N.
  a) What acceleration is produced?
  b) How far will the body travel in 10 s?
  c) What will be its velocity at the end of 10 s?

4-8 A body of mass 0.5 kg is at rest at the origin, $x = 0$, on a horizontal frictionless surface. At time $t = 0$ a force of 0.001 N is applied to the body parallel to the x-axis, and 5 s later this force is removed.
  a) What are the position and velocity of the body at $t = 5$ s?
  b) If the same force is again applied at $t = 15$ s, what are the position and velocity of the body at $t = 20$ s?

4-9 A body of mass 10 kg is moving with a constant velocity of 5 m·s^{-1} on a horizontal surface. The coefficient of sliding friction between body and surface is 0.20.
  a) What horizontal force is required to maintain the motion?
  b) If the force is removed, how soon will the body come to rest?

4-10 A hockey puck leaves a player's stick with a velocity of 10 m·s^{-1} and slides 40 m before coming to rest. Find the coefficient of friction between the puck and the ice.

4-11 An electron (mass = 9×10^{-31} kg) leaves the cathode of a radio tube with zero initial velocity and travels in a straight line to the anode, which is 1 cm away. It reaches the anode with a velocity of 6×10^6 m·s^{-1}. If the accelerating force was constant, compute
  a) the accelerating force, in newtons,
  b) the time to reach the anode,
  c) the acceleration.

(The gravitational force on the electron may be neglected.)

4-12 The moon is 0.38×10^9 m from the earth and has a mass of 7.36×10^{22} kg. Find the gravitational force it exerts on a 1-kg body on earth; express your result also as a fraction of the body's weight (the earth's gravitational force on it.)

4-13 A communications satellite of mass 200 kg is in a circular orbit of radius 40,000 km around the earth (mass 5.97×10^{24} kg). What is the gravitational force on the satellite, and what fraction is this of its weight on earth?

4-14 A spaceship travels from earth directly toward the sun. At what distance from the center of the earth do the gravitational forces of the sun and the earth on the ship exactly cancel? Use the data in Problem 4-18 or in Appendix G.

4-15 In an experiment using the Cavendish balance to measure the gravitational constant G, it is found that a sphere of mass 0.8 kg attracts another sphere of mass 0.004 kg with a force of 13×10^{-11} N when the distance between the centers of the spheres is 0.04 m. The acceleration of gravity at the earth's surface is 9.80 m·s^{-2}, and the radius of the earth is 6400 km. Compute the mass of the earth from these data.

4-16 Two spheres, each of mass 6.4 kg, are fixed at points A and B (Fig. 4-10). Find the magnitude and direction of the initial acceleration of a sphere of mass 0.010 kg if released from rest at point P and acted on only by forces of gravitational attraction of the spheres at A and B.

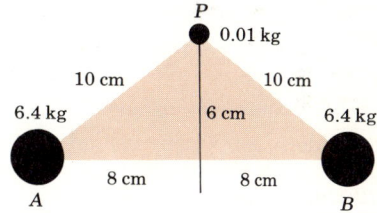

Figure 4-10

4-17 The mass of the moon is about one eighty-first, and its radius one-fourth, that of the earth. What is the acceleration due to gravity on the surface of the moon?

4-18 In round numbers, the distance from the earth to the moon is 250,000 mi, the distance from the earth to the sun is 93 million mi, the mass of the earth is 6×10^{27} g, and the mass of the sun is 2×10^{33} g. Approximately what is the ratio of the gravitational pull of the sun on the moon to that of the earth on the moon?

4-19 An elevator with mass 2000 kg rises with an acceleration of 1 m·s^{-2}. What is the tension in the supporting cable?

4-20 A 4-kg block is accelerated upward by a cord whose breaking strength is 60 N. Find the maximum acceleration that can be given to the block without breaking the cord.

4-21 An elevator weighing 4000 lb rises with an acceleration of 4 ft·s^{-2}. What is the tension in the supporting cable?

4-22 A block weighing 8 lb is accelerated upward by a cord whose breaking strength is 12 lb. Find the maximum acceleration that can be given to the block without breaking the cord.

Good test question

4–23 A 5-kg block is supported by a cord and pulled upward with an acceleration of 2 m·s^{-2}.

a) What is the tension in the cord?

b) After the block has been set in motion, the tension in the cord is reduced to 49 N. What sort of motion will the block have then?

c) If the cord is now slackened completely, the block is observed to rise 2 m farther before coming to rest. With what velocity was it traveling when the cord became slack?

4–24 A block weighing 10 N is supported by a string that can be moved up or down. What conclusions can you draw regarding magnitude and direction of the acceleration and velocity of the upper end of the string when the tension in the string is

a) 5 N?   b) 10 N?   c) 15 N?

4–25 A body hangs from a spring balance supported from the roof of an elevator.

a) If the elevator has an upward acceleration of 2.45 m·s^{-2} and the balance reads 50 N, what is the true weight of the body?

b) Under what circumstances will the balance read 30 N?

c) What will the balance read if the elevator cable breaks?

4–26 A transport plane is to take off from a level landing field with two gliders in tow, one behind the other. Each glider has a mass of 1200 kg, and the friction force or drag on each may be assumed constant and equal to 2000 N. The tension in the towrope between the transport plane and the first glider is not to exceed 10,000 N.

a) If a velocity of 40 m·s^{-1} is required for the take-off, what minimum length of runway is needed?

b) What is the tension in the towrope between the two gliders while they are accelerating for the take-off?

4–27 If the coefficient of friction between tires and road is 0.5, what is the shortest distance in which an automobile can be stopped when traveling at 60 mi·hr^{-1}?

4–28 A 40-kg packing case is on the floor of a truck. The coefficient of static friction between the case and the truck floor is 0.30, and the coefficient of sliding friction is 0.20. Find the magnitude and direction of the friction force acting on the case

a) when the truck is accelerating at 2 m·s^{-2};

b) when it is accelerating at -3 m·s^{-2}.

4–29 A balloon is descending with a constant acceleration a, less than the acceleration due to gravity g. The weight of the balloon, with its basket and contents, is w. What weight, W, of ballast should be released so that the balloon will begin to accelerate upward with constant acceleration a? Neglect air resistance.

4–30 A 5-kg block slides down a plane inclined at 30° to the horizontal. Find the acceleration of the block

a) if the plane is frictionless;

b) if the coefficient of kinetic friction is 0.2.

4–31 A 20-kg box rests on the flat floor of a truck. The coefficient of friction between box and floor is 0.1. The truck stops at a stop sign and then starts, with an acceleration of 2 m·s^{-2}. If the box is 5 m from the rear of the truck when it starts, how much time elapses before it falls off the rear of the truck? How far does the truck travel in this time?

4–32 A 160-lb man stands on a bathroom scale in an elevator. As the elevator starts moving, the scale reads 200 lb.

a) Find the acceleration of the elevator (magnitude and direction).

b) What is the acceleration if the scale reads 120 lb?

c) If the scale reads zero, should the man worry? Explain.

4–33 A short commuter train consists of a locomotive and two cars. The mass of the locomotive is 6000 kg and that of each car 2000 kg. The train pulls away from a station with an acceleration of 0.5 m·s^{-2}.

a) Find the tension in the coupler joining the locomotive to the first car, and in the coupler joining the two cars.

b) What total horizontal force must the locomotive wheels exert on the track?

4–34 A loaded elevator with very worn cables has a total mass of 2000 kg, and the cables can withstand a maximum tension of 24,000 N.

a) What is the maximum acceleration for the elevator if the cables are not to break?

b) What is the answer for part (a) if the elevator is taken to the moon, where $g = 1.67$ m·s^{-2}?

4–35 A block rests on an inclined plane that makes an angle θ with the horizontal. The coefficient of sliding friction is 0.50, and the coefficient of static friction is 0.75.

a) As the angle θ is increased, find the minimum angle at which the block starts to slip.

b) At this angle, find the acceleration once the block has begun to move.

c) How much time is required for the block to slip 20 ft along the inclined plane?

4–36

a) What constant horizontal force is required to drag a 16-lb block along a horizontal surface with an acceleration of 4 ft·s^{-2} if the coefficient of sliding friction between block and surface is 0.5?

b) What weight, hanging from a cord attached to the 16-lb block and passing over a small frictionless pulley, will produce this acceleration?

4–37 A block of mass 5 kg resting on a horizontal surface is connected by a cord passing over a light frictionless pulley to a hanging block of mass 5 kg. The coefficient of friction between the block and the horizontal surface is 0.5. Find

a) the tension in the cord, and
b) the acceleration of each block.

4-38 A block having a mass of 2 kg is projected up a long 30° incline with an initial velocity of 22 m·s^{-1}. The coefficient of friction between the block and the plane is 0.3.

a) Find the friction force acting on the block as it moves up the plane.
b) How long does the block move up the plane?
c) How *far* does the block move up the plane?
d) How long does it take the block to slide from its position in part (c) to its starting point?
e) With what velocity does it arrive at this point?
f) If the mass of the block had been 5 kg instead of 2 kg, would the answers in the preceding parts be changed?

4-39 A 20-kg box is dragged across a rough level floor having a coefficient of kinetic friction of 0.3 by a rope which is pulled upward at an angle of 30° to the horizontal with a force of magnitude 80 N.

a) What is the normal force?
b) What is the frictional force?
c) What is the acceleration of the box?
d) If the force is reduced until the acceleration becomes zero, what is the tension in the rope?

4-40 A box of mass m is dragged across a rough level floor having a coefficient of kinetic friction μ_k by a rope which is pulled upward at an angle θ to the horizontal with a force of magnitude F. In terms of m, μ_k, θ, F, and g, obtain expressions for:

a) The normal force;
b) The frictional force;
c) The acceleration of the box;
d) The magnitude of force required to move the box with constant speed.

4-41 A 20-kg box is pushed up a rough sloping ramp, inclined at 30° to the horizontal, having a coefficient of kinetic friction of 0.3, by a horizontal force of magnitude 300 N.

a) What is the normal force?
b) What is the frictional force?
c) What is the acceleration of the box?
d) If the force is reduced until the acceleration becomes zero, what is the magnitude of the force?

4-42 A box of mass m is pushed up a rough sloping ramp, inclined at angle θ to the horizontal, having a coefficient of kinetic friction of μ_k, by a horizontal force of magnitude F. In terms of m, θ, F, and g, obtain expressions for the following:

a) What is the normal force?

b) What is the frictional force?
c) What is the acceleration of the box?
d) What force is required to push the box with constant velocity?
e) Show that if θ is too large, it is impossible to push the box up the ramp with a horizontal force, no matter how large. Derive an expression for the critical value of θ.

4-43 A 20-kg block on a level frictionless surface is attached by a cord passing over a small frictionless pulley to a hanging block originally at rest 1 m above the floor. The hanging block strikes the floor in 2 s after the system is released from rest. Find

a) the weight of the hanging block;
b) the tension in the string while both blocks were in motion.

4-44 Two blocks, each having mass 20 kg, rest on frictionless surfaces, as shown in Fig. 4–11. Assuming the pulleys to be light and frictionless, compute

a) the time required for block A to move 1 m down the plane, starting from rest;
b) the tension in the cord connecting the blocks.

Figure 4–11

4-45 A block of mass 0.2 kg rests on top of a block of mass 0.8 kg. The combination is dragged along a horizontal surface at constant velocity by a hanging block of mass 0.2 kg as in Fig. 4–12a.

a) The first 0.2-kg block is removed from the 0.8-kg block

Figure 4–12

and attached to the hanging block, as in Fig. 4–12b. What is now the acceleration of the system?

b) What is the tension in the cord attached to the 0.8-kg block in part (a)?

4–46 Block A in Fig. 4–13 has a mass of 2 kg and block B 20 kg. The coefficient of kinetic friction between B and the horizontal surface is 0.1.

a) What is the mass of block C if the acceleration of B is 2 m·s^{-2} toward the right?

b) What is the tension in each cord when B has the acceleration stated above?

Figure 4–13

4–47 Two blocks connected by a cord passing over a small frictionless pulley rest on frictionless planes, as shown in Fig. 4–14.

a) Which way will the system move?

b) What is the acceleration of the blocks?

c) What is the tension in the cord?

Figure 4–14

4–48 Two 0.2-kg blocks hang at the ends of a light flexible cord passing over a small frictionless pulley, as in Fig. 4–15. A 0.1-kg block is placed on the right, and then removed after 2 s.

a) How far will each block move in the first second after the 0.1-kg block is removed?

b) What was the tension in the cord before the 0.1-kg block was removed? After it was removed?

c) What was the tension in the cord supporting the pulley before the 0.1-kg block was removed? Neglect the weight of the pulley.

4–49 In terms of m_1, m_2, and g, find the acceleration of both blocks in Fig. 4–16.

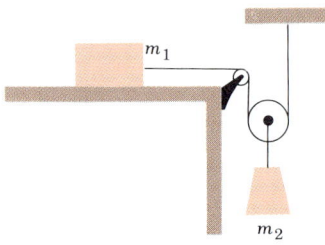

Figure 4–16

4–50 The masses of bodies A and B in Fig. 4–17 are 20 kg and 10 kg, respectively. They are initially at rest on the floor and are connected by a weightless string passing over a weightless and frictionless pulley. An upward force F is applied to the pulley. Find the accelerations a_1 of body A and a_2 of body B when F is

a) 98 N, b) 196 N, c) 394 N, d) 788 N.

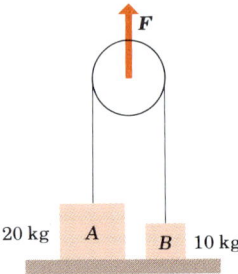

Figure 4–17

4–51 The two blocks in Fig. 4–18 are connected by a heavy uniform rope of mass 4 kg. An upward force of 200 N is applied as shown.

a) What is the acceleration of the system?

b) What is the tension at the top of the heavy rope?

c) What is the tension at the midpoint of the rope?

Figure 4–18

4–52 Two blocks with masses of 4 kg and 8 kg, respectively, are connected by a string and slide down a 30° inclined plane, as in Fig. 4–19. The coefficient of sliding friction between the 4-kg block and the plane is 0.25, and between the 8-kg block and the plane it is 0.50.

a) Calculate the acceleration of each block.

Figure 4–15

b) Calculate the tension in the string.

Figure 4-19

4-53 Two bodies with masses of 5 kg and 2 kg, respectively, hang 1 m above the floor from the ends of a cord 3 m long passing over a frictionless pulley. Both bodies start from rest. Find the maximum height reached by the 2-kg body.

4-54 A man of mass 80 kg stands on a platform of mass 40 kg. He pulls a rope that is fastened to the platform and runs over a pulley on the ceiling. With what force does he have to pull in order to give himself and the platform an upward acceleration of $1 \text{ m} \cdot \text{s}^{-2}$?

4-55 What acceleration must the cart in Fig. 4-20 have in order that the block A will not fall? The coefficient of static friction between the block and the cart is μ_s. How would the behavior of the block be described by an observer on the cart?

Figure 4-20

4-56 Which way will the accelerometer in Fig. 4-8 deflect under the following conditions?

a) The cart is moving toward the right and traveling faster.
b) The cart is moving toward the right and traveling slower.
c) The cart is moving toward the left and traveling faster.
d) The cart is moving toward the left and traveling slower.
e) The cart is at rest on a sloping surface.
f) The cart is given an upward velocity on a frictionless inclined plane. It first moves up, then stops, and then moves down. What is the deflection in each stage of the motion?

4-57 In Problem 3-38, suppose the particle has a mass of 0.20 kg.

a) Derive an expression for the force on the particle, as a function of time.
b) What is the maximum force in the positive direction?

When does it occur? Where is the particle at this time?

4-58 A body of mass $m = 5$ kg falls from rest in a viscous medium. The body is acted on by a net constant downward force of 20 N, and by a viscous retarding force proportional to its speed and equal to $5v$, where v is the speed in meters per second.

a) Find the initial acceleration, a_0.
b) Find the acceleration when the speed is $3 \text{ m} \cdot \text{s}^{-1}$.
c) Find the speed when the acceleration equals $0.1a_0$.
d) Find the terminal velocity, v_t.
e) Find the coordinate, velocity, and acceleration 2 s after the start of the motion.
f) Find the time required to reach a speed $0.9v_t$.
g) Construct a graph of v versus t, for the first three seconds of the motion.

4-59 A body falls from rest through a medium which exerts a resisting force that varies directly with the square of the velocity ($R = -kv^2$).

a) Draw a diagram showing the direction of motion and indicate with the aid of vectors all of the forces acting on the body.
b) Apply Newton's second law and infer from the resulting equation the general properties of the motion.
c) Show that the body acquires a terminal velocity and calculate it.
d) Derive the equation for the velocity at any time.

4-60 A particle of mass m, originally at rest, is subjected to a force whose direction is constant but whose magnitude varies with the time according to the relation

$$F = F_0 \left[1 - \left(\frac{t-T}{T} \right)^2 \right],$$

where F_0 and T are constants. The force acts only for the time interval $2T$.

a) Make a rough graph of F versus t.
b) Prove that the speed v of the particle after a time $2T$ has elapsed is equal to $4F_0 T/3\, m$.
c) Choose numbers for v, T, and m that might be appropriate to a batted baseball, and calculate the force F_0. Judge whether the answer is sensible.

4-61 In Problem 3-44, suppose the mass of the motorboat is 2000 kg. Find the force accelerating the boat when its speed is

a) $6 \text{ m} \cdot \text{s}^{-1}$, and
b) $3 \text{ m} \cdot \text{s}^{-1}$.
c) If the same boat were towed at a constant speed of $3 \text{ m} \cdot \text{s}^{-1}$, what would be the tension in the towline?

4-62 A 20-kg monkey has a firm hold on a light rope that passes over a frictionless pulley and is attached to a 20-kg

82 NEWTON'S SECOND LAW. GRAVITATION

Figure 4-21

bunch of bananas, as shown in Fig. 4-21. The monkey looks upward, sees the bananas, and starts to climb the rope to get at the bananas.

a) As he climbs, do the bananas move up, down, or remain at rest?

b) As he climbs, does the distance between him and the bananas decrease, increase, or remain constant?

c) The monkey releases his hold on the rope. What about the distance between him and the bananas while he is falling?

d) Before he reaches the ground, he grabs the rope to stop his fall. What do the bananas do?

MOTION IN A PLANE 5

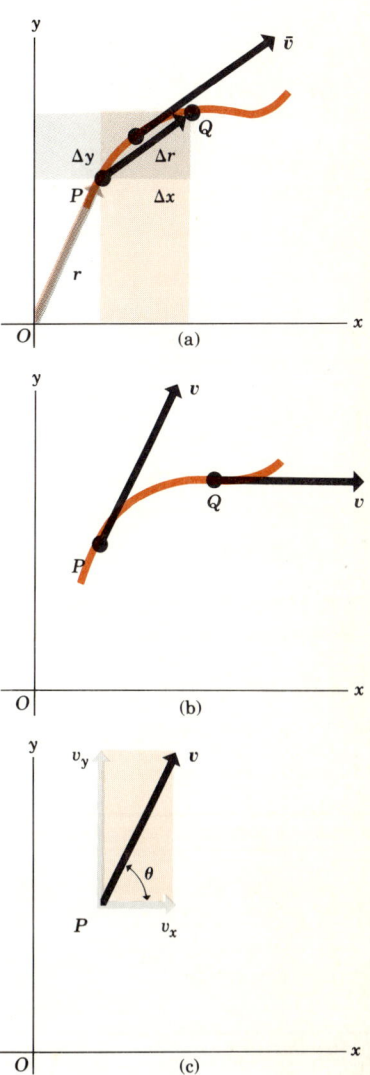

5-1 Motion in a plane

Thus far we have discussed only motion along a straight line, or *rectilinear motion*. In this chapter we shall consider *plane motion*, that is, motion in a path that lies in a fixed plane. Examples of such motion are the flight of a thrown or batted baseball, a projectile shot from a gun, a ball whirled at the end of a cord, the motion of the moon or of a satellite around the earth, and the motion of the planets around the sun.

We shall consider several kinds of problems in plane motion. In some the path of the particle is known and we wish to determine its velocity and acceleration and the resultant force acting on it. An example is that of a ball attached to a cord and whirled in a circle at constant speed. What is the tension in the cord? In others, the force acting on a particle may be known at every point of space and we may wish to find the equation of motion of the particle. Examples are the orbit of a planet around the sun, or the path followed by a rocket.

For these problems the kinematic language developed in Chapter 4 must be generalized; velocity and acceleration are in general *vector* quantities, subject to the methods of treatment introduced in Chapter 1.

5-2 Average and instantaneous velocity

In Fig. 5–1a, a particle moves along a curved path in the xy-plane; points P and Q represent the positions of the particle at two different times. The position of the particle at P can be described by the displacement vector r from the origin O to P; r is called the *position vector* of the particle at this point. The components of r are the coordinates x and y, and we may write

$$r = xi + yj. \qquad (5\text{-}1)$$

The particle's displacement as it moves from P to Q is the change Δr in the position vector r, as shown in the figure. The x- and y-components of Δr are the quantities Δx and Δy, as shown.

5-1 (a) The displacement Δr between points P and Q has components Δx and Δy. The average velocity v has the same direction as Δr. (b) The instantaneous velocity v at each point is always tangent to the path at that point. (c) Components of the instantaneous velocity at P are shown.

Let Δt be the time interval during which the particle moves from P to Q. The average velocity during this interval is defined to be a vector quantity equal to the displacement divided by the time interval:

$$\text{Average velocity } \bar{v} = \Delta r/\Delta t. \tag{5-2}$$

That is, the average velocity is a vector quantity having the same direction as Δr and a magnitude equal to the magnitude of Δr divided by Δt. The magnitude of Δr is always the straight-line distance from P to Q regardless of the actual shape of the path taken by the particle. Thus the average velocity would be the same for any path that would take the particle from P to Q in the time interval Δt. Because the vector quantity $\bar{v}$ is associated with the entire displacement Δr, it has been drawn in Fig. 5-1a at a point intermediate between P and Q.

The *instantaneous* velocity v at the point P is defined in magnitude and direction as the *limit approached by the average velocity* when point Q is taken closer and closer to point P:

$$\text{Instantaneous velocity } v = \lim_{\Delta t \to 0} \frac{\Delta r}{\Delta t} = \frac{dr}{dt}. \tag{5-3}$$

As point Q approaches point P, the direction of the vector Δr approaches that of the tangent to the path at P, so that the instantaneous velocity vector at any point is *tangent* to the path at that point. The instantaneous velocities at points P and Q are shown in Fig. 5-1b.

It follows from Eq. (5-1) that the components $\bar{v}_x$ and $\bar{v}_y$ of average velocity are the corresponding components of displacement, divided by Δt. That is,

$$\bar{v}_x = \frac{\Delta x}{\Delta t}, \qquad \bar{v}_y = \frac{\Delta y}{\Delta t}.$$

Similarly, the instantaneous velocities of these projections, denoted by v_x and v_y, are equal respectively to the x- and y-components of the instantaneous velocity v, and we may write

$$v_x = \lim_{\Delta t \to 0} \frac{\Delta x}{\Delta t} = \frac{dx}{dt}, \qquad v_y = \lim_{\Delta t \to 0} \frac{\Delta y}{\Delta t} = \frac{dy}{dt}. \tag{5-4}$$

The magnitude of the instantaneous velocity is given by

$$|v| = \sqrt{v_x^2 + v_y^2},$$

and the angle θ in Fig. 5-1c by

$$\tan \theta = \frac{v_y}{v_x}.$$

In terms of unit vectors,

$$v = \frac{dr}{dt} = \frac{d}{dt}(xi + yj) = \frac{dx}{dt}i + \frac{dy}{dt}j. \tag{5-5}$$

If x and y are known functions of time, then the derivatives in Eqs. (5-4) and (5-5) can be computed and the velocity determined at any time. In many of the problems of this chapter our objective will be to obtain x and y as functions of time; these functions then provide a complete description of the motion.

Thus we may represent velocity, a vector quantity, in terms of its components or in terms of its magnitude and direction, just as with other vector quantities such as displacement and force. The *direction* of the instantaneous velocity of a particle at any point is *always* in a direction tangent to the path at that point, no matter how complicated the motion.

5-3 Average and instantaneous acceleration

In Fig. 5-2a the vectors v_1 and v_2 represent the instantaneous velocities, at points P and Q, of a particle moving in a curved path. The velocity v_2 differs in *direction* from the velocity v_1. The diagram has been constructed for a case in which it also differs in *magnitude,* although in some special cases the magnitude of the velocity may remain constant.

The *average acceleration* $\bar{a}$ of the particle as it moves from P to Q is defined as the *vector change in velocity,* Δv, divided by the time interval Δt:

$$\text{Average acceleration } \bar{a} = \frac{\Delta v}{\Delta t}. \tag{5-6}$$

5-2 (a) The vector $\bar{a} = \Delta v/\Delta t$ represents the average acceleration between P and Q. (b) Construction for obtaining $\Delta v = v_2 - v_1$. (c) Instantaneous acceleration a at point P. Vector v is tangent to the path; vector a points toward the concave side of the path.

Average acceleration is a vector quantity, in the same direction as the vector Δv.

The vector change in velocity, Δv, means the vector difference $v_2 - v_1$:

$$\Delta v = v_2 - v_1,$$

or

$$v_2 = v_1 + \Delta v.$$

Thus v_2 is the vector sum of v_1 and Δv. This relationship is shown in Fig. 5-2b.

The average acceleration vector, $\bar{a} = \Delta v/\Delta t$, is shown in Fig. 5-2a at a point midway between P and Q.

The *instantaneous acceleration* a at point P is defined in magnitude and direction as the limit approached by the average acceleration when point Q approaches point P and Δv and Δt both approach zero:

$$\text{Instantaneous acceleration } a = \lim_{\Delta t \to 0} \frac{\Delta v}{\Delta t} = \frac{dv}{dt}. \tag{5-7}$$

The instantaneous acceleration vector at point P is shown in Fig. 5-2c. Note that it does *not* have the same direction as the velocity vector. Reference to the construction in Fig. 5-2c shows that the acceleration vector must always lie on the *concave* side of the curved path.

5-4 Components of acceleration

Figure 5-3 again shows the motion of a particle referred to a rectangular coordinate system. The x- and y-components, a_x and a_y, of the instantaneous acceleration a of the particle are

$$a_x = \lim_{\Delta t \to 0} \frac{\Delta v_x}{\Delta t} = \frac{dv_x}{dt}, \qquad a_y = \lim_{\Delta t \to 0} \frac{\Delta v_y}{\Delta t} = \frac{dv_y}{dt}. \tag{5-8}$$

5-3 The acceleration a is resolved into its components a_x and a_y.

In terms of unit vectors,

$$a = \frac{dv_x}{dt}i + \frac{dv_y}{dt}j. \qquad (5\text{-}9)$$

By using Eqs. (5-4) and (5-5), we may also express the acceleration as

$$a_x = \frac{d^2x}{dt^2}, \qquad a_y = \frac{d^2y}{dt^2}, \qquad (5\text{-}10)$$

or

$$a = \frac{d^2x}{dt^2}i + \frac{d^2y}{dt^2}j. \qquad (5\text{-}11)$$

If the components a_x and a_y are known, the magnitude and direction of the acceleration a can be obtained, just as with velocity. When the acceleration is known, the force on the particle can be found from Newton's second law,

$$F = ma.$$

Conversely, if the force F is known at every point, the acceleration a and its components a_x and a_y can be found from Newton's second law.

Example The coordinates of a particle moving in the xy-plane are given as functions of time by

$$x = 1\text{ m} + (2\text{ m}\cdot\text{s}^{-2})t^2, \qquad y = (2\text{ m}\cdot\text{s}^{-1})t + (1\text{ m}\cdot\text{s}^{-3})t^3.$$

Find the particle's position, velocity, and acceleration at time $t = 2$ s.

Solution At time $t = 2$ s,

$$x = 1\text{ m} + (2\text{ m}\cdot\text{s}^{-2})(2\text{ s})^2 = 9\text{ m},$$
$$y = (2\text{ m}\cdot\text{s}^{-1})(2\text{ s}) + (1\text{ m}\cdot\text{s}^{-3})(2\text{ s})^3 = 12\text{ m}.$$

The particle's distance from the origin at this time is

$$r = \sqrt{x^2 + y^2} = \sqrt{(9\text{ m})^2 + (12\text{ m})^2} = 15\text{ m}.$$

The velocity at any time t is given by

$$v_x = \frac{dx}{dt} = 2(2\text{ m}\cdot\text{s}^{-2})t, \qquad v_y = \frac{dx}{dt} = (2\text{ m}\cdot\text{s}^{-1}) + 3(1\text{ m}\cdot\text{s}^{-3})t^2.$$

The velocity at time $t = 2$ s is given by

$$v_x = 2(2\text{ m}\cdot\text{s}^{-2})(2\text{ s}) = 8\text{ m}\cdot\text{s}^{-1},$$
$$v_y = (2\text{ m}\cdot\text{s}^{-1}) + 3(1\text{ m}\cdot\text{s}^{-3})(2\text{ s})^2 = 14\text{ m}\cdot\text{s}^{-1},$$

or

$$v = (8\text{ m}\cdot\text{s}^{-1})i + (14\text{ m}\cdot\text{s}^{-1})j.$$

The magnitude of the velocity at this time is

$$v = \sqrt{v_x^2 + v_y^2} = \sqrt{(8\text{ m}\cdot\text{s}^{-1})^2 + (14\text{ m}\cdot\text{s}^{-1})^2} = 16.1\text{ m}\cdot\text{s}^{-1},$$

and its direction with respect to the positive x-axis is

$$\theta = \arctan\frac{14\text{ m}\cdot\text{s}^{-1}}{8\text{ m}\cdot\text{s}^{-1}} = 60.3°.$$

The acceleration at any time is given by

$$a_x = \frac{dv_x}{dt} = \frac{d^2x}{dt^2} = 2(2 \text{ m}\cdot\text{s}^{-2}), \qquad a_y = \frac{dv_y}{dt} = \frac{d^2y}{dt^2} = 6(1 \text{ m}\cdot\text{s}^{-3})t.$$

The acceleration at time $t = 2$ s is

$$a_x = 4 \text{ m}\cdot\text{s}^{-2}, \qquad a_y = 6(1 \text{ m}\cdot\text{s}^{-3})(2 \text{ s}) = 12 \text{ m}\cdot\text{s}^{-2},$$

or

$$\boldsymbol{a} = (4 \text{ m}\cdot\text{s}^{-2})\boldsymbol{i} + (12 \text{ m}\cdot\text{s}^{-2})\boldsymbol{j}.$$

The magnitude of the acceleration at this time is

$$a = \sqrt{a_x^2 + a_y^2} = \sqrt{(4 \text{ m}\cdot\text{s}^{-2})^2 + (12 \text{ m}\cdot\text{s}^{-2})^2} = 12.6 \text{ m}\cdot\text{s}^{-2},$$

and its direction with respect to the positive x-axis is

$$\theta = \arctan \frac{a_y}{a_x} = \arctan \frac{12 \text{ m}\cdot\text{s}^{-2}}{4 \text{ m}\cdot\text{s}^{-2}} = 71.6°.$$

We note that the direction of the acceleration is different from that of the velocity, and that it is *not* tangent to the particle's path.

The acceleration of a particle moving in a curved path can also be resolved into rectangular components $a_\perp$ and $a_\parallel$, in directions *normal* and *parallel* (or tangential) to the path, as shown in Fig. 5-4a. Unlike the rectangular components referred to a set of fixed axes, the normal and tangential components do not have fixed directions in space. They do, however, have a direct physical significance. The parallel component $a_\parallel$ arises from a change in the *magnitude* of the velocity vector $\boldsymbol{v}$, while the normal component $a_\perp$ arises from a change in the *direction* of the velocity.

To understand these statements more fully, we consider the two special cases shown in Fig. 5-4. In Fig. 5-4b, the acceleration is at some instant *parallel* to the velocity $\boldsymbol{v}_1$ at that instant. Then because $\boldsymbol{a}$ gives the rate of change of velocity, the change in $\boldsymbol{v}$ during a small time interval Δt is a vector $\Delta \boldsymbol{v}$ having the same direction as $\boldsymbol{a}$ and hence the same direction as $\boldsymbol{v}_1$. Thus the velocity $\boldsymbol{v}_2$ at the end of Δt, being equal to $\boldsymbol{v}_1 + \Delta \boldsymbol{v}$, is a vector having the same direction as $\boldsymbol{v}_1$ but somewhat greater magnitude.

In Fig. 5-4c, the acceleration is perpendicular to the instantaneous velocity. Then in an interval Δt, the change $\Delta \boldsymbol{v}$ is a vector perpendicular to $\boldsymbol{v}_1$, as shown. Again $\boldsymbol{v}_2 = \boldsymbol{v}_1 + \Delta \boldsymbol{v}$, but in this case $\boldsymbol{v}_1$ and $\boldsymbol{v}_2$ have different directions. As the time interval Δt approaches zero, the angle θ in the figure also approaches zero, $\Delta \boldsymbol{v}$ approaches perpendicularity to both $\boldsymbol{v}_1$ and $\boldsymbol{v}_2$, and $\boldsymbol{v}_1$ and $\boldsymbol{v}_2$ have the same magnitude.

Thus when $\boldsymbol{a}$ is *parallel* to $\boldsymbol{v}$, its effect is to change the magnitude of $\boldsymbol{v}$ but not its direction; when $\boldsymbol{a}$ is *perpendicular* to $\boldsymbol{v}$, its effect is to change the direction of $\boldsymbol{v}$ but not its magnitude. In general $\boldsymbol{a}$ may have components *both* parallel and perpendicular to $\boldsymbol{v}$, but the above statements are still valid for the individual components. In particular, a particle traveling along a curved path with constant speed has a nonzero acceleration, even though the magnitude of $\boldsymbol{v}$ does not change. In this case the acceleration is always perpendicular to $\boldsymbol{v}$ at each point. When a

5-4 (a) The acceleration is resolved into a component $a_\parallel$ parallel or *tangent* to the path and a component $a_\perp$ perpendicular or *normal* to the path. (b) When $\boldsymbol{a}$ is parallel to $\boldsymbol{v}$, the magnitude of $\boldsymbol{v}$ increases but its direction does not change. (c) When $\boldsymbol{a}$ is perpendicular to $\boldsymbol{v}$, the direction of $\boldsymbol{v}$ changes but its magnitude is constant.

particle moves in a circle with constant speed, the acceleration is at each instant directed toward the center of the circle. This special case will be considered in detail in Sec. 5-6.

When a body moves in a curved path, a transverse force must be exerted on it to deviate it sidewise. The ratio of transverse (or normal) force to normal acceleration is the mass of the body, and this is also equal to the ratio of tangential force to tangential acceleration. That is, if $F_\perp$ and $F_\parallel$ are the normal and tangential components of force on a body moving in a curved path, Newton's second law (with the same value of m) applies to each of these components:

$$F_\perp = ma_\perp, \quad F_\parallel = ma_\parallel.$$

If the force on a particle is *always* normal to the path, then $F_\parallel = 0$, $a_\parallel = 0$, and the particle has no tangential component of acceleration. The *magnitude* of the velocity then remains constant and the only effect of the force is to change the *direction* of motion, that is, to deviate the particle sidewise.

If the force has no normal component, then $F_\perp = 0$, $a_\perp = 0$, and there is no change in the *direction* of the velocity; that is, the particle moves in a straight line.

5-5 Motion of a projectile

Any object that is given an initial velocity and that subsequently follows a path determined by the gravitational force acting on it and by the frictional resistance of the atmosphere is called a *projectile*. Thus the term applies to a missile shot from a gun, a rocket after its fuel is exhausted, an object dropped from an airplane, or a thrown or batted baseball. The motion of a freely falling body discussed in Chapter 3 is a special case of projectile motion. The path followed by a projectile is called its *trajectory*.

We shall consider only trajectories that are of sufficiently short range so that the gravitational force can be considered *constant* in both magnitude and direction. The motion will be referred to axes fixed with respect to the earth. This is not precisely an inertial system, but for trajectories of short range, the error incurred by regarding it as an inertial system is very small. Finally, all effects of air resistance will be ignored; thus our results are precise only for motion in a vacuum on a flat, nonrotating earth. These simplifying assumptions form the basis of an idealized *model* of the physical problem, in which we neglect unimportant details and focus attention on the most important aspects of the phenomenon.

Since the only force on a projectile in this idealized case is its weight, considered constant in magnitude and direction, the motion is best referred to a set of rectangular coordinate axes. We shall take the x-axis horizontal and the y-axis vertically upward. The x-component of the force on the projectile is then zero and the y-component is the weight of the projectile, $-mg$. Then from Newton's second law,

$$a_x = \frac{F_x}{m} = 0, \quad a_y = \frac{F_y}{m} = \frac{-mg}{m} = -g$$

or

$$a = -gj.$$

That is, the horizontal component of acceleration is zero and the vertical component is downward and equal to that of a freely falling body. Since zero acceleration means constant velocity, the motion can be described as a combination of *horizontal motion with constant velocity* and *vertical motion with constant acceleration*.

The key to analysis of projectile motion is the fact that all the needed vector relationships, including Newton's second law and the definitions of velocity and acceleration, can be expressed in terms of separate equations for the x- and y-components of these vector quantities. It was seen in Chapter 4 that the single vector equation $F = ma$ is equivalent to the two component equations

$$F_x = ma_x \quad \text{and} \quad F_y = ma_y.$$

Similarly, each component of velocity is the rate of change of the corresponding coordinate, and each component of acceleration is the rate of change of the corresponding velocity component. In this sense the x and y motions are independent and may be analyzed separately. The actual motion is then the superposition of these separate motions.

Suppose that at time $t = 0$ our particle is at the point (x_0, y_0) and has velocity components v_{0x} and v_{0y}. As seen above, the components of acceleration are $a_x = 0$, $a_y = -g$. The variation of each coordinate with time is a case of motion with constant acceleration, and Eqs. (3-12) and (3-16) can be used directly. Substituting v_{0x} for v_0 and 0 for a_x, we find for x

$$v_x = v_{0x}, \tag{5-12}$$

$$x = x_0 + v_{0x}t. \tag{5-13}$$

Similarly, substituting v_{0y} for v_0 and $-g$ for a,

$$v_y = v_{0y} - gt, \tag{5-14}$$

$$y = y_0 + v_{0y}t - \tfrac{1}{2}gt^2. \tag{5-15}$$

The content of Eqs. (5-12) through (5-15) can also be represented by the vector equations

$$v = v_0 - gtj, \tag{5-16}$$

$$r = r_0 + v_0 t - \tfrac{1}{2}gt^2 j, \tag{5-17}$$

where r_0 is the position vector at time $t = 0$.

Usually it is convenient to take the initial position as the origin; in this case, $x_0 = y_0 = 0$ or $r_0 = 0$. This might be, for example, the position of a ball at the instant it leaves the thrower's hand, or the position of a bullet at the instant it leaves the gun barrel.

Figure 5-5 shows the path of a projectile that passes through the origin at time $t = 0$. The position, velocity, and velocity components of the projectile are shown at a series of times separated by equal intervals. As the figure shows, v_x does not change, but v_y changes by equal amounts in successive intervals, corresponding to constant y-acceleration.

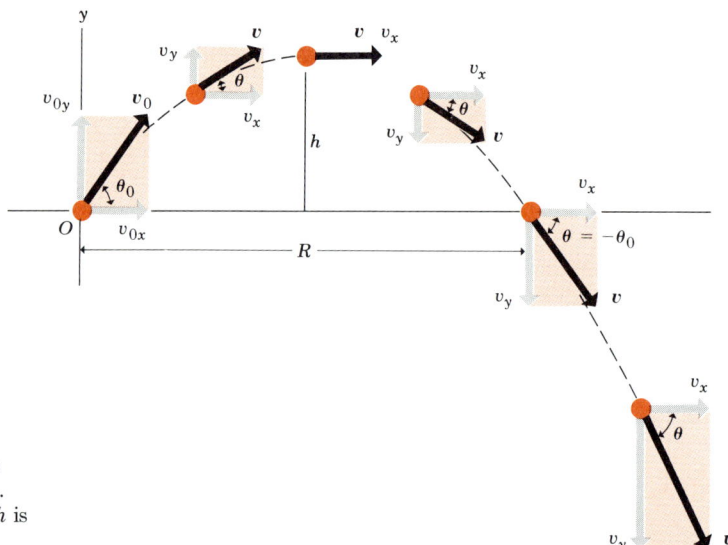

5-5 Trajectory of a body projected with an initial velocity v_0 at an angle of departure θ_0. The distance R is the horizontal range, and h is the maximum height.

The initial velocity v_0 may be represented by its magnitude v_0 (the initial speed) and the angle θ_0 it makes with the positive x-axis. In terms of these quantities, the *components* v_{0x} and v_{0y} of initial velocity are

$$v_{0x} = v_0 \cos \theta_0,$$
$$v_{0y} = v_0 \sin \theta_0. \tag{5-18}$$

Using these relations in Eqs. (5-12) through (5-15) and setting $x_0 = y_0 = 0$, we obtain

$$x = (v_0 \cos \theta_0)t, \tag{5-19}$$
$$y = (v_0 \sin \theta_0)t - \tfrac{1}{2}gt^2, \tag{5-20}$$
$$v_x = v_0 \cos \theta_0, \tag{5-21}$$
$$v_y = v_0 \sin \theta_0 - gt. \tag{5-22}$$

These equations describe the position and velocity of the projectile in Fig. 5-5 at any time t.

A variety of additional information can be obtained from Eqs. (5-19) through (5-22). For example, the distance r of the projectile from the origin at any time (the magnitude of the position vector $\mathbf{r}$) is given by

$$r = \sqrt{x^2 + y^2}. \tag{5-23}$$

The projectile's speed (the magnitude of its resultant velocity) is

$$v = \sqrt{v_x^2 + v_y^2}. \tag{5-24}$$

The *direction* of the velocity, in terms of the angle θ it makes with the positive x-axis, is given by

$$\tan \theta = \frac{v_y}{v_x}. \tag{5-25}$$

The velocity vector $\mathbf{v}$ is tangent to the trajectory, so its direction is the same as that of the trajectory.

Equations (5-19) and (5-20) give the position of the particle in terms of the parameter t. The equation in terms of x and y can be obtained by eliminating t. We find $t = x/v_0 \cos\theta_0$ and

$$y = (\tan\theta_0)x - \frac{g}{2v_0^2 \cos^2\theta_0} x^2. \tag{5-26}$$

The quantities v_0, $\tan\theta_0$, $\cos\theta_0$, and g are constants, so the equation has the form

$$y = ax - bx^2,$$

where a and b are constants. This is the equation of a *parabola*.

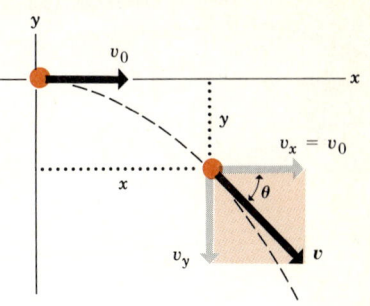

5-6 Trajectory of a body projected horizontally.

Example 1 A ball is projected horizontally with a velocity v_0 of magnitude $5 \text{ m}\cdot\text{s}^{-1}$. Find its position and velocity after $\frac{1}{4}$ s (see Fig. 5-6).

Solution In this case, the departure angle is zero. The initial vertical velocity component is therefore zero. The horizontal velocity component equals the initial velocity and is constant.

The x- and y-coordinates, when $t = \frac{1}{4}$ s, are, from Eqs. (5-19) and (5-20),

$$x = v_0 t = (5 \text{ m}\cdot\text{s}^{-1})(\tfrac{1}{4} \text{ s}) = 1.25 \text{ m},$$

$$y = -\tfrac{1}{2}gt^2 = -\tfrac{1}{2}(9.8 \text{ m}\cdot\text{s}^{-2})(\tfrac{1}{4} \text{ s})^2 = -0.306 \text{ m}.$$

The distance of the projectile from the origin at this time is

$$r = \sqrt{x^2 + y^2} = \sqrt{(1.25 \text{ m})^2 + (-0.306 \text{ m})^2} = 1.29 \text{ m}.$$

The components of velocity are, from Eqs. (5-21) and (5-22),

$$v_x = v_0 = 5 \text{ m}\cdot\text{s}^{-1},$$

$$v_y = -gt = (-9.8 \text{ m}\cdot\text{s}^{-2})(\tfrac{1}{4} \text{ s}) = -2.45 \text{ m}\cdot\text{s}^{-1}.$$

The resultant velocity has magnitude

$$v = \sqrt{v_x^2 + v_y^2} = 5.57 \text{ m}\cdot\text{s}^{-1}.$$

The angle θ is

$$\theta = \arctan\frac{v_y}{v_x} = \arctan\frac{-2.45 \text{ m}\cdot\text{s}^{-1}}{5.00 \text{ m}\cdot\text{s}^{-1}} = -26.1°.$$

That is, the velocity is 26.1° *below* the horizontal.

Example 2 In Fig. 5-5, let $v_0 = 160 \text{ ft}\cdot\text{s}^{-1}$, $\theta_0 = 53.1°$. Then

$$v_{0x} = v_0 \cos\theta_0 = (160 \text{ ft}\cdot\text{s}^{-1})(0.60) = 96 \text{ ft}\cdot\text{s}^{-1},$$

$$v_{0y} = v_0 \sin\theta_0 = (160 \text{ ft}\cdot\text{s}^{-1})(0.80) = 128 \text{ ft}\cdot\text{s}^{-1}.$$

a) Find the position of the projectile, and the magnitude and direction of its velocity, when $t = 2.0$ s.

$$x = (96 \text{ ft}\cdot\text{s}^{-1})(2.0 \text{ s}) = 192 \text{ ft},$$

$$y = (128 \text{ ft}\cdot\text{s}^{-1})(2.0 \text{ s}) - \tfrac{1}{2}(32 \text{ ft}\cdot\text{s}^{-2})(2.0 \text{ s})^2$$
$$= 192 \text{ ft}.$$

$$v_x = 96 \text{ ft} \cdot \text{s}^{-1},$$
$$v_y = 128 \text{ ft} \cdot \text{s}^{-1} - (32 \text{ ft} \cdot \text{s}^{-2})(2.0 \text{ s}) = 64 \text{ ft} \cdot \text{s}^{-1},$$
$$v = \sqrt{v_x^2 + v_y^2} = 115.4 \text{ ft} \cdot \text{s}^{-1},$$
$$\theta = \arctan \frac{64 \text{ ft} \cdot \text{s}^{-1}}{96 \text{ ft} \cdot \text{s}^{-1}} = \arctan 0.667 = 33.7°.$$

b) Find the time at which the projectile reaches the highest point of its flight, and find the elevation of this point.

At the highest point, the vertical velocity v_y is zero. If t_1 is the time at which this point is reached,

$$v_y = 0 = 128 \text{ ft} \cdot \text{s}^{-1} - (32 \text{ ft} \cdot \text{s}^{-2})t_1,$$
$$t_1 = 4 \text{ s}.$$

The elevation h of the point is the value of y when $t = 4$ s.

$$h = (128 \text{ ft} \cdot \text{s}^{-1})(4 \text{ s}) - \tfrac{1}{2}(32 \text{ ft} \cdot \text{s}^{-2})(4 \text{ s})^2 = 256 \text{ ft}.$$

c) Find the *horizontal range R,* that is, the horizontal distance from the starting point to the point at which the projectile returns to its original elevation and at which, therefore, $y = 0$. Let t_2 be the time at which this point is reached. Then

$$y = 0 = (128 \text{ ft} \cdot \text{s}^{-1})t_2 - \tfrac{1}{2}(32 \text{ ft} \cdot \text{s}^{-2})t_2^2.$$

This quadratic equation has two roots

$$t_2 = 0 \quad \text{and} \quad t_2 = 8 \text{ s},$$

corresponding to the two times at which $y = 0$. Evidently the time desired is the second root, $t_2 = 8$ s, which is just twice the time to reach the highest point. The time of descent therefore equals the time to rise.

The horizontal range R is the value of x when $t = 8$ s:

$$R = v_x t_2 = (96 \text{ ft} \cdot \text{s}^{-1})(8 \text{ s}) = 768 \text{ ft}.$$

The vertical component of velocity at this point is

$$v_y = 128 \text{ ft} \cdot \text{s}^{-1} - (32 \text{ ft} \cdot \text{s}^{-2})(8 \text{ s}) = -128 \text{ ft} \cdot \text{s}^{-1}.$$

That is, the vertical velocity has the same magnitude as the initial vertical velocity, but the opposite direction. Since v_x is constant, the angle below the horizontal at this point equals the angle of departure.

d) If unimpeded, the projectile continues to travel beyond its horizontal range. For example, the projectile might have been fired from the edge of a high cliff, so that negative values of y are possible. It is left as an exercise to compute the position and velocity at a time 10 s after the start, corresponding to the last position shown in Fig. 5–5. The results are:

$$x = 960 \text{ ft}, \quad y = -320 \text{ ft},$$
$$v_x = 96 \text{ ft} \cdot \text{s}^{-1}, \quad v_y = -192 \text{ ft} \cdot \text{s}^{-1}.$$

Example 3 Figure 5–7 illustrates an interesting experimental demonstration of the properties of projectile motion. A ball (shown by the light-colored circle) is projected directly toward a second ball (the dark

5-7 Trajectory of a ball thrown directly at another ball that is freely falling.

circle). The second ball is released from rest at the instant the first is projected, and it is found that the balls collide as shown regardless of the value of the initial velocity. To show that this must happen, we note that the initial elevation of the second ball is $x \tan \theta_0$, and that in time t it falls a distance $\tfrac{1}{2}gt^2$. Its elevation at the instant of collision is therefore

$$y = x \tan \theta_0 - \tfrac{1}{2}gt^2.$$

In this same time the first ball travels the distance x with constant x-component of velocity $v_0 \cos \theta$, so $x = v_0 \cos \theta_0 t$. Solving this for t and substituting in the preceding equation, we obtain

$$y = x \tan \theta_0 - \frac{1}{2}g \left(\frac{x}{v_0 \cos \theta_0} \right)^2.$$

But this is the same expression as Eq. (5-26) for the path of the first ball. Thus at the instant the first ball reaches the line along which the second is falling, the two heights are the same, and the balls must collide at this point.

(monkey/hunter problem)

Example 4 For a projectile launched with speed v_0 at initial angle θ_0, derive general expressions for the maximum height h and range R shown in Fig. 5-5. For a given v_0, what value of θ_0 gives maximum range?

Solution We follow the same pattern as in Example 2b. First, *when* does the projectile reach its maximum height? At this point $v_y = 0$, so the time t_1 at maximum height is given by

$$v_y = v_0 \sin \theta_0 - gt_1 = 0,$$

$$t_1 = \frac{v_0 \sin \theta_0}{g}.$$

Next, what is the value of y at this time? From Eq. (5-20),

$$h = v_0 \sin \theta_0 \left(\frac{v_0 \sin \theta_0}{g} \right) - \frac{1}{2} g \left(\frac{v_0 \sin \theta_0}{g} \right)^2$$

$$= \frac{v_0^2 \sin^2 \theta_0}{2g}. \tag{5-27}$$

As expected, the maximum value of h occurs when the projectile is launched straight up, for then $\theta_0 = 90°$, $\sin \theta_0 = 1$, and $h = v_0^2/2g$. If it is launched horizontally, $\theta_0 = 0$ and the maximum height is zero!

To find the range, we first find the time t_2 when the projectile returns to the ground. At that time $y = 0$ and, from Eq. (5–20),

$$t_2(v_0 \sin \theta_0 - \tfrac{1}{2}gt_2) = 0.$$

The two roots of this quadratic equation for t_2 are $t_2 = 0$ and $t_2 = 2v_0 \sin \theta_0/g$. The first is obviously the time the projectile *leaves* the ground; the range R is the value of x at the second time. From Eq. (5–19),

$$R = (v_0 \cos \theta_0)\left(\frac{2v_0 \sin \theta_0}{g}\right).$$

According to a familiar trigonometric identity, $2 \sin \theta_0 \cos \theta_0 = \sin 2\theta_0$, so

$$R = \frac{v_0^2 \sin 2\theta_0}{g}. \tag{5–28}$$

The maximum value of $\sin 2\theta_0$, namely, unity, occurs when $2\theta_0 = 90°$, or $\theta_0 = 45°$, and this angle gives the maximum range for a given initial speed.

In some problems we want to know what the departure angle θ_0 should be for a given v_0 to produce a certain range R. Suppose a football player wants to throw a football at 20 m·s^{-1} to a receiver 30 m away; at what angle should he throw it? From Eq. (5–28),

$$\theta_0 = \frac{1}{2} \arcsin \frac{Rg}{v_0^2}$$

$$= \frac{1}{2} \arcsin \frac{(30 \text{ m})(9.80 \text{ m}\cdot\text{s}^{-2})}{(20 \text{ m}\cdot\text{s}^{-1})^2}$$

$$= \frac{1}{2} \arcsin 0.735.$$

There are *two* values of θ_0 between 0 and 90° satisfying this requirement, for arcsin 0.735 = 47.3° or 132.7°, giving $\theta_0 = 23.7°$ or 66.3°. Both these angles give the same range; the time of flight and the maximum height are greater for the higher-angle trajectory. Incidentally, the two possible values of θ_0 add to exactly 90°; is this a coincidence?

Figure 5–8 is a drawing made from a multiflash photograph of the trajectory of a ball; x- and y-axes and the initial velocity vector have been added. The horizontal distances between consecutive positions are all equal, showing that the horizontal velocity component is constant. The vertical distances first decrease and then increase, showing that the vertical motion is accelerated.

If the departure angle θ_0 is *below* the horizontal as, for instance, in the motion of a ball after it rolls off a sloping roof, exactly the same principles apply. The horizontal velocity component remains constant and equal to $v_0 \cos \theta_0$. The vertical motion is the same as that of a body projected *downward* with an initial y-component of velocity $v_0 \sin \theta_0$, which is now a negative quantity.

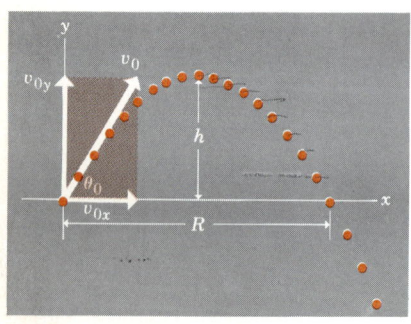

5–8 Trajectory of a body projected at an angle with the horizontal. (Reproduction of a multiflash photograph.)

5-6 Circular motion

It was shown in Sec. 5-3 that when a particle moves along a curved path it must have a component of acceleration perpendicular to the path, even if its speed is constant. For a *circular* path, there is a simple relation between the normal component of acceleration, the particle's speed, and the radius of the path. We now derive this relation for a particle moving in a circular path with constant speed. Such motion is called *uniform circular motion*.

Figure 5-9a represents a particle moving in a circular path of radius R with center at O. Vectors v_1 and v_2 represent its velocities at points P and Q. The vector change in velocity, Δv, is shown in Fig. 5-9b. The particle moves from P to Q in a time Δt.

The triangles OPQ and opq in Fig. 5-9 are *similar*, since both are isosceles triangles and the angles labeled $\Delta \theta$ are the same. Hence

$$\frac{\Delta v}{v_1} = \frac{\Delta s}{R} \quad \text{or} \quad \Delta v = \frac{v_1}{R} \Delta s.$$

The magnitude of the average normal acceleration $\bar{a}_\perp$ during Δt is $\Delta v / \Delta t$; from the above equation, this is equal to

$$\bar{a}_\perp = \frac{\Delta v}{\Delta t} = \frac{v_1}{R} \frac{\Delta s}{\Delta t}.$$

The *instantaneous* acceleration $a_\perp$ at point P is the limiting value of this expression, as point Q is taken closer and closer to point P, and as $\Delta t \to 0$:

$$a_\perp = \lim_{\Delta t \to 0} \frac{v_1}{R} \frac{\Delta s}{\Delta t} = \frac{v_1}{R} \lim_{\Delta t \to 0} \frac{\Delta s}{\Delta t}.$$

But the limiting value of $\Delta s / \Delta t$ is the speed v_1 at point P, and since P can be any point of the path, we can drop the subscript from v_1 and let v represent the speed at any point. Then

$$a_\perp = \frac{v^2}{R}. \tag{5-29}$$

The magnitude of the instantaneous normal acceleration is therefore equal to the square of the speed divided by the radius. The direction is perpendicular to v and inward along the radius, toward the center of the circle. Because of this it is called a *central*, a *centripetal*, or a *radial acceleration*. (The term "centripetal" means "seeking a center.")

An alternative derivation of Eq. (5-29), more analytic and less geometric than the above discussion, is sketched in Problem 5-45.

Figure 5-10 shows the directions of the velocity and acceleration vectors at a number of points, for a particle moving in a circle with constant speed.

In the above discussion the particle's speed was assumed constant. If the speed varies, Eq. (5-29) still gives the normal component of acceleration, but in that case there is also a *tangential* component of acceleration $a_\parallel$, equal to the rate of change of speed:

$$a_\parallel = \lim_{\Delta t \to 0} \frac{\Delta v_\parallel}{\Delta t}.$$

5-9 Construction for finding change in velocity, Δv, of a particle moving in a circle.

5-10 Velocity and acceleration vectors of a particle in uniform circular motion.

If the speed is constant, there is no tangential component of acceleration and the acceleration is purely normal, resulting only from the continuous change in *direction* of the velocity.

Example 1 A car traveling at a constant speed of $20 \text{ m} \cdot \text{s}^{-1}$ rounds a curve of radius 100 m. What is its acceleration?

Solution The magnitude of the acceleration is given by Eq. (5–29):

$$a_\perp = \frac{v^2}{R} = \frac{(20 \text{ m} \cdot \text{s}^{-1})^2}{100 \text{ m}} = 4.0 \text{ m} \cdot \text{s}^{-2}.$$

The direction of $\boldsymbol{a}$ at each instant is perpendicular to the velocity and directed toward the center of the circle.

Example 2 In a carnival ride, the passengers travel in a circle of radius 5.0 m, making one complete circle in 4.0 s. What is the acceleration?

Solution The speed is the circumference of the circle divided by the time τ for one revolution:

$$v = \frac{2\pi R}{\tau} = \frac{2\pi(5.0 \text{ m})}{4.0 \text{ s}} = 7.85 \text{ m} \cdot \text{s}^{-1}.$$

The centripetal acceleration is

$$a_\perp = \frac{v^2}{R} = \frac{(7.85 \text{ m} \cdot \text{s}^{-1})^2}{5.0 \text{ m}} = 12.3 \text{ m} \cdot \text{s}^{-2}.$$

As in the preceding example, the direction of $\boldsymbol{a}$ is always toward the center of the circle. The magnitude of $\boldsymbol{a}$ is greater than g, the acceleration due to gravity, so this is quite a wild ride!

5–7 Centripetal force

Newton's second law governs circular motion as well as all other motion of a particle. The acceleration toward the center of the circle, for a particle in uniform circular motion, must be caused by a *force* also directed toward the center. Since the magnitude of the radial acceleration equals v^2/R, and its direction is toward the center, the magnitude of the radial force on a particle of mass m is

$$F = ma_\perp = m\frac{v^2}{R}. \qquad (5\text{--}30)$$

A familiar example of such a force occurs when one ties an object to a string and whirls it in a circle. The string must constantly pull in toward the center; if it breaks, then the inward force no longer acts, and the object flies off along a tangent to the circle.

More generally, there may be several forces acting on a body in uniform circular motion. In this case the *vector sum* of all forces on the body must have the magnitude given by Eq. (5–30) and must be directed toward the center of the circle. For such problems the problem-solving techniques developed in Sec. 4–6 are applicable, as the following examples will show.

The force in Eq. (5–30) is sometimes called **centripetal force.** The usage is unfortunate because it implies that this force is somehow different from ordinary forces, or that the fact of circular motion somehow generates an additional force; neither is correct. *Centripetal* refers to the *effect* of the force, namely to the fact that it causes circular motion in which the direction of the velocity changes but not its magnitude. In the equation $\Sigma F = ma$, the sum of forces must include only the real physical forces, pushes or pulls exerted by strings, rods, or other agencies; the quantity mv^2/R does *not* appear in ΣF but belongs on the ma side of the equation.

Example 1 A small body of mass 0.2 kg revolves uniformly in a circle on a horizontal frictionless surface, attached by a cord 0.2 m long to a pin set in the surface. If the body makes two complete revolutions per second, find the force P exerted on it by the cord. (See Fig. 5-11.)

5-11 Body revolving uniformly in a circle on a horizontal frictionless surface.

Solution The circumference of the circle is

$$2\pi(0.2 \text{ m}) = 0.4\pi \text{ m}$$

so the speed is 0.8π m·s^{-1}. The magnitude of the centripetal acceleration is

$$a = \frac{v^2}{R} = \frac{(0.8\pi \text{ m}\cdot\text{s}^{-1})^2}{0.2 \text{ m}}$$
$$= 31.6 \text{ m}\cdot\text{s}^{-2}.$$

Since the body has no vertical acceleration, the forces n and w are equal and opposite and the force P is the resultant force. Therefore

$$P = ma = (0.2 \text{ kg})(31.6 \text{ m}\cdot\text{s}^{-2})$$
$$= 6.32 \text{ N}.$$

Example 2 Figure 5-12 represents a small body of mass m revolving in a horizontal circle with velocity v of constant magnitude at the end of a cord of length L. As the body swings around its path, the cord sweeps over the surface of a cone. The cord makes an angle θ with the vertical, so the radius of the circle in which the body moves is $R = L \sin \theta$ and the magnitude of the velocity v equals $2\pi L \sin \theta / \tau$, where τ is the time for one complete revolution.

The forces exerted on the body when it is in the position shown are its weight w and the tension T in the cord. (Note that the force diagram in Fig. 5-12 is exactly like that in Fig. 4-8b. The only difference is that in this case the acceleration a is the *radial* acceleration, v^2/R.) Let T be resolved into a horizontal component $T \sin \theta$ and a vertical component $T \cos \theta$. The body has no vertical acceleration, so the vertical forces $T \cos \theta$ and w are equal in magnitude. The resultant inward radial or *centripetal* force is the horizontal component $T \sin \theta$, which is equal to the mass m times the radial (or centripetal) acceleration:

$$T \sin \theta = m\frac{v^2}{R}, \qquad T \cos \theta = w.$$

When the first of these equations is divided by the second, and w is replaced by mg, the result is

$$\tan \theta = \frac{v^2}{Rg}. \tag{5-31}$$

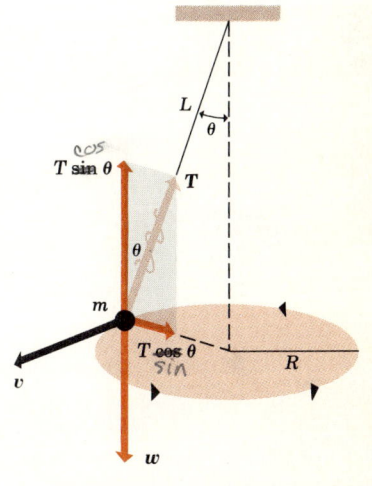

5-12 The conical pendulum.

When we make use of the relations $R = L \sin \theta$ and $v = 2\pi L \sin \theta / \tau$, Eq. (5–31) becomes

$$\cos \theta = \frac{g\tau^2}{4\pi^2 L}, \qquad (5\text{–}32)$$

or

$$\tau = 2\pi \sqrt{L(\cos \theta)/g}. \qquad (5\text{–}33)$$

Equation (5–32) indicates how the angle θ depends on the time of revolution τ and the length L of the cord. For a given length L, $\cos \theta$ decreases as the time is made shorter, and the angle θ increases. The angle never becomes 90°, however, since this requires that $\tau = 0$ or $v = \infty$. The *largest possible* value of τ, equal to $2\pi (L/g)^{1/2}$, occurs for very small θ, for which $\cos \theta = 1$.

Equation (5–33) is similar in form to the expression for the time of swing of a simple pendulum, which will be derived in Chapter 11. Because of this similarity, the present device is called a *conical pendulum*.

Some readers may wish to add to the forces shown in Fig. 5–12 an outward, "centrifugal" force, to "keep the body out there," or to "keep it in equilibrium." ("Centrifugal" means "fleeing a center.") Let us examine this point of view. First, the body *doesn't* stay there! A moment later it will be at a different position on its circular path. At the instant shown it is moving in the direction of the velocity vector v, and unless a resultant force acts on it, it will, according to Newton's first law, continue to move in this direction. If an outward force *were* acting on it, equal and opposite to the inward component of the force T, there would be *no* resultant inward force to deviate it sidewise from its present direction of motion.

Those who wish to add a force to "keep the body in equilibrium" forget that the term *equilibrium* refers to a state of rest, or of motion *in a straight line* with constant speed. Here, the body is *not* moving in a straight line, but in a circle. It is *not* in equilibrium, but has an acceleration toward the center of the circle and must be acted on by a resultant or *un*balanced force to produce this acceleration. In this example there is no outward force on the body!

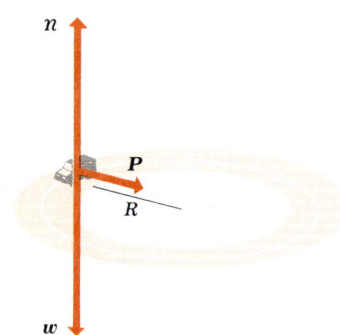

Example 3 Figure 5–13a represents an automobile or a railway car rounding a flat curve of radius R, on a level road or track. The forces acting on it are its weight w, the normal force n, and the force P, which must be provided by friction, in the case of an automobile, or by a force exerted by the rails against the flanges on the wheels of a railway car.

In order not to have to rely on friction, or to reduce wear on the rails and flanges, the road or the track may be *banked*, as shown in Fig. 5–13b. The normal force n then has a vertical component of magnitude $n \cos \theta$, and a horizontal component of magnitude $n \sin \theta$ toward the center, which provides the centripetal force. The banking angle θ can be computed as follows. If v is the speed and R the radius, then

$$n \sin \theta = \frac{mv^2}{R}.$$

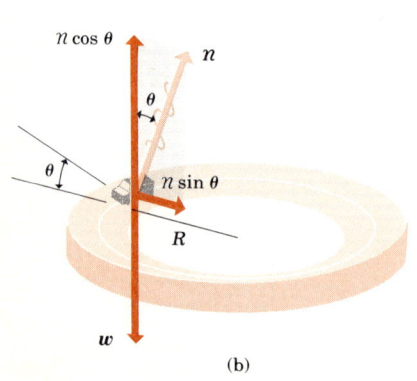

5–13 (a) Forces on a vehicle rounding a curve on a level track. (b) Forces when the track is banked.

Since there is no vertical acceleration,

$$n \cos \theta = w = mg.$$

Dividing the first equation by the second, we obtain

$$\tan \theta = \frac{v^2}{Rg}. \quad (5\text{-}34)$$

The tangent of the angle of banking is proportional to the square of the speed and inversely proportional to the radius. For a given radius no one angle is correct for all speeds. Hence in the design of highways and railroads, curves are banked for the *average speed* of the traffic over them. The same considerations apply to the correct banking angle of a plane when it makes a turn in level flight.

Note that the banking angle is given by the same expression as that for the angle with the vertical made by the cord of a conical pendulum. In fact, the force diagrams in Figs. 5-12 and 5-13 are identical.

5-8 Motion in a vertical circle

Figure 5-14 represents a small body attached to a cord of length R and whirling in a *vertical* circle about a fixed point O to which the other end of the cord is attached. The motion, while circular, is not *uniform,* since the speed increases on the way down and decreases on the way up. However, the normal component of acceleration is still given by $a_\perp = v^2/R$, as discussed above. In addition there is now a *tangential* component of acceleration.

The forces on the body at any point are its weight $\mathbf{w} = m\mathbf{g}$ and the tension $\mathbf{T}$ in the cord. Let the weight be resolved into a normal component, of magnitude $mg \cos \theta$, and a tangential component of magnitude $mg \sin \theta$, as in Fig. 5-14. The resultant tangential and normal forces are then

$$F_\parallel = mg \sin \theta \quad \text{and} \quad F_\perp = T - mg \cos \theta.$$

The tangential acceleration, from Newton's second law, is

$$a_\parallel = \frac{F_\parallel}{m} = g \sin \theta,$$

and is the same as that of a body sliding on a frictionless inclined plane of slope angle θ. The normal (radial) acceleration $a_\perp = v^2/R$ is

$$a_\perp = \frac{F_\perp}{m} = \frac{T - mg \cos \theta}{m} = \frac{v^2}{R},$$

and the tension in the cord is therefore

$$T = m\left(\frac{v^2}{R} + g \cos \theta\right). \quad (5\text{-}35)$$

At the lowest point of the path, $\theta = 0$, $\sin \theta = 0$, $\cos \theta = 1$. Hence at this point $F_\parallel = 0$, $a_\parallel = 0$, and the acceleration is purely radial (upward). The magnitude of the tension, from Eq. (5-35), is

$$T = m\left(\frac{v^2}{R} + g\right).$$

At the highest point, $\theta = 180°$, $\sin \theta = 0$, $\cos \theta = -1$, and the acceleration is again purely radial (downward). The tension is

$$T = m\left(\frac{v^2}{R} - g\right). \quad (5\text{-}36)$$

5-14 Forces on a body whirling in a vertical circle with center at O.

With motion of this sort, it is a familiar fact that there is a certain critical speed v_c at the highest point, below which the cord becomes slack and the path is no longer circular. To find this speed, set $T = 0$ in Eq. (5–36):

$$0 = m\left(\frac{v_c^2}{R} - g\right), \qquad v_c = \sqrt{Rg}.$$

The multiflash photographs of Fig. 5–15 illustrate another case of motion in a vertical circle, a small ball "looping the loop" on the inside of

5–15 Reproductions of multiflash photographs of a ball looping the loop in a vertical circle.

a vertical circular track. The inward normal force exerted on the ball by the track takes the place of the tension T in Fig. 5-14.

In Fig. 5-15a, the ball is released from an elevation such that its speed at the top of the track is greater than the critical speed, $\sqrt{Rg}$. In Fig. 5-14b, the ball starts from a lower elevation and reaches the top of the circle with a speed such that its own weight is slightly greater than the requisite centripetal force. In other words, the track would have to pull *outward* to maintain the circular motion. Since this is impossible, the ball leaves the track and moves for a short distance in a parabola. This parabola soon intersects the circle, however, and the remainder of the trip is completed successfully. In Fig. 5-15c, the start is made from a still lower elevation, the ball leaves the track sooner, and the parabolic path is clearly evident.

Example In Fig. 5-16, a small body of mass $m = 0.10$ kg swings in a vertical circle at the end of a cord of length $R = 1.0$ m. If its speed $v = 2.0$ m·s^{-1} when the cord makes an angle $\theta = 30°$ with the vertical, find (a) the radial and tangential components of its acceleration at this instant, (b) the magnitude and direction of the resultant acceleration, and (c) the tension T in the cord.

Solution
a) The radial component of acceleration is

$$a_\perp = \frac{v^2}{R} = \frac{(2.0 \text{ m·s}^{-1})^2}{1.0 \text{ m}} = 4.0 \text{ m·s}^{-2}.$$

The tangential component of acceleration, due to the tangential force $mg \sin \theta$, is

$$a_\parallel = g \sin \theta = (9.8 \text{ m·s}^{-2})(0.50) = 4.9 \text{ m·s}^{-2}.$$

b) The magnitude of the resultant acceleration (see Fig. 5-16b) is

$$a = \sqrt{a_\perp^2 + a_\parallel^2} = 6.33 \text{ m·s}^{-2}.$$

The angle ϕ is

$$\phi = \tan^{-1}\frac{a_\parallel}{a_\perp} = 50.8°.$$

c) The tension in the cord is given by $F_\perp = ma_\perp$: $T - mg \cos \theta = mv^2/R$, so

$$T = m\left(\frac{v^2}{R} + g \cos \theta\right) = 1.25 \text{ N}.$$

Note that the magnitude of the tangential acceleration is not constant but is proportional to the sine of the angle θ. Hence the equations of motion with constant acceleration *cannot* be used to find the speed at other points of the path. We shall show in the next chapter, however, how the speed at any point can be found from *energy* considerations.

5-16 (a) Forces on a body swinging in a vertical circle. (b) The radial and tangential components of acceleration are combined to obtain the resultant acceleration a.

5-9 Motion of a satellite

In discussing the trajectory of a projectile in Sec. 5-5, we assumed that the gravitational force on the projectile (its weight $\boldsymbol{w}$) had the same

102 MOTION IN A PLANE

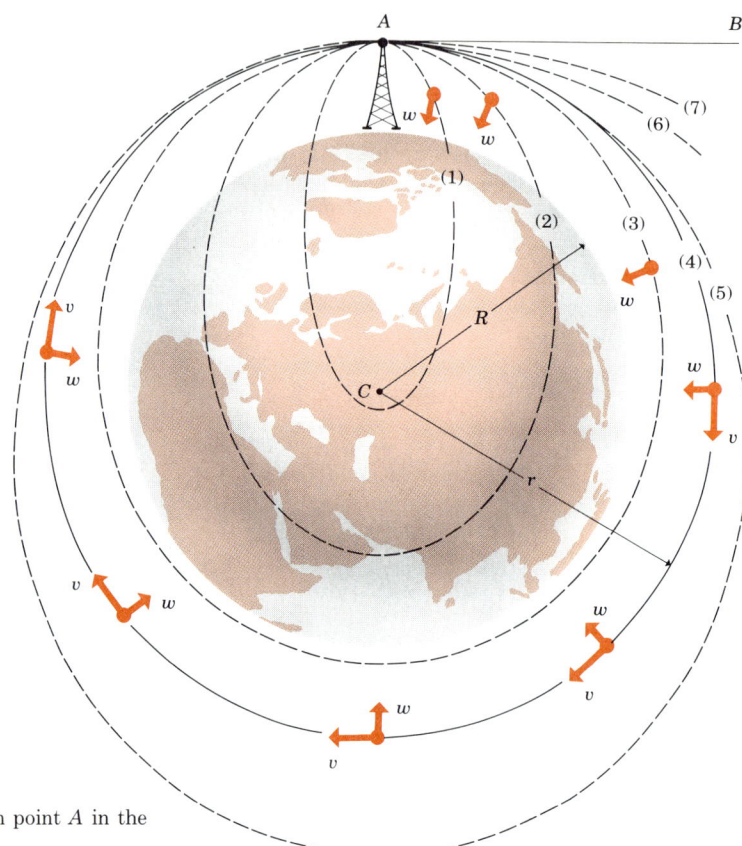

5–17 Trajectories of a body projected from point A in the direction AB with different initial velocities.

direction and magnitude at all points of the trajectory. These conditions are approximately satisfied if the projectile remains near the earth's surface and if its trajectory is small compared to the earth's radius. Under these conditions (if air resistance is neglected) the trajectory is a parabola.

In reality, the gravitational force is directed toward the center of the earth and is inversely proportional to the square of the distance from the earth's center, so it is *not* constant in either magnitude or direction. It can be shown that under an inverse-square force directed toward a fixed point the trajectory must always be a *conic section* (ellipse, circle, parabola, or hyperbola).

Suppose that a very tall tower could be constructed as in Fig. 5–17, and that a projectile were launched from point A at the top of the tower in the "horizontal" direction AB. If the initial velocity is not too great, the trajectory will be like that numbered (1), which is a portion of an ellipse with the earth's center C at one focus. If the trajectory is so short that changes in the magnitude and direction of w can be neglected, the ellipse *approximates* a parabola.

Trajectories (2) through (7) illustrate the effect of increasing the initial velocity. Trajectory (2) is again a portion of an ellipse. Trajectory (3), which just misses the earth, is a *complete* ellipse, and the projectile has become an earth satellite. Its velocity when it returns to point A is the same as its initial velocity, and in the absence of retarding forces it will repeat its motion indefinitely. The earth's rotation will have moved the tower to a different point by the time the satellite returns to point A, but this does not affect the orbit.

Trajectory (4) is a special case in which the orbit is a circle. Trajectory (5) is again an ellipse, (6) is a parabola, and (7) is a hyperbola. Trajectories (6) and (7) are not closed orbits; for these the projectile never returns to its starting point.

For simplicity we shall consider only circular orbits; many manmade satellites have very nearly circular orbits. The force that provides the necessary centripetal acceleration is the gravitational attraction of the earth (mass m_E) for the satellite (mass m). If the radius of the circular orbit (measured from the center of the earth) is r, then the gravitational force is given by $Gm_E m/r^2$, as discussed in Sec. 4-4. This is to be equated to the product of the satellite mass m and its acceleration v^2/r. Thus for a circular satellite orbit, Newton's second law requires

$$\frac{Gm_E m}{r^2} = \frac{mv^2}{r}.$$

When this is solved for v, the result is

$$v = \sqrt{\frac{Gm_E}{r}}. \qquad (5\text{-}37)$$

The relation shows that once the orbit radius is specified the satellite speed cannot be chosen arbitrarily but must be given by Eq. (5-37).

The orbit radius can also be related to the time required for one revolution, called the *period* τ of the satellite, or to the number of revolutions per unit time, called the *frequency* f. We first note that τ and f are related reciprocally; $\tau = 1/f$. Next, the speed is equal to the distance traveled in one revolution (the circumference $2\pi r$) divided by the time for one revolution. Thus

$$v = \frac{2\pi r}{\tau} = 2\pi r f. \qquad (5\text{-}38)$$

An expression for τ can be obtained by solving this for τ and combining with Eq. (5-37):

$$\tau = \frac{2\pi r}{v} = 2\pi r \sqrt{\frac{r}{Gm_E}} = \frac{2\pi r^{3/2}}{\sqrt{Gm_E}}. \qquad (5\text{-}39)$$

These quantities can also be expressed in terms of the radius R of the earth and the acceleration of gravity g at its surface. From Eq. (4-8), $g = Gm_E/R^2$. Using this to eliminate the quantity $(Gm_E)^{1/2}$ in Eqs. (5-37) and (5-39) yields

$$v = R\sqrt{\frac{g}{r}}, \qquad (5\text{-}40)$$

$$\tau = \frac{2\pi r}{v} = \frac{2\pi r}{R\sqrt{g/r}} = \frac{2\pi}{R\sqrt{g}} r^{3/2}. \qquad (5\text{-}41)$$

These relations show that larger orbits correspond to longer periods and slower speeds.

The acceleration, given by $a = v^2/r$ (*not* v^2/R!) can also be expressed in terms of g:

$$a = \frac{R^2}{r^2} g. \qquad (5\text{-}42)$$

This is the acceleration of gravity at radius r; the satellite, like any projectile, is a freely falling body. As expected, the acceleration is less than g at the earth's surface, in the ratio of the squares of the radii.

Example An earth satellite revolves in a circular orbit at a height of 300 km (about 200 miles) above the earth's surface.

a) What is the speed of the satellite, assuming the earth's radius to be 6380 km and g to be 9.80 m·s^{-2}? We have

$$R = 6380 \text{ km} = 6.38 \times 10^6 \text{ m}$$

and

$$r = 6380 \text{ km} + 300 \text{ km} = 6680 \text{ km} = 6.68 \times 10^6 \text{ m}.$$

From Eq. (5-40),

$$v = R\sqrt{\frac{g}{r}} = (6.38 \times 10^6 \text{ m}) \left(\frac{9.80 \text{ m} \cdot \text{s}^{-2}}{6.68 \times 10^6 \text{ m}} \right)^{1/2}$$

$$= 7730 \text{ m} \cdot \text{s}^{-1} = 25{,}350 \text{ ft} \cdot \text{s}^{-1}$$

$$= 17{,}300 \text{ mi} \cdot \text{hr}^{-1}.$$

Alternatively, Eq. (5-37) may be used. The mass of the earth is $m_E = 5.98 \times 10^{24}$ kg, so we have:

$$v = \sqrt{\frac{(6.67 \times 10^{-11} \text{ N} \cdot \text{m}^2 \cdot \text{kg}^{-2})(5.98 \times 10^{24} \text{ kg})}{6.68 \times 10^6 \text{ m}}} = 7730 \text{ m} \cdot \text{s}^{-1}.$$

b) What is the period τ?

$$\tau = \frac{2\pi r}{v} = 5430 \text{ s} = 90.5 \text{ min} = 1.51 \text{ hr}.$$

c) What is the radial acceleration of the satellite?

$$a = \frac{v^2}{r} = 8.94 \text{ m} \cdot \text{s}^{-2}.$$

This is equal to free-fall acceleration at a height of 300 km above the earth.

Our discussion has centered around the motion of manmade earth satellites, but it should be clear that it is also applicable to the circular motion of *any* body under its gravitational attraction to a stationary body. Other familiar examples are the motion of our moon around earth, that of the moons of Jupiter around Jupiter, and the motions of the planets around the sun, all in nearly circular orbits.

Indeed, the study of planetary motion historically played a pivotal role in the development of physics. Johannes Kepler (1571–1630) spent several painstaking years analyzing the motions of the planets, basing his work on precise measurements made by the Danish astronomer

Tycho Brahe (1546–1601). Kepler discovered that the orbits of the planets are (nearly circular) ellipses, and that the period of a planet in its orbit is proportional to the three-halves power of the orbit radius, as Eq. (5–39) shows. But it remained for Newton (1642–1727) to show, with the aid of his laws of motion and law of gravitation, that this behavior of the planets could be understood on the basis of the very same physical principles he had developed to analyze *terrestrial* motion. From our historical perspective three hundred years later, there is absolutely no doubt that this *Newtonian synthesis* (as it has come to be called) is one of the greatest achievements in the entire history of science, certainly comparable in significance to the development of quantum mechanics, the theory of relativity, and the understanding of DNA, in our own century.

*5-10 Effect of the earth's rotation on g

Because the earth rotates, it is not precisely an inertial frame of reference, and the apparent weight of an object on earth is not precisely equal to the earth's gravitational attraction. Figure 5–18 is a cutaway view of the earth, showing three observers, each holding a body of mass m hanging from a string. Each observer, unaware of the earth's rotation, *thinks* the body is in equilibrium under the action of the tension T in his string and an equal and opposite force w, which we shall call the *apparent* weight. This is different at different points on the earth, as we shall see.

If we assume the earth to be spherically symmetric, the earth's gravitational attraction, that is, the *true* weight, which we call w_0, has the

5-18 The resultant of the forces T and w_0 is the centripetal force F, equal in magnitude to mv^2/r.

same magnitude $F_g = Gmm_E/R^2$ at all points on the earth's surface. We also suppose that the center of the earth can be taken as the origin of an inertial coordinate system; this ignores the orbital motion of the earth, which is a much smaller effect than that of rotation about its axis. Then the body at the north pole really *is* in equilibrium in an inertial system, and the tension in that observer's string is equal to w_0. But the body at the equator is moving in a circle of radius R with speed v, and there must be a net inward force equal to the mass times the centripetal acceleration:

$$w_0 - T = mv^2/R. \tag{5-43}$$

Thus the magnitude of the apparent weight (equal to T at this location) is

$$w = w_0 - \frac{mv^2}{R}, \tag{5-44}$$

We also note that if the earth were not rotating, the body when released would have a free-fall acceleration g_0 given by $g_0 = w_0/m$, but that its actual acceleration relative to the observer at the equator is given by $g = w/m$. Dividing Eq. (5-44) by m and using these relations, we find

$$g = g_0 - \frac{v^2}{R}. \tag{5-45}$$

To evaluate v^2/R, we note that in one day (86,400 s) a point at the equator moves a distance equal to the earth's circumference, $2\pi R = 2\pi(6.38 \times 10^6 \text{ m})$. Thus

$$v = \frac{2\pi(6.38 \times 10^6 \text{ m})}{86,400 \text{ s}} = 464 \text{ m} \cdot \text{s}^{-1},$$

and

$$\frac{v^2}{R} = \frac{(464 \text{ m} \cdot \text{s}^{-1})^2}{6.38 \times 10^6 \text{ m}} = 0.0337 \text{ m} \cdot \text{s}^{-2}.$$

Thus for a spherically symmetric earth the acceleration due to gravity should be about 0.03 m·s^{-2} less at the equator than at the poles.

Table 5-1 gives the values of g at various locations, showing variations of approximately this magnitude. There are also small variations due to lack of perfectly spherical symmetry of the earth, local variations in density, and differences in elevation.

Table 5-1 Variations of g with latitude and elevation

Station	North latitude	Elevation, m	g, m·s^{-2}	g, ft·s^{-2}
Canal Zone	9°	0	9.78243	32.0944
Jamaica	18°	0	9.78591	32.1059
Bermuda	32°	0	9.79806	32.1548
Denver	40°	1638	9.79609	32.1393
Cambridge, Mass.	42°	0	9.80398	32.1652
Standard station			9.80665	32.1740
Greenland	70°	0	9.82534	32.2353

At points intermediate between the equator and the poles, the force relations are as shown for the observer at latitude θ. The difference in the magnitudes of g and g_0 is intermediate between zero and $0.0337 \text{ m} \cdot \text{s}^{-2}$, and as shown the *direction* of action of the apparent weight differs from the direction toward the center, by an angle α as shown. Working out the relationships for this more general case is left as an exercise; the values of g and α are given approximately by

$$g = g_0 - \frac{v^2 \cos \theta}{R}, \qquad \alpha = \frac{v^2 \sin \theta}{g_0 R}, \tag{5-46}$$

where v is the speed of a point on the rotating earth at latitude θ (less than the speed at the equator by a factor $\cos \theta$). At $\theta = 45°$, one finds

$$g = g_0 - 0.0049 \text{ m} \cdot \text{s}^{-2}, \qquad \alpha = 0.00035 \text{ rad} = 0.0020°.$$

A similar discussion can be applied to the phenomenon of "weightlessness" in satellites. A space vehicle in orbit is a freely falling body, with an acceleration $\boldsymbol{a}$ toward the earth's center equal to the value of the acceleration of gravity $\boldsymbol{g}$ at its orbit radius. The apparent weight w is given as before by

$$w = w_0 - ma = mg - a.$$

But, in this case,

$$\boldsymbol{g} = \boldsymbol{a},$$

so

$$w = 0.$$

It is in this sense that an astronaut in the vehicle is said to be "weightless" or in a state of "zero g." The vehicle is like the freely falling elevator discussed in Sec. 4-6, Example 7, except that it has a large constant tangential speed along its orbit.

5-11 Relative velocity

The concept of relative velocity, introduced in Sec. 3-8 for motion along a straight line, is easily extended to include motion in a plane or in space. Suppose that, in the example preceding Eq. (3-25), the automobile is traveling at some instant not in the same direction as the train but perpendicular to this direction, *across* the flatcars at $20 \text{ m} \cdot \text{s}^{-1}$. If the flatcars are moving at $15 \text{ m} \cdot \text{s}^{-1}$ relative to the earth, then the velocity of the automobile *relative to the earth* is the vector sum of these two velocities. That is, Eq. (3-25) is a special case of the more general *vector equation*:

$$v_{\text{AE}} = v_{\text{AF}} + v_{\text{FE}}. \tag{5-47}$$

Thus the automobile's velocity relative to the earth would be $25 \text{ m} \cdot \text{s}^{-1}$ at an angle of 53° to the direction of the train's motion.

Equation (5-47) can be extended to include any number of relative velocities. For example, if a bug B crawls along the floor of the automobile with a velocity relative to the automobile of v_{BA}, his velocity relative to the earth is the vector sum of his velocity relative to the automobile and that of the velocity of the automobile relative to the earth:

$$v_{\text{BE}} = v_{\text{BA}} + v_{\text{AE}}.$$

When this is combined with Eq. (5-47), we find

$$v_{BE} = v_{BA} + v_{AF} + v_{FE}. \qquad (5\text{-}48)$$

This equation illustrates the general rule for combining relative velocities.

1. Write each velocity with a double subscript in the *proper order*, meaning "velocity of (first subscript) relative to (second subscript)."
2. When relative velocities are added, the first letter of any subscript is to be the same as the last letter of the preceding subscript.
3. The first letter of the subscript of the first velocity in the sum, and the second letter of the subscript of the last velocity, are the subscripts, in that order, of the relative velocity represented by the sum. This somewhat lengthy statement should be clear when it is compared with Eq. (5-47) or (5-48).

Any of the relative velocities in an equation such as Eq. (5-47) can be transferred from one side of the equation to the other, with sign reversed. Thus Eq. (5-47) can be written

$$v_{AF} = v_{AE} - v_{FE}.$$

The velocity of the automobile relative to the flatcar equals the *vector difference* between the velocities of automobile and flatcar, each relative to the earth.

One more point should be noted. The velocity of body A relative to body B, v_{AB}, is the negative of the velocity of B relative to A, v_{BA}:

$$v_{AB} = -v_{BA}.$$

That is, v_{AB} is equal in magnitude and opposite in direction to v_{BA}. If the automobile is traveling to the *right* at 20 m·s^{-1}, relative to the flatcars, the flatcars are traveling to the *left* at 20 m·s^{-1}, relative to the automobile.

Example 1 The compass of an aircraft indicates that it is headed due north, and its airspeed indicator shows that it is moving through the air at 120 m·hr^{-1}. If there is a wind of 50 mi·hr^{-1} from west to east, what is the velocity of the aircraft relative to the earth?

Solution Let subscript A refer to the aircraft, and subscript F to the moving air (which now corresponds to the flatcar in Fig. 3-10). Subscript E refers to the earth. We have given

$$v_{AF} = 120 \text{ m·hr}^{-1}, \quad \text{due north,}$$
$$v_{FE} = 50 \text{ mi·hr}^{-1}, \quad \text{due east,}$$

and we wish to find the magnitude and direction of v_{AE}:

$$v_{AE} = v_{AF} + v_{FE}.$$

The three relative velocities are shown in Fig. 5-19. It follows from this diagram that

$$v_{AE} = 130 \text{ mi·hr}^{-1}, \quad 22.6° \text{ E of N.}$$

5-19 Vector diagram for aircraft flying north, wind blowing east, and resultant velocity vector of the plane.

Example 2 In what direction should the pilot head in order to travel due north? What will then be his velocity relative to the earth? (The magnitude of his airspeed, and the wind velocity, are the same as in the preceding example.)

Solution We now have given

$$v_{AF} = 120 \text{ mi} \cdot \text{hr}^{-1}, \quad \text{direction unknown,}$$
$$v_{FE} = 50 \text{ mi} \cdot \text{hr}^{-1}, \quad \text{due east,}$$

and we wish to find v_{AE}, whose magnitude is unknown but whose direction is due north. (Note that both this and the preceding example require us to determine two unknown quantities. In the former example, these were the *magnitude and direction* of v_{AE}. In this example, the unknowns are the *direction* of v_{AF} and the *magnitude* of v_{AE}.)

The three relative velocities must still satisfy the *vector equation*

$$v_{AE} = v_{AF} + v_{FE}.$$

The appropriate vector diagram is shown in Fig. 5–20. We find

$$v_{AE} = \sqrt{(120 \text{ mi} \cdot \text{hr}^{-1})^2 - (50 \text{ mi} \cdot \text{hr}^{-1})^2} = 109 \text{ mi} \cdot \text{hr}^{-1},$$
$$\theta = \sin^{-1} \frac{50 \text{ mi} \cdot \text{hr}^{-1}}{120 \text{ mi} \cdot \text{hr}^{-1}} = 24.6°.$$

That is, the pilot should head 24.6° W of N, and his ground speed will be 109 mi·hr⁻¹.

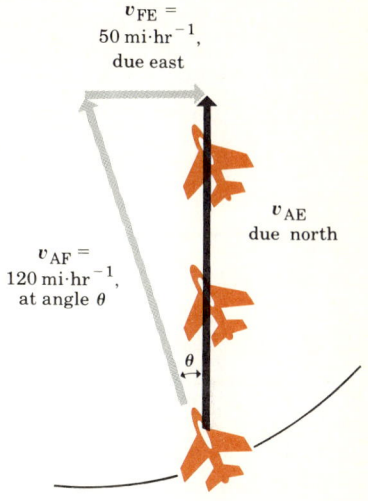

5–20 The vector v_{AF} shows the direction in which the pilot should point the plane in order to have the plane travel due north.

Questions

5-1 A simple pendulum (a mass swinging at the end of a string) swings back and forth in a circular arc. What is the direction of its acceleration at the ends of the swing? At the midpoint?

5-2 A football is thrown in a parabolic path. Is there any point at which the acceleration is parallel to the velocity? Perpendicular to the velocity?

5-3 When a rifle is fired at a distant target, the barrel is not lined up exactly on the target. Why not? Does the angle of correction depend on the distance of the target?

5-4 The acceleration of a falling body is measured in an elevator traveling upward at a constant speed of 9.8 m·s⁻¹. What result is obtained?

5-5 One can play catch with a softball in an airplane in level flight just as though the plane were at rest. Is this still possible when the plane is making a turn?

5-6 A package is dropped out of an airplane in level flight. If air resistance could be neglected, how would the motion of the package look to the pilot? To an observer on the ground?

5-7 No matter what the initial velocity, the motion of a projectile (neglecting air resistance) is *always* confined to a single plane. Why?

5-8 If a jumper can give himself the same initial speed regardless of the direction he jumps (forward or straight up), how is his maximum vertical jump (high jump) related to his maximum horizontal jump (broad jump)?

5-9 A manned space-flight projectile is launched in a parabolic trajectory. A man in the capsule feels weightless. Why? In what sense is he weightless?

5-10 The maximum range of a projectile occurs when it is aimed at 45°, if air resistance is neglected. Is this still true if air resistance is included? If not, is the optimum angle greater or less than 45°? Why?

5-11 A passenger in a car rounding a sharp curve is thrown toward the outside of the curve. What force throws him in this direction?

5-12 In uniform circular motion, what is the *average* velocity during one revolution? The average acceleration?

5-13 In uniform circular motion the acceleration is perpendicular to the velocity at every instant, even though both change continuously in direction. Is there any other motion having this property, or is uniform circular motion unique?

5-14 A certain centrifuge is claimed to operate at 100,000g. What does this mean?

5-15 If an artificial earth satellite has a period of exactly one day, how does its motion look to an observer on the rotating earth?

5-16 How could the motion of a satellite be used to determine the mass of the earth? Could a similar procedure be used to obtain the mass of the sun?

5-17 Raindrops hitting the side windows of a car in motion often leave diagonal streaks. Why? What about diagonal streaks on the windshield? Is the explanation the same or different?

5-18 In a rainstorm with a strong wind, what determines the best position in which to hold an umbrella?

5-19 In uninformed discussions of satellites, one hears questions such as "What keeps the satellite moving in its orbit?" and "What keeps the satellite up?" How do you answer these questions? Are your answers applicable to the moon?

Problems

5-1 A ball rolls off the edge of a tabletop 1 m above the floor, and strikes the floor at a point 1.5 m horizontally from the edge of the table.
a) Find the time of flight.
b) Find the initial velocity.
c) Find the magnitude and direction of the velocity of the ball just before it strikes the floor. Draw a diagram to scale.

5-2 A block slides off a horizontal tabletop 1 m high with a speed of $3 \text{ m} \cdot \text{s}^{-1}$. Find
a) the horizontal distance from the edge of the table at which the block strikes the floor, and
b) the horizontal and vertical components of its velocity when it reaches the floor.

5-3 An airplane flying at $100 \text{ m} \cdot \text{s}^{-1}$ drops a box at an elevation of 2000 m.
a) How much time is required for the box to reach the earth?
b) How far does it travel horizontally while falling?
c) Find the horizontal and vertical components of its velocity when it strikes the earth.

5-4 A block sliding toward the edge of a horizontal tabletop has a speed of $4 \text{ m} \cdot \text{s}^{-1}$ at a point 3 m from the edge. It slides off the edge of the table, which is 1 m high, and strikes the floor 1 m from the edge of the table. What was the coefficient of sliding friction between block and table?

5-5 A golf ball is driven horizontally from an elevated tee with a speed of $25 \text{ m} \cdot \text{s}^{-1}$. It strikes the fairway 2.5 s later.
a) How far has it fallen vertically?
b) How far has it traveled horizontally?
c) Find the horizontal and vertical components of its velocity, and the magnitude and direction of its resultant velocity, just before it strikes.

5-6 A level-flight plane, flying at an altitude of 1024 ft with a speed of $240 \text{ ft} \cdot \text{s}^{-1}$, is overtaking a motor boat traveling at $80 \text{ ft} \cdot \text{s}^{-1}$ in the same direction as the plane. At what distance astern of the boat should a bag be dropped from the plane in order to hit the boat?

5-7 A .22 calibre rifle bullet is fired in a horizontal direction with a muzzle velocity of $900 \text{ ft} \cdot \text{s}^{-1}$. In the absence of air resistance, how far will it have dropped in traveling a horizontal distance of
a) 50 yd?   b) 100 yd?   c) 150 yd?
d) How far will it drop in one second?

5-8 A ball is projected with an initial upward velocity component of $20 \text{ m} \cdot \text{s}^{-1}$ and a horizontal velocity component of $25 \text{ m} \cdot \text{s}^{-1}$.
a) Find the position and velocity of the ball after 2 s, 3 s, 4 s.
b) How much time is required to reach the highest point of the trajectory?
c) How high is this point?
d) How much time (after launch) is required for the ball to return to its original level?
e) How far has it traveled horizontally during this time? Show your results in a neat sketch, large enough to show all features clearly.

5-9 A batted baseball leaves the bat at an angle of 30° above the horizontal, and is caught by an outfielder 400 ft from the plate.
a) What was the initial speed of the ball?
b) How high did it rise?
c) How long was it in the air?

5-10 A spring gun projects a golf ball at an angle of 45° above the horizontal. The horizontal range is 10 m.
a) What is the maximum height to which the ball rises?
b) For the same initial speed, what are the two angles of departure for which the range is 6 m?
c) Sketch all three trajectories to scale in the same diagram.

5-11 Suppose the departure angle θ_0 in Fig. 5-7 is 15° and the distance x is 5 m. Where will the balls collide if the initial speed of the first is
a) $20 \text{ m} \cdot \text{s}^{-1}$?   b) $5 \text{ m} \cdot \text{s}^{-1}$?

Sketch both trajectories.

c) Will a collision take place if the departure angle is below horizontal?

5-12 If a baseball player can throw a ball a maximum

distance of 60 m over the ground, what is the maximum vertical height to which he can throw it? Assume the ball to have the same initial speed in each case.

5-13 A player kicks a football at an angle of 37° with the horizontal and with an initial speed of 48 ft·s^{-1}. A second player standing at a distance of 100 ft from the first in the direction of the kick starts running to meet the ball at the instant it is kicked. How fast must he run in order to catch the ball before it hits the ground?

5-14 A baseball leaves the bat at a height of 4 ft above the ground, traveling at an angle of 45° with the horizontal, and with a speed such that the horizontal range would be 400 ft. At a distance of 360 ft from home plate is a fence 30 ft high. Will the ball be a home run?

5-15 A projectile shot at an angle of 60° above the horizontal strikes a building 30 m away at a point 15 m above the point of projection.

a) Find the speed of projection.
b) Find the magnitude and direction of the velocity of the projectile when it strikes the building.

5-16

a) What must be the velocity of a projectile fired vertically upward to reach an altitude of 60 km?
b) What velocity is required to reach the same height if the gun makes an angle of 45° with the vertical?
c) Compute the time required to reach the highest point in each trajectory.
d) How far would a plane traveling at 500 km·hr^{-1} move in this time?

5-17 The angle of elevation of an antiaircraft gun is 70° and the muzzle velocity is 900 m·s^{-1}. For what time after firing should the fuse be set if the shell is to explode at an altitude of 1500 m?

5-18 A boy throws a water-filled balloon at an angle of 53.1° with a speed of 10 m·s^{-1}. A car is advancing toward the boy at a constant speed of 5 m·s^{-1}. If the balloon is to hit the car, how far away should the car be when the balloon is thrown?

5-19 Prove that a projectile launched at angle θ_0 has the same range as one launched with the same speed at angle $(90° - \theta_0)$.

5-20 An airplane diving at an angle of 36.9° with the horizontal drops a bag of sand from an altitude of 800 m. The bag is observed to strike the ground 5 s after its release.

a) What is the speed of the plane?
b) How far does the bag travel horizontally during its fall?
c) What are the horizontal and vertical components of its velocity just before it strikes the ground?

5-21 A 10-kg stone is dropped from a cliff in a high wind. The wind exerts a steady horizontal 50-N force on the stone as it falls. Is the path of the stone a straight line, a parabola, or some more complicated path? Explain.

5-22 A man is riding on a flatcar traveling with a constant speed of 30 ft·s^{-1} (Fig. 5-21). He wishes to throw a ball through a stationary hoop 16 ft above the height of his hands in such a manner that the ball will move horizontally as it passes through the hoop. He throws the ball with a speed of 40 ft·s^{-1} with respect to himself.

a) What must be the vertical component of the initial velocity of the ball?
b) How many seconds after he releases the ball will it pass through the hoop?
c) At what horizontal distance in front of the hoop must he release the ball?

Figure 5-21

5-23 In the example of Sec. 5-4, suppose the particle has mass 2 kg.

a) Derive an expression for the force on the particle, as a function of time.
b) Find the magnitude and direction of the force on the particle at time $t = 2$ s.
c) Find the times at which the particle's acceleration is parallel to its velocity.
d) At what times is the particle's acceleration parallel to the force?

5-24 The coordinates of a particle moving in the xy-plane are given as functions of time by

$$x = (2 \text{ m·s}^{-1})t, \qquad y = 19 \text{ m} - (2 \text{ m·s}^{-2})t^2.$$

a) What is the particle's distance from the origin at time $t = 2$ s?
b) What is the particle's velocity (magnitude and direction) at time $t = 2$ s?
c) What is the particle's acceleration (magnitude and direction) at time $t = 2$ s?
d) At what times is the particle's velocity perpendicular to its acceleration?
e) At what times is the particle's velocity perpendicular to its position vector? What are the locations of the particle at these times?

f) What is the particle's minimum distance from the origin? At what times does this minimum occur?

g) Sketch the path of the particle.

5-25 A particle having a mass of 3 kg moves in the xy-plane under the action of a force given by

$$F = (6\text{ N}\cdot\text{s}^{-2})t^2 i + (4\text{ N}\cdot\text{s}^{-1})t j.$$

a) Assuming that the particle is at rest at the origin at time $t = 0$, derive expressions for acceleration, velocity, and position vectors as functions of time.

b) Sketch the path of the particle.

c) Find the magnitude and direction of the velocity at $t = 3$ s.

5-26 An automobile travels around a circular track 2000 m in circumference at a constant speed of $30\text{ m}\cdot\text{s}^{-1}$.

a) Show in a scale diagram the velocity vectors of the automobile at the beginning and end of a time interval $\Delta t = 5$ s. Let 1 cm = $5\text{ m}\cdot\text{s}^{-1}$.

b) Find graphically the change in velocity, Δv, in this time interval.

c) Find the magnitude of the average acceleration during this interval, $\Delta v/\Delta t$.

d) What is the radial acceleration, v^2/R?

e) If the track is 10 m wide, what should be the elevation of the outer circumference above the inner circumference, for the speed above?

5-27 The radius of the earth's orbit around the sun (assumed circular) is 1.49×10^{11} m and the earth travels around this orbit in 365 days.

a) What is the magnitude of the orbital velocity of the earth, in meters per second?

b) What is the radial acceleration of the earth toward the sun, in meters per second squared?

5-28 A model of a helicopter rotor has four blades, each 2 m in length, and is rotated in a wind tunnel at 1500 rev·min^{-1}.

a) What is the linear speed of the blade tip, in meters per second?

b) What is the radial acceleration of the blade tip, expressed as a multiple of the acceleration of gravity, g?

c) A pressure-measuring device with a mass of 0.1 kg is mounted at the blade tip. Find the centripetal force on it, and compare with its weight.

5-29 A highway curve of radius 1600 ft is to be banked so that a car traveling 50 mi·hr^{-1} will have no tendency to skid sideways. At what angle should it be banked?

5-30 A flat (unbanked) curve on a highway has a radius of 800 ft, and a car rounds the curve at a speed of 50 mi·hr^{-1}. What must be the minimum coefficient of friction to prevent sliding?

5-31 A stone of mass 1 kg is attached to one end of a string 1 m long, of breaking strength 500 N, and is whirled in a horizontal circle on a frictionless tabletop. The other end of the string is kept fixed. Find the maximum velocity the stone can attain without breaking the string.

5-32 An unbanked circular highway curve on level ground makes a turn of 90°. The highway carries traffic at 60 mi·hr^{-1}, and the centripetal force on a vehicle is not to exceed $\frac{1}{10}$ of its weight. What is the minimum length of the curve, in miles?

5-33 A coin placed on a 12-in. record will revolve with the record when it is brought up to a speed of $33\frac{1}{3}$ rev·min^{-1}, provided the coin is not more than 4 in. from the axis.

a) What is the coefficient of static friction between coin and record?

b) How far from the axis can the coin be placed, without slipping, if the turntable rotates at 45 rev·min^{-1}?

5-34

a) At how many revolutions per second must the apparatus of Fig. 5-22 rotate about the vertical axis in order that the cord shall make an angle of 45° with the vertical?

b) What is then the tension in the cord?

c) Find the angle θ which the cord makes with the vertical if the system is rotating at 1.5 rev·s^{-1}. (Set up the general equation relating the angle θ to the number of revolutions per second, n, the lengths a and L, and the acceleration of gravity, g. Then find by trial the angle θ that satisfies this equation.)

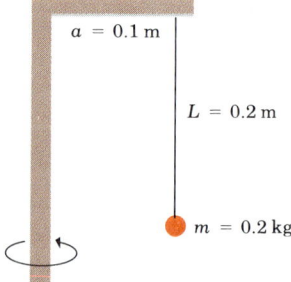

Figure 5-22

5-35 The "Giant Swing" at a county fair consists of a vertical central shaft with a number of horizontal arms attached at its upper end. Each arm supports a seat suspended from a cable 5 m long, the upper end of the cable being fastened to the arm at a point 4 m from the central shaft.

a) Find the time of one revolution of the swing if the cable supporting a seat makes an angle of 30° with the vertical.

b) Does the angle depend on the weight of the passenger for a given rate of revolution?

5-36 The 4-kg block in Fig. 5-23 is attached to a vertical rod by means of two strings. When the system rotates about the axis of the rod, the strings are extended as shown in the diagram.

a) How many revolutions per minute must the system make in order that the tension in the upper cord shall be 60 N?
b) What is then the tension in the lower cord?

Figure 5–23

5–37 A bead can slide without friction on a circular hoop of radius 0.1 m in a vertical plane. The hoop rotates at a constant rate of 2 rev·s⁻¹ about a vertical diameter, as in Fig. 5–24.

a) Find the angle θ at which the bead is in vertical equilibrium. (Of course it has a radial acceleration toward the axis.)
b) Is it possible for the bead to "ride" at the same elevation as the center of the hoop?
c) What will happen if the hoop rotates at 1 rev·s⁻¹?

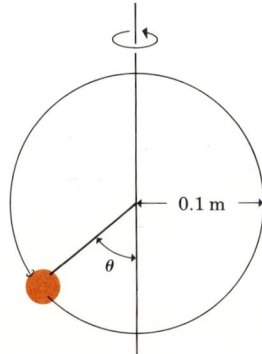

Figure 5–24

5–38 A curve of 200-m radius on a level road is banked at the correct angle for a velocity of 15 m·s⁻¹. If an automobile rounds this curve at 30 m·s⁻¹, what is the minimum coefficient of friction between tires and road so that the automobile will not skid? Assume all forces to act at the center of gravity.

5–39 An airplane in level flight is said to make a *standard turn* when it makes a complete circular turn in 2 min.

a) What is the banking angle of a standard turn if the speed of the airplane is 100 m·s⁻¹?
b) What is the radius of the circle in which it turns?
c) What is the centripetal force on the airplane, expressed as a fraction (or multiple) of its weight?

5–40 The radius of the circular track in Fig. 5–15 is 0.4 m and the mass of the ball is 0.1 kg.

a) Find the critical velocity at the highest point of the track.
b) If the actual velocity at the highest point of the track is twice the critical velocity, find the force exerted by the ball against the track.

5–41 A stunt pilot who has been diving vertically at a velocity of 400 mi·hr⁻¹ pulls out of the dive by changing his course to a circle in a vertical plane.

a) What is the minimum radius of the circle in order that the acceleration at the lowest point shall not exceed "$7g$"?
b) How much does a 180-lb pilot apparently weigh at the lowest point of the pullout?

5–42 A cord is tied to a pail of water and the pail is swung in a vertical circle of radius 1 m. What must be the minimum velocity of the pail at the highest point of the circle if no water is to spill from the pail?

5–43 The radius of a Ferris wheel is 5 m and it makes one revolution in 10 s.

a) Find the difference between the apparent weight of a passenger at the highest and lowest points, expressed as a fraction of his weight. (That is, find the difference between the upward force exerted on the passenger by the seat at these two points.)
b) What would the time of one revolution be if his apparent weight at the highest point were zero?
c) What would then be his apparent weight at the lowest point?
d) What would happen, at this rate of revolution, if his seat belt broke at the highest point, and if he did not hang onto his seat?

5–44 A ball is held at rest in position A in Fig. 5–25 by two light cords. The horizontal cord is cut and the ball starts swinging as a pendulum. What is the ratio of the tension in the supporting cord, in position B, to that in position A?

Figure 5–25

5-45 A particle moves in the xy-plane; its coordinates are given as functions of time by
$$x = R \cos \omega t, \qquad y = R \sin \omega t,$$
where R and ω are constants.

a) Show that the particle's distance from the origin is constant and equal to R, i.e., that its path is a circle of radius R.

b) Show that at every point the particle's velocity is perpendicular to its position vector. (*Hint:* Show that the scalar product $\mathbf{r} \cdot \mathbf{v}$ is always zero.)

c) Show that the particle's acceleration is always opposite in direction to its position vector and has magnitude $\omega^2 R$.

d) Show that the magnitude of the particle's velocity is constant and equal to ωR.

e) Combine the results of parts (c) and (d) to show that the particle's acceleration has constant magnitude v^2/R and that the motion is uniform circular motion.

5-46 The asteroid Toro was discovered in 1964. Its radius is about 5 km.

a) Assuming the density of Toro is the same as that of earth, find its total mass, and find the acceleration due to gravity at its surface.

b) Suppose a body is to be placed in a circular orbit around Toro, with radius just slightly larger than the asteroid's radius. What is the speed of the body? Could you launch yourself into orbit around Toro by running?

5-47 What is the period of revolution of a manmade satellite of mass m that is orbiting the earth in a circular path of radius 8000 km? (Mass of earth = 5.98×10^{24} kg.)

5-48 There exist several satellites moving in a circle in the earth's equatorial plane and at such a height above the earth's surface that they remain always above the same point. Find the height of such a satellite.

5-49 It is desired to launch a satellite in a circular orbit 1000 m above the earth. What orbital speed must be imparted to the satellite?

5-50 An earth satellite rotates in a circular orbit of radius 4400 mi (about 400 mi above the earth's surface) with an orbital speed of 17,000 mi·hr^{-1}.

a) Find the time of one revolution.

b) Find the acceleration of gravity at the orbit.

5-51 What would be the length of a day if the rate of revolution of the earth were such that $g = 0$ at the equator?

5-52 The weight of a man as determined by a spring balance at the equator is 180 lb. By how many ounces does this differ from the true force of gravitational attraction at the same point?

5-53 A passenger on a ship traveling due east with a speed of 18 knots observes that the stream of smoke from the ship's funnels makes an angle of 20° with the ship's wake. The wind is blowing from south to north. Assume that the smoke acquires a velocity (with respect to the earth) equal to the velocity of the wind, as soon as it leaves the funnels. Find the velocity of the wind.

5-54 An airplane pilot wishes to fly due north. A wind of 60 mi·hr^{-1} is blowing toward the west. If the flying speed of the plane (its speed in still air) is 180 mi·hr^{-1}, in what direction should the pilot head? What is the speed of the plane over the ground? Illustrate with a vector diagram.

5-55 An airplane pilot sets a compass course due west and maintains an air speed of 240 km·hr^{-1}. After flying for $\frac{1}{2}$ hr, he finds himself over a town that is 150 km west and 40 km south of his starting point.

a) Find the wind velocity, in magnitude and direction.

b) If the wind velocity were 120 km·hr^{-1} due south, in what direction should the pilot set his course in order to travel due west? Take the same air speed of 240 km·hr^{-1}.

5-56 When a train has a speed of 10 m·s^{-1} eastward, raindrops that are falling vertically with respect to the earth make traces that are inclined 30° to the vertical on the windows of the train.

a) What is the horizontal component of a drop's velocity with respect to the earth? With respect to the train?

b) What is the velocity of the raindrop with respect to the earth? With respect to the train?

5-57 A river flows due north with a velocity of 2 m·s^{-1}. A man rows a boat across the river, his velocity relative to the water being 3 m·s^{-1} due east.

a) What is his velocity relative to the earth?

b) If the river is 1000 m wide, how far north of his starting point will he reach the opposite bank?

c) How much time is required to cross the river?

5-58

a) In what direction should the rowboat in Problem 5-57 be headed in order to reach a point on the opposite bank directly east from the starting point?

b) What will be the velocity of the boat relative to the earth?

c) How much time is required to cross the river?

5-59 A motorboat is observed to travel 10 mi·hr^{-1} relative to the earth in the direction 37° north of east. If the velocity of the boat due to the wind is 2 mi·hr^{-1} eastward and that due to the current is 4 mi·hr^{-1} southward, what is the magnitude and direction of the velocity of the boat due to its own power?

WORK AND ENERGY

6-1 Work

In everyday life the term *work* is applied to any activity requiring the exertion of muscular or mental effort. In physics the term is used in a much more specific sense, which involves the application of a *force* to a body and *displacement* of that body. When a body moves a distance d along a straight line, while acted on by a constant force of magnitude F in the same direction as the motion, the work W done by the force is defined as

$$W = Fd. \tag{6-1}$$

This definition may seem quite arbitrary but, as we shall see in this chapter, work is directly related to quantities characterizing the *motion* of the body and is therefore a useful quantity.

More generally, the force need not have the same direction as the displacement. In Fig. 6-1, the force F, assumed constant, makes an angle θ with the displacement. The work W of this force (or the work done by the force), when its point of application undergoes a displacement s, is defined as the product of the magnitude of the displacement and the *component* of the force in the direction of the displacement.

The component of F in the direction of s is $F\cos\theta$. Then

$$W = (F\cos\theta)s. \tag{6-2}$$

6-1 The work done by the force F in a displacement s is $(F\cos\theta)\cdot s$.

115

An alternative interpretation of Eq. (6–2) is that $s \cos \theta$ is the component of displacement in the direction of $\boldsymbol{F}$. Thus the work may also be obtained by finding the component of displacement in the direction of $\boldsymbol{F}$ and multiplying by the magnitude of $\boldsymbol{F}$.

Comparing Eq. (6–2) with the definition of the *scalar product* of two vectors, Eq. (1–19), we see that work can also be expressed compactly as

$$W = \boldsymbol{F} \cdot \boldsymbol{s}. \tag{6–3}$$

Although it is calculated from two vector quantities, work itself is a *scalar* quantity. A 5-N force toward the east acting on a body that is displaced 6 m to the east does exactly the same amount of work as a 5-N force toward the north acting on a body displaced 6 m to the north. Work is, however, an *algebraic* quantity; it can be positive or negative. When the component of the force is in the *same direction* as the displacement, the work W is *positive*. When it is *opposite* to the displacement, the work is *negative*. If the force is at *right angles* to the displacement, it has no component in the direction of the displacement and the work is *zero*.

Thus, when a body is lifted, the work done by the lifting force is positive; when a spring is stretched, the work done by the stretching force is positive; when a gas is compressed in a cylinder, the work of the compressing force is positive. The work of the gravitational force on a body being lifted is negative, since the (downward) gravitational force is opposite to the (upward) displacement. When a body slides on a stationary surface, the work of the frictional force exerted *on the body* is negative, since this force is always opposite to the displacement of the body. *No* work is done by the frictional force exerted *by* the body *on the stationary surface* because there is no motion of this surface. Also, although it is considered "hard work" to hold a heavy object stationary at arm's length, no work is done in the technical sense because there is no motion. Even if one were to walk along a level floor while carrying the object, no work would be done, since the (vertical) supporting force has no component in the direction of the (horizontal) motion. Similarly, the work of the normal force exerted on a body by a surface on which it moves is zero, and the work of the centripetal force on a body moving in a circle is also zero.

The *unit* of work in any system is the unit of force multiplied by the unit of distance. In SI (mks) units the unit of force is the newton and the unit of distance is the meter; thus in this system the unit of work is one *newton meter* (1 N·m). This combination of units appears so frequently in mechanics that it is given a special name, the *joule* (abbreviated J).

$$1 \text{ joule} = (1 \text{ newton})(1 \text{ meter}) \quad \text{or} \quad 1 \text{ J} = 1 \text{ N·m}.$$

In the cgs system, the unit of work is one dyne-centimeter, also called one *erg*. Because 1 dyn = 10^{-5} N and 1 cm = 10^{-2} m,

$$1 \text{ erg} = 10^{-7} \text{ J}.$$

In the British engineering system the unit of work is one *foot pound* (1 ft-lb); no special name is given this unit. The following conversions are useful:

$$1 \text{ J} = 0.7376 \text{ ft·lb},$$

$$1 \text{ ft·lb} = 1.356 \text{ J}.$$

When several external forces act on a body, it is useful to consider the works of the separate forces. Each of these may be computed from the definition of work in Eq. (6–2). Then, since work is a scalar quantity, the *total* work is the *algebraic sum* of the individual works.

Example 1 Figure 6–2 shows a box being dragged along a horizontal surface by a constant force P making a constant angle θ with the direction of motion. The other forces on the box are its weight w, the upward normal force n exerted by the surface, and the friction force $\mathcal{F}$. What is the work of each force when the box moves a distance s along the surface to the right?

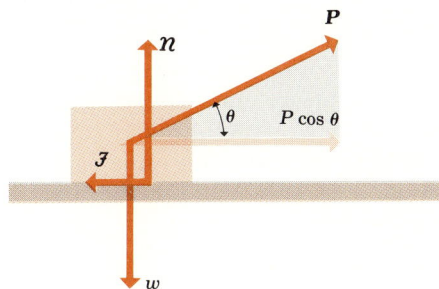

6–2 An object on a rough horizontal surface moving to the right under the action of a force P inclined at an angle θ.

Solution The component of P in the direction of motion is $P\cos\theta$. The work of the force P is therefore

$$W_P = (P\cos\theta)s.$$

The forces w and n are both at right angles to the displacement. Hence,

$$W_w = 0, \qquad W_n = 0.$$

The friction force $\mathcal{F}$ is opposite to the displacement, so the work of the friction force is

$$W_{\mathcal{F}} = -\mathcal{F}s.$$

Since work is a scalar quantity, the total work W of all forces on the body is the algebraic sum (*not* the vector sum) of the individual works:

$$\begin{aligned} W &= W_P + W_w + W_n + W_{\mathcal{F}} \\ &= (P\cos\theta)s + 0 + 0 - \mathcal{F}s \\ &= (P\cos\theta - \mathcal{F})s. \end{aligned}$$

But $(P\cos\theta - \mathcal{F})$ is the *resultant* force on the body. Hence *the total work of all forces is equal to the work of the resultant force.*

Suppose that $P = 50$ N, $\mathcal{F} = 15$ N, $\theta = 36.9°$, and $s = 20$ m. Then

$$\begin{aligned} W_P &= (P\cos\theta)s \\ &= (50\text{ N})(0.800)(20\text{ m}) = 800\text{ N}\cdot\text{m}, \\ W_{\mathcal{F}} &= -\mathcal{F}s = (-15\text{ N})(20\text{ m}) = -300\text{ N}\cdot\text{m}, \\ W &= W_P + W_{\mathcal{F}} = 500\text{ N}\cdot\text{m}. \end{aligned}$$

As a check, the total work may be expressed as

$$\begin{aligned} W &= (P\cos\theta - \mathcal{F})s \\ &= (40\text{ N} - 15\text{ N})(20\text{ m}) = 500\text{ N}\cdot\text{m}. \end{aligned}$$

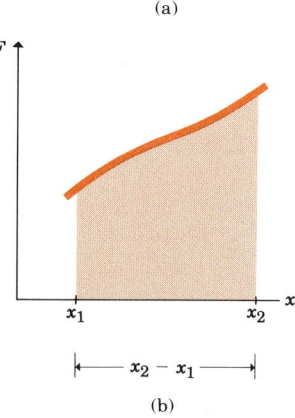

6-3 (a) Curve showing how F varies with x. (b) If the area is partitioned into small rectangles, their sum approximates the total work done during the displacement; the greater the number of rectangles used, the closer is the approximation.

6-2 Work done by a varying force

In the preceding section we defined the work of a *constant* force. In many important cases, however, the work is done by a force that varies in magnitude or direction during the displacement of the body on which it acts. Thus when a spring is stretched slowly, the force required to stretch it increases steadily as the spring elongates; when a body is projected vertically upward, the gravitational force exerted on it by the earth decreases inversely with the square of its distance from the earth's center.

Suppose a particle moves along a line under the action of a force directed along the line but varying with the particle's position. In Fig. 6-3a the force magnitude is shown as a function of the particle's coordinate x. To find the work of this force, we divide the displacement into short segments Δx_1, Δx_2, etc., and in each segment approximate the varying force by one that is constant in each segment. The force then has approximately the value F_1 in segment Δx_1, F_2 in segment Δx_2, and so on. The work done in the first segment is then $F_1 \Delta x_1$, that in the second is $F_2 \Delta x_2$, and so on, and the *total* work is

$$W = F_1 \Delta x_1 + F_2 \Delta x_2 + \cdots \tag{6-4}$$

As the number of segments becomes very large and the size of each very small, this becomes (in the limit) the *integral* of F from x_1 to x_2:

$$W = \int_{x_1}^{x_2} F\, dx. \tag{6-5}$$

We also note that $F_1 \Delta x_1$ represents the *area* of the first vertical strip in Fig. 6-3a, and that the integral in Eq. (6-5) represents the area under the curve in Fig. 6-3b.

If the force also varies in *direction* during the displacement, then F in Eq. (6-5) must be replaced by the *component* of force in the direction of displacement; this is given by $F \cos \theta$, where θ is the angle between F and the x-axis at each point. In general θ may vary from point to point. Then we have

$$W = \int_{x_1}^{x_2} F \cos \theta\, dx. \tag{6-6}$$

Example 1 Compute the work required to stretch a spring a distance X from its unstretched length, if the force required increases in direct proportion to the amount of elongation.

To keep a spring stretched at an elongation x beyond its unstretched length, a force $\boldsymbol{F}$ must be exerted at one end and a force equal in magnitude but opposite in direction at the other end, as shown in Fig. 6-4. If

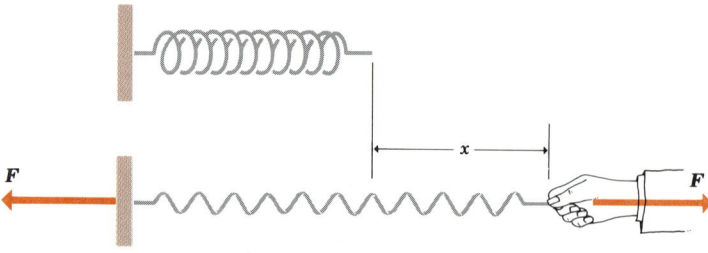

6-4 The force to stretch a spring is proportional to its elongation; $F = kx$.

the elongation is not too great, F is directly proportional to x:

$$F = kx, \qquad (6\text{-}7)$$

where k is a constant called the *force constant* or the *stiffness* of the spring. This direct proportion between force and elongation, for elongations that are not too great, was discovered by Robert Hooke in 1678 and is known as Hooke's law. It will be discussed more fully in a later chapter.

Suppose that forces equal in magnitude and opposite in direction are exerted on the ends of the spring, and that the forces are gradually increased from zero. Let one end of the spring be kept fixed. The work of the force at the fixed end is zero, but work is done by the varying force F at the moving end. This force is in the same direction as the displacement. In Fig. 6-5, the force F is plotted vertically and the displacement x of the moving end, representing the elongation of the spring, is plotted horizontally.

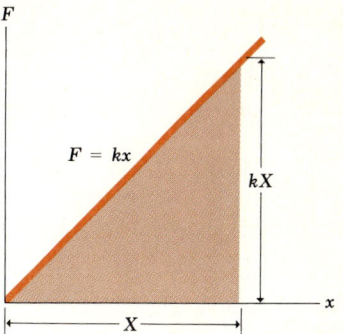

6-5 The work done in stretching a spring is equal to the area of the shaded triangle.

The work required to stretch the spring from $x = 0$ (no elongation) to $x = X$ is

$$W = \int_0^X F\,dx = \int_0^X kx\,dx = \tfrac{1}{2}kX^2. \qquad (6\text{-}8)$$

This result can also be obtained graphically; the area of the shaded triangle in Fig. 6-5, representing the total work, is equal to half the product of base and altitude, or

$$W = \tfrac{1}{2}(X)(kX) = \tfrac{1}{2}kX^2$$

in agreement with the above result.

The work is therefore equal to one-half the product of the force constant k and the *square* of the final elongation X. When the elongation is doubled, the work increases by a factor of 4.

The reader should convince himself that the work done in changing the elongation of a spring from x_1 to x_2 is

$$W = \tfrac{1}{2}kx_2^2 - \tfrac{1}{2}kx_1^2.$$

The definition of work can be generalized further to include motion along a curved path. Suppose the particle moves from point P_1 to P_2 along the curve; we imagine dividing the portion of the curve between these points into many infinitesimal vector displacements, and we call a typical one of these $d\mathbf{l}$. Each $d\mathbf{l}$ is tangent to the path at its position. Let $\mathbf{F}$ be the force at a typical point along the path, and θ the angle between $\mathbf{F}$ and $d\mathbf{l}$ at this point. Then the small element of work dW done on the particle during the displacement $d\mathbf{l}$ may be written as

$$dW = F\cos\theta\, dl = F_\parallel\, dl = \mathbf{F}\cdot d\mathbf{l},$$

where as before $F_\parallel = F\cos\theta$ is the component of $\mathbf{F}$ in the direction of $d\mathbf{l}$. The total work done on the particle by $\mathbf{F}$ as the particle moves from P_1 to P_2 is then represented symbolically as

$$W = \int_{P_1}^{P_2} F\cos\theta\, dl = \int_{P_1}^{P_2} F_\parallel\, dl = \int_{P_1}^{P_2} \mathbf{F}\cdot d\mathbf{l}. \qquad (6\text{-}9)$$

This integral is called a *line integral*; we shall have several other oppor-

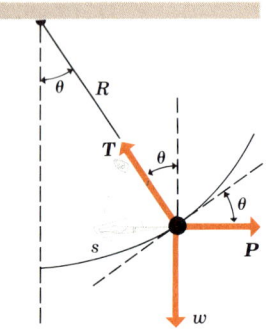

6-6 A variable horizontal force P acts on a small object while the displacement varies from zero to s.

tunities to use line integrals, particularly in the area of electricity and magnetism. To actually evaluate Eq. (6-9) in a specific problem we must have some sort of detailed description of the path and of the variation of F along the path.

Example 2 A small object of weight w hangs from a string of length R, as shown in Fig. 6-6. A *variable* horizontal force P, which starts at zero and gradually increases, is used to pull the object very slowly (so that equilibrium exists at all times) until the string makes an angle θ_0 with the vertical. Calculate the work of the force P.

Since the object is in equilibrium, the sum of the horizontal forces equals zero:

$$P = T \sin \theta.$$

Equating the sum of the vertical forces to zero, we find

$$w = T \cos \theta.$$

Dividing these two equations, we get

$$P = w \tan \theta.$$

The point of application of P swings through the arc s. Since $s = R\theta$, $dl = R\, d\theta$, and

$$W = \int \boldsymbol{P} \cdot d\boldsymbol{l} = \int P \cos \theta \, dl$$
$$= \int_0^{\theta_0} w \tan \theta \cos \theta R \, d\theta$$
$$= wR \int_0^{\theta_0} \sin \theta \, d\theta = wR(1 - \cos \theta_0). \quad (6\text{-}10)$$

If $\theta_0 = 0$, there is no displacement; in that case $\cos \theta = 1$ and $W = 0$, as expected, but if $\theta_0 = 90°$, then $\cos \theta = 0$ and $W = wR$. In that case the work is the same as though the body had been lifted straight up a distance R by a force equal to its weight w. In fact, as the reader is invited to verify, the quantity $R(1 - \cos \theta)$ is the increase in the height of the body during the displacement, so for any value of θ_0 the work of force P is the change in height multiplied by the weight. This result will be proved more generally in Sec. 6-4.

6-3 Work and kinetic energy

The work done on a body for a force is related very directly to the change in the body's motion that results. To develop this relationship, we consider first a body of mass m moving along a straight line under the action of a constant resultant force of magnitude F directed along the line. The body's acceleration is given by Newton's second law, $F = ma$. Suppose the speed increases from v_1 to v_2 while the body undergoes a displacement s. Then, from the analysis of motion with constant acceleration, and from Eq. (3-17), we have

$$v_2{}^2 = v_1{}^2 + 2as$$

$$a = \frac{v_2{}^2 - v_1{}^2}{2s}.$$

Hence
$$F = m\frac{v_2^2 - v_1^2}{2s},$$
and
$$Fs = \tfrac{1}{2}mv_2^2 - \tfrac{1}{2}mv_1^2. \qquad (6\text{-}11)$$

The product Fs is the work W of the resultant force F. The quantity $\tfrac{1}{2}mv^2$, one-half the product of the mass of the body and the square of its speed, is called its *kinetic energy, K*.

$$K = \tfrac{1}{2}mv^2. \qquad (6\text{-}12)$$

The first term on the right side of Eq. (6-11), containing the final speed v_2, is the final kinetic energy of the body, K_2, and the second term is the initial kinetic energy, K_1. The difference between these terms is the *change* in kinetic energy, and we have the important result that *the work of the resultant external force on a body is equal to the change in kinetic energy of the body.*

$$W = K_2 - K_1 = \Delta K \qquad (6\text{-}13)$$

Kinetic energy, like work, is a *scalar* quantity. The kinetic energy of a moving body depends only on its speed (the *magnitude* of its velocity) but not on the *direction* in which it is moving. The *change* in kinetic energy depends only on the work $W = Fs$ and not on the individual values of F and s. That is, the force F could have been large and the displacement s small, or the reverse might have been true. If the mass m and the speeds v_1 and v_2 are known, the *work* of the resultant force can be found without any knowledge of the force F and the displacement s.

If the work W is *positive*, the final kinetic energy is greater than the initial kinetic energy and the kinetic energy *increases*. If the work is *negative*, the kinetic energy *decreases*. In the special case in which the work is *zero*, the kinetic energy remains *constant*.

In Eq. (6-13), W is the work of the *resultant* force, that is, of the *vector sum* of *all* the forces acting on the body. Alternatively, one may calculate the work of each force separately; W is then the *algebraic sum* of all these quantities of work. The example in Sec. 6-1 illustrates these two alternatives.

Although Eq. (6-13) was derived for the special case of a *constant* resultant force, it is true even when the force varies in any arbitrary way. The work of *any* resultant force on a rigid body equals the change in kinetic energy of the body. To prove this statement, we divide the total displacement s into a large number of small segments Δs, just as in the calculation of work done by a varying force. The change of kinetic energy in segment Δs_1 is equal to the work $F_1 \Delta s_1$, and so on; the total kinetic energy is the sum of the changes in the individual segments and is thus equal to the total work done.

When SI units are used, m is measured in kilograms and v in meters per second. But from Eq. (6-13) kinetic energy must have the same units as work. To verify this we recall that $1\,\text{N} = 1\,\text{kg}\cdot\text{m}\cdot\text{s}^{-2}$. Thus

$$1\,\text{J} = 1\,\text{N}\cdot\text{m} = \left(1\,\text{kg}\cdot\text{m}\cdot\text{s}^{-2}\right)\text{m} = 1\,\text{kg}\cdot\text{m}^2\cdot\text{s}^{-2}.$$

The joule is thus the SI unit of both work and kinetic energy and, as we shall see later, of all kinds of energy. Similarly, in the cgs system, with m in grams and v in centimeters per second, we have

$$1\,\text{erg} = 1\,\text{dyn}\cdot\text{cm} = 1\,\text{g}\cdot\text{cm}\cdot\text{s}^{-2}\cdot\text{cm} = 1\,\text{g}\cdot\text{cm}^2\cdot\text{s}^{-2}.$$

In the British engineering system,

$$1 \text{ ft} \cdot \text{lb} = 1 \text{ ft} \cdot \text{slug} \cdot \text{ft} \cdot \text{s}^{-2} = 1 \text{ slug} \cdot \text{ft}^2 \cdot \text{s}^{-2}.$$

Example Refer again to the body in Fig. 6-2 and the numerical values given in the example of Sec. 6-1. The *total* work of the external forces was shown to be 500 N·m. Hence the kinetic energy of the body increases by 500 N·m. To verify this, suppose the initial speed v_1 is 4 m·s^{-1} and the mass of the body is 10 kg. The initial kinetic energy is

$$K_1 = \tfrac{1}{2}mv_1^2 = \tfrac{1}{2}(10 \text{ kg})(4 \text{ m} \cdot \text{s}^{-1})^2 = 80 \text{ J}.$$

To find the final kinetic energy we must first find the acceleration:

$$a = \frac{F}{m} = \frac{40 \text{ N} - 15 \text{ N}}{10 \text{ kg}} = 2.5 \text{ m} \cdot \text{s}^{-2}.$$

Then,

$$v_2^2 = v_1^2 + 2as = (4 \text{ m} \cdot \text{s}^{-1})^2 + 2(2.5 \text{ m} \cdot \text{s}^{-2})(20 \text{ m}) = 116 \text{ m}^2 \cdot \text{s}^{-2},$$

and

$$K_2 = \tfrac{1}{2}(10 \text{ kg})(116 \text{ m}^2 \cdot \text{s}^{-2}) = 580 \text{ J}.$$

The increase in kinetic energy is therefore 500 J.

6-4 Gravitational potential energy

When a gravitational force acts on a body undergoing a displacement, the force does work on the body, and we shall see that this work can be expressed conveniently in terms of the initial and final *position* of the body. In Fig. 6-7a, a body of mass m (and of weight $\boldsymbol{w} = m\boldsymbol{g}$) moves vertically from a height y_1 above an arbitrary reference plane to a height y_2. For the present, we consider only displacements near the earth's surface, so that variations of gravitational force with distance from the earth's center can be neglected. The downward gravitational force on the body is then constant and equal to $\boldsymbol{w}$. Let $\boldsymbol{P}$ represent the resultant of all other forces acting on the body. The direction of the gravitational force $\boldsymbol{w}$ is opposite to the upward displacement, and the work of this force is

$$W_{\text{grav}} = -w(y_2 - y_1) = -(mgy_2 - mgy_1). \tag{6-14}$$

The reader should check that this expression also has the correct sign when the body moves *downward* (so y_2 is less than y_1), and that it still gives a correct result when y_1, or y_2, or both, are negative, corresponding to a position *below* the reference plane.

Thus W_{grav} can be determined from the values of the quantity mgy at the beginning and end of the displacement. This quantity, the product of the weight mg and the height y above the origin of coordinates, is called the *gravitational potential energy*, U:

$$U(\text{gravitational}) = mgy. \tag{6-15}$$

Then the initial value of gravitational potential energy is $U_1 = mgy_1$ and the final value $U_2 = mgy_2$. The work W_{grav} done by the gravitational force during the displacement from y_1 to y_2 can then be written as

$$W_{\text{grav}} = U_1 - U_2 = -\Delta U.$$

6-7 Work of the gravitational force w during the motion of an object from one point in a gravitational field to another.

Thus when gravity does positive work, corresponding to downward displacement and decreasing y, the potential energy decreases; when the body moves upward the work done by gravity is negative and the potential energy increases.

Now let W' represent the total work done by P, that is, by all forces *other than* the gravitational force. The total work done by *all* forces is then $W_{\text{grav}} + W'$. Since the *total* work equals the change in kinetic energy,

$$W' + W_{\text{grav}} = K_2 - K_1,$$
$$W' - (mgy_2 - mgy_1) = (\tfrac{1}{2}mv_2^2 - \tfrac{1}{2}mv_1^2). \quad (6\text{-}16)$$

The quantities $\tfrac{1}{2}mv_2^2$ and $\tfrac{1}{2}mv_1^2$ depend only on the final and initial *speeds;* the quantities mgy_2 and mgy_1 depend only on the final and initial *elevations*. Let us therefore rearrange this equation, transferring the quantities mgy_2 and mgy_1 from the "work" side of the equation to the "energy" side:

$$W' = (\tfrac{1}{2}mv_2^2 - \tfrac{1}{2}mv_1^2) + (mgy_2 - mgy_1). \quad (6\text{-}17)$$

The left side of Eq. (6-17) contains only the work of the force P. The terms on the right depend only on the final and initial states of the body (its speed and elevation) and not specifically on the way in which it moved.

The first expression in parentheses on the right of Eq. (6-17) is the change in kinetic energy of the body, and the second is the change in its gravitational potential energy. Equation (6-17) can also be written

$$W' = (\tfrac{1}{2}mv_2^2 + mgy_2) - (\tfrac{1}{2}mv_1^2 + mgy_1). \quad (6\text{-}18)$$

The sum of kinetic and potential energies of the body is called its *total mechanical energy*. The first expression in parentheses on the right of Eq. (6-18) is the *final* value of the total mechanical energy, and the second is the *initial* value. Hence, *the work of all forces acting on the body,* **with the exception of the gravitational force,** *equals the change in the total mechanical energy of the body*. If the work W' is positive, the mechanical energy increases. If W' is negative, the energy decreases.

In the special case in which the *only* force on the body is the gravitational force, the work W' is zero. Equation (6-18) can then be written

$$\tfrac{1}{2}mv_2^2 + mgy_2 = \tfrac{1}{2}mv_1^2 + mgy_1. \tag{6-19}$$

Under these conditions, then, the *total mechanical energy remains constant*, or is *conserved*. This is a special case of the principle of *conservation of mechanical energy*.

Example 1 A man holds a ball of mass $m = 0.2$ kg at rest in his hand. He then throws the ball vertically upward. In this process, his hand moves up 0.5 m before the ball leaves his hand with an upward velocity of 20 m·s^{-1}. Discuss the motion of the ball from the work–energy standpoint, assuming $g = 10$ m·s^{-2}.

Solution First, consider the throwing process. Take the reference level ($y = 0$) at the initial position of the ball. Then $K_1 = 0$, $U_1 = 0$. Take point 2 at the point where the ball leaves the thrower's hand. Then

$$U_2 = mgy_2 = (0.2 \text{ kg})(10 \text{ m·s}^{-2})(0.5 \text{ m}) = 1.0 \text{ J},$$
$$K_2 = \tfrac{1}{2}mv_2^2 = \tfrac{1}{2}(0.2 \text{ kg})(20 \text{ m·s}^{-1})^2 = 40 \text{ J}.$$

Let P represent the upward force exerted on the ball by the man in the throwing process. The work W' is then the work of this force and is equal to the sum of the changes in kinetic and potential energy of the ball.

The kinetic energy of the ball increases by 40 J and its potential energy by 1 J. The work W' of the upward force P is therefore 41 J.

If the force P is constant, the work of this force is given by

$$W' = P(y_2 - y_1),$$

and the force P is then

$$P = \frac{W'}{y_2 - y_1} = \frac{41 \text{ J}}{0.5 \text{ m}} = 82 \text{ N}.$$

However, the *work* of the force P is 41 J whether or not the force is constant.

Now consider the flight of the ball *after* it leaves the thrower's hand. In the absence of air resistance, the only force on the ball is then its weight $w = mg$. Hence the total mechanical energy of the ball remains constant. The calculations will be simplified if we take a new reference level at the point where the ball leaves the thrower's hand. Calling this point 1, we have

$$K_1 = 40 \text{ J}, \qquad U_1 = 0,$$
$$K + U = 40 \text{ J},$$

and the total mechanical energy at *any* point of the path equals 40 J.

Suppose we wish to find the speed of the ball at a height of 15 m above the reference level. Its potential energy at this elevation is 30 J. (Why?) Its kinetic energy is therefore 10 J. To find its speed, we have

$$\tfrac{1}{2}mv^2 = K, \qquad v = \pm\sqrt{2K/m} = \pm 10 \text{ m·s}^{-1}.$$

The significance of the $\pm$ sign is that the ball passes this point *twice*, once on the way up and again on the way down. Its *potential* energy at this point is the same whether it is moving up or down. Hence its kinetic energy is the same and its *speed* is the same. The algebraic sign of the speed is + when the ball is moving up and − when it is moving down.

Next, let us find the height of the highest point reached. At this point, $v = 0$ and $K = 0$. Therefore $U = 40$ J, and the ball rises to a height h above the point where it leaves the thrower's hand, given by

$$mgh = (0.2 \text{ kg})(10 \text{ m} \cdot \text{s}^{-2})h = 40 \text{ J}.$$

Thus

$$h = 20 \text{ m}.$$

Finally, suppose we were asked to find the speed at a point 30 m above the reference level. The potential energy at this point would be 60 J. But the *total* energy is only 40 J, so the ball never reaches a height of 30 m.

A body may travel from an initial elevation y_1 to a final elevation y_2 along a curved path, as shown in Fig. 6–7b. The work done by gravity during this displacement is the same as when the body travels straight up, as in Fig. 6–7a. To prove this we divide the path into a large number of small segments Δs; a typical example is shown in Fig. 6–7c. According to the remark following Eq. (6–2), the work done during this displacement may be obtained by multiplying the component of displacement in the direction of the force by the magnitude of the force. As shown in Fig. 6–7c, the vertical component of displacement has magnitude $\Delta s \cos \theta = \Delta y$, but its direction is opposite to that of the force w. Thus the work done by w is

$$-mg \, \Delta s \cos \theta = -mg \, \Delta y$$

and is the same as though the body had been displaced straight up a distance Δy. Similarly, the *total* work done by gravity depends only on the *total* vertical displacement $(y_2 - y_1)$ and is independent of any horizontal motion that may occur.

Example 2 A body slides down a curved track that is one quadrant of a circle of radius R, as in Fig. 6–8. If it starts from rest and there is no friction, find its speed at the bottom of the track. (The motion of this body is exactly the same as that of a body attached to one end of a string of length R, the other end of which is held at point 0.)

Solution The equations of motion with constant acceleration cannot be used, since the acceleration decreases during the motion. (The slope angle of the track becomes smaller and smaller as the body descends.) However, if there is no friction, the only force on the body other than its weight is the normal force n exerted on it by the track. The work of this force is zero because at each point it is perpendicular to the small element of displacement near that point. Thus $W' = 0$ and mechanical energy is conserved. Take point 1 at the starting point and point 2 at the bottom of the track. Take the reference level at point 2. Then $y_1 = R$, $y_2 = 0$, and

$$K_2 + U_2 = K_1 + U_1,$$
$$\tfrac{1}{2}mv_2^2 + 0 = 0 + mgR,$$
$$v_2 = \pm \sqrt{2gR}.$$

6–8 An object sliding down a frictionless curved track.

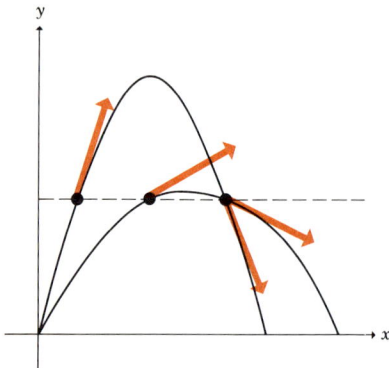

6-9 For the same initial speed, the speed is the same at all points at the same elevation.

The speed is therefore the same as if the body had fallen *vertically* through a height R. (What is now the significance of the $\pm$ sign?)

As a numerical example, let $R = 1$ m. Then
$$v = \pm\sqrt{2(9.8 \text{ m}\cdot\text{s}^{-2})(1 \text{ m})} = \pm 4.43 \text{ m}\cdot\text{s}^{-1}.$$

Example 3 Suppose a body of mass 0.5 kg slides down a track of radius $R = 1$ m, like that in Fig. 6-8, but its speed at the bottom is only 3 m·s^{-1}. What was the work of the frictional force acting on the body?

Solution In this case, $W' = W_{\mathcal{F}}$, and
$$W_{\mathcal{F}} = (\tfrac{1}{2}mv_2^2 - \tfrac{1}{2}mv_1^2) + (mgy_2 - mgy_1)$$
$$= \tfrac{1}{2}(0.5 \text{ kg})(9 \text{ m}^2\cdot\text{s}^{-2}) - 0 + 0 - (0.5 \text{ kg})(9.8 \text{ m}\cdot\text{s}^{-2})(1 \text{ m})$$
$$= 2.25 \text{ J} - 4.9 \text{ J} = -2.65 \text{ J}.$$

The frictional work was therefore -2.65 J, and the total mechanical energy *decreased* by 2.65 J. The mechanical energy of a body is *not* conserved when friction forces act on it.

Example 4 In the absence of air resistance, the only force on a projectile is its weight, and the mechanical energy of the projectile remains constant. Figure 6-9 shows two trajectories of a projectile with the same initial speed (and hence the same total energy) but with different angles of departure. At all points at the same elevation the potential energy is the same; hence the kinetic energy is the same and the speed is the same.

Example 5 A small object of weight w hangs from a string of length R, as shown in Fig. 6-10. A *variable* horizontal force P that starts at zero and gradually increases is used to pull the object very slowly until the string makes an angle θ with the vertical. Calculate the work of the force P.

Solution The sum of the works W' of all the forces other than the gravitational force must equal the change of kinetic energy plus the change of gravitational potential energy. Hence
$$W' = W_P + W_T = \Delta K + \Delta U.$$

Since T is perpendicular to the path of its point of application, $W_T = 0$; and since the body was pulled very slowly at all times, the change of kinetic energy is also zero. Hence
$$W_P = \Delta U = w\,\Delta y,$$
where Δy is the distance that the object has been raised. From Fig. 6-10 Δy is seen to be $R(1 - \cos\theta)$. Therefore
$$W_P = wR(1 - \cos\theta).$$

This same result was obtained in Sec. 6-2, Example 2, but the potential-energy approach is a much simpler calculation.

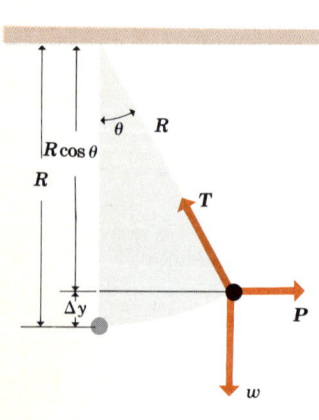

6-10 $\Delta y = R(1 - \cos\theta)$.

Thus far in this section it has been assumed that the changes in elevation are small enough so that the gravitational force on a body can be considered constant. But as discussed in Sec. 4-5, the gravitational force is given in general by

$$F_g = \frac{Gmm_E}{r^2},$$

where m_E is the mass of the earth and r is the distance of the body from the earth's center. When the change in r is sufficiently large, we may no longer consider the gravitational force constant; it decreases as $1/r^2$. When r increases from r_1 to r_2, the work of the gravitational force is

$$W_{grav} = -Gmm_E \int_{r_1}^{r_2} \frac{dr}{r^2} = \frac{Gmm_E}{r_2} - \frac{Gmm_E}{r_1}. \quad (6\text{-}20)$$

If the gravitational force is the only force on the body, then this is equal to the change in kinetic energy from r_1 to r_2:

$$Gmm_E \left(\frac{1}{r_2} - \frac{1}{r_1}\right) = \tfrac{1}{2}mv_2^2 - \tfrac{1}{2}mv_1^2$$

or

$$\tfrac{1}{2}mv_1^2 - \frac{Gmm_E}{r_1} = \tfrac{1}{2}mv_2^2 - \frac{Gmm_E}{r_2}. \quad (6\text{-}21)$$

The quantity $-G(mm_E/r)$ is therefore the general expression for the gravitational potential energy of a body attracted by the earth:

$$U(\text{gravitational}) = -G\frac{mm_E}{r}. \quad (6\text{-}22)$$

The *total mechanical energy* of the body, the sum of its kinetic energy and potential energy, is

$$E = K + U = \tfrac{1}{2}mv^2 - G\frac{mm_E}{r}. \quad (6\text{-}23)$$

If the only force on the body is the gravitational force, the total mechanical energy remains constant, or is *conserved*.

Example 6 Use the principle of conservation of mechanical energy to find the velocity with which a body must be projected vertically upward, in the absence of air resistance, (a) to rise to a height above the earth's surface equal to the earth's radius, R, and (b) to escape from the earth.

Solution

a) Let v_1 be the initial velocity. Then

$$r_1 = R, \quad r_2 = 2R, \quad v_2 = 0,$$

and

$$\tfrac{1}{2}mv_1^2 - G\frac{mm_E}{R} = 0 - G\frac{mm_E}{2R},$$

or

$$v_1^2 = \frac{Gm_E}{R}.$$

b) When v_1 is the escape velocity, $r_1 = R$, $r_2 = \infty$, $v_2 = 0$. Then

$$\tfrac{1}{2}mv_1^2 - G\frac{mm_E}{R} = 0 \quad \text{or} \quad v_1^2 = \frac{2Gm_E}{R}.$$

It may seem strange that the general expression for gravitational potential energy, Eq. (6–22), should contain a minus sign. The reason for this lies in the choice of a *reference state* or reference level in which the potential energy is considered zero. If we set $U = 0$ in Eq. (6–22) and solve for r, we find

$$r = \infty.$$

That is, the *gravitational potential energy of a body is zero when the body is at an infinite distance from the earth.* Since the potential energy *decreases* as the body approaches the earth, it must be *negative* at any finite distance from the earth. The *change* in potential energy of a body as it moves from one point to another is the same, whatever the choice of reference level, and it is only changes in potential energy that are significant.

6–5 Elastic potential energy

Figure 6–11 shows a body of mass m on a level frictionless surface. One end of a spring is attached to the body and the other end of the spring is fixed. Take the origin of coordinates at the position of the body when the spring is unstretched (Fig. 6–11a). An outside agent exerts a force P of sufficient magnitude to cause the spring to stretch. As soon as the slightest extension takes place, a force F is created within the spring that acts in a direction opposite that of P. The force F is called an *elastic force*. If the force P is reduced to zero, the elastic force restores the spring to its original unstretched condition. It may therefore be referred to as a *restoring force*.

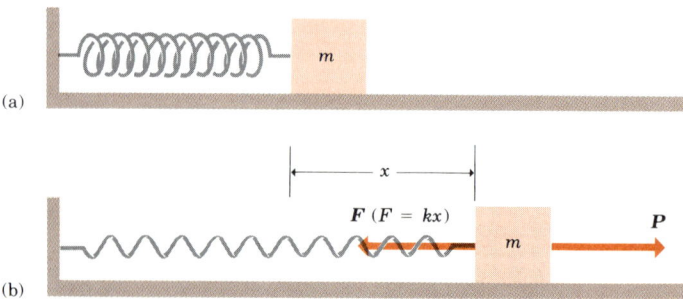

6–11 When an applied force P produces an extension x of a spring, an elastic restoring force F is created within the spring, where $F = kx$.

As shown in Sec. 6–2, for a force obeying Hooke's law, stretching the string to a final elongation x requires an amount of work $\tfrac{1}{2}kx^2$. Thus the work done *by the spring* (the restoring force) during this elongation is $-\tfrac{1}{2}kx^2$. Similarly, in a displacement from an initial elongation x_1 to a final elongation x_2, the elastic restoring force does an amount of work W_{el} given by

$$W_{\text{el}} = -\tfrac{1}{2}kx_2^2 - (-\tfrac{1}{2}kx_1^2).$$

The quantity $\frac{1}{2}kx^2$, one-half the product of the force constant and the square of the coordinate of the body, is called the *elastic potential energy* of the body, U. (The symbol U is used for any form of potential energy.)

$$U(\text{elastic}) = \tfrac{1}{2}kx^2. \tag{6-24}$$

The work done by the elastic force can thus be expressed in terms of a change in potential energy, just as was done for gravitational work:

$$W_{el} = \tfrac{1}{2}kx_1^2 - \tfrac{1}{2}kx_2^2 = U_1 - U_2. \tag{6-25}$$

When x increases, the elastic work is negative and the potential energy increases; when x decreases, the elastic force does positive work and the potential energy decreases.

Let W' stand for the work of the applied force $\mathbf{P}$. Setting the total work equal to the change in kinetic energy of the body, we have

$$W' + W_{el} = \Delta K,$$

$$W' - (\tfrac{1}{2}kx_2^2 - \tfrac{1}{2}kx_1^2) = (\tfrac{1}{2}mv_2^2 - \tfrac{1}{2}mv_1^2).$$

The quantities $\tfrac{1}{2}kx_2^2$ and $\tfrac{1}{2}kx_1^2$ depend only on the initial and final positions of the body and not specifically on the way in which it moved. Let us therefore transfer them from the "work" side of the equation to the "energy" side. Then

$$W' = (\tfrac{1}{2}mv_2^2 - \tfrac{1}{2}mv_1^2) + (\tfrac{1}{2}kx_2^2 - \tfrac{1}{2}kx_1^2) = \Delta K + \Delta U. \tag{6-26}$$

Hence the work W' of the force $\mathbf{P}$ equals the sum of the change in the kinetic energy of the body and the change in its elastic potential energy.

Equation (6-26) can also be written

$$W' = (\tfrac{1}{2}mv_2^2 + \tfrac{1}{2}kx_2^2) - (\tfrac{1}{2}mv_1^2 + \tfrac{1}{2}kx_1^2) = E_2 - E_1. \tag{6-27}$$

The sum of the kinetic and potential energies of the body is its total mechanical energy; and *the work of all forces acting on the body,* **with the exception of the elastic force,** *equals the change in the total mechanical energy of the body.*

If the work W' is positive, the mechanical energy increases. If W' is negative, it decreases. In the special case in which W' is zero, the mechanical energy remains constant or is *conserved*.

In that special case Eq. (6-27) may be rewritten

$$\tfrac{1}{2}mv_1^2 + \tfrac{1}{2}kx_1^2 = \tfrac{1}{2}mv_2^2 + \tfrac{1}{2}kx_2^2. \tag{6-28}$$

Example 1 Let the force constant k of the spring in Fig. 6-11 be 24 N·m^{-1}, and let the mass of the body be 4 kg. The body is initially at rest, and the spring is initially stretched 0.5 m. Then the body is released and moves back toward its equilibrium position. What is its speed when the spring is stretched 0.3 m? Assume that the body rests on a frictionless surface.

Solution The equations of motion with constant acceleration cannot be used, since the spring force varies with position, but the speed can be found from energy considerations. The spring force is the only force acting on the body that does work; thus in Eq. (6-27), $W' = 0$. The total mechanical energy is constant, and Eq. (6-28) is applicable. We have

$v_1 = 0$, $x_1 = 0.5$ m, $x_2 = 0.3$ m, and v_2 is to be found. From Eq. (6-28),

$$\tfrac{1}{2}(4 \text{ kg})(0)^2 + \tfrac{1}{2}(24 \text{ N} \cdot \text{m}^{-1})(0.5 \text{ m})^2$$
$$= \tfrac{1}{2}(4 \text{ kg})(v_2)^2 + \tfrac{1}{2}(24 \text{ N} \cdot \text{m}^{-1})(0.3 \text{ m})^2.$$

Evaluating the left side, we find that the total mechanical energy is 3 J, and solving for v_2 yields the result

$$v_2 = \pm 0.98 \text{ m} \cdot \text{s}^{-1}.$$

What is the physical significance of the $\pm$ sign?

Example 2 For the system of Example 1, suppose the body is initially at rest, and the spring is initially unstretched. Suppose that a constant force $\boldsymbol{P}$ of magnitude 10 N is exerted on the body, and that there is no friction. What will be the speed of the body when it has moved 0.5 m?

Solution The equations of motion with constant acceleration cannot be used, since the resultant force on the body varies as the spring is stretched. However, the speed can be found from energy considerations:

$$W' = \Delta K + \Delta U.$$
$$(10 \text{ N})(0.5 \text{ m}) = \tfrac{1}{2}(4 \text{ kg})v_2{}^2 - 0 + \tfrac{1}{2}(24 \text{ N} \cdot \text{m}^{-1})(0.50 \text{ m})^2 - 0,$$
$$v_2 = 1 \text{ m} \cdot \text{s}^{-1}.$$

Example 3 Suppose the force $\boldsymbol{P}$ ceases to act when the body has moved 0.5 m. *How much farther* will the body move before coming to rest?

Solution The elastic force is now the only force, and mechanical energy is conserved. The kinetic energy is $\tfrac{1}{2}mv^2 = 2$ J, and the potential energy is $\tfrac{1}{2}kx^2 = 3$ J. The total energy is therefore 5 J (equal to the work of the force $\boldsymbol{P}$). When the body comes to rest, its kinetic energy is zero and its potential energy is therefore 5 J. Hence

$$\tfrac{1}{2}kx_{\text{max}}{}^2 = \tfrac{1}{2}(24 \text{ N} \cdot \text{m}^{-1})(x_{\text{max}})^2 = 5 \text{ J}, \quad x_{\text{max}} = 0.645 \text{ m}.$$

Example 4 A block of mass m, initially at rest, is dropped from a height h onto a massless spring whose force constant is k. Find the maximum distance y that the spring will be compressed. (See Fig. 6-12.)

Solution This is a process for which the principle of the conservation of mechanical energy holds. At the moment of release, the kinetic energy is zero. At the moment when maximum compression occurs, there is also no kinetic energy. Hence, the loss of gravitational potential energy of the block equals the gain of elastic potential energy of the spring. As shown in Fig. 6-12, the total fall of the block is $h + y$, whence

$$mg(h + y) = \tfrac{1}{2}ky^2, \quad \text{or} \quad y^2 - \frac{2mg}{k}y - \frac{2mgh}{k} = 0.$$

Therefore, from the quadratic formula,

$$y = \frac{1}{2}\left[\frac{2mg}{k} \pm \sqrt{\left(\frac{2mg}{k}\right)^2 + \frac{8mgh}{k}}\right].$$

The positive root is the desired result; the negative root corresponds to

Release   Maximum compression

6-12 The total fall of the block is $h + y$.

the height to which the block and spring would rebound if they were fastened together after contact.

6-6 Conservative and dissipative forces

We have seen that when an object acted on by gravity moves from any position above a zero reference level to any other position, the work of the gravitational force is independent of the path and equal to the difference between the final and the initial values of a function called the *gravitational potential energy*. If the gravitational force *alone* acts on the object, the total mechanical energy (the sum of the kinetic and gravitational potential energies) is constant or *conserved,* and therefore the gravitational force is called a *conservative force*. Thus, if the object is ascending, the work of the gravitational force is accomplished at the expense of the kinetic energy. When the object descends to its original level, the work of the gravitational force serves to increase the kinetic energy back to its original value. Thus the work done by gravity on ascent is completely recovered on descent. Complete recoverability is an important aspect of the work of a conservative force.

When an object attached to a spring is moved from one value of the spring extension to any other value, the work of the elastic force is also independent of the path and equal to the difference between the final and initial values of a function called the *elastic potential energy*. If the elastic force *alone* acts on the object, the sum of the kinetic and elastic potential energies is conserved, and therefore the elastic force is also a conservative force. If the object moves so as to increase the extension of the spring, the work of the elastic force is accomplished at the expense of the kinetic energy. If, however, the extension is decreasing, then the work of the elastic force serves to increase the kinetic energy so that this work is completely recovered also.

To summarize, we see that the work of a conservative force has the following properties:

1. It is independent of the path.
2. It is equal to the difference between the final and initial values of a *potential energy* function.
3. It is completely recoverable.

Contrast a conservative force with a friction force exerted on a moving object by a fixed surface. The work of the friction force *does depend* on the path; the longer the path between two given points, the greater the work. There is *no* function the difference of two values of which equals the work of the friction force. When we slide an object on a rough fixed surface back to its original position, the friction force reverses, and instead of recovering the work done in the first displacement, we must again do work on the return trip. In other words, frictional work is *not* completely recoverable. When the friction force acts alone, the total mechanical energy is *not* conserved. The friction force is therefore called a *nonconservative* or a *dissipative force. The mechanical energy of a body is conserved only when no dissipative forces act on it.*

We find that when friction forces act on a moving body, another form of energy is involved. The more general principle of conservation of energy includes this other form of energy, along with kinetic and potential energy; and, when it is included, the *total* energy of any system remains constant. We shall study this general conservation principle more fully in a later chapter.

Example Example 3 in Sec. 6-4 illustrates the motion of a body acted on by a dissipative friction force. The initial mechanical energy of the body is its initial potential energy of 4.9 J. Its final mechanical energy is its final kinetic energy of 2.25 J. The frictional work $W_{\mathcal{F}}$ is -2.65 J. A quantity of thermal energy equivalent to 2.65 J is developed as the body slides down the track. The sum of this energy and the final mechanical energy equals the initial mechanical energy, and the total energy of the system is conserved.

*6-7 Internal work

Figure 6–13a shows a man standing on frictionless roller skates on a level surface, facing a rigid wall. Suppose that he sets himself in motion backward by pushing against the wall. The forces acting *on the man* are his weight w, the upward forces n_1 and n_2 exerted by the ground, and the horizontal force P exerted by the wall. (The latter is the reaction to the force with which the man pushes against the wall.) The works of w and of n_1 and n_2 are zero because they are perpendicular to the motion. The force P is the unbalanced horizontal force that imparts to the system a horizontal acceleration. *The work of P, however, is zero because there is no motion of its point of application.* We are therefore confronted with a curious situation in which a force is responsible for acceleration, but its work, being zero, is not equal to the increase in kinetic energy of the system!

The new feature in this situation is that the man is a composite system with several parts that can move in relation to each other and thus can do work on each other, even in the absence of any interaction with externally applied forces. Such work is called *internal work*. Although internal forces play no role in accelerating the composite system, their points of application can move so that work is done; thus the man's kinetic energy can change even though the *external* forces do no work.

When both external and internal forces act on the particles of a system, the total work W of *all* forces, external and internal, is the sum of the works W_e and W_i and is equal to the change in the total kinetic energy of the system:

$$W = W_e + W_i = \Delta K. \tag{6-29}$$

The same principle holds in the case of an accelerated automobile. The portions of the rubber tires that are in contact with the rough roadway push back on the ground and the reactions to these forces, designated by Q_1 and Q_2 in Fig. 6–13b, are the external horizontal forces acting *on* the automobile, which are responsible for imparting the horizontal acceleration to the system. Since, however, the portions of the tires momentarily in contact with the road are at rest with respect to the road, the works of the forces Q_1 and Q_2 are zero. As a result of the expanding gases in the cylinders of the automobile engine there are

6-13 (a) External forces acting on a man who is pushing against a wall. The work of these forces is zero. (b) External forces on an automobile. The work of these forces is zero. In both cases, the work of the internal force is responsible for the increase in kinetic energy.

many internal forces, some of which do work; these internal forces are responsible for the increase in kinetic energy.

*6-8 Internal potential energy

If any of the *external* forces on the particles of a system are conservative (gravitational or elastic), the work of these forces can be transferred to the energy side of the work-energy equation and called the change in *external potential energy of the system*. In many instances, the *internal* forces depend only on the distances between *pairs* of particles. The internal work then depends only on the initial and final distances between the particles, and not on the particular way in which they moved. The work of these internal forces can then also be transferred to the energy side of the work-energy equation and called the change in *internal potential energy of the system*. Internal potential energy is a property of the system as a whole and cannot be assigned to any specific particle.

Let W' represent the sum of the works of all external and internal forces that have *not* been transferred to the energy side of the work-energy equation and called changes in external and internal potential energies. Let U_e and U_i represent the external and internal potential energies. The work-energy equation then takes the form

$$W' = \Delta K + \Delta U_e + \Delta U_i, \qquad (6\text{-}30)$$

where the total mechanical energy now includes the kinetic energy of the system and both its external and internal potential energies. If the work W' is zero, the total mechanical energy is conserved.

Example 1 Consider a system consisting of two particles in outer space, very far from all other matter. No external forces act on the system, no external work is done when the bodies move, and the system has no external potential energy. A gravitational force of attraction acts between the bodies, which depends only on the distance between them. We can therefore say that the system has an internal potential energy. If the particles start from rest and accelerate toward each other, the total kinetic energy of the system increases and its internal potential energy decreases. The sum of its kinetic energy and internal potential energy remains constant.

Energy considerations *alone* do not suffice to tell us how much kinetic energy is gained by each body separately. This can be determined, however, from *momentum* considerations, as will be explained in the next chapter.

Example 2 In Sec. 6-5 we computed the elastic potential energy of a body acted on by a force exerted by a spring. This force was an *external* force acting on the body, and the elastic potential energy $\frac{1}{2}kx^2$ should properly be called the *external* elastic potential energy *of the body*.

Let us now consider an alternative viewpoint. The spring itself is a *system* composed of an enormous number of molecules that exert internal forces on one another. When the spring is stretched, the distances between its molecules change. The intermolecular forces depend only on the distances between pairs of molecules, so that a stretched spring has internal elastic potential energy. To calculate this, suppose the spring is

134 WORK AND ENERGY

slowly stretched from its no-load length by equal and opposite forces applied at its ends. The work of these forces was shown to equal $\frac{1}{2}kx^2$, where x is the elongation of the spring above its no-load length. Let us retain this work on the left side of the work–energy equation so that it becomes the work W' in Eq. (6–30). There is no change in the kinetic energy of the spring and no change in its external potential energy. If we call the internal potential energy zero when $x = 0$, then the *internal elastic potential energy of a stretched spring* is equal to $\frac{1}{2}kx^2$. Note that the internal potential energy can be calculated without any detailed information regarding the intermolecular forces.

6-9 Power

Time considerations are not involved in the definition of work. The same amount of work is done in raising a given weight through a given height whether the work is done in 1 s, 1 hr, or 1 yr. In many instances, however, it is necessary to consider the *rate* at which work is done as well as the total amount of work accomplished. The rate at which work is done by a working agent is called the *power* developed by that agent.

If a quantity of work ΔW is done in a time interval Δt, the average power $\bar{P}$ is defined as

$$\text{Average power} = \frac{\text{work done}}{\text{time interval}},$$

$$\bar{P} = \frac{\Delta W}{\Delta t}.$$

If the rate at which work is done is not constant, this ratio may vary; in this case we may define an *instantaneous* power P as the limit of this quotient as $\Delta t \to 0$:

$$P = \lim_{\Delta t \to 0} \frac{\Delta W}{\Delta t} = \frac{dW}{dt}. \tag{6-31}$$

The SI unit of power is one joule per second (1 J·s⁻¹), which is called one *watt* (1 W). The kilowatt (1 kW = 10³ W) and the megawatt (1 MW = 10⁶ W) are also commonly used. The cgs power unit is one erg per second (1 erg·s⁻¹). No single term is assigned to this unit.

In the British system, where work is expressed in foot-pounds and time in seconds, the unit of power is one foot-pound per second. A larger unit called the *horsepower* (hp) is in common use:

$$1 \text{ hp} = 550 \text{ ft·lb·s}^{-1} = 33{,}000 \text{ ft·lb·min}^{-1}.$$

That is, a 1-hp motor running at full load is doing 33,000 ft-lb of work every minute it runs.

A common misconception is that there is something inherently *electrical* about a watt or a kilowatt. This is not the case. It is true that electrical power is usually expressed in watts or kilowatts, but the power consumption of an incandescent lamp could equally well be expressed in horsepower, or an automobile engine rated in kilowatts.

From the relations between the newton, pound, meter, and foot, we can show that

$$1 \text{ hp} = 746 \text{ W} = 0.746 \text{ kW},$$

or about $\frac{3}{4}$ of a kilowatt, a useful figure to remember.

Since power is energy or work per unit time, the units of power may be used to define new units of work or energy. One such unit in common use is the kilowatt-hour, a unit of energy commonly used for electrical energy.

One kilowatt-hour is the work done in one hour by an agent working at the constant rate of one kilowatt.

Since such an agent does 1000 J of work each second, the work done in 1 hr is $3600 \times 1000 = 3{,}600{,}000$ J:

$$1 \text{ kWh} = 3.6 \times 10^6 \text{ J} = 3.6 \text{ MJ}.$$

Note that the horsepower-hour and the kilowatt-hour are units of *work* or *energy*, not power. Although energy is an abstract physical quantity, it nevertheless has a monetary value. A newton of force or a meter per second of velocity are not things that are bought and sold as such, but a kilowatt-hour of energy is a quantity offered for sale at a definite market rate. In the form of electrical energy, a kilowatt-hour can be purchased at a price varying from a few tenths of a cent to a few cents, depending on the locality and the quantity purchased.

6-10 Power and velocity

Suppose a force $\boldsymbol{F}$ acts on a body while it undergoes a displacement Δs along its path. If $F_\parallel$ is the component of $\boldsymbol{F}$ tangent to the path, then the work is given by $\Delta W = F_\parallel \Delta s$, and the average power is

$$\bar{P} = \frac{\Delta W}{\Delta t} = F_\parallel \frac{\Delta s}{\Delta t} = F_\parallel \bar{v}.$$

In the limit as $\Delta t \to 0$ we obtain the instantaneous power:

$$P = F_\parallel v, \qquad (6\text{-}32)$$

where v is the instantaneous velocity. This relation may also be expressed in terms of the scalar product:

$$P = \boldsymbol{F} \cdot \boldsymbol{v}. \qquad (6\text{-}33)$$

Example A jet airplane engine develops a thrust of 15,000 N (roughly 3000 lb). When the plane is flying at 300 m·s⁻¹ (roughly 600 mph), what horsepower is developed?

Solution

$$P = Fv = (1.5 \times 10^4 \text{ N})(300 \text{ m} \cdot \text{s}^{-1}) = 4.5 \times 10^6 \text{ W}$$

$$= (4.5 \times 10^6 \text{ W}) \left(\frac{1 \text{ hp}}{746 \text{ W}}\right) = 6030 \text{ hp}.$$

*6-11 Mass and energy

At sufficiently high speeds (comparable to the speed of light) the laws of newtonian mechanics are no longer precisely correct but must be replaced by the more general relations predicted by the special relativity, to be studied in Chapter 43. One aspect of this generalization is that kinetic energy is no longer precisely $\frac{1}{2}mv^2$, but is given instead by

$$K = \frac{mc^2}{\sqrt{1 - v^2/c^2}} - mc^2, \qquad (6\text{-}34)$$

where c is the speed of light, $c = 3.00 \times 10^8$ m·s^{-1}.

This appears quite different from $\frac{1}{2}mv^2$, but when v is small it reduces to $\frac{1}{2}mv^2$. To see this, we apply the binomial theorem to the radical in Eq. (6-34):

$$(1 - a)^{-1/2} = 1 + \tfrac{1}{2}a + \tfrac{3}{8}a^2 + \cdots,$$

where, in this case, $a = v^2/c^2$. Equation (6-34) becomes

$$K = mc^2\left(1 + \frac{v^2}{2c^2} + \frac{3v^4}{8c^4} + \cdots\right) - mc^2$$

$$= \tfrac{1}{2}mv^2 + \tfrac{3}{8}m\left(\frac{v^4}{c^2}\right) + \cdots.$$

When v is much smaller than c, powers of v^2/c^2 become very small and may be neglected, and what remains is simply $\frac{1}{2}mv^2$.

The relativistic kinetic energy expression was predicted theoretically by Lorentz and Einstein, and has been verified experimentally with a variety of observations on rapidly moving particles such as electrons. One interesting feature is that the expression becomes very large as v becomes nearly equal to c, and predicts an infinitely large energy at $v = c$. This suggests physically that no particle may have a speed as large as c, since this would require an infinite quantity of energy. The relation of Eq. (6-34) to the nonrelativistic expression $\frac{1}{2}mv^2$ is shown in Fig. 6-14.

Equation (6-34) may be reinterpreted in an interesting way. Suppose a particle with mass m has an intrinsic energy equal to mc^2, not related directly to its speed. Then the total energy of a moving particle is the sum

$$E = K + mc^2 = \frac{mc^2}{\sqrt{1 - v^2/c^2}}. \qquad (6\text{-}35)$$

We may call mc^2 the "rest energy" of the particle. The validity of this concept may be tested by considering situations in which the mass of a particle changes; we may investigate whether, when mass changes, there is a corresponding change in kinetic energy, so that the *total* energy is conserved.

Example When the nucleus of a lithium atom is struck by a rapidly moving proton, a momentary union of the two nuclei takes place, after which the compound nucleus breaks up into two alpha particles (helium nuclei). The alpha particles recoil in almost opposite directions and move initially with very high velocities. Their combined kinetic energy is much *greater* than the kinetic energy of the original proton. The source of this kinetic energy is the so-called "binding energy" of the nuclear particles, which is a form of internal potential energy. That is, the internal potential energy of the assemblage of protons and neutrons that makes up the unstable composite nucleus is larger than the potential energy when the same number of particles is combined in the form of two helium nuclei. A crude analogy is that of two masses forced apart by

6-14 Relation between the speed of a particle and its kinetic energy. The speed never exceeds c, no matter how great the kinetic energy. At speeds small compared to c, the kinetic energy is approximately $(1/2)mv^2$, but at greater speeds this expression deviates more and more from the relativistically correct equation.

a compressed spring, but tied together by a cord. If the cord is cut, the internal potential energy of the spring is transformed into kinetic energy of the recoiling masses.

The mass of a proton is 1.6727×10^{-27} kg. The mass of a lithium nucleus is 11.6478×10^{-27} kg, and that of an alpha particle is 6.6448×10^{-27} kg. Hence the mass of the original system is

$$1.6727 + 11.6478 \times 10^{-27} \text{ kg} = 13.3205 \times 10^{-27} \text{ kg}.$$

The mass of the two alpha particles is

$$2(6.6448 \times 10^{-27} \text{ kg}) = 13.2896 \times 10^{-27} \text{ kg}.$$

Thus, the final mass is less than the initial by 0.0309×10^{-27} kg, corresponding to an energy difference

$$(0.0309 \times 10^{-27} \text{ kg})(3.00 \times 10^8 \text{ m} \cdot \text{s}^{-1})^2 = 0.278 \times 10^{-11} \text{ J}.$$

In order for energy to be conserved, the final total kinetic energy must be greater than the initial by this amount. Experiments have confirmed this prediction and hence also confirmed the validity of the "rest energy" concept.

Mass changes occur also in ordinary chemical reactions, but they are so small as not to be directly observable. For example, when two molecules of hydrogen H_2 and one of oxygen O_2 combine to form two molecules of water (H_2O), an energy of about 4×10^{-19} J is released for each water molecule produced. This corresponds to a mass change of $4 \times 10^{-19} \text{ J}/c^2 = 0.44 \times 10^{-35}$ kg. Since the mass of a water molecule is about 3×10^{-26} kg, this is a mass change on the order of 1 part in 10^{10}; such small mass changes in chemical reactions have never been observed, but in nuclear reactions the fractional mass changes may be as large as 10^{-4} and are readily observable.

Thus the concepts of *mass* and *energy* emerge as two aspects of a single conserved entity that may be called *mass–energy*. Equation (6-35) is derived and discussed in greater detail in Chapter 43.

Questions

6-1 An elevator is hoisted by its cables at constant speed. Is the total work done on the elevator positive, negative, or zero?

6-2 A rope tied to a body is pulled, causing the body to accelerate. But according to Newton's third law the body pulls back on the rope with an equal and opposite force. Is the total work done then zero? If so, how can the body's kinetic energy change?

6-3 In Fig. 6-9, the projectile has the same initial kinetic energy in each case. Why does it not then rise to the same maximum height in each case?

6-4 Are there any cases where a frictional force can *increase* the mechanical energy of a system? If so, give examples.

6-5 An automobile jack is used to lift a heavy weight by exerting a force much smaller in magnitude than the weight. Does this mean that less work is done than if the weight had been lifted directly?

6-6 A compressed spring is clamped in its compressed position and is then dissolved in acid. What becomes of its potential energy?

6-7 A child standing on a playground swing can increase the amplitude of his motion by "pumping up," pulling back on the swing ropes at the appropriate points during the motion. Where does the added energy come from?

6-8 In a siphon, water is lifted above its original level during its flow from one container to another. Where does it get the needed potential energy?

6-9 A man bounces on a trampoline, going a little higher with each bounce. Explain how he increases his total mechanical energy.

6-10 Is it possible for the second hill on a roller-coaster track to be higher than the first? What would happen if it were higher?

6-11 Does the kinetic energy of a car change more when it speeds up from 10 to 15 m·s^{-1} or from 15 to 20 m·s^{-1}?

6-12 A car accelerates from an initial speed to a greater final speed while the engine develops constant power. Is the acceleration greater at the beginning of this process or at the end?

6-13 Time yourself while running up a flight of steps, and compute your maximum power, in horsepower. Are you stronger than a horse?

6-14 When a constant force is applied to a body moving with constant acceleration, is the power of the force constant? If not, how would the force have to vary with speed for the power to be constant?

6-15 An advertisement for a power saw states: "This power tool uses the energy of the motor to stop the blade rotation within 5 seconds after the switch is turned off." Is this an accurate statement? Please explain.

Problems

6-1 The locomotive of a freight train exerts a constant force of 60,000 N on the train while drawing it at 50 km·hr^{-1} on a level track. How much work does it do in a distance of 1 km?

6-2 A 40-kg block is pushed a distance of 5 m along a level floor at a constant speed by a force at an angle of 30° below the horizontal. The coefficient of friction between block and floor is 0.25. How much work is done by the force? By friction?

6-3 A horse is towing a canal boat, the towrope making an angle of 10° with the towpath. If the tension in the rope is 20 N, how many joules of work are done while moving 50 m along the towpath?

6-4 A block is pushed 2 m along a fixed horizontal surface by a horizontal force of 2 N. The opposing force of friction is 0.4 N.

a) How much work is done by the 2-N force?

b) What is the work of the friction force?

6-5 A body is attracted toward the origin with a force given by $F = -6x^3$ where F is in newtons and x in meters.

a) What force is required to hold the body at point a, 1 m from the origin?

b) At point b, 2 m from the origin?

c) How much work must be done to move the body from point a to point b, if it is at rest at the start and at the end?

6-6 The force exerted by a gas in a cylinder on a piston whose area is A is given by $F = pA$, where p is the force per unit area, or *pressure*. The work W in a displacement of the piston from x_1 to x_2 is

$$W = \int_{x_1}^{x_2} F\, dx = \int_{x_1}^{x_2} pA\, dx = \int_{V_1}^{V_2} p\, dV,$$

where dV is the accompanying infinitesimal change of volume of the gas.

a) During an expansion of a gas at constant temperature (isothermal) the pressure depends on the volume according to the relation

$$p = \frac{nRT}{V},$$

where n and R are constants and T is the constant temperature. Calculate the work in expanding isothermally from volume V_1 to volume V_2.

b) During an expansion of a gas at constant entropy (adiabatic) the pressure depends on the volume according to the relation

$$p = \frac{K}{V^\gamma},$$

where K and γ are constants. Calculate the work in expanding adiabatically from V_1 to V_2.

6-7

a) Compute the kinetic energy of a 1200-kg automobile traveling at 20 km·hr^{-1}.

b) How many times as great is the kinetic energy if the velocity is doubled?

6-8 Compute the kinetic energy, in joules, of a 2-g rifle bullet traveling at 500 m·s^{-1}.

6-9 An electron strikes the screen of a cathode-ray tube with a velocity of 10^7 m·s^{-1}. Compute its kinetic energy in joules. The mass of an electron is 9.11×10^{-31} kg.

6-10 A body of mass 2 kg is initially at rest on a horizontal frictionless plane. It is then pulled 4 m by a horizontal force of magnitude 25 N. Use the work–energy relation to find its final speed.

6-11 In Problem 6-10, suppose the body has an initial speed of 10 m·s^{-1} and is then pulled 4 m by a force of magnitude 25 N in the direction of the initial velocity. What is its final speed?

6-12 A body of mass 8 kg moves in a straight line on a horizontal frictionless surface. At one point in its path its speed is 4 m·s^{-1}, and after it has traveled 3 m its speed is 5 m·s^{-1} in the same direction. Use the work–energy relation to find the force acting on the body, assuming it is constant.

6-13 A force of magnitude 5 N acts on a 2-kg body moving initially in the direction of the force with a speed of $4 \text{ m} \cdot \text{s}^{-1}$. Over what distance must the force act in order to change the body's speed to $6 \text{ m} \cdot \text{s}^{-1}$?

6-14 What is the potential energy of an 800-kg elevator at the top of the Empire State Building 380 m above street level? Assume the potential energy at street level to be zero.

6-15 What is the increase in potential energy of a 1-kg body when lifted from the floor to a table 1 m high?

6-16 A meter stick whose mass is 0.3 kg is pivoted at one end, as in Fig. 6–15, and displaced through an angle of 60° from the vertical. What is the increase in its potential energy?

Figure 6–15

6-17 The force in newtons required to stretch a certain spring a distance of x m beyond its unstretched length is given by $F = 100x$.

a) What force will stretch the spring 0.1 m? 0.2 m? 0.4 m?

b) How much work is required to stretch the spring 0.1 m? 0.2 m? 0.4 m?

6-18 The scale of a certain spring balance reads from zero to 1200 N and is 0.1 m long.

a) What is the potential energy of the spring when it is stretched 0.1 m? 0.05 m?

b) When a 600-kg mass hangs from the spring?

6-19 A body moves a distance of 10 m under the action of a force that has the constant value of 5.5 N for the first 6 m and then decreases to a value of 2 N, as shown by the graph in Fig. 6–16.

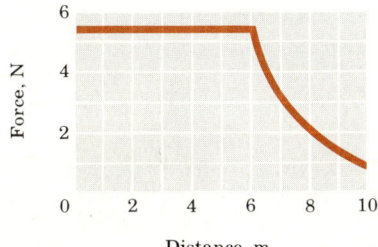

Figure 6–16

a) How much work is done in the first 6 m of the motion?

b) Approximately how much work is done in the last 4 m?

6-20 A block of mass 8 kg is pushed 5 m along a horizontal frictionless surface by a horizontal force of 40 N. The block starts from rest.

a) How much work is done? What becomes of this work?

b) Check your answer by computing the acceleration of the block, its final velocity, and its kinetic energy.

6-21 In the preceding problem, suppose the block had an initial speed of $4 \text{ m} \cdot \text{s}^{-1}$, other quantities remaining the same.

a) How much work is done?

b) Check by computing the final velocity and the increase in kinetic energy.

6-22 A 5-kg block is lifted vertically at a constant velocity of $4 \text{ m} \cdot \text{s}^{-1}$ through a height of 12 m.

a) How great a force is required?

b) How much work is done? What becomes of this work?

6-23 A 12-kg block is pushed 20 m up the sloping surface of a plane inclined at an angle of 37° to the horizontal, by a constant force F of 120 N acting parallel to the plane. The coefficient of friction between the block and plane is 0.25.

a) What is the work of the force F?

b) Compute the increase in kinetic energy of the block.

c) Compute the increase in potential energy of the block.

d) Compute the work done against friction. What becomes of this work?

e) What can you say about the sum of (b), (c), and (d)?

6-24 A man of mass 80 kg sits on a platform suspended from a movable pulley and raises himself at constant speed by a rope passing over a fixed pulley (Fig. 6–17). Assuming no friction losses, find

a) the force he must exert,

b) the increase in his energy when he raises himself 1 m.

Figure 6–17

Answer part (b) by calculating his increase in potential energy, and also by computing the product of the force on the rope and the length of rope passing through his hands.

6-25 A barrel of mass 120 kg is suspended by a rope 10 m long.

a) What horizontal force is necessary to hold the barrel in a position displaced sideways 2 m from the vertical?

b) How much work is done in moving it to this position?

6-26 The system in Fig. 6-18 is released from rest with the 12-kg block 3 m above the floor. Use the principle of conservation of energy to find the velocity with which the block strikes the floor. Neglect friction and inertia of the pulley.

Figure 6-18

6-27 The spring of a spring gun has a force constant of 500 N·m^{-1}. It is compressed 0.05 m and a ball of mass 0.01 kg is placed in the barrel against the compressed spring.

a) Compute the maximum velocity with which the ball leaves the gun when released.

b) Determine the maximum velocity if a constant resisting force of 10 N acts on the ball.

6-28 A block of mass 1 kg is forced against a horizontal spring of negligible mass, compressing the spring an amount $x_1 = 0.2$ m. When released, the block moves on a horizontal tabletop a distance $x_2 = 1.0$ m before coming to rest. The spring constant k is 100 N·m^{-1} (Fig. 6-19). What is the coefficient of friction, μ, between the block and the table?

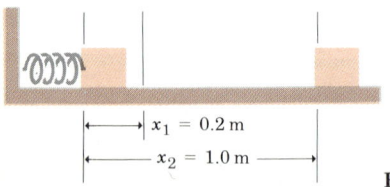

Figure 6-19

6-29 A 2-kg block is dropped from a height of 0.4 m onto a spring whose force constant k is 1960 N·m^{-1}. Find the maximum distance the spring will be compressed.

6-30 An 8-kg projectile is fired from a gun with a muzzle velocity of 300 m·s^{-1} at an angle of departure of 45°. The angle is then increased to 90° and a similar projectile is fired with the same muzzle velocity. Neglecting air resistance,

a) Find the maximum height attained by each projectile.

b) Show that the total energy at the top of the trajectory is the same in the two cases.

c) Using the energy principle, find the height attained by a similar projectile if fired at an angle of 30°.

6-31 A block of mass 2 kg is released from rest at point A on a track that is one quadrant of a circle of radius 1 m (Fig. 6-20). It slides down the track and reaches point B with a speed of 4 m·s^{-1}. From point B it slides on a level surface a distance of 3 m to point C, where it comes to rest.

a) What was the coefficient of sliding friction on the horizontal surface?

b) How much work was done against friction as the body slid down the circular arc from A to B?

Figure 6-20

6-32 A small sphere of mass m is fastened to a weightless string of length 0.5 m to form a pendulum. The pendulum is swinging so as to make a maximum angle of 60° with the vertical.

a) What is the velocity of the sphere when it passes through the vertical position?

b) What is the instantaneous acceleration when the pendulum is at its maximum deflection?

6-33 A ball is tied to a cord and set in rotation in a vertical circle. Prove that the tension in the cord at the lowest point exceeds that at the highest point by six times the weight of the ball.

6-34 A small body of mass m slides without friction around the loop-the-loop apparatus shown in Fig. 6-21. It starts from rest at point A at a height $3R$ above the bottom of the loop. When it reaches point B at the end of a horizontal diameter of the loop, compute

a) its radial acceleration,

b) its tangential acceleration, and

c) its resultant acceleration. Show these accelerations in a diagram, approximately to scale.

Figure 6-21

6-35 A meter stick, pivoted about a horizontal axis through its center, has a body of mass 2 kg attached to one end and a body of mass 1 kg attached to the other. The mass of the meter stick can be neglected. The system is released from rest with the stick horizontal. What is the velocity of each body as the stick swings through a vertical position?

6-36 A variable force P is maintained tangent to a frictionless cylindrical surface of radius a, as shown in Fig. 6-22. By slowly varying this force, a block of weight w is moved and the spring to which it is attached is stretched from position 1 to position 2. Calculate the work of the force P.

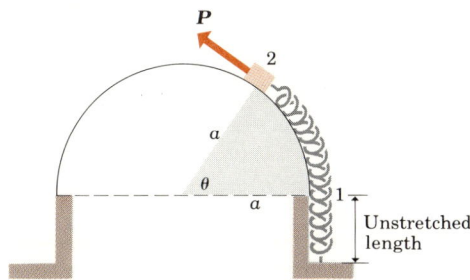

Figure 6-22

6-37 A 5-kg block is pushed up a frictionless plane inclined at 30° to the horizontal. It is pushed 2.0 m along the plane by a constant 100-N force parallel to the plane. If its speed at the bottom is $2 \text{ m} \cdot \text{s}^{-1}$, what is its speed at the top?

6-38 In Problem 6-37, suppose the plane is not frictionless but has $\mu = 0.1$.

a) What is the speed of the block at the top?

b) What fraction of the work done by the force is dissipated by friction?

6-39 An 80-kg man jumps from a height of 2 m onto a platform mounted on springs. As the springs compress, the platform is pushed down a maximum distance of 0.2 m below its initial position, and it then rebounds.

a) What is the man's speed at the instant the platform is depressed 0.1 m?

b) If the man had just stepped gently onto the platform how far would it have been pushed down?

6-40 In Problem 6-34, if point A is at a height h (not equal to $3R$) above the bottom of the loop, what is the minimum value of h (in terms of R) so that the body moves around the loop without falling off at the top?

6-41 A ball of mass 0.5 kg is tied to a string of length 1.0 m, and the other end of the string is tied to a rigid support. The ball is held straight out horizontally from the point of support, with the string pulled taut, and is then released.

a) What is the speed of the ball at the lowest point of its motion?

b) What is the tension in the string at this point?

6-42 A skier starts at the top of a large frictionless spherical snowball, with very small initial velocity, and skis straight down the side. At what point does he lose contact with the snowball and fly off at a tangent? That is, at the instant he loses contact with the snowball, what angle does a radial line from the center to the skier make with the vertical?

6-43 In Sec. 6-4, Example 6, calculate the escape velocity, in meters per second, for an object at the surface of the earth. Why is this velocity independent of the object's mass?

6-44 Calculate the escape velocity for an object at the surface of the moon.

6-45 There is a point on the line joining the centers of the earth and moon where the gravitational forces add to zero.

a) Determine the location of this point. The appropriate data are given in Appendix G.

b) With what velocity would a body have to be launched at the surface of the earth in order to just reach this point with zero velocity?

c) If a body is given a velocity just slightly greater than that found in (b), what is its velocity just as it reaches the surface of the moon?

6-46 Show that the escape velocity of a body at the surface of the earth is greater than the kinetic energy of the same body in a circular orbit around the earth (with orbit radius only slightly greater than earth's radius) by a factor of $\sqrt{2}$.

6-47 The asteroid Toro, discovered in 1964, has a radius of about 5 km and a mass of about 2×10^{15} kg. Calculate the escape velocity for a body at the surface of Toro. Could a person *jump* off Toro, i.e., jump with a velocity greater than the escape velocity?

6-48 If the potential energy of a 1-kg body, associated with the gravitational attraction of the earth and the moon, is zero far away from both bodies, what is it at the "zero-gravity" point described in Problem 6-45?

6-49 What average horsepower is developed by an 80-kg man while climbing in 10 s a flight of stairs that rises 6 m vertically? Express this power in watts and kilowatts.

6-50 The hammer of a pile driver has a mass of 500 kg and must be lifted a vertical distance of 2 m in 3 s. What horsepower engine is required?

6-51 A ski tow is to be operated on a 37° slope 300 m long. The rope is to move at $12 \text{ km} \cdot \text{hr}^{-1}$ and power must be provided for 80 riders at one time, with average mass per rider of 70 kg. Estimate the horsepower required to operate the tow.

6-52

a) If energy costs 5 cents per kWh, how much is one horsepower-hour worth?

b) How many ft-lb can be purchased for one cent?

6-53 Compute the monetary value of the kinetic energy of the projectile of a 14-in. naval gun, at the rate of 2 cents per kWh. The projectile has a mass of 600 kg and its muzzle velocity is 800 m·s^{-1}.

6-54 At 5 cents per kWh, what does it cost to operate a 10-hp motor for 8 hr?

6-55 The engine of an automobile develops 20 hp when the automobile is traveling at 50 km·hr^{-1}.

a) What is the resisting force acting on the automobile, in newtons?

b) If the resisting force is proportional to the velocity, what horsepower will drive the car at 25 km·hr^{-1}? At 100 km·hr^{-1}?

6-56 The engine of a motorboat delivers 40 hp to the propeller while the boat is moving at 10 m·s^{-1}. What would be the tension in the towline if the boat were being towed at the same speed?

6-57 A man whose mass is 70 kg walks up to the third floor of a building. This is a vertical height of 12 m above the street level.

a) How many joules of work has he done?

b) By how much has he increased his potential energy?

c) If he climbs the stairs in 20 s, what was his rate of working, in horsepower?

6-58 A pump is required to lift 800 kg (about 200 gallons) of water per minute from a well 10 m deep and eject it with a speed of 20 m·s^{-1}.

a) How much work is done per minute in lifting the water?

b) How much in giving it kinetic energy?

c) What horsepower engine is needed?

6-59 A 2000-kg elevator starts from rest and is pulled upward with a constant acceleration of 4 m·s^{-1}.

a) Find the tension in the supporting cable.

b) What is the velocity of the elevator after it has risen 15 m?

c) Find the kinetic energy of the elevator 3 s after it starts.

d) How much is its potential energy increased in the first 3 s?

e) What horsepower is required when the elevator is traveling 8 m·s^{-1}?

6-60 An automobile of mass 1200 kg has a speed of 30 m·s^{-1} on a horizontal road when the engine is developing 50 hp. What is its speed, with the same horsepower, if the road rises 1 m in 20 m? Assume all friction forces to be constant.

6-61

a) If 20 hp are required to drive a 1200-kg automobile at 50 km·hr^{-1} on a level road, what is the total retarding force due to friction, windage, etc.?

b) What power is necessary to drive the car at 50 km·hr^{-1} up a 10-percent grade (i.e., one rising 10 m vertically in 100 m horizontally)?

c) What power is necessary to drive the car at 50 km·hr^{-1} *down* a 2-percent grade?

d) Down what percent grade would the car coast at 50 km·hr^{-1}?

6-62 The total consumption of electrical energy in the United States is of the order of 10^{19} joules per year.

a) What is the average rate of energy consumption, in watts?

b) If the population of the United States is 200 million, what is the average rate of energy consumption per person?

6-63 The sun transfers energy to the earth by radiation, at a rate of approximately 1.4 kW per square meter of surface. If this energy could be collected and converted to electrical energy with 100-percent efficiency, how great an area would be required to collect the energy cited in the preceding problem?

6-64 An automobile of mass m accelerates, starting from rest, while the engine supplies constant power P.

a) Show that the velocity is given as a function of time by $v = (2Pt/m)^{1/2}$.

b) Show that the position is given as a function of time by $x = (8P/9m)^{1/2} t^{3/2}$.

6-65 A nuclear bomb containing 20 kg of plutonium explodes. The rest mass of the products of the explosion is less than the original rest mass by one ten-thousandth of the original rest mass.

a) How much energy is released in the explosion?

b) If the explosion takes place in 1 μs, what is the average power developed by the bomb?

c) How much water could the released energy lift to a height of 1 km?

6-66 In a hypothetical nuclear-fusion reactor two deuterium nuclei combine or "fuse" to form one helium nucleus. The mass of a deuterium nucleus, expressed in atomic mass units (u) is 2.0147 u; that of a helium nucleus is 4.0039 u (1 u = 1.66 × 10^{-27} kg).

a) How much energy is released when 1 kg of deuterium undergoes fusion?

b) The annual consumption of electrical energy in the United States is of the order of 10^{19} J. How much deuterium must react to produce this much energy?

6-67 Compute the kinetic energy of an electron (mass 9.11 × 10^{-31} kg) using both the nonrelativistic and relativistic expressions, and compare the two results, for speeds of

a) 1.0 × 10^7 m·s^{-1};   b) 2.0 × 10^8 m·s^{-1}.

IMPULSE AND MOMENTUM 7

7-1 Impulse and momentum

In the preceding chapter, the concepts of work and energy were developed from Newton's laws of motion. We shall see next how two similar concepts, those of *impulse* and *momentum*, also arise from these laws.

Let us consider again a particle of mass m moving along a straight line. For the moment we assume that the force F on this particle is constant and directed along the line of motion. If the particle's velocity at some initial time $t = 0$ is v_0, the velocity v at a later time t is given by

$$v = v_0 + at,$$

where the constant acceleration a is determined from the relation $F = ma$. When the above equation is multiplied through by m and ma is replaced by F, the result is

$$mv = mv_0 + Ft,$$

or

$$mv - mv_0 = Ft. \tag{7-1}$$

The right side of this equation, the product of the force and the time during which it acts, is called the *impulse* of the force, denoted by J. More generally, if a constant force F acts during a time interval from t_1 to t_2, the impulse of the force is defined to be

$$\text{Impulse} = J = F(t_2 - t_1). \tag{7-2}$$

The left side of Eq. (7-1) contains the product of mass and velocity of the particle at two different times. This product is also given a special name, *momentum*. It is also sometimes called *linear* momentum to distinguish it from a similar quantity called *angular momentum*, to be discussed later. The symbol p is often used for momentum:

$$\text{Momentum} = p = mv. \tag{7-3}$$

In terms of these newly defined quantities, the meaning of Eq. (7-1) is that the impulse of the force from time zero to time t is equal to the change of momentum during that interval, that is, the momentum at the end of the interval minus that at the beginning. We see that there is nothing special about the two times zero and t, so we may also say that if the particle's velocity at time t_1 is v_1, and its velocity at time t_2 is v_2, then

$$F(t_2 - t_1) = mv_2 - mv_1. \tag{7-4}$$

Straight-line motion is, of course, a special case. Strictly speaking, the quantities F, v, and a should have been called the *components* along the straight line of the vector quantities **F, v,** and **a,** respectively. The definitions of momentum and impulse and the relationship expressed by Eq. (7-4) are easily generalized for motion not confined to a straight line. We define impulse and momentum to be *vector quantities,* as follows:

$$\text{\textbf{Impulse}} = \boldsymbol{J} = \boldsymbol{F}(t_2 - t_1). \tag{7-5}$$

$$\text{\textbf{Momentum}} = \boldsymbol{p} = m\boldsymbol{v}. \tag{7-6}$$

The corresponding generalization of Eq. (7-4) is the vector equation

$$\boldsymbol{F}(t_2 - t_1) = m\boldsymbol{v}_2 - m\boldsymbol{v}_1. \tag{7-7}$$

This relation between impulse and momentum change has the same form as the relation between work and kinetic-energy change developed in Chapter 6. There are important differences, however. First, impulse is a product of a force and a *time* interval, while work is a product of a force and a *distance,* and depends on the angle between force and displacement. Furthermore, both force and velocity are *vector* quantities, so impulse and momentum are also vector quantities, unlike work and kinetic energy, which are scalars. In motion along a straight line, where only one component of a vector is involved, the force and velocity may have components along this line that are either positive or negative.

Example 1 A particle of mass 2 kg moves along the x-axis with an initial velocity of $3 \text{ m} \cdot \text{s}^{-1}$. A force $F = -6$ N (i.e., in the negative x-direction) is applied for a period of 3 s. Find the final velocity.

Solution From Eq. (7-4),

$$(-6 \text{ N})(3 \text{ s}) = (2 \text{ kg})v_2 - (2 \text{ kg})(3 \text{ m} \cdot \text{s}^{-1})$$

or

$$v_2 = -6 \text{ m} \cdot \text{s}^{-1}.$$

The particle's final velocity is in the negative x-direction.

The unit of impulse, in any system, equals the product of the units of force and time in that system. Thus, in SI (mks) units, the unit is one newton second (1 N·s), in the cgs system it is one dyne second (1 dyn·s), and in the engineering system it is one pound second (1 lb·s).

The unit of momentum in SI (mks) units is one kilogram meter per second (1 kg·m·s^{-1}), in the cgs system it is one gram centimeter per second (1 g·cm·s^{-1}), and in the engineering system it is one slug foot per second (1 slug·ft·s^{-1}). Since

$$1 \text{ kg} \cdot \text{m} \cdot \text{s}^{-1} = (1 \text{ kg} \cdot \text{m} \cdot \text{s}^{-2})\text{s} = 1 \text{ N} \cdot \text{s},$$

it follows that the units of momentum and impulse are equivalent.

The concept of impulse can be generalized to include forces that vary with time; both the magnitude and the direction of the force may vary. We consider a particle of mass m moving in the xy-plane, or in space, acted on by a varying resultant force F. Newton's second law states that

$$F = m\frac{dv}{dt},$$

or

$$F\, dt = m\, dv.$$

If v_1 is the velocity when $t = t_1$, and v_2 the velocity when $t = t_2$, it follows that

$$\int_{t_1}^{t_2} F\, dt = \int_{v_1}^{v_2} m\, dv. \qquad (7\text{--}8)$$

The integral on the left is the *impulse J of the force F* in the time interval $t_2 - t_1$, and is a *vector quantity*:

$$\textbf{Impulse} = \boldsymbol{J} = \int_{t_1}^{t_2} \boldsymbol{F}\, dt.$$

This integral can be evaluated, of course, only when the force is known as a function of the time. The integral on the right, however, is just m multiplied by the total change in v:

$$\int_{v_1}^{v_2} m\, dv = m \int_{v_1}^{v_2} dv = m(v_2 - v_1).$$

Thus Eq. (7–8) may be written as

$$\int_{t_1}^{t_2} F\, dt = mv_2 - mv_1, \qquad (7\text{--}9)$$

In the special case when F is constant, this reduces to Eq. (7–7).

Equation (7–9) is also equivalent (for forces and velocities in the xy-plane) to the two scalar equations

$$\int_{t_1}^{t_2} F_x\, dt = mv_{x2} - mv_{x1},$$
$$\int_{t_1}^{t_2} F_y\, dt = mv_{y2} - mv_{y1}, \qquad (7\text{--}10)$$

The impulse of any force component, or any force whose direction is constant, can be represented graphically by plotting the force vertically and the time horizontally, as in Fig. 7-1. The *area* under the curve, between vertical lines at t_1 and t_2, is equal to the impulse of the force in this time interval.

If the component in a given direction of the impulse of a force is *positive*, the corresponding component of momentum of the body on which the impulse acts *increases* algebraically; if negative, it decreases. If the impulse is zero, there is no change in momentum.

Example 2 Consider the changes in momentum produced by the following forces:

a) A body moving on the x-axis is acted on for 2 s by a constant force of 10 N toward the right.

7-1 The area of the small rectangle is an approximation of the total change in momentum in the interval Δt.

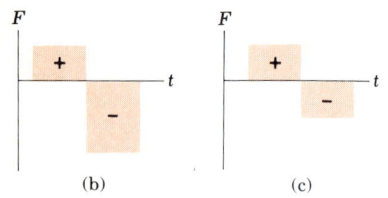

7-2 Graphic representation of the forces in Example 2.

b) The body is acted on for 2 s by a constant force of 10 N toward the right and then for 2 s by a constant force of 20 N toward the left.

c) The body is acted on for 2 s by a constant force of 10 N toward the right, then for 1 s by a constant force of 20 N toward the left. The forces for each case are shown graphically in Fig. 7-2.

Solution

a) The impulse of the force is $(+10\text{ N})(2\text{ s}) = +20\text{ N}\cdot\text{s}$. Hence the momentum of *any* body on which the force acts increases by $20\text{ kg}\cdot\text{m}\cdot\text{s}^{-1}$. This change is the same whatever the mass of the body and whatever the magnitude and direction of its initial velocity.

Suppose the mass of the body is 2 kg and that it is initially at rest. Its *final* momentum then equals its *change* in momentum and its final velocity is $10\text{ m}\cdot\text{s}^{-1}$ toward the right. (The reader should verify by computing the acceleration.)

Had the body been initially moving toward the right at $5\text{ m}\cdot\text{s}^{-1}$, its initial momentum would have been $10\text{ kg}\cdot\text{m}\cdot\text{s}^{-1}$ toward the right, its final momentum $30\text{ kg}\cdot\text{m}\cdot\text{s}^{-1}$, and its final velocity $15\text{ m}\cdot\text{s}^{-1}$ toward the right.

Had the body been moving initially toward the *left* at $5\text{ m}\cdot\text{s}^{-1}$, its initial momentum would have been $-10\text{ kg}\cdot\text{m}\cdot\text{s}^{-1}$, its final momentum $+10\text{ kg}\cdot\text{m}\cdot\text{s}^{-1}$, and its final velocity $5\text{ m}\cdot\text{s}^{-1}$ toward the right. That is, the constant force of 10 N toward the right would first have brought the body to rest and then given it a velocity in the direction opposite to its initial velocity.

b) The impulse of this force is

$$(+10\text{ N})(2\text{ s}) - (20\text{ N})(2\text{ s}) = -20\text{ N}\cdot\text{s}.$$

The momentum of *any* body on which this impulse acts is changed by $-20\text{ kg}\cdot\text{m}\cdot\text{s}^{-1}$. The reader should examine various possibilities, as in the preceding example.

c) The impulse of this force is

$$(+10\text{ N})(2\text{ s}) - (20\text{ N})(1\text{ s}) = 0.$$

Hence the momentum of any body on which it acts is not changed. Of course, the momentum of the body is increased during the first 2 s but it is *decreased* by an equal amount in the next second. As an exercise, describe the motion of a body of mass 2 kg, moving initially to the left at $5\text{ m}\cdot\text{s}^{-1}$, and acted on by this force. It will help to construct a graph of velocity versus time.

Example 3 A ball of mass 0.4 kg is thrown against a brick wall. When it strikes the wall it is moving horizontally to the left at $30\text{ m}\cdot\text{s}^{-1}$, and it rebounds horizontally to the right at $20\text{ m}\cdot\text{s}^{-1}$. Find the impulse of the force exerted on the ball by the wall.

Solution The initial momentum of the ball is

$$(0.4\text{ kg})(-30\text{ m}\cdot\text{s}^{-1}) = -12\text{ kg}\cdot\text{m}\cdot\text{s}^{-1}.$$

The final momentum is $+8\text{ kg}\cdot\text{m}\cdot\text{s}^{-1}$. The *change* in momentum is

$$mv_2 - mv_1 = 8\text{ kg}\cdot\text{m}\cdot\text{s}^{-1} - (-12\text{ kg}\cdot\text{m}\cdot\text{s}^{-1})$$
$$= 20\text{ kg}\cdot\text{m}\cdot\text{s}^{-1}.$$

Hence, the impulse of the force exerted on the ball was 20 N·s. Since the impulse is *positive,* the force must be toward the right.

Note that the *force* exerted on the ball cannot be found without further information regarding the collision. The general nature of the force-time graph is shown by one of the curves in Fig. 7–3. The force is zero before impact, rises to a maximum, and decreases to zero when the ball leaves the wall. If the ball is relatively rigid, like a baseball, the time of collision is small and the maximum force is large, as in curve (a). If the ball is more yielding, like a tennis ball, the collision time is larger and the maximum force is less, as in curve (b). In any event, the *area* under the force-time graph must equal 20 N·s.

For an idealized case in which the force is constant and the collision time is 1 ms (10^{-3} s), as represented by the horizontal straight line, the force is 20,000 N.

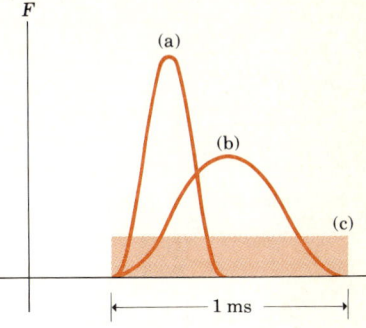

7–3 Force-time graph for Example 3.

7–2 Conservation of momentum

Let us consider now a system consisting of two bodies that interact with each other but not with anything else. Because each body exerts a force on the other, the momentum of each body changes. According to Newton's third law, the forces on the two bodies are always equal in magnitude and opposite in direction. Thus the two *impulses* in any given time interval are also equal and opposite. It follows that *in any time interval, the vector change in momentum of either particle is equal in magnitude and opposite in direction to the vector change in momentum of the other.*

This principle can be stated more simply by defining the *total momentum* of the system as the vector sum of the momenta of the separate bodies. If the change of momentum of one body is exactly the negative of that of the other, then the change in the *total* momentum must be zero. Thus *when two bodies interact only with each other, their total momentum is constant.*

A force that one part of a system exerts on another is called an *internal* force, and a force exerted on a part of a system by some body or agency outside the system is called an *external* force. When no external force acts on a system, or when the vector sum (resultant) of the external forces is zero, the total momentum of the system is constant, in both magnitude and direction. This is the *principle of conservation of linear momentum:*

When no resultant external force acts on a system, the total momentum of the system remains constant in magnitude and direction.

The principle of conservation of momentum is one of the most fundamental and important principles of mechanics. Note that it is more general than the principle of conservation of mechanical energy; mechanical energy is conserved *only* when the internal forces are *conservative.* The principle of conservation of momentum holds whatever the nature of the internal forces, conservative or not.

Example 1 Figure 7–4 shows two bodies moving toward each other on a frictionless surface or a level air track. There is no friction, and the resultant vertical force on each body is zero. Thus the resultant force on each body is the force exerted on it by the other body, and the total momentum of the system is constant, in magnitude and direction.

7–4 Momentum diagram for Example 1.

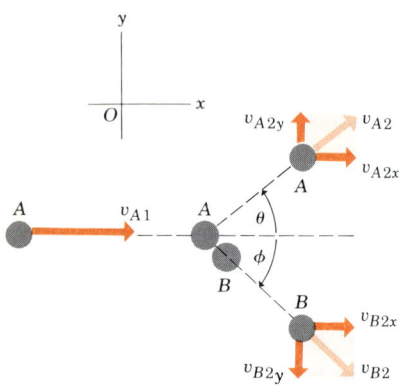

7-5 Velocity vector diagram for Example 2.

Suppose that, after the bodies collide, body B moves away with a final velocity of $+2$ m·s^{-1}. What is the final velocity of body A?

Solution Let the final velocity of A be v_{A2}. Then we write an equation showing the equality of the total momentum before and after the collision:

$$(5 \text{ kg})(2 \text{ m·s}^{-1}) + (3 \text{ kg})(-2 \text{ m·s}^{-1}) = (5 \text{ kg})v_{A2} + (3 \text{ kg})(+2 \text{ m·s}^{-1}).$$

Solving this equation for v_{A2}, we find $v_{A2} = -0.4$ m·s^{-1}.

Example 2 In Fig. 7–5, body A, having mass $m_A = 5$ kg, moves with initial velocity $v_{A1} = 2$ m·s^{-1} parallel to the x-axis, and collides with body B, having mass $m_B = 3$ kg, initially at rest. After the collision, the velocity of A is found to be $v_{A2} = 1$ m·s^{-1} in a direction making an angle $\theta = 30°$ with the initial direction. What is the final velocity of B?

Solution Again the total momentum of the system is the same before and after the collision. In this example, however, the velocities are not all along a single line, and the *vector* nature of momentum must be recognized. This is most easily accomplished by finding the components of each momentum in the x- and y-directions. Then momentum conservation requires that the sum of the x-components before the collision must equal that after the collision, and similarly for the y-components. Just as with force equilibrium problems, we write a separate equation for each component. For the x-components, we have

$$(5 \text{ kg})(2 \text{ m·s}^{-1}) + (3 \text{ kg})(0) = (5 \text{ kg})(1 \text{ m·s}^{-1})(\cos 30°) + (3 \text{ kg})v_{B2x}.$$

From this we find $v_{B2x} = 1.89$ m·s^{-1}.

Similarly, conservation of the y-component of total momentum gives

$$(5 \text{ kg})(0) + (3 \text{ kg})(0) = (5 \text{ kg})(1 \text{ m·s}^{-1})(\sin 30°) + (3 \text{ kg})v_{B2y},$$

from which $v_{B2y} = -0.83$ m·s^{-1}.

The magnitude of $\mathbf{v}_{B2}$ is

$$v_{B2} = \sqrt{(1.89 \text{ m·s}^{-1})^2 + (-0.83 \text{ m·s}^{-1})^2} = 2.06 \text{ m·s}^{-1},$$

and the angle of its direction from the positive x-axis is

$$\phi = \tan^{-1} \frac{-0.83 \text{ m·s}^{-1}}{1.89 \text{ m·s}^{-1}} = -23.7°.$$

7-3 Collisions

Suppose we are given the masses and initial velocities of two colliding bodies, and we wish to compute their velocities after the collision. If the collision is "head-on," as in the first example in Sec. 7–2, then momentum conservation provides the equation

$$m_A v_{A1} + m_B v_{B1} = m_A v_{A2} + m_B v_{B2} \tag{7-11}$$

for the unknowns v_{A2} and v_{B2}. If it is of the more general type of the second example in Sec. 7–2, then momentum conservation gives two equations for the four unknowns v_{A2x}, v_{A2y}, v_{B2x}, and v_{B2y}:

$$m_A v_{A1x} + m_B v_{B1x} = m_A v_{A2x} + m_B v_{B2x}, \tag{7-12}$$

$$m_A v_{A1y} + m_B v_{B1y} = m_A v_{A2y} + m_B v_{B2y}. \tag{7-13}$$

In either case, we have more unknowns than equations. Hence momentum considerations *alone* do not suffice to completely determine the final velocities; we must have more information regarding the collision process.

If the forces of interaction between the bodies are *conservative*, the total kinetic energy is the same before and after the collision and the collision is said to be *completely elastic*. Such a collision is closely approximated if one end of a steel spring is attached to one of the bodies, as in Fig. 7-6. When the bodies collide, the spring is momentarily compressed and some of the original kinetic energy is momentarily converted to elastic potential energy. The spring then expands, and when the bodies separate, this potential energy is reconverted to kinetic energy.

At the opposite extreme from a completely elastic collision is one in which the colliding bodies stick together and move *as a unit* after the collision. Such a collision is called a *completely inelastic collision*, and will result if the bodies in Fig. 7-4 are provided with a coupling mechanism such as that between two freight cars, or if the bodies in Fig. 7-6 are locked together at the instant when their velocities become equal and the spring is fully compressed.

7-6 Body A (with spring) and body B approach each other on a frictionless surface.

7-4 Inelastic collisions

For the special case of a completely inelastic collision between two bodies A and B, we have, from the definition of such a collision,

$$v_{A2} = v_{B2} = v_2.$$

When this is combined with the principle of conservation of momentum, we obtain

$$m_A v_{A1} + m_B v_{B1} = (m_A + m_B)v_2, \tag{7-14}$$

and the final velocity can be computed if the initial velocities and the masses are known.

It is of interest to compare the initial and final *kinetic energies* in such a collision. The kinetic energy of the system before collision is

$$K_1 = \tfrac{1}{2}m_A v_{A1}^2 + \tfrac{1}{2}m_B v_{B1}^2.$$

The final kinetic energy is

$$K_2 = \tfrac{1}{2}(m_A + m_B)v_2^2.$$

For the special case in which body B is *initially* at rest ($v_{B1} = 0$), the ratio of the final to the initial kinetic energy is

$$\frac{K_2}{K_1} = \frac{(m_A + m_B)v_2^2}{m_A v_{A1}^2}.$$

Inserting the expression for v_2 from Eq. (7-14), we find that this reduces to

$$\frac{K_2}{K_1} = \frac{m_A}{m_A + m_B}.$$

The right side is necessarily less than unity, so *the total kinetic energy always decreases in an inelastic collision*. Even when v_{B1} is not zero, it is not difficult to show that the kinetic energy after an inelastic collision is always less than before.

Example 1 Suppose that the collision in Fig. 7-4 is completely inelastic and that the masses and initial velocities have the values shown. The common velocity after the collision is then

$$v_2 = \frac{m_A v_{A1} + m_B v_{B1}}{m_A + m_B} = 0.5 \text{ m} \cdot \text{s}^{-1}.$$

Since v_2 is positive, the system moves to the right after the collision.

The kinetic energy of body A before the collision is

$$\tfrac{1}{2} m_A v_{A1}^2 = \tfrac{1}{2}(5 \text{ kg})(2 \text{ m} \cdot \text{s}^{-1})^2 = 10 \text{ J},$$

and that of body B is

$$\tfrac{1}{2} m_B v_{B1}^2 = \tfrac{1}{2}(3 \text{ kg})(-2 \text{ m} \cdot \text{s}^{-1})^2 = 6 \text{ J}.$$

The total kinetic energy before collision is therefore 16 J. Note that the kinetic energy of body B is positive, although its velocity v_{B1} and its momentum $m v_{B1}$ are both negative.

The kinetic energy after the collision is $\tfrac{1}{2}(m_A + m_B) v_2^2 = 1$ J. Hence, far from remaining constant, the final kinetic energy is only $\tfrac{1}{16}$ of the original, and $\tfrac{15}{16}$ is "lost" in the collision. If the bodies couple together like two freight cars, most of this energy is converted to elastic waves, which are eventually absorbed.

If there is a spring between the bodies, as in Fig. 7-6, and the bodies are locked together when their velocities become equal, the energy is trapped as potential energy in the compressed spring. If all these forms of energy are taken into account, the *total* energy of the system is conserved although its *kinetic* energy is not. However, momentum is always conserved in a collision, whether or not the collision is elastic.

Example 2 The *ballistic pendulum* is a device for measuring the velocity of a bullet. The bullet is allowed to make a completely inelastic collision with a body of much greater mass. The momentum of the system immediately after the collision equals the original momentum of the bullet, but since the *velocity* is very much smaller, it can be determined more easily. Although the ballistic pendulum has now been superseded by other devices, it is still an important laboratory experiment for illustrating the concepts of momentum and energy.

In Fig. 7-7, the pendulum, consisting perhaps of a large wooden block of mass M, hangs vertically by two cords. A bullet of mass m, traveling with a velocity v, strikes the pendulum and remains embedded in it. If the collision time is very small compared with the time of swing of the pendulum, the supporting cords remain practically vertical during this time. Hence, no external horizontal forces act on the system during the collision, and the horizontal momentum is conserved. Then if V represents the velocity of bullet and block immediately after the collision,

$$mv = (m + M)V, \qquad v = \frac{m + M}{m} V.$$

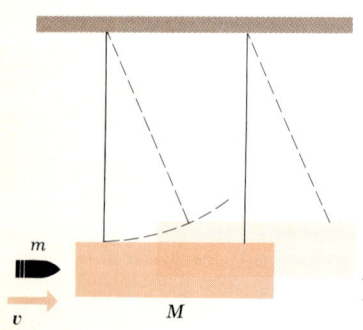

7-7 The ballistic pendulum.

The kinetic energy of the system, immediately after the collision, is $K = \frac{1}{2}(m + M)V^2$.

The pendulum now swings to the right and upward until its kinetic energy is converted to gravitational potential energy. (Small frictional effects can be neglected.) Hence,

$$\frac{1}{2}(m + M)V^2 = (m + M)gy,$$
$$V = \sqrt{2gy},$$

and

$$v = \frac{m + M}{m}(\sqrt{2gy}).$$

By measuring m, M, and y, we can compute the original velocity v of the bullet.

It is important to remember that kinetic energy is *not* conserved *during the collision*. The ratio of the kinetic energy of bullet and pendulum, after the collision, to the original kinetic energy of the bullet, is

$$\frac{\frac{1}{2}(m + M)V^2}{\frac{1}{2}mv^2} = \frac{m}{m + M}.$$

Thus, if $M = 1000$ g, and $m = 1$ g, only about 0.1 percent of the original energy remains as kinetic energy; 99.9 percent is converted to internal energy.

7–5 Elastic collisions

Consider next a perfectly elastic "head-on" or *central* collision between two bodies A and B. The bodies separate after the collision and have different velocities, v_{A2} and v_{B2}. Since kinetic energy and momentum are *both* conserved, we have:

Conservation of kinetic energy

$$\tfrac{1}{2}m_A v_{A1}^2 + \tfrac{1}{2}m_B v_{B1}^2 = \tfrac{1}{2}m_A v_{A2}^2 + \tfrac{1}{2}m_B v_{B2}^2.$$

Conservation of momentum

$$m_A v_{A1} + m_B v_{B1} = m_A v_{A2} + m_B v_{B2}.$$

Hence if the masses and initial velocities are known, we have two independent equations, from which the final velocities can be found. Simultaneous solution of these equations is somewhat involved; the results are:

$$(v_{B2} - v_{A2}) = -(v_{B1} - v_{A1}), \qquad (7\text{–}15)$$

$$v_{A2} = \frac{2m_B v_{B1} + v_{A1}(m_A - m_B)}{m_A + m_B}, \qquad (7\text{–}16)$$

$$v_{B2} = \frac{2m_A v_{A1} - v_{B1}(m_A - m_B)}{m_A + m_B}. \qquad (7\text{–}17)$$

The difference $(v_{B2} - v_{A2})$ is the velocity of B relative to A after the collision, while $(v_{B1} - v_{A1})$ is its relative velocity after the collision. Thus, from Eq. (7–15), *the relative velocity of two particles in a central, completely elastic collision is unchanged in magnitude but reversed in direction.*

For the special case in which body B is at rest before the collision, $v_{B1} = 0$ and Eqs. (7-16) and (7-17) simplify to

$$v_{A2} = \frac{m_A - m_B}{m_A + m_B} v_{A1}, \qquad v_{B2} = \frac{2m_A}{m_A + m_B} v_{A1}.$$

If the masses of A and B are equal, $v_{A2} = 0$ and $v_{B2} = v_{A1}$. That is, the first body comes to rest and the second moves off with a velocity equal to the original velocity of the first. Both the momentum and kinetic energy of the first are completely transferred to the second.

When the masses are unequal, the kinetic energies after the collisions (still assuming $v_{B1} = 0$) are

$$(K_2)_A = \tfrac{1}{2} m_A v_{A2}^2 = \left(\frac{m_A - m_B}{m_A + m_B}\right)^2 (K_1)_A,$$

$$(K_2)_B = \tfrac{1}{2} m_B v_{B2}^2 = \left(\frac{4 m_A m_B}{(m_A + m_B)^2}\right)(K_1)_A.$$

It is of interest to derive the expression for the *fractional decrease* in kinetic energy of body A, that is, the ratio of its loss of kinetic energy to its original kinetic energy. Since, in an elastic collision, the energy lost by A equals the energy gained by B, this ratio is

$$\frac{(K_2)_B}{(K_1)_A} = \frac{4 m_A m_B}{(m_A + m_B)^2} = 4\left(\frac{m_A}{m_B}\right)\frac{1}{[1 + (m_A/m_B)]^2}.$$

The fractional decrease in kinetic energy of A is plotted in Fig. 7-8 as a function of the ratio m_A/m_B (note the logarithmic scale). The energy loss approaches zero as m_A/m_B approaches zero (a very small body colliding with a very large one) and also as m_A/m_B approaches infinity (a very large body colliding with a very small one). In the former case, the first body rebounds with practically its original velocity. In the latter, the first body continues to move with very nearly its original velocity. The maximum fractional energy decrease occurs when $m_A = m_B$, and for this ratio of masses the fractional energy loss equals unity, as shown above.

This fact has an important bearing on the problem of slowing down the rapidly moving neutrons in the moderator of a nuclear reactor. As neutrons pass through matter, they make occasional elastic collisions with nuclei. In a head-on collision, the greatest loss of kinetic energy results when the mass of the nucleus equals that of the neutron, that is, when the nucleus is that of ordinary hydrogen. The greater the nuclear mass, the less of the neutron's energy is transferred to the nucleus.

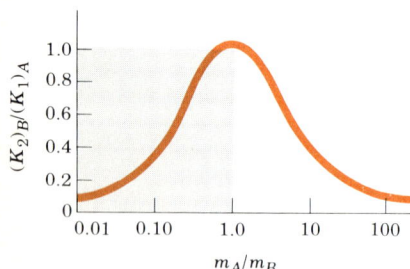

7-8 Fractional loss in energy in a head-on elastic collision, plotted as a function of the ratio of the masses of the colliding bodies.

Example 1 Suppose the collision illustrated in Fig. 7-4 is completely elastic. What are the velocities of A and B after the collision?

Solution From the principle of conservation of momentum,

$$(5 \text{ kg})(2 \text{ m·s}^{-1}) + (3 \text{ kg})(-2 \text{ m·s}^{-1}) = (5 \text{ kg}) v_{A2} + (3 \text{ kg}) v_{B2},$$

$$5 v_{A2} + 3 v_{B2} = 4 \text{ m·s}^{-1}.$$

Since the collision is completely elastic,

$$v_{B2} - v_{A2} = -(v_{B1} - v_{A1}) = -(-2 \text{ m·s}^{-1} - 2 \text{ m·s}^{-1}) = 4 \text{ m·s}^{-1}.$$

Solving these equations simultaneously, we obtain

$$v_{A2} = -1 \text{ m} \cdot \text{s}^{-1}, \qquad v_{B2} = 3 \text{ m} \cdot \text{s}^{-1}.$$

Both bodies therefore reverse their directions of motion, A traveling to the left at $1 \text{ m} \cdot \text{s}^{-1}$ and B to the right at $3 \text{ m} \cdot \text{s}^{-1}$.

The total kinetic energy after the collision is

$$\tfrac{1}{2}(5 \text{ kg})(-1 \text{ m} \cdot \text{s}^{-1})^2 + \tfrac{1}{2}(3 \text{ kg})(3 \text{ m} \cdot \text{s}^{-1})^2 = 16 \text{ J},$$

which equals the total kinetic energy before the collision.

If an elastic collision is not head-on, the velocities do not all lie on a single line. Each final velocity then has two unknown components and there are four unknowns in all. Conservation of energy and of the x- and y-components of momentum give only three equations, so additional information is needed. This may take the form of the direction of one of the final velocities, perhaps obtained from geometrical considerations, or of one of the final velocity magnitudes.

Example 2 In Fig. 7-5 suppose the 5-kg mass m_A has an initial velocity of $4 \text{ m} \cdot \text{s}^{-1}$ in the positive direction and a final velocity of $2 \text{ m} \cdot \text{s}^{-1}$ in an unknown direction. Find the final speed v_{B2} of the 3-kg mass and the angles θ and ϕ in the figure, assuming the collision is perfectly elastic.

Solution Because the collision is elastic, the total final kinetic energy is the same as the initial kinetic energy. Thus

$$\tfrac{1}{2}(5 \text{ kg})(4 \text{ m} \cdot \text{s}^{-1})^2 = \tfrac{1}{2}(5 \text{ kg})(2 \text{ m} \cdot \text{s}^{-1})^2 + \tfrac{1}{2}(3 \text{ kg})v_{B2}^2,$$

from which

$$v_{B2} = 4.47 \text{ m} \cdot \text{s}^{-1}.$$

Conservation of the x- and y-components of total momentum gives, respectively,

$$(5 \text{ kg})(4 \text{ m} \cdot \text{s}^{-1}) = (5 \text{ kg})(2 \text{ m} \cdot \text{s}^{-1}) \cos \theta + (3 \text{ kg})(4.47 \text{ m} \cdot \text{s}^{-1}) \cos \phi,$$

$$0 = (5 \text{ kg})(2 \text{ m} \cdot \text{s}^{-1}) \sin \theta - (3 \text{ kg})(4.47 \text{ m} \cdot \text{s}^{-1}) \sin \phi.$$

These are two simultaneous equations for θ and ϕ; the simplest solution is to eliminate ϕ as follows: We solve the first equation for $\cos \phi$ and the second for $\sin \phi$; we then square each equation and add; since $\sin^2 \phi + \cos^2 \phi = 1$, this eliminates ϕ and leaves an equation which may be solved for $\cos \theta$ and hence for θ. This value may then be substituted back into either of the two equations and the result solved for ϕ. The details are left as an exercise; the results are

$$\theta = 36.9°, \qquad \phi = 53.5°.$$

The examples in this section and the preceding one show that collisions can be classified according to energy considerations. A collision in which kinetic energy is conserved is called *perfectly elastic* or simply *elastic*. A collision in which the total kinetic energy decreases and the two bodies have a common final velocity is called *completely inelastic* or simply *inelastic*. There are intermediate cases in which some kinetic energy is lost but not enough for the two bodies to reach a common final velocity; such collisions are called *semielastic*. There are also cases in which the final kinetic energy is *greater* than the initial value. This occurs, for example, when an explosive between the two bodies is detonated during the collision, blowing them apart with great speed. Such a collision might be called *superelastic,* although this term is not in common use. A special case of a superelastic collision is *recoil,* discussed in the next section.

Finally, momentum conservation can be applied even to systems that are not really isolated. It may happen that there are some external forces acting on the colliding bodies, but that the internal forces during the collision are so much larger than the external that the latter may be neglected during the actual collision. This is certainly the case, for example, when two cars collide at an icy intersection.

7-6 Recoil

Figure 7-9 shows two blocks A and B, between which there is a compressed spring. When the system is released from rest, the spring exerts equal and opposite forces on the blocks until it has expanded to its natural unstressed length. It then drops to the surface, while the blocks continue to move. The original momentum of the system is zero, and if frictional forces can be neglected, the resultant external force on the system is zero. The momentum of the system therefore remains constant and equal to zero. Then if v_A and v_B are the velocities acquired by A and B, we have

$$m_A v_A + m_B v_B = 0, \qquad \frac{v_A}{v_B} = -\frac{m_B}{m_A}.$$

The velocities are of opposite sign and their magnitudes are inversely proportional to the corresponding masses.

7-9 Conservation of momentum in recoil.

The original kinetic energy of the system is also zero. The final kinetic energy is

$$K = \tfrac{1}{2} m_A v_A{}^2 + \tfrac{1}{2} m_B v_B{}^2.$$

The source of this energy is the original elastic potential energy of the system. The ratio of kinetic energies is

$$\frac{\tfrac{1}{2} m_A v_A{}^2}{\tfrac{1}{2} m_B v_B{}^2} = \frac{m_A}{m_B} \left(\frac{v_A}{v_B}\right)^2 = \frac{m_B}{m_A}.$$

Thus, although the final momenta are equal in magnitude, the final kinetic energies are inversely proportional to the corresponding masses, the body of smaller mass receiving the larger share of the original potential energy. The reason is that the change in *momentum* of a body equals the *impulse* of the force acting on it, while the change in *kinetic energy* equals the *work* of the force. The forces on the two bodies are equal in magnitude and act for equal *times,* so they produce equal and opposite changes in momentum. The points of application of the forces, however, *do not* move through equal *distances* (except when $m_A = m_B$), since the acceleration, velocity, and displacement of the smaller body are greater than those of the larger. Hence more *work* is done on the body of smaller mass.

These considerations may be applied to the firing of a rifle. The initial momentum of the system is zero. When the rifle is fired, the bullet and the powder gases acquire a forward momentum, and the rifle (together with any system to which it is attached) acquires a rearward momentum of the same magnitude. Because of the relatively large mass of the rifle compared with that of the bullet and powder charge, the velocity and kinetic energy of the rifle are much *smaller* than those of the bullet and powder gases.

7-7 Center of mass

The principle of conservation of momentum for an isolated system can be restated in a useful way using the concept of *center of mass,* which we now introduce. First, we define the *total momentum* P of a collection of any number of particles as the vector sum of the individual momenta:

$$P = p_1 + p_2 + p_3 + \cdots = m_1 v_1 + m_2 v_2 + m_3 v_3 + \cdots. \quad (7\text{-}18)$$

The total mass, which may be denoted by M, is just the sum of the individual masses:

$$M = m_1 + m_2 + m_3 + \cdots. \quad (7\text{-}19)$$

Now suppose there is a point called the center of mass, whose motion is representative of the motion of the whole system, in the sense that when its velocity, which we shall call V, is multiplied by M, the result is equal to the total momentum P. That is

$$MV = P = m_1 v_1 + m_2 v_2 + m_3 v_3 + \cdots$$

or

$$V = \frac{m_1 v_1 + m_2 v_2 + m_3 v_3 + \cdots}{m_1 + m_2 + m_3 + \cdots}. \quad (7\text{-}20)$$

This defines the velocity V of the center of mass, and it also suggests a natural way to define the *position* of the center of mass. We recall that the particle velocities $v_1, v_2, \ldots,$ are the time derivatives of the position vectors $r_1, r_2, r_3, \ldots$ of the particles, as defined in Sec. 5-2. Letting R be the position vector of the center of mass, we define R as

$$R = \frac{m_1 r_1 + m_2 r_2 + m_3 r_3 + \cdots}{m_1 + m_2 + m_3 + \cdots}. \quad (7\text{-}21)$$

By definition, $V = dR/dt$, and so the time derivative of Eq. (7-21) is Eq. (7-20), as it must be.

In component form, Eq. (7-21) becomes

$$X = \frac{m_1 x_1 + m_2 x_2 + m_3 x_3 + \cdots}{m_1 + m_2 + m_3 + \cdots},$$

$$Y = \frac{m_1 y_1 + m_2 y_2 + m_3 y_3 + \cdots}{m_1 + m_2 + m_3 + \cdots}, \quad (7\text{-}22)$$

where (x_1, y_1) are the coordinates of the first particle, and so on.

It follows directly from Eq. (7-20) that when the total momentum is constant, the velocity of the center of mass is also constant; this occurs whenever the vector sum of external forces is zero. When this is not the case, the center of mass has an acceleration, which we may call $A = dV/dt$. Taking the time derivative of Eq. (7-20) and multiplying by the total mass, we obtain

$$MA = m_1 a_1 + m_2 a_2 + m_3 a_3 + \cdots. \quad (7\text{-}23)$$

Now $m_1 a_1$ is equal to the vector sum of forces on the first particle, and so on, and the right side of Eq. (7-23) is equal to the vector sum of *all* the forces on *all* the particles. Just as in Sec. 7-2, we may classify each force as internal or external; because of Newton's third law, the internal forces all cancel in pairs, and what survives on the right side is the sum of the *external* forces only. Thus

$$\sum F_{\text{ext}} = MA. \quad (7\text{-}24)$$

That is, the center of mass moves just as though all the mass were concentrated at that point and it were acted on by a resultant force equal to the sum of the external forces on the system. This result is of central importance in the analysis of the motion of rigid bodies, where the motion is described as a combination of motion of the center of mass and rotational motion about an axis through the center of mass. We return to this topic in Chapter 9.

Example A 2-kg body and a 3-kg body are moving along the x-axis. At a particular instant the 2-kg body is 1 m from the origin and has a velocity of $3 \text{ m} \cdot \text{s}^{-1}$, and the 3-kg body is 2 m from the origin and has a velocity of $-1 \text{ m} \cdot \text{s}^{-1}$. Find the position and velocity of the center of mass, and also find the total momentum.

Solution From Eq. (7-22),

$$X = \frac{(2 \text{ kg})(1 \text{ m}) + (3 \text{ kg})(2 \text{ m})}{2 \text{ kg} + 3 \text{ kg}} = 1.6 \text{ m},$$

and from Eq. (7-20),

$$V = \frac{(2 \text{ kg})(3 \text{ m} \cdot \text{s}^{-1}) + (3 \text{ kg})(-1 \text{ m} \cdot \text{s}^{-1})}{2 \text{ kg} + 3 \text{ kg}} = 0.60 \text{ m} \cdot \text{s}^{-1}.$$

Strictly speaking, this is the x-component of velocity of the center of mass. The total momentum (strictly speaking, its x-component) is

$$P = (2 \text{ kg})(3 \text{ m} \cdot \text{s}^{-1}) + (3 \text{ kg})(-1 \text{ m} \cdot \text{s}^{-1}) = 3 \text{ kg} \cdot \text{m} \cdot \text{s}^{-1}).$$

Alternatively, from Eq. (7-20),

$$P = (5 \text{ kg})(0.6 \text{ m} \cdot \text{s}^{-1}) = 3 \text{ kg} \cdot \text{m} \cdot \text{s}^{-1}.$$

In principle the position of the center of mass of a body can always be calculated by regarding it as a collection of particles and using Eqs. (7-22). In practice these calculations can become quite complex. Usually we shall deal with symmetric bodies in which the center of mass lies at the geometrical center. Other calculational techniques will be discussed in the following chapter, in connection with the related concept of *center of gravity*.

*7-8 Rocket propulsion

A rocket is propelled by the ejection of a portion of its mass to the rear. The forward force on the rocket is the reaction to the backward force on the ejected material, and as more material is ejected, the mass of the rocket decreases. The analysis of this process is best handled by impulse–momentum considerations. In order not to bring in too many complicating forces, we shall consider a rocket fired vertically upward, and neglect air resistance and variations in g.

Figure 7–10a represents the rocket at a time t after take-off, when its mass is m and its upward velocity is v. The total momentum at this instant is thus mv. In a short time interval dt, a mass dm of gas is ejected from the rocket. Let v_r represent the downward speed of this gas *relative to the rocket*. The velocity v' of the gas relative to the earth is then

$$v' = v - v_r,$$

and its momentum is

$$dm\, v' = dm(v - v_r).$$

At the end of the time interval dt, the mass of rocket and unburned fuel has decreased to $m - dm$, and its velocity has increased to $v + dv$. Its momentum is therefore

$$(m - dm)(v + dv).$$

Thus the *total* momentum at time $t + dt$ is

$$(m - dm)(v + dv) + dm(v - v_r).$$

Figure 7–10b represents rocket and ejected gas at this time.

We now make use of the impulse–momentum relation; the product of the resultant external force F on a system, and the time interval dt during which it acts, is equal to the change in momentum of the system. If air resistance is neglected, the external force F on the rocket is its weight, $-mg$. (We take the upward direction as positive.) The change in momentum, in time dt, is the difference between the momentum of the system at the end and at the beginning of the time interval. Hence

$$-mg\, dt = (m - dm)(v + dv) + dm(v - v_r) - mv$$
$$= m\, dv - dm\, v_r - dm\, dv.$$

The term $dm\, dv$ may be dropped because it is a product of two small quantities and thus is much smaller than the other terms. Dropping this term, dividing by dt, and rearranging, we obtain

$$m\left(\frac{dv}{dt}\right) = v_r\left(\frac{dm}{dt}\right) - mg. \tag{7-25}$$

Now dv/dt is the acceleration of the rocket, so the left side of this equation (mass times acceleration) equals the resultant force on the rocket. The first term on the right equals the upward thrust on the rocket, and the resultant force equals the difference between this thrust

7-10 (a) Rocket at time t after take-off, with mass m and upward velocity v. Its momentum is mv. (b) At time $t + dt$, the mass of the rocket (and unburned fuel) is $m - dm$, its velocity is $v + dv$, and its momentum is $(m - dm)(v + dv)$. The ejected gas has momentum $dm(v - v_r)$.

and the weight of the rocket, *mg*. It will be seen that the upward thrust is proportional both to the relative velocity v_r of the ejected gas and to the mass of gas ejected per unit time, dm/dt.

The acceleration is

$$\frac{dv}{dt} = \frac{v_r}{m}\left(\frac{dm}{dt}\right) - g. \tag{7-26}$$

As the rocket rises, the value of g decreases, according to Newton's law of gravitation. (In "outer space," far from all other bodies, g becomes negligibly small.) If the values of v_r and dm/dt remain approximately constant while the fuel is being consumed and m continually decreases, the acceleration *increases* until all the fuel is burned.

Equation (7-26) can be integrated to find a relation between the velocity at any time and the remaining mass. From Eq. (7-26),

$$dv = v_r \frac{dm}{m} - g\,dt.$$

Now dm is a positive quantity, representing the mass ejected in time dt, so the change in mass of the rocket in that time is $-dm$. Thus, in computing the total mass change *in the rocket* we must change the sign of the term containing dm.

Let the mass and velocity at time $t = 0$ be m_0 and v_0, respectively; then

$$\int_{v_0}^{v} dv = -\int_{m_0}^{m} v_r \frac{dm}{m} - \int_{0}^{t} g\,dt,$$

and

$$v - v_0 = -v_r \ln \frac{m}{m_0} - gt,$$

$$v = v_0 + v_r \ln \frac{m_0}{m} - gt. \tag{7-27}$$

Example 1 In the first second of its flight, a rocket ejects 1/60 of its mass with a relative velocity of 2400 m·s^{-1}. What is the acceleration of the rocket?

We have: $dm = m/60$, $dt = 1$ s. From Eq. (7-26),

$$\frac{dv}{dt} = \frac{2400 \text{ m·s}^{-1}}{(60)(1 \text{ s})} - 9.8 \text{ m·s}^{-2}$$

$$= 31.2 \text{ m·s}^{-2}.$$

Example 2 Suppose the ratio of initial mass to final mass for the rocket above is 4 and that the fuel is consumed in a time $t = 60$ s. The velocity at the end of this time, from Eq. (7-27), is

$$v = (2400 \text{ m·s}^{-1})(\ln 4) - (9.8 \text{ m·s}^{-2})(60 \text{ s})$$

$$= 2740 \text{ m·s}^{-1}.$$

At the start of the flight, when the velocity of the rocket is zero, the ejected gases are moving downward, relative to the earth, with a velocity equal to the relative velocity v_r. When the velocity of the rocket has increased to v_r, the ejected gases have a velocity zero relative to the earth. When the rocket velocity becomes greater than v_r, the velocity of the ejected gases is in the same direction as that of the rocket. Thus the velocity acquired by the rocket can be greater (and is often much greater) than the relative velocity v_r. In the example above, where the final velocity of the rocket was 2740 m·s^{-1} and the relative velocity was 2400 m·s^{-1}, the last portion of the ejected fuel had an upward velocity of $(2740 - 2400)$ m·s^{-1} = 340 m·s^{-1}.

*7-9 Generalizations

The momentum–impulse relations and the principle of conservation of momentum have been developed from Newton's laws of motion, which we have observed are strictly valid only for bodies moving with speeds small compared to the speed of light. Thus it should not be surprising that when v is *not* small compared to c, modifications are needed. Relativistic momentum relations are discussed in detail in Chapter 43, but we quote a few important results here, without detailed derivation.

The correct relativistic generalization, as predicted theoretically by Lorentz and Einstein and also verified experimentally, is to define momentum not as mv, but as

$$\text{Momentum} = p = \frac{mv}{\sqrt{1 - v^2/c^2}}. \tag{7-28}$$

When v is much smaller than c, this reduces to mv, but in general the momentum of a particle is larger than mv. Equation (7-28) can be interpreted as saying that the mass of the particle increases by the factor $1/(1 - v^2/c^2)^{1/2}$ when it is moving, but an equally valid and, in some ways, less misleading view is simply that the definition of momentum is no longer mv, but instead is given by Eq. (7-28). In either case, it is found that in collisions where there is no resultant external force, *the principle of conservation of momentum is still obeyed,* provided the generalized definition of momentum is used.

A further generalization of the concept of momentum is needed when we deal with particles that have no mass in the usual sense. One such particle is the *photon,* or *quantum* of electromagnetic wave energy, which will be discussed more fully in later chapters. An electromagnetic wave, such as light or a radio wave, transfers energy in packages of definite size, with the amount in each package proportional to the frequency of the radiation. For a wave of frequency f, the energy E in each package or *photon* is given by

$$E = hf, \tag{7-29}$$

where h is a fundamental physical constant called *Planck's constant.* In SI units, $h = 6.62 \times 10^{-34}$ J·s.

The theory of relativity predicts that any particle with energy must also have momentum, whether it has mass or not. For a massless particle the momentum turns out to be the energy divided by the speed of light. Thus, from Eq. (7-29) we obtain, for the momentum of a massless particle,

$$p = \frac{hf}{c} = \frac{h}{\lambda}, \tag{7-30}$$

where $\lambda = c/f$ is the wavelength of the radiation.

Example A sodium atom initially at rest emits a photon of yellow light, recoiling in the opposite direction to the direction of emission. What is the atom's recoil velocity?

Solution The wavelength of the light is about $\lambda = 589 \times 10^{-9}$ m, and the momentum of a photon is

$$p = \frac{h}{\lambda} = \frac{6.62 \times 10^{-34} \text{ J} \cdot \text{s}}{589 \times 10^{-9} \text{ m}} = 1.12 \times 10^{-27} \text{ kg} \cdot \text{m} \cdot \text{s}^{-1}.$$

For conservation of momentum, the atom must recoil with a momentum of this magnitude in the opposite direction. The mass of a sodium atom is about 3.8×10^{-26} kg, so its speed after emission is

$$v = \frac{1.12 \times 10^{-27} \text{ kg} \cdot \text{m} \cdot \text{s}^{-1}}{3.8 \times 10^{-26} \text{ kg}} = 0.029 \text{ m} \cdot \text{s}^{-1}.$$

In conclusion, we emphasize again that the principle of conservation of momentum, like that of conservation of energy, has much greater generality than the newtonian context in which it was originally developed. As indicated briefly above, the definition of momentum can be generalized to relativistic mechanics and to quantum mechanics. With these generalizations, conservation of momentum is believed to be a *universal* conservation law; *no exception has ever been found.*

Questions

7-1 Suppose you catch a baseball, and then someone invites you to catch a bullet with the same momentum or with the same kinetic energy. Which would you choose?

7-2 In splitting logs with a hammer and wedge, is a heavy hammer more effective than a lighter hammer? Why?

7-3 When a large, heavy truck collides with a passenger car, the occupants of the car are much more likely to be hurt than the truck driver. Why?

7-4 A glass dropped on the floor is more likely to break if the floor is concrete than if it is wood. Why?

7-5 "It ain't the fall that hurts you; it's the sudden stop at the bottom." Discuss.

7-6 When a person fires a rifle or shotgun, it is advisable to hold the butt firmly against the shoulder rather than a bit away from it, to minimize the impact on the shoulder. Why?

7-7 When a catcher in a baseball game catches a fast ball, he does not hold his arms rigid, but relaxes them so the mitt moves several inches while the ball is being caught. Why is this important?

7-8 When rain falls from the sky, what becomes of its momentum as it hits the ground? Is your answer also valid for Newton's famous apple?

7-9 A man stands in the middle of a perfectly smooth, frictionless frozen lake. He can set himelf in motion by throwing things, but suppose he has nothing to throw. Can he propel himself to shore *without* throwing anything?

7-10 A machine gun is fired at a steel plate. Is the average force on the plate from the bullet impact greater if the bullets bounce off or if they are squashed and stick to the plate?

7-11 How do Mexican jumping beans work? Do they violate conservation of momentum? Of energy?

7-12 Early critics of Robert Goddard, a pioneer in the use of rocket propulsion, claimed that rocket engines could not be used in outer space where there is no air for the rocket to push against. How would you answer such criticism? Would the criticism be valid for a jet engine in an ordinary airplane?

7-13 In a zero-gravity environment, can a rocket-propelled spaceship ever attain a speed greater than the relative speed with which the burnt fuel is exhausted?

7-14 Theoretically, a flashlight emitting a beam of light should tend to recoil in the opposite direction. Why is this recoil not actually observed?

7-15 Because light carries momentum, a beam of light striking a surface exerts a force on the surface. Is the force greater when the surface absorbs the light or when it reflects it back toward its source?

7-16 It has been proposed to propel spaceships using "ion engines," whose operation is similar to rockets but which eject mass in the form of subatomic particles at very great speeds. What advantages would such an engine have compared to a more conventional rocket?

Problems

7-1
a) What is the momentum of a 10,000-kg truck whose velocity is 20 m·s^{-1}? What velocity must a 5,000-kg truck attain in order to have (b) the same momentum, (c) the same kinetic energy?

7-2 A baseball has a mass of about 0.2 kg.
a) If the velocity of a pitched ball is 30 m·s^{-1}, and after being batted it is 50 m·s^{-1} in the opposite direction, find the change in momentum of the ball and the impulse of the blow.
b) If the ball remains in contact with the bat for 0.002 s, find the average force of the blow.

7-3 A bullet having a mass of 0.05 kg, moving with a velocity of 400 m·s^{-1}, penetrates a distance of 0.1 m into a wooden block firmly attached to the earth. Assume the accelerating force constant. Compute
a) the acceleration of the bullet,
b) the accelerating force,
c) the time of acceleration,
d) the impulse of the collision.

Compare the answer to part (d) with the initial momentum of the bullet.

7-4 A bullet emerges from the muzzle of a gun with a velocity of 300 m·s^{-1}. The resultant force on the bullet, while it is in the gun barrel, is given by

$$F = 400 \text{ N} - \frac{(4 \times 10^5 \text{ N·s}^{-1})t}{3}.$$

a) Construct a graph of F versus t.
b) Compute the time required for the bullet to travel the length of the barrel, assuming the force becomes zero just at the end of the barrel.
c) Find the impulse of the force.
d) Find the mass of the bullet.

7-5 A box, initially sliding on the floor of a room, is eventually brought to rest by friction. Is the momentum of the box conserved? If not, does this process contradict the principle of conservation of momentum? What becomes of the original momentum of the box?

7-6 A golf ball of mass 0.10 kg is initially at rest when it is struck by a club and given a speed of 50 m·s^{-1}. If the club and ball are in contact for 0.002 s, what average force acted on the ball? Is the effect of the ball's weight during the time of contact significant?

7-7 A tennis ball approaches a player's racket horizontally at 10 m·s^{-1}. After it is struck its velocity is horizontal, in the opposite direction, with magnitude 20 m·s^{-1}. The ball has mass 0.060 kg, and it is in contact with the racket for 0.01 s. What average force acted on the ball?

7-8 A steel ball of mass 0.5 kg is dropped from a height of 4 m onto a horizontal steel slab. The collision is elastic, and the ball rebounds to its original height.
a) Calculate the impulse delivered to the ball during impact.
b) If the ball is in contact with the slab for 0.002 s, find the average force on the ball during impact.

7-9 A baseball of mass 0.25 kg is struck by a bat. Just before impact, the ball is traveling horizontally at 40 m·s^{-1}, and it leaves the bat at an angle of 30° above horizontal with a speed of 60 m·s^{-1}. If the ball and bat were in contact for 0.005 s, find the horizontal and vertical components of the average force on the ball.

7-10 Compare the damage to an automobile (and its occupants) in the following circumstances:
a) The automobile makes a completely inelastic head-on collision with an identical automobile traveling with the same speed in the opposite direction, and
b) it makes a completely inelastic head-on collision with a vertical rock cliff.
c) Which would be worse (for the occupants of a light car), to collide head-on with a truck traveling in the opposite direction with a momentum of equal magnitude, or to collide head-on with a truck having the same kinetic energy? (A reasonable criterion to use is the change in velocity of the car in each case.)

7-11
a) An empty freight car of mass 10,000 kg rolls at 2 m·s^{-1} along a level track and collides with a loaded car of mass 20,000 kg, standing at rest with brakes released. If the cars couple together, find their speed after the collision.
b) Find the decrease in kinetic energy as a result of the collision.
c) With what speed should the loaded car be rolling toward the empty car, in order that both shall be brought to rest by the collision?

7-12 When a bullet of mass 20 g strikes a ballistic pendulum of mass 10 kg, the center of gravity of the pendulum is observed to rise a vertical distance of 7 cm. The bullet remains embedded in the pendulum.
a) Calculate the original velocity of the bullet.
b) What fraction of the original kinetic energy of the bullet remains as kinetic energy of the system immediately after the collision?
c) What fraction of the original momentum remains as momentum of the system?

7-13 A bullet of mass 5 g is shot *through* a 1-kg wood block suspended on a string 2 m long. The center of gravity of the block is observed to rise a distance of 0.50 cm. Find the speed of the bullet as it emerges from the block if the initial speed is 300 m·s⁻¹.

7–14 A frame of mass 200 g, when suspended from a certain coil spring, is found to stretch the spring 10 cm. A lump of putty of mass 200 g is dropped from rest onto the frame from a height of 30 cm (Fig. 7–11). Find the maximum distance the frame moves downward.

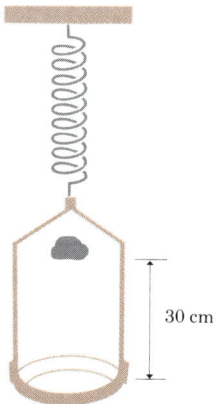

Figure 7–11

7–15 A bullet of mass 2 g, traveling in a horizontal direction with a velocity of 500 m·s⁻¹, is fired into a wooden block of mass 1 kg, initially at rest on a level surface. The bullet passes through the block and emerges with its velocity reduced to 100 m·s⁻¹. The block slides a distance of 20 cm along the surface from its initial position.

a) What was the coefficient of sliding friction between block and surface?
b) What was the decrease in kinetic energy of the bullet?
c) What was the kinetic energy of the block at the instant after the bullet passed through it?

7–16 A rifle bullet of mass 10 g is fired with a velocity of 800 m·s⁻¹ into a ballistic pendulum of mass 5 kg, suspended from a cord 1 m long. Compute (a) the vertical height through which the pendulum rises, (b) the initial kinetic energy of the bullet, (c) the kinetic energy of bullet and pendulum after the bullet is embedded in the pendulum.

7–17 A 5-g bullet is fired horizontally into a 3-kg wooden block resting on a horizontal surface. The coefficient of sliding friction between block and surface is 0.20. The bullet remains embedded in the block, which is observed to slide 25 cm along the surface before stopping. What was the velocity of the bullet?

7–18 A rifle bullet of mass 0.01 kg strikes and embeds itself in a block of mass 0.99 kg which rests on a horizontal frictionless surface and is attached to a coil spring, as shown in Fig. 7–12. The impact compresses the spring 10 cm. Calibration of the spring shows that a force of 1.0 N is required to compress the spring 1 cm.

a) Find the maximum potential energy of the spring.
b) Find the velocity of the block just after the impact.
c) What was the initial velocity of the bullet?

Figure 7–12

7–19 A 2000-kg automobile going eastward on Chestnut Street at 60 km·hr⁻¹ collides with a 4000-kg truck which is going southward *across* Chestnut Street at 20 km·hr⁻¹. If they become coupled on collision, what is the magnitude and direction of their velocity immediately after colliding?

7–20 On a frictionless table, a 3-kg block moving 4 m·s⁻¹ to the right collides with an 8-kg block moving 1.5 m·s⁻¹ to the left.

a) If the two blocks stick together, what is the final velocity?
b) If the two blocks make a completely elastic head-on collision, what are their final velocities?
c) How much mechanical energy is dissipated in the collision of part (a)?

7–21 Two blocks of mass 300 g and 200 g are moving toward each other along a horizontal frictionless surface with velocities of 50 cm·s⁻¹ and 100 cm·s⁻¹, respectively.

a) If the blocks collide and stick together, find their final velocity.
b) Find the loss of kinetic energy during the collision.
c) Find the final velocity of each block if the collision is completely elastic.

7–22
a) Prove that when a moving body makes a perfectly inelastic collision with a second body of equal mass, initially at rest, one-half of the original kinetic energy is "lost."
b) Prove that when a very heavy particle makes a perfectly elastic head-on collision with a very light particle that is at rest, the light one goes off with twice the velocity of the heavy one.

7–23 A 10-g block slides with a velocity of 20 cm·s⁻¹ on a smooth level surface and makes a head-on collision with a 30-g block moving in the opposite direction with a velocity of 10 cm·s⁻¹. If the collision is perfectly elastic, find the velocity of each block after the collision.

7–24 A block of mass 200 g, sliding with a velocity of 12 cm·s⁻¹ on a smooth, level surface, makes a perfectly elastic head-on collision with a block of mass m, initially at rest. After the collision, the velocity of the 200-g block is 4 cm·s⁻¹ in the same direction as its initial velocity. Find (a) the mass m, and (b) its velocity after the collision.

7-25 A body of mass 600 g is initially at rest. It is struck by a second body of mass 400 g initially moving with a velocity of 125 cm·s⁻¹ toward the right along the x-axis. After the collision, the 400-g body has a velocity of 100 cm·s⁻¹ at an angle of 37° above the x-axis in the first quadrant. Both bodies move on a horizontal frictionless plane.

a) What is the magnitude and direction of the velocity of the 600-g body after the collision?
b) What is the loss of kinetic energy during the collision?

7-26 A small steel ball moving with speed v_0 in the positive x-direction makes a perfectly elastic, noncentral collision with an identical ball originally at rest. After impact, the first ball moves with speed v_1 in the first quadrant at an angle θ_1 with the x-axis and the second with speed v_2 in the fourth quadrant at an angle θ_2 with the x-axis.

a) Write the equations expressing conservation of linear momentum in the x-direction, and in the y-direction.
b) Square these equations and add them.
c) At this point, introduce the fact that the collision is perfectly elastic.
d) Prove that $\theta_1 + \theta_2 = \pi/2$.

7-27 A jet of liquid of cross-sectional area A and density ρ moves with speed v_J in the positive x-direction and impinges against a perfectly smooth blade B, which deflects the stream at right angles but does not slow it down, as shown in Fig. 7-13.

Figure 7-13

a) If the blade is *stationary*, prove that the rate of arrival of mass at the blade is $dm/dt = \rho A v_J$.
b) If the impulse–momentum theorem is applied to a small mass dm, prove that the x-component of the force acting on this mass for the time interval dt is given by
$$F_x = -\frac{dm}{dt} v_J.$$
c) Prove that the *steady* force exerted *on* the blade in the x-direction is
$$F_x = \rho A v_J^2.$$
If the blade moves to the right with a speed v_B ($v_B < v_J$), derive the equations for:
d) the rate of arrival of mass at the moving blade,
e) the force F_x on the blade, and
f) the power delivered to the blade.

7-28 A stone whose mass is 100 g rests on a horizontal frictionless surface. A bullet of mass 2.5 g, traveling horizontally at 400 m·s⁻¹, strikes the stone and rebounds horizontally at right angles to its original direction with a speed of 300 m·s⁻¹.

a) Compute the magnitude and direction of the velocity of the stone after it is struck.
b) Is the collision perfectly elastic?

7-29 A hockey puck B rests on a smooth ice surface and is struck by a second puck, A, which was originally traveling at 30 m·s⁻¹ and which is deflected 30° from its original direction (Fig. 7-14). Puck B acquires a velocity at 45° with the original velocity of A.

a) Compute the speed of each puck after the collision.
b) Is the collision perfectly elastic? If not, what fraction of the original kinetic energy of puck A is "lost"?

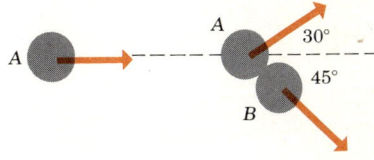

Figure 7-14

7-30 In Problem 7-29, suppose the collision is perfectly elastic, and A is deflected 30° from its initial direction. Find the final speed of each puck and the direction of B's velocity.

7-31 A neutron of mass m collides elastically with a nucleus of mass M, which is initially at rest. Show that if the neutron's initial kinetic energy is K_0, the maximum kinetic energy that it can *lose* during the collision is $4mMK_0/(M+m)^2$. [*Hint.* The maximum energy loss occurs in a head-on collision.]

7-32 Imagine that a ball of mass 200 g rolls back and forth between opposite sides of a billiard table 1 m wide, with a velocity that remains constant in magnitude but reverses in direction at each collision with the cushions. The magnitude of the velocity is 4 m·s⁻¹.

a) What is the change in momentum of the ball at each collision?
b) How many collisions per unit time are made by the ball with one of the cushions?

c) What is the average time rate of change of momentum of the ball as a result of these collisions?

d) What is the average force exerted by the ball on a cushion?

e) Sketch a graph of the force exerted on the ball as a function of time, for a time interval of 5 s. [*Note.* This problem illustrates how one computes the average force exerted by the molecules of a gas on the walls of the containing vessel.]

7-33 A 20-kg projectile is fired at an angle of 60° above the horizontal and with a muzzle velocity of 400 m·s^{-1}. At the highest point of its trajectory the projectile explodes into two fragments of equal mass, one of which falls vertically with zero initial speed.

a) How far from the point of firing does the other fragment strike if the terrain is level?

b) How much energy was released during the explosion?

7-34 A railroad handcar is moving along straight frictionless tracks. In each of the following cases, the car initially has a total mass (car and contents) of 200 kg and is traveling with a velocity of 4 m·s^{-1}. Find the *final velocity* of the car in each of the three cases.

a) A 20-kg mass is thrown sideways out of the car with a velocity of 2 m·s^{-1} relative to the car.

b) A 20-kg mass is thrown backwards out of the car with a velocity of 4 m·s^{-1} relative to the car.

c) A 20-kg mass is thrown into the car with a velocity of 6 m·s^{-1} relative to the ground and opposite in direction to the velocity of the car.

7-35 A bullet weighing 0.02 lb is fired with a muzzle velocity of 2700 ft·s^{-1} from a rifle weighing 7.5 lb.

a) Compute the recoil velocity of the rifle, assuming it free to recoil.

b) Find the ratio of the kinetic energy of the bullet to that of the rifle.

7-36 Block A in Fig. 7-15 has a mass of 1 kg, and block B has a mass of 2 kg. The blocks are forced together, compressing a spring S between them, and the system is released from rest on a level frictionless surface. The spring is not fastened to either block and drops to the surface after it has expanded. Block B acquires a speed of 0.5 m·s^{-1}. How much potential energy was stored in the compressed spring?

Figure 7-15

7-37 An open-topped freight car of mass 10,000 kg is coasting without friction along a level track. It is raining very hard, with the rain falling vertically downward. The car is originally empty and moving with a velocity of 1 m·s^{-1}. What is the velocity of the car after it has traveled long enough to collect 1000 kg of rain water?

7-38 A neutron of mass 1.67×10^{-27} kg, moving with a velocity of 2×10^4 m·s^{-1}, makes a head-on collision with a boron nucleus of mass 17.0×10^{-27} kg, originally at rest.

a) If the collision is completely inelastic, what is the final kinetic energy of the system, expressed as a fraction of the original kinetic energy?

b) If the collision is perfectly elastic, what fraction of its original kinetic energy does the neutron transfer to the boron nucleus?

7-39 A nucleus, originally at rest, decays radioactively by emitting an electron of momentum 9.22×10^{-21} kg·m·s^{-1}, and at right angles to the direction of the electron a neutrino with momentum 5.33×10^{-21} kg·m·s^{-1}.

a) In what direction does the residual nucleus recoil?

b) What is its momentum?

c) If the mass of the residual nucleus is 3.90×10^{-25} kg, what is its kinetic energy?

7-40 An 80-kg man standing on ice throws a 0.2-kg ball horizontally with a speed of 30 m·s^{-1}.

a) With what speed and in what direction will the man begin to move?

b) If the man throws 4 such balls every 3 s, what is the average force acting on him? [*Hint.* Average force equals average rate of change of momentum.]

7-41 Find the average recoil force on a machine gun firing 120 shots per minute. The mass of each bullet is 10 g, and the muzzle velocity is 800 m·s^{-1}.

7-42 A rifleman, who together with his rifle has a mass of 100 kg, stands on roller skates and fires 10 shots horizontally from an automatic rifle. Each bullet has a mass of 10 g and a muzzle velocity of 800 m·s^{-1}.

a) If the rifleman moves back without friction, what is his velocity at the end of the 10 shots?

b) If the shots were fired in 10 s, what was the average force exerted on him?

c) Compare his kinetic energy with that of the 10 bullets.

7-43 A 1000-kg automobile is moving along a straight highway at 10 m·s^{-1}. Another car, with mass 2000 kg and speed 20 m·s^{-1} is 30 m ahead of the first.

a) Find the position of the center of mass of the two automobiles.

b) Find the total momentum from the above data.

c) Find the velocity of the center of mass.

d) Find the total momentum, using the velocity of the center of mass. Compare your result with that of part (b).

7-44 Find the position of the center of mass of the earth-moon system. Use the data in Appendix G.

7-45 A uniform steel rod 1 m in length is bent in a 90° angle at its midpoint. Determine the position of its center of mass. [*Hint.* The mass of each side of the angle may be assumed to be concentrated at its center.]

7-46 Three particles have the following masses and coordinates: (1) 2 kg, (3 m, 2 m); (2) 3 kg, (1 m, −4 m); (3) 4 kg, (−3 m, 5 m). Find the coordinates of the center of mass of the system.

7-47 A canoeist of mass 80 kg stands up in a 30-kg canoe of length 5 m. He walks from a point 1 m from one end to a point 1 m from the other end. If resistance to motion of the canoe in the water can be neglected, how far does the canoe move during this process?

7-48 A rocket burns 0.05 kg of fuel per second, ejecting it as a gas with a velocity relative to the rocket of 5000 m·s^{-1}.

a) What force does this gas exert on the rocket?
b) Would the rocket operate in free space?
c) If it would operate in free space, how would you steer it? Could you brake it?

7-49 If a single-stage rocket, fired vertically from rest at the earth's surface, burns its fuel in a time of 30 s, and the relative velocity $v_r = 3000$ m·s^{-1}, what must be the mass ratio m_0/m for a final velocity v of 8 km·s^{-1} (about equal to the orbital velocity of an earth satellite)?

7-50

a) Show that the acceleration of a rocket fired vertically upward is given by

$$a = -\frac{v_r}{m}\frac{dm}{dt} - g.$$

b) Suppose that the rate of ejection of mass by the rocket is constant, that is, the rate of decrease of mass is $dm/dt = -km_0$, where k is a positive constant and m_0 is the initial mass. What is the numerical value of k, and in what units is it expressed, for the rocket in Problem 7–49?

c) Show that the mass at any time t is given by the equation

$$m = m_0(1 - kt).$$

d) Show that the acceleration at any time is equal to

$$\frac{v_r k}{1 - kt} - g.$$

e) Find the initial acceleration of the rocket in Problem 7–49, in terms of the acceleration of gravity, g.
f) Find the acceleration 15 s after the motion starts.
g) Sketch the acceleration–time graph.

7-51 This problem illustrates the advantage of using a multi-stage rather than a single-stage rocket. Suppose that the first stage of a two-stage rocket has a total mass of 12,000 kg, of which 9000 kg is fuel. The total mass of the second stage is 1000 kg, of which 750 kg is fuel. Assume that the relative velocity v_r of ejected material is constant, and neglect any effect of gravity. (The latter effect is small during the firing period if the rate of fuel consumption is large.)

a) Suppose that the entire fuel supply carried by the two-stage rocket were utilized in a single-stage rocket of the same total mass of 13,000 kg. What would be the velocity of the rocket, starting from rest, when its fuel was exhausted?
b) What is the velocity when the fuel of the first stage is exhausted, if the first stage carries the second stage with it to this point? This velocity then becomes the initial velocity of the second stage.
c) What is the final velocity of the second stage?
d) What value of v_r would be required to give the second stage of the above rocket a velocity of 8 km·s^{-1}?

7-52 At what speed does the momentum of a particle differ by 1 percent from the value obtained using the nonrelativistic expression mv? Is the correct relativistic value greater or less than that obtained from the nonrelativistic expression?

7-53 By combining the relativistic energy and momentum expressions, show that these two quantities are related by

$$E^2 = (mc^2)^2 + (pc)^2,$$

where p is the relativistic momentum and E the *total* energy, including rest energy. Hence, show that when the total energy of a particle is much greater than its rest energy, the energy–momentum relation is approximately the same as for a massless particle. Note that for this to occur the speed of the particle must be very close to c.

7-54 Construct a right triangle in which one of the angles is α, where $\sin \alpha = v/c$ (v is the speed of a particle, c the speed of light). If the base of the triangle (the side adjacent to α) is the rest energy mc^2, show that

a) the hypotenuse is the *total energy* and
b) the side opposite α is c times the relativistic momentum.
c) Describe a simple graphical procedure for finding the kinetic energy K.

7-55 The mass of an electron is 9.11×10^{-31} kg. Comparing the classical definition of momentum with its relativistic generalization, by how much is the classical expression in error if (a) $v = 0.01c$; (b) $v = 0.5c$; (c) $v = 0.9c$?

7-56 A radioactive isotope of cobalt, Co^{60}, emits an electromagnetic photon (γ ray) of wavelength 0.932×10^{-12} m. The cobalt nucleus contains 27 protons and 33 neutrons, each with a mass of about 1.66×10^{-27} kg. If the nucleus is at rest before emission, what is the speed afterward? Is it necessary to use the relativistic generalization of momentum?

7-57 In Problem 7–56, suppose the cobalt atom is in a metallic crystal containing 0.01 mol of cobalt (about 6.02×10^{21} atoms) and that the entire crystal recoils as a unit, rather than just the single nucleus. Find the recoil velocity. (This recoil of the entire crystal rather than a single nucleus is called the *Mössbauer effect,* in honor of its discoverer, who first observed it in 1958.)

EQUILIBRIUM. MOMENT OF A FORCE

8

8-1

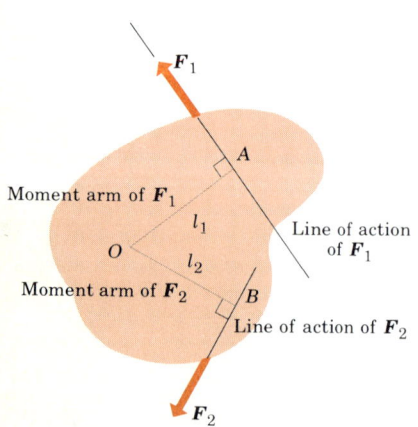

8-2 The moment of a force about an axis is the product of the force and its moment arm.

8-1 Moment or torque of a force

The discussion of equilibrium problems in Chapter 2 was concerned mostly with bodies that could be represented as points. It was mentioned, however, that, in order for an extended body to be in equilibrium, the body must also have no tendency to *rotate*. We now return to this more general problem of equilibrium of a rigid body. In this context we introduce the concept of *moment* or *torque* of a force, which also plays an essential role in motion of a rigid body, to be discussed in Chapter 9.

The tendency of a force to cause rotation depends on the line along which the force acts as well as the magnitude of the force. For example, when one pushes a door open, a force applied near the knob is more effective than the same force applied near the hinge side. Similarly, in Fig. 8–1, the force F_1, if applied by itself to the body shown, would cause a *counterclockwise* rotation and a translation to the right, while force F_2 applied by itself would cause a *clockwise* rotation (and translation to the right), even though it has the same magnitude and direction as F_1.

It is useful to think of a body that is pivoted or hinged so it can rotate about some stationary axis. Suppose such a body is acted on by several forces, which act along lines lying in a plane perpendicular to the axis. In Fig. 8–2 this plane is the plane of the figure, and the axis of rotation is perpendicular to this plane at point O.

The tendency of force F_1 in Fig. 8–2 to cause a rotation about the axis through O is measured jointly by the magnitude of the force and the perpendicular distance l_1 between the line of action of the force and the axis. Clearly if $l_1 = 0$ there is *no* tendency to cause rotation. The role of distance l_1 is analogous to that of a wrench handle; a tight bolt is turned more easily by using a long-handled wrench than by a short-handled one. Distance l_1 is called the *moment arm* of force F_1 about the axis O, and the product of F_1 and l_1 is called the *moment* or *torque* of the force about point O. These two terms are used interchangeably. We shall use the Greek letter Γ (capital "gamma") for moment or torque:

$$\Gamma = Fl. \tag{8-1}$$

In Fig. 8-2 the body is acted on by the forces F_1 and F_2, lying in the plane of the diagram. The moment arm of F_1 is the perpendicular distance OA of length l_1, and the moment arm of F_2 is the perpendicular distance OB of length l_2.

The effect of the force F_1 is to produce *counterclockwise* rotation about the axis, while that of F_2 is to produce *clockwise* rotation. To distinguish between these directions of rotation, we shall adopt the convention that *counterclockwise moments are positive, and clockwise moments are negative*. Hence the moment Γ_1 of the force F_1 about the axis through O is

$$\Gamma_1 = +F_1 l_1,$$

and the moment Γ_2 of F_2 is

$$\Gamma_2 = -F_2 l_2.$$

In SI (mks) units, where the unit of force is the newton and the unit of length the meter, the unit of torque is the newton-meter. If forces are expressed in pounds and lengths in feet, torques are expressed in pound-feet.

8-2 The second condition for equilibrium

We saw in Sec. 2-3 that, when a body is in equilibrium under the action of several coplanar forces, the vector sum of the forces must be zero. This is the *first condition for equilibrium*,

$$\sum F_x = 0, \quad \sum F_y = 0. \tag{8-2}$$

The second condition for equilibrium of a rigid body corresponds to the absence of any tendency of the body to rotate: *The sum of the moments of all forces acting on the body, calculated with respect to some specified axis, must be zero.* Furthermore, the body need not actually be pivoted about the chosen axis. In order to be in equilibrium the body must have no tendency to rotate about *any* axis, so the sum of moments must be zero, no matter what axis is chosen.

$$\sum \Gamma = 0 \quad \text{(about any arbitrary axis).} \tag{8-3}$$

Although the choice of axis is arbitrary, one must, of course, use the *same* axis for *all* the moments.

For the present we regard the second condition for equilibrium of a rigid body as an *empirical* relation; we use it because it has been found experimentally to be correct. In Chapter 9 this condition will emerge as a special case of the general dynamical principle governing rotational motion of a rigid body, a principle which is derived from Newton's laws.

An additional observation that is often useful in solving problems is the following. When a force is represented in terms of its components, the moment of that force with respect to a specified axis can be obtained by computing separately the moments of the components, each with its appropriate moment arm, and adding the results. This is often easier than determining the moment arm of the original force. This technique is illustrated in some of the following examples.

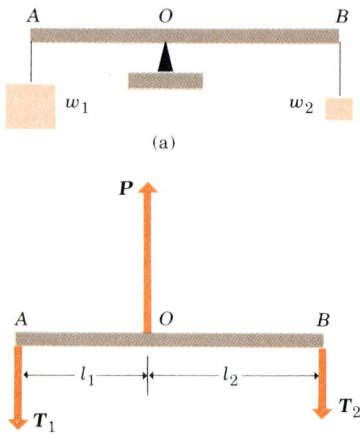

8-3 A rod in equilibrium under three parallel forces.

Example 1 A rigid rod whose own weight is negligible (Fig. 8-3) is pivoted at point O and carries a body of weight w_1 at end A. Find the weight w_2 of a second body that must be attached at end B if the rod is to be in equilibrium, and find the force P exerted on the rod by the pivot at Q.

Solution Figure 8-3b is a free-body diagram for the rod. The forces T_1 and T_2 are equal respectively to w_1 and w_2. The conditions of equilibrium, taking moments about an axis through O, perpendicular to the diagram, give

$$\sum F_y = P - T_1 - T_2 = 0, \quad \text{(First condition)}$$

$$\sum \Gamma_O = T_1 l_1 - T_2 l_2 = 0. \quad \text{(Second condition)}$$

Let $l_1 = 1.2$ m, $l_2 = 1.6$ m, $w_1 = 20$ N. Then, from the equations above,

$$P = 35 \text{ N}, \quad T_2 = w_2 = 15 \text{ N}.$$

To illustrate that the resultant moment about *any* axis is zero, let us compute moments about an axis through point A:

$$\sum \Gamma_A = P l_1 - T_2 (l_1 + l_2)$$
$$= (35 \text{ N})(1.2 \text{ m}) - (15 \text{ N})(2.8 \text{ m}) = 0.$$

The point about which moments are computed need not lie on the rod. To verify this, the reader may calculate the resultant moment about a point 0.4 m to the left of A and 0.4 m above it.

Example 2 In Fig. 8-4 a ladder 10 m long, of weight 400 N considered as concentrated at its center, leans in equilibrium against a vertical frictionless wall and makes an angle of 53.1° with the horizontal (forming a 3-4-5 right triangle). We wish to find the magnitudes and directions of the forces $\mathbf{F}_1$ and $\mathbf{F}_2$.

Solution If the wall is frictionless, $\mathbf{F}_1$ is horizontal. The direction of $\mathbf{F}_2$ is unknown; except in special cases, its direction does *not* lie along the ladder. Instead of considering its magnitude and direction as unknowns, it is simpler to resolve the force $\mathbf{F}_2$ into (unknown) x- and y-components and solve for these. The magnitude and direction of $\mathbf{F}_2$ may then be computed. The first condition of equilibrium therefore provides the equations

$$\left. \begin{array}{l} \sum F_x = F_{2x} - F_1 = 0, \\ \sum F_y = F_{2y} - 400 \text{ N} = 0. \end{array} \right\} \quad \text{(First condition)}$$

In writing the second condition, moments may be computed about an axis through any point. The resulting equation is simplest if one selects a point through which the lines of action of two or more forces pass, since these forces then do not appear in the equation. Let us therefore take moments about an axis through point A.

$$\sum \Gamma_A = F_1(8 \text{ m}) - (400 \text{ N})(3 \text{ m}) = 0. \quad \text{(Second condition)}$$

8-4 Forces on a ladder in equilibrium, leaning against a frictionless wall.

From the second equation, $F_{2y} = 400$ N, and from the third,

$$F_1 = \frac{1200 \text{ N} \cdot \text{m}}{8 \text{ m}} = 150 \text{ N}.$$

Then from the first equation,

$$F_{2x} = 150 \text{ N}.$$

Hence

$$F_2 = \sqrt{(400 \text{ N})^2 + (150 \text{ N})^2} = 427.2 \text{ N},$$

$$\theta = \tan^{-1} \frac{400 \text{ N}}{150 \text{ N}} = 69.4°.$$

Example 3 Figure 8-5a shows a human arm lifting a dumbbell, and Fig. 8-5b shows a free-body diagram for the forearm, showing the forces involved. The forearm is in equilibrium under the action of the weight w, the tension T in the tendon connected to the biceps muscle, and the forces exerted on it by the upper arm, through the elbow joint. We want to find the tendon tension and the components of force at the elbow, three unknown scalar quantities in all.

Solution First we represent the tendon force in terms of its components T_x and T_y, using the given angle θ and the unknown magnitude T:

$$T_x = T \cos \theta, \qquad T_y = T \sin \theta.$$

Next we note that if moments about the elbow joint are taken, the resulting moment equation does not contain the unknowns E_x and E_y, since their lines of action pass through this point, as also does T_x. The moment equation is then simply

$$lw - dT_y = 0.$$

From this we immediately find

$$T_y = \frac{lw}{d} \quad \text{and} \quad T = \frac{lw}{d \sin \theta}.$$

To find E_x and E_y we could now use the first conditions for equilibrium, $\Sigma F_x = 0$ and $\Sigma F_y = 0$. Instead, for added practice in using moments, we take moments about the point A where the tendon is attached:

$$(l-d)w + dE_y = 0 \quad \text{and} \quad E_y = -\frac{(l-d)w}{d}.$$

The negative sign of this result shows that our initial guess for the direction of E_y was wrong; it is actually vertically *downward*.

Finally, we take moments about point B in the figure:

$$lw - hE_x = 0 \quad \text{and} \quad E_x = \frac{lw}{h}.$$

We note that each stage of this calculation is simplified by choosing the point for calculating moments so as to eliminate one or more of the unknown quantities. In the last step, the force T has no moment about point B; thus when the moments of T_x and T_y are computed separately, they must add to zero. The reader is invited to verify this statement in detail.

8-5

As a specific example, suppose $w = 50$ N, $d = 0.1$ m, $l = 0.5$ m, and $\theta = 80°$. Since $\tan \theta = h/d$, we find

$$h = d \tan \theta = (0.1 \text{ m})(5.67) = 0.567 \text{ m}.$$

From the previous general results, we find

$$T = \frac{(0.5 \text{ m})(50 \text{ N})}{(0.1 \text{ m})(0.985)} = 254 \text{ N},$$

$$E_y = -\frac{(0.5 \text{ m} - 0.1 \text{ m})(50 \text{ N})}{0.1 \text{ m}} = -200 \text{ N},$$

$$E_x = \frac{(0.5 \text{ m})(50 \text{ N})}{0.567 \text{ m}} = 44.1 \text{ N}.$$

The magnitude of the force at the elbow is

$$E = \sqrt{E_x^2 + E_y^2} = 205 \text{ N}.$$

As mentioned above, we have not explicitly used the first condition for equilibrium, that the vector sum of the forces be zero. As a check, the reader should verify that this condition *is* satisfied by the above results.

Units of meters have been used in the above example; centimeters could have been used. In fact, *any* unit of distance may be used in a moment equation, provided the *same* unit is used throughout a given equation.

8-3 Center of gravity

In equilibrium problems where one of the forces acting on a body is its own weight, it is necessary to know the moment of this force about a specified point. It turns out that the moment can always be calculated correctly by assuming the entire force of gravity to be applied at a single point called the *center of gravity*. Furthermore, this point is identical with the *center of mass* introduced in Sec. 7-7. We now discuss the center of gravity.

Every particle of matter in a body is attracted by the earth, and the single force that we call the *weight* of the body is the resultant of all these forces of attraction. The direction of the force on each particle is toward the center of the earth, but the distance to the earth's center is so great that for all practical purposes the forces can be considered parallel to one another. Hence the weight of a body is the resultant of a large number of parallel forces.

Figure 8-6a shows a flat object of arbitrary shape in the xy-plane. Let the body be subdivided into a large number of small particles of weights w_1, w_2, etc., and let the coordinates of these particles be (x_1, y_1), (x_2, y_2), etc. The total weight W of the object is

$$W = w_1 + w_2 + \cdots = \sum w. \tag{8-4}$$

Each particle's weight also contributes to the total moment acting on the body. Computing moments about point O, we see that the moment associated with particle 1 is $w_1 x_1$, that for particle 2 is $w_2 x_2$, and so

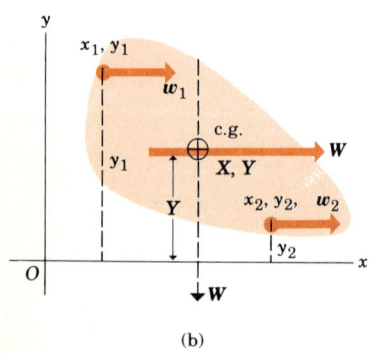

8-6 The body's weight W is the resultant of a large number of parallel forces. The line of action of W always passes through the center of gravity.

on, and that the total torque is

$$w_1 x_1 + w_2 x_2 + \cdots = \sum wx.$$

For convenience in the present discussion we are taking *clockwise* torques to be positive, the opposite choice from that used in preceding sections.

Along what line must the total weight W act in order for the total torque to equal the above expression? Suppose it acts along a line a distance X to the right of the origin. Then it must be true that $w_1 x_1 + w_2 x_2 + \cdots = WX$, or

$$X = \frac{w_1 x_1 + w_2 x_2 + \cdots}{W} = \frac{\sum wx}{W} = \frac{\sum wx}{\sum w}. \tag{8-5}$$

That is, the moment due to the weight of the object can always be obtained by assuming the total weight to be acting along a line determined by Eq. (8-5). Stated another way, in computing the moment due to the weight of an object, we must consider the weight as acting at a point whose x-coordinate X is given by Eq. (8-5). (See Fig. 8-6a.)

Now let the object and the reference axes be rotated 90° clockwise or, which amounts to the same thing, let us consider the gravitational forces to be rotated 90° counterclockwise, as in Fig. 8-6b. The total weight W is unaltered, but now it must act along a line a distance Y above the origin, such that $WY = \sum wy$, or

$$Y = \frac{w_1 y_1 + w_2 y_2 + \cdots}{w_1 + w_2 + \cdots} = \frac{\sum wy}{\sum w} = \frac{\sum wy}{W}. \tag{8-6}$$

The point of intersection of the lines of action of W in the two parts of Fig. 8-6 has the coordinates X and Y and is called the *center of gravity* of an object. By considering some arbitrary orientation of the object, one can show that the line of action of W *always* passes through the center of gravity, and thus that the moment of W can always be obtained correctly by taking W to act at the center of gravity.

If we divide numerator and denominator in Eqs. (8-5) and (8-6) by g and use the fact that $w_1 = m_1 g$, and so on, the right sides of these equations become identical to those in the definitions of the center of mass in Sec. 7-7, Eqs. (7-22). Thus <u>the center of gravity of any body is identical to its center of mass</u>. We also note that the definition of center of gravity makes sense only in a uniform gravitational field, where g has the same value everywhere; otherwise the w's would depend on position. The center of *mass*, conversely, is defined independently of any gravitational effect.

If the centers of gravity of each of a number of bodies have been determined, the coordinates of the center of gravity of the combination can be computed from Eqs. (8-5) and (8-6), letting w_1, w_2, etc., be the weights of the bodies and x_1 and y_1, x_2 and y_2, etc., be the coordinates of the center of gravity of each.

Symmetry considerations are often useful in finding the position of the center of gravity. Thus the center of gravity of a homogeneous sphere, cube, circular disk, or rectangular plate is at its geometrical center. That of a cylinder or right circular cone is on the axis of symmetry, and so on.

Example Locate the center of gravity of the machine part in Fig. 8-7, consisting of a disk 4 cm in diameter and 2 cm long, and a rod 2 cm in diameter and 12 cm long, constructed of a homogeneous material.

8-7

Solution By symmetry, the center of gravity lies on the axis and the center of gravity of each part is midway between its ends. The volume of the disk is 8π cm^3 and that of the rod is 12π cm^3. Since the weights of the two parts are proportional to their volumes,

$$\frac{w(\text{disk})}{w(\text{rod})} = \frac{w_1}{w_2} = \frac{8\pi}{12\pi} = \frac{2}{3}.$$

Take the origin O at the left face of the disk, on the axis. Then

$$x_1 = 1 \text{ cm}, \qquad x_2 = 8 \text{ cm},$$

and

$$X = \frac{w_1(1 \text{ cm}) + \tfrac{3}{2}w_1(8 \text{ cm})}{w_1 + \tfrac{3}{2}w_1} = 5.2 \text{ cm}.$$

The center of gravity is on the axis, 5.2 cm to the right of O.

The center of gravity (center of mass) has several other important properties. First, when a body is suspended in a gravitational field from a single point, the center of gravity always hangs directly below the point of suspension; otherwise the body could not be in rotational equilibrium. This fact can be used for an experimental determination of the location of the center of gravity of an irregular body.

Second, a force applied to a body at its center of gravity does not tend to cause the body to rotate, while a force applied at any other point tends to cause both rotational and translational motion. Third, as discussed in Sec. 7-7, the total momentum of a body is always given by the product of its total mass and its center-of-mass velocity, and the acceleration of the center of mass is always the same as that of a point mass equal to the total mass of the body, acted on by the resultant external force. The second and third points will be considered in greater detail in Chapter 9, where we study the dynamics of rigid-body motion.

8-4 Couples

It sometimes happens that the forces on a body include two forces of equal magnitude and opposite direction, having lines of action that are parallel but do not coincide. Such a pair of forces is called a *couple*. A common example is a compass needle in the earth's magnetic field, as shown in Fig. 8-8. The north and south poles of the needle are acted on by equal forces, one toward the north and the other toward the south.

8-8 Forces on the poles of a compass needle.

Except when the needle points in the N–S direction, the two forces do not have the same line of action.

Figure 8–9 shows a couple consisting of two forces, each of magnitude F, separated by a perpendicular distance l. The resultant R of the forces is

$$R = F - F = 0.$$

The fact that the resultant force is zero means that a couple has no effect in producing translation (as a whole) of the body on which it acts. *The only effect of a couple is to produce rotation.*

The resultant moment of the couple in Fig. 8–9 about an arbitrary point O is

$$\begin{aligned}\sum \Gamma_0 &= x_2 F - x_1 F \\ &= (x_1 + l)F - x_1 F \\ &= lF.\end{aligned} \qquad (8\text{-}7)$$

Since the distances x_1 and x_2 do not appear in the result, we conclude that *the moment of a couple is the same about all points in the plane of the forces and is equal to the product of the magnitude of either force and the perpendicular distance between their lines of action.*

8–9 Two equal and opposite forces having different lines of action are called a *couple*. The moment of the couple is the same about all points and is equal to Fl.

*8–5 Vector torque

The concept of torque can be generalized using the vector product, introduced in Sec. 1–8. This generalization is not necessary for most of the problems in this book, but it is essential for a more general treatment of the dynamics of rotational motion.

In Fig. 8–10, a force F is applied to a rigid body at a point P whose position relative to the origin O is given by the position vector r. The angle between the directions of r and F is denoted as θ. The moment arm for the force, with respect to O, is the distance l shown. This is given by $l = r \sin \theta$, where r, the magnitude of the vector r, is the distance OP. Thus the torque Γ of the force with respect to O may be written as

$$\Gamma = rF \sin \theta. \qquad (8\text{-}8)$$

We recognize this as the magnitude of the vector product $r \times F$. This suggests defining a vector torque Γ by

$$\Gamma = r \times F. \qquad (8\text{-}9)$$

In Fig. 8–10, the direction of Γ is out of the plane of the figure. Thus torques which we have previously called positive are represented as vectors pointing out of the plane, negative ones as vectors into the plane. The relation between the sense of rotation implied by a given torque and the direction of the associated vector is exactly the same as that described by the right-hand rule in Sec. 1–8, Fig. 1–12.

The corresponding generalization of the second condition for equilibrium is that the *vector sum* of torques on a body, with respect to a given point, must be zero.

8–10 The point of application of force F is described by the position vector r of point P. The vector torque or moment of this force, with respect to the origin O, is defined as $\Gamma = r \times F$.

Questions

8-1 Mechanics sometimes extend the length of a wrench handle by slipping a section of pipe over the handle. Why is this a dangerous procedure?

8-2 Manuals for car-engine repair always specify the *torque* (moment) to be applied when tightening the cylinder-head bolts. Why is torque specified, rather than the *force* to be applied to the wrench handle?

8-3 Car tires are sometimes "balanced" on a machine that pivots the tire and wheel about the center, by laying weights around the wheel rim until it does not tip from the horizontal plane. Discuss this procedure in terms of the center of gravity.

8-4 Is it possible for a solid body to have no matter at its center of gravity? Consider, for example, a disk with a hole in the center, an empty bottle, and a hollow cylinder.

8-5 A man climbs a tall, old stepladder that has a tendency to sway. He feels much more unstable when standing near the top than when near the bottom. Why?

8-6 What determines whether a body in equilibrium is stable or unstable? As an example, discuss a cone resting on its base on a horizontal floor; balanced on its point; and lying on its side.

8-7 When a heavy load is placed in the back end of a truck, behind the rear wheels, the effect is to *raise* the front end of the truck. Why?

8-8 When a tall heavy object such as a refrigerator is pushed across a rough floor, what determines whether it slides or tips?

8-9 Why is a tapered water glass with a narrow base easier to tip over than one with straight sides? Does it matter whether the glass is full or empty?

8-10 People sometimes prop open a door by wedging some object in the space between the hinged side of the door and the frame. Explain why this often results in the hinge screws being ripped out.

8-11 Discuss the action of a claw hammer in pulling out nails, in terms of moments.

8-12 In trying to lift an object that is much too heavy to be lifted directly, one often speaks of the importance of "leverage." Discuss this term in relation to moments, and explain how it is related to the concept of leverage in trading stock options and other securities.

8-13 Why is it easier to hold a 10-kg body in your hand at your side than to hold it with your arm extended horizontally?

8-14 In pioneer days when a Conestoga wagon was stuck in the mud, a man would grasp a wheel spoke and try to turn the wheel, rather than simply pushing the wagon. Why?

8-15 Does the center of gravity of a solid body always lie within the body? If not, give a counterexample.

Problems

8-1 A heavy electric motor is to be carried by two men by placing it on a light board 2.0 m long. To lift the board with the motor on it, one man must lift at one end with a force of 600 N and the other at the opposite end with a force of 400 N. What is the weight of the motor, and where is it located?

8-2 In Problem 8-1, suppose the board is not light but weighs 200 N. What is the weight of the motor, and where is it located?

8-3 A child weighing 400 N sits on one end of a seesaw that is 3.0 m long and is pivoted 1.4 m from the child. If another child sitting at the opposite end just balances, what is his weight? Neglect the weight of the seesaw. What is the total force on the pivot?

8-4 A certain automobile has a wheelbase (distance between front and rear axles) of 3.0 m. If 60 percent of the weight rests on the front wheels, how far behind the front wheels is the center of gravity?

8-5 Two men are carrying a ladder 20 ft long, weighing 200 lb. If one man can lift a maximum of 80 lb and lifts at one end, at what point should the other man lift?

8-6 You are given (a) a meter stick through which a number of holes have been bored so that its center of gravity is not at its midpoint, (b) a knife edge, on which the meter stick can be pivoted, (c) a body whose weight is known to be w, and (d) a spool of thread. Using this equipment only, explain, with the aid of a diagram, how you would determine the weight of the meter stick.

8-7 A uniform plank 15 m long, weighing 400 N, rests symmetrically on two supports 8 m apart, as shown in Fig. 8-11. A boy weighing 640 N starts at point A and walks toward the right.

a) Construct in the same diagram two graphs showing the upward forces F_A and F_B exerted on the plank at points A and B, as functions of the coordinate x of the boy. Let 1 in. = 250 N vertically, and 1 in. = 2.5 m horizontally.

b) Find from your diagram how far beyond point B the boy can walk before the plank tips.

c) How far from the right end of the plank should support B be placed in order that the boy walk just to the end of the plank without causing it to tip?

Figure 8-11

8-8 The strut in Fig. 8-12 weighs 200 N and its center of gravity is at its center. Find

a) the tension in the cable and

b) the horizontal and vertical components of the force exerted on the strut at the wall.

Figure 8-12

8-9 Find the tension in the cable BD in Fig. 8-13, and the horizontal and vertical components of the force exerted on the strut AB at pin A, using

a) the first and second conditions of equilibrium ($\Sigma F_x = 0$, $\Sigma F_y = 0$, $\Sigma \Gamma = 0$), taking moments about an axis through point A perpendicular to the plane of the diagram;

b) the second condition of equilibrium only, taking moments first about an axis through A, then about an axis through B, and finally about an axis through D. The weight of the strut can be neglected.

c) Represent the computed forces by vectors in a diagram drawn to scale, and show that the lines of action of the forces exerted on the strut at points A, B, and C intersect at a common point.

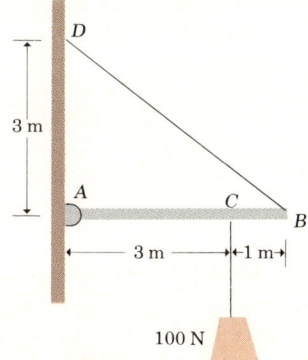

Figure 8-13

8-10 Find the tension T in each cable, and the magnitude and direction of the force exerted on the strut by the pivot, in each of the arrangements in Fig. 8-14. Let the weight of the suspended object in each case be 1000 N. Neglect the weight of the strut.

Figure 8-14

8-11 A horizontal boom 4 m long is hinged to a vertical wall at one end, and a 500-N body hangs from its outer end. The boom is supported by a guy wire from its outer end to a point on the wall directly above the boom.

a) If the tension in this wire is not to exceed 1000 N, what is the minimum height above the boom at which it may be fastened to the wall?

b) By how many newtons would the tension be increased if the wire were fastened 0.5 m below this point, the boom remaining horizontal? Neglect the weight of the boom.

8-12 Find the largest weight w that can be supported by the structure in Fig. 8-15 if the maximum tension the

Figure 8-15

upper rope can withstand is 1000 N and the maximum compression the strut can withstand is 2000 N. The vertical rope is strong enough to carry any load required. Neglect the weight of the strut.

8-13 A ball of mass 1 kg is attached by a light rod 0.4 m in length to a second ball of mass 3 kg. Where is the center of gravity of this system?

8-14 In Problem 8-13, suppose the rod is uniform and has mass 2 kg. Where is the center of gravity of the system?

8-15 Three small objects of equal mass are located in the x, y-plane, at points having coordinates (0.1 m, 0), (0, 0.1 m), (0.1 m, 0.1 m). Determine the coordinates of the center of gravity. Can you use the symmetry of the situation to simplify your calculation?

8-16 A person of mass 80 kg stands 1 m from one end of a floating canoe 5 m long, having mass 40 kg, initially at rest. He walks toward the opposite end until he is 1 m from it. Assuming that the center of gravity of the system does not move, how far does the canoe move?

8-17 In a molecule of carbon monoxide (CO), the nuclei of the two atoms are about 1.13×10^{-10} m apart. Where is the center of mass of the molecule?

8-18 A solid cube 5 cm on a side is glued to a solid cube of the same material, 10 cm on a side, so that one corner and two edges of the two cubes coincide. Define a suitable coordinate system and determine the position of the center of gravity of the system.

8-19 A station wagon of mass 2000 kg has a wheelbase (distance between front and rear axles) of 3.0 m. Ordinarily 1100 kg rests on the front wheels and 900 kg on the rear wheels. A box of mass 200 kg is now placed on the tailgate, 1.0 m behind the rear axle. How much total weight now rests on the front wheels? On the rear wheels?

8-20 End A of the bar AB in Fig. 8-16 rests on a frictionless horizontal surface, while end B is hinged. A horizontal force P of 60 N is exerted on end A. Neglect the weight of the bar. What are the horizontal and vertical components of the force exerted by the bar on the hinge at B?

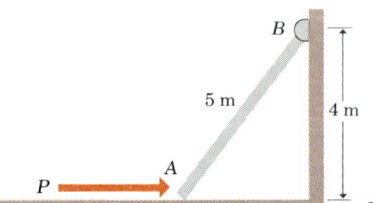

Figure 8-16

8-21 A single additional force is to be applied to the bar in Fig. 8-17 to maintain it in equilibrium in the position shown. The weight of the bar can be neglected.

a) What are the x- and y-components of the required force?

b) What is the tangent of the angle that the force must make with the bar?

Figure 8-17

c) What is the magnitude of the required force?

d) Where should the force be applied?

8-22 A circular disk 0.5 m in diameter, pivoted about a horizontal axis through its center, has a cord wrapped around its rim. The cord passes over a frictionless pulley P and is attached to a body of weight 240 N. A uniform rod 2 m long is fastened to the disk, with one end at the center of the disk. The apparatus is in equilibrium, with the rod horizontal, as shown in Fig. 8-18.

a) What is the weight of the rod?

b) What is the new equilibrium direction of the rod when a second body weighing 20 N is suspended from the outer end of the rod, as shown by the broken line?

Figure 8-18

8-23 A roller whose diameter is 1.0 m weighs 360 N. What horizontal force is necessary to pull the roller over a brick 0.1 m high when the force is applied (a) at the center, (b) at the top of the roller?

8-24 The boom in Fig. 8-19 is uniform and weighs 2500 N. (a) Find the tension in the guy wire, and the horizontal and vertical components of the force exerted on the boom at its lower end. (b) Does the line of action of this force lie along the boom?

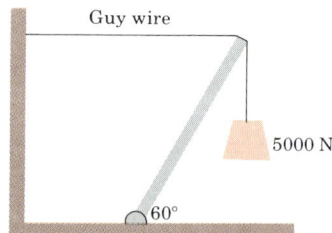

Figure 8-19

8-25 Two ladders, 4 m and 3 m long, respectively, are hinged at point A and tied together by a horizontal rope 0.6 m above the floor, as in Fig. 8-20. The ladders weigh 400 N and 300 N, respectively, and the center of gravity of each is at its center. If the floor is frictionless, find

a) the upward force at the bottom of each ladder,
b) the tension in the rope, and
c) the force that one ladder exerts on the other at point A.
d) If a load of 1000 N is now suspended from point A, find the tension in the rope.

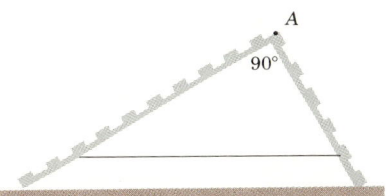

Figure 8-20

8-26 A uniform ladder 10 m long rests against a vertical frictionless wall with its lower end 6 m from the wall. The ladder weighs 400 N. The coefficient of static friction between the foot of the ladder and the ground is 0.40. A man weighing 800 N climbs slowly up the ladder.

a) What is the maximum frictional force that the ground can exert on the ladder at its lower end?
b) What is the actual frictional force when the man has climbed 3 m along the ladder?
c) How far along the ladder can the man climb before the ladder starts to slip?

8-27 One end of a meter stick is placed against a vertical wall, as in Fig. 8-21. The other end is held by a light cord making an angle θ with the stick. The coefficient of static friction between the end of the meter stick and the wall is 0.30.

a) What is the maximum value the angle θ can have if the stick is to remain in equilibrium?
b) Let the angle θ be 10°. A body of the same weight as the meter stick is suspended from the stick as shown, at a distance x from the wall. What is the minimum value of x for which the stick will remain in equilibrium?
c) When $\theta = 10°$, how large must the coefficient of static friction be so that the body can be attached at the left end of the stick without causing it to slip?

Figure 8-21

8-28 One end of a post weighing 500 N rests on a rough horizontal surface with $\mu_s = 0.3$. The upper end is held by a rope fastened to the surface and making an angle of 37° with the post, as in Fig. 8-22. A horizontal force F is exerted on the post as shown.

a) If the force F is applied at the midpoint of the post, what is the largest value it can have without causing the post to slip?
b) How large can the force be, without causing the post to slip, if its point of application is $\frac{7}{10}$ of the way from the ground to the top of the post?
c) Show that if the point of application of the force is too high, the post cannot be made to slip, no matter how great the force. Find the critical height for the point of application.

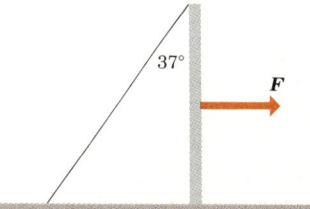

Figure 8-22

8-29 A door 1.0 m wide and 2.5 m high weighs 200 N and is supported by two hinges, one 0.5 m from the top and the other 0.5 m from the bottom. Each hinge supports half the total weight of the door. Assuming the door's center of gravity is at its center, find

a) the components of force exerted on the door by each hinge;
b) the magnitude and direction of the force exerted by each hinge.

8-30 A gate 4 m long and 2 m high weighs 400 N. Its center of gravity is at its center, and it is hinged at A and B. To relieve the strain on the top hinge, a wire CD is connected as shown in Fig. 8-23. The tension in CD is increased until the horizontal force at hinge A is zero.

a) What is the tension in the wire CD?
b) What is the magnitude of the horizontal component of force at hinge B?
c) What is the combined vertical force exerted by hinges A and B?

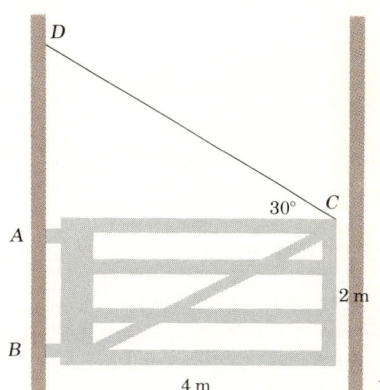

Figure 8-23

8-31 A uniform rectangular block, 0.5 m high and 0.25 m wide, rests on a plank AB, as in Fig. 8-24. The coefficient of static friction between block and plank is 0.40.

a) In a diagram drawn to scale, show the line of action of the resultant normal force exerted on the block by the plank when the angle $\theta = 15°$.

b) If end B of the plank is slowly raised, will the block start to slide down the plank before it tips over? Find the angle θ at which it starts to slide, or at which it tips.

c) What would be the answer to part (b) if the coefficient of static friction were 0.60? If it were 0.50?

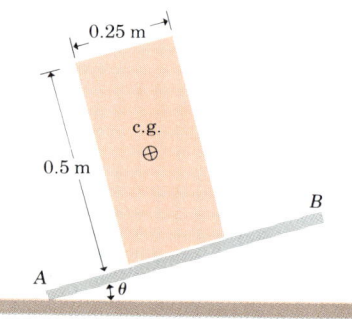

Figure 8-24

8-32 A rectangular block 0.25 m wide and 0.5 m high is dragged to the right along a level surface at constant speed by a horizontal force P, as shown in Fig. 8-25. The coefficient of sliding friction is 0.40, the block weighs 25 N, and its center of gravity is at its center.

a) Find the magnitude of the force P.

b) Find the position of the line of action of the normal force N exerted on the block by the surface, if the height $h = 0.125$ m.

c) Find the value of h at which the block just starts to tip.

Figure 8-25

8-33 A garage door is mounted on an overhead rail, as in Fig. 8-26. The wheels at A and B have rusted so that they do not roll, but slide along the track. The coefficient of sliding friction is 0.5. The distance between the wheels is 2 m, and each is 0.5 m in from the vertical sides of the door. The door is symmetrical and weighs 800 N. It is pushed to the left at constant velocity by a horizontal force P.

a) If the distance h is 1.5 m, what is the vertical component of the force exerted on each wheel by the track?

b) Find the maximum value h can have without causing one wheel to leave the track.

Figure 8-26

8-34 The objects in Fig. 8-27 are constructed of uniform wire bent into the shapes shown. Find the position of the center of gravity of each.

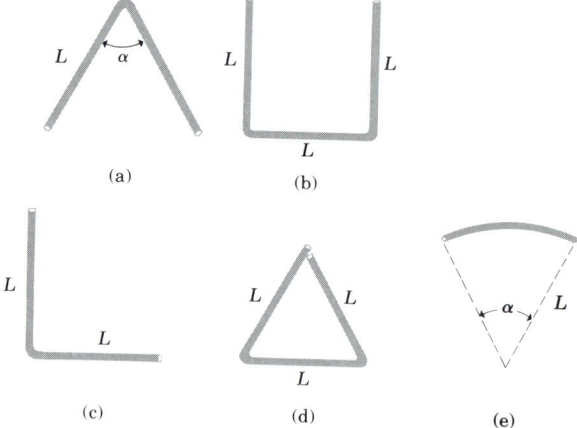

Figure 8-27

ROTATION 9

9-1 Introduction

Up to this point our discussion of principles of dynamics has been concerned primarily with *particles*. Even in problems where the size of a body is not negligible, we have assumed that its motion can be described in terms of the motion of a *point*. That is, the point mass serves as a *model* to represent a body whose size and shape are not relevant in the particular problem under discussion.

There are many problems in which this model is inadequate. One important class of problems involves *rigid bodies*, which can undergo both translational and rotational motion. The rigid body, a body with a perfectly definite and unchanging shape, is itself an idealized model, since real materials always deform somewhat when forces are exerted, but it is a useful model in cases where such deformation may be neglected.

In the preceding chapter *equilibrium* of an extended body was considered. We found that equilibrium requires both the absence of any tendency for change in *translational* motion and the absence of any tendency for the body to change its *rotational* motion. Similarly, motion of a rigid body can always be represented as a combination of translational motion of some point in the body and rotational motion about an axis through that point. For simplicity we begin our discussion with rotation about a stationary axis. Just as with dynamics of a particle, we first develop *kinematic* language for describing rotational motion.

9-2 Angular velocity

Figure 9-1 represents a rigid body of arbitrary shape rotating about a fixed axis through point O and perpendicular to the plane of the diagram. Line OP is a line fixed with respect to the body and rotating with it. The position of the entire body is evidently completely specified by the point O and the angle θ which the line OP makes with some reference line fixed in space, such as Ox. The motion of the body is therefore

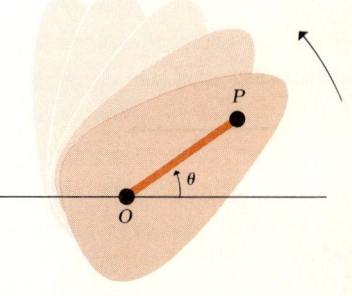

9-1 Body rotating about a fixed axis through point O.

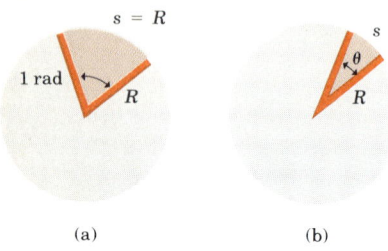

9-2 An angle θ in radians is defined as the ratio of the arc s to the radius R.

analogous to the rectilinear motion of a particle whose position is completely specified by a single coordinate such as x or y. The equations of motion are greatly simplified if the angle θ is expressed in *radians*.

One radian (1 rad) is the angle subtended at the center of a circle by an arc of length equal to the radius of the circle, as shown in Fig. 9–2a. Since the radius is contained 2π times ($2\pi = 6.28\ldots$) in the circumference, there are 2π or $6.28\ldots$ rad in one complete revolution or $360°$. Hence,

$$1 \text{ rad} = \frac{360°}{2\pi} = 57.3 \text{ degrees,}$$
$$360° = 2\pi \text{ rad} = 6.28 \text{ rad,}$$
$$180° = \pi \text{ rad} = 3.14 \text{ rad,}$$
$$90° = \pi/2 \text{ rad} = 1.57 \text{ rad,}$$
$$60° = \pi/3 \text{ rad} = 1.05 \text{ rad,}$$
$$45° = \pi/4 \text{ rad} = 0.79 \text{ rad,}$$

and so on.

In general (Fig. 9–2b), if θ represents any arbitrary angle subtended by an arc of length s on the circumference of a circle of radius R, then θ (in radians) is equal to the length of the arc s divided by the radius R:

$$\theta = \frac{s}{R}, \qquad s = R\theta. \tag{9-1}$$

For example, if $\theta = 2\pi$ rad, then s is the circumference of the circle.

An angle in radians, being defined as the ratio of a length to a length, is a pure number. If $s = 1.5$ m and $R = 1$ m, the angle is usually described as $\theta = 1.5$ rad, but it would be equally correct to say simply $\theta = 1.5$.

In Fig. 9–3, a reference line OP in a rotating body makes an angle θ_1 with the reference line Ox, at a time t_1. At a later time t_2 the angle has increased to θ_2. The *average angular velocity* of the body, $\bar{\omega}$, in the time interval between t_1 and t_2, is defined as the ratio of the *angular displacement* $\theta_2 - \theta_1$, or $\Delta\theta$, to the elapsed time $t_2 - t_1$ or Δt:

$$\bar{\omega} = \frac{\Delta\theta}{\Delta t}.$$

The *instantaneous angular velocity* ω is defined as the limit approached by this ratio as Δt approaches zero, that is, the derivative of θ with respect to t:

$$\omega = \lim_{\Delta t \to 0} \frac{\Delta\theta}{\Delta t} = \frac{d\theta}{dt}. \tag{9-2}$$

Since the body is rigid, *all* lines in it rotate through the same angle in the same time, and the angular velocity is characteristic of the body as a whole. If the angle θ is in radians, the unit of angular velocity is one radian per second (1 rad·s^{-1}). Other units, such as the revolution per minute (rev·min^{-1}), are in common use. We note that 1 rev·s$^{-1} = 2\pi$ rad·s^{-1}.

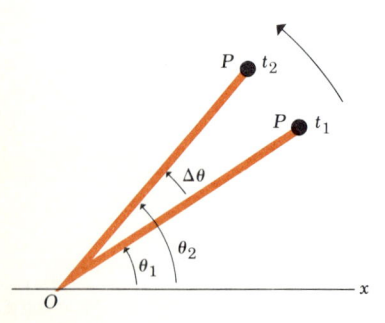

9-3 Angular displacement $\Delta\theta$ of a rotating body.

9-3 Angular acceleration

If the angular velocity of a body changes, it is said to have an angular acceleration. If ω_1 and ω_2 are the instantaneous angular velocities at times t_1 and t_2, the *average angular acceleration* $\bar{\alpha}$ is defined as

$$\bar{\alpha} = \frac{\omega_2 - \omega_1}{t_2 - t_1} = \frac{\Delta\omega}{\Delta t},$$

and the *instantaneous angular acceleration* α is defined as the limit of this ratio when Δt approaches zero:

$$\alpha = \lim_{\Delta t \to 0} \frac{\Delta\omega}{\Delta t} = \frac{d\omega}{dt}. \tag{9-3}$$

The unit of angular acceleration is $1 \text{ rad} \cdot \text{s}^{-2}$ or 1 s^{-2}. Angular velocity and angular acceleration are exactly analogous to linear velocity and acceleration.

Since $\omega = d\theta/dt$, the angular acceleration can be written

$$\alpha = \frac{d}{dt}\frac{d\theta}{dt} = \frac{d^2\theta}{dt^2}.$$

Also, by the chain rule,

$$\alpha = \frac{d\omega}{d\theta}\frac{d\theta}{dt} = \omega\frac{d\omega}{d\theta}. \tag{9-4}$$

9-4 Rotation with constant angular acceleration

The simplest type of accelerated rotational motion is that in which the angular acceleration is *constant*. When this is the case, expressions for the angular velocity and angular coordinate can readily be derived by integration. We have

$$d\omega/dt = \alpha = \text{constant},$$

$$\int d\omega = \int \alpha \, dt,$$

$$\omega = \alpha t + C_1.$$

If ω_0 is the angular velocity when $t = 0$, the integration constant C_1 is equal to ω_0 and

$$\omega = \omega_0 + \alpha t. \tag{9-5}$$

Then, since $\omega = d\theta/dt$,

$$\int d\theta = \int \omega_0 \, dt + \int \alpha t \, dt, \qquad \theta = \omega_0 t + \tfrac{1}{2}\alpha t^2 + C_2.$$

The integration constant C_2 is the value of θ when $t = 0$, which we denote as θ_0.

$$\theta = \theta_0 + \omega_0 t + \tfrac{1}{2}\alpha t^2. \tag{9-6}$$

If we write the angular acceleration as

$$\alpha = \omega\frac{d\omega}{d\theta},$$

then

$$\int \alpha \, d\theta = \int \omega \, d\omega + C_3, \qquad \alpha\theta = \tfrac{1}{2}\omega^2 + C_3.$$

If the angle θ again has the value θ_0 when $t = 0$, and if the initial angular velocity is ω_0, then $C_3 = \alpha\theta_0 - \tfrac{1}{2}\omega_0^2$ and

$$\omega^2 = \omega_0^2 + 2\alpha(\theta - \theta_0). \tag{9-7}$$

Table 9-1 shows the similarity between Eqs. (9-5), (9-6), and (9-7) for motion with constant angular acceleration and the equations for motion with constant linear acceleration.

Table 9-1

analogies

Motion with constant linear acceleration	Motion with constant angular acceleration
a = constant	α = constant
$v = v_0 + at$	$\omega = \omega_0 + \alpha t$
$x = x_0 + v_0 t + \tfrac{1}{2}at^2$	$\theta = \theta_0 + \omega_0 t + \tfrac{1}{2}\alpha t^2$
$v^2 = v_0^2 + 2a(x - x_0)$	$\omega^2 = \omega_0^2 + 2\alpha(\theta - \theta_0)$

Example The angular velocity of a body is 4 rad·s^{-1} at time $t = 0$, and its angular acceleration is constant and equal to 2 rad·s^{-2}. A line OP in the body is horizontal at time $t = 0$.

a) What angle does this line make with the horizontal at time $t = 3$ s?
b) What is the angular velocity at this time?

Solution

a) $\theta = \theta_0 + \omega_0 t + \tfrac{1}{2}\alpha t^2$

$\quad = 0 + (4 \text{ rad·s}^{-1})(3 \text{ s}) + \tfrac{1}{2}(2 \text{ rad·s}^{-2})(3 \text{ s})^2$

$\quad = 21 \text{ rad} = \dfrac{21}{2\pi} \text{ rev} = 3.34 \text{ rev}.$

The body turns through three complete revolutions plus an additional 0.34 rev or $(0.34 \text{ rev})(2\pi \text{ rad·rev}^{-1}) = 2.15 \text{ rad} = 123°$. The line OP thus turns through 123° and makes an angle of 57° with the horizontal.

b) $\omega = \omega_0 + \alpha t = 4 \text{ rad·s}^{-1} + (2 \text{ rad·s}^{-2})(3 \text{ s}) = 10 \text{ rad·s}^{-1}.$

Alternatively, from Eq. (9-7),

$$\omega^2 = \omega_0^2 + 2\alpha(\theta - \theta_0)$$
$$= (4 \text{ rad·s}^{-1})^2 + 2(2 \text{ rad·s}^{-2})(21 \text{ rad})$$
$$= 100 \text{ rad}^2\text{·s}^{-2},$$
$$\omega = 10 \text{ rad·s}^{-1}.$$

9-5 Relation between angular and linear velocity and acceleration

In Sec. 5-6 we discussed the linear velocity and acceleration of a *particle* revolving in a circle. When a *rigid body* rotates about a fixed axis, every point in the body moves in a circle whose center is on the axis and which lies in a plane perpendicular to the axis. There are some useful and simple relations between the angular velocity and acceleration of the rotating body and the linear velocity and acceleration of points within it.

Let r be the distance from the axis to some point P in the body, so that the point moves in a circle of radius r, as in Fig. 9-4. When the radius makes an angle θ with the reference axis, the distance s to the point P, measured along the circular path, is

$$s = r\theta. \tag{9-8}$$

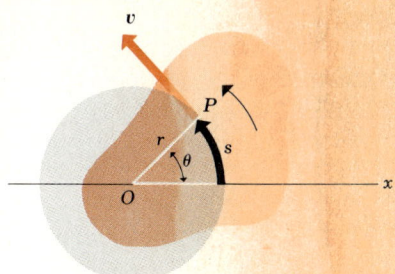

9-4 The distance s moved through by point P equals $r\theta$.

When the angle is very small, say $\Delta\theta$, the length of the arc Δs is also small, and the time interval Δt for the body to rotate through the angle $\Delta\theta$ is also small. We have

$$\Delta s = r\,\Delta\theta, \qquad \frac{\Delta s}{\Delta t} = r\frac{\Delta\theta}{\Delta t}.$$

That is, the average speed $\Delta s/\Delta t$ is r times the average angular velocity $\Delta\theta/\Delta t$, and in the limit, as $\Delta t \to 0$,

$$v = r\omega. \tag{9-9}$$

If the angular velocity about the given axis should change *in magnitude only* (the axis remaining fixed) by $\Delta\omega$, the linear velocity in a direction tangent to a circle of radius r will change by Δv, where

$$\Delta v = r\,\Delta\omega.$$

If these changes take place in a short time interval Δt, then

$$\frac{\Delta v}{\Delta t} = r\frac{\Delta\omega}{\Delta t},$$

and, in the limit, as $\Delta t \to 0$,

$$a_\parallel = r\alpha, \quad \text{(tangential)} \tag{9-10}$$

where $a_\parallel$ is the tangential component of the linear acceleration of a point a distance r from the axis.

The *radial* component of acceleration v^2/r of the point P can also be expressed in terms of the angular velocity,

$$a_\perp = \frac{v^2}{r} = \omega^2 r. \quad \text{(radial)} \tag{9-11}$$

This is true at each instant *even when ω and v are not constant*.

The tangential and radial components of acceleration of a point P in a rotating body are shown in Fig. 9-5. Their vector sum is the acceleration $\boldsymbol{a}$, as shown.

9-6 Kinetic energy of rotation. Moment of inertia

We have shown that the magnitude of the velocity of a particle in a rigid body rotating about a fixed axis is

$$v = r\omega,$$

where r is the distance of the particle from the axis and ω is the angular velocity of the body. The kinetic energy of a particle of mass m is there-

9-5 Nonuniform rotation about a fixed axis through point O. The tangential component of acceleration of point P equals $r\alpha$; the radial component equals $\omega^2 r$.

fore
$$\tfrac{1}{2}mv^2 = \tfrac{1}{2}mr^2\omega^2.$$

The total kinetic energy of the body is the *sum* of the kinetic energies of all particles in the body,

$$K = \tfrac{1}{2}m_1 r_1^2 \omega^2 + \tfrac{1}{2}m_2 r_2^2 \omega^2 + \cdots$$
$$= \sum \tfrac{1}{2} mr^2 \omega^2,$$

where the symbol Σ stands for *sum*, indicating that the quantity $\tfrac{1}{2}mr^2\omega^2$ is to be computed for each particle making up the body, and the results added. Since ω is the same for all particles in a rigid body,

$$K = \tfrac{1}{2}\left[\sum mr^2\right]\omega^2.$$

To obtain the sum Σmr^2, the body is subdivided (in imagination) into a large number of particles, the mass of each particle is multiplied by the square of its distance from the axis, and these products are summed over all particles. The result is called the *moment of inertia* of the body, about the axis of rotation, and is represented by I:

$$I = \sum mr^2. \tag{9-12}$$

In SI (mks) units the unit of moment of inertia is $1\ \mathrm{kg\cdot m^2}$.

The rotational kinetic energy of a rigid body can now be written as

$$K = \tfrac{1}{2}I\omega^2. \tag{9-13}$$

The form of this expression is analogous to that of translational kinetic energy

$$K = \tfrac{1}{2}mv^2.$$

That is, for rotation about a fixed axis, this moment of inertia I is analogous to mass m (or inertia), and angular velocity ω is analogous to linear velocity v. However, the moment of inertia of a body depends on the location of the axis of rotation, and this must always be specified when a moment of inertia is described.

Example 1 Three small bodies, which can be considered as particles, are connected by light rigid rods, as in Fig. 9-6. What is the moment of inertia of the system (a) about an axis through point A, perpendicular to the plane of the diagram, and (b) about an axis coinciding with the rod BC?

Solution

a) The particle at point A lies on the axis. Its distance *from* the axis is zero and it contributes nothing to the moment of inertia. Therefore,

$$I = \sum mr^2 = (0.1\ \mathrm{kg})(0.5\ \mathrm{m})^2 + (0.2\ \mathrm{kg})(0.4\ \mathrm{m})^2$$
$$= 0.057\ \mathrm{kg\cdot m^2}.$$

b) The particles at B and C both lie on the axis. The moment of inertia is

$$I = \sum mr^2 = (0.3\ \mathrm{kg})(0.4\ \mathrm{m})^2 = 0.048\ \mathrm{kg\cdot m^2}.$$

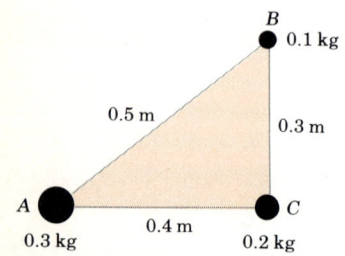

9-6

This illustrates the important fact that the moment of inertia of a body, unlike its mass, is *not* a unique property of the body but depends on the position of the axis about which it is computed.

c) If the body rotates about an axis through A and perpendicular to the plane of the diagram, with an angular velocity $\omega = 4$ rad·s^{-1}, what is the rotational kinetic energy?

$$K = \tfrac{1}{2}I\omega^2 = \tfrac{1}{2}(0.057 \text{ kg·m}^2)(4 \text{ rad·s}^{-1})^2 = 0.456 \text{ J}.$$

For a body that is *not* composed of discrete point masses but is a continuous distribution of matter, the summation expressed in the definition of moment of inertia, $I = \Sigma mr^2$, is usually evaluated by means of integration; several examples are given in Sec. 9–7. The moments of inertia of a few simple but important bodies are listed in Fig. 9–7 for convenience.

It should be noted that it is *not* correct in general to compute the moment of inertia of a body by assuming all the mass to be concentrated at the center of mass (center of gravity). For example, when a uniform thin rod of length L and mass M is pivoted about an axis through one end, perpendicular to the rod, the moment of inertia is $I = ML^2/3$ while, if we take the mass as concentrated at the center, distance $L/2$ from the axis, we obtain the *incorrect* result $I = M(L/2)^2 = ML^2/4$.

The concept of moment of inertia, together with the general work-energy principle, is very useful in problems involving the rotation of a rigid body. The following examples show some applications of these ideas.

Example 2 A light flexible rope is wrapped several times around a solid cylinder of mass 50 kg and diameter 0.12 m, which rotates without friction about a fixed horizontal axis. The free end of the rope is pulled with a constant force of magnitude 9 N for a distance of 2 m. If the cylinder is initially at rest, find its final angular velocity and the final speed of the rope.

Solution Because no energy is lost in friction, the final kinetic energy $\tfrac{1}{2}I\omega^2$ of the cylinder is equal to the work Fd done by the force, which is $(9 \text{ N})(2 \text{ m}) = 18$ J. From Fig. 9–7, the moment of inertia is

$$I = \tfrac{1}{2}MR^2 = \tfrac{1}{2}(50 \text{ kg})(0.06 \text{ m})^2 = 0.09 \text{ kg·m}^2.$$

The work–energy relation then gives

$$\tfrac{1}{2}(0.09 \text{ kg·m}^2)\omega^2 = 18 \text{ J},$$
$$\omega = 20 \text{ rad·s}^{-1}.$$

The final speed of the rope is equal to the tangential speed of the cylinder, which is given by Eq. (9–9):

$$v = r\omega = (0.06 \text{ m})(20 \text{ rad·s}^{-1}) = 1.2 \text{ m·s}^{-1}.$$

Example 3 A light flexible rope is wrapped several times around a solid cylinder of mass M and radius R, which rotates, with no friction, about a fixed horizontal axis, as in Fig. 9–8. The free end of the rope is tied to a mass m, which is released from rest a distance h above the floor. Find its speed and the angular velocity of the cylinder just as mass m strikes the floor.

(a) Slender rod, axis through center — $\tfrac{1}{12}Ml^2$

(b) Rectangular plate, axis through center — $\tfrac{1}{12}M(a^2 + b^2)$

(c) Hollow cylinder — $\tfrac{1}{2}M(R_1^2 + R_2^2)$

(d) Solid cylinder — $\tfrac{1}{2}MR^2$

(e) Thin-walled hollow cylinder — MR^2

(f) Solid sphere — $\tfrac{2}{5}MR^2$

9–7 Moments of inertia.

9–8

Solution Initially the system has no kinetic energy, but it has potential energy mgh. Just as mass m strikes the floor, the potential energy is zero, but both this mass and the cylinder have kinetic energy. The total kinetic energy is

$$K = \tfrac{1}{2}mv^2 + \tfrac{1}{2}I\omega^2. \qquad (9\text{-}14)$$

Now, according to Fig. 9-7, the moment of inertia of the cylinder is $I = \tfrac{1}{2}MR^2$. Furthermore, v and ω are related by $v = R\omega$, since the speed of mass m must be equal to the tangential velocity at the outer surface of the cylinder. Using these relations and setting K equal to mgh, we obtain

$$mgh = \tfrac{1}{2}mv^2 + \tfrac{1}{2}(\tfrac{1}{2}MR^2)\left(\frac{v}{R}\right)^2 = \tfrac{1}{2}(m + \tfrac{1}{2}M)v^2,$$

$$v = \sqrt{\frac{2gh}{1 + M/2m}}.$$

When M is much larger than m, v is very small, as might be expected.

*9-7 Calculation of moments of inertia

Although the moment of inertia of a solid body is defined in principle by Eq. (9-12), this equation can be applied directly only in cases where the body consists of a few point masses, as was the case in Sec. 9-6, Example 1. When the body consists of a *continuous* distribution of matter, the sum is expressed in terms of an integral, as we now discuss.

We imagine dividing the entire volume of the body into small volume elements dV in such a way that all points in a particular element are very nearly the same distance from the axis of rotation; we call this distance r as before. Let dm be the mass in the volume element dV. The moment of inertia may then be expressed as

$$I = \lim_{\Delta m \to 0} \sum r^2 \,\Delta m = \int r^2 \,dm. \qquad (9\text{-}15)$$

Since the density ρ of the element is its mass per unit volume, $\rho = dm/dV$, one may also write

$$I = \int r^2 \rho \,dV.$$

If the body is homogeneous, that is, if the density is uniform, then ρ may be taken outside the integral:

$$I = \rho \int r^2 \,dV. \qquad (9\text{-}16)$$

In using this equation, the volume element dV is expressed in terms of differentials of the integration variables, usually coordinates of the volume element, and dV must be chosen so that all points within it are at very nearly the same distance from the axis of rotation. For regularly shaped bodies this integration can often be carried out quite easily. Three examples are given below.

Example 1 *Uniform slender rod, axis perpendicular to length.* Figure 9-9 shows a slender uniform rod of mass m and length l. We wish to

9-9

compute its moment of inertia about an axis through O, at an arbitrary distance h from one end. Select as an element of volume a short section of length dx and cross-sectional area A, at a distance x from point O. Then

$$dm = \rho\, dV = \rho A\, dx = \frac{\rho A l}{l} dx = \frac{M}{l} dx.$$

Now

$$I_O = \int x^2\, dm = \frac{M}{l} \int_{-h}^{l-h} x^2\, dx$$

$$= \frac{M}{l} \frac{x^3}{3}\bigg]_{-h}^{l-h} = \frac{1}{3} M(l^2 - 3lh + 3h^2).$$

From this general expression, we can find the moment of inertia about an axis through any point on the rod. For example, if the axis is at the left end, $h = 0$ and

$$I = \frac{1}{3} M l^2. \tag{9-17}$$

If the axis is at the right end, $h = l$ and

$$I = \frac{1}{3} M l^2,$$

as would be expected. If the axis passes through the center,

$$h = \frac{l}{2} \quad \text{and} \quad I = \frac{1}{12} M l^2, \tag{9-18}$$

as shown also in Fig. 9-7.

Example 2 *Hollow or solid cylinder, axis of symmetry.* Figure 9-10 shows a hollow cylinder of length l and inner and outer radii R_1 and R_2. We choose as the most convenient volume element a thin cylindrical shell of radius r, thickness dr, and length l. The volume of this sheet is very nearly equal to that of a flat sheet of thickness dr, length l, and width $2\pi r$ (the circumference of the shell). Then

$$dm = \rho\, dV = 2\pi \rho l r\, dr.$$

The moment of inertia is given by

$$I = \rho \int r^2\, dV$$

$$= 2\pi \rho l \int_{R_1}^{R_2} r^3\, dr$$

$$= \frac{\pi \rho l}{2} (R_2^4 - R_1^4)$$

$$= \frac{\pi \rho l}{2} (R_2^2 - R_1^2)(R_2^2 + R_1^2). \tag{9-19}$$

9-10

It is usually more convenient to express this in terms of the total mass M

of the body, which is the density multiplied by the total volume. The volume is given by

$$\pi l(R_2{}^2 - R_1{}^2).$$

Hence

$$M = \pi l \rho (R_2{}^2 - R_1{}^2),$$

and the moment of inertia is therefore

$$I = \tfrac{1}{2} M(R_1{}^2 + R_2{}^2), \tag{9-20}$$

as shown also in Fig. 9–7.

If the cylinder is solid, $R_1 = 0$, and if we let R represent the outer radius,

$$I = \tfrac{1}{2} MR^2. \tag{9-21}$$

If the cylinder is very thin-walled (like a stovepipe), $R_1 = R_2$ (very nearly) and if R represents this common radius,

$$I = MR^2.$$

Note that the moment of inertia of a cylinder about an axis coinciding with its axis of symmetry does not depend on the length l. Two hollow cylinders of the same inner and outer radii, one of wood and one of brass, but having the same mass M, have equal moments of inertia even though the length of the former is much greater. Moment of inertia depends only on the *radial* distribution of mass, not on its distribution along the axis. Thus Eq. (9-20) holds also for a very short cylinder such as a washer, and Eq. (9-21) for a thin disk.

Example 3 *Uniform sphere of radius R, axis through center.* Divide the sphere into thin disks, as indicated in Fig. 9–11. The radius r of the disk shown is

$$r = \sqrt{R^2 - x^2}.$$

Its volume is

$$dV = \pi r^2 \, dx = \pi (R^2 - x^2) \, dx$$

and its mass is

$$dm = \rho \, dV.$$

Hence from Eq. (9-21) its moment of inertia is

$$dI = \frac{\pi \rho}{2} (R^2 - x^2)^2 \, dx,$$

and for the whole sphere,

$$I = (2) \frac{\pi \rho}{2} \int_0^R (R^2 - x^2)^2 \, dx,$$

since by symmetry the right hemisphere has the same moment of inertia as the left. Carrying out the integration, we obtain

$$I = \frac{8\pi\rho}{15} R^5.$$

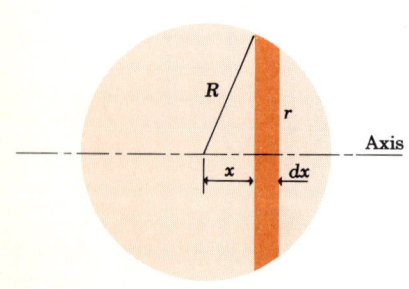

9–11

The mass M of the sphere is

$$M = \rho V = \frac{4\pi \rho R^3}{3}.$$

Hence

$$I = \tfrac{2}{5}MR^2.$$

9-8 Work and power in rotational motion

A force applied to a rotating body does *work* on the body; this work may be expressed in terms of the torque of the force and the angular displacement.

Suppose a force $\boldsymbol{F}$ acts as shown in Fig. 9-12 at the rim of a pivoted wheel of radius R while the wheel rotates through a small angle $d\theta$. If this angle is small enough, the force may be regarded as constant during the correspondingly small time interval. By definition, the work of the force $\boldsymbol{F}$ is

$$dW = F\,ds.$$

But $ds = R\,d\theta$, so that

$$dW = FR\,d\theta.$$

Now FR is the *torque*, Γ, due to the force $\boldsymbol{F}$, so we have finally

$$dW = \Gamma\,d\theta. \tag{9-22}$$

If the torque is constant while the angle changes by a finite amount from θ_1 to θ_2,

$$W = \Gamma(\theta_2 - \theta_1) = \Gamma\,\Delta\theta. \tag{9-23}$$

That is, *the work done by a constant torque equals the product of the torque and the angular displacement.* If Γ is expressed in meters·newton, the work is in joules.

The force in Fig. 9-12 has no component along the *radial* direction. Such a component, if it existed, would do no work, since the displacement of the point of application has no radial component. Similarly, a radial component of force would make no contribution to the torque. Hence Eqs. (9-22) and (9-23) are still correct even when $\boldsymbol{F}$ does have a radial component.

When both sides of Eq. (9-22) are divided by the time interval dt during which the displacement occurs, we obtain

$$\frac{dW}{dt} = \Gamma\left(\frac{d\theta}{dt}\right).$$

But dW/dt is the rate of doing work, or the *power*, and $d\theta/dt$ is the angular velocity. Hence

$$P = \Gamma\omega. \tag{9-24}$$

That is, *the instantaneous power developed by an agent exerting a torque equals the product of the torque and the instantaneous angular velocity.* This is the analog of $P = Fv$ for linear motion.

9-12 A force applied to a rotating body does work on the body.

Example The driveshaft of an automobile rotates at 3600 rpm and transmits 80 hp from the engine to the rear wheels. Compute the torque developed by the engine.

$$\omega = \frac{(3600 \text{ rev} \cdot \text{min}^{-1})(2\pi \text{ rad} \cdot \text{rev}^{-1})}{(60 \text{ s} \cdot \text{min}^{-1})} = 120\,\pi \text{ rad} \cdot \text{s}^{-1},$$

$$80 \text{ hp} = (80 \text{ hp})(746 \text{ W} \cdot \text{hp}^{-1}) = 59,700 \text{ W},$$

$$\Gamma = \frac{P}{\omega} = \frac{59,700 \text{ W}}{120\pi \text{ rad} \cdot \text{s}^{-1}} = 158 \text{ m} \cdot \text{N}.$$

9-9 Torque and angular acceleration

We now consider more generally the *dynamics* of rotation of a rigid body about a fixed axis, that is, the relation between the forces on a rotating body and its motion. This is best expressed in terms of the torque Γ applied to the body and its angular acceleration α. This formulation exhibits a very close parallel to the relation $\Sigma \boldsymbol{F} = m\boldsymbol{a}$ for a point mass.

The relation of Γ to α is obtained most simply from the work-energy theorem of Sec. 6-3. The rate at which the torque does work on the body (that is, its power $\Gamma\omega$) must equal the rate of change of the body's kinetic energy. Thus

$$\Gamma\omega = \frac{d}{dt}(\tfrac{1}{2}I\omega^2) = I\omega\frac{d\omega}{dt} = I\omega\alpha,$$

and

$$\Gamma = I\alpha.$$

(We have used the fact that, for a given axis, I is constant.)

If there are several torques acting on the body, each one does work, and the total rate of change of kinetic energy is the sum of the separate powers of the individual torques, which is ω multiplied by the sum of the torques. Thus, in general,

$$\sum \Gamma = I\alpha. \tag{9-25}$$

This is the fundamental dynamical principle for rotation of a rigid body. It is the rotational analog of $\Sigma \boldsymbol{F} = m\boldsymbol{a}.$

Equation (9-25) can also be derived directly from Newton's second law without direct use of energy considerations. Though not difficult, that derivation is somewhat more involved than the one above, and will not be given here.

Example 1 A rope is wrapped several times around a uniform solid cylinder of radius 0.1 m and mass 50 kg, pivoted so it can rotate about its axis. What is the angular acceleration when the rope is pulled with a force of 20 N?

Solution The torque is $\Gamma = (0.1 \text{ m})(20 \text{ N}) = 2.0 \text{ m} \cdot \text{N}$, and the angular acceleration is

$$\alpha = \frac{\Gamma}{I} = \frac{2.0 \text{ m} \cdot \text{N}}{\tfrac{1}{2}(50 \text{ kg})(0.1 \text{ m})^2} = 8 \text{ rad} \cdot \text{s}^{-2}.$$

Example 2 In Example 3, Sec. 9-6, find the acceleration of mass m and the angular acceleration of the cylinder.

Solution We treat the two bodies separately. Figure 9-13 shows the forces acting on each body. We take the positive direction of motion of mass m to be downward, and the positive sense of rotation for the cylinder to be counterclockwise. Applying Newton's second law to mass m yields the relation

$$mg - T = ma.$$

Applying Eq. (9-25) to the cylinder gives

$$RT = I\alpha = \tfrac{1}{2}MR^2\alpha.$$

Now the acceleration a of mass m must equal the tangential acceleration of a point on the surface of the cylinder, which, according to Eq. (9-10), is given by $a_\parallel = R\alpha$. We use this in the cylinder equation, divide by R, and substitute the resulting expression for T in the equation for m, obtaining

$$mg - \tfrac{1}{2}Ma = ma, \qquad a = \frac{mg}{m + M/2} = \frac{g}{1 + M/2m}.$$

The tension in the rope is *not* equal to the weight mg of mass m; if it were, m could not accelerate. From the above relations,

$$T = mg - ma = \frac{mg}{1 + 2m/M}.$$

When M is much larger than m, the tension is nearly equal to mg, and the acceleration is correspondingly much less than g. When M is zero, $T = 0$ and $a = g$; the mass then falls freely.

If mass m starts from rest at a height h above the floor, its velocity v when it strikes the ground is given by $v^2 - v_0^2 = 2ah$. In this case $v_0 = 0$ and

$$v = \sqrt{2ah} = \sqrt{\frac{2gh}{1 + M/2m}},$$

in agreement with the result obtained from energy considerations in Sec. 9-6.

9-13

It should be noted that the sequence of steps in the solution of this problem is exactly the same as that used in Sec. 4-6 for applications of Newton's laws. The reader should review the discussion at the beginning of Sec. 4-6 and then observe its application to the following example.

Example 3 In Fig. 9-14a, mass m_1 slides without friction on the horizontal surface, the frictionless pulley is in the form of a thin cylindrical shell of mass M and radius R, and the string turns the pulley without slipping. Find the acceleration of each mass, the angular acceleration of the pulley, and the tension in each part of the string.

Solution Figure 9-14b shows free-body diagrams for the three bodies involved, and also shows a choice of positive directions for the various coordinates. We note that the two tensions T_1 and T_2 *cannot* be equal; if they were, the pulley could not have an angular acceleration. Hence, to

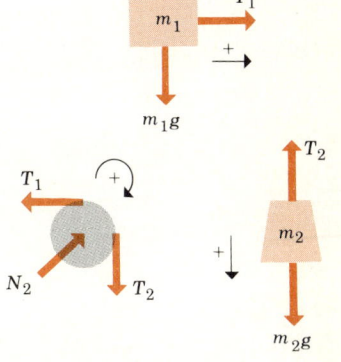

9-14

label the tension in both parts of the string as simply T would be a serious error.

The equations of motion for masses m_1 and m_2 are

$$T_1 = m_1 a_1, \tag{9-26}$$

$$m_2 g - T_2 = m_2 a_1. \tag{9-27}$$

The unknown normal force N_2 acting on the axis of the pulley has no torque with respect to the axis of rotation, and the equation of motion of the pulley is

$$T_2 R - T_1 R = I\alpha = (MR^2)\alpha. \tag{9-28}$$

Assuming the string does not stretch or slip, we have the additional *kinematic* relations

$$a_1 = a_2 = R\alpha. \tag{9-29}$$

(The accelerations of m_1 and m_2 have different directions but the same magnitude.)

Equation (9-29) can be used to eliminate a_2 and α from Eqs. (9-26) through (9-28). The result is the following set of equations for the three unknowns T_1, T_2, and a_1:

$$T_1 = m_1 a_1,$$
$$m_2 g - T_2 = m_2 a_1,$$
$$T_2 - T_1 = M a_1.$$

These may be solved simultaneously; the simplest procedure is simply to add the three equations to eliminate T_1 and T_2, and then to solve for a_1. The result is

$$a_1 = \frac{m_2 g}{m_1 + m_2 + M}.$$

This result may then be substituted back into Eqs. (9-26) and (9-27) to find the tensions. The results are:

$$T_1 = \frac{m_1 m_2 g}{m_1 + m_2 + M},$$

$$T_2 = \frac{(m_1 + M) m_2 g}{m_1 + m_2 + M}.$$

We note that if either m_1 or M is much larger than m_2, the accelerations are very small and T_2 is approximately $m_2 g$, while if m_2 is much larger than either of m_1 or M, the acceleration is approximately g, as should be expected.

*9-10 Rotation about a moving axis

The foregoing analysis of the dynamics of rotational motion of a rigid body can be extended to some cases where the axis of rotation moves, that is, where both translational and rotational motion occur at once. Familiar examples of such motion include a ball rolling down a hill and a yo-yo unwinding at the end of a string. The key to this more general analysis, which we shall not derive in detail, is that Eq. (9-25) remains valid when the axis of rotation moves, provided the axis passes through the center of mass (center of gravity) of the body and does not change its direction.

Analysis of motion that includes both translation and rotation of a body requires two separate equations of motion for the body, one based on $\Sigma F = ma$ for the translational motion, the other based on $\Sigma \Gamma = I\alpha$ for the rotational motion. In addition, there is often a kinematic relation between the two motions, such as a wheel that rolls without slipping or a string that unwinds from a pulley while turning it.

Example 1 A string is wrapped several times around a solid cylinder, and then the end of the string is held stationary while the cylinder is released from rest with no initial motion. Find the acceleration of the cylinder and the tension in the string.

Solution Figure 9–15 shows a free-body diagram for the situation, including the choice of positive coordinate directions. The equation for the translational motion of the center of mass is

$$Mg - T = Ma, \qquad (9\text{-}30)$$

and the equation for rotational motion about the axis through the center of mass is

$$TR = I\alpha = (\tfrac{1}{2}MR^2)\alpha. \qquad (9\text{-}31)$$

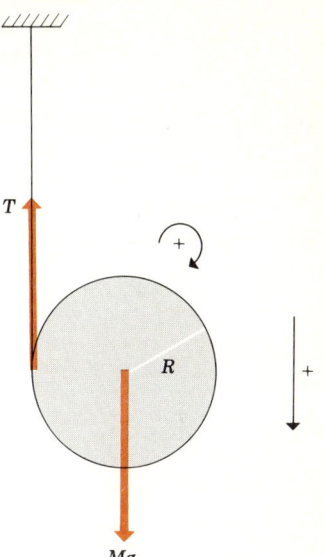

9–15

In addition, if the string unwinds without slipping there is the kinematic relation

$$a = R\alpha. \qquad (9\text{-}32)$$

This may be used to eliminate α from Eq. (9–31), and then Eqs. (9–30) and (9–31) may be solved simultaneously for T and a. The results are

$$a = \tfrac{2}{3}g, \qquad T = \tfrac{1}{3}Mg.$$

Example 2 A solid sphere rolls without slipping down a plane inclined at angle θ to the horizontal. What is the acceleration of its center?

Solution Figure 9–16 shows a free-body diagram, with positive coordinate directions indicated. The equations of motion for translational and rotational motion, respectively, are

$$mg \sin \theta - \mathcal{F} = ma,$$

$$\mathcal{F}R = I\alpha = (\tfrac{2}{5}mR^2)\alpha. \quad \text{Friction causes the torque}$$

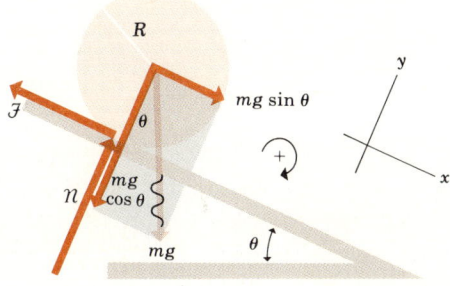

9–16

If the sphere rolls without slipping, then $a = R\alpha$. We use this to eliminate α and then solve for a and $\mathcal{F}$ to obtain

$$a = \tfrac{5}{7} g \sin\theta, \qquad \mathcal{F} = \tfrac{2}{7} mg \sin\theta.$$

We note that the acceleration is just $\tfrac{5}{7}$ as large as it would be if the sphere could slide without friction down the slope. Also, the friction force is essential to prevent slipping and thus to cause the angular acceleration of the sphere. An expression for the minimum coefficient of friction can be obtained by noting that the normal force is $\mathcal{N} = mg \cos\theta$. To prevent slipping, the coefficient of (static) friction must be at least as great as

$$\mu_s = \frac{\mathcal{F}}{\mathcal{N}} = \frac{\tfrac{2}{7} mg \sin\theta}{mg \cos\theta} = \tfrac{2}{7} \tan\theta.$$

If the plane is tilted only slightly, θ is small and only a small value of μ_s is needed to prevent slipping; but as the angle increases the required value of μ_s increases, as might be expected intuitively.

*9-11 Parallel-axis theorem

If the moment of inertia I_{cm} of a body about an axis through its center of mass is known, then the moment of inertia I_P about any other axis parallel to the original one but displaced from it by distance d is easily obtained, by means of a relation called the *parallel-axis theorem,* which states

$$I_P = I_{cm} + Md^2. \tag{9-33}$$

To prove this theorem, we consider the body shown in Fig. 9–17. The origin of coordinates has been chosen to coincide with the center of mass, and we wish to compute the moment of inertia about an axis through point P, perpendicular to the plane of the figure. Let m_i be a typical mass element having coordinates x_i and y_i. Then the moment of inertia about

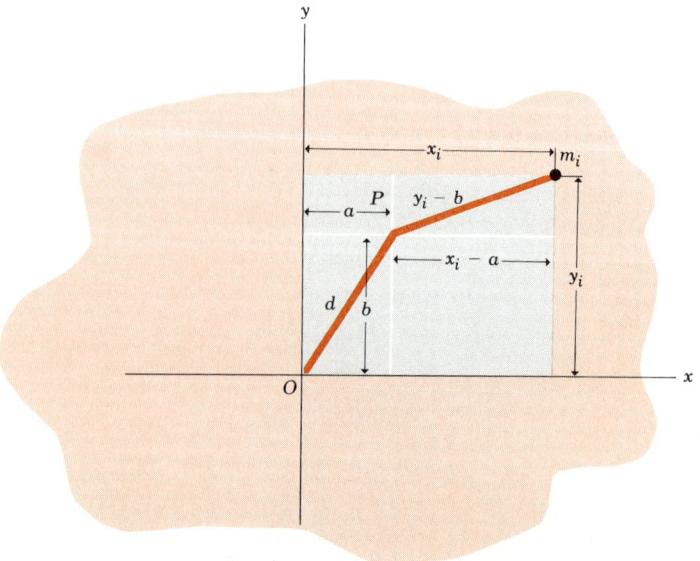

9–17

an axis through O perpendicular to the figure is

$$I_{cm} = \sum m_i(x_i^2 + y_i^2),$$

and the moment of inertia about the axis through P is

$$I_P = \sum m_i \left[(x_i - a)^2 + (y_i - b)^2\right].$$

We expand the squared terms and regroup, obtaining

$$I_P = \sum m_i(x_i^2 + y_i^2) - 2a \sum m_i x_i - 2b \sum m_i y_i + (a^2 + b^2) \sum m_i.$$

The first sum is recognized as I_{cm}; the second and third are zero because they represent the product of the total mass and the x- and y-coordinates of the center of mass, which are zero because we have taken our origin to be the center of mass. The final term is d^2 multiplied by the total mass, so the theorem is proved.

Example 1 The moment of inertia of a thin rod about an axis through its midpoint, perpendicular to the rod, is $I = ML^2/12$ (as shown in Fig. 9-7). Find its moment of inertia about an axis perpendicular to the rod at one end.

Solution We have $I_{cm} = ML^2/12$ and $d = L/2$; Eq. (9-33) yields

$$I_P = \frac{ML^2}{12} + M\left(\frac{L^2}{2}\right) = \frac{ML^2}{3}.$$

Example 2 Find the moment of inertia of a thin uniform disk about an axis perpendicular to its plane at the edge.

Solution From Fig. 9-7 we find $I_{cm} = MR^2/2$, and in this case $d = R$. Thus

$$I_P = \frac{MR^2}{2} + MR^2 = \frac{3 MR^2}{2}.$$

9-12 Angular momentum and angular impulse

Figure 9-18a represents a small body of mass m moving in the plane of the diagram with a speed v and a momentum mv. We define its *angular momentum* about an axis through O perpendicular to the plane of the diagram, as the product of its linear momentum and the perpendicular distance from the axis to its line of motion. Angular momentum will be denoted by L. Thus, for a particle,

$$\text{Angular momentum} = L = mvr. \qquad (9\text{-}34)$$

This definition of angular momentum is analogous to the definition of moment or torque of a force, and indeed angular momentum is sometimes called *moment of momentum*.

Figure 9-18b shows a body of finite size rotating in the plane of the diagram about an axis through O. The velocity v of a small element of the body is related to the angular velocity ω of the body by $v = \omega r$. The angular momentum of the element is therefore $mvr = \omega mr^2$, and the

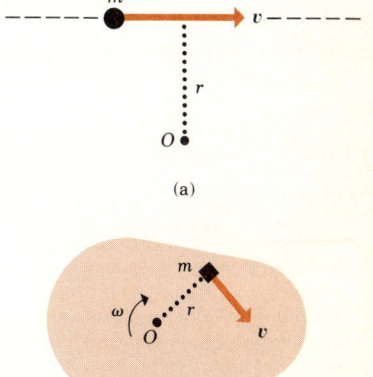

9-18 Angular momentum.

total angular momentum of the body is $\Sigma \omega m r^2 = \omega \Sigma m r^2$. But $\Sigma m r^2$ is the moment of inertia of the body about its axis of rotation. Hence the angular momentum can be written as $I\omega$. Again denoting angular momentum by the symbol L, we have

$$L = I\omega. \tag{9-35}$$

This is analogous to the definition of linear momentum mv for a particle.

When a constant torque Γ acts on a body of moment of inertia I, for a time interval from t_1 to t_2, the angular velocity changes from ω_1 to ω_2, according to the relation

$$\Gamma = I\alpha = I\left(\frac{\omega_2 - \omega_1}{t_2 - t_1}\right)$$

Rearranging this equation, we obtain

$$\Gamma(t_2 - t_1) = I\omega_2 - I\omega_1 = L_2 - L_1. \tag{9-36}$$

The product of the torque and the time interval during which it acts is called the *angular impulse* of the torque; we denote this quantity by J_θ.

$$\text{Angular impulse} = J_\theta = \Gamma(t_2 - t_1). \tag{9-37}$$

This quantity is the rotational analog of the impulse of a force, defined in Sec. 7-1. Equation (9-36) states that *the angular impulse acting on a body equals the change of angular momentum of the body about the same axis*.

To include the possibility that the torque may vary with time, we generalize the definition of angular impulse to

$$J_\theta = \int_{t_1}^{t_2} \Gamma \, dt. \tag{9-38}$$

Thus the general relationship between angular impulse and angular momentum is

$$J_\theta = I\omega_2 - I\omega_1 = L_2 - L_1, \tag{9-39}$$

with J_θ given by Eq. (9-37) if Γ is constant, and by Eq. (9-38) otherwise. A torque that is large but acts only during a short time interval is often called an *impulsive torque*.

The basic dynamic relation for rigid-body rotation, given by Eq. (9-25), may be restated in terms of angular momentum. Taking the time derivative of Eq. (9-35), we find

$$\frac{dL}{dt} = I\frac{d\omega}{dt} = I\alpha.$$

Combining this with Eq. (9-25) yields the relation

$$\sum \Gamma = \frac{dL}{dt}. \tag{9-40}$$

9-13 Conservation of angular momentum

Here is an example of the usefulness of the concepts of angular impulse and angular momentum. Figure 9-19 shows two disks with moments of inertia I and I', rotating with initial constant angular velocities ω_0 and

ω'_0, respectively. The disks are then pushed together by a force directed along the axis, without any torque on either disk. After a short time the disks reach a common final angular velocity ω.

During this time the larger disk exerts a torque Γ' on the smaller, and the smaller exerts a torque Γ on the larger. Both Γ and Γ' vary during the contact, becoming zero after the common final angular velocity is reached. At each instant the two torques are equal in magnitude and opposite in direction, because of Newton's third law. That is, at each instant $\Gamma = -\Gamma'$; the total impulses on the two bodies are thus related by $J_\theta = -J'_\theta$.

To see why this must be so, we consider a point of contact between the two bodies. The two forces which the bodies exert on each other at this common point have equal magnitude and opposite direction, according to Newton's third law. Because they act at the same point, they also have the same moment arm with respect to any axis. The two forces lie along the same line, and the torque of one is the negative of the torque of the other. The same thing is true of the pairs of forces at all the other contact points. Thus at any instant $\Gamma = -\Gamma'$, and the total impulses on the two bodies are related by $J_\theta = -J'_\theta$.

Now according to Eq. (9–39), the impulse on each disk equals the change of angular momentum of that disk:

$$J_\theta = I\omega - I\omega_0,$$
$$J'_\theta = I'\omega - I'\omega'_0.$$

But because $J_\theta = -J'_\theta$, these are equal and opposite:

$$I\omega - I\omega_0 = -(I'\omega - I'\omega'_0)$$

or

$$I\omega_0 + I'\omega'_0 = (I + I')\omega. \qquad (9\text{–}41)$$

9–19 An impulsive torque acts when two rotating disks engage.

The left side of Eq. (9–41) is the total angular momentum of the system (consisting of both disks) before contact, and the right side is the total angular momentum after contact. We have therefore derived the important result that the total angular momentum of the whole system is unaltered. When both disks are regarded as one system, then the torques Γ and Γ' are *internal* torques; during the engaging process no *external* torque acts. *When the resultant external torque on a system is zero, the angular momentum of the system remains constant*; hence any internal interaction between the parts of a system cannot alter its total angular momentum. This is the *principle of conservation of angular momentum,* and it ranks with the principles of conservation of linear momentum and conservation of energy as one of the most fundamental of physical laws.

This principle may also be derived using Eq. (9–40) as a starting point. We first define the total angular momentum of a composite system in the obvious way, as the sum of angular momenta of the parts. The time derivative of angular momentum of each part is given by Eq. (9–40), and the time derivative of the *total* angular momentum is thus equal to the sum of all the torques acting on all the parts of the system. These may be classified as *internal* (exerted by one part of the system on another) or external (exerted by something outside the system). According to the above argument, the internal torques add to zero, assum-

9-20 Conservation of angular momentum.

ing each action–reaction pair of forces acts along a single line. When the sum of the external torques is zero, the rate of change of total angular momentum is zero, and so the total angular momentum is constant.

A circus acrobat, a diver, or a skater performing a pirouette on the toe of one skate, all take advantage of the principle. Suppose an acrobat has just left a swing as in Fig. 9–20, with arms and legs extended and with a counterclockwise angular momentum. When he pulls his arms and legs in, his moment of inertia I becomes much smaller. Since his angular momentum $I\omega$ remains constant and I decreases, his angular velocity ω increases.

Example 1 In the situation of Fig. 9–19, suppose the large disk has mass 2 kg, radius 0.2 m, and initial angular velocity 50 rad·s^{-1}, and the small disk has mass 4 kg, radius 0.1 m, and initial angular velocity 200 rad·s^{-1} (about 1900 rpm). Find the common final angular velocity after the disks are pushed into contact. Is kinetic energy conserved during this process?

Solution The moments of inertia of the two disks are

$$I = \tfrac{1}{2}(2 \text{ kg})(0.2 \text{ m})^2 = 0.04 \text{ kg·m}^2,$$

and

$$I' = \tfrac{1}{2}(4 \text{ kg})(0.1 \text{ m})^2 = 0.02 \text{ kg·m}^2.$$

From conservation of angular momentum, we have

$$(0.04 \text{ kg·m}^2)(50 \text{ rad·s}^{-1}) + (0.02 \text{ kg·m}^2)(200 \text{ rad·s}^{-1})$$
$$= (0.04 \text{ kg·m}^2 + 0.02 \text{ kg·m}^2)\omega,$$
$$\omega = 100 \text{ rad·s}^{-1}.$$

The initial kinetic energy is

$$K_0 = \tfrac{1}{2}(0.04 \text{ kg·m}^2)(50 \text{ rad·s}^{-1})^2 + \tfrac{1}{2}(0.02 \text{ kg·m}^2)(200 \text{ rad·s}^{-1})^2$$
$$= 450 \text{ J}.$$

The final kinetic energy is

$$K = \tfrac{1}{2}(0.04 \text{ kg·m}^2 + 0.02 \text{ kg·m}^2)(100 \text{ rad·s}^{-1})^2 = 300 \text{ J}.$$

One-third of the kinetic energy was lost during this "angular collision," which is the rotational analog of an inelastic collision. We should not expect kinetic energy to be conserved, even though the resultant external force and torque are zero, since nonconservative (frictional) internal forces act during the contact.

Example 2 A man stands at the center of a turntable, holding his arms extended horizontally, with a 5-kg mass in each hand, as in Fig. 9–21. He is set rotating about a vertical axis with an angular velocity of one revolution in 2 s. Find his new angular velocity if he drops his hands to his sides. The moment of inertia of the man may be assumed constant and equal to 6 kg·m^2. The original distance of the weights from the axis is 1 m, and their final distance is 0.2 m.

Solution If friction in the turntable is neglected, no external torques act about a vertical axis and the angular momentum about this axis is

9-21 Conservation of angular momentum about a fixed axis.

constant. That is,
$$I\omega = (I\omega)_0 = I_0\omega_0,$$
where I and ω are the final moment of inertia and angular velocity, and I_0 and ω_0 are the initial values of these quantities:

$$I = I_{\text{man}} + I_{\text{weights}},$$
$$I = 6 \text{ kg} \cdot \text{m}^2 + 2(5 \text{ kg})(0.2 \text{ m})^2 = 6.4 \text{ kg} \cdot \text{m}^2,$$
$$I_0 = 6 \text{ kg} \cdot \text{m}^2 + 2(5 \text{ kg})(1.0 \text{ m})^2 = 16 \text{ kg} \cdot \text{m}^2,$$
$$\omega_0 = 2\pi \left(\frac{1}{2}\right) \text{ rad} \cdot \text{s}^{-1},$$
$$\omega = \omega_0 \left(\frac{I_0}{I}\right) = \pi \frac{16 \text{ kg} \cdot \text{m}^2}{6.4 \text{ kg} \cdot \text{m}^2} = 2.5\pi \text{ rad} \cdot \text{s}^{-1} = 1.25 \text{ rev} \cdot \text{s}^{-1}.$$

That is, the angular velocity is more than doubled.
The initial kinetic energy is
$$K_0 = \tfrac{1}{2}(16 \text{ kg} \cdot \text{m}^2)(\pi \text{ rad} \cdot \text{s}^{-1})^2 = 79 \text{ J}.$$
The final kinetic energy is
$$K = \tfrac{1}{2}(6.4 \text{ kg} \cdot \text{m}^2)(2.5\pi \text{ rad} \cdot \text{s}^{-1})^2 = 197 \text{ J}.$$
Where did the extra energy come from?

Example 3 A door 1.0 m wide, having a mass of 15 kg, is hinged at one side so it can rotate without friction about a vertical axis. A bullet having mass 10 g and speed 400 m·s^{-1} is fired into the door, in a direction perpendicular to the plane of the door, and imbeds itself at the exact center of the door. Find the angular velocity of the door just after the bullet imbeds itself. Is kinetic energy conserved?

Solution There is no external torque about the axis defined by the hinges, so angular momentum about this axis is conserved. The initial angular momentum of the bullet is given by Eq. (9–34):
$$L = mvr = (0.01 \text{ kg})(400 \text{ m} \cdot \text{s}^{-1})(0.5 \text{ m}) = 2.0 \text{ kg} \cdot \text{m}^2 \cdot \text{s}^{-1}.$$
This is equal to the final angular momentum $I\omega$, where I is the total moment of inertia of door and bullet. To find I for the door alone, we note that dimensions parallel to the axis are irrelevant, and that its moment of inertia is the same as that of a rod 1 m long, pivoted at one end. As discussed in Example 1 of Sec. 9–11, this is given by
$$\frac{ML^2}{3} = \frac{(15 \text{ kg})(1.0 \text{ m})^2}{3} = 5.0 \text{ kg} \cdot \text{m}^2.$$
The moment of inertia of the bullet is
$$mr^2 = (0.01 \text{ kg})(0.5 \text{ m})^2 = 0.0025 \text{ kg} \cdot \text{m}^2.$$
Conservation of angular momentum requires that
$$2.0 \text{ kg} \cdot \text{m}^2 \cdot \text{s}^{-1} = (5.0 \text{ kg} \cdot \text{m}^2 + 0.0025 \text{ kg} \cdot \text{m}^2)\omega,$$
$$\omega = 0.4 \text{ rad} \cdot \text{s}^{-1}.$$

The collision of bullet and door is inelastic, so we do not expect energy to be conserved. To check, we calculate initial and final kinetic energies:

$$K_i = \tfrac{1}{2}mv^2 = \tfrac{1}{2}(0.01 \text{ kg})(400 \text{ m}\cdot\text{s}^{-1})^2 = 800 \text{ J};$$
$$K_f = \tfrac{1}{2}I\omega^2 = \tfrac{1}{2}(5.0025 \text{ kg}\cdot\text{m}^2)(0.4 \text{ rad}\cdot\text{s}^{-1})^2 = 0.4 \text{ J}.$$

The final kinetic energy is only 1/2000 of the initial value.

*9-14 Vector representation of angular quantities

A rotational quantity associated with an axis, such as angular velocity, angular momentum, or torque, can be represented by a *vector* along the axis. Thus angular velocity can be defined as a vector quantity $\boldsymbol{\omega}$ with magnitude equal to the number of radians through which the body turns in unit time (our original definition) and direction along the axis of rotation.

A vector can have either of two opposite directions along a given line (corresponding to an arrowhead on either end of the arrow representing the vector); it is customary to define angular velocity as having the direction in which a righthand-thread screw would advance if turned with the body, as shown in Fig. 9-22. Alternatively, when the fingers of the right hand are wrapped around the axis with the fingers pointing in the direction of rotation, the thumb points in the direction of the vector angular velocity, as shown in Fig. 9-22.

These definitions are consistent with the definition of the vector product in Sec. 1-8. As we shall see, the vector product plays an important role in the vector representation of angular quantities. Thus far we have considered only problems in which the axis of rotation has a fixed direction. For motion where the direction of the axis *changes,* the vector nature of angular velocity is an essential feature of the description of motion.

The vector generalization of torque has already been introduced in Sec. 8-5; when a force $\boldsymbol{F}$ is applied at a point whose position vector with respect to the origin is $\boldsymbol{r}$, the torque $\boldsymbol{\Gamma}$ of that force with respect to the origin is

$$\boldsymbol{\Gamma} = \boldsymbol{r} \times \boldsymbol{F}. \tag{9-42}$$

Angular momentum is generalized similarly. A particle of mass m, at a point with position vector $\boldsymbol{r}$ with respect to the origin, moving with velocity $\boldsymbol{v}$ and momentum $\boldsymbol{p} = m\boldsymbol{v}$, has angular momentum $\boldsymbol{L}$ defined as

$$\boldsymbol{L} = \boldsymbol{r} \times \boldsymbol{p} = \boldsymbol{r} \times m\boldsymbol{v}. \tag{9-43}$$

As an example, if a particle of mass m moves counterclockwise in a circle of radius r, in the plane of this page, with speed v, its angular momentum is a vector pointing out of the page, with magnitude mvr, in agreement with Eq. (9-34). This can also be expressed in terms of the angular velocity; since $v = r\omega$, we have

$$L = mvr = m(r\omega)r = mr^2\omega = I\omega.$$

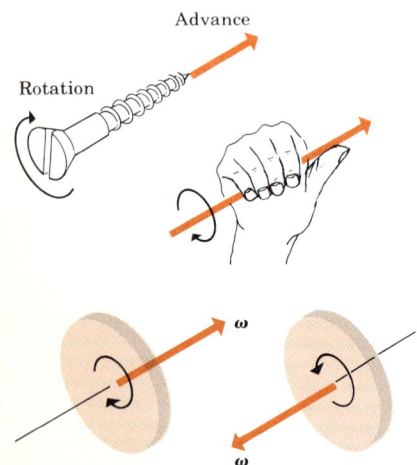

9-22 Vector angular velocity of a rotating body.

We have used the fact that the moment of inertia I of the particle about an axis through the center of the circle is mr^2.

The total angular momentum of a rigid body is the vector sum of the angular momenta of the particles composing it. If the body lies in a single plane, as in Fig. 9-23, and rotates about an axis perpendicular to this plane, then all these vectors lie along the axis. Then, the *vector* sum becomes an *arithmetic* sum and the magnitude of the angular momentum of the plate is

$$L = \sum rmv.$$

But $v = r\omega$, and ω has the same value for all particles. Hence

$$L = \left(\sum mr^2\right)\omega = I\omega,$$

and this definition of angular momentum leads to the same expression for L as that stated in Sec. 9-12.

The basic dynamic principle for rotation about an axis with fixed direction, that is, Eq. (9-25), can be restated in terms of angular momentum:

$$\sum \Gamma = I\alpha = I\frac{d\omega}{dt} = \frac{d}{dt}(I\omega) = \frac{dL}{dt}. \quad (9\text{-}44)$$

The corresponding vector generalization, which we shall not derive, is

$$\sum \boldsymbol{\Gamma} = \frac{d\boldsymbol{L}}{dt}. \quad (9\text{-}45)$$

This form of the relationship includes the possibility that $\boldsymbol{\Gamma}$ and $\boldsymbol{L}$ may have different directions, in which case the direction of the axis of rotation may change.

Here is an application of some of these concepts. Figure 9-24 shows a familiar toy gyroscope. When the flywheel is set spinning by wrapping a

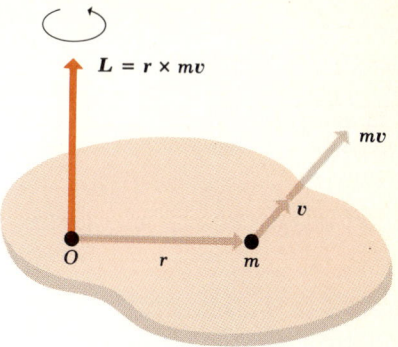

9-23 Vector $\boldsymbol{L}$ represents the moment of momentum, or angular momentum, of a particle of mass m about point O: $\boldsymbol{L} = \boldsymbol{r} \times m\boldsymbol{v}$.

9-24 Vector $\Delta \boldsymbol{L}$ is the change in angular momentum produced in time Δt by the moment $\boldsymbol{\Gamma}$ of the force $\boldsymbol{w}$. Vectors $\Delta \boldsymbol{L}$ and $\boldsymbol{\Gamma}$ are in the same direction.

string around its shaft and pulling, one possible motion when the shaft is supported at only one end is a steady circular motion of the axis in a horizontal plane. This is an interesting phenomenon, inasmuch as intuition suggests the end of the axis opposite O should drop if it is not supported. Vector angular momentum considerations permit an understanding of this behavior.

The forces acting on the gyroscope are its weight w, acting downward at the center of gravity, and the upward force P at the pivot point O. These forces acting on a body at rest would tend to cause a rotation about an axis in the horizontal plane, perpendicular to the gyroscope axis, and the associated torque Γ is in the direction shown. However, the body is *not* at rest, and the effect of this torque is not to *initiate* a rotational motion but to *change* a motion that already exists. At the instant shown the body has angular momentum described by the vector L, and the *change* ΔL in angular momentum in a short time interval is given, from Eq. (9–45), by $\Delta L = \Gamma \, \Delta t$. Thus ΔL has the same direction as Γ. After time Δt the angular momentum is $L + \Delta L$ and, as the vector diagram shows, this means that the gyroscope axis has turned through a small angle $\Delta \phi$ given by $\Delta \phi = |\Delta L|/|L|$. Thus the motion of the axis is consistent with the torque–angular momentum relationship. This motion of the axis is called *precession*.

The rate at which the axis moves, $\Delta \phi / \Delta t$, is called the *precession angular velocity;* denoting this quantity by Ω, we find

$$\Omega = \frac{\Delta \phi}{\Delta t} = \frac{|\Delta L|/|L|}{\Delta t} = \frac{\Gamma}{L}. \tag{9-46}$$

Thus the precession angular velocity is *inversely* proportional to the angular velocity of spin about the axis. A rapidly spinning gyroscope precesses slowly, and as it slows down the precession angular velocity *increases!*

We may now return to the question of why the unsupported end of the axis does not fall. The reason is that the upward force P exerted on the gyroscope by the pivot is just equal in magnitude to its weight w, so that the resultant vertical *force* is zero and the vertical acceleration of the center of gravity is zero. In other words, the vertical component of its linear momentum remains zero, since there is no resultant vertical force. However, the resultant *torque* of these forces is *not* zero and the angular momentum changes.

If the top were not rotating, it would have no angular momentum L initially. Its angular momentum ΔL after a time Δt would be that acquired from the torque acting on it and would have the same direction as the torque. In other words, the top would rotate about an axis through O in the direction of the vector Γ. But if the top is originally rotating, the *change* in its angular momentum produced by the torque adds vectorially to the large angular momentum it already has, and since ΔL is horizontal and perpendicular to L, the result is a motion of precession, with both the angular momentum vector and the axis remaining horizontal.

9-15 Comparison of linear and angular motion

At various points in this chapter we have compared rotational motion of a rigid body to straight-line motion of a particle. There is in fact a very complete parallel between these two classes of motion. Table 9-2 summarizes the pairs of analogous quantities for the two.

Table 9-2 Analogy between translational and rotational motion

Concept	Translation	Rotation
Displacement	x	θ
Velocity	$v = dx/dt$	$\omega = d\theta/dt$
Acceleration	$a = dv/dt$	$\alpha = d\omega/dt$
Force, moment	F	Γ
Equilibrium	$\Sigma F = 0$	$\Sigma \Gamma = 0$
Constant acceleration	$v = v_0 + at$ $x = x_0 + v_0 t + \tfrac{1}{2}at^2$ $v^2 = v_0^2 + 2a(x - x_0)$	$\omega = \omega_0 + \alpha t$ $\theta = \theta_0 + \omega_0 t + \tfrac{1}{2}\alpha t^2$ $\omega^2 = \omega_0^2 + 2\alpha(\theta - \theta_0)$
Mass, moment of inertia	m	I
Newton's second law	$\Sigma F = ma$	$\Sigma \Gamma = I\alpha$
Work	$W = \int F\, dx$	$W = \int \Gamma\, d\theta$
Power	$P = Fv$	$P = \Gamma \omega$
Kinetic energy	$K = \tfrac{1}{2}mv^2$	$K = \tfrac{1}{2}I\omega^2$
Impulse	$\int F\, dt$	$\int \Gamma\, dt$
Momentum	mv	$L = I\omega$

Questions

9-1 What is the difference between tangential and radial acceleration, for a point on a rotating body?

9-2 A flywheel rotates with constant angular velocity. Does a point on its rim have a tangential acceleration? A radial acceleration? Are these accelerations constant? In magnitude? In direction?

9-3 A flywheel rotates with constant angular acceleration. Does a point on its rim have a tangential acceleration? A radial acceleration? Are these accelerations constant? In magnitude? In direction?

9-4 Can a single force applied to a body change both its translational and rotational motion?

9-5 How might you determine experimentally the moment of inertia of an irregularly shaped body?

9-6 Can you think of a body that has the same moment of inertia for all possible axes? For all axes passing through a certain point? What point?

9-7 A cylindrical body has mass M and radius R. Is it ever possible for its moment of inertia to be greater than MR^2?

9-8 In order to maximize the moment of inertia of a flywheel while minimizing its weight, what shape should it have?

9-9 In tightening cylinder-head bolts in an automobile engine, the critical quantity is the *torque* applied to the bolts. Why is this more important than the actual *force* applied to the wrench handle?

9-10 The flywheel of an automobile engine is included to increase the moment of inertia of the engine crankshaft. Why is this desirable?

9-11 A solid ball and a hollow ball with the same mass and radius roll down a slope. Which one reaches the bottom first?

9-12 A solid ball, a solid cylinder, and a hollow cylinder roll down a slope. Which reaches the bottom first? Last? Does it matter whether the radii are the same?

9-13 When an electric motor is turned on, it takes longer to come up to final speed if there is a grinding wheel attached to the shaft. Why?

9-14 Experienced cooks can tell whether an egg is raw or hard-boiled by rolling it down a slope (taking care to catch it at the bottom). How is this possible?

9-15 Consider the idea of an automobile powered by energy stored in a rotating flywheel, which can be "recharged" using an electric motor. What advantages and disadvantages would such a scheme have, compared to more conventional drive mechanisms? Could as much energy be stored as in a tank of gasoline? What factors would limit the maximum energy storage?

9-16 An electric grinder coasts for a minute or more after the power is turned off, while an electric drill coasts for only a few seconds. Why is there a difference?

9-17 Part of the kinetic energy of a moving automobile is in rotational motion of its wheels. When the brakes are applied hard on an icy street, the wheels "lock" and the car starts to slide. What becomes of the rotational kinetic energy?

Problems

9-1

a) What angle in radians is subtended by an arc 3 m in length, on the circumference of a circle whose radius is 2 m?

b) What angle in radians is subtended by an arc of length 78.54 cm on the circumference of a circle of diameter 100 cm? What is this angle in degrees?

c) The angle between two radii of a circle is 0.60 radian. What length of arc is intercepted on the circumference of a circle of radius 200 cm? of radius 200 ft?

9-2 Compute the angular velocity, in rad·s^{-1}, of the crankshaft of an automobile engine that is rotating at 4800 rev·min^{-1}.

9-3

a) A cylinder 6 in. in diameter rotates in a lathe at 750 rev·min^{-1}. What is the tangential velocity of the surface of the cylinder?

b) The proper tangential velocity for machining cast iron is about 2 ft·s^{-1}. At how many rev·min^{-1} should a piece of stock 2 inches in diameter be rotated in a lathe?

9-4 An electric motor running at 1800 rev·min^{-1} has on its shaft three pulleys, of diameters 5, 10, and 15 cm, respectively. Find the linear velocity of the surface of each pulley in cm·s^{-1}. The pulleys may be connected by a belt to a similar set on a countershaft, the 5-cm to the 15-cm, the 10-cm to the 10-cm, or the 15-cm to the 5-cm. Find the three possible angular velocities of the countershaft in rev·min^{-1}.

9-5 A wheel 0.6 m in diameter starts from rest and accelerates with constant angular acceleration to an angular velocity of 100 rad·s^{-1} in 20 s. Find the angular acceleration and the angle turned through. $\frac{\partial \omega}{\partial t} = \alpha \quad \theta = \theta_0 + \omega t + \frac{1}{2}\alpha t^2$

9-6 The angular velocity of a flywheel decreases uniformly from 1000 rev·min^{-1} to 400 rev·min^{-1} in 5 s. Find the angular acceleration and the number of revolutions made by the wheel in the 5-s interval. How many more seconds are required for the wheel to come to rest?

9-7 A flywheel requires 3 s to rotate through 234 radians. Its angular velocity at the end of this time is 108 rad·s^{-1}. Find its constant angular acceleration.

9-8 A flywheel whose angular acceleration is constant and equal to 2 rad·s^{-1} rotates through an angle of 100 radians in 5 s. How long had it been in motion at the beginning of the 5-s interval if it started from rest?

9-9 A record-player turntable starts from rest and turns with constant angular acceleration, reaching its final angular velocity of $33\frac{1}{3}$ rev·min^{-1} after 3 s. Find the components of acceleration of a point at the edge of a 12-inch record 2 s after the turntable is turned on. Also find the magnitude and direction of the resultant acceleration.

9-10 A wheel 1.0 m in diameter is rotating about a fixed axis with an initial angular velocity of 2 rev·s^{-1}. The angular acceleration is 3 rev·s^{-2}.

a) Compute the angular velocity after 6 s.

b) Through what angle has the wheel turned in this time interval?

c) What is the tangential velocity of a point on the rim of the wheel at $t = 6$ s?

d) What is the resultant acceleration of a point on the rim of the wheel at $t = 6$ s?

9-11 A wheel having a diameter of 0.2 m starts from rest and accelerates with constant angular acceleration to an angular velocity of 900 rev·min^{-1} in 5 s.

a) Find the position at the end of 1 s of a point originally at the top of the wheel.

b) Compute and show in a diagram the magnitude and direction of the acceleration of this point at the end of 1 s.

9-12 A flywheel of radius 30 cm starts from rest and accelerates with constant angular acceleration of 0.50 rad·s^{-2}. Compute the tangential acceleration, the radial acceleration, and the resultant acceleration of a point on its rim

a) at the start,

b) after it has turned through 120°,

c) after it has turned through 240°.

9-13 A wheel starts from rest and accelerates with constant angular acceleration to an angular velocity of 900 rev·min^{-1} in 20 s. At the end of 1 s,

a) find the angle through which the wheel has rotated; and

b) compute and show in a diagram the magnitude and direction of the tangential and radial components of acceleration of a point 0.2 m from the axis.

9-14 A wheel rotates with a constant angular velocity of 10 rad·s^{-1}.

a) Compute the radial acceleration of a point 0.5 m from the axis, from the relation $a_\perp = \omega^2 R$.

b) Find the tangential velocity of the point, and compute its radial acceleration from the relation $a_\perp = v^2/R$.

9-15 Find the required angular velocity of an ultracentrifuge, in rpm, in order that the radial acceleration of a point 1 cm from the axis shall equal 300,000g (i.e., 300,000 times the acceleration of gravity).

9-16 An automobile engine is idling at 500 rev·min^{-1}. When the accelerator is depressed, the angular velocity increases to 3000 rev·min^{-1} in 5 s. Assume a constant angular acceleration.

a) What are the initial and final angular velocities, expressed in radians per second?

b) What was the angular acceleration, in radians per second squared?

c) How many revolutions did the engine make during the acceleration period?

d) The flywheel of the engine is 0.5 m in diameter. What is the linear speed of a point at its rim when the angular speed is 3000 rev·min^{-1}?

e) What was the tangential acceleration of the point during the acceleration period?

f) What is the radial acceleration of the point when the angular speed is 3000 rev·min^{-1}?

9-17

a) Prove that when a body starts from rest and rotates about a fixed axis with constant angular acceleration, the radial acceleration of a point in the body is directly proportional to its angular displacement.

b) Through what angle has the body turned at the instant when the resultant acceleration of a point makes an angle of 60° with the radial direction?

9-18 Four small bodies, each of mass 3 kg, are arranged in a square 0.5 m on a side, and connected by light rods. Find the moment of inertia of the system about an axis

a) through the center of the square, perpendicular to its plane;

b) bisecting two opposite sides of the square.

9-19 A thin rectangular sheet of steel is 0.3 m by 0.4 m and has mass 24 kg. Find the moment of inertia about an axis

a) through the center, parallel to the long sides;

b) through the center, parallel to the short sides;

c) through the center, perpendicular to the plane.

Do you notice anything strange about your results?

9-20 A bicycle wheel of radius 0.3 m has a rim of mass 1.0 kg and 50 spokes, each of mass 0.01 kg. What is its moment of inertia about its axis of rotation?

9-21 Find the moment of inertia of a rod 4 cm in diameter and 2 m long, of mass 8 kg,

a) about an axis perpendicular to the rod and passing through its center,

b) about an axis perpendicular to the rod and passing through one end,

c) about a longitudinal axis passing through the center of the rod.

9-22 The four bodies shown in Fig. 9-25 have equal masses m. Body A is a solid cylinder of radius R. Body B is a hollow thin cylinder of radius R. Body C is a solid square with length of side $= 2R$. Body D is the same size as C, but hollow (i.e., made up of four thin sticks). The bodies have axes of rotation perpendicular to the page and through the center of gravity of each body.

a) Which body has the smallest moment of inertia?

b) Which body has the largest moment of inertia?

Figure 9-25

9-23 Small blocks, each of mass m, are clamped at the ends and at the center of a light rigid rod of length L. Compute the moment of inertia of the system about an axis perpendicular to the rod and passing through a point one-quarter of the length from one end. Neglect the moment of inertia of the rod.

9-24 The inner radius of a hollow cylinder is 10 cm, the outer radius is 20 cm, and the length is 30 cm. What is the radius of a thin cylindrical shell of equal mass that would have the same moment of inertia as the hollow cylinder?

9-25 A flywheel consists of a solid disk 0.5 m in diameter and 0.02 m thick, and two projecting hubs 0.1 m in diameter and 0.1 m long. If the material of which it is constructed has a density of 6000 kg·m^{-3}, find its moment of inertia about the axis of rotation.

9-26 A grinding wheel 0.2 m in diameter, of mass 3 kg, is rotating at 3600 rev·min^{-1}.

a) What is its kinetic energy?

b) How far would it have to drop in free fall to acquire the same kinetic energy?

9-27 A solid uniform disk of mass m and radius R is pivoted about a horizontal axis through its center, and a small body of mass m is attached to the rim of the disk. If the disk is released from rest with the small body at the end of a horizontal radius, find the angular velocity when the small body is at the bottom.

9-28 The flywheel of a gasoline engine is required to give up 300 J of kinetic energy while its angular velocity decreases from 600 rev·min^{-1} to 540 rev·min^{-1}. What moment of inertia is required?

9-29 The flywheel of a punch press has a moment of inertia of 25 kg·m^2 and it runs at 300 rev·min^{-1}. The flywheel supplies all the energy needed in a quick punching operation.

a) Find the speed in rev·min^{-1} to which the flywheel will be reduced by a sudden punching operation requiring 4000 J of work.

b) What must be the constant power supply to the flywheel in horsepower to bring it back to its initial speed in 5 s?

9-30 A magazine article described a passenger bus in Zurich, Switzerland, which derived its motive power from the energy stored in a large flywheel. The wheel was brought up to speed periodically, when the bus stopped at a station, by an electric motor, which could then be attached to the electric power lines. The flywheel was a solid cylinder of mass 1000 kg, diameter 1.8 m, and its top speed was 3000 rev·min^{-1}.

a) At this speed, what is the kinetic energy of the flywheel?

b) If the average power required to operate the bus is 25 hp, how long can it operate between stops?

9-31 A grindstone in the form of a solid cylinder has a radius of 0.5 m and a mass of 50 kg.

a) What torque will bring it from rest to an angular velocity of 300 rev·min^{-1} in 10 s?

b) What is its kinetic energy when it is rotating at 300 rev·min^{-1}?

9-32 The flywheel of a motor has a mass of 300 kg and a moment of inertia of 675 kg·min^2. The motor develops a constant torque of 2000 N·m and the flywheel starts from rest.

a) What is the angular acceleration of the flywheel?

b) What will be its angular velocity after making 4 revolutions?

c) How much work is done by the motor during the first 4 revolutions?

9-33 The flywheel of a stationary engine has a moment of inertia of 30 kg·m^2.

a) What constant torque is required to bring it up to an angular velocity of 900 rev·min^{-1} in 10 s, starting from rest?

b) What is its final kinetic energy?

9-34 A grindstone 1.0 m in diameter, of mass 50 kg, is rotating at 900 rev·min^{-1}. A tool is pressed normally against the rim with a force of 200 N, and the grindstone comes to rest in 10 s. Find the coefficient of friction between the tool and the grindstone. Neglect friction in the bearings.

9-35 A 60-kg grindstone is 1 m in diameter and has a moment of inertia of 3.75 kg·m^2. A tool is pressed down on the rim with a normal force of 50 N. The coefficient of sliding friction between the tool and stone is 0.6, and there is a constant friction torque of 5 N·m between the axle of the stone and its bearings.

a) How much force must be applied normally at the end of a crank handle 0.5 m long to bring the stone from rest to 120 rev·min^{-1} in 9 s?

b) After attaining a speed of 120 rev·min^{-1}, what must the normal force at the end of the handle become to maintain a constant speed of 120 rev·min^{-1}?

c) How long will it take the grindstone to come from 120 rev·min^{-1} to rest if it is acted on by the axle friction alone?

9-36 A constant torque of 20 N·m is exerted on a pivoted wheel for 10 s, during which time the angular velocity of the wheel increases from zero to 100 rev·min^{-1}. The external torque is then removed and the wheel is brought to rest by friction in its bearings in 100 s. Compute

a) the moment of inertia of the wheel,

b) the friction torque,

c) the total number of revolutions made by the wheel.

9-37 A cord is wrapped around the rim of a flywheel 0.5 m in radius, and a steady pull of 50 N is exerted on the cord, as in Fig. 9-26a. The wheel is mounted in frictionless bearings on a horizontal shaft through its center. The moment of inertia of the wheel is 4 kg·m^2.

a) Compute the angular acceleration of the wheel.

b) Show that the work done in unwinding 5 m of cord equals the gain in kinetic energy of the wheel.

c) If a mass having a weight of 50 N hangs from the cord, as in Fig. 9-26b, compute the angular acceleration of the wheel. Why is this not the same as in part (a)?

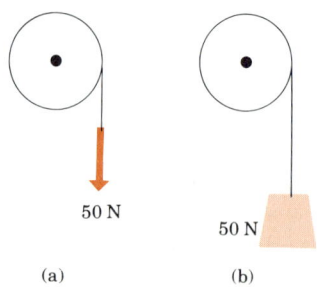

(a)   (b)   **Figure 9-26**

9-38 A solid cylinder of mass 15 kg, 0.3 m in diameter, is pivoted about a horizontal axis through its center, and a rope wrapped around the surface of the cylinder carries at its end a block of mass 8 kg.

a) How far does the block descend in 5 s, starting from rest?
b) What is the tension in the rope?
c) What is the force exerted on the cylinder by its bearings?

9-39 A bucket of water of mass 20 kg is suspended by a rope wrapped around a windlass in the form of a solid cylinder 0.2 m in diameter, also of mass 20 kg. The bucket is released from rest at the top of a well and falls 20 m to the water.

a) What is the tension in the rope while the bucket is falling?
b) With what velocity does the bucket strike the water?
c) What was the time of fall? Neglect the weight of the rope.

9-40 A 5-kg block rests on a horizontal frictionless surface. A cord attached to the block passes over a pulley, whose diameter is 0.2 m, to a hanging block also of mass 5 kg. The system is released from rest, and the blocks are observed to move 4 m in 2 s.

a) What was the moment of inertia of the pulley?
b) What was the tension in each part of the cord?

9-41 Figure 9-27 represents an *Atwood's machine*. Find the linear accelerations of blocks A and B, the angular acceleration of the wheel C, and the tension in each side of the cord (a) if the surface of the wheel is frictionless; (b) if there is no slipping between the cord and the surface of the wheel. Let the masses of blocks A and B be 4 kg and 2 kg, respectively, the moment of inertia of the wheel about its axis be 0.2 kg·m², and the radius of the wheel be 0.1 m.

Figure 9-27

9-42 A flywheel 1.0 m in diameter is pivoted on a horizontal axis. A rope is wrapped around the outside of the flywheel, and a steady pull of 50 N is exerted on the rope. It is found that 10 m of rope are unwound in 4 s.

a) What was the angular acceleration of the flywheel?
b) What is its final angular velocity?
c) What is its final kinetic energy?
d) What is its moment of inertia?

9-43 A block of mass $m = 5$ kg slides down a surface inclined 37° to the horizontal, as shown in Fig. 9-28. The coefficient of sliding friction is 0.25. A string attached to the block is wrapped around a flywheel on a fixed axis at O. The flywheel has a mass $M = 20$ kg, an outer radius $R = 0.2$ m, and a moment of inertia with respect to the axis of 0.2 kg·m².

a) What is the acceleration of the block down the plane?
b) What is the tension in the string?

Figure 9-28

9-44 A body of mass m is attached to a light cord wound around the shaft of a wheel, as in Fig. 9-29. The radius of the shaft is r, and the shaft is supported in fixed, frictionless bearings. When released from rest, the body descends a distance of 1.75 m in 5 s. Find the moment of inertia of the wheel and shaft, in terms of m and r.

Figure 9-29

9-45 A solid cylinder rolls without slipping down a 30° slope. Find the acceleration, the frictional force, and the minimum coefficient of friction needed to prevent slipping.

9-46 A yo-yo is made from two uniform disks of radius a connected by a light axle of radius b. A string is wound several times around the axle and then held stationary while the yo-yo is released from rest, dropping as the string unwinds. Find the acceleration of the yo-yo and the tension in the string.

9-47 A lawn roller in the form of a hollow cylinder of mass M is pulled horizontally with a constant force F applied by a handle attached to the axle. If it rolls without slipping, find the acceleration and the frictional force.

9-48 Find the moment of inertia of a thin rod of mass M and length L, about an axis perpendicular to the rod, a distance $L/4$ from one end.

9-49 Find the moment of inertia of a thin square sheet of side length a and mass M, about an axis perpendicular to the sheet at a corner.

9-50 A uniform thin rod is bent into a square of side length a. If the total mass is M, find the moment of inertia about an axis through the center, perpendicular to the plane of the square.

9-51 A light rigid rod 1.0 m long has a small block of mass 0.05 kg attached at one end. The other end is pivoted, and the rod rotates in a vertical circle. At a certain instant, the rod is 36.9° above the horizontal, and the tangential speed of the block is 4 m·s⁻¹.

a) What are the horizontal and vertical components of the velocity of the block?
b) What is the moment of inertia of the system?
c) What is the radial acceleration of the block?
d) What is the tangential acceleration of the block?
e) What is the tension or compression in the rod?

9-52

a) Compute the torque developed by an airplane engine whose output is 2000 hp at an angular velocity of 2400 rev·min⁻¹.
b) If a drum of negligible mass, 0.5 m in diameter, were attached to the motor shaft and the power output of the motor were used to raise a weight hanging from a rope wrapped around the drum, how large a weight could be lifted?
c) With what velocity would it rise?

9-53 A solid wood door 1.0 m wide and 2.0 m high is hinged along one side and has a total mass of 50 kg. Initially open and at rest, the door is struck with a hammer at its center; during the blow, an average force of 2000 N acts for 0.01 s. Find the angular velocity of the door after the impact.

9-54 The door of Problem 9-53 is struck at its center by a handful of sticky mud, of mass 0.5 kg, traveling 10 m·s⁻¹ just before impact. Find the final angular velocity of the door. Is the moment of inertia of the mud significant?

9-55 A target in a shooting gallery consists of a vertical square wooden board, 0.2 m on a side, of mass 2.0 kg, pivoted on an axis along its top edge. It is struck at the center by a bullet of mass 5 g, traveling 300 m·s⁻¹.

a) What is the angular velocity of the board just after the bullet's impact?
b) What maximum height above the equilibrium position does the center of the board reach before starting to swing down again?
c) What bullet speed would be required for the board to swing all the way over after impact?

9-56 A large wooden turntable of radius 2.0 m and total mass 120 kg is rotating about a vertical axis through its center, with angular velocity 3.0 rad·s⁻¹. A bag of sand of mass 100 kg is dropped vertically onto it, at a point near the outer edge.

a) Find the angular velocity of the turntable after the sandbag is dropped.
b) Compute the kinetic energies before and after the sandbag is dropped. Why are they not equal?

9-57 In Fig. 9-30, the steel balls A and B have a mass of 500 g each, and are rotating about the vertical axis with an angular velocity of 4 rad·s⁻¹ at a distance of 15 cm from the axis. Collar C is now forced down until the balls are at a distance of 5 cm from the axis. How much work must be done to move the collar down?

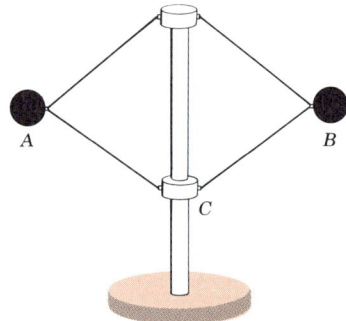

Figure 9-30

9-58 A man sits on a piano stool holding a pair of dumbbells at a distance of 0.6 m from the axis of rotation of the stool. He is given an angular velocity of 5 rad·s⁻¹, after which he pulls the dumbbells in until they are only 0.2 m distant from the axis. The moment of inertia of the man about the axis of rotation is 5 kg·m² and may be considered constant. Each dumbbell has a mass of 5 kg and may be considered a point mass. Neglect friction.

a) What is the initial angular momentum of the system?
b) What is the angular velocity of the system after the dumbbells are pulled in toward the axis?
c) Compute the kinetic energy of the system before and after the dumbbells are pulled in. Account for the difference, if any.

9-59 A block of mass 0.05 kg is attached to a cord passing through a hole in a horizontal frictionless surface, as in Fig. 9-31. The block is originally revolving at a distance of 0.2 m from the hole, with an angular velocity of 3 rad·s⁻¹. The cord is then pulled from below, shortening the radius of the circle in which the block revolves to 0.1 m. The block may be considered a point mass.

a) What is the new angular velocity?
b) Find the change in kinetic energy of the block.

9-60 A small block of mass 4 kg is attached to a cord passing through a hole in a horizontal frictionless surface. The block is originally revolving in a circle of radius 0.5 m about the hole, with a tangential velocity of 4 m·s⁻¹. The

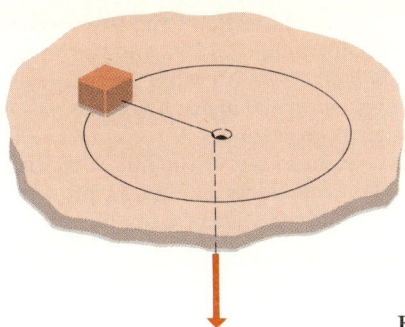

Figure 9-31

cord is then pulled slowly from below, shortening the radius of the circle in which the block revolves. The breaking strength of the cord is 600 N. What will be the radius of the circle when the cord breaks?

9-61 A block of mass M rests on a turntable that is rotating at constant angular velocity ω. A smooth cord runs from the block through a hole in the center of the table down to a hanging block of mass m. The coefficient of friction between the first block and the turntable is μ. (See Fig. 9-32). Find the largest and smallest values of the radius r for which the first block will remain at rest relative to the turntable.

Figure 9-32

9-62 A uniform rod of mass 0.03 kg and 0.2 m long rotates in a horizontal plane about a fixed vertical axis through its center. Two small bodies, each of mass 0.02 kg, are mounted so that they can slide along the rod. They are initially held by catches at positions 0.05 m on each side of the center of the rod, and the system is rotating at 15 rev·min^{-1}. Without otherwise changing the system, the catches are released and the masses slide outward along the rod and fly off at the ends.

a) What is the angular velocity of the system at the instant when the small masses reach the ends of the rod?

b) What is the angular velocity of the rod after the small masses leave it?

9-63 A turntable rotates about a fixed vertical axis, making one revolution in 10 s. The moment of inertia of the turntable about this axis is 1200 kg·m^2. A man of mass 80 kg, initially standing at the center of the turntable, runs out along a radius. What is the angular velocity of the turntable when the man is 2 m from the center?

9-64 Disks A and B are mounted on a shaft SS and may be connected or disconnected by a clutch C, as in Fig. 9-33. The moment of inertia of disk A is one-half that of disk B. With the clutch disconnected, A is brought up to an angular velocity ω_0. The accelerating torque is then removed from A and it is coupled to disk B by the clutch. (Bearing friction may be neglected.) It is found that 2000 J of heat are developed in the clutch when the connection is made. What was the original kinetic energy of disk A?

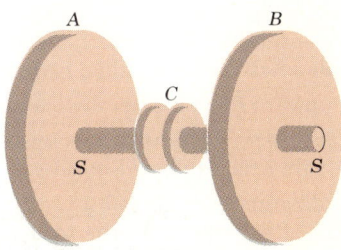

Figure 9-33

9-65 A man of mass 100 kg stands at the rim of a turntable of radius 2 m and moment of inertia 4000 kg·m^2, mounted on a vertical frictionless shaft at its center. The whole system is initially at rest. The man now walks along the outer edge of the turntable with a velocity of 1 m·s^{-1}, relative to the earth.

a) With what angular velocity and in what direction does the turntable rotate?

b) Through what angle will it have rotated when the man reaches his initial position on the turntable?

c) Through what angle will it have rotated when he reaches his initial position relative to the earth?

9-66 A man of mass 60 kg runs around the edge of a horizontal turntable mounted on a vertical frictionless axis through its center. The velocity of the man, relative to the earth, is 1 m·s^{-1}. The turntable is rotating in the opposite direction with an angular velocity of 0.2 rad·s^{-1}. The radius of the turntable is 2 m and its moment of inertia about the axis of rotation is 400 kg·m^2. Find the final angular velocity of the system if the man comes to rest, relative to the turntable.

9-67 Two flywheels, A and B, are mounted on shafts that can be connected or disengaged by a friction clutch C (see Fig. 9-33). The moment of inertia of wheel A is 8 kg·m^2. With the clutch disengaged, wheel A is brought up to an angular velocity of 600 rev·min^{-1}. Wheel B is initially at rest. The clutch is now engaged, accelerating B and decelerating A, until both wheels have the same angular velocity. The final angular velocity of the system is 400 rev·min^{-1}.

a) What was the moment of inertia of wheel B?

b) How much mechanical energy was lost in the process? (Neglect all bearing friction.)

9-68 The stabilizing gyroscope of a ship is a solid disk of mass 50,000 kg, its radius is 2 m, and it rotates about a vertical axis with an angular velocity of 900 rev·min^{-1}.

a) How long a time is required to bring it up to speed, starting from rest, with a constant power input of 100 hp?

b) Find the torque needed to cause the axis to precess in a vertical fore-and-aft plane at the rate of 1 degree·s^{-1}.

9-69 The mass of the rotor of a toy gyroscope is 150 g and its moment of inertia about its axis is 1500 g·cm^2. The mass of the frame is 30 g. The gyroscope is supported on a single pivot, as in Fig. 9-34, with its center of gravity distant 4 cm horizontally from the pivot, and is precessing in a horizontal plane at the rate of one revolution in 6 s.

a) Find the upward force exerted by the pivot.

b) Find the angular velocity with which the rotor is spinning about its axis, expressed in rev·min^{-1}.

c) Copy the diagram, and show by vectors the angular momentum of the rotor and the torque acting on it.

Figure 9-34

9-70 The moment of inertia of the front wheel of a bicycle is 0.4 kg·m^2, its radius is 0.5 m, and the forward speed of the bicycle is 5 m·s^{-1}. With what angular velocity must the front wheel be turned about a vertical axis to counteract the capsizing torque due to a mass of 60 kg, 0.02 m horizontally to the right or left of the line of contact of wheels and ground? (Bicycle riders: compare your own experience and see if your answer seems reasonable.)

9-71 The rotor of a small control gyro can be accelerated from rest to an angular velocity of 50,000 rev·min^{-1} in 0.2 s. The moment of inertia of the rotor about its spin axis is 385 g·cm^2.

a) What is the angular acceleration (assumed constant) as the rotor is brought up to speed?

b) What torque is needed to produce this angular acceleration?

9-72 A demonstration gyro wheel is constructed by removing the tire from a bicycle wheel 1.0 m in diameter, wrapping lead wire around the rim, and taping it in place. The shaft projects 0.2 m at each side of the wheel and a man holds the ends of the shaft in his hands. The mass of the system is 5 kg and its entire mass may be assumed to be located at its rim. The shaft is horizontal and the wheel is spinning about the shaft at 5 rev·s^{-1}. Find the magnitude and direction of the force each hand exerts on the shaft under the following conditions:

a) The shaft is at rest.

b) The shaft is rotating in a horizontal plane about its center at 0.04 rev·s^{-1}.

c) The shaft is rotating in a horizontal plane about its center at 0.20 rev·s^{-1}.

d) At what rate must the shaft rotate in order that it may be supported at one end only?

ELASTICITY 10

10-1 Stress

The preceding chapter dealt with the motion of a rigid body, an idealized model used to represent the motion of a body when rotational motion may not be neglected but when the changes in the *shape* of the body are so small as to be negligible. Real materials always yield to some extent under the influence of applied forces; we shall now develop models to describe the relation between deformation of real materials and the forces responsible for these deformations.

Ultimately, the change in shape or volume of a body when outside forces act on it is determined by the forces between its molecules. Molecular theory is at present not sufficiently advanced to enable us to calculate the elastic properties of, say, a block of copper, starting from the properties of a copper atom. However, the study of the solid state is an active subject in many research laboratories, and our knowledge of it is steadily increasing. In this chapter we shall confine ourselves to quantities that are directly measurable, and shall not attempt detailed *molecular* explanation of the observed behavior.

Figure 10–1a shows a bar of uniform cross-sectional area A subjected to equal and opposite pulls F at its ends. The bar is said to be in *tension*. Consider a section through the bar at right angles to its length, as indicated by the broken line. Since every portion of the bar is in equilibrium, that portion at the right of the section must be pulling on the portion at the left with a force F, and vice versa. If the section is not too near the ends of the bar, these pulls are uniformly distributed over the cross-sectional area A, as indicated by the short arrows in Fig. 10–1b. We define the *stress* S at the section as the ratio of the force F to the area A:

$$\text{Stress} = \frac{F}{A}. \tag{10-1}$$

The stress is called a *tensile* stress, meaning that each portion *pulls* on the other, and it is also a *normal* stress because the distributed force is perpendicular to the area.

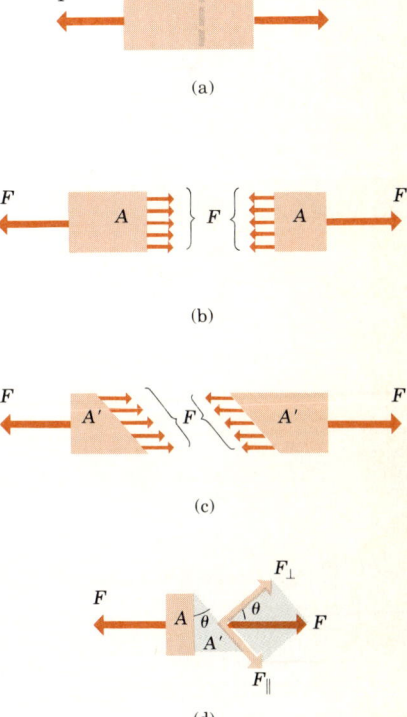

10-1 (a) A bar in tension. (b) The stress at a perpendicular section equals F/A. (c) and (d) The stress at an inclined section can be resolved into a normal stress $F_\perp/A'$, and a tangential or shear stress $F_\parallel \cdot A'$.

The SI unit of stress is the newton per square meter (N·m^{-2}). This unit is also given the special name, the *pascal* (abbreviated Pa).

One pascal = 1 Pa = 1 N·m^{-2}.

Other units of stress are the dyne per square centimeter (dyn·cm^{-2}) and the pound per square foot (lb·ft^{-2}). In the British system the hybrid unit, the pound per square inch (lb·in^{-2} or psi), is commonly used.

Example A human biceps (upper arm muscle) may exert a force of the order of 600 N on the bones to which it is attached. If the muscle has a cross-sectional area at its center of 50 cm^2 = 0.005 m^2, and the tendon attaching its lower end to the bones below the elbow joint has a cross section of 0.5 cm^2 = 5 × 10^{-5} m^2, find the tensile stress in each of these.

Solution In each case the stress is the force per unit area. For the muscle,

$$\text{Tensile stress} = \frac{600 \text{ N}}{0.005 \text{ m}^2} = 120{,}000 \text{ N·m}^{-2} = 1.2 \times 10^5 \text{ Pa}.$$

For the tendon,

$$\text{Tensile stress} = \frac{600 \text{ N}}{5 \times 10^{-5} \text{ m}^2} = 1.2 \times 10^7 \text{ Pa}.$$

The maximum force a muscle can exert depends on its cross-sectional area, but the maximum *stress* is nearly the same for a wide variety of muscles.

Returning to Fig. 10–1, we consider next a section through the bar in some arbitrary direction, as in Fig. 10–1c. The resultant force exerted on the portion at either side of this section, by the portion at the other, is equal and opposite to the force F at the end of the section. Now, however, the force is distributed over a larger area A' and is not at right angles to the area. If we represent the resultant of the distributed forces by a single vector of magnitude F, as in Fig. 10–1d, this vector can be resolved into a component $F_\perp$ normal to the area A', and a component $F_\parallel$ tangent to the area. The *normal* stress is defined, as before, as the ratio of the component $F_\perp$ to the area A'. The ratio of the component $F_\parallel$ to the area A' is called the *tangential* stress or, more commonly, the *shear* stress at the section:

$$\text{Normal stress} = \frac{F_\perp}{A'}$$
$$\text{Tangential (shear) stress} = \frac{F_\parallel}{A'}. \tag{10-2}$$

Stress is not a vector quantity since, unlike a force, we cannot assign to it a specific direction. The *force* acting on the portion of the body on a specified side of a section has a definite direction, but this depends on the orientation of the section. Stress is one of a class of physical quantities called *tensors*.

A bar subjected to *pushes* at its ends, as in Fig. 10–2, is said to be in *compression*. The stress on the broken section, illustrated in part (b), is also a normal stress but is now a *compressive* stress, since each portion

(a)

(b)

10–2 A bar in compression.

pushes on the other. If we again take a section in some arbitrary direction, it is subject to both a tangential (shear) and a normal stress, the latter now being a compression.

As another example of a body under stress, consider the block of square cross section in Fig. 10–3a, acted on by two equal and opposite couples produced by the pairs of forces F_x and F_y distributed over its surfaces. The block is in equilibrium, and any portion of it is in equilibrium also. Thus the distributed forces over the diagonal face in part (b) must have a resultant F whose components are equal to F_x and F_y. The stress at this section is therefore a pure compression, although the stresses at the right face and the bottom are both shear stresses. Similarly, we see from Fig. 10–3c that the other diagonal face is in pure tension.

Consider next a fluid under pressure. The term "fluid" means a substance that can flow; hence the term applies to both liquids and gases. If there is a shearing stress at any point in a fluid, the fluid slips sidewise so long as the stress is maintained. Hence in a fluid at rest, the shear stress is everywhere zero. Figure 10–4 represents a fluid in a cylinder provided with a piston, on which is exerted a downward force. The triangle is a side view of a wedge-shaped portion of the fluid. If for the moment we neglect the weight of the fluid, the only forces on this portion are those exerted by the rest of the fluid, and since these forces can have no shear (or tangential) component, they must be normal to the surfaces of the wedge. Let F_x, F_y, and F represent the forces against the three faces. Since the fluid is in equilibrium, it follows that

$$F \sin \theta = F_x, \qquad F \cos \theta = F_y.$$

Also,

$$A \sin \theta = A_x, \qquad A \cos \theta = A_y.$$

Dividing the upper equations by the lower, we find

$$\frac{F}{A} = \frac{F_x}{A_x} = \frac{F_y}{A_y}.$$

Hence the force per unit area is the *same*, regardless of the direction of the section, and is always a compression. Any one of the preceding ratios defines the *hydrostatic pressure p* in the fluid,

$$p = \frac{F}{A}, \qquad F = pA. \tag{10-3}$$

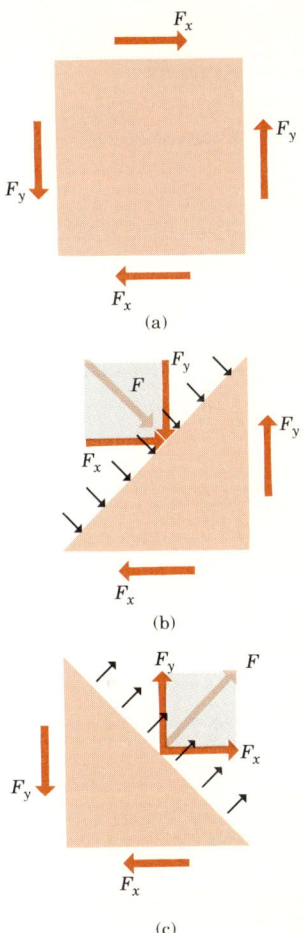

10–3 (a) A body in shear. The stress on one diagonal, part (b), is a pure compression; that on the other, part (c), is a pure tension.

Commonly used units of pressure include 1 Pa ($= 1\ \mathrm{N\cdot m^{-2}}$), 1 dyn·cm^{-2}, 1 lb·ft^{-2}, and 1 lb·in^{-2}. Also in common use is the *atmosphere*, abbreviated atm. One atmosphere is defined to be the average pressure of the earth's atmosphere at sea level.

One atmosphere = 1 atm = 1.013×10^5 Pa = 14.7 lb·in^{-2}.

Like other types of stress, pressure is not a vector quantity; no direction can be assigned to it. *The force against any area within (or bounding) a fluid at rest and under pressure is normal to the area, regardless of the orientation of the area.* This is what is meant by the common statement that "the pressure in a fluid is the same in all directions."

The stress within a solid can also be a hydrostatic pressure, provided the stress at all points of the surface of the solid is of this nature. That is,

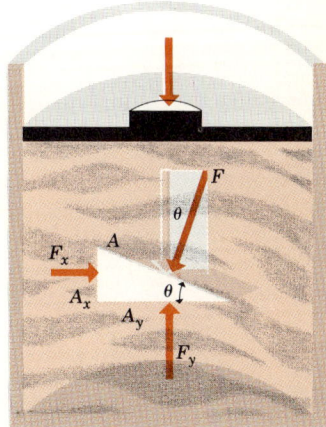

10–4 A fluid under hydrostatic pressure. The force on a surface in any direction is normal to the surface.

10-5 The longitudinal strain is defined as $\Delta l/l_0$.

the force per unit area must be the same at *all* points of the surface, and the force must be normal to the surface and directed inward. This is *not* the case in Fig. 10-2, where forces are applied at the ends of the bar only, but it is automatically the case if a solid is immersed in a fluid under pressure.

10-2 Strain

The term *strain* refers to the relative change in dimensions or shape of a body that is subjected to stress. Associated with each type of stress described in the preceding section is a corresponding type of strain.

Figure 10-5 shows a bar whose natural length is l_0 and which elongates to a length $l = l_0 + \Delta l$ when equal and opposite pulls F are exerted at its ends. The elongation, Δl, of course, does not occur at the ends only; every element of the bar stretches in the same proportion as does the bar as a whole. The *tensile strain* in the bar is defined as the ratio of the increase in length to the original length:

$$\text{Tensile strain} = \frac{l - l_0}{l_0} = \frac{\Delta l}{l_0}. \tag{10-4}$$

The *compressive strain* of a bar in compression is defined in the same way, as the ratio of the decrease in length to the original length.

Figure 10-6a illustrates the nature of the deformation when *shear* stresses act on the faces of a block, as in Fig. 10-3. The dotted outline *abcd* represents the unstressed block, and the solid lines $a'b'c'd'$ represent the block under stress. The centers of the stressed and unstressed block coincide in part (a). In part (b), the deformation is the same as in (a), but the edges *ad* and $a'd'$ coincide. The lengths of the faces under shear remain very nearly constant; all dimensions parallel to the diagonal *ac* increase in length, and those parallel to the diagonal *bd* decrease in length. Note that this is to be expected in view of the nature of the corresponding internal stresses (see Fig. 10-3). This type of strain is called a *shear strain,* and is defined as the ratio of the displacement x of corner b to the transverse dimension h:

$$\text{Shear strain} = \frac{x}{h} = \tan \phi. \tag{10-5}$$

In practice x is nearly always much smaller than h, so $\tan \phi$ is very nearly equal to ϕ, and the strain is simply the angle ϕ (measured in radians, of course). Like all strains, shear strain is a pure number with no units because it is a ratio of two lengths.

The strain produced by a hydrostatic pressure, called a *volume strain,* is defined as the ratio of the change in volume, ΔV, to the original volume V. It also is a pure number:

$$\text{Volume strain} = \frac{\Delta V}{V}. \tag{10-6}$$

10-3 Elasticity and plasticity

The relation between each of the three kinds of stress and its corresponding strain plays an important role in the branch of physics called the *theory of elasticity,* or its engineering counterpart, *strength of mate-*

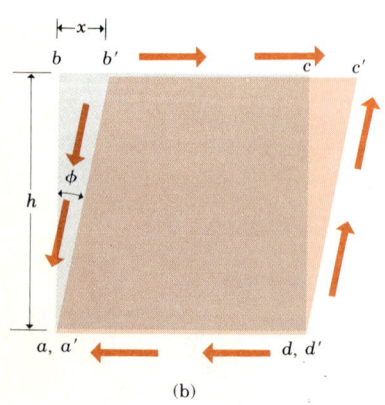

10-6 Change in shape of a block in shear. The shear strain is defined as x/h.

rials. When any stress is plotted against the appropriate strain, the resulting stress–strain diagram is found to have one of several different shapes, depending on the kind of material.

A typical stress–strain diagram for a ductile metal is shown in Fig. 10-7. The stress is a simple tensile stress and the strain is the percentage elongation. During the first portion of the curve (up to a strain of less than 1 percent), the stress and strain are proportional until the point *a*, the *proportional limit,* is reached. The fact that there is a region in which stress and strain are proportional is called *Hooke's Law.* (Robert Hooke (1635–1703) was a contemporary of Newton.)

From *a* to *b*, stress and strain are not proportional, but nevertheless, if the load is removed at any point between *O* and *b*, the curve will be retraced and the material will return to its original length. In the region *Ob*, the material is said to be *elastic* or to exhibit *elastic behavior,* and the point *b* is called the *elastic limit,* or the *yield point.* Up to this point, the forces exerted by the material are *conservative;* when the material returns to its original shape, work done in producing the deformation is recovered. The deformation is said to be *reversible.*

If the material is loaded further, the strain increases rapidly, but when the load is removed at some point beyond *b*, say *c*, the material does not come back to its original length but traverses the thin line in Fig. 10-7. The length at zero stress is now *greater* than the original length, and the material is said to have a *permanent set.* Further increase of load beyond *c* produces a large increase in strain (even if the stress decreases) until a point *d* is reached at which *fracture* takes place. From *b* to *d*, the metal is said to undergo *plastic flow* or *plastic deformation.* A plastic deformation is *irreversible.* If large plastic deformation takes place between the elastic limit and the fracture point, the metal is said to be *ductile.* If, however, fracture occurs soon after the elastic limit is passed, the metal is said to be *brittle.*

Figure 10-8 shows a stress–strain curve for a typical sample of vulcanized rubber that has been stretched to over seven times its original length. During *no* portion of this curve is the stress proportional to the strain! The substance, however, is elastic, in the sense that, when the load is removed, the rubber is restored to its original length. On decreasing the load, the stress–strain curve is *not* retraced but follows the thin curve of Fig. 10-8.

The lack of coincidence of the curves for increasing and decreasing stress is known as *elastic hysteresis* (accent on the third syllable). (An analogous phenomenon observed with magnetic materials is called *magnetic hysteresis.*) When the stress–strain relation exhibits this behavior, the associated forces are *not* conservative, since the work done by the material in returning to its original shape is *less* than the work required to deform it. It can be shown that the area bounded by the two curves, that is, the area of the *hysteresis loop,* is proportional to the energy (per unit volume) dissipated within the elastic or magnetic material.

The large elastic hysteresis of some types of rubber makes these materials very valuable as vibration absorbers. If a block of such material is placed between a piece of vibrating machinery and the floor, elastic hysteresis takes place during every cycle of vibration. Mechanical energy is converted to a form known as internal energy, which evidences itself by a rise in temperature. As a result, only a small amount of energy of vibration is transmitted to the floor.

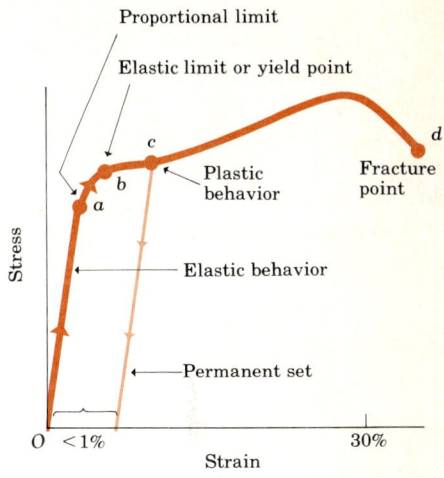

10-7 Typical stress–strain diagram for a ductile metal under tension.

10-8 Typical stress–strain diagram for vulcanized rubber, showing elastic hysteresis.

The stress required to cause actual fracture of a material is called the *breaking stress* or the *ultimate strength*. Two materials, such as two steels, may have very similar elastic constants but vastly different breaking stresses.

10-4 Elastic modulus

The stress required to produce a given strain depends on the nature of the material under stress. The ratio of stress to strain, or the *stress per unit strain*, is called an *elastic modulus* of the material. The larger the elastic modulus, the greater the stress needed for a given strain.

Consider first longitudinal (tensile or compressive) stresses and strains. Experiment shows that up to the proportional limit, *a given longitudinal stress produces a strain of the same magnitude whether the stress is a tension or a compression.* Hence, the ratio of tensile stress to tensile strain, for a given material, equals the ratio of compressive stress to compressive strain. This ratio is called *Young's modulus* for the material, and will be denoted by Y:

$$Y = \frac{\text{tensile stress}}{\text{tensile strain}} = \frac{\text{compressive stress}}{\text{compressive strain}}$$

$$= \frac{F_\perp/A}{\Delta l/l_0} = \frac{l_0}{A}\frac{F_\perp}{\Delta l}. \tag{10-7}$$

If the proportional limit is not exceeded, the ratio of stress to strain is constant, and Hooke's law is therefore equivalent to the statement that *within the proportional limit, the elastic modulus of a given material is constant,* depending only on the nature of the material.

Since a strain is a pure number, the units of Young's modulus are the same as those of stress, namely, force per unit area. Some typical values are listed in Table 10-1.

When a material elongates under tensile stress, the dimensions *perpendicular* to the direction of stress become *shorter* by an amount proportional to the fractional change in length. If w_0 is the original width and Δw the change in width, then it is found that

$$\frac{\Delta w}{w_0} = -\sigma \frac{\Delta l}{l_0}, \tag{10-8}$$

Table 10-1 Approximate elastic constants

Material	Young's modulus, Y		Shear modulus, S		Bulk modulus, B		Poisson's ratio, σ
	10^{11} Pa	10^6 lb·in^{-2}	10^{11} Pa	10^6 lb·in^{-2}	10^{11} Pa	10^6 lb·in^{-2}	
Aluminum	0.70	10	0.30	3.4	0.70	10	0.16
Brass	0.91	13	0.36	5.1	0.61	8.5	0.26
Copper	1.1	16	0.42	6.0	1.4	20	0.32
Glass	0.55	7.8	0.23	3.3	0.37	5.2	0.19
Iron	1.9	26	0.70	10	1.0	14	0.27
Lead	0.16	2.3	0.056	0.8	0.077	1.1	0.43
Nickel	2.1	30	0.77	11	2.6	34	0.36
Steel	2.0	29	0.84	12	1.6	23	0.19
Tungsten	3.6	51	1.5	21	2.0	29	0.20

where σ is a dimensionless constant characteristic of the material, called *Poisson's ratio*. For many common materials σ has a value between 0.1 and 0.3. Similarly, a material under compressive stress "bulges" at the sides, and again the fractional change in width is given by Eq. (10-8).

The *shear modulus S* of a material, within the Hooke's law region, is defined as the ratio of the shear stress to the shear strain it produces:

$$S = \frac{\text{shear stress}}{\text{shear strain}} \qquad (10\text{-}9)$$

$$= \frac{F_\parallel/A}{x/h} = \frac{h}{A}\frac{F_\parallel}{x} = \frac{F_\parallel/A}{\phi}.$$

(Refer to Fig. 10-6 for the meaning of x and of h.)

The shear modulus of a material is also expressed as force per unit area. For most materials it is one-half to one-third as great as Young's modulus. The shear modulus is also called the *modulus of rigidity* or the *torsion modulus*.

The shear modulus has a significance for *solid* materials only. A liquid or gas flows under the influence of a shearing stress, and cannot *permanently* support such a stress.

The modulus relating a hydrostatic pressure to the volume strain it produces is called the *bulk modulus,* and we shall represent it by B. The general definition of bulk modulus is the (negative) ratio of a change in pressure dp to the volume strain dV/V_0 (fractional change in volume) produced by it:

$$B = -\frac{dp}{dV/V_0} = -V\frac{dp}{dV}. \qquad (10\text{-}10)$$

The minus sign is included in the definition of B because an *increase* of pressure always causes a *decrease* in volume. That is, if dp is positive, dV is negative. By including a minus sign in its definition, we make the bulk modulus itself a positive quantity.

The change in volume of a *solid* or *liquid* under pressure is so small that the value of B is independent of the initial volume V_0. Provided the pressure is not too great, the ratio dp/dV is constant also, the bulk modulus is constant, and we can replace dp and dV by finite changes Δp and ΔV in pressure and volume. The volume of a *gas,* however, changes markedly with pressure, and the general definition of B must be used for gases.

The reciprocal of the bulk modulus is called the *compressibility k*. From its definition,

$$k = \frac{1}{B} = -\frac{dV/V_0}{dp} = -\frac{1}{V_0}\frac{dV}{dp}. \qquad (10\text{-}11)$$

The compressibility of a material thus equals the *fractional decrease in volume,* $-dV/V$, per unit increase dp in pressure.

The units of bulk modulus are the same as those of pressure, and the units of compressibility are those of *reciprocal pressure*. Thus the statement that the compressibility of water (see Table 10-2) is 46.4×10^{-6} atm^{-1} means that the volume decreases by 46.4 one-millionths of the original volume for each atmosphere increase in pressure.

The various types of stress, strain, and elastic modulus are summarized in Table 10-3.

Table 10-2 Compressibilities of liquids at 20°C

	Compressibility, $k = \frac{1}{B}$		
	10^{-11} Pa^{-1}	10^{-7} (lb·in^{-2})$^{-1}$	10^{-6} atm^{-1}
Carbon disulfide	93	64	94
Ethyl alcohol	110	76	111
Glycerine	21	14	21
Mercury	3.7	2.6	3.8
Water	45.8	31.6	46.4

Table 10-3 Stresses and strains

Type of stress	Stress	Strain	Elastic modulus	Name of modulus
Tension or compression	$\dfrac{F_\perp}{A}$	$\dfrac{\Delta l}{l_0}$	$Y = \dfrac{F_\perp/A}{\Delta l/l_0}$	Young's modulus
Shear	$\dfrac{F_\parallel}{A}$	$\tan\phi \approx \phi$	$S = \dfrac{F_\parallel/A}{\phi}$	Shear modulus
Hydrostatic pressure	$p\left(=\dfrac{F_\perp}{A}\right)$	$\dfrac{\Delta V}{V_0}$	$B = -V_0\dfrac{dp}{dV}$	Bulk modulus

Example 1 In an experiment to measure Young's modulus, a load of 500 kg, hanging from a steel wire of length 3 m and cross section 0.20 cm², was found to stretch the wire 0.4 cm above its no-load length. What were the stress, the strain, and the value of Young's modulus for the steel of which the wire was composed?

Solution

$$\text{Stress} = \frac{F_\perp}{A} = \frac{(500 \text{ kg})(9.8 \text{ m·s}^{-2})}{2.0 \times 10^{-5} \text{ m}^2}$$
$$= 2.45 \times 10^8 \text{ Pa};$$

$$\text{Strain} = \frac{\Delta l}{l_0} = \frac{0.004 \text{ m}}{3 \text{ m}} = 0.00133;$$

$$Y = \frac{\text{stress}}{\text{strain}} = \frac{2.45 \times 10^8 \text{ Pa}}{0.00133} = 1.84 \times 10^{11} \text{ Pa}.$$

Example 2 Suppose the object in Fig. 10-6 is a brass plate 1.0 m square and 0.5 cm thick. How large a force F must be exerted on each of its edges if the displacement x in Fig. 10-6b is 0.02 cm? The shear modulus of brass is 0.36×10^{11} Pa.

Solution The shear stress on each edge is

$$\text{Shear stress} = \frac{F_\parallel}{A} = \frac{F}{(1.0 \text{ m})(0.005 \text{ m})} = (200 \text{ m}^{-2})F.$$

The shear strain is

$$\text{Shear strain} = \frac{x}{h} = \frac{2 \times 10^{-4}}{1.0 \text{ m}} = 2.0 \times 10^{-4}.$$

The shear modulus S is

$$S = \frac{\text{stress}}{\text{strain}} = 0.36 \times 10^{11} \text{ Pa} = \frac{(200 \text{ m}^{-2})F}{2.0 \times 10^{-4}},$$

$$F = 3.6 \times 10^4 \text{ N}.$$

Example 3 The volume of oil contained in a certain hydraulic press is 0.2 m^3. Find the decrease in volume of the oil when subjected to a pressure increase of 2.04×10^7 Pa. The compressibility of the oil is $20 \times 10^{-6} \text{ atm}^{-1}$.

Solution We first convert the pressure to atmospheres;

$$2.04 \times 10^7 \text{ Pa} = 201 \text{ atm}.$$

From Eq. (10-11),

$$\Delta V = -kV_0 \Delta p = -(20 \times 10^{-6} \text{ atm}^{-1})(0.2 \text{ m}^3)(201 \text{ atm})$$
$$= -8.04 \times 10^{-4} \text{ m}^3.$$

Although the three elastic moduli and Poisson's ratio have been discussed separately, they are not completely independent. For materials having no distinction between various directions (i.e., *isotropic* materials), only two of these are really independent. For example, the bulk and shear moduli may be expressed in terms of Young's modulus and Poisson's ratio:

$$B = \frac{Y}{3(1 - 2\sigma)}, \qquad S = \frac{Y}{2(1 + \sigma)}. \qquad (10\text{-}12)$$

For materials having directional properties, such as wood (which has a grain direction) and single crystals of materials, these relations do not hold, and the elastic behavior is more complex.

10-5 The force constant

The various elastic moduli are quantities that describe the elastic properties of a particular *material* and do not directly indicate how much a given rod, cable, or spring constructed of the material will distort under load. When Eq. (10-7) is solved for $F_\perp$, one obtains

$$F_\perp = \frac{YA}{l_0} \Delta l;$$

or, if YA/l_0 is replaced by a single constant k and the elongation Δl is represented by x, then

$$F_\perp = kx. \qquad (10\text{-}13)$$

In other words, the elongation of a body in tension beyond its no-load length is directly proportional to the stretching force. Hooke's law was originally stated in this form, rather than in terms of stress and strain.

When a helical wire spring is stretched, the stress in the wire is practically pure *shear*. The elongation of the spring as a whole is directly proportional to the stretching force. That is, an equation of the form $F = kx$ still applies, the constant k depending on the shear modulus of the wire, its radius, the radius of the coils, and the number of coils.

The constant k, or the ratio of the force to the elongation, is called the *force constant* or the *stiffness* of the spring, and is expressed in newtons per meter or pounds per foot. It is equal numerically to the force required to produce unit elongation.

The ratio of elongation to the force, or the elongation per unit force, is called the *compliance* of the spring. The compliance is the reciprocal of the force constant and is expressed in meters per newton or feet per pound. It is numerically equal to the elongation produced by unit force.

Questions

10-1 When a wire is bent back and forth, it becomes hot. Why?

10-2 When a wire is stretched, the cross section decreases somewhat from the undeformed value. How does this affect the definition of tensile stress? Should the original or the decreased value be used?

10-3 Is the work required to stretch a metal rod proportional to the amount of stretch? Explain.

10-4 Why is concrete with steel reinforcing rods imbedded in it stronger than plain concrete?

10-5 Is the shear modulus of steel greater or less than that of jelly? By roughly what factor?

10-6 Is the bulk modulus for steel greater or less than that of air? By roughly what factor?

10-7 Looking at the molecular structure of matter, discuss why gases are generally more compressible than liquids and solids.

10-8 Climbing ropes used by mountaineers are usually made of nylon. Would a steel cable of equal strength be just as good? What advantages and disadvantages would it have, compared to nylon?

10-9 How could you measure the force constants of the springs in an automobile?

10-10 In a nylon mountaineering rope, is a lot of mechanical hysteresis desirable or undesirable?

10-11 A spring scale for measuring weight has a spring and a scale that indicates how much the spring stretches under a given weight. Such scales are often illegal in commerce. (Cf. the inscription "no springs, honest weight" found on some grocery-store scales.) Why? (What property of a metal spring could affect its accuracy?)

10-12 Coil springs found in automobile suspension systems are sometimes designed *not* to obey Hooke's law. How can a spring be made so as to achieve this result? Why is it desirable?

10-13 Compare the mechanical properties of a steel cable, made by twisting many thin wires together, with those of a solid steel wire of the same diameter. What advantages does each have?

10-14 Electric power lines are sometimes made using wires with steel core and copper jacket, or strands of copper and steel twisted together. Why?

10-15 A spring is compressed, clamped in its compressed position, and then dissolved in acid. What becomes of the elastic potential energy?

10-16 When rubber mounting blocks are used to absorb machine vibrations through mechanical hysteresis, as discussed in Sec. 10-3, what becomes of the energy associated with the vibrations?

Problems

10-1 A steel rod 4 m long and 0.5 cm² in cross section is found to stretch 0.2 cm under a tension of 12,000 N. What is Young's modulus for this steel?

10-2 The elastic limit of a steel elevator cable is 40,000 lb·in^{-2}. Find the maximum upward acceleration that can be given a 2-ton elevator when supported by a cable whose cross section is $\frac{1}{2}$ in², if the stress is not to exceed $\frac{1}{4}$ of the elastic limit.

10-3 A nylon rope used by mountaineers elongates 1.5 m under the weight of an 80-kg climber. If the rope is 50 m in length and 9 mm in diameter, what is Young's modulus for this material? If Poisson's ratio for nylon is 0.2, find the change in diameter under this stress.

10-4 A relaxed biceps muscle requires a force of 25 N for an elongation of 5 cm, and the same muscle under maximum tension requires a force of 500 N for the same elongation. Regarding the muscle as a uniform cylinder of length 0.2 m and cross-sectional area 50 cm², find Young's modulus for the muscle tissue under each of these conditions.

10-5 A copper wire 4 m long and 1.0 mm in diameter was given the test below. A load of 20 N was originally hung from the wire to keep it taut. The position of the lower end of the wire was read on a scale.

Added load, N	Scale reading, cm
0	3.02
10	3.07
20	3.12
30	3.17
40	3.22
50	3.27
60	3.32
70	4.27

a) Make a graph of these values, plotting the increase in length horizontally and the added load vertically.
b) Calculate the value of Young's modulus.
c) What was the stress at the proportional limit?

10-6 A steel wire has the following properties:

Length = 5 m
Cross section = 0.05 cm²
Young's modulus = 1.8×10^{11} Pa
Shear modulus = 0.6×10^{11} Pa
Proportional limit = 3.6×10^8 Pa
Breaking stress = 7.2×10^8 Pa

The wire is fastened at its upper end and hangs vertically.

a) How great a load can be supported without exceeding the proportional limit?
b) How much will the wire stretch under this load?
c) What is the maximum load that can be supported?

10-7

a) What is the maximum load that can be supported by an aluminum wire 0.1 cm in diameter without exceeding the proportional limit of 8×10^7 Pa?
b) If the wire was originally 5 m long, how much will it elongate under this load?
c) How much does the diameter change under this load?

10-8 A 5-kg mass hangs on a vertical steel wire 0.5 m long and 0.004 cm² in cross section. Hanging from the bottom of this mass is a similar steel wire which supports a 10-kg mass. Compute

a) the longitudinal strain, and
b) the elongation of each wire.

10-9 A 15-kg mass, fastened to the end of a steel wire of unstretched length 0.5 m, is whirled in a vertical circle with an angular velocity of 2 rev·s⁻¹ at the bottom of the circle. The cross section of the wire is 0.02 cm². Calculate the elongation of the wire when the weight is at the lowest point of its path.

10-10 A copper wire 8 m long and a steel wire 4 m long, each of cross section 0.5 cm², are fastened end to end and stretched with a tension of 500 N.

a) What is the change in length of each wire?
b) What is the elastic potential energy of the system?

10-11 A copper rod of length 2 m and cross-sectional area 2.0 cm² is fastened end to end to a steel rod of length L and cross-sectional area 1.0 cm². The compound rod is subjected to equal and opposite pulls of magnitude 3×10^4 N at its ends.

a) Find the length L of the steel rod if the elongations of the two rods are equal.
b) What is the stress in each rod?
c) What is the strain in each rod?

10-12 A rod 1.05 m long, whose weight is negligible, is supported at its ends by wires A and B of equal length, as shown in Fig. 10-9. The cross section of A is 1 mm², that of B is 2 mm². Young's modulus for wire A is 2.4×10^{11} Pa, and for B it is 1.6×10^{11} Pa. At what point along the bar should a weight w be suspended in order to produce (a) equal stresses in A and B, (b) equal strains in A and B?

Figure 10-9

10-13 A bar of length L, cross-sectional area A, and Young's modulus Y, is subjected to a tension F. Represent the stress in the bar by Q and the strain by P. Derive the expression for the elastic potential energy, per unit volume, of the bar in terms of Q and P.

10-14 A certain elevator cable is to have a maximum stress of 10,000 lb·in⁻², to allow for appropriate safety factors. It if is to support a loaded elevator of total weight 4000 lb that has a maximum upward acceleration of 5 ft·s⁻², what should be the diameter of the cable?

10-15 A steel bar 0.2 cm square and 5 m long is stretched with a force of 400 N at each end. Find the stress, the strain, the total elongation, and the fractional change in thickness of the bar.

10-16 Two round rods, one of steel, the other of brass, are joined end to end. Each rod is 0.5 m long and 2 cm in diameter. The combination is subjected to tensile forces of 5000 N.

a) What is the strain in each rod?
b) What is the elongation of each rod?
c) What is the change in diameter of each rod?

10-17 A specimen of oil having an initial volume of 1000 cm³ is subjected to a pressure increase of 12×10^5 Pa, and the volume is found to decrease by 0.3 cm³. What is the bulk modulus for the material? The compressibility?

10-18 The compressibility of sodium is to be measured by observing the displacement of the piston in Fig. 10-4 when a force is applied. The sodium is immersed in an oil that fills the cylinder below the piston. Assume that the piston and walls of the cylinder are perfectly rigid, that there is no friction, and no oil leak. Compute the compressibility of the sodium in terms of the applied force F, the piston displacement x, the piston area A, the initial volume of the oil V_0, the initial volume of the sodium v_0, and the compressibility of the oil k_0.

10-19 Two strips of metal are riveted together at their ends by four rivets, each of diameter 0.5 cm. What is the maximum tension that can be exerted by the riveted strip if the shearing stress on the rivets is not to exceed 6×10^8 Pa? Assume each rivet to carry one-quarter of the load.

10-20 Find the mass per cubic meter of ocean water at a depth of 500 m where the pressure is about 5.0×10^6 Pa. The density at the surface is 1.03×10^3 kg·m⁻³.

10-21 Compute the compressibility of steel, in reciprocal atmospheres, and compare with that of water. Which material is the more readily compressed?

10-22 A steel post 15 cm in diameter and 3 m long is placed vertically and is required to support a load of 10,000 kg.

a) What is the stress in the post?
b) What is the strain in the post?
c) What is the change in the length of the post?

10-23 A hollow cylindrical steel column 5 m high shortens 0.03 cm under a compression load of 40,000 kg. If the inner radius of the cylinder is 0.80 of the outer one, what is the outer radius?

10-24 A bar of cross section A is subjected to equal and opposite tensile forces F at its ends. Consider a plane through the bar making an angle θ with a plane at right angles to the bar (Fig. 10-10).

a) What is the tensile (normal) stress at this plane, in terms of F, A, and θ?

Figure 10-10

b) What is the shear (tangential) stress at the plane, in terms of F, A, and θ?
c) For what value of θ is the tensile stress a maximum?
d) For what value of θ is the shear stress a maximum?

10-25 Suppose the block in Fig. 10-3 is rectangular instead of square, but is in equilibrium under the action of shear stresses on those outside faces perpendicular to the plane of the diagram. (Note that then $F_x \neq F_y$.)

a) Show that the shear stress is the same on all outside faces perpendicular to the plane of the diagram.
b) Show that on all sections perpendicular to the plane of the diagram and making an angle of 45° with an end face, the stress is still a pure tension or compression.

10-26 The equation of state (the equation relating pressure, volume, and temperature) for an ideal gas is $pV = nRT$, where n and R are constants. Show that if the gas is compressed while the temperature T is held constant, the bulk modulus is equal to the pressure.

10-27 When an ideal gas is compressed without the transfer of any heat in or out of it, the pressure and volume are related by $pV^\gamma =$ constant, where γ is a constant having different values for different gases. Show that in this case the bulk modulus is given by $B = \gamma p$.

10-28 In Fig. 10-6, suppose the body is a square steel plate, 10 cm on a side and 1 cm thick. Find the magnitude of force required on each of the four sides to cause a shear strain of 0.01.

10-29 In Fig. 10-6, suppose the body is a square aluminum plate 0.2 m on a side. It is desired that when forces are applied to the four edges, of equal magnitude 1.0×10^6 N each, the resulting shear strain is no greater than 0.01. What minimum thickness of plate is required?

PERIODIC MOTION 11

11-1 Introduction

The motion of a body when acted upon by a constant force was considered in detail in Chapters 3 and 4. The motion is one of constant acceleration, and it was found useful to derive expressions for the position and velocity of the body at any time, and for its velocity at any position. In the present chapter, we study the motion of a body when the resultant force on it is *not* constant, but varies during the motion. Naturally, there is an infinite number of ways in which a force may vary; hence no general expressions can be given for the motion of a body when acted on by a variable force, except that the acceleration at each instant must equal the force at that instant divided by the mass of the body.

There is however a particular class of problems that occur so often in practical situations that it is useful to develop formulas for this special case. In these problems, a body has an *equilibrium* position, and when displaced from equilibrium it experiences a *restoring* force, which causes it to undergo a back-and-forth motion past the equilibrium position. A familiar example is the pendulum, a mass suspended from a string. In the equilibrium position, the mass hangs straight down; when displaced from this position the mass does not simply return to the equilibrium position, but instead swings back and forth in a regular, repetitive manner.

Such a repetitive motion is said to be *oscillatory* or *periodic*. Many other examples might be cited: the balance wheel of a watch, the pistons in a gasoline engine, the strings in a musical instrument. The molecules of a solid body vibrate with oscillatory motion about their equilibrium positions in the crystal lattice, although of course this motion cannot be observed directly. The beating of the human heart is a periodic motion; Galileo is alleged to have used his own heartbeat for timing observations of motion.

In many kinds of *wave motion,* the particles of the medium in which the wave is traveling oscillate with periodic motion. This is true even for light waves and radio waves in empty space, except that instead of material particles the quantities that oscillate are the electric and magnetic fields associated with the wave. As a final example, an electrical circuit in which there is an alternating current is described in terms of voltages, currents, and electric charges that oscillate with time. Thus a study of periodic motion will lay the foundation for future work in many different fields of physics.

11-2 Elastic restoring forces

It was shown in Chapter 10 that, when a body is deformed, the deformation is proportional to the deforming force, provided the proportional limit is not exceeded. The change may be an increase in length, as of a wire or a coil spring, or a decrease in length, or a bending as of a flat spring, or a twisting of a rod about its axis, or many other forms. The term "force" is to be interpreted liberally as the force, or torque, or pressure, or whatever may be producing the distortion. If we restrict the discussion to the case of a push or a pull, where the distortion is simply the displacement of the point of application of the force, the force and displacement are related by Hooke's law; then

$$F = kx,$$

where F is the force exerted *on the spring,* k is a proportionality constant called the *force constant,* and x is the displacement from the equilibrium position.

11-3 Basic concepts

To illustrate basic concepts, let us consider the system shown in Fig. 11-1. A body of mass m can move without friction along a straight line. It might be, for example, a glider on a linear track, as described in Sec. 2-8. The body is attached to one end of a spring, the other end of which is held stationary. We describe the position of the body with the coordinate x, taking the origin ($x = 0$) as the equilibrium position, where the spring is neither stretched nor compressed. When the body is displaced to the right, x is positive, the spring is stretched, and it exerts a force on the body toward the left (the negative x-direction), toward the equilibrium position. When it is displaced to the left, x is negative, the spring is compressed, and it exerts a force on the body toward the right (the positive x-direction), again toward equilibrium. Thus the sign of the x-component of force *on the body* is always opposite to that of x itself. If, in addition, the spring obeys Hooke's law, then the force on the body is given by

$$F = -kx, \qquad (11-1)$$

where k is the *force constant* for the spring, discussed in Sec. 10-5. Equation (11-1) is valid for both positive and negative x.

Now suppose the body is displaced to the right a distance A and released. The spring exerts a restoring force; the body accelerates in the direction of this force, and moves toward the equilibrium position with increasing speed. The *rate* of increase (i.e., the acceleration) is not constant, however, since the accelerating force becomes smaller as the body approaches the equilibrium position.

11-1 Model for periodic motion. If the spring obeys Hooke's law, the motion is simple harmonic motion.

When the body reaches the center, the restoring force has decreased to zero; but because of the velocity that has been acquired, the body "overshoots" the equilibrium position and continues to move toward the left. As soon as the equilibrium position is passed, the restoring force again comes into play, directed now toward the right. The body's speed thus decreases, at a rate that increases with increasing distance from O. It therefore comes to rest at some point to the left of O, and repeats its motion in the opposite direction.

As will be shown in the next section, the motion is confined to a range $\pm A$ on either side of the equilibrium position, each to-and-fro movement taking place in the same interval of time. If there were no loss of energy by friction, the motion would continue indefinitely, once it had been started. This type of motion, under the influence of an elastic restoring force proportional to displacement and in the absence of all friction, is called *simple harmonic motion,* abbreviated SHM.

A *complete vibration* or *complete cycle* means one round trip, say from A to $-A$ and back to A, or from O to A to O to $-A$ and back to O.

The *periodic time,* or simply the *period* of the motion, represented by τ, is the time required for one complete vibration.

The *frequency, f,* is the number of complete vibrations per unit time. Evidently the frequency is the reciprocal of the period, or $\tau = 1/f$. The SI unit, one cycle per second, is called one *hertz* ($1 \text{ Hz} = 1 \text{ s}^{-1}$).

The *amplitude, A,* is the maximum displacement from equilibrium, that is, the maximum value of $|x|$. The total range of the motion is therefore $2A$.

11-4 Equations of simple harmonic motion

The motion of a point mass m under a restoring force given by $-kx$, where k is the force constant and x the displacement from equilibrium, is an idealized model constituting the simplest possible example of periodic motion. In more complex examples the force may depend on displacement in a more complicated way, but when the force is *directly proportional* to displacement, the resulting motion is called *simple harmonic motion*. Since many more complex motions are *approximately* simple harmonic and may therefore be described approximately by this model, it is useful to analyze simple harmonic motion in some detail. We shall obtain expressions for the coordinate, velocity, and acceleration of a body moving with simple harmonic motion, just as we found those for a body moving with constant acceleration. It must be emphasized that the equations of motion with *constant* acceleration cannot be applied, since the acceleration is continually changing.

Figure 11-2 represents the vibrating body of Fig. 11-1 at some instant when its displacement from the equilibrium position O is described

11-2

by the coordinate x. The particle's mass is m, and the resultant force acting on it is simply the elastic restoring force $-kx$. From Newton's second law,

$$F = -kx = ma,$$

or

$$a = -\frac{k}{m}x. \qquad (11\text{-}2)$$

Thus, an alternative description of the essential feature of simple harmonic motion is that *the acceleration at each instant is proportional to the negative of the displacement at that instant*. When x has its maximum positive value A, the acceleration has its maximum negative value $-kA/m$; and at the instant the particle passes the equilibrium position ($x = 0$), the acceleration is zero. Its velocity is, of course, *not* zero at this point.

The energy principle provides a convenient basis for analysis of some aspects of simple harmonic motion. The elastic restoring force is a *conservative* force, and the work done by this force can be represented in terms of a potential energy $U = \frac{1}{2}kx^2$, just as in Sec. 6-5. The kinetic energy is $K = \frac{1}{2}mv^2$ and, according to the principle of conservation of energy, the total energy $E = K + U$ is constant. That is,

$$E = \tfrac{1}{2}mv^2 + \tfrac{1}{2}kx^2 = \text{constant.} \qquad (11\text{-}3)$$

The total energy E is also closely related to the amplitude A of the motion. When the particle reaches its maximum displacement $\pm A$, it stops and turns back toward equilibrium. At this instant, $v = 0$, so there is no kinetic energy; the (constant) total energy is thus equal to the potential energy at this point, which is $\frac{1}{2}kA^2 = E$. Thus,

$$\tfrac{1}{2}mv^2 + \tfrac{1}{2}kx^2 = \tfrac{1}{2}kA^2,$$

or

$$v = \pm\sqrt{\frac{k}{m}}\sqrt{A^2 - x^2}. \qquad (11\text{-}4)$$

This relation permits us to obtain the velocity (apart from a sign) for any given position, and Eq. (11-4) is analogous to the relation $v^2 - v_0^2 = 2a(x - x_0)$ for straight-line motion with constant acceleration.

The significance of Eq. (11-3) is shown by the graph in Fig. 11-3, in which energy is plotted vertically and the coordinate x horizontally. The curve represents the potential energy, $U = \frac{1}{2}kx^2$; this curve is a parabola. A horizontal line is drawn at a height equal to the total energy E. We see that the motion is restricted to values of x lying between the points at which the horizontal line intersects the parabola since, if x were outside this range, the potential energy would exceed the total energy, and that is impossible. The motion of the vibrating body is analogous to that of a particle released at a height E on a frictionless track shaped like the potential energy curve, and the motion is said to take place in a "potential energy well."

If a vertical line is constructed at any value of x within the permitted range, the length of the segment between the x-axis and the parabola represents the potential energy U at that value of x, and the length of the segment between the parabola and the horizontal line at height E

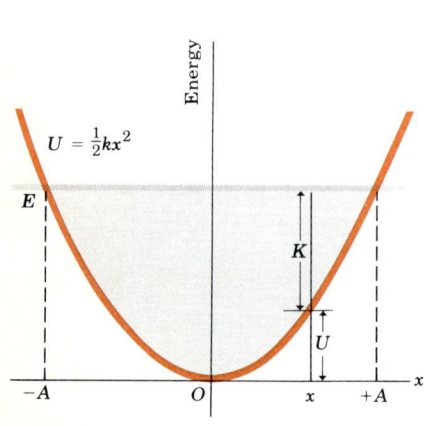

11-3 Relation between total energy E, potential energy U, and kinetic energy K, for a body oscillating with SHM.

represents the corresponding kinetic energy K. At the endpoints, therefore, the energy is all potential and at the midpoint it is all kinetic. The speed has its maximum value $v_{\max}$ at the midpoint:

$$\tfrac{1}{2}mv_{\max}^2 = E, \qquad v_{\max} = \sqrt{2E/m}. \tag{11-5}$$

The fact that both the maximum potential energy and maximum kinetic energy are equal to the total energy (and thus to each other) may be used to relate the maximum velocity to the amplitude:

$$\tfrac{1}{2}kA^2 = \tfrac{1}{2}mv_{\max}^2, \qquad v_{\max} = \sqrt{\frac{k}{m}}\,A. \tag{11-6}$$

The position–velocity relation given by Eq. (11-4) is very useful, but it does not tell us where the particle is at any given *time*. To have a complete description of the motion we need to know the position, velocity, and acceleration for any given time. The appropriate relations may be obtained either by integrating the energy equation, Eq. (11-4), or by working directly with Newton's second law, Eq. (11-2). We take the energy approach first.

In Eq. (11-4) we replace v by dx/dt, temporarily disregarding the $\pm$ sign:

$$\frac{dx}{dt} = \sqrt{\frac{k}{m}}\,\sqrt{A^2 - x^2}. \tag{11-7}$$

The next step is called *separation of variables;* we write all factors containing x on one side, and all those containing t on the other, so that each side can be integrated:

$$\int \frac{dx}{\sqrt{A^2 - x^2}} = \sqrt{\frac{k}{m}} \int dt.$$

Carrying out the integration, we find

$$\arcsin \frac{x}{A} = \sqrt{\frac{k}{m}}\,t + C,$$

where C is an integration constant. Solving for x, we obtain

$$x = A \sin\!\left(\sqrt{\frac{k}{m}}\,t + C\right). \tag{11-8}$$

The quantity in parentheses plays the role of an angle (measured always in radians.) Because the sine function is *periodic,* this shows that x is a periodic function of time, as anticipated, and specifically it is a *sinusoidal* function.

The *period* of motion, denoted by τ, is the time for one complete cycle or oscillation. The sine function repeats itself whenever the quantity in parentheses increases by 2π. Thus if we start at time $t = 0$, the time τ at which one cycle has been completed is given by

$$\sqrt{\frac{k}{m}}\,\tau = 2\pi,$$

or

$$\tau = 2\pi \sqrt{\frac{m}{k}}. \tag{11-9}$$

Thus the period of the motion is determined by the mass m and the force constant k. It *does not* depend on the amplitude or on the total energy. For given values of m and k, the time of one complete oscillation is the same whether the amplitude is large or small. A motion that has this property is said to be *isochronous* (equal time).

The frequency f, or the number of complete oscillations per unit time, is the reciprocal of the period τ:

$$f = \frac{1}{\tau} = \frac{1}{2\pi}\sqrt{\frac{k}{m}}. \qquad (11\text{-}10)$$

The basic unit for f is "cycles per second" or simply s^{-1}. In SI units one cycle per second is given the special name *one hertz* (1 Hz) in honor of Heinrich Hertz, one of the pioneers in investigating electromagnetic waves during the latter part of the nineteenth century. Thus

$$\text{one hertz} = 1 \text{ Hz} = 1 \text{ cycle}\cdot\text{s}^{-1}.$$

The *angular frequency* ω is defined as $\omega = 2\pi f$ and is expressed in radians per second. It follows from either of the two preceding equations that

$$\omega = \sqrt{k/m}, \qquad (11\text{-}11)$$

and Eqs. (11-8) and (11-4) can be written more compactly as

$$x = A \sin(\omega t + C), \qquad (11\text{-}12)$$

$$v = \omega\sqrt{A^2 - x^2}. \qquad (11\text{-}13)$$

Relationships in simple harmonic motion are often expressed in terms of ω rather than f because it saves writing factors of 2π. By convention, the hertz is *not* used as a unit of angular frequency, which is usually given the unit rad·s^{-1} or simply s^{-1}.

In SI units, m is expressed in kilograms and k in newtons per meter. The frequency f is then in vibrations per second, or hertz, and the period τ in seconds per vibration.

The general form of Eqs. (11-9) and (11-10) is as expected. When m is large, we expect the motion to be slow and ponderous, corresponding to small f and large τ. A large value of k means a very stiff spring, corresponding to large f and small τ. The most noteworthy feature of these equations, however, is that they *do not* contain the amplitude A of the motion; this may be surprising. If a given spring–mass system is given an initial displacement and released and its frequency measured, and then it is stopped, given a *different* displacement, and released, the two frequencies are the same! To be sure, the maximum displacement, maximum speed, and maximum acceleration are all different in the two cases, but *not* the frequency. Thus an additional important characteristic of simple harmonic motion is that *the frequency is not dependent on the amplitude of the motion*.

In Eq. (11-8), the sine function can never be greater than $+1$ or less than -1, so x can never be greater than A or less than $-A$; this confirms again that A is the *amplitude* of the motion. The integration constant C determines the position of the body at time $t = 0$. If $C = 0$, then at time $t = 0$, $x = A \sin 0 = 0$, and the body starts at the origin. But if $C = \pi/2$, then at time $t = 0$, $x = A \sin \pi/2 = A$, and the body starts at its maximum positive displacement. In that case one can use the identity

$\sin(\theta + \pi/2) = \cos\theta$ to rewrite Eq. (11–12) as

$$x = A \cos \omega t. \tag{11–14}$$

Before exploring the meaning of Eq. (11–12) further, we return to an alternative derivation starting with Eq. (11–2). Replacing a by d^2x/dt^2 and using Eq. (11–11), we find

$$\frac{d^2x}{dt^2} = -\omega^2 x. \tag{11–15}$$

This is a *differential equation;* it does not give x as a function of t, but it says that x must be a function such that when its second derivative is calculated, the result is a negative constant multiplied by the original function. What functions have this property? Sines and cosines immediately come to mind; all of the following functions have the required property.

$$x = A \sin \omega t, \tag{11–16a}$$

$$x = A \cos \omega t, \tag{11–16b}$$

$$x = A \cos(\omega t + \theta_0). \tag{11–16c}$$

The reader is invited to calculate the second derivative of each of these, substitute it in the left side of Eq. (11–15), substitute the function itself into the right side, and verify that the left and right sides are indeed equal, thus verifying that each of Eqs. (11–16) is a *solution* of Eq. (11–15). This in turn shows that each function represents a motion that is consistent with Newton's second law and is thus a physically possible motion.

The three functions in Eqs. (11–16) are plotted as functions of time in Fig. 11–4, showing that the essential difference among them is the position of the body at the instant of time we choose to call $t = 0$.

Equations (11–16) will form the basis of our further description of simple harmonic motion. For given values of A and ω, they differ in the position of the particle at time $t = 0$, that is, in the particular point in the cycle at which $t = 0$. If the body is given an initial displacement A at time $t = 0$ and released with no initial velocity, the motion is described by Eq. (11–16b). If it is given an initial velocity v_0 at the equilibrium position ($x = 0$) at time $t = 0$, the appropriate equation is Eq. (11–16a);

11–4 Graphs of Eqs. (11–16), showing phase relationships.

in that case, reference to Eq. (11–6) shows that v_0 and A are related by

$$v_0 = \omega A,$$

since the velocity at $x = 0$ is the maximum velocity.

When the body is given *both* an initial displacement x_0 and an initial velocity v_0 at time $t = 0$, Eq. (11–16c) must be used. If x_0 and v_0 are given, A and θ_0 can be determined; θ_0 is called a *phase angle*, and its purpose is to describe the point in the cycle at which $t = 0$. To determine A and θ, we first note that when $t = 0$ is substituted into Eq. (11–16c), the result must equal x_0; thus

$$x_0 = A \cos \theta_0. \tag{11-17}$$

Also, the velocity v at any time t is given by

$$v = \frac{dx}{dt} = -\omega A \sin(\omega t + \theta_0). \tag{11-18}$$

Again substituting $t = 0$ and equating the result to v_0, we find

$$v_0 = -\omega A \sin \theta_0. \tag{11-19}$$

This and Eq. (11–17) may be solved for A and θ_0. To eliminate A, we divide Eq. (11–19) by Eq. (11–17), obtaining

$$\tan \theta_0 = -\frac{v_0}{\omega x_0}. \tag{11-20}$$

To eliminate θ_0, we divide Eq. (11–19) by ω, square it, square Eq. (11–17), and add the two, using the identity $\sin^2 \theta_0 + \cos^2 \theta_0 = 1$, to obtain

$$A^2 = x_0^2 + \frac{v_0^2}{\omega^2}. \tag{11-21}$$

The amplitude is then no longer equal to the initial displacement. This is reasonable: If at time $t = 0$ the particle has an initial displacement in the positive direction and also a positive velocity in that direction, then it will move *farther* in that direction before returning; hence A must be greater than x_0. The frequency and period relations, Eqs. (11–9) and (11–10), are unchanged.

Example 1 A spring is mounted as in Fig. 11–1. By attaching a spring balance to the free end and pulling sideways, we determine that the force is proportional to the displacement, a force of 4 N causing a displacement of 0.02 m. We attach a 2-kg body to the end, pull it aside a distance of 0.04 m, and release it.

a) Find the force constant of the spring.

$$k = \frac{F}{x} = \frac{4 \text{ N}}{0.02 \text{ m}} = 200 \text{ N} \cdot \text{m}^{-1}.$$

b) Find the period and frequency of vibration.

$$\tau = 2\pi \sqrt{\frac{m}{k}} = 2\pi \sqrt{\frac{2 \text{ kg}}{200 \text{ N} \cdot \text{m}^{-1}}} = \frac{\pi}{5} \text{ s} = 0.628 \text{ s};$$

$$f = \frac{1}{\tau} = \frac{5}{\pi} \text{ s}^{-1} = 1.59 \text{ s}^{-1} = 1.59 \text{ Hz};$$

$$\omega = 2\pi f = \sqrt{\frac{k}{m}} = 10 \text{ s}^{-1}.$$

c) Compute the maximum velocity attained by the vibrating body.

The maximum velocity occurs at the equilibrium position, where the coordinate is zero. Since

$$v = \pm \omega \sqrt{A^2 - x^2},$$

then when $x = 0$,

$$v = v_{\max} = \pm \omega A.$$

Also,

$$A = 0.04 \text{ m},$$

$$v_{\max} = \pm(10 \text{ s}^{-1})(0.04 \text{ m}) = \pm 0.4 \text{ m} \cdot \text{s}^{-1}.$$

d) Compute the maximum acceleration.

From Eq. (11-2) or (11-15),

$$a = -\frac{k}{m}x = -\omega^2 x.$$

The maximum acceleration occurs at the ends of the path, where $x = \pm A$. Therefore

$$a_{\max} = \mp \omega^2 A = \mp(10 \text{ s}^{-1})^2(0.04 \text{ m}) = \mp 4.0 \text{ m} \cdot \text{s}^{-2}.$$

e) Compute the velocity and acceleration when the body has moved halfway in toward the center from its initial position.

At this point, from Eq. (11-4),

$$x = \frac{A}{2} = 0.02 \text{ m},$$

$$v = -(10 \text{ s}^{-1})\sqrt{(0.04 \text{ m})^2 - (0.02 \text{ m})^2}$$

$$= -\left(\frac{2\sqrt{3}}{10}\right) \text{m} \cdot \text{s}^{-1} = -0.346 \text{ m} \cdot \text{s}^{-1},$$

$$a = -\omega^2 x = -(10 \text{ s}^{-1})^2(0.02 \text{ m}) = -2.0 \text{ m} \cdot \text{s}^{-2}.$$

f) How much time is required for the body to move halfway in to the center from its initial position?

The position at any time is given by $x = A \cos \omega t$. From this,

$$A/2 = A \cos(10 \text{ s}^{-1})t,$$

$$\cos(10 \text{ s}^{-1})t = \frac{1}{2},$$

$$(10 \text{ s}^{-1})t = \arccos \frac{1}{2} = \frac{\pi}{3},$$

$$t = \frac{\pi}{30} \text{s}.$$

Example 2 The system in Example 1 is given an initial displacement of 0.05 m and an initial velocity of 2 m·s⁻¹. Find the amplitude, the phase angle, and the total energy of the motion, and write an equation for the position as a function of time.

Solution From Eq. (11-21),
$$A = \sqrt{x_0^2 + (v_0/\omega)^2}$$
$$= \sqrt{(0.05 \text{ m})^2 + (2 \text{ m}\cdot\text{s}^{-1}/10 \text{ s}^{-1})^2}$$
$$= 0.206 \text{ m}.$$

From Eq. (11-20),
$$\theta_0 = \arctan\frac{-v_0}{\omega x_0}$$
$$= \arctan\frac{-2 \text{ m}\cdot\text{s}^{-1}}{(10 \text{ s}^{-1})(0.05 \text{ m})}$$
$$= -76.0°$$
$$= -1.33 \text{ rad}.$$

From Eq. (11-3) and the following discussion,
$$E = \tfrac{1}{2}(200 \text{ N}\cdot\text{m}^{-1})(0.206 \text{ m})^2$$
$$= 4.25 \text{ J}.$$

Alternatively, from the initial conditions,
$$E = \tfrac{1}{2}mv_0^2 + \tfrac{1}{2}kx_0^2$$
$$= \tfrac{1}{2}(2 \text{ kg})(2 \text{ m}\cdot\text{s}^{-1})^2 + \tfrac{1}{2}(200 \text{ N}\cdot\text{m}^{-1})(0.05 \text{ m})^2$$
$$= 4.25 \text{ J}.$$

The position is given by Eq. (11-16c):
$$x = (0.206 \text{ m})\cos[(10 \text{ s}^{-1})t - 1.33 \text{ rad}].$$

11-5 Circle of reference

Understanding the various relationships of simple harmonic motion is facilitated by a graphical representation called the *circle of reference*, which we now introduce.

In Fig. 11-5, point Q moves counterclockwise around a circle of radius A with constant angular velocity ω (measured in rad·s^{-1}). The vector from O to Q represents the position of point Q relative to O; θ is the angle this vector makes with the positive x-axis. This vector, whose horizontal component represents the actual motion, is called a *phasor;* this representation is also useful in many other areas of physics, including ac circuit analysis and optics.

Let P be a point on the horizontal diameter of the circle, directly below Q. Point P is called the projection of Q onto the diameter. Point Q is referred to as the *reference point,* and the circle in which it moves as the *reference circle.* As the reference point revolves, the point P moves back and forth along a horizontal line, keeping always directly below (or above) Q. We shall show that the motion of P is the same as that of a body moving under the influence of an elastic restoring force in the absence of friction, i.e., that it is *simple harmonic motion*.

The displacement of P at any time t is the distance OP or x; from the figure,
$$x = A \cos\theta.$$

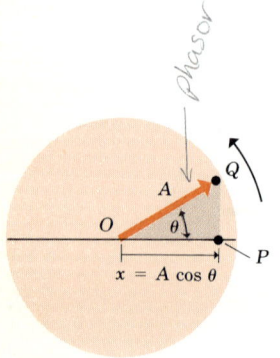

11-5 Coordinate of a body in simple harmonic motion.

If point Q is at the extreme right end of the diameter at time $t = 0$, the angle θ may be written

$$\theta = \omega t.$$

Hence,

$$x = A \cos \omega t. \qquad (11\text{-}22)$$

Now ω, the angular velocity of Q in radians per second, is related to f, the number of complete revolutions of Q per second, by

$$\omega = 2\pi f,$$

since there are 2π radians in one complete revolution. Furthermore, the point P makes one complete vibration for each revolution of Q. Hence, f may also be interpreted as the number of vibrations per second or the *frequency* of vibration of point P. Thus, Eq. (11-22) may also be written

$$x = A \cos 2\pi f t. \qquad (11\text{-}23)$$

The instantaneous velocity of P may be found with the aid of Fig. 11-6. The reference point Q moves with a tangential velocity given by Eq. (9-9):

$$v_\| = \omega A = 2\pi f A.$$

Since point P is always directly below or above the reference point, the velocity of P at each instant must equal the x-component of the velocity of Q. That is, from Fig. 11-6,

$$v = -v_\| \sin \theta = -\omega A \sin \theta;$$

$$v = -\omega A \sin \omega t. \qquad (11\text{-}24)$$

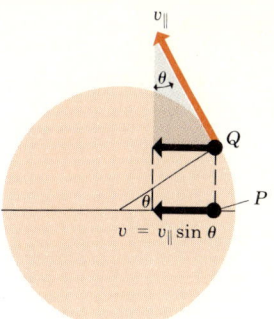

11-6 Velocity in simple harmonic motion.

The minus sign is introduced because the direction of the velocity is toward the left. When Q is below the horizontal diameter, the velocity of P is toward the right, but since $\sin \theta$ is negative at such points, the minus sign is still needed. Equation (11-24) gives the velocity of point P at any time.

Finally, the acceleration of point P may be found by making use again of the fact that, since P is always directly below or above Q, its acceleration must equal the x-component of the acceleration of Q. Point Q, since it moves in a circular path with a constant angular velocity ω, has at each instant an acceleration toward the center given by Eq. (9-11):

$$a_\perp = \frac{v_\|^2}{A} = \omega^2 A.$$

From Fig. 11-7, the x-component of this acceleration is

$$a = -a_\perp \cos \theta;$$

$$a = -\omega^2 A \cos \omega t. \qquad (11\text{-}25)$$

The minus sign is introduced because the acceleration is toward the left. When Q is to the left of the center, the acceleration of P is toward the right but, since $\cos \theta$ is negative at such points, the minus sign is still required. Equation (11-25) gives the acceleration of P at any time.

Now comes the crucial step in showing that the motion of P is simple harmonic. We combine Eqs. (11-22) and (11-25), obtaining

$$a = -\omega^2 x. \qquad (11\text{-}26)$$

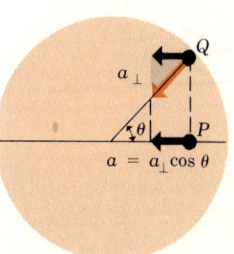

11-7 Acceleration in simple harmonic motion.

Since ω is constant, the acceleration a at each instant equals a negative constant times the displacement x at that instant. But this is just the essential feature of simple harmonic motion, as given by Eq. (11-2). Hence the motion of P is indeed simple harmonic!

In order to make Eqs. (11-2) and (11-26) agree precisely, we must choose an angular velocity ω for the reference point Q such that

$$\omega^2 = \frac{k}{m}.$$

Hence the angular velocity of point Q is identical with the angular frequency of motion defined by Eq. (11-11).

Throughout the above discussion we have assumed that the initial position of the particle (at time $t = 0$) is its maximum positive displacement A, but this is not an essential restriction. Different initial positions correspond to different initial positions of the reference point Q. For example, if at time $t = 0$ the phasor OQ makes an angle θ_0 with the positive x-axis, then the angle θ at time t is given not by $\theta = \omega t$ as before, but by

$$\theta = \theta_0 + \omega t. \tag{11-27}$$

The only change in the above discussion is to replace (ωt) in Eqs. (11-22), (11-24), and (11-25) by $(\omega t + \theta_0)$. These equations then become

$$\begin{aligned} x &= A\cos(\omega t + \theta_0), \\ v &= -\omega A \sin(\omega t + \theta_0), \\ a &= -\omega^2 A \cos(\omega t + \theta_0) = -\omega^2 x. \end{aligned} \tag{11-28}$$

The initial position x_0 and initial velocity v_0 (at time $t = 0$) are then given by

$$x_0 = A\cos\theta_0, \qquad v_0 = -\omega A \sin\theta_0,$$

in agreement with Eqs. (11-17) and (11-19).

Example In Example 1 of Sec. 11-4, use the circle of reference to find the time for the displacement from $x = A$ to $x = A/2$, requested in part (f).

Solution While the body moves halfway in, the reference point revolves through an angle of 60° (Fig. 11-8). Since the reference point moves with constant angular velocity and in this example makes one complete revolution during one period ($\tau = \pi/5$ s), the time to rotate through 60° is

$$\left(\frac{1}{6}\right)\left(\frac{\pi}{5}\text{ s}\right) = \frac{\pi}{30}\text{ s} = 0.105\text{ s}.$$

11-6 Motion of a body suspended from a coil spring

Figure 11-9a shows a coil spring of force constant k and no-load length l. When a body of mass m is attached to the spring as in part (b), it hangs in equilibrium with the spring extended by an amount Δl such that the upward force P exerted by the spring is equal to the weight of the body, mg. But $P = k\Delta l$, so

$$k\,\Delta l = mg.$$

11-8

11-9 The restoring force on a body suspended by a spring is proportional to the coordinate measured from the equilibrium position.

Now suppose the body is at a distance x *above* its equilibrium position, as in Fig. 11-9c. The extension of the spring is now $\Delta l - x$, the upward force it exerts on the body is $k(\Delta l - x)$, and the resultant force F on the body is

$$F = k(\Delta l - x) - mg = -kx.$$

The resultant force is therefore proportional to the displacement of the body *from its equilibrium position;* and, if set in vertical motion, the body oscillates with an angular frequency

$$\omega = \sqrt{k/m}.$$

Example A body of mass 5 kg is suspended by a spring, which stretches 0.1 m when the body is attached. It is then displaced downward an additional 0.05 m and released. Find the amplitude, period, and frequency of the resulting simple harmonic motion.

Solution Since the initial position is 0.05 m from equilibrium and there is no initial velocity, $A = 0.05$ m. To find the period we first find the force constant of the spring. The spring is stretched 0.1 m by a force of $(5 \text{ kg})(9.8 \text{ m} \cdot \text{s}^{-2})$, so

$$k = \frac{mg}{\Delta l} = \frac{(5 \text{ kg})(9.8 \text{ m} \cdot \text{s}^{-2})}{0.1 \text{ m}} = 490 \text{ N} \cdot \text{m}^{-1},$$

$$\tau = 2\pi \sqrt{\frac{m}{k}} = 2\pi \sqrt{\frac{5 \text{ kg}}{490 \text{ N} \cdot \text{m}^{-1}}} = 0.635 \text{ s},$$

$$f = \frac{1}{\tau} = 1.57 \text{ Hz}.$$

11-7 Angular harmonic motion

Angular harmonic motion results when a body that is pivoted about an axis experiences a restoring torque proportional to the angular displacement from its equilibrium position. This type of vibration is very similar to linear harmonic motion, and the corresponding equations may be written down immediately from the analogies between linear and angular quantities.

A restoring torque proportional to angular displacement is expressed by

$$\Gamma = -k'\theta, \tag{11-29}$$

where k' is a proportionality constant called the *torque constant*. The moment of inertia of the pivoted body corresponds to the mass of a body in linear motion. Hence the formula for the period of angular harmonic motion is

$$\tau = 2\pi \sqrt{\frac{I}{k'}}, \tag{11-30}$$

where k' is the constant in Eq. (11-29). The angular frequency is now given by

$$\omega = \sqrt{\frac{k'}{I}}. \tag{11-31}$$

Caution is required in the choice of symbols; ω must not be used for the angular velocity of the body (a quantity that varies with time) because it has already been used for the angular frequency of the motion (a constant for any given system). A natural choice for the body's angular velocity is Ω. Then the appropriate formulas are obtained by replacing x everywhere by θ, v by Ω, and a by α.

The balance wheel of a watch is a familiar example of angular harmonic motion. If the hairspring behaves according to Eq. (11-29), the motion is *isochronous;* the period is constant even though the amplitude decreases somewhat as the mainspring unwinds.

11-8 The simple pendulum

A simple pendulum, an idealized model for a more complex system, consists of a point mass suspended by an inextensible weightless string in a uniform gravitational field. When pulled to one side of its equilibrium position and released, the pendulum bob vibrates about this position. We wish to analyze this motion, asking in particular whether it is simple harmonic.

The necessary condition for simple harmonic motion is that the restoring force F shall be directly proportional to the coordinate x and oppositely directed. The path of the bob is not a straight line, but the arc of a circle of radius L, where L is the length of the supporting cord. The coordinate x refers to distances measured *along this arc*. (See Fig. 11-10.) Hence, if $F = -kx$, the motion will be simple harmonic.

Figure 11-10 shows the forces on the bob at an instant when its coordinate is x. We choose axes tangent to the circle and along the radius, and resolve the weight into components. The restoring force F is

$$F = -mg \sin \theta. \tag{11-32}$$

The restoring force is therefore proportional *not* to θ but to $\sin \theta$, so the motion is *not* simple harmonic. However, *if the angle θ is small*, $\sin \theta$ is very nearly equal to θ. For example, when $\theta = 0.1$ rad (about 6°),

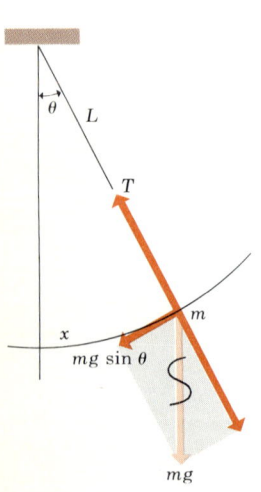

11-10 Forces on the bob of a simple pendulum.

$\sin\theta = 0.0998$, a difference of only 0.2 percent. With this approximation, Eq. (11–32) becomes

$$F = -mg\theta = -mg\frac{x}{L},$$

or

$$F = -\frac{mg}{L}x. \qquad (11\text{–}33)$$

The restoring force is then proportional to the coordinate *for small displacements*, and the constant mg/L represents the force constant k. The period of a simple pendulum when its amplitude is small is therefore

$$\tau = 2\pi\sqrt{\frac{m}{k}} = 2\pi\sqrt{\frac{m}{mg/L}},$$

or

$$\tau = 2\pi\sqrt{\frac{L}{g}}. \qquad (11\text{–}34)$$

The corresponding frequency relations are *pendulum*

$$f = \frac{1}{2\pi}\sqrt{\frac{g}{L}}, \qquad \omega = \sqrt{\frac{g}{L}}. \qquad (11\text{–}35)$$

We note that these expressions do not contain the *mass* of the particle; this is because the restoring force, a component of the particle's weight, is proportional to m. Thus the mass appears on both sides of $F = ma$ and may be canceled out. For small oscillations the period of a pendulum for a given value of g is determined entirely by its length.

We emphasize again that the motion of a pendulum is only *approximately* simple harmonic, and when the amplitude is not small the departures from simple harmonic motion can be substantial. What constitutes a "small" amplitude? It can be shown that the general equation for the time of swing, when the maximum angular displacement is α, is

$$\tau = 2\pi\sqrt{\frac{L}{g}}\left(1 + \frac{1^2}{2^2}\sin^2\frac{\alpha}{2} + \frac{1^2 \cdot 3^2}{2^2 \cdot 4^2}\sin^4\frac{\alpha}{2} + \cdots\right). \qquad (11\text{–}36)$$

The time may be computed to any desired degree of precision by taking enough terms in the infinite series. When $\alpha = 15°$ (on either side of the central position), the true period differs from that given by the approximate Eq. (11–34) by less than 0.5%.

The utility of the pendulum as a timekeeper is based on the fact that the period is practically independent of the amplitude. Thus, as a clock runs down and the amplitude of the swings becomes slightly smaller, the clock will still keep very nearly correct time.

The simple pendulum is also a precise and convenient method of measuring the acceleration of gravity g without actually resorting to free fall, since L and τ may readily be measured. More complicated pendulums find considerable application in the field of geophysics. Local deposits of ore or oil, if their density differs from that of their surroundings, affect the local value of g, and precise measurements of this quantity over an area that is being prospected often furnish valuable information regarding the nature of underlying deposits.

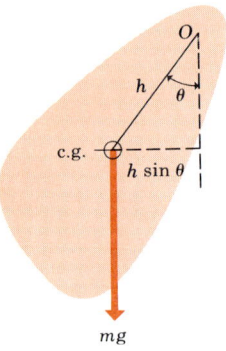

11-11 A physical pendulum.

*11-9 The physical pendulum

A so-called "physical" pendulum is any real pendulum, as contrasted with a simple pendulum in which all the mass is assumed to be concentrated at a point. In Fig. 11-11, a body of irregular shape is pivoted about a horizontal frictionless axis O and displaced from the vertical by an angle θ. The distance from the pivot to the center of gravity is h, the moment of inertia of the pendulum about an axis through the pivot is I, and the mass of the pendulum is m. The weight mg causes a restoring torque

$$\Gamma = -(mg)(h \sin \theta). \tag{11-37}$$

When released, the body oscillates about its equilibrium position; as in the case of the simple pendulum, the motion is not simple harmonic, since the torque Γ is proportional not to θ but to $\sin \theta$. However, if θ is small, we can again approximate $\sin \theta$ by θ, and the motion is approximately harmonic. With this approximation,

$$\Gamma \approx -(mgh)\theta,$$

and the effective torque constant is

$$k' = -\frac{\Gamma}{\theta} = mgh.$$

The angular frequency is

$$\omega = \sqrt{\frac{k'}{I}} = \sqrt{\frac{mgh}{I}}, \tag{11-38}$$

and the period is

$$\tau = 2\pi \sqrt{\frac{I}{k'}} = 2\pi \sqrt{\frac{I}{mgh}}. \tag{11-39}$$

Example 1 Let the body in Fig. 11-11 be a meterstick pivoted at one end, and let L be the total length (1 m).

Solution From Example 1 of Sec. 9-11,

$$I = \tfrac{1}{3}mL^2, \qquad h = \frac{L}{2}, \qquad g = 9.8 \text{ m} \cdot \text{s}^{-2},$$

$$\tau = 2\pi \sqrt{\frac{\tfrac{1}{3}mL^2}{mgL/2}} = 2\pi \sqrt{\frac{2L}{3g}}$$

$$= 2\pi \sqrt{(2/3)(1 \text{ m})/9.8 \text{ m} \cdot \text{s}^{-2}} = 1.64 \text{ s}.$$

Example 2 Equation (11-39) may be solved for the moment of inertia I, giving

$$I = \frac{\tau^2 mgh}{4\pi^2}.$$

The quantities on the right of the equation are all directly measurable. Hence the moment of inertia of a body of any complex shape may be found by suspending the body as a physical pendulum and measuring its period of vibration. The location of the center of gravity can be found by balancing. Since τ, m, g, and h are known, I can be computed. For example, Fig. 11-12 illustrates a connecting rod pivoted about a horizontal knife-edge. The connecting rod has a mass of 2 kg, and its center of gravity has been found by balancing to be 0.2 m below the knife-edge. When set into oscillation, it is found to make 100 complete vibrations in 120 s, so that $\tau = 120 \text{ s}/100 = 1.2$ s. Therefore,

$$I = \frac{(1.2 \text{ s})^2 (2 \text{ kg})(9.8 \text{ m} \cdot \text{s}^{-2})(0.2 \text{ m})}{4\pi^2} = 0.143 \text{ kg} \cdot \text{m}^2.$$

11-12

Questions

11-1 Think of several examples in everyday life of motion that is at least approximately simple harmonic. In what respects does each differ from SHM?

11-2 Does a tuning fork or similar tuning instrument undergo simple harmonic motion? Why is this a crucial question to musicians?

11-3 If a spring is cut in half, what is the force constant of each half? How would the frequency of SHM using a half-spring differ from that using the same mass and the entire spring?

11-4 The analysis of SHM in this chapter neglected the mass of the spring. How does the spring mass change the characteristics of the motion?

11-5 The system shown in Fig. 11-9 is mounted in an elevator, which accelerates upward with constant acceleration. Does the period increase, decrease, or remain the same?

11-6 A highly elastic "superball" bouncing on a hard floor has a motion that is approximately periodic. In what ways is the motion similar to SHM? In what ways is it different?

11-7 How could one determine the force constants of a car's springs by bouncing each end up and down?

11-8 Do the pistons in an automobile engine undergo simple harmonic motion?

11-9 Why is the "springiness" of a diving board adjusted for different dives and different weights of divers? How is the adjustment made?

11-10 In Fig. 11-1, suppose the stationary end of the spring is connected instead to another mass, equal to the original mass and free to slide along the same line. Could such a system undergo SHM? How would the period compare with that of the original system?

11-11 For the mass-spring system of Fig. 11-1, is there any point during the motion at which the mass is in equilibrium?

11-12 In any periodic motion, unavoidable friction always causes the amplitude to decrease with time. Does friction also affect the *period* of the motion? Give a qualitative argument to support your answer.

11-13 If a pendulum clock is taken to a mountaintop, does it gain or lose time, assuming it is correct at a lower elevation?

11-14 When the amplitude of a simple pendulum increases, should its period increase or decrease? Give a qualitative argument; do not rely on Eq. (11-36). Is your argument also valid for a physical pendulum?

11-15 A pendulum is mounted in an elevator that accelerates upward with constant acceleration. Does the period increase, decrease, or remain the same?

11-16 At what point in the motion of a simple pendulum is the string tension greatest? Least?

11-17 A child on a swing can increase his amplitude by "pumping up." Where does the extra energy come from? (The answer is not "from the child.")

11-18 Could a standard of time be based on the period of a certain standard pendulum? What advantages and disadvantages would such a standard have, compared to the actual present-day standard discussed in Sec. 1-2?

Problems

11-1 The general equation of simple harmonic motion,
$$x = A \cos(\omega t + \theta_0),$$
can be written in the equivalent form
$$x = B \sin \omega t + C \cos \omega t.$$

a) Find the expressions for the amplitudes B and C in terms of the amplitude A and the initial phase angle θ_0.
b) Interpret these expressions in terms of a phasor diagram.

11-2 A body of mass 0.25 kg is acted on by an elastic restoring force of force constant $k = 25$ N·m^{-1}.

a) Construct the graph of elastic potential energy U as a function of displacement x, over a range of x from -0.3 m to $+0.3$ m. Let 1 cm = 0.1 J vertically, and 1 cm = 0.05 m horizontally.

The body is set into oscillation with an initial potential energy of 0.6 J and an initial kinetic energy of 0.2 J. Answer the following questions by reference to the graph.

b) What is the amplitude of oscillation?
c) What is the potential energy when the displacement is one-half the amplitude?
d) At what displacement are the kinetic and potential energies equal?
e) What is the speed of the body at the midpoint of its path?

Find
f) the period τ,
g) the frequency f, and
h) the angular frequency ω.
i) What is the initial phase angle θ_0 if the amplitude $A = 15$ cm, the initial displacement $x_0 = 7.5$ cm, and the initial velocity v_0 is negative?

11-3 A body is vibrating with simple harmonic motion of amplitude 15 cm and frequency 4 Hz. Compute

a) the maximum values of the acceleration and velocity,
b) the acceleration and velocity when the coordinate is 9 cm, and
c) the time required to move from the equilibrium position to a point 12 cm distant from it.

11-4 A body of mass 10 g moves with simple harmonic motion of amplitude 24 cm and period 4 s. The coordinate is $+24$ cm when $t = 0$. Compute

a) the position of the body when $t = 0.5$ s,
b) the magnitude and direction of the force acting on the body when $t = 0.5$ s,
c) the minimum time required for the body to move from its initial position to the point where $x = -12$ cm,
d) the velocity of the body when $x = -12$ cm.

11-5 The motion of the piston of an automobile engine is approximately simple harmonic.

a) If the stroke of an engine (twice the amplitude) is 10 cm, and the engine runs at 3600 rev·min^{-1}, compute the acceleration of the piston at the end of its stroke.
b) If the piston has a mass of 0.5 kg, what resultant force must be exerted on it at this point?
c) What is the velocity of the piston, in kilometers per hour, at the midpoint of its stroke?

11-6 A body of mass 2 kg is suspended from a spring of negligible mass, and is found to stretch the spring 20 cm.

a) What is the force constant of the spring?
b) What is the period of oscillation of the body, if pulled down and released?
c) What would be the period of a body of mass 4 kg hanging from the same spring?

11-7 The scale of a spring balance reading from zero to 180 N is 9 cm long. A body suspended from the balance is observed to oscillate vertically at 1.5 Hz. What is the mass of the body? Neglect the mass of the spring.

11-8 A body of mass 4 kg is attached to a coil spring and oscillates vertically in simple harmonic motion. The amplitude is 0.5 m, and at the highest point of the motion the spring has its natural unstretched length. Calculate the elastic potential energy of the spring, the kinetic energy of the body, its gravitational potential energy relative to the lowest point of the motion, and the sum of these three energies, when the body is

a) at its lowest point,
b) at its equilibrium position, and
c) at its highest point.

11-9 A mass of 100 kg suspended from a wire whose unstretched length l_0 is 4 m is found to stretch the wire by 0.004 m. The cross-sectional area of the wire, which can be assumed constant, is 0.1 cm^2.

a) If the load is pulled down a small additional distance and released, find the frequency at which it will vibrate.
b) Compute Young's modulus for the wire.

11-10 A small block is executing simple harmonic motion in a horizontal plane with an amplitude of 10 cm. At a point 6 cm away from equilibrium, the velocity is 24 cm·s^{-1}.

a) What is the period?
b) What is the displacement when the velocity is ± 12 cm·s^{-1}?
c) If a small object placed on the oscillating block is just on the verge of slipping at the endpoint of the path, what is the coefficient of friction?

11-11 A force of 30 N stretches a vertical spring 15 cm.

a) What mass must be suspended from the spring so that the system will oscillate with a period of $(\pi/4)$ s?
b) If the amplitude of the motion is 5 cm, where is the body and in what direction is it moving $(\pi/12)$ s after it has passed the equilibrium position, moving downward?
c) What force does the spring exert on the body when it is 3 cm below the equilibrium position, moving upward?

11-12 A body of mass m is suspended from a coil spring and the time for 100 complete oscillations is measured for the following values of m:

m (g)	100	200	400	1000
Time of 100 oscillations (s)	23.4	30.6	41.8	64.7

Plot graphs of the measured values of (a) τ vs. m, and (b) τ^2 vs. m.

c) Are the experimental results in agreement with theory?
d) Is either graph a straight line?
e) Does the straight line pass through the origin?
f) What is the force constant of the spring?
g) What is the mass of the spring?

11-13 A body of mass 100 g hangs from a long spiral spring. When pulled down 10 cm below its equilibrium position and released, it vibrates with a period of 2 s.

a) What is its velocity as it passes through the equilibrium position?
b) What is its acceleration when it is 5 cm above the equilibrium position?
c) When it is moving upward, how much time is required for it to move from a point 5 cm below its equilibrium position to a point 5 cm above it?
d) How much will the spring shorten if the body is removed?

11-14 A body of mass 5 kg hangs from a spring and oscillates with a period of 0.5 s. How much will the spring shorten when the body is removed?

11-15 Four passengers whose combined mass is 300 kg are observed to compress the springs of an automobile by 5 cm when they enter the automobile. If the total load supported by the springs is 900 kg, find the period of vibration of the loaded automobile.

11-16

a) A block suspended from a spring vibrates with simple harmonic motion. At an instant when the displacement of the block is equal to one-half the amplitude, what fraction of the total energy of the system is kinetic and what fraction is potential?
b) When the block is in equilibrium, the length of the spring is an amount s greater than in the unstretched state. Prove that $\tau = 2\pi \sqrt{s/g}$.

11-17

a) With what additional force must a vertical spring carrying a 4-kg body in equilibrium be stretched so that, when released, it will perform 48 complete oscillations in 32 s with an amplitude of 5 cm?
b) What force is exerted by the spring on the body when it is at the lowest point, the middle, and the highest point of the path?
c) What is the kinetic energy of the system when the body is 2 cm below the middle of the path? its potential energy?

11-18 A force of 60 N stretches a certain spring 30 cm. A body of mass 4 kg is hung from this spring and allowed to come to rest. It is then pulled down 10 cm and released.

a) What is the period of the motion?
b) What are the magnitude and direction of the acceleration of the body when it is 5 cm above the equilibrium position, moving upward?
c) What is the tension in the spring when the body is 5 cm above the equilibrium position?
d) What is the shortest time required to go from the equilibrium position to the point 5 cm above?
e) If a small object were placed on the oscillating body, would it remain in contact with the body or leave it?
f) If a small object were placed on the oscillating body and its amplitude doubled, where would the object and the oscillating body begin to separate?

11-19 Two springs with the same unstretched length but different force constants k_1 and k_2 are attached to a block of mass m on a level frictionless surface. Calculate the effective force constant in each of the three cases. (a), (b), and (c) depicted in Fig. 11-13.

d) A body of mass m, suspended from a spring with a force constant k, vibrates with a frequency f_1. When the spring is cut in half and the same body is suspended from one of the halves, the frequency is f_2. What is the ratio f_2/f_1?

Figure 11-13

11-20 Two springs, each of unstretched length 0.2 m but having different force constants k_1 and k_2, are attached to opposite ends of a block of mass m on a level frictionless surface. The outer ends of the springs are now attached to two pins P_1 and P_2, 10 cm from the original positions of the ends of the springs. Let

$$k_1 = 1 \text{ N} \cdot \text{m}^{-1}, \quad k_2 = 3 \text{ N} \cdot \text{m}^{-1}, \quad m = 0.1 \text{ kg}.$$

(See Fig. 11-14.)

a) Find the length of each spring when the block is in its new equilibrium position, after the springs have been attached to the pins.

b) Find the period of vibration of the block if it is slightly displaced from its new equilibrium position and released.

Figure 11-14

11-21 The block in Problem 11-20 and Fig. 11-14 is oscillating with an amplitude of 0.05 m. At the instant it passes through its equilibrium position, a lump of putty of mass 0.1 kg is dropped vertically onto the block and sticks to it.

a) Find the new period and amplitude.

b) Was there a loss of mechanical energy? If so, where did it go?

c) Would the answers be the same if the putty had been dropped on the block when it was at one end of its path?

11-22 A simple pendulum 4 m long swings with an amplitude of 0.2 m.

a) Compute the velocity of the pendulum at its lowest point.

b) Compute its acceleration at the ends of its path.

11-23 Find the length of a simple pendulum whose period is exactly 1 s at a point where $g = 9.80 \text{ m} \cdot \text{s}^{-2}$.

11-24

a) What is the change $\Delta\tau$ in the period of a simple pendulum when the acceleration of gravity g changes by Δg? [Hint. The new period $\tau + \Delta\tau$ is obtained by substituting $g + \Delta g$ for g:

$$\tau + \Delta\tau = 2\pi \sqrt{\frac{L}{g + \Delta g}}.$$

To obtain an approximate expression, expand the factor $(g + \Delta g)^{-1/2}$ using the binomial theorem and keeping only the first two terms:

$$(g + \Delta g)^{-1/2} = g^{-1/2} - \tfrac{1}{2}g^{-3/2} \Delta g + \cdots.$$

The other terms contain higher powers of Δg and are very small if Δg is small.]

b) Find the *fractional* change in period $\Delta\tau/\tau$ in terms of the fractional change $\Delta g/g$.

c) A pendulum clock, which keeps correct time at a point where $g = 9.8000 \text{ m} \cdot \text{s}^{-2}$, is found to lose 10 seconds each day at a higher elevation. Use the result of part (a) or (b) above to find approximately the value of g at this new location.

11-25 A certain simple pendulum has a period on earth of 2.0 s. What is its period on the surface of the moon, where $g = 1.7 \text{ m} \cdot \text{s}^{-2}$? [Note. The approximate method of Problem 11-24 is *not* adequate here because Δg is not small compared to g.]

11-26 The balance wheel of a watch vibrates with an angular amplitude of π radians and with a period of 0.5 s.

a) Find its maximum angular velocity.

b) Find its angular velocity when its displacement is one-half its amplitude.

c) Find its angular acceleration when its displacement is 45°.

11-27 A certain alarm clock ticks four times each second, each tick representing half a period. The balance wheel consists of a thin rim of radius 1.5 cm, connected to the balance staff by thin spokes of negligible mass. The total mass is 0.8 kg.

a) What is the moment of inertia of the balance wheel?

b) What is the torque constant of the hairspring?

11-28 At one time it was proposed to define the second as the time required for a simple pendulum exactly one meter long to complete one side-to-side swing (one-half period).

a) By how much is this definition in error if $g = 9.8000 \text{ m} \cdot \text{s}^{-2}$?

b) For what value of g would this definition be exactly correct?

11-29

a) Find the moment of inertia of a thin uniform rod of length L about a perpendicular axis through the rod at a distance $L/4$ from the center of mass.

b) If the rod is pivoted about this axis as a simple pendulum, what is the period?

11-30 A thin uniform rod is pivoted about a perpendicular axis through the rod, and its period as a physical pendulum is measured. By trial and error a second pivot point is found so that the period is the same as for the first point. Show that in this case the period depends only on the distance L between the two points and is given by $\tau = 2\pi(L/g)^{1/2}$.

11-31 A monkey wrench is pivoted at one end and allowed to swing as a physical pendulum. The period is 0.9 s and the pivot is 20 cm from the center of gravity.

a) What is the ratio of moment of inertia to mass for the wrench, about an axis through the pivot?

b) If the wrench was initially displaced 0.1 rad from its equilibrium position, what is the angular velocity of the wrench as it passes through the equilibrium position?

11-32 A solid disk of radius $R = 12$ cm oscillates as a physical pendulum about an axis perpendicular to the plane of the disk at a distance r from its center. (See Fig. 11-15.)

a) Calculate the period of oscillation (for small amplitudes) for the following values of r: 0, $R/4$, $R/2$, $3R/4$, R. [*Hint*. Use the parallel-axis theorem, Sec. 9-11.]

b) Let τ_0 represent the period when $r = R$, and τ the period at any other value of r. Construct a graph of the dimensionless ratio τ/τ_0 as a function of the dimensionless ratio r/R. (Note that the graph then describes the behavior of *any* solid disk, whatever its radius.)

11-33 It is desired to construct a pendulum of period 10 s.

a) What is the length of a *simple* pendulum having this period?

b) Suppose the pendulum must be mounted in a case not over 0.5 m high. Can you devise a pendulum, having a period of 10 s, that will satisfy this requirement?

11-34 A meterstick hangs from a horizontal axis at one end and oscillates as a physical pendulum. A body of small dimensions, and of mass equal to that of the meterstick, can be clamped to the stick at a distance h below the axis. Let τ represent the period of the system with the body attached, and τ_0 the period of the meterstick alone. Find the ratio τ/τ_0

a) when $h = 0.5$ m,

b) when $h = 1.0$ m.

c) Is there any value of h for which $\tau = \tau_0$? If so, find it and explain why the period is unchanged when h has this value.

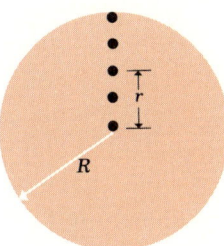

Figure 11-15

12 FLUID STATICS

12-1 Density

The term "fluid statics" is applied to the study of fluids at rest, and "fluid dynamics" to fluids in motion. The special branch of fluid dynamics relating to the flow of gases and of air in particular is called "aerodynamics."

A fluid is a substance that can flow. Hence the term includes both liquids and gases. Liquids and gases differ markedly in their compressibilities; a gas is easily compressed, while a liquid is practically incompressible. The small volume changes of a liquid under pressure can usually be neglected in this part of the subject.

The *density* of a homogeneous material is defined as its mass per unit volume. Units of density are one kilogram per cubic meter (1 kg·m^{-3}) in SI (mks) units, and one gram per cubic centimeter (1 g·cm^{-3}) in the cgs system. We shall represent density by the Greek letter ρ (rho):

$$\rho = \frac{m}{V}, \qquad m = \rho V. \tag{12-1}$$

Typical values of density at room temperature are given in Table 12-1.

Table 12-1 Densities

Material	Density, g·cm^{-3}	Material	Density, g·cm^{-3}
Aluminum	2.7	Silver	10.5
Brass	8.6	Steel	7.8
Copper	8.9	Mercury	13.6
Gold	19.3	Ethyl alcohol	0.81
Ice	0.92	Benzene	0.90
Iron	7.8	Glycerin	1.26
Lead	11.3	Water	1.00
Platinum	21.4	Sea water	1.03

The conversion factor

$$1 \text{ g} \cdot \text{cm}^{-3} = 1000 \text{ kg} \cdot \text{m}^{-3}$$

is useful in connection with this table.

The *specific gravity* of a material is the *ratio* of its density to that of water and is therefore a pure number. "Specific gravity" is a poor term, since it has nothing to do with gravity. "Relative density" would describe the concept more precisely.

Density measurements are an important analytical technique in a wide variety of circumstances. The condition of an automobile storage battery can be judged by measuring the density of the electrolyte, a sulfuric acid solution. As the battery discharges, the sulfuric acid (H_2SO_4) combines with lead in the battery plates to form insoluble lead sulfate ($PbSO_4$), decreasing the concentration of the solution. The density varies from about $1.30 \text{ g} \cdot \text{cm}^{-3}$ for a fully charged battery to $1.15 \text{ g} \cdot \text{cm}^{-3}$ for a discharged battery. Similarly, permanent-type antifreeze is usually a solution of ethylene glycol (density $1.12 \text{ g} \cdot \text{cm}^{-3}$) in water, with small quantities of additives to retard corrosion. The glycol concentration, which determines the freezing point of the solution, can be determined from a simple density measurement. Both these measurements are performed routinely in service stations with the aid of a simple hydrometer, which measures density by observation of the level at which a calibrated body floats in a sample of the solution. The hydrometer is discussed in Sec. 12-5.

Medical science finds many uses for density measurements; examples include tests on body fluids such as blood and urine. The density of normal human blood is about 1.04 to $1.06 \text{ g} \cdot \text{cm}^{-3}$. Since density increases with the concentration of red cells, an unusually low density may indicate anemia. Similarly, the usual density of urine is about $1.02 \text{ g} \cdot \text{cm}^{-3}$. Certain diseases lead to increased excretion of salts and a corresponding increase in urine density.

12-2 Pressure in a fluid

When the concept of hydrostatic pressure was introduced in Sec. 10-1, the *weight* of the fluid was neglected and the pressure was assumed the same at all points. It is a familiar fact, however, that atmospheric pressure is greater at sea level than at mountain altitudes and that the pressure in a lake or in the ocean increases with increasing depth below the surface. We therefore generalize the definition of pressure and define the pressure at *any point* as the ratio of the normal force dF exerted on a small area dA including the point, to the area dA:

$$p = \frac{dF}{dA}, \qquad dF = p\, dA. \qquad (12\text{-}2)$$

If the pressure is the same at all points of a finite plane surface of area A, these equations reduce to Eq. (10-3):

$$p = \frac{F}{A}, \qquad F = pA.$$

Let us find the general relation between the pressure p at any point in a fluid in a gravitational field and the elevation of the point y. If the

246 FLUID STATICS

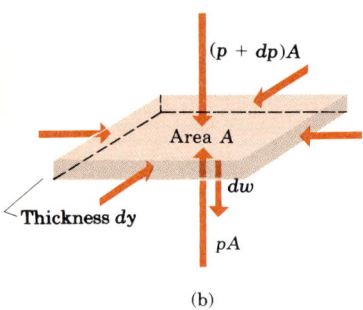

12-1 Forces on an element of fluid in equilibrium.

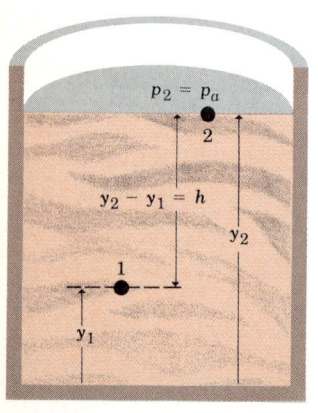

12-2

fluid is in equilibrium, every volume element is in equilibrium. Consider an element in the form of a thin slab, shown in Fig. 12-1, whose thickness is dy and whose faces have an area A. If ρ is the density of the fluid, the mass of the element is $\rho A\,dy$ and its weight dw is $\rho g A\,dy$. The force exerted on the element by the surrounding fluid is everywhere normal to its surface. By symmetry, the resultant horizontal force on its vertical sides is zero. The upward force on its lower face is pA, and the downward force on its upper face is $(p+dp)A$. Since it is in equilibrium,

$$\sum F_y = 0,$$

$$pA - (p+dp)A - \rho g A\,dy = 0,$$

and therefore

$$\frac{dp}{dy} = -\rho g \qquad (12\text{-}3)$$

Since ρ and g are both positive quantities, it follows that a positive dy (an increase of elevation) is accompanied by a negative dp (decrease of pressure). If p_1 and p_2 are the pressures at elevations y_1 and y_2 above some reference level, then

$$p_2 - p_1 = -\rho g(y_2 - y_1). \qquad (12\text{-}4)$$

Let us apply this equation to a liquid in an open vessel, such as that shown in Fig. 12-2. Take point 1 at any level and let p represent the pressure at this point. Take point 2 at the top where the pressure is atmospheric pressure, p_a. Then

$$p_a - p = -\rho g(y_2 - y_1),$$

$$p = p_a + \rho g h. \qquad (12\text{-}5)$$

Note that the shape of the containing vessel does not affect the pressure, and that the pressure depends only on depth. It also follows from Eq. (12-5) that if the pressure p_a is increased in any way, say by inserting a piston on the top surface and pressing down on it, the pressure p at any depth must increase by exactly the same amount. This fact was recognized in 1653 by the French scientist Blaise Pascal (1623–1662) and is called "Pascal's law." It is often stated: *"Pressure applied to an enclosed fluid is transmitted undiminished to every portion of the fluid and the walls of the containing vessel."* We can see now that it is not an independent principle but a necessary consequence of the laws of mechanics.

Pascal's law is illustrated by the operation of a hydraulic press or jack, shown in Fig. 12-3. A piston of small cross-sectional area a is used to exert a small force f directly on a liquid such as oil. The pressure $p = f/a$ is transmitted through the connecting pipe to a larger cylinder equipped with a larger piston of area A. Since the pressure is the same in both cylinders,

$$p = \frac{f}{a} = \frac{F}{A} \qquad \text{and} \qquad F = \frac{A}{a}f.$$

It follows that the hydraulic press is a force-multiplying device with a multiplication factor equal to the ratio of the areas of the two pistons. Barber chairs, dentist chairs, car lifts and jacks, and hydraulic brakes are all devices that make use of the principle of the hydraulic press.

In Eq. (12-4) it is assumed that the density ρ is constant; this is a reasonable assumption for liquids, which are relatively incompressible, but it is not realistic for gases. In calculating the pressure variation with altitude in the earth's atmosphere, for example, the variation in density must be included. This calculation is carried out in Sec. 17-2, Example 4.

12-3

12-3 Pressure gauges

The analysis in Sec. 12-2 can be applied directly to various devices used to measure pressure in a gas; these are usually called pressure *gauges*. The simplest type of pressure gauge is the open-tube manometer, illustrated in Fig. 12-4a. It consists of a U-shaped tube containing a liquid, one end of the tube being at the pressure p which it is desired to measure, while the other end is open to the atmosphere at pressure p_a.

The pressure at the bottom of the left column is $p + \rho g y_1$, while that at the bottom of the right column (the same point) is $p_a + \rho g y_2$, where ρ is the density of the manometric liquid. Since these pressures must be equal,

$$p + \rho g y_1 = p_a + \rho g y_2,$$

and

$$p - p_a = \rho g(y_2 - y_1) = \rho g h. \quad (12\text{-}6)$$

The pressure p is called the *absolute pressure,* whereas the difference $p - p_a$ between this and atmospheric pressure is called the *gauge pressure.* Thus the gauge pressure is proportional to the difference in height of the liquid columns.

The mercury barometer is a long glass tube that has been filled with mercury and then inverted in a dish of mercury, as shown in Fig. 12-4b. The space above the mercury column contains only mercury vapor, whose pressure, at room temperature, is so small that it may be neglected. It is easily seen that

$$p_a = \rho g(y_2 - y_1) = \rho g h. \quad (12\text{-}7)$$

As discussed in Chapter 10, the SI unit of pressure is *one pascal* (1 Pa), equal to one newton per square meter (1 N·m^{-2}). A related unit is *one bar,* defined as 10^5 Pa. Because mercury manometers and barometers are commonly used laboratory instruments, pressures are also sometimes described in terms of the height of the corresponding mercury column, as so many "inches of mercury" or "millimeters of mercury" (abbreviated mm Hg). The pressure due to a column of mercury one millimeter high has even been given a special name, *one Torr* (after Torricelli, inventor of the mercury barometer). Such units depend on the density of mercury, which varies with temperature, and on the value of g, which varies with location. For these and other reasons, they are gradually passing out of common use in favor of the pascal.

One type of blood-pressure gauge commonly used by physicians, called a *sphygmomanometer,* includes a form of manometer similar to Fig. 12-4a. Blood-pressure readings, such as 130/80, refer to the maxi-

12-4 (a) The open-tube manometer. (b) The barometer.

mum and minimum pressures, measured in millimeters of mercury or Torrs. Because of height differences, the hydrostatic pressure varies at different points in the body; the standard reference point is the upper arm, level with the heart. Pressure is also affected by the viscous nature of blood flow, by valves throughout the vascular system, and by changes in diameters of blood vessels to regulate pressure.

Example Compute the atmospheric pressure on a day when the height of the barometer is 76.0 cm.

Solution The height of the mercury column depends on ρ and g as well as on the atmospheric pressure. As mentioned, ρ varies with temperature, and g with latitude and elevation above sea level. If we assume $g = 9.8 \text{ m} \cdot \text{s}^{-2}$ and $\rho = 13.6 \times 10^3 \text{ kg} \cdot \text{m}^{-3}$,

$$p_a = \rho g h = (13.6 \times 10^3 \text{ kg} \cdot \text{m}^{-3})(9.8 \text{ m} \cdot \text{s}^{-2})(0.76 \text{ m})$$
$$= 101{,}300 \text{ N} \cdot \text{m}^{-2} = 1.013 \times 10^5 \text{ Pa}.$$

In English units,

$$0.76 \text{ m} = 30 \text{ in.} = 2.5 \text{ ft},$$
$$\rho g = 850 \text{ lb} \cdot \text{ft}^{-3},$$
$$p_a = 2120 \text{ lb} \cdot \text{ft}^{-2} = 14.7 \text{ lb} \cdot \text{in}^{-2}.$$

A pressure of 1.013×10^5 Pa $= 14.7$ lb$\cdot$in^{-2} is called one *atmosphere* (1 atm). A pressure of exactly 10^5 N$\cdot$m^{-2} is called one *bar*, and a pressure one one-thousandth as great is one *millibar*. Atmospheric pressures are of the order of 1000 millibars, and are now stated in terms of this unit by the National Weather Service.

The Bourdon-type pressure gauge is more convenient for most purposes than a liquid manometer. It consists of a flattened brass tube closed at one end and bent into a circular form. The closed end of the tube is connected by a gear and pinion to a pointer that moves over a scale. The open end of the tube is connected to the apparatus, the pressure within which is to be measured. When pressure is exerted within the flattened tube, it straightens slightly, just as a bent rubber hose straightens when water is admitted. The resulting motion of the closed end of the tube is transmitted to the pointer.

12-4 Pumps

Many devices of modern life include glass or metal containers from which nearly all the air has been exhausted; examples include electric light bulbs, TV picture tubes, thermos bottles, and many others. Successful operation of these devices requires air pressures as low as 10^{-6} Pa, and special pumps are used to produce these "high" vacuums. Two commonly used types are the *rotary oil pump* for pressures as low as 10^{-2} Pa, and the *mercury* or *oil diffusion pump* for pressures as low as 10^{-6} Pa.

Figure 12-5 depicts schematically a common type of rotary oil pump. The vessel to be exhausted is connected to the tube A, which communicates directly with the space marked B. As an eccentric cylinder C rotates in the direction shown, the point of contact between it and the inner walls of the stationary cylinder moves around in a clockwise direction, thereby trapping some air in the space marked E. The sliding

12-5 The rotary oil pump.

vane D is kept in contact with the rotating cylinder by the pressure of the rod F. When the air in E is compressed enough to increase the pressure slightly above atmospheric, the valve G opens and the air bubbles through the oil and leaves through an opening H in the upper plate. The cylinder is turned by a small electric motor.

For pressures below about 10^{-2} Pa, a diffusion pump is usually employed. In this pump a rapidly moving jet of mercury, or of a special oil of low vapor pressure, sweeps or pushes the air molecules away from the vessel to be exhausted. Air molecules from the vessel keep diffusing into the jet. A common type of diffusion pump is depicted in Fig. 12-6. A rotary oil pump is used to reduce the pressure within the diffusion pump to the low value necessary to ensure a well-defined jet to prevent air molecules from diffusing from the jet to the vessel. The air molecules that diffuse into the jet are removed by the oil pump (called in this case the *forepump*), and the mercury or oil condenses on the cool walls of the pump and returns to the well at the bottom.

Pumps are also used to compress gases and circulate liquids. Compressors for home refrigerators and air conditioners frequently use rotary pumps similar in principle to that shown in Fig. 12-5; reciprocating-piston pumps, equipped with pistons similar to those in gasoline engines, are also used for refrigerator compressors, and as air compressors. In these pumps, the essential principle is a varying volume and a valve arrangement to ensure one-way flow of the fluid being pumped.

In situations where it is important to prevent contamination of the fluid, the varying volume is often provided by a flexible diaphragm, as shown in Fig. 12-7. The center portion of the diaphragm moves up and down, driven by a motor. When it moves up, fluid is pushed into the chamber from the left by gravity or external air pressure; when it moves down, the fluid in the chamber is forced out the right side. Backflow is prevented by the two valves. Many automobile-engine fuel pumps operate on this principle, and the human heart comprises four such pumps, with the expansion and contraction of the four chambers by the heart muscles playing the role of the diaphragm in Fig. 12-7. Some artificial heart pumps use diaphragm pumps.

12-6 The diffusion pump.

12-7 A diaphragm pump.

12-5 Archimedes' principle

Buoyancy is a familiar phenomenon; a body immersed in water seems to have less weight than when immersed in air, and a body whose average density is less than that of the fluid in which it is immersed can float in that fluid. Examples are the human body in water, or a helium-filled balloon in air.

Archimedes' principle states that *when a body is immersed in a fluid, the fluid exerts an upward force on the body equal to the weight of the fluid that is displaced by the body.* To prove this principle, we consider an arbitrary portion of fluid at rest. In Fig. 12-8, the irregular outline is the surface bounding this portion of fluid. The short arrows represent the forces exerted by the surrounding fluid against small elements of the boundary surface.

Since the entire fluid is at rest, the x-component of the resultant of these surface forces is zero. The y-component of the resultant, F_y, must

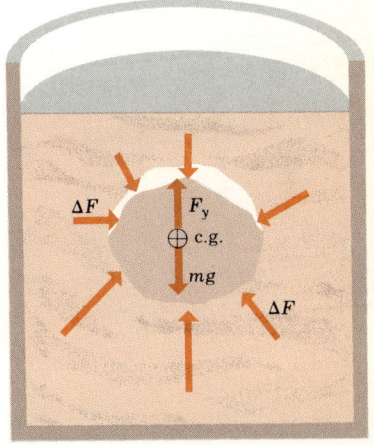

12-8 Archimedes' principle. The buoyant force F_y equals the weight of the displaced fluid.

equal the weight mg of the fluid inside the arbitrary surface, and its line of action must pass through the center of gravity of this fluid.

Now suppose that the fluid inside the surface is removed and replaced by a solid body having exactly the same shape. The pressure at every point is exactly the same as before; the force exerted on the body by the surrounding fluid is unaltered, and is therefore equal to the weight mg of fluid displaced. The line of action of this force again passes through the center of gravity of the displaced fluid.

The submerged body need not be in equilibrium. Its weight may be greater or less than F_y, and if it is not homogeneous, its center of gravity may not lie on the line of F_y. Therefore, in general, it will be acted on by a resultant force through its own center of gravity and by a torque, and will rise or fall and also rotate.

The weight of a balloon floating in air, or of a submarine floating at some depth below the surface of the water, is just equal to the weight of a volume of air, or water, that is equal to the volume of the balloon or submarine. That is, the average density of the balloon equals that of air, and the average density of the submarine equals the density of water.

A body whose average density is less than that of a liquid can float partially submerged at the free upper surface of the liquid. A familiar example is the hydrometer, shown in Fig. 12-9. The instrument sinks in the fluid until the weight of fluid it displaces is exactly equal to its own weight. In a fluid of greater density, less fluid is displaced for the same weight, so less of the instrument is immersed; i.e., in denser fluids it floats *higher*. This corresponds to the familiar fact that when swimming in sea water (density $1.03 \text{ g} \cdot \text{cm}^{-3}$), one's body floats higher than in fresh water.

The hydrometer is weighted at its bottom end so the upright position is stable, and a scale in the stem at the top permits direct density readings. Figure 12-9b shows a hydrometer commonly used to measure density of battery acid or antifreeze. When the tube at the bottom is immersed in the fluid and the bulb is squeezed to expel air and then released (like a giant medicine dropper), the resulting pressure difference causes the fluid to rise into the outer tube, and the hydrometer floats in this sample of the fluid.

Example 1 A block of brass has a mass of 0.5 kg and a density of $8.0 \times 10^3 \text{ kg} \cdot \text{m}^{-3}$. It is suspended from a string. Find the tension in the string if the block is in air, and if it is completely immersed in water.

Solution If the very small buoyant force of air can be neglected, the tension when the block is in air is just equal to its weight:

$$mg = (0.5 \text{ kg})(9.8 \text{ m} \cdot \text{s}^{-2}) = 4.9 \text{ N}.$$

When immersed in water the block experiences an upward buoyant force equal to the weight of water displaced. To find this, we can first find the volume of brass:

$$V = \frac{m}{\rho} = \frac{0.5 \text{ kg}}{8.0 \times 10^3 \text{ kg} \cdot \text{m}^{-3}} = 6.25 \times 10^{-5} \text{ m}^3.$$

Next, the weight of this volume of water is

$$w = mg = (\rho V)g = (1.0 \times 10^3 \text{ kg} \cdot \text{m}^{-3})(6.25 \times 10^{-5} \text{ m}^3)(9.8 \text{ m} \cdot \text{s}^{-2})$$
$$= 0.612 \text{ N}.$$

12-9 (a) A simple hydrometer. (b) Hydrometer used as a tester for battery acid or antifreeze.

The tension in the string is the actual weight less the upward buoyant force, or $4.9\text{ N} - 0.612\text{ N} = 4.29\text{ N}$.

A shortcut leading to the same result is as follows: The density of water is less than that of brass by a factor

$$\frac{1.0 \times 10^3 \text{ kg} \cdot \text{m}^{-3}}{8.0 \times 10^3 \text{ kg} \cdot \text{m}^{-3}} = \frac{1}{8}.$$

Therefore the weight of an amount of water of volume equal to that of the brass will be less than the weight of the brass by the same factor. Thus the weight of the displaced water is $(\frac{1}{8})(4.9\text{ N}) = 0.612\text{ N}$.

Example 2 If the water container in the above example sits on a scale, how does the scale reading change when the brass is immersed in the water?

Solution Taking the water and the brass together as a system, we note that its total weight does not change when the brass is immersed. Thus the sum of the two supporting forces, the string tension and the upward force of the scale and the container, must be the same in both cases. But the string tension was found to decrease by 0.612 N upon immersion of the brass, so the scale reading must *increase* by 0.612 N. Qualitatively we can see that the scale force must increase because when the brass is immersed, the water level in the container rises, increasing the pressure of water on the bottom of the container.

Example 3 When a sensitive analytical balance is used for precise weight determinations, corrections must be made for buoyancy effects in air if the object being weighed has a density appreciably different from that of the weighing standards or "weights," which are often brass. For example, suppose a block of wood having density $0.5 \times 10^3 \text{ kg} \cdot \text{m}^{-3}$ is just balanced on an equal-arm balance by brass weights of total mass 0.0500 kg. Because the wood is less dense than brass, the upward buoyant force of air on the wood is greater than that on the brass weights. Hence the true mass of wood is greater than 0.0500 kg by an amount corresponding to the difference of buoyant forces. The density of air is about $1.3 \text{ kg} \cdot \text{m}^{-3}$.

The apparent weight of each body is the difference between its true weight and the buoyant force of the air. If ρ_w, ρ_b, and ρ_a are the densities of the wood, brass, and air, and V_w and V_b are the volumes of the wood and brass, the apparent weights, which are equal, are

$$\rho_w V_w g - \rho_a V_w g = \rho_b V_b g - \rho_a V_b g.$$

The true mass of the wood is $\rho_w V_w$, and the true mass of the standard is $\rho_b V_b$. Hence,

True mass $= \rho_w V_w = \rho_b V_b + \rho_a(V_w - V_b) = \begin{pmatrix}\text{mass of} \\ \text{standard}\end{pmatrix} + \rho_a(V_w - V_b).$

In the specific example cited,

$$V_w = \frac{0.05 \text{ kg}}{0.5 \times 10^3 \text{ kg} \cdot \text{m}^{-3}} = 1.0 \times 10^{-4} \text{ m}^3 \text{ (approximately)}$$

$$V_b = \frac{0.05 \text{ kg}}{8.0 \times 10^3 \text{ kg} \cdot \text{m}^{-3}} = 6.25 \times 10^{-6} \text{ m}^3,$$

$$\rho_a = 1.3 \text{ kg} \cdot \text{m}^{-3}.$$

Hence,

$$\rho_a(V_w - V_b) = (1.3 \text{ kg}\cdot\text{m}^{-3})(1.0 \times 10^{-4} \text{ m}^3 - 6.25 \times 10^{-6} \text{ m}^3)$$
$$= 1.22 \times 10^{-4} \text{ kg}.$$

Therefore,

$$\text{True mass} = 0.050122 \text{ kg}.$$

Since analytical balances have a precision of the order of 1 mg (10^{-6} kg), this is a significant correction, of the order of 0.2%.

An interesting medical application of buoyancy is found in the technique of *hydrotherapy*. A patient may be unable to lift a limb because of disease or injury in the associated muscles or joints. When the body is immersed in water it becomes, in effect, nearly weightless because its average density is only a little greater than that of water. As a result, the forces required to move the limb are greatly reduced, and therapeutic exercise becomes possible.

*12-6 Forces against a dam

Suppose water stands at a depth H behind the vertical upstream face of a dam (Fig. 12-10). It exerts a certain resultant horizontal force on the dam, tending to slide it along its foundation, and a certain torque, tending to overturn the dam about the point O. We wish to find the total horizontal force. The *width* of the dam is L.

Figure 12-10b is a view of the upstream face of the dam. The pressure at an elevation y is

$$p = \rho g(H - y).$$

Atmospheric pressure can be omitted, since it also acts upstream against the other face of the dam. The force dF against the shaded strip is

$$dF = p \, dA = \rho g(H - y)L \, dy,$$

and the total force is

$$F = \int dF = \int_0^H \rho g L(H - y) \, dy$$
$$= \tfrac{1}{2}\rho g L H^2. \tag{12-8}$$

12-10 Forces on a dam.

The moment of the force dF about an axis through O is
$$d\Gamma = y\, dF = \rho g L y (H - y)\, dy.$$
The total torque about O is
$$\Gamma = \int d\Gamma = \int_0^H \rho g L y (H - y)\, dy$$
$$= \tfrac{1}{6}\rho g L H^3. \tag{12-9}$$

If $\bar{H}$ is the height above O at which the total force F would have to act to produce this torque,
$$F\bar{H} = \tfrac{1}{2}\rho g L H^2 \times \bar{H} = \tfrac{1}{6}\rho g L H^3,$$
$$\bar{H} = \tfrac{1}{3}H.$$

Hence the line of action of the resultant force is at $\tfrac{1}{3}$ of the depth above O, or at $\tfrac{2}{3}$ of the depth below the surface.

12-7 Surface tension

A liquid flowing slowly from the tip of a medicine dropper emerges not as a continuous stream but as a succession of drops. A sewing needle, if placed carefully on a water surface, makes a small depression in the surface and rests there without sinking, even though its density may be as much as 10 times that of water. When a clean glass tube of small bore is dipped into water, the water rises in the tube, but if the tube is dipped in mercury, the mercury is depressed. All these phenomena, and many others of a similar nature, are associated with the existence of a *boundary surface* between a liquid and some other substance.

All surface phenomena indicate that the surface of a liquid can be considered to be in a state of stress such that, for any line lying in or bounding the surface, the material on either side of the line exerts a *pull* on the material on the other side. This pull lies in the plane of the surface and is perpendicular to the line. In Fig. 12-11, a wire ring has a loop of thread attached to it as shown. When the ring and thread are dipped in a soap solution and removed, a thin film of liquid is formed in which the thread "floats" freely, as shown in part (a). If the film inside

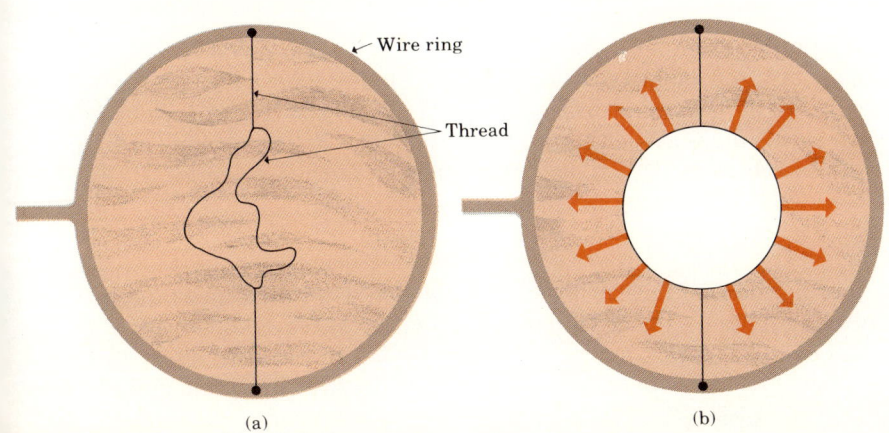

12-11 A wire ring with a flexible loop of thread, dipped in a soap solution, (a) before and (b) after puncturing the surface films inside the loop.

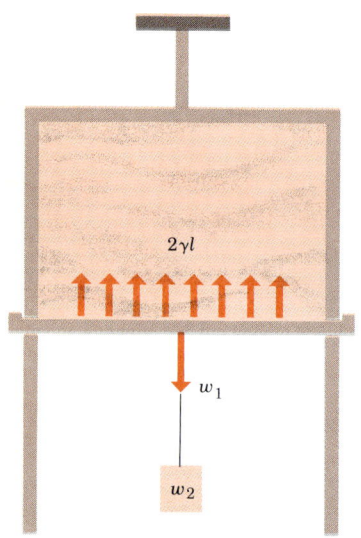

12-12 The horizontal slide wire is in equilibrium under the action of the upward surface force $2\gamma l$ and downward pull $w_1 + w_2$.

the loop of thread is punctured, the thread springs out into a circular shape as in part (b), as if the surfaces of the liquid were pulling radially outward on it, as shown by the arrows. Presumably, the same forces were acting before the film was punctured, but since there was film on *both* sides of the thread the net force exerted by the film on every portion of the thread was zero.

Another simple apparatus for demonstrating surface effects is shown in Fig. 12–12. A piece of wire is bent into the shape of a U and a second piece of wire is used as a slider. When the apparatus is dipped in a soap solution and removed, the slider (if its weight w_1 is not too great) is quickly pulled up to the top of the U. It may be held in equilibrium by adding a second weight w_2. Surprisingly, the same total force $F = w_1 + w_2$ will hold the slider at rest in *any* position, regardless of the area of the liquid film, provided the film remains at constant temperature. That is, the force does not increase as the surface is stretched farther. This is very different from the elastic behavior of a sheet of rubber, for which the force would be greater as the sheet was stretched.

Although a soap film like that in Fig. 12–12 is very thin, its thickness is still enormous compared with the size of a molecule. Hence it can be considered as made up chiefly of bulk liquid, bounded by two surface layers a few molecules thick. When the crossbar in Fig. 12–12 is pulled down and the area of the film is increased, molecules formerly in the main body of the liquid move into the surface layers. That is, these layers are not "stretched" as a rubber sheet would be; more surface is created by molecules moving from the bulk liquid.

Let l be the length of the wire slider. Since the film has two surfaces, the total length along which the surface force acts on the slider is $2l$. The *surface tension* γ in the film is defined as *the ratio of the surface force to the length* (perpendicular to the force) *along which the force acts.* Hence, in this case,

$$\gamma = \frac{F}{2l}. \tag{12-10}$$

The SI unit of surface tension is the newton per meter ($1 \text{ N} \cdot \text{m}^{-1}$). This unit is not in common use; the usual unit is the cgs unit, the dyne per centimeter ($1 \text{ dyn} \cdot \text{cm}^{-1}$). The conversion factor is

$$1 \text{ N} \cdot \text{m}^{-1} = 1000 \text{ dyn} \cdot \text{cm}^{-1}.$$

Another example is shown in Fig. 12–13. A circular wire of circumference l is lifted out from the body of a liquid. The additional force F needed to balance the surface forces $2\gamma l$ due to the two surface films on each side is measured either by the stretch of a delicate spring or by the twist of a torsion wire. The surface tension is then given by

$$\gamma = \frac{F}{2l}.$$

12-13 Lifting a circular wire of length l out of a liquid requires an additional force F to balance the surface forces $2\gamma l$. This method is commonly used to measure surface tension.

Other methods of measuring surface tension will be apparent in what is to follow. Some typical values are shown in Table 12–2.

The surface tension of a liquid surface in contact with its own vapor or with air is found to depend *only* on the nature of the liquid and on the temperature. Surface tension usually decreases as temperature increases; Table 12-2 illustrates this behavior for water.

Table 12-2 Experimental values of surface tension

Liquid in contact with air	T, °C	Surface tension, dyn·cm^{-1}
Benzene	20	28.9
Carbon tetrachloride	20	26.8
Ethyl alcohol	20	22.3
Glycerin	20	63.1
Mercury	20	465
Olive oil	20	32.0
Soap solution	20	25.0
Water	0	75.6
Water	20	72.8
Water	60	66.2
Water	100	58.9
Oxygen	−193	15.7
Neon	−247	5.15
Helium	−269	0.12

A surface under tension tends to contract until it has the minimum area consistent with any fixed boundaries that may be present and with pressure differences that may be present on opposite sides of the surface. For example, a drop of liquid under no external forces, or in free fall in vacuum, is always spherical in shape because the sphere has smaller surface area for a given volume than has any other geometric shape. Figure 12-14 is a beautiful example of formation of spherical droplets in a very complex phenomenon, the impact of a drop on a rigid surface. This was taken by Dr. Edgerton of M.I.T., one of the pioneers in the development of high-speed photographic technique.

12-14 A drop of milk splashes on a hard surface. (Reproduced from *Flash*, courtesy of Ralph S. Hale & Co.)

12-8 Pressure difference across a surface film

A soap bubble consists of two spherical surface films very close together, with liquid between. Surface tension makes the films tend to contract, but as the bubble contracts it compresses the inside air, increasing the interior pressure to a point that prevents further contraction. To obtain a relation between surface tension and this excess pressure, we may apply the principles of statics to one-half of the bubble.

As a preliminary, consider first a small element ΔA of the surface, shown in Fig. 12–15. Suppose the air pressure above this element is p and that below is p_a. The force normal to the element is therefore $(p - p_a)\Delta A$. The component of this force in the x-direction is

$$(p - p_a)\Delta A \cos \theta.$$

But $\Delta A \cos \theta$ is the area projected on a plane perpendicular to the y-axis. The force in the y-direction is therefore the difference of pressure multiplied by the *projected area* in the y-direction.

Now consider the half-bubble shown in Fig. 12–16. The other half (not shown) exerts an upward force equal to twice the surface tension times the perimeter or

$$F(\text{up}) = (2\gamma)(2\pi R).$$

The downward force on this half-bubble is equal to the pressure difference $p - p_a$ multiplied by the area obtained by projecting the half-bubble on a plane perpendicular to the direction in question. Since this projected area is πR^2,

$$F(\text{down}) = (p - p_a)\pi R^2.$$

Since the half-bubble is in equilibrium,

$$(p - p_a)\pi R^2 = 4\pi R\gamma,$$

or

$$p - p_a = \frac{4\gamma}{R} \quad \text{(soap bubble)}. \tag{12-11}$$

If the surface tension remains constant (this means constant temperature), the pressure difference is greater for small bubbles than for large ones. If two bubbles are blown at opposite ends of a pipe, the smaller of the two will force air into the larger; the smaller one will get still smaller, and the larger will increase in size.

For a liquid drop, which has only *one* surface film, the difference between pressure of the liquid and that of the outside air is half that for a soap bubble:

$$p - p_a = \frac{2\gamma}{R} \quad \text{(liquid drop)}. \tag{12-12}$$

Example Calculate the excess pressure inside a drop of mercury of diameter 4 mm at 20°C.

Solution

$$p - p_a = \frac{2\gamma}{R} = \frac{(2)(465 \times 10^{-3} \text{ N} \cdot \text{m}^{-1})}{0.002 \text{ m}}$$
$$= 465 \text{ N} \cdot \text{m}^{-2} \cong 0.005 \text{ atm}.$$

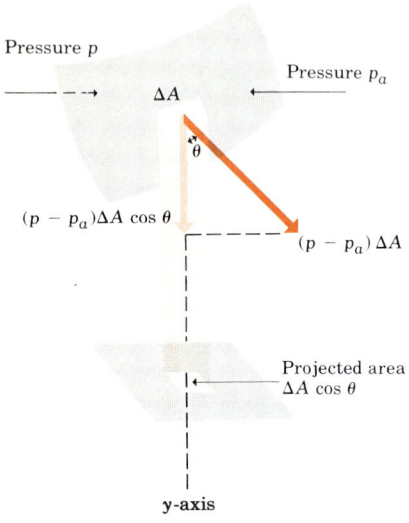

12–15 The force in the y-direction is the difference of pressure multiplied by the projected area in the y-direction.

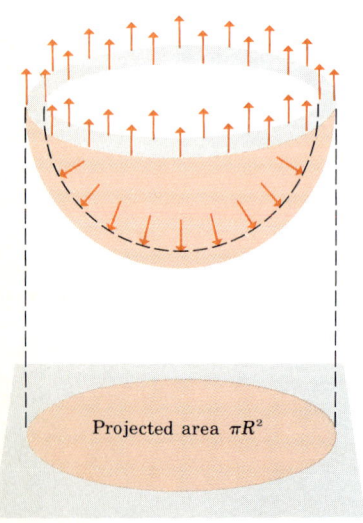

12–16 Equilibrium of half a soap bubble. The force exerted by the other half is $(2\gamma)(2\pi R)$, and the net force exerted by the air inside and outside the bubble is the pressure difference times projected area, or $(p - p_0)R^2$.

*12-9 Contact angle and capillarity

The preceding sections have been concerned with surface films lying in the boundary between a liquid and a gas. Surface films also exist between a solid wall and a liquid, and between a solid and a vapor. The three boundaries and their accompanying films are shown schematically in Fig. 12-17. The films are only a few molecules thick. Associated with each film is an appropriate surface tension. Thus,

γ_{SL} = surface tension of the solid-liquid film,
γ_{SV} = surface tension of the solid-vapor film,
γ_{LV} = surface tension of the liquid-vapor film.

12-17 Surface films exist at the solid-vapor boundary as well as at the liquid-vapor boundary.

The symbol γ without subscripts, defined and used in the preceding sections, now appears as γ_{LV}.

The curvature of the surface of a liquid near a solid wall depends upon the *difference* between γ_{SV} and γ_{SL}. Consider a portion of a glass wall in contact with methylene iodide, as shown in Fig. 12-18a. When γ_{SV} is greater than γ_{SL}, the line along which the three films meet is pulled upward, as in Figs. 12-17 and 12-18a. Angle θ in Fig. 12-18 is called the *contact angle,* and we see that when γ_{SV} is greater than γ_{SL}, the angle of contact is between zero and 90°. Figure 12-18b shows a case where γ_{SV} is *less* than γ_{SL}, the line of intersection of the films is pulled *downward,* and the contact angle is between 90° and 180°. In Fig. 12-18c, γ_{SV} and γ_{SL} are very nearly equal and the contact angle is 90°. Table 12-3 gives contact angles of representative liquids in containers of several materials.

12-18 The surface of a liquid near a solid wall is curved if the solid-vapor surface tension γ_{SV} differs from the solid-liquid surface tension γ_{SL}.

Impurities and adulterants present in or added to a liquid may alter the contact angle considerably. Wetting agents or detergents change the contact angle from a large value, greater than 90°, to a value much *smaller* than 90°. Conversely, waterproofing agents applied to cloth cause the contact angle of water in contact with the cloth to be larger than 90°. The effect of a detergent on a drop of water resting on a block of paraffin is shown in Fig. 12-19.

An important phenomenon resulting from surface tension is the elevation of a liquid in an open tube of small cross section. The term

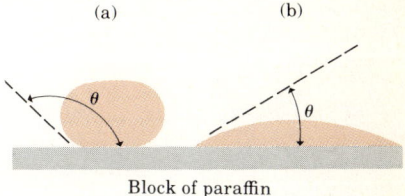

12-19 Effect of decreasing contact angle by a wetting agent.

258 FLUID STATICS

Table 12-3 Contact angles

Liquid	Wall	Contact angle
α-Bromonaphthalene ($C_{10}H_7Br$)	Soda-lime glass	5°
	Lead glass	6°45′
	Pyrex	20°30′
	Fused quartz	21°
Methylene iodide (CH_2I_2)	Soda-lime glass	29°
	Lead glass	30°
	Pyrex	29°
	Fused quartz	33°
Water	Paraffin	107°
Mercury	Soda-lime glass	140°

capillarity, used to describe effects of this sort, originates from the description of such tubes as "capillary" or "hairlike." In the case of a liquid that wets the tube, the contact angle is less than 90° and the liquid rises until an equilibrium height y is reached, as shown in Fig. 12-20a. The curved liquid surface in the tube is called a *meniscus.*

If the tube is a cylinder of radius r, the liquid makes contact with the tube along a line of length $2\pi r$. When we isolate the cylinder of liquid of height y and radius r, along with its liquid–vapor film, the total upward force is

$$F = 2\pi r\, \gamma_{LV} \cos \theta.$$

The downward force is the weight w of the cylinder; this is equal to the weight-density ρg times the volume, which is approximately $\pi r^2 y$, neglecting the small volume of the meniscus itself. Thus

$$w = \rho g \pi r^2 y.$$

Since the cylinder is in equilibrium,

$$\rho g \pi r^2 y = 2\pi r\, \gamma_{LV} \cos \theta,$$

or

$$y = \frac{2 \gamma_{LV} \cos \theta}{\rho g r}. \qquad (12\text{-}13)$$

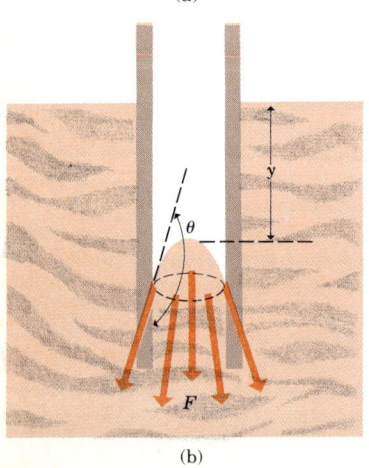

12-20 Surface tension forces on a liquid in a capillary tube. The liquid rises if $\theta < 90°$ and is depressed if $\theta > 90°$.

The same equation holds for capillary depression, shown in Fig. 12-20b. Capillarity accounts for the rise of ink in blotting paper, the rise of lighter fluid in the wick of a cigarette lighter, and many other common phenomena.

Capillarity is very important in a variety of life processes. A familiar example is the rising of water (actually a dilute aqueous solution) from the roots of a plant to its foliage, due partly to capillarity and osmotic pressure developed in the roots. In the higher animals, including humans, blood is pumped through the arteries and veins, but capillarity is still important in the smallest blood vessels, which indeed are called capillaries.

A related phenomenon is that of *negative pressure.* Although ordinarily the stress in a liquid is of a compressive nature, there are circumstances in which liquids can sustain actual *tensile* stresses. Imagine a cylindrical tube, closed at one end and having a tight-fitting piston at the other. Suppose the tube is completely filled with a liquid that wets both the inner surface of the tube and the piston face; that is, the molecules of liquid adhere to the surfaces. If we pull the piston out, we expect that the liquid will pull away from the piston, leaving an empty space, but if the surfaces are very clean and the liquid very pure, an actual tensile stress, accompanied by an increase in volume, can be observed. It is as though we are stretching the liquid, and it is prevented from pulling away from the walls of the container by adhesive forces.

With water, tensile stresses as large as 300 atm have been observed. In the laboratory this situation is a highly unstable one; a liquid under tension tends to break up into many small droplets. In tall trees, however, negative pressures are a regular occurrence. It is generally believed that negative pressure is an important mechanism for transport of water and nutrients from the roots to the leaves in the small xylem tubes (diameter of the order of 0.1 mm) in the growing layers of the tree.

Questions

12-1 Why are very narrow heels found on some women's shoes more likely to make dents in a soft wood floor than wider heels, even though both may support the same weight?

12-2 A rubber hose is attached to a funnel, and the free end is bent around to point upward. When water is poured into the funnel, it rises in the hose to the same level as in the funnel, despite the fact that the funnel has a lot more water in it than the hose. Why?

12-3 If you drive a car into a lake 10 m deep, do you have a better chance of escaping from the car if you try to open a door or if you open a window and crawl out? Why?

12-4 Is the pressure inside a siphon tube greater or less than atmospheric pressure? Does this suggest a limitation on the maximum height over which a siphon can be used?

12-5 How can one instrument function as both a barometer and an altimeter?

12-6 A pan of water sits on a kitchen scale. You stick your finger into the water, without touching the pan. Does the scale reading change? How?

12-7 In describing the size of a large ship, one uses such expressions as "it displaces 20,000 tons." What does this mean? Can the weight of the ship be obtained from this information?

12-8 A popular, though expensive, sport is hot-air ballooning, in which a large nylon balloon is filled with air heated by a gas burner at the bottom. Why must the air be heated? How does the balloonist control his ascent and decent?

12-9 A submarine is 50 m below the surface of the ocean, with its weight exactly balanced by the buoyant force. If it descends to 100 m, does the buoyant force increase, decrease, or remain the same?

12-10 A rigid lighter-than-air balloon filled with helium cannot continue to rise indefinitely. Why not? What determines the maximum height it can attain?

12-11 Does the buoyant force on an object immersed in water change if the object and the container of water are placed in an elevator that accelerates upward?

12-12 Some people can float in fresh water only by taking a deep breath; when they exhale they sink. Why?

12-13 If a person can barely float in sea water, will he or she float more easily in fresh water, or be unable to float at all? Why?

12-14 The purity of gold can be tested by weighing it in air and in water. How? Why is it not feasible to make a fake gold brick by gold-plating some cheaper material?

12-15 Explain how the admonition "go soak your head" can be taken literally to measure the weight of your head without removing it from your body.

12-16 A possible new technology made feasible by the "zero-gravity" environment of an orbiting space station is the making of metal foam. Why is it easier to make metal foam under zero-g conditions than on earth?

12-17 Hot water running into the sink doesn't splash as much as cold water, and it sounds different. Why?

12-18 Why can't a skin diver obtain an air supply at any desired depth by breathing through a "snorkel," a tube connected to his face mask and having its upper end above the water surface?

12-19 Suppose the door of a room makes an airtight but frictionless fit in its frame. Do you think you could open the door if the air pressure on one side were standard atmospheric pressure and that on the other side differed from standard by 1%?

Problems

12-1 The piston of a hydraulic automobile lift is 30 cm in diameter. What pressure, in pascals, is required to lift a car of mass 1500 kg? Also express this pressure in atmospheres.

12-2 The expansion tank of a household hot-water heating system is open to the atmosphere and is 10 m above a pressure gauge attached to the furnace. What is the gauge pressure at the furnace, in pascals? In atmospheres?

12-3 The liquid in the open-tube manometer in Fig. 12–4a is mercury, and $y_1 = 3$ cm, $y_2 = 8$ cm. Atmospheric pressure is 970 millibars.

a) What is the absolute pressure at the bottom of the U-tube?

b) What is the absolute pressure in the open tube, at a depth of 5 cm below the free surface?

c) What is the absolute pressure of the gas in the tank?

d) What is the gauge pressure of the gas, in "cm of mercury"?

e) What is the gauge pressure in "cm of water"?

12-4 What gauge pressure must a pump produce in order to pump water from the bottom of Grand Canyon (730 m elevation) to Indian Gardens (1370 m)? Express your results in pascals and in atmospheres.

12-5 A diving bell is to be designed to withstand the pressure of water at a depth of 2000 ft.

a) If sea water weighs 64 lb per cubic foot, what is the pressure at this depth?

b) What is the total force on a circular glass window 6 inches in diameter?

12-6 A common method for testing gold for purity is to measure its density by weighing it in air and then in water, as discussed in Sec. 12–5. In an era of rising gold prices, a swindler proposed to make a fake gold ingot by using a hollow slab of iridium (density 22.5 g·cm^{-3}) and plating it with a thin layer of gold (density 19.3 g·cm^{-3}).

a) To make a fake ingot of total mass 0.5 kg, what should be the total volume?

b) What is the volume of the interior air space?

c) Is iridium really any cheaper than gold?

12-7 According to advertising claims, a certain small car will float in water.

a) If the mass of the car is 1000 kg and its interior volume 4.0 m^3, what fraction of the car is immersed when it floats? The buoyancy of steel and other materials may be neglected.

b) As water gradually leaks in and displaces the air in the car, what fraction of the interior volume is water when the car sinks?

12-8

a) A small test tube, partially filled with water, is inverted in a large jar of water and floats, as shown in Fig. 12–21. The lower end of the test tube is open and the top of the large jar is covered by a tightly fitting rubber membrane. When the membrane is pressed down the test tube sinks, and when the membrane is released it rises again. Explain. (A hollow glass figure in human form is often used instead of the test tube and is called a "Cartesian diver.")

b) A torpedoed ship sinks below the surface of the ocean. If the depth is sufficiently great, is it possible for the ship to remain suspended in equilibrium at some point above the ocean bottom?

Figure 12–21

12-9 When an iceberg floats in sea water, what fraction of its volume is submerged? (The ice, being of glacial origin, is fresh water.)

12-10 A slab of ice floats on a fresh-water lake. What minimum volume must the slab have in order for an 80-kg man to be able to stand on it without getting his feet wet?

12-11 A tube 1 cm^2 in cross section is attached to the top of a vessel 1 cm high and of cross section 100 cm^2. Water is poured into the system, filling it to a depth of 100 cm above the bottom of the vessel, as in Fig. 12–22.

a) What is the force exerted by the water against the bottom of the vessel?

b) What is the weight of the water in the system?

c) Explain why (a) and (b) are not equal.

Figure 12–22

12-12 A piece of gold–aluminum alloy weighs 10 lb. When suspended from a spring balance and submerged in water, the balance reads 8 lb. What is the weight of gold in the alloy if the specific gravity of gold is 19.3 and the specific gravity of aluminum is 2.5?

12-13 What is the area of the smallest block of ice 0.5 m thick that will just support a man of mass 100 kg? The specific gravity of the ice is 0.917, and it is floating in fresh water.

12-14 A cubical block of wood 10 cm on a side floats at the interface between oil and water as in Fig. 12–23, with its lower surface 2 cm below the interface. The density of the oil is 0.6 g·cm^{-3}.

a) What is the mass of the block?
b) What is the gauge pressure at the lower face of the block?

Figure 12-23

12-15 The densities of air, helium, and hydrogen (at standard conditions) are, respectively, 1.29 kg·m^{-3}, 0.178 kg·m^{-3}, and 0.0899 kg·m^{-3}. What is the volume in cubic meters displaced by a hydrogen-filled dirigible that has a total "lift" of 10,000 kg? What would be the "lift" if helium were used instead of hydrogen?

12-16 A piece of wood is 0.5 m long, 0.2 m wide, and 0.02 m thick. Its specific gravity is 0.6. What volume of lead must be fastened underneath to sink the wood in calm water so that its top is just even with the water level?

12-17 A cubical block of wood 10 cm on a side and of density 0.5 g·cm^{-3} floats in a jar of water. Oil of density 0.8 g·cm^{-3} is poured on the water until the top of the oil layer is 4 cm below the top of the block.

a) How deep is the oil layer?
b) What is the gauge pressure at the lower face of the block?

12-18 A cubical block of steel (density = 7.8 g·cm^{-3}) floats on mercury (density = 13.6 g·cm^{-3}).

a) What fraction of the block is above the mercury surface?
b) If water is poured on the mercury surface, how deep must the water layer be so that the water surface just rises to the top of the steel block?

12-19 Block A in Fig. 12–24 hangs by a cord from spring balance D and is submerged in a liquid C contained in beaker B. The mass of the beaker is 1 kg; the mass of the liquid is 1.5 kg. Balance D reads 2.5 kg and balance E reads 7.5 kg. The volume of block A is 0.003 m^3.

a) What is the mass per unit volume of the liquid?
b) What will each balance read if block A is pulled up out of the liquid?

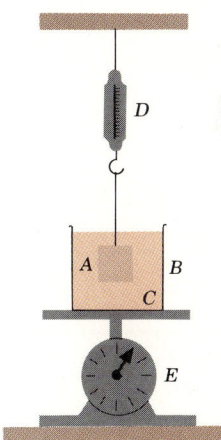

Figure 12-24

12-20 A hollow sphere of inner radius 9 cm and outer radius 10 cm floats half submerged in a liquid of specific gravity 0.8.

a) Calculate the density of the material of which the sphere is made.
b) What would be the density of a liquid in which the hollow sphere would just float completely submerged?

12-21 Two spherical bodies having the same diameter are released simultaneously from the same height. If the mass of one is 10 times that of the other and if the air resistance on each is the same, show that the heavier body will arrive at the ground first.

12-22 When a life preserver having a volume of 0.03 m^3 is immersed in sea water (specific gravity 1.03), it will just support an 80-kg man (specific gravity 1.2) with $\frac{2}{10}$ of his volume above water. What is the mass per unit volume of the material composing the life preserver?

12-23 A block of balsa wood placed in one scale pan of an equal-arm balance is found to be exactly balanced by a 100-g brass "weight" in the other scale pan. Find the true mass of the balsa wood, if its specific gravity is 0.15.

12-24 A 1500-kg cylindrical can buoy floats vertically in salt water (specific gravity = 1.03). The diameter of the buoy is 1.0 m. Calculate:

a) the additional distance the buoy will sink when a 100-kg man stands on top,
b) the period of the resulting vertical simple harmonic motion when the man dives off.

12-25 A cubical block of wood 1 ft on a side is weighted so that its center of gravity is at the point shown in Fig. 12-25a, and it floats in water with one-half its volume submerged. Compute the restoring torque when the block is "heeled" at an angle of 45° as in Fig. 12-25b.

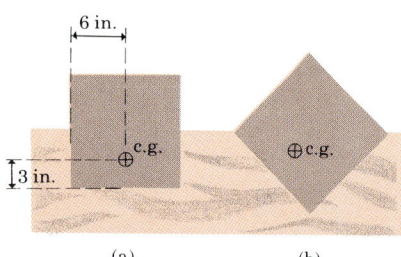

Figure 12-25

12-26 A hydrometer consists of a spherical bulb and a cylindrical stem of cross section 0.4 cm². The total volume of bulb and stem is 13.2 cm³. When immersed in water the hydrometer floats with 8 cm of the stem above the water surface. In alcohol, 1 cm of the stem is above the surface. Find the density of the alcohol.

12-27 The following is quoted from a letter. How would you reply?

It is the practice of carpenters hereabouts, when laying out and leveling up the foundations of relatively long buildings, to use a garden hose filled with water, into the ends of the hose being thrust glass tubes 10 to 12 inches long.

The theory is that the water, seeking a common level, will be of the same height in both the tubes and thus effect a level. Now the question rises as to what happens if a bubble of air is left in the hose. Our greybeards contend the air will not affect the reading from one end to the other. Others say that it will cause important inaccuracies.

Can you give a relatively simple answer to this question, together with an explanation? I include a rough sketch [Fig. 12-26] of the situation that caused the dispute.

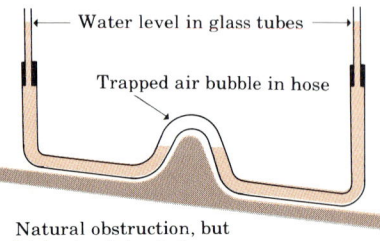

Figure 12-26

12-28 A swimming pool measures $25 \times 8 \times 3$ m deep. Compute the force exerted by the water against either end and against the bottom.

12-29 The upper edge of a vertical gate in a dam lies along the water surface. The gate is 2 m wide and is hinged along the bottom edge, which is 3 m below the water surface. What is the torque about the hinge?

12-30 The upper edge of a gate in a dam runs along the water surface. The gate is 2 m high and 3 m wide and is hinged along a horizontal line through its center. Calculate the torque about the hinge.

12-31 Figure 12-27 is a cross-sectional view of a masonry dam whose length perpendicular to the diagram is 30 m. The depth of water behind the dam is 10 m. The masonry of which the dam is constructed has a density of 3000 kg·m⁻³.

a) Find the dimensions x and $2x$, if the weight of the dam is to be 10 times as great as the horizontal force exerted on it by the water.

b) Is the dam then stable with respect to its overturning about the edge through point O?

c) How does the size of the reservoir behind the dam affect the answers above?

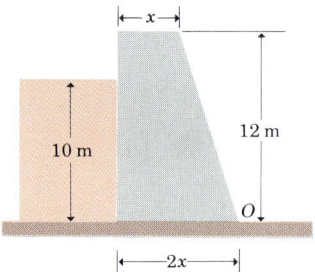

Figure 12-27

12-32 A U-tube of length l (see Fig. 12-28) contains a liquid. What is the difference in height between the liquid columns in the vertical arms,

a) if the tube has an acceleration a toward the right, and

b) if the tube is mounted on a horizontal turntable rotating with an angular velocity ω, with one of the vertical arms on the axis of rotation?

c) Explain why the difference in height does not depend on the density of the liquid, or on the cross-sectional area of the tube. Would it be the same if the vertical tubes did not have equal cross sections? Would it be the same if the horizontal portion were tapered from one end to the other?

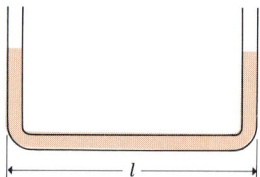

Figure 12-28

12-33 Compare the tension of a soap bubble with that of a rubber balloon in the following respects:

a) Has each a surface tension?

b) Does the surface tension depend on area?

c) Is Hooke's law applicable?

12-34 Water can rise to a height y in a certain capillary. Suppose that this tube is immersed in water so that only a length $y/2$ is above the surface. Will you have a fountain or not? Explain your reasoning.

12-35 A capillary tube is dipped in water with its lower end 10 cm below the water surface. Water rises in the tube to a height of 4 cm above that of the surrounding liquid, and the angle of contact is zero. What gauge pressure is required to blow a hemispherical bubble at the lower end of the tube?

12-36 A glass tube of inside diameter 1 mm is dipped vertically into a container of mercury, with its lower end 1 cm below the mercury surface.

a) What must be the gauge pressure of air in the tube to blow a hemispherical bubble at its lower end?

b) To what height will mercury rise in the tube if the air pressure in the tube is 3000 Pa below atmospheric? The angle of contact between mercury and glass is 140°.

12-37 Find the excess pressure inside a water drop of radius 2 mm, and inside one of radius 0.01 mm, typical of water droplets in fog. Assume $T = 20°C$.

12-38 What radius must a water drop have for the difference between inside and outside pressures to be 0.1 atm?

12-39 On a day when the atmospheric pressure is 950 millibars,

a) what would be the height of the mercury column in a barometric tube of inside diameter 2 mm?

b) What would be the height in the absence of any surface-tension effects?

c) What is the minimum diameter a barometric tube may have in order that the correction for capillary depression shall be less than 0.01 cm of mercury?

12-40

a) Derive the expression for the height of capillary rise in the space between two parallel plates dipping in a liquid.

b) Two glass plates, parallel to each other and separated by 0.5 mm, are dipped in water. To what height will the water rise between them? Assume zero angle of contact.

12-41 A tube of circular cross section and outer radius 0.14 cm is closed at one end. This end is weighted and the tube floats vertically in water, heavy end down. The total mass of the tube and weights is 0.20 g. If the angle of contact is zero, how far below the water surface is the bottom of the tube? Assume a water temperature of 20°C.

12-42 Find the gauge pressure, in dynes per square centimeter, in a soap bubble 5 cm in diameter. The surface tension is $25 \text{ dyn} \cdot \text{cm}^{-1} = 25 \times 10^{-3} \text{ N} \cdot \text{m}^{-1}$.

12-43 Two large glass plates are clamped together along one edge and separated by spacers a few millimeters thick along the opposite edge to form a wedge-shaped air film. These plates are then placed vertically in a dish of colored liquid. Show that the edge of the liquid forms an equilateral hyperbola.

12-44 When mercury is poured onto a flat glass surface which it does not wet, it spreads out into a pool of uniform thickness regardless of the size of the pool. Find the thickness of the pool.

12-45 Suppose the xylem tubes in the actively growing outer layer of a tree are uniform cylinders, and that the rising of sap is due entirely to capillarity, with a contact angle of 45° and surface tension of $0.05 \text{ N} \cdot \text{m}^{-1}$. What is the maximum radius of the tubes for a tree 20 m tall? Assume the density of sap to be the same as that of water.

12-46 A tree 50 m tall has xylem tubes (sap-carrying pipes) in the form of uniform cylinders of radius 2×10^{-4} mm. If the surface tension is $0.05 \text{ N} \cdot \text{m}^{-1}$ and the contact angle 45°, what minimum pressure must be developed in the roots in order for the sap to reach the top of the tree, if the pressure at the top is one atmosphere? (See Problem 12-45.)

13 FLUID DYNAMICS

13-1 Introduction

Fluid dynamics is the study of fluids in motion. It is one of the most complex branches of mechanics, as illustrated by such familiar examples of fluid flow as a river in flood or a swirling cloud of cigarette smoke. While each drop of water or each smoke particle is governed by Newton's laws of motion, the resulting equations can be exceedingly complex. Fortunately, many situations of practical importance can be represented by idealized models that are simple enough to permit detailed analysis.

To begin, we shall consider only a so-called *ideal fluid,* that is, one that is incompressible and has no internal friction or viscosity. The assumption of incompressibility is usually a good approximation for liquids. A gas can also be treated as incompressible provided the flow is such that pressure differences are not too great. Internal friction in a fluid gives rise to shear stresses when two adjacent layers of fluid move relative to each other, or when the fluid flows inside a tube or around an obstacle. In some cases these shear forces can be neglected in comparison with gravitational forces and forces arising from pressure differences.

The path followed by an element of a moving fluid is called a *line of flow*. In general, the velocity of the element changes in both magnitude and direction along its line of flow. If every element passing through a given point follows the same line of flow as that of preceding elements, the flow is said to be *steady* or *stationary*. When any given flow is first started, it passes through a nonsteady state, but in many instances the flow becomes steady after a certain period of time has elapsed. In steady flow, the velocity at each point of space remains constant in time, although the velocity of a particular particle of the fluid may change as it moves from one point to another.

A *streamline* is defined as a curve whose tangent, at any point, is in the direction of the fluid velocity at that point. In steady flow, the streamlines coincide with the lines of flow.

If we construct all of the streamlines passing through the periphery of an element of area, such as the area A in Fig. 13–1, these lines enclose

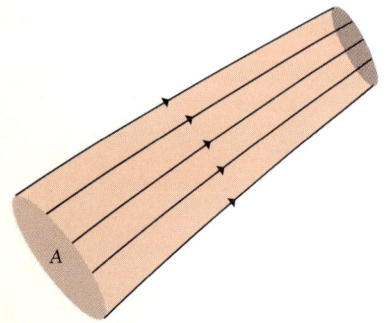

13-1 A flow tube bounded by streamlines.

(a)   (b)   (c)   (d)

13-2 (a), (b), (c): Streamline flow around obstacles of various shapes. (d) Flow in a channel of varying cross-sectional area.

a tube called a *flow tube* or *tube of flow*. From the definition of a streamline, no fluid can cross the side walls of a tube of flow; in steady flow there can be no mixing of the fluids in different flow tubes.

Figure 13-2 illustrates the nature of the flow around a number of obstacles, and in a channel of varying cross section. The photographs were made using an apparatus designed by Pohl, in which alternate streams of clear and colored water flow between two closely spaced glass plates. The obstacles and the channel walls are opaque flat plates that fit between the glass plates. It will be noted that each obstacle is completely surrounded by a tube of flow. The tube splits into two portions at a so-called *stagnation point* on the upstream side of the obstacle. These portions rejoin at a second stagnation point on the downstream side. The velocity at the stagnation points is zero. It will also be noted that the cross sections of all flow tubes decrease at a constriction and increase again when the channel widens.

The patterns of Fig. 13-2 are typical of *streamline* or *laminar* flow, in which adjacent layers of fluid slide smoothly past each other. At sufficiently high flow rates, or when boundary surfaces cause abrupt changes in velocity, the flow becomes irregular and much more complex and is called *turbulent* flow. In turbulent flow there is, strictly speaking, no steady-state pattern, since the flow pattern continuously changes.

13-2 The equation of continuity

Let us consider any stationary, closed surface in a moving fluid; in general, fluid flows into the volume enclosed by the surface at some points and flows out at other points. The *equation of continuity* is a mathematical statement of the fact that the *net* rate of flow of mass *inward* across any closed surface is equal to the rate of increase of the mass within the surface.

For an incompressible fluid in steady flow, the equation takes the following form. Figure 13-3 represents a portion of a tube of flow, between two fixed cross sections of areas A_1 and A_2. Let v_1 and v_2 be the speeds at these sections. There is no flow across the side wall of the tube because at every point on the wall the velocity is tangent to the wall. The volume of fluid that flows into the tube across A_1 in a time interval Δt is that contained in the short cylindrical element of base A_1 and height $v_1 \Delta t$, that is, $A_1 v_1 \Delta t$. If the density of the fluid is ρ, the *mass*

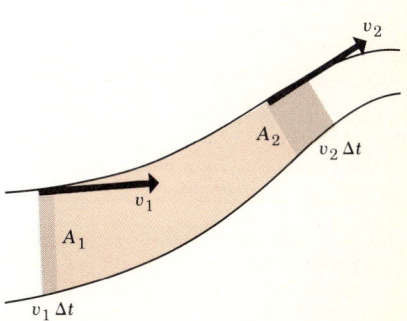

13-3 Flow into and out of a portion of a tube of flow.

flowing in is $\rho A_1 v_1 \Delta t$. Similarly, the mass that flows out across A_2 in the same time is $\rho A_2 v_2 \Delta t$. The volume between A_1 and A_2 is constant, and since the flow is steady, the mass flowing out equals that flowing in. Hence

$$\rho A_1 v_1 \Delta t = \rho A_2 v_2 \Delta t,$$

or,

$$A_1 v_1 = A_2 v_2, \tag{13-1}$$

and the product Av is constant along any given tube of flow. It follows that when the cross section of a flow tube decreases, as in the constriction in Fig. 13-2d, the velocity increases. This can readily be shown by introducing small particles into the fluid and observing their motion.

13-3 Bernoulli's equation

When an incompressible fluid flows along a horizontal flow tube of varying cross section, its velocity must change, as Eq. (13-1) shows. A force is required to produce this acceleration, and for this force to be caused by the fluid surrounding a particular element of fluid, the pressure must be different in different regions. If the pressure were the same everywhere, the net force on any fluid element would be zero. Thus when the cross section of a flow tube varies, the pressure *must* vary along the tube, even when there is no difference in elevation. If the elevation also changes, there is an additional pressure difference. Bernoulli's equation is a general expression that relates the pressure difference between two points in a flow tube to both velocity changes and elevation changes. Daniel Bernoulli developed this relationship in 1738.

To derive the Bernoulli equation, we apply the work–energy theorem to the fluid in a section of a flow tube. In Fig. 13-4, we consider the fluid which at some initial time lies between the two cross sections a and c. In a time interval Δt, the fluid that was initially at a moves to b, a distance $\Delta s_1 = v_1 \Delta t$, where v_1 is the speed at this end of the tube. In the same interval the fluid initially at c moves to d, a distance $\Delta s_2 = v_2 \Delta t$. The cross-sectional areas at the two ends are A_1 and A_2, as shown. Because of the continuity relation, the volume of fluid ΔV passing any cross section in time Δt is $\Delta V = A_1 \Delta s_1 = A_2 \Delta s_2$.

We can compute the *work* done on this fluid during Δt. The force on the cross section at a is $p_1 A_1$, and that at c is $p_2 A_2$, where p_1 and p_2 are the pressures at the two ends. The net *work* done on the element during this displacement is therefore

$$W = p_1 A_1 \Delta s_1 - p_2 A_2 \Delta s_2 = (p_1 - p_2) \Delta V. \tag{13-2}$$

The negative sign in the second term arises because the force at c is opposite in direction to the displacement.

We now equate this work to the total change in energy, kinetic and potential, of the element. The kinetic energy of the fluid between sections b and c does not change. At the beginning of Δt, the fluid between a and b has volume $A_1 \Delta s_1$, mass $\rho A_1 \Delta s_1$, and kinetic energy $\frac{1}{2}\rho(A_1 \Delta s_1) v_1^2 = \frac{1}{2}\rho \Delta V v_1^2$. Similarly, at the end of Δt, the fluid between c and d has kinetic energy $\frac{1}{2}\rho(A_2 \Delta s_2) v_2^2 = \frac{1}{2}\rho \Delta V v_2^2$. Thus the net *change* in kinetic energy is

$$\Delta K = \tfrac{1}{2}\rho\, \Delta V (v_2^2 - v_1^2). \tag{13-3}$$

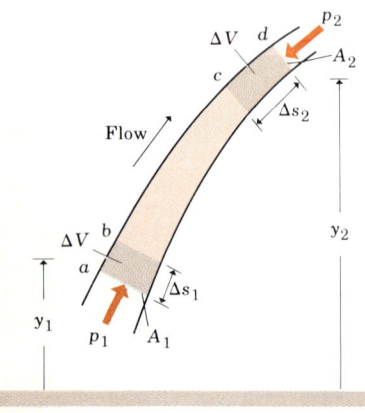

13-4 The net work done on the portion of the fluid initially between sections a and c equals the increase in its kinetic and potential energy.

The change in potential energy is obtained similarly. The potential energy of the mass entering at a in time Δt is $\Delta m\, gy_1 = \rho\, \Delta V\, gy_1$, and that of the mass leaving at c is $\Delta m\, gy_2 = \rho\, \Delta V\, gy_2$. The net *change* in potential energy is

$$\Delta U = \rho\, \Delta V g(y_2 - y_1). \tag{13-4}$$

Combining Eqs. (13-2), (13-3), and (13-4) in the work–energy theorem $W = \Delta K + \Delta U$, we obtain

$$(p_1 - p_2)\, \Delta V = \tfrac{1}{2}\rho\, \Delta V(v_2{}^2 - v_1{}^2) + \rho\, \Delta V g(y_2 - y_1)$$

or

$$p_1 - p_2 = \tfrac{1}{2}\rho(v_2{}^2 - v_1{}^2) + \rho g(y_2 - y_1). \tag{13-5}$$

In this form, Bernoulli's equation represents the equality of the work per unit volume of fluid $(p_1 - p_2)$ to the sum of the changes in kinetic and potential energies per unit volume that occur during the flow. Or we may interpret Eq. (13-5) in terms of pressures. The second term on the right is the pressure difference arising from the weight of the fluid and the difference in elevation of the two ends of the fluid element. The first term on the right is the additional pressure difference associated with the change of velocity of the fluid. Equation (13-5) can also be written

$$p_1 + \rho g y_1 + \tfrac{1}{2}\rho v_1{}^2 = p_2 + \rho g y_2 + \tfrac{1}{2}\rho v_2{}^2 \tag{13-6}$$

and since the subscripts 1 and 2 refer to *any* two points along the tube of flow, Bernoulli's equation may also be written

$$p + \rho g y + \tfrac{1}{2}\rho v^2 = \text{constant.} \tag{13-7}$$

Note carefully that p is the *absolute* (not gauge) pressure, and that a consistent set of units must be used. In SI units, pressure is expressed in Pa, density in kg·m^{-3}, and velocity in m·s^{-1}.

Example Water enters a house through a pipe 2.0 cm in inside diameter, at an absolute pressure of 4×10^5 Pa (about 4 atm). The pipe leading to the second-floor bathroom 5 m above is 1.0 cm in diameter. When the flow velocity at the inlet pipe is 4 m·s^{-1}, find the flow velocity and pressure in the bathroom.

Solution The flow velocity is obtained from the continuity equation:

$$v_2 = \frac{A_1}{A_2} v_1 = \frac{\pi(1.0\text{ cm})^2}{\pi(0.5\text{ cm})^2}(4\text{ m·s}^{-1}) = 16\text{ m·s}^{-1}.$$

The pressure is now obtained from Bernoulli's equation:

$$\begin{aligned}
p_2 &= p_1 - \tfrac{1}{2}\rho(v_2{}^2 - v_1{}^2) - \rho g(y_2 - y_1)\\
&= 4 \times 10^5 \text{ Pa} - \tfrac{1}{2}(1.0 \times 10^3 \text{ kg·m}^{-3})(256 \text{ m}^2\text{·s}^{-2} - 16 \text{ m}^2\text{·s}^{-2})\\
&\quad - (1.0 \times 10^3 \text{ kg·m}^{-3})(9.8 \text{ m·s}^{-2})(5\text{ m})\\
&= 2.3 \times 10^5 \text{ Pa} \simeq 2.3 \text{ atm.}
\end{aligned}$$

We note that when the water is turned off, the second term on the right vanishes and the pressure rises to 3.5×10^5 Pa.

13-4 Applications of Bernoulli's equation

1. The equations of hydrostatics are special cases of Bernoulli's equation, when the velocity is everywhere zero. Thus, when v_1 and v_2 are zero,

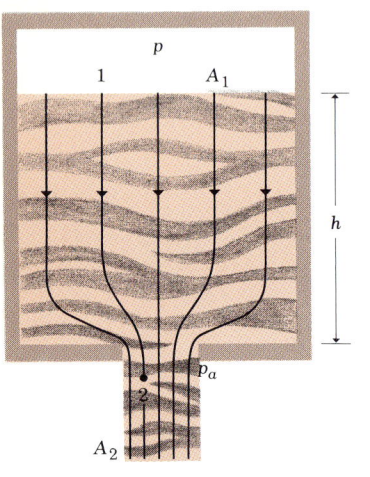

13-5 Flow of a liquid out of an orifice.

Eq. (13-5) reduces to

$$p_1 - p_2 = \rho g(y_2 - y_1),$$

which is the same as Eq. (12-4).

2. *Speed of efflux. Torricelli's theorem.* Figure 13-5 represents a tank of cross-sectional area A_1, filled to a depth h with a liquid of density ρ. The space above the top of the liquid contains air at pressure p, and the liquid flows out of an orifice of area A_2. Let us consider the entire volume of moving fluid as a single tube of flow, and let v_1 and v_2 be the speeds at points 1 and 2. The quantity v_2 is called the *speed of efflux*. The pressure at point 2 is atmospheric, p_a. Applying Bernoulli's equation to points 1 and 2, and taking the bottom of the tank as our reference level, we get

$$p + \tfrac{1}{2}\rho v_1^2 + \rho g h = p_a + \tfrac{1}{2}\rho v_2^2,$$

or

$$v_2^2 = v_1^2 + 2\frac{p - p_a}{\rho} + 2gh. \tag{13-8}$$

From the equation of continuity,

$$v_2 = \frac{A_1}{A_2} v_1. \tag{13-9}$$

Because of the converging of the streamlines as they approach the orifice, the cross section of the stream continues to diminish for a short distance outside the tank. It is the area of smallest cross section, known as the *vena contracta*, that should be used in Eq. (13-9). For a sharp-edged circular opening, the area of the *vena contracta* is about 65 percent as great as the area of the orifice.

Now let us consider some special cases. Suppose the tank is open to the atmosphere, so that

$$p = p_a \quad \text{and} \quad p - p_a = 0.$$

Suppose also that $A_1 \gg A_2$. Then v_1^2 is very much less than v_2^2 and can be neglected, and from Eq. (13-8),

$$v_2 = \sqrt{2gh}. \tag{13-10}$$

That is, *the speed of efflux is the same as the speed a body would acquire in falling freely through a height h*. This is *Torricelli's theorem*. It is not restricted to an opening in the bottom of a vessel, but applies also to a hole in the side walls at a depth h below the surface.

Now suppose again that the ratio of areas is such that v_1^2 is negligible and that the pressure p (in a closed vessel) is so large that the term $2gh$ in Eq. (13-8) can be neglected, compared with $2(p - p_a)/\rho$. The speed of efflux is then

$$v_2 = \sqrt{2(p - p_a)/\rho}. \tag{13-11}$$

The density ρ is that of the fluid escaping from the orifice. If the vessel is partly filled with a liquid, as in Fig. 13-5, ρ is the density of the liquid. On the other hand, if the vessel contains only a gas, ρ is the density of the gas. The efflux speed of a gas may be very great, even for small pressures, since its density is small. However, if the pressure is too great, our idealized model is no longer adequate. Compressibility of the gas must be considered, and if the speed is too great, the motion may become turbulent. Bernoulli's equation can no longer be applied to the motion under these conditions.

A flow of fluid out of an orifice in a vessel gives rise to a *thrust* or *reaction force* on the remainder of the system. The mechanics of the problem are the same as those involved in rocket propulsion. The thrust can be computed as follows, provided conditions are such that Bernoulli's equation is applicable. If A is the area of the orifice, ρ the density of the escaping fluid, and v the speed of efflux, the mass of fluid flowing out in time Δt is $\rho A v \, \Delta t$, and its momentum (mass × velocity) is $\rho A v^2 \, \Delta t$. Since we are neglecting the relatively small velocity of the fluid in the container, we can say that the escaping fluid started from rest and acquired the momentum above in time Δt. Its *rate of change* of momentum was therefore $\rho A v^2$, and from Newton's second law this equals the force acting on it. By Newton's third law, an equal and opposite reaction force acts on the remainder of the system. Taking the expression for v^2 from Eq. (13–11), we can write the reaction force as

$$F = \rho A v^2 = \rho A \frac{2(p - p_a)}{\rho},$$

or

$$F = 2A(p - p_a). \tag{13-12}$$

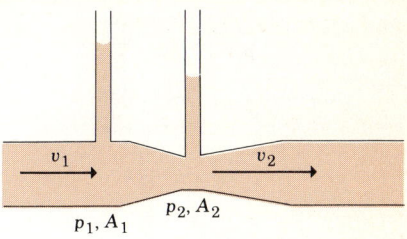

13-6 The Venturi tube.

Thus, while the *speed* of efflux is inversely proportional to the density, the *thrust* is independent of the density and depends only on the area of the orifice and the gauge pressure $p - p_a$.

3. The *Venturi tube*, illustrated in Fig. 13–6, consists of a constriction or throat inserted in a pipeline and having properly designed tapers at inlet and outlet to avoid turbulence. Bernoulli's equation, applied to the wide and to the constricted portions of the pipe, becomes

$$p_1 + \tfrac{1}{2}\rho v_1^2 = p_2 + \tfrac{1}{2}\rho v_2^2.$$

From the equation of continuity, the speed v_2 is greater than the speed v_1, and hence the pressure p_2 in the throat is *less* than the pressure p_1. Thus a net force to the right acts to accelerate the fluid as it enters the throat, and a net force to the left decelerates it as it leaves. The pressures p_1 and p_2 can be measured by attaching vertical side tubes as shown in the diagram. From a knowledge of these pressures and of the cross-sectional areas A_1 and A_2, the velocities and the mass rate of flow can be computed. When used for this purpose, the device is called a *Venturi meter*.

The reduced pressure at a constriction finds a number of technical applications. Gasoline vapor rushes into the manifold of an internal combustion engine because of the low pressure produced in a Venturi throat to which the carburetor is connected. The *aspirator pump* is a Venturi throat through which water is forced. Air rushes into the low-pressure water flowing through the constricted portion.

4. *Measurement of pressure in a moving fluid.* The pressure p in a fluid flowing in an enclosed channel can be measured with an open-tube manometer, as shown in Fig. 13–7. In (a), one arm of the manometer is connected to an opening in the channel wall. In (b), a *probe* is inserted in the stream. The probe should be small enough so that the flow is not

13-7 Pressure gauges for measuring the static pressure p in a fluid flowing in an enclosed channel.

13-8 The Pitot tube.

13-9 The Prandtl tube.

13-10 Flow lines around an airfoil.

appreciably disturbed and should be shaped so as to avoid turbulence. The difference h_1 in height of the liquid in the arms of the manometer is proportional to the difference between atmospheric pressure p_a and the fluid pressure p. That is,

$$p_a - p = \rho_m g h_1,$$
$$p = p_a - \rho_m g h_1, \qquad (13\text{-}13)$$

where ρ_m is the density of the manometer liquid.

The *Pitot tube*, shown in Fig. 13-8, is a probe with an opening at its upstream end. A stagnation point forms at the opening, where the pressure is p_2 and the speed is zero. Applying Bernoulli's equation to the stagnation point, and to a point at a large distance from the probe where the pressure is p and the speed is v, we get

$$p_2 = p + \tfrac{1}{2}\rho v^2 \qquad (13\text{-}14)$$

(ρ is the density of the flowing fluid). The pressure p_2 at the stagnation point is therefore the sum of the pressure p and the quantity $\tfrac{1}{2}\rho v^2$.

The quantity $\tfrac{1}{2}\rho v^2$ is sometimes called the *dynamic pressure* and p the *static pressure*. These terms are somewhat misleading, inasmuch as p is always the true pressure (force per unit area) in both static and dynamic situations.

The instrument shown in Fig. 13-9 combines in a single device the functions of the instruments in Figs. 13-7 and 13-8. This device is called a *Prandtl tube;* the term Pitot tube is also used. The pressure at opening 1 corresponds to the pressure p measured in Fig. 13-7, and that at opening 2, a stagnation point, is the quantity $p_2 = p + \tfrac{1}{2}\rho v^2$ measured in Fig. 13-8. The manometric height h_3 is proportional to the difference of these, or to $\tfrac{1}{2}\rho v^2$. Hence

$$\tfrac{1}{2}\rho v^2 = \rho_m g h_3. \qquad (13\text{-}15)$$

This instrument is self-contained, and its reading does not depend on atmospheric pressure. If held at rest, it can be used to measure the velocity of a stream of fluid flowing past it. If mounted on an aircraft, it indicates the velocity of the aircraft relative to the surrounding air and is known as an *airspeed indicator.*

5. *Lift on an aircraft wing.* Figure 13-10 shows flow lines around a section of an aircraft wing or an airfoil. The orientation of the wing relative to the flow direction causes the flow lines to crowd together above the wing, corresponding to increased flow velocity in this region, much as in the throat of a Venturi. Hence the region above the wing is one of increased velocity and reduced pressure, while below the wing the pressure remains nearly atmospheric. Because the upward force on the underside of the wing is greater than the downward force on the top side, there is a net upward force or *lift.*

This phenomenon can also be understood qualitatively on the basis of Newton's laws. In order for the fluid to exert a net upward force on the wing, the wing must exert a *downward* reaction force on the fluid, deflecting the stream downward. This effect is shown in Fig. 13-10; the fluid suffers a net change in the vertical component of momentum as it passes the wing, corresponding to the downward force the wing exerts on it. The force *on* the wing is thus *upward,* in agreement with the above discussion.

As the angle of the wing relative to the flow increases, turbulent flow occurs in a larger and larger region above the wing, and the pressure drop is no longer as great as that predicted by Bernoulli's principle. The lift on the wing decreases, and in extreme cases the airplane stalls.

6. *The curved flight of a spinning ball.* This effect is somewhat more subtle than the lift on an airplane wing, but the same basic principles are involved. Figure 13–11a represents a stationary ball in a blast of air moving from right to left. The motion of the air stream around and past the ball is the same as though the ball were moving through still air from left to right. Because of the large velocities ordinarily involved, there is a region of turbulent flow behind the ball, as shown.

When the ball is spinning, as in Fig. 13–11b, the viscosity of air causes layers of air near the ball's surface to be pulled around in the direction of spin. The velocity of air relative to the ball surface is greater at the top than at the bottom. The region of turbulence becomes asymmetric, with turbulence occurring farther forward on the top side than on the bottom. This asymmetry causes a pressure difference; the average pressure at the top becomes greater than that at the bottom. The corresponding net downward force deflects the ball as shown. In a baseball curve pitch, the actual deflection is sideways, and in that case Fig. 13–11 shows a *top* view of the situation.

Corresponding to the downward force on the ball in Fig. 13–11b, the ball must exert an equal and opposite reaction force *upward* on the air. This deflects the air stream upward near the ball, as the figure shows, just as the lift force on an airplane wing corresponds to a downwad deflection of the air stream.

A similar effect occurs with golf balls, which always have "backspin" from impact with the slanted club face. The resulting pressure difference between top and bottom of the ball causes a "lift" force which keeps the ball in the air considerably longer than would be possible without spin. A well-hit drive appears from the tee to "float" or even curve *upward* during the initial portion of its flight, and this is a real effect, not an illusion. The dimples on the ball play an essential role; an undimpled ball has a much shorter trajectory than a dimpled one given the same initial velocity and spin. Some manufacturers even claim that the *shape* of the dimples is significant; one feels that polygonal dimples are superior to round ones!

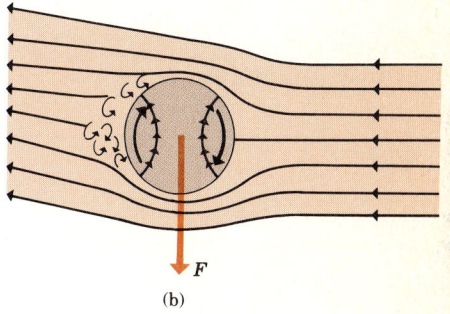

13–11 (a) Air flow past a stationary ball, showing symmetric region of turbulence behind ball. (b) Air flow past a spinning ball, showing asymmetric region of turbulence and deflection of airstream. The net force on the ball is in the direction shown. Motion of the air from right to left relative to the ball corresponds to motion of a ball through still air from left to right.

13–5 Viscosity

Viscosity may be thought of as the internal friction of a fluid. Because of viscosity, a force must be exerted to cause one layer of a fluid to slide past another, or to cause one surface to slide past another if there is a layer of fluid between the surfaces. Both liquids and gases exhibit viscosity, although liquids are much more viscous than gases. The problem of viscous flow is similar to that of shear stress and strain in a solid.

The simplest example of flow of a viscous fluid is flow between two parallel plates, as shown in Fig. 13–12. The bottom plate is stationary, while the top plate moves with constant speed v. It is found that the fluid in contact with each surface has the same speed as that surface; thus at the top surface the fluid has speed v, while the fluid adjacent to the bottom surface is at rest. The speeds of intermediate layers of fluid increase uniformly from one surface to the other, as shown by the arrows.

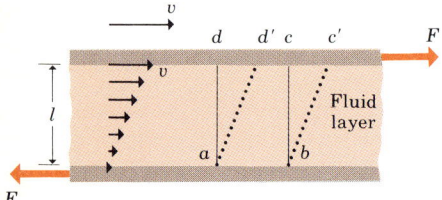

13-12 Laminar flow of a viscous fluid.

Flow of this type is called *laminar*. (A lamina is a thin sheet.) The layers of liquid slide over one another much as do the leaves of a book when it is placed flat on a table and a horizontal force applied to the top cover. As a consequence of this motion, a portion of the liquid which at some instant has the shape *abcd*, will a moment later take the shape *abc'd'*, and will become more and more distorted as the motion continues. That is, the liquid is in a state of continually increasing shear strain.

In order to maintain the motion, a constant force must be exerted to the right on the upper, moving plate, and hence indirectly on the upper liquid surface. This force tends to drag the fluid, and the lower plate as well, to the right. Therefore an equal force must be exerted toward the left on the lower plate to hold it stationary. These forces are lettered F in Fig. 13-12. If A is the area of the fluid over which these forces are applied (i.e., the surface area of the plates), the ratio F/A is the shear *stress* exerted on the fluid.

When a shear stress is applied to a *solid*, the effect of the stress is to produce a certain displacement of the solid, such as dd'. The shear strain is defined as the ratio of this displacement to the transverse dimension l, and within the elastic limit the shear stress is proportional to the shear strain. With a *fluid*, on the other hand, the shear strain increases without limit so long as the stress is applied, and the stress is found by experiment to depend not on the shear strain, but on its *rate of change*. The strain in Fig. 13-12 (at the instant when the volume of fluid has the shape $abc'd'$) is dd'/ad, or dd'/l. Since l is constant, the *rate of change* of strain equals $1/l$ times the rate of change of dd'. But the rate of change of dd' is simply the speed of point d', that is, the speed v of the moving wall. Hence

$$\text{Rate of change of shear strain} = \frac{v}{l}.$$

The rate of change of shear strain is also referred to simply as the *strain rate*.

The *coefficient of viscosity* of the fluid, or simply its *viscosity* η, is defined as the ratio of the shear stress, F/A, to the rate of change of shear strain:

$$\eta = \frac{\text{shear stress}}{\text{rate of change of shear strain}} = \frac{F/A}{v/l},$$

or

$$F = \eta A \frac{v}{l}. \tag{13-16}$$

For a liquid that flows readily, such as water or kerosene, the shear stress is relatively small for a given strain rate, and the viscosity also is relatively small. For a liquid such as molasses or glycerine, a greater shear stress is necessary for the same strain rate, and the viscosity is correspondingly greater. Viscosities of gases at ordinary pressures and temperatures are very much *smaller* than those of common liquids. Viscosities of all fluids are markedly dependent on temperature, increasing for gases and decreasing for liquids as the temperature is increased, hence the expression "as slow as molasses in January." An important consideration in the design of oils for engine lubrication is to *reduce* the temperature variation of viscosity as much as possible.

From Eq. (13–16), the unit of viscosity is that of force times distance, divided by area times velocity. In SI units it is

$$1\text{ N}\cdot\text{m}\cdot\text{m}^{-2}(\text{m}\cdot\text{s}^{-1})^{-1} = 1\text{ N}\cdot\text{s}\cdot\text{m}^{-2}.$$

The corresponding cgs unit, $1\text{ dyn}\cdot\text{s}\cdot\text{cm}^{-2}$, is the only viscosity unit in common use, and is called 1 *poise* in honor of the French scientist Poiseuille. Thus,

$$1\text{ poise} = 1\text{ dyn}\cdot\text{s}\cdot\text{cm}^{-2} = 10^{-1}\text{ N}\cdot\text{s}\cdot\text{m}^{-2}.$$

Small viscosities are expressed in *centipoises* ($1\text{ cp} = 10^{-2}$ poise) or *micropoises* ($1\text{ }\mu\text{p} = 10^{-6}$ poise). A few typical values are given in Table 13–1.

Table 13-1 Typical values of viscosity

Temperature, °C	Viscosity of castor oil, poise	Viscosity of water, centipoise	Viscosity of air, micropoise
0	53	1.792	171
20	9.86	1.005	181
40	2.31	0.656	190
60	0.80	0.469	200
80	0.30	0.357	209
100	0.17	0.284	218

Not all fluids behave according to the direct proportionality of force and velocity predicted by Eq. (13–16). An interesting exception is blood, for which velocity increases more rapidly than force. Thus doubling the force in Fig. 13–13 produces *more* than a twofold increase in velocity. This behavior may be understood on the basis that, on a microscopic scale, blood is not a homogeneous fluid but rather a suspension of solid particles in a liquid. The suspended particles have characteristic shapes; for example, red cells are roughly disk-shaped. At small velocities their orientations are random, but as velocity increases they tend to become oriented so as to facilitate flow. The fluids that provide lubrication in human joints exhibit similar behavior.

Fluids for which Eq. (13–16) holds are called *newtonian fluids;* we see that this description is an idealized model which not all fluids obey. In general, fluids that are suspensions or dispersions are often nonnewtonian in their viscous behavior. Nevertheless, Eq. (13–16) provides a useful model to describe approximately the properties of many pure substances.

13-6 Poiseuille's law

When a viscous fluid flows in a tube, the flow velocity is different at different points of a cross section. The outermost layer of fluid clings to the walls of the tube, and its velocity is zero. The tube walls exert a backward drag on this layer, which in turn drags backward on the next layer beyond it, and so on. If the velocity is not too great, the flow is *laminar,* with a velocity that is greatest at the center of the tube and decreases to zero at the walls. The flow is like that of a number of tele-

13-13 (a) Forces on a cylindrical element of a viscous fluid. (b) Velocity distribution for viscous flow.

scoping tubes sliding relative to one another, the central tube advancing most rapidly and the outer tube remaining at rest.

Let us consider the variation of velocity with radius for a cylindrical pipe of inner radius R. We consider the flow of a cylindrical element of fluid coaxial with the pipe, of radius r and length L, as shown in Fig. 13-13a. The force on the left end is $p_1 \pi r^2$, and that on the right end $p_2 \pi r^2$, as shown. The net force is thus

$$F = (p_1 - p_2)\pi r^2.$$

Since the element does not accelerate, this force must just balance the viscous retarding force at the surface of this element. This force is given by Eq. (13–16), but since the velocity does not vary uniformly with distance from the center, we must replace v/l in this expression with dv/dr, where dv is the small change of velocity when we go from distance r to $r + dr$ from the axis. The area over which the viscous force acts is $A = 2\pi rL$. Thus the viscous force is

$$F = \eta 2\pi r L \frac{dv}{dr}.$$

Equating this to the net force due to pressure on the ends and rearranging, we find that

$$\frac{dv}{dr} = -\frac{(p_1 - p_2)r}{2\eta L}.$$

This shows that the velocity changes more and more rapidly as we go from the center ($r = 0$) to the pipe wall ($r = R$). The negative sign must be introduced because v decreases as r increases. Integrating, we find

$$-\int_v^0 dv = \frac{p_1 - p_2}{2\eta L}\int_r^R r\,dr,$$

and

$$v = \frac{p_1 - p_2}{4\eta L}(R^2 - r^2). \tag{13-17}$$

The velocity decreases from a maximum value $(p_1 - p_2)R^2/4\eta L$ at the center to zero at the wall. Thus the maximum velocity is proportional to the *square* of the pipe radius and is also proportional to the pressure change per unit length $(p_1 - p_2)/L$, called the *pressure gradient*. The curve in Fig. 13-13b is a graph of Eq. (13-17) with the v-axis horizontal and the r-axis vertical.

Equation (13-17) may be used to find the total rate of flow of fluid through the pipe. The velocity at each point is proportional to the pressure gradient $(p_1 - p_2)/L$, so the *total* flow rate must also be proportional to this quantity. Let us consider the thin-walled element in Fig. 13-13c. The volume of fluid dV crossing the ends of this element in a time dt is $v\,dA\,dt$, where v is the velocity at the radius r and dA is the shaded area, equal to $2\pi r\,dr$. Taking the expression for v from Eq. (13-17), we get

$$dV = \frac{p_1 - p_2}{4\eta L}(R^2 - r^2)\,2\pi r\,dr\,dt.$$

The volume flowing across the entire cross section is obtained by integrating over all elements between $r = 0$ and $r = R$.

$$dV = \frac{\pi(p_1 - p_2)}{2\eta L} \int_0^R (R^2 - r^2) r \, dr \, dt$$

$$= \frac{\pi}{8} \frac{R^4}{\eta} \frac{p_1 - p_2}{L} dt.$$

The total volume of flow per unit time, dV/dt, is given by

$$\frac{dV}{dt} = \frac{\pi}{8} \frac{R^4}{\eta} \frac{p_1 - p_2}{L}. \qquad (13\text{–}18)$$

This relation was first derived by Poiseuille* and is called *Poiseuille's law*. The volume rate of flow is inversely proportional to viscosity, as might be expected. It is proportional to the pressure gradient along the pipe, and it varies as the fourth power of the radius. For example, if the radius is halved, the flow rate is reduced by a factor of 16. This relation is familiar to physicians in connection with the selection of needles for hypodermic syringes. Needle size is much more important than thumb pressure in determining the flow rate from the needle; doubling the needle diameter has the same effect as increasing the thumb force sixteenfold. Similarly, blood flow in arteries and veins can be controlled over a wide range by relatively small changes in diameter, an important temperature-control mechanism in warm-blooded animals.

The difference between the flow of an ideal nonviscous fluid and one having viscosity is illustrated in Fig. 13–14, where fluid is flowing along a horizontal tube of varying cross section. The height of the fluid in the small vertical tubes is proportional to the gauge pressure.

In part (a), the fluid is assumed to have no viscosity. The pressure at b is very nearly the static pressure $\rho g y$, since the velocity is small in the large tank. The pressure at c is less than at b because the fluid must accelerate between these points. The pressures at c and d are equal, however, since the velocity and elevation at these points are the same. There is a further pressure drop between d and e, and between f and g. The pressure at g is atmospheric, and the gauge pressure at this point is zero.

Part (b) of the diagram illustrates the effect of viscosity. Again, the pressure at b is nearly the static pressure $\rho g y$. There is a pressure drop from b to c, due now in part to viscous effects, and also a further drop from c to d. The pressure gradient in this part of the tube is represented by the slope of the dotted line. The drop from d to e results in part from acceleration and in part from viscosity. The pressure gradient between e and f is greater than between c and d because of the smaller radius in this portion. Finally, the pressure at g is somewhat above atmospheric, since there is now a pressure gradient between this point and the end of the tube.

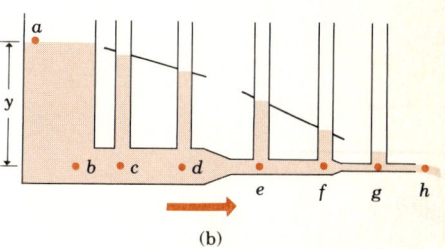

13–14 Pressures along a horizontal tube in which is flowing (a) an ideal fluid, (b) a viscous fluid.

---

*Approximate pronunciation: Pwah-zoy'.

13-7 Stokes' law

When an ideal fluid of zero viscosity flows past a sphere, or when a sphere moves through a stationary fluid, the streamlines form a perfectly symmetrical pattern around the sphere, as shown in Fig. 13-2a. The pressure at any point on the upstream hemispherical surface is exactly the same as that at the corresponding point on the downstream face, and the resultant force on the sphere is zero. If the fluid has viscosity, however, there will be a viscous drag on the sphere. (A viscous drag is experienced by a body of *any* shape, but only for a sphere is the drag readily calculable.)

We shall not attempt to derive the expression for the viscous force directly from the laws of flow of a viscous fluid. The only quantities on which the force can depend are the viscosity η of the fluid, the radius r of the sphere, and its velocity v relative to the fluid. A complete analysis shows that the force F is given by

$$F = 6\pi\eta rv. \tag{13-19}$$

This equation was first deduced by Sir George Stokes in 1845 and is called *Stokes' law*. We have already used it in Sec. 4-6 (Example 9) to study the motion of a sphere falling in a viscous fluid. At that point it was necessary to know only that the viscous force on a given sphere in a given fluid is proportional to the relative velocity.

A sphere falling in a viscous fluid reaches a *terminal velocity* v_T at which the viscous retarding force plus the buoyant force equals the weight of the sphere. Let ρ be the density of the sphere and ρ' the density of the fluid. The weight of the sphere is then $(4/3)\pi r^3\rho g$, and the buoyant force is $(4/3)\pi r^3\rho'g$; when the terminal velocity is reached, the total force is zero and

$$\tfrac{4}{3}\pi r^3\rho'g + 6\pi\eta rv_\mathrm{T} = \tfrac{4}{3}\pi r^3\rho g,$$

or

$$v_\mathrm{T} = \frac{2}{9}\frac{r^2 g}{\eta}(\rho - \rho'). \tag{13-20}$$

When the terminal velocity of a sphere of known radius and density is measured, the viscosity of the fluid in which it is falling can be found from the equation above. Conversely, if the viscosity is known, the radius of the sphere can be determined by measuring the terminal velocity. This method was used by Millikan to determine the radius of very small electrically charged oil drops (used to measure the electrical charge of an inidividual electron) by observing free fall in air.

Even for nonspherical bodies, a relation of the form of Eq. (13-19) holds, with a different numerical coefficient. Biologists call the terminal velocity the *sedimentation velocity,* and experiments with sedimentation can give useful information concerning very small particles. It is often useful to increase the terminal velocity by spinning the sample in a centrifuge, which greatly increases the effective acceleration of gravity.

13-8 Reynolds number

When the velocity of a fluid flowing in a tube exceeds a certain critical value (which depends on the properties of the fluid and the diameter of the tube), the nature of the flow becomes extremely complicated. Within a very thin layer adjacent to the tube walls, called the *boundary layer,* the flow is still laminar. The flow velocity in the boundary layer is zero at the tube walls and increases uniformly throughout the layer. The properties of the boundary layer are of the greatest importance in determining the resistance to flow, and the transfer of heat to or from the moving fluid.

Beyond the boundary layer, the motion is highly irregular. Random local circular currents called *vortices* develop within the fluid, with a large increase in the resistance to flow. Flow of this sort is called *turbulent.*

Experiment indicates that a combination of four factors determines whether the flow of a fluid through a tube or pipe is laminar or turbulent. This combination is known as the *Reynolds number,* N_R, and is defined as

$$N_R = \frac{\rho v D}{\eta}, \qquad (13\text{-}21)$$

where ρ is the density of the fluid, v the average forward velocity, η the viscosity, and D the diameter of the tube. (The average velocity is defined as the uniform velocity over the entire cross section of the tube, which would result in the same volume rate of flow.) The Reynolds number, $\rho v D/\eta$, is a *dimensionless* quantity and has the same numerical value in any consistent system of units. For example, for water at 20°C flowing in a tube of diameter 1 cm with an average velocity of 10 cm·s^{-1}, the Reynolds number is

$$N_R = \frac{\rho v D}{\eta} = \frac{(1 \text{ g} \cdot \text{cm}^{-3})(10 \text{ cm} \cdot \text{s}^{-1})(1 \text{ cm})}{0.01 \text{ dyn} \cdot \text{s} \cdot \text{cm}^{-2}} = 1000.$$

Had the four quantities been expressed originally in SI units, the same value of 1000 would have been obtained.

A variety of experiments have shown that when the Reynolds number is less than about 2000 the flow is laminar, whereas above about 3000 the flow is turbulent. In the transition region between 2000 and 3000 the flow is unstable and may change from one type to the other. Thus for water at 20°C flowing in a tube 1 cm in diameter, the flow is laminar when

$$\frac{\rho v D}{\eta} \leq 2000,$$

or when

$$v \leq \frac{(2000)(0.01 \text{ dyn} \cdot \text{s} \cdot \text{cm}^{-2})}{(1 \text{ g} \cdot \text{cm}^{-3})(1 \text{ cm})} = 20 \text{ cm} \cdot \text{s}^{-1}.$$

Above about 30 cm·s^{-1} the flow is turbulent. If air at the same temperature were flowing at 30 cm·s^{-1} in the same tube, the Reynolds number would be

$$N_R = \frac{(0.0013 \text{ g} \cdot \text{cm}^{-3})(30 \text{ cm} \cdot \text{s}^{-1})(1 \text{ cm})}{181 \times 10^{-6} \text{ dyn} \cdot \text{s} \cdot \text{cm}^{-2}} = 215.$$

(a)   (b)   (c)

13–15 (a) Laminar flow. (b) Turbulent flow. (c) First laminar, then turbulent.

Since this is much less than 3000, the flow would be laminar and would not become turbulent unless the velocity were as great as 420 cm·s^{-1}.

The distinction between laminar and turbulent flow is shown in the photographs of Fig. 13–15. In (a) and (b) the fluid is water and in (c) air and smoke particles.

The Reynolds number of a system forms the basis for the study of the behavior of real systems through the use of small scale models. A common example is the wind tunnel, in which one measures the aerodynamic forces on a scale model of an aircraft wing. The forces on a full-size wing are then deduced from these measurements.

Two systems are said to be *dynamically similar* if the Reynolds number, $\rho v D/\eta$, is the same for both. The letter D may refer, in general, to any dimension of a system, such as the span or chord of an aircraft wing. Thus the flow of a fluid of given density ρ and viscosity η, about a half-scale model, is dynamically similar to that around the full-size object if the velocity v is twice as great.

Questions

13–1 Is the continuity relation, Eq. (13–1), valid for compressible fluids? If not, is there a similar relation that *is* valid?

13–2 If the velocity at each point in space in steady-state fluid flow is constant, how can a fluid particle accelerate?

13–3 Whenever possible, airplanes take off and land heading into the wind. Why?

13–4 Does the "lift" of an airplane wing depend on altitude?

13–5 How does a baseball pitcher give the ball the spin that makes it curve? Can he make it curve in either direction? Does it matter whether he is righthanded or lefthanded? What is a spitball? Why is it illegal?

13–6 When a car on a highway is passed by a large truck, the car is sometimes pulled *toward* the truck. What does Bernoulli's relation have to say about this?

13-7 In a store-window vacuum cleaner display, a ping-pong ball is suspended in midair in a jet of air blown from the outlet hose of a tank-type vacuum cleaner. The ball bounces around a little but always returns to the center of the jet, even if it is tilted. How does this behavior illustrate the Bernoulli relation?

13-8 Why do jet airplanes usually fly at altitudes of about 30,000 feet even though it takes a lot of fuel to climb to that altitude?

13-9 What causes the sharp hammering sound sometimes heard from a water pipe when an open faucet is suddenly turned off?

13-10 When a smooth-flowing stream of water comes out of a faucet, it narrows as it falls, and if it falls far enough it eventually breaks up into drops. Why does it narrow? Why does it break up?

13-11 A tornado consists of a rapidly whirling air vortex. Why is the pressure always much lower in the center than at the outside? How does this condition account for the destructive power of a tornado?

13-12 When paddling a canoe, one can attain a certain critical speed with relatively little effort, and then a much greater effort is required to make the canoe go even a little faster. Why?

13-13 Why does air escaping from the open end of a small pipe make a lot more noise than air escaping at the same volume rate from a large pipe?

13-14 Hot water running into a sink makes less noise and splashes less than cold water. Part of the difference is caused by varying surface tension, but part is due to something else. What?

13-15 When water is running in the shower, the shower curtain seems to be pulled in toward the falling water, rather than being blown outward. Why?

Problems

13-1 A circular hole 2 cm in diameter is cut in the side of a large standpipe, 10 m below the water level in the standpipe. Find

a) the velocity of efflux, and

b) the volume discharged per unit time.

Neglect the contraction of the streamlines after emerging from the hole.

13-2 Water stands at a depth H in a large open tank whose side walls are vertical (Fig. 13-16). A hole is made in one of the walls at a depth h below the water surface.

a) At what distance R from the foot of the wall does the emerging stream of water strike the floor?

b) At what height above the bottom of the tank could a second hole be cut so that the stream emerging from it would have the same range?

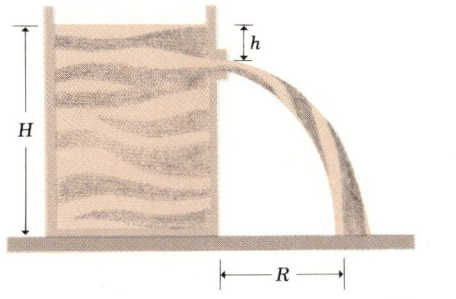

Figure 13-16

13-3 A cylindrical vessel, open at the top, is 20 cm high and 10 cm in diameter. A circular hole whose cross-sectional area is 1 cm² is cut in the center of the bottom of the vessel. Water flows into the vessel from a tube above it at the rate of 140 cm³·s⁻¹. How high will the water in the vessel rise?

13-4 At a certain point in a horizontal pipeline the velocity is 2 m·s⁻¹ and the gauge pressure is 1.0×10^4 Pa above atmospheric. Find the gauge pressure at a second point in the line if the cross-sectional area at the second point is one-half that at the first. The liquid in the pipe is water.

13-5 Water in an enclosed tank is subjected to a gauge pressure of 2×10^4 Pa, applied by compressed air introduced into the top of the tank. There is a small hole in the side of the tank 5 m below the level of the water. Calculate the speed with which water escapes from this hole.

13-6 What gauge pressure is required in the city mains in order that a stream from a fire hose connected to the mains may reach a vertical height of 20 m?

13-7 A tank of large area is filled with water to a depth of 0.3 m. A hole of 5 cm² cross section in the bottom allows water to drain out in a continuous stream.

a) What is the rate at which water flows out of the tank, in m³·s⁻¹?

b) At what distance below the bottom of the tank is the cross-sectional area of the stream equal to one-half the area of the hole?

13-8 A sealed tank containing sea water to a height of 2 m also contains air above the water at a gauge pressure of 40 atm. Water flows out from a hole at the bottom. The cross-sectional area of the hole is 10 cm².

a) Calculate the efflux velocity of the water.

b) Calculate the reaction force on the tank exerted by the water in the emergent stream.

13-9 A pipeline 0.2 m in diameter, flowing full of water, has a constriction of diameter 0.1 m. If the velocity in the 0.2-m portion is $2 \text{ m} \cdot \text{s}^{-1}$, find

a) the velocity in the constriction, and

b) the discharge rate in cubic meters per second.

13-10 A horizontal pipe of 0.2 m^2 cross section tapers to a cross section of 0.1 m^2. If water is flowing with a velocity of $50 \text{ m} \cdot \text{min}^{-1}$ in the large pipe where a pressure gauge reads 0.8×10^5 Pa, what is the gauge pressure in the adjoining part of the small pipe? The barometer reads 76 cm of mercury.

13-11 At a certain point in a pipeline the velocity is $1 \text{ m} \cdot \text{s}^{-1}$ and the gauge pressure is 3×10^5 Pa. Find the gauge pressure at a second point in the line 20 m lower than the first, if the cross section at the second point is one-half that at the first. The liquid in the pipe is water.

13-12 Water stands at a depth of 1 m in an enclosed tank whose side walls are vertical. The space above the water surface contains air at a gauge pressure of 8×10^5 Pa. The tank rests on a platform 2 m above the floor. A hole of cross-sectional area 1 cm^2 is made in one of the side walls just above the bottom of the tank.

a) Where does the stream of water from the hole strike the floor?

b) What is the vertical force exerted on the floor by the stream?

c) What is the horizontal force exerted on the tank? Assume the water level and the pressure in the tank to remain constant, and neglect any effect of viscosity.

13-13 Water flows steadily from a reservoir, as in Fig. 13-17. The elevation of point 1 is 10 m; of points 2 and 3 it is 1 m. The cross section at point 2 is 0.04 m^2 and at point 3 it is 0.02 m^2. The area of the reservoir is very large compared with the cross sections of the pipe.

a) Compute the gauge pressure at point 2.

b) Compute the discharge rate in cubic meters per second.

Figure 13-17

13-14 Water flows steadily in a pipeline of constant cross section leading out of an elevated tank. At a point 2 m below the water level in the tank the gauge pressure in the flowing stream is 10^4 Pa.

a) What is the velocity of the water at this point?

b) If the pipe rises to a point 3 m above the level of the water in the tank, what are the velocity and the pressure at the latter point?

13-15 Water flows through a horizontal pipe of cross-sectional area 10 cm^2. At one section the cross-sectional area is 5 cm^2. The pressure difference between the two sections is 300 Pa. How many cubic meters of water will flow out of the pipe in 1 min?

13-16 Two very large open tanks, A and F (Fig. 13-18), both contain the same liquid. A horizontal pipe BCD, having a constriction at C, leads out of the bottom of tank A, and a vertical pipe E opens into the constriction at C and dips into the liquid in tank F. Assume streamline flow and no viscosity. If the cross section at C is one-half that at D, and if D is at distance h_1 below the level of the liquid in A, to what height h_2 will liquid rise in pipe E? Express your answer in terms of h_1. Neglect changes in atmospheric pressure with elevation.

Figure 13-18

13-17 At a certain point in a horizontal pipeline the gauge pressure is 0.5×10^5 Pa. At another point the gauge pressure is 0.3×10^5 Pa. If the areas of the pipe at these two points are 20 cm^2 and 10 cm^2, respectively, compute the number of cubic meters of water that flow across any cross section of the pipe per minute.

13-18 Water flowing in a horizontal pipe discharges at the rate of $0.004 \text{ m}^3 \cdot \text{s}^{-1}$. At a point in the pipe where the cross section is 0.001 m^2, the absolute pressure is 1.2×10^5 Pa. What must be the cross section of a constriction in the pipe such that the pressure there is reduced to 1.0×10^5 Pa?

13-19 The pressure difference between the main pipeline and the throat of a Venturi meter is 10^5 Pa. The areas of the pipe and the constriction are 0.1 m^2 and 0.05 m^2, respectively. How many cubic meters per second are flowing through the pipe? The liquid in the pipe is water.

13-20 The section of pipe shown in Fig. 13-19 has a cross section of 40 cm^2 at the wider portions and 10 cm^2 at the constriction. The discharge from the pipe is $3000 \text{ cm}^3 \cdot \text{s}^{-1}$.

a) Find the velocities at the wide and the narrow portions.

Figure 13-19

b) Find the pressure difference between these portions.

c) Find the difference in height between the mercury columns in the U-tube.

13-21 Water is used as the manometric liquid in a Prandtl tube mounted in an aircraft to measure airspeed. If the maximum difference in height between the liquid columns is 0.1 m, what is the maximum airspeed that can be measured? The density of air is 1.3 kg·m^{-3}.

13-22 In Problem 13-21, suppose the manometric liquid is mercury. What is the maximum airspeed that can be measured?

13-23 In a wind-tunnel experiment the pressure on the top surface of an airplane wing was 0.90×10^5 Pa, and the pressure on the bottom surface 0.91×10^5 Pa. If the area of each surface is 40 m^2, what is the net lift force on the wing?

13-24 An airplane of mass 6000 kg has a wing area of 60 m^2. If the pressure on the lower wing surface is 0.60×10^5 Pa during level flight at an elevation of 4000 m, what is the pressure on the upper wing surface?

13-25 Water at 20°C flows through a pipe of radius 1.0 cm. If the flow velocity at the center is 10 cm·s^{-1}, find the pressure drop along a 2-m section of pipe due to viscosity.

13-26 An aluminum ball of radius 1.0 cm falls through water at 20°C.

a) What is the terminal velocity, assuming laminar flow and including buoyancy?

b) What is the Reynolds number? Is the flow really laminar?

13-27 A viscous liquid flows through a tube with laminar flow, as in Fig. 13-13b. Prove that the volume rate of flow is the same as if the velocity were uniform at all points of a cross section and equal to half the velocity at the axis.

13-28

a) With what terminal velocity will an air bubble 1 mm in diameter rise in a liquid of viscosity 150 cp and density 0.90 g·cm^{-3}?

b) What is the terminal velocity of the same bubble in water?

13-29

a) With what velocity is a steel ball 1 mm in radius falling in a tank of glycerine at an instant when its acceleration is one-half that of a freely falling body?

b) What is the terminal velocity of the ball? The densities of steel and of glycerine are 8.5 g·cm^{-3} and 1.32 g·cm^{-3}, respectively, and the viscosity of glycerine is 8.3 poise.

13-30 Assume that air is streaming horizontally past an aircraft wing such that the velocity is 40 m·s^{-1} over the top surface and 30 m·s^{-1} past the bottom surface. If the wing has a mass of 300 kg and an area of 5 m^2, what is the net force on the wing? The density of air is 1.3 kg·m^{-3}.

13-31 Modern airplane design calls for a "lift" of about 1000 N per square meter of wing area. Assume that air flows past the wing of an aircraft with streamline flow. If the velocity of flow past the lower wing surface is 100 m·s^{-1}, what is the required velocity over the upper surface to give a "lift" of 1000 N·m^{-2}? The density of air is 1.3 kg·m^{-3}.

13-32 Water at 20°C flows with a speed of 50 cm·s^{-1} through a pipe of diameter 3 mm.

a) What is the Reynolds number?

b) What is the nature of the flow?

13-33 Water at 20°C is pumped through a horizontal smooth pipe 15 cm in diameter and discharges into the air. If the pump maintains a flow velocity of 30 cm·s^{-1},

a) What is the nature of the flow?

b) What is the discharge rate in liters per second?

13-34 The tank at the left in Fig. 13-14a has a very large cross section and is open to the atmosphere. The depth $y = 40$ cm. The cross sections of the horizontal tubes leading out of the tank are respectively 1 cm^2, 0.5 cm^2, and 0.2 cm^2. The liquid is ideal, having zero viscosity.

a) What is the volume rate of flow out of the tank?

b) What is the velocity in each portion of the horizontal tube?

c) What are the heights of the liquid in the vertical side tubes?

Suppose that the liquid in Fig. 13-14b has a viscosity of 0.5 poise, a density of 0.8 g·cm^{-3}, and that the depth of liquid in the large tank is such that the volume rate of flow is the same as in part (a) above. The distance between the side tubes at c and d, and between those at e and f, is 20 cm. The cross sections of the horizontal tubes are the same in both diagrams.

d) What is the difference in level between the tops of the liquid columns in tubes c and d?

e) In tubes e and f?

f) What is the flow velocity on the axis of each part of the horizontal tube?

13-35

a) Is it reasonable to assume that the flow in the second part of Problem 13-34 is laminar?

b) Would the flow be laminar if the liquid were water?

13-36 Oil having a viscosity of 300 centipoise and a density of 0.90 g·cm^{-3} is to be pumped from one large open tank to another through 1 km of smooth steel pipe 15 cm in diameter. The line discharges into the air at a point 30 m above the level of the oil in the supply tank.

a) What gauge pressure, in atmospheres, must the pump exert in order to maintain a flow of 0.05 m^3·s^{-1}?

b) What is the power consumed by the pump?

13-37 In the Reynolds number example following Eq. (13-21), convert the given quantities to SI units; compute the Reynolds number and verify that it has the same value as when the original cgs units are used.

TEMPERATURE AND EXPANSION 14

14-1 Concept of temperature

In our study of mechanical systems thus far, only three fundamental quantities (in the sense of Sec. 1-1) have been needed: length, time, and mass. All other mechanical quantities such as force, energy, and momentum can be expressed in terms of these three quantities. We now begin to consider a class of phenomena, called *thermal* or *heat* phenomena, that requires a fourth fundamental quantity, *temperature*.

The familiar sensations of hotness and coldness are described with adjectives such as cold, cool, tepid, warm, hot, etc. When we touch an object, we use our *temperature sense* to ascribe *to the object* a property called *temperature*, which determines whether it will feel hot or cold to the touch. The hotter it feels, the higher the temperature. To define temperature *quantitatively*, however, we must define it in terms of operations that are independent of our sense perceptions of hotness or coldness, and that involve quantities that can be *measured* objectively. How this is done will be explained in the following paragraphs.

Even before treating the concept of *temperature* in a precise, quantitative manner, we can note that there are numerous simple systems in which a measurable property of the system varies with the hotness or coldness of the system. A simple example is a liquid such as mercury or alcohol in a bulb attached to a very thin tube, as in Fig. 14-1a. The significant quantity characterizing the state of this system is the length L of the liquid column, measured from some arbitrary fixed point. When the system becomes hotter, the liquid rises in the tube and L increases. Another simple system is a quantity of gas in a constant-volume container, shown in Fig. 14-1b. The pressure p measured by the gauge, or perhaps by a manometer as described in Sec. 12-3, increases or decreases as the gas becomes hotter or colder. A third example is the electrical resistance R of a wire, which also varies with hotness or coldness. In each of these examples, the quantity describing the varying state of the system, such as the length L, the pressure p, or the resistance R, is called a *state coordinate* for the system.

(a)

(b)

14-1 (a) A system whose state is specified by the value of L. (b) A system whose state is given by the value of p.

284 TEMPERATURE AND EXPANSION

Adiabatic wall
(a)

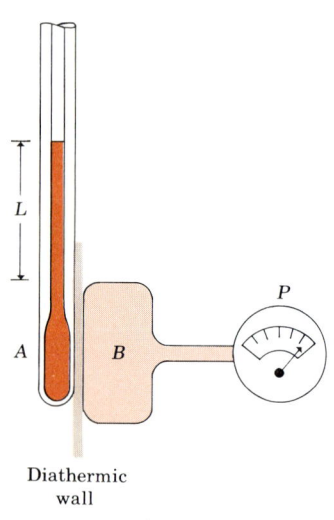

Diathermic wall
(b)

14-2 System A, a liquid column, and system B, a gas at constant volume, separated by (a) an adiabatic wall, p and L independent, and (b) a diathermic wall, p and L dependent.

Let us refer to the tube-and-liquid set-up as system A and the gas in a container as system B. Then system A has the state coordinate L, and B has the state coordinate p. When systems A and B are brought into contact, in general their state coordinates change. Qualitatively, we say that initially one is hotter than the other, and each system changes the state of the other. If, however, the two systems are separated by an insulating material such as wood, plastic foam, or fiber glass, they influence each other much more slowly. Generalizing from this observation, we postulate the existence of an *ideal* insulating wall. Such a wall is called an *adiabatic wall*. Two systems on opposite sides of an adiabatic wall do not influence each other's state coordinates *at all*, and so the state coordinates of the two systems can vary independently of each other. Such a wall is an idealized *model*, which can be approximately realized in the real world. In Fig. 14–2a an adiabatic wall is represented by a thick darkly-shaded region.

The opposite of an adiabatic wall is a partition that *does* allow the two systems on opposite sides to influence each other; such a wall is called a *diathermic wall*. A sheet of a heat-conducting material such as copper or aluminum can be used as a diathermic wall. In Fig. 14–2b, a diathermic wall is shown as a thin colored region.

When systems A and B are first put into actual contact or are separated by a diathermic wall, their state coordinates may or may not change. Eventually, a state is reached when no further change in the coordinates of A and B takes place. The joint state of both systems that exists when all changes in the coordinates have ceased is called **thermal equilibrium**.

Now suppose two systems A and B are separated from each other by an adiabatic wall but each is in contact with a third system C through diathermic walls. The whole assembly is surrounded by an adiabatic wall, as shown in Fig. 14–3. Experiment shows that the two systems will come to thermal equilibrium with the third and that no further change will occur if the adiabatic wall separating them is then replaced by a diathermic wall, as in Fig. 14–3b. If, instead of allowing both systems A and B to come to equilibrium with C at the same time, we first have equilibrium between A and C and then equilibrium between B and C (the state of system C being the same in both cases), then, when A and B are brought into communication through a diathermic wall, they will be

(a)

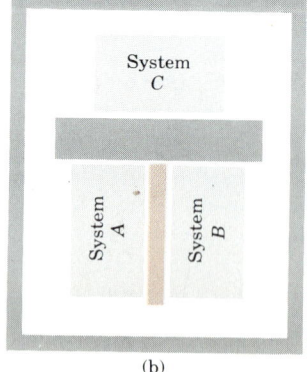

(b)

14-3 The zeroth law of thermodynamics. (a) If A and B are each in thermal equilibrium with C, then (b) A and B are in thermal equilibrium with each other.

found to be in thermal equilibrium. We shall use the expression "two systems are in thermal equilibrium" to mean that the two systems are in states such that if the two *were* to be connected through a diathermic wall, the combined system *would be* in thermal equilibrium.

These experimental facts may then be stated concisely in the following form: *Two systems in thermal equilibrium with a third are in thermal equilibrium with each other.* This principle is called the zeroth law of thermodynamics. This rather odd-sounding name refers to the fact that the most important principles of thermodynamics are called the first, second, and third laws of thermodynamics. The principle we have just discussed is fundamental to all of these, so it deserves to be called the zeroth law. Although it may seem obvious, its truth must be verified by experiment.

But what is it about two systems A and B that determines whether or not they are in thermal equilibrium? It is a property called *temperature*. *The temperature of a system is that property that determines whether or not it will be in thermal equilibrium with other systems.* When two or more systems are in thermal equilibrium, they are said to have the same temperature.

The temperature of a system may be represented by a number. The establishment of a temperature scale is merely the adoption of a set of rules for assigning numbers to temperatures. Once this is done, the condition for thermal equilibrium between two systems is that they have the same temperature. When the temperatures of two systems are different, we may be sure that they are *not* in thermal equilibrium.

14-2 Thermometers

In defining a temperature scale, the simplest procedure is to choose a system such as one of those described above, arbitrarily assigning a numerical value of temperature to each value of the state coordinate of the system. This then defines quantitatively the temperature of this system, and of all systems in thermal equilibrium with it.

Although the system of Fig. 14–1a was one of the earliest thermometers, various other systems are now used. Important characteristics of a thermometer include sensitivity (an appreciable change in the state coordinate produced by a small change in temperature), accuracy in the measurement of the state coordinate, and reproducibility. Another often desirable property is speed in coming to thermal equilibrium with other systems. The thermometers that satisfy these requirements best will be described in the following paragraphs.

A thermometer widely used in research and engineering laboratories is the *thermocouple,* which makes use of the fact that junctions between dissimilar metals or alloys in an electrical circuit give rise to an electromotive force or "voltage" if they are at different temperatures. In Fig. 14–4, the *test junction* is placed in contact with the system whose temperature is to be measured, and the two *reference junctions* are kept at a constant reference temperature, in this case the temperature of melting ice. The test junction is often embedded in the material whose temperature is to be measured. Since the test junction is small and has a small mass, it can follow temperature changes rapidly and come to equilibrium quickly.

14-4 Thermocouple, showing the test junction and the reference junction.

14-5 Principle of the optical pyrometer.

A different electrical thermometer, the *resistance thermometer*, consists of a coil of fine wire, often enclosed in a thin-walled silver tube for protection. Copper wires lead from the thermometer unit to a resistance-measuring device. Since resistance may be measured with great precision, the resistance thermometer is one of the most precise instruments for the measurement of temperature. In the region of extremely low temperatures, a small carbon cylinder or a small piece of germanium crystal is used instead of a coil of wire.

To measure temperatures above the range of thermocouples and resistance thermometers, an *optical pyrometer* is used. As shown in Fig. 14-5, it consists essentially of a telescope T, with a filter F of red glass and a small electric lamp bulb L mounted in the tube. When the pyrometer is directed toward a furnace, an observer looking through the telescope sees the dark lamp filament against the bright background of the furnace. The lamp filament is connected to a battery B and a variable resistor R. By varying R we may gradually increase the current in the filament, and hence its brightness, until the brightness of the filament just matches the brightness of the background. From previous calibration of the instrument at known temperatures, the scale of the ammeter A in the circuit may be marked to read the unknown temperature directly. Since no part of the instrument needs to come into contact with the hot body, the optical pyrometer may be used at very high temperatures, above the melting points of metals.

Among all the state coordinates or, as they are often called, *thermometric properties,* the pressure of a gas in a constant-volume container is outstanding in its sensitivity, accuracy of measurement, and reproducibility. The constant-volume gas thermometer is illustrated schematically in Fig. 14-6. The gas, usually helium, is contained in bulb C, and the pressure exerted by it can be measured by the open-tube mercury manometer. As the temperature of the gas increases, the gas expands, forcing the mercury down in tube B and up in tube A. Tubes A and B are connected through a rubber tube D to a mercury reservoir R. By raising R, the mercury level in B may be brought back to a reference mark E. The gas is thus kept at constant volume.

14-6 The constant-volume gas thermometer.

Gas thermometers are used mainly in bureaus of standards and in some university research laboratories. They are usually large, bulky, and slow in coming to thermal equilibrium.

14-3 Defining a temperature scale

If the temperature of a system remains constant, the state coordinate of any thermometer in contact with it must also remain constant. By this means it has been found that only at a certain definite temperature is it possible for the solid and liquid phases of a material to exist together without the liquid changing to solid or the solid to liquid. This condition is called *phase equilibrium*. The temperature at which phase equilibrium can exist for a particular material varies somewhat with pressure. Similarly, a liquid will remain in phase equilibrium with its vapor at only one definite temperature when the pressure is maintained constant.

The temperature at which a solid and liquid of the same material coexist in phase equilibrium *at atmospheric pressure* is called the *normal melting point*. The temperature at which a liquid and its vapor exist in phase equilibrium at atmospheric pressure is called the *normal boiling point*. Phase equilibrium between a solid and its vapor is possible at atmospheric pressure for some materials. The temperature at which this takes place is the *normal sublimation point*.

It is possible for *all three* phases—solid, liquid, and vapor—to coexist in equilibrium, but only at one definite pressure and temperature known as the *triple point*. The triple-point pressure of water is 4.58 mm of mercury, or 610 Pa. The triple-point temperature is 0.01°C. Any of these phase-equilibrium conditions for any material can be chosen as a standard reference point for the purpose of setting up a temperature scale. Any temperature so chosen is called a *fixed point*. *The standard fixed point used in modern thermometry is the triple point of water,* to which is given the arbitrary* number

$$273.16 \text{ K},$$

read "273.16 kelvins." The triple point is discussed in greater detail in Sec. 17-6.

To illustrate the process of establishing a temperature scale, we consider the constant-volume gas thermometer, for which the thermometric property is the gas pressure p. We define the ratio of two temperatures to be the ratio of the two corresponding values of p. Thus if a gas thermometer is put into thermal equilibrium with a certain system and registers the value p, and is then put into thermal equilibrium with another system and registers the value p_0, the ratio of the two temperatures, T and T_0, respectively, is

$$\frac{T}{T_0} = \frac{p}{p_0}. \tag{14-1}$$

If now we let p_0 be the pressure at the standard fixed point, the triple point of water, then by definition $T_0 = 273.16$ K. Hence the temperature

---

*By extrapolating backwards with the aid of the gas-law formulas, the "absolute zero" point was "determined" to be -273.15°C. Since the kelvin has the same magnitude as one degree on the Celsius scale, the triple point on the Kelvin scale becomes 273.16 K.

T corresponding to any other pressure p is

$$T = (273.16 \text{ K})\frac{p}{p_0}. \tag{14-2}$$

Example Suppose a constant-volume gas thermometer has a pressure of 1.50×10^4 Pa at the triple point of water and a pressure of 2.05×10^4 Pa at the normal boiling point. What is the temperature T_B at the normal boiling point?

Solution From Eq. (14-2),

$$T_B = (273.16 \text{ K})\frac{2.05 \times 10^4 \text{ Pa}}{1.50 \times 10^4 \text{ Pa}} = 373 \text{ K}.$$

The remaining problem in temperature measurement is that when different thermometers are used to measure the same temperature, the results do not always agree. Best agreement is found with gas thermometers; there is some variation with pressure for any gas, but it is found that, in the limit as pressure becomes smaller and smaller, all gas thermometers at the same temperature approach the same reading, independent of the nature of the gas. Thus the usual procedure is to use the constant-volume thermometer (in the limit of very low pressure) to *define* a temperature scale, and then use this device to calibrate the other thermometers in terms of this scale.

The lowest temperature that can be measured with a gas thermometer is about 1 K, provided low-pressure helium is used as the gas. In Chapter 19 we shall discuss the Kelvin temperature scale, which is independent of the properties of any particular substance. It can be shown that in the temperature region in which a gas thermometer can be used, the gas scale and the Kelvin scale are identical. In anticipation of this result, we write K after a gas temperature. It will also be shown in Chapter 19 how the absolute zero of temperature is defined on the Kelvin scale.

The body temperatures of warm-blooded animals are held constant to within a few tenths of one Celsius degree by an elaborate temperature-control system. The thermometric property that senses the temperature of the blood is provided by a chemical equilibrium condition in a part of the brain called the hypothalamus; this activates appropriate temperature-controlling mechanisms. The most important mechanisms are dilation or contraction of blood vessels near the surface, increasing or decreasing loss of body heat by conduction, and activation or deactivation of sweat glands to increase or decrease evaporative cooling.

14-4 The Celsius, Rankine, and Fahrenheit scales*

The Celsius temperature scale employs a degree of the same magnitude as that of the Kelvin scale, but its zero point is shifted so that *the Celsius temperature of the triple point of water is* 0.01 *degree Celsius,* abbreviated 0.01°C. Thus, if T_C denotes the Celsius temperature

$$T_C = T - 273.15 \text{ K.} \tag{14-3}$$

---

* Named after Anders Celsius (1701–1744), William John MacQuorn Rankine (1820–1872), and Gabriel Fahrenheit (1686–1736).

The Kelvin temperature at which steam condenses at 1 atm pressure is found to be 373.15 K, so on the Celsius scale this is

$$373.15° - 273.15° = 100.00°C.$$

There are two other scales in common use in engineering and in everyday life in the United States and in Great Britain. The Rankine temperature T_R (written °R) is proportional to the Kelvin temperature according to the relation

$$T_R = \frac{9}{5}T. \tag{14-4}$$

A degree of the same size is used in the Fahrenheit scale T_F (written °F), but with the zero point shifted according to the relation

$$T_F = T_R - 459.67°R. \tag{14-5}$$

Substituting Eqs. (14-3) and (14-4) into Eq. (14-5), we find

$$T_F = \frac{9}{5}T_C + 32\ F°, \tag{14-6}$$

from which it follows that the Fahrenheit temperature of the ice point ($T_C = 0°C$) is 32°F and of the steam point ($T_C = 100°C$) is 212°F. The 100 Celsius or Kelvin degrees between the ice point and the steam point correspond to 180 Fahrenheit or Rankine degrees, as shown in Fig. 14-7, where the four scales are compared.

14-7 Relations among Kelvin, Celsius, Rankine, and Fahrenheit temperature scales. Temperatures have been rounded off to the nearest degree.

We use the phrase "10 degrees Celsius," or "10°C," when referring to an *actual temperature* on the Celsius scale, and "10 Celsius degrees," or "10 C°" to mean a temperature *interval*. Thus there is an interval of 10 Celsius degrees between 20 degrees Celsius and 30 degrees Celsius.

14-5 Thermal expansion

Most solid materials expand when heated. Suppose a rod of material has a length L_0 at some initial temperature, and that, when the temperature increases by an amount ΔT, the length increases by ΔL. Experiment shows that if ΔT is not too large, ΔL is *directly proportional* to ΔT. Of course, ΔL is also proportional to L_0; if two rods of the same material have the same temperature change, but one is initially twice as long as the other, then the *change* in its length is also twice as great. Introducing a proportionality constant α (which is different for different materials), we may summarize this relation as follows:

$$\Delta L = \alpha L_0 \Delta T. \tag{14-7}$$

The constant α, which characterizes the thermal expansion properties of a particular material, is called the *temperature coefficient of linear expansion* or, more briefly, the *coefficient of linear expansion*.

For materials having no preferential directions, every linear dimension changes according to Eq. (14-7). Thus L could equally well represent the thickness of the rod, the side length of a square sheet, or the diameter of a hole in the material. There are some exceptional cases. Wood, for example, has different expansion properties along the grain and across the grain; and single crystals of material can have different properties along different crystal-axis directions. We exclude these exceptional cases from the present discussion.

It must be emphasized that the direct proportionality expressed by Eq. (14-7) is not exact, but that it is *approximately* correct for sufficiently small temperature changes. For any temperature one can define a coefficient of thermal expansion by the equation

$$\alpha = \frac{1}{L}\frac{dL}{dT}. \tag{14-8}$$

In this case, α for a given material will be found to vary somewhat with the initial temperature and the size of the temperature interval. Because Eq. (14-7) is at best an approximation, we shall ignore this variation. Average values of α for several materials are listed in Table 14-1.

Example A steel measuring tape 5 m long is calibrated at a temperature of 20°C. What is its length on a hot summer day when the temperature is 35°C?

Solution From Eq. (14-7),

$$\Delta L = (5 \text{ m})(1.2 \times 10^{-5} (\text{C}°)^{-1})(35°\text{C} - 20°\text{C})$$
$$= 0.9 \times 10^{-3} \text{ m} = 0.9 \text{ mm} = 0.0009 \text{ m}.$$

Thus the length at 35°C is 5.0009 m.

Increasing temperature usually causes increases in *volume,* for both solid and liquid materials; experiment shows that if the temperature change ΔT is not too great, the increase in volume ΔV is approximately *proportional* to the temperature change. It is also proportional to the initial volume V_0, just as with linear expansion. The relationship can be expressed as follows:

$$\Delta V = \beta V_0 \Delta T. \tag{14-9}$$

The constant β, which characterizes the volume expansion properties of a particular material, is called the *temperature coefficient of volume expansion,* or the *coefficient of volume expansion*.

Like the coefficient of linear expansion, β varies somewhat with temperature, and Eq. (14-9) should be regarded as an approximate relation, valid for sufficiently small temperature changes. For many substances β decreases as the temperature is lowered, approaching zero as the Kelvin temperature approaches zero. It is also noteworthy that the higher the

Table 14-1 Coefficient of linear expansion

Material	α $(\text{C}°)^{-1}$
Aluminum	2.4×10^{-5}
Brass	2.0×10^{-5}
Copper	1.7×10^{-5}
Glass	$0.4–0.9 \times 10^{-5}$
Steel	1.2×10^{-5}
Invar	0.09×10^{-5}
Quartz (fused)	0.04×10^{-5}

melting temperature of a metal, the lower the coefficient of volume expansion.

Some values of β in the neighborhood of room temperature are listed in Table 14–2. Note that the values for liquids are much larger than those for solids.

Table 14–2 Coefficient of volume expansion

Solids	β, $(C°)^{-1}$	Liquids	β, $(C°)^{-1}$
Aluminum	7.2×10^{-5}	Alcohol, ethyl	75×10^{-5}
Brass	6.0×10^{-5}	Carbon disulfide	115×10^{-5}
Copper	5.1×10^{-5}	Glycerin	49×10^{-5}
Glass	1.2–2.7×10^{-5}	Mercury	18×10^{-5}
Steel	3.6×10^{-5}	Water, 20°C	20×10^{-5}
Invar	0.27×10^{-5}	Water, 50°C	60×10^{-5}
Quartz (fused)	0.12×10^{-5}		

If there is a hole in a solid body, the volume of the hole increases when the body expands, just as if the hole were a solid of the same material as the body. This remains true even if the hole becomes so large that the surrounding body is reduced to a thin shell. Thus the volume enclosed by a thin-walled glass flask or thermometer bulb increases just as would a solid body of glass of the same size.

Example A glass flask of volume 200 cm³ is just filled with mercury at 20°C. How much mercury overflows when the temperature of the system is raised to 100°C? The coefficient of volume expansion of the glass is $1.2 \times 10^{-5}(C°)^{-1}$.

Solution The increase in the volume of the flask is

$$\Delta V = (1.2 \times 10^{-5}(C°)^{-1})(200 \text{ cm}^3)(100° - 20°) = 0.192 \text{ cm}^3.$$

The increase in the volume of the mercury is

$$\Delta V = (18 \times 10^{-5}(C°)^{-1})(200 \text{ cm}^3)(100° - 20°) = 2.88 \text{ cm}^3.$$

The volume of mercury overflowing is therefore

$$2.88 \text{ cm}^3 - 0.19 \text{ cm}^3 = 2.69 \text{ cm}^3.$$

The volume expansion coefficient for a solid material is related to the linear coefficient. To obtain the relation, we consider a solid body in the form of a rectangular parallelepiped with dimensions L_1, L_2, and L_3. Then the volume is

$$V_0 = L_1 L_2 L_3.$$

According to Eq. (14–7), when the temperature increases by ΔT, each linear dimension changes, and the new volume is

$$V_0 + \Delta V = L_1 L_2 L_3 (1 + \alpha \Delta T)^3 = V_0 (1 + \alpha \Delta T)^3$$
$$= L_1 L_2 L_3 [1 + 3\alpha \Delta T + 3\alpha^2 (\Delta T)^2 + \alpha^3 (\Delta T)^3].$$

If ΔT is small we may neglect the terms containing $(\Delta T)^2$ and $(\Delta T)^3$.

Dropping these and subtracting $V_0 = L_1 L_2 L_3$ from both sides, we find
$$\Delta V = (3\alpha) V_0 \Delta T.$$
Comparing this with Eq. (14–9), we obtain
$$\beta = 3\alpha. \qquad (14\text{–}10)$$

Water, in the temperature range from 0°C to 4°C, *decreases* in volume with increasing temperature, contrary to the behavior of most substances. That is, between 0°C and 4°C the coefficient of expansion of water is *negative*. Above 4°C, water expands when heated. Since the volume of a given mass of water is smaller at 4°C than at any other temperature, the density of water is maximum at 4°C. Water also expands upon freezing, unlike most materials.

This anomalous behavior has an important effect on plant and animal life in lakes. When a lake cools, the cooled water at the surface flows to the bottom because of its greater density. But when the temperature reaches 4°C, this flow ceases, and the water near the surface remains colder (and less dense) than that at the bottom. As the surface freezes, the ice floats because it is less dense than water. The water at the bottom remains at 4°C until nearly the entire body is frozen. If water behaved like most substances, contracting continuously on cooling and freezing, lakes would freeze from the bottom up, circulation due to density differences would continuously carry warmer water to the surface for efficient cooling, and lakes would freeze solid much more easily, thus destroying all plant and animal life that can withstand cold water but not freezing.

The anomalous expansion of water in the temperature range 0°C to 10°C is shown in Fig. 14–8. Table 14–3 covers a wider range of temperatures.

The *milliliter* (ml) was formerly defined to be the volume of 1 g of water at maximum density (at 4°C). As Table 14–3 shows, this made the milliliter slightly larger than the cubic centimeter. In 1964 the milliliter was officially redefined to be *exactly* one cubic centimeter.

Table 14–3 Density and volume of water

T, °C	Density, g·cm^{-3}	Volume of 1 gram, cm^3
0	0.99984	1.00016
4	0.99997	1.00003
10	0.99970	1.00030
20	0.99820	1.00180
50	0.98804	1.01210
75	0.97486	1.02579
100	0.95835	1.04346

*14–6 Thermal stresses

If the ends of a rod are rigidly fixed so as to prevent expansion or contraction and the temperature of the rod is changed, tensile or compressive stresses, called *thermal stresses,* will be set up in the rod. These stresses may become very large, sufficiently so to stress the rod beyond its elastic limit or even beyond its breaking strength. Hence in the design

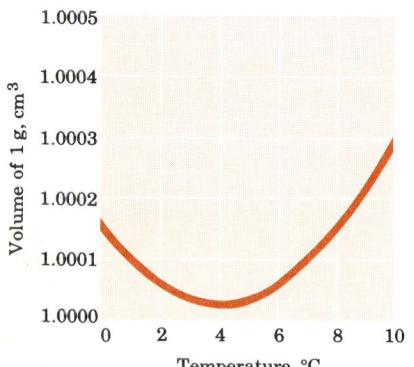

14–8 Volume of one gram of water in the temperature range from 0 to 10°C, showing anomalous thermal expansion behavior. Between 0 and 4°C the coefficient of volume expansion is negative.

of any structure that is subject to changes in temperature, some provision must, in general, be made for expansion. In a long steam pipe this is accomplished by the insertion of expansion joints or a section of pipe in the form of a U. In bridges, one end may be rigidly fastened to its abutment while the other rests on rollers.

It is a simple matter to compute the thermal stress set up in a rod that is not free to expand or contract. Suppose that a rod of length L_0 and cross-section area A has its ends rigidly fastened while the temperature is reduced by an amount ΔT.

The fractional change in length if the rod were free to contract would be

$$\frac{\Delta L}{L_0} = \alpha \, \Delta T, \tag{14-11}$$

where ΔL and ΔT are both negative. Since the rod is *not* free to contract, the tension must increase by a sufficient amount to produce an equal and opposite fractional change in length. But from the definition of Young's modulus,

$$Y = \frac{F/A}{\Delta L/L_0}, \qquad \frac{\Delta L}{L_0} = \frac{F}{AY}, \tag{14-12}$$

where this ΔL is positive. The tensile force F is determined by the requirement that the *total* fractional change in length, thermal expansion plus elastic strain, must be zero:

$$\alpha \, \Delta T + \frac{F}{AY} = 0,$$

$$F = -AY\alpha \, \Delta T. \tag{14-13}$$

Since ΔT represents a decrease in temperature, it is negative, so F is positive. The tensile *stress* in the rod is

$$\frac{F}{A} = -Y\alpha \, \Delta T. \tag{14-14}$$

If, instead, ΔT represents an *increase* in temperature, then F and F/A become negative, corresponding to compressive force and stress, respectively.

Thermal stresses can also be induced by nonuniform heating. Even if a solid body at uniform temperature has no internal stresses, stress may be induced by nonuniform expansion if the body is heated unevenly. The breaking of a thick glass container when hot water is poured into it is a familiar phenomenon. Heat-resistant glasses such as Pyrex have exceptionally low expansion coefficients, and usually also have high strength to permit thin-wall construction to minimize temperature differences.

Similar phenomena occur with volume expansion. If a bottle is completely filled with water, tightly capped, and then heated, it will break because the thermal expansion coefficient for water is greater than that for glass. If a material is enclosed in a very rigid container so that its volume cannot change, then a rise in temperature ΔT is accompanied by an increase in pressure Δp. An analysis similar to that leading to Eq. (14-14) shows that the pressure increase is given by

$$\Delta p = B\beta \, \Delta T, \tag{14-15}$$

where B is the bulk modulus and β the coefficient of volume expansion.

Questions

14-1 Does it make sense to say that one body is twice as hot as another?

14-2 A student claimed that thermometers are useless because a thermometer always registers *its own* temperature. How would you respond?

14-3 What other properties of matter, in addition to those mentioned in the text, might be used as thermometric properties? How could they be used to make a thermometer?

14-4 A thermometer is laid out in direct sunlight. Does it measure the temperature of the air, or of the sun, or what?

14-5 Thermometers sometimes contain red or blue liquid, which is often ethanol. What advantages and disadvantages does this have compared with mercury?

14-6 Could a thermometer similar to that shown in Fig. 14-1a be made using water as the liquid? What difficulties would this thermometer present?

14-7 What is the temperature of a vacuum?

14-8 Is there any particular reason for constructing a temperature scale with higher numbers corresponding to hotter bodies, rather than the reverse?

14-9 If a brass pin is a little too large to fit in a hole in a steel block, should you heat the pin and cool the block, or the reverse?

14-10 When a block with a hole in it is heated, why doesn't the material around the hole expand into the hole and make it smaller?

14-11 Many automobile engines have cast-iron cylinders and aluminum pistons. What kinds of problems could occur if the engine gets too hot?

14-12 When a hot-water faucet is turned on, the flow often decreases gradually before it settles down. This is an annoyance in the shower; why does it happen?

14-13 Two bodies made of the same material have the same external dimensions and appearance, but one is solid and the other is hollow. When they are heated, is the overall volume expansion the same or different?

14-14 A thermostat for controlling household heating systems often contains a bimetallic element consisting of two strips of different metals, welded together face to face. When the temperature changes, this composite strip bends one way or the other. Why?

14-15 Why is it sometimes possible to loosen caps on screw-top bottles by dipping the cap briefly in hot water?

14-16 The rate at which a pendulum clock runs depends on the length of the pendulum. Would a pendulum clock gain time in hot weather and lose in cold, or the reverse? Could one design a pendulum, perhaps using two different metals, that would *not* change length with temperature?

14-17 When a rod is cooled but prevented from contracting, as in Sec. 14-6, thermal tension develops. Under these circumstances, does the *thickness* of the rod change? If so, how would the change be calculated?

Problems

14-1 The ratio of the pressures of a gas at the melting point of lead and at the triple point of water, when the gas is kept at constant volume, is found to be 2.19816. What is the Kelvin temperature of the melting point of lead?

14-2

a) If you feel sick in France and are told you have a temperature of 40°C, should you be concerned?

b) What is normal body temperature on the Celsius scale?

c) The normal boiling point of liquid oxygen is −182.97°C. What is this temperature on the Kelvin and Rankine scales?

d) At what temperature do the Fahrenheit and Celsius scales coincide?

14-3 When the United States finally converts officially to metric units, the Celsius temperature scale will replace the Fahrenheit scale for everyday use. As a familiarization exercise, find the Celsius temperatures corresponding to

a) a cool room (68°F);

b) a hot summer day (95°F);

c) a cold winter day (5°F).

14-4 In a rather primitive experiment with a constant-volume gas thermometer, the pressure at the triple point of water was found to be 4.0×10^4 Pa, and the pressure at the normal boiling point 5.4×10^4 Pa. According to these data, what is the temperature of absolute zero on the Celsius scale?

14-5 A gas thermometer of the type shown in Fig. 14-6 registered a pressure corresponding to 5.0 cm of mercury when in contact with water at the triple point. What pressure will it read when in contact with water at the normal boiling point?

14-6 The electrical resistance of some metals varies with temperature (as measured by a gas thermometer) approximately according to $R = R_0[1 + \beta(T - T_0)]$, where R_0 is the resistance at temperature T_0. For a certain metal, the constant β is found to be 0.004 K^{-1}.

a) If the resistance at 0°C is 100 ohms, what is the resistance at 20°C?

b) At what temperature is the resistance 200 ohms?

14-7 The pendulum shaft of a clock is of aluminum. What is the fractional change in length of the shaft when it is cooled from 25°C to 10°C?

14-8 A surveyor's 100-ft steel tape is correct at a temperature of 20°C. The distance between two points, as measured by this tape on a day when the temperature is 35°C, is 86.57 ft. What is the true distance between the points?

14-9 To ensure a tight fit, the aluminum rivets used in airplane construction are made slightly larger than the rivet holes and cooled by "dry ice" (solid CO_2) before being driven. If the diameter of a hole is 0.2500 in., what should be the diameter of a rivet at 20°C if its diameter is to equal that of the hole when the rivet is cooled to $-78°C$, the temperature of dry ice? Assume the expansion coefficient to remain constant at the value given in Table 14-1.

14-10 A steel ring of 3.000 inches inside diameter at 20°C is to be heated and slipped over a brass shaft measuring 3.002 inches in diameter at 20°C.

a) To what temperature should the ring be heated?

b) If the ring and shaft together are cooled by some means such as liquid air, at what temperature will the ring just slip off the shaft?

14-11 A metal rod 30.0 cm long expands by 0.075 cm when its temperature is raised from 0°C to 100°C. A rod of a different metal and of the same length expands by 0.045 cm for the same rise in temperature. A third rod, also 30.0 cm long, is made up of pieces of each of the above metals placed end to end, and expands 0.065 cm between 0°C and 100°C. Find the length of each portion of the composite bar.

14-12 A hole 1.000 in. in diameter is bored in a brass plate at a temperature of 20°C. What is the diameter of the hole when the temperature of the plate is increased to 200°C? Assume the expansion coefficient to remain constant.

14-13 Suppose that a steel hoop could be constructed around the earth's equator, just fitting it at a temperature of 20°C. What would be the thickness of space between the hoop and the earth if the temperature of the hoop were increased by 1 C°?

14-14 A clock whose pendulum makes one vibration in 2 s is correct at 25°C. The pendulum shaft is of steel and its mass may be neglected compared with that of the bob.

a) What is the fractional change in length of the shaft when it is cooled to 15°C?

b) How many seconds per day will the clock gain or lose at 15°C? The calculational technique outlined for Problem 11-24 may be used.

14-15 A clock with a brass pendulum shaft keeps correct time at a certain temperature.

a) How closely must the temperature be controlled if the clock is not to gain or lose more than 1 second a day? Does the answer depend on the period of the pendulum?

b) Will an increase of temperature cause the clock to gain or lose?

14-16 A thermometer similar to that shown in Fig. 14-1a has a spherical bulb of radius 0.2 cm and a capillary tube of radius 0.05 mm. The liquid is mercury, and β for the bulb is $1.2 \times 10^{-5} (C°)^{-1}$. What distance on the scale is covered by a temperature range from 0 to 100°C?

14-17 A 250-cm³ glass bottle is completely filled with water at 50°C. The bottle and water are then heated to 60°C. How much water runs over

a) if the expansion of the bottle is neglected;

b) if the expansion of the bottle is included? Take $\beta = 1.2 \times 10^{-5} (C°)^{-1}$ for glass.

14-18 An area is measured on the surface of a solid body. If the area is A_0 at some initial temperature and then changes by ΔA when the temperature changes by ΔT, show that

$$\Delta A = (2\alpha) A_0 \Delta T.$$

14-19 An aluminum cube 10 cm on a side is heated from 10°C to 30°C. What is the change in its volume? In its density?

14-20 A brass ball 6 cm in radius is cooled from 100°C to 20°C. Find its change in volume

a) by first finding the change in radius and then computing the new volume from the new radius;

b) by using the coefficient of volume expansion.

14-21 A glass flask whose volume is exactly 1000 cm³ at 0°C is filled level full of mercury at this temperature. When flask and mercury are heated to 100°C, 15.2 cm³ of mercury overflow. If the coefficient of volume expansion of mercury is 0.000182 per Celsius degree, compute the coefficient of linear expansion of the glass.

14-22 At a temperature of 20°C, the volume of a certain glass flask, up to a reference mark on the stem of the flask, is exactly 100 cm³. The flask is filled to this point with a liquid whose coefficient of volume expansion is $120 \times 10^{-5} (C°)^{-1}$, with both flask and liquid at 20°C. The coefficient of linear expansion of the glass is $8 \times 10^{-6} (C°)^{-1}$. The cross section of the stem is 1 mm² and can be considered constant. How far will the liquid rise or fall in the stem when the temperature is raised to 40°C?

14-23 The pressure p, volume V, number of moles n, and Kelvin temperature T of an ideal gas are related by the equation $pV = nRT$. Prove that the coefficient of volume expansion is equal to the reciprocal of the Kelvin temperature.

14-24 The length of a steel bridge is 2000 ft.

a) If it were a continuous span, fixed at one end and free

to move at the other, about what would be the range of motion of the free end between a cold winter day ($-20°$F) and a hot summer day ($100°$F)?

b) If both ends were rigidly fixed on the summer day, what would be the stress on the winter day?

14-25 The cross section of a steel rod is 10 cm². What is the least force that will prevent it from contracting while cooling from $520°$C to $20°$C?

14-26

a) A steel wire that is 3 m long at $20°$C is found to increase in length by 1.5 cm when heated to $520°$C. Compute its average coefficient of linear expansion.

b) Find the stress in the wire if it is stretched taut at $520°$C and cooled to $20°$C without being allowed to contract.

14-27 A steel rod of length 40 cm and a copper rod of length 36 cm, both of the same diameter, are placed end to end between two rigid supports, with no initial stress in the rods. The temperature of the rods is now raised by $50°$C. What is the stress in each rod?

14-28 A heavy brass bar has projections at its ends, as in Fig. 14-9. Two fine steel wires fastened between the projections are just taut (zero tension) when the whole system is at $0°$C. What is the tensile stress in the steel wires when the temperature of the system is raised to $300°$C? Make any simplifying assumptions you think are justified, but state what they are.

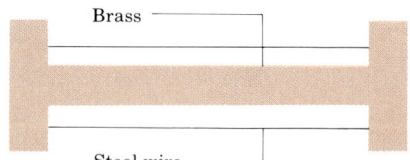

Figure 14-9

14-29 Steel railroad rails 60 ft long are laid on a winter day when the temperature is $20°$F.

a) How much space must be left between rails if they are to just touch on a summer day when the temperature is $110°$F?

b) If the rails were originally laid in contact, what would be the stress in them on a summer day when the temperature is $110°$F?

14-30 Prove that if a body under hydrostatic pressure is raised in temperature but not allowed to expand, the increase in pressure is

$$\Delta p = B\beta \, \Delta T,$$

where the bulk modulus B and the average coefficient of volume expansion β are both assumed positive and constant.

14-31

a) A block of steel at an initial pressure of 1 atm and a temperature of $20°$C is kept at constant volume. If the temperature is raised to $32°$C, what is the final pressure?

b) If the block is maintained at constant volume by rigid walls that can withstand a maximum pressure of 1200 atm, what is the highest temperature to which the system may be raised? Assume B and β to remain practically constant at the values 1.5×10^{11} Pa and $5.0 \times 10^{-5}(C°)^{-1}$, respectively.

14-32 What hydrostatic pressure is necessary to prevent a copper block from expanding when its temperature is increased from $20°$C to $30°$C?

14-33 Table 14-3 lists the density of water, and the volume of 1 g, at atmospheric pressure. A steel bomb is filled with water at $10°$C and atmospheric pressure, and the system is heated to $75°$C. What is then the pressure in the bomb? Assume the bomb to be sufficiently rigid so that its volume is not affected by the increased pressure.

14-34 A liquid is enclosed in a metal cylinder provided with a piston of the same metal. The system is originally at atmospheric pressure and at a temperature of $80°$C. The piston is forced down until the pressure on the liquid is increased by 100 atm, and it is then clamped in this position. Find the new temperature at which the pressure of the liquid is again 1 atm. Assume that the cylinder is sufficiently strong so that its volume is not altered by changes in pressure, but only by changes in temperature.

Compressibility of liquid $\kappa = 50 \times 10^{-6}$ atm^{-1}.
Coefficient of volume
expansion of liquid $\beta = 5.3 \times 10^{-4}(C°)^{-1}$.
Coefficient of linear
expansion of metal $\alpha = 10 \times 10^{-6}(C°)^{-1}$.

14-35

a) For any material, density ρ, mass m, and volume V are related by $\rho = m/V$. Prove that

$$\beta = -\frac{1}{\rho}\frac{\partial \rho}{\partial T}.$$

b) The density of rock salt between $-193°$C and $-13°$C is given by the empirical formula

$$\rho = 2.1680(1 - 11.2 \times 10^{-5}T - 0.5 \times 10^{-7}T^2),$$

with T measured on the Celsius scale. Calculate β at $-100°$C.

QUANTITY OF HEAT 15

15-1 Heat transfer

In the preceding chapter we discussed the concept of temperature in connection with thermal equilibrium. When two bodies *not* initially in thermal equilibrium are placed in contact or are separated by a diathermic wall, their temperatures change until they reach thermal equilibrium. We may now examine the nature of the interaction that takes place between the bodies during their approach to thermal equilibrium. A quantitative discussion of this interaction leads to the concept of *heat*, the subject of the present chapter.

Suppose that system A, at a higher temperature than system B, is put in contact with B. When thermal equilibrium has been reached, A will be found to have undergone a temperature decrease and B a temperature increase. It was therefore natural for early investigators in this field to assume that A lost something and that this "something" flowed into B. While the temperature changes are taking place, it is customary to refer to a *heat flow* or a *heat transfer* from A to B.

At one time the process of heat transfer was thought to be a flow of an invisible weightless fluid called *caloric*, but the work of Count Rumford (1753-1814) and Sir James Prescott Joule (1818-1889), established firmly that heat flow is an *energy transfer*. When an energy transfer takes place by virtue of a temperature difference exclusively, it is called a *heat flow*. For example, water is heated and converted to steam in a steam boiler by contact with a metal pipe or container kept at a high temperature by a hot flame from burning coal or gas. The steam has a greater ability to do *work* (by pushing against a turbine blade or the piston in a steam engine) than it had when it was in the form of cold water, and therefore the water must have received *energy* in a process involving a heat flow from the hot flame (to the container) to the cooler water.

Of course, energy transfer can also occur without heat flow. In an air compressor, a moving piston pushes against a mass of air, doing work on the air as it compresses it to a smaller volume. In this compressed state the gas is capable of doing more work than before, and hence has acquired energy.

Finally, if in the air compressor, the air and the piston are at different temperatures, heat flow between the piston and the air can occur. This is an example of a process involving two kinds of energy transfer simultaneously, flow of *heat* and performance of *work*.

15-2 Quantity of heat

Let us consider various kinds of energy transfer in the specific system shown in Fig. 15-1, consisting of a quantity of water and a piece of resistance wire. In (a) the system is caused to undergo a temperature rise ΔT in a process involving a heat flow through a diathermic wall (Sec. 14-1) at the bottom of the enclosure. In (b) the system is surrounded by adiabatic walls and there is no heat flow. The same temperature change ΔT is produced by permitting a falling body to turn an electric generator, causing a current in the resistance wire. Thus in (b) the method of energy transfer to the system is the performance of mechanical work, with no heat flow into the system.

Suppose, however, that we regard the water alone as our system in (b). Then the flow of energy *into the water* is due to the presence of a resistance wire whose temperature is maintained a trifle higher than that of the water by means of the electric current in the wire. This process involves a heat flow from the wire to the water.

Two conclusions may be drawn from these considerations:

1. Whether a process is regarded as involving a flow of heat or a performance of work depends on the *choice of the system*.
2. Whether the energy transfer shown in Fig. 15-1b is regarded as a flow of heat or as a performance of work has no bearing on the quantity of energy delivered to the water.

We may describe the results of the two processes in Fig. 15-1 by saying that a heat flow and a performance of work are *equivalent*. Once either process is completed, the energy of the system is greater than before, and no experiment can tell whether this energy increase has been caused by a flow of heat or a performance of work. They are both methods of energy transfer; in each case the amount of energy transferred is expressed in energy units such as joules or foot-pounds, and the time rate of energy transfer is expressed in power units, such as watts (joules per second) or foot-pounds per second.

Up to this point, the word *heat* has been used only in conjunction with *flow* or *transfer*. A heat flow is an energy transfer brought about solely by a temperature difference. Ideally, the word *heat* should be used *only* to refer to a *method* of energy *transfer*, and when the transfer is completed to refer to the total amount of energy so transferred. The expression "quantity of heat," however, has played such a huge role in the development of the subject and is to be found in so many books and tables, that it is almost impossible to avoid. It is essential to bear in mind that the concept "quantity of heat" has meaning *only* in the context of an interaction in which energy is *transferred* from one system to another as a result of a temperature difference. As we shall see later, to say that a given system *contains* a certain quantity of heat is meaningless. A careful discussion of this point leads to the concept of *internal energy*, to be discussed in detail in Chapter 18.

(a)

(b)

15-1 The same temperature change of the same system may be accomplished by either (a) a heat flow or (b) the performance of work.

In the eighteenth century, a unit of quantity of heat, the *calorie*, was defined as that amount of heat required to raise the temperature of one gram of water one kelvin or one Celsius degree. It was found later that more heat was required to raise the temperature of 1 g of water from, say, 90°C to 91°C than from 30°C to 31°C. The definition was then refined, and the chosen calorie became known as the "15° calorie," that is, the quantity of heat required to change the temperature of 1 g of water from 14.5°C to 15.5°C. A corresponding unit defined in terms of Fahrenheit degrees and British units is the *British thermal unit,* or Btu. By definition, 1 Btu, is the quantity of heat required to raise the temperature of one pound of water from 63°F to 64°F. A third unit in common use, especially in measuring food energy, is the kilocalorie (kcal). The relations among these three units are

$$1 \text{ Btu} = 252 \text{ cal} = 0.252 \text{ kcal}.$$

It should also be noted that the unit of quantity of heat is fundamentally a unit of energy; thus there must be a relation between the above units and the familiar mechanical energy units such as the joule. It has been found experimentally that

$$1 \text{ cal} = 1 \text{ kcal} = 4.186 \text{ J},$$
$$1 \text{ Btu} = 778 \text{ ft-lb}.$$

The International Committee on Weights and Measures no longer recognizes the calorie as a fundamental unit, but recommends instead that the joule be used for quantity of heat as well as for all other forms of energy. Although the calorie is convenient in problems involving water, it is awkward when both heat and other forms of energy are involved. There seems little doubt that ultimately the joule will become the universal unit of energy; many of the examples and problems of this and the following chapters will anticipate this usage. The relation between heat and mechanical energy will be explored in detail in Chapters 18 through 20.

15-3 Heat capacity

Suppose that a small quantity of heat dQ is transferred between a system of mass m and its surroundings. If the system undergoes a small temperature change dT, the *specific heat capacity c of the system is defined as*

$$c = \frac{1}{m} \frac{dQ}{dT}. \qquad (15\text{-}1)$$

That is, the heat dQ needed to increase the temperature of a mass m of material by an amount dT is

$$dQ = mc\, dT. \qquad (15\text{-}2)$$

The specific heat capacity of water is approximately

$$4.19 \text{ J} \cdot \text{g}^{-1} \cdot (\text{C}°)^{-1}, \quad 4190 \text{ J} \cdot \text{kg}^{-1} \cdot (\text{C}°)^{-1},$$
$$1 \text{ cal} \cdot \text{g}^{-1} \cdot (\text{C}°)^{-1}, \quad \text{or} \quad 1 \text{ Btu} \cdot \text{lb}^{-1} \cdot (\text{F}°)^{-1}.$$

It is often convenient to use the *mole* to describe the amount of substance. By definition, one mole (1 mol) of any substance is a quantity of matter such that its mass in grams is numerically equal to the *molecular mass M* (often called *molecular weight*).* To calculate the number of moles n, we divide the mass in grams by the molecular mass; thus $n = m/M$. Replacing the mass m in Eq. (15-1) by the product nM, we get

$$Mc = \frac{1}{n}\frac{dQ}{dT}.$$

The product Mc is called the <u>molar heat capacity</u> and is represented by the symbol C. Hence, by definition

$$C = Mc = \frac{1}{n}\frac{dQ}{dt}, \qquad (15\text{-}3)$$

and

$$dQ = nC\,dT. \qquad (15\text{-}4)$$

The molar heat capacity of water is approximately

$$75.3 \text{ J}\cdot\text{mol}^{-1}\cdot(\text{C}°)^{-1} \quad \text{or} \quad 18 \text{ cal}\cdot\text{mol}^{-1}\cdot(\text{C}°)^{-1}.$$

The quantity defined by Eq. (15-1) is sometimes called simply *specific heat*, and the molar heat capacity defined by Eq. (15-3) is often called the *molar specific heat*. However, the terms *specific heat capacity* and *molar heat capacity* will be used throughout this book. Representative values of specific and molar heat capacity are given in Table 15-1.

Table 15-1 Mean specific and molar heat capacities of metals (constant pressure)

Metal	(Specific) $\text{J}\cdot\text{g}^{-1}\cdot(\text{C}°)^{-1}$	Temperature range, °C	M, $\text{g}\cdot\text{mol}^{-1}$	(Molar) $C = Mc$ $\text{J}\cdot\text{mol}^{-1}\cdot(\text{C}°)^{-1}$
Beryllium	1.97	20–100	9.01	17.7
Aluminum	0.91	17–100	27.0	24.6
Iron	0.47	18–100	55.9	26.3
Copper	0.39	15–100	63.5	24.8
Silver	0.234	15–100	108	25.3
Mercury	0.138	0–100	201	27.7
Lead	0.130	20–100	207	26.9

If the specific heat capacity of a material is constant over a temperature interval from T_1 to T_2, then from Eq. (15-2), the total quantity of heat Q must be supplied to a body of mass m to change its temperature from T_1 to T_2 is

$$Q = mc(T_2 - T_1). \qquad (15\text{-}5)$$

If T_2 is less than T_1, Q is negative, indicating transfer of heat *out of* the body rather than *into* it.

---

*Although the term molecular *weight* is in common use, the term measures the *mass,* not the *weight* of a molecule. Thus the term *molecular mass* is preferable. This term is coming into more common use and will be used throughout this book.

Similarly, when the molar heat capacity may be treated as constant, Eq. (15-4) gives

$$Q = nC(T_2 - T_1). \qquad (15\text{-}6)$$

In fact, the heat capacities of all materials vary somewhat with temperature and also with pressure. When this variation may not be neglected, the total heat needed for a temperature change from T_1 to T_2 is obtained by integrating Eq. (15-2):

$$Q = m \int_{T_1}^{T_2} c \, dT. \qquad (15\text{-}7)$$

Of course, c must be expressed as a function of T in order to carry out the integration.

Precise measurements of heat capacities require great experimental skill, partly because of the difficulty of avoiding (and compensating for) unwanted heat transfer between the sample and its surroundings. Still, such measurements are very important. The temperature variation of specific heats provides the most direct approach to the understanding of molecular energies in matter. Low-temperature thermal properties in particular are of great interest in contemporary physics.

Figure 15-2 shows the variation of the specific heat capacity of water with temperature. The quantity of heat necessary to raise the temperature of 1 g of water from 14.5°C to 15.5°C is

$$\text{one } 15° \text{ cal} = 4.1858 \text{ J}.$$

Two other calories are frequently used. The *international table calorie* (IT cal) is *defined* to be

$$\text{one IT cal} = 4.1868 \text{ J},$$

and is almost identical with the 15° cal. The *thermochemical calorie*, however, is equal to 4.1840 J and corresponds to about a 17° calorie. There is also a *mean* calorie, equal to 4.190 J. This variety of definitions of the calorie is an additional argument in favor of *eliminating* the calorie completely and using the joule as the fundamental unit of quantity of heat as well as of all other forms of energy.

The specific or molar heat capacity of a substance is not the only physical property whose experimental determination requires the measurement of a quantity of heat. Heat conductivity, heat of fusion, heat of vaporization, heat of combustion, heat of solution, and heat of reaction

15-2 Specific heat capacity of water as a function of temperature.

are examples of other such properties that are called *thermal properties* of matter. The field of physics and physical chemistry concerned with the measurement of thermal properties is called *calorimetry*.

15-4 Experimental values of heat capacities

The amount of heat transferred to or from a system depends on the manner in which the system is controlled or constrained during the transfer. The most common constraints are for the system to be kept at *constant pressure* or at *constant volume*. The two corresponding specific (or molar) heat capacities are denoted by the respective symbols c_p (or C_p) and c_v (or C_v). In the electrical method of measuring heat capacities, the sample under investigation is usually maintained at constant pressure. The two heat capacities differ because if the system is permitted to expand during heating, there is additional energy exchange through performance of mechanical work by the system on its surroundings. If the volume is held constant there is no such work.

A few average values of heat capacities of metals are listed in Table 15-1. The specific heat capacities are seen to be *less* than that of water and to decrease with increasing molecular mass. The last column reveals an interesting regularity, first noted in 1819 by two French physicists, Dulong and Petit. The average molar heat capacities (at constant pressure) for all metals except the very lightest are approximately the same and equal to about 25 J·mol^{-1}·(C°)$^{-1}$. This is known as the *Dulong and Petit law;* although only an approximate rule, it contains the germ of a very important idea. We know from chemistry that the number of molecules in one mole is the same for all substances. It follows that (nearly) the same amount of heat is required *per molecule* to raise the temperature of each of these metals by a given amount, even though the *mass* of a molecule of lead (for example) is nearly 10 times as great as that of a molecule of aluminum. To put it in a different way, the heat required to raise the temperature of a sample of metal depends only on *how many* molecules the sample contains, and not on the mass of an individual molecule. This is the first time in our study of physics that we have met a property of matter so directly related to its molecular structure.

The really important theoretical interpretation of heat-capacity measurements can be made only when the theoretician has at his disposal the complete temperature dependence, from the lowest to the highest possible temperatures, of the *molar heat capacity at constant volume,* C_v. It is this quantity that is directly connected with the molecular energy, which, in turn, may be calculated by statistical methods. Fortunately, there are relations that permit obtaining values of C_v from experimental values of C_p. Figure 15-3 shows the resulting values of C_v and their temperature variation from about 4 K to 1000 K. These curves represent the extremes of behavior. All other metals and nonmetals lie within the boundaries of lead and diamond. Except for certain anomalies that are too complicated to dwell upon, all such curves approach about 25 J·mol^{-1}·K^{-1} as the temperature approaches infinity. The Dulong and Petit value, therefore, is approached by all solids, lead arriving at this value at about 200 K (below room temperature) and diamond requiring well over 2000 K.

Each curve in Fig. 15-3 has exactly the same shape; it starts at zero, rises rapidly at first, then bends over and approaches the value

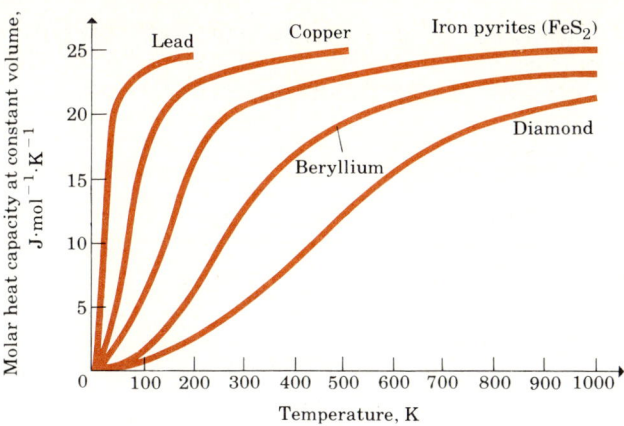

15-3 Temperature variation of C_v of solids.

25 J·mol⁻¹·K⁻¹. It was shown by Debye that the behavior of the molar heat capacity of *nonmetals* over the entire temperature range can be accounted for by the *vibrational motion of the molecules occupying regular positions in the crystal lattice*. There are, however, many substances whose temperature variation of heat capacity cannot be explained entirely on the basis of lattice vibrations. The low-temperature behavior of four "abnormal" types of substance is depicted in Fig. 15–4. There are many more different types of heat-capacity curves; each of them indicates some special molecular or atomic or ionic property of the particles that occupy the regular positions within the crystal lattice. The theoretical understanding of heat capacity curves is one of the most interesting areas of modern solid-state physics.

15-5 Change of phase

The term *phase* as used here refers to the fact that matter exists either as a solid, liquid, or gas. Thus the chemical substance H_2O exists in the *solid phase* as ice, in the *liquid phase* as water, and in the *gaseous phase* as steam. Provided they do not decompose at high temperatures, all substances can exist in any of the three phases under the proper conditions of temperature and pressure. Transitions from one phase to another are accompanied by the absorption or liberation of heat and usually by a change in volume, even when the transition occurs at constant temperature. The temperature at which a phase change occurs also depends on pressure.

As an example, suppose that ice is taken from a freezer at $-25°C$. It is then crushed and placed in a container surrounded by a heating coil which supplies heat at a uniform rate. The container is insulated so that no other heat reaches the ice, and a thermometer is inserted into the ice. The temperature is observed to increase steadily, as shown by the segment *a* to *b* in Fig. 15–5, until the temperature has risen to $0°C$. In this temperature range the specific heat capacity of ice is approximately

$$2.00 \text{ J} \cdot \text{g}^{-1}(\text{C}°)^{-1} \quad \text{or} \quad 0.48 \text{ cal} \cdot \text{g}^{-1} \cdot (\text{C}°)^{-1}.$$

As soon as this temperature is reached, some liquid water begins to appear in the container. In other words, the ice begins to *melt*. The melting process is a *change of phase*, from the solid phase to the liquid phase.

15-4 Different ways in which the molar heat capacity varies with the temperature at low temperatures, indicating widely different atomic processes.

304 QUANTITY OF HEAT

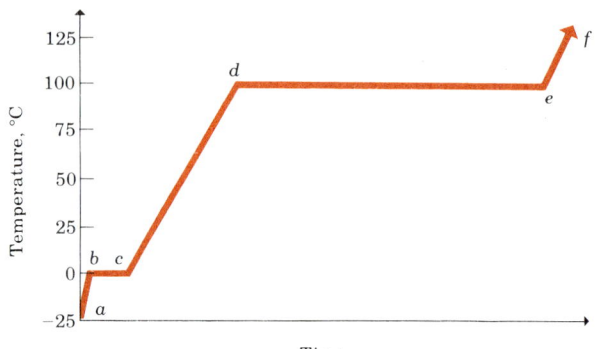

15-5 The temperature remains constant during each change of phase, provided the pressure remains constant.

The thermometer, however, shows no *increase in temperature,* and even though heat is being supplied at the same rate as before, the temperature remains at 0°C until *all* the ice is melted (point c) if the pressure is maintained constant at one atmosphere.

As soon as the last of the ice has melted, the temperature begins to rise again at a uniform rate (from c to d) although this rate is *slower* than that from a to b because the specific heat capacity of water is greater than that of ice. When a temperature of 100°C is reached (point d), bubbles of steam (gaseous water or water vapor) start to escape from the liquid surface; the water begins to *boil*. The temperature remains constant at 100°C (at constant atmospheric pressure) until all the water has boiled away. Another change of phase has therefore taken place, from the liquid phase to the gaseous phase.

If all the water vapor had been trapped and not allowed to diffuse away (a very large container would be needed), the heating process could be continued as from e to f. The gas would now be called "superheated steam."

An essential point in this discussion is that when heat is added slowly (in order to maintain thermal equilibrium) to a substance which can exist in different phases, *either* the temperature rises *or* some of the substance undergoes a phase change, but *never* both at the same time. Once the temperature for a phase change (e.g., the melting or boiling temperature) has been reached, no further temperature change occurs until *all* the substance has undergone the phase change.

Although water was chosen as an example in the process just described, the same type of curve as in Fig. 15-5 is obtained for many other substances. Some, of course, decompose before reaching a melting or boiling point, and others, such as glass or pitch, do not undergo a phase change, with a corresponding definite melting temperature, but become gradually softer as their temperature is raised. Crystalline substances, such as ice or a metal, always *do* undergo a phase change and melt at a definite temperature. Glass and pitch behave like supercooled liquids of very high viscosity.

The quantity of heat per unit mass that must be supplied to a material at its melting point to convert it completely to a liquid *at the same temperature* is called the *heat of fusion* of the material. The quantity of heat per unit mass that must be supplied to a material at its boiling point to convert it completely to a gas at the same temperature is called the *heat of vaporization* of the material. Heats of fusion and vaporization are expressed in units of energy per unit mass, such as joules per kilogram, calories per gram, or Btu per pound-mass. Thus, the heat of fusion of ice is about 3.34×10^5 J·kg^{-1}, 79.7 cal·g^{-1}, or 143 Btu·lb^{-1}. The heat of vaporization of water at 100°C is 2.26×10^6 J·kg^{-1}, 540 cal·g^{-1}, or 972 Btu·lb^{-1}.

The temperatures at which phase transitions occur depend on pressure, and the accompanying heats also depend on pressure. For example, water at normal atmospheric pressure boils at 100°C with a heat of vaporization of 2259 J·g^{-1}; but if the pressure is reduced to 0.5 atm (corresponding to an elevation above sea level of about 6000 m), the boiling temperature is 82°C and the heat of vaporization 2310 J·g^{-1}. Similar effects occur with fusion, although the pressure dependence is usually less pronounced.

Some heats of fusion and vaporization at normal atmospheric pressure are listed in Table 15-2. These quantities of heat are sometimes called *latent* heats because they change the phase of the material but not its temperature. This term is somewhat redundant and will not be used here.

Table 15-2 Heats of fusion and vaporization

Substance	Normal melting point		Heat of fusion, J·g^{-1}	Normal boiling point		Heat of vaporization, J·g^{-1}
	K	°C		K	°C	
Hydrogen	13.84	−259.31	58.6	20.26	−252.89	452
Nitrogen	63.18	−209.97	25.5	77.34	−195.81	201
Oxygen	54.36	−218.79	13.8	90.18	−182.97	213
Ethyl alcohol	159	−114	104.2	351	78	854
Mercury	234	−39	11.8	630	357	272
Water	273.15	0.00	335	373.15	100.00	2256
Sulfur	392	119	38.1	717.75	444.60	326
Lead	600.5	327.3	24.5	2023	1750	871
Antimony	903.65	630.50	165	1713	1440	561
Silver	1233.95	960.80	88.3	2466	2193	2336
Gold	1336.15	1063.00	64.5	2933	2660	1578
Copper	1356	1083	134	1460	1187	5069

When heat is removed from a gas, its temperature falls; at the same temperature at which it boiled, it returns to the liquid phase, or *condenses*. In so doing it gives up to its surroundings the same quantity of heat that was required to vaporize it. The heat so given up, per unit mass, is equal to the heat of vaporization. Similarly, when a liquid is cooled to the temperature at which it melted, it returns to the solid phase or freezes, and gives up heat per unit mass exactly equal to the heat of fusion.

Ordinarily melting and freezing occur at the same temperature, as do boiling and condensing (as stated above), but under some conditions a material can be heated above or cooled below the normal phase-change temperature without a phase change occurring. The resulting state is unstable and is described as *superheated* or *supercooled*. Very pure water can be cooled several degrees below the normal freezing point under ideal conditions; when a small ice crystal is dropped in or the water is agitated, it crystallizes very quickly. Similarly, supercooled water vapor condenses quickly into fog droplets when a disturbance, such as dust particles or ionizing radiation, is introduced. Bubbles of vapor may form in a superheated liquid in a similar way. These phenomena are used in the bubble chamber and the cloud chamber, to be discussed in Sec. 17-9.

Whether a substance, at its melting point, is melting or freezing depends on whether heat is being supplied or removed. That is, if heat is supplied to a beaker containing both ice and water at 0°C, some of the ice melts; if heat is removed, some of the water freezes; the temperature in either case remains at 0°C so long as both ice and water are present. If heat is *neither supplied nor removed,* no change at all takes place and the relative amounts of ice and water, and the temperature, all remain constant. Such a system is said to be in *phase equilibrium*. This furnishes another point of view regarding the melting point. That is, the melting (or freezing) point of a substance is *that temperature at which both the liquid and solid phases can exist together*. At any higher temperature, the substance can be only a liquid; at any lower temperature, it can be only a solid.

The general term *heat of transformation* is applied to both heats of fusion and heats of vaporization, and both are designated by the letter L. Since L represents the heat absorbed or liberated in the change of phase of unit mass, the heat Q absorbed or liberated in the change of phase of a mass m is

$$Q = mL. \tag{15-8}$$

When a solid melts or a liquid boils, heat is added to the material and Q is positive; when a vapor condenses or a liquid freezes, heat is released by the material and Q is negative.

The household steam-heating system makes use of a boiling–condensing process to transfer heat from the furnace to the radiators. Each kilogram of water that is turned to steam in the furnace absorbs 2.26×10^6 J (the heat of vaporization of water) from the furnace, and gives up this same amount when it condenses in the radiators. (This figure is correct if the steam pressure is 1 atm. It is slightly smaller at higher pressures.) Thus the steam-heating system does not need to circulate as much water as a hot-water heating system. If water leaves a hot-water furnace at 60°C and returns at 40°C (a temperature drop of 20°C), about 27 kg of water must circulate to transfer the same heat as is transferred in the form of heat of vaporization by 1 kg of steam.

The temperature-control mechanisms of many warm-blooded animals operate on a similar principle. When the hypothalamus detects a slight rise in blood temperature, the sweat glands are activated. As sweat (chiefly water) evaporates from the surface of the body, it removes heat from the body as heat of vaporization. The heat is brought to the surface of the skin by the blood in the vascular system. This plays the same role as the pipes connecting the furnace to the radiators, with the heart playing the role of the circulating pump in a forced-circulation hot-water or steam system.

Under proper conditions of temperature and pressure, a substance can change directly from the solid to the gaseous phase without passing through the liquid phase. The transfer from solid to vapor is called *subli-

mation, and the solid is said to *sublime.* "Dry ice" (solid carbon dioxide) sublimes at atmospheric pressure. Liquid carbon dioxide cannot exist at a pressure lower than about 5×10^5 Pa (about 5 atm). Heat is absorbed in the process of sublimation, and liberated in the reverse process. The quantity of heat per unit mass is called the *heat of sublimation.*

Definite quantities of heat are involved in chemical reactions. Perhaps the most familiar are those associated with combustion; complete combustion of one gram of gasoline produces about 46,000 J or about 11,000 cal, and the *heat of combustion* is said to be

$$46{,}000 \text{ J} \cdot \text{g}^{-1} \quad \text{or} \quad 46 \times 10^6 \text{ J} \cdot \text{kg}^{-1}.$$

Energy values of foods are defined similarly; the unit of food energy, although called a calorie, is really a *kilocalorie,* equal to 1000 cal or 4186 J. When we say that a gram of peanut butter "contains" 12 calories, we mean that, when it reacts with oxygen, with the help of enzymes, to convert the carbon and hydrogen completely to CO_2 and H_2O, the total energy liberated as heat is 12,000 cal or 50,200 J. Not all of this energy is directly useful for mechanical work; the matter of efficiency of utilization of energy will be discussed in detail in Chapter 19.

15-6 Examples

As indicated above, the basic principle in calculations involving quantity of heat is that when heat flow occurs between two bodies in thermal contact, the amount of heat lost by one body must equal that gained by the other. To avoid confusion about algebraic signs when implementing this principle, the student should use Eqs. (15–5) and (15–8) consistently for each body, noting that each Q is positive when heat enters a body and negative when it leaves. Then the *algebraic sum* of all the Q's must be zero. The following examples illustrate this principle in the context of phenomena discussed in this chapter.

Example 1 A copper cup of mass 0.1 kg, initially at 20°C, is filled with 0.2 kg of coffee initially at 70°C. What is the final temperature after the coffee and cup attain thermal equilibrium?

Solution Let the final temperature be T. The (negative) heat added to the coffee (which is assumed to have a specific heat capacity equal to that of water) is

$$Q = (0.2 \text{ kg})(4186 \text{ J} \cdot \text{kg}^{-1} \cdot (\text{C}°)^{-1})(T - 70°\text{C}).$$

Similarly, the heat gained by the copper cup is

$$Q = (0.1 \text{ kg})(390 \text{ J} \cdot \text{kg}^{-1} \cdot (\text{C}°)^{-1})(T - 20°\text{C}).$$

Equating the sum of these two quantities of heat to zero yields an algebraic equation for T; solution of this equation gives $T = 67.8°\text{C}$. The final temperature is much closer to the initial temperature of the coffee than that of the cup because of the much larger specific heat capacity of water.

Example 2 How much ice at $-20°\text{C}$ must be dropped into 0.25 kg of water, initially at 20°C, in order for the final temperature to be 0°C with the ice all melted? The heat capacity of the container may be neglected.

Solution The (negative) heat added to the water is
$$Q = (0.25 \text{ kg})(4186 \text{ J} \cdot \text{kg}^{-1} \cdot (\text{C}°)^{-1})(0°\text{C} - 20°\text{C})$$
$$= -2.09 \times 10^4 \text{ J}.$$

The specific heat capacity of ice is $2000 \text{ J} \cdot \text{kg}^{-1} \cdot (\text{C}°)^{-1}$. Let the mass of ice be m; then the heat needed to heat it from $-20°\text{C}$ to $0°\text{C}$ is
$$Q = m(2.00 \times 10^3 \text{ J} \cdot \text{kg}^{-1} \cdot (\text{C}°)^{-1})(0°\text{C} - (-20°\text{C}))$$
$$= m(4.00 \times 10^4 \text{ J} \cdot \text{kg}^{-1}).$$

The additional heat needed to melt the ice is the latent heat of fusion times the mass:
$$Q = m(33.4 \times 10^4 \text{ J} \cdot \text{kg}^{-1}).$$

The sum of these three quantities must equal zero:
$$-2.09 \times 10^4 \text{ J} + m(4.00 \times 10^4 \text{ J} \cdot \text{kg}^{-1}) + m(33.4 \times 10^4 \text{ J} \cdot \text{kg}^{-1}) = 0,$$

from which $m = 0.056 \text{ kg} = 56 \text{ g}$. This is roughly equivalent to two medium-size ice cubes.

Example 3 A certain gasoline camping lantern emits as much light as a 25-watt electric light bulb. Assuming that the efficiency of conversion of heat into light is the same for the lantern and the bulb (which is not actually correct), how much gasoline does the lantern burn in 10 hours?

Solution The rate of energy conversion in the lantern is 25 W, so in 10 hours (36,000 s), the total energy needed is
$$(25 \text{ J} \cdot \text{s}^{-1})(36{,}000 \text{ s}) = 0.9 \times 10^6 \text{ J}.$$

As mentioned in Sec. 15–5, combustion of one gram of gasoline produces 46,000 J, so the mass of gasoline required is
$$\frac{0.9 \times 10^6 \text{ J}}{46{,}000 \text{ J} \cdot \text{g}^{-1}} = 19.6 \text{ g}.$$

Actual lanterns are much less efficient than this, and typically require of the order of 300 to 400 g of gasoline (roughly one pint) to operate for ten hours.

Questions

15–1 Suppose you have a Thermos bottle half full of cold coffee. Can you heat it to drinking temperature by shaking it? Is this possible *in principle*? Is it feasible in practice? Are you adding heat to the coffee?

15–2 When the oil in an automatic transmission is churned up by the turbine blades, it becomes hot, and usually an oil-cooling system is required. Is the engine adding heat to the oil?

15–3 A student asserted that a suitable unit for specific heat capacity is $1 \text{ m}^2 \cdot \text{s}^{-2} \cdot (\text{C}°)^{-1}$. Is this correct?

15–4 The specific heat capacity of water has the same numerical value when expressed in $\text{cal} \cdot \text{g}^{-1} \cdot (\text{C}°)^{-1}$ as when expressed in $\text{Btu} \cdot \text{lb}^{-1} \cdot (\text{F}°)^{-1}$. Is this a coincidence? Does the same relation hold for heat capacities of other materials?

15–5 A student claimed that when two bodies not initially in thermal equilibrium are placed in contact, the temperature rise of the cooler body must always equal the temperature drop of the warmer. Do you agree? Is there a principle of conservation of temperature, or something like that?

15–6 In choosing a fluid to circulate inside a gasoline engine to cool it (such as water or antifreeze), should you choose a material with a large or small specific heat capacity? Why? What other considerations are important?

15-7 Any latent heat has a numerical value that is $\frac{5}{9}$ as great when expressed in cal·g^{-1} as when expressed in Btu·lb^{-1}. Why is the conversion so simple?

15-8 Why do you think the heat of vaporization for water is so much larger than the heat of fusion?

15-9 Some household air conditioners used in dry climates cool air by blowing it through a water-soaked filter, evaporating some of the water. How does this work? Would such a system work well in a high-humidity climate?

15-10 Why does food cook faster in a pressure cooker than in boiling water?

15-11 How does the human body maintain a temperature of 37.0°C (98.6°F) in the desert where the temperature is 50°C (122°F)?

15-12 Desert travelers sometimes keep water in a canvas bag. Some water seeps through the bag and evaporates. How does this cool the water inside?

15-13 When water is placed in ice-cube trays in a freezer, why does the water not freeze all at once when the temperature has reached 0°C? In fact, it freezes first in a layer adjacent to the sides of the tray. Why?

15-14 When an automobile engine overheats and the radiator water begins to boil, the car can still be driven some distance before catastrophic engine damage occurs. Why? What determines the onset of really disastrous overheating?

15-15 Why do automobile manufacturers recommend that antifreeze (typically a 50% solution of ethylene glycol in water) be kept in the engine in summer as well as winter?

15-16 When you step out of the shower you feel cold, but as soon as you are dry you feel warmer, even though the room temperature is the same. Why?

15-17 Suppose the heat of fusion of ice were only 10 J·g^{-1} instead of 334 J·g^{-1}. Would this change the way we make iced tea? Martinis? Lemonade?

15-18 Why is the climate of regions adjacent to large bodies of water usually more moderate than that of regions far from large bodies of water?

Problems

15-1 A combustion experiment is performed by burning a mixture of fuel and oxygen in a constant-volume "bomb" surrounded by a water bath. During the experiment the temperature of the water is observed to rise. Considering the mixture of fuel and oxygen as the system,

a) has heat been transferred?

b) has work been done?

15-2 A liquid is irregularly stirred in a well-insulated container and thereby undergoes a rise in temperature. Considering the liquid as the system,

a) has heat been transferred?

b) has work been done?

15-3 An automobile of mass 1500 kg is traveling at 5 m·s^{-1}. How many calories of heat are transferred in the brake mechanism when it is brought to rest?

15-4 A copper vessel of mass 200 g contains 400 g of water. The water is heated by a friction device that dissipates mechanical energy, and it is observed that the temperature of the system rises at the rate of 3 C°·min^{-1}. Neglect heat losses to surroundings. What power in watts is dissipated in the water?

15-5 How long could a 2000-hp motor be operated on the heat energy liberated by 1 mi^3 of ocean water when the temperature of the water is lowered by 1 C° if all this heat were converted to mechanical energy? Why do we not utilize this tremendous reservoir of energy?

15-6

a) A certain house burns 10 tons of coal in a heating season. The heat of combustion of the coal is 11,000 Btu·lb^{-1}. If stack losses are 15%, how many Btu were actually used to heat the house?

b) In some localities large tanks of water are heated by solar radiation during the summer and the stored energy is used for heating during the winter. Find the required dimensions of the storage tank, assuming it to be a cube, to store a quantity of energy equal to that computed in part (a). Assume that the water is raised to 120°F in the summer and cooled to 80°F in the winter.

15-7 An artificial satellite, constructed of aluminum, encircles the earth at a speed of 9000 m·s^{-1}.

a) Find the ratio of its kinetic energy to the energy required to raise its temperature by 600 C°. (The melting point of aluminum is 660°C.) Assume a constant specific heat capacity of 0.20 cal·g^{-1}·(C°)$^{-1}$.

b) Discuss the bearing of your answer on the problem of the re-entry of a satellite into the earth's atmosphere.

15-8 A calorimeter contains 100 g of water at 0°C. A 1000-g copper cylinder and a 1000-g lead cylinder, both at 100°C, are placed in the calorimeter. Find the final temperature if there is no loss of heat to the surroundings.

15-9

a) Compare the heat capacities (heat capacity is the heat per unit temperature change) of equal *masses* of water, copper, and lead.

b) Compare the heat capacities of equal *volumes* of water, copper, and lead.

15-10 An aluminum can of mass 500 g contains 117.5 g of water at a temperature of 20°C. A 200-g block of iron at 75°C is dropped into the can. Find the final temperature, assuming no heat loss to the surroundings.

15-11 A casting weighing 100 lb is taken from an annealing furnace where its temperature was 900°F and plunged into a tank containing 800 lb of oil at a temperature of 80°F. The final temperature is 100°F, and the specific heat capacity of the oil is 0.5 Btu·lb^{-1}·(F°)$^{-1}$. What was the specific heat capacity of the casting? Neglect the heat capacity of the tank itself and any heat losses.

15-12 A lead bullet, traveling at 350 m·s^{-1}, strikes a target and is brought to rest. What would be the rise in temperature of the bullet if there were no heat loss to the surroundings? Does the bullet melt?

15-13 A copper calorimeter (mass 300 g) contains 500 g of water at a temperature of 15°C. A 560-g block of copper, at a temperature of 100°C, is dropped into the calorimeter and the temperature is observed to increase to 22.5°C. Neglect heat losses to the surroundings. Find the specific heat capacity of copper.

15-14 A 50-g sample of a material, at a temperature of 100°C, is dropped into a calorimeter containing 200 g of water initially at 20°C. The calorimeter is of copper and its mass is 100 g. The final temperature of the calorimeter is 22°C. Compute the specific heat capacity of the sample.

15-15 In Fig. 15–6 an electric heater is shown whose purpose is to provide a continuous supply of hot water. Water is flowing at the rate of 300 g·min^{-1}, the inlet thermometer registers 15°C, the voltmeter reads 120 V and the ammeter 10 A (corresponding to a power input of (120 V)(10 A) = 1200 W).

a) When a steady state is finally reached, what is the reading of the outlet thermometer?

b) Why is it unnecessary to take into account the heat capacity (mc or nC) of the apparatus itself?

15-16 The molar heat capacity at constant pressure of a certain substance varies with temperature according to the empirical equation

$$C_p = 27.2 \text{ J mol}^{-1}\text{ K}^{-1} + (4 \times 10^{-3} \text{ J mol}^{-1}\text{ K}^{-2})T.$$

How much heat is necessary to change the temperature of 10 moles from 27°C to 527°C?

15-17 At very low temperatures, the molar heat capacity of rock salt varies with the temperature according to "Debye's T^3 law"; thus

$$C = k\frac{T^3}{\Theta^3},$$

where $k = 1940 \text{ J·mol}^{-1}\text{·K}^{-1}$ and $\Theta = 281$ K.

a) How much heat is required to raise the temperature of 2 moles of rock salt from 10 K to 50 K?

b) What is the mean molar heat capacity in this range?

c) What is the true molar heat capacity at 50 K?

15-18 An automobile engine whose output is 40 hp uses 4.5 gallons of gasoline per hour. The heat of combustion is 12×10^7 joules per gallon. What is the efficiency of the engine? That is, what fraction of the heat of combustion is converted to mechanical work?

15-19 The electric power input to a certain electric motor is 0.5 kW, and the mechanical power output is 0.54 hp.

a) What is the efficiency of the motor? (That is, what fraction of the electrical power is converted to mechanical power?)

b) How many calories of heat are developed in the motor in one hour of operation, assuming all electrical energy not converted to mechanical energy is converted to heat?

15-20 Evaporation of perspiration is an important mechanism for temperature control in warm-blooded animals. What mass of water must evaporate from the surface of an 80-kg human body to cool it 1 C°? The specific heat capacity of the human body is approximately 1 cal·g^{-1}·(C°)$^{-1}$, and the latent heat of vaporization of water at body temperature (37°C) is 577 cal·g^{-1}.

15-21 If a person's food-energy intake is 2400 calories per day, and if all this energy eventually is given off as heat, what is the average rate of heat output, in watts? How many people in a room, giving off heat at this rate, would be required in order to have the same heat output as a 1500-W electric heater?

15-22 A piece of ice at 0°C falls from rest into a lake at 0°C, and one-half of one percent of the ice melts. Compute the minimum height from which the ice falls.

15-23 What must be the initial velocity of a lead bullet at a temperature of 25°C, so that the heat developed when it is brought to rest shall be just sufficient to melt it?

Figure 15–6

Figure 15-7

15-24 Figure 15-7 shows a sketch of a continuous-flow calorimeter used to measure the heat of combustion of a gaseous fuel. Water is supplied at the rate of 12.5 lb·min^{-1} and natural gas at 0.020 ft^3·min^{-1}. In the steady state, the inlet and outlet thermometers register 60°F and 76°F, respectively. What is the heat of combustion of natural gas in Btu·ft^{-3}? Why should the gas flow be made as small as possible?

15-25 An electrical resistor is immersed in a liquid and electrical energy is dissipated for 100 s at a constant rate of 50 W. The mass of the liquid is 530 g, and its temperature increases from 17.64°C to 20.77°C. Find the mean specific heat capacity of the liquid in this temperature range.

15-26 How much heat is required to convert 1 g of ice at −10°C to steam at 100°C?

15-27 A beaker of very small mass contains 500 g of water at a temperature of 80°C. How many grams of ice at a temperature of −20°C must be dropped in the water so that the final temperature of the system will be 50°C?

15-28 An open vessel contains 500 g of ice at −20°C. The mass of the container can be neglected. Heat is supplied to the vessel at the constant rate of 100 cal·min^{-1} for 100 min. Plot a curve showing the elapsed time as abscissa and the temperature as ordinate.

15-29 A copper calorimeter of mass 100 g contains 150 g of water and 8 g of ice in thermal equilibrium at atmospheric pressure. 100 g of lead at a temperature of 200°C are dropped into the calorimeter. Find the final temperature if no heat is lost to the surroundings.

15-30 500 g of ice at −16°C are dropped into a calorimeter containing 1000 g of water at 20°C. The calorimeter is of copper and has a mass of 278 g. Compute the final temperature of the system, assuming no heat losses.

15-31 A tube leads from a flask in which water is boiling under atmospheric pressure to a calorimeter. The mass of the calorimeter is 150 g, its specific heat capacity is 0.1 cal·g^{-1}·(C°)$^{-1}$, and it contains originally 340 g of water at 15°C. Steam is allowed to condense in the calorimeter until its temperature increases to 71°C, after which the total mass of calorimeter and contents is found to be 525 g. Compute the heat of condensation of steam from these data.

15-32 An aluminum canteen whose mass is 500 g contains 750 g of water and 100 g of ice. The canteen is dropped from an aircraft to the ground. After landing, the temperature of the canteen is found to be 25°C. Assuming that no energy is given to the ground in the impact, what was the velocity of the canteen just before it landed? State any additional assumptions needed.

15-33 A calorimeter contains 500 g of water and 300 g of ice, all at a temperature of 0°C. A block of metal of mass 1000 g is taken from a furnace where its temperature was 240°C and is dropped quickly into the calorimeter. As a result, all the ice is just melted. What would the final temperature of the system have been if the mass of the block had been twice as great? Neglect heat loss from and the heat capacity of the calorimeter.

15-34 An ice cube whose mass is 50 g is taken from a refrigerator where its temperature was −10°C and dropped into a glass of water at 0°C. If no heat is gained or lost from outside, how much water will freeze onto the cube?

15-35 A copper calorimeter can (mc = 30 cal·(C°)$^{-1}$) contains 50 g of ice. The system is initially at 0°C. 12 g of steam at 100°C and 1 atm pressure are admitted into the calorimeter. What is the final temperature of the calorimeter and its contents?

15-36 A vessel whose walls are thermally insulated contains 2100 g of water and 200 g of ice, all at a temperature of 0°C. The outlet of a tube leading from a boiler, in which water is boiling at atmospheric pressure, is inserted into the water. How many grams of steam must condense to raise the temperature of the system to 20°C? Neglect the heat capacity of the container.

15-37 A 2-kg iron block is taken from a furnace where its temperature was 650°C and placed on a large block of ice at 0°C. Assuming that all the heat given up by the iron is used to melt the ice, how much ice is melted?

15-38 In a household hot-water heating system, water is delivered to the radiators at 140°F and leaves at 100°F. The system is to be replaced by a steam system in which steam at atmospheric pressure condenses in the radiators, the condensed steam leaving the radiators at 180°F. How many pounds of steam will supply the same heat as was supplied by 1 lb of hot water in the first installation?

15-39 A "solar house" has storage facilities for 4 million Btu. Compare the space requirements for this storage on the assumption

a) that the heat is stored in water heated from a minimum temperature of 80°F to a maximum at 120°F, and

b) that the heat is stored in Glauber salt ($Na_2SO_4 \cdot 10H_2O$) heated in the same temperature range.

Properties of Glauber Salt

Specific heat capacity	
Solid	0.46 Btu·lb^{-1}·(F°)$^{-1}$
Liquid	0.68 Btu·lb^{-1}·(F°)$^{-1}$
Specific gravity	1.6
Melting point	90°F
Heat of fusion	104 Btu·lb^{-1}

15-40 The nominal food-energy value of butter is about 6 kcal·g^{-1}, where 1 kcal = 1000 cal = 4186 J. If all this energy could be converted completely to mechanical energy, how much butter would be required to power an 80-kg mountaineer on his journey from Lupine Meadows (elevation 2070 m) to the summit of Grand Teton (4196 m)?

15-41 The capacity of air conditioners is sometimes expressed in Btu·hr^{-1} or "tons," the latter being the number of tons of ice that can be frozen from water at 0°C in 24 hr by the unit.

a) What is the rating in Btu/hr of a one-ton air conditioner?

b) Express the capacity of a one-ton air conditioner in watts.

15-42 A 6000 Btu·hr^{-1} air conditioner typically consumes about 800 W of electrical power. What is the ratio of the rate of removal of heat to the rate of consumption of electrical energy? Express the two rates in the same units to obtain a dimensionless ratio.

15-43 Suppose 1 liter of gasoline will propel a car a distance of 10 km. The density of gasoline is about 0.7 g·cm^{-3}, and its heat of combustion is about 4.6×10^4 J·g^{-1}.

a) If the engine is 25% efficient, that is, if $\frac{1}{4}$ of the heat of combustion is converted into useful mechanical work, what total work does the engine do during the 10-km displacement?

b) If this work is assumed to be done in opposing a constant resisting force F, find the magnitude of F.

HEAT TRANSFER 16

16-1 Conduction

If one end of a metal rod is placed in a flame while the other is held in the hand, that part of the rod one is holding becomes hotter and hotter, although it is not itself in direct contact with the flame. Heat is said to reach the cooler end of the rod by *conduction* through the material of the rod. The molecules at the hot end of the rod increase the energy of their vibration as the temperature increases. Then, as they collide with their more slowly moving neighbors farther out on the rod, some of their energy of motion is shared with these neighbors, and they in turn pass it along to those still farther from the flame. Hence energy of thermal motion is passed along from one molecule to the next, while each individual molecule remains at its original position.

Most metals are good conductors of electricity and also good conductors of heat. The ability of a metal to conduct an electric current is due to the fact that there are within it so-called "free" electrons, that is, electrons that have become detached from their parent molecules. The free electrons also play a part in the conduction of heat, and the reason metals are such good heat conductors is that the free electrons provide an effective mechanism for carrying thermal energy from the hotter to the cooler portions of the metal.

Conduction of heat can take place in a body only when different parts of the body are at different temperatures, and the direction of heat flow is always from points of higher to points of lower temperature. Figure 16–1 represents a rod of material of cross-sectional area A and length L. Let the left end of the rod be kept at a temperature T_2, and the right end at a lower temperature T_1. The direction of the heat current is then from left to right through the rod. The sides of the rod are covered by an insulating material, so no heat transfer occurs at the sides. This is of course an idealized model; all real materials, even the best thermal insulators, conduct heat to some extent.

After the ends of the rod have been kept at the temperatures T_1 and T_2 for a sufficiently long time, the temperature at points within the rod is

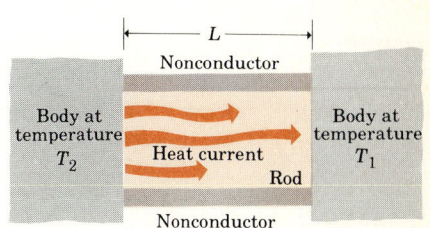

16–1 Steady-state heat flow in a uniform rod.

found to decrease uniformly with distance from the hot to the cold face. At each point, however, the temperature remains constant with time. This condition is called "steady-state" heat flow.

Experiment shows that the rate of flow of heat through the rod in the steady state is proportional to the area A, proportional to the temperature difference $(T_2 - T_1)$, and inversely proportional to the length L. These proportions may be converted to an equation by introducing a constant k whose numerical value depends on the material of the rod. The quantity k is called the *thermal conductivity* of the material,

$$H = \frac{kA(T_2 - T_1)}{L}, \qquad (16\text{-}1)$$

where H is the quantity of heat flowing through the rod per unit time, also called the *heat current*. The units of H are energy per unit time, or power. Thus in SI units the unit of H is 1 J·s^{-1} or 1 W.

Equation (16-1) may also be used to compute the rate of heat flow through a slab, or *any* homogeneous body having a uniform cross section perpendicular to the direction of flow, provided the flow has attained steady-state conditions and the ends are kept at constant temperatures.

When the cross section is not uniform, or when steady-state conditions are not present, the temperature does not necessarily change uniformly along the direction of heat flow. Equation (16-1) may still be applied to a thin layer of material perpendicular to the direction of flow. If x is the coordinate measured along the flow path, dx the thickness of the layer, and A the cross section area perpendicular to that path, then Eq. (16-1) may be rewritten

$$H = -kA \frac{dT}{dx}, \qquad (16\text{-}2)$$

where dT is the temperature change across the distance dx. The negative sign is included because, if the temperature *increases* in the direction of increasing x (dx and dT both positive), the direction of heat flow is the direction of *decreasing x,* and conversely. The temperature change per unit length dT/dx is called the *temperature gradient;* Eq. (16-1) refers to the special case where the temperature gradient is constant, and equal to $(T_2 - T_1)/L$.

The SI unit of rate of heat flow is the joule per second; other units such as the calorie per second or Btu per second are also sometimes used. The units of k are obtained by solving Eq. (16-2) for k:

$$k = -\frac{H}{A}\frac{dx}{dT}.$$

This shows that in SI units, the unit of k is

$$\frac{(1\ \text{J}\cdot\text{s}^{-1})(1\ \text{m})}{(1\ \text{m})^2(1\ \text{C}°)} = 1\ \text{J}\cdot(\text{s}\cdot\text{m}\cdot\text{C}°)^{-1}.$$

Values of thermal conductivity are sometimes tabulated using cgs units, with the calorie as the energy unit. The unit of k then is 1 cal·(s·cm·C°)$^{-1}$. The conversion is

$$1\ \text{cal}\cdot(\text{s}\cdot\text{cm}\cdot\text{C}°)^{-1} = 419\ \text{J}\cdot(\text{s}\cdot\text{m}\cdot\text{C}°)^{-1}.$$

16-1

For most materials, the thermal conductivity increases slightly with increasing temperature, but the variation is small and can often be neglected. Some numerical values of k, at temperatures near room temperature, are given in Table 16–1. The properties of materials used commercially as heat insulators are expressed in a mixed system in which the unit of heat current is 1 Btu·hr^{-1}, the unit of area is 1 ft^2, and the unit of temperature gradient is one Fahrenheit degree per *inch* (1 F°·in^{-1})!

Table 16–1 Thermal conductivities (k)

	$J \cdot s^{-1} \cdot m^{-1} \cdot (C°)^{-1}$	$cal \cdot s^{-1} \cdot cm^{-1} \cdot (C°)^{-1}$
Metals		
Aluminum	205	0.49
Brass	109	0.26
Copper	385	0.92
Lead	34.7	0.083
Mercury	8.3	0.020
Silver	406	0.97
Steel	50.2	0.12
Various solids		
(Representative values)		
Insulating brick	0.15	0.00035
Red brick	0.6	0.0015
Concrete	0.8	0.002
Cork	0.04	0.0001
Felt	0.04	0.0001
Glass	0.8	0.002
Ice	1.6	0.004
Rock wool	0.04	0.0001
Styrofoam	0.01	0.00002
Wood	0.12–0.04	0.0003–0.0001
Gases		
Air	0.024	0.000057
Argon	0.016	0.000039
Helium	0.14	0.00034
Hydrogen	0.14	0.00033
Oxygen	0.023	0.000056

Equation (16–1) shows that the larger the thermal conductivity k, the larger the heat current, other factors being equal. A material for which k is large is therefore a good heat conductor, while if k is small, the material is a poor conductor or a good insulator. In ordinary circumstances a "perfect heat conductor" ($k = \infty$) or a "perfect heat insulator" ($k = 0$) does not exist. However, Table 16–1 shows that the metals as a group have much greater thermal conductivities than the nonmetals, and that those of gases are extremely small.

Example 1 A styrofoam box is used to keep beer cold at a picnic. It has total wall area (including the lid) of 0.8 m^2 and wall thickness 2.0 cm. The thermal conductivity of styrofoam is about 0.01 J·m^{-1}·s^{-1}·(C°)$^{-1}$.

It is filled with ice and beer at 0°C. What is the rate of heat flow into the box if the outside temperature is 30°C? How much ice melts in one day?

Solution We assume that the total heat flow is approximately the same as it would be through a flat slab of area 0.8 m² and thickness 2.0 cm = 0.02 m. The rate of heat flow, from Eq. (16–1), is

$$H = (0.01 \text{ J} \cdot \text{m}^{-1} \cdot \text{s}^{-1} \cdot (\text{C}°)^{-1})(0.8 \text{ m}^2)\frac{30 \text{ C}°}{0.02 \text{ m}} = 12 \text{ J} \cdot \text{s}^{-1}.$$

There are 86,400 s in one day, so the total heat flow in one day is

$$(12 \text{ J})(86{,}400 \text{ s}) = 1.04 \times 10^6 \text{ J}.$$

The latent heat of fusion of ice is 334 J·g⁻¹, so the quantity of ice melted by this quantity of heat is

$$\frac{1.04 \times 10^6 \text{ J}}{334 \text{ J} \cdot \text{g}^{-1}} = 3110 \text{ g} \quad \text{or about 3.1 kg.}$$

Example 2 A steel bar 10 cm long is welded end-to-end to a copper bar 20 cm long. Each bar has a square cross section, 2 cm on a side. The free end of the steel bar is placed in contact with steam at 100°C, and the free end of the copper bar with ice at 0°C. Find the temperature at the junction of the two bars and the total rate of heat flow, when steady-state conditions have been reached.

Solution The key to the solution is the fact that the rates of heat flow in the two bars must be equal; otherwise some sections would have more heat flowing in than out, or the reverse, and steady-state conditions could not exist. Let T be the unknown junction temperature; we use Eq. (16–1) for each bar and equate the two expressions:

$$\frac{k_s A(100°\text{C} - T)}{L_s} = \frac{k_c A(T - 0°\text{C})}{L_c}.$$

The areas A are equal and may be divided out. Substituting numerical values, we find

$$\frac{(50.2 \text{ J} \cdot \text{s}^{-1} \cdot \text{m}^{-1} \cdot (\text{C}°)^{-1})(100°\text{C} - T)}{0.1 \text{ m}} = \frac{(385 \text{ J} \cdot \text{s}^{-1} \cdot \text{m}^{-1} \cdot (\text{C}°)^{-1})(T - 0°\text{C})}{0.2 \text{ m}}.$$

Rearranging and solving for T, we find

$$T = 20.7°\text{C}.$$

Even though the steel bar is shorter, the temperature drop across it is much greater than across the copper bar, because steel is a much poorer conductor.

The total heat current is now obtained by substituting this value for T back into either of the above expressions:

$$H = \frac{(50.2 \text{ J} \cdot \text{s}^{-1} \cdot \text{m}^{-1} \cdot (\text{C}°)^{-1})(0.02 \text{ m})^2(100°\text{C} - 20.7°\text{C})}{0.1 \text{ m}}$$

$$= 15.9 \text{ J} \cdot \text{s}^{-1} = 15.9 \text{ W},$$

or
$$H = \frac{(385 \text{ J} \cdot \text{s}^{-1} \cdot \text{m}^{-1} \cdot (\text{C}°)^{-1})(0.02 \text{ m})^2(20.7°\text{C} - 0°\text{C})}{0.2 \text{ m}}$$
$$= 15.9 \text{ J} \cdot \text{s}^{-1} = 15.9 \text{ W}.$$

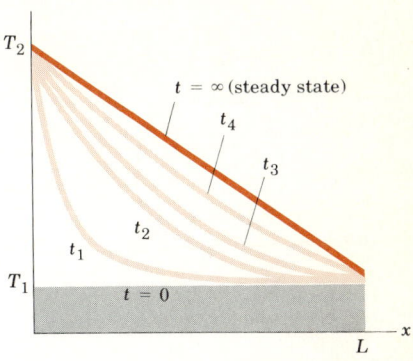

16-2 Transient and steady-state temperature distribution along a rod initially at temperature T_1. The transient distributions were computed for $t_2 = 5t_1$, $t_3 = 10t_1$, $t_4 = 20t_1$.

Example 3 In Example 2 above, suppose the two bars are separated; one end of each bar is placed in contact with steam at 100°C, and the other end of each with ice at 0°C. What is the total rate of heat flow in the two bars?

Solution In this case the bars are in parallel rather than in series. The total heat flow is the sum of the flows in the two bars, and for each bar $T_2 - T_1 = 100°\text{C} - 0°\text{C} = 100 \text{ C}°$. Thus

$$H = \frac{(50.2 \text{ J} \cdot \text{s}^{-1} \cdot \text{m}^{-1} \cdot (\text{C}°)^{-1})(0.02 \text{ m})^2(100 \text{ C}°)}{0.1 \text{ m}}$$
$$+ \frac{(385 \text{ J} \cdot \text{s}^{-1} \cdot \text{m}^{-1} \cdot (\text{C}°)^{-1})(0.02 \text{ m})^2(100 \text{ C}°)}{0.2 \text{ m}}$$
$$= 20.1 \text{ J} \cdot \text{s}^{-1} + 77.0 \text{ J} \cdot \text{s}^{-1} = 97.1 \text{ J} \cdot \text{s}^{-1} = 97.1 \text{ W}.$$

The heat flow in the copper bar is much greater than in the steel bar, even though it is longer, because its thermal conductivity is much larger. The total heat flow is much larger than in Example 2 because the full 100 C° temperature difference appears across each bar.

Before steady-state conditions are reached, the temperature at each point may change with time. For example, suppose the rod in Fig. 16-1 is initially at the temperature T_1 throughout, and then at time $t = 0$ the ends are placed in contact with the bodies at temperatures T_1 and T_2. The resulting temperature distributions at various times are shown in Fig. 16-2. At time t_1 the temperature changes much more rapidly with x near the left end than near the right (i.e., the temperature gradient is greater), corresponding to the fact that the heat flow is much greater near that end to provide a net flow of heat into the middle region and raise its temperature. With increasing time the differences become smaller and smaller, until after a very long time ($t = \infty$) the steady-state uniform temperature distribution is attained. The final steady state is independent of the initial conditions at $t = 0$; it depends only on the temperatures at the ends.

In the steady state the heat current in the rod is the same in all cross sections; otherwise there would be a net flow of heat *into* certain regions and *out of* others and their temperatures would change, contradicting the assumption of a steady state.

*16-2 Radial heat flow

We consider next two examples of heat flow that is not in a single direction and in which the temperature gradient is not uniform. Figure 16-3 represents a steam pipe of radius a, surrounded by a layer of insulating material of outer radius b. We let T_2 and T_1 be the inner and outer temperatures; if $T_2 > T_1$, heat flows radially outward. In the steady

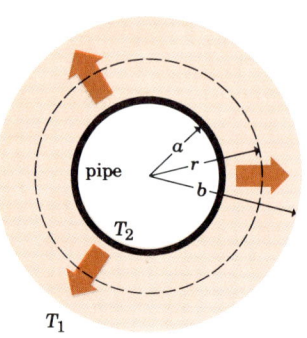

16–3 Radial heat flow in a cylinder.

state the total heat current H is the same across every surface enclosing the pipe, such as the imaginary cylinder of radius r shown in the figure. If the length of the pipe is L, the area of this cylinder is $2\pi rL$, and the heat current is given by

$$H = -kA\frac{dT}{dr} = -2\pi krL\frac{dT}{dr} = \text{constant}.$$

We separate variables and integrate:

$$\int \frac{dr}{r} = -\int \frac{2\pi kL}{H}\,dT,$$

$$\ln r = -\frac{2\pi kL}{H}T + C, \tag{16-3}$$

where C is an integration constant. The constants C and H are obtained from the two simultaneous equations which result when we substitute the values ($r = a$, $T = T_2$) and ($r = b$, $T = T_1$), respectively, into Eq. (16–3). We omit the intermediate steps in the calculation; the final results are

$$H = \frac{2\pi kL(T_2 - T_1)}{\ln(b/a)}, \tag{16-4}$$

$$T = T_2 - \frac{\ln(r/a)}{\ln(b/a)}(T_2 - T_1). \tag{16-5}$$

This same calculation can be carried out for heat flow between two concentric spheres. In that case the area is $4\pi r^2$, and the final results are

$$H = \frac{4\pi kab(T_2 - T_1)}{b - a}, \tag{16-6}$$

$$T = T_2 - \frac{b}{r}\cdot\frac{r - a}{b - a}(T_2 - T_1). \tag{16-7}$$

Derivation of these results is left as an exercise.

16–3 Convection

Convection is transfer of heat by actual motion of material. The hot-air furnace, the hot-water heating system, and the flow of blood in the body are examples. If the material is forced to move by a blower or pump, the process is called *forced convection*; if the material flows due to differences in density (caused by thermal expansion), the process is called *natural* or *free convection*. To understand the latter, consider a U-tube, as illustrated in Fig. 16–4.

In (a), the water is at the same temperature in both arms of the U and hence stands at the same level in each. In (b), the right side of the U has been heated. The water in this side expands and therefore has smaller density; thus a longer column is needed to balance the pressure produced by the cold water in the left column. When the stopcock is opened, water flows from the top of the warmer column into the colder column. This increases the pressure at the bottom of the U produced by the cold column, and decreases the pressure at this point due to the hot column. Hence at the bottom of the U, water is forced from the cold to the hot side. If heat is continuously applied to the hot side and removed from the cold side, the circulation continues indefinitely. The net result is a continuous transfer of heat from the hot to the cold side of the column. In the common household hot-water heating system, the "cold" side corresponds to the radiators and the "hot" side to the furnace.

The anomalous expansion of water, mentioned in Sec. 14-5, has an important effect on the way in which lakes and ponds freeze in winter. Suppose a pond is at a temperature of 20°C throughout, and suppose the air temperature at its surface falls to −10°C. The water at the surface becomes cooled; it contracts, becomes more dense than the warmer water below it, and sinks in this less dense water, its place being taken by water at 20°C. The sinking of the cooled water causes a mixing process, which continues until all of the water has been cooled to 4°C. Now, however, when the surface water cools to 3°C, it *expands*, is *less* dense than the water below it, and hence floats on the surface. Convection and mixing then cease, and the remainder of the water can lose heat only by *conduction*. Since water is a poor heat conductor, cooling takes place very slowly after 4°C is reached, with the result that the pond freezes first at its surface. Then, since the density of ice is even smaller than that of water at 0°C, the ice floats on the water below it, and further freezing can result only from heat flow upward by conduction.

There is no simple equation for convective heat transfer as there is for conduction. The heat lost or gained by a surface at one temperature in contact with a fluid at another temperature depends on many factors, such as the shape and orientation of the surface, the mechanical and thermal properties of the fluid, and the nature of the fluid flow, laminar or turbulent.

For practical calculations the usual procedure is to define a convection coefficient h by means of the equation

$$H = hA\,\Delta T, \qquad (16\text{-}8)$$

where H is the heat current due to convection, A is the surface area, and ΔT is the temperature difference between the surface and the main body of the fluid. Values of h are determined experimentally; it is found not to be constant but to depend on ΔT. The results are published as tables or graphs from which a physicist or engineer may obtain convection coefficients appropriate for specific situations.

A situation of practical importance is that of natural convection from a wall or a pipe that is at a constant temperature and is surrounded by air at atmospheric pressure whose temperature is less than that of the wall or pipe by an amount ΔT. The convection coefficients applicable in this situation are given in Table 16-2. If the air is *hotter* than a horizontal plate, the first two values of h should be interchanged.

Example The air in a room is at a temperature of 25°C, and the outside air is at −15°C. How much heat is transferred per unit area of a glass windowpane of thermal conductivity 2.5×10^{-3} cal·cm^{-1}·s^{-1}·(C°)$^{-1}$ and of thickness 2 mm?

Solution To assume that the inner surface of the glass is at 25°C and the outer surface is at −15°C is erroneous, as anyone can verify by touching the inner surface of a glass windowpane on a cold day. One

16-4 Convection is brought about by differences in density.

Table 16-2 Coefficients of natural convection in air at atmospheric pressure

Equipment	Convection coefficient h, cal·s^{-1}·cm^{-2}·(C°)$^{-1}$
Horizontal plate, facing upward	$0.595 \times 10^{-4}(\Delta T)^{1/4}$
Horizontal plate, facing downward	$0.314 \times 10^{-4}(\Delta T)^{1/4}$
Vertical plate	$0.424 \times 10^{-4}(\Delta T)^{1/4}$
Horizontal or vertical pipe $\left(\dfrac{\text{diameter}}{D}\right)$	$1.00 \times 10^{-4}\left(\dfrac{\Delta T}{D}\right)^{1/4}$

must expect a much *smaller* temperature difference across the windowpane, so that in the steady state the rates of transfer of heat (1) by convection in the room, (2) by conduction through the glass, and (3) by convection in the outside air, are all equal.

As a first approximation in the solution of this problem, let us assume that the window is at a uniform temperature T. If $T = 5°C$, then the temperature difference between the inside air and the glass is the same as that between the glass and the outside air, or 20°C. Hence the convection coefficient in both cases is

$$h = 0.424 \times 10^{-4}(20)^{1/4} \text{ cal·s}^{-1}\text{·cm}^{-2}\text{·(C°)}^{-1}$$
$$= 0.897 \times 10^{-4} \text{ cal·s}^{-1}\text{·cm}^{-2}\text{·(C°)}^{-1},$$

and, from Eq. (16-8), the heat transferred per unit area is

$$\frac{H}{A} = (0.897 \times 10^{-4} \text{ cal·s}^{-1}\text{·cm}^{-2}\text{·(C°)}^{-1})(20 \text{ C°})$$
$$= 17.9 \times 10^{-4} \text{ cal·s}^{-1}\text{·cm}^{-2}.$$

The glass, however, is *not* at uniform temperature; there must be a temperature difference ΔT across the glass sufficient to provide heat conduction at the rate of 17.9×10^{-4} cal·s^{-1}·cm^{-2}. Using the conduction equation, Eq. (16-1), we obtain

$$\Delta T = \frac{LH}{kA} = \frac{(0.2 \text{ cm})(17.9 \times 10^{-4} \text{ cal·s}^{-1}\text{·cm}^{-2})}{2.5 \times 10^{-3} \text{ cal·cm}^{-1}\text{·s}^{-1}\text{·(C°)}^{-1}} = 0.14 \text{ C°}.$$

Thus the inner surface is at about 5.07°C and the outer at 4.93°C.

Heat transfer in the human body involves a combination of mechanisms, which together maintain a remarkably constant and uniform temperature despite large changes in environmental conditions. As mentioned above, the chief *internal* mechanism is forced convection, with the heart serving as the pump and the blood as the circulating fluid. Heat transfer with the surroundings involves conduction, convection, and radiation, in proportions depending on circumstances. Total heat loss from the body is on the order of 2000 to 5000 kcal per day, depending on the amount of activity. A dry nude body in still air loses about half its heat by radiation, but under conditions of vigorous activity and copious perspiration, the dominant mechanism is evaporative cooling. Radiation will be discussed in the following sections.

16-4 Radiation

When one's hand is placed in direct contact with the surface of a hot-water or steam radiator, heat reaches the hand by *conduction* through the radiator walls. If the hand is held above the radiator but not in contact with it, heat reaches the hand by way of the upward-moving *convection* currents of warm air. If the hand is held at one side of the radiator it still becomes warm, even though conduction through the air is negligible and the hand is not in the path of the convection currents. Energy now reaches the hand by *radiation*.

The term radiation refers to the continual emission of energy from the surface of all bodies. This energy is called *radiant energy* and is in the form of electromagnetic waves. These waves travel with the speed of light and are transmitted through vacuum as well as through air. (Better, in fact, since they are absorbed by air to some extent.) When they fall on a body that is not transparent to them, such as the surface of one's hand or the walls of the room, they are absorbed, resulting in a transfer of heat to the absorbing material.

The radiant energy emitted by a surface depends on the nature of the surface and on its temperature. At low temperatures the rate of radiation is small and the radiant energy is chiefly of relatively long wavelength. As the temperature is increased, the rate of radiation increases very rapidly, in proportion to the *fourth power* of the absolute temperature. For example, a copper block at a temparature of 100°C (373 K) radiates about 0.03 J·s^{-1} or 0.03 W from each square centimeter of its surface. At a temperature of 500°C (773 K), it radiates about 0.54 W from each square centimeter, and at 1000°C (1273 K), it radiates about 4 W per square centimeter. This rate is 130 times as great as that at a temperature of 100°C.

At any temperature, the radiant energy emitted is a mixture of waves of different wavelengths. Wavelengths of light in the visible spectrum range from 0.4×10^{-6} m (violet) to 0.7×10^{-6} m (red). At a temperature of 300°C, practically all of the radiant energy emitted by a body is carried by waves of *longer* wavelength than these. Such waves are called *infrared,* meaning "beyond the red." As the temperature rises, the wavelengths shift to shorter values. At 800°C a body emits enough visible radiant energy to be self-luminous and appears "red hot." By far the larger part of the energy emitted, however, is still carried by infrared waves. At 3000°C, which is about the temperature of an incandescent lamp filament, the radiant energy contains enough of the shorter wavelengths so that the body appears "white hot."

16-5 Stefan-Boltzmann Law

Experiment shows that the rate of radiation of energy from a surface is proportional to the surface area and to the fourth power of the *absolute temperature T*. It also depends on the nature of the surface, described by a dimensionless number e, which is between 0 and 1. Thus the relation may be expressed as

$$H = Ae\sigma T^4, \qquad (16\text{-}9)$$

where σ is a universal physical constant called the Stefan–Boltzmann constant. This relation was deduced by Josef Stefan (1835–1893) on the basis of experimental measurements made by John Tyndall (1820–1893) and was later derived from theoretical considerations by Ludwig Boltzmann (1844–1906).

In Eq. (16–9), H has units of power (energy per unit time). Thus in SI units σ has the units $\text{W} \cdot \text{m}^{-2} \cdot \text{K}^{-4}$. The numerical value is found experimentally to be

$$\sigma = 5.6699 \times 10^{-8} \text{ W} \cdot \text{m}^{-2} \cdot \text{K}^{-4}.$$

The number e characterizing the emitting properties of a particular surface is called the *emissivity*. It is generally larger for dark, rough surfaces than for light, smooth ones. The emissivity of a smooth copper surface is about 0.3.

Example A thin square steel plate, 10 cm on a side, is heated in a blacksmith's forge to a temperature of 800°C. If the emissivity is unity, what is the total rate of radiation of energy?

Solution The total surface area, including both sides, is $2(0.1 \text{ m})^2 = 0.02 \text{ m}^2$. The temperature used in Eq. (16–9) must be *absolute* temperature; 800°C = 1073 K. Then Eq. (16–9) gives

$$H = (0.02 \text{ m}^2)(1)(5.67 \times 10^{-8} \text{ W} \cdot \text{m}^{-2} \cdot \text{K}^{-4})(1073 \text{ K})^4 = 1503 \text{ W}.$$

If the plate were heated by an electric heater instead, an electric power input of 1503 W would be required to maintain it at constant temperature.

If the surfaces of all bodies continuously emit radiant energy according to Eq. (16–9), then why do they not eventually radiate away *all* their energy and cool down to a temperature of absolute zero? The answer is that they *would* do so if energy were not supplied to them in some way. In the case of an electric heater element or the filament of an electric lamp, energy is supplied electrically to make up for the energy radiated. As soon as this energy supply is cut off, these bodies do, in fact, cool down very quickly to the temperature of their surroundings. The reason they do not cool further is that the surroundings are *also* radiating. Some of this radiated energy is intercepted and absorbed.

The rate at which a body *radiates* energy is determined by the temperature of the *body*, but the rate at which it *absorbs* energy by radiation depends on the temperature of its *surroundings*. When a body is hotter than its surroundings, the rate of emission exceeds the rate of absorption; there is a net loss of energy, and the body cools down, unless it is heated by some other means. When a body is cooler than its surroundings, the rate of absorption is greater than the rate of emission, and its temperature rises. At thermal equilibrium the two rates are equal.

Thus when a body at temperature T is surrounded by walls also at temperature T, maintenance of thermal equilibrium requires that the body's rate of *absorption* of radiant energy from the walls must be equal to the rate of radiation from its surface, which is

$$H = Ae\sigma T^4.$$

Hence, for such a body at a temperature T_1, surrounded by walls at a temperature T_2, (as in Fig. 16-5), the *net* rate of loss (or gain) of energy per unit area by radiation is

$$H_{\text{net}} = Ae\sigma T_1^4 - Ae\sigma T_2^4 = Ae\sigma(T_1^4 - T_2^4). \quad (16\text{-}10)$$

Infrared emission from a body can be studied using a camera equipped with infrared-sensitive film, or better with a special device similar in principle to a television camera, sensitive to infrared radiation. The resulting picture is called a *thermograph;* since emission depends on temperature, thermography permits detailed study of temperature distributions. Some present-day instrumentation is sensitive to differences as small as 0.1 C°.

Thermography has a variety of important medical applications. Local temperature variations in the body are associated with various tumors, such as breast cancers, and growths as small as 1 cm in diameter can be detected. Vascular disorders leading to local temperature anomalies can be studied, and many other examples could be cited.

16-5 In thermal equilibrium, the rate of emission of radiant energy equals the rate of absorption. Hence a good absorber is a good emitter.

16-6 The ideal radiator

In Fig. 16-5, a body with initial temperature T_1 is suspended inside an enclosure whose walls are kept at temperature T_2. It is found that eventually the body attains the temperature T_2, irrespective of its emissivity or other characteristics. That is, the body reaches *thermal equilibrium* with its surroundings. When in thermal equilibrium, the body must *emit* radiant energy at the same rate at which it *absorbs* it. But the rate of emission depends on the emissivity; thus a body that is a good emitter (*e* close to unity) must also be a good absorber, and a poor emitter is a poor absorber. Furthermore, since the radiation striking the body must be either absorbed or reflected, a *good* absorber is a *poor* reflector, and a *poor* absorber is a *good* reflector. Finally we conclude that *a good reflector is a poor emitter.*

This is the reason for silvering the walls of vacuum ("Thermos") bottles. A vacuum bottle is constructed with double glass walls; the space between the walls is evacuated so that heat flow by conduction and convection is practically eliminated. To reduce the radiant emission to as low a value as possible, the walls are covered with a coating of silver, which is highly reflecting and hence is a very poor emitter.

Since a good absorber is a good emitter, the *best* emitter is that surface that is the best absorber. But no surface can absorb more than *all* of the radiant energy which strikes it. Any surface which *does* absorb all of the incident energy is the best emitting surface possible. Such a surface would reflect no radiant energy, and hence would appear black in color (provided its temperature is not so high that it is self-luminous). It is called an *ideally black surface.* A body having such a surface is called an ideal blackbody, an ideal radiator, or simply a *blackbody.*

No actual surface is ideally black; the closest approach is lampblack, which reflects only about 1%. Blackbody conditions can be closely approximated, however, by a small opening in the walls of a closed container. Radiant energy entering the opening is in part absorbed by the interior walls. Of the part reflected, only a very little escapes through the opening, the remainder being eventually absorbed by the walls. Hence the *opening* behaves like an ideal absorber.

Conversely, the radiant energy emitted by the walls (or by any body within the enclosure) and escaping through the opening, is, if the walls are of uniform temperature, of the same nature as that emitted by an ideal radiator. This fact is of importance when using an optical pyrometer, described in Section 14–2. The readings of such an instrument are correct only when it is sighted on a blackbody. If used to measure the temperature of a red-hot ingot of iron in the open, its readings will be too low, since iron is a poorer emitter than a blackbody. If, however, the pyrometer is sighted on the iron while still in the furnace, where it is surrounded by walls at the same temperature, "blackbody conditions" are fulfilled and the reading is correct. The failure of the iron to emit as effectively as a blackbody is then just compensated for by the radiant energy that it reflects.

The emissivity e of an ideally black surface is equal to unity. For all real surfaces it is less than unity.

Example Assuming the total surface of the human body is 1.2 m² and the surface temperature is 30°C = 303 K, find the total rate of radiation of energy from the body.

Solution Surprisingly, for infrared radiation the body is a very good approximation to an ideal blackbody, irrespective of skin pigmentation. The rate of energy transfer per unit area is given by Eq. (16–9). Taking $e = 1$, we find

$$H = (1.2 \text{ m}^2)(1)(5.67 \times 10^{-8} \text{ W}\cdot\text{m}^{-2}\cdot\text{K}^{-4})(303 \text{ K})^4 = 574 \text{ W}.$$

Of course this loss is partially balanced by *absorption* of radiation, which depends on the temperature of the surroundings. The *net* rate of radiative energy transfer is given by Eq. (16–10).

Questions

16-1 Why does a marble floor feel cooler to the feet than a carpet at the same temperature?

16-2 A beaker of boiling water can be picked up with bare fingers without burning if it is grasped only at the thin turned-out rim at the top. Why?

16-3 Old-time kitchen lore suggests that things cook better (evenly and without burning) in heavy cast-iron pots. What desirable characteristics would such pots have?

16-4 If you have wet hands and pick up a piece of metal that is below freezing, you may stick to it. This doesn't happen with wood. Why not?

16-5 A cold block of metal feels colder than a block of wood at the same temperature, but a hot block of metal feels hotter than a block of wood at the same temperature. Is there any temperature at which they feel equally hot or cold?

16-6 A person pours a cup of hot coffee, intending to drink it five minutes later. To keep it as hot as possible, should he put cream in it now or wait until just before he drinks it?

16-7 In late afternoon when the sun is about to set, swarms of insects such as mosquitoes are sometimes seen in vertical "plumes" above a tree or above the hood of a car with a hot engine. Why?

16-8 Mountaineers caught in a storm sometimes survive by digging a cave in snow or ice. How in the world can you keep warm in an ice cave?

16-9 Old-time pioneers sometimes kept themselves warm on cold winter nights by heating bricks in the fire, wrapping them in thick layers of cloth, and taking them to bed. Discuss the roles of conduction, convection, and radiation in this technique.

16-10 It is well known that a potato bakes faster if a large nail is stuck through it. Why? Is aluminum better than steel? There is also a gadget on the market to hasten roasting of meat, consisting of a hollow metal tube containing a wick and some water; this is claimed to be much better than a solid metal rod. How does this work?

16-11 The temperature in outer space, far from any solid body, is believed to be about 3 K. If you leave a spaceship and go for a space walk, do you get cold very quickly?

16-12 Aluminum foil used for food cooking and storage sometimes has one shiny surface and one dull surface. When food is wrapped for baking, should the shiny side be in or out? Which side should be out when it is to be frozen?

16-13 Some cooks claim that the bottom crust of a pie gets browner if the pie pan is glass than if it is metal. Why should this be?

16-14 Glider pilots in the midwest know that thermal updrafts are likely to occur above freshly plowed fields. Why?

16-15 We're lucky the earth isn't in thermal equilibrium with the sun. But why isn't it?

16-16 On a chilly fall morning, the grass is covered with frost but the concrete sidewalk isn't. Why?

16-17 Why can a microwave oven cook massive objects such as potatoes and beef roasts much faster than a conventional oven? What disadvantages are associated with the reasons for this greater speed?

Problems

16-1 A slab of a thermal insulator is 100 cm² in cross section and 2 cm thick. Its thermal conductivity is $0.1 \text{ J} \cdot \text{s}^{-1} \cdot \text{m}^{-1} \cdot (\text{C}°)^{-1}$. If the temperature difference between opposite faces is 100 C°, how much heat flows through the slab in one day?

16-2 Suppose that the rod in Fig. 16-1 is of copper, of length 10 cm and of cross-sectional area 1 cm². Let $T_2 = 100°\text{C}$ and $T_1 = 0°\text{C}$.

a) What is the final steady-state temperature gradient along the rod?

b) What is the heat current in the rod, in the final steady state?

c) What is the final steady-state temperature at a point in the rod 2 cm from its left end?

16-3 A long rod, insulated to prevent heat losses, has one end immersed in boiling water (at atmospheric pressure) and the other end in a water–ice mixture. The rod consists of 100 cm of copper (one end in steam) and a length, L_2, of steel (one end in ice). Both rods are of cross-sectional area 5 cm². The temperature of the copper–iron junction is 60°C, after a steady state has been set up.

a) How much heat per second flows from the steam bath to the ice-water mixture?

b) How long is L_2?

16-4 A rod is initially at a uniform temperature of 0°C throughout. One end is kept at 0°C and the other is brought into contact with a steam bath at 100°C. The surface of the rod is insulated so that heat can flow only lengthwise along the rod. The cross-sectional area of the rod is 2 cm², its length is 100 cm, its thermal conductivity is $0.8 \text{ cal} \cdot \text{s}^{-1} \cdot \text{cm}^{-1} \cdot (\text{C}°)^{-1}$, its density is $10 \text{ g} \cdot \text{cm}^{-3}$, and its specific heat capacity is $0.1 \text{ cal} \cdot \text{g}^{-1} \cdot (\text{C}°)^{-1}$. Consider a short cylindrical element of the rod 1 cm in length.

a) If the temperature gradient at one end of this element is 200 C°·cm⁻¹, how many calories flow across this end per second?

b) If the average temperature of the element is increasing at the rate of 5 C°·s⁻¹, what is the temperature gradient at the other end of the element?

16-5 One experimental method of measuring the thermal conductivity of an insulating material is to construct a box of the material and measure the power input to an electric heater, inside the box, which maintains the interior at a measured temperature above the outside surface. Suppose that in such an apparatus a power input of 120 W is required to keep the interior surface of the box 120 F° above the temperature of the outer surface. The total area of the box is 25 ft² and the wall thickness is 1.5 in. Find the thermal conductivity of the material in the commercial system of units.

16-6 A container of wall area 5000 cm² and thickness 2 cm is filled with water in which there is a stirrer. The outer surface of the walls is kept at a constant temperature of 0°C. The thermal conductivity of the walls is 0.000478 cgs units, and the effect of edges and corners can be neglected. The power required to run the stirrer at an angular velocity of 1800 rpm is found to be 100 W. What will be the final steady-state temperature of the water in the container? Assume that the stirrer keeps the entire mass of water at a uniform temperature.

16-7 A boiler with a steel bottom 1.5 cm thick rests on a hot stove. The area of the bottom of the boiler is 1500 cm². The water inside the boiler is at 100°C, and 750 g are evaporated every 5 min. Find the temperature of the lower surface of the boiler, which is in contact with the stove.

16-8 A camping icebox, having wall area of 2 m² and thickness 5 cm, is constructed of insulating material having a thermal conductivity of $0.05 \text{ J} \cdot \text{s}^{-1} \cdot \text{m}^{-1} \cdot (\text{C}°)^{-1}$. The outside temperature is 20°C, and the inside of the box is to be maintained at 5°C by ice. The melted ice drains from the box at a temperature of 15°C. If ice costs 10 cents per kilogram, what will it cost to run the icebox for 1 hr?

16-9 An outer house wall has a layer of wood 3 cm thick on the outside, and a layer of styrofoam insulation 3 cm thick inside. If the wood has $k = 0.04 \text{ J} \cdot \text{s}^{-1} \cdot \text{m}^{-1} \cdot (\text{C}°)^{-1}$, the interior temperature is 20°C, and the exterior temperature is −10°C,

a) what is the temperature at the plane where the wood meets the styrofoam, and

b) what is the rate of heat flow, per square meter, through this wall?

16-10 A layer of ice 10 cm thick has formed on the surface of a lake, where the temperature is $-10°C$. The water under the slab is at $0°C$. At what rate does the thickness of the slab increase if the heat of fusion of the water freezing on the underside of the slab is conducted through the slab?

16-11 Three rods of equal length and cross section are placed end to end, one copper, one steel, one aluminum, in that order. The free end of the copper rod is in contact with ice water at $0°C$, and the free end of the aluminum rod is in contact with steam at $100°C$. If no heat transfer occurs at the sides, find the temperature at each junction.

16-12 A solid wood door has dimensions $2.0 \text{ m} \times 0.8 \text{ m} \times 4 \text{ cm}$. Its thermal conductivity is

$$k = 0.04 \text{ J}\cdot\text{s}^{-1}\cdot\text{m}^{-1}\cdot(\text{C}°)^{-1}.$$

The inside air temperature is $20°$ and the outside air temperature is $-10°C$.

a) What is the rate of heat flow through the door, assuming the surface temperatures are those of the surrounding air?

b) By what factor is the heat flow increased if a window 0.5 m square is inserted, assuming the glass is 0.5 cm thick and the surface temperatures are again those of the surrounding air?

16-13 A steam pipe of radius 2 cm, carrying steam at $120°C$, is surrounded by a cylindrical jacket of cork with inner and outer radii 2 cm and 4 cm, and this in turn is surrounded by a cylindrical jacket of styrofoam having inner and outer radii 4 cm and 6 cm. The thermal conductivity of styrofoam is about $0.01 \text{ J}\cdot\text{s}^{-1}\cdot\text{m}^{-1}\cdot(\text{C}°)^{-1}$. The outer surface of the styrofoam is in contact with air at $20°C$.

a) What is the temperature at a radius of 4 cm, where the two insulating layers meet?

b) What is the total rate of transfer of heat out of a 2-m length of pipe?

16-14 A spherical shell has inner and outer radii a and b, respectively, and the temperatures at the inner and outer surfaces are T_a and T_b.

a) Derive Eq. (16-6) for the total heat current through the shell.

b) Derive Eq. (16-7) for the temperature in the shell as a function of radius.

16-15 Rods of copper, brass, and steel are welded together to form a Y-shaped figure. The cross-sectional area of each rod is 2 cm^2. The end of the copper rod is maintained at $100°C$ and the ends of the brass and steel rods at $0°C$. Assume there is no heat loss from the surfaces of the rods. The lengths of the rods are: copper, 46 cm; brass, 13 cm; steel, 12 cm.

a) What is the temperature of the junction point?

b) What is the heat current in the copper rod?

16-16 A compound bar 2 meters long is constructed of a solid steel core 1 cm in diameter surrounded by a copper casing whose outside diameter is 2 cm. The outer surface of the bar is thermally insulated and one end is maintained at $100°C$, the other at $0°C$.

a) Find the total heat current in the bar.

b) What fraction of this total is carried by each material?

16-17 An electric transformer is in a cylindrical tank 60 cm in diameter and 1 m high, with flat top and bottom. If the tank transfers heat to the air only by natural convection, and the electrical losses are to be dissipated at the rate of 1 kW, how many degrees will the tank surface rise above room temperature?

16-18

a) What would be the difference in height between the columns in the U-tube in Fig. 16-4 if the liquid is water and the left arm is 1 m high at $4°C$ while the other is at $75°C$?

b) What is the difference between the pressures at the foot of two columns of water each 10 m high if the temperature of one is $4°C$ and that of the other is $75°C$?

16-19 A flat wall is maintained at constant temperature of $100°C$, and the air on both sides is at atmospheric pressure and at $20°C$. How much heat is lost by natural convection from 1 m^2 of wall (both sides) in 1 hr if (a) the wall is vertical and (b) the wall is horizontal?

16-20 A vertical steam pipe of outside diameter 7.5 cm and height 4 m has its outer surface at the constant temperature of $95°C$. The surrounding air is at atmospheric pressure and at $20°C$. How much heat is delivered to the air by natural convection in 1 hr?

16-21 What is the rate of energy radiation per unit area of a blackbody at a temperature of (a) 300 K, (b) 3000 K?

16-22 The emissivity of tungsten is approximately 0.35. A tungsten sphere 1 cm in radius is suspended within a large evacuated enclosure whose walls are at 300 K. What power input is required to maintain the sphere at a temperature of 3000 K if heat conduction along the supports is neglected?

16-23 A small blackened solid copper sphere of radius 2 cm is placed in an evacuated enclosure whose walls are kept at $100°C$. At what rate must energy be supplied to the sphere to keep its temperature constant at $127°C$?

16-24 A cylindrical metal can 10 cm high and 5 cm in diameter contains liquid helium at 4 K, at which temperature its heat of vaporization is $20 \text{ J}\cdot\text{g}^{-1}$. Completely surrounding the helium can are walls maintained at the temperature of liquid nitrogen, 80 K, the intervening space being evacuated. How much helium is lost per hour? Assume the radiant emittance of the helium can to be 0.2 that of a blackbody at 4 K.

16-25 A solid cylindrical copper rod 10 cm long has one end maintained at a temperature of 20.00 K. The other end is blackened and exposed to thermal radiation from a body at 300 K, no energy being lost or gained elsewhere. When equilibrium is reached, what is the temperature of the blackened end? [*Hint.* Since copper is a very good conductor of heat at low temperature, $k = 4$ cal·s^{-1}·cm^{-1}·(C°)$^{-1}$, the temperature of the blackened end is only slightly greater than 20 K.]

16-26 The operating temperature of a tungsten filament in an incandescent lamp is 2450 K and its emissivity is 0.30. Find the surface area of the filament of a 25-watt lamp.

16-27 A certain incandescent lightbulb operates at 3000 K. The total surface area of the filament is 0.05 cm², and the emissivity is 0.3. What electric power must be supplied to the filament?

16-28 The rate at which radiant energy reaches the surface of the earth from the sun is about 1.4 kW·m^{-2}. The distance from earth to sun is about 1.5×10^{11} m, and the radius of the sun is about 0.7×10^9 m.

a) What is the rate of radiation of energy, per unit area, from the sun's surface?

b) If the sun radiates as an ideal blackbody, what is the temperature of its surface?

16-29 Suppose that both ends of the rod in Fig. 16-1 are kept at a temperature of 0°C, and that the initial temperature distribution along the rod is given by $T = 100 \sin \pi x/L$, where T is in °C. Let the rod be of copper, of length $L = 10$ cm and of cross section 1 cm².

a) Show the initial temperature distribution in a diagram.

b) What is the final temperature distribution after a very long time has elapsed?

c) Sketch curves which you think would represent the temperature distribution at intermediate times.

d) What is the initial temperature gradient at the ends of the rod?

e) What is the initial heat current from the ends of the rod into the bodies making contact with its ends?

f) What is the initial heat current at the center of the rod? Explain. What is the heat current at this point at any later time?

g) What is the value of $k/\rho c$ for copper, and in what unit is it expressed? (This quantity is called the *thermal diffusivity*.)

h) What is the initial rate of change of temperature at the center of the rod?

i) How long a time would be required for the rod to reach its final temperature, if the temperature continued to decrease at this rate? (This time can be described as the *relaxation time* of the rod.)

j) From the graphs in part (c), would you expect the rate of change of temperature at the midpoint to remain constant, increase, or decrease?

k) What is the initial rate of change of temperature at a point in the rod, 2.5 cm from its left end?

THERMAL PROPERTIES OF MATTER

17-1 Equations of state

The *state* of a certain mass m of material is described by its pressure p, volume V, and temperature T. In general these cannot all vary independently. In preceding chapters we have studied the variation of volume with pressure (elasticity) and with temperature (thermal expansion). Mathematically speaking, there is a functional relationship among these quantities; symbolically this relationship can be represented as

$$V = f(p, T, m).$$

In words, the volume of a certain mass of material depends on the pressure, temperature, and mass. In more general problems, other state variables such as magnetization or electric polarization may be needed, but such generalizations will not be considered here.

Any such relationship is called an *equation of state*. To express this relationship in specific mathematical form for a particular substance is usually very complex. It is often useful to know how some *one* of the quantities changes when some other is varied, the rest being kept constant. Thus the (isothermal) compressibility k describes the change in volume when the pressure is changed, for a constant mass at a constant temperature. Similarly, the coefficient of volume expansion β gives the change in volume when the temperature is changed, for a constant mass at a constant pressure.

The term *state* as used here implies an *equilibrium* state. This means that the temperature and pressure are the same at all points. Hence if heat is added at some point to a system in an equilibrium state, we must wait until the processes of heat transfer within the system have brought about a new uniform temperature before the system is again in an equilibrium state. Similarly, when expansion or compression occurs, there is mass in motion, requiring acceleration and nonuniform pressure, and only when mechanical equilibrium is reestablished is the system's state described by a single pressure.

17-2 The ideal gas

Gases at low pressures have particularly simple equations of state. We may study the behavior of a gas by means of a cylinder with a movable piston, equipped with a pressure gauge and a thermometer. We may vary the pressure, volume, and temperature, and pump any desired mass of any gas into the cylinder, in order to explore the relationships between pressure, volume, temperature, and quantity of substance. It is often convenient to measure the amount of gas in terms of the number of *moles n* rather than the mass m. Since the molecular mass M is the mass per mole, the total mass m is given by

$$m = nM.$$

From measurements of the pressure, volume, temperature, and number of moles of various gases, a number of conclusions emerge. First, the volume is proportional to the number of moles. If we double the quantity of gas, keeping pressure and temperature constant, the volume doubles. Second, the volume varies *inversely* with pressure; if we double the pressure, holding the temperature and quantity of material constant, the gas is compressed to one-half its initial volume. Third, the pressure is proportional to the *absolute* temperature; if we double the absolute temperature, holding the volume and quantity of material constant, the pressure doubles.

These conclusions can be summarized neatly in a single equation of state:

$$pV = nRT. \tag{17-1}$$

The constant of proportionality R might be expected to have different values for different gases, but instead it turns out to have the same value for *all* gases, at least at sufficiently high temperature and low pressure. This quantity is called the *universal gas constant*. The numerical value of R depends, of course, on the units in which p, V, n, and T are expressed. The adjective "universal" means that in any one system of units R has the same value for *all* gases. In SI (mks) units, where the unit of p is 1 Pa or 1 N·m^{-2} and the unit of V is 1 m^3, the numerical value of R is found to be

$$R = 8.314 \, (\text{N} \cdot \text{m}^{-2}) \cdot \text{m}^3 \cdot \text{mol}^{-1} \cdot \text{K}^{-1} = 8.314 \, \text{J} \cdot \text{mol}^{-1} \cdot \text{K}^{-1}.$$

In the cgs system, where the unit of p is 1 dyn·cm^{-2} and the unit of V is 1 cm^3,

$$R = 8.314 \times 10^7 \, (\text{dyn} \cdot \text{cm}^{-2}) \cdot \text{cm}^3 \cdot \text{mol}^{-1} \cdot \text{K}^{-1}$$
$$= 8.314 \times 10^7 \, \text{erg} \cdot \text{mol}^{-1} \cdot \text{K}^{-1}.$$

We note that the units of pressure times volume are the same as units of energy, so, in *all* systems of units, R has units of energy per mole, per unit of absolute temperature. In terms of calories,

$$R = 1.99 \, \text{cal} \cdot \text{mol}^{-1} \cdot \text{K}^{-1}.$$

In chemical calculations volumes are commonly expressed in liters (l), pressures in atmospheres, and temperatures in kelvins. In this system,

$$R = 0.08207 \, \text{l} \cdot \text{atm} \cdot \text{mol}^{-1} \cdot \text{K}^{-1}.$$

We now define an *ideal gas* as one for which Eq. (17–1) holds precisely for *all* pressures and temperatures. As the term suggests, the ideal gas is an idealized model, which represents the behavior of gases very well in some circumstances, less well in others. Generally, gas behavior approximates the ideal-gas model most closely at very low pressures, when the gas molecules are far apart. However, the deviations are not very great at moderate pressures and at temperatures not too near those at which the gas liquefies.

Deviations from ideal-gas behavior are shown in Fig. 17–1, where the ratio pV/nT is plotted as a function of p and T. For an ideal gas this quantity is constant, but for real gases it varies increasingly at lower and lower temperatures. At sufficiently high temperature and low pressure this ratio approaches the value R for an ideal gas.

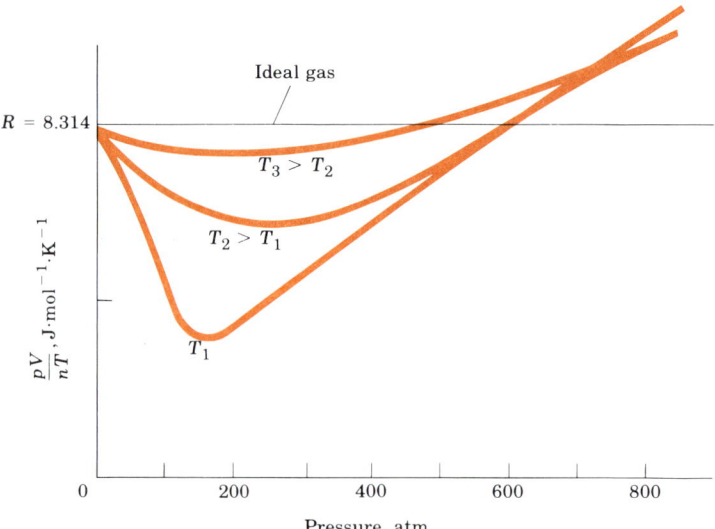

17–1 The limiting value of pV/nT is independent of T for all gases. For an ideal gas, pV/nT is constant.

For a *fixed mass* (or fixed number of moles) of an ideal gas, the product nR is constant and hence pV/T is constant also. Thus if the subscripts 1 and 2 refer to two states of the same mass of a gas, but at different pressures, volumes, and temperatures,

$$\frac{p_1 V_1}{T_1} = \frac{p_2 V_2}{T_2} = \text{constant.} \tag{17-2}$$

If the temperatures T_1 and T_2 are the same, then

$$p_1 V_1 = p_2 V_2 = \text{constant.} \tag{17-3}$$

The fact that at constant temperature the product of the pressure and volume of a fixed mass of gas is very nearly constant was discovered experimentally by Robert Boyle in 1660, and Eq. (17–3) is called *Boyle's law*. Although exactly true (by definition) for an *ideal gas*, it is obeyed only approximately by real gases and is not a fundamental law like Newton's laws or the law of conservation of energy.

Example 1 Find the volume of one mole of any ideal gas at "standard temperature and pressure" (STP), that is, 1 atm pressure and 0°C.

Solution In SI units, $p = 1 \text{ atm} = 1.013 \times 10^5$ Pa and $T = 273$ K. Thus, from Eq. (17–1),

$$V = \frac{nRT}{p} = \frac{(1 \text{ mol})(8.314 \text{ J}\cdot\text{mol}^{-1}\cdot\text{K}^{-1})(273 \text{ K})}{1.013 \times 10^5 \text{ Pa}}$$
$$= 0.0224 \text{ m}^3.$$

Alternatively, using liters and atmospheres,

$$V = \frac{(1 \text{ mol})(0.08207 \text{ l}\cdot\text{atm}\cdot\text{mol}^{-1}\cdot\text{K}^{-1})(273 \text{ K})}{1 \text{ atm}}$$
$$= 22.4 \text{ l} = 0.0224 \text{ m}^3.$$

Example 2 A tank attached to an air compressor contains 20 l of air at a temperature of 30°C and gauge pressure 4.0×10^5 Pa (about 60 lb·in^{-2}). What is the mass of air, and what volume would it occupy at normal atmospheric pressure and 0°C?

Solution Air is a mixture of gases, consisting of about 78% nitrogen and 21% oxygen, with small percentages of other gases. The *average* molecular mass, which we may use in the ideal-gas equation of state, is $M = 28.8$ g·mol^{-1}. Gauge pressure is the amount in excess of atmospheric pressure, so the total pressure is

$$p = 4.0 \times 10^5 \text{ Pa} + 1.01 \times 10^5 \text{ Pa} = 5.01 \times 10^5 \text{ Pa}.$$

The temperatures must be expressed in kelvins: 30°C = 303 K. We express the volume in cubic meters: 20 l = 20×10^{-3} m^3. To find the mass of gas, we first use Eq. (17–1) to find the number of moles:

$$n = \frac{pV}{RT} = \frac{(5.01 \times 10^5 \text{ Pa})(20 \times 10^{-3} \text{ m}^3)}{(8.314 \text{ J}\cdot\text{mol}^{-1}\cdot\text{K}^{-1})(303 \text{ K})} = 3.98 \text{ mol};$$
$$m = nM = (3.98 \text{ mol})(28.8 \text{ g}\cdot\text{mol}^{-1}) = 115 \text{ g} = 0.115 \text{ kg}.$$

The volume at 1 atm and 0°C (273 K) is

$$V = \frac{nRT}{p} = \frac{(3.98 \text{ mol})(8.314 \text{ J}\cdot\text{mol}^{-1}\cdot\text{K}^{-1})(273 \text{ K})}{1.01 \times 10^5 \text{ Pa}}$$
$$= 89.4 \times 10^{-3} \text{ m}^3 = 89.4 \text{ l}.$$

Example 3 The volume of an oxygen tank is 50 l. As oxygen is withdrawn from the tank, the reading of a pressure gauge drops from 300 lb·in^{-2} to 100 lb·in^{-2}, and the temperature of the gas remaining in the tank drops from 30°C to 10°C.

a) How many kilograms of oxygen were in the tank originally?
b) How many kilograms were withdrawn?
c) What volume would be occupied by the oxygen withdrawn from the tank at a pressure of 1 atm and a temperature of 20°C?

Solution

a) Let us express pressures in atmospheres, volumes in liters, and temperatures in kelvins. Then $R = 0.0821$ l·atm·mol^{-1}·K^{-1}. The initial and final *gauge* pressures are

$$300 \text{ lb} \cdot \text{in}^{-2} = \frac{300}{14.7} = 20.4 \text{ atm},$$

$$100 \text{ lb} \cdot \text{in}^{-2} = 6.80 \text{ atm}.$$

The corresponding *absolute* pressures are 21.4 atm and 7.80 atm. Initially,

$$n_1 = \frac{p_1 V}{RT_1} = \frac{(21.4 \text{ atm})(50 \text{ l})}{(0.0821 \text{ l} \cdot \text{atm} \cdot \text{mol}^{-1} \cdot \text{K}^{-1})(303 \text{ K})} = 43.0 \text{ mol}.$$

The original mass was therefore

$$m_1 = (43.0 \text{ mol})(32 \text{ g} \cdot \text{mol}^{-1}) = 1376 \text{ g} = 1.376 \text{ kg}.$$

b) The number of moles remaining in the tank is

$$n_2 = \frac{p_2 V}{RT_2} = \frac{(7.8 \text{ atm})(50 \text{ l})}{(0.0821 \text{ l} \cdot \text{atm} \cdot \text{mol}^{-1} \cdot \text{K}^{-1})(283 \text{ K})} = 16.8 \text{ mol};$$

$$m_2 = (16.8 \text{ mol})(32 \text{ g} \cdot \text{mol}^{-1}) = 538 \text{ g} = 0.538 \text{ kg}.$$

The number of kilograms withdrawn is

$$m_1 - m_2 = 1.376 \text{ kg} - 0.538 \text{ kg} = 0.838 \text{ kg}.$$

c) The number of moles withdrawn is $43.0 - 16.8 = 26.2$ mol. The volume occupied would be

$$V = \frac{nRT}{p} = \frac{(26.2 \text{ mol})(0.0821 \text{ l} \cdot \text{atm} \cdot \text{mol}^{-1} \cdot \text{K}^{-1})(293 \text{ K})}{1 \text{ atm}}$$

$$= 630 \text{ l}.$$

Example 4 Find the variation of atmospheric pressure with elevation in the earth's atmosphere, assuming the temperature to be uniform throughout (which is not actually the case).

Solution We begin with the hydrostatic relation of Eq. (12-3), $dp/dy = -\rho g$. The density ρ is not constant but varies with pressure. Replacing n in the ideal-gas equation by m/M, we obtain

$$pV = \frac{m}{M} RT \quad \text{and} \quad \rho = \frac{m}{V} = \frac{pM}{RT}.$$

We substitute this expression for ρ into Eq. (12-3), separate variables, and integrate, letting p_1 be the pressure at elevation y_1 and p_2 that at y_2.

$$\frac{dp}{dy} = -\frac{pMg}{RT},$$

$$\int_{p_1}^{p_2} \frac{dp}{p} = -\frac{Mg}{RT} \int_{y_1}^{y_2} dy,$$

$$\ln \frac{p_2}{p_1} = -\frac{Mg}{RT}(y_2 - y_1).$$

If the pressure at $y = 0$ is p_0, then the pressure p at any y is given by

$$p = p_0 e^{-Mgy/RT}. \tag{17-4}$$

At the summit of Mt. Everest, where $y = 8882$ m,

$$\frac{Mgy}{RT} = \frac{(28.8 \times 10^{-3} \text{ kg} \cdot \text{mol}^{-1})(9.8 \text{ m} \cdot \text{s}^{-2})(8882 \text{ m})}{(8.314 \text{ J} \cdot \text{mol}^{-1} \cdot \text{K}^{-1})(273 \text{ K})} = 1.10,$$

$$p = (1.01 \times 10^5 \text{ Pa})e^{-1.10} = 0.336 \times 10^5 \text{ Pa}$$
$$= 0.333 \text{ atm}.$$

We have used a constant temperature of $0°\text{C} = 273$ K, which is not completely realistic, but this example shows why mountaineers have so much difficulty breathing on Mt. Everest; air pressure is only one-third of its sea-level value. For similar reasons, jet airplanes, which typically fly at altitudes of 8,000 to 10,000 m, must have pressurized cabins for passenger comfort and health.

17-3 pVT-surface for an ideal gas

Since the equation of state for a fixed mass of a substance is a relation among the three variables p, V, and T, it defines a *surface* in a rectangular coordinate system in which p, V, and T are plotted along the three axes. Figure 17-2 shows the pVT-surface of an ideal gas. The dark color lines on the surface show the relation between p and V when T is constant (Boyle's law), the black lines show the relation between V and T when p is constant, and the light color lines show the relation between p and T when V is constant. When viewed perpendicular to the pT-plane, the surface appears as at the left, and its appearance when viewed perpendicular to the pV-plane is shown at the right.

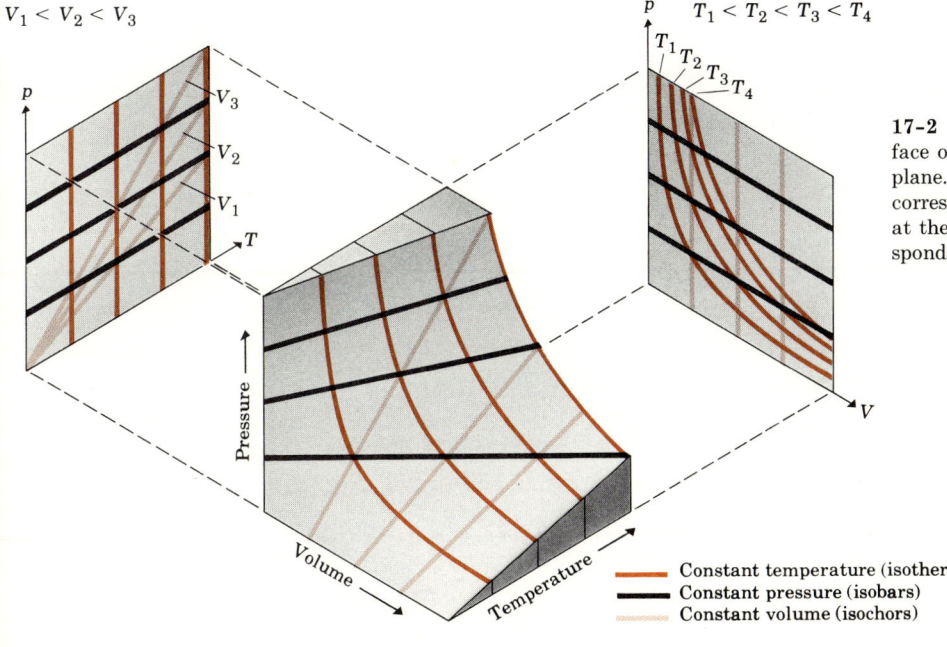

17-2 Projections of ideal-gas pVT-surface on the pT-plane and on the pV-plane. At the left, each light color line corresponds to a certain constant volume; at the right, each dark color line corresponds to a certain constant temperature.

Every possible equilibrium state of a given quantity of gas corresponds to a point on the surface, and every point of the surface corresponds to a possible equilibrium state. The gas cannot exist in a state

334　THERMAL PROPERTIES OF MATTER

that is not on the surface. For example, if the volume and temperature are given, thus locating a point in the base plane of Fig. 17–2, the pressure is then determined by the nature of the gas, and it can have only the value represented by the height of the surface above this point.

In any process in which the gas passes through a succession of equilibrium states, the point representing its state moves along a curve lying on the pVT-surface. Such a process must be carried out very slowly to give the temperature and pressure time to remain uniform at all points of the gas.

*17–4 pVT-surface for a real substance

While all real substances approximate ideal gases at sufficiently low pressures, their behavior departs more and more from that of an ideal gas at high pressures and low temperatures. As the temperature is lowered and the pressure increased, all substances change from the *gas phase* to the *liquid phase* or the *solid phase*. For a fixed mass of a substance, however, there is still a definite relation between the pressure, temperature, and total volume. In other words, the substance has an equation of state in *any* circumstances, and although the general form of the equation is much too complicated to express mathematically, we can represent it *graphically* by a pVT-surface. Figure 17–3 shows a pVT-surface for a substance that expands on melting. The substance can exist in either the solid, liquid, or gas phase, or in two phases simultaneously, or, along the triple line, in all three phases. (The distinction between the terms "gas" and "vapor" will be explained later. For the present they may be considered synonymous.)

17–3 pVT-surface for a substance that expands on melting. Projections of the surface on the pT- and pV-planes are also shown.

In order that the diagram shall represent the properties of a particular *substance,* but not depend on *how much* of the substance is present, we plot along the volume axis not the *total* volume V but the *specific volume v,* the volume per unit mass. Thus for a system of mass m,

$$v = \frac{V}{m}.$$

The specific volume is the reciprocal of the density ρ,

$$\rho = \frac{m}{V} = \frac{1}{v}.$$

To correlate these diagrams with our familiar experiences regarding the behavior of solids, liquids, and gases, let us start with a substance in the solid phase at point a in Fig. 17-3. Let the substance be contained in a cylinder with a movable piston; a constant force is exerted on the piston so that the pressure remains constant as the substance expands or contracts. A constant-pressure process is called an *isobaric* process, and lines representing constant-pressure processes are called *isobars.* We now bring the cylinder in contact with a heat source.

At the start of the process, the temperature of the substance rises as heat is added, at a rate determined by the specific heat capacity of the solid, and the volume increases slightly at a rate determined by the coefficient of volume expansion. When point b is reached the substance starts to melt, that is, to change from the solid to the liquid phase. The temperature ceases to rise even though heat is being continuously supplied. The volume increases as the substance melts.

When point c is reached, the substance is wholly in the liquid phase. The temperature now starts to rise again, at a rate determined by the specific heat capacity of the *liquid,* and the volume increases at a rate determined by its coefficient of volume expansion.

When point d is reached, the temperature again ceases to rise, although heat is still being added, and bubbles of vapor start to form in the cylinder and rise to its upper surface. Since the density of the vapor phase is smaller than that of the liquid, the volume increases.

At point e, the substance is wholly in the vapor phase. The relative volume increase from d to e, except at very high pressures, is enormously greater than the volume change between liquid and solid, so Fig. 17-3 is not constructed on a uniform scale. With further addition of heat the temperature rises again, the rate now being determined by the specific heat capacity of the *vapor* phase (at constant pressure). The volume also increases much more rapidly than did that of the solid or liquid.

In the process just described, the pressure was kept constant as heat was added and the temperature increased. Such a process is called an *isobaric* process. As a second example of how the pVT-surface describes the behavior of a material, we consider a process in which the pressure is increased and the temperature held constant. Such a process is called an *isothermal* process. We begin with the state described by point g in Fig. 17-3. As the pressure is increased, the volume decreases along the line gh; simultaneously, heat flows from the gas to the body with which it is in contact; otherwise the temperature of the gas would rise. When point h is reached, drops of liquid begin to form in the cylinder, and the volume continues to decrease without further increase of pressure.

At point j, all of the substance has condensed to the liquid phase. With further increase of pressure the volume decreases, but only slightly because of the small compressibility of liquids. Another break in the curve takes place at point k. Crystals of the solid begin to appear, and the volume again decreases without an increase of external pressure. At point l, the substance has been completely converted to the solid phase. Further increase of pressure reduces the volume of the solid only slightly, and unless the substance can exist in more than one modification of the solid phase, no further changes in phase result.

As a final example, suppose we again start with a gas in the cylinder but at a lower temperature, corresponding to point n in Fig. 17–3. If the pressure is increased isothermally, crystals of the solid begin to appear at point o and the gas changes directly to a solid without passing through the liquid phase. The pressure remains constant along the line op, and at point p the substance is all in the solid phase.

In all the processes described above, heat must be removed or added continuously in order to keep the pressure or temperature constant. If at any stage of a process the system is thermally insulated so that there can be no flow of heat in or out, and if the external pressure is kept constant, the system remains in equilibrium. Thus at any point on the surfaces lettered solid–liquid, solid–vapor, or liquid–vapor, *two* phases can coexist in equilibrium, and along the *triple line* all three phases can coexist. A vapor at the pressure and temperature at which it can exist in equilibrium with its liquid is called a *saturated vapor,* and the liquid is called a *saturated liquid*. Thus points e and h represent saturated vapor, and points j and d, saturated liquid. (The term "saturated" is poorly chosen. It does not have the same meaning as a "saturated solution" in chemistry. There is no dissolving of one substance in another.)

17–5 Phase diagrams

It is difficult to draw and reproduce pVT-surfaces, and the information represented by such a surface is often represented by means of *projections* of the surface on the pT- or pV-plane, as shown in Fig. 17–3. The pT projection is called a *phase diagram*. Each phase of a material can exist for certain ranges of temperature and pressure, and the pT-phase diagram shows what phase is present for each pair of values of p and T. A typical phase diagram is shown in Fig. 17–4. At each point on the diagram only a single phase can exist, except for points on the lines, where two phases can coexist in *phase equilibrium*. Points on the lines therefore represent conditions under which phase changes occur.

If a substance is heated at constant pressure, it passes through a sequence of states represented by points on the broken horizontal line labeled (a). The melting and boiling temperatures at this pressure are the temperatures at which the broken line crosses the fusion and vaporization curves, respectively. Similarly, if material is compressed by increasing the pressure while we hold the temperature constant, as represented by the vertical broken line labeled (b), the material passes from vapor to liquid and then to solid at the points where the broken line crosses the vaporization curve and fusion curve, respectively.

When the pressure is sufficiently low, a material may be transformed by constant-pressure heating from solid directly to vapor, as represented

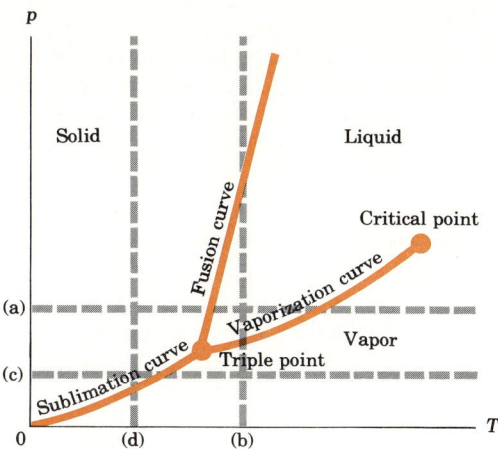

17-4 A pT phase diagram, showing regions of temperature and pressure at which the various phases exist and where phase changes occur.

by line (c). This process, called *sublimation*, occurs where the line crosses the sublimation curve. At atmospheric pressure solid carbon dioxide undergoes sublimation; there is no liquid phase at this pressure. Similarly, when a substance with initial temperature lower than the triple point is compressed by increasing its pressure, it goes directly from the vapor to the solid phase, as shown by line (d).

Figures 17-3 and 17-4 are characteristic of materials that expand on melting. There are a few materials that *contract* on melting; the most familiar examples are water and the metal antimony. For such a substance the fusion curve always slopes upward to the left, as in Fig. 17-5, rather than to the right. The melting temperature decreases when the pressure increases, so solid material can be melted by an increase of

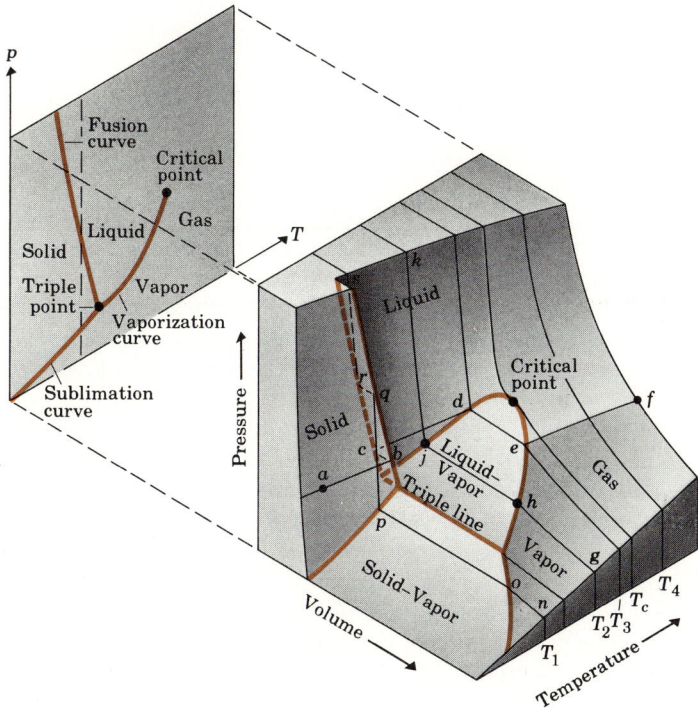

17-5 pVT-surface and phase diagram for a material that contracts on melting. Such a material can be melted by increasing its pressure, as shown by the broken vertical line.

pressure. Ice-skating is possible because the pressure of the skate blades against the ice (which may be as much as several hundred atmospheres) melts a thin film of water and lubricates the surfaces.

17-6 Triple point and critical point

The intersection point of the equilibrium curves on the left sides of Figs. 17-3 and 17-5 is called the *triple point*. In each figure this point is an end view of the *triple line* on the pVT-surface. For any substance there is only one pressure and temperature at which all three phases can coexist, namely, the triple-point pressure and temperature. Triple-point data for a few substances are given in Table 17-1.

Table 17-1 Triple-point data

Substance	Temperature, K	Pressure, 10^5 Pa
Hydrogen (normal)	13.84	0.0704
Deuterium (normal)	18.63	0.171
Neon	24.57	0.432
Nitrogen	63.18	0.125
Oxygen	54.36	0.00152
Ammonia	195.40	0.0607
Carbon dioxide	216.55	5.17
Sulfur dioxide	197.68	0.00167
Water	273.16	0.00610

Figures 17-3, 17-4, and 17-5 show that the liquid and gas (or vapor) phases can exist together only when the temperature and pressure are less than those at the point lying at the top of the tongue-shaped surface lettered liquid–vapor. This point is called the *critical point*, and the corresponding values of T, p, and v are the *critical temperature, pressure, and specific volume*. A gas at a temperature *above* the critical temperature, such as T_4 in Fig. 17-3, does not separate into two phases when compressed isothermally, but its properties change gradually and continuously from those we ordinarily associate with a gas (low density, large compressibility) to those of a liquid (high density, small compressibility). Table 17-2 lists critical constants for a few substances. The very low critical temperatures of hydrogen and helium make it evident why these gases defied attempts to liquefy them for many years.

The term *vapor* is sometimes used to mean a gas at any temperature below its critical temperature, and sometimes is restricted to mean a gas in equilibrium with the liquid phase, that is, a saturated vapor. The term is really unnecessary; no sudden change takes place in the properties of a substance when the critical isotherm is crossed either on the portion of the surface "gas and vapor," or on the portion "liquid."

Many substances can exist in more than one modification of the solid phase. Transitions from one modification to another occur at definite temperatures and pressures, just as with phase changes from liquid to solid, etc. Water is one such substance; at least eight types of ice have been observed at very high pressures.

Table 17-2 Critical constants

Substance	Critical temperature, K	Critical pressure, 10^5 Pa	Critical volume, 10^{-6} m$^3 \cdot$ mol^{-1}	Critical density, kg $\cdot$ m^{-3}
Helium (4)	5.3	2.29	57.8	69.3
Helium (3)	3.34	1.17	72.6	41.3
Hydrogen (normal)	33.3	13.0	65.0	31.0
Deuterium (normal)	38.4	16.6	60.3	66.3
Nitrogen	126.2	33.9	90.1	311
Oxygen	154.8	50.8	78	410
Ammonia	405.5	112.8	72.5	235
Freon 12	384.7	40.1	218	555
Carbon dioxide	304.2	73.9	94.0	468
Sulfur dioxide	430.7	78.8	122	524
Water	647.4	221.2	56	320
Carbon disulfide	552	79.0	170	440

As numerical examples, consider the pT-diagrams of water and carbon dioxide, in Fig. 17-6. In (a), a horizontal line at a pressure of 1 atm intersects the freezing-point curve at 0°C and the boiling-point curve at 100°C. The boiling point increases with increasing pressure up to the critical temperature of 374°C. Solid, liquid, and vapor can remain in equilibrium only at the triple point, where the vapor pressure is 0.006 atm and the temperature is 0.01°C.

For carbon dioxide, the triple-point temperature is −56.6°C and the corresponding pressure is 5.11 atm. Hence at atmospheric pressure CO_2 can exist only as a solid or vapor. When heat is supplied to solid CO_2, open to the atmosphere, it changes directly to a vapor, without passing through the liquid phase, whence the name "dry ice." Liquid CO_2 can exist only at a pressure greater than 5.11 atm. The steel tanks in which CO_2 is commonly stored contain liquid and vapor (both saturated). The pressure in these tanks is the vapor pressure of CO_2 at the temperature of the tank. If the latter is 20°C the vapor pressure is 56 atm or about 830 lb·in^{-2}.

The pVT-surface for ordinary helium (of mass number 4) is shown in Fig. 17-7. Two remarkable properties may be seen. First, helium has no triple point at which solid, liquid, and gas coexist in phase equilibrium. Second, there are *two* liquid phases; this is a property unique to helium. Liquid II is the *superfluid* phase; it has several unusual properties, including essentially zero viscosity, very large thermal conductivity, and the ability to form a surface film which creeps up the sides of a container.

The phase transition between Liquid I and Liquid II is unusual; there is no heat of phase change for this transition. The two phases cannot coexist side by side; the material must be entirely in one phase or entirely in the other, and the corresponding line on the pT diagram shows conditions of temperature and pressure at which the transition can occur. There are two points that would ordinarily be called triple points, but at neither is it possible to have phase equilibrium with three phases.

(a) Water

(b) Carbon dioxide

17-6 Pressure–temperature diagrams (not to a uniform scale).

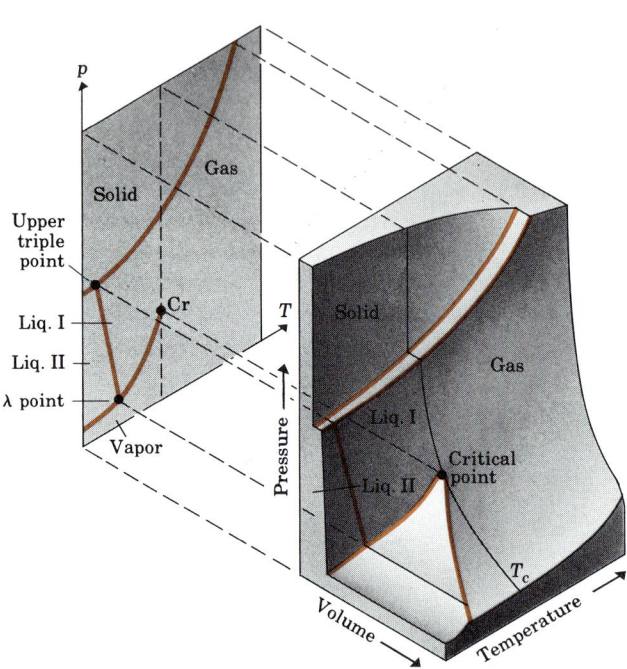

17-7 pVT-surface for helium, and pT phase diagram.

Finally, at any pressure less than about 25 atm, helium does not solidify as the temperature is lowered, but remains liquid all the way to absolute zero, or at least to the lowest temperature presently attainable, of the order of 10^{-6} K. To obtain solid helium the pressure must be raised above 25 atm, and then at sufficiently low temperature the helium atoms form a crystal lattice.

17-7 Vapor pressure

The pressure of a vapor in equilibrium with liquid or solid at any temperature is called the *vapor pressure* of the substance at that temperature. Thus the sublimation and vaporization curves in Figs. 17-4 and 17-5 are graphs of vapor pressure versus temperature. The vapor pressure of a substance is a function of *temperature only,* not of volume. That is, in a vessel containing a liquid (or solid) and vapor in equilibrium at a *fixed temperature,* the pressure does not depend on the relative amounts of liquid and vapor present. If the volume is decreased some of the vapor condenses, and vice versa; but if the temperature is kept constant by removing or adding heat the pressure does not change.

The *boiling-point* temperature of a liquid is the temperature at which its vapor pressure equals the external pressure. Table 17-3 gives the vapor pressure of water as a function of temperature; we see that the vapor pressure is 1 atm (1.01×10^5 Pa) at a temperature of 100°C. If the external pressure is reduced to 0.0233×10^5 Pa, water will boil at room temperature (20°C) while, under a pressure of 90 lb·in^{-2} (about 6 atm), the boiling temperature is 160°C.

*17-8 Humidity

Atmospheric air is a mixture of gases consisting of about 78 percent nitrogen, 21 percent oxygen, and small amounts of carbon dioxide, water

Table 17-3 Vapor pressure of water

T, °C	Vapor pressure, 10^5 Pa	Vapor pressure, lb·in^{-2}	T, °F
0	0.00610	0.0886	32
5	0.00868	0.126	41
10	0.0119	0.173	50
15	0.0169	0.245	59
20	0.0233	0.339	68
40	0.0734	1.07	104
60	0.199	2.89	140
80	0.473	6.87	176
100	1.01	14.7	212
120	1.99	28.8	248
140	3.61	52.4	284
160	6.17	89.6	320
180	10.0	145	356
200	15.5	225	392
220	23.2	336	428

vapor, and other gases. The mass of water vapor per unit volume is called the *absolute humidity*. The total pressure exerted by the atmosphere is the sum of the pressures exerted by its component gases. These pressures are called the *partial pressures* of the components. It is found that the partial pressure of each of the component gases of a gas mixture is very nearly the same as would be the actual pressure of that component alone if it occupied the same volume as does the mixture, a fact known as *Dalton's law*. That is, each of the gases of a gas mixture behaves independently of the others. The partial pressure of water vapor in the atmosphere is ordinarily of the order of 0.005 to 0.05×10^5 Pa.

The partial pressure of water vapor at any given air temperature can never exceed the vapor pressure of water at that particular temperature. Thus, from Table 17-3, the partial pressure at 10°C cannot exceed 0.0119×10^5 Pa, and at 15°C it cannot exceed 0.0169×10^5 Pa. If the concentration of water vapor, or the absolute humidity, is such that the partial pressure *equals* the vapor pressure, the vapor is *saturated*. If the partial pressure is *less* than the vapor pressure, the vapor is *unsaturated*. The ratio of the partial pressure to the vapor pressure at the same temperature is called the *relative humidity,* and is usually expressed as a percentage:

$$\text{Relative humidity (\%)} = (100\%) \frac{\text{partial pressure of water vapor}}{\text{vapor pressure at same temperature}}.$$

The relative humidity is 100% if the vapor is saturated and zero if no water vapor at all is present.

Example The partial pressure of water vapor in the atmosphere is 0.012×10^5 Pa and the temperature is 20°C. Find the relative humidity.

Solution From Table 17-3, the vapor pressure at 20°C is 0.0233×10^5 Pa. Hence,

$$\text{Relative humidity} = \frac{0.012 \times 10^5 \text{ Pa}}{0.0233 \times 10^5 \text{ Pa}} (100\%) = 52\%.$$

Since the water vapor in the atmosphere is saturated when its partial pressure equals the vapor pressure at the air temperature, saturation can be brought about either by increasing the water vapor content or by lowering the temperature. For example, let the partial pressure of water vapor be 0.012×10^5 Pa when the air temperature is 20°C, as in the preceding example. Saturation or 100% relative humidity could be attained either by introducing enough more water vapor (keeping the temperature constant) to increase the partial pressure to 0.0233×10^5 Pa, *or by lowering the temperature* to 10°C, at which, from Table 17–3, the vapor pressure is 0.012×10^5 Pa.

If the temperature were to be lowered *below* 10°C, the vapor pressure would be less than 0.012×10^5 Pa. The partial pressure would then be higher than the vapor pressure and enough vapor would condense to reduce the partial pressure to the vapor pressure at the lower temperature. It is this process that brings about the formation of clouds, fog, and rain. The phenomenon is also of frequent occurrence at night when the earth's surface becomes cooled by radiation. The condensed moisture is called *dew*. If the partial pressure is so low that the temperature must fall below 0°C before saturation occurs, the vapor condenses into ice crystals in the form of frost or snow.

The temperature at which the water vapor in a given sample of air becomes saturated is called the *dew point*. Measuring the temperature of the dew point is the most accurate method of determining relative humidity. The usual method is to cool a metal container having a bright polished surface, and to observe its temperature when the surface becomes clouded with condensed moisture. In the above example, the dew point would be found to be 10°C.

The rate of evaporation of water from a water–air surface depends on the relative humidity. Evaporation is most rapid when the vapor pressure is low, but when the relative humidity is high evaporation is slower. At 100% relative humidity, *no* further evaporation can take place. The added discomfort that high relative humidity brings to a hot summer day is a familiar effect; the reason is that one of the body's important temperature-control mechanisms, cooling by evaporation of sweat from the skin, is inhibited by high relative humidity.

17–9 The bubble chamber

The bubble chamber is an instrument for studying the interactions of fundamental particles, especially in high-energy collisions, by making their paths visible. As shown in Fig. 17–8, it consists in principle of an enclosure having glass windows and a movable piston. The enclosure contains a liquefied gas, usually hydrogen but occasionally propane, at pressure and temperature just below the boiling point. When the pressure is suddenly decreased slightly by pulling down the piston, the boiling temperature drops below the actual temperature. However, the liquid does not necessarily begin to boil immediately; it may persist in the liquid phase for some time in an unstable *superheated* state represented by a point *below* the pVT-surface.

Bubbles of vapor then tend to form first around ionized molecules or other inhomogeneities in the liquid. Hence when a high-energy charged particle passes through, leaving a string of ions along its path, a string of

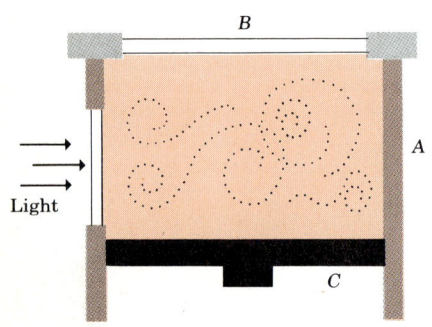

17–8 The bubble chamber.

bubbles forms on the ions. This trail is visible and can be recorded by stereoscopic photography; analysis of this photographic record then determines the precise path and speed of the particle.

A historical forerunner of the bubble chamber was the *cloud chamber,* in which a supercooled vapor is used instead of a superheated liquid. A charged particle passing through leaves a trail of ions, and droplets of liquid condense around these ions, marking the particle's path. In present-day research the cloud chamber has been supplanted by the bubble chamber and other more recent inventions such as the spark chamber, but in the early days of high-energy physics it was of great importance.

Questions

17-1 If the ideal-gas model were perfectly valid at all temperatures, what would be the volume of a gas when the temperature is absolute zero?

17-2 In the ideal-gas equation of state, could Celsius temperature be used instead of Kelvin if an appropriate numerical value of the gas constant R is used?

17-3 Comment on this statement: Equal masses of two different gases, placed in containers of equal volume at equal temperature, must exert equal pressures.

17-4 When a car is driven some distance, the air pressure in the tires increases. Why? Should you let some air out to reduce the pressure?

17-5 Sec. 17-1 states that pressure, volume, and temperature cannot change individually without one affecting the others. Yet when a liquid evaporates its volume changes, even though its pressure and temperature are constant. Is this inconsistent?

17-6 At low pressure or high temperature, gas behavior deviates from the ideal-gas model. One reason is the finite size of gas molecules. From these deviations, how could one estimate what fraction of the total volume in a gas is occupied by the molecules themselves?

17-7 Radiator caps in automobile engines will release coolant when the pressure reaches a certain value, typically 15 psi or so. Why is it desirable to have the coolant under pressure? Why not just seal the system completely?

17-8 People speak of being angry enough to make their blood boil. Could your blood boil if you go to a high enough altitude? Is the pressure of your blood the same as atmospheric pressure? Does this matter?

17-9 Can it be too cold for good ice skating? What about skiing? Does a thin layer of ice under the skis melt?

17-10 Why does the air cool off more on a clear night than a cloudy night? Desert areas often have exceptionally large day–night temperature variations. Why?

17-11 On a chilly evening you can "see your breath." Can you really? What are you really seeing? Does this phenomenon depend on the temperature of the air, the humidity, or both? How is it related to the dew point?

17-12 Why do frozen water pipes burst? Would a mercury thermometer break if the temperature goes below the freezing temperature of mercury?

17-13 Why does moisture condense on the insides of windows on cold days? If it is cold enough outside, frost will form, but only when the outside temperature is far below freezing. Why? Why are the bottoms frostier than the tops? How do storm windows change the situation?

17-14 "It's not the heat; it's the humidity!" Why are you more uncomfortable on a hot day when the humidity is high than when it is low?

17-15 Why does a shower room get steamy or foggy when the hot water runs for a long time?

17-16 The critical volumes of several substances are listed in Table 17-2, and with a few exceptions they are all in the range of 50 to $100 \times 10^{-6}\,\text{m}^3 \cdot \text{mol}^{-1}$. Why is there not more variation?

17-17 Why do flames from a fire always go up rather than down?

17-18 Unwrapped food placed in a freezer experiences dehydration, known to housewives as "freezer burn." Why? The same process, carried out intentionally, is called "freeze-drying." For freeze-drying the food is usually frozen first and then placed in a partial vacuum. Why? What advantages might freeze-drying have, compared to ordinary drying?

17-19 An old or burned-out incandescent lightbulb usually has a dark gray area on part of the inside of the bulb. What is this? Why is it not a uniform deposit all over the inside of the bulb?

17-20 Why do glasses containing cold liquids sweat in warm weather? Glasses containing alcoholic drinks sometimes form frost on the outside. How can this happen?

Problems

17-1 What is the mass of one cubic meter of nitrogen at 1 atm and 20°C?

17-2 What is the ratio of the density of oxygen to that of water at 1 atm and 0°C?

17-3 A tank contains $0.5\,\text{m}^3$ of nitrogen at 27°C and 1.5×10^5 Pa (absolute pressure). What is the pressure if the volume is increased to $5.0\,\text{m}^3$ and the temperature is increased to 327°C?

17-4 A tank having a volume of $0.10\,\text{m}^3$ is filled with oxygen at a gauge pressure of 4.0×10^5 Pa and temperature of 47°C. Later it is found that, because of a leak, the gauge pressure has dropped to 3.0×10^5 Pa and the temperature has decreased to 27°C. Find

a) the initial mass of oxygen and

b) the mass that has leaked out.

17-5 A helium storage tank has a capacity of $0.05\,\text{m}^3$. If the gas pressure is 100 atm at 27°C, determine

a) the number of moles of helium and

b) the mass of helium.

17-6 A hot-air balloon makes use of the fact that hot air at atmospheric pressure is less dense than cooler air at the same pressure; the buoyant force is the difference between the weight of hot air and that of an equal volume of the cooler surrounding air. If the volume of the balloon is $500\,\text{m}^3$ and the surrounding air is at 0°C, what must be the temperature of the air in the balloon in order for it to lift a total mass of 200 kg?

17-7 A flask of volume 2 l, provided with a stopcock, contains oxygen at 300 K and atmospheric pressure. The system is heated to a temperature of 400 K, with the stopcock open to the atmosphere. The stopcock is then closed and the flask cooled to its original temperature.

a) What is the final pressure of the oxygen in the flask?

b) How many grams of oxygen remain in the flask?

17-8 A balloon whose volume is $500\,\text{m}^3$ is to be filled with hydrogen at atmospheric pressure.

a) If the hydrogen is stored in cylinders of volume $0.05\,\text{m}^3$ at an absolute pressure of 15×10^5 Pa, how many cylinders are required?

b) What is the total weight that can be supported by the balloon, in air at standard conditions?

c) What weight could be supported if the balloon were filled with helium instead of hydrogen?

17-9 Derive from the equation of state of an ideal gas an equation for the density of an ideal gas in terms of pressure, temperature, and appropriate constants.

17-10 At the beginning of the compression stroke, a cylinder of a diesel engine contains $800\,\text{cm}^3$ of air at atmospheric pressure and a temperature of 27°C. At the end of the stroke, the air has been compressed to a volume of $50\,\text{cm}^3$ and the gauge pressure has increased to 40×10^5 Pa. Compute the temperature.

17-11 A bubble of air rises from the bottom of a lake, where the pressure is 3 atm, to the surface, where the pressure is 1 atm. The temperature at the bottom of the lake is 7°C and the temperature at the surface is 27°C. What is the ratio of the volume of the bubble as it reaches the surface to its volume at the bottom?

17-12 A liter of helium under a pressure of 2 atm and at a temperature of 27°C is heated until both pressure and volume are doubled.

a) What is the final temperature?

b) How many grams of helium are there?

17-13 A flask contains 1 g of oxygen at an absolute pressure of 10 atm and at a temperature of 47°C. At a later time it is found that because of a leak the pressure has dropped to $\tfrac{5}{8}$ of its original value and the temperature has decreased to 27°C.

a) What is the volume of the flask?

b) How many grams of oxygen leaked out between the two observations?

17-14 The submarine *Squalus* sank at a point where the depth of water was 240 ft. The temperature at the surface is 27°C and at the bottom it is 7°C. The density of sea water may be taken as $2\,\text{slugs}\cdot\text{ft}^{-3}$.

a) If a diving bell in the form of a circular cylinder 8 ft high, open at the bottom and closed at the top, is lowered to this depth, to what height will the water rise within it when it reaches the bottom?

b) At what gauge pressure must compressed air be supplied to the bell while on the bottom to expel all the water from it?

17-15 A bicycle pump is full of air at an absolute pressure of $15\,\text{lb}\cdot\text{in}^{-2}$. The length of stroke of the pump is 18 in. At what part of the stroke does air begin to enter a tire in which the gauge pressure is $40\,\text{lb}\cdot\text{in}^{-2}$? Assume the compression to be isothermal.

17-16 A vertical cylindrical tank 1 m high has its top end closed by a tightly fitting frictionless piston of negligible weight. The air inside the cylinder is at an absolute pressure of 1 atm. The piston is depressed by pouring mercury on it slowly. How far will the piston descend before mercury spills over the top of the cylinder? The temperature of the air is maintained constant.

17-17 A barometer is made of a tube 90 cm long and of cross section $1.5\,\text{cm}^2$. Mercury stands in this tube to a height of 75 cm. The room temperature is 27°C. A small amount of nitrogen is introduced into the evacuated space above the mercury and the column drops to a height of 70 cm. How many grams of nitrogen were introduced?

17-18 A large tank of water has a hose connected to it, as shown in Fig. 17-9. The tank is sealed at the top and has compressed air between the water surface and the top. When the water height h_2 is 3 m, the gauge pressure p_1 is 1.0×10^5 N·m^{-2}. Assume that the air above the water surface expands isothermally.

a) What is the velocity of flow out of the hose when $h_2 = 3$ m?

b) What is the velocity of flow when h_2 has decreased to 2 m? Neglect friction.

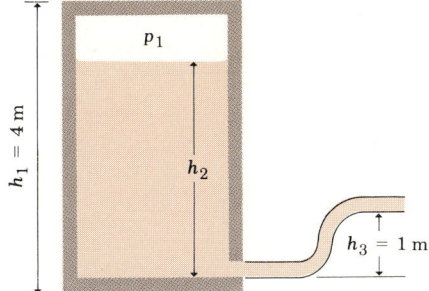

Figure 17-9

17-19 What is atmospheric pressure at an altitude of 5000 m, if the air temperature is a uniform 0°C? This is roughly the maximum altitude usually attained by aircraft with nonpressurized cabins.

17-20 In the lower part of the atmosphere (the troposphere) the temperature is not uniform but decreases with increasing elevation. Show that if the temperature variation is approximated by the linear relation

$$T = T_0 - \alpha y,$$

where T_0 is the temperature at the earth's surface and T is the temperature at a height y, the pressure is given by

$$\ln \frac{p_0}{p} = \frac{Mg}{R\alpha} \ln \frac{T_0}{T_0 - \alpha y}$$

where p_0 is the pressure at the earth's surface and M is the molecular mass.

The coefficient α is called the "lapse rate of temperature." While it varies with atmospheric conditions, an average value is about 0.6 C°/100 m.

17-21 Construct two graphs for a real substance, one showing pressure as a function of volume, and the other showing pressure as a function of temperature. Show on each graph the region in which the substance exists as

a) a gas or vapor,
b) a liquid,
c) a solid.

Show also the triple point and the critical point.

17-22 A small amount of liquid is introduced into a glass tube, all air is removed, and the tube is sealed off. Describe the behavior of the meniscus when the temperature of the system is raised:

a) If the volume of the tube is much greater than the critical volume;

b) If the volume of the tube is much less than the critical volume;

c) If the volume of the tube is only slightly different from the critical volume.

17-23 A piece of ice at 0°C is placed alongside a beaker of water at 0°C in a glass vessel, from which all air has been removed. If the ice, water, and vessel are all maintained at a temperature of 0°C by a suitable thermostat, describe the final equilibrium state inside the vessel.

17-24 Compare the density of carbon dioxide at the critical point to the density of solid carbon dioxide ("dry ice") at 1 atm, which is about 1.56 g cm^{-3}.

17-25 Determine the density of water at the critical point, and express it as a multiple of the density at 0°C and 1 atm.

17-26 Find the ratio of the density of oxygen to that of water (a) at 0°C and 1 atm, and (b) at their respective critical points.

17-27

a) What is the relative humidity on a day when the temperature is 20°C and the dew point is 5°C?

b) What is the partial pressure of water vapor in the atmosphere?

c) What is the absolute humidity, in grams per cubic meter?

17-28 The temperature in a room is 40°C. A can is gradually cooled by adding cold water. At 10°C the surface of the can clouds over. What is the relative humidity in the room?

17-29 A pan of water is placed in a sealed room of volume 60 m^3 at a temperature of 27°C and initial relative humidity of 60 percent.

a) How many grams of water will evaporate?

b) What is the absolute humidity in kg·m^{-3} after equilibrium has been reached?

c) If the temperature of the room is then increased 1 C°, how many more grams of water will evaporate?

17-30

a) What is the dew-point temperature on a day when the air temperature is 20°C and the relative humidity is 60 percent?

b) What is the absolute humidity, expressed in grams per cubic meter?

17-31 An air-conditioning system is required to increase the relative humidity of 0.5 m^3 of air per second from 30 percent to 65 percent. The air temperature is 20°C. How many kilograms of water are needed by the system per hour?

17-32 The volume of a closed room, kept at a constant temperature of 20°C, is 60 m^3. The relative humidity in the room is initially 10%. If a pan of water is brought into the room, how many grams will evaporate?

18 THE FIRST LAW OF THERMODYNAMICS

18-1 Energy and work in thermodynamics

Thermodynamics is concerned with energy relationships involving heat, mechanical energy, and other aspects of energy and energy transfer. The principles of thermodynamics are usually stated with reference to some well-defined *system,* usually a specified quantity of matter. A *thermodynamic system* is one that can interact with its surroundings in at least two ways, one of which must be transfer of heat. A familiar example is a quantity of a gas confined in a cylinder with a piston. Energy can be added to the system by conduction of heat, and it is also possible to do mechanical *work* on the system, since the piston exerts a force that can move through a displacement.

Thermodynamics has its roots in intensely practical problems. A steam engine or steam turbine, for example, makes use of the heat of combustion of coal or other fuel to perform mechanical work to drive an electric generator, pull a train, or perform some other useful function. The gasoline engine in an automobile has a similar function. Because of these practical applications, it is customary to discuss energy relations for a thermodynamic system not in terms of the work done *on* the system *by* its surroundings, but rather in terms of the work done *by the system* on its surroundings. In view of Newton's third law, this is the negative of the work done *on* the system. This convention will be clarified further in the next section.

Heat can be understood on the basis of microscopic mechanical energy, i.e., the kinetic and potential energies of individual molecules in a material; and it is also possible to develop the principles of thermodynamics from a microscopic viewpoint. In the present chapter that development is deliberately avoided, in order to emphasize that the central principles and concepts of thermodynamics can be treated in a wholly *macroscopic* way, without reference to microscopic models. Indeed, part of the great power and generality of thermodynamics springs from the fact that it is *not* dependent on details of the structure of matter. In Chapter 20 we return to microscopic considerations and examine their relation to the principles of thermodynamics.

18-2 Work in volume changes

18-1 Force exerted *by* a system during a small expansion.

When a gas expands, it pushes out on its boundary surfaces as they move outward; hence an expanding gas always does a positive quantity of work. To calculate the work done by a thermodynamic system during a volume change, let us consider a solid or fluid contained in a cylinder equipped with a movable piston, as shown in Fig. 18-1. Suppose that the cylinder has a cross-sectional area A and that the pressure exerted by the system at the piston face is p. The force exerted by the system is therefore pA. If the piston moves out a small distance dx, the work dW of this force is

$$dW = pA\, dx.$$

But

$$A\, dx = dV,$$

where dV is the change of volume of the system. Therefore

$$dW = p\, dV, \qquad (18\text{-}1)$$

and in a finite change of volume from V_1 to V_2,

$$W = \int_{V_1}^{V_2} p\, dV. \qquad (18\text{-}2)$$

In general the pressure of the system may vary during the volume change, and the integral can be evaluated only when the pressure is known as a function of volume. It is customary to represent this relationship graphically by plotting p as a function of V, as in Fig. 18-2. Then Eq. (18-2) may be interpreted graphically as the *area* under the curve between the limits V_1 and V_2.

According to the convention stated above, the work is *positive* when a system *expands*. Thus, if a system expands from 1 to 2 in Fig. 18-2, the area is regarded as positive. A *compression* from 2 to 1 would give rise to a *negative* area; when a system is compressed, its volume decreases and it does *negative* work on its surroundings.

If the pressure remains constant while the volume changes, then the work is

$$W = p(V_2 - V_1) \qquad \text{(constant pressure only)}. \qquad (18\text{-}3)$$

Example An ideal gas is kept at constant temperature T and undergoes an *isothermal expansion* in which its volume changes from V_1 to V_2. How much work is done?

Solution From Eq. (18-2)

$$W = \int_{V_1}^{V_2} p\, dV.$$

For an ideal gas

$$p = \frac{nRT}{V}.$$

Since n, R, and T are constant,

$$W = nRT \int_{V_1}^{V_2} \frac{dV}{V} = nRT \ln \frac{V_2}{V_1}. \qquad (18\text{-}4)$$

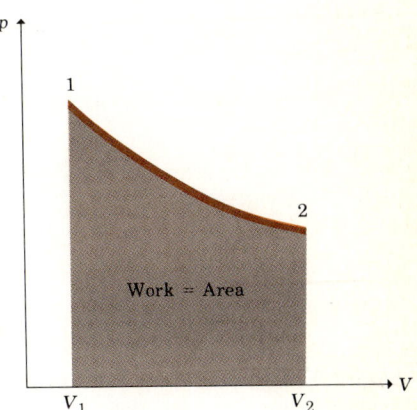

18-2 The work done when a substance changes its volume and pressure is the area under the curve on a pV-diagram.

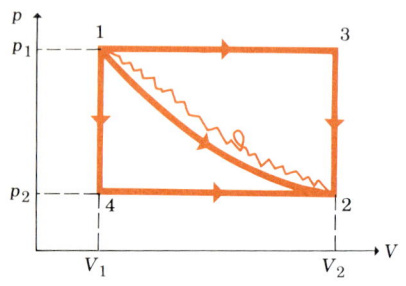

18-3 Work depends on the path.

In an expansion, $V_2 > V_1$ and W is positive. At constant T,

$$p_1 V_1 = p_2 V_2, \quad \text{or} \quad \frac{V_2}{V_1} = \frac{p_1}{p_2},$$

and the isothermal work may be expressed also in the form

$$W = nRT \ln \frac{p_1}{p_2}. \tag{18-5}$$

On the p-V diagram in Fig. 18–3 an initial state 1 (characterized by pressure p_1 and volume V_1) and a final state 2 (characterized by pressure p_2 and volume V_2) are represented by the two points 1 and 2. There are many ways in which the system may be taken from 1 to 2. For example, the pressure may be kept constant from 1 to 3 and then the volume kept constant from 3 to 2, in which case the work done is equal to the area under the line $1 \to 3$. Another possibility is the path $1 \to 4 \to 2$, in which case the work is the area under the line $4 \to 2$. The jagged line and the smooth curve from 1 to 2 represent other possibilities, and the work done is different in each. We can see, therefore, that *the work depends not only on the initial and final states but also on the intermediate states*, i.e., *on the path*.

18-3 Heat in volume changes

As we have seen in Chapter 15, heat is energy transferred to or from a system by virtue of a temperature difference between the system and its surroundings. Heat is regarded as positive when it enters a system and negative when it leaves. The performance of work and the transfer of heat are *means of energy transfer,* that is, processes whereby the energy of a system may be increased or decreased.

An important aspect of energy relations is illustrated by the situation of Fig. 18–4. In Fig. 18–4a, a quantity of gas is contained in a cylinder with a piston, with initial volume of two liters, and maintained at a temperature of 300 K by an electric stove. If the pressure exerted on the piston is only a trifle smaller than the gas pressure, the gas expands slowly, and heat flows from the stove to the gas, maintaining the temperature of the gas at the constant value 300 K. Suppose that the gas expands in this slow, controlled, isothermal manner until its volume

18-4 (a) Slow, controlled, isothermal expansion of a gas from an initial state 1 to a final state 2. (b) Rapid, uncontrolled expansion of the same gas starting at the same state 1 and ending at the same state 2.

becomes, say, seven liters. A finite amount of heat is absorbed by the gas during this process.

Figure 18–4b shows a vessel surrounded by adiabatic walls and divided into two compartments (the lower one of volume two liters, the upper, five liters) by a thin, breakable partition. In the lower compartment the same gas has the same initial volume and temperature as in (a). Suppose the partition is broken and the gas in (b) undergoes a rapid, uncontrolled expansion into the vacuum above, where it encounters no movable piston and therefore exerts no external force. Not only is no work done in this expansion, but also no heat passes through the adiabatic walls. The final volume is seven liters as in (a). The rapid, uncontrolled expansion of a gas into a vacuum is called a *free expansion* and will be discussed later in this chapter. Experiments have indicated that very little temperature change takes place, provided the initial gas pressure is less than a few atmospheres, and therefore the final *state* of the gas is the same as the final state in (a). But the states traversed by the gas (the pressures and volumes) while proceeding from state 1 to state 2 are entirely different, so that (a) and (b) represent *two different paths* connecting the *same states* 1 and 2. During path (a) heat is transferred and work is done; during path (b) *no* heat is transferred and no work is done. Heat, like work, *depends not only on the initial and final states but also on the path*.

Thus it would not be correct to say that a body *contains* a certain amount of heat. For suppose we assigned an arbitrary value to "the heat in a body" in some standard reference state. The "heat in the body" in some other state would then equal the "heat" in the reference state plus the heat added when the body is carried to the second state. But the heat added depends entirely on the path by which we go from one state to the other, and since there are an infinite number of paths that might be followed, there are an infinite number of values that might equally well be assigned to the "heat in the body" in the second state. Since it is not possible to assign any one value to the "heat in the body," we conclude that this concept is meaningless.

18–4 Internal energy

The transfer of heat and the performance of work constitute two means of adding energy to or subtracting energy from a system. Once the transfer of energy has occurred, the system is said to have undergone a change in *internal energy*.

Suppose a system changes from state 1 to state 2 along a definite path and that the heat Q absorbed by the system and the work W done by it are measured. Expressing both Q and W in the same units (either thermal or mechanical), we may then calculate the difference $Q - W$. If now we do the same thing over again for many different paths (between the same states 1 and 2), we find experimentally that $Q - W$ *is the same for all paths connecting 1 and 2*.

But Q is the energy that has been added to a system by the transfer of heat and W is equal to the energy that has been extracted from the system by the performance of work. The difference $Q - W$, therefore, must represent the internal energy change of the system. It follows that *the internal energy change of a system is independent of the path,* and is therefore equal to the energy of the system in state 2 minus the energy in state 1. Internal energy is usually denoted by the symbol U. Thus if U_1 is the internal energy in state 1 and U_2 the internal energy in state 2, then

$$U_2 - U_1 = \Delta U = Q - W. \qquad (18\text{-}6)$$

If some arbitrary value is assigned to the internal energy in some standard reference state, its value in any other state is uniquely defined, since $Q - W$ is the same for all processes connecting the states. Equation (18-6) is known as *the first law of thermodynamics*. In applications of the law in this form, it must be remembered that (1) all quantities must be expressed in the same units, (2) Q is positive when the heat goes *into* the system, and (3) W is positive when the system expands and does positive work on its surroundings.

Rewriting Eq. (18-6) in the form

$$Q = U_2 - U_1 + W = \Delta U + W, \qquad (18\text{-}7)$$

we may say that when a quantity of heat Q enters a system during a process, some of it (ΔU) remains in the system as increased internal energy, while the remainder (W) leaves the system again in the form of work done by the system against its surroundings.

Internal energy can be interpreted in terms of microscopic mechanical energy, that is, kinetic and potential energies of individual molecules in a material. From the thermodynamic standpoint, however, this is not necessary. Equation (18-6) is the *definition* of the internal energy of a system or, more precisely, of the *change* in its internal energy in any process. As with other forms of energy, only *differences* in internal energy are defined, and not absolute values.

If a system is carried through a process that eventually returns it to its initial state (a *cyclic* process), then

$$U_2 = U_1 \quad \text{and} \quad Q = W.$$

Thus, although net work W may be done by the system in the process, energy has not been created, since an equal amount of energy must have flowed into the system as heat Q.

An *isolated* system is one that does no external work and into which there is no flow of heat. Then, for any process taking place in such a system,

$$W = Q = 0$$

and

$$U_2 - U_1 = 0 \quad \text{or} \quad \Delta U = 0.$$

That is, *the internal energy of an isolated system remains constant*. This is the most general statement of the *principle of conservation of energy*. The internal energy of an *isolated* system cannot be changed by any process (mechanical, electrical, chemical, nuclear, or biological) taking place *within* the system. The energy of a system can be changed only by a flow of heat across its boundary, or by the performance of work. If either of these takes place, the system is no longer isolated. The increase in energy of the system is then equal to the energy flowing in as heat, minus the energy flowing out as work.

18-5

Example A thermodynamic process is shown in Fig. 18-5. In process ab, 600 J of heat are added, and in process bd 200 J of heat are added. Find (a) the internal energy change in process ab; (b) the internal energy change in process abd; and (c) the total heat added in process acd.

Solution

a) There is no volume change in process ab, so $W = 0$ and $\Delta U = Q = 600$ J.

b) Process bd occurs at constant pressure, so the work done by the system is
$$W = p(V_2 - V_1) = (8 \times 10^4 \text{ Pa})(5 \times 10^{-3} \text{ m}^3 - 2 \times 10^{-3} \text{ m}^3)$$
$$= 240 \text{ J.}$$
Thus the total work for abd is $W = 240$ J and the total heat is $Q = 800$ J. Equation (18-6) gives
$$\Delta U = Q - W = 800 \text{ J} - 240 \text{ J} = 560 \text{ J.}$$

c) Because ΔU is independent of path, the internal energy change is the same for path acd as for abd, that is, 560 J. The total work for path acd is
$$W = (3 \times 10^4 \text{ Pa})(5 \times 10^{-3} \text{ m}^3 - 2 \times 10^{-3} \text{ m}^3) = 90 \text{ J.}$$
Equation (18-6) or (18-7) then gives
$$Q = \Delta U + W = 560 \text{ J} + 90 \text{ J} = 650 \text{ J.}$$

We see that although ΔU is the same for abd and acd, W and Q are quite different for the two processes.

In the next several sections, several general classes of thermodynamic processes are introduced, and the heat, work, and internal energy considerations appropriate for each are discussed.

18-5 Adiabatic process

A process in which no heat enters or leaves a system is called an *adiabatic process*. For every adiabatic process, $Q = 0$. This prevention of heat flow may be accomplished either by surrounding the system with a thick layer of heat-insulating material (such as cork, asbestos, firebrick, or styrofoam), or by performing the process quickly. The flow of heat requires finite time, so any process performed quickly enough will be practically adiabatic. Applying the first law to an adiabatic process, we get

$$U_2 - U_1 = \Delta U = -W \quad \text{(adiabatic process)}. \tag{18-8}$$

Thus, the change in the internal energy of a system, in an adiabatic process, is equal *in magnitude* to the work done by the system. If the work W is negative, as when a system is compressed, then $-W$ is positive, U_2 is greater than U_1, and the internal energy of the system increases. If W is positive, as when a system expands, the internal energy of the system decreases. An increase of internal energy is usually (but not always) accompanied by a rise in temperature, and a decrease in internal energy by a temperature drop.

The compression of the mixture of gasoline vapor and air that takes place during the compression stroke of a gasoline engine is an example of an approximately adiabatic process involving a temperature rise. The expansion of the combustion products during the power stroke of the engine is an approximately adiabatic process involving a temperature decrease. Adiabatic processes, therefore, play a very important role in mechanical engineering.

18-6 Isochoric process

When a substance undergoes a process in which the volume remains unchanged ($\Delta V = 0$), the process is called *isochoric*. The rise of pressure and temperature produced by a flow of heat into a substance contained in a rigid container of fixed volume is an example of an isochoric process. When the volume does not change, no work is done; $W = 0$, and therefore, from the first law,

$$U_2 - U_1 = \Delta U = Q \quad \text{(isochoric process)}. \tag{18-9}$$

All the added heat has served to increase the internal energy. The very sudden increase of temperature and pressure accompanying the explosion of gasoline vapor and air in a gasoline engine may be treated mathematically as though it were an isochoric addition of heat.

18-7 Isothermal process

A process taking place at constant temperature is said to be *isothermal*. In order for the temperature to remain constant the changes in the pressure and volume must be carried out very slowly so that the state approximates thermal equilibrium very closely throughout the process. In general, *none* of the quantities Q, W, or $U_2 - U_1$ is zero.

There are a few special cases in which the internal energy of the system depends only on temperature, not on pressure or volume. An ideal gas and an ideal paramagnetic crystal are two examples. When such a system undergoes an isothermal process, its internal energy does not change, and therefore, $Q = W$. This relation does *not* hold for systems other than those mentioned.

18-8 Isobaric process

A process taking place at constant pressure is called an *isobaric process*. When water enters the boiler of a steam engine and is heated to its boiling point and vaporized, and then the steam is superheated, all these processes take place isobarically. Such processes play an important role in mechanical engineering and also in chemistry.

As discussed in Sec. 18-2,

$$W = p(V_2 - V_1) \quad \text{(isobaric process)}. \tag{18-10}$$

A simple example is the vaporization of a mass m of liquid at constant pressure and temperature. If V_L is the volume of liquid and V_V is the volume of vapor, the work done by the system in expanding from V_L to V_V at constant pressure p is

$$W = p(V_V - V_L).$$

The heat absorbed per unit mass is the heat of vaporization L. Hence,

$$Q = mL.$$

From the first law,

$$U_V - U_L = mL - p(V_V - V_L). \tag{18-11}$$

Example One gram of water (1 cm³) becomes 1671 cm³ of steam when boiled at a pressure of 1 atm. The heat of vaporization at this pressure is 2256 J·g⁻¹. Compute the external work and the increase in internal energy.

Solution

$$\begin{aligned} W = p(V_V - V_L) &= (1.013 \times 10^5 \text{ Pa})(1671 \times 10^{-6} \text{ m}^3 \\ &\quad - 1 \times 10^{-6} \text{ m}^3) \\ &= 169 \text{ J}. \end{aligned}$$

From Eq. (18-11),

$$\begin{aligned} U_V - U_L = mL - W &= 2256 \text{ J} - 169 \text{ J} \\ &= 2087 \text{ J}. \end{aligned}$$

That is, when one gram of water vaporizes, 2256 J of heat are added. Of this added energy, 2087 J remains in the system as an increase in internal energy, and the remaining 169 J leaves the system again as it does work against the surroundings.

18-9 Throttling process

A *throttling process* is one in which a fluid, originally at a constant high pressure, seeps through a porous wall or a narrow opening (needle valve or "throttling" valve) into a region of constant lower pressure without a transfer of heat taking place. In Fig. 18-6a, a fluid is discharged from a

(a) System with two pressures

(b) Fluid at constant high pressure p_1

(c) Fluid at constant lower pressure p_2

18-6 Throttling process.

pump at a high pressure, then passes through a throttling valve into a pipe that leads directly to the intake or low-pressure side of the pump. Every successive element of fluid undergoes the throttling process in a continuous stream.

Consider any element of fluid enclosed between the piston and throttling valve of Fig. 18-6b. Suppose this piston moves toward the right and another piston on the other side of the valve moves to the right also, at such rates that the pressure on the left remains at a constant high value, p_1, and that on the right at a constant lower value p_2. After all the fluid has been forced through the valve, the final state is that of Fig. 18-6c.

The net work done in this process is the difference between the work done in forcing the righthand piston out and the work done in forcing the lefthand piston in. Let V_1 and V_2 be the initial and final volumes. Since the low-pressure fluid changes in volume from zero to V_2 at the constant pressure p_2, the work is

$$p_2(V_2 - 0),$$

and since the high-pressure fluid changes in volume from V_1 to zero at the constant high pressure p_1, the work is

$$p_1(0 - V_1).$$

The net work W done by the system is, therefore,

$$W = p_2 V_2 - p_1 V_1.$$

Since the process is adiabatic, $Q = 0$; hence, from the first law,

$$U_2 - U_1 = 0 - (p_2 V_2 - p_1 V_1)$$

or

$$U_2 + p_2 V_2 = U_1 + p_1 V_1. \tag{18-12}$$

This result is of great importance in steam engineering and in refrigeration. The sum $U + pV$, called the *enthalpy*, is tabulated for steam and for many refrigerants. The throttling process plays an essential role in the action of a refrigerator, since this is the process that brings about the drop in temperature needed for refrigeration. Liquids that are about to evaporate (saturated liquids) always undergo a drop in temperature and partial vaporization as a result of a throttling process. Gases, however, may undergo either a temperature rise or a drop, depending on the initial temperature and pressure and on the final pressure.

18-10 Differential form of the first law

Up to this point we have used the first law of thermodynamics only in its *finite* form

$$\Delta U = U_2 - U_1 = Q - W,$$

referring to a process in which states 1 and 2 differ in pressure, volume, and temperature by a finite amount. Now suppose states 1 and 2 differ only infinitesimally. Then a small amount of heat dQ is transferred, a small amount of work dW is done, and the corresponding small internal energy change is dU. In this case the first law becomes

$$dU = dQ - dW. \tag{18-13}$$

In systems such as we have been considering, the work is given by $dW = p\,dV$, and the first law in differential form becomes

$$dU = dQ - p\,dV. \tag{18-14}$$

This form will be used several times in the following sections.

18-11 Internal energy of an ideal gas

Imagine a thermally insulated vessel with rigid walls, divided into two compartments by a partition. Suppose that there is a gas in one compartment and that the other is empty. If the partition is removed, the gas undergoes what is known as a *free expansion,* in which no work is done and no heat is transferred. From the first law, since both Q and W are zero, it follows that the internal energy remains unchanged during a free expansion.

Whether or not the *temperature* of a gas changes during a free expansion is an important question, for it provides information concerning the dependence of internal energy on other quantities. If the temperature should change while the internal energy stays the same, one would have to conclude that the internal energy depends on *both* the temperature and the volume, or both the temperature and the pressure, but certainly not on the temperature alone. If, on the other hand, T remains unchanged during a free expansion in which we know U remains unchanged then the only conclusion that is admissible is that *U is a function of T only,* not of p or V.

Experiment has shown that the internal energy of a real gas does depend to some extent on the pressure or volume as well as on the temperature. At sufficiently low pressures and high temperatures, however, the behavior approaches more and more closely that of an ideal gas, and in this limit the internal energy is found to depend *only* on temperature. In fact, it can be proved that for any system whose equation of state is the ideal-gas equation, the internal energy *must* depend only on temperature. The proof requires the second law of thermodynamics, to be introduced in the next chapter, and it will not be given here. But the fact that the internal energy of an ideal gas depends only on its temperature is an important aspect of the ideal-gas model, and we shall use this property several times in the following sections.

18-12 Heat capacities of an ideal gas

The temperature of a substance may be changed under a variety of conditions. The volume may be kept constant, or the pressure may be kept constant, or both may be allowed to vary in some definite manner. The amount of heat per mole necessary to cause unit rise of temperature is different in each case. In other words, a substance has many different molar heat capacities. Two, however, are particularly useful, namely, those at *constant volume* and at *constant pressure*. We shall denote these by C_v and C_p, respectively.

In a constant-volume temperature change, no work is done, and the change in internal energy equals the heat added. In a constant-pressure temperature change, on the other hand, the volume must increase—otherwise the pressure could not remain constant—and so the gas does work. Thus the heat input for the constant-pressure process must be

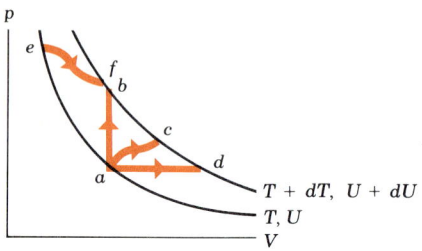

18–7 The change in internal energy of an ideal gas is the same in all processes between two given temperatures.

greater, for a given temperature change, than for the constant-volume process since, in the former, additional energy must be supplied to account for the work done in the expansion. There is in fact a simple relation between the two molar heat capacities of an ideal gas, at constant volume and at constant pressure, and it will be developed below.

Figure 18–7 shows two isotherms (lines of constant temperature) for an ideal gas, one representing possible states of the system at temperature T, the other at $T + dT$. Since the internal energy of an ideal gas depends only on the temperature, it is constant if the temperature is constant and the isotherms are also curves of *constant internal energy*. The internal energy therefore has a constant value U at every point on the isotherm at temperature T, and a constant value $U + dU$ at every point on the isotherm at $T + dT$. It follows that the *change* in internal energy, dU, is the same in all processes in which the gas is taken from *any* point on one isotherm to *any* point on the other. Thus dU is the same for *all* the processes ab, ac, ad, and ef, in Fig. 18–7.

Consider first the process ab, at constant volume. To carry out such a process, the gas at temperature T is enclosed in a rigid container and brought in contact with a body at a slightly higher temperature $T + dT$. There is a flow of heat dQ into the gas, and by definition of the molar heat capacity at constant volume, C_v,

$$dQ = nC_v \, dT \quad \text{(constant volume)}. \tag{18-15}$$

The pressure of the gas increases during this process, but no work is done, since the volume is constant. Hence $dW = 0$ and, from the first law (in differential form), $dU = dQ - dW$, we have

$$dU = nC_v \, dT. \tag{18-16}$$

But dU is the same for *all* processes between the isotherms in Fig. 18–7, so Eq. (18–16) gives the change in internal energy in *all* the processes in Fig. 18–7, *even if they are not at constant volume*.

Now consider the process ad in Fig. 18–7, at constant *pressure*. To carry out such a process the gas might be enclosed in a cylinder with a piston, acted on by a constant external pressure, and brought in contact with a body at a temperature $T + dT$. As heat flows into the gas it expands at constant pressure and does work. By definition of the molar heat capacity at constant pressure, C_p, the heat dQ flowing into the gas is

$$dQ = nC_p \, dT \quad \text{(constant pressure)}. \tag{18-17}$$

The work dW is

$$dW = p \, dV.$$

But from the equation of state, since p is constant,

$$pV = nRT, \quad p \, dV = nR \, dT,$$

so

$$dW = nR \, dT.$$

Then from the first law,

$$dU = dQ - dW,$$

$$nC_v \, dT = nC_p \, dT - nR \, dT,$$

or

$$C_p - C_v = R. \qquad (18\text{-}18)$$

As predicted, the molar heat capacity of an ideal gas at constant pressure is *greater* than that at constant volume; the difference is the universal gas constant R. Of course, R must be expressed in the same units as C_p and C_v, such as J·mol^{-1}·K^{-1}.

Although Eq. (18-18) was derived for an ideal gas, it is very nearly true for many real gases at moderate pressures. Measured values of C_p and C_v are given in Table 18-1 for some real gases at low pressures; the difference in most cases is approximately 8.31 J·mol^{-1}·K^{-1}.

Table 18-1 Molar heat capacities of gases at low pressure

Type of gas	Gas	C_p, J·mol^{-1}·K^{-1}	C_v, J·mol^{-1}·K^{-1}	$C_p - C_v$	$\gamma = \dfrac{C_p}{C_v}$
Monatomic	He	20.78	12.47	8.31	1.67
	A	20.78	12.47	8.31	1.67
Diatomic	H_2	28.74	20.42	8.32	1.41
	N_2	29.07	20.76	8.31	1.40
	O_2	29.41	21.10	8.31	1.40
	CO	29.16	20.85	8.31	1.40
Polyatomic	CO_2	36.94	28.46	8.48	1.30
	SO_2	40.37	31.39	8.98	1.29
	H_2S	34.60	25.95	8.65	1.33

In the last column of Table 18-1 are listed the values of the ratio C_p/C_v, denoted by the Greek letter γ (gamma).

$$\gamma = \frac{C_p}{C_v}. \qquad (18\text{-}19)$$

It is seen that γ is 1.67 for monatomic gases, and is about 1.40 for diatomic gases. There is no simple regularity for polyatomic gases. Table 18-1 also shows that the molar heat capacity of a gas seems to be related to its molecular structure; all the *monatomic* gases have approximately equal values of C_v, and all the *diatomic* gases have about equal values. This is not an accident; in Chapter 20 we shall return to a more detailed theoretical study of heat capacities of gases.

Solids and liquids also expand when heated, if free to do so, and hence perform work. The coefficients of volume expansion of solids and liquids are, however, so much smaller than those of gases that the work is small. Usually for solids and liquids C_p is greater than C_v, but the difference is small and is not expressible as simply as that for a gas. Because of the large stresses set up when solids or liquids are heated and *not* allowed to expand, most heating processes involving them take place at constant pressure, and hence C_p (rather than C_v) is the quantity usually measured for a solid or liquid.

Finally, we emphasize once more that for an ideal gas the internal energy change in *any* process is given by Eq. (18-16),

$$dU = nC_v\, dT,$$

whether the volume is constant or not, because the internal energy of an ideal gas depends only on its temperature. If the volume is *not* constant, then work is done, and the heat added is *not* equal to $nC_v\,dT$; but this expression is *still* equal to the internal energy change dU. This statement is, however, *not* true for any system other than an ideal gas and a few other very specific systems.

18-13 Adiabatic process for an ideal gas

Any process in which there is no flow of heat into or out of a system is called *adiabatic*. For a truly adiabatic process the system would have to be surrounded by a perfect heat insulator, or the surroundings of the system kept always at the same temperature as the system. However, if a process such as compression or expansion of a gas is carried out very rapidly, it is *nearly* adiabatic, since the flow of heat into or out of the system requires finite time. Thus the compression stroke of a gasoline or diesel engine is approximately adiabatic.

For an adiabatic process, $Q = 0$ and the first law becomes

$$\Delta U = -W. \tag{18-20}$$

Hence when an ideal gas expands adiabatically it does *positive* work, the internal energy *decreases,* and the temperature also decreases. Conversely, compressing a gas adiabatically increases the internal energy and the temperature.

In Fig. 18-8, the black curves represent adiabatic processes on a pV diagram. The colored curves represent isotherms for comparison. An adiabatic curve, at any point, has a somewhat steeper slope than the isothermal curve passing through the same point. That is, as one follows along an adiabatic from right to left (compression process), the curve continually cuts across isotherms of higher and higher temperatures, in agreement with the fact that the temperature continually increases in an adiabatic compression.

It is useful to have relations between volume and temperature, or volume and pressure, for an adiabatic process. We consider first a *small* change of state in which the temperature changes by dT and the volume by dV. Equation (18-16) gives the internal energy change for *any* process for an ideal gas, adiabatic or not, so we have $dU = nC_v\,dT$. Also, the work done by the gas during the process is given by $W = p\,dV$. Then, since $dU = -W$ for an adiabatic process, we have

$$nC_v\,dT = -p\,dV. \tag{18-21}$$

To obtain a relation containing only the variables T and V, we may eliminate p using the ideal-gas equation in the form $p = nRT/V$. Substituting this in Eq. (18-21) and rearranging, we obtain

$$\frac{dT}{T} + \frac{R}{C_v}\frac{dV}{V} = 0,$$

or, since $R/C_v = (C_p - C_v)/C_v = C_p/C_v - 1 = \gamma - 1$,

$$\frac{dT}{T} + (\gamma - 1)\frac{dV}{V} = 0. \tag{18-22}$$

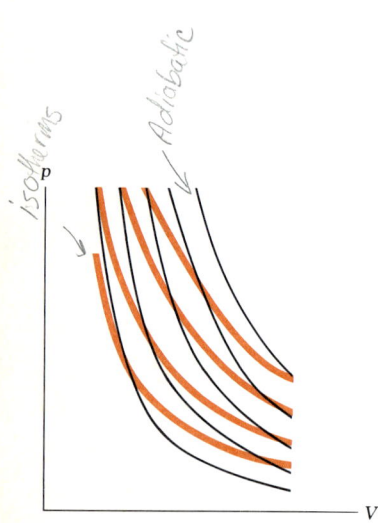

18-8 Adiabatic curves (black lines) and isothermal curves (colored lines).

This gives the relation between dT and dV for an infinitesimal adiabatic process; for a finite change we integrate both sides of Eq. (18-22), obtaining

$$\ln T + (\gamma - 1) \ln V = \text{constant},$$
$$\ln T + \ln V^{\gamma-1} = \text{constant},$$
$$\ln [TV^{\gamma-1}] = \text{constant},$$

and finally

$$TV^{\gamma-1} = \text{constant}. \qquad (18\text{-}23)$$

Thus for an initial state (T_1, V_1) and a final state (T_2, V_2),

$$T_1 V_1^{\gamma-1} = T_2 V_2^{\gamma-1}. \qquad (18\text{-}24)$$

In application of Eqs. (18-23) and (18-24) it is important to remember that the T's must be *absolute* temperatures (Kelvin or Rankine).

Equation (18-23) may also be converted into a relation between pressure and volume by eliminating T, using the ideal-gas equation in the form $T = pV/nR$. Since n and R are constant, we obtain

$$\left(\frac{pV}{nR}\right) V^{\gamma-1} = \text{constant},$$

or, since n and R are constants,

$$pV^\gamma = \text{constant}. \qquad (18\text{-}25)$$

Thus for an initial state (p_1, V_1) and a final state (p_2, V_2),

$$p_1 V_1^\gamma = p_2 V_2^\gamma. \qquad (18\text{-}26)$$

Example The compression ratio of a certain diesel engine is 15. This means that air in the cylinders is compressed to 1/15 of its initial volume. If the initial pressure is 1.0×10^5 Pa and the initial temperature is 27°C (= 300 K), find the final pressure and temperature after compression. Air is mostly a mixture of oxygen and nitrogen, and $\gamma = 1.40$.

Solution From Eq. (18-24),

$$T_2 = T_1 \left(\frac{V_1}{V_2}\right)^{\gamma-1} = (300 \text{ K})(15)^{0.40} = 886 \text{ K} = 613°\text{C}.$$

From Eq. (18-26),

$$p_2 = p_1 \left(\frac{V_1}{V_2}\right)^\gamma = (1.0 \times 10^5 \text{ Pa})(15)^{1.40}$$
$$= 44.3 \times 10^5 \text{ Pa} \simeq 44 \text{ atm}.$$

If the compression had been isothermal, the final pressure would have been 15 atm, but since the temperature also increases, the final pressure is much greater. The high temperature attained during compression causes the fuel to ignite spontaneously without the need for spark plugs, when it is injected into the cylinders at the end of the compression stroke.

Because $W = -\Delta U$ for an adiabatic process, it is easy to calculate the *work* done by an ideal gas during an adiabatic process. If the initial and final temperatures are known, we have simply

$$W = nC_v(T_1 - T_2); \qquad (18\text{-}27)$$

or, if the pressures or volumes are known, we may use $pV = nRT$ to obtain

$$W = \frac{C_v}{R}(p_1V_1 - p_2V_2) = \frac{1}{\gamma - 1}(p_1V_1 - p_2V_2). \qquad (18\text{-}28)$$

We note that if the process is an expansion, the temperature drops, T_1 is greater than T_2, p_1V_1 is greater than p_2V_2, and the work is *positive,* as we should expect. If it is a compression, the work is negative.

Example 2 In the above example, how much work does the gas do during the compression if the initial volume of the cylinder is $1.01 = 1.0 \times 10^{-3}$ m³?

Solution We may determine the number of moles and then use Eq. (18-27), or we may use Eq. (18-28). In the first case we have

$$n = \frac{p_1V_1}{RT_1} = \frac{(1.0 \times 10^5 \text{ Pa})(1.0 \times 10^{-3} \text{ m}^3)}{(8.314 \text{ J}\cdot\text{mol}^{-1}\cdot\text{K}^{-1})(300 \text{ K})} = 0.040 \text{ mol}.$$

C_v for air is 20.8 J·mol⁻¹·K⁻¹;

$$W = nC_v(T_1 - T_2)$$
$$= (0.040 \text{ mol})(20.8 \text{ J}\cdot\text{mol}\cdot\text{K}^{-1})(300 \text{ K} - 886 \text{ K})$$
$$= -488 \text{ J}.$$

In the second,

$$W = \frac{1}{1.40 - 1}\left[(1.0 \times 10^5 \text{ Pa})(1.0 \times 10^{-3} \text{ m}^3)\right.$$
$$\left. - (44.3 \times 10^5 \text{ Pa})\left(\frac{1.0}{15} \times 10^{-3} \text{ m}^3\right)\right]$$
$$= -488 \text{ J}.$$

The work is negative because the gas is compressed.

In the above analysis we have used the ideal-gas equation of state, which holds only when the state of the gas changes slowly enough so that at each step the pressure and temperature are *uniform* throughout the gas. Thus the validity of our results is limited to situations where the process is rapid enough to prevent appreciable heat exchange with the surroundings, yet slow enough so the system does not depart very much from thermal and mechanical equilibrium.

Questions

18-1 It isn't correct to say that a body contains a certain amount of heat, yet a body can transfer heat to another body. How can a body give away something it doesn't have in the first place?

18-2 Discuss the application of the first law of thermodynamics to a mountaineer who eats food, gets warm and sweats a lot during the climb, and does a lot of mechanical work raising himself to the summit. What about the descent? One also gets warm during the descent. Is the source of this energy the same as during the ascent?

18-3 How can you cool a room (i.e., take heat out of it) by adding energy to it in the form of electric energy supplied to an air conditioner?

18-4 If you are told the initial and final states of a system and the associated change in internal energy, can you determine whether the internal energy change was due to work or to heat transfer?

18-5 Household refrigerators always have arrays or coils of tubing, usually at the back or bottom. When the refrigerator is running, the tubing becomes quite hot. Where does the heat come from?

18-6 There are a few materials that contract when heated, such as water between 0°C and 4°C. Would you expect C_p for such a material to be greater or less than C_v?

18-7 When water melts (decreasing its volume), is the internal energy change greater or less than the heat added?

18-8 When one drives a car in cool, foggy weather, ice sometimes forms in the throat of the carburetor, even though the outside air temperature is above freezing. Why?

18-9 On a warm summer day a large cylinder of compressed gas (propane or butane) was used to supply several large gas burners at a cookout. After a while, frost formed on the outside of the tank. Why?

18-10 Air escaping from an air hose at a gas station always feels cold. Why?

18-11 The prevailing winds blow across the central valley of California and up the western slopes of the Sierra Nevada mountains. They cool as they reach the slopes, and the precipitation there is much greater than in the valley. But what makes them cool?

18-12 Applying the same considerations as in Question 18-11, can you explain why Death Valley, on the opposite side of the Sierra from the central valley, is so hot and dry?

18-13 In the situation of Question 18-11, during certain seasons the wind blows in the opposite direction. Although the mountains are cool, the wind in the valley (called the "Santa Anna," after the notorious Mexican general) is always very hot. What heats it? A similar phenomenon in the Alps is called the "Foehn"; by local legend it is blamed for irrational behavior in man and beast, and has even been used as a defense in murder trials.

18-14 When a gas expands adiabatically, it does work on its surroundings. But if there is no heat input to the gas, where does the energy come from?

18-15 In a constant-volume process, $dU = nC_v \, dT$, but in a constant-pressure process it is *not* true that $dU = nC_p \, dT$. Why not?

18-16 Since C_v is defined with specific reference to a constant-volume process, how can it be correct that for an ideal gas $dU = nC_v \, dT$ even when the volume is not constant?

Problems

18-1 A combustion experiment is performed by burning a mixture of fuel and oxygen in a constant-volume "bomb" surrounded by a water bath. During the experiment the temperature of the water is observed to rise. Regarding the mixture of fuel and oxygen as the system:

a) Has heat been transferred?
b) Has work been done?
c) What is the sign of ΔU?

18-2 A liquid is irregularly stirred in a well-insulated container and thereby undergoes a rise in temperature. Regarding the liquid as the system:

a) Has heat been transferred?
b) Has work been done?
c) What is the sign of ΔU?

18-3 A resistor immersed in running water carries an electric current. Consider the resistor as the system under consideration.

a) Is there a flow of heat into the resistor?
b) Is there a flow of heat into the water?
c) Is work done?
d) Assuming the state of the resistor to remain unchanged, apply the first law to this process.

18-4 In a certain process, 500 cal of heat are supplied to a system, and at the same time 100 J of work are done on the system. What is the increase in the internal energy of the system?

18-5 A piece of ice falls from rest into a lake at 0°C, and one-half of one percent of the ice melts. Compute the minimum height from which the ice falls.

18-6 What must be the initial velocity of a lead bullet at a temperature of 25°C, so that the heat developed when it is brought to rest shall be just sufficient to melt it?

18-7 In a certain process, 200 Btu are supplied to a system and at the same time the system expands against a constant external pressure of 100 lb·in^{-2}. The internal energy of the system is the same at the beginning and end of the process. Find the increase in volume of the system.

18-8 An inventor claims to have developed an engine which takes in 100,000 Btu from its fuel supply, wastes 25,000 Btu in the exhaust, and delivers 25 kWh of mechanical work. Do you advise investing money to put this engine on the market?

18-9 A vessel with rigid walls and covered with asbestos is divided into two parts by an insulating partition. One part contains a gas at temperature T and pressure p. The other part contains a gas at temperature T' and pressure p'. The partition is removed. What conclusion may be drawn by applying the first law of thermodynamics?

18-10 A mixture of hydrogen and oxygen is enclosed in a rigid insulating container and exploded by a spark. The temperature and pressure both increase considerably. Neglecting the small amount of energy provided by the spark itself, what conclusion may be drawn by applying the first law of thermodynamics?

18-11 When water is boiled under a pressure of 2 atm, the heat of vaporization is 2.20×10^6 J·kg^{-1} and the boiling point is 120°C. At this pressure, one kilogram of water has a volume of 10^{-3} m^3, and one kilogram of steam a volume of 0.824 m^3.

a) Compute the work done when one kilogram of steam is formed at this temperature.

b) Compute the increase in internal energy.

18-12 When a system is taken from state a to state b, in Fig. 18-9, along the path acb, 80 J of heat flow into the system, and 30 J of work are done.

a) How much heat flows into the system along path adb if the work is 10 J?

b) When the system is returned from b to a along the curved path, the work is 20 J. Does the system absorb or liberate heat, and how much?

c) If $U_a = 0$ and $U_d = 40$ J, find the heat absorbed in the processes ad and db.

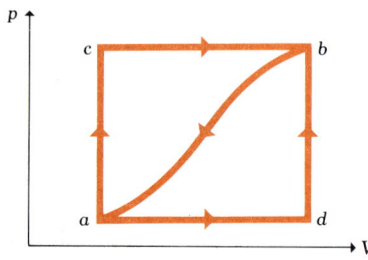

Figure 18-9

18-13 The van der Waals equation of state, an approximate representation of the behavior of gases at high pressure, is

$$\left(p + \frac{an^2}{V^2}\right)(V - nb) = nRT,$$

where a and b are constants having different values for different gases. (If $a = b = 0$, this is the ideal-gas equation.) Calculate the work done by a van der Waals gas in an expansion from V_1 to V_2.

18-14 An ideal gas undergoes an isothermal expansion at 0°C from 0.010 m^3 to 0.200 m^3. For 5 mol of gas, compute the work done, the heat added, and the change in internal energy.

18-15 Consider the isothermal compression of 0.10 mol of an ideal gas at $T = 0$°C. The initial pressure is 1 atm and the final volume is $\frac{1}{5}$ the initial volume.

a) Determine the work required.

b) Does the gas exchange heat with its surroundings? If so, how much and in which direction?

c) What is the change in internal energy?

18-16 In a cylinder, 2.0 mol of an ideal monatomic gas initially at 1.0×10^6 Pa and 300 K expands until its volume doubles. Compute the work done if the expansion is

a) isothermal,

b) adiabatic,

c) isobaric.

d) Show each process on a pV-diagram.

e) In which case is the heat transfer greatest? Least?

f) In which case is the internal energy change greatest? Least?

18-17 A steel cylinder of cross-sectional area 0.01 m^2 contains 1×10^{-2} m^3 of glycerin. The cylinder is equipped with a tightly fitting piston that supports a load of 3×10^4 N. The temperature of the system is increased from 20°C to 70°C. Neglect the expansion of the steel cylinder. Find

a) the increase in volume of the glycerin,

b) the mechanical work of the 3×10^4 N force,

c) the amount of heat added to the glycerin [C_p for glycerin is 0.58 cal·g^{-1}·(C°)$^{-1}$], and

d) the change in internal energy of the glycerin.

18-18 The volume of an ideal gas is 4 l, the pressure is 2 atm, and the temperature is 300 K. The gas first expands at constant pressure to twice its original volume; it is then compressed isothermally to its original volume, and finally cooled at constant volume to its original pressure.

a) Show the process in a pV-diagram.

b) Compute the temperature during the isothermal compression.

c) Compute the work done by the gas during the expansion.

d) Compute the maximum pressure attained in the process.

18-19 A cylinder contains 1 mol of oxygen gas at a temperature of 27°C. The cylinder is provided with a frictionless piston, which maintains a constant pressure of 1 atm on the gas. The gas is heated until its temperature increases to 127°C.

a) Draw a diagram representing the process in the pV-plane.
b) How much work is done by the gas in this process?
c) On what is this work done?
d) What is the change in internal energy of the gas?
e) How much heat was supplied to the gas?
f) How much work would have been done if the pressure had been 0.5 atm?

18-20

a) Compare the quantity of heat required to raise the temperature of 1 g of hydrogen through 1 C°, at constant pressure, with that required to raise the temperature of 1 g of water by the same amount.
b) Of the substances listed in Table 18-1, which has the largest *specific* heat capacity?

18-21 Ten liters of air at atmospheric pressure are compressed isothermally to a volume of 2 l, and are then allowed to expand adiabatically to a volume of 10 l. Show the process in a pV-diagram.

18-22 An ideal gas is contained in a cylinder closed with a movable piston. The initial pressure is 1 atm and the initial volume is 1 l. The gas is heated at constant pressure until the volume is doubled, then heated at constant volume until the pressure is doubled, and finally expanded adiabatically until the temperature drops to its initial value. Show the process in a pV-diagram.

18-23 An ideal gas initially at 10 atm and 300 K is permitted to expand adiabatically until its volume doubles. Find the final pressure and temperature if the gas is (a) monatomic; (b) diatomic.

18-24 A gasoline engine takes in air at 20°C and 1 atm, and compresses it adiabatically to $\frac{1}{10}$ the original volume. Find the final temperature and pressure.

18-25 During an adiabatic expansion, the temperature of 0.1 mol of oxygen drops from 30°C to 10°C. How much work does the gas do? How much heat is added?

18-26 A certain ideal gas has $\gamma = 1.33$. Determine the molar heat capacities at constant volume and at constant pressure.

18-27 Compressed air at a pressure of 2×10^6 Pa is used to drive an air engine, which exhausts at a pressure of 2×10^5 Pa. What must be the temperature of the compressed air in order that there may be no possibility of frost forming in the exhaust ports of the engine? Assume the expansion to be adiabatic. [*Note.* Frost frequently forms in the exhaust ports of an air-driven engine. This happens when the moist air is cooled below 0°C by the expansion that takes place in the engine.]

18-28 Initially at a temperature of 140°F, 10 ft³ of air expands at a constant gauge pressure of 20 lb·in⁻² to a volume of 50 ft³, and then expands further adiabatically to a final volume of 80 ft³ and a final gauge pressure of 3 lb·in⁻². Sketch the process in the pV-plane and compute the work done by the air.

18-29 The cylinder of a pump compressing air from atmospheric pressure into a very large tank at 60 lb·in⁻² gauge pressure is 10 in. long.

a) At what position in the stroke will air begin to enter the tank? Assume the compression to be adiabatic.
b) If the air is taken into the pump at 27°C, what is the temperature of the compressed air?

18-30 Two moles of helium are initially at a temperature of 27°C and occupy a volume of 20 l. The helium is first expanded at constant pressure until the volume has doubled, and then adiabatically until the temperature returns to its initial value.

a) Draw a diagram of the process in the pV-plane.
b) What is the total heat supplied in the process?
c) What is the total change in internal energy of the helium?
d) What is the total work done by the helium?
e) What is the final volume?

THE SECOND LAW OF THERMODYNAMICS

19-1 Heat engines

A dominant aspect of a technological society is its ability to use sources of energy other than the muscles of men and animals. Sometimes, as with water power, mechanical energy is directly available. Most of our energy, however, comes from the burning of fossil fuels and from nuclear reactions; both of these supply energy that is transferred as *heat*. Some heat can be used directly for heating buildings, for cooking, and for chemical and metallurgical processing, but to operate a machine or propel a vehicle we require *mechanical* energy. Thus a crucial problem is that of withdrawing heat from a source and converting as much of it as possible into mechanical energy. Such conversion is also fundamental to the functioning of various biological systems in which energy from chemical reactions is converted into mechanical energy of motion or position.

Any device for converting heat to mechanical energy is called a *heat engine*. Most of this chapter is concerned with various aspects of the analysis of heat engines. During this analysis we shall consider a physical principle not previously encountered, the Second Law of Thermodynamics.

In a heat engine a certain quantity of matter undergoes various thermal and mechanical processes such as addition or subtraction of heat, expansion, compression, and change of phase. This material is called the *working substance* of the engine. We consider for simplicity an engine in which the working substance is carried through a *cyclic* process, that is, a sequence of processes in which it eventually returns to its original state. In the condensing type of steam engine used in marine propulsion, the working substance, in this case pure water, is actually used over and over again. Water is evaporated in the boilers at high pressure and temperature, does work in expanding against a piston or in a turbine, is condensed by cooling water from the ocean, and pumped back into the boilers. The refrigerant in a household refrigerator also undergoes a cyclic process. Internal combustion engines and steam locomotives do not carry a system through a closed cycle, but they can be analyzed in terms of cyclic processes that approximate their actual operations.

All these devices absorb heat from a source at a high temperature, perform some mechanical work, and discard some heat at a lower temperature. When a system is carried through a cyclic process, its initial and final internal energies are equal; from the first law, for any number of complete cycles,

$$U_2 - U_1 = 0 = Q - W,$$
$$Q = W.$$

That is, the *net heat flowing into the engine in a cyclic process equals the net work done by the engine*.

In Sec. 18-2 the representation of thermodynamic processes on a pV diagram was discussed, and it was shown that the area under a curve on a pV diagram represents the work done by the system during the corresponding process. A *cyclic* process must be represented by a *closed* curve on a pV diagram, and a little thought will show that the area enclosed by the curve corresponds to the net work done by the system during one cycle. If the curve is traversed in a clockwise sense, the net work is positive; if counterclockwise, negative.

In the operation of a heat engine there are always two bodies capable of either providing or absorbing large quantities of heat without appreciable changes of temperature. Thus the flames and hot gases surrounding the boilers of a steam-turbine system can give up large quantities of heat at a high temperature, and constitute therefore what may be called the *hot reservoir*. Heat transferred between the hot reservoir and the working substance in a heat engine will be represented by the symbol Q_H, where it is understood that a positive value of Q_H means heat entering the working substance. The cooler water or air used to cool and condense the steam after it has passed through the turbine functions as the *cold reservoir*. (The words *hot* and *cold* are, of course, relative.) The heat transferred between the working substance and the cold reservoir will be denoted by Q_C. A negative value of Q_C means that energy is transferred *out of* the working substance, to the cold reservoir.

The energy transformations in a heat engine are conveniently represented schematically by the *flow diagram* of Fig. 19-1. The engine itself is represented by the circle. The heat Q_H supplied to the engine by the hot reservoir is proportional to the cross section of the incoming "pipeline" at the top of the diagram. The cross section of the outgoing pipeline at the bottom is proportional to the magnitude $|Q_C|$ of the heat that is rejected as heat in the exhaust. The branch line to the right represents that portion of the heat supplied which the engine converts to mechanical work, W.

Consider a heat engine operating in a cycle over and over again and let Q_H and Q_C stand for the heats absorbed and rejected by the working substance *per cycle*. The net heat absorbed is

$$Q = Q_H + Q_C = Q_H - |Q_C|, \quad (19\text{-}1)$$

where Q_C is a negative number. The useful output of the engine is the net work W done by the working substance, and from the first law,

$$W = Q = Q_H + Q_C. \quad (19\text{-}2)$$

19-1 Schematic flow diagram of a heat engine.

The heat *absorbed* is usually obtained from the combustion of fuel. The heat *rejected* ordinarily has no economic value. The *thermal efficiency* of a cycle, denoted by *e*, is defined as the ratio of the useful work to the heat absorbed ("what you get" divided by "what you pay for"):

$$\text{Thermal efficiency} = e = \frac{W}{Q_H} = \frac{Q_H + Q_C}{Q_H} = 1 + \frac{Q_C}{Q_H}. \quad (19\text{-}3)$$

Because of friction losses, the *useful* work delivered by an engine is less than the work W, and the overall efficiency is less than the thermal efficiency.

In terms of the flow diagram in Fig. 19–1, the most efficient engine is the one for which the branch pipeline representing the work obtained is as *large* as possible, and the exhaust pipeline representing the heat rejected is as *small* as possible, for a given incoming pipeline or quantity of heat supplied.

As an example of a heat engine and a calculation of thermal efficiency, we consider next an idealized model to describe the thermodynamic processes in a gasoline engine of the type found in automobiles.

19-2 Internal-combustion engines

In the common gasoline engine a mixture of air and gasoline vapor is drawn into a cylinder through an open intake valve while the piston descends, increasing the volume of the cylinder from a minimum of V (when the piston is all the way up) to a maximum of RV (when it is all the way down). The ratio R is called the *compression ratio*, and for present-day automobile engines it is typically about 8. At the end of this *intake stroke* the intake valve closes and the mixture is compressed to volume V during the *compression stroke*. The mixture is then ignited by the spark plug, and the heated gas expands back to volume RV, pushing on the piston and doing work; this is the *power stroke*. Finally, the exhaust valve opens and the combustion products are pushed out (during the *exhaust stroke*) to prepare the cylinder for the next intake stroke.

Figure 19–2 shows an idealized model of the corresponding thermodynamic processes. At a the gasoline–air mixture has entered the cylinder. It is compressed adiabatically along line ab, and is then ignited. Heat Q_H is added to the system by the burning gasoline, and the power stroke is the adiabatic expansion cd. The gas is cooled to the temperature of the outside air (da); during this process heat Q_C is exchanged. In practice, of course, this same air does not enter the engine again but, since an equivalent amount does enter, we may consider the process to be cyclic.

It is not difficult to calculate the efficiency of this idealized cycle. Processes bc and da are constant-volume, so the heats Q_H and Q_C are related simply to the temperatures:

$$Q_H = nC_v(T_c - T_b),$$
$$Q_C = nC_v(T_a - T_d).$$

The thermal efficiency is given by Eq. (19–3); inserting the above expressions and canceling out the common factors nC_v, we obtain

$$e = \frac{T_c - T_b + T_a - T_d}{T_c - T_b}, \quad (19\text{-}4)$$

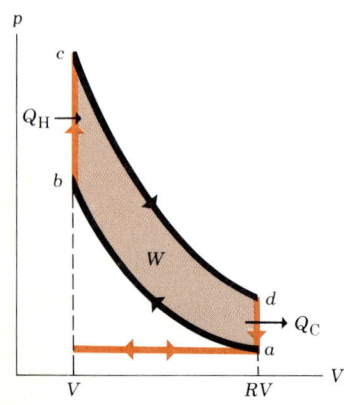

19-2 pV-diagram of the Otto cycle.

This may be further simplified by using the temperature–volume relation for adiabatic processes, Eq. (18–24). For the two adiabatic processes ab and cd, we find

$$T_a(RV)^{\gamma-1} = T_b V^{\gamma-1},$$
$$T_d(RV)^{\gamma-1} = T_c V^{\gamma-1}.$$

We divide out the common factor $V^{\gamma-1}$ and substitute the resulting expressions for T_b and T_c back into Eq. (19–4). The result is

$$e = \frac{T_d R^{\gamma-1} - T_a R^{\gamma-1} + T_a - T_d}{T_d R^{\gamma-1} - T_a R^{\gamma-1}} = \frac{(T_d - T_a)(R^{\gamma-1} - 1)}{(T_d - T_a)R^{\gamma-1}}.$$

Finally, we divide out the common factor $(T_d - T_a)$, obtaining the simple result

$$e = 1 - \frac{1}{R^{\gamma-1}}. \tag{19-5}$$

The thermal efficiency given by Eq. (19–5) is always less than unity, even for this idealized model. Using $R = 8$ and $\gamma = 1.4$ (the value for air), we find $e = 0.56$, or 56 percent. The efficiency can be increased by increasing R. However, this also increases the temperature at the end of the adiabatic compression of the gas–air mixture. If it is too high, the mixture explodes spontaneously and prematurely, instead of burning evenly after ignition by the spark plug. The mechanical strength and wear of engine parts also pose problems; the maximum practical compression ratio for ordinary gasoline is about 10. Higher ratios can be used with more exotic fuels.

The cycle just described, called the *Otto cycle*, is of course a highly idealized model; it neglects friction, turbulence, loss of heat to cylinder walls, and many other effects that combine to reduce the efficiency of a real engine. Another source of inefficiency is incomplete combustion. A mixture of gasoline vapor with just enough air for complete combustion of the hydrocarbons to H_2O and CO_2 does not ignite readily. Reliable ignition requires a mixture "richer" in gasoline, but this leads to CO and unburned hydrocarbons in the exhaust. The heat obtained from the gasoline is then *less* than the total heat of combustion; the difference is wasted, and the exhaust contributes to air pollution. One attack on this problem is the stratified-charge engine, in which the concentration of gasoline vapor near the spark plug is greater than in the remainder of the combustion chamber.

The operation of the Diesel engine is similar to that of the gasoline engine; the principal difference is that there is no fuel in the cylinder during compression. At the beginning of the power stroke fuel is injected into the cylinder just rapidly enough to keep the pressure approximately constant during the first part of the power stroke. The fuel ignites spontaneously because of the high temperature developed during the adiabatic compression; no spark plugs are needed.

The idealized Diesel cycle is shown in Fig. 19–3. Starting at point a, air is compressed adiabatically to point b, heated at constant pressure to point c, expanded adiabatically to point d, and cooled at constant volume to point a.

Since there is no fuel in the cylinder of a diesel engine during the compression stroke, pre-ignition cannot occur, and the compression ratio

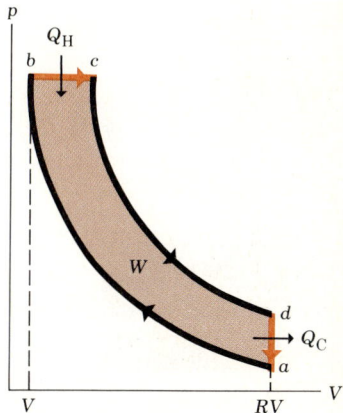

19-3 pV-diagram of the Diesel cycle.

R may be much higher than for a gasoline engine. Values of 15 to 20 are typical; with these values and $\gamma = 1.4$, the efficiency of the idealized Diesel cycle is about 0.65 to 0.70. As with the Otto cycle, the efficiency of any actual engine is less than this.

19-3 Steam engines

A type of heat engine widely used in electric power generating plants is the steam turbine. In a typical modern plant, water is boiled at very high pressure and temperature (of the order of 500°C and 200 atm) in a boiler heated by burning coal. It expands and does work against the rotating blades of a turbine, which then drives the electrical generator. The steam leaves the turbine at a lower temperature, of the order of 200°C, and is then further cooled and condensed so it can be cycled through the boiler again. The thermal efficiency of such an engine is of the order of 35 to 40 percent. That is, 60 to 65 percent of the energy supplied by the burning coal is wasted, some of it in the stack gases but most of it in the energy thrown away when the used steam is condensed.

19-4 The refrigerator

A refrigerator may be considered to be a heat engine operated in reverse. A heat engine takes in heat from a hot reservoir, converts a part of the heat into mechanical work output, and rejects the difference as heat to a cold reservoir. A refrigerator, however, takes in heat from a cold reservoir, the compressor supplies mechanical work *input,* and heat is rejected to a hot reservoir. With reference to the ordinary home refrigerator, the food and ice cubes constitute the cold reservoir, work is done by the electric motor, and the hot reservoir is the air in the kitchen.

The flow diagram of a refrigerator is given in Fig. 19-4. In one cycle, heat Q_C enters the refrigerator at low temperature T_C, work W is done on the refrigerator and heat Q_H leaves at a higher temperature T_H. Both W and Q_H are negative quantities, while Q_C is positive. It follows from the first law that

$$-Q_H = Q_C - W,$$

or

$$|Q_H| = Q_C + |W|,$$

and the heat rejected to the hot reservoir is the sum of the heat taken from the cold reservoir and the heat equivalent of the work done by the motor.

From an economic point of view, the best refrigeration cycle is one that removes the greatest amount of heat Q_C from the refrigerator, for the least expenditure of mechanical work W. We therefore define *the performance coefficient* (rather than the efficiency) of a refrigerator as the ratio $K = -Q_C/W$. Since $W = Q_H + Q_C$,

$$\text{Performance coefficient} = K = -\frac{Q_C}{W} = -\frac{Q_C}{Q_H + Q_C}. \quad (19\text{-}6)$$

The principles of the common refrigeration cycle are illustrated schematically in Fig. 19-5. Compressor A delivers gas, typically CCl_2F_2

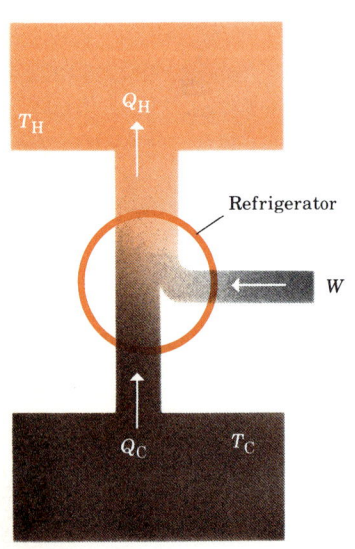

19-4 Schematic flow diagram of a refrigerator.

or another member of the "Freon" family, at high temperature and pressure to coils B. Heat is removed from the gas in B by water or air cooling, resulting in condensation of the gas to a liquid, still under high pressure. The liquid passes through the throttling valve or expansion valve C, emerging as a mixture of liquid and vapor at a much lower temperature. In coils D, heat is supplied that converts the remaining liquid into vapor, which enters compressor A to repeat the cycle. In a refrigerator, coils D are placed in the cold compartment, where they cool the refrigerator directly. In an air-conditioner, air is cooled as it is blown through these coils, while coils B are cooled by warmer outside air.

19-5 Principle of the mechanical refrigeration cycle.

The operating principle of the refrigerator is also used in the *heat pump,* a device which heats a house by cooling the outside air. It is in effect a refrigerator turned inside out. For the heat pump the appropriate definition of the performance coefficient is the ratio

$$K' = \frac{-Q_\mathrm{H}}{W}.$$

The reader may show that this is related to Eq. (19-6) by

$$K' = K + 1.$$

If no work were needed to operate a refrigerator, the coefficient of performance (heat extracted divided by work done) would be infinite. Experience shows that work is *always* needed to transfer heat from a colder to a hotter body. Heat flows spontaneously from a hotter to a colder body, and to reverse this flow requires the addition of work from the outside. No one has ever succeeded in making a refrigerator that transports heat from a cold body to a hotter body without the addition of work.

19-5 The second law of thermodynamics

None of the heat engines described above has a thermal efficiency of 100%. That is, none of them absorbs heat and converts it *completely* into work. There is nothing in the *first* law of thermodynamics that precludes this possibility. The first law requires only that the energy output of an engine, in the form of mechanical work, shall equal the difference between the energies absorbed and rejected in the form of heat. An engine that rejected no heat and converted all the heat absorbed to mechanical work would therefore be perfectly consistent with the first law.

There is however another principle, independent of the first law and not derivable from it, which determines the maximum fraction of the energy absorbed by an engine as heat that can be converted to mechanical work. The basis of this principle lies in the difference between the nature of internal energy and that of mechanical energy. The former is the energy of *random* molecular motion, while the latter represents *ordered* molecular motion. Superposed on their random motion, the molecules of a moving body have an ordered motion in the direction of the velocity of the body. The total molecular kinetic energy associated with the ordered motion is what we call in mechanics the kinetic energy of the moving body. The kinetic (and potential) energy associated with the *random* motion constitutes the internal energy. When the moving body makes an inelastic collision and comes to rest, the *ordered* portion of the molecular kinetic energy becomes converted to *random* motion. Since we cannot control the motions of individual molecules, it is impossible to reconvert the random motion completely to ordered motion. We can, however, convert a portion of it; this is what is accomplished by a heat engine.

The impossibility of converting heat *completely* into mechanical energy forms the basis of one form of the *second law of thermodynamics*. Although we have introduced it with a discussion of molecular motion, the second law (as with all of thermodynamics) deals with directly measurable quantities such as heat and work, and can be stated in terms of macroscopic quantities, independently of any molecular model. One statement of the second law is as follows:

> It is impossible for any system to undergo a process in which it absorbs heat from a reservoir at a single temperature and converts it completely into mechanical work, while ending in the same state in which it began.

That is, no heat engine can have a thermal efficiency of unity (100%).

If the second law were *not* true, it would be possible to drive a steamship across the ocean by extracting heat from the ocean, or to run a power plant by extracting heat from the surrounding air. It should be noted again that neither of these "impossibilities" violates the *first* law of thermodynamics. After all, both the ocean and the surrounding air contain an enormous store of internal energy which, in principle, may be extracted in the form of a flow of heat. The second law, therefore, is not a deduction from the first but stands by itself as a separate law of nature, referring to an aspect of nature different from that contemplated by the first law. The first law denies the possibility of creating or destroying energy; the second denies the possibility of utilizing energy in a particular way.

The fact that work may be dissipated completely into heat, whereas heat may *not* be converted entirely into work, expresses an essential one-sidedness of nature. All natural, spontaneous processes may be studied in the light of the second law, and in all such cases, this peculiar one-sidedness is found. Thus, heat always flows spontaneously from a hotter to a colder body; gases always seep through an opening spontaneously from a region of high pressure to a region of low pressure; gases and liquids left by themselves always tend to mix, not to unmix. Salt dissolves in water but a salt solution does not separate by itself into pure salt and pure water. Rocks weather and crumble; iron rusts; people grow old. These are all examples of *irreversible* processes that take place naturally in only one direction and, by their one-sidedness, express the second law of thermodynamics. We also note that in the thermodynamic sense an irreversible process is always a *nonequilibrium* process. Irreversible heat flow accompanies departures from thermal equilibrium, a free expansion of a gas involves states that are not in mechanical equilibrium, and so on. In each case the process tends to move the system toward an equilibrium state.

The analysis of refrigerators forms the basis for an alternative statement of the second law of thermodynamics. Heat flows spontaneously from hotter to colder bodies, never the reverse. A refrigerator does take heat from a colder to a hotter body, but its operation depends on input of mechanical energy or work. Generalizing this observation, we state:

It is impossible for any process to have as its sole result the transfer of heat from a cooler to a hotter body.

This statement may seem not to be very closely related to the previous statement about the impossibility of converting heat completely into mechanical work. In fact, though, the two statements are completely equivalent. For example, if we could build a workless refrigerator, violating the second or "refrigerator" statement of the second law, we could use it in conjunction with a heat engine, pumping the heat rejected by the engine back to the hot reservoir to be reused, thereby obtaining a composite machine (Fig. 19–6a) that violates the first or "engine" statement of the second law.

19-6 (a) A workless refrigerator (left), if it existed, could be used in combination with an ordinary heat engine (right) to form an engine with 100-percent efficiency. (b) An engine with 100-percent efficiency (left), if it existed, could be used in combination with an ordinary refrigerator (right) to form a workless refrigerator. Thus if either of these is impossible, the other must be also.

Or if we could make an engine with 100% thermal efficiency, in violation of the first statement, we could run it using heat from the hot reservoir, and use the work output to drive a refrigerator (Fig. 19–6b) that pumps heat from the cold reservoir to the hot. This would violate the "refrigerator" statement. Thus any device that violates one form of the second law can also be used to violate the other form; we conclude that if violations of the first form are impossible, so are violations of the second.

Heat flow across a finite temperature gradient is an *irreversible* process, in the sense that no minor change in conditions can make it proceed in the reverse direction. Similarly, conversion of mechanical energy to heat, as in friction or turbulent fluid flow, is an irreversible process. The second law of thermodynamics is an expression of the inherent one-way aspect of these processes and the resulting limitations on heat engines and refrigerators.

19-6 The Carnot cycle

According to the second law, no heat engine can have 100% efficiency. But what is the *maximum* possible efficiency of an engine, given two heat reservoirs at temperatures T_H and T_C? This question was answered in 1824 by the French engineer Sadi Carnot, who developed a hypothetical, idealized heat engine that has the maximum possible efficiency con-

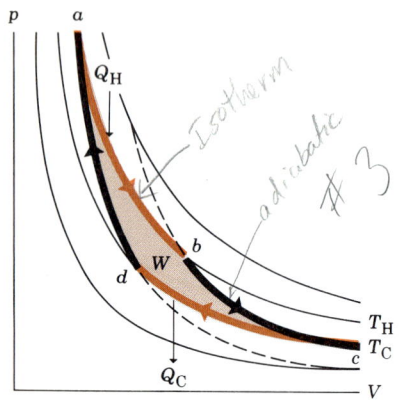

19-7 The Carnot cycle. Color lines are isothermals; black lines are adiabatics.

sistent with the second law. His engine is called the *Carnot engine,* and its cycle is called the *Carnot cycle.*

To understand the rationale of the Carnot cycle we return to the matter of reversibility, discussed at the end of Sec. 19-5. Heat flow through a finite temperature drop and conversion of work into heat are irreversible processes. In maximizing the efficiency of a heat engine (which converts work into heat), we should therefore *avoid* all irreversible processes. During heat transfer there must be no finite temperature difference; when the engine takes heat from the hot reservoir at T_H, the engine itself must be at T_H; otherwise irreversible heat flow would occur. Similarly, when heat is rejected to the cold reservoir at T_C, the engine itself must be at T_C. That is, every process involving heat transfer must be *isothermal.* Conversely, there must be *no* heat transfer in any process where the temperature of the engine changes, for it could not be reversible. In short, every process in our idealized cycle must be either *isothermal* or *adiabatic.* In addition, not only thermal but mechanical equilibrium must be maintained at all times, so that each process is completely reversible.

The Carnot cycle consists of two isothermal and two adiabatic processes. A Carnot cycle using an ideal gas as the working substance is shown on a pV diagram in Fig. 19-7. It consists of the following steps:

1. The gas expands isothermally at temperature T_H, absorbing heat Q_H (*ab*);
2. It expands adiabatically until its temperature drops to T_C (*bc*);
3. It is compressed isothermally at T_C, rejecting heat Q_C (*cd*);
4. It is compressed adiabatically back to its initial state (*da*).

The heat and work in each of these steps can be calculated in terms of the volumes and temperatures, and from these the thermal efficiency of the Carnot engine may be determined.

It is not difficult to calculate the efficiency of a Carnot engine using an ideal gas. From Eq. (19-3), we need to find the ratio Q_C/Q_H of the quantities of heat transferred in the two isothermal processes. For an ideal gas the internal energy depends only on temperature and is thus constant in an isothermal process. Thus, for example, Q_H is equal to the work W_{ab} done by the gas during its isothermal expansion at temperature T_H. This is in turn given by Eq. (18-4), and so we find

$$W_{ab} = Q_H = nRT_H \ln \frac{V_b}{V_a}.$$

Similarly,

$$Q_C = nRT_C \ln \frac{V_d}{V_c} = -nRT_C \ln \frac{V_c}{V_d}.$$

This quantity is negative because V_d is less than V_c. The ratio of the two quantities of heat is thus

$$\frac{Q_C}{Q_H} = -\frac{T_C}{T_H} \frac{\ln(V_c/V_d)}{\ln(V_b/V_a)}. \qquad (19\text{-}7)$$

This can be simplified further by use of the temperature–volume relation for an adiabatic process, Eq. (18–24). We find for the two adiabatic processes:

$$T_H V_b^{\gamma-1} = T_C V_c^{\gamma-1}$$

and

$$T_H V_a^{\gamma-1} = T_C V_d^{\gamma-1}.$$

Dividing the first of these by the second, we find

$$\frac{V_b^{\gamma-1}}{V_a^{\gamma-1}} = \frac{V_c^{\gamma-1}}{V_d^{\gamma-1}} \quad \text{and} \quad \frac{V_b}{V_a} = \frac{V_c}{V_d}.$$

Thus the two logarithms in Eq. (19–7) are equal, and that equation reduces to

$$\frac{Q_C}{Q_H} = -\frac{T_C}{T_H}. \tag{19-8}$$

Then from Eq. (19–3) the efficiency of the Carnot engine is

$$e = 1 - \frac{T_C}{T_H} = \frac{T_H - T_C}{T_H}. \tag{19-9}$$

This surprisingly simple result says that the efficiency of a Carnot engine depends only on the temperatures of the two heat reservoirs. When the difference of the two temperatures is large, the efficiency is nearly unity; when it is small, e is must less than unity.

Example 1 Suppose 0.2 mol of an ideal diatomic gas ($\gamma = 1.40$) undergoes a Carnot cycle with temperatures $T_H = 400$ K and $T_C = 300$ K. The initial pressure is $p_a = 10.0 \times 10^5$ Pa, and during the isothermal expansion at temperature T_H the volume doubles.

a) Find the pressure and volume at each of points a, b, c, and d in Fig. 19–7.
b) Find Q, W, and ΔU for each step in the cycle, and for the entire cycle.
c) Determine the efficiency directly from the results of (b), and compare with the result from Eq. (19–9).

Solution

a) First,

$$V_a = \frac{nRT_H}{p_a} = \frac{(0.2 \text{ mol})(8.314 \text{ J} \cdot \text{mol}^{-1} \cdot \text{K}^{-1})(400 \text{ K})}{10.0 \times 10^5 \text{ Pa}}$$

$$= 6.65 \times 10^{-4} \text{ m}^3,$$

$$V_b = 2V_a = 2(6.65 \times 10^{-4} \text{ m}^3) = 13.3 \times 10^{-4} \text{ m}^3.$$

For the isothermal expansion $a \to b$, $p_a V_a = p_b V_b$, so

$$p_b = \frac{p_a V_a}{V_b} = 5.0 \times 10^5 \text{ Pa}.$$

For the adiabatic expansion $b \to c$, $T_H V_b^{\gamma-1} = T_C V_c^{\gamma-1}$, so

$$V_c = V_b \left(\frac{T_H}{T_C}\right)^{1/\gamma-1} = (13.3 \times 10^{-4} \text{ m}^3)\left(\frac{4}{3}\right)^{2.5} = 27.3 \times 10^{-4} \text{ m}^3.$$

So
$$p_c = \frac{nRT_C}{V_c} = \frac{(0.2 \text{ mol})(8.314 \text{ J} \cdot \text{mol}^{-1} \cdot \text{K}^{-1})(300 \text{ K})}{27.3 \times 10^{-4} \text{ m}^3}$$
$$= 1.83 \times 10^5 \text{ Pa}.$$

For the adiabatic compression $d \to a$, $T_C V_d^{\gamma-1} = T_H V_a^{\gamma-1}$, and
$$V_d = V_a \left(\frac{T_H}{T_C}\right)^{1/\gamma-1} = (6.65 \times 10^{-4} \text{ m}^3)\left(\frac{4}{3}\right)^{2.5} = 13.65 \times 10^{-4} \text{ m}^3;$$
$$p_d = \frac{nRT_C}{V_d} = \frac{(0.2 \text{ mol})(8.314 \text{ J} \cdot \text{mol}^{-1} \cdot \text{K}^{-1})(300 \text{ K})}{13.65 \times 10^{-4} \text{ m}^3}$$
$$= 3.65 \times 10^5 \text{ Pa}.$$

b) For the isothermal expansion $a \to b$, $\Delta U = 0$ and, from Eq. (18-4),
$$W = Q_H = nRT_H \ln \frac{V_b}{V_a}$$
$$= (0.2 \text{ mol})(8.314 \text{ J} \cdot \text{mol}^{-1} \cdot \text{K}^{-1})(400 \text{ K}) \ln 2$$
$$= 461 \text{ J}.$$

For the adiabatic expansion $b \to c$, $Q = 0$; from Eq. (18-27),
$$W = -\Delta U = nC_v(T_H - T_C)$$
$$= (0.2 \text{ mol})(20.78 \text{ J} \cdot \text{mol}^{-1} \cdot \text{K}^{-1})(400 \text{ K} - 300 \text{ K})$$
$$= 415.7 \text{ J}.$$

For the isothermal compression $c \to d$, $\Delta U = 0$, and
$$W = Q_C = nRT_C \ln \frac{V_d}{V_c}$$
$$= (0.2 \text{ mol})(8.314 \text{ J} \cdot \text{mol}^{-1} \cdot \text{K}^{-1})(300 \text{ K}) \ln \frac{13.65 \times 10^{-4} \text{ m}^3}{27.3 \times 10^{-4} \text{ m}^3}$$
$$= -345.8 \text{ J}.$$

For the adiabatic compression $d \to a$, $Q = 0$;
$$W = -\Delta U = nC_v(T_C - T_H)$$
$$= (0.2 \text{ mol})(8.314 \text{ J} \cdot \text{mol}^{-1} \cdot \text{K}^{-1})(300 \text{ K} - 400 \text{ K})$$
$$= -415.7 \text{ J}.$$

Alternatively Eq. (18-28) may be used for either adiabatic process. For $d \to a$,
$$W = \frac{1}{\gamma - 1}(p_d V_d - p_a V_a) = (2.5)[(3.65 \times 10^5 \text{ Pa})(13.65 \times 10^{-4} \text{ m}^3)$$
$$- (10.0 \times 10^5 \text{ Pa})(6.65 \times 10^{-4} \text{ m}^3)]$$
$$= -415.7 \text{ J}.$$

The results may be tabulated as follows:

	Q	W	ΔU
$a \to b$	461 J	461 J	0
$b \to c$	0	415.7 J	−415.7 J
$c \to d$	−345.8 J	−345.8 J	0
$d \to a$	0	−415.7 J	415.7 J
total	115.2 J	115.2 J	0

We note that, for the entire cycle, $Q = W$ and $\Delta U = 0$. Also the quantities of work in the two adiabatic processes are negatives of each other; it is easy to prove from the analysis leading to Eq. (19-9) that this must always be the case.

c) From the table, the total work is 115.2 J, and $Q_H = 461$ J. Thus

$$e = \frac{W}{Q_H} = \frac{115.2 \text{ J}}{461 \text{ J}} = 0.250.$$

From Eq. (19-9),

$$e = \frac{T_H - T_C}{T_H} = \frac{400 \text{ K} - 300 \text{ K}}{200 \text{ K}} = 0.250.$$

Because each step in the Carnot cycle is reversible, the entire cycle may be reversed, converting the engine into a refrigerator. The performance coefficient of the Carnot refrigerator is obtained by combining Eqs. (19-6) and (19-8), first rewriting Eq. (19-6) as

$$K = -\left(\frac{Q_H + Q_C}{Q_C}\right)^{-1}.$$

We find

$$K = \frac{T_C}{T_H - T_C}. \tag{19-10}$$

When the temperature difference is small, K is much larger than unity; in that case a lot of heat can be "pumped" from the lower to the higher temperature with only a little expenditure of work. But the greater the temperature difference, the smaller is K, and the more work is required to transfer a given quantity of heat.

Example 2 If the cycle described in Example 1 is run backwards as a refrigerator, the performance coefficient is given by Eq. (19-6):

$$K = -\frac{Q_C}{W} = -\frac{(-345.8 \text{ J})}{115.2 \text{ J}} = 3.00.$$

Because the cycle is a Carnot cycle, we may also use Eq. (19-10):

$$K = \frac{T_C}{T_H - T_C} = \frac{300 \text{ K}}{400 \text{ K} - 300 \text{ K}} = 3.00.$$

For a Carnot cycle, e and K depend only on the temperatures, and it is not necessary to calculate Q and W. For cycles containing irreversible processes, however, more detailed calculations are necessary.

It is easy to prove that no engine can be more efficient than a Carnot engine operating between the same two temperatures. The key to the proof is the above observation that since each step in the Carnot cycle is reversible, the *entire cycle* may be reversed. Run backward, the engine becomes a refrigerator. If any engine is more efficient than a Carnot engine, its work output may be used to drive a Carnot refrigerator and pump the rejected heat back to the hot reservoir, thus violating the engine statement of the second law. Thus still another equivalent statement of the second law is:

> No engine operating between two given temperatures can be more efficient than a Carnot engine operating between the same two temperatures.

It also follows directly that

> All Carnot engines operating between the same two temperatures have the same efficiency, irrespective of the nature of the working substance.

Thus, although Eqs. (19–9) and (19–10) were derived for a Carnot engine using an ideal gas as its working substance, they are in fact valid for *any* Carnot engine, no matter what its working substance.

Equation (19–9) points the way to the conditions that a real engine, such as a steam turbine, must fulfill to approach as closely as possible the maximum attainable efficiency. These conditions are that the intake temperature T_H must be made as high as possible and the exhaust temperature T_C as low as possible.

The exhaust temperature cannot be lower than the lowest temperature available for cooling the exhaust. This is usually the temperature of the air, or perhaps of river water if this is available at the plant. The only recourse then is to raise the boiler temperature T_H. Since the vapor pressure of all liquids increases rapidly with increasing temperature, a limit is set by the mechanical strength of the boiler. At 500°C, the vapor pressure of water is about 240×10^5 Pa, 235 atm, or 3450 lb·in^{-2}, and this is about the maximum practical pressure in large present-day steam boilers.

The unavoidable exhaust heat loss in electric power plants creates a serious environmental problem. When a lake or river is used for cooling, the temperature of the body of water may be raised several degrees. Such a temperature change has a severely disruptive effect on the overall ecological balance, inasmuch as relatively small temperature changes can have significant effects on metabolic rates in plants and animals. Since *thermal pollution,* as this effect is called, is an inevitable consequence of the second law of thermodynamics, careful planning is essential to minimize the ecological impact of new power plants.

*19–7 The Kelvin temperature scale

The Carnot cycle can be used to define a temperature scale that is completely independent of the properties of any particular material. As we have seen, the efficiency of a Carnot engine operating between reservoirs at two given temperatures *is independent of the nature of the working substance and is a function only of the temperatures.* If we consider a number of Carnot engines using different working substances and absorbing and rejecting heat to the same two reservoirs, the thermal efficiency is the same for all:

$$e = \frac{Q_H + Q_C}{Q_H} = 1 + \frac{Q_C}{Q_H}$$
$$= \text{constant.}$$

Hence, the ratio Q_H/Q_C is a constant for all of the Carnot engines. Kelvin proposed that the ratio of the temperatures of the reservoirs be *defined* as equal to this constant ratio of the absolute magnitudes of the quantities of heat absorbed and rejected or, since Q_C is a negative quantity, as equal to the negative of the ratio Q_H/Q_C. Thus,

$$\frac{T_H}{T_C} = \frac{|Q_H|}{|Q_C|} = -\frac{Q_H}{Q_C}. \tag{19-11}$$

Equation (19-11) appears identical to Eq. (19-8), but there is a subtle and crucial difference. The temperatures in Eq. (19-8) are those based on an ideal-gas thermometer, as defined in Sec. (14-3), while Eq. (19-11) defines a temperature scale, based on the Carnot cycle and the second law of thermodynamics, which is completely independent of the behavior of any particular substance. Thus the Kelvin temperature scale is truly *absolute*.

To complete the definition of the Kelvin scale we proceed, as in Chapter 14, to assign the arbitrary value of 273.16 K to the temperature of the triple point of water, T_0. For a Carnot engine operating between reservoirs at the temperatures T and T_0, we have

$$\frac{|Q|}{|Q_0|} = \frac{T}{T_0}$$

or

$$T = 273.16 \text{ K} \frac{|Q|}{|Q_0|}. \tag{19-12}$$

Comparing this with the corresponding expression for the ideal gas temperature, namely,

$$273.16 \text{ K} \left[\lim_{p_0 \to 0} \left(\frac{p}{p_0} \right)_{\text{constant volume}} \right],$$

we see that in the Kelvin scale Q plays the role of a "thermometric property." This does not, however, have the objection attached to a coordinate of an arbitrarily chosen thermometer, inasmuch as the behavior of a Carnot engine is independent of the nature of the working substance.

As mentioned in Sec. 19-6, if an ideal gas is taken around a Carnot cycle, the ratio of the heats absorbed and rejected, $|Q_H|/|Q_C|$, is equal to the ratio of the temperatures of the reservoirs *as expressed on the gas scale,* defined in Chapter 14. Since, in both scales, the triple point of water is chosen to be 273.16 K, it follows that *the Kelvin and the ideal gas scales are identical.*

*19-8 Absolute zero

It follows from Eq. (19-11) or (19-12) that the heat transferred isothermally between two given adiabatics decreases as the temperature decreases. Conversely, the smaller the value of Q, the lower the corresponding T. The smallest possible value of Q is zero, and the corresponding T is absolute zero. Thus, *if a system undergoes a reversible isothermal process without transfer of heat, the temperature at which this process takes place is called absolute zero*. In other words, at absolute zero, an isotherm and an adiabatic are identical.

It should be noted that the definition of absolute zero holds for all substances and is therefore independent of the peculiar properties of any one arbitrarily chosen substance. Furthermore, the definition is in terms of purely large-scale concepts. No reference is made to molecules or to molecular energy. Whether absolute zero may be achieved experimentally is a question of some interest and importance. To achieve temperatures below 4.2 K, at which ordinary helium (mass number 4) liquefies, it is necessary to lower the vapor pressure by pumping away the vapor as fast as possible. The lowest temperature that has ever been reached in this way is 0.7 K, and this required larger pumps and larger pumping tubes than are usually employed in low-temperature laboratories. With the aid of the light isotope of helium (mass number 3), which liquefies at 3.2 K, vigorous pumping yields a temperature of about 0.3 K. Still lower temperatures, as low as 10^{-6} K, may be achieved magnetically, but it becomes apparent that the closer one approaches absolute zero, the more difficult it is to go further. It is generally accepted as a law of nature that, although absolute zero may be approached as closely as we please, it is impossible actually to reach the zero of temperature. This is known as the "*unattainability statement of the third law of thermodynamics.*"

Absolute zero can also be interpreted on a molecular level, although this must be done with some discretion. Because of quantum effects, it is *not* correct to say that at $T = 0$ all molecular motion ceases. Rather, at absolute zero the system has its *minimum* possible total energy (kinetic plus potential), although this minimum amount is in general not zero.

*19-9 Entropy

The second law can be restated in quantitative form using the concept of *entropy*. The first law of thermodynamics is the law of energy, the second law of thermodynamics is the law of entropy, and every process that takes place in nature, whether it be mechanical, electrical, chemical, or biological, must proceed in conformity with these two laws.

Entropy is directly related to irreversibility and to the directionality of natural processes such as heat flow and conversion of work to heat. As we shall see, any irreversible process must be accompanied by an increase in entropy. In defining entropy, we first recall the discussion of Sec. 18-4, in which internal energy was defined. When a system is carried from one state to another it is found by experiment that the difference between the heat added and the work done by the system, $Q - W$, has the same value for all paths. This makes it possible to introduce the concept of internal energy, with the change in internal energy in any process defined and measured by the quantity $Q - W$.

Entropy, or rather changes in entropy, can be defined in a similar way. We consider two states of a system and any number of processes leading from one to the other. For the present we consider only reversible (i.e., equilibrium) processes. (This restriction was not necessary when defining internal energy.) Denoting entropy by S, we define the change in entropy ΔS in any process leading from state 1 to state 2 as

$$\Delta S = \int_1^2 \frac{dQ}{T} \quad \text{(reversible processes)}. \qquad (19\text{-}13)$$

That is, the process is represented as a series of infinitesimal steps, each involving a heat transfer dQ at absolute temperature T, and the quotients dQ/T are summed for the entire process.

It may be proved, using the second law of thermodynamics, that ΔS as defined by Eq. (19-13) has the same value for *all possible* reversible processes from state 1 to state 2. Thus we are dealing with a quantity (the entropy of the system) that has a definite value for any given state of the system, that depends *only* on this state, and that, in any process, changes by an amount determined only by the initial and final states, not the path taken between them.

If the process happens to be isothermal, T may be taken outside the integral in Eq. (19-13); in this case

$$\Delta S = \frac{Q}{T} \quad \text{(reversible isothermal process)}. \quad (19\text{-}14)$$

The unit of entropy is 1 J·K^{-1}, 1 cal·K^{-1}, 1 Btu·(R°)$^{-1}$, etc.

Example 1 One kilogram of ice at 0°C is melted and converted to water at 0°C. Compute its change in entropy.

Solution The temperature remains constant at 273 K, and

$$S_2 - S_1 = \frac{Q}{T}.$$

But Q is simply the total heat that must be supplied to melt the ice, or 334×10^3 J. Hence

$$S_2 - S_1 = \frac{334 \times 10^3 \text{ J}}{273 \text{ K}} = 1223 \text{ J·K}^{-1},$$

and the increase in entropy of the system is 1223 J·K^{-1}. In any *isothermal* reversible process, the entropy change equals the heat added divided by the absolute temperature.

Example 2 One kilogram of water at 0°C is heated to 100°C. Compute its change in entropy.

Solution The temperature is not constant and dQ and T must be expressed in terms of a single variable in order to carry out the integration. This is easily done, since

$$dQ = mc\, dT.$$

Hence

$$\Delta S = S_2 - S_1 = \int_{T_1}^{T_2} mc\, \frac{dT}{T} = mc \ln \frac{T_2}{T_1}$$

$$= (1000 \text{ g})(4.19 \text{ J·g}^{-1}\text{·K}^{-1}) \ln \frac{373 \text{ K}}{273 \text{ K}}$$

$$= 1308 \text{ J·K}^{-1}.$$

Example 3 A gas is allowed to expand adiabatically and reversibly. What is its change in entropy?

Solution In an adiabatic process no heat is allowed to enter or leave the system. Hence $Q = 0$ and there is *no* change in entropy. It follows that every *reversible* adiabatic process is one of constant entropy and may be described as *isentropic*.

Example 4 A thermally insulated box is divided by a partition into two compartments, each having volume V. Initially one compartment contains n moles of an ideal gas at temperature T, and the other is evacuated. The partition is then broken, and the gas expands to fill both compartments. What is its entropy change?

Solution For this process, $Q = 0$, $W = 0$, $\Delta U = 0$, and therefore (since it is an ideal gas) $\Delta T = 0$. One might think that the entropy change is zero because there is no heat exchange. But Eq. (19–13) is valid only for *reversible* processes; this free expansion is *not* reversible, and there *is* an entropy change. To calculate it we can use the fact that the entropy change depends only on the initial and final states and not on the intermediate process. Thus we can devise a *reversible* process having the same endpoints, use Eq. (19–13) to calculate its entropy change, and thus obtain the entropy change in the original process. The appropriate reversible process in this case is an isothermal expansion from V to $2V$ at temperature T. The gas does work during this expansion, so heat must be supplied in order to keep the internal energy constant; the total heat equals the total work, which is given by Eq. (18–4):

$$W = Q = nRT \ln 2.$$

Thus the entropy change is

$$\Delta S = \frac{Q}{T} = nR \ln 2,$$

and this is also the entropy change for the free expansion. For one mole,

$$\Delta S = (1 \text{ mol})(8.314 \text{ J} \cdot \text{mol}^{-1} \cdot \text{K}^{-1})(0.693) = 5.76 \text{ J} \cdot \text{K}^{-1}.$$

Entropy can be interpreted microscopically in terms of *disorder* or *randomness* of a system. Adding heat to a substance increases its disorder because it increases the randomness of molecular motion. The entropy of a gas increases during a free expansion because of the increase in randomness of *position* of gas molecules. In each of the above examples, there is a correlation between microscopic disorder, or randomness, and entropy. This correlation is also significant in connection with a restatement of the second law of thermodynamics in terms of entropy, discussed in the next section.

*19–10 Entropy and the second law

One of the features that distinguish entropy from such quantities as energy, momentum, and angular momentum is that there is *no* principle of conservation of entropy. In fact, the reverse is true; the entropy of an isolated system *can* change but, as we shall see, it *can never* decrease. In fact, there is an increase in entropy in every natural process, if all systems taking part in the process are included. In an idealized, completely reversible process involving only equilibrium states, there is no change in entropy, but all natural (i.e., irreversible) processes take place with an increase in entropy.

Example Suppose 1 kg of water at 100°C is placed in thermal contact with 1 kg of water at 0°C. What is the total change in entropy?

Solution We assume the specific heat capacity of water is constant in this temperature range. Then the first 1000 cal transferred cools the hot water to 99°C and warms the cold water from 0°C to 1°C. The net change of entropy is approximately

$$\Delta S = -\frac{1000 \text{ cal}}{373 \text{ K}} + \frac{1000 \text{ cal}}{273 \text{ K}} = 0.98 \text{ cal} \cdot \text{K}^{-1}.$$

Further increases in entropy occur as the system approaches thermal equilibrium. The *total* increase in entropy can be calculated using the same method as in Example 2 of Sec. 19-9. The entropy change of the hot water is

$$\Delta S = (1 \text{ kg})(1000 \text{ cal} \cdot \text{kg}^{-1} \cdot \text{K}^{-1}) \int_{373 \text{ K}}^{323 \text{ K}} \frac{dT}{T}$$

$$= (1000 \text{ cal} \cdot \text{K}^{-1}) \ln \frac{323 \text{ K}}{373 \text{ K}} = -144 \text{ cal} \cdot \text{K}^{-1};$$

the entropy change of the cold water is

$$\Delta S = (1000 \text{ cal} \cdot \text{K}^{-1}) \ln \frac{323 \text{ K}}{273 \text{ K}} = +168 \text{ cal} \cdot \text{K}^{-1},$$

and the *total* entropy change of the system is

$$\Delta S = +168 \text{ cal} \cdot \text{K}^{-1} - 144 \text{ cal} \cdot \text{K}^{-1} = +24 \text{ cal} \cdot \text{K}^{-1}.$$

Thus an irreversible heat flow is accompanied by an increase in entropy.

Of course, the same result could be attained by simply mixing the hot and cold water. This too is an irreversible process, and the entropy increase is the same. These examples of the mixing of substances at different temperatures, or the flow of heat from a higher to a lower temperature, are characteristic of *all* natural (i.e., irreversible) processes. When all the entropy changes in the process are included, the increases in entropy are always greater than the decreases. In the special case of a reversible process, the increases and decreases are equal. Hence we can formulate the general principle, which is considered a part of the second law of thermodynamics, that *when all systems taking part in a process are included, the entropy either remains constant or increases*. In other words, no process is possible in which the entropy decreases, when all systems taking part in the process are included.

What is the significance of the increase of entropy that accompanies every natural process? The answer, or one answer, is that it represents the extent to which the universe becomes more disordered, or random, in that process. Consider again the example of the mixing of hot and cold water. We *might* have used the hot and cold water as the high- and low-temperature reservoirs of a heat engine, and in the course of removing heat from the hot water and giving heat to the cold water we could have obtained some mechanical work. But once the hot and cold water have been mixed and have come to a uniform temperature, this opportunity of converting heat to mechanical work is lost and, moreover, it is lost irretrievably. The lukewarm water will never *unmix* itself and separate into a hotter and colder portion.* Of course, there is no decrease in

---

*The branch of physics called "statistical mechanics" would modify this statement to read, "It is highly improbable that the water will separate spontaneously into a hotter and a colder portion, but it is not impossible."

energy when the hot and cold water are mixed, and what has been "lost" in the mixing process is not *energy,* but *opportunity*—the opportunity to convert a portion of the heat flowing out of the hot water to mechanical work. Hence when entropy increases, energy becomes more unavailable, and we say that the universe has become more random or "run down" to that extent. This is the true significance of the term "irreversible."

The tendency of all natural processes such as heat flow, mixing, diffusion, etc., is to bring about a uniformity of temperature, pressure, composition, etc., at all points. One may visualize a distant future in which, as a consequence of these processes, the entire universe has attained a state of absolute uniformity throughout. When and if such a state is reached, although there would have been no change in the energy of the universe, all physical, chemical, and presumably biological processes would have to cease. This fate toward which we appear headed has been described as the "heat death" of the universe!

19-11 Energy conversion

We have seen in this chapter that the laws of thermodynamics place very general limitations on conversion of energy from one form to another. In this day of increasing energy demand and diminishing resources, these matters are of the utmost practical importance. We conclude this chapter with a brief discussion of a few energy-conversion systems, present and proposed.

Most (over 80 percent) of the electric power generated in the United States is obtained from coal-fired steam-turbine generating plants. Modern boilers can transfer about 80 to 90 percent of the heat of combustion of coal into steam. The theoretical thermodynamic efficiency of the turbine, given by Eq. (19–9), is usually limited to about 0.55, and the actual efficiency is typically 90 percent of the theoretical value, or about 0.50. The efficiency of large electrical generators in converting mechanical power to electrical is very large, typically 99 percent. Thus the overall thermal efficiency of such a plant is roughly (0.85)(0.50)(0.99), or about 40 percent.

In 1970, a generator with a capacity of about 1 gigawatt ($= 1000$ MW $= 10^9$ W) went into operation at the Tennessee Valley Authority's Paradise power plant. The steam is heated to 1003°F, at a pressure of 3650 lb·in^{-2}. The plant consumes 10,500 tons of coal per day, and the overall thermal efficiency is 39.3 percent.

Nuclear power plants have the same theoretical efficiency limit as coal-fired plants. Because at present it is not practical to run nuclear reactors at as high temperatures as coal boilers, the theoretical thermal efficiency is usually lower. The overall thermal efficiency of a nuclear plant is typically 30 percent.

In both coal-fired and nuclear plants, the energy not converted to electrical energy is wasted and must be disposed of. A common practice is to locate such a plant near a lake or river and use the water for disposal of excess heat; this can raise the water temperature several degrees, often with serious ecological consequences.

Solar energy is an inviting possibility. The power in the sun's radiation is about 1.4 kW per square meter, above the earth's atmosphere. A maximum of about 1.0 kW·m^{-2} reaches the surface of the earth on a clear day, and the time average over a 24-hour period is about 0.2 kW·m^{-2}. This radiation could be collected, and focused with mirrors and used to generate steam for a heat engine. A more exotic proposal is to use large banks of photocells for direct conversion to electricity. Such a process is not a heat engine in the usual sense and is not limited by the Carnot efficiency. There are other fundamental limitations on photocell efficiency, but 50% seems attainable in multilayer p-n junction photocells.

An indirect scheme for collection and conversion of solar energy would use the temperature gradient in the ocean. In the Caribbean, for example, the water temperature near the surface is about 25°C, while at a depth of a few hundred meters it may be 10°C. While the second law of thermodynamics forbids taking heat from the ocean and converting it completely into work, there is nothing to forbid running a heat engine between these two temperatures. The thermodynamic efficiency would be very low, but with such a vast reservoir of energy available this would not be a serious problem.

This is only a small sample of present-day activity in energy-conversion research. Many other processes are under discussion or development, and the principles of thermodynamics outlined in this chapter are of central importance in all of them.

Questions

19-1 Suppose you want to increase the efficiency of a heat engine. Would it be better to increase T_H or to decrease T_C by an equal amount?

19-2 If an energy-conversion process involves two steps, each wth its own efficiency, such as heat to work and work to electrical energy, is the efficiency of the composite process equal to the product of the two efficiencies, or the sum, or the difference, or what?

19-3 In some climates it is practical to heat a house using a heat pump, which acts as an air-conditioner in reverse, cooling the outside air and heating the inside air. Can the heat delivered to the house ever exceed the electric energy input to the pump?

19-4 What irreversible processes occur in a gasoline engine?

19-5 A housewife tries to cool her kitchen on a hot day by leaving the refrigerator door open. What happens? Would the result be different if an old-fashioned ice box were used?

19-6 Is it a violation of the second law to convert mechanical energy completely into heat?

19-7 A growing plant creates a highly complex and organized structure out of simple materials, such as air, water, and trace minerals. Does this violate the second law of thermodynamics? What is the plant's ultimate source of energy?

19-8 An electric motor has its shaft coupled to that of a generator. The motor drives the generator, and the current from the generator is used to run the motor. The excess current is used to power a television set. What is wrong with this scheme?

19-9 Think of some reversible and some irreversible processes in purely mechanical systems, such as blocks sliding on planes, springs, pulleys, strings, and all that stuff.

19-10 Why must a room air-conditioner be placed in a window? Why can't it just be set on the floor and plugged in?

19-11 Discuss the following examples of increasing disorder or randomness: mixing of hot and cold water; free expansion of a gas; irreversible heat flow; development of heat through mechanical friction. Are entropy increases involved in all these?

19-12 When the sun shines on a glass-roofed greenhouse, the temperature becomes higher inside than outside. Does this violate the second law?

19-13 When a wet cloth is hung up in a hot wind in the desert, it is cooled by evaporation to a temperature that may be 20°C or so below that of the air. Discuss this process in the light of the second law.

19-14 Are the earth and the sun in thermal equilibrium? Are there entropy changes associated with the transmission of energy from the sun to the earth? Does radiation differ from other modes of heat transfer with respect to entropy changes?

19-15 Discuss the entropy changes involved in the preparation and consumption of a hot-fudge sundae.

Problems

19-1 A heat engine takes 0.1 mole of an ideal gas around the cycle shown in the pV diagram of Fig. 19-8. Process 1–2 is at constant volume, process 2–3 is adiabatic, and process 3–1 is at a constant pressure of 1 atm. The value of γ for this gas is $\tfrac{5}{3}$.

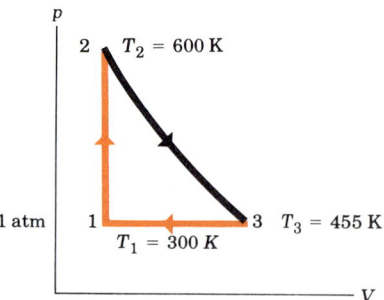

Figure 19-8

a) Find the pressure and volume at points 1, 2, and 3.

b) Find the net work done by the gas in the cycle.

19-2 A cylinder contains oxygen at a pressure of 2 atm. The volume is 3 l and the temperature is 300 K. The oxygen is carried through the following processes:

1. Heated at constant pressure to 500 K.
2. Cooled at constant volume to 250 K.
3. Cooled at constant pressure to 150 K.
4. Heated at constant volume to 300 K.

a) Show the four processes above in a pV-diagram, giving the numerical values of p and V at the end of each process.

b) Calculate the net work done by the oxygen.

c) Find the net heat flowing into the oxygen.

d) What is the efficiency of this device as a heat engine?

19-3 What is the thermal efficiency of an engine that operates by taking an ideal gas through the following cycle? Let $C_v = 12 \text{ J} \cdot \text{mol}^{-1} \cdot \text{K}^{-1}$.

1. Start with n moles at P_0, V_0, T_0.
2. Change to $2P_0$, V_0, at constant volume.
3. Change to $2P_0$, $2V_0$, at constant pressure.
4. Change to P_0, $2V_0$, at constant volume.
5. Change to P_0, V_0, at constant pressure.

19-4 A Carnot engine whose low-temeprature reservoir is at 280 K has an efficiency of 40%. It is desired to increase this to 50%.

a) By how many degrees must the temperature of the high-temperature reservoir be increased if the temperature of the low-temperature reservoir remains constant?

b) By how many degrees must the temperature of the low-temperature reservoir be decreased if that of the high-temperature reservoir remains constant?

19-5 A Carnot engine whose high-temperature reservoir is at 400 K takes in 100 cal of heat at this temperature in each cycle, and gives up 80 cal to the low-temperature reservoir.

a) What is the temperature of the latter reservoir?

b) What is the thermal efficiency of the cycle?

19-6 A Carnot refrigerator takes heat from water at 0°C and rejects heat to a room at 27°C. Suppose that 50 kg of water at 0°C are converted to ice at 0°C.

a) How much heat is rejected to the room?

b) How much energy must be supplied to the refrigerator?

19-7 A Carnot engine is operated between two heat reservoirs at temperatures of 400 K and 300 K.

a) If the engine receives 1200 cal from the reservoir at 400 K in each cycle, how many calories per cycle does it reject to the reservoir at 300 K?

b) If the engine is operated in reverse, as a refrigerator, and receives 1200 cal from the reservoir at 300 K, how many calories does it deliver to the reservoir at 400 K?

c) How many calories would be produced if the mechanical work required to operate the refrigerator in part (b) were converted directly to heat?

19-8 An ice-cube maker freezes ice cubes at the rate of 5 g every second, starting with water at the freezing point. Heat is given off to the room at 30°C. If the system uses an ideal Carnot refrigerator,

a) What electric power input is required?

b) At what rate is heat given off to the room?

19-9 Show that the efficiency e of a Carnot engine and the performance coefficient K of a Carnot refrigerator are related by $K = (1 - e)/e$.

19-10 A Carnot engine operates between two heat reservoirs at temperatures T_H and T_C. An inventor proposes to increase the efficiency by running one engine between T_H and an intermediate temperature T', and a second engine between T' and T_C using the heat expelled by the first engine. Compute the efficiency of this composite system, and compare it to that of the original engine.

19-11 Heat is added to 0.5 kg of ice at 0°C until it is all melted.

a) What is the change in entropy of the water?

b) If the source of heat is a very massive body at a temperature of 20°C, what is the change in entropy of this body?

c) What is the total change in entropy of the water and the heat source?

19-12 A block of aluminum of mass 1 kg, initially at 100°C, is dropped into 1 kg of water initially at 0°C.

a) What is the final temperature?

b) What is the total change in entropy of the system?

19-13 Two moles of an ideal gas undergo a reversible isothermal expansion from 0.02 m³ to 0.04 m³ at a temperature of 300 K. What is the change in entropy of the gas?

19-14 One end of a copper rod is immersed in boiling water at 100°C, the other end in an ice–water mixture at 0°C. The sides of the rod are insulated. During a certain time interval, 0.5 kg of ice melts. Find:

a) the entropy change of the boiling water;

b) the entropy change of the ice–water mixture;

c) the entropy change of the copper rod;

d) the total entropy change of the entire system.

19-15 Consider the entropy of water to be zero when it is in the liquid phase at 0°C and atmospheric pressure.

a) How much heat must be supplied to 1 kg of water to raise its temperature from 0°C to 100°C?

b) What is the entropy of 1 kg of water at 100°C?

c) 1 kg of water at 0°C is mixed with 1 kg at 100°C. What is the final temperature?

d) Find the entropy of the hot and cold water before mixing, and the entropy of the system after mixing. (This is another example of the increase of entropy in an irreversible process.)

19-16

a) Draw a graph of a Carnot cycle, plotting Kelvin temperature vertically and entropy horizontally (a temperature–entropy or TS diagram).

b) Show that the area under any curve in a temperature–entropy diagram represents the heat absorbed by the system.

c) Derive from your diagram the expression for the thermal efficiency of a Carnot cycle.

19-17 An engine is to be built to extract power from the temperature gradient of the ocean. If the surface and deep-water temperatures are 25°C and 10°C, respectively, what is the maximum theoretical efficiency of such an engine?

19-18 A coal-fired steam-turbine power plant that produces 1000 MW of electrical power is located beside a river. The overall thermal efficiency of the plant is 40 percent.

a) What is the total thermal power input to the plant?

b) At what rate is waste heat discharged from the plant?

c) If the waste heat is delivered to the river, and if the temperature rise must be no greater than 5°C, how much water must be available per second?

d) In part (c), if the river is 100 m wide and 5 m deep, what must be the minimum flow velocity of the water?

19-19 A solar power plant is to be built with a power output capacity of 1000 MW. What land area must the solar energy collectors occupy if they are:

a) photocells with 60-percent efficiency?

b) mirrors that generate steam for a turbine–generator unit with overall efficiency of 30 percent?

MOLECULAR PROPERTIES OF MATTER

20-1 Molecular theory of matter

An abundance of physical and chemical evidence has shown that matter in all phases is composed of particles called *molecules*. The molecules of any substance which is a specific chemical compound are identical. Many of the properties of matter in bulk (discussed in previous chapters) such as elasticity, surface tension, phase changes, heat capacities, and others, can be understood more completely on the basis of a microscopic or molecular view of matter.

The smallest molecules are of the order of 10^{-10} m in size; the largest are at least 10,000 times this large. One essential characteristic of a molecule is the *force* describing its interaction with a neighboring molecule. Intermolecular forces which hold molecules of a liquid or solid together are chiefly *electrical* in nature, arising from electrical interactions of the charged fundamental particles of which molecules are composed. Intermolecular forces do not, however, follow a simple inverse-square law. *Gravitational* attraction between molecules is so weak compared with the electrical forces that it is completely negligible.

When molecules are far apart, as in a gas, the intermolecular force is very small and attractive. As a gas is compressed and its molecules are brought closer together the force increases. In liquids, relatively large pressures are required to compress the substance appreciably, and we conclude that at separations between molecules that are only slightly *less* than their normal spacing the force becomes *repulsive* and relatively large.

Thus the force must vary with the distance r between molecules somewhat as shown in Fig. 20–1. At large distances the force is small and attractive. As the molecules are brought closer together, the force of attraction becomes larger, passes through a maximum, and then decreases to zero at an equilibrium separation r_0. When the distance is less than r_0 the force becomes repulsive and increases quite rapidly. The figure also shows the potential energy as a function of r. It is related to the force by $F = -dU/dr$; the potential energy has a *minimum* at r_0,

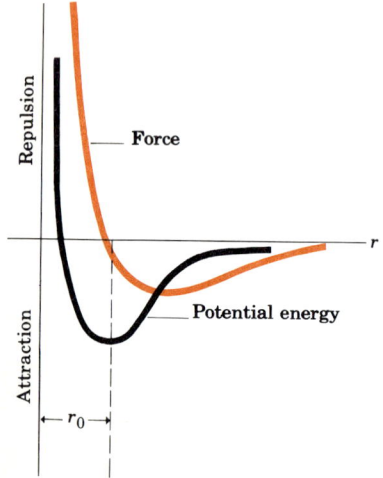

20-1 The force between two molecules (color curve) changes from an attraction when the separation is large, to a repulsion when the separation is small. The potential energy (black curve) is minimum at r_0, where the force is zero.

where the force is zero. Such a potential-energy function is often called a *potential well*.

In view of the attractive intermolecular forces, why do not all molecules eventually coalesce into matter in the liquid or solid phase? The answer is that the molecules are always in *motion*, and have kinetic energy associated with this motion. Generally speaking, the kinetic energy increases with temperature. At very low temperatures the average kinetic energy of a molecule may be much *less* than the maximum magnitude of potential energy, the "depth" of the potential well in Fig. 20-1. The molecules then condense into a liquid or solid phase with average intermolecular spacing of about r_0. But at higher temperatures the average kinetic energy becomes larger than the depth of the potential well; molecules can then escape the intermolecular force and become free to move independently, as in the gaseous phase of matter.

In solids the molecules execute vibratory motion about more or less fixed centers; if the potential well is approximately parabolic in shape near its minimum, the motion is approximately simple harmonic. The amplitudes of the vibratory motions are relatively small, and the fixed centers comprise a space lattice, which gives rise to the repeated spatial patterns and symmetry of *crystals*. A photograph of the individual, very large molecules of *necrosis virus protein* is shown in Fig. 20-2. It was taken with an electron microscope at a magnification of about 80,000. The molecules look like neatly stacked oranges. Each molecule is about 1.2×10^{-8} m in diameter.

20-2 Crystal of necrosis virus protein. The actual size of the entire crystal is about two thousandths of a millimeter (0.002 mm). (Courtesy of Ralph Wyckoff. Reprinted with the permission of Educational Services, Inc., from *Physics*, D. C. Heath and Co., Boston, 1960.)

In liquids the intermolecular distances are usually only a trifle greater than those of a solid. The molecules execute vibratory motion of greater energy about centers that are free to move but remain at approximately the same distances from one another. Liquids show a certain regularity of structure only in the immediate neighborhood of a few molecules. This is called *short-range order,* in contrast with the *long-range order* of a solid crystal.

The molecules of a gas have greater kinetic energy than those of liquids and solids; on the average, they are far away from one another, where only very small attractive forces exist. A molecule of a gas therefore moves with linear motion until a collision takes place either with another molecule or with a wall. In molecular terms, an *ideal gas* is a gas whose molecules exert *no* forces of attraction on each other. The mathematical analysis of a collection of such idealized molecules in random linear motion between collisions is called the *kinetic theory of gases.* The analysis of an ideal gas given in this chapter is a simple example of kinetic theory.

20-2 Avogadro's number

In 1811 John Dalton, often called the father of modern chemistry, first suggested that the molecules of a given chemical substance are all alike. This hypothesis, since confirmed by an overwhelming variety of evidence, makes it easy to understand why, in chemical reactions, compounds combine with each other in definite proportions by mass. It also leads directly to the concepts of *molecular mass* and *number of moles.* Specifically, *one mole* of any pure chemical compound contains a definite number of molecules, the same number for all compounds. The number of molecules in a mole is called *Avogadro's number,* denoted by N_A. (Avogadro was a contemporary of Dalton.)

Although the existence of Avogadro's number is implied by Dalton's hypothesis of identical molecules, obtaining a numerical value is not easy. One early attempt consisted of permitting a drop of oil to spread out over the surface of water. On the assumption that the oil continues to spread until the layer is only one molecule thick, and can then spread no farther, one can estimate the volume of a single molecule, and from that its mass. Thus the number of molecules in a given mass can be determined. More recent and much more precise measurements have been obtained by using x-rays to measure the distances between layers of molecules in a crystal. The presently accepted value of N_A is

$$N_A = 6.022 \times 10^{23} \text{ molecules} \cdot \text{mol}^{-1}.$$

The molecular mass M of a compound is the mass of one mole; the mass m of a single molecule is thus given by

$$M = N_A m. \tag{20-1}$$

Once Avogadro's number has been determined, it can be used to compute the mass of a molecule. The mass of 1 mol of atomic hydrogen (i.e., the atomic mass) is 1.008 g. Since by definition the number of molecules in 1 mol is Avogadro's number, it follows that the mass of a single atom of hydrogen is

$$m_H = \frac{1.008 \text{ g} \cdot \text{mol}^{-1}}{6.022 \times 10^{23} \text{ molecules} \cdot \text{mol}^{-1}} = 1.674 \times 10^{-24} \text{ g} \cdot \text{molecule}^{-1}.$$

For an oxygen molecule of molecular mass 32 g·mol^{-1},

$$m_{O_2} = \frac{32 \text{ g} \cdot \text{mol}^{-1}}{6.022 \times 10^{23} \text{ molecules} \cdot \text{mol}^{-1}} = 53.12 \times 10^{-24} \text{ g} \cdot \text{molecule}^{-1}.$$

At standard conditions, 1 mol of an ideal gas occupies 22,400 cm³. The number of molecules per cubic centimeter in an ideal gas at standard conditions is therefore

$$\frac{6.022 \times 10^{23} \text{ molecules} \cdot \text{mol}^{-1}}{22,400 \text{ cm}^3 \cdot \text{mol}^{-1}} = 2.69 \times 10^{19} \text{ molecules} \cdot \text{cm}^{-3}.$$

This is known as *Loschmidt's number,* and of course it is the same for *all* gases.

Numbers of this enormous magnitude are very difficult to comprehend. Sir James Jeans, the eminent British astrophysicist, offered a striking illustration. Each human breath contains the order of 400 cm³, and thus the order of 10^{22} molecules. Let us assume that George Washington's dying breath has by now become uniformly distributed through the atmosphere, which Jeans estimated to contain 10^{44} molecules. Thus the number of molecules in a breath is roughly equal to the number of breaths in the atmosphere, and since the total volume of the lungs (typically 2000 to 3000 cm³) is several times the volume of a single breath, each of us at this moment probably has several molecules of Washington's last breath in his lungs!

20-3 Properties of matter

Any molecular theory of matter attempts to account for the complex *macroscopic* properties of matter in bulk in terms of simple microscopic behavior and properties of its molecules. For example, by assuming that a monatomic gas consists of a large number of particles having no properties other than mass and velocity, we can explain the observed equation of state of a gas at low pressure and can then show that C_v for the gas should equal 12.5 J·mol⁻¹·K⁻¹ (see Table 18-1). If we add the hypothesis that the "particles" are not simply geometrical points but have a finite size, then the general features of the viscosity, thermal conductivity, and coefficient of diffusion of a gas can be understood, as well as the fact that polyatomic gases have larger values of C_v than monatomic gases. By assuming that there are forces between the particles, we can refine the equation of state to bring it into better agreement with that of a real gas, and we can explain the phenomena of liquefaction and solidification at low temperatures.

The *electrical* and *magnetic* properties of matter, and the emission and absorption of light by matter, call for a molecular model that is itself an aggregate of subatomic particles. Some of these are electrically charged, and the forces between molecules originate in these electric charges. This development of molecular theory will take us into the area of atomic and nuclear structure in Chapter 47. In the next section we return to the starting point and see what properties of a gas at low pressure can be explained by the simplest possible molecular model, an aggregate of particles having mass and velocity.

20-4 Kinetic theory of an ideal gas

We are now ready to develop in detail the relation between the kinetic-molecular model of an ideal gas and the ideal-gas equation of state, discussed in Sec. 17-2. We consider a container of volume V, containing N identical molecules each of mass m. We assume that each collision of

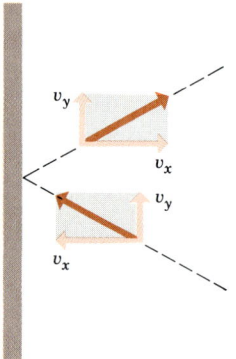

20-3 Elastic collision of a molecule with container wall. The component v_y parallel to the wall does not change; the component v_x perpendicular to the wall reverses direction. The speed v does not change.

a molecule with a wall of the container is perfectly elastic, as shown in Fig. 20–3. In each collision the component of velocity parallel to the wall is unchanged and the component perpendicular to the wall is reversed.

Our program will be to determine for a given wall area A the number of collisions per unit time, the associated momentum change, and the force needed for the momentum change. Then we can obtain an expression for the pressure, which is force per unit area. We let v_x be the *magnitude* of the x-component of velocity of a molecule. At first we shall assume that all molecules have the same v_x. This assumption, though unrealistic, helps to clarify basic ideas; and we shall show soon that it is not really necessary.

In each collision the change in x-component of momentum is $2mv_x$. To find the number of collisions in a time interval Δt, with a given wall area A, we note that, in order to experience a collision during Δt, a molecule must be within a distance $v_x \Delta t$ at the beginning of Δt, as shown in Fig. 20–4, and must be headed toward the wall. Thus to collide with A during Δt, a molecule must, at the beginning of Δt, be within a cylinder of base area A and length $v_x \Delta t$. The volume of such a cylinder is $Av_x \Delta t$.

Assuming the number of molecules per unit volume (N/V) is uniform, the *number* of molecules in this cylinder is $(N/V)(Av_x \Delta t)$. But, on the average, half of these molecules are moving *away from* the wall. Thus the number of collisions with A during Δt is

$$\frac{1}{2}\left(\frac{N}{V}\right)(Av_x \Delta t). \tag{20-2}$$

The total momentum change ΔP_x due to all these collisions is $2mv_x$ times the *number* of collisions:

$$\Delta P_x = \frac{1}{2}\left(\frac{N}{V}\right)(Av_x \Delta t)(2mv_x) = \frac{NAmv_x^2 \Delta t}{V}, \tag{20-3}$$

and the *rate* of change of momentum is

$$\frac{\Delta P_x}{\Delta t} = \frac{NAmv_x^2}{V}. \tag{20-4}$$

According to Newton's second law, this equals the average force exerted by the wall area A on the molecules; from the *third* law, this is the negative of the force exerted *on* the wall *by* the molecules. Finally, pressure p is force per unit area, and we obtain

$$p = \frac{F}{A} = \frac{Nmv_x^2}{V}. \tag{20-5}$$

Now in fact v_x is *not* the same for all molecules. But we could have sorted them into groups having the same v_x within each group, and added up the resulting contributions to the pressure. The net effect of this is simply to replace v_x^2 in Eq. (20–5) by the *average* value of v_x^2, which we denote by $\overline{v_x^2}$. Furthermore, $\overline{v_x^2}$ is related simply to the *speeds* of the molecules. The speed v (magnitude of velocity) of any molecule is related to the velocity components v_x, v_y, and v_z, by

$$v^2 = v_x^2 + v_y^2 + v_z^2.$$

20-4 A molecule moving toward the wall with speed v_x collides with the area A during the time interval Δt only if it is within a distance $v_x \Delta t$ of the wall at the beginning of the interval. All such molecules are contained within a volume $Av_x \Delta t$.

This may be averaged over all molecules:
$$\overline{v^2} = \overline{v_x^2} + \overline{v_y^2} + \overline{v_z^2}.$$
But since the x-, y-, and z-directions are all equivalent,
$$\overline{v_x^2} = \overline{v_y^2} = \overline{v_z^2}.$$
Hence
$$\overline{v_x^2} = \tfrac{1}{3}\overline{v^2},$$
and Eq. (20–5) becomes
$$pV = \tfrac{1}{3}Nm\overline{v^2} = \tfrac{2}{3}(\tfrac{1}{2}Nm\overline{v^2}). \tag{20-6}$$
But $\tfrac{1}{2}m\overline{v^2}$ is the average kinetic energy of a single molecule, and the product of this and the total number of molecules N equals the total random kinetic energy, or internal energy U. Hence the product pV equals two-thirds of the internal energy:
$$pV = \tfrac{2}{3}U. \tag{20-7}$$

By experiment, at low pressures, the equation of state of a gas is
$$pV = nRT.$$

The theoretical and experimental laws will therefore be in complete agreement if we set
$$U = \tfrac{3}{2}nRT. \tag{20-8}$$
The average kinetic energy of a single molecule is then
$$\frac{U}{N} = \tfrac{1}{2}m\overline{v^2} = \frac{3nRT}{2N}.$$

But the number of moles, n, equals the total number of molecules, N, divided by Avogadro's number, N_A, the number of molecules per mole:
$$n = \frac{N}{N_A}, \qquad \frac{n}{N} = \frac{1}{N_A}.$$
Hence
$$\tfrac{1}{2}m\overline{v^2} = \frac{3}{2}\frac{R}{N_A}T. \tag{20-9}$$

The ratio R/N_A occurs frequently in molecular theory. It is called the *Boltzmann constant, k*:
$$k = \frac{R}{N_A} = \frac{8.31 \text{ J} \cdot \text{mol}^{-1} \cdot \text{K}^{-1}}{6.02 \times 10^{23} \text{ molecules} \cdot \text{mol}^{-1}}$$
$$= 1.38 \times 10^{-23} \text{ J} \cdot \text{molecule}^{-1} \cdot \text{K}^{-1}.$$

Since R and N_A are universal constants, the same is true of k. Then
$$\tfrac{1}{2}m\overline{v^2} = \tfrac{3}{2}kT. \tag{20-10}$$

Thus the average kinetic energy per molecule depends only on the temperature, not on the pressure, volume, or molecular species. An equivalent statement can also be obtained from Eq. (20–9) by using the relation $M = N_A m$:

$$N_A(\tfrac{1}{2}m\overline{v^2}) = \tfrac{1}{2}M\overline{v^2} = \tfrac{3}{2}RT. \tag{20-11}$$

That is, the kinetic energy of a *mole* of molecules depends only on T.

From Eqs. (20–10) and (20–11) we can obtain expressions for the square root of $\overline{v^2}$, called the *root-mean-square speed* v_{rms}:

$$v_{\text{rms}} = \sqrt{\overline{v^2}} = \sqrt{\frac{3kT}{m}} = \sqrt{\frac{3RT}{M}}. \tag{20-12}$$

Finally, it is sometimes convenient to rewrite the ideal-gas equation on a molecular basis. Since $N = N_A n$ and $R = N_A k$, an alternative form of the gas equation is

$$pV = NkT. \tag{20-13}$$

Thus k may be regarded as a gas constant on a "per molecule" basis instead of the usual "per mole" basis for R.

Example 1 What is the average kinetic energy of a molecule of a gas at a temperature of 300 K?

Solution

$$\tfrac{1}{2}m\overline{v^2} = \tfrac{3}{2}kT = (\tfrac{3}{2})(1.38 \times 10^{-23} \text{ J} \cdot \text{K}^{-1})(300 \text{ K})$$
$$= 6.21 \times 10^{-21} \text{ J}.$$

Example 2 What is the total random kinetic energy of the molecules in 1 mole of a gas at a temperature of 300 K?

Solution

$$U = \tfrac{3}{2}nRT$$
$$= (\tfrac{3}{2})(1 \text{ mol})(8.314 \text{ J} \cdot \text{mol}^{-1} \cdot \text{K}^{-1})(300 \text{ K})$$
$$= 3741 \text{ J} = 894 \text{ cal}.$$

Example 3 What is the root-mean-square speed of a hydrogen molecule at 300 K?

Solution The mass of a hydrogen molecule (see Sec. 20–2) is

$$m_{\text{H}_2} = (2)(1.674 \times 10^{-27} \text{ kg}) = 3.347 \times 10^{-27} \text{ kg}.$$

Hence, from Eq. (20–12),

$$v_{\text{rms}} = \sqrt{\frac{3kT}{m}} = \sqrt{\frac{3(1.38 \times 10^{-23} \text{ J} \cdot \text{K}^{-1})(300 \text{ K})}{3.347 \times 10^{-27} \text{ kg}}}$$
$$= 1927 \text{ m} \cdot \text{s}^{-1}.$$

Alternatively,

$$v_{\text{rms}} = \sqrt{\frac{3RT}{M}} = \sqrt{\frac{3(8.314 \text{ J} \cdot \text{mol}^{-1} \cdot \text{K}^{-1})(300 \text{ K})}{2(1.008 \times 10^{-3} \text{ kg} \cdot \text{mol}^{-1})}}$$
$$= 1927 \text{ m} \cdot \text{s}^{-1}.$$

We note that in applying Eq. (20–12) using the value of R in SI units, M must be expressed in *kilograms* per mole, not grams per mole. Thus in this example $M = 2.016 \times 10^{-3} \text{ kg} \cdot \text{mol}^{-1}$, not $2.016 \text{ g} \cdot \text{mol}^{-1}$.

Example 4 Five gas molecules chosen at random are found to have speeds of 500, 600, 700, 800, and 900 m·s^{-1}. Find the rms speed; is it the same as the *average* speed?

Solution There are five molecules; the average value of v^2 is

$$v^2 = \frac{(500 \text{ m·s}^{-1})^2 + (600 \text{ m·s}^{-1})^2 + (700 \text{ m·s}^{-1})^2 + (800 \text{ m·s}^{-1})^2 + (900 \text{ m·s}^{-1})^2}{5}$$

$$= 510{,}000 \text{ m}^2\text{·s}^{-2},$$

and v_{rms} is the square root of this:

$$v_{\text{rms}} = 714 \text{ m·s}^{-1}.$$

The *average* speed v is given by

$$v = \frac{500 \text{ m·s}^{-1} + 600 \text{ m·s}^{-1} + 700 \text{ m·s}^{-1} + 800 \text{ m·s}^{-1} + 900 \text{ m·s}^{-1}}{5}$$

$$= 700 \text{ m·s}^{-1}.$$

Clearly, v_{rms} and v are not, in general, the same.

Example 5 Find the number of molecules in one cubic meter of air at atmospheric pressure and 0°C.

Solution From Eq. (20–13),

$$N = \frac{pV}{kT} = \frac{(1.013 \times 10^5 \text{ Pa})(1 \text{ m}^3)}{(1.38 \times 10^{-23} \text{ J·K}^{-1})(273 \text{ K})} = 2.69 \times 10^{25}.$$

When a gas expands against a moving piston it does work. This work is accompanied by a decrease in the random kinetic energy of the gas molecules. Conversely, when work is done in compressing a gas, the random kinetic energy of its molecules increases. But if the collisions with the walls are perfectly elastic, as we have assumed, how can a molecule gain or lose energy in a collision with a piston? To understand this, we must consider the collision of a molecule with a moving wall.

When a molecule makes a collision with a *stationary* wall, it exerts a momentary force on the wall but does no work, since the wall does not move. But if the wall is in motion, work *is* done in a collision. Thus, if the wall in Fig. 20–3 is moving to the left, work is done on it by the molecules that strike it, and their velocities (and kinetic energies) after colliding are smaller than they were before a collision. The collision is still completely elastic, since the work done on the moving piston is just equal to the decrease in the kinetic energy of the molecule. Similarly, if the piston is moving toward the right, the kinetic energy of a colliding molecule is *increased,* the increase in kinetic energy being equal to the work done on the molecule.

The assumption that individual molecules undergo *elastic* collisions with the container wall is not strictly correct. More detailed investigation has shown that in most cases molecules actually adhere to the wall for a short time, and then leave again with speeds characteristic of the temperature *of the wall*. However, the gas and the wall are ordinarily in thermal equilibrium, and the validity of our conclusions is not altered by this discovery.

20-5 Molar heat capacity of a gas

When heat flows into a system in a process at constant volume, no work is done and all of the energy inflow goes into an increase in the internal energy U of the system. From the molecular viewpoint, the internal energy of a system is the sum of the kinetic and potential energies of its molecules. If this sum can be computed as a function of temperature, then, from its rate of change with temperature, we can derive a theoretical expression for the molar heat capacity.

The simplest system is a *monatomic* ideal gas; in this case the molecular energy is wholly kinetic and is given by Eq. (20-8):

$$\text{Random kinetic energy} = U = \tfrac{3}{2}nRT.$$

If the temperature increases by dT, the random kinetic energy increases by

$$dU = \tfrac{3}{2}nR\, dT.$$

In a constant-volume process in which the temperature increases by dT, the energy flowing into a system is, by definition of C_v,

$$dQ = nC_v\, dT = dU.$$

Hence,

$$C_v = \tfrac{3}{2}R. \tag{20-14}$$

In SI units,

$$C_v = \tfrac{3}{2}(8.314\text{ J}\cdot\text{mol}^{-1}\cdot\text{K}^{-1}) = 12.47\text{ J}\cdot\text{mol}^{-1}\cdot\text{K}^{-1}.$$

The experimental values of C_v for monatomic gases in Table 18-1 are, in fact, almost exactly equal to $\tfrac{3}{2}R$. This agreement is a striking confirmation of the basic correctness of the kinetic model of a gas and did much to establish this theory at a time when many scientists still refused to accept it.

Since $C_p = C_v + R$, it follows that the theoretical ratio of molar heat capacities for a monatomic ideal gas is

$$\frac{C_p}{C_v} = \gamma = \frac{\tfrac{3}{2}R + R}{\tfrac{3}{2}R} = \frac{5}{3} = 1.67.$$

This also is in good agreement with the experimental values in Table 18-1.

Polyatomic gases present a more complicated problem. We might expect that a molecule composed of two or more atoms could have additional energy associated with *vibrational* motion of the atoms relative to each other and with *rotational* motion of the molecule about its center of mass. Also, since every atom consists of a positive nucleus and one or more electrons, there may be additional energy associated with position and motion of these electric charges.

The *temperature* of a gas depends only on the average random *translational* kinetic energy of its molecules. When heat flows into a *monatomic* gas, at constant volume, all of this energy goes into an increase in random *translational* molecular kinetic energy, as evidenced by the agreement between the measured values of C_v and the values computed from the increase in translational kinetic energy. But when heat flows into a *diatomic* or *polyatomic* gas, part of the energy may go toward increasing rotational and vibrational motion. Hence for a given temperature change a *greater* total amount of energy is required, and polyatomic gases have *larger* molar heat capacities than monatomic gases. The experimental values in Table 18-1 show this effect.

To make the discussion somewhat more quantitative, we introduce the principle of *equipartition of energy*. According to this principle, each velocity component has, on the average, an associated kinetic energy $\frac{1}{2}kT$. The number of velocity components needed to describe the motion of a molecule completely is called the number of *degrees of freedom*. For a monatomic gas it is three; for a diatomic molecule there are two possible axes of rotation, perpendicular to each other and to the molecule's axis. (Rotations about the molecule's own axis are not considered because, in ordinary collisions, there is no way for this rotational motion to change.) Thus if five degrees of freedom are assigned to a diatomic molecule, the average total kinetic energy per molecule is $5kT/2$ instead of $3kT/2$. The total internal energy of n moles is $U = 5nRT/2$, and the molar heat capacity (at constant volume) is

$$C_v = \frac{5}{2}R. \tag{20-15}$$

In SI units,

$$C_v = (5/2)(8.314 \text{ J} \cdot \text{mol}^{-1} \cdot \text{K}^{-1}) = 20.78 \text{ J} \cdot \text{mol}^{-1} \cdot \text{K}^{-1}.$$

Reference to Table 18-1 shows that this is in approximate agreement with measured values for diatomic gases. The corresponding value of γ is

$$\gamma = \frac{C_p}{C_v} = \frac{\frac{5}{2}R + R}{\frac{5}{2}R} = \frac{7}{5} = 1.40,$$

which is again in reasonable agreement with experimental values.

Additional contributions to heat capacities of gases can arise from *vibrational* motion. Molecules are never perfectly rigid; the molecular bonds can stretch and bend, permitting internal vibrations to occur. There are additional degrees of freedom and energies associated with vibrational motion. For most diatomic gases, however, vibrational motion does *not* contribute appreciably to heat capacity, for reasons that require quantum mechanics for their understanding. Briefly, the energy of a vibrational motion can change only in finite steps. If the energy change of the first step is much larger than the energy possessed by most molecules, then nearly all the molecules will remain in the minimum-energy state of motion. In this case changing the temperature does not change their average vibrational energy appreciably, and the vibrational degrees of freedom are said to be "frozen out." But in more complex molecules the gaps between permitted energy levels are sometimes much *smaller*, and then vibration *does* contribute to heat capacity. In Table 18-1, the large values of C_v for some polyatomic molecules show the contributions of vibrational energy. In addition, a molecule with three or more atoms not in a straight line has three, not two, rotational degrees of freedom.

*20-6 Distribution of molecular speeds

As mentioned above, the molecules in a gas do not all have the same speed. Direct measurements of the distribution of molecular speeds have been made by a number of methods. Figure 20-5 is a diagram of the

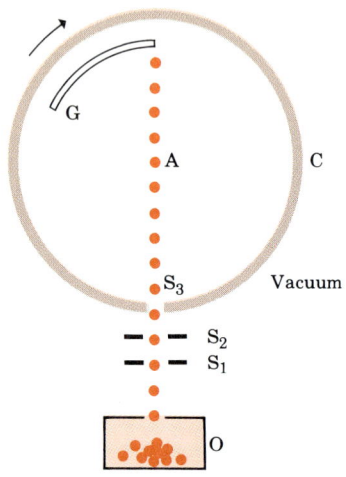

20-5 Apparatus used by Zartman and Ko in studying distribution of velocities.

apparatus used by Zartman and Ko in 1930–1934, a modification of a technique developed by Stern in 1920. Metallic silver is melted and evaporated in the oven O. A beam of silver atoms escapes through a small opening in the oven and passes through the slits S_1 and S_2 into an evacuated region. The cylinder C can be rotated at approximately 6000 rpm about the axis A. If the cylinder is at rest, the molecular beam enters the cylinder through a slit S_3 and strikes a curved glass plate G. The molecules stick to the glass plate, and the number arriving at any portion can be determined by removing the plate and measuring with a recording microphotometer the darkening that has resulted.

Now suppose the cylinder is rotated. Molecules can enter it only during the short time intervals during which the slit S_3 crosses the molecular beam. If the rotation is clockwise, as indicated, the glass plate moves toward the right while the molecules cross the diameter of the cylinder. They therefore strike the plate to the left of the point of impact when the cylinder is at rest, and the more slowly they travel, the farther to the left is their point of impact. The blackening of the plate is therefore a measure of the "velocity spectrum" of the molecular beam.

Several other techniques have been devised in more recent years for measuring distributions of molecular speeds with greater precision. Results of these experiments can be compared with a predicted speed distribution derived from a more detailed kinetic theory of gases and called the *Maxwell–Boltzmann distribution*. This theoretical distribution is shown in Fig. 20-6 for several different temperatures. At any temperature the curve shows a peak corresponding to the most probable speed; very small and very large values of v occur infrequently. As the temperature increases, the peak shifts to higher and higher speeds, corresponding to the increase in average molecular kinetic energy with temperature. Measured molecular speed distributions are generally in excellent agreement with the theoretical prediction of the Maxwell–Boltzmann distribution.

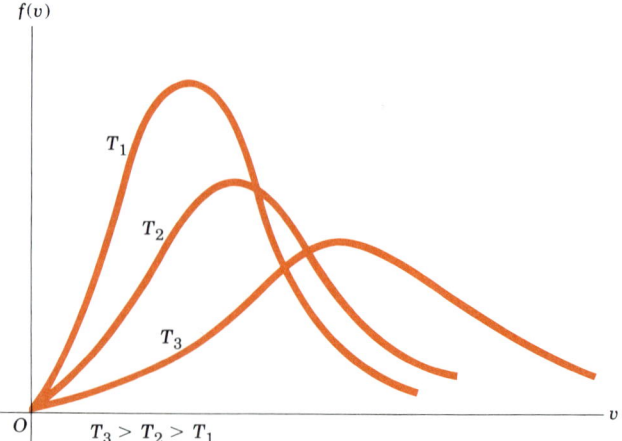

20-6 Maxwell–Boltzmann distribution curves for various temperatures. As the temperature increases, the curve becomes flatter, and its maximum shifts to higher temperature.

20-7 Crystals

Many materials can exist in a variety of solid forms; a familiar example is the element *carbon*. The black soot deposited on a kettle by a smoky

20-7 Close packing of the sodium ions (black spheres) and chlorine ions (white spheres) in a sodium chloride crystal. (Courtesy of Alan Holden, reprinted with permission of Educational Services, Inc., from *Crystals and Crystal Growing*. Doubleday and Co., New York, 1960.)

campfire is nearly pure carbon. The so-called "lead" in a pencil is not lead at all, but chiefly a different form of carbon called graphite. Diamond is a third form of solid carbon.

The remarkably dissimilar mechanical, thermal, electrical, and optical properties of these three forms of carbon can be understood in terms of the arrangement of the carbon atoms in the solid structure. In soot the atoms have no regular arrangement and are said to form an *amorphous* solid. Graphite and diamond are both crystals, but the orderly arrangement of atoms is different in the two crystal lattices. The study of the spatial arrangements of atoms (or molecules or ions) in the various types of crystal lattices is a fascinating area of physics. We shall be concerned here with only a few basic ideas that are needed to understand some of the mechanical and thermal properties of single crystals or solids composed of an aggregate of crystals, that is, *polycrystalline solids*.

The particles comprising a crystal lattice are *closely packed,* as seen in Fig. 20-2 and as suggested by the model in Fig. 20-7, where the sodium and chlorine ions of a sodium chloride crystal are represented as spheres touching one another. Actually, each ion consists of a nucleus surrounded by electrons in motion, so the boundary of an ion must be thought of as the average positions of the outermost electrons. It follows that the outer electrons of one particle in a crystal lattice come close to and even at times interpenetrate with the outer electrons of a neighboring particle. In drawings it is customary to exaggerate the distances between neighboring particles, as in the *face-centered cubic* lattice structure shown in Fig. 20-8.

If the chemical composition and density of a very small volume element of a substance are measured at many different places, and are found to be the same at all points, the substance is said to be *homogeneous*. If the physical properties determining the transport of heat, electricity, light, etc., are *the same in all directions,* the substance is said to

20-8 Symbolic representation of a sodium chloride crystal, with exaggerated distances between ions. Sodium ions are shown as black spheres, chlorine ions as colored spheres.

be *isotropic*. Cubic crystals are both homogeneous and isotropic, whereas all other crystals, although homogeneous, are not isotropic (i.e., they are *anisotropic*.)

The *anisotropy* of noncubic crystals can be understood in terms of the forces between neighboring particles in a lattice. A convenient way of visualizing these forces is shown in Fig. 20-9, an idealized model that represents a crystal lattice as a large number of spheres, each connected to its neighbors by springs. The force constants of parallel springs are imagined equal, but the force constant of each spring pointing in one direction may be different from that of springs pointing in different directions. If each of the particles represented by spheres in Fig. 20-9 is imagined to be vibrating about its equilibrium position, one obtains a rough picture of the dynamic character of a crystal lattice.

The *elastic* properties of crystals can be partially understood from a diagram such as that in Fig. 20-9. Young's modulus is simply a measure of the stiffness of the springs in the direction in which the crystal is pulled or pushed. The shear modulus depends on the springs that are stretched and those that are compressed when the crystal is twisted. One of the first tests of the correctness of a lattice structure representing a particular material is to compare a calculated value of an elastic modulus with a measured value.

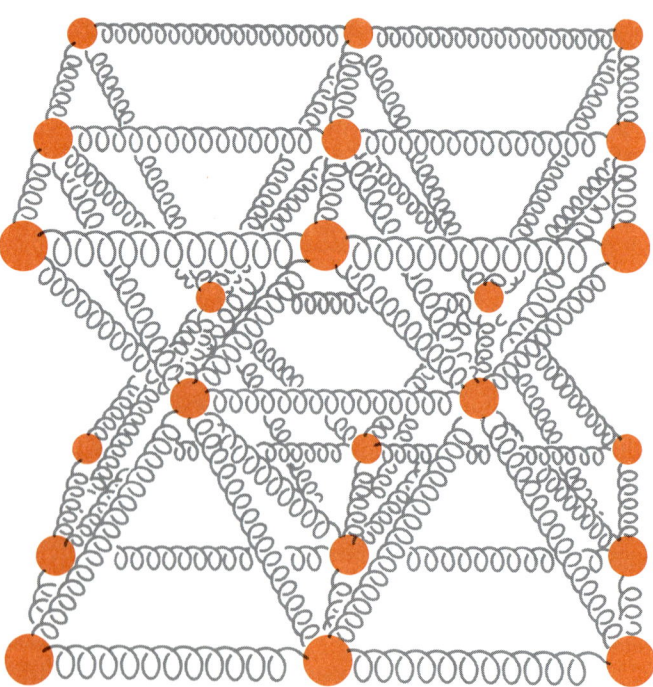

20-9 The forces between neighboring particles in a crystal may be visualized by imagining every particle to be connected to its neighbors by springs. In the case of a cubic crystal, all springs are assumed to have the same spring constant. Anisotropy is connected with differing spring constants in different directions.

*20-8 Heat capacity of a crystal

In Sec. 20-5 we derived some simple results concerning heat capacities of ideal gases from a kinetic-molecular model. Similar considerations can be applied to crystalline solids. Let us consider a crystal consisting of N

identical atoms. Each atom is bound to an equilibrium position by forces that may be pictured as springs, in the same spirit as Fig. 20-9. Thus each atom can vibrate about its equilibrium position; according to the equipartition principle introduced in Sec. 20-5, each atom should have an average kinetic energy of $\frac{1}{2}kT$ for each of its three degrees of freedom (corresponding to three components of velocity). In addition each atom has on the average some *potential* energy associated with the elastic deformation. Now, for a simple harmonic oscillator (discussed in Chapter 11), it is not hard to show that the average kinetic energy is *equal* to the average potential energy. In our crystal lattice model, each atom is essentially a three-dimensional harmonic oscillator; and it can be shown that the equality of average kinetic and potential energies also holds here, provided the "spring" forces are proportional to the displacement from the equilibrium position.

Thus we expect each atom to have an average kinetic energy $3kT/2$ and an average potential energy $3kT/2$, or an average total energy $3kT$. Thus the total energy of the whole crystal, which may be expressed in either molecular or molar terms, is

$$U = 3NkT = 3nRT. \qquad (20\text{-}16)$$

From this we conclude that the molar heat capacity of a crystal of N identical atoms should be

$$C_v = 3R. \qquad (20\text{-}17)$$

In SI units,

$$C_v = (3)(8.314 \text{ J}\cdot\text{mol}^{-1}\cdot\text{K}^{-1}) = 24.9 \text{ J}\cdot\text{mol}^{-1}\cdot\text{K}^{-1}.$$

But this is just the law of Dulong and Petit, which we encountered in Sec. 15-4; there it was regarded as an *empirical* rule, but now we have *derived* it from kinetic theory. The agreement is only approximate, to be sure, but, considering the very simple nature of the model, it is quite significant.

At low temperatures the heat capacities of most solids *decrease* with decreasing temperature. As with vibrational energy of molecules, this phenomenon requires quantum-mechanical concepts for its understanding. The vibrating atoms in the crystal lattice can gain or lose energy only in certain finite increments. At very low temperatures the quantity kT is much *smaller* than the smallest energy increment the vibrators can accommodate. As a result, at low T most of the systems remain in their *lowest* energy states because the energy required to make transitions to the next higher energy levels is not available. Thus, at low temperatures the average vibrational energy per vibrator is *less* than kT, and correspondingly the heat capacity per molecule is *less* than k. But in the other limit, where kT is *large* compared to the minimum energy increment, the classical equipartition theorem holds, and the total heat capacity is $3k$ per molecule, or $3R$ per mole, as the Dulong and Petit relation predicts. This behavior is illustrated by Fig. 15-3. Quantitative understanding of the temperature variation of specific heats was one of the triumphs of quantum mechanics during the early years of its development, early in the twentieth century.

Questions

20-1 Which has more atoms, a kilogram of hydrogen or a kilogram of lead? Which has more mass?

20-2 Chlorine is a mixture of two isotopes, one having molecular mass 35 g·mol^{-1}, the other 37 g·mol^{-1}. Which molecules move faster, on the average?

20-3 The proportion of various gases in the earth's atmosphere changes somewhat with altitude. Would you expect the proportion of oxygen at high altitude to be greater or less than at sea level?

20-4 *Comment on this statement:* When two gases are mixed, if they are to be in thermal equilibrium they must have the same average molecular speed.

20-5 In deriving the ideal-gas equation from the kinetic-molecular model, we ignored potential energy due to the earth's gravity. Is this omission justified?

20-6 In deriving the ideal-gas equation, we assumed the number of molecules to be very large, so that we could compute the average force due to many collisions; however the ideal-gas equation holds accurately only at low pressures, where the molecules are few and far between. Is this inconsistent?

20-7 Some elements of solid crystalline form have molar heat capacities *larger* than $3R$. What effects could account for this?

20-8 Considering molecular speeds, is v_{rms} equal to the *most probable* speed, as indicated by the peak on one of the curves in Fig. 20-6?

20-9 A gas storage tank has a small leak. The pressure in the tank drops more quickly if the gas is hydrogen or helium than if it is oxygen. Why?

20-10 Consider two specimens of gas at the same temperature, both having the same total mass but having different molecular masses. Which has the greater internal energy? Does your answer depend on the molecular structure of the gases?

20-11 A student claims that the enthalpy H of an ideal gas depends only on its temperature, and that the change in enthalpy in any process is given by $\Delta H = nC_p \Delta T$, irrespective of any pressure or volume changes. Is this correct?

Problems

Molecular data

N_A = Avogadro's number
 = 6.02×10^{23} molecules·mol^{-1},
Mass of a hydrogen atom = 1.67×10^{-27} kg,
Mass of a nitrogen molecule = $28 \times 1.66 \times 10^{-27}$ kg,
Mass of an oxygen molecule = $32 \times 1.66 \times 10^{-27}$ kg.

20-1 Consider an ideal gas at 0°C and at 1 atm pressure. Imagine each molecule to be, on the average, at the center of a small cube.

a) What is the length of an edge of this small cube?

b) How does this distance compare with the diameter of a molecule?

20-2 A mole of liquid water occupies a volume of 18 cm^3. Imagine each molecule to be, on the average, at the center of a small cube.

a) What is the length of an edge of this small cube?

b) How does this distance compare with the diameter of a molecule?

20-3 What is the length of the side of a cube, in a gas at standard conditions, that contains a number of molecules equal to the population of the United States (about 200 million)?

20-4 The lowest pressures readily attainable in the laboratory are of the order of 10^{-13} atm. At this pressure and ordinary temperature (say $T = 300$ K), how many molecules are present in a volume of 1 cm^3?

20-5 Experiment shows that the size of an oxygen molecule is of the order of 2×10^{-10} m. Make a rough estimate of the pressure at which the finite volume of the molecules should cause noticeable deviations from ideal-gas behavior at ordinary temperatures ($T = 300$ K).

20-6 How many moles are there in a glass of water (0.2 kg)? How many molecules?

20-7 The density of crystalline sodium chloride is 2.16 g·cm^{-3}, and its crystal structure is shown in Fig. 20-8. Find the spacing of adjacent atoms in the crystal lattice. The atomic mass of sodium is 23.0 g·mol^{-1} and that of chlorine is 35.5 g·mol^{-1}.

20-8 The density of liquid oxygen at 1 atm and -185°C is about 1.14 g·cm^{-3}. From this, estimate the volume and size of a single oxygen molecule.

20-9

a) At what temperature is the rms speed of hydrogen molecules equal to the speed of the first earth satellite (about 18,000 mi hr^{-1})?

b) At what temperature is the rms speed equal to the escape speed from the gravitational field of the earth?

20-10 At what temperature is the rms speed of oxygen molecules equal to the rms speed of hydrogen molecules at 0°C?

20-11 Isotopes of uranium are sometimes separated by gaseous diffusion, using the fact that the rms speeds of their molecules in vapor are slightly different, and hence the vapors diffuse at slightly different rates. Assuming the atomic masses for ^{235}U and ^{238}U are 235 g mol^{-1} and 238 g mol^{-1}, respectively, what is the ratio of the rms speed of the ^{235}U atoms in the vapor to that of the ^{238}U atoms, assuming the temperature is uniform?

20-12 A flask contains a mixture of mercury vapor, neon, and helium. Compare

a) the average kinetic energies of the three types of atoms, and

b) the root-mean-square speeds.

20-13 The speed of propagation of a sound wave in air at 27°C is about 350 m·s^{-1}. Compare this with

a) v_{rms} for nitrogen molecules;

b) the root-mean-square value of v_x at this temperature.

(If sound propagation were an isothermal process, which it ordinarily is not, the speed of sound would be equal to $(v_x)_{\text{rms}}$. See Sec. 21-5.)

20-14

a) What is the average translational kinetic energy of a molecule of oxygen at a temperature of 300 K?

b) What is the average value of the square of its speed?

c) What is the root-mean-square speed?

d) What is the momentum of an oxygen molecule traveling at this speed?

e) Suppose a molecule traveling at this speed bounces back and forth between opposite sides of a cubical vessel 10 cm on a side. What is the average force it exerts on the walls of the container? (Assume that the molecule's velocity is perpendicular to the two sides that it strikes.)

f) What is the average force per unit area?

g) How many molecules traveling at this speed are necessary to produce an average pressure of 1 atm?

h) Compare with the number of oxygen molecules actually contained in a vessel of this size, at 300 K and atmospheric pressure.

i) Your answer for (h) should be three times as large as the answer for (g). Where does this discrepancy arise?

20-15 The surface of the sun has a temperature of about 6000 K and consists largely of hydrogen.

a) Find the rms speed and average kinetic energy of a hydrogen molecule.

b) The escape velocity for a particle to leave the gravitational influence of the sun is given by $(2GM/R)^{1/2}$, where M is the sun's mass, R its radius, and G the gravitational constant (Sec. 6-4, Example 6). Can appreciable quantities of hydrogen escape from the sun's gravitational field?

20-16 What is the total random kinetic energy of the molecules in 1 mole of helium at a temperature of (a) 300 K? (b) 301 K? (c) Compare the difference between these with the change in internal energy of 1 mole of helium when its temperature is increased by 1 K, as computed from the relation $\Delta U = nC_v \Delta T$.

20-17 Smoke particles in the air typically have masses of the order of 10^{-16} kg. The Brownian motion of these particles resulting from collisions with air molecules can be observed with a microscope.

a) Find the root-mean-square speed of Brownian motion for such a particle in air at 300 K.

b) Would the speed be different if the particle were in hydrogen gas at the same temperature? Explain.

20-18 It can be shown that a projectile of mass m can "escape" from the earth's gravitational field if it is launched vertically upward with a kinetic energy greater than mgR, where g is the acceleration due to gravity at the earth's surface, and R is the earth's radius.

a) At what temperature would an average oxygen molecule have this much energy? An average hydrogen molecule?

b) Can you explain on this basis why the earth's atmosphere contains very little hydrogen?

20-19 Compute the increase in gravitational potential energy of an oxygen molecule for an increase in elevation of 1 m near the earth's surface. At what temperature is this equal to the average kinetic energy of oxygen molecules?

20-20

a) For what mass of molecule or particle is v_{rms} equal to 1 m·s^{-1} at a temperature of 300 K?

b) If the particle is an ice crystal, how many molecules does it contain?

c) Estimate the size of the particle. Would it be visible to the naked eye? Through a microscope?

20-21 The oven in Fig. 20-5 contains bismuth at a temperature of 840 K; the drum is 10 cm in diameter and rotates at 6000 rev·min^{-1}. Find the displacement on the glass plate G, measured from a point directly opposite the slit, of the points of impact of the molecules Bi and Bi$_2$. Assume that all molecules of each species travel with the rms speed appropriate to that species.

20-22 Suppose the two atoms in an oxygen molecule are represented as two point masses 2×10^{-10} m apart.

a) Calculate the moment of inertia of a molecule about an axis through the midpoint, perpendicular to the line joining the atoms.

b) If the rotational kinetic energy of the molecule is $\frac{1}{2}kT$, what is the root-mean-square angular velocity of rotation at $T = 300$ K?

20-23 Compute the specific heat capacity at constant volume of hydrogen gas, and compare it with the specific heat capacity of water.

20-24 Calculate the molar heat capacity of water vapor, assuming the triatomic molecule has three translational and three rotational degrees of freedom and that vibrational motion does not contribute. The actual specific heat capacity of water vapor at low pressures is about 0.48 cal·g^{-1}·K^{-1}. Compare this with your calculation, and comment on the actual role of vibrational motion.

20-25 For each of the triatomic gases in Table 18-1, compute the value of C_v on the assumption that there is no vibrational energy. Compare with the measured values in the table, and compute the fraction of the total heat capacity which is due to vibration for each of the three gases.

MECHANICAL WAVES 21

(a) Transverse displacement

(b) Longitudinal displacement

(c) Longitudinal and transverse displacement

21-1 Introduction

A *wave* is any disturbance from an equilibrium condition that travels or *propagates* with time from one region of space to another. There are many examples of waves in everyday experience, and examples of wave phenomena are found in all branches of physics. The concept of waves, in fact, is one of the most important unifying elements in all of physics.

Waves on the surface of a body of water, produced by the wind or by some other disturbance, are a familiar sight. A sound is carried by means of traveling waves in the intervening atmosphere between source and listener. Many of the observed properties of light are best explained by a wave theory, and light waves are of the same fundamental nature as radio waves, infrared and ultraviolet waves, x-rays, and gamma rays. One of the outstanding developments of twentieth-century physics has been the discovery that all matter is endowed with wave properties and that a beam of electrons, for example, is reflected by a crystal in much the same way as is a beam of x-rays.

The waves most easily understood are *mechanical waves,* in which the wave travels through some material medium as it is displaced from an equilibrium state. Imagine a medium consisting of a large number of particles, each connected or coupled to its neighbors by *elastic* material. If one end of the medium is disturbed or displaced in any way, the displacement does not occur immediately at all other parts of the medium. The original displacement gives rise to an elastic force in the material adjacent to it; then the next particle is displaced, and then the next, and so on. In other words, *the displacement propagates through the medium with a definite speed.*

In Fig. 21-1a the medium is a spring, such as a "Slinky," or even just a wire under tension. If the left end is given a small displacement in a direction perpendicular to the medium, this transverse displacement will occur at successive intervals of time at each coil of the spring, resulting in the propagation of a *transverse pulse* along the spring.

In Fig. 21-1b the medium is a liquid or a gas contained in a tube closed at the right end with a rigid wall and at the left end with a

(d) Torsional displacement

21-1 Propagation of disturbances.

movable piston. If the piston is moved slightly toward the right, a *longitudinal pulse* propagates through the medium in the tube. Longitudinal waves can also travel in solid materials.

In Fig. 21–1c the medium is a liquid contained in a trough. The horizontal motion of a flat piece of wood at the left end will create a displacement of the liquid that is *both* longitudinal and transverse, and this disturbance travels along the medium.

In Fig. 21–1d the medium is a set of "dumbbells" connected to a steel strip. A slight rotation of the lowest dumbbell constitutes a *torsional displacement,* which propagates up the medium with a finite speed.

21–2 Periodic waves

When the motions of the particles are at right angles to the direction of travel of the wave, the wave is called *transverse*. If the particles move back and forth along the direction of propagation, the wave is called *longitudinal*. Suppose that one end of a medium is forced to vibrate periodically, the displacement y (either transverse or longitudinal) varying with time according to the equation of simple harmonic motion:

$$y = \begin{cases} A \sin \omega t, \\ \quad \text{or} \\ A \cos \omega t. \end{cases}$$

The resulting continuous train of disturbances, traveling with a speed depending on the properties of the medium, is called a *wave*.

As a specific example, suppose that one end of a stretched string is forced to vibrate periodically in a transverse direction with simple harmonic motion of amplitude A, frequency f, and period $\tau = 1/f$. For the present we shall assume the string to be long enough so that any effects at the far end need not be considered. A *continuous train* of transverse sinusoidal waves then advances along the string. The shape of a portion of the string near the end, at intervals of $\frac{1}{8}$ of a period, is shown in Fig. 21–2 for a total time of one period. The string is assumed to have been vibrating for a sufficiently long time so that the shape of the string is sinusoidal for an indefinite distance from the driven end. It will be seen from the figure that the waveform advances steadily toward the right, as indicated by the short arrow pointing to one particular wave crest, while any one point on the string (see the black dot) oscillates about its equilibrium position with simple harmonic motion. It is important to distinguish between the motion of the *waveform,* which moves with constant speed c *along* the string, and the motion of *a particle of the string,* which is simple harmonic and *transverse* to the string.

The distance between two successive maxima (or between any two successive points in the same phase) is the *wavelength* of the wave and is denoted by λ. Since the waveform, traveling with constant speed c, advances a distance of one wavelength in a time interval of one period, it follows that $c = \lambda/\tau$, or

$$c = f\lambda. \tag{21–1}$$

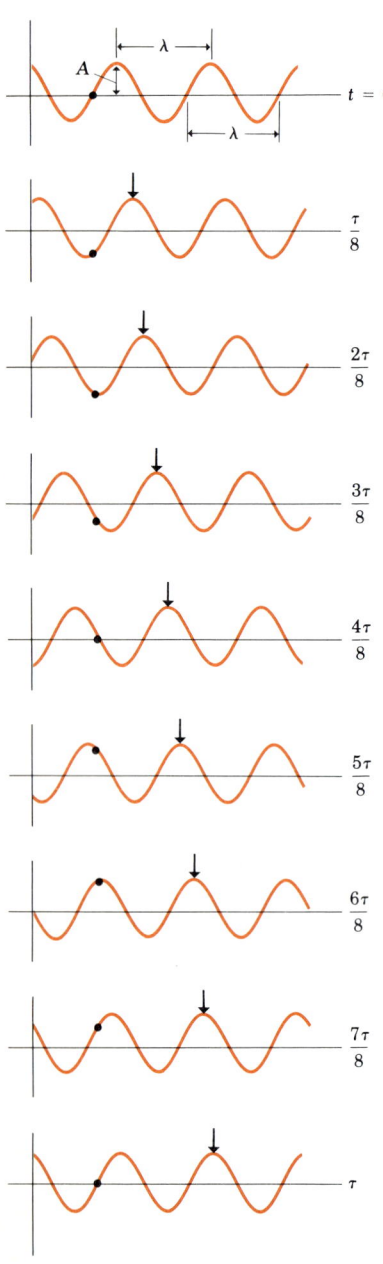

21–2 A sinusoidal transverse wave traveling toward the right, shown at intervals of one-eighth of a period.

That is, *the speed of propagation equals the product of frequency and wavelength.*

21-3 A sinusoidal longitudinal wave traveling toward the right, shown at intervals of one-eighth of a period.

To understand the mechanics of a *longitudinal* wave, consider a long tube filled with a fluid and provided with a plunger at the left end, as shown in Fig. 21–3. The dots represent particles of the fluid. Suppose the plunger is forced to undergo a simple harmonic vibration parallel to the direction of the tube. During a part of each oscillation, a region whose pressure is above the equilibrium pressure is formed. Such a region is called a *condensation* and is represented by a darkly shaded area. Following the production of a condensation, a region is formed in which the pressure is lower than the equilibrium value. This is called a *rarefaction* and is represented by a lightly shaded area. The condensations and rarefactions move to the right with constant speed c, as indicated by successive positions of the small vertical arrow. The speed of longitudinal waves in air (sound) at 20°C is 344 m·s^{-1} or 1130 ft·s^{-1}. The motion of a single particle of the medium, shown by a heavy black dot, is simple harmonic, parallel to the direction of propagation.

The wavelength is the distance between two successive condensations or two successive rarefactions, and the same fundamental equation, $c = f\lambda$, holds in this, as in all types of periodic or repetitive waves.

Example What is the wavelength of a sound wave having a frequency of 262 Hz (the approximate frequency of the note "middle C" on the piano)?

Solution At 20°C the speed of sound in air is 344 m·s^{-1}, as mentioned above. From Eq. (21–1),

$$\lambda = \frac{c}{f} = \frac{344 \text{ m·s}^{-1}}{262 \text{ s}^{-1}} = 1.31 \text{ m}.$$

21-3 Mathematical description of a wave

For detailed analysis of wave motion we need a mathematical language for describing waves. Transverse waves on a string will be considered first, but the methods introduced are applicable to any wave motion.

We consider a stretched string which in its equilibrium position lies along the x-axis of a coordinate system. During wave motion each particle, labeled by its equilibrium position x, is displaced some distance y in the transverse direction. The value of y depends on which particle of string is being described (that is, on x) and also on the time t at which we look at it. Thus y is a *function* of x and t: $y = f(x, t)$. If this function is known for a particular wave motion, it can be used to predict the position of any particle at any time. From this the velocity and acceleration of any point, the shape of the string, its slope at any point, and any other kinematic information, can be obtained.

Thus the function $y = f(x, t)$, once it is known, constitutes a complete description of the motion. Such a function is called a *wave function;* in this section we study wave functions for sinusoidal waves.

As a sinusoidal wave travels from left to right on a string, each particle undergoes simple harmonic motion, but successive particles have successively larger delays, or phase lags, compared to the motion at the left end (at $x = 0$) where the motion originates. Let the displacement of a particle at the origin ($x = 0$) be given by

$$y = A \sin \omega t = A \sin 2\pi f t. \qquad (21\text{-}2)$$

The time required for the wave disturbance to travel from $x = 0$ to some point x to the right of the origin is given by x/c, where c is the wave speed. The motion of point x at time t is the same as the motion of point $x = 0$ at the earlier time $t - x/c$. Thus the displacement of point x at time t is obtained simply by replacing t in Eq. (21-2) by $(t - (x/c))$, and we find

$$y(x, t) = A \sin \omega \left(t - \frac{x}{c}\right)$$
$$= A \sin 2\pi f \left(t - \frac{x}{c}\right). \qquad (21\text{-}3)$$

The notation $y(x, t)$ is a reminder that the displacement y is a function of both the location x of the point and the time t.

Equation (21-3) can be rewritten in several alternative forms, conveying the same information in different ways. In terms of the period τ and wavelength λ, we find, using Eq. (21-1),

$$y(x, t) = A \sin 2\pi \left(\frac{t}{\tau} - \frac{x}{\lambda}\right). \qquad (21\text{-}4)$$

Another convenient form is obtained by defining a quantity k, called the *propagation constant* or the *wave number:*

$$k = \frac{2\pi}{\lambda}. \qquad (21\text{-}5)$$

In terms of k and ω, the wavelength–frequency relation $c = \lambda f$ becomes

$$\omega = ck, \qquad (21\text{-}6)$$

and Eq. (21-4) can be written

$$y(x, t) = A \sin(\omega t - kx). \qquad (21\text{-}7)$$

Which of these various forms is used is a matter of convenience in a specific problem; fundamentally they all say the same thing. It should be mentioned in passing that some authors define the wave number as $1/\lambda$ rather than $2\pi/\lambda$. In that case, our k is still called the propagation constant.

At any given *time t*, Eq. (21-3), (21-4), or (21-7) gives the displacement y of a particle from its equilibrium position, as a function of the *coordinate x* of the particle. If the wave is a transverse wave in a string, the equation represents the *shape* of the string at that instant, as if we had taken a photograph of the string. Thus at time $t = 0$,

$$y = A \sin(-kx) = -A \sin kx = -A \sin 2\pi \frac{x}{\lambda}.$$

This curve is plotted in Fig. 21-4.

At any given *coordinate x,* Eq. (21-3), (21-4), or (21-7) gives the displacement y of the particle at that coordinate, as a function of *time*. Thus at the position $x = 0$,

$$y = A \sin \omega t = A \sin 2\pi \frac{t}{\tau}.$$

This curve is plotted in Fig. 21-5.

The above formulas may be used to represent a wave traveling in the *negative x*-direction by making a simple modification. In this case the displacement of point x at time t is the same as the motion of point $x = 0$ at the *later* time $(t + (x/c))$, and in Eq. (21-2) we must replace t by $(t + (x/c))$. Thus, for a wave traveling in the negative x-direction,

$$y = A \sin 2\pi f \left(t + \frac{x}{c} \right) = A \sin 2\pi \left(\frac{t}{\tau} + \frac{x}{\lambda} \right)$$

$$= A \sin(\omega t + kx). \qquad (21\text{-}8)$$

It is essential to distinguish carefully between the *speed of propagation c* of the waveform and the *particle speed v* of a particle of the medium in which the wave is traveling. The wave speed c is given by

$$c = \lambda f = \frac{\omega}{k}. \qquad (21\text{-}9)$$

The particle speed v for any point in a transverse wave (that is, at a fixed value of x) is obtained by taking the derivative of y with respect to t, holding x constant. Such a derivative is called a *partial derivative* and is written $\partial y/\partial t$. Thus for a sinusoidal wave given by

$$y = A \sin(\omega t - kx), \qquad (21\text{-}10)$$

we have

$$v = \frac{\partial y}{\partial t} = \omega A \cos(\omega t - kx). \qquad (21\text{-}11)$$

The *acceleration* of the point is the *second* partial derivative:

$$a = \frac{\partial^2 y}{\partial t^2} = -\omega^2 A \sin(\omega t - kx). \qquad (21\text{-}12)$$

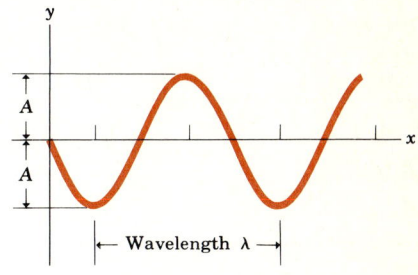

21-4 Wave form at $t = 0$.

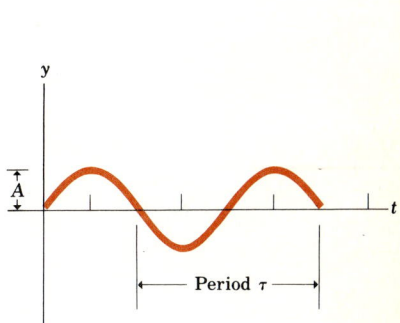

21-5 Vibration at $x = 0$.

One may also compute partial derivatives with respect to x, holding t constant. The first derivative $\partial y/\partial x$ is the *slope* of the string at any point. The second partial derivative with respect to x is

$$\frac{\partial^2 y}{\partial x^2} = -k^2 A \sin(\omega t - kx). \qquad (21\text{-}13)$$

It follows from Eqs. (21-12) and (21-13) that

$$\frac{\partial^2 y/\partial t^2}{\partial^2 y/\partial x^2} = \frac{\omega^2}{k^2} = c^2,$$

since $c = \omega/k$. The *partial differential equation*

$$\frac{\partial^2 y}{\partial t^2} = c^2 \frac{\partial^2 y}{\partial x^2} \qquad (21\text{-}14)$$

is one of the most important in all of physics. It is called the *wave equation,* and whenever it occurs, the conclusion is made immediately that y propagates as a traveling wave along the x-axis with a wave speed c.

An example of application of this wave-function formulation is given in the next section.

21-4 Speed of a transverse wave

The speed of propagation c of a transverse wave is determined by two mechanical quantities, the *mass* per unit length of the string, and the *tension*. We shall now develop that relationship by analyzing a particularly simple type of wave motion.

Consider a perfectly flexible string, as shown in Fig. 21-6, under a tension S and with linear density (mass per unit length) μ. In Fig. 21-6a the string is at rest. At time $t = 0$, a constant transverse force F is applied at the left end of the string. It might be expected that the end would move with constant acceleration; this would certainly occur if the force were applied to a point mass. But here the effect of the force is to set successively more and more mass in motion. Force is the rate of change of momentum mv; in this case mv changes because m changes, not v. Hence the end of the string moves upward with constant *velocity* v.

Figure 21-6b shows the shape of the string after a time t has elapsed. All points of the string at the left of the point P are moving with speed v,

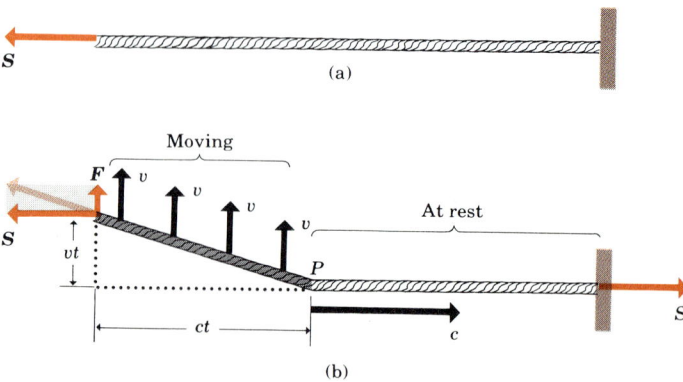

21-6 Propagation of a transverse disturbance in a string.

whereas all points at the right of P are still at rest. The boundary point P between the moving and the stationary portions is traveling to the right with the *speed of propagation* c. The left end of the string has moved up a distance vt and the boundary point P has advanced a distance ct.

The tension at the left end of the string is the vector sum of the forces $\mathbf{S}$ and $\mathbf{F}$. Because there is no motion along the length of the string, there is no unbalanced horizontal force, so S, the magnitude of the horizontal component, does not change when the string is displaced. As a result of the increased tension, the string clearly stretches somewhat. It can be shown that, for small displacements, the amount of stretch is approximately proportional to the increase in tension, as would be expected from Hooke's Law.

The speed of propagation c can be calculated by applying the impulse–momentum relation to the portion of the string that is in motion at time t, that is, the darkly shaded portion in Fig. 21-6. We set the transverse impulse (transverse force × time) equal to the change of transverse momentum of the moving portion (mass × transverse velocity). The *impulse* of the transverse force F in time t is Ft. By similar triangles,

$$\frac{F}{S} = \frac{vt}{ct}, \qquad F = S\frac{v}{c}.$$

Hence,

$$\text{Transverse impulse} = S\frac{v}{c}t.$$

The mass of the moving portion is the product of the mass per unit length μ and the length ct. (In its displaced position, the section of string is stretched somewhat, making its mass per unit length somewhat *less* than μ and its length somewhat *greater* than ct. But the mass of the section is still μct, the same as in the undisplaced position.) Hence,

$$\text{Transverse momentum} = \mu ctv.$$

We note that the momentum increases with time *not* because mass is moving faster, as was usually the case in Chapter 7, but because *more mass* is brought into motion. Nevertheless, the impulse of the force F is still equal to the total change in momentum of the system. Applying this relation, we obtain

$$\underbrace{S\frac{v}{c}t}_{\text{Impulse}=Ft} = \underbrace{\mu ctv}_{\text{Transverse momentum}},$$

and therefore

$$c = \sqrt{\frac{S}{\mu}} \quad \text{(transverse wave)}. \qquad (21\text{-}15)$$

[← Force; ↙ mass/unit length]

Thus the speed of propagation of a transverse pulse in a string depends only on the tension (a force) and the mass per unit length. Although the above calculation of wave speed considered only a very special kind of pulse, a little thought shows that *any* shape of wave disturbance can be considered as a series of pulses with different rates of transverse displacement. Thus, although derived for a special case, Eq.

(21–15) is valid for *any* transverse wave motion on a string, including, in particular, the sinusoidal and other periodic waves discussed in Sections 21–2 and 21–3.

Example A certain string has a linear mass density of 0.25 kg·m^{-1} and is stretched with a tension of 25 N. One end is given a sinusoidal motion with frequency 5 Hz and amplitude 0.01 m. At time $t = 0$ the end has zero displacement and is moving in the $+y$-direction.

a) Find the wave speed, amplitude, angular frequency, period, wavelength, and wave number.
b) Write a wave function describing the wave.
c) Find the position of the point at $x = 0.25$ m at time $t = 0.1$ s.
d) Find the transverse velocity of the point at $x = 0.25$ m at time $t = 0.1$ s.
e) Find the slope of the string at the point $x = 0.25$ m at time $t = 0.1$ s.

Solution
a) From Eq. (21–15), the wave speed is

$$c = \sqrt{\frac{S}{\mu}} = \sqrt{\frac{25 \text{ N}}{0.25 \text{ kg}\cdot\text{m}^{-1}}} = 10 \text{ m}\cdot\text{s}^{-1}.$$

The amplitude A is just the amplitude of the motion of the endpoint, $A = 0.01$ m. The angular frequency is

$$\omega = 2\pi f = 2\pi(5 \text{ s}^{-1}) = 31.4 \text{ s}^{-1}.$$

The period is $\tau = 1/f = 0.2$ s. The wavelength is obtained from Eq. (21–1):

$$\lambda = \frac{c}{f} = \frac{10 \text{ m}\cdot\text{s}^{-1}}{5 \text{ s}^{-1}} = 2 \text{ m}.$$

The wave number is obtained from Eq. (21–5) or Eq. (21–6):

$$k = \frac{2\pi}{\lambda} = \frac{2\pi}{2 \text{ m}} = 3.14 \text{ m}^{-1}, \quad \text{or} \quad k = \frac{\omega}{c} = \frac{31.4 \text{ s}^{-1}}{10 \text{ m}\cdot\text{s}^{-1}} = 3.14 \text{ m}^{-1}.$$

b) The wave function, using the form of Eq. (21–4), is

$$y = (0.01 \text{ m}) \sin 2\pi \left(\frac{t}{0.2 \text{ s}} - \frac{x}{2 \text{ m}}\right)$$
$$= (0.01 \text{ m}) \sin[(31.4 \text{ s}^{-1})t - (3.14 \text{ m}^{-1})x].$$

Alternatively, using the form of Eq. (21–7),

$$y = (0.01 \text{ m}) \sin[(31.4 \text{ s}^{-1})t - (3.14 \text{ m}^{-1})x].$$

c) The displacement of the point $x = 0.25$ m at time $t = 0.1$ s is given by substituting these values into either of the above wave equations:

$$y = (0.01 \text{ m})\sin 2\pi\left(\frac{0.1 \text{ s}}{0.2 \text{ s}} - \frac{0.25 \text{ m}}{2 \text{ m}}\right)$$
$$= (0.01 \text{ m})\sin 2\pi(0.375) = 0.00707 \text{ m}.$$

d) The transverse velocity at any point at any time is given by

$$v = \frac{\partial y}{\partial t} = \omega A \cos(\omega t - kx).$$

In the present problem,

$$v = (31.4 \text{ s}^{-1})(0.01 \text{ m}) \cos[(31.4 \text{ s}^{-1})(0.1 \text{ s}) - (3.14 \text{ m}^{-1})(0.25 \text{ m})]$$
$$= -0.22 \text{ m} \cdot \text{s}^{-1}.$$

e) The slope at any point at any time is given by

$$\frac{\partial y}{\partial x} = -kA \cos(\omega t - kx).$$

In the present problem,

$$\frac{\partial y}{\partial x} = -(3.14 \text{ m}^{-1})(0.01 \text{ m}) \cos[(31.4 \text{ s}^{-1})(0.1 \text{ s}) - (3.14 \text{ m}^{-1})(0.25)]$$
$$= 0.022.$$

For a wave traveling in the $+x$-direction, the transverse velocity is positive at every point where the slope is negative, and negative where the slope is positive. Can you prove this?

21-5 Speed of a longitudinal wave

The propagation speed of longitudinal as well as transverse waves is determined by the mechanical properties of the medium. Figure 21-7 shows a fluid (liquid or gas) of density ρ in a tube of cross-sectional area A and under a pressure p. In Fig. 21-7a the fluid is at rest. At time $t = 0$, the piston at the left end of the tube is set in motion toward the right with a speed v. Figure 21-7b shows the fluid after a time t has elapsed. All portions of the fluid at the left of point P are moving with speed v, whereas all portions at the right of P are still at rest. The boundary between the moving and the stationary portions travels to the right with the speed of propagation c. At time t the piston has moved a distance vt and the boundary has advanced a distance ct. As in the case of a transverse disturbance in a string, the speed of propagation can be computed from the impulse–momentum theorem.

The quantity of fluid set in motion in time t is the amount that originally occupied a volume of length ct and of cross-sectional area A. The mass of this fluid is therefore ρctA and the longitudinal momentum it has acquired is

$$\text{Longitudinal momentum} = \rho ctAv.$$

We next compute the increase of pressure, Δp, in the moving fluid. The original volume of the moving fluid, Act, has been decreased by an amount Avt. From the definition of bulk modulus B (see Chapter 10),

$$B = \frac{\text{Change in pressure}}{\text{Fractional change in volume}} = \frac{\Delta p}{Avt/Act}.$$

Therefore,

$$\Delta p = B\frac{v}{c}.$$

The pressure in the moving fluid is therefore $p + \Delta p$, and the force exerted on it by the piston is $(p + \Delta p)A$. The net force on the moving fluid (see Fig. 21-7b) is ΔpA, and the longitudinal impulse is

$$\text{Longitudinal impulse} = \Delta pAt = B\frac{v}{c}At.$$

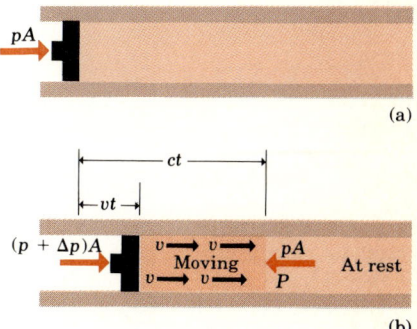

21-7 Propagation of a longitudinal disturbance in a fluid confined in a tube.

Applying the impulse–momentum theorem, we find

$$B\frac{v}{c}At = \rho ctAv,$$

and therefore

$$c = \sqrt{\frac{B}{\rho}} \quad \text{(longitudinal wave)}. \tag{21-16}$$

The speed of propagation of a longitudinal pulse in a fluid therefore depends only on the bulk modulus and the density of the medium.

When a longitudinal wave propagates in a solid bar, the situation is somewhat different from that of a fluid confined in a tube of constant cross section, since the bar expands slightly sidewise when it is compressed longitudinally. It can be shown, by the same type of reasoning as that just given, that the velocity of a longitudinal pulse in the bar is given by

$$c = \sqrt{\frac{Y}{\rho}} \quad \text{(longitudinal wave)}, \tag{21-17}$$

where Y is Young's modulus, defined in Chapter 10.

As with the calculation for the transverse wave on a string, Eqs. (21-16) and (21-17) are valid for all wave motions, not just the special case discussed here. In particular, they are valid for sinusoidal waves as discussed in Sec. 21-3.

Since the *shape* of a fluid does not change when a longitudinal wave passes through it, it is not as easy to visualize the relation between particle motion and wave motion as it is for transverse waves in a string. Figure 21-8 may help in correlating these motions. To use this figure, cut a slit about $\frac{1}{16}$ in. wide and $3\frac{1}{4}$ in. long in a 3-in. × 5-in. card (or fasten two cards edge-to-edge with a $\frac{1}{16}$-in. gap), place the card over the figure with the slit at the top of the diagram, and move the card downward with constant velocity. The portions of the sine curves that are visible through the slit correspond to a row of particles along which there is traveling a longitudinal, sinusoidal wave. Note that each particle executes simple harmonic motion about its equilibrium position, with a phase that increases continuously along the slit, while the regions of maximum condensation and rarefaction move from left to right with constant speed. Moving the card upward simulates a wave traveling from right to left.

Example 1 Determine the speed of sound waves in water, and find the wavelength of a wave having a frequency of 262 Hz.

Solution We use Eq. (21-16) to find the wave speed. From Table 10-2, we find that the compressibility of water, which is the reciprocal of the bulk modulus, is $k = 45.8 \times 10^{-11}$ Pa^{-1}. Thus $B = (1/49) \times 10^{11}$ Pa. The density of water is $\rho = 1.00 \times 10^3$ kg·m^{-3}. We obtain

$$c = \sqrt{\frac{B}{\rho}} = \sqrt{\frac{(1/45.8) \times 10^{11} \text{ Pa}}{1.00 \times 10^3 \text{ kg·m}^{-3}}} = 1478 \text{ m·s}^{-1}.$$

The wavelength is given by

$$\lambda = \frac{c}{f} = \frac{1478 \text{ m·s}^{-1}}{262 \text{ s}^{-1}} = 5.64 \text{ m}.$$

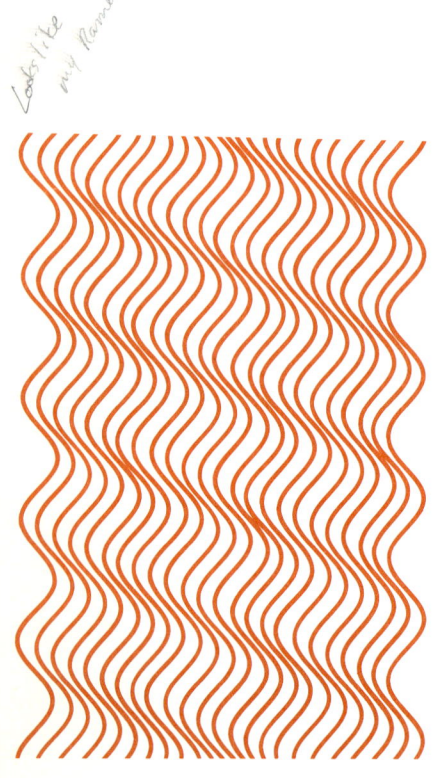

21-8 Diagram for illustrating longitudinal traveling waves.

For comparison, we note that a wave of this frequency in air has a wavelength of 1.31 m, as obtained in the example of Sec. 21-2.

Example 2 What is the speed of longitudinal sound waves in a steel rod?

Solution We use Eq. (21-17). From Table 10-1, $Y = 2.0 \times 10^{11}$ Pa, and from Table 12-1, $\rho = 7.8$ g·cm^{-3} = 7.8×10^3 kg·m^{-3}. From these,

$$c = \sqrt{\frac{Y}{\rho}} = \sqrt{\frac{2.0 \times 10^{11} \text{ Pa}}{7.8 \times 10^3 \text{ kg·m}^{-3}}} = 5064 \text{ m·s}^{-1}.$$

21-6 Adiabatic character of a longitudinal wave

It is a familiar fact that compression of a fluid causes a rise in its temperature unless heat is withdrawn in some way. Conversely, an expansion is accompanied by a temperature decrease unless heat is added. As a longitudinal wave advances through a fluid, the regions that are compressed at any instant are slightly warmer than those that are expanded. The condition is present, therefore, for the conduction of heat from a condensation to a rarefaction. The quantity of heat conducted per unit time and per unit area depends on the thermal conductivity of the fluid and upon the distance between a condensation and its adjacent rarefaction (half a wavelength). For ordinary frequencies, say from 20 to 20,000 Hz (the frequency range in which the human ear is sensitive) and for even the best known heat conductors, the wavelength is too large and the thermal conductivity too small for an appreciable amount of heat to flow. The compressions and rarefactions are therefore *adiabatic* rather than isothermal.

The change in volume produced by a given change of pressure depends upon whether the compression (or expansion) is adiabatic or isothermal. Thus there are two bulk moduli, the *adiabatic* bulk modulus B_{ad} and the *isothermal* bulk modulus. The expression for the speed of a longitudinal wave should therefore be written

$$c = \sqrt{\frac{B_{ad}}{\rho}}. \qquad (21\text{-}18)$$

In the case of an ideal gas, the relation between pressure p and volume V during an adiabatic process is given (as discussed in Sec. 18-13) by

$$pV^\gamma = \text{constant}, \qquad (21\text{-}19)$$

where γ is the ratio of the heat capacity at constant pressure to that at constant volume. The bulk modulus B is defined in general as

$$B = -V\frac{dp}{dV}.$$

Taking the derivative of Eq. (21-19) with respect to V yields

$$\frac{dp}{dV}V^\gamma + \gamma p V^{\gamma-1} = 0$$

Dividing by $V^{\gamma-1}$ and rearranging, we find that the adiabatic bulk modulus for an ideal gas is simply

$$B_{ad} = \gamma p. \qquad (21\text{-}20)$$

For an isothermal process, on the other hand, $pV = \text{constant}$, and the isothermal bulk modulus is

$$B_{\text{iso}} = p. \tag{21-21}$$

In each case the bulk modulus (characterizing the material's resistance to being compressed) is proportional to the pressure, but the adiabatic modulus is *larger* than the isothermal by a factor γ.

Combining Eqs. (21-18) and (21-20), we obtain

$$c = \sqrt{\frac{\gamma p}{\rho}} \quad \text{(ideal gas)}. \tag{21-22}$$

But, for an ideal gas,

$$\frac{p}{\rho} = \frac{RT}{M},$$

where R is the gas constant, M the molecular mass, and T the absolute temperature. Therefore

$$c = \sqrt{\frac{\gamma RT}{M}} \quad \text{(ideal gas)}, \tag{21-23}$$

and since, for a given gas, γ, R, and M are constants, we see that the speed of propagation is proportional to the square root of the absolute temperature.

Example Compute the speed of longitudinal waves in air.

Solution The mean molecular mass of air is

$$28.8 \text{ g}\cdot\text{mol}^{-1} = 28.8 \times 10^{-3} \text{ kg}\cdot\text{mol}^{-1}.$$

Also, $\gamma = 1.40$ and $R = 8.314 \text{ J}\cdot\text{mol}^{-1}\cdot\text{K}^{-1}$. At $T = 300$ K we obtain

$$c = \sqrt{\frac{(1.40)(8.314 \text{ J}\cdot\text{mol}^{-1}\cdot\text{K}^{-1})(300 \text{ K})}{28.8 \times 10^{-3} \text{ kg}\cdot\text{mol}^{-1}}}$$

$$= 348 \text{ m}\cdot\text{s}^{-1} = 1142 \text{ ft}\cdot\text{s}^{-1} = 779 \text{ mi}\cdot\text{hr}^{-1}.$$

This agrees with the measured speed at this temperature to within 0.3 percent.

Longitudinal waves in air give rise to the sensation of *sound*. The ear is sensitive to a range of sound frequencies from about 20 to about 20,000 Hz. From the relation $c = f\lambda$, the corresponding wavelength range is from about 17 m, corresponding to a 20-Hz note, to about 1.7 cm, corresponding to 20,000 Hz.

The *molecular* nature of a gas has been ignored in the preceding discussion, and a gas has been treated as though it were a continuous medium. Actually, we know that a gas is composed of molecules in random motion, separated by distances that are large compared with their diameters. The vibrations that constitute a wave in a gas are superposed on the random thermal motion. At atmospheric pressure, the average distance traveled between collisions is about 10^{-5} cm, while the displacement amplitude of a faint sound may be only one ten-thousandth of this amount. An element of gas in which a sound wave is traveling can be compared to a swarm of gnats, where the swarm as a whole can be seen to oscillate slightly while individual insects move about through the swarm, apparently at random.

*21-7 Water waves

Probably the most familiar type of wave motion is that observed at the surface of a body of water and produced by the winds or some other disturbance. The oscillations of the water particles in these waves are not confined to the surface, however, but extend with diminishing amplitude to the very bottom. Furthermore, the oscillations have *both* a longitudinal and a transverse component. For this reason, and also because the motion is governed by the laws of hydrodynamics, a complete analysis of this wave motion calls for mathematical methods beyond the scope of this book. The basic physical approach, however, is perfectly straightforward.

Consider a long canal of rectangular cross section and with frictionless walls, containing a depth h of an ideal incompressible liquid of density ρ and surface tension γ. When a wave travels along the canal, each element of the liquid is displaced from its equilibrium position both horizontally and vertically. The restoring force on the element results in part from pressure differences brought about by the variations in depth from point to point, and in part from surface tension effects arising from the curvature of the free surface. The solution of the resulting equations of motion must satisfy the conditions that the pressure at the upper surface is constant and equal to atmospheric pressure, and that the vertical displacement at the bottom of the canal is always zero. In addition, the motion of the fluid must satisfy the equation of continuity.

For simple harmonic waves, in which the x- and y-components of displacement are sine or cosine functions, the speed c is found to be given by

$$c^2 = \left(\frac{g\lambda}{2\pi} + \frac{2\pi\gamma}{\rho\lambda}\right)\tanh\frac{2\pi h}{\lambda}, \qquad (21\text{-}24)$$

where g is the acceleration of gravity and λ is the wavelength.

We immediately recognize one very significant difference between this expression and that for the speed of propagation of transverse waves in a string, or compressional waves in a fluid or solid. Here, the speed depends not only on the properties of the medium *but on the wavelength λ as well*. When this is the case, the medium is said to be *dispersive*. Such a dependence of speed on wavelength occurs in many other types of waves. For example, a prism separates a parallel beam of white light into a spectrum, with different wavelengths emerging in different directions, because glass is a dispersive medium for light waves.

In general, the particles of liquid move in ellipses in vertical planes parallel to the length of the canal, with the major axis of the ellipse horizontal. This motion can be considered as the superposition of two simple harmonic oscillations of the same frequency and different amplitudes, one in the horizontal and one in the vertical direction, and with a *phase difference* of 90°. The wave can therefore be considered as the superposition of a longitudinal and a transverse wave, 90° out of phase with each other, having different amplitudes. If the wavelength is as small as or smaller than the depth, the two amplitudes are very nearly

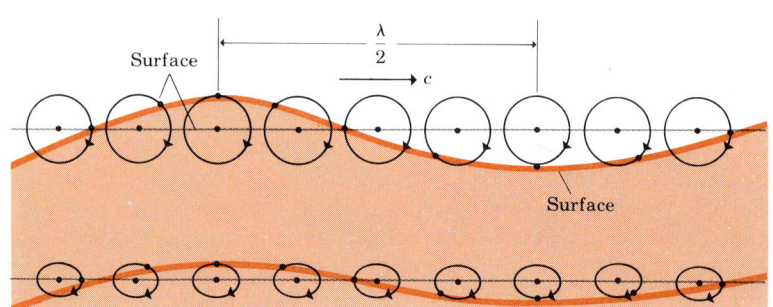

21-9 Particle motion and wave shape for waves in a canal.

equal at the surface and the surface particles move in circles. The amplitudes of both components decrease with increasing depth, but the vertical component decreases more rapidly than does the horizontal. At the bottom, the vertical component becomes zero and the oscillation is wholly longitudinal.

Figure 21-9 illustrates the paths of motion of particles in the surface layer and at some depth below the surface. The upper horizontal black line represents the free surface of the liquid at rest. The circles are the paths of particles whose equilibrium position is at the center of the circle. When a wave travels from left to right, the particles revolve clockwise in these paths. The full line gives the shape of the surface at the instant shown.

The lower black line passes through the equilibrium positions of particles at some depth below the surface. Their paths are elliptical, as indicated, and the lower colored line is their locus at the instant when the free surface has the shape shown by the upper colored line.

Similar waves can occur at the surfaces of elastic solids. The rigidity of the material is an added complication in this case, with the result that the phase difference between the longitudinal and transverse components of displacement changes with depth beneath the surface. Such waves, called *Rayleigh waves,* promise to become important in connection with mechanical devices for processing radar, television, and radio signals. For example, it is possible to make *mechanical* band-pass filters for the megahertz frequency range that are simpler and more compact than their electrical analogs.

Wave phenomena may be demonstrated effectively by means of a "ripple tank" consisting of a horizontal sheet of glass or plastic forming the bottom of a shallow tank that is filled with water about 1 cm deep. A strong source of light beneath the tank sends light through the water waves, which cast a shadow on a translucent screen above the tank. A convenient source of water waves is provided by a smooth sphere about the size of a pencil eraser, attached to a rod and periodically dipped into the water. At low frequencies the wavelength λ may be made as large as, say, 10 cm. The surface tension of water is about $0.070 \text{ N} \cdot \text{m}^{-1}$, and the density of water is $10^3 \text{ kg} \cdot \text{m}^{-3}$. Under these conditions, the second term in the parentheses of Eq. (21-24) turns out to be negligible compared with the first; also, $\tanh 2\pi h/\lambda$ is very nearly equal to $2\pi h/\lambda$. Then

$$c \approx \sqrt{\frac{g\lambda}{2\pi} \cdot \frac{2\pi h}{\lambda}} \approx \sqrt{gh}. \tag{21-25}$$

Thus for waves of sufficiently long wavelength, the speed depends only on the acceleration due to gravity, g, and the depth h, and is independent of the wavelength (no dispersion). Since, according to Eq. (21–25), the shallower the water, the slower the speed of the waves, it is possible to eliminate reflections from the edges of the ripple tank by beveling the edges of the tank. A wave approaching a "shore" is therefore slowed down to zero speed.

Questions

21-1 The term *wave* has a variety of meanings in ordinary language. Think of several, and discuss their relation to the precise physical sense in which the term is used in this chapter.

21-2 What kinds of energy are associated with waves on a stretched string? How could such energy be detected experimentally?

21-3 Are torsional waves (Fig. 21–1d) longitudinal or transverse?

21-4 Is it possible to have a longitudinal wave on a stretched string? A transverse wave on a steel rod?

21-5 The speed of sound waves in air depends on temperature, but the speed of light waves does not. Why?

21-6 For the wave motions discussed in this chapter, does the speed of propagation depend on the amplitude?

21-7 Children make toy telephones by sticking each end of a long string through a hole in the bottom of a paper cup and knotting it so it won't pull out. When the string is pulled taut, sound can be transmitted from one cup to the other. How does this work? Why is the transmitted sound louder than the sound traveling through air for the same distance?

21-8 An echo is sound reflected from a distant object such as a wall or a cliff. Explain how you can determine how far away the object is by timing the echo.

21-9 Why do you see lightning before you hear the thunder? A familiar rule of thumb is to start counting slowly, once per second, when you see the lightning; when you hear thunder, divide the number you have reached by five, to obtain your distance (in miles) from the lightning. Why does this work? Or does it?

21-10 When ocean waves approach a beach, the crests are always nearly parallel to the shore, despite the fact that they must come from various directions. Why?

21-11 The amplitudes of ocean waves increase as they approach the shore, and often the crests bend over and "break." Why?

21-12 When a rock is thrown into a pond and the resulting ripples spread in ever-widening circles, the amplitude decreases with increasing distance from the center. Why?

21-13 In speaker systems designed for high-fidelity music reproduction, the "tweeters" that reproduce the high frequencies are always much smaller than the "woofers" used for low frequencies. Why?

21-14 When sound travels from air into water, does the frequency of the wave change? The wavelength? The speed?

21-15 Which of the quantities describing a sinusoidal wave is most closely related to musical pitch? To loudness?

21-16 Musical notes produced by different instruments (such as a flute and an oboe) may have the same pitch and loudness and yet sound different. What is the difference, in physical terms?

Problems

21-1 The speed of sound in air at 20°C is 344 m·s^{-1}.

a) What is the wavelength of a sound wave of frequency 32 Hz, the lowest pedal note on medium-size pipe organs?

b) What is the frequency of a wave having a wavelength of 4 ft, corresponding approximately to the D above middle C on the piano?

21-2 The speed of radio waves in vacuum (equal to the speed of light) is 3.00×10^8 m·s^{-1}. Find the wavelength for

a) an AM radio station with frequency 1000 kHz;

b) an FM radio station with frequency 100 MHz.

21-3 The speed of sound in water is about 1480 m·s^{-1}. Find the frequency of a sound wave such that its wavelength in water is the same as the wavelength in air at 20°C of a sound wave of frequency 1000 Hz.

21-4 Middle C on the piano corresponds to a frequency of 262 Hz = 262 s^{-1}.

a) Find the wavelength and wave number of the corresponding sound wave in air at 20°C.
b) What is the angular frequency?
c) What is the wavelength of the wave in water at 20°C, if the wavespeed is 1480 m·s^{-1}?

21-5 The equation of a certain traveling transverse wave is

$$y = 2 \sin 2\pi \left(\frac{t}{0.01} - \frac{x}{30} \right),$$

where x and y are in centimeters and t is in seconds. What are

a) the amplitude,
b) the wavelength,
c) the frequency, and
d) the speed of propagation of the wave?

21-6 Show that Eq. (21-4) may be written

$$y = -A \sin \frac{2\pi}{\lambda} (x - ct).$$

21-7 A traveling transverse wave on a string is represented by the equation in Problem 21-6. Let $A = 8$ cm, $\lambda = 16$ cm, and $c = 2$ cm·s^{-1}.

a) At time $t = 0$, compute the transverse displacement y at 2-cm intervals of x (that is, at $x = 0$, $x = 2$ cm, $x = 4$ cm, etc.) from $x = 0$ to $x = 32$ cm. Show the results in a graph. This is the shape of the string at time $t = 0$.
b) Repeat the calculations, for the same values of x, at times $t = 1$ s, $t = 2$ s, $t = 3$ s, and $t = 4$ s. Show on the same graph the shape of the string at these instants. In what direction is the wave traveling?

21-8 The equation of a transverse traveling wave on a string is

$$y = 2 \cos[\pi(0.5x - 200t)],$$

where x and y are in cm and t is in seconds.

a) Find the amplitude, wavelength, frequency, period, and velocity of propagation.
b) Sketch the shape of the string at the following values of t: 0, 0.0025 s, and 0.005 s.
c) If the mass per unit length of the string is 5 g·cm^{-1}, find the tension.

21-9 A transverse sine wave of amplitude 10 cm and wavelength 200 cm travels from left to right along a long horizontal stretched string with a speed of 100 cm·s^{-1}. Take the origin at the left end of the undisturbed string. At time $t = 0$, the left end of the string is at the origin and is moving downward.

a) What is the frequency of the wave?
b) What is the angular frequency?
c) What is the propagation constant?
d) What is the equation of the wave?
e) What is the equation of motion of the left end of the string?
f) What is the equation of motion of a particle 150 cm to the right of the origin?
g) What is the (absolute) maximum transverse velocity of any particle of the string?
h) Find the transverse displacement and the transverse velocity of a particle 150 cm to the right of the origin, at time $t = 3.25$ s.
i) Make a sketch of the shape of the string, for a length of 400 cm, at time $t = 3.25$ s.

21-10 A steel wire is stretched with a tension of 50 N. What is its mass per unit length if the speed of transverse waves on the wire is the same as the speed of sound in air at 20°C?

21-11 A steel wire 6 m long has a mass of 60 g and is stretched with a tension of 1000 N. What is the speed of propagation of a transverse wave in the wire?

21-12 One end of a horizontal string is attached to a prong of an electrically driven tuning fork whose frequency of vibration is 240 Hz. The other end passes over a pulley and supports a mass of 5 kg. The linear mass density of the string is 0.02 kg·m^{-1}.

a) What is the speed of a transverse wave in the string?
b) What is the wavelength?

21-13 One end of a stretched rope is given a periodic transverse motion with a frequency of 10 Hz. The rope is 50 m long, has total mass 0.5 kg, and is stretched with a tension of 400 N.

a) Find the wave speed and the wavelength.
b) If the tension is doubled, how must the frequency be changed to maintain the same wavelength?

21-14 One end of a rubber tube 20 m long, of total mass 1 kg, is fastened to a fixed support. A cord attached to the other end passes over a pulley and supports a body of mass 10 kg. The tube is struck a transverse blow at one end. Find the time required for the pulse to reach the other end.

21-15 A metal wire has these properties:

Coefficient of linear expansion = 1.5×10^{-5} (C°)$^{-1}$,
Young's modulus = 2.0×10^{11} Pa,
Density = 9.0×10^3 kg·m^{-3}.

At each end are rigid supports. If the tension is zero at 20°C, what will be the speed of a transverse wave at 8°C?

21-16 A metal wire, of density 5×10^3 kg·m^{-3}, with a Young's modulus equal to 2.0×10^{11} Pa, is stretched between rigid supports. At one temperature the speed of a transverse wave is found to be 200 m·s^{-1}. When the temperature is raised 100 C°, the speed decreases to 160 m·s^{-1}. Determine the coefficient of linear expansion.

21-17 The measured value of the speed of longitudinal waves in water is approximately 1486 m·s^{-1} at 20°C. Compute the adiabatic compressibility ($1/B_{ad}$) of water and compare with the isothermal compressibility in Table 10-2.

21-18 Provided the amplitude is sufficiently great, the human ear can respond to longitudinal waves over a range of frequencies from about 20 Hz to about 20,000 Hz. Compute the wavelengths corresponding to these frequencies

a) for waves in air,

b) for waves in water.

(See Problem 21-17). Assume that in air $c = 345$ m·s^{-1}.

21-19 At a temperature of 27°C, what is the speed of longitudinal waves in

a) argon,

b) hydrogen?

c) Compare with the speed in air at the same temperature.

21-20 What is the difference between the speeds of longitudinal waves in air at $-3°$C and at $57°$C?

21-21 The sound waves from a loudspeaker spread out nearly uniformly in all directions when their wavelength is large compared with the diameter of the speaker. When the wavelength is small compared with the diameter of the speaker, much of the sound energy is concentrated in the forward direction. For a speaker of diameter 25 cm, compute the frequency for which the wavelength of the sound waves, in air, is

a) 10 times the diameter of the speaker,

b) equal to the diameter of the speaker,

c) $\frac{1}{10}$ the diameter of the speaker.

21-22 A steel pipe 100 m long is struck at one end. A person at the other end hears two sounds as a result of two longitudinal waves, one in the pipe and the other in the air. What is the time interval between the two sounds? Take Young's modulus of steel to be 2×10^{11} Pa, and the speed of sound in air to be 345 m·s^{-1}.

21-23 What must be the stress (F/A) in a stretched wire of a material whose Young's modulus is Y, in order that the speed of longitudinal waves shall equal 10 times the speed of transverse waves?

21-24 What is the ratio of the speed of sound in a diatomic gas to the rms speed of gas molecules at the same temperature?

21-25

a) If the propagation of sound waves in gases were characterized by isothermal rather than adiabatic expansions and compressions, show that the speed of sound would be given by $(RT/M)^{1/2}$.

b) What would be the speed of sound in air at 20°C in this case?

c) Under what circumstances might the wave propagation be expected to be isothermal?

21-26 A longitudinal wave propagates in a steel bar having density 7.0 g·m^{-3} and Young's modulus 2×10^{11} Pa.

a) What is the wave speed?

b) By what factor is this speed greater than the speed of sound in air at 20°C?

21-27

a) By how many meters per second, at a temperature of 27°C, does the speed of sound in air increase per Celsius degree rise in temperature? [*Hint:* Compute dc/dT.]

b) Is the rate of change of speed with temperature the same at all temperatures?

21-28 A loose coil of flexible rope of length L and mass M rests on a frictionless table. A force S is applied at one end of the rope and more and more rope is pulled from the coil with a constant velocity v.

a) What is the relation between S and v?

b) A transverse pulse is produced while the rope is being pulled from the coil. How does the pulse behave?

c) What is the total work of the force S at the moment when the last bit of rope leaves the coil?

d) What is the kinetic energy of the rope?

e) Is there a discrepancy between (c) and (d)? Explain.

21-29 The ends of a flexible rope of length L and mass M are joined so as to form a circular loop of radius R. The loop rotates about an axis through its center on a horizontal frictionless surface with a constant linear speed v.

a) Apply Newton's second law to a short piece of rope and calculate the tension S in the rope.

b) Explain the effect of imparting to the rope, while it is spinning, a small rapid radial displacement.

VIBRATING 22 BODIES

22-1 Boundary conditions for a string

In Chapter 21 we studied the propagation of wave in *unbounded* media. That is, we were not concerned with the possibility that a wave may arrive at an end or boundary of the elastic medium in which it propagates. When such boundaries do play a significant role, interesting and important new aspects of wave motion emerge.

As an example, let us consider what happens when a wave pulse or wave train, advancing along a stretched string, arrives at the end of the string. If the end is fastened to a rigid support, it must remain at rest. The arriving wave exerts a force on the support; the reaction to this force "kicks back" on the string and sets up a *reflected* pulse or wave train traveling in the reverse direction.

At the opposite extreme from a rigidly fixed end is one that is perfectly free to move in the direction transverse to the length of the string. For example, the string might be tied to a light ring, which slides on a smooth rod perpendicular to the length of the string; the ring and rod maintain the tension of the string but exert no transverse force. At a free end the arriving pulse or wave train causes the end to "overshoot," and again a reflected wave is set up. The conditions imposed on the motion of the end of the string, such as attachment to a rigid support or the complete absence of transverse force, are called *boundary conditions*.

The multiflash photograph of Fig. 22-1 shows the reflection of a pulse at a fixed end of a string. (The camera was tipped vertically while the photographs were taken so that successive images lie one under the other. The "string" is a rubber tube, and it sags somewhat.) The pulse is reflected with both its displacement and its direction of propagation reversed. When reflection takes place at a *free* end, the direction of propagation is reversed but the direction of the displacement is unchanged.

It is helpful to think of the process of reflection in the following way. Imagine the string to be extended indefinitely beyond its actual end. The

22-1 A pulse starts in the upper left corner, travels to the right, and is reflected from the fixed end of the string at the right.

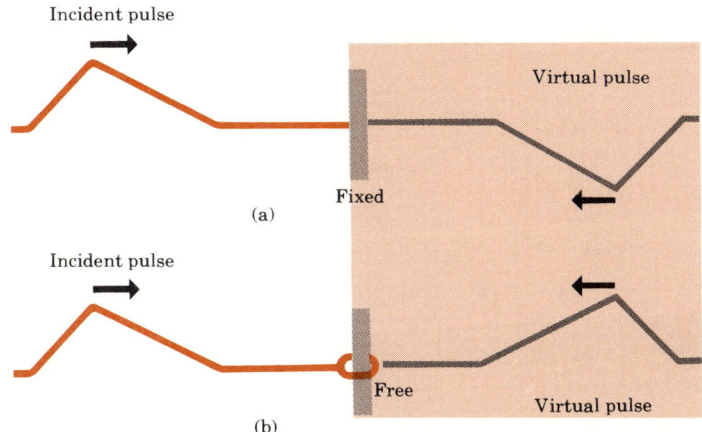

22-2 Description of reflection of a pulse (a) at a fixed end of a string, and (b) at a free end, in terms of an imaginary "virtual" pulse.

actual pulse can be considered to continue on into the imaginary portion as though the support were not there, while at the same time a "virtual" pulse, which has been traveling in the imaginary portion, moves out into the real string and forms the reflected pulse. The nature of the reflected pulse depends on whether the end is fixed or free. The two cases are shown in Fig. 22-2.

The displacement at a point where the actual and virtual pulses cross each other is the *algebraic sum* of the displacements in the individual pulses. Figures 22-3 and 22-4 show the shape of the region near the end of the string for both types of reflected pulses, Fig. 22-3 for a free end and Fig. 22-4 for a fixed end. In the latter case, the incident and reflected pulses must combine in such a way that the displacement of the end of the string is always zero.

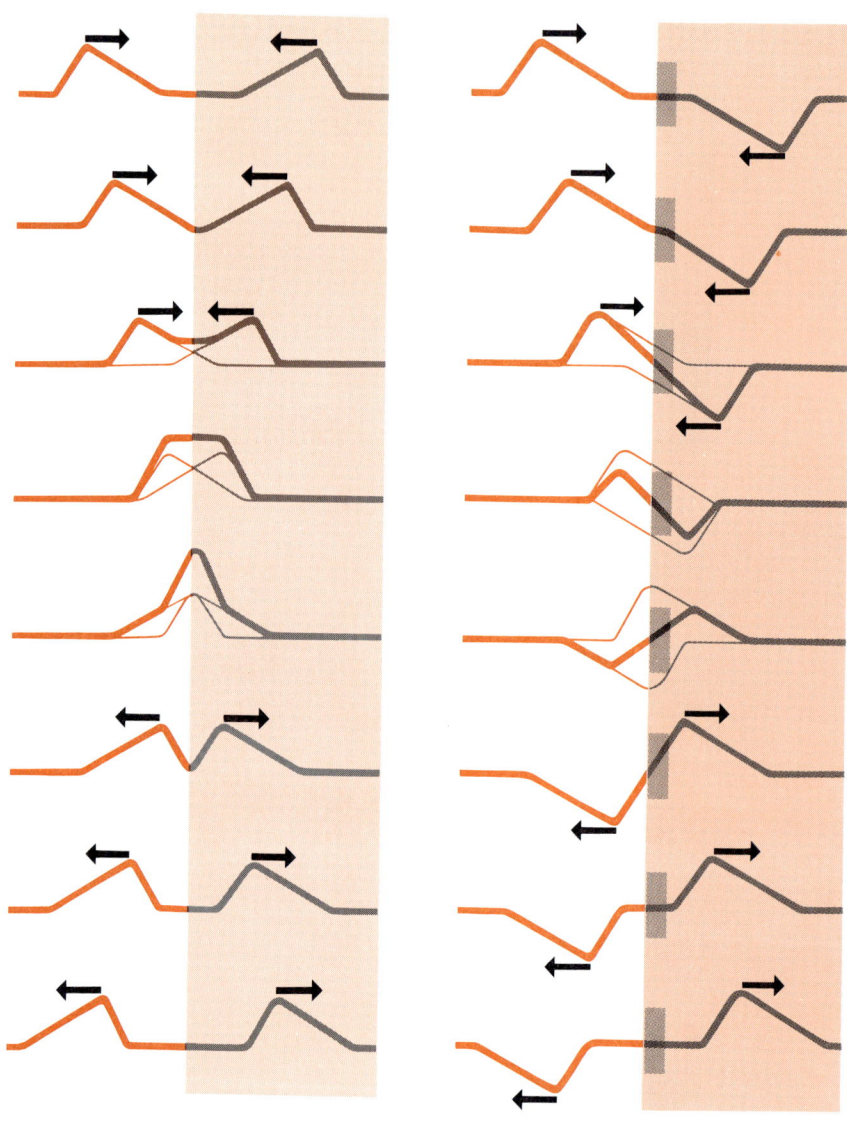

22-3 Reflection at a free end.

22-4 Reflection at a fixed end.

22-2 Superposition and standing waves

When a continuous succession of waves, such as a sinusoidal wave, arrives at a fixed end of a string, a corresponding continuous succession of *reflected* waves originates at this end and travels in the opposite direction. The resulting motion of the string is determined by an extremely important principle known as the *principle of superposition,* which states that the actual displacement of any point on the string, at any time, is obtained by adding the displacement that point would have if only the first wave were present, and the displacement it would have with only the second wave.

Mathematically speaking, the principle of superposition states that the wave function describing the resulting motion in the above situation is obtained by adding the two wave functions for the two separate waves.

This additive property of wave functions depends in turn on the form of the wave equation, Eq. (21–14), which must be satisfied by any physically possible wave function. Specifically, the wave equation is *linear;* that is, it contains derivatives of the function y only to the first power. As a result, if any two functions $y_1(x, t)$ and $y_2(x, t)$ each satisfy the equation separately, their *sum* $y_1 + y_2$ automatically satisfies the wave equation and hence is also a physically possible motion.

The principle of superposition is of central importance in all types of wave motion; it applies not only to waves on a string, but also to sound waves, electromagnetic waves (such as light), and all other wave phenomena in which the wave equation is linear. The general term *interference* is used to describe phenomena which result from two or more waves passing through the same region at the same time.

When a sinusoidal wave is reflected by a fixed point on a string, the appearance of the resulting motion gives no evidence that there are two waves traveling in opposite directions. If the frequency is sufficiently great so that the eye cannot follow the motion, the string appears subdivided into a number of segments, as in the time exposure photograph of Fig. 22–5a. A multiflash photograph of the same string, in Fig. 22–5b, indicates a few of the instantaneous shapes of the string. At every instant its shape is a sine curve, but whereas in a traveling wave the amplitude remains constant while the wave progresses, here the waveform remains fixed in position (longitudinally) while the amplitude fluctuates. Certain points known as the *nodes* remain always at rest. Midway between these points, at the *loops* or *antinodes,* the fluctuations are a maximum. The vibration as a whole is called a *standing wave.*

To understand the formation of a standing wave, consider the separate graphs of the waveform at four instants one-tenth of a period apart, shown in Fig. 22–6. The gray curves represent a wave traveling to the

22–5 (a) Standing waves in a stretched string (time exposure). (b) Multiflash photograph of a standing wave, with nodes at the center and at the ends.

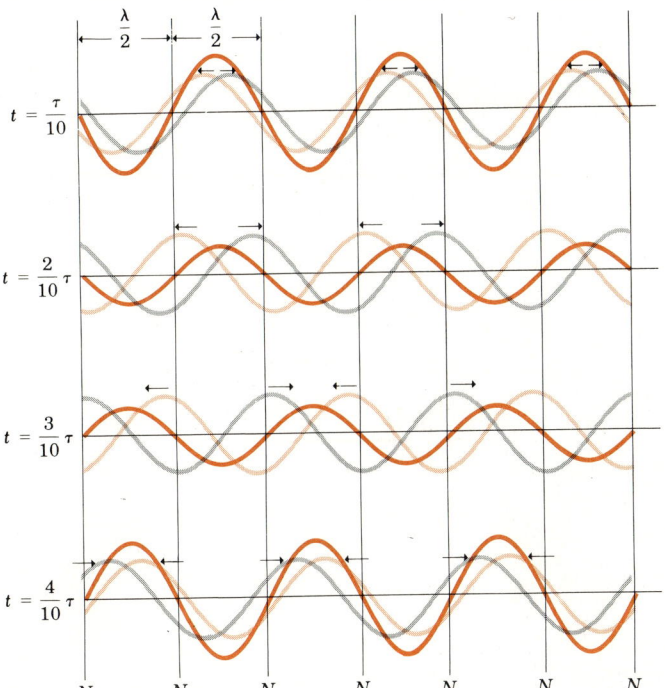

22–6 The formation of a standing wave. A wave traveling to the right (gray lines) combines with a wave traveling to the left (light color lines) to form a standing wave (dark color line).

right. The light color curves represent a wave of the same propagation speed, wavelength, and amplitude traveling to the *left*. The dark color curves represent the resultant waveform, obtained by applying the principle of superposition, that is, by adding displacements. At those places marked N at the bottom of Fig. 22-6, the resultant displacements are *always* zero. These are the **nodes**. Midway between the nodes, the vibrations have the *largest* amplitude. These are the **antinodes**. It is evident from the figure that

$$\left\{ \begin{array}{c} \text{Distance between adjacent nodes} \\ \text{or} \\ \text{Distance between adjacent antinodes} \end{array} \right\} = \frac{\lambda}{2}.$$

The wave function for the standing wave of Fig. 22-6 may be obtained by adding the displacements of two waves of equal amplitude, period, and wavelength, traveling in opposite directions. Thus, if

$$y_1 = A \sin(\omega t - kx) \quad \text{(positive } x\text{-direction)},$$
$$y_2 = -A \sin(\omega t + kx) \quad \text{(negative } x\text{-direction)},$$

then

$$y_1 + y_2 = A[\sin(\omega t - kx) - \sin(\omega t + kx)].$$

Introducing the expressions for the sine of the sum and difference of two angles and combining terms, we obtain

$$y_1 + y_2 = -[2A \cos \omega t] \sin kx. \qquad (22\text{-}1)$$

The shape of the string at each instant is, therefore, a sine curve whose amplitude (the expression in brackets) varies with time. The appearance is not that of a traveling wave shape, but of a sinusoidal displacement in one position which grows larger and smaller with time. Each point in the string still undergoes simple harmonic motion, but instead of the progressively increasing phase differences between motions of adjacent points, all points move *in phase* (or 180° out of phase).

The positions of the nodes can be obtained from Eq. (22-1). Wherever $\sin kx = 0$, the displacement is always zero. This occurs when $kx = 0, \pi, 2\pi, 3\pi, \ldots$, or

$$\begin{aligned} x &= 0, \quad \pi/k, \quad 2\pi/k, \quad 3\pi/k, \ldots \\ &= 0, \quad \lambda/2, \quad \lambda, \quad 3\lambda/2, \ldots \end{aligned} \qquad (22\text{-}2)$$

22-3 Vibration of a string fixed at both ends

Thus far we have discussed a long string fixed at one end and have considered the standing waves set up near that end by interference between the incident and reflected waves. Let us next consider the more usual case, that of a string fixed at *both* ends. A sinusoidal wave is reflected and re-reflected; and since the string is fixed at both ends, both ends must be nodes. Since adjacent nodes are one-half wavelength apart, the length of the string may be

$$\frac{\lambda}{2}, \quad \frac{2\lambda}{2}, \quad \frac{3\lambda}{2},$$

or, in general, *any* integer number of half-wavelengths. Or (to put it differently) if one considers a particular string of length L, standing waves may be set up in the string by vibrations of a number of different frequencies, namely, those that give rise to waves of wavelengths

$$\lambda = 2L, \quad \frac{2L}{2}, \quad \frac{2L}{3}, \quad \ldots = \frac{2L}{n} \qquad (n = 1, 2, 3, \ldots). \qquad (22\text{-}3)$$

The wave speed c is the same for all frequencies. From the relation $f = c/\lambda$, the possible frequencies are

$$f = \frac{c}{2L}, \quad \frac{2c}{2L}, \quad \frac{3c}{2L}, \quad \ldots = \frac{nc}{2L} \qquad (n = 1, 2, 3, \ldots). \qquad (22\text{-}4)$$

The lowest frequency, $c/2L$, is called the *fundamental* frequency f_1 and the others are the *overtones*. The frequencies of the latter are therefore $2f_1$, $3f_1$, $4f_1$, and so on. Overtones whose frequencies are integral multiples of the fundamental are said to form a *harmonic series*. The fundamental is the *first harmonic*. The frequency $2f_1$ is the *first overtone* or the *second harmonic*, the frequency $3f_1$ is the *second overtone* or the *third harmonic*, and so on.

The above results may also be obtained directly from Eq. (22-1). The boundary conditions require that $y_1 + y_2 = 0$ at the ends of the string, that is at $x = 0$ and $x = L$. Since the sine of zero is zero, the first condition is satisfied automatically. The second requires that $\sin kL = 0$, and this is true only when k has certain special values. The sine of an angle is zero only when the angle is zero or an integer multiple of π (180°). Thus we must have

$$kL = n\pi \qquad (n = 1, 2, 3, \ldots). \qquad (22\text{-}5)$$

We do not include the possibility $n = 0$ because that gives $k = 0$, i.e., a wave with zero displacement *everywhere* (a possible case, to be sure, but not a very interesting one!).

Replacing k above by $2\pi/\lambda$, we obtain

$$\frac{2\pi L}{\lambda} = n\pi \quad \text{or} \quad \lambda = \frac{2L}{n}, \qquad (22\text{-}6)$$

in agreement with Eq. (22-3).

Each of the frequencies given by Eq. (22-4) corresponds to a possible *normal mode* of motion, that is, a motion in which each particle of the string moves sinusoidally, all with the same frequency. As the above analysis shows, there is an infinite number of normal modes, each with its characteristic frequency. This situation is in striking contrast with the simpler harmonic oscillator system, a single mass and a spring. The harmonic oscillator has only *one* normal mode and one characteristic frequency, while the vibrating string has an infinite number.

When a body suspended from a spring is pulled down and released, only one frequency of vibration will ensue. If a string is initially distorted so that its shape is the same as *any one* of the possible harmonics, it will vibrate, when released, at the frequency of that particular harmonic. But when a piano string is struck or a guitar string is plucked, not only the fundamental but many of the overtones are present in the resulting vibration. This motion is therefore a combination or *superposition* of normal modes. Several frequencies and motions are present simultaneously, and the displacement of any point on the string is the sum (or superposition) of displacements associated with the individual modes.

The fundamental frequency of the vibrating string is $f_1 = c/2L$, where, from Eq. (21-15), $c = \sqrt{S/\mu}$. It follows that

$$f_1 = \frac{1}{2L}\sqrt{\frac{S}{\mu}}. \qquad (22\text{-}7)$$

Stringed instruments afford many examples of the implications of this equation. For example, all such instruments are "tuned" by varying the tension S, an increase of tension increasing the frequency or pitch, and vice versa. The inverse dependence of frequency on length L is illustrated by the long strings of the bass section of the piano or the bass viol compared with the shorter strings of the piano treble or the violin. Pressing the strings against the fingerboard of a violin or guitar with the fingers to change the length of the vibrating portion of the string is the usual means of varying pitch when playing these instruments. One reason for winding the bass strings of a piano with wire is to increase the mass per unit length μ, so as to obtain the desired low frequency without resorting to a string that is inconveniently long.

22-4 Longitudinal standing waves

Longitudinal waves traveling along a tube of finite length are reflected at the ends of the tube in much the same way that transverse waves in a string are reflected at its ends. Interference between the waves traveling in opposite directions again gives rise to standing waves.

When reflection takes place at a closed end, the displacement of the particles there must always be zero. Hence a closed end is a *displacement node*. If the end of the tube is open, the nature of the reflection is more complex and depends on whether the tube is wide or narrow compared with the wavelength. If the tube is narrow compared with the wavelength, which is the case in most musical instruments, the reflection is such that the open end is a *displacement antinode,* for reasons discussed below. Thus the longitudinal waves in a column of fluid are reflected at the closed and open ends of a tube in the same way that transverse waves in a string are reflected at fixed and free ends, respectively.

Standing longitudinal waves in a column of gas may be demonstrated conveniently with the aid of the apparatus shown in Fig. 22-7, known as Kundt's tube. A glass tube a meter or so long is closed at one

22-7 Kundt's tube for determining the velocity of sound in a gas. The shading represents the density of the gas molecules at an instant when the pressure at the displacement nodes is a maximum or a minimum.

end with glass and at the other with a flexible diaphragm. The gas to be studied is admitted to the tube at a known temperature and at atmospheric pressure. A source of longitudinal waves, S, whose frequency may be varied, causes vibration of the flexible diaphragm. A small amount of light powder or cork dust is sprinkled uniformly along the tube.

When a frequency is found at which the air column is in resonance, the amplitude of the standing waves becomes large enough for the gas particles to sweep the cork dust along the tube, at all points where the gas is in motion. The powder therefore collects at the displacement nodes, where the gas remains at rest. Sometimes a wire, running along the axis of the tube, is maintained at a dull red heat by an electric current. The nodes show themselves as hot points, compared with the antinodes, which are cooled by moving gas.

In a standing wave, the distance between two adjacent nodes is one-half (not one!) wavelength; the wavelength λ is obtained by measuring the distance between alternate clumps of powder. When the frequency f is known, the velocity c is then

$$c = f\lambda.$$

At a displacement node, the pressure variations above and below the average are *maximum,* whereas at a displacement antinode, pressure does not vary. This may be understood easily when it is realized that two small masses of gas on opposite sides of a displacement node vibrate in *opposite phase*. When they approach each other, the pressure at the node is maximum, and when they recede from each other, the pressure at the node is minimum. Two small masses of gas, however, on opposite sides of a displacement *antinode* vibrate *in phase,* and hence cause no pressure variations at the antinode.

We may speak of *pressure nodes,* the points where the pressure does not vary, and *pressure antinodes,* the points where its variation is maximum. With this language, a pressure node always is a displacement antinode, and a pressure antinode is a displacement node. An open end of a tube or a pipe is a pressure node because such an end is open to the atmosphere and is thus at constant pressure. But for this reason also an open end is always a displacement *antinode,* as stated above.

Figure 22-8 may be helpful in visualizing a longitudinal standing wave. To use this figure, cut a slit about $\frac{1}{16}$ in. wide and $3\frac{1}{2}$ in. long in a card. Place the card over the diagram with the slit horizontal and move it vertically with constant velocity. The portions of the curves that appear in the slit will correspond to the oscillations of the particles in a longitudinal standing wave.

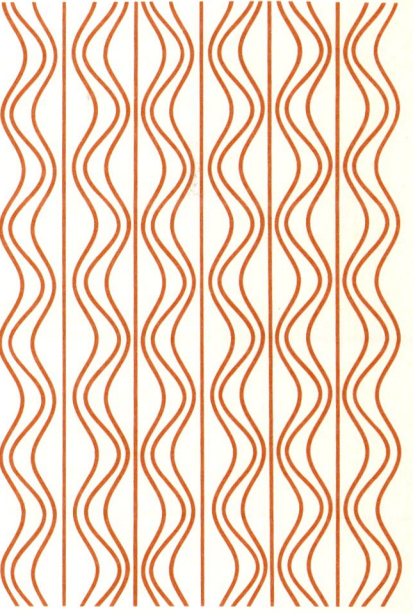

22-8

22-5 Vibrations of organ pipes

If one end of a pipe is open and a stream of air is directed against an edge, vibrations are set up and the tube resonates at its natural frequencies. As in the case of a plucked string, the fundamental and overtones exist at the same time. In the case of a pipe open at both ends, the fundamental frequency f_1 corresponds to an antinode at each end and a node in the middle, as shown at the top of Fig. 22-9. Succeeding diagrams of Fig.

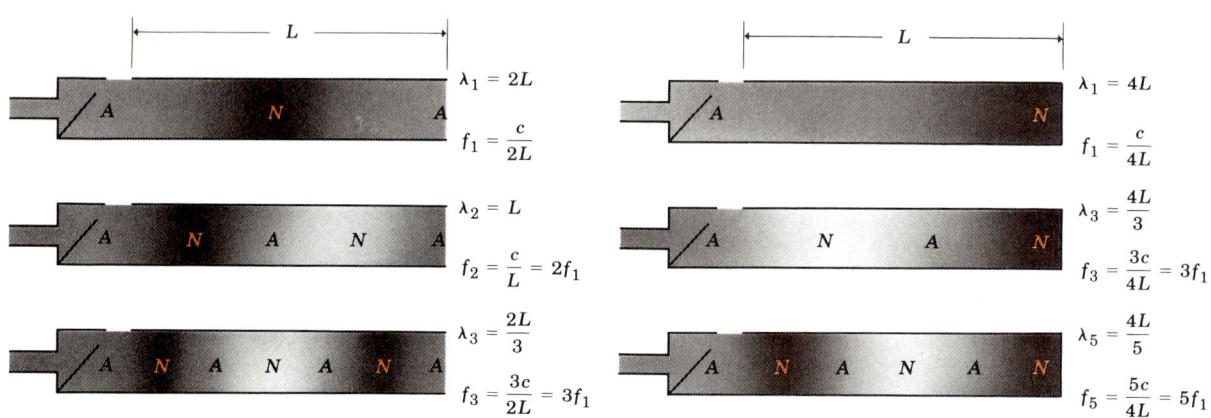

22-9 Modes of vibration of an open organ pipe.

22-10 Modes of vibration of a stopped organ pipe.

22-9 show two of the overtones, which are seen to be the second and third harmonics. *In an open pipe the fundamental frequency is $c/2L$ and all harmonics are present.*

The properties of a *stopped* pipe (open at one end, closed at the other) are shown in the diagrams of Fig. 22-10. The fundamental frequency is seen to be $c/4L$, which is one-half that of an open pipe of the same length. In the language of music, the *pitch* of a closed pipe is one octave lower (a factor of two in frequency) than that of an open pipe of equal length. From the remaining diagrams of Fig. 22-10, it may be seen that the second, fourth, etc., harmonics are missing. Hence, *in a stopped pipe, the fundamental frequency is $c/4L$ and only the odd harmonics are present.*

In an organ pipe several modes are almost always present at once. The extent to which modes higher than the fundamental are present depends on the cross section of the pipe, the ratio of length to width, whether the pipe is straight or tapered, and to a slight extent on the material. The harmonic content of the tone is an important factor in determining the "tone quality" or timbre. A very thin pipe produces a tone rich in higher harmonics, which the ear perceives as thin and "stringy," while a fatter pipe produces principally the fundamental tone, perceived as a softer, more flute-like tone.

*22-6 Vibrations of rods and plates

A rod may be set in longitudinal vibration by clamping it at some point and applying a time-varying force of appropriate frequency. In Fig. 22-11a the rod is clamped in the middle; consequently, a standing wave is set up with a node in the middle and antinodes at each end, exactly the same as the fundamental mode of an open organ pipe. The fundamental frequency of the rod is then $c/2L$, where c is the velocity of a longitudinal wave in the rod. Since the velocity of a longitudinal wave in a solid is much greater than that in air, a rod has a higher fundamental frequency than an open organ pipe of the same length.

When the rod is clamped at a point $\frac{1}{4}$ of its length from one end, as shown in Fig. 22-11b, the second harmonic may be produced, and so on.

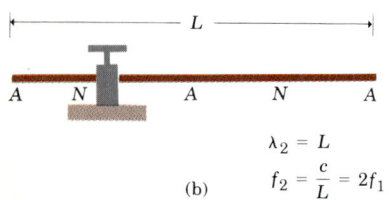

22-11 Modes of vibration of a rod.

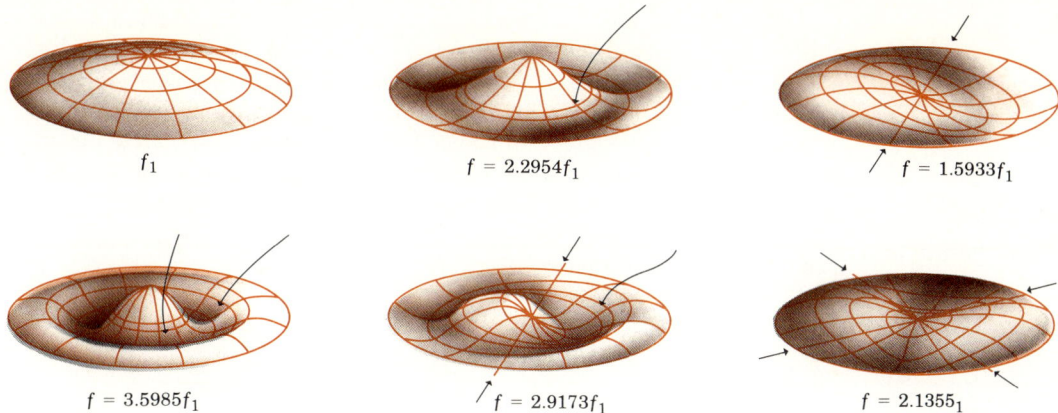

f_1

$f = 2.2954f_1$

$f = 1.5933f_1$

$f = 3.5985f_1$

$f = 2.9173f_1$

$f = 2.1355_1$

When a stretched flexible membrane, such as a drumhead, is struck a blow, a two-dimensional pulse travels outward from the struck point and is reflected and re-reflected at the boundary of the membrane. If some point of the membrane is forced to vibrate periodically, continuous trains of waves travel along the membrane. Just as with the stretched string, standing waves can be set up in the membrane; each of these waves has a certain natural frequency. The lowest frequency is the fundamental and the others are overtones. In general, when the membrane is vibrating, a number of overtones are present.

The nodes of a vibrating membrane are *lines* (nodal lines) rather than points. The boundary of the membrane is evidently one such line. Some of the other possible nodal lines of a circular membrane are indicated by arrows in Fig. 22-12. The natural frequency of each mode is given in terms of the fundamental f_1. It will be noted that the frequencies of the overtones are *not* integer multiples of f_1. That is, they are not harmonics.

The restoring force in a vibrating flexible membrane arises from the tension with which it is stretched. A metal plate, if sufficiently thick, vibrates in a similar way, the restoring force being produced by bending stresses in the plate. The study of vibrations of membranes and plates is of importance in connection with the design of loudspeaker diaphragms, the diaphragms of telephone receivers and microphones, and many other acoustical devices.

22-12 Modes of vibration of a membrane, showing nodal lines. The frequency of each mode is given in terms of fundamental frequency, f_1. (Adapted from *Vibration and Sound,* by Philip M. Morse, second edition, McGraw-Hill Book Company, Inc., 1948. By permission of the publishers.)

22-7 Interference of longitudinal waves

The phenomenon of *interference* between two longitudinal waves in air may be demonstrated with the aid of the apparatus depicted in Fig. 22-13. A wave emitted by a source S is sent into a tube, where it divides into two waves, one following the constant path SAR, the other the path SBR, which may be varied by sliding the tube B to the right. Suppose the frequency of the source is 350 Hz. Then the wavelength is $\lambda = c/f = 1.0$ m. If both paths are of equal length, the two waves arrive at R at the same time and the vibrations set up by both waves are *in phase*. The resulting vibration has an amplitude equal to the sum of the two individual amplitudes and the phenomenon of *reinforcement* or *constructive interference* may be detected either with the ear at R or with the aid of a microphone, amplifier, and loudspeaker.

22-13 Apparatus for demonstrating interference of longitudinal waves.

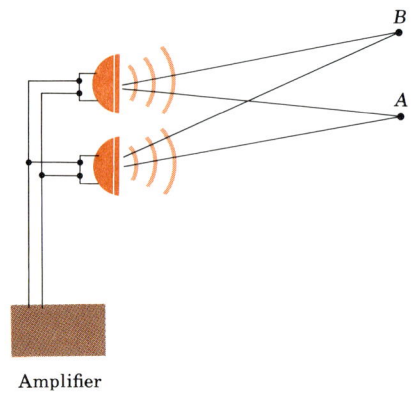

22-14
Amplifier

Now suppose the tube B is moved out a distance of 0.25 m, thereby making the path SBR 0.5 m longer than the path SAR. The righthand wave travels a distance $\lambda/2$ greater than the lefthand wave and the vibration set up at R by the righthand wave is therefore in opposite phase to that set up by the lefthand wave. The consequent *destructive interference* or *cancellation* is shown by a marked *reduction* in sound at R.

If the tube B is now pulled out another 0.25 m, so that the *path difference,* SBR minus SAR, is 1.0 m (1 wavelength), the two vibrations at R again reinforce each other. Thus

$$\left\{\begin{array}{l}\text{Reinforcement takes place}\\ \text{when the path difference}\end{array}\right\} = 0, \quad \lambda, \quad 2\lambda, \quad \text{etc.}$$

$$\left\{\begin{array}{l}\text{Cancellation takes place}\\ \text{when the path difference}\end{array}\right\} = \frac{\lambda}{2}, \quad \frac{3\lambda}{2}, \quad \frac{5\lambda}{2}, \quad \text{etc.}$$

A similar phenomenon can be demonstrated with two loudspeakers driven by the same amplifier, as shown in Fig. 22–14. Suppose the speakers both emit a pure sinusoidal sound wave of constant frequency. When a microphone is placed at point A in the figure, equidistant from the speakers, it receives a strong acoustic signal, corresponding to the fact that two waves emitted from the speakers in phase also arrive at point A in phase and add to each other. But at point B the signal is much *weaker* than when only one speaker is present, corresponding to the fact that the wave from one speaker travels a half-wavelength farther than that from the other; the two arrive a half-cycle out of phase and cancel each other out almost completely. An experiment closely analogous to this one but using light waves, provided the first conclusive evidence of the wave nature of light. This is discussed in detail in Chapter 41.

22–8 Resonance

In previous sections we have discussed several examples of mechanical systems having normal modes of oscillation. In each mode, every particle of the system oscillates with simple harmonic motion of the same frequency as the frequency of this mode. The systems discussed have an infinite series of normal modes, but the basic concept is closely related to motion of a simple harmonic oscillator, which has only a single normal mode.

Now suppose a periodically varying force is applied to a system having normal modes. The system vibrates with a frequency equal to that of the *force;* this motion is called a *forced oscillation*. In general the amplitude of this motion is relatively small, but if the frequency of the force is close to one of the normal-mode frequencies, the amplitude can become quite large.

If the frequency of the force were precisely *equal* to a normal-mode frequency, and if there were no friction or other energy-dissipating mechanism, then the force would continue to add energy to the system, and the amplitude would increase indefinitely. In any real system there is always some dissipation of energy, but the "response" of the system (that is, the amplitude of the forced oscillation) is greatest when the force frequency is equal to one of the normal-mode frequencies. This behavior is called *resonance*.

A common example of mechanical resonance is provided by pushing a swing. The swing is a pendulum with a single natural frequency depending on its length. If a series of regularly spaced pushes is given to the swing, with a frequency equal to that of the swing, the motion may be made quite large. If the frequency of the pushes differs from the natural frequency of the swing, or if the pushes occur at irregular intervals, the swing will hardly execute any vibration at all.

Unlike a simple pendulum, which has only one natural frequency, a stretched string (and other systems discussed in this chapter) has a large number of natural frequencies. Suppose that one end of a stretched string is fixed while the other is moved back and forth in a transverse direction. The amplitude at the driven end is fixed by the driving mechanism. Standing waves are set up in the string, whatever the value of the frequency f. If the frequency is *not* equal to one of the natural frequencies of the string, the amplitude at the antinodes will be fairly small. However, if the frequency is equal to *any one* of the natural frequencies, the string is in resonance and the amplitude at the antinodes will be very much *larger* than that at the driven end. In other words, although the driven end is not a node, it lies much closer to a node than to an antinode when the string is in resonance. In Fig. 22–5a, the right end of the string was held fixed and the left end was forced to oscillate vertically with small amplitude. The photographs show the standing waves of relatively large amplitude that resulted when the frequency of oscillation of the left end was equal to the fundamental frequency or to one of the first three overtones.

A steel bridge or, for that matter, any elastic structure, is capable of vibrating with certain natural frequencies. If the regular footsteps of a marching band were to have a frequency equal to one of the natural frequencies of a bridge which the band is crossing, a vibration of dangerously large amplitude might result. Therefore, in crossing a bridge, the group always "breaks step."

Tuning a radio or television receiver is an example of *electrical* resonance. By turning a dial, one adjusts the natural frequency of an alternating current in the receiving circuit to be equal to the frequency of the waves broadcast by the desired station. *Optical* resonance may also take place between atoms in a gas at low pressure and light waves from a lamp containing the same atoms. Thus, light from a sodium lamp may cause the sodium atoms in a glass bulb to glow with characteristic yellow sodium light.

The phenomenon of resonance may be demonstrated with the aid of the longitudinal waves set up in air by a vibrating plate or tuning fork. If two identical tuning forks are placed some distance apart and one is struck, the other will be heard when the first is suddenly damped. Should a small piece of wax or modeling clay be put on one of the forks, the frequency of that fork will be altered enough to destroy the resonance.

A similar phenomenon can be demonstrated with a piano. With the damper pedal depressed so the dampers are lifted and the strings free to vibrate, one sings a steady tone into the piano. When the singing stops the piano seems to continue to sing the same note. The sound waves from the voice excite vibrations in those strings with natural frequencies close to those (fundamental or harmonics) of the note sung initially.

Questions

22-1 When you inhale helium, your voice becomes high and squeaky. Why? (Don't try this with hydrogen; you might explode like the Hindenburg.) What happens when you inhale carbon dioxide?

22-2 A musical interval of an octave corresponds to a factor of two in frequency. By what factor must the tension in a piano or violin string be increased to raise its pitch one octave?

22-3 The pitch (or frequency) of an organ pipe changes with temperature. Does it increase or decrease with increasing temperature? Why?

22-4 By touching a string lightly at its center while bowing, a violinist can produce a note exactly one octave above the note to which the string is tuned, i.e., a note with exactly twice the frequency. Why is this possible?

22-5 In most modern wind instruments the pitch is changed by changing the length of the vibrating air column using keys or valves. The bugle, however, has no valves or keys, yet it can play many notes. How? Are there restrictions on what notes it can play?

22-6 Kettledrums (tympani) have a pitch, determined by the frequency of the fundamental mode. How can a kettledrum be tuned?

22-7 Energy can be transferred along a string by wave motion. However, in a standing wave no energy can ever be transferred past a node. Why?

22-8 Can a standing wave be produced on a string by superposing two waves with the same frequency but different amplitudes, traveling in opposite directions?

22-9 In discussing standing longitudinal waves in an organ pipe, we spoke of pressure nodes and antinodes. What physical quantity is analogous to pressure for standing transverse waves on a stretched string?

22-10 A glass of water sits on a kitchen counter just above a dishwasher that is running. The surface of the water has a set of concentric stationary circular ripples. What is happening?

22-11 Consider the peculiar sound of a large bell. Do you think the overtones are harmonic?

22-12 What is the difference between music and noise?

22-13 Is the mass per unit length the same for all strings on a piano? On a guitar? Why?

22-14 When a heavy truck drives up a steep hill, the windows in nearby houses sometimes vibrate. Why?

22-15 A popular children's pastime is lining up a row of dominoes, standing on edge, so that when the first one is pushed over the whole row falls, one after another. Is this a wave motion? In what respects is it similar to waves on a stretched string? In what respects is it different?

22-16 A series of traffic lights along a street is timed so that a car going at the right speed can hit a succession of green lights and not have to stop. Considering several cars on the same street, does this situation have wavelike properties?

22-17 Some opera singers are reputed to be able to break a glass by singing the appropriate note. What physical phenomenon could account for this?

Problems

22-1 A steel piano wire 50 cm long, of mass 5 g, is stretched with a tension of 400 N.
  a) What is the frequency of its fundamental mode of vibration?
  b) What is the number of the highest overtone that could be heard by a person who is capable of hearing frequencies up to 10,000 Hz?

22-2 A steel wire of length $L = 100$ cm and density $\rho = 8$ g·cm^{-3} is stretched tightly between two rigid supports. Vibrating in its fundamental mode, the frequency is $f = 200$ Hz.
  a) What is the speed of transverse waves on this wire?
  b) What is the longitudinal stress (force per unit area) in the wire?
  c) If the maximum acceleration at the midpoint of the wire is 80,000 cm·s^{-2}, what is the amplitude of vibration at the midpoint?

22-3 A stretched string is observed to vibrate with a frequency of 30 Hz in its fundamental mode when the supports are 60 cm apart. The amplitude at the antinode is 3 cm. The string has a mass of 30 g.
  a) What is the speed of propagation of a transverse wave in the string?
  b) Compute the tension in the string.

22-4 The fundamental frequency of the A-string on a cello is 220 Hz. The vibrating portion of the string is 68 cm long and has a mass of 1.29 g. With what tension must it be stretched?

22-5 A standing wave of frequency 1100 Hz in a column of methane at 20°C produces nodes that are 20 cm apart. What is the ratio of the heat capacity at constant pressure to that at constant volume.

22-6 An aluminum object is hung from a steel wire. The fundamental frequency for transverse standing waves on the wire is 300 Hz. The object is then immersed in water so that one-half of its volume is submerged. What is the new fundamental frequency?

22-7 Standing waves are set up in a Kundt's tube by the longitudinal vibration of an iron rod 1 m long, clamped at the center. If the frequency of the iron rod is 2480 Hz and the powder heaps within the tube are 6.9 cm apart, what is the speed of the waves

a) in the iron rod, and

b) in the gas?

22-8 The speed of a longitudinal wave in a mixture of helium and neon at 300 K was found to be 758 m·s^{-1}. What is the composition of the mixture?

22-9 The atomic mass of iodine is 127 g·mol^{-1}. A standing wave in iodine vapor at 400 K produces nodes that are 6.77 cm apart when the frequency is 1000 Hz. Is iodine vapor monatomic or diatomic?

22-10 A copper rod 1 m long, clamped at the $\frac{1}{4}$ point, is set in longitudinal vibration and is used to produce standing waves in a Kundt's tube containing air at 300 K. Heaps of cork dust within the tube are found to be 4.95 cm apart. What is the speed of the longitudinal waves in copper?

22-11 Find the fundamental frequency and the first four overtones of a 6-in. pipe

a) if the pipe is open at both ends,

b) if the pipe is closed at one end.

c) How many overtones may be heard by a person having normal hearing for each of the above cases?

Assume $c = 1130$ ft·s^{-1}.

22-12 A long tube contains air at a pressure of 1 atm and temperature 77°C. The tube is open at one end and closed at the other by a movable piston. A tuning fork near the open end is vibrating with a frequency of 500 Hz. Resonance is produced when the piston is at distances 18.0, 55.5, and 93.0 cm from the open end.

a) From these measurements, what is the speed of sound in air at 77°C?

b) From the above result, what is the ratio γ of the heat capacities of air at constant pressure and at constant volume?

22-13 An organ pipe A of length 2 ft, closed at one end, is vibrating in the first overtone. Another organ pipe B of length 1.35 ft, open at both ends, is vibrating in its fundamental mode. Take the speed of sound in air as 1120 ft·s^{-1}. Neglect end corrections.

a) What is the frequency of the tone from A?

b) What is the frequency of the tone from B?

22-14 A plate cut from a quartz crystal is often used to control the frequency of an oscillating electrical circuit. Longitudinal standing waves are set up in the plate with displacement antinodes at opposite faces. The fundamental frequency of vibration is given by the equation

$$f_1 = \frac{2.87 \times 10^5}{s},$$

where f_1 is in hertz and s is the thickness of the plate in centimeters.

a) Compute Young's modulus for the quartz plate.

b) Compute the thickness of plate required for a frequency of 1200 kHz. (1 kHz = 1000 Hz.) The density of quartz is 2.66 g·cm^{-3}.

22-15 In Problem 22-4, what percent increase in tension is needed to increase the frequency from 220 Hz to 233 Hz, corresponding to a rise in pitch from A to A-sharp?

22-16 The longest pipe found in most medium-size pipe organs is 16 feet long. What is the frequency of the corresponding note if the pipe is (a) open at both ends; (b) open at one end, closed at the other? Assume $T = 293$ K.

22-17 The frequency of middle C is 262 Hz.

a) If an organ pipe is open at both ends, what length must it have to produce this note at 20°C?

b) At what temperature will the frequency be 6 percent higher, corresponding to a rise in pitch from C to C-sharp?

22-18 A certain organ pipe produces a frequency of 440 Hz in air. If the pipe is filled with helium at the same temperature, what frequency does it produce?

23 ACOUSTIC PHENOMENA

23-1 Sound waves

In this chapter we are concerned primarily with longitudinal waves in air which, when striking the ear, give rise to the sensation of *sound*. The human ear is sensitive to waves in the frequency range from about 20 to 20,000 Hz, although the term *sound* is sometimes applied also to similar waves with frequencies outside the range of human audibility.

The simplest sound waves are sinusoidal waves with definite frequency, amplitude, and wavelength. When such a wave arrives at the ear, it causes a vibration of the air particles at the eardrum with a definite frequency and amplitude. This vibration may also be described in terms of the variation of *air pressure* at the same point. The air pressure rises above atmospheric pressure, and then drops below atmospheric pressure, with simple harmonic motion of the same frequency as that of an air particle.

The simplest periodic sound wave in an elastic medium is described by a wave function of the form introduced in Sec. 21-3, Eq. (21-7):

$$y = A \sin(\omega t - kx), \qquad (23\text{-}1)$$

where y is the displacement from equilibrium of a point in the medium, and the amplitude A is as usual the *maximum* displacement from equilibrium.

From a practical standpoint it is nearly always easier to measure the *pressure* variations in a sound wave than the displacements, so it is worthwhile to develop a relation between the two. Let p be the instantaneous pressure fluctuation at any point, that is, the amount by which the pressure *differs* from normal atmospheric pressure. If the displacements of two neighboring points x and $x + \Delta x$ are the same, the gas between these points is neither compressed nor rarefied, there is no volume change, and consequently $p = 0$. Only when y varies from one point to a neighboring point is there a change of volume and therefore of pressure.

The fractional volume change $\Delta V/V$ in an element near point x turns out to be given simply by $\partial y/\partial x$, which is the rate of change of y and x as we go from one point to a neighboring point. To see why this is so, we note that $\Delta V/V$ is proportional to the change in length of an element which has length Δx when no wave disturbance is present, divided by Δx. The change in length is the value of y at the point $x + \Delta x$, minus the value at the point x. If Δx is very small, this is approximately Δx multiplied by the *derivative* of y with respect to x. Thus

$$y(x + \Delta x, t) - y(x, t) = \frac{\partial y}{\partial x} \Delta x$$

and

$$\frac{\Delta V}{V} = \frac{y(x + \Delta x, t) - y(x, t)}{\Delta x} = \frac{\partial y}{\partial x}. \qquad (23\text{-}2)$$

Now from the definition of the bulk modulus B, Eq. (10-10), $p = -B\,\Delta V/V$, and we find

$$p = -B\left(\frac{\partial y}{\partial x}\right). \qquad (23\text{-}3)$$

The negative sign arises because, when $\partial y/\partial x$ is positive, the displacement is greater at $x + \Delta x$ than at x, corresponding to an increase in volume and a *decrease* in pressure. For the sinusoidal wave of Eq. (23-1) we find

$$p = BkA \cos(\omega t - kx). \qquad (23\text{-}4)$$

The *maximum* amount by which the pressure differs from atmospheric, that is, the maximum value of p, is called the *pressure amplitude*, denoted P. Equation (23-4) shows that

$$P = BkA. \qquad (23\text{-}5)$$

We note that the pressure amplitude is directly proportional to the displacement amplitude A, as might be expected, and that it also depends on the wavelength. Waves of shorter wavelength (larger k) have greater pressure variations for a given amplitude because the maxima and minima are squeezed closer together.

We may also develop a relation between p and the particle velocity v of points in the medium (not to be confused with the wave propagation speed c). The particle velocity is given by $v = \partial y/\partial t$, and for Eq. (23-1) this is

$$v = \frac{\partial y}{\partial t} = \omega A \cos(\omega t - kx). \qquad (23\text{-}6)$$

The maximum velocity, called the velocity amplitude V, is given by $V = \omega A$. Combining this with Eq. (23-5), we find

$$P = \frac{BkV}{\omega} = \frac{BV}{c}. \qquad (23\text{-}7)$$

Thus the relation between pressure amplitude and velocity amplitude does not depend on frequency. Recalling that the propagation speed c is

given by Eq. (21–16),
$$c = \sqrt{B/\rho}, \tag{23-8}$$
we may rewrite Eq. (23–7) as follows:
$$P = \sqrt{B\rho}\, V. \tag{23-9}$$

The quantity $(B\rho)^{\frac{1}{2}}$ is called the *mechanical impedance* of the system, in reference to an analogy with electric circuits, in which pressure is analogous to voltage, velocity to current, and mechanical impedance to resistance or impedance. In terms of this analogy, Eq. (23–9) is the mechanical analog of Ohm's law, discussed in Sec. 28–3.

Example Measurements of sound waves show that the maximum pressure variations in the loudest sounds that the ear can tolerate without pain are of the order of 30 Pa (above and below atmospheric pressure, which is about 100,000 Pa). Find the corresponding maximum displacement, if the frequency is 1000 Hz and $c = 350$ m·s^{-1}.

Solution We have $\omega = (2\pi)(1000 \text{ Hz}) = 6283 \text{ s}^{-1}$ and
$$k = \frac{\omega}{c} = \frac{6283 \text{ s}^{-1}}{350 \text{ m}\cdot\text{s}^{-1}} = 18.0 \text{ m}^{-1}.$$

The adiabatic bulk modulus for air is
$$B = \gamma p = (1.4)(1.013 \times 10^5 \text{ Pa}) = 1.42 \times 10^5 \text{ Pa}.$$

From Eq. (23–5) we find
$$A = \frac{P}{Bk} = \frac{(30 \text{ Pa})}{(1.42 \times 10^5 \text{ Pa})(18.0 \text{ m}^{-1})}$$
$$= 1.18 \times 10^{-5} \text{ m} = 0.0118 \text{ mm}.$$

Thus the displacement amplitude of even the loudest sound is *extremely* small. The maximum pressure variation in the *faintest* audible sound of frequency 1000 Hz is only about 3×10^{-5} Pa. The corresponding displacement amplitude is about 10^{-9} cm. For comparison, the wavelength of yellow light is 6×10^{-5} cm, and the diameter of a molecule is about 10^{-8} cm. The ear is an extremely sensitive organ!

23–2 Intensity

An essential aspect of wave propagation of all sorts is transfer of *energy*. A familiar example is the energy supply of the earth, which reaches us from the sun via electromagnetic waves. The intensity I of a traveling wave is defined as *the time average rate at which energy is transported by the wave, per unit area*, across a surface perpendicular to the direction of propagation. More briefly, the intensity is the average *power* transported per unit area.

We have seen that the power developed by a force equals the product of force and velocity. Hence the power *per unit area* in a sound wave equals the product of the excess pressure (force per unit area) and the *particle* velocity, given by Eqs. (23–4) and (23–6), respectively. We find
$$pv = \omega Bk A^2 \cos^2(\omega t - kx). \tag{23-10}$$

The intensity is, by definition, the average value of this quantity. The average value of the function $\cos^2 z$ is $\frac{1}{2}$, so we find

$$I = \tfrac{1}{2}\omega BkA^2. \tag{23-11}$$

It is usually more convenient to express I in terms of the pressure amplitude P. Using Eq. (23-5), we find

$$I = \tfrac{1}{2}\omega Bk\left(\frac{P}{Bk}\right)^2 = \frac{cP^2}{2B}. \tag{23-12}$$

By use of the wave speed relation $c = (B/\rho)^{1/2}$, this may also be written in the alternative forms

$$I = \frac{P^2}{2\rho c} = \frac{P^2}{2\sqrt{\rho B}}. \tag{23-13}$$

Derivation of these expressions is left as an exercise.

The intensity of a sound wave of the largest amplitude tolerable to the human ear (about $P = 30$ Pa) is

$$\begin{aligned} I &= \frac{(30 \text{ Pa})^2}{2(1.22 \text{ kg}\cdot\text{m}^{-3})(346 \text{ m}\cdot\text{s}^{-1})} \\ &= 1.07 \text{ J}\cdot\text{s}^{-1}\cdot\text{m}^{-2} = 1.06 \text{ W}\cdot\text{m}^{-2} \\ &= 1.07 \times 10^{-4} \text{ W}\cdot\text{cm}^{-2}. \end{aligned}$$

The unit 1 W·cm^{-2} is a mixed unit, neither cgs nor mks. We mention it here because it is unfortunately in general use among acousticians.

The pressure amplitude of the *faintest* sound wave that can be heard is about 3×10^{-5} Pa, and the corresponding intensity is about 10^{-12} W·m^{-2} or 10^{-16} W·cm^{-2}.

The *total* power carried across a surface by a sound wave equals the product of the intensity at the surface and the surface area, if the intensity over the surface is uniform. The average total power developed as sound waves by a person speaking in an ordinary conversational tone is about 10^{-5} W, while a loud shout corresponds to about 3×10^{-2} W. Since the population of the city of New York is about eight million persons, the acoustical power developed, if all were to speak at the same time, would be about 80 W, or enough to operate a moderate-sized electric light. On the other hand, the power required to fill a large auditorium with loud sound is considerable. Suppose the intensity over the surface of a hemisphere 20 m in radius is 1 W·m^{-2}. The area of the surface is about 2500 m^2. Hence the acoustic power output of a speaker at the center of the sphere would have to be

$$(1 \text{ W}\cdot\text{m}^{-2})(2500 \text{ m}^2) = 2500 \text{ W},$$

or 2.5 kW. The electrical power input to the speaker would need to be considerably larger, since the efficiency of such devices is not very high.

23-3 Intensity level and loudness

Because of the extremely large range of intensities over which the ear is sensitive, a *logarithmic* rather than an arithmetic intensity scale is convenient. Accordingly, the *intensity level* β of a sound wave is defined by the equation

$$\beta = 10 \log \frac{I}{I_0}, \qquad (23\text{-}14)$$

where I_0 is an arbitrary reference intensity, taken as 10^{-12} W·m^{-2}, which corresponds roughly to the faintest sound that can be heard. Intensity levels are expressed in *decibels,* abbreviated db.*

If the intensity of a sound wave equals I_0 or 10^{-12} W·m^{-2}, its intensity level is 0 db. The maximum intensity that the ear can tolerate, about 1 W·m^{-2}, corresponds to an intensity level of 120 db. Table 23-1 gives the intensity levels in decibels of a number of familiar noises. It is taken from a survey made by the New York City Noise Abatement Commission.

Table 23-1 Noise levels due to various sources (representative values)

Source or description of noise	Noise level, db	Intensity, W·m^{-2}
Threshold of pain	120	1
Riveter	95	3.2×10^3
Elevated train	90	10^{-3}
Busy street traffic	70	10^{-5}
Ordinary conversation	65	3.2×10^{-6}
Quiet automobile	50	10^{-7}
Quiet radio in home	40	10^{-8}
Average whisper	20	10^{-10}
Rustle of leaves	10	10^{-11}
Threshold of hearing	0	10^{-12}

Within the range of audibility, the sensitivity of the ear varies with frequency. The *threshold of audibility* at any frequency is the minimum intensity of sound at that frequency that can be detected. For a young adult with normal hearing, the threshold of audibility at 1000 Hz is about 0 db; at 200 and 15,000 Hz it is about 20 db, and at 50 and 18,000 Hz it is about 50 db. Thus the ear's sensitivity drops off at the low and high ends of the frequency scale.

The term *loudness* refers to the listener's subjective perception of the magnitude of a sound sensation. Loudness usually increases with intensity, but because of the varying sensitivity of the ear, there is no direct relation between the two. It is possible, however, to define a unit of loudness, and then by experimenting with human subjects, compare the perceived loudness of various sounds and work out a correlation between intensity and loudness for sounds of various frequencies and combinations of frequencies.

---

* Originally, a scale of intensity levels in *bels* was defined by the relation

Intensity level = $\log (I/I_0)$.

This unit proved inconveniently large, and hence the *decibel,* one-tenth of a bel, has come into general use. The unit was named in honor of Alexander Graham Bell.

The standard of perceived loudness is the *sone*. One sone is defined as the loudness experienced by a listener with normal hearing when a 1000-Hz tone of intensity 40 db is presented to both ears. A 2-sone sound is perceived as twice as loud as a 1-sone sound, and so on. One sone is found to correspond to about 70 db at 60 Hz, 53 db at 200 Hz, 40 db at 1000 Hz, or 35 db at 3000 Hz. In many situations the loudness from two combined sources is found to be *additive;* if one source causes a loudness at a certain location of 3 sones, and another 5 sones, then the combined loudness from both sources is found to be 8 sones. The sone is coming into common use as a measure of noise levels; it takes into account the subjective, psychological aspects of sound, as the intensity (measured in decibels) does not.

23-4 Quality and pitch

A string that has been plucked or the column of air in a wind instrument vibrates with many frequencies at the same time. It is unusual for a body to vibrate with only one frequency. A carefully made tuning fork struck lightly on a rubber block may vibrate with only one frequency, but in the case of musical instruments, the fundamental and many harmonics are usually present at the same time. The impulses that are sent from the ear to the brain give rise to one net effect that is characteristic of the instrument. Suppose, for example, that the sound spectrum of a tone consisted of a fundamental of 200 Hz and harmonics 2, 3, 4, and 5, each with a different intensity, whereas the sound spectrum of another tone consisted of exactly the same frequencies but with a different intensity distribution. The two tones would sound different; they are said to differ in *quality* or *timbre*.

Adjectives used to describe the quality of musical tones are purely subjective in character, such as reedy, golden, round, mellow, tinny, etc. The quality of a sound is determined in part by the number of overtones present and their respective intensity-versus-time curves. The sound spectra of several musical instruments are shown in Fig. 23-1.

Another important factor in determining tone quality is the behavior at the beginning and end of a tone. A piano tone begins percussively with a thump and then dies away gradually; a harpsichord tone, in addition to having different harmonic content, begins much more quickly and incisively with a click, the higher harmonics beginning before the lower ones. The cessation of tone when the key is released is also much

23-1 Sound spectra of some musical instruments. (Courtesy of Dr. Harvey Fletcher.)

more prompt with the harpsichord than the piano. Similar effects are present in other musical instruments; with wind and string instruments the player has considerable control over the attack and decay of the tone, and these characteristics help define the unique characteristics of each instrument.

The term *pitch* refers to the attribute of a sound sensation that enables one to classify a note as "high" or "low." Like loudness, it is a subjective quantity and cannot be measured with instruments. Pitch is related to the objective quantity *frequency,* but there is no one-to-one correspondence. For a pure tone of constant intensity, the pitch becomes higher as the frequency is increased, but the pitch of a pure tone of constant frequency becomes lower as the intensity level is raised.

Many of the notes played on musical instruments are rich in harmonics, some of which may be more prominent than the fundamental. Presented with an array of frequencies constituting a harmonic series, the ear assigns a characteristic pitch to the combination, this pitch being that associated with the fundamental frequency of the series. So definite is this pitch sensation that it is possible to eliminate the fundamental frequency entirely, by means of filters, without changing the pitch! The ear supplies the fundamental, provided the correct harmonics are present. It is this rather surprising property of the ear that enables a small loudspeaker that does not radiate low frequencies well to give nevertheless the impression of good radiation in the low-frequency region. Because the speaker is a fairly efficient radiator for the frequencies of the harmonics, the listener believes he is actually hearing the low frequencies, when instead he is hearing only multiples of these frequencies and his ear is supplying the fundamental.

The frequency spectrum of musical tones is often even more complex than is suggested by Fig. 23-1. Rather than having a precisely defined fundamental frequency, a tone usually has a continuous distribution of frequencies, with intensity peaks at frequencies which are identified as the fundamental and the various overtones. The perception of pitch of a particular tone is therefore related to its frequency spectrum in an even more complex way than would be the case with a sharply defined fundamental frequency.

*23-5 Musical intervals and scales

When certain musical tones are produced in succession or together, even an untrained listener recognizes a relationship among them. Such relations are described in musical language by words such as octave, major third, minor third, etc. The listener recognizes something basic in these combinations of tones; and experimental measurement of the fundamental frequencies of the separate tones discloses that their relationships, one to the other, may be approximated by simple whole-number ratios. For example, the fundamental frequencies of middle C of the piano and its *octave* above, C′, have a ratio of 1 to 2. Another basic set of tones is obtained by playing C, E, G. These constitute what is known as a *major triad,* and the frequencies are found to be approximately proportional to 4, 5, and 6.

Starting at middle C of the piano, and playing only the white keys toward the right, it is possible to find, within only nine notes, three major triads. These are shown in Table 23-2; they enable one to calculate the frequencies of all the notes, once one of them has been chosen arbitrarily. By international agreement, the frequency of the A above middle C is chosen to be 440 vibrations per second. The frequencies of all the notes, shown in row 5 of Table 23-2, are those that correspond to the *just diatonic scale* of the key of C. The frequency ratios of adjacent notes are seen to be either 9/8, 10/9, or 16/15. The interval between two tones whose frequencies bear the ratio 9/8 or 10/9 are called *whole tones*, whereas the interval between two notes of frequency ratio 16/15 is a *half tone*.

Table 23-2 Frequency relations in the just diatonic scale and in the equally tempered scale

Frequency relations	Do Middle C	Re D	Mi E	Fa F	Sol G	La A	Ti B	Do' C'	Re' D'
Octave, key of C	1							2	
Major triad, key of C	4		5		6			(8)	
Major triad, key of F				4		5		6	
Major triad, key of G		(3)			4		5		6
Just scale, key of C	264	297	330	352	396	440	495	528	594
Intervals in just scale		9/8 whole	10/9 whole	16/15 half	9/8 whole	10/9 whole	9/8 whole	16/15 half	9/8 whole
Equally tempered scale suitable for all keys	261.6	293.7	329.6	349.2	392.0	440	493.9	523.3	587.4
Intervals in equally tempered scale		$\sqrt[6]{2}$	$\sqrt[6]{2}$	$\sqrt[12]{2}$	$\sqrt[6]{2}$	$\sqrt[6]{2}$	$\sqrt[6]{2}$	$\sqrt[12]{2}$	$\sqrt[6]{2}$

If a major diatonic scale were constructed starting at D instead of at C, four new notes would be needed for a perfect diatonic scale. If all possible musical keys were to be provided for, many additional notes would be needed for each octave. To avoid this tremendous complication, what is known as the *equally tempered scale* has been devised. In this scheme there are 12 half-tone intervals in every range of an octave, adjacent notes a half tone apart bearing the constant ratio of the twelfth root of 2, i.e., 1.05946.

Simple as this scheme is, it results in no one scale being exactly diatonic. Since the ratios of the diatonic scale were originally selected to suit the preferences of the ear (being made up of three sets of major triads, each of which constitutes a harmonious combination), this means that an instrument tuned to the equally tempered scale is not quite so pleasant to the ear. Thus a piano whose white keys are tuned to the "just" scale sounds better (i.e., more "in tune") than one tuned to equal temperament when played in the key of C, but in some other keys it sounds worse. Hence an instrument to be used for compositions in all keys is usually tuned to equal temperament.

In the Baroque period, however, keys with more than three sharps or flats were used relatively infrequently, and various compromise temperaments were devised by Kirnberger, Werckmeister, and others, to favor the commonly used keys. Organs and harpsichords intended primarily for music of this period are sometimes tuned to one of these temperaments. There is evidence that J. S. Bach favored these rather than equal temperament. His composition "The Well-Tempered Clavier" contains preludes and fugues in all the major and minor keys, but it is important not to misconstrue *well-tempered* as *equally tempered!*

23-6 Beats

Standing waves in an air column have been cited as one example of *interference*. They arise when two waves of the same amplitude and frequency are traveling through the same region in opposite directions. We now wish to consider another type of interference that results when two waves of equal amplitude but slightly different frequency travel through the same region. Such a condition exists when two tuning forks of slightly different frequency are sounded simultaneously or when two piano strings struck by the same key are slightly "out of tune."

Let us consider a particular point in space through which the two waves pass simultaneously. The individual displacements are plotted as functions of time in Fig. 23-2a. If the total length of the time axis represents one second, the frequencies of the two waves are 16 and 18 Hz. Applying the principle of superposition, we add the two displacements at each time to find the total displacement at that time, obtaining the graph of Fig. 23-2b. The appearance is that of a single sinusoidal wave with a varying amplitude; in this example, the amplitude goes through two maxima and two minima in one second. This varying amplitude causes variations of loudness that are called *beats*. The amplitude varies with a frequency called the *beat frequency*, which is different from that of either wave. If this frequency is a few hertz, it is perceived as a waver or pulsation in the tone.

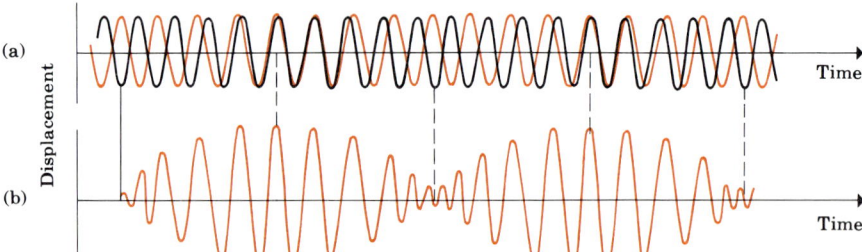

23-2 Beats are fluctuations in amplitude produced by two sound waves of slightly different frequency.

The production of beats may be analyzed as follows. At a particular point, the displacements due to two waves of equal amplitude A but different frequency f_1 and f_2 may be written

$$y_1 = A \cos 2\pi f_1 t, \qquad y_2 = A \cos 2\pi f_2 t.$$

By the principle of superposition, the resultant displacement is

$$y = y_1 + y_2 = A[\cos 2\pi f_1 t + \cos 2\pi f_2 t],$$

and, since

$$\cos a + \cos b = 2 \cos\left(\frac{a+b}{2}\right) \cos\left(\frac{a-b}{2}\right),$$

this may be written

$$y = \left[2A \cos 2\pi \left(\frac{f_1 - f_2}{2}\right) t\right] \cos 2\pi \left(\frac{f_1 + f_2}{2}\right) t. \qquad (23\text{-}15)$$

The resulting vibration can then be considered to have frequency $(f_1 + f_2)/2$ (the average frequency of the two tones) and amplitude given by the expression in brackets. The amplitude therefore varies with time at a frequency $(f_1 - f_2)/2$. If f_1 and f_2 are nearly equal, this term is small and the amplitude fluctuates very slowly. When the amplitude is large, the sound is loud, and vice versa. A beat, or a maximum of amplitude, will occur when $\cos 2\pi t(f_1 - f_2)/2$ equals 1 or -1. Since *each* of these values occurs once in each cycle, the number of beats per second is *twice* the frequency $(f_1 - f_2)/2$; *the number of beats per second equals the difference of the frequencies.*

Beats between two tones can be detected by the ear up to a beat frequency of 6 or 7 Hz. Two piano strings or two organ pipes differing in frequency by two or three hertz sound wavery and "out of tune," although some organ stops contain two sets of pipes deliberately tuned to beat frequencies of about 1 to 2 Hz for a gently undulating effect. Listening for beats is an important technique in tuning all musical instruments.

At higher frequencies, individual beats can no longer be distinguished and the sensation merges into one of *consonance* or *dissonance* depending on the frequency ratio of the tones. A beat frequency, even though it lies within the frequency range of the ear, is not necessarily interpreted by the ear as a tone of that frequency. Nevertheless, a tone can be heard of frequency equal to the frequency difference between two others sounded simultaneously. Such a tone is called a *difference* tone. Although not as easy to recognize, a frequency called a *summation* tone and equal to the sum of the frequencies of the two tones can also be heard. The general term applied to both difference tones and summation tones is *combination* tones.

*23-7 The Doppler effect

When a source of sound, or a listener, or both, are in motion relative to the air, the pitch of the sound, as heard by the listener, is in general not the same as when source and listener are at rest. The most common example is the sudden drop in pitch of the sound from an automobile horn as one meets and passes a car proceeding in the opposite direction. This phenomenon is called the *Doppler effect.*

Let v_L and v_S represent the velocities of a listener and a source, relative to the air. We shall consider only the special case in which the velocities lie along the line joining listener and source. Since these velocities may be in the same or opposite directions, and the listener may be either ahead of or behind the source, a convention of signs is required. We shall take the positive directions of v_L and v_S as that *from* the position of the listener, toward the position of the source. The speed of propagation of sound waves, c, will always be considered positive.

We consider first a listener L moving with velocity v_L toward a stationary source S, as in Fig. 23–3. The source emits a wave with frequency f_S and wavelength $\lambda = c/f_S$. The figure shows several wave crests, separated by equal distances λ. The waves approaching the moving listener have a speed of propagation *relative to him* of $(c + v_L)$. Thus the frequency f_L with which the listener encounters wave crests, i.e., the frequency he *hears,* is

$$f_L = \frac{c + v_L}{\lambda} = \frac{c + v_L}{c/f_S}, \tag{23-16}$$

or

$$f_L = f_S\left(\frac{c + v_L}{c}\right) = f_S\left(1 + \frac{v_L}{c}\right). \tag{23-17}$$

Thus an observer moving toward a source hears a larger frequency and higher pitch than a stationary observer. Similarly, a listener moving away from the source ($v_L < 0$) hears a lower pitch.

Now suppose the source is also moving, with velocity v_S, as in Fig. 23–4. The wave speed relative to the air is still c; the speed of propagation of a wave is not altered by the motion of the source, but is a property of the wave medium alone. But the wavelength is no longer given by

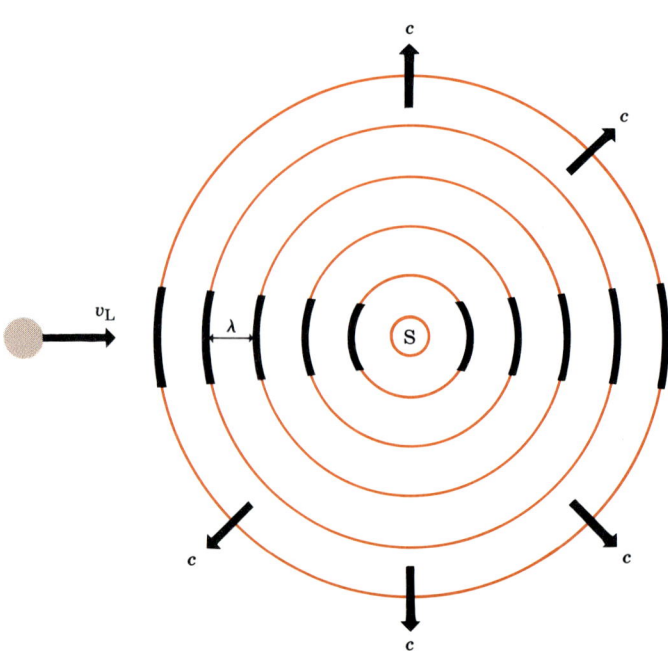

23-3 A listener moving toward a stationary source hears a frequency higher than the source frequency because the relative velocity of listener and wave is greater than c.

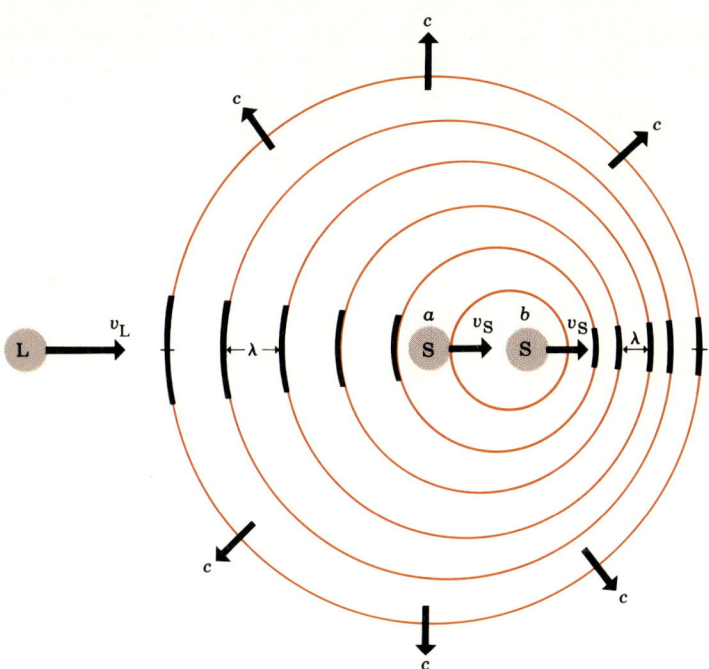

23-4 Wave surfaces emitted by a moving source are crowded together in front of the source and stretched out behind it.

c/f_S. The time for emission of one cycle of the wave is the period $\tau = 1/f_S$. During this time the wave travels a distance $c\tau = c/f_S$, and the source moves a distance $v_S\tau = v_S/f_S$. The wavelength is the distance between successive wave crests, and this is determined by the *relative* displacement of source and wave. In the region to the right of the source, the wavelength is

$$\lambda = \frac{c}{f_S} - \frac{v_S}{f_S} = \frac{c - v_S}{f_S}, \qquad (23\text{-}18)$$

and in the region to the left it is

$$\lambda = \frac{c + v_S}{f_S}. \qquad (23\text{-}19)$$

The waves are, respectively, compressed and stretched out by the motion of the source.

The frequency measured by the listener is now given by substituting Eq. (23-19) into the first form of Eq. (23-16):

$$f_L = \frac{c + v_L}{\lambda} = \frac{c + v_L}{(c + v_S)/f_S},$$

or

$$\frac{f_L}{c + v_L} = \frac{f_S}{c + v_S}, \qquad (23\text{-}20)$$

which expresses the frequency f_L heard by the listener in terms of the frequency f_S of the source.

This general relation includes all possibilities for collinear motion of source and listener relative to the medium. If one happens to be at rest in the medium, the corresponding velocity is zero, and of course when both are at rest or have the same velocity relative to the medium, then $f_L = f_S$. Whenever source or listener is moving in the opposite direction to that which we have designated as positive, the corresponding velocity to be used in Eq. (23-18) is negative. The examples illustrate these sign conventions.

Examples Let $f_S = 300$ Hz, and $c = 300$ m·s^{-1}. The wavelength of the waves emitted by a stationary source is then $c/f_S = 1.00$ m.

a) What are the wavelengths ahead of and behind the moving source in Fig. 23-4 if its velocity is 30 m·s^{-1}?

In front of the source,

$$\lambda = \frac{c - v_S}{f_S} = \frac{300 \text{ m·s}^{-1} - 30 \text{ m·s}^{-1}}{300 \text{ Hz}} = 0.90 \text{ m}.$$

Behind the source,

$$\lambda = \frac{c + v_S}{f_S} = \frac{300 \text{ m·s}^{-1} + 30 \text{ m·s}^{-1}}{300 \text{ Hz}} = 1.10 \text{ m}.$$

b) If the listener L in Fig. 23-4 is at rest and the source is moving away from him at 30 m·s^{-1}, what is the frequency as heard by the listener?

Since

$$v_L = 0 \quad \text{and} \quad v_S = 30 \text{ m·s}^{-1},$$

we have

$$f_L = f_S \frac{c}{c + v_S}$$
$$= 300 \text{ Hz} \left(\frac{300 \text{ m·s}^{-1}}{300 \text{ m·s}^{-1} + 30 \text{ m·s}^{-1}} \right) = 273 \text{ Hz}.$$

c) If the source in Fig. 23-4 is at rest and the listener is moving toward the left at 30 m·s^{-1}, what is the frequency as heard by the listener?

The positive direction (from listener to source) is still from left to right, so

$$v_L = -30 \text{ m·s}^{-1}, \quad v_S = 0,$$
$$f_L = f_S \frac{c + v_L}{c}$$
$$= 300 \text{ Hz} \left(\frac{300 \text{ m·s}^{-1} - 30 \text{ m·s}^{-1}}{300 \text{ m·s}^{-1}} \right) = 270 \text{ Hz}.$$

Thus, while the frequency f_L heard by the listener is less than the frequency f_S both when the source moves away from the listener and when the listener moves away from the source, the decrease in frequency is not the same for the same speed of recession.

In the preceding equations, the velocities v_L, v_S, and c are all *relative to the air*, or more generally, to the medium in which the waves are traveling. The Doppler effect exists also for electromagnetic waves in empty space, such as light waves or radio waves. In this case, there is no "medium" relative to which a velocity can be defined, and we can speak only of the *relative* velocity v of source and receiver.

To derive the expression for the Doppler frequency shift for light requires the use of relativistic kinematic relations. These will be derived in Chapter 43, but meanwhile we quote the result without derivation. The wave speed c is the speed of light and is the same for both source and listener. In the frame of reference in which the listener is at rest, the source is moving away from the listener with velocity v. (If, instead, the source is *approaching* the listener, v is negative.) The source frequency is again f_S. The frequency f_L measured by the listener (i.e., the frequency of arrival of the waves at L) is then given by

$$f_L = \left(\sqrt{\frac{c-v}{c+v}}\right) f_S. \quad (23\text{-}21)$$

23-5 Change of velocity component along the line of sight of a satellite passing a tracking station.

When v is positive, the source moves *away* from the listener and f_L is always *less* than f_S; when v is negative, the source moves *toward* the listener and f_L is *greater* than f_S. Thus, the qualitative effect is the same as for sound, although the quantitative relationship is different.

The Doppler effect provides a convenient means of tracking a satellite that is emitting a radio signal of constant frequency f_S. The frequency f_L of the signal received on the earth decreases as the satellite is passing, since the velocity component *toward* the earth decreases from position 1 to position 2 in Fig. 23-5, and then points *away* from the earth from 2 to 3. If the received signal is combined with a constant signal generated in the receiver to give rise to *beats,* then the beat frequency may be such as to produce an audible note whose pitch decreases as the satellite passes overhead.

A similar technique is used by law-enforcement officers to measure automobile speeds. An electromagnetic wave is emitted by a source at the side of the road, typically attached to a police car. The wave is reflected from a moving car, which thus acts as a moving source; the reflected wave is Doppler-shifted in frequency. Measurement of the frequency shift using beats, as with satellite tracking, permits simple measurement of the speed.

The Doppler effect for *light* is important in astronomy. Analysis of the spectra of light from distant stars shows shifts in wavelength compared to spectra from the same elements on earth. These can be interpreted as Doppler shifts due to motion of the stars. The shift is nearly always toward the longer wavelength or red end of the spectrum, and is therefore called the *red shift*. Such observations have provided practically all the evidence for the "exploding universe" cosmological theories, which represent the universe as having evolved from a great explosion several billion years ago in a relatively small region of space.

*23-8 Radiation from a piston

As we have seen, sound waves can be produced in a wide variety of ways. One of the simplest to understand is the moving cone of a loudspeaker, and it is instructive to examine the action of such a system. For simplicity, we consider an idealized model in which the cone is represented as a plane moving perpendicularly to its surface. A single surface oscillating this way produces sound waves radiating away from both sides. To simplify the problem further, let us assume that we have an oscillating piston fitting closely in a large wall or *baffle,* as in Fig. 23-6a. Then only

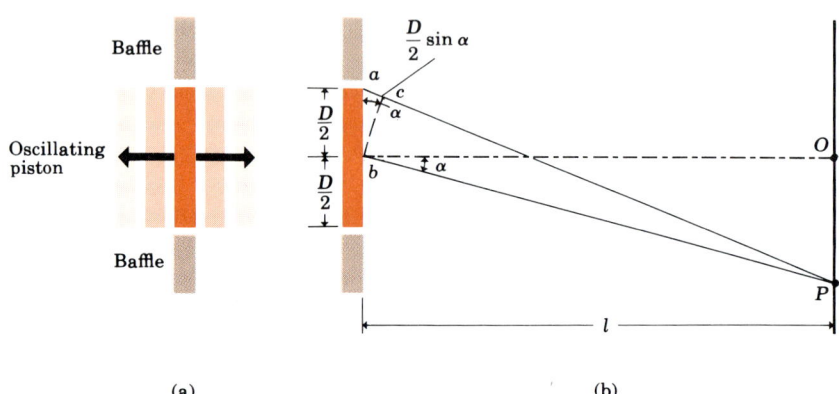

23-6 (a) An oscillating piston set in a baffle. (b) Construction for locating diffraction minima in radiation pattern from a long rectangular source.

the waves radiated from the right side of the piston need be considered in the space to the right of the baffle.

The directional distribution of radiated energy depends on the shape of the piston and on its dimensions relative to the wavelength of the emitted waves. The mathematics becomes very complicated for any shapes except circles and rectangles. The simplest case is that of a very long rectangle such as a long plank, and Fig. 23-6b is an end view of such a piston. Imagine the right face of the piston, of width D, to be subdivided into a large number of very narrow strips parallel to the length of the piston and perpendicular to the plane of the diagram. Each of these strips can be treated as a *line source,* sending out waves having *cylindrical* wave surfaces coaxial with the source. We now apply the principle of superposition to these cylindrical waves.

Point P is a point in space at a distance l large compared with the width D of a source, and with the wavelength λ. Consider the two elementary line sources at a and b, one at the top edge of the piston and the other just below its centerline. With P as a center and Pb as radius, draw the arc shown by the broken line, intersecting the line Pa at c. The distance ac is then the path difference between the waves reaching P from a and b. Since l is large compared with D, the arc bc is very nearly a straight line and abc is very nearly a right triangle with the angle α equal to the angle between Pb and the normal Ob. The path difference ac is then

$$ac = \frac{D}{2}\sin\alpha. \tag{23-22}$$

If the angle α has such a value that ac equals a half-wavelength, the waves from a and b reach P 180° out of phase and very nearly cancel one another.

Consider next the pair of line sources just below a and b. A diagram similar to Fig. 23-6b can be constructed for them, and the waves from these two sources *also* cancel at P. Proceeding in this way over the entire surface of the piston, we see that in a direction making an angle α with the normal, nearly complete cancellation occurs, provided that

$$ac = \frac{\lambda}{2},$$

or, from Eq. (23–22),

$$\sin \alpha = \frac{\lambda}{D}. \tag{23-23}$$

As an example, if $D = 0.3$ m and $\lambda = 0.15$ m (corresponding to a frequency of about 2000 Hz),

$$\sin \alpha = \frac{0.15 \text{ m}}{0.3 \text{ m}} = 0.50,$$

$$\alpha = 30°,$$

and no energy is radiated in a direction at 30° on either side of the normal.

Other minima occur in directions for which $\sin \alpha = 2\lambda/D, 3\lambda/D$, etc. This can be shown by dividing the surface of the piston into quarters, sixths, etc., and pairing off one element against another, as in Fig. 23–6b. Between the minima in the radiation pattern are regions of maximum intensity, and there is a maximum straight out to the side (at point O) where radiation from all regions of the piston travels the same distance and arrives in phase. A completely analogous phenomenon occurs with light, and in that context it is called *diffraction*. This is discussed in detail in Chapter 41.

The analysis of the radiation pattern from a *circular* piston can be carried out in the same way as that for a long rectangle. The piston is subdivided into narrow circular zones instead of long strips, and the effect of the waves from all the zones is summed at a distant point. There is a maximum of intensity along the axis of the piston, as would be expected. The angle α at which the first minimum occurs is given by

$$\sin \alpha = 1.22 \frac{\lambda}{D}, \tag{23-22}$$

where D is now the piston diameter. If $D = 0.3$ m, $\lambda = 0.15$ m,

$$\sin \alpha = 1.22 \left(\frac{0.15 \text{ m}}{0.3 \text{ m}} \right) = 0.61,$$

$$\alpha \approx 38°.$$

About 85 percent of the radiated energy is concentrated within a cone of this half-angle. Other minima and maxima of rapidly decreasing intensity surround the central maximum. Figure 23–7 is a space diagram of the intensity distribution in front of an oscillating circular piston set in a baffle.

This *directivity* of the sound radiated by a piston has a number of important applications. In a motion picture theater, for example, we wish the sound waves radiated by a speaker behind the screen to spread out over a large angle. If a loudspeaker 0.3 m in diameter, approximated by a plane circular piston, is the sound source, then at a frequency of 2000 Hz the directly radiated wave is concentrated mainly in a "beam" of half-angle 38°, centered on the speaker axis. For a frequency of 10,000 Hz, or a wavelength of about 0.03 m, the half-angle is

$$\sin \alpha = 1.22 \left(\frac{0.03 \text{ m}}{0.3 \text{ m}} \right) = 0.122, \quad \alpha \approx 7°,$$

23-7 Space diagram of intensity distribution in front of an oscillating circular piston set in a baffle. A vector drawn from the center of the piston to any point on the surface has a length proportional to the sound intensity in that direction, as observed at a large fixed distance from the piston.

while at a frequency of 1200 Hz, or a wavelength of about 0.25 m,

$$\sin \alpha = 1.22 \left(\frac{0.25 \text{ m}}{0.30 \text{ m}} \right) \approx 1.0. \qquad \alpha \approx 90°.$$

This means that for frequencies of 1200 Hz or less the radiated energy spreads out fairly uniformly, while at 10,000 Hz it is concentrated in a narrow beam, like a searchlight beam.

Since the intelligibility of speech depends largely on the high-frequency components, these are usually channeled into a number of speakers directed toward different parts of the auditorium; the low frequencies can be handled by a single speaker, since the low-frequency radiation pattern has a much wider angular spread. On the other hand, it is sometimes desirable, as in underwater sound signaling or when leading the cheering section, to produce a beam having only a small angular divergence. To accomplish this, the diameter of the source (approximating it by a circular piston) must be large compared with the wavelength of the radiated sound.

If instead of a vibrating piston set in a wall there is merely an *aperture* in the wall and a plane wave is incident from the left, the waves transmitted through the aperture will propagate beyond it in the same way as waves *originating* at a piston set in the aperture. If the wavelength is small compared with the dimensions of the aperture, the spreading of the waves is small, while if the wavelength is large, the waves spread out in all directions.

When there is an obstacle in the path of a train of waves, the resultant effect at the far side of the obstacle is due to those portions of the advancing wave surface that are *not* obstructed. In very general terms, if the wavelength is relatively small, the spreading is small and the obstacle casts a sharp "shadow." The larger the wavelength, the greater the spreading or bending of the waves. Thus one *can* hear around the corner of a building, but the effect is greater for waves of long wavelength (or low frequency) than for waves of short wavelength (or high frequency). The general term for the phenomena described above, in which one is concerned with the resultant effect of a large number of waves from different parts of a source, is *diffraction*. Diffraction phenomena with *light* are very important and will be discussed in detail in Chapter 41.

23-9 Applications of acoustic phenomena

The relevance of acoustical principles is by no means limited to sound and hearing. Analysis of mechanical waves is a powerful investigative tool in a wide variety of fields. Elastic waves in water are reflected by solid bodies immersed in the water, such as schools of fish, submarines, or submerged wrecked ships. *Sonar* systems make use of reflections of sound waves in water to determine position and motion of such bodies.

Analysis of elastic waves in the earth provides important information about its structure. The interior of the earth may be pictured crudely as made of concentric spherical shells, with mechanical properties such as density and elastic moduli that are different in different shells. Waves produced by explosions or earthquakes are reflected and refracted at the interfaces between these shells, and analysis of these waves permits partial determination of the dimensions and properties of the shells.

Some of the most interesting recent applications of acoustics lie in the field of medicine, where sound waves are used for both diagnosis and therapy. In diagnosis, *ultrasonic* (i.e., above 20 kHz) frequencies are usually used because their short wavelengths permit study of smaller-scale phenomena than with audible sound. Reflection of such waves from regions in the interior of the body can be used to detect a wide variety of anomalous conditions such as tumors, and to study various phenomena such as heart-valve action. Ultrasound is more sensitive than x-rays in distinguishing various kinds of tissues; it is believed to be less hazardous than x-rays, although possible hazards of ultrasound have not yet been thoroughly explored. At much higher power levels, ultrasound appears to have promise as a selective destroyer of pathological tissue, and may find usefulness in the treatment of certain cancers as well as arthritis and related diseases.

In medical and other applications of ultrasonics, a variety of electronic instrumentation is used. The sound is always produced by first generating an electrical wave, and using this to drive a *transducer* which converts electrical to mechanical (i.e., sound) waves. Thus the *transducer* is similar in action to a loudspeaker except that the frequency ranges may be different; frequencies as high as several megahertz (i.e., several million cycles per second) are routinely used. The detecting instruments include transducers, functioning in the role of microphones, as well as amplifiers and devices for displaying reflected signals. A common technique is to send out pulses, and observe the transmitted and reflected pulse "blips" on an oscilloscope, which permits a direct measurement of the time interval between transmission and reflection.

The applications of acoustic principles to environmental problems are obvious. It is now recognized that noise is an important aspect of environmental degradation. The design of quiet mass-transit vehicles, for example, involves detailed study of sound generation and propagation in the motors, wheels, and supporting structures of vehicles. Excessive noise levels can lead to permanent hearing impairment; recent studies have shown that many young "rock" musicians have suffered hearing losses typical of persons 65 years of age, and even prolonged listening to high-level rock music (90 to 100 db) can lead to permanent damage.

Questions

23-1 Ultrasonic cleaners use ultrasonic waves of high intensity in water or a solvent to clean dirt from dishes, engine parts, and so on. How do they work? What advantages and disadvantages does this process have, compared with other cleaning methods?

23-2 Lane dividers on highways sometimes have regularly spaced ridges or ripples. When the tires of a moving car roll along such a divider, a musical note is produced. Why? Could this phenomenon be used to measure the car's speed? How?

23-3 Two tuning forks have identical frequencies, but one is stationary while the other is mounted on a rotating record turntable. What does a listener hear?

23-4 The organist in a cathedral plays a loud chord and then releases it. The sound persists for a few seconds and gradually dies away. Why does it persist? What happens to the energy when it dies away?

23-5 Why do foghorns always have very low pitches?

23-6 Some stereo amplifiers intended for home use have a maximum power output of 600 W or more. What would happen if you had 600 W of actual sound power in a moderate-sized room?

23-7 Why does your voice sound different over the telephone than in person?

23-8 Why does a guitar have a sharper, more metallic sound when played with a hard pick than when plucked with the bare fingertips?

23-9 When a record is being played with the volume control turned down all the way, a faint sound can be heard coming directly from the stylus. It seems to be all treble and no bass. How is the sound produced, and why are the frequencies so unbalanced?

23-10 The tone quality of a violin is different when the bow is near the bridge (the ends of the strings) than when it is nearer the centers of the strings. Why?

23-11 An engineer who likes to express everything in technical terms remarked that the price of gasoline rose by 2 db in 1979. What do you think he meant?

23-12 When you are shouting at someone a fair distance away, it is easier for him to hear you if the wind is blowing from you to him, than if it is in the opposite direction. What is the physical basis for this difference?

23-13 Two notes an octave apart on a piano have a frequency ratio of 2:1. When one is slightly out of tune, beats are heard. How are they produced?

23-14 A large church has part of the organ in front and part in back. When both divisions are playing at once and a person walks rapidly down the center aisle, the two divisions sound out of tune. Why?

23-15 Can you think of circumstances in which a Doppler effect would be observed for surface waves in water? For elastic waves propagating in a body of water?

23-16 How does a person perceive the direction from which a sound comes? Is our directional perception more acute for some frequency ranges than for others? Why?

23-17 A sound source and a listener are both at rest on the earth, but a strong wind is blowing. Is there a Doppler effect?

Problems

Unless otherwise indicated, assume the speed of sound in air to be $c = 345 \text{ m} \cdot \text{s}^{-1}$.

23-1

a) If the pressure amplitude in a sound wave is tripled, by what factor is the intensity of the wave increased?

b) By what factor must the pressure amplitude of a sound wave be increased in order to increase the intensity by a factor of 16?

23-2

a) Two sound waves of the same frequency, one in air and one in water, are equal in intensity. What is the ratio of the pressure amplitude of the wave in water to that of the wave in air?

b) If the pressure amplitudes of the waves are equal, what is the ratio of their intensities?

c) What is the difference between their intensity levels? (The speed of sound in water may be taken as $1490 \text{ m} \cdot \text{s}^{-1}$.)

23-3

a) Relative to the arbitrary reference intensity of $10^{-12} \text{ W} \cdot \text{m}^{-2}$, what is the intensity level in decibels of a sound wave whose intensity is $10^{-6} \text{ W} \cdot \text{m}^{-2}$?

b) What is the intensity level of a sound wave in air whose pressure amplitude is 0.2 Pa?

23-4

a) Show that if β_1 and β_2 are the intensity levels in db of sounds of intensities I_1 and I_2, respectively, the difference in intensity levels of the sounds is

$$\beta_2 - \beta_1 = 10 \log \frac{I_2}{I_1}.$$

b) Show that if P_1 and P_2 are the pressure amplitudes of two sound waves, the difference in intensity levels of the waves is

$$\beta_2 - \beta_1 = 20 \log \frac{P_2}{P_1}.$$

23-5

a) Show that if the reference level of intensity is $I_0 = 10^{-12}$ W·m^{-2}, the intensity level of a sound of intensity I (in W·m^{-2}) is

$$\beta = 120 + 10 \log I.$$

b) Is this relation valid if I is expressed in W·cm^{-2} instead of W·m^{-2}? If not, how should it be modified?

23-6 The intensity due to a number of independent sound sources is the sum of the individual intensities. How many decibels greater is the intensity level when all five quintuplets cry simultaneously than when a single one cries? How many more crying babies would be required to produce a further increase in the intensity level of an equal number of decibels?

23-7 A window whose area is 1 m^2 opens on a street where the street noises result in an intensity level, at the window, of 60 db. How much "acoustic power" enters the window via the sound waves?

23-8 A source of sound emits a total power of 10 W, uniformly in all directions. At what distance from the source is the sound level

a) 100 db?

b) 60 db?

23-9 A certain sound source radiates uniformly in all directions in air. At a distance of 5 m the sound level is 80 db. The frequency is 440 Hz.

a) What is the displacement amplitude at this distance?

b) What is the pressure amplitude?

c) At what distance is the sound level 60 db?

23-10 Derive Eq. (23-13) from the relations preceding it.

23-11 Two loudspeakers, A and B, radiate sound uniformly in all directions. The output of acoustic power from A is 8×10^{-4} W, and from B it is 13.5×10^{-4} W. Both loudspeakers are vibrating in phase at a frequency of 173 Hz.

a) Determine the difference in phase of the two signals at a point C along the line joining A and B, 3 m from B and 4 m from A.

b) Determine the intensity at C from speaker A if speaker B is turned off, and the intensity at C from speaker B if speaker A is turned off.

c) With both speakers on, what are the intensity and intensity level at C?

23-12 What should be the diameter of a sound source in the form of a circular piston set in a wall, if the central lobe of the diffraction pattern is to have a half-angle of 45°, for a frequency of 10,000 Hz?

23-13 The sound source of a sonar system operates at a frequency of 50,000 Hz. Approximate the source by a circular disk set in the hull of a ship. The velocity of sound in water can be taken as 1450 m·s^{-1}.

a) What is the wavelength of the waves emitted by the source?

b) What must be the diameter of the source if the half-angular divergence of the main beam is not to be more than 10°.

c) What is the difference in frequency between the directly radiated waves and the waves reflected from a whale traveling directly away from the ship at 25 km·hr^{-1}?

23-14 Two identical piano strings when stretched with the same tension have a fundamental frequency of 400 Hz. By what fractional amount must the tension in one string be increased in order that 4 beats·s^{-1} shall occur when both strings vibrate simultaneously?

23-15 In Table 23-2, the interval from C to G is called a *fifth*. Compare the frequency ratio for this interval in the just scale to that in the equally tempered scale. Which is larger, and by what percent?

23-16 In Table 23-2, compare the frequency ratio for the interval C to G with the frequency ratio for the interval D to A, in the just scale.

a) By what percent do the two ratios differ?

b) Show that the two ratios are equal in the equally tempered scale.

23-17 The frequency ratio of a half-tone interval on the equally tempered scale is 1.059. Find the speed of an automobile passing a listener at rest in still air, if the pitch of the car's horn drops a half-tone between the times when the car is coming directly toward him and when it is moving directly away from him.

23-18

a) Refer to Fig. 23-4 and the examples in Section 23-7. Suppose that a wind of speed 15 m·s^{-1} is blowing in the same direction as that in which the source is moving. Find the wavelengths ahead of and behind the source.

b) Find the frequency heard by a listener at rest when the source is moving away from him.

23-19 A railroad train is traveling at 30 m·s^{-1} in still air. The frequency of the note emitted by the locomotive whis-

tle is 500 Hz. What is the wavelength of the sound waves
a) in front of and
b) behind the locomotive?
What would be the frequency of the sound heard by a stationary listener
c) in front of and
d) behind the locomotive?
What frequency would be heard by a passenger on a train traveling at 15 m·s⁻¹ and
e) approaching the first,
f) receding from the first?
g) How would each of the preceding answers be altered if a wind of speed 10 m·s⁻¹ were blowing in the same direction as that in which the locomotive was traveling?

23-20 A plane sound wave of frequency f_0 and wavelength λ_0 travels horizontally toward the right. It strikes and is reflected from a large, rigid, vertical plane surface, perpendicular to the direction of propagation of the wave and moving toward the left with a speed v.
a) How many positive wave crests strike the surface in a time interval t?
b) At the end of this time interval, how far to the left of the surface is the wave that was reflected at the beginning of the time interval?
c) What is the wavelength of the reflected waves, in terms of λ_0?
d) What is the frequency, in terms of f_0?
e) A listener is at rest at the left of the moving surface. Describe the sensation of sound that he hears as a result of the combined effect of the incident and reflected wave trains.

23-21 Two whistles, A and B, each have a frequency of 500 Hz. A is stationary and B is moving toward the right (away from A) at a speed of 50 m·s⁻¹. An observer is between the two whistles, moving toward the right with a speed of 25 m·s⁻¹. Take the speed of sound in air as 350 m·s⁻¹.

a) What is the frequency from A as heard by the observer?
b) What is the frequency from B as heard by the observer?
c) What is the beat frequency heard by the observer?

23-22 A man stands at rest in front of a large smooth wall. Directly in front of him, between him and the wall, he holds a vibrating tuning fork of frequency 400 Hz. He now moves the fork toward the wall with a speed of 1 m·s⁻¹. How many beats per second will he hear between the sound waves reaching him directly from the fork, and those reaching him after being reflected from the wall?

23-23 A source of sound waves, S, emitting waves of frequency 1000 Hz, is traveling toward the right in still air with a speed of 30 m·s⁻¹. At the right of the source is a large, smooth, reflecting surface moving toward the left with a speed of 120 m·s⁻¹.
a) How far does an emitted wave travel in 0.01 s?
b) What is the wavelength of the emitted waves in front of (i.e., at the right of) the source?
c) How many waves strike the reflecting surface in 0.001 s?
d) What is the speed of the reflected waves?
e) What is the wavelength of the reflected waves?

23-24
a) Show that Eq. (23-21) can be written
$$f_L = f_S\left(1 - \frac{v}{c}\right)^{1/2}\left(1 + \frac{v}{c}\right)^{-1/2}.$$
b) Use the binomial theorem to show that if $v \ll c$, this is approximately equal to
$$f_L = f_S\left(1 - \frac{v}{c}\right).$$
c) An earth satellite emits a radio signal of frequency 10^8 Hz. An observer on the ground detects beats between the received signal and a local signal also of frequency 10^8 Hz. At a particular moment, the beat frequency is 2400 Hz. What is the component of the satellite's velocity directed toward the earth at this moment?

ELECTRICITY AND MAGNETISM, LIGHT, AND ATOMIC PHYSICS
PART II

COULOMB'S LAW 24

24-1 Electric charges

It was known to the ancient Greeks as long ago as 600 B.C. that amber, rubbed with wool, acquires the property of attracting light objects. In describing this property today, we say that the amber is *electrified,* or possesses an *electric charge,* or is *electrically charged*. These terms are derived from the Greek word *elektron,* meaning amber. It is possible to impart an electric charge to any solid material by rubbing it with any other material. Thus a person becomes electrified by scuffing his shoes across a nylon carpet, a comb is electrified in passing through dry hair, an electric charge is developed on a sheet of paper moving through a printing press, and so on.

Plastic rods and fur are often used in demonstrations. Suppose a plastic rod is electrified by rubbing it with fur and is then touched to two small, light balls of cork or pith, suspended by thin silk or nylon threads. The balls are found to be *repelled* by the plastic rod, and they also repel each other.

A similar experiment performed with a glass rod that has been rubbed with silk gives rise to the same result; pith balls electrified by contact with such a glass rod are repelled not only by the rod but by each other. On the other hand, when a pith ball that has been in contact with electrified plastic is placed near one that has been in contact with electrified glass, the pith balls *attract* each other. We are therefore led to the conclusion that there are *two kinds* of electric charge—that possessed by plastic after being rubbed with fur, called a *negative* charge, and that possessed by glass after being rubbed with silk, called a *positive* charge. The experiments on pith balls described above lead to the fundamental results that (1) *like charges repel,* (2) *unlike charges attract*.

These repulsive or attractive forces of electrical origin are distinct from *gravitational* attraction, and usually so much larger that the gravitational interaction between two electrically charged bodies may be neglected. Of course, a charged body may be acted on by the gravitational field of a large body such as the earth, and in such cases the electrical and gravitational forces may be of comparable magnitude.

Two bodies may also interact by means of magnetic interaction; the most familiar example is attraction of iron objects to a permanent magnet. Although magnetic forces were once believed to be a fundamentally different type of interaction, it is now known that magnetic interactions are really one aspect of the interactions between charged particles in motion. Indeed, an electromagnet manifests magnetic interactions when an electric current passes through its coils. Magnetic interactions will be discussed in later chapters; for the present we concentrate on charges *at rest,* that is, on *electrostatics.*

Suppose a plastic rod is rubbed with fur and then touched to a suspended pith ball. Both the rod and the pith ball are then negatively charged. If the *fur* is now brought near the pith ball, the ball is *attracted,* indicating that the fur is *positively* charged. It follows that when plastic is rubbed with fur, opposite charges appear on the two materials. This is found to happen whenever any substance is rubbed with any other substance. Thus, glass becomes positive, while the silk with which the glass was rubbed becomes negative. This suggests strongly that electric charges are not generated or created, but that the process of acquiring an electric charge consists of *transferring* something from one body to another, so that one body has an excess and the other a deficiency of that something. It was not until the end of the nineteenth century that this "something" was found to consist of very small, negatively charged particles, known today as *electrons.*

24-2 Atomic structure

The interactions responsible for the structure of atoms and molecules, and hence of all matter, are primarily electrical interactions between electrically charged particles. The fundamental building blocks are three kinds of particles, the negatively charged *electron,* the positively charged *proton,* and the neutral *neutron.* The negative charge of the electron is of the same magnitude as the positive charge of the proton, and no charges of smaller magnitude have ever been observed. The charge of a proton or an electron is the ultimate, natural unit of charge. Measurements of the charge of the electron are discussed in Sec. 26-6.

The particles are arranged in the same general way in all atoms. The protons and neutrons always form a closely packed group called the *nucleus,* which has a net positive charge due to the protons. The diameter of the nucleus, if we think of it as roughly spherical, is of the order of 10^{-14} m. Outside the nucleus, but at relatively large distances from it, are the electrons, whose number is equal to the number of protons within the nucleus. If the atom is undisturbed, and no electrons are removed from or added to the space around the nucleus, the atom as a whole is electrically *neutral.* That is, *the algebraic sum of the positive charges of the nucleus and the negative charges of the electrons is zero,* just as equal positive and negative numbers sum to zero. If one or more electrons are removed, the remaining positively charged structure is called a positive *ion.* A negative ion is an atom that has gained one or more extra electrons. The process of losing or gaining electrons is called *ionization.*

In the atomic model proposed by the Danish physicist Niels Bohr in 1913, the electrons were pictured as whirling about the nucleus in circular or elliptical orbits. More recent research has shown that the electrons are more accurately represented as spread-out distributions of electric charge, governed by the principles of quantum mechanics, which will be discussed in Chapter 45. Nevertheless, the Bohr model is still useful for visualizing the structure of an atom. The diameters of the electron charge distributions, which the Bohr model pictures as orbits, determine the overall size of the atom as a whole, and these are of the order of 10^{-10} m, or about ten thousand times as great as the diameter of the nucleus. A Bohr atom is analogous to a solar system in miniature, with electrical forces taking the place of gravitational forces. The massive, positively charged central nucleus corresponds to the sun, while the electrons, moving around the nucleus under the electrical force of its attraction, correspond to the planets moving around the sun under the influence of its gravitational attraction.

The masses of the proton and neutron are nearly equal, and the mass of the proton is about 1836 times that of the electron. Nearly all the mass of an atom, therefore, is concentrated in its nucleus. Since one mole of monatomic hydrogen consists of 6.022×10^{23} particles (Avogadro's number) and its mass is 1.008 g, the mass of a single hydrogen atom is

$$\frac{1.008 \text{ g}}{6.022 \times 10^{23}} = 1.674 \times 10^{-24} \text{ g} = 1.674 \times 10^{-27} \text{ kg}.$$

The nucleus of a hydrogen atom is a single proton, and around it there is a single electron. Hence, of the total mass of the hydrogen atom, $\frac{1}{1837}$ part is the mass of the electron and the remainder is the mass of a proton. To four significant figures,

$$\text{Mass of electron} = 9.110 \times 10^{-31} \text{ kg},$$
$$\text{Mass of proton} = 1.673 \times 10^{-27} \text{ kg},$$
$$\text{Mass of neutron} = 1.675 \times 10^{-27} \text{ kg}$$

After hydrogen, the atom with the next simplest structure is that of helium. Its nucleus consists of two protons and two neutrons, and it has two extranuclear electrons. When these two electrons are removed, the doubly charged helium ion (which is the helium nucleus itself) is often called an *alpha particle,* or α-particle. The next element, lithium, has three protons in its nucleus and has thus a nuclear charge of three units. In the un-ionized state the lithium atom has three extranuclear electrons. Each element has a different number of nuclear protons and therefore a different positive nuclear charge. In the table of elements listed at the end of this book, known as the *periodic table,* each element occupies a box with which is associated a number, called the *atomic number.*

The atomic number is the number of nuclear protons; in the un-ionized state, this is equal to the number of extranuclear electrons.

Every material body contains a tremendous number of charged particles, positively charged protons in the nuclei of its atoms and negatively charged electrons outside the nuclei. When the total number of protons equals the total number of electrons, the body as a whole is electrically neutral.

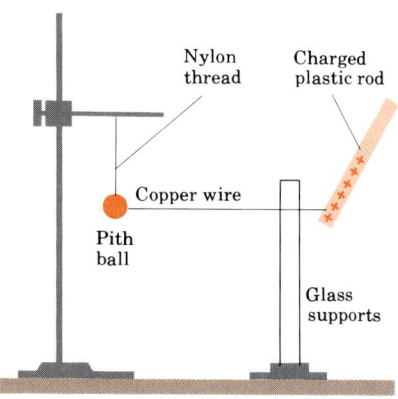

24-1 Copper is a conductor of electricity.

To give a body an excess negative charge, we may either *add* a number of *negative* charges to a neutral body, or *remove* a number of *positive* charges from the body. Similarly, either an *addition* of *positive* charge or a *removal* of *negative* charge results in an excess positive charge. In most instances, it is negative charges (electrons) that are added or removed, and a "positively charged body" is one that has lost some of its normal complement of electrons.

The "charge" of a body refers to its *excess* charge only. The excess charge is always a very small fraction of the total positive or negative charge in the body.

Implicit in the above statements is the *principle of conservation of charge*. This principle states that the algebraic sum of all the electric charges in any closed system is constant. Charge can be transferred from one body to another, but it cannot be created or destroyed. Conservation of charge is believed to be a *universal* conservation law; there is no experimental evidence for any violation of this principle.

24-3 Conductors and insulators

Some materials permit electric charge to move from one region of the material to another, while other materials do not. Suppose we touch one end of a copper wire to an electrified plastic rod and the other end to an initially uncharged pith ball, as in Fig. 24-1. The pith ball is found (by its subsequent interaction with other charged pith balls) to become charged. Thus charge has moved along the wire from the rod to the ball. The copper wire is called a *conductor* of electricity. If the experiment is repeated using a rubber band or nylon thread in place of the wire, *no* transfer of charge is observed; and these materials are called *insulators*. The motion of charge through a material substance will be studied in more detail in Chapter 28, but for the present it is sufficient to state that most substances fall into one or the other of the two classes above. Conductors permit the passage of charge through them, while insulators do not.

Metals in general are good conductors, while most nonmetals are insulators. The positive valency of metals and the fact that they form positive ions in solution indicate that the atoms of a metal will part readily with one or more of their outer electrons. Within a metallic conductor such as a copper wire, a few outer electrons become detached from each atom and can move freely throughout the metal in much the same way that the molecules of a gas can move through the spaces between grains of sand in a sand-filled container. In fact, these free electrons are often referred to as an "electron gas." The positive nuclei and the remainder of the electrons remain fixed in position. Within an insulator, on the other hand, there are no (or at most very few) free electrons.

24-4 Charging by induction

In charging a pith ball by contact with, say, a plastic rod that has been rubbed with fur, some of the extra electrons on the plastic are transferred to the ball, leaving the plastic with a smaller negative charge. There is, however, another way to use the plastic rod to charge other

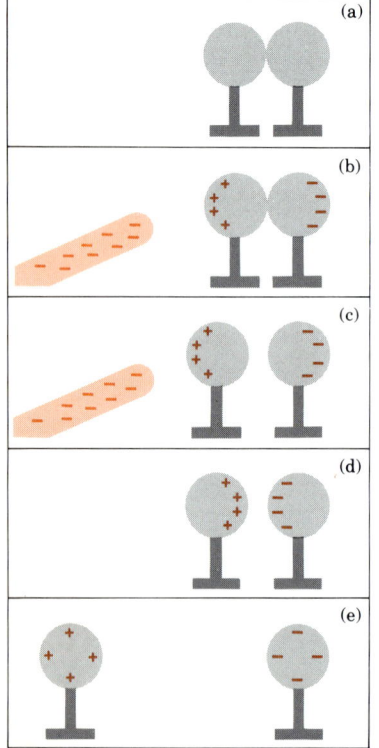

24-2 Two metal spheres are oppositely charged by induction.

bodies, in which the plastic may impart a charge of *opposite* sign and lose none of its own charge. This process, called *charging by induction,* is illustrated in Fig. 24-2.

In Fig. 24-2a, two neutral metal spheres are in contact, both supported on insulating stands. When a negatively charged rod is brought near one of the spheres but without touching it, as in (b), the free electrons in the metal spheres are repelled and drift slightly away from the rod, toward the right. Since the electrons cannot escape from the spheres, an excess negative charge accumulates at the right surface of the right sphere. This leaves a deficiency of negative charge, or an excess positive charge, at the left surface of the left sphere. These excess charges are called *induced* charges.

It should not be inferred that *all* of the free electrons in the spheres are driven to the surface of the right sphere. As soon as any induced charges develop, they also exert forces on the free electrons within the spheres. In this case this force is toward the left (a repulsion by the negative induced charge and an attraction by the positive induced charge). Within an extremely short time the system reaches an equilibrium state in which, at every point in the interior of the spheres, the force on an electron toward the right, exerted by the charged rod, is just balanced by a force toward the left exerted by the induced charges.

The induced charges remain on the surfaces of the spheres as long as the rod is held nearby. When the rod is removed, the electron cloud in the spheres moves to the left and the original neutral condition is restored.

Suppose that the spheres are separated slightly, as shown in (c), while the plastic rod is nearby. If the rod is now removed, as in (d), we are left with two oppositely charged metal spheres whose charges attract each other. When the two spheres are separated by a great distance, as in (e), each of the two charges becomes uniformly distributed over its sphere. It should be noticed that the negatively charged rod has lost none of its charge in the steps from (a) to (e).

The steps from (a) to (e) in Fig. 24-3 should be self-explanatory. In this figure, a single metal sphere (on an insulating stand) is charged by induction. The symbol lettered "ground" in part (c) simply means that the sphere is connected to the earth (a conductor). The earth thus takes the place of the second sphere in Fig. 24-2. In step (c), electrons are repelled to ground either through a conducting wire, or along the moist skin of a person who touches the sphere with his finger. The earth thus acquires a negative charge equal to the induced positive charge remaining on the sphere.

The process taking place in Figs. 24-2 and 24-3 could be explained equally well if the mobile charges in the spheres were *positive* or, in fact, if *both* positive and negative charges were mobile. Although we now know that in a metallic conductor it is actually the *negative* charges that move, it is often convenient to describe a process *as if* the positive charges moved.

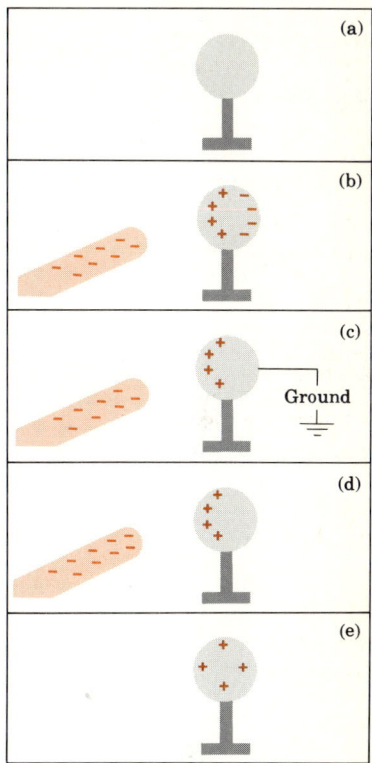

24-3 Charging a single metal sphere by induction.

24-5 Coulomb's law

The electrical interaction between two charged particles is described in terms of the *forces* they exert on each other. The first quantitative investigation of the behavior of this force was carried out by Augustin de Coulomb (1736–1806) in 1784, using for force measurements a torsion balance of the type used 13 years later by Cavendish to study (much weaker) gravitational forces (Sec. 4-4). Coulomb studied the force of attraction or repulsion between two "point charges," that is, charged bodies whose dimensions are small compared with the distance between them.

Coulomb found that the force grows weaker with increasing separation between the bodies. When the distance doubles, the force decreases to $\frac{1}{4}$ of its initial value. Thus it varies inversely with the square of the distance. If r is the distance between the particles, then the force is proportional to $1/r^2$.

The force also depends on the quantity of charge on each body, which is usually denoted by q or Q. In Coulomb's time, no *unit* of charge had been defined, nor had any method been developed for comparing a given charge with a unit. Despite this, Coulomb devised an ingenious method of showing how the force exerted on or by a charged body depended on its charge. He reasoned that if a charged spherical conductor were brought in contact with a second identical conductor, originally *uncharged,* the charge on the first would, by symmetry, be shared equally between the conductors. He thus had a method for obtaining one-half, one-quarter, and so on, of any given charge. The results of his experiments were consistent with the conclusion that the force between two point charges q and q' is proportional to each charge and hence proportional to the *product* of the charges. The complete expression for the magnitude of the force between two point charges is therefore

$$F = k \frac{|qq'|}{r^2}, \tag{24-1}$$

where k is a proportionality constant whose magnitude depends on the units in which F, q, q', and r are expressed. Equation (24-1) is the mathematical statement of what is known today as *Coulomb's law:*

The force of attraction or repulsion between two point charges is directly proportional to the product of the charges and inversely proportional to the square of the distance between them.

The *direction* of the force on each particle is always along the line joining the two particles, pulling each particle toward the other in the case of attractive forces on unlike charges, and pushing them apart in the case of repulsive forces on like charges.

The form of Eq. (24-1) is the same as that of the law of gravitation, discussed in Sec. 4-4, but electrical and gravitational interactions are two distinct classes of phenomena. The proportionality of the force to $1/r^2$ has been verified with great precision. There is no reason to suspect, for example, that the electrical force might vary as $1/r^{2.0001}$.

The charges q and q' are *algebraic* quantities, corresponding to the existence of two kinds of charge (positive and negative), but Eq. (24-1) gives the magnitude of the interaction force in all cases. When the charges are of like sign the forces are repulsive; when they are unlike, the forces are attractive. In either case, the forces obey Newton's third law; the force that q exerts on q' is the negative of the force q' exerts on q.

The "absolute value" bars in Eq. (24–1) are needed because F, the magnitude of a vector quantity, is by definition always positive, while the product qq' is negative whenever the two charges have opposite signs.

When two or more charges exert forces simultaneously on a given charge, the total force experienced by that charge is found to be the *vector sum* of the forces that the various charges would exert individually. This important property, called the *principle of superposition,* permits the application of Coulomb's law to arrays of charge of any degree of complexity, although the computational problems can be very great. Example 4 below illustrates the application of the superposition principle.

If there is matter in the space between the charges, the *net* force acting on each is altered because charges are induced in the molecules of the intervening material. This effect will be described later. As a practical matter, the law can be used as stated for point charges in air, since even at atmospheric pressure the effect of the air is to alter the force from its value in vacuum by only about one part in two thousand.

In the chapters of this book dealing with electrical phenomena, we shall use the SI (mks) system of units exclusively. The SI electrical units include all the familiar electrical units such as the volt, the ampere, the ohm, and the watt. The cgs system is also used, more so in scientific work than in commerce and industry, but there is *no* British system of electrical units. This is one of many reasons for abandoning the British system and adopting metric units universally, and there seems little doubt that SI units will eventually receive worldwide adoption.

To the three basic SI units (the meter, kilogram, and second) we now add a fourth, the unit of electric charge. This unit is called one *coulomb* (1 C). The electrical constant k in Eq. (24–1) is, in this system,

$$k = 8.98755 \times 10^9 \text{ N} \cdot \text{m}^2 \cdot \text{C}^{-2}$$
$$\approx 9.0 \times 10^9 \text{ N} \cdot \text{m}^2 \cdot \text{C}^{-2}.$$

Later, in connection with the study of electromagnetic radiation, we shall show that k is closely related to the speed of light in vacuum,

$$c = 2.998 \times 10^8 \text{ m} \cdot \text{s}^{-1}.$$

Specifically,

$$k = 10^{-7} c^2.$$

The relationship is not accidental, but results from the definition of the unit of current, which in turn is related to the interaction of electric and magnetic fields, to be studied later.

In the cgs system of electrical units (not used in this book) the constant k is defined to be unity, without units. This defines a unit of electric charge called the *statcoulomb* or the *esu* (electrostatic unit). The conversion factor is

$$1 \text{ C} = 2.998 \times 10^9 \text{ esu}.$$

In SI units the constant k in Eq. (24–1) is usually written not as k but as $1/4\pi\epsilon_0$, where ϵ_0 is another constant. This appears to complicate matters, but it actually simplifies some formulas to be encountered later. Thus Coulomb's law is usually written as

$$F = \frac{1}{4\pi\epsilon_0} \frac{|qq'|}{r^2}, \qquad (24\text{-}2)$$

with

$$\frac{1}{4\pi\epsilon_0} = 8.987554 \times 10^9 \text{ N} \cdot \text{m}^2 \cdot \text{C}^{-2}$$

and

$$\epsilon_0 = 8.854185 \times 10^{-12} \text{ C}^2 \cdot \text{N}^{-1} \cdot \text{m}^{-2}.$$

The "natural" unit of charge is the magnitude of charge of an electron or a proton. This quantity is denoted by e; the most precise measurements to date yield the value

$$e = 1.602192 \times 10^{-19} \text{ C} \approx 1.60 \times 10^{-19} \text{ C}.$$

One coulomb therefore represents the negative of the total charge carried by about 6×10^{18} electrons. For comparison, the population of the earth is estimated to be about 3×10^9 persons, while a cube of copper 1 cm on a side contains about 2.4×10^{24} electrons.

In electrostatics problems, charges as large as one coulomb are unusual; a more typical range of magnitude is 10^{-9} to 10^{-6} C. The microcoulomb (1 μC = 10^{-6} C) is often used as a practical unit of charge.

Example 1 Two charges are located on the positive x-axis of a coordinate system, as shown in Fig. 24-4. Charge $q_1 = 2 \times 10^{-9}$ C is 2 cm from the origin, and charge $q_2 = -3 \times 10^{-9}$ C is 4 cm from the origin. What is the total force exerted by these two charges on a charge $q_3 = 5 \times 10^{-9}$ C located at the origin?

Solution The total force on q_3 is the vector sum of the forces due to q_1 and q_2 individually. Converting distances to meters, we use Eq. (24-2) to find the magnitude F_1 of the force on q_3 due to q_1:

$$F_1 = \frac{(9.0 \times 10^9 \text{ N} \cdot \text{m}^2 \cdot \text{C}^{-2})(2 \times 10^{-9} \text{ C})(5 \times 10^{-9} \text{ C})}{(0.02 \text{ m})^2} = 2.25 \times 10^{-4} \text{ N}.$$

This force has a negative x-component because q_3 is repelled (i.e., pushed in the $-x$-direction) by q_1, which has the same sign. Similarly, the force due to q_2 is found to have magnitude

$$F_2 = \frac{(9.0 \times 10^9 \text{ N} \cdot \text{m}^2 \cdot \text{C}^{-2})(3 \times 10^{-9} \text{ C})(5 \times 10^{-9} \text{ C})}{(0.04 \text{ m})^2} = 0.84 \times 10^{-4} \text{ N}.$$

This force has a positive x-component because q_3 is attracted (i.e., pulled in the positive x-direction) by the opposite charge q_2. The sum of the x-components is

$$\sum F_x = -2.25 \times 10^{-4} \text{ N} + 0.84 \times 10^{-4} \text{ N} = -1.41 \times 10^{-4} \text{ N}.$$

There are no y- or z-components. Thus the total force on q_3 is directed to the left, with magnitude 1.41×10^{-4} N.

24-4

Example 2 An α-particle is a nucleus of doubly ionized helium. It has a mass m of 6.68×10^{-27} kg and a charge q of $+2e$ or 3.2×10^{-19} C. Compare the force of electrostatic repulsion between two α-particles with the force of gravitational attraction between them.

Solution The electrostatic force F_e is

$$F_e = \frac{1}{4\pi\epsilon_0} \frac{q^2}{r^2},$$

and the gravitational force F_g is

$$F_g = G \frac{m^2}{r^2}.$$

The ratio of the electrostatic to the gravitational force is

$$\frac{F_e}{F_g} = \frac{1}{4\pi\epsilon_0 G} \frac{q^2}{m^2} = 3.1 \times 10^{35}.$$

The gravitational force is evidently negligible compared with the electrostatic force.

Example 3 The Bohr model of the hydrogen atom (discussed in Chapter 45) consists of a single electron of charge $-e$ revolving in a circular orbit about a single proton of charge $+e$. The electrostatic force of attraction between electron and proton provides the centripetal force that retains the electron in its orbit. Hence if v is the orbital velocity, Newton's second law gives

$$\frac{1}{4\pi\epsilon_0} \frac{e^2}{r^2} = m\left(\frac{v^2}{r}\right).$$

In Bohr's theory, the electron may revolve only in some one of a number of specified orbits. The orbit of smallest radius is that for which the angular momentum L of the electron is $h/2\pi$, where h is a universal constant called *Planck's constant*, equal to 6.625×10^{-34} J·s. Then,

$$L = mvr = \frac{h}{2\pi}. \qquad (24\text{--}3)$$

When v is eliminated between the preceding equations, we find

$$r = \frac{\epsilon_0 h^2}{\pi k m e^2},$$

and when numerical values are inserted, we find, for the radius of the *first Bohr orbit*,

$$r = 5.29 \times 10^{-11} \text{ m} = 0.529 \times 10^{-8} \text{ cm}.$$

This result corresponds reasonably well with other estimates of the "size" of a hydrogen atom obtained from deviations from ideal gas behavior, the density of hydrogen in the liquid and solid states, and other observations.

Example 4 In Fig. 24–5, two equal positive charges $q = 2.0 \times 10^{-6}$ C interact with a third charge $Q = 4.0 \times 10^{-6}$ C. Find the magnitude and direction of the total (resultant) force on Q.

Solution The key word is *total;* we must compute the force each charge exerts on Q, and then obtain the *vector sum* of the forces. This is most easily accomplished using components. The figure shows the force on Q due to the upper charge q. From Coulomb's law,

$$F = (9.0 \times 10^9 \text{ N} \cdot \text{C}^{-2} \cdot \text{m}^{-2}) \frac{(4.0 \times 10^{-6} \text{ C})(2.0 \times 10^{-6} \text{ C})}{(0.5 \text{ m})^2}$$

$$= 0.29 \text{ N}.$$

The components of this force are given by:

$$F_x = F \cos \theta = (0.29 \text{ N}) \left(\frac{0.4 \text{ m}}{0.5 \text{ m}} \right) = 0.23 \text{ N},$$

$$F_y = -F \sin \theta = -(0.29 \text{ N}) \left(\frac{0.3 \text{ m}}{0.5 \text{ m}} \right) = -0.17 \text{ N}.$$

The lower charge q exerts a force of the same magnitude, but in a different direction. From symmetry we see that its x-component is the same as that due to the upper charge, but its y-component is opposite. Hence,

$$\sum F_x = 2(0.23 \text{ N}) = 0.46 \text{ N},$$

$$\sum F_y = 0.$$

The total force on Q is horizontal, with magnitude 0.46 N. How would this solution differ if the lower charge were *negative*?

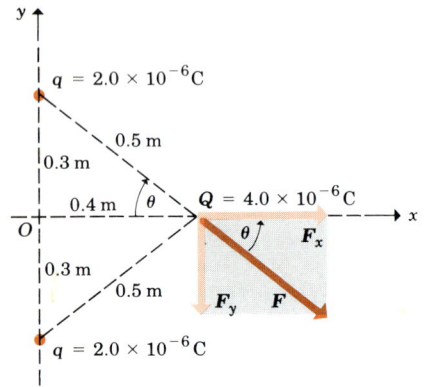

24-5 F is the force on Q due to the upper charge q.

24-6 Electrical interactions

Because matter is made up of charged particles, it is not surprising that electrical interactions play a central and dominant role in all aspects of the structure of matter. The forces that hold atoms together in a molecule or in a solid crystal lattice, the adhesive force of glue, the forces associated with surface tension—all these are basically electrical in nature, arising from the electrical forces between the charged particles making up the interacting atoms. A complete description of the detailed behavior of these forces requires new *mechanical* principles and the introduction of quantum-mechanical concepts, to be discussed in Chapter 45. Nevertheless, Coulomb's law and the additional effects resulting from relative motion of the charges still describe the basic electrical interactions involved.

Electrical interactions alone are *not* sufficient to understand the structure of the *nuclei* of atoms, however. A nucleus is made up of protons, which repel each other, and neutrons, which have no electrical charge. For nuclei to be stable there must be additional forces, attractive in nature, to hold their constituent parts together despite the electrical repulsion. This new kind of interaction, not seen outside the nucleus, is called the *nuclear force;* many phenomena associated with the stability or instability of nuclei pivot around the competition between the repulsive electrical forces and the attractive nuclear forces. These matters will be discussed in greater detail in Chapter 47.

Questions

24-1 Plastic food wraps can be used to cover a container by simply stretching the material across the top and pressing the overhanging material against the sides. What makes it stick? Does it stick to itself with equal tenacity? Why? Does it matter whether the container is metallic or not?

24-2 Bits of paper are attracted to an electrified comb or rod, even though they have no net charge. How is this possible?

24-3 How do we know that the magnitudes of electron and proton charge are *exactly* equal? With what precision is this really known?

24-4 When you walk across a nylon rug and then touch a large metallic object, you may get a spark and a shock. Why does this tend to happen more in the winter than the summer? Why do you not get a spark when you touch a *small* metal object?

24-5 The free electrons in a metal have mass and therefore weight and are gravitationally attracted toward the earth. Why, then, do they not all settle to the bottom of the conductor, as sediment settles to the bottom of a river?

24-6 Simple electrostatics experiments, such as picking up bits of paper with an electrified comb, never work as well on rainy days as on dry days. Why?

24-7 High-speed printing presses sometimes use gas flames to reduce electric charge build-up on the paper passing through the press. Why does this help? (An added benefit is rapid drying of the ink.)

24-8 What similarities do electrical forces have to gravitational forces? What are the most significant differences?

24-9 Given two identical metal objects mounted on insulating stands, describe a procedure for placing charges of equal magnitude and opposite sign on the two objects.

24-10 How do we know that protons have positive charge and electrons negative charge, rather than the reverse?

24-11 Gasoline transport trucks sometimes have chains that hang down and drag on the ground at the rear end. What are these for?

24-12 When a nylon sleeping bag is dragged across a rubberized cloth air mattress in a dark tent, small sparks are sometimes seen. What causes them?

24-13 Atomic nuclei are made of protons and neutrons. This fact by itself shows that there must be another kind of interaction in addition to the electrical forces. Explain.

24-14 When transparent plastic tape is pulled off a roll and one tries to position it precisely on a piece of paper, it often jumps over and sticks where it isn't wanted. Why does it do this?

Problems

24-1 How many excess electrons must be placed on each of two small spheres spaced 3 cm apart if the force of repulsion between the spheres is to be 10^{-19} N?

24-2 Each of two small spheres is positively charged, the combined charge totaling 4×10^{-8} C. What is the charge on each sphere if they are repelled with a force of 27×10^{-5} N when placed 0.1 m apart?

24-3 6.02×10^{23} atoms of monatomic hydrogen have a mass of one gram. How far would the electron of a hydrogen atom have to be removed from the nucleus for the force of attraction to equal the weight of the atom?

24-4 What is the total positive charge, in coulombs, of all the protons in 1 mol of hydrogen atoms?

24-5 If all the positive charges in a mole of hydrogen atoms were lumped into a single charge, and all the negative charges into a single charge, what force would the two lumped charges exert on each other at a distance of

a) 1 m;

b) 10^7 m (comparable to the diameter of the earth)?

24-6 An alpha particle consists of two protons and two neutrons bound together. What is the repulsive force between two alpha particles at a distance of 10^{-15} m, comparable to the sizes of nuclei?

24-7 Two copper spheres, each having mass 1 kg, are separated by 1 m.

a) How many electrons does each sphere contain?

b) How many electrons would have to be removed from one sphere and added to the other to cause an attractive force of 10^4 N (roughly one ton)?

c) What fraction of all the electrons on a sphere does this represent?

24-8 Point charges of 2×10^{-9} C are situated at each of three corners of a square whose side is 0.20 m. What would be the magnitude and direction of the resultant force on a point charge of -1×10^{-9} C if it were placed (a) at the center of the square? (b) at the vacant corner of the square?

24-9 Two charges of $+10^{-9}$ C each are 8 cm apart in air. Find the magnitude and direction of the force exerted by these charges on a third charge of $+5 \times 10^{-11}$ C that is 5 cm distant from each of the first two charges.

24-10 Two positive point charges, each of magnitude q, are located on the y-axis at points $y = +a$ and $y = -a$. A third positive charge of the same magnitude is located at some point on the x-axis.

a) What is the force exerted on the third charge when it is at the origin?

b) What is the magnitude and direction of the force on the third charge when its coordinate is x?

c) Sketch a graph of the force on the third charge as a function of x, for values of x between $+4a$ and $-4a$. Plot forces to the right upward, forces to the left downward.

24-11 A negative point charge of magnitude q is located on the y-axis at the point $y = +a$, and a positive charge of the same magnitude is located at $y = -a$. A third positive charge of the same magnitude is located at some point on the x-axis.

a) What is the magnitude and direction of the force exerted on the third charge when it is at the origin?

b) What is the force on the third charge when its coordinate is x?

c) Sketch a graph of the force on the third charge as a function of x, for values of x between $+4a$ and $-4a$.

24-12 Two small balls, each of mass 10 g, are attached to silk threads 1 m long and hung from a common point. When the balls are given equal quantities of negative charge, each thread makes an angle of 4° with the vertical.

a) Draw a diagram showing all of the forces on each ball.

b) Find the magnitude of the charge on each ball.

24-13 A certain metal sphere of volume 1 cm³ has a mass of 7.5 g and contains 8.2×10^{22} free electrons.

a) How many electrons must be removed from each of two such spheres so that the electrostatic force of repulsion between them just balances the force of gravitational attraction? Assume the distance between the spheres is great enough so that the charges on them can be treated as point charges.

b) Express the number of electrons removed as a fraction of the total number of free electrons.

24-14 In the Bohr model of atomic hydrogen, an electron of mass 9.11×10^{-31} kg revolves about a proton in a circular orbit of radius 5.29×10^{-11} m. The proton has a positive charge equal in magnitude to the negative charge on the electron and its mass is 1.67×10^{-27} kg.

a) What is the radial acceleration of the electron?

b) What is its velocity?

c) What is its angular velocity?

24-15 One gram of monatomic hydrogen contains 6.02×10^{23} atoms, each consisting of an electron with charge -1.60×10^{-19} C and a proton with charge $+1.60 \times 10^{-19}$ C.

a) Suppose all these electrons could be located at the north pole of the earth and all the protons at the south pole. What would be the total force of attraction exerted on each group of charges by the other? The diameter of the earth is 12,800 km.

b) What would be the magnitude and direction of the force exerted by the charges in part (a) on a third positive charge, equal in magnitude to the total charge at one of the poles and located at a point on the equator? Draw a diagram.

24-16 The dimensions of atomic nuclei are of the order of 10^{-14} m. Suppose that two α-particles are separated by this distance.

a) What is the force exerted on each α-particle by the other?

b) What is the acceleration of each? (See Example 2 in Sec. 24–5 for numerical data.)

24-17 The pair of equal and opposite charges in Problem 24-11 is called an *electric dipole*.

a) Show that when the x-coordinate of the third charge in Problem 24-11 is large compared with the distance a, the force on it is inversely proportional to the *cube* of its distance from the midpoint of the dipole.

b) Show that if the third charge is located on the y-axis, at a y-coordinate large compared with the distance a, the force on it is also inversely proportional to the cube of its distance from the midpoint of the dipole.

24-18 A small ball having a positive charge q_1 hangs by an insulating thread. A second ball with a negative charge $q_2 = -q_1$ is kept at a horizontal distance a to the right of the first. (The distance a is large compared with the diameter of the ball.)

a) Show in a diagram all of the forces on the hanging ball in its final equilibrium position.

b) You are given a third ball having a positive charge $q_3 = 2q_1$. Find at least two points at which this ball can be placed so that the first ball will hang vertically.

24-19 Two point charges are located in the xy-plane, as follows: A charge 2.0×10^{-9} C is at the point $(x = 0, y = 4$ cm$)$, and a charge -3.0×10^{-9} C is at the point $(x = 3$ cm$, y = 4$ cm$)$.

a) If a third charge of 4.0×10^{-9} C is placed at the origin, find the x- and y-components of the total force on this third charge.

b) Find the magnitude and direction of the total force on the charge at the origin in (a).

24-20 A charge -3×10^{-9} C is placed at the origin of an xy-coordinate system, and a charge 2×10^{-9} C is placed on the positive y-axis, at $y = 4$ cm. If a third charge 4×10^{-9} C is now placed at the point $(x = 3$ cm$, y = 4$ cm$)$, find the components of the total force exerted on this charge by the other two. Also find the magnitude and direction of this force.

THE ELECTRIC FIELD; GAUSS'S LAW 25

25-1 The electric field

The electrical interaction between charged particles can be reformulated using the concept of *electric field*. This is not only an aid in calculations but also an important concept with fundamental theoretical significance. To introduce the concept, we consider the mutual repulsion of two positively charged bodies A and B, as shown in Fig. 25–1a. In particular, the force on B is labeled F in the figure. This is an "action-at-a-distance" force; it can act across empty space and does not need any matter in the intervening space to transmit the force.

Now let us think of body A as having the effect of modifying some of the properties of the space in its vicinity. We remove body B and label its former position as point P. The charged body A is said to produce or cause an *electric field* at point P (and at all other points in its vicinity). Then when body B is placed at point P and experiences the force F, we take the point of view that the force is exerted on B *by the field*, rather than directly by A. Since B would experience a force at any point in space around A, the electric field exists at all points in the region around A. (One could equally well consider that body B sets up an electric field, and that the force on body A is exerted by the field due to B.)

The experimental test for the existence of an electric field at any point is simply to place a small charged body, which will be called a *test charge*, at the point. If a force (of electrical origin) is exerted on the test charge, then an electric field exists at the point.

An electric field is said to exist at a point if a force of electrical origin is exerted on a stationary charged body placed at the point.

Since force is a vector quantity, the electric field is a *vector quantity* whose properties are determined when both the magnitude and the direction of an electric force are specified. We define the *electric field E* at a point as the quotient obtained when the force F, acting on a positive test charge, is divided by the magnitude q' of the test charge. Thus,

$$E = \frac{F}{q'} \quad (25\text{-}1)$$

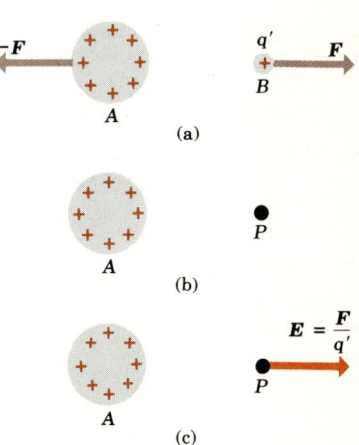

25-1 A charged body creates an electric field in the space around it.

and the direction of **E** is the direction of **F**. It follows that

$$F = q'E.$$

The force on a *negative* charge, such as an electron, is *opposite* to the direction of the electric field.

The electric field is sometimes called *electric intensity* or *electric field intensity*. In SI (mks) units, where the unit of force is 1 N and the unit of charge is 1 C, the unit of electric field is 1 newton per coulomb (1 N·C^{-1}). Electric field may also be expressed in other units, to be defined later.

The force experienced by the test charge q' varies from point to point, and so the electric field is also different at different points. Thus in general **E** is not a single vector quantity, but an infinite set of vector quantities, one associated with each point in space. It is an example of a *vector field*. Another example of a vector field is the description of motion of a flowing fluid. In general, different points in the fluid have different velocities, so the velocity is a vector field. If a rectangular coordinate system is used, then in principle each component of **E** can be expressed as a *function* of the coordinates (x, y, z) of a point in space. Vector fields are an important part of the mathematical language used in many areas of physics, particularly in electricity and magnetism.

One difficulty with our definition of electric field is that in Fig. 25–1 the force exerted by the test charge q' may change the charge distribution A, especially if the body is a conductor on which charge is free to move, so that the electric field around A when q' is present is not the same as when it is absent. However, when q' is very small, the redistribution of charge on body A is also very small; thus the difficulty can be avoided by refining the definition of electric field to be *the limiting value of the force per unit charge on a test charge q' at the point, as the charge q' approaches zero*:

$$E = \lim_{q' \to 0} \frac{F}{q'}.$$

If an electric field exists within a *conductor,* a force is exerted on every charge in the conductor. The motion of the free charges brought about by this force is called a *current*. Conversely, if there is *no* current in a conductor, and hence no motion of its free charges, *the electric field in the conductor must be zero.*

In most instances, the magnitude and direction of an electric field vary from point to point. If the magnitude and direction are constant throughout a certain region, the field is said to be *uniform* in this region.

Example 1 What is the electric field 30 cm from a charge $q = 4 \times 10^{-9}$ C?

Solution From Coulomb's law, the *force* on a test charge q' 30 cm from q is

$$F = \frac{1}{4\pi\epsilon_0}\frac{|qq'|}{r^2} = \frac{(9 \times 10^9 \text{ N·m}^2\text{·C}^{-2})(4 \times 10^{-9} \text{ C})(q')}{(0.3 \text{ m})^2}$$
$$= (400 \text{ N·C}^{-1})(q').$$

Then from Eq. (25–1), the magnitude of **E** is

$$E = \frac{F}{q'} = 400 \text{ N·C}^{-1}.$$

The *direction* of **E** at this point is along the line joining q and q', away from q.

Example 2 When the terminals of a 100-V battery are connected to two large parallel plates 1 cm apart, the field in the region between the plates is very nearly uniform and the electric field magnitude E is 10^4 N·C^{-1}. Suppose the direction of **E** is vertically upward. Compute the force on an electron in this field and compare with the weight of the electron.

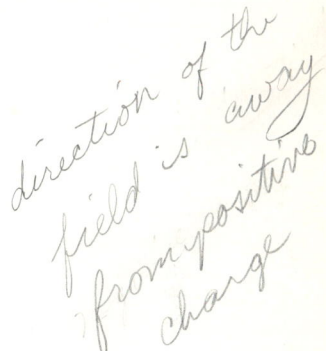

direction of the field is away from positive charge

Solution

$$\text{Electron charge } e = 1.60 \times 10^{-19} \text{ C},$$
$$\text{Electron mass } m = 9.1 \times 10^{-31} \text{ kg}.$$
$$F_{\text{elec}} = eE = (1.60 \times 10^{-19} \text{ C})(10^4 \text{ N·C}^{-1})$$
$$= 1.60 \times 10^{-15} \text{ N};$$
$$F_{\text{grav}} = mg = (9.1 \times 10^{-31} \text{ kg})(9.8 \text{ m·s}^{-2})$$
$$= 8.9 \times 10^{-30} \text{ N}.$$

The ratio of the electrical to the gravitational force is therefore

$$\frac{1.60 \times 10^{-15} \text{ N}}{8.9 \times 10^{-30} \text{ N}} = 1.8 \times 10^{14}.$$

The gravitational force is negligibly small compared to the electrical force.

Example 3 If released from rest, what speed does the electron of Example 2 acquire while traveling 1 cm? What is then its kinetic energy? How much time is required?

Solution The force is constant, so the electron moves with a constant acceleration of

$$a = \frac{F}{m} = \frac{eE}{m} = \frac{1.60 \times 10^{-15} \text{ N}}{9.1 \times 10^{-31} \text{ kg}}$$
$$= 1.8 \times 10^{15} \text{ m·s}^{-2}.$$

Its speed after traveling 1 cm, or 10^{-2} m, is

$$v = \sqrt{2ax} = 6.0 \times 10^6 \text{ m·s}^{-1}.$$

Its kinetic energy is

$$\tfrac{1}{2}mv^2 = 1.6 \times 10^{-17} \text{ J}.$$

The time required is

$$t = \frac{v}{a} = 3.3 \times 10^{-9} \text{ s}.$$

Example 4 If the electron of Example 2 is projected into the field with a horizontal velocity, as in Fig. 25-2, find the equation of its trajectory.

Solution The direction of the field is upward in Fig. 25-2, so the force on the electron is downward. The initial velocity is along the positive x-axis. The x-acceleration is zero, the y-acceleration is $-(eE/m)$. Hence,

$-F/m$

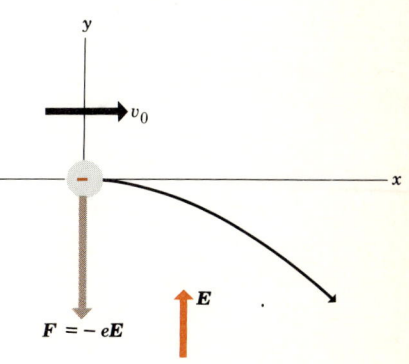

25-2 Trajectory of an electron in an electric field.

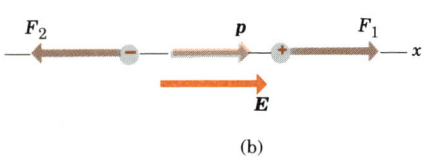

25-3 (a) The torque on the dipole is $\Gamma = pE \sin \theta$. (b) The dipole is in equilibrium in a uniform field when p and E are parallel. If the field is not uniform, the net force on the dipole is $p(\Delta E/\Delta x)$.

after a time t,

$$x = v_0 t,$$
$$y = -\frac{1}{2}\left(\frac{eE}{m}\right)t^2.$$

Elimination of t gives

$$y = -\left(\frac{eE}{2mv_0^2}\right)x^2,$$

which is the equation of a parabola. The motion is the same as that of a body projected horizontally in the earth's gravitational field. The deflection of electrons by an electric field is used to control the direction of an electron stream in many electronic devices, such as the cathode-ray oscilloscope.

Example 5 Figure 25-3 represents two point charges of equal magnitude q but of opposite sign, separated by a distance l. Such a pair of charges is called an *electric dipole*. The dipole is in a *uniform* electric field E, whose direction makes an angle θ with the line joining the two charges, called the *dipole axis*. A force F_1, of magnitude qE, in the direction of the field, is exerted on the positive charge, and a force F_2, of the same magnitude but in the opposite direction, is exerted on the negative charge. The resultant *force* on the dipole is zero, but since the two forces do not have the same line of action, they constitute a *couple* (see Sec. 8-4). The moment of the couple is

$$\Gamma = (qE)(l \sin \theta),$$

since $l \sin \theta$ is the perpendicular distance between the action lines of the forces.

The product ql of the charge q and the distance l is called the *electric dipole moment*, and is represented by p:

$$p = ql.$$

The torque exerted by the couple is therefore

$$\Gamma = pE \sin \theta. \tag{25-2}$$

The *vector dipole moment* of the dipole, p, is defined as a vector of magnitude p lying along the dipole axis and pointing from the negative toward the positive charge. With this definition and the concept of vector torque introduced in Sec. 8-5, the torque on an electric dipole in an electric field may be written as

$$\Gamma = p \times E. \tag{25-3}$$

The effect of the torque is to tend to rotate the dipole to a position in which the dipole moment p is parallel to the electric vector E, as in Fig. 25-3b. If the field is *uniform*, the dipole is in equilibrium in this position.

Suppose, however, that the field at each charge has the direction of the vector E but that it is *not* uniform and its magnitude is greater at the position of the $+$ charge than at the position of the $-$ charge. The force F_1 is then *greater* than the force F_2, and there is a net force on the dipole toward the right, urging it toward a region of *stronger* field.

Suppose the field increases uniformly in magnitude along the x-axis, and that it changes by an amount dE in an interval dx. Then, over the length l of the dipole, when it is parallel to the x-axis, the total change of E is $l(dE/dx)$. The difference of the two forces is given by

$$F = F_1 - F_2 = ql\left(\frac{dE}{dx}\right) = p\left(\frac{dE}{dx}\right).$$

The quantity dE/dx, representing the rate of change of E with position, is called the *electric field gradient;* the net force on a dipole oriented parallel to a nonuniform field is equal to the product of the dipole moment and the field gradient.

25-2 Calculation of electric field

The preceding section has described an experimental way to measure the electric field at a point. The method consists of placing a small test charge at the point, *measuring* the force on it, and taking the ratio of the force to the charge. The electric field at a point may also be *computed* from Coulomb's law if the magnitudes and positions of all charges contributing to the field are known. Thus, to find the magnitude of the electric field at a point P, at a distance r from a point charge q, we imagine a test charge q' to be placed at P. The force on the test charge, by Coulomb's law, has magnitude

$$F = \frac{1}{4\pi\epsilon_0}\left(\frac{qq'}{r^2}\right), \quad = k\left(\frac{qq'}{r^2}\right)$$

and hence the electric field at P has magnitude

$$E = \frac{F}{q'} = \frac{1}{4\pi\epsilon_0}\left(\frac{q}{r^2}\right). \quad = k\left(\frac{q}{r^2}\right)$$

The direction of the field is away from the charge q if the latter is positive (as in Fig. 25-4a), toward q if it is negative.

The *direction* of the electric field caused by a point charge is represented conveniently by use of *unit vectors,* introduced in Sec. 1-7. Let $\hat{r}$ be a unit vector directed along the line from the charge q to the point P at which the field is to be determined. Then for either a positive or negative charge q,

$$E = \frac{1}{4\pi\epsilon_0}\frac{q\hat{r}}{r^2}, \qquad (25\text{-}4)$$

where r is the distance from the charge to point P. When q is negative, the direction of E is toward q, opposite to $\hat{r}$.

If a number of point charges q_1, q_2, etc., are at distances r_1, r_2, etc., from a given point P, as in Fig. 25-4b, each exerts a force on a test charge q' placed at the point, and the resultant force on the test charge is the *vector sum* of these forces. The resultant electric field is the *vector sum* of the individual electric fields, and

$$E = E_1 + E_2 + \cdots$$

$$= \frac{1}{4\pi\epsilon_0}\left(\frac{q_1\hat{r}_1}{r_1^2} + \frac{q_2\hat{r}_2}{r_2^2} + \cdots\right). \qquad (25\text{-}5)$$

$$E = k\sum \frac{q}{r^2}$$

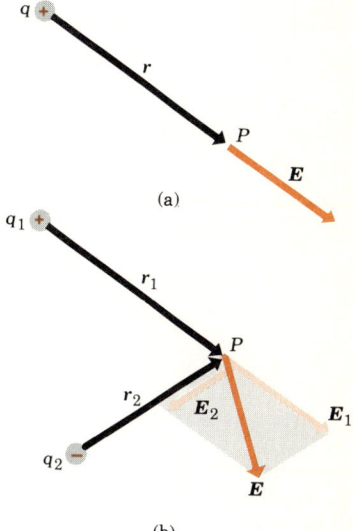

25-4 (a) The electric field E is in the same direction as the vector r when q is positive. (b) The resultant electric field at point P is the vector sum of E_1 and E_2.

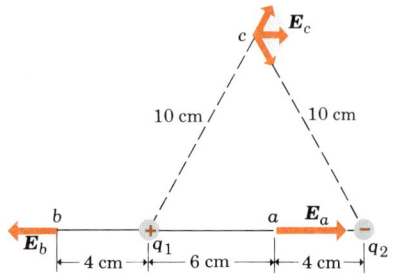

25-5 Electric field at three points, a, b, and c, in the field set up by charges q_1 and q_2.

Because each term in the sum is a vector quantity, the sum is a *vector sum*. The fact that the total field is the sum of the separate fields that would be caused by the individual charges is a direct result of the *principle of superposition*, discussed in Sec. 24-5.

Example 1 Point charges q_1 and q_2 of $+12 \times 10^{-9}$ C and -12×10^{-9} C, respectively, are placed 0.1 m apart, as in Fig. 25-5. Compute the electric fields due to these charges at points a, b, and c.

Solution At point a, the vector due to the positive charge q_1 is directed toward the right, and its magnitude is

$$E_1 = (9 \times 10^9 \text{ N} \cdot \text{m}^2 \cdot \text{C}^{-2}) \frac{(12 \times 10^{-9} \text{ C})}{(0.06 \text{ m})^2}$$
$$= 3.00 \times 10^4 \text{ N} \cdot \text{C}^{-1}.$$

The vector due to the negative charge q_2 is also directed toward the right; its magnitude is

$$E_2 = (9 \times 10^9 \text{ N} \cdot \text{m}^2 \cdot \text{C}^{-2}) \frac{(12 \times 10^{-9} \text{ C})}{(0.04 \text{ m})^2}$$
$$= 6.75 \times 10^4 \text{ N} \cdot \text{C}^{-1}.$$

Hence, at point a,

$$E_a = (3.00 + 6.75) \times 10^4 \text{ N} \cdot \text{C}^{-1}$$
$$= 9.75 \times 10^4 \text{ N} \cdot \text{C}^{-1}, \quad \text{toward the right.}$$

At point b, the vector due to q_1 is directed toward the left, with magnitude

$$E_1 = (9 \times 10^9 \text{ N} \cdot \text{m}^2 \cdot \text{C}^{-2}) \frac{(12 \times 10^{-9} \text{ C})}{(0.04 \text{ m})^2}$$
$$= 6.75 \times 10^4 \text{ N} \cdot \text{C}^{-1}.$$

The vector due to q_2 is directed toward the right, with magnitude

$$E_2 = (9 \times 10^9 \text{ N} \cdot \text{m}^2 \cdot \text{C}^{-2}) \frac{(12 \times 10^{-9} \text{ C})}{(0.14 \text{ m})^2}$$
$$= 0.55 \times 10^4 \text{ N} \cdot \text{C}^{-1}.$$

Hence, at point b

$$E_b = (6.75 - 0.55) \times 10^4 \text{ N} \cdot \text{C}^{-1}$$
$$= 6.20 \times 10^4 \text{ N} \cdot \text{C}^{-1}, \quad \text{toward the left.}$$

At point c, the magnitude of each vector is

$$E = (9 \times 10^9 \text{ N} \cdot \text{m}^2 \cdot \text{C}^{-2}) \frac{(12 \times 10^{-9} \text{ C})}{(0.1 \text{ m})^2}$$
$$= 1.08 \times 10^4 \text{ N} \cdot \text{C}^{-1}.$$

The directions of these vectors are shown in the figure; their resultant is easily seen to be

$$E_c = 1.08 \times 10^4 \text{ N} \cdot \text{C}^{-1}, \quad \text{toward the right.}$$

In practical situations, electric fields are often set up by charges distributed over the surfaces of conductors of finite size, rather than by point charges. The electric field must then be calculated by imagining the charge distribution to be subdivided into many small elements of charge Δq, at varying distances from the point P at which the field is to be calculated, and adding the separate contributions of these elements to the total field. Not all of the charge in each element will be at the same distance from the point P, but if the elements are small compared with the distance to the point, and r represents the distance from any point within the element to the point P, then approximately

$$\mathbf{E} \approx \frac{1}{4\pi\epsilon_0} \sum \frac{\Delta q \hat{\mathbf{r}}}{r^2}.$$

The finer the subdivision, the better the approximation, and in the limit as $\Delta q \to 0$,

$$\mathbf{E} = \frac{1}{4\pi\epsilon_0} \lim_{\Delta q \to 0} \sum \frac{\Delta q \hat{\mathbf{r}}}{r^2}.$$

The limit of the vector sum, however, is the *vector integral*

$$\mathbf{E} = \frac{1}{4\pi\epsilon_0} \int \frac{\hat{\mathbf{r}} \, dq}{r^2}. \quad = k \int \frac{\hat{r} \, dq}{r^2} \quad (25\text{-}6)$$

$$E = k \sum \frac{q}{r^2}$$

The limits of integration must be assigned so as to include all charges contributing to the field. Like any vector equation, Eq. (25-6) implies three scalar equations, one for each component of the vectors $\mathbf{E}$ and $\hat{\mathbf{r}}$. To evaluate the vector integral, we evaluate each of the three scalar integrals.

Example 2 A ring-shaped conductor of radius a carries a total charge Q. Find the electric field at a point a distance x from the center, along the line perpendicular to the plane of the ring, through its center.

Solution The situation is shown in Fig. 25-6. We represent the ring as made up of small segments ds, as shown. The charge of the segment ds is dQ. At point P the element of charge dQ causes an electric field contribution $d\mathbf{E}$ having magnitude dE given by

$$dE = \frac{1}{4\pi\epsilon_0} \frac{dQ}{x^2 + a^2}. \quad = k \frac{dQ}{(\sqrt{x^2+a^2})^2}$$

The component dE_x of this field along the x-axis is

$$dE_x = dE \cos\theta = \left(\frac{1}{4\pi\epsilon_0}\right)\left(\frac{dQ}{x^2 + a^2}\right)\left(\frac{x}{\sqrt{x^2 + a^2}}\right)$$

$$= \frac{1}{4\pi\epsilon_0} \frac{x \, dQ}{(x^2 + a^2)^{3/2}}.$$

To find the total x-component of field, E_x, we integrate this expression:

$$E_x = \int \frac{1}{4\pi\epsilon_0} \frac{x \, dQ}{(x^2 + a^2)^{3/2}}.$$

On the right side, everything is constant except dQ; x does not vary as we move from point to point on the ring. Thus everything except dQ

25-6 Electric field due to ring of charge.

may be taken outside the integral. The integral of dQ is simply the total charge Q, and we finally obtain

$$E_x = \frac{1}{4\pi\epsilon_0} \frac{Qx}{(x^2 + a^2)^{3/2}}. \qquad (25\text{-}7)$$

In principle, this calculation should also be performed for the components perpendicular to the x-axis, but it is easy to see from symmetry that these add to zero.

Equation (25-7) shows that, at the center of the ring ($x = 0$), the total field is zero, as might be expected; charges on opposite sides pull in opposite directions on a test charge at that point, and their fields cancel. When x is much larger than a, Eq. (25-7) becomes approximately equal to $Q/4\pi\epsilon_0 x^2$, corresponding to the fact that at distances much greater than the dimensions of the ring it appears as a point charge.

Example 3 *Long charged wire.* In Fig. 25-7 a long thin wire of length $2L$ lies along the y-axis and has an electric charge per unit length of λ. We wish to find the electric field caused by this charge at point P, a distance r from the midpoint of the wire, as shown.

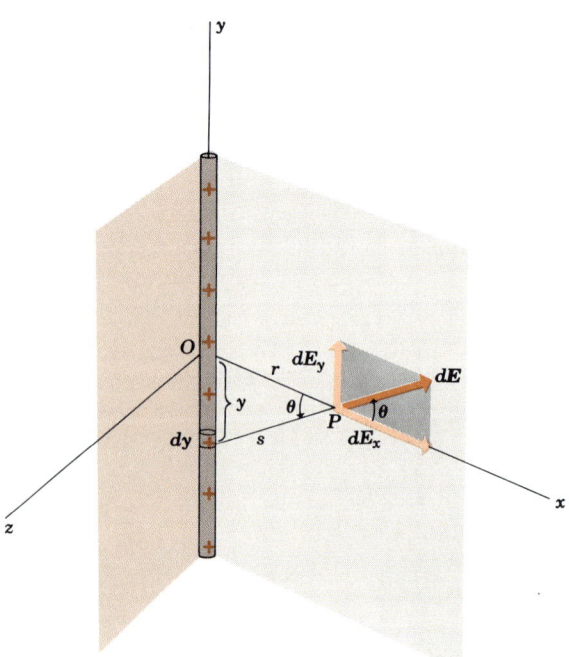

25-7 Electric field of a long thin charged wire.

Solution We imagine dividing the wire into infinitesimal segments dy. The charge dq of a segment dy is then given by $dq = \lambda \, dy$. The distance s from a segment at position y to point P is $s = (y^2 + r^2)^{1/2}$, and the contribution $d\mathbf{E}$ to the field at P due to this segment has magnitude dE given by

$$dE = \frac{1}{4\pi\epsilon_0} \frac{\lambda \, dy}{y^2 + r^2}.$$

To find the total field at P we need to add (integrate) the contributions from *all* the segments. We first express $d\mathbf{E}$ in terms of its x- and y-components. Reference to Fig. 25-7 shows that

$$\cos\theta = \frac{r}{\sqrt{y^2 + r^2}}, \quad \sin\theta = -\frac{y}{\sqrt{y^2 + r^2}},$$

$$dE_x = dE\cos\theta = \frac{1}{4\pi\epsilon_0}\frac{\lambda r\, dy}{(y^2+r^2)^{3/2}},$$

$$dE_y = dE\sin\theta = -\frac{1}{4\pi\epsilon_0}\frac{\lambda y\, dy}{(y^2+r^2)^{3/2}}.$$

The total E_x and E_y caused by the charge on the entire wire are then given by

$$E_x = \int_{-L}^{L} \frac{1}{4\pi\epsilon_0}\frac{\lambda r\, dy}{(y^2+r^2)^{3/2}},$$

$$E_y = -\int_{-L}^{L} \frac{1}{4\pi\epsilon_0}\frac{\lambda y\, dy}{(y^2+r^2)^{3/2}}. \tag{25-8}$$

These integrals may be evaluated by trigonometric substitution or by use of a table of integrals. The details are left as an exercise; the final results are

$$E_x = \frac{1}{4\pi\epsilon_0}\frac{2\lambda L}{r\sqrt{L^2+r^2}},$$

$$E_y = 0. \tag{25-9}$$

The result that $E_y = 0$ could have been predicted from symmetry; for every charge element dq at a given position y there is a corresponding element at $-y$, and the y-components of field caused by these two charges are opposite. Thus the total y-component of field *must* be zero.

If the length $2L$ of the wire is much longer than the distance r (i.e., if the wire is infinitely long), then in Eq. (25-9) r becomes negligible compared to L in the radical, L is canceled by $(L^2+r^2)^{1/2}$, and we obtain simply

$$E_x = \frac{1}{2\pi\epsilon_0}\frac{\lambda}{r} \quad \text{(long straight line of charge)}. \tag{25-10}$$

The total field is proportional to λ, as expected, and it is inversely proportional to the *first power* of the distance r rather than the *inverse-square* relation for a single point charge.

If the point P had been on the z-axis instead of the x-axis, the only nonzero component of E would have been E_z. The electric field at any point in the xz-plane is directed radially outward from the wire. For an infinitely long wire, the field at *every point* in space is perpendicular to the wire and radially outward from it.

Example 4 *Infinite plane sheet of charge.* In Fig. 25-8, positive charge is distributed uniformly over the entire xy-plane, with a *charge per unit area*, or *surface density of charge, σ*. We wish to calculate the electric field at the point P, at a distance a from the plane.

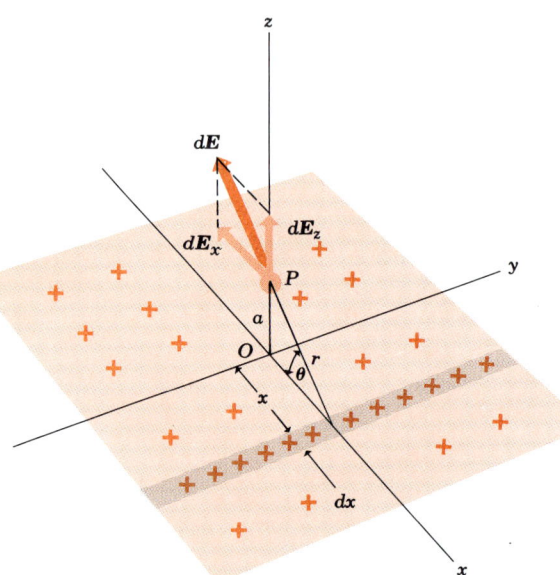

25-8

We divide the charge into narrow strips of width dx, parallel to the y-axis. Each strip can be considered a *line* charge, and we can use the result of the preceding example.

The area of a portion of a strip of length L is $L\,dx$, and the charge dq on the strip is

$$dq = \sigma L\,dx.$$

The charge per unit length, λ, is therefore

$$\lambda = \frac{dq}{L} = \sigma\,dx.$$

From Eq. (25–10), the strip sets up at point P a field $d\mathbf{E}$, lying in the xz-plane, of magnitude

$$dE = \frac{\sigma}{2\pi\epsilon_0}\frac{dx}{r}.$$

The field can be resolved into components $d\mathbf{E}_x$ and $d\mathbf{E}_z$. By symmetry, the components $d\mathbf{E}_x$ will sum to zero when the entire sheet of charge is considered. (Be sure that you understand why.) The resultant field at P is therefore in the z-direction, perpendicular to the sheet of charge. From the diagram,

$$dE_z = dE\sin\theta$$

and hence

$$E = \int dE_z = \frac{\sigma}{2\pi\epsilon_0}\int_{-\infty}^{+\infty}\frac{\sin\theta\,dx}{r}.$$

But

$$\sin\theta = \frac{a}{r}, \qquad r^2 = a^2 + x^2,$$

and therefore

$$E = \frac{\sigma a}{2\pi\epsilon_0} \int_{-\infty}^{+\infty} \frac{dx}{a^2 + x^2} = \frac{\sigma a}{2\pi\epsilon_0} \left[\frac{1}{a} \tan^{-1} \frac{x}{a} \right]_{-\infty}^{+\infty},$$

$$E = \frac{\sigma}{2\epsilon_0}. \tag{25-11}$$

Note that the distance a from the plane to the point P *does not* appear in the final result. This means that the intensity of the field set up by an infinite plane sheet of charge is *independent of the distance from the charge*. In other words, the field is *uniform* and *normal* to the plane of charge.

The same result would have been obtained if point P, in Fig. 25-8, had been taken *below* the xy-plane. That is, a field of the same magnitude but in the opposite sense is set up on the opposite side of the plane.

25-3 Field lines

The concept of field lines was introduced by Michael Faraday (1791–1867) as an aid in visualizing electric (and magnetic) fields. A *field line* (in an electric field) is *an imaginary line drawn in such a way that its direction at any point* (i.e., the direction of its tangent) *is the same as the direction of the field at that point.* (See Fig. 25-9.) Since, in general, the direction of a field varies from point to point, field lines are usually curves. Faraday called these lines "lines of force", but the term "field line" is preferable.

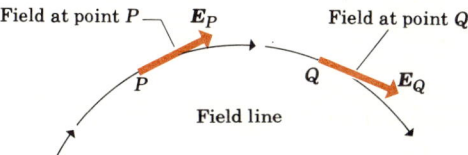

25-9 The direction of the electric field at any point is tangent to the field line through that point.

Figure 25-10 shows some of the field lines in two planes containing (a) a single positive charge; (b) two equal charges, one positive and one negative (an electric dipole); and (c) two equal positive charges. The direction of the resultant field at every point in each diagram is along the tangent to the field line passing through the point. Arrowheads on the field lines indicate the direction in which the tangent is to be drawn.

No field lines originate or terminate in the space surrounding a charge. Every field line in an *electrostatic* field is a continuous line terminated by a positive charge at one end and a negative charge at the other.* While sometimes for convenience we speak of an "isolated" charge and draw its field as in Fig. 25-10a, this simply means that the charges on which the lines terminate are at large distances from the charge under consideration. For example, if the charged body in Fig. 25-10a is a small sphere suspended by a thread from the laboratory ceiling, the negative charges on which its field lines terminate would be

---

* We shall see in a larger chapter that a *changing magnetic field* sets up an electric field whose lines *do not* terminate on electric charges, but close on themselves.

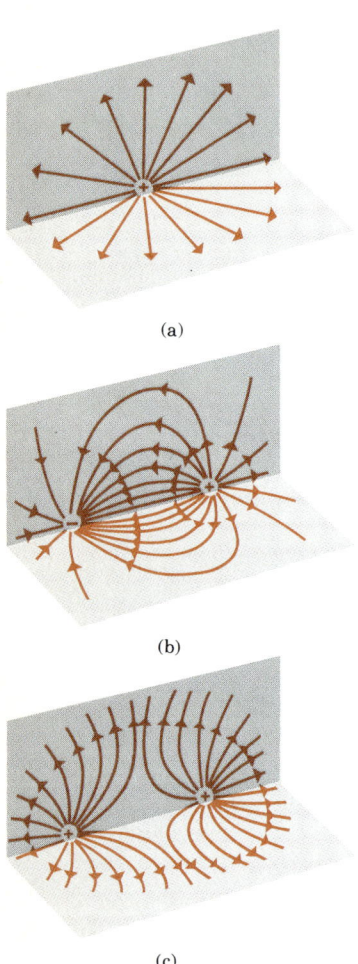

25-10 The mapping of an electric field with the aid of field lines.

found on the walls, floor, or ceiling, as well as on other objects in the laboratory.

At any one point, the resultant field can have but one direction. Hence, only one field line can pass through each point of the field. In other words, field lines never intersect.

If a field line were to be drawn through every point of an electric field, all of space and the entire surface of a diagram would be filled with lines, and no individual line could be distinguished. By suitably limiting the number of field lines one draws to represent a field, the lines can be used to indicate the *magnitude* of a field as well as its *direction*. This is accomplished by spacing the lines in such a way that *the number per unit area crossing a surface at right angles to the direction of the field is at every point proportional to the electric field*. In a region where the field is large, such as that between the positive and negative charges of Fig. 25-10b, the field lines are closely spaced, whereas in a region where the field is small, such as that between the two positive charges of Fig. 25-10c, the lines are widely separated. In a *uniform* field, the field lines are straight, parallel, and uniformly spaced.

25-4 Gauss's Law

Karl Friedrich Gauss (1777–1855) was a German scientist and mathematician who made many contributions to experimental and theoretical physics and to mathematics. The relation known as *Gauss's law* is a statement of an important property of electrostatic fields.

The content of Gauss's law is suggested by consideration of field lines, discussed in Sec. 25-3. The field of an isolated positive point charge q is represented by lines radiating out in all directions. Suppose we imagine this charge as surrounded by a spherical surface of radius R, with the charge at its center. The area of this imaginary surface is $4\pi R^2$, so if the total number of field lines emanating from q is N, then the number of lines *per unit surface area* on the spherical surface is $N/4\pi R^2$. We imagine a second sphere concentric with the first, but with radius $2R$. Its area is $4\pi(2R)^2 = 16\pi R^2$, and the number of lines per unit area on this sphere is $N/16\pi R^2$, one-fourth the density of lines on the first sphere. This corresponds to the fact that, at distance $2R$, the field has only one-fourth the magnitude it has at distance R, and verifies our qualitative statement in Sec. 25-3 that the *density* of lines is proportional to the magnitude of the field.

The fact that the *total* number of lines at distance $2R$ is the same as at R can be expressed another way. The field is inversely proportional to R^2, but the *area* of the sphere is proportional to R^2, so the *product* of the two is independent of R. For a sphere of arbitrary radius r, the magnitude of E on the surface is

$$E = \frac{1}{4\pi\epsilon_0} \frac{q}{r^2},$$

the surface area is

$$A = 4\pi r^2,$$

and the product of the two is

$$EA = \frac{q}{\epsilon_0}. \qquad (25\text{-}12)$$

This is independent of r and depends *only* on the charge q. As we shall see, this result is of crucial importance in the following development.

What is true of the entire sphere is also true of any portion of its surface. In the construction of Fig. 25-11, an area ΔA is outlined on a sphere of radius R and then projected onto the sphere of radius $2R$ by drawing lines from the center through points on the boundary of ΔA. The area projected on the larger sphere is clearly $4\,\Delta A$; thus again the product $E\,\Delta A$ is independent of the radius of the sphere.

This projection technique shows how this discussion may be extended to nonspherical surfaces. Instead of a second sphere, let us surround the sphere of radius R by a surface of irregular shape, as in Fig. 25-12a. Consider a small areal element ΔA; we note that this area is *larger* than the corresponding element on a spherical surface at the same distance from q. If a normal to the surface makes an angle θ with a radial line from q, two sides of the area projected on the spherical surface are foreshortened by a factor $\cos\theta$, as shown in Fig. 25-12b. Thus, the quantity corresponding to $E\,\Delta A$ for the spherical surface is $E\,\Delta A\,\cos\theta$ for the irregular surface.

Now we may divide the entire irregular surface into small elements ΔA, compute the quantity $E\,\Delta A\,\cos\theta$ for each, and sum the results. Each of these projects onto a corresponding element of area on the sphere, so summing the quantities $E\,\Delta A\,\cos\theta$ over the irregular surface must yield the same result as summing the quantities $E\,\Delta A$ over the sphere. But we have already performed that calculation; the result, given by Eq. (25-12), depends only on the charge q. Thus, for the irregular surface the result is

$$\sum E\,\Delta A\,\cos\theta = \frac{q}{\epsilon_0}, \qquad (25\text{-}13)$$

no matter what the shape of the surface, provided only that it is a *closed* surface enclosing the charge q. Correspondingly, for a closed surface en-

25-11 Projection of an element of area ΔA on a sphere of radius R, onto a sphere of radius $2R$. The projection multiplies each linear dimension by two, so the area element on the larger sphere is $4\,\Delta A$.

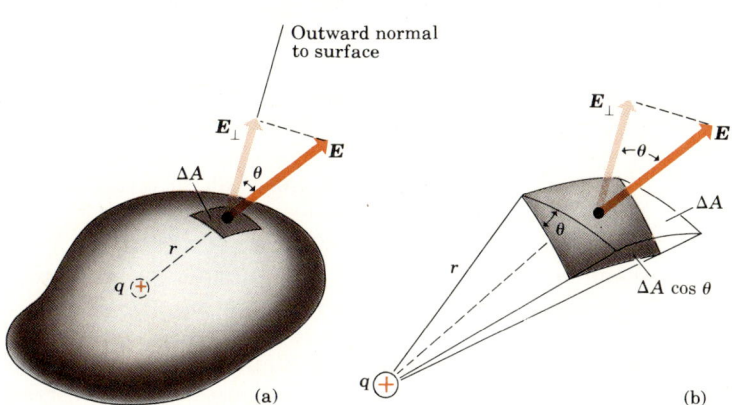

25-12

closing *no* charge,

$$\sum E \, \Delta A \cos \theta = 0.$$

This is a mathematical statement of the fact that when a region contains no charge, any field line that enters on one side must leave again at some other point on the boundary surface. Field lines can begin or end inside a region of space only when there is charge in that region.

Because the field varies from point to point on the irregular surface, Eq. (25-13) is strictly true only in the limit when the area elements become very small. In this limit, the sum becomes an integral called the *surface integral* of $E \cos \theta$, written

$$\oint E \cos \theta \, dA = \frac{q}{\epsilon_0}. \qquad (25\text{-}14)$$

The circle on the integral sign reminds us that the integral is always taken over a *closed* surface enclosing the charge q.

Since $E \cos \theta$ is the component of $\mathbf{E}$ perpendicular to the surface at each point, we may use the notation $E_\perp = E \cos \theta$, and write

$$\oint E_\perp \, dA = \frac{q}{\epsilon_0}. \qquad (25\text{-}15)$$

The quantity $E_\perp \, dA = E \cos \theta \, dA$ is also called the *electric flux* through the area dA. Electric flux may be denoted by Ψ, and an element of flux corresponding to a small area dA by $d\Psi$. Then

$$d\Psi = E_\perp \, dA = E \cos \theta \, dA = \mathbf{E} \cdot d\mathbf{A}. \qquad (25\text{-}16)$$

The total flux Ψ through a finite surface area is

$$\Psi = \int E_\perp \, dA = \int E \cos \theta \, dA = \int \mathbf{E} \cdot d\mathbf{A}. \qquad (25\text{-}17)$$

(The notation $\mathbf{E} \cdot d\mathbf{A}$ is explained below.) Thus Eqs. (25-14) and (25-15) state that the total electric flux out of a closed surface is proportional to the charge enclosed:

$$\Psi = \frac{q}{\epsilon_0}. \qquad (25\text{-}18)$$

If the point charge in Fig. 25-12 is negative, the $\mathbf{E}$ field is directed radially *inward;* the angle θ is then greater than 90°, its cosine is negative, and the integral in Eq. (25-15) is negative. But since q is also negative, Eq. (25-15) still holds.

If a point charge lies *outside* a closed surface (the reader may construct his own diagram), the electric field is *outward* at some points of the surface and *inward* at others. It is not difficult to show that the positive and negative contributions to the integral over the surface exactly cancel, and the sum is zero. But the charge *inside* the closed surface is also zero, so again Eq. (25-15) is obeyed.

Although we have been concerned only with a single point charge, it is easy to generalize the above results to *any* charge distribution. The total electric field $\mathbf{E}$ at a point on the surface is the vector sum of the fields produced by the individual charges, and the quantity $E\,dA\cos\theta$ is therefore the sum of the contributions from these charges. Since Eq. (25-15) holds for each point charge, a corresponding relation holds for the *total* E field and the *total* charge enclosed by the surface. That is,

$$\oint E_\perp\, dA = \frac{1}{\epsilon_0} \sum q, \qquad (25\text{--}19)$$

where E is now the total electric field and Σq represents the *algebraic* sum of all charges enclosed by the surface.

Equation (25-19) is the mathematical statement of *Gauss's law*. It states that when we multiply each element of area of a closed surface by the normal component of E at the element, and sum over the entire surface, the result is a constant times the total charge inside the surface.

The notation can be simplified by use of the *vector area d**A***, defined as a vector whose magnitude equals dA and whose direction is that of the *outward* normal at dA. The product $E_\perp\, dA = E\cos\theta\, dA$ can then be written as the *scalar product* or *dot product* of the vectors E and $d\mathbf{A}$:

$$E_\perp\, dA = \mathbf{E}\cdot d\mathbf{A}.$$

Denoting the total charge enclosed by $Q = \Sigma q$, we may write Gauss's law more compactly as

$$\oint \mathbf{E}\cdot d\mathbf{A} = Q/\epsilon_0. \qquad (25\text{--}20)$$

As mentioned above, the flux of E across a surface, as well as Gauss's law, can be interpreted graphically in terms of field lines. If the number of lines per unit area at right angles to their direction is proportional to E, the surface integral of $E_\perp$ over a closed surface is proportional to the total number of lines crossing the surface in an outward direction, and the net charge within the surface is proportional to this number. As an example, consider the field of two equal and opposite point charges shown in Fig. 25-13. Surface A encloses the positive charge only, and 18 lines cross it in an outward direction. Surface B encloses the negative charge only, and it also is crossed by 18 lines, but in an inward direction. Surface C enclosed *both* charges. It is intersected by lines at 16 points, at 8 of which the intersections are outward and at 8 of which they are inward. The *net* number of lines crossing in an outward direction is zero, and the net charge inside the surface is also zero. Surface D is intersected at 6 points, at 3 of which the intersections are outward while at the other 3 they are inward. The net number of lines crossing in an outward direction, and the enclosed charge, are both zero.

In evaluating the surface integral of $E_\perp$ over a closed surface, it is often necessary to divide the surface, in imagination, into a number of portions. The integral over the entire surface is the *sum* of the integrals over each portion. There are many cases of practical importance where symmetry considerations simplify the evaluation of the integral enough that only simple algebraic manipulations are necessary. Several examples will be discussed in the next section. The following observations are also useful:

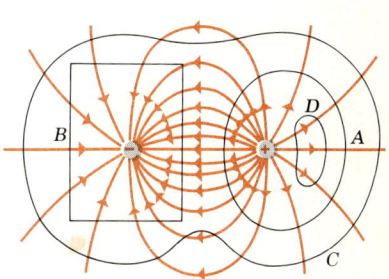

25-13

1. If E is at right angles to a surface of area A at all points, and has the

same *magnitude* at all points of the surface, then $E_\perp = E =$ constant, and

$$\int E_\perp \, dA = EA.$$

2. If E is *parallel* to a surface at all points, $E_\perp = 0$ and the integral is zero.
3. If $E = 0$ at all points of a surface, the integral is zero.
4. The surface to which Gauss's law is applied need not be a real physical surface, such as the surface of a solid body. Indeed, in most applications of this law, one considers an imaginary or geometrical surface that may be in empty space, embedded in a solid body, or partly in space and partly within a body.
5. In the integral $\int \mathbf{E} \cdot d\mathbf{A}$ or $\int E_\perp \, dA$, $\mathbf{E}$ is always the *total* electric field at each point on the surface. In general this field is caused partly be charges within the volume and partly by charges outside. Thus even when there is *no* charge within the volume, the field at points on the surface need not be zero. In that case, however, $\int \mathbf{E} \cdot d\mathbf{A}$ is always zero.
6. In applications of Gauss's law to field calculations, some judgment is required in choosing a surface. Two useful guiding principles are that the point or points at which the field is to be determined must lie on the surface, and that the surface must have enough symmetry so that it is possible to evaluate the integral. Thus if the problem has spherical or cylindrical symmetry, the gaussian surface will usually be spherical or cylindrical, respectively.

25-5 Applications of Gauss's Law

1. Location of excess charge on a conductor. It has been explained that the electric field $\mathbf{E}$ is zero at all points within a conductor when the charges in the conductor are at rest. (If $\mathbf{E}$ were *not* zero, the charges would move, and we would not have an electro*static* situation.) We may construct an imaginary surface in the interior of a conductor, such as surface A in Fig. 25–14a. (In applications of Gauss's law, such a surface is often called a *gaussian surface*.) Because $\mathbf{E} = \mathbf{0}$ everywhere on this surface, Eq. (25-17) requires that the net charge inside the surface be zero.

If we now imagine the surface to shrink to zero like a collapsing balloon, as suggested in Fig. 25–14a, until it essentially encloses a point, the charge at the point must be zero. Since this process can take place at *any* point in the conductor, *there can be no net charge at any point within the conductor*. It follows that the entire excess charge on the conductor must be located on the *outer surface* of the conductor, as shown.

Now suppose there is a cavity in the conductor, as in Fig. 25–14b, and there are no charges within the cavity. Again a gaussian surface such as A may be used to show that there is no charge within the material of the conductor. Furthermore, consideration of a surface such as B shows that the net charge on the surface of the cavity must be zero. This does not necessarily prove that the entire cavity wall is uncharged, but we shall

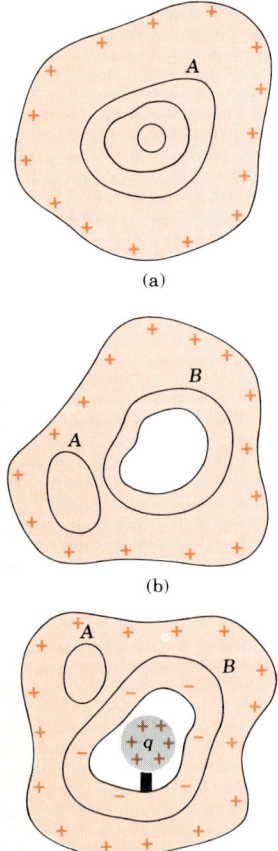

25-14

prove later that this is in fact the case. Thus the entire charge on the conductor lies on its *outer* surface, not on the cavity wall.

Next, suppose that there is a conductor inside the cavity but insulated from it, as in Fig. 25–14c, and that the inner conductor has a charge q. Application of Gauss's law to surface B shows again that the *net* charge inside this surface is zero, so there must be a charge on the cavity wall, equal and opposite in sign to the charge q. If the outer conductor is initially *uncharged* before the charge q is inserted (and is insulated so that the total charge on it cannot change), there must be a charge on its outer surface, equal and opposite to the charge on the cavity wall and therefore equal to, and of the *same* sign as, the charge q. If the outer conductor originally had a charge q', the charge on its outer surface becomes $q + q'$.

It follows that insertion of a charge into a cavity in a hollow conductor results in the appearance of an exactly equal charge on the outer surface of the hollow conductor, whether or not this conductor was originally charged.

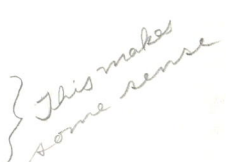

2. *Coulomb's law.* We have considered Coulomb's law as the fundamental equation of electrostatics and have derived Gauss's law from it. An alternative procedure is to consider Gauss's law as a fundamental experimental relation. Coulomb's law can then be derived from Gauss's law, by using this law to obtain the expression for the electric field E due to a point charge.

Consider the electric field of a single positive point charge q, shown in Fig. 25–15. By *symmetry*, the field is everywhere radial (there is no reason why it should deviate to one side of a radial direction rather than to another) and its magnitude is the same at all points at the same distance r from the charge (any point at this distance is like any other). Hence, if we select as a gaussian surface a spherical surface of radius r, $E_\perp = E = $ constant at all points of the surface. Then

$$\oint E_\perp \, dA = E_\perp \oint dA = EA = 4\pi r^2 E.$$

From Gauss's law

$$4\pi r^2 E = \frac{q}{\epsilon_0} \quad \text{and} \quad E = \frac{1}{4\pi\epsilon_0} \frac{q}{r^2}.$$

The force on a point charge q' at a distance r from the charge q is then

$$F = q'E = \frac{1}{4\pi\epsilon_0} \frac{qq'}{r^2},$$

which is Coulomb's law.

3. *Field of a charged conducting sphere.* Any excess charge on an isolated solid conducting sphere is, by symmetry, distributed *uniformly* over its outer surface. The electric field at any point can be calculated (at least in principle) by summing the contributions from elements of charge on the surface, but it is much simpler to use Gauss's law. At points outside the sphere the field is radial everywhere, for the same reason as for a point charge; because of the spherical symmetry, there is no reason for the field at a given point to deviate one way rather than

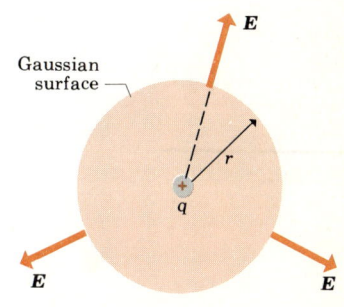

25–15

another from the radial direction. Furthermore, the magnitude of the field is again uniform over a spherical surface of radius r, concentric with the conductor.

Thus if we construct a gaussian surface of radius r, where r is greater than the radius R of the sphere, and if q is the total charge on the sphere,

$$4\pi r^2 E = \frac{q}{\epsilon_0},$$

$$E = \frac{1}{4\pi\epsilon_0} \frac{q}{r^2}.$$

(25-21)

The field *outside* the sphere is, therefore, the same as though the entire charge were concentrated at a point at its center. Just outside the surface of the sphere, where $r = R$,

$$E = \frac{1}{4\pi\epsilon_0} \frac{q}{R^2}.$$

Inside the sphere, as in the interior of any conductor where no charge is in motion, the field is zero. Thus when r is less than R, $E = 0$.

The same argument may be applied to a conducting, hollow, spherical *shell*, that is, a spherical conductor with a concentric spherical hole in the center, if there is no charge in the hole. This time we take a spherical gaussian surface of radius r less than the radius of the hole. If there *is* a field inside the hole, it must be spherically symmetric (radial) as before, so again $E = q/4\pi\epsilon_0 r^2$. But this time $q = 0$, so E must also be zero.

Because of the relation of Gauss's law to Coulomb's law, this result holds only because the electric field obeys an inverse *square* law. If the field were inversely proportional to r^3, or to $r^{2.147}$, Gauss's law would not hold. Very precise measurements have shown that the internal field of a charged sphere is, in fact, so small that the exponent of r cannot differ from exactly 2 by more than 1 part in 10^7. There seems no reason to doubt that it is exactly 2.

It is left as a problem to find, from Gauss's law, the electric field in the interspace between a charged sphere and a concentric hollow sphere surrounding it.

4. Field of a line charge and of a charged cylindrical conductor. We consider next the electric field set up by a long thin uniformly charged wire. We solved this problem in Sec. 25-2, Example 3, by a straightforward application of Coulomb's law, requiring a somewhat involved integration. Gauss's law makes it possible to find the field by an almost trivial calculation.

If the wire is very long and we are not too near either end, then, by symmetry, the field lines outside the wire are *radial* and lie in planes perpendicular to the wire. Also, the field has the same magnitude at all points at the same radial distance from the wire. This suggests that we use as a gaussian surface a *cylinder* of arbitrary radius r and arbitrary length l, with its ends perpendicular to the wire, as in Fig. 25-16. If λ is the charge *per unit length* on the wire, the charge within the gaussian surface is λl. Since $\mathbf{E}$ is at right angles to the wire, the component of $\mathbf{E}$ normal to the end faces is zero. Thus the end faces make no contribution

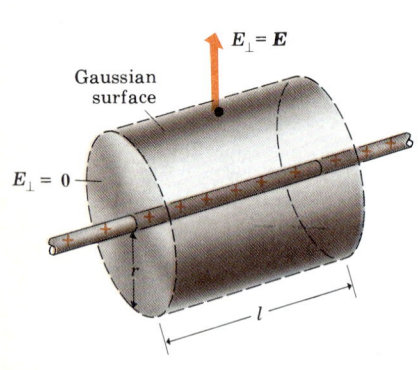

25-16 Cylindrical gaussian surface for calculating the electric field due to a long charged wire.

to the integral in Gauss's law. At all points of the curved surface, $E_\perp = E = $ constant, and since the area of this surface is $2\pi r l$, we have

$$(E)(2\pi r l) = \frac{\lambda l}{\epsilon_0}, \qquad (25\text{-}22)$$

$$E = \frac{1}{2\pi\epsilon_0}\frac{\lambda}{r}.$$

It should be noted that although the *entire* charge on the wire contributes to the field E, only that portion of the total charge lying within the gaussian surface is used when we apply Gauss's law. This feature of the law is puzzling at first; it appears as though we had somehow obtained the right answer by ignoring a part of the charge, and that the field of a *short* wire of length l would be the same as that of a very long wire. The existence of the entire charge on the wire *is*, however, taken into account when we consider the *symmetry* of the problem. Suppose the wire had been a short one, of length l. Then we could *not* conclude by symmetry that the field at one end of the cylinder, say, would equal that at the center, or that the field lines would everywhere be perpendicular to the wire. So the entire charge on the wire actually *is* taken into account, but in an indirect way.

It is left as a problem (1) to show that the field outside a long charged cylinder is the same as though the charge on the cylinder were concentrated in a line along its axis, and (2) to calculate the electic field in the interspace between a charged cylinder and a coaxial hollow cylinder that surrounds it.

5. *Field of an infinite plane sheet of charge.* To solve this problem, construct the gaussian surface shown by the shaded area in Fig. 25–17, consisting of a cylinder whose ends have an area A and whose walls are perpendicular to the sheet of charge. By symmetry, since the sheet is infinite, the electric field E has the same magnitude E on both sides of the surface, is uniform, and is directed normally away from the sheet of charge. No field lines cross the *side* walls of the cylinder; that is, the component of E normal to these walls is zero. At the ends of the cylinder the normal component of E is equal to E. The integral $\int E_\perp\, dA$, calculated over the entire surface of the cylinder, therefore reduces to $2EA$. If σ is the charge *per unit area* in the plane sheet, the net charge within the gaussian surface is σA. Hence

$$2EA = \frac{\sigma A}{\epsilon_0}, \qquad E = \frac{\sigma}{2\epsilon_0}. \qquad (25\text{-}23)$$

Note that the magnitude of the field is <u>*independent*</u> of the distance from the <u>sheet and does *not*</u> decrease inversely with the square of the distance. The field lines remain everywhere straight, parallel, and uniformly spaced. This is because the sheet was assumed infinitely large.

Of course, nothing in nature can really be infinitely large; this infinitely large plane sheet is an idealization. Our result is still useful, however; the real content of Eq. (25–23) is that it is a very good approximation to the behavior of the electric field caused by a large but finite sheet of charge, at points that are not near its edge and are close to the plane compared to its dimensions. In such cases the field is very nearly uniform and perpendicular to the plane.

25–17 Gaussian surface in the form of a cylinder for finding the field of an infinite plane sheet of charge.

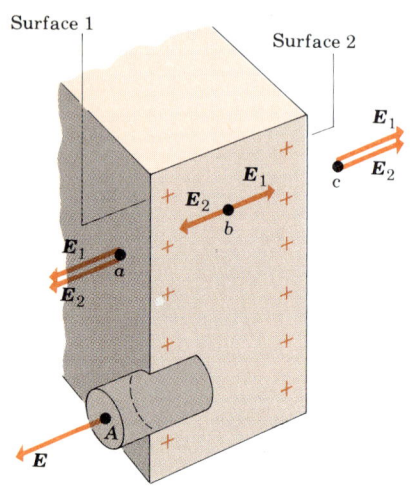

25-18 Electric field inside and outside a charged conducting plate.

6. Field of an infinite plane charged conducting plate. When a flat metal plate is given a net charge, this charge distributes itself over the entire outer surface of the plate and, if the plate is of uniform thickness and is infinitely large (or if we are not too near the edges of a finite plate), the charge per unit area is uniform and is the same on both surfaces. Hence, the field of such a charged plate arises from the superposition of the fields of *two* sheets of charge, one on each surface of the plate. By symmetry, the field is perpendicular to the plate, directed away from it (if the plate has a positive charge) and is uniform. The magnitude of the electric field at any point can be found from Gauss's law, or by using the results already derived for a sheet of charge.

Figure 25-18 shows a portion of a large charged conducting plate. Let σ represent the charge per unit area in the sheet of charge on *each* surface (i.e., the total charge per unit area on both surfaces together is 2σ). At point a, outside the plate at the left, the component of electric field $\mathbf{E}_1$, due to the sheet of charge on the left face of the plate, is directed toward the left, and its magnitude is $\sigma/2\epsilon_0$. The component $\mathbf{E}_2$ due to the sheet of charge on the right face of the plate is also toward the left and its magnitude is also $\sigma/2\epsilon_0$. The magnitude of the *resultant* intensity E is therefore

$$E = E_1 + E_2 = \frac{\sigma}{2\epsilon_0} + \frac{\sigma}{2\epsilon_0} = \frac{\sigma}{\epsilon_0}. \qquad (25\text{-}24)$$

At point b, inside the plate, the two components of electric field are in opposite directions and their resultant is zero, as it must be in any conductor in which the charges are at rest. At point c, the components again add and the magnitude of the resultant is σ/ϵ_0, directed toward the right.

To derive these results directly from Gauss's law, consider the cylinder shown in the figure. Its end faces are of area A; one end lies inside and one outside the plate. The field inside the conductor is zero. The field outside, by symmetry, is perpendicular to the plate, so the normal component of $\mathbf{E}$ is zero over the walls of the cylinder and is equal to E over the outside end face. Hence, from Gauss's law,

$$EA = \frac{\sigma A}{\epsilon_0}, \qquad E = \frac{\sigma}{\epsilon_0}. \qquad (25\text{-}25)$$

7. Field between oppositely charged parallel conducting plates. When two plane parallel conducting plates, having the size and spacing shown in Fig. 25-19, are given equal and opposite charges, the field between and around them is approximately as shown in Fig. 25-19a. While most of the charge accumulates at the opposing faces of the plates and the field is essentially uniform in the space between them, there is a small quantity of charge on the outer surfaces of the plates and a certain spreading or "fringing" of the field at the edges of the plates.

As the plates are made larger and the distance between them diminished, the fringing becomes relatively less. Such an arrangement, two oppositely charged plates separated by a distance small compared with their linear dimensions, is encountered in many pieces of electrical equipment, notably in capacitors. Often the fringing is entirely negligible; even if it is not, neglecting it often provides useful approximations in cases where the work of more detailed calculations is not warranted. We shall therefore assume that the field between two oppositely charged plates is uniform, as in Fig. 25-19b, and that the charges are distributed uniformly over the opposing surfaces.

The electric field at any point can be considered as the resultant of that due to two sheets of charge of opposite sign, or it may be found from Gauss's law. Thus at points a and c in Fig. 25-19b, the components $\mathbf{E}_1$ and $\mathbf{E}_2$ are each of magnitude $\sigma/2\epsilon_0$ but are oppositely directed, so their resultant is zero. At any point b between the plates, the components are in the same direction and their resultant is σ/ϵ_0. It is left as an exercise to show that the same results follow from applying Gauss's law to the surfaces shown by dotted lines.

8. *Field just outside any charged conductor.* Figure 25-20 represents a portion of the surface of a charged conductor of irregular shape. In general, the surface density of charge will vary from point to point of the surface. Let σ represent the surface density at a small area A. We shall show in the next chapter that the electric field *just outside* the surface of any charged conductor is at right angles to the surface.

Let us construct a gaussian surface in the form of a small cylinder, one of whose end faces, of area A, lies within the conductor, while the other lies just outside. The charge within the gaussian surface is σA. The electric field is zero at all points within the conductor. Outside the conductor, the normal component of $\mathbf{E}$ is zero at the side walls of the cylinder (since $\mathbf{E}$ is normal to the conductor), while over the end face the normal component is equal to E. Hence from Gauss's law,

$$EA = \frac{\sigma A}{\epsilon_0}, \qquad E = \frac{\sigma}{\epsilon_0}. \qquad (25\text{-}26)$$

This agrees with the results already obtained for spherical, cylindrical, and plane surfaces. Just outside the surface of a sphere of radius R carrying a charge q, for example, the electric field is

$$E = \frac{1}{4\pi\epsilon_0} \frac{q}{R^2}.$$

But the surface density of charge on the sphere is $q/4\pi R^2$, so $E = \sigma/\epsilon_0$.

The field outside an infinite charged conducting plate was also shown to equal σ/ϵ_0. In this case, the field is the same at *all* distances from the plate, but in general it decreases with increasing distance from the surface.

The expressions that we have derived for the electric fields set up by a number of simple charge distributions are summarized in Table 25-1.

Finally, we note that, in each of the above examples, Gauss's law could be used to obtain an expression for the field because it enabled us to make effective use of the *symmetry* of the situation. There are many situations where such symmetry is lacking and where Gauss's law is *not* helpful as a tool for practical calculations. To cite a simple example, we consider the electric field produced by a dipole. The field does have axial symmetry about the dipole axis, but there is no simple surface over which $E_\perp$ is constant. Thus there is no practical way to apply Gauss's law to such a configuration, even though the law is still valid in principle in this situation.

25-19 Electric field between oppositely charged parallel plates.

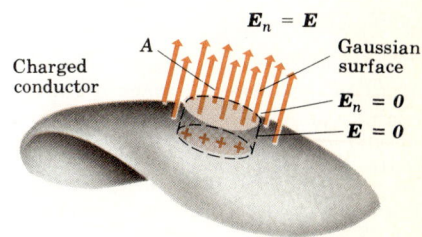

25-20 The field just outside a charged conductor is perpendicular to the surface and is equal to σ/ϵ_0.

Table 25-1 Electric fields around simple charge distributions

Charge distribution responsible for the electric field	Arbitrary point in the electric field	Electric field at this point
Single point charge q	Distance r from q	$E = \dfrac{1}{4\pi\epsilon_0}\dfrac{q\hat{r}}{r^2}$
Several point charges, $q_1, q_2, \ldots$	Distance r_1 from q_1, r_2 from $q_2 \ldots$	$E = \dfrac{1}{4\pi\epsilon_0}\left(\dfrac{q_1\hat{r}_1}{r_1^2} + \dfrac{q_2\hat{r}_2}{r_2^2} + \cdots\right)$
Charge q uniformly distributed on the surface of a solid conducting sphere of radius R	(a) Outside, $r \geq R$ (b) Inside, $r < R$	(a) $E = \dfrac{1}{4\pi\epsilon_0}\dfrac{q\hat{r}}{r^2}$ (b) $E = 0$
Long cylinder of radius R, with charge per unit length λ	(a) Outside, $r \geq R$ (b) Inside, $r < R$	(a) $E = \dfrac{1}{2\pi\epsilon_0}\dfrac{\lambda}{r}$ (b) $E = 0$
Two oppositely charged conducting plates with charge per unit area σ	Any point between plates	$E = \dfrac{\sigma}{\epsilon_0}$
Any charged conductor	Just outside the surface	$E = \dfrac{\sigma}{\epsilon_0}$

Questions

25-1 It was shown in the text that the electric field inside a spherical hole in a conductor is zero. Is this also true for a cubical hole? Can the same argument be used?

25-2 Coulomb's law and Newton's law of gravitation have the same *form*. Can Gauss's law be applied to gravitational fields as well as electric fields? If so, what modifications are needed?

25-3 By considering how the lines of force must look near a conducting surface, can you see why the charge density and electric field magnitude at the surface of an irregularly shaped solid conductor must be greatest in regions where the surface curves most sharply, and least in flat regions?

25-4 The electric field and the velocity field in a moving fluid are two examples of vector fields. Think of several other examples. There are also *scalar* fields, which associate a single number with each point in space. Temperature is an example; think of several others.

25-5 Consider the electric field caused by two point charges separated by some distance. Suppose there is a point where the field is zero; what does this tell you about the *signs* of the charges?

25-6 Does an electric charge experience a force due to the field that the charge itself produces?

25-7 A particle having electric charge and mass moves in an electric field. If it starts from rest, does it always move along the field line that passes through its starting point? Explain.

25-8 If the exponent 2 in the r^2 of Coulomb's law were 3 instead, would Gauss's law still be valid?

25-9 A student claimed that an appropriate unit for electric field magnitude is $1\,\text{J}\cdot\text{C}^{-1}\cdot\text{m}^{-1}$. Is this correct?

25-10 A certain region of space bounded by an imaginary closed surface contains no charge. Is the electric field always zero everywhere on the surface? If not, under what circumstances is it zero on the surface?

25-11 A student claimed that the electric field produced by a dipole is represented by field lines that cross each other. Is this correct? Is there any simple rule governing field lines for a superposition of fields due to point charges, if the field lines due to the separate charges are known? Do field lines *ever* cross?

25-12 Nineteenth-century physicists liked to give everything mechanical attributes. Faraday and his contemporaries thought of field lines as elastic strings that repelled each other and arranged themselves in equilibrium under the action of their elastic tension and mutual repulsion. Try this picture on several examples and decide whether it makes any sense. (These mechanical properties are of course now known to be completely fictitious.)

25-13 Are Coulomb's law and Gauss's law *completely* equivalent? Are there any situations in electrostatics where one is valid and the other isn't

25-14 The text states that, in an electrostatic field, every field line must start on a positive charge and terminate on a negative charge. But suppose the field is that of a single positive point charge. Then what?

25-15 Is the total (net) electric charge in the universe positive, negative, or zero?

Problems

25-1 A small object carrying a charge of -5×10^{-9} C experiences a downward force of 20×10^{-9} N when placed at a certain point in an electric field.

a) What is the electric field at the point?
b) What would be the magnitude and direction of the force acting on an electron placed at the point?

25-2 What must be the charge on a particle of mass 2 g for it to remain stationary in the laboratory when placed in a downward-directed electric field of 500 N·C^{-1}?

25-3 A uniform electric field exists in the region between two oppositely charged plane parallel plates. An electron is released from rest at the surface of the negatively charged plate and strikes the surface of the opposite plate, 2 cm distant from the first, in a time interval of 1.5×10^{-8} s.

a) Find the electric field.
b) Find the velocity of the electron when it strikes the second plate.

25-4 An electron is projected with an initial velocity $v_0 = 10^7$ m·s^{-1} into the uniform field between the parallel plates in Fig. 25-21. The direction of the field is vertically downward, and the field is zero except in the space between the plates. The electron enters the field at a point midway between the plates. If the electron just misses the upper plate as it emerges from the field, find the magnitude of the electric field.

Figure 25-21

25-5 An electron is projected into a uniform electric field of 5000 N·C^{-1}. The direction of the field is vertically upward. The initial velocity of the electron is 10^7 m·s^{-1}, at an angle of 30° above the horizontal.

a) Find the maximum distance the electron rises vertically above its initial elevation.
b) After what horizontal distance does the electron return to its original elevation?
c) Sketch the trajectory of the electron.

25-6 In a rectangular coordinate system a charge of 25×10^{-9} C is placed at the origin of coordinates, and a charge of -25×10^{-9} C is placed at the point $x = 6$ m, $y = 0$. What is the electric field at

a) $x = 3$ m, $y = 0$?
b) $x = 3$ m, $y = 4$ m?

25-7 A charge of 16×10^{-9} C is fixed at the origin of coordinates, a second charge of unknown magnitude is at $x = 3$ m, $y = 0$, and a third charge of 12×10^{-9} C is at $x = 6$ m, $y = 0$. What is the magnitude of the unknown charge if the resultant field at $x = 8$ m, $y = 0$ is 20.25 N·C^{-1} directed to the right?

25-8 In a rectangular coordinate system, two positive point charges of 10^{-8} C each are fixed at the points $x = +0.1$ m, $y = 0$, and $x = -0.1$ m, $y = 0$. Find the magnitude and direction of the electric field at the following points:

a) the origin;
b) $x = 0.2$ m, $y = 0$;
c) $x = 0.1$ m, $y = 0.15$ m;
d) $x = 0$, $y = 0.1$ m.

25-9 Same as Problem 25-8, except that one of the point charges is positive and the other negative.

25-10

a) What is the electric field of a gold nucleus, at a distance of 10^{-12} cm from the nucleus?
b) What is the electric field of a proton, at a distance of 5.28×10^{-9} cm from the proton?

25-11 A small sphere whose mass is 0.1 g carries a charge of 3×10^{-10} C and is attached to one end of a silk fiber 5 cm long. The other end of the fiber is attached to a large vertical conducting plate, which has a surface charge of 25×10^{-6} C·m^{-2} on each side. Find the angle the fiber makes with the vertical.

25-12 How many excess electrons must be added to an isolated spherical conductor 10 cm in diameter to produce a field of intensity 1300 N·C^{-1} just outside the surface?

25-13 What is the magnitude of an electric field in which the force on an electron is equal in magnitude to the weight of the electron?

25-14 Electric charge is distributed uniformly over a disk of radius a, with total charge Q. Find the electric field at a point on the axis of the disk, distance x from its center. [*Hint:* Divide the disk into concentric rings, use the result of Example 2, Sec. 25-2, to find the field due to each ring, and integrate to find the total field.]

25-15 In Example 3 of Sec. 25-2, suppose that the wire is not infinitely long but has total length $2r$, and that point P lies on the perpendicular bisector of the wire. What is the electric field at point P? Compare this with the field of an infinitely long wire.

25–16 Electric charge is uniformly distributed around a semicircle of radius a, with total charge Q. What is the electric field at the center of curvature?

25–17 The electric field in the region between a pair of oppositely charged plane parallel plates, each 100 cm^2 in area, is 10^4 N·C^{-1}. What is the charge on each plate? Neglect edge effects.

25–18 A wire is bent into a ring of radius R and given a charge q.

a) What is the magnitude of the electric field at the center of the ring?

b) Derive the expression for the electric field at a point on a line perpendicular to the plane of the ring and passing through its center, at a distance r from the center of the ring. What is the direction of the **E**-vector at points on this line?

c) Sketch a graph of the magnitude of **E** as a function of r, from $r = 0$ to $r = 2R$.

25–19 The electric field **E** in Fig. 25–22 is everywhere parallel to the x-axis. The field has the same magnitude at all points in any given plane perpendicular to the x-axis (parallel to the yz-plane), but the magnitude is different for various planes. That is, E_x depends on x but not on y and z, and E_y and E_z are zero. At points *in* the yz-plane, $E_x = 400$ N·C^{-1}. (The volume shown could be a section of a large insulating slab 1 m thick, with its faces parallel to the yz-plane and with a uniform volume charge distribution imbedded in it.)

a) What is the value of $\int E_\perp \, dA$ over surface I in the diagram?

b) What is the value of the surface integral of **E** over surface II?

c) There is a positive charge of 26.6×10^{-9} C within the volume. What is the magnitude and direction of **E** at the face opposite I?

25–20 Apply Gauss's law to the dotted gaussian surfaces in Fig. 25–19b to calculate the electric field between and outside the plates.

25–21 A small conducting sphere of radius a, mounted on an insulating handle and having a positive charge q, is inserted through a hole in the walls of a hollow conducting sphere of inner radius b and outer radius c. The hollow sphere is supported on an insulating stand and is initially uncharged, and the small sphere is placed at the center of the hollow sphere. Neglect any effect of the hole.

a) Show that the electric field at a point in the region between the spheres, at a distance r from the center, is equal to

$$E = \frac{1}{4\pi\epsilon_0}\frac{q}{r^2}.$$

b) What is the electric field at a point outside the hollow sphere?

c) Sketch a graph of the magnitude of **E** as a function of r, from $r = 0$ to $r = 2c$.

d) Represent the charge on the small sphere by four + signs. Sketch the field lines of the system, within a spherical volume of radius $2c$.

e) The small sphere is moved to a point near the inner wall of the hollow sphere. Sketch the field lines.

25–22

a) If the charge per unit length λ on the wire in Fig. 25–16 is finite, and the wire is infinitely long, the *total* charge on the wire is infinite. Explain why this infinite charge does not give rise to an infinite electrical field.

b) Draw a diagram showing an end view of an infinitely long charged wire, and the field lines in a plane perpendicular to the wire, far from either end. Explain, in terms of field lines, why the field decreases with $1/r$, although the field of a *point* charge decreases with $1/r^2$.

25–23 Prove that the electric field outside an infinitely long cylindrical conductor with a uniform surface charge is the same as if all the charge were on the axis.

25–24 A long coaxial cable consists of an inner cylindrical conductor of radius a and an outer coaxial cylinder of inner radius b and outer radius c. The outer cylinder is mounted on insulating supports and has no net charge. The inner cylinder has a uniform positive charge λ per unit length. Calculate the electric field

a) at any point between the cylinders, and

b) at any external point.

c) Sketch a graph of the magnitude of **E** as a function of the distance r from the axis of the cable, from $r = 0$ to $r = 2c$.

d) Find the charge per unit length on the inner surface of the outer cylinder, and on the outer surface.

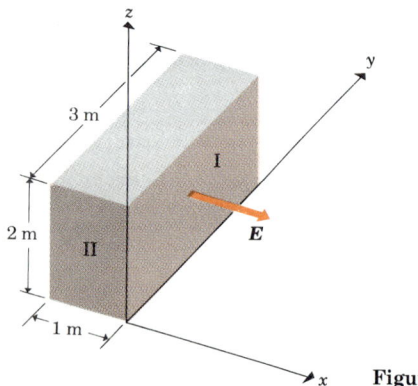

Figure 25–22

25-25 Suppose that positive charge is uniformly distributed throughout a spherical volume of radius R, the charge per unit volume being ρ.

a) Use Gauss's law to prove that the magnitude of the electric field inside the volume, at a distance r from the center, is

$$E = \frac{\rho r}{3\epsilon_0}.$$

b) What is the electric field at a point outside the spherical volume at a distance r from the center? Express your answer in terms of the total charge q within the spherical volume.

c) Compare the answers to (a) and (b) when $r = R$.

d) Sketch a graph of the magnitude of E as a function of r, from $r = 0$ to $r = 3R$.

25-26 Suppose that positive charge is uniformly distributed throughout a very long cylindrical volume of radius R, the charge per unit volume being ρ.

a) Derive the expression for the electric field inside the volume at a distance r from the axis of the cylinder, in terms of the charge density ρ.

b) What is the electric field at a point outside the volume, in terms of the charge per unit length λ in the cylinder?

c) Compare the answers to (a) and (b) when $r = R$.

d) Sketch a graph of the magnitude of E as a function of r, from $r = 0$ to $r = 3R$.

25-27 A conducting spherical shell of inner radius a and outer radius b has a positive point charge Q located at its center. The total charge on the shell is zero, and it is insulated from its surroundings.

a) Derive expressions for the electric field magnitude in terms of the distance r from the center, for the regions $r < a$, $a < r < b$, and $r > b$.

b) What is the surface charge density on the inner surface of the conducting shell?

c) Draw a sketch showing electric field lines and the location of all charges.

d) Draw a graph of E as a function of r.

25-28 The earth has net electric charge that causes a field at points near its surface of the order of 100 N·C⁻¹. If the earth is regarded as a conducting sphere of radius 6.38×10^6 m, what is the magnitude of its charge?

26 POTENTIAL

26-1 Electrical potential energy

In Chapter 6 the concepts of *work* and *energy* were introduced, and we saw that they provide a very important method of analysis for many mechanical problems. Energy plays an equally fundamental role in electricity and magnetism; in this chapter we shall apply work and energy considerations to the electric field. When a charged particle moves in an electric field, the field does *work* on the particle. We shall see that the work can always be expressed in terms of a potential energy, which in turn is associated with a new concept called *electrical potential* or simply *potential*.

In Sec. 6-6 the concept of a *conservative force field* was introduced; the reader would do well to review that discussion. Whenever the work done on a body undergoing a displacement can be expressed in terms of a potential energy function, the corresponding force is said to be *conservative*. If the potential energy function U has the value U_a at point a and the value U_b at point b, then the work $W_{a \to b}$ done by the force during any displacement from a to b, along any path, is given by

$$W_{a \to b} = U_a - U_b. \tag{26-1}$$

That is, the work done on the body equals its *loss* in potential energy. In order for this statement to have meaning, this work must be the same for all possible paths from a to b. If all the forces acting on the body are conservative, then the total mechanical energy of the body, kinetic and potential, is *conserved;* as we have seen, this principle simplifies the analysis of many mechanical systems.

We shall now show that the force on a charged particle in an electric field, due to other charges at rest, is a conservative force field. To begin, we consider only the interaction of two point charges. Specifically, consider a particle with a positive charge q' moving in the field produced by a stationary point charge q, as shown in Fig. 26-1. We shall first calculate the work done on the charge q' during a displacement along a radial line from point a to point b.

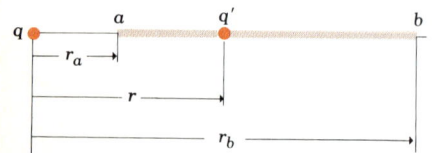

26-1 Charge q' moves along a straight line extending radially from charge q. As it moves from a to b, the distance varies from r_a to r_b.

The magnitude of the force on q' is given by Coulomb's law:

$$F = \frac{1}{4\pi\epsilon_0} \frac{qq'}{r^2}. \tag{26-2}$$

The work done by this force on q' as it moves from r_a to r_b is given by

$$W_{a \to b} = \int_{r_a}^{r_b} F\, dr = \int_{r_a}^{r_b} \frac{1}{4\pi\epsilon_0} \frac{qq'}{r^2}\, dr$$

$$= \frac{qq'}{4\pi\epsilon_0}\left(\frac{1}{r_a} - \frac{1}{r_b}\right). \tag{26-3}$$

Thus the work for this particular path depends only on the endpoints. We have not yet proved that the work is the same for *all possible* paths from a to b. To do this, we consider a more general displacement, in which a and b do *not* lie on the same radial line, as in Fig. 26-2. To find the work in the displacement from point a to point b we must now use the more general definition of work given by Eq. (6-9):

$$W = \int_a^b F \cos\theta\, dl = \int_a^b \mathbf{F} \cdot d\mathbf{l}. \tag{26-4}$$

But from the figure, $\cos\theta\, dl = dr$. That is, the work done during a small displacement $d\mathbf{l}$ depends only on the change dr in the distance r between the charges, which is the *radial component* of the displacement. In any displacement in which r does not change, no work is done because $\mathbf{F}$ and $d\mathbf{l}$ are perpendicular, $\cos\theta = 0$, and $\mathbf{F}\cdot d\mathbf{l} = 0$.

Thus Eq. (26-3) is equal to the work even for this more general displacement. This result shows that the work done on q' by the $\mathbf{E}$ field depends only on the initial and final distances r_a and r_b and not on the

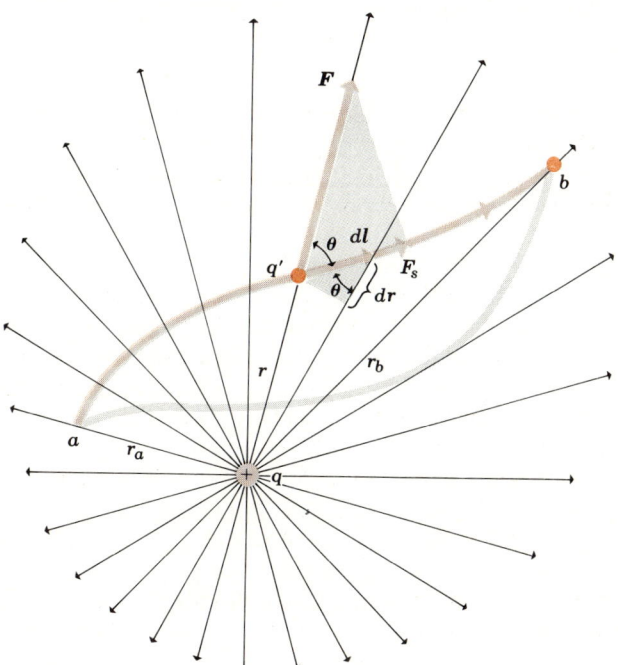

26-2 The work done by the electric-field force on charge q' depends only on the distances r_a and r_b.

path connecting these points. The work is the same for *any* path between these points, or between any pair of points at distances r_a and r_b from the charge q. Also, if the test charge is returned from b to a along any path, the work done by the **E** field is the *negative* of that done in the displacement from a to b. Thus, the total work done during a *loop* displacement returning to the starting point is always zero.

Comparing Eqs. (26-1) and (26-3), we see that the term $qq'/4\pi\epsilon_0 r_a$ is the potential energy U_a when q' is at point a, at distance r_a from q, and $qq'/4\pi\epsilon_0 r_b$ the potential energy U_b when it is at point b, at distance r_b from q. Thus the potential energy U of the test charge q' at *any* distance r from charge q is given by

$$U = \frac{1}{4\pi\epsilon_0} \frac{qq'}{r}. \tag{26-5}$$

If the field in which charge q' moves is not that of a single point charge but is due to a more general charge distribution, we may always divide the charge distribution into small elements and treat each element as a point charge. Then, since the total field is the vector sum of the fields due to the individual elements, and since the total work on q' is the sum of the contributions from the individual charge elements, we may conclude that *every* electric field due to a static charge distribution is a conservative force field. Furthermore, the potential energy of a test charge q' at point a in Fig. 26-3, due to a collection of charges q_1, q_2, q_3, etc., at distances r_1, r_2, r_3, etc., from the test charge q', is given by

$$U = \frac{q'}{4\pi\epsilon_0}\left(\frac{q_1}{r_1} + \frac{q_2}{r_2} + \frac{q_3}{r_3} + \cdots\right),$$

$$U = \frac{q'}{4\pi\epsilon_0} \sum \frac{q_i}{r_i}. \tag{26-6}$$

At a second point b, the potential energy is given by the same expression except that $r_1, r_2, \ldots$ now represent the distances from the respective charges to point b. The work of the electric force in moving the test charge from a to b along any path is equal to the difference $U_a - U_b$ between its potential energies at a and at b.

In the discussion of potential energy in Chapter 6, it was observed that it is always possible to add an arbitrary constant C to a potential-energy function U, since in Eq. (26-1) adding the same constant C to both U_a and U_b does not change the physically significant quantity, namely the difference $U_a - U_b$. Thus we are always free to choose the constant C so that U is zero at some convenient reference position. In Chapter 6, the potential energy of a body in a uniform gravitational field could be taken to be zero at some convenient reference level, often the surface of the earth. When the body is above this reference level, its potential energy is positive; when below, negative.

In Eqs. (26-5) and (26-6), the reference position for electrical potential energy, the position at which $U = 0$, has been chosen implicitly to be that at which all the distances $r_1, r_2, \ldots$ are *infinite*. That is, the potential energy of the test charge is zero when it is very far removed from all the charges setting up the field. This is the most convenient reference level for most electrostatic problems. When dealing with electrical *circuits*, other reference levels are more convenient, which simply means

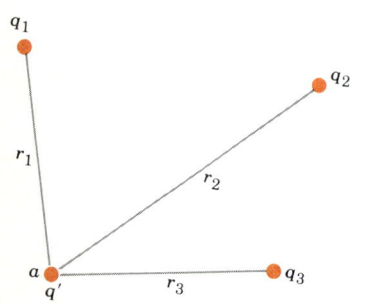

26-3 Potential energy of a charge q' at point a depends on charges $q_1, q_2,$ and q_3, and on their distances r_1, r_2, r_3 from point a.

that a constant term is added to the potential energy. This is not significant, since it is only *differences* in potential energy that are of practical significance or even measurable.

From the above definitions, *the potential energy of a test charge at any point in an electric field is equal to the work of the electric force when the test charge is brought from the point in question to a zero reference level, often taken at infinity.*

26-2 Potential

Instead of dealing directly with the potential energy U of a charged particle, it is useful to introduce the more general concept of *potential energy per unit charge*. This quantity is called the *potential*, and *the potential at any point of an electrostatic field is defined as the potential energy per unit charge* at the point. Potential is represented by the letter V:

$$V = \frac{U}{q'}, \quad \text{or} \quad U = q'V. \qquad (26\text{-}7)$$

Potential energy and charge are both *scalars*, so potential is a scalar quantity. Its basic SI unit is 1 *joule per coulomb* ($1 \text{ J} \cdot \text{C}^{-1}$). For brevity, a potential of $1 \text{ J} \cdot \text{C}^{-1}$ is called 1 *volt* (1 V). The unit is named in honor of the Italian scientist Alessandro Volta (1745–1827). Commonly used multiples of the volt are the kilovolt ($1 \text{ kV} = 10^3 \text{ V}$), the megavolt ($1 \text{ MV} = 10^6 \text{ volts}$), and the gigavolt ($1 \text{ GV} = 10^9 \text{ V}$). Submultiples include the millivolt ($1 \text{ mV} = 10^{-3} \text{ V}$) and the microvolt ($1 \text{ } \mu\text{V} = 10^{-6} \text{ V}$).

To put Eq. (26-1) on a "work per unit charge" basis, we divide both sides by q', obtaining

$$\frac{W_{a \rightarrow b}}{q'} = \frac{U_a}{q'} - \frac{U_b}{q'} = V_a - V_b, \qquad (26\text{-}8)$$

where $V_a = U_a/q'$ is the potential energy per unit charge at point a, and similarly for V_b. V_a and V_b are called the *potential at point a* and *potential at point b*, respectively. From Eq. (26-6), the potential V at a point due to an arbitrary collection of point charges is given by

$$V = \frac{U}{q'} = \frac{1}{4\pi\epsilon_0} \sum \frac{q_i}{r_i} \qquad (26\text{-}9)$$

The difference $V_a - V_b$ is called the *potential of a with respect to b*, and is often abbreviated V_{ab}. It should be noted that potential, like electric field, is independent of the test charge q' used to define it.

The potential due to a collection of point charges is usually obtained most easily by use of Eq. (26-9), but in some problems where the $\boldsymbol{E}$ field is known it is easier to work with it directly. The force $\boldsymbol{F}$ on the test charge q' can be written as $\boldsymbol{F} = q'\boldsymbol{E}$. When this relation is used in Eq. (26-4) and the result combined with Eq. (26-8), one obtains the useful result

$$V_{ab} = V_a - V_b = \int_a^b \boldsymbol{E} \cdot d\boldsymbol{l} = \int_a^b E \cos \theta \, dl. \qquad (26\text{-}10)$$

Either integral on the right side is called the *line integral* of $\boldsymbol{E}$. It represents the conceptual process of dividing a path into small elements $d\mathbf{l}$, multiplying each magnitude dl by the component of $\mathbf{E}$ parallel to $d\mathbf{l}$ at that point, and summing the results for the entire path. The concept of line integral is very useful in formulating principles of physics, particularly in electricity and magnetism.

Equation (26–10) states that when a positive test charge moves from a region of high potential to one of lower potential (that is, $V_b < V_a$), the electric field does positive work on it. Thus a positive charge tends to "fall" from a high-potential region to a lower-potential one. The opposite is true for a negative charge.

A *voltmeter* is an instrument that measures the potential *difference* between the points to which its terminals are connected. The principle of the common type of moving-coil voltmeter will be described later. There are also much more sensitive potential-measuring devices that make use of electronic amplification, and devices that can measure a potential difference of $1\,\mu V$ are used routinely.

Example 1 A particle having a charge $q = 3 \times 10^{-9}\,\text{C}$ moves from point a to point b along a straight line, a total distance $d = 0.5$ m. The electric field is uniform along this line, in the direction from a to b, with magnitude $E = 200\,\text{N}\cdot\text{C}^{-1}$. Determine the force on q, the work done on it by the field, and the potential difference $V_a - V_b$.

Solution The force is in the same direction as the electric field, and its magnitude is given by

$$F = qE = (3 \times 10^{-9}\,\text{C})(200\,\text{N}\cdot\text{C}^{-1}) = 600 \times 10^{-9}\,\text{N}.$$

The work done by this force is

$$W = Fd = (600 \times 10^{-9}\,\text{N})(0.5\,\text{m}) = 300 \times 10^{-9}\,\text{J}.$$

The potential difference is the work per unit charge, which is

$$V_a - V_b = \frac{W}{q} = \frac{300 \times 10^{-9}\,\text{J}}{3 \times 10^{-9}\,\text{C}} = 100\,\text{J}\cdot\text{C}^{-1} = 100\,\text{V}.$$

Alternatively, since E is force per unit charge, the work per unit charge is obtained by multiplying E by the distance d:

$$V_a - V_b = Ed = (200\,\text{N}\cdot\text{C}^{-1})(0.5\,\text{m}) = 100\,\text{J}\cdot\text{C}^{-1} = 100\,\text{V}.$$

Example 2 Point charges of $+12 \times 10^{-9}\,\text{C}$ and $-12 \times 10^{-9}\,\text{C}$ are placed 10 cm apart, as in Fig. 26–4. Compute the potentials at points a, b, and c.

Solution We must evaluate the *algebraic* sum $(1/4\pi\epsilon_0)\,\Sigma(q_i/r_i)$ at each point. At point a, the potential due to the positive charge is

$$(9 \times 10^9\,\text{N}\cdot\text{m}^2\cdot\text{C}^{-2})\frac{12 \times 10^{-9}\,\text{C}}{0.06\,\text{m}} = 1800\,\text{N}\cdot\text{m}\cdot\text{C}^{-1}$$

$$= 1800\,\text{J}\cdot\text{C}^{-1} = 1800\,\text{V},$$

and the potential due to the negative charge is

$$(9 \times 10^9\,\text{N}\cdot\text{m}^2\cdot\text{C}^{-2})\frac{-12 \times 10^{-9}\,\text{C}}{0.04\,\text{m}} = -2700\,\text{J}\cdot\text{C}^{-1} = -2700\,\text{V}.$$

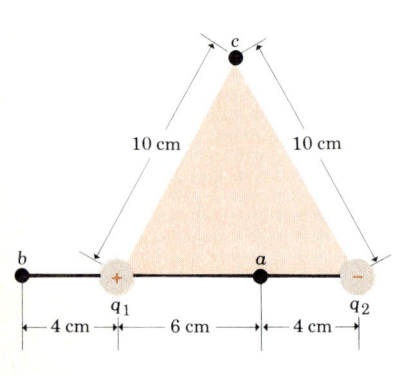

26–4

Hence
$$V_a = 1800 \text{ V} - 2700 \text{ V} = -900 \text{ V}$$
$$= -900 \text{ J} \cdot \text{C}^{-1}.$$

At point b, the potential due to the positive charge is $+2700$ V and that due to the negative charge is -770 V. Hence,
$$V_b = 2700 \text{ V} - 770 \text{ V} = 1930 \text{ V}$$
$$= 1930 \text{ J} \cdot \text{C}^{-1}.$$

At point c the potential is
$$V_c = 1080 \text{ V} - 1080 \text{ V} = 0.$$

Example 3 Compute the potential energy of a point charge of $+4 \times 10^{-9}$ C if placed at points a, b, and c in Fig. 26-4.

Solution First,
$$U = qV.$$
Hence at point a,
$$U = qV_a = (4 \times 10^{-9} \text{ C})(-900 \text{ J} \cdot \text{C}^{-1})$$
$$= -36 \times 10^{-7} \text{ J}.$$

At point b,
$$U = qV_b = (4 \times 10^{-9} \text{ C})(1930 \text{ J} \cdot \text{C}^{-1})$$
$$= 77 \times 10^{-7} \text{ J}.$$

At point c,
$$U = qV_c = 0.$$
(All relative to a point at infinity.)

Example 4 In Fig. 26-5, a particle having mass $m = 5$ g and charge $q' = 2 \times 10^{-9}$ C starts from rest at point a and moves in a straight line to point b. What is its speed v at point b?

Solution Conservation of energy gives
$$K_a + U_a = K_b + U_b.$$
For this situation, $K_a = 0$ and $K_b = \frac{1}{2}mv^2$. The potential energies are given in terms of the potentials by Eq. (26-6): $U_a = q'V_a$ and $U_b = q'V_b$. Thus the energy equation becomes
$$0 + q'V_a = \frac{1}{2}mv^2 + q'V_b.$$
When solved for v, this yields
$$v = \sqrt{\frac{2q'(V_a - V_b)}{m}}.$$
We obtain the potentials just as we did in the preceding examples:
$$V_a = (9.0 \times 10^9 \text{ N} \cdot \text{m}^2 \cdot \text{C}^{-2})\left(\frac{3 \times 10^{-9} \text{ C}}{0.01 \text{ m}} + \frac{-3 \times 10^{-9} \text{ C}}{0.02 \text{ m}}\right) = 1350 \text{ V},$$
$$V_b = (9.0 \times 10^9 \text{ N} \cdot \text{m}^2 \cdot \text{C}^{-2})\left(\frac{3 \times 10^{-9} \text{ C}}{0.02 \text{ m}} + \frac{-3 \times 10^{-9} \text{ C}}{0.01 \text{ m}}\right) = -1350 \text{ V}.$$

26-5

Finally,

$$v = \sqrt{\frac{2(2 \times 10^{-9}\,\text{C})(2700\,\text{V})}{5 \times 10^{-3}\,\text{kg}}} = 4.65 \times 10^{-2}\,\text{m}\cdot\text{s}^{-1} = 4.65\,\text{cm}\cdot\text{s}^{-1}.$$

Consistency of units may be checked by noting that $1\,\text{V} = 1\,\text{J}\cdot\text{C}^{-1}$, and thus the numerator under the radical has units of J or $\text{kg}\cdot\text{m}^2\cdot\text{s}^{-2}$.

26-3 Calculation of potential differences

In principle the potential difference between any two points can be calculated by using Eq. (26-9) to find the potential at each point, but in many cases, especially where the electric field is easily obtained, it is easier to use Eq. (26-10), calculating from the known electric field the work done on a test charge during a displacement from one point to the other. In some problems a combination of these two approaches is useful. The following problems illustrate these remarks.

Example 1 *Charged spherical conductor.* We consider first a solid conducting sphere of radius R, with total charge q. From Gauss's law, as discussed in Chapter 25, we conclude that, at all points *outside* the sphere, the field is the same as that of a point charge q at the center of the sphere. *Inside* the sphere the field is zero everywhere; otherwise charge would move within the sphere.

As mentioned above, it is convenient to take the reference level of potential (the point where $V = 0$) at a very large distance from all charges. Since the field at any point r greater than R is the same as for a point charge, the work on a test charge moving from a finite value of r to infinity is also the same as for a point charge, so the potential V at a radial distance r, relative to a point at infinity, is

$$V = \frac{1}{4\pi\epsilon_0}\frac{q}{r}. \tag{26-11}$$

The potential is positive if q is positive, negative if q is negative.

Equation (26-11) applies to the field of a charged spherical conductor only when r is greater than or equal to the radius R of the sphere. The potential *at* the surface is

$$V = \frac{1}{4\pi\epsilon_0}\frac{q}{R}. \tag{26-12}$$

Inside the sphere the field is zero everywhere, and no work is done on a test charge displaced from any point to any other in this region. Thus the potential is the same at all points inside the sphere, and is equal to its value $q/4\pi\epsilon_0 R$ at the surface. The field and potential are shown as functions of r in Fig. 26-6.

The electric field E at the surface has a magnitude

$$E = \frac{1}{4\pi\epsilon_0}\frac{q}{R^2}. \tag{26-13}$$

The maximum potential to which a conductor in air can be raised is limited by the fact that air molecules become ionized, and hence the air becomes a conductor, at an electric field of about $3 \times 10^6\,\text{N}\cdot\text{C}^{-1}$. Com-

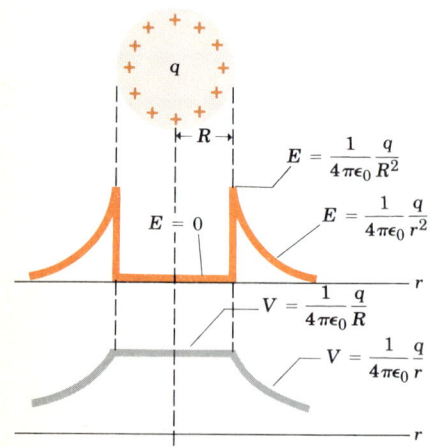

26-6 Electric field magnitude E and potential V at points inside and outside a charged spherical conductor.

paring Eqs. (26-12) and (26-13), we note that at the surface the field and potential are related by $V = ER$. Thus if E_m represents the upper limit of electric field, known as the *dielectric strength,* the maximum potential to which a spherical conductor can be raised is

$$V_m = RE_m.$$

For a sphere 1 cm in radius, in air,

$$V_m = (10^{-2} \text{ m})(3 \times 10^6 \text{ N} \cdot \text{C}^{-1}) = 30{,}000 \text{ V},$$

and no amount of "charging" could raise the potential of a sphere of this size, in air, higher than about 30,000 V. It is this fact that necessitates the use of large spherical terminals on high-voltage machines. If we make $R = 2$ m, then

$$V_m = (2 \text{ m})(3 \times 10^6 \text{ N} \cdot \text{C}^{-1}) = 6 \times 10^6 \text{ V} = 6 \text{ MV}.$$

At the other extreme is the effect produced by sharp points, a "point" being a surface of very *small* radius of curvature. Since the maximum potential is proportional to the radius, even relatively small potentials applied to sharp points in air will produce sufficiently high fields just outside the point to result in ionization of the surrounding air.

Example 2 *Parallel plates.* The electric field between oppositely charged parallel plates, as derived in Sec. 25-5, is

$$E = \frac{\sigma}{\epsilon_0}, \quad (26\text{-}14)$$

where σ is the magnitude of the surface charge density (charge per unit area) on either plate. To obtain the potential difference between points a and x in Fig. 26-7, we note that the force on a test charge q' is $F = Eq'$, and the work done during a displacement from a to x is

$$W_{ax} = Fx = Eq'x.$$

Thus, the potential difference, the work per unit charge, also called the potential at a with respect to x, is

$$V_a - V_x = \frac{W_{ax}}{q'} = Ex,$$

or

$$V_x = V_a - Ex = V_a - \frac{\sigma}{\epsilon_0}x. \quad (26\text{-}15)$$

The potential therefore decreases *linearly* with x. At point b, where $x = l$ and $V_x = V_b$,

$$V_b = V_a - El,$$

and hence

$$E = \frac{V_a - V_b}{l} = \frac{V_{ab}}{l}. \quad (26\text{-}16)$$

That is, *the electric field equals the potential difference between the plates divided by the distance between them.* This is a more useful ex-

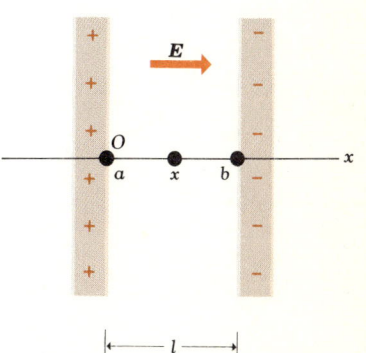

26-7

pression for E than Eq. (26–14) is, because the potential difference V_{ab} can readily be measured with a voltmeter, while there are no instruments that read surface density of charge directly. It must be understood, however, that Eqs. (26–15) and (26–16) are valid only when E is *uniform*.

Equation (26–16) also shows that the unit of electric field can be expressed as 1 *volt per meter* (1 V·m^{-1}), as well as 1 N·C^{-1}. In practice, the volt per meter is most commonly used as the unit of E.

Example 3 A simple type of vacuum tube known as a *diode* consists essentially of two electrodes within a highly evacuated enclosure. One electrode, the cathode, is maintained at a high temperature and emits electrons from its surface. A potential difference of a few hundred volts is maintained between the cathode and the other electrode, known as the *anode,* with the anode at the higher potential. Suppose that, in a certain diode, the anode potential V_a is 250 V above that (V_c) of the cathode, and an electron is emitted from the cathode with no initial velocity. What is its velocity when it reaches the anode?

Solution Let V_c and V_a represent the cathode and anode potentials, respectively, and let the charge on the electron be $-e$. Since this charge is negative, the electron goes from the cathode, where the potential is lower, to the anode, where the potential is higher. Hence the work done on the elctron by the field is its charge ($-e$), where e is the (positive) magnitude of the electron charge, multiplied by the (negative) potential difference $V_c - V_a = V_{ca}$. This work in turn equals the change of kinetic energy of the electron.

Letting v_c and v_a be the speeds at cathode and anode, respectively, we have

$$(-e)(V_c - V_a) = e(V_a - V_c) = eV_{ac} = \tfrac{1}{2}mv_a^2 - \tfrac{1}{2}mv_c^2.$$

Since $v_c = 0$,

$$v_a = \sqrt{2eV_{ac}/m}$$
$$= \sqrt{\frac{2(1.6 \times 10^{-19}\text{ C})(250\text{ J·C}^{-1})}{9.1 \times 10^{-31}\text{ kg}}}$$
$$= 9.4 \times 10^6 (\text{J·kg}^{-1})^{1/2} = 9.4 \times 10^6 \text{ m·s}^{-1}.$$

Note that the shape or separation of the electrodes need not be known. The final velocity depends only on the difference of potential between cathode and anode. Of course, the time of transit from cathode to anode depends on the geometry of the tube.

Example 4 *Line charge and charged conducting cylinder.* The field of a line charge, and the field outside a charged conducting cylinder, are both given by

$$E = \frac{1}{2\pi\epsilon_0} \frac{\lambda}{r},$$

where λ is the charge per unit length.

The potential difference between any two points a and b at radial distances r_a and r_b is

$$V_a - V_b = \frac{\lambda}{2\pi\epsilon_0} \int_{r_a}^{r_b} \frac{dr}{r} = \frac{\lambda}{2\pi\epsilon_0} \ln \frac{r_b}{r_a}. \tag{26-17}$$

If we take point b at infinity and set $V_b = 0$, we find for the potential V_a,

$$V_a = \frac{\lambda}{2\pi\epsilon_0} \ln \frac{\infty}{r_a} = \infty.$$

Hence a reference point at infinity is not suitable for this field! We can, however, set $V = 0$ at some arbitrary radius r_0. Then at any radius r,

$$V = \frac{\lambda}{2\pi\epsilon_0} \ln \frac{r_0}{r}. \qquad (26\text{-}18)$$

Equations (26-17) and (26-18) give the potential in the field of a cylinder only for values of r equal to or greater than the radius R of the cylinder. If r_0 is taken as the cylinder radius R, so that the potential of the cylinder is considered zero, the potential at any external point, relative to that of the cylinder, is

$$V = \frac{\lambda}{2\pi\epsilon_0} \ln \frac{R}{r}. \qquad (26\text{-}19)$$

26-4 Equipotential surfaces

The potential distribution in an electric field may be represented graphically by *equipotential surfaces*. An equipotential surface is a surface such that the potential has the same value at all points on the surface. While an equipotential surface may be constructed through every point of an electric field, it is customary to show only a few of the equipotentials in a diagram.

Since the potential energy of a charged body is the same at all points of a given equipotential surface, it follows that no (electrical) work is needed to move a charged body over such a surface. Hence, the equipotential surface through any point must be at right angles to the direction of the field at that point. If this were not so, the field would have a component lying in the surface and work would have to be done against electrical forces to move a charge in the direction of this component. The field lines and the equipotential surfaces thus form a mutually perpendicular network. In general, the field lines of a field are curves and the equipotentials are curved surfaces. For the special case of a uniform field, where the field lines are straight and parallel, the equipotentials are parallel planes perpendicular to the field lines.

Figure 26-8 shows several arrangements of charges. The field lines are represented by colored lines, and cross sections of the equipotential surfaces as black lines. The actual field is of course three-dimensional. At each crossing of an equipotential and a field line, the two are perpendicular.

When charges at rest reside on the surface of a conductor, the electric field just outside the conductor must be everywhere perpendicular to the surface. To prove this, we imagine transporting a test charge q' around the loop $abcda$ in Fig. 26-9a. The segments bc and da (and the associated work) may be made arbitrarily small, and no work is done in the segment cd because, as already shown, the field is zero everywhere inside the conductor. If the field just outside the conductor has a component $E_\parallel$ parallel to the surface, this component does work equal to $q'E_\parallel l$.

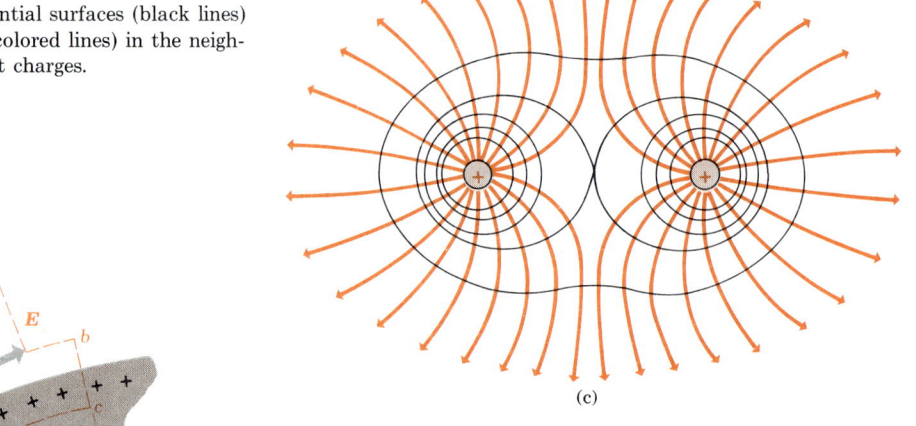

26-8 Equipotential surfaces (black lines) and field lines (colored lines) in the neighborhood of point charges.

26-9 (a) Construction for finding the direction of E outside the surface of a conductor. (b) Field lines always meet charged conducting surfaces at right angles.

But then the net work done in the displacement ab is different from that in the displacement $adcb$, and the force field is not conservative. To avoid this contradiction, we must conclude that there *cannot* be a component of E parallel to the surface, and that E is therefore perpendicular to the surface. It also follows that, when all charges are at rest, *the surface of a conductor is always an equipotential surface*. Figure 26-9b illustrates these conclusions. Field lines (color) and equipotentials (black) are shown.

26-5 Potential gradient

The electric field E and the potential V in a region of space can always be regarded as functions of the coordinates of the point under consideration. These functions are of course closely related, and Eq. (26–10) is one form of this relation. A different form which is often useful relates the components of E to *derivatives* of V.

In Eq. (26–10), $V_{ab} = V_a - V_b$ is the change of potential when the point moves from b to a. This may be written as

$$V_{ab} = \int_b^a dV = -\int_a^b dV,$$

where dV is the infinitesimal change of potential accompanying a small displacement dl. As usual, $E \cos \theta$ is the component of E parallel to dl.

For brevity we denote this by $E_\parallel$; then, using the above expression, for V_{ab}, we may rewrite Eq. (26-10) as

$$-\int_a^b dV = \int_a^b E_\parallel \, dl.$$

Since these two integrals are equal for any pair of limits a and b, the *integrands* must be equal. Thus for an *infinitesimal* displacement dl,

$$-dV = E_\parallel \, dl,$$

or (26-20)

$$E_\parallel = -\frac{dV}{dl}.$$

The derivative dV/dl is the rate of change of V for a displacement in the direction of dl. In particular, if dl is parallel to the x-axis, then the component of $\boldsymbol{E}$ parallel to dl is just the x-component of $\boldsymbol{E}$, that is, E_x. Thus $E_x = -dV/dx$. Because V is in general also a function of y and z, the appropriate notation for a derivative in which only x varies is $\partial V/\partial x$. The y- and z-components of $\boldsymbol{E}$ are related to the corresponding derivatives of V in the same way, so we have

$$E_x = -\frac{\partial V}{\partial x},$$

$$E_y = -\frac{\partial V}{\partial y}, \qquad (26\text{-}21)$$

$$E_z = -\frac{\partial V}{\partial z}.$$

In terms of unit vectors, $\boldsymbol{E}$ may be written as

$$\boldsymbol{E} = -\left(\boldsymbol{i}\frac{\partial V}{\partial x} + \boldsymbol{j}\frac{\partial V}{\partial y} + \boldsymbol{k}\frac{\partial V}{\partial z}\right)$$

(26-22)

$$= -\left(\boldsymbol{i}\frac{\partial}{\partial x} + \boldsymbol{j}\frac{\partial}{\partial y} + \boldsymbol{k}\frac{\partial}{\partial z}\right)V.$$

In vector notation the operation

$$\boldsymbol{i}\frac{\partial}{\partial x} + \boldsymbol{j}\frac{\partial}{\partial y} + \boldsymbol{k}\frac{\partial}{\partial z}$$

is called the *gradient* and is denoted by the symbol $\boldsymbol{\nabla}$. Thus in vector notation, Eqs. (26-21) are summarized compactly as

$$\boldsymbol{E} = -\boldsymbol{\nabla} V. \qquad (26\text{-}23)$$

This is read "$\boldsymbol{E}$ is the negative of the gradient of V."

Clearly, the units of potential gradient (volts per meter) must be the same as those of electric field (newtons per coulomb). This may be verified by recalling that potential is potential energy per unit charge and that $1\text{ V} = 1\text{ J}\cdot\text{C}^{-1}$. Then

$$1\text{ V}\cdot\text{m}^{-1} = 1\text{ J}\cdot\text{C}^{-1}\cdot\text{m}^{-1} = 1\text{ N}\cdot\text{m}\cdot\text{C}^{-1}\cdot\text{m}^{-1}$$
$$= 1\text{ N}\cdot\text{C}^{-1}.$$

Example 1 We have shown that the potential at a radial distance r from a point charge q is

$$V = \frac{1}{4\pi\epsilon_0} \frac{q}{r}.$$

By symmetry, the electric field is in the radial direction, so

$$E = E_r = -\frac{dV}{dr} = -\frac{d}{dr}\left(\frac{1}{4\pi\epsilon_0} \frac{q}{r}\right) = \frac{1}{4\pi\epsilon_0} \frac{q}{r^2},$$

in agreement with Coulomb's law.

Example 2 The potential outside a charged conducting cylinder of radius R and charge per unit length λ (Example 4 in Sec. 26-3) is

$$V = \frac{\lambda}{2\pi\epsilon_0} \ln\frac{R}{r} = \frac{\lambda}{2\pi\epsilon_0}(\ln R - \ln r).$$

The electric field is radial, and its magnitude is given by

$$E = -\frac{dV}{dr} = \frac{\lambda}{2\pi\epsilon_0 r},$$

in agreement with our previous result.

26-6 The Millikan oil-drop experiment

We have now developed the theory of electrostatics to a point where one of the classical physical experiments of all time can be described. In a brilliant series of investigations carried out at the University of Chicago in the period 1909–1913, Robert Andrews Millikan not only demonstrated conclusively the discrete nature of electric charge, but actually measured the charge of an individual electron.

Millikan's apparatus is shown schematically in Fig. 26–10a. Two accurately parallel horizontal metal plates A and B, are insulated from each other and separated by a few millimeters. Oil is sprayed in fine droplets from an atomizer above the upper plate and a few of the droplets are allowed to fall through a small hole in this plate. A beam of light is directed horizontally between the plates, and a telescope is set up with its axis at right angles to the light beam. The oil drops, illuminated by the light beam and viewed through the telescope, appear like tiny bright stars, falling slowly with a terminal velocity determined by their weight and by the viscous air-resistance force opposing their motion.

It is found that some of the oil droplets are electrically charged, presumably because of frictional effects. Charges can also be given the drops if the air in the apparatus is ionized by x-rays or a bit of radioactive material. Some of the electrons or ions then collide with the drops and stick to them. The drops are usually negatively charged, but occasionally one with a positive charge is found.

The simplest method, in principle, for measuring the charge on a drop is as follows. Suppose a drop has a negative charge and the plates are maintained at a potential difference such that a downward electric field of intensity $E\,(= V_{AB}/l)$ is set up between them. The forces on the drop are then its weight mg and the upward force qE. By adjusting the

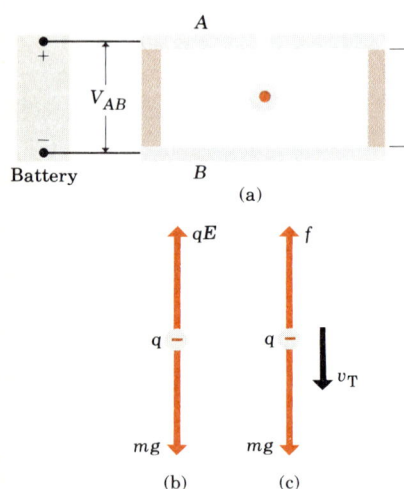

26-10 (a) Schematic diagram of Millikan apparatus. (b) Forces on a drop at rest. (c) Forces on a drop falling with its terminal velocity v_T.

field E, qE can be made just equal to mg, so that the drop remains at rest, as indicated in Fig. 26-10b. Under these circumstances,

$$q = \frac{mg}{E}.$$

The mass of the drop equals the product of its density ρ and its volume, $4\pi r^3/3$, and $E = V_{AB}/l$, so

$$q = \frac{4\pi}{3} \frac{\rho r^3 g l}{V_{AB}}. \qquad (26\text{-}24)$$

All the quantities on the right are readily measurable with the exception of the drop radius r, which is of the order of 10^{-5} cm and is much too small to be measured directly. It can be calculated, however, by cutting off the electric field and measuring the terminal velocity v_T of the drop as it falls through a known distance d defined by reference lines in the ocular of the telescope.

The terminal velocity is that at which the weight mg is just balanced by the viscous force f. The viscous force on a sphere of radius r, moving with a velocity v through a fluid of viscosity η, is given by Stokes' law, discussed in Sec. 13-7:

$$f = 6\pi \eta r v.$$

If Stokes' law applies, and the drop is falling with its terminal velocity v_T,

$$mg = f,$$
$$\frac{4}{3}\pi r^3 \rho g = 6\pi \eta r v_T,$$

and

$$r = 3\sqrt{\eta v_T/2\rho g}.$$

When this expression for r is inserted in Eq. (26-24), we have

$$q = 18\pi \frac{l}{V_{AB}} \sqrt{\frac{\eta^3 v_T^3}{2\rho g}}, \qquad (26\text{-}25)$$

which expresses the charge q in terms of measurable quantities.

In actual practice this procedure is modified somewhat. To correct for the buoyant force of the air through which the drop falls, the density ρ of the oil should be replaced by $(\rho - \rho_g)$, where ρ_g is the density of air. A correction to Stokes' law is also required, because air is not a continuous fluid but a collection of molecules separated by distances that are of the same order of magnitude as the dimensions of the drops.

Millikan and his co-workers measured the charges of thousands of drops, and found that, within the limits of their experimental error, every drop had a charge equal to some small integer multiple of a basic charge e. That is, drops were observed with charges of e, $2e$, $3e$, etc., but never with such values as $0.76e$ or $2.49e$. The evidence is conclusive that electric charge is not something that can be divided indefinitely, but that it exists in nature only in units of magnitude e. When a drop is observed with charge e, we conclude it has acquired one extra electron; if its charge is $2e$, it has two extra electrons, and so on.

As stated in Sec. 24-5, the best experimental value of the charge e is

$$e = 1.602192 \times 10^{-19} \text{ C} \simeq 1.60 \times 10^{-19} \text{ C}.$$

The quark theory of fundamental particle structure, discussed in Sec. 47-11, incorporates particles called *quarks* having fractional charges $\pm e/3$ and $\pm 2e/3$. There are conflicting views among physicists as to whether properties (such as charge) of an individual quark should be directly observable. In 1979 the American physicist William Fairbank reported an experiment in which fractionally charged particles were observed. His results have not been duplicated in other laboratories, and their significance is still uncertain.

26-7 The electronvolt

The change in potential energy of a particle having a charge q, when it moves from a point where the potential is V_a to a point where the potential is V_b, is

$$\Delta U = q(V_a - V_b) = qV_{ab}.$$

In particular, if the charge q is the electronic charge $e = 1.60 \times 10^{-19}$ C, and the potential difference $V_{ab} = 1$ V, the change in energy is

$$\Delta U = (1.60 \times 10^{-19} \text{ C})(1 \text{ V}) = 1.60 \times 10^{-19} \text{ J}.$$

This quantity of energy is called 1 *electronvolt* (1 eV):

$$1 \text{ eV} = 1.60 \times 10^{-19} \text{ J}.$$

Other commonly used units are:

$$1 \text{ keV} = 10^3 \text{ eV},$$
$$1 \text{ MeV} = 10^6 \text{ eV},$$
$$1 \text{ GeV} = 10^9 \text{ eV}.$$

The electronvolt is a convenient energy unit when one is dealing with the motions of electrons and ions in electric fields, because the change in potential energy between two points on the path of a particle having a charge e, when expressed in electronvolts, is *numerically* equal to the potential difference between the points, in volts. If the charge is some multiple of e, say Ne, the change in potential energy in electronvolts is, numerically, N times the potential difference in volts. For example, if a particle having a charge $2e$ moves between two points for which the potential difference is 1000 V, the change in its potential energy is

$$\Delta U = qV_{ab} = (2)(1.6 \times 10^{-19} \text{ C})(10^3 \text{ V})$$
$$= 3.2 \times 10^{-16} \text{ J}$$
$$= 2000 \text{ eV}.$$

Although the electronvolt was defined above in terms of *potential* energy, energy of *any* form, such as the kinetic energy of a moving particle, can be expressed in terms of the electronvolt. Thus one may speak of a "one-million-volt electron," meaning an electron having a kinetic energy of one million electronvolts (1 MeV).

One of the principles of the special theory of relativity, to be developed in Chapter 43, is that the mass m of a particle is equivalent to a quantity of energy mc^2, where c is the speed of light. The rest mass of an electron is 9.108×10^{-31} kg, and the energy equivalent to this is

$$E_0 = mc^2 = (9.108 \times 10^{-31} \text{ kg})(3 \times 10^8 \text{ m} \cdot \text{s}^{-1})^2$$
$$= 82 \times 10^{-15} \text{ J}.$$

Since $1 \text{ eV} = 1.60 \times 10^{-19}$ J, this is equivalent to

$$E_0 = 511{,}000 \text{ eV} = 0.511 \text{ MeV}.$$

When the kinetic energy of a particle becomes of comparable magnitude to its rest energy, Newton's laws of motion are not strictly valid and must be replaced by the more general relations of relativistic mechanics, to be developed in detail in Chapter 43. For example, an electron accelerated through a potential difference of 500 kV acquires a kinetic energy approximately *equal* to its rest energy, and a correct analysis of the motion of the particle requires the use of relativistic mechanics.

*26–8 The cathode-ray oscilloscope

Figure 26–11 is a schematic diagram of the elements of a cathode-ray oscilloscope tube. The interior of the tube is highly evacuated. The *cathode* at the bottom is raised to a high temperature by the *heater*, and electrons evaporate from its surface. (Before the nature of this process of electron emission was fully understood, these electrons were given the name "cathode rays.") The *accelerating anode*, which has a small hole at its center, is maintained at a high positive potential V_1 relative to the cathode, so that there is an electric field, directed downward between the anode and cathode. This field is confined to the cathode-anode region, and electrons passing through the hole in the anode travel with *constant* vertical velocity from the anode to the *fluorescent screen*.

The function of the *control grid* is to regulate the number of electrons that reach the anode (and hence the brightness of the spot on the screen). The *focusing anode* ensures that electrons leaving the cathode in slightly different directions all arrive at the same spot on the screen. These two electrodes need not be considered in the following analysis. The complete assembly of cathode, control grid, focusing anode, and accelerating electrode is referred to as an *electron gun*.

The accelerated electrons pass between two pairs of *deflecting plates*. An electric field between the first pair of plates deflects them to the front or back, and a field between the second pair deflects them right or left. In the absence of such fields the electrons travel in a straight line from the hole in the accelerating anode to the *fluorescent screen* and produce a bright spot on the screen where they strike it.

We first calculate the speed v imparted to the electrons by the electron gun, just as in Example 4 of Sec. 26–2. We assume the electrons leave the cathode with zero initial speed; in fact, the electrons do have some small initial speed when they evaporate from the cathode, but it is very small compared to the final speed and can be neglected. As in the example, we find

$$v = \sqrt{\frac{2eV_1}{m}}. \quad (26\text{–}26)$$

26–11 Basic elements of a cathode-ray tube.

As a numerical example, if $V_1 = 2000$ V,

$$v = \sqrt{\frac{2(1.6 \times 10^{-19}\text{ C})(2 \times 10^3\text{ V})}{9.11 \times 10^{-31}\text{ kg}}}$$
$$= 2.65 \times 10^7 \text{ m} \cdot \text{s}^{-1}.$$

We note that the kinetic energy of an electron at the anode depends only on the *potential difference* between anode and cathode, and not at all on the details of the fields within the electron gun or on the shape of the electron trajectory within the gun.

If there is no electric field between the front-back deflection plates, the electrons enter the region between the other plates with a speed equal to v and represented by v_x in Fig. 26-12. (This view is turned 90° from that of Fig. 26-11.) If there is a potential difference V_2 between the plates, and the upper plate is positive, a downward electric field of magnitude $E = V_2/l$ is set up between the plates. A constant upward force eE then acts on the electrons, and their upward acceleration is

$$a_y = \frac{eE}{m} = \frac{eV_2}{ml}. \tag{26-27}$$

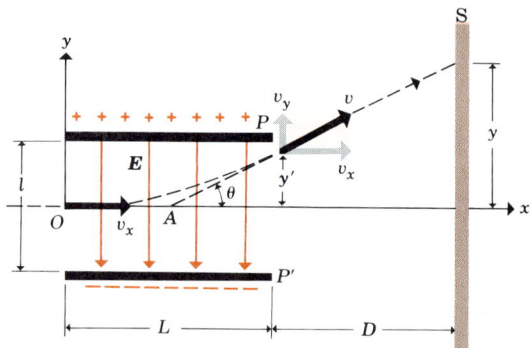

26-12 Electrostatic deflection of cathode rays.

The *horizontal* component of velocity v_x remains constant, so the time required for the electrons to travel the length L of the plates is

$$t = \frac{L}{v_x}. \tag{26-28}$$

In this time, they acquire an upward velocity component given by

$$v_y = a_y t, \tag{26-29}$$

and are displaced upward by an amount

$$y' = \frac{1}{2} a_y t^2.$$

On emerging from the deflecting field, their velocity v makes an angle θ with the x-axis, where

$$\tan \theta = \frac{v_y}{v_x};$$

and from this point on they travel in a straight line to the screen. It is not difficult to show that this straight line, if projected backward, intersects the x-axis at a point A that is midway between the ends of the plates. Then if y is the vertical coordinate of the point of impact with screen S,

$$\tan \theta = \frac{y}{D + (L/2)}.$$

When this is combined with Eqs. (26-27), (26-28), and (26-29), we get finally

$$y = \left[\frac{L}{2l}\left(D + \frac{L}{2}\right)\right]\frac{V_2}{V_1}. \qquad (26\text{-}30)$$

The term in brackets is a purely geometrical factor. If the accelerating voltage V_1 is held constant, *the deflection y is proportional to the deflecting voltage V_2.*

If a field is also set up between the *horizontal* deflecting plates, the beam is also deflected in the horizontal direction, perpendicular to the plane of Fig. 26-12. The coordinates of the luminous spot on the screen are then proportional, respectively, to the horizontal and vertical deflecting voltages. This device, with its associated circuitry, forms the basis of the *oscilloscope,* a laboratory instrument widely used for the observation and measurement of rapidly varying voltages.

The picture tube in a television set is similar in its operation to the oscilloscope tube, except that the beam is deflected by magnetic fields (to be discussed in later chapters) rather than by electric fields. Graphic display devices on some computer terminals also are similar, using a deflected electron beam and a fluorescent screen. The accelerating voltage in TV picture tubes (V_1 in the above discussion) is typically 20 to 25 kV.

Questions

26-1 Are there cases in electrostatics where a conducting surface is *not* an equipotential surface? If so, give an example.

26-2 If the electric potential at a single point is known, can the electric field at that point be determined?

26-3 If two points are at the same potential, is the electric field necessarily zero everywhere between them?

26-4 If the electric field is zero throughout a certain region of space, is the potential also zero in this region? If not, what *can* be said about the potential?

26-5 A student said: "Since electrical potential is always proportional to potential energy, why bother with the concept of potential at all?" How would you respond?

26-6 Is potential gradient a scalar or vector quantity?

26-7 A conducting sphere is to be charged by bringing in positive charge, a little at a time, until the total charge is Q. The total work required for this process is alleged to be proportional to Q^2. Is this correct? Why or why not?

26-8 The potential (relative to a point at infinity) midway between two charges of equal magnitude and opposite sign is zero. Can you think of a way to bring a test charge from infinity to this midpoint in such a way that no work is done in any part of the displacement?

26-9 A high-voltage dc power line falls on a car, so the entire metal body of the car is at a potential of 10,000 V with respect to the ground. What happens to the occupants

a) when they are sitting in the car;

b) when they step out of the car?

26-10 In electronics it is customary to define the potential of ground (thinking of the earth as a large conductor) as zero. Is this consistent with the fact that the earth has a net electric charge that is not zero? (Cf. Problem 25-28.)

26-11 A positive point charge is placed near a very large conducting plane. A professor of physics asserted that the field caused by this configuration is the same as would be obtained by removing the plane and placing a negative point charge of equal magnitude in the mirror-image position behind the initial position of the plane. Is this correct?

26-12 It is easy to produce a potential of several thousand volts on your body by scuffing your shoes across a nylon carpet, yet contact with a power line of comparable voltage would probably be fatal. What's the difference?

Problems

26-1 Two large parallel metal sheets carrying equal and opposite electric charges are separated by a distance of 0.05 m. The electric field between them is approximately uniform and has magnitude 600 N·C^{-1}. What is the potential difference between the plates? Which plate is at higher potential?

26-2 Two large parallel metal plates carry opposite charges. They are separated by 0.1 m, and the potential difference between them is 500 V.

a) What is the magnitude of the electric field, if it is uniform, in the region between the plates?

b) Compute the work done by this field on a charge of 2.0×10^{-9} C as it moves from the higher-potential plate to the lower.

c) Compare the result of (b) to the change of potential energy of the same charge, computed from the electrical potential.

26-3 A total electric charge of 4×10^{-9} C is distributed uniformly over the surface of a sphere of radius 0.20 m. If the potential is zero at a point at infinity, what is the value of the potential

a) at a point on the surface of the sphere;

b) at a point inside the sphere, 0.1 m from the center?

26-4 A particle of charge $+3 \times 10^{-9}$ C is in a uniform electric field directed to the left. It is released from rest and moves a distance of 5 cm, after which its kinetic energy is found to be $+4.5 \times 10^{-5}$ J.

a) What work was done by the electrical force?

b) What is the magnitude of the electric field?

c) What is the potential of the starting point with respect to the endpoint?

26-5 In Problem 26-4, suppose that another force in addition to the electrical force acts on the particle, so that when it is released from rest it moves to the right. After it has moved 5 cm, the additional force has done 9×10^{-5} J of work and the particle has 4.5×10^{-5} J of kinetic energy.

a) What work was done by the electrical force?

b) What is the magnitude of the electric field?

c) What is the potential of the starting point with respect to the endpoint?

26-6 A charge of 2.5×10^{-8} C is placed in an upwardly directed uniform electric field having magnitude 5×10^4 N·C^{-1}. What is the work of the electrical force when the charge is moved

a) 45 cm to the right?

b) 80 cm downward?

c) 260 cm at an angle of 45° upward from the horizontal?

26-7

a) Show that 1 N·C$^{-1} = 1$ V·m^{-1}.

b) A potential difference of 2000 V is established between parallel plates in air. If the air becomes electrically conducting when the electric field exceeds 3×10^6 N·C^{-1}, what is the minimum separation of the plates?

26-8 A small sphere of mass 0.2 g hangs by a thread between two parallel vertical plates 5 cm apart. The charge on the sphere is 6×10^{-9} C. What potential difference between the plates will cause the thread to assume an angle of 30° with the vertical?

26-9 Two point charges $q_1 = +40 \times 10^{-9}$ C and $q_2 = -30 \times 10^{-9}$ C are 10 cm apart. Point A is midway between them, point B is 8 cm from q_1 and 6 cm from q_2. Find

a) the potential at point A;

b) the potential at point B;

c) the work required to carry a charge of 25×10^{-9} C from point B to point A.

26-10 Three equal point charges of 3×10^{-7} C are placed at the corners of an equilateral triangle whose side is one meter. What is the potential energy of the system? Take as zero potential energy the energy of the three charges when they are infinitely far apart.

26-11 The potential at a certain distance from a point charge is 600 V, and the electric field is 200 N·C^{-1}.

a) What is the distance to the point charge?

b) What is the magnitude of the charge?

26-12 Two point charges whose magnitudes are $+20 \times 10^{-9}$ C and -12×10^{-9} C are separated by a distance of 5 cm. An electron is released from rest between the two charges, 1 cm from the negative charge, and moves along the line connecting the two charges. What is its velocity when it is 1 cm from the positive charge?

26-13 Two positive point charges, each of magnitude q, are fixed on the y-axis at the points $y = +a$ and $y = -a$.

a) Draw a diagram showing the positions of the charges.

b) What is the potential V_0 at the origin?

c) Show that the potential at any point on the x-axis is
$$V = \frac{1}{4\pi\epsilon_0} \frac{2q}{\sqrt{a^2 + x^2}}.$$

d) Sketch a graph of the potential on the x-axis as a function of x over the range from $x = +4a$ to $x = -4a$.

e) At what value of x is the potential one-half that at the origin?

26-14 Consider the same distribution of charges as in Problem 26-13.

a) Sketch a graph of the potential on the y-axis as a function of y, over the range from $y = +4a$ to $y = -4a$.

b) Discuss the physical meaning of the graph at the points $+a$ and $-a$.

c) At what point or points on the y-axis is the potential equal to that at the origin?

d) At what points on the y-axis is the potential equal to half its value at the origin?

26-15 Consider the same charge distribution as in Problem 26-13.

a) Suppose a positively charged particle of charge q' and mass m is placed precisely at the origin and released from rest. What happens?

b) What will happen if the charge in part (a) is displaced slightly in the direction of the y-axis?

c) What will happen if it is displaced slightly in the direction of the x-axis?

26-16 Again consider the charge distribution in Problem 26-13. Suppose a positively charged particle of charge q' and mass m is displaced slightly from the origin in the direction of the x-axis.

a) What is its speed at infinity?

b) Sketch a graph of the velocity of the particle as a function of x.

c) If the particle is projected toward the left along the x-axis from a point at a large distance to the right of the origin, with a velocity half that acquired in part (a), at what distance from the origin will it come to rest?

d) If a negatively charged particle were released from rest on the x-axis, at a very large distance to the left of the origin, what would be its velocity as it passed the origin?

26-17 A positive charge $+q$ is located at the point $x = -a, y = -a$, and an equal negative charge $-q$ is located at the point $x = +a, y = -a$.

a) Draw a diagram showing the positions of the charges.

b) What is the potential at the origin?

c) What is the expression for the potential at a point on the x-axis, as a function of x?

d) Sketch a graph of the potential as a function of x, in the range from $x = +4a$ to $x = -4a$. Plot positive potentials upward, negative potentials downward.

26-18 A potential difference of 1600 V is established between two parallel plates 4 cm apart. An electron is released from the negative plate at the same instant that a proton is released from the positive plate.

a) How far from the positive plate will they pass each other?

b) How do their velocities compare when they strike the opposite plates?

c) How do their energies compare when they strike the opposite plates?

26-19 Consider the same charge distribution as in Problem 26-13.

a) Construct a graph of the potential energy of a positive point charge on the x-axis, as a function of x.

b) Construct a graph of the potential energy of a negative point charge on the axis, as a function of x.

c) What is the x-component of the potential gradient at the origin?

26-20 In the Bohr model of the hydrogen atom, a single electron revolves around a single proton in a circle of radius R.

a) By equating the electrical force to the electron mass times its acceleration, derive an expression for the electron's speed.

b) Obtain an expression for the electron's kinetic energy, and show that its magnitude is just half that of the electrical potential energy.

c) Obtain an expression for the total energy, and evaluate it using $R = 0.528 \times 10^{-10}$ m.

26-21 A vacuum diode consists of a cylindrical cathode of radius 0.05 cm, mounted coaxially within a cylindrical anode 0.45 cm in radius. The potential of the anode is 300 V higher than that of the cathode. An electron leaves the surface of the cathode with zero initial speed. Find its speed when it strikes the anode.

26-22 A vacuum triode may be idealized as follows. A plane surface (the cathode) emits electrons with negligible initial velocities. Parallel to the cathode and 3 mm away from it is an open grid of fine wire at a potential of 18 V above the cathode. A second plane surface (the anode) is 12 mm beyond the grid and is at a potential of 15 V above the cathode. Assume that the plane of the grid is an equipotential surface, and that the potential gradients between cathode and grid, and between grid and anode, are uniform. Assume also that the structure of the grid is sufficiently open for electrons to pass through it freely.

a) Draw a diagram of potential vs. distance, along a line from cathode to anode.

b) With what speed will electrons strike the anode?

26-23 A ring-shaped conductor of radius a carries a total charge Q; it is placed with its plane perpendicular to the x-axis and its center at the origin. (Cf. Sec. 25-2, Example 2.)

a) Derive an expression for the potential at any point on the x-axis.

b) Using Eq. (26-21) and the result of (a), derive an expression for the x-component of electric field at any point on the x-axis. Compare your result to that of the example cited above.

c) Construct graphs of the potential V and the x-component of field E_x as functions of x, and show the geometric relationship between the two graphs.

26-24 A metal sphere of radius r_a is supported on an insulating stand at the center of a hollow metal sphere of inner radius r_b. There is a charge $+q$ on the inner sphere and a charge $-q$ on the outer. (See Problem 25-21.)

a) Show that the potential of the inner sphere with respect to the outer is

$$V_{ab} = \frac{q}{4\pi\epsilon_0}\left(\frac{1}{r_a} - \frac{1}{r_b}\right).$$

b) Show that the electric field at any point between the spheres has magnitude

$$E = \frac{V_{ab}}{(1/r_a - 1/r_b)} \cdot \frac{1}{r^2}.$$

c) Find the electric field at a point outside the larger sphere, at a distance r from the center, where $r > r_b$.

d) Suppose the charge on the outer sphere is not $-q$ but a negative charge of different magnitude, say $-Q$. Show that the answers for (a) and (b) are the same as before but (c) is different.

26-25 A long metal cylinder of radius r_a is supported on an insulating stand on the axis of a long hollow metal cylinder of inner radius r_b. The positive charge per unit length on the inner cylinder is λ and there is an equal negative charge per unit length on the outer cylinder. (See Problem 25-24.)

a) Show that the potential of the inner cylinder with respect to the outer is

$$V_{ab} = \frac{\lambda}{2\pi\epsilon_0} \ln \frac{r_b}{r_a}.$$

b) Show that the electric field at any point between the cylinders has magnitude

$$E = \frac{V_{ab}}{\ln(r_b/r_a)} \cdot \frac{1}{r}.$$

c) What is the potential difference if the outer cylinder has no net charge?

26-26 Refer to Problem 25-25.

a) Find the expression for the potential V as a function of r, both inside and outside the sphere, relative to a point at infinity.

b) Sketch graphs of V and E as functions of r from $r = 0$ to $r = 3R$, and compare with Fig. 26-6.

26-27 Refer to Problem 25-26.

a) Find the expressions for the potential V as a function of r, both inside and outside the cylinder. Let $V = 0$ at the surface of the cylinder.

b) Sketch graphs of V and E as functions of r, from $r = 0$ to $r = 3R$.

26-28 Refer to Problem 24-19.

a) Calculate the potential at the origin, and at the point (3 cm, 0) due to the first two point charges.

b) Calculate the work the electric field would do on the third charge if it moved from the origin to the point (3 cm, 0).

26-29 Refer to Problem 24-20.

a) Calculate the potential at the point (3 cm, 0) and at the point (3 cm, 4 cm) due to the first two charges.

b) If the third charge moves from the point (3 m, 0) to the point (3 cm, 4 cm), calculate the work done on it by the field of the first two charges. Comment on the *sign* of this work; is your result reasonable?

26-30 In an apparatus for measuring the electronic charge e by Millikan's method, an electric intensity of 6.34×10^4 V·m^{-1} is required to maintain a certain charged oil drop at rest. If the plates are 1.5 cm apart, what potential difference between them is required?

26-31 An oil droplet of mass 3×10^{-11} g and of radius 2×10^{-4} cm carries 10 excess electrons. What is its terminal velocity

a) when falling in a region in which there is no electric field?

b) When falling in an electric field of magnitude 3×10^5 N·C^{-1} directed downward?

The viscosity of air is 180×10^{-7} N·s·m^{-2}. (Neglect the buoyant force of the air.)

26-32 A charged oil drop, in a Millikan oil-drop apparatus, is observed to fall through a distance of 1 mm in a time of 27.4 s, in the absence of any external field. The same drop can be held stationary in a field of 2.37×10^4 N·C^{-1}. How many excess electrons has the drop acquired? The viscosity of air is 180×10^{-7} N·s·m^{-2}. The density of oil is 824 kg·m^{-3}, and the density of air is 1.29 kg·m^{-3}.

26-33 Find the energy equivalent of the rest mass of the proton; express your result in MeV.

26-34 Find the potential energy of the interaction of two protons at a distance of 10^{-15} m, typical of the dimensions of atomic nuclei. Express your result in MeV.

26-35 An alpha particle with kinetic energy 10 MeV makes a head-on collision with a gold nucleus at rest.

What is the distance of closest approach of the two particles?

26-36

a) Prove that when a particle of constant mass and charge is accelerated from rest in an electric field, its final velocity is proportional to the square root of the potential difference through which it is accelerated.

b) Find the magnitude of the proportionality constant if the particle is an electron, the velocity is in meters per second, and the potential difference is in volts.

c) What is the final velocity of an electron accelerated through a potential difference of 1136 V if it has an initial velocity of 10^7 m·s^{-1}?

26-37

a) What is the maximum potential difference through which an electron can be accelerated if its kinetic energy is not to exceed 1% of the rest energy?

b) What is the speed of such an electron, expressed as a fraction of the speed of light, c?

c) Make the same calculations for a *proton*.

26-38 The electric field in the region between the deflecting plates of a certain cathode-ray oscilloscope is 30,000 N·C^{-1}.

a) What is the force on an electron in this region?

b) What is the acceleration of an electron when acted on by this force?

26-39 In Fig. 26-13, an electron is projected along the axis midway between the plates of a cathode-ray tube with an initial velocity of 2×10^7 m·s^{-1}. The uniform electric field between the plates has an intensity of 20,000 N·C^{-1} and is upward.

a) How far below the axis has the electron moved when it reaches the end of the plates?

b) At what angle with the axis is it moving as it leaves the plates?

c) How far below the axis will it strike the fluorescent screen S?

Figure 26-13

26-40 An electron with kinetic energy 100 MeV collides head-on with a gold nucleus at rest. Assuming that the gold nucleus can be treated as a uniform distribution of charge through a sphere of radius 7×10^{-15} m and that the electron can penetrate into the nucleus, what is its kinetic energy when it reaches the center of the nucleus?

27 CAPACITANCE. PROPERTIES OF DIELECTRICS

27-1 Capacitors

Any two conductors separated by an insulator are said to form a *capacitor*. In most cases of practical interest the conductors usually have charges of equal magnitude and opposite sign, so that the net charge on the capacitor as a whole is zero. The electric field in the region between the conductors is then proportional to the magnitude of this charge, and it follows that the *potential difference* V_{ab} between the conductors is also proportional to the charge magnitude Q.

The *capacitance C* of a capacitor is defined as the ratio of the magnitude of the charge Q on *either* conductor to the magnitude of the potential difference V_{ab} between the conductors:

$$C = \frac{Q}{V_{ab}}. \qquad (27\text{-}1)$$

It follows from its definition that the unit of capacitance is one *coulomb per volt* ($1 \text{ C} \cdot \text{V}^{-1}$). A capacitance of one coulomb per volt is called one *farad* (1 F) in honor of Michael Faraday. A capacitor is represented by the symbol

When one speaks of a capacitor as having charge Q, what is really meant is that the conductor at higher potential has a charge Q, and the conductor at lower potential a charge $-Q$ (assuming Q is a positive quantity). This interpretation should be kept in mind in the following discussion and examples.

Capacitors find many applications in electrical circuits. Capacitors are used for tuning radio circuits and for "smoothing" the rectified current delivered by a power supply. A capacitor is used to eliminate sparking when a circuit containing inductance is suddenly opened. The ignition system of every automobile engine contains a capacitor to eliminate sparking of the "points" when they open and close. The efficiency of

516

alternating-current power transmission can often be increased by the use of large capacitors.

27-2 The parallel-plate capacitor

The most common type of capacitor consists in principle of two conducting plates parallel to each other and separated by a distance that is small compared with the linear dimensions of the plates (see Fig. 27-1). Practically the entire field of such a capacitor is localized in the region between the plates, as shown. There is a slight "fringing" of the field at its outer boundary (bending outward of field lines near the edges), but if the plates are sufficiently close, this may be neglected. The field between the plates is then uniform, and the charges on the plates are uniformly distributed over their opposing surfaces. This arrangement is known as a *parallel-plate capacitor*.

Let us assume first that the plates are in vacuum. It has been shown that the electric field magnitude between a pair of closely spaced parallel plates in vacuum is

$$E = \frac{\sigma}{\epsilon_0} = \frac{Q}{\epsilon_0 A},$$

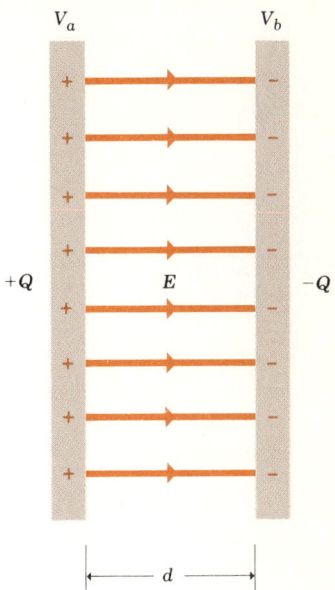

27-1 Parallel-plate capacitor.

where σ is the magnitude of surface charge density on either plate, A is the area of each plate, and Q is the magnitude of total charge on each plate. Since the electric field (potential gradient) between the plates is uniform, the potential difference between the plates is

$$V_{ab} = Ed = \frac{1}{\epsilon_0} \frac{Qd}{A},$$

where d is the separation of the plates. Hence the capacitance of a parallel-plate capacitor in vacuum is

$$C = \frac{Q}{V_{ab}} = \epsilon_0 \frac{A}{d}. \tag{27-2}$$

Since ϵ_0, A, and d are constants for a given capacitor, the capacitance is a constant independent of the charge on the capacitor, and is directly proportional to the area of the plates and inversely proportional to their separation. If SI units are used, A is expressed in square meters and d in meters. The capacitance C is then in farads.

As an example, let us compute the area of the plates of a 1-F parallel-plate capacitor if the separation of the plates is 1 mm and the plates are in vacuum:

$$C = \epsilon_0 \frac{A}{d},$$

$$A = \frac{Cd}{\epsilon_0} = \frac{(1 \text{ F})(10^{-3} \text{ m})}{8.85 \times 10^{-12} \text{ C}^2 \cdot \text{N}^{-1} \cdot \text{m}^{-2}}$$

$$= 1.13 \times 10^8 \text{ m}^2.$$

This corresponds to a square 10,600 m, or 34,800 ft, or about $6\frac{1}{2}$ miles on a side!

27-2 Variable air capacitor. (Courtesy of General Radio Company.)

Since the farad is such a large unit of capacitance, units of more convenient size are the *microfarad* (1 μF = 10^{-6} F), and the *picofarad* (1 pF = 10^{-12} F). For example, a common radio set contains in its power supply several capacitors whose capacitances are of the order of ten or more microfarads, while the capacitances of the tuning capacitors are of the order of ten to one hundred picofarads.

Variable capacitors whose capacitance may be varied (between limits) are widely used in the tuning circuits of radio receivers. These usually have a number of fixed parallel metal plates connected together and constituting one "plate" of the capacitor, while a second set of movable plates (also connected together) forms the other "plate" (Fig. 27-2). The movable plates are mounted on a shaft and may be caused to interleave the fixed plates to a greater or lesser extent. The effective area of the capacitor is that of the interleaved portion of the plates.

A variable capacitor is represented by the symbol

Capacitors consisting of concentric spheres and of coaxial cylinders are sometimes used in standards laboratories, since the corrections for the "fringing" fields can be made more readily and the capacitance can be accurately calculated from the dimensions of the apparatus. Problems involving spherical and cylindrical capacitors will be found at the end of the chapter.

Example The plates of a parallel-plate capacitor are 5 mm apart and 2 m² in area. The plates are in vacuum. A potential difference of 10,000 V is applied across the capacitor. Compute (a) the capacitance, (b) the charge on each plate, and (c) the electric field in the space between them.

Solution

a)
$$C = \epsilon_0 \frac{A}{d} = \frac{(8.85 \times 10^{-12} \, \text{C}^2 \cdot \text{N}^{-1} \cdot \text{m}^{-2})(2 \, \text{m}^2)}{5 \times 10^{-3} \, \text{m}}$$
$$= 3.54 \times 10^{-9} \, \text{C}^2 \cdot \text{N}^{-1} \cdot \text{m}^{-1}.$$

But
$$1 \, \text{C}^2 \cdot \text{N}^{-1} \cdot \text{m}^{-1} = 1 \, \text{C}^2 \cdot \text{J}^{-1} = 1 \, \text{C} \left(\frac{\text{J}}{\text{C}}\right)^{-1}$$
$$= 1 \, \text{C} \cdot \text{V}^{-1} = 1 \, \text{F},$$

so
$$C = 3.54 \times 10^{-9} \, \text{F} = 0.00354 \, \mu\text{F}.$$

b) The charge on the capacitor is
$$Q = CV_{ab} = (3.54 \times 10^{-9} \, \text{C} \cdot \text{V}^{-1})(10^4 \, \text{V})$$
$$= 3.54 \times 10^{-5} \, \text{C}.$$

That is, the plate at higher potential has charge $+3.54 \times 10^{-5}$ C, and the other plate has charge -3.54×10^{-5} C.

c) The electric field is
$$E = \frac{\sigma}{\epsilon_0} = \frac{Q}{\epsilon_0 A} = \frac{3.54 \times 10^{-5} \, \text{C}}{(8.85 \times 10^{-12} \, \text{C}^2 \cdot \text{N}^{-1} \cdot \text{m}^{-2})(2 \, \text{m}^2)}$$
$$= 20 \times 10^5 \, \text{N} \cdot \text{C}^{-1};$$

or, since the electric field equals the potential gradient,

$$E = \frac{V_{ab}}{d} = \frac{10^4 \text{ V}}{5 \times 10^{-3} \text{ m}}$$
$$= 20 \times 10^5 \text{ V} \cdot \text{m}^{-1}.$$

Of course, the newton per coulomb and the volt per meter are equivalent units.

27-3 Capacitors in series and parallel

In Fig. 27-3a, two capacitors are connected in *series* between points a and b, maintained at a constant potential difference V_{ab}. The capacitors are both initially uncharged. In this connection, both capacitors always have the same charge Q. One might ask whether the lower plate of C_1 and the upper plate of C_2 might have charges different from those on the remaining two plates, but in this case the net charge on each capacitor would not be zero, and the resulting electric field in the conductor connecting the two capacitors would cause a current to flow until the total charge on each capacitor is zero. Hence, *in a series connection the magnitude of charge on all plates is the same.*

Referring again to Fig. 27-3a, we have

$$V_{ac} \equiv V_1 = \frac{Q}{C_1}, \qquad V_{cb} \equiv V_2 = \frac{Q}{C_2},$$

$$V_{ab} \equiv V = V_1 + V_2 = Q\left(\frac{1}{C_1} + \frac{1}{C_2}\right),$$

and

$$\frac{V}{Q} = \frac{1}{C_1} + \frac{1}{C_2}. \tag{27-3}$$

The *equivalent* capacitance C of the series combination is defined as the capacitance of a *single* capacitor for which the charge Q is the same as for the combination, when the potential difference V is the same. For such a capacitor, shown in Fig. 27-3b,

$$Q = CV, \qquad \frac{V}{Q} = \frac{1}{C}. \tag{27-4}$$

Hence, from Eqs. (27-3) and (27-4),

$$\frac{1}{C} = \frac{1}{C_1} + \frac{1}{C_2}.$$

Similarly, for any number of capacitors in *series*,

$$\frac{1}{C} = \frac{1}{C_1} + \frac{1}{C_2} + \frac{1}{C_3} + \cdots \tag{27-5}$$

The reciprocal of the equivalent capacitance equals the sum of the reciprocals of the individual capacitances.

In Fig. 27-4a, two capacitors are connected in *parallel* between points a and b. In this case, the potential difference $V_{ab} \equiv V$ is the same for both, and the charges Q_1 and Q_2, not necessarily equal, are

$$Q_1 = C_1 V, \qquad Q_2 = C_2 V.$$

27-3 (a) Two capacitors in series, and (b) their equivalent.

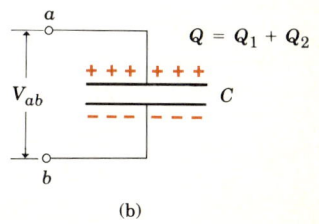

27-4 (a) Two capacitors in parallel, and (b) their equivalent.

The *total* charge Q supplied by the source was
$$Q = Q_1 + Q_2 = V(C_1 + C_2),$$
and
$$\frac{Q}{V} = C_1 + C_2. \qquad (27\text{-}6)$$

The *equivalent* capacitance C of the parallel combination is defined as that of a single capacitor, shown in Fig. 27-4b, for which the total charge is the same as in part (a). For this capacitor,
$$\frac{Q}{V} = C,$$
and hence
$$C = C_1 + C_2.$$

In the same way, for any number of capacitors in *parallel*
$$C = C_1 + C_2 + C_3 + \cdots \qquad (27\text{-}7)$$

The equivalent capacitance equals the *sum* of the individual capacitances.

Example In Figs. 27-3 and 27-4, let $C_1 = 6\,\mu\text{F}$, $C_2 = 3\,\mu\text{F}$, $V_{ab} = 18$ V.

The equivalent capacitance of the series combination in Fig. 27-3a is given by
$$\frac{1}{C} = \frac{1}{6\,\mu\text{F}} + \frac{1}{3\,\mu\text{F}}, \qquad C = 2\,\mu\text{F}.$$

The charge Q is
$$Q = CV = (2\,\mu\text{F})(18\text{ V}) = 36\,\mu\text{C}.$$

The potential differences across the capacitors are
$$V_{ac} \equiv V_1 = \frac{Q}{C_1} = \frac{36\,\mu\text{C}}{6\,\mu\text{F}} = 6\text{ V}, \qquad V_{cb} \equiv V_2 = \frac{Q}{C_2} = \frac{36\,\mu\text{C}}{3\,\mu\text{F}} = 12\text{ V}.$$

The *larger* potential difference appears across the *smaller* capacitor.

The equivalent capacitance of the parallel combination in Fig. 27-4a is
$$C = C_1 + C_2 = 9\,\mu\text{F}.$$

The charges Q_1 and Q_2 are
$$Q_1 = C_1 V = (6\,\mu\text{F})(18\text{ V}) = 108\,\mu\text{C},$$
$$Q_2 = C_2 V = (3\,\mu\text{F})(18\text{ V}) = 54\,\mu\text{C}.$$

27-4 Energy of a charged capacitor

The process of charging a capacitor consists of transferring charge from the plate at lower potential to the plate at higher potential. The charging process therefore requires the expenditure of *energy*. We may imagine charging the capacitor by starting with both plates completely uncharged, and then repeatedly removing small positive charges dq from one plate and transferring them to the other plate. The final charge Q

and the final potential difference V are related by

$$Q = CV.$$

At a stage of the process at which the magnitude of the net charge on either plate is q, the potential difference v between the plates is $v = q/C$. The work dW required to transfer the next charge dq is

$$dW = v\,dq = \frac{q\,dq}{C}.$$

The total work W to increase the charge from zero to a final value Q is

$$W = \int dW = \frac{1}{C}\int_0^Q q\,dq = \frac{Q^2}{2C}.$$

The final potential difference V between the plates is $V = Q/C$ and we can also write

$$W = \frac{Q^2}{2C} = \frac{1}{2}CV^2 = \frac{1}{2}QV. \tag{27-8}$$

When Q is expressed in coulombs and V in volts (joules per coulomb), W is expressed in joules.

The last form of Eq. (27-8) also shows that the total work is equal to the *average* potential $V/2$ during the charging process, multiplied by the total charge Q transferred.

A charged capacitor is the electrical analog of a stretched spring, whose elastic potential energy equals $\frac{1}{2}kx^2$. The charge Q is analogous to the elongation x, and the *reciprocal* of the capacitance, $1/C$, is analogous to the force constant k. The energy supplied to a capacitor in the charging process is stored by the capacitor and is released when the capacitor discharges.

It is often useful to consider the stored energy to be localized in the *electric field* between the capacitor plates. The capacitance of a parallel-plate capacitor in vacuum is

$$C = \epsilon_0 \frac{A}{d}.$$

The electric field fills the space between the plates, of volume Ad, and is given by

$$E = \frac{V}{d}.$$

The energy per unit volume, or the *energy density*, denoted by u, is

$$u = \text{Energy density} = \frac{\frac{1}{2}CV^2}{Ad}.$$

Making use of the preceding equations, we can express this as

$$u = \frac{1}{2}\epsilon_0 E^2. \tag{27-9}$$

Example In Fig. 27-5 capacitor C_1 is initially charged by connecting it to a source of potential difference V_0 (not shown in the figure). Let $C_1 = 8\,\mu\text{F}$ and $V_0 = 120$ V. The charge Q_0 is

$$Q_0 = C_1 V_0 = 960\,\mu\text{C},$$

27-5 When the switch S is closed, the charged capacitor C_1 is connected to an uncharged capacitor C_2.

and the energy of the capacitor is

$$\frac{1}{2}Q_0 V_0 = 0.0576 \text{ J}.$$

After the switch S is closed, the positive charge Q_0 is distributed over the upper plates of both capacitors, and the negative charge $-Q_0$ is distributed over the lower plates of both. Let Q_1 and Q_2 represent the magnitudes of the final charges on the respective capacitors. Then

$$Q_1 + Q_2 = Q_0.$$

When the motion of charges has ceased, both upper plates are at the same potential and both lower plates are at the same potential (different from that of the upper plates). The final potential difference between the plates, V, is therefore the same for both capacitors, and

$$Q_1 = C_1 V, \qquad Q_2 = C_2 V.$$

When this is combined with the preceding equation, we find that

$$V = \frac{Q_0}{C_1 + C_2} = \frac{960 \ \mu\text{C}}{12 \ \mu\text{F}} = 80 \text{ V},$$
$$Q_1 = 640 \ \mu\text{C}, \qquad Q_2 = 320 \ \mu\text{C}.$$

The final energy of the system is

$$\frac{1}{2}Q_0 V = 0.0384 \text{ J}.$$

This is less than the original energy of 0.0576 J; the difference is converted to energy of some other form. If the resistance of the connecting wires is large, most of the energy is converted to heat. If the resistance is small, most of the energy is radiated in the form of electromagnetic waves.

This process is exactly analogous to an inelastic collision of a moving car with a stationary car. In the electrical case, the charge $Q = CV$ is conserved. In the mechanical case, the momentum $p = mv$ is conserved. The electrical energy $\frac{1}{2}CV^2$ is *not* conserved, and the mechanical energy $\frac{1}{2}mv^2$ is *not* conserved.

* 27-5 Effect of a dielectric

Most capacitors have a solid, nonconducting material or *dielectric* between their plates. A common type of capacitor incorporates strips of metal foil, forming the plates, separated by strips of wax-impregnated paper or plastic sheet such as Mylar, which serves as the dielectric. A sandwich of these materials is rolled up, forming a compact unit that can provide a capacitance of several microfarads in a relatively small volume.

Electrolytic capacitors have as their dielectric an extremely thin layer of nonconducting oxide between a metal plate and a conducting solution. Because of the thinness of the dielectric, electrolytic capacitors of relatively small dimensions may have a capacitance of the order of 100 to 1000 μF.

The function of a solid dielectric between the plates of a capacitor is threefold. First, it solves the mechanical problem of maintaining two large metal sheets at an extremely small separation but without actual contact.

Second, any dielectric material, when subjected to a sufficiently large electric field, experiences *dielectric breakdown*, a partial ionization, which permits conduction through a material that is supposed to insulate. Many insulating materials can tolerate stronger electric fields without breakdown than can air.

Third, the capacitance of a capacitor of given dimensions is *larger* when there is dielectric material between the plates than when the plates are separated only by air or vacuum. This effect can be demonstrated with the aid of a sensitive electrometer, a device that can measure the potential difference between two conductors without permitting any charge to flow from one to the other. In Fig. 27-6a, a capacitor is charged, with magnitude of charge Q on each plate and potential difference V_0. When a sheet of dielectric, such as glass, paraffin, or polystyrene, is inserted between the plates, the potential difference is found to *decrease* to a smaller value V. When the dielectric is removed, the potential difference returns to its original value, showing that the original charges on the plates have not been affected by insertion of the dielectric.

27-6 Effect of a dielectric between the plates of a parallel-plate capacitor. (a) With a given charge, the potential difference is V_0. (b) With the same charge, the potential difference V is smaller than V_0.

The original capacitance of the capacitor, C_0, was

$$C_0 = \frac{Q}{V_0}.$$

Since Q does not change and V is observed to be less than V_0, it follows that C is *greater* than C_0. The ratio of C to C_0 is called the *dielectric constant* of the material, K.

$$K = \frac{C}{C_0}. \quad \begin{array}{l}\rightarrow C \text{ w/dielectric} \\ \rightarrow C \text{ w/air or vacuum}\end{array} \quad (27\text{-}10)$$

Since C is always greater than C_0, the dielectric constants of all dielectrics are greater than unity. Some representative values of K are given in Table 27-1. For vacuum, of course, $K = 1$ by definition, and K for air is so nearly equal to 1 that for most purposes an air capacitor is equivalent to one in vacuum; the original measurement of V_0 in Fig. 27-6a could have been made with the plates in air instead of in vacuum.

Table 27-1 Dielectric constant K at 20°C

Material	K	Material	K
Vacuum	1	Strontium titanate	310
Glass	5–10	Titanium dioxide (rutile)	173($\perp$),
Mica	3–6		86($\parallel$)
Mylar	3.1	Water	80.4
Neoprene	6.70	Glycerin	42.5
Plexiglas	3.40	Liquid ammonia ($-78°$C)	25
Polyethylene	2.25	Benzene	2.284
Polyvinyl chloride	3.18	Air (1 atm)	1.00059
Teflon	2.1	Air (100 atm)	1.0548
Germanium	16		

27-7 Induced charges on the faces of a dielectric in an external field.

With vacuum (or air) between the plates, the electric field E_0 in the region between the plates of a parallel-plate capacitor is

$$E_0 = \frac{V_0}{d} = \frac{\sigma}{\epsilon_0}.$$

The observed reduction in potential difference, when a dielectric is inserted between the plates, implies a reduction in the electric field, which in turn implies a reduction in the charge per unit area. Since no charge has leaked off the plates, such a reduction could be caused only by induced charges of opposite sign appearing on the two surfaces of the *dielectric*. That is, the dielectric surface adjacent to the positive plate must have an *induced negative charge* and that adjacent to the negative plate an *induced positive charge of equal magnitude*, as shown in Fig. 27-7.

If σ_i is the magnitude of induced charge per unit area on the surfaces of the dielectric, then the *net* surface charge on each side that contributes to the electric field within the dielectric has magnitude $(\sigma - \sigma_i)$, and the electric field in the dielectric is

$$E = \frac{V}{d} = \frac{\sigma - \sigma_i}{\epsilon_0}. \tag{27-11}$$

But

$$K = \frac{C}{C_0} = \frac{Q/V}{Q/V_0} = \frac{V_0}{V} = \frac{E_0}{E} = \frac{\sigma}{\sigma - \sigma_i}; \tag{27-12}$$

and therefore

$$\sigma - \sigma_i = \frac{\sigma}{K}. \tag{27-13}$$

Substituting Eq. (27-13) into Eq. (27-11), we get

$$E = \frac{\sigma}{K\epsilon_0}. \tag{27-14}$$

The product $K\epsilon_0$ is called the *permittivity* of the dielectric and is represented by ϵ.

$$\epsilon = K\epsilon_0. \tag{27-15}$$

The electric field within the dielectric may therefore be written

$$E = \frac{\sigma}{\epsilon}. \tag{27-16}$$

Also,

$$C = KC_0 = K\epsilon_0 \frac{A}{d},$$

and the capacitance of a parallel-plate capacitor with a dielectric between its plates is therefore

$$C = \epsilon \frac{A}{d}. \tag{27-17}$$

In empty space, where $K = 1$, $\epsilon = \epsilon_0$, and therefore ϵ_0 may be described as the "permittivity of empty space" or the "permittivity of vacuum." Since K is a pure number, the units of ϵ and ϵ_0 are evidently the same, $C^2 \cdot N^{-1} \cdot m^{-2}$.

Example The parallel plates in Fig. 27-7 have an area of 2000 cm² or 2×10^{-1} m², and are 1 cm or 10^{-2} m apart. The original potential difference between them, V_0, is 3000 V, and it decreases to 1000 V when a sheet of dielectric is inserted between the plates. Compute (a) the original capacitance C_0, (b) the charge Q on each plate, (c) the capacitance C after insertion of the dielectric, (d) the dielectric constant K of the dielectric, (e) the permittivity ϵ of the dielectric, (f) the induced charge Q_i on each face of the dielectric, (g) the original electric field E_0 between the plates, and (h) the electric field E after insertion of the dielectric.

$\epsilon_0 = 8.85 \times 10^{-12} \frac{C^2}{N \cdot m^2}$

Solution

a) $C_0 = \epsilon_0 \dfrac{A}{d} = (8.85 \times 10^{-12} \, C^2 \cdot N^{-1} \cdot m^{-2}) \dfrac{2 \times 10^{-1} \, m^2}{10^{-2} \, m}$

$\qquad = 17.7 \times 10^{-11} \, F = 177 \, pF.$

b) $Q = C_0 V_0 = (17.7 \times 10^{-11} \, F)(3 \times 10^3 \, V) = 53.1 \times 10^{-8} \, C.$

c) $C = \dfrac{Q}{V} = \dfrac{53.1 \times 10^{-8} \, C}{10^3 \, V} = 53.1 \times 10^{-11} \, F = 531 \, pF.$

d) $K = \dfrac{C}{C_0} = \dfrac{53.1 \times 10^{-11} \, F}{17.7 \times 10^{-11} \, F} = 3.$

The dielectric constant could also be found from Eq. (27–12),

$$K = \dfrac{V_0}{V} = \dfrac{3000 \, V}{1000 \, V} = 3.$$

e) $\epsilon = K\epsilon_0 = (3)(8.85 \times 10^{-12} \, C^2 \cdot N^{-1} \cdot m^{-2})$

$\qquad = 26.6 \times 10^{-12} \, C^2 \cdot N^{-1} \cdot m^{-2}.$

f) $\qquad Q_i = A\sigma_i, \qquad Q = A\sigma,$

$\sigma - \sigma_i = \dfrac{\sigma}{K}, \qquad \sigma_i = \sigma\left(1 - \dfrac{1}{K}\right),$

$\qquad Q_i = Q\left(1 - \dfrac{1}{K}\right) = (53.1 \times 10^{-8} \, C)\left(1 - \dfrac{1}{3}\right)$

$\qquad\qquad = 35.4 \times 10^{-8} \, C.$

g) $E_0 = \dfrac{V_0}{d} = \dfrac{3000 \, V}{10^{-2} \, m} = 3 \times 10^5 \, V \cdot m^{-1}.$

h) $E = \dfrac{V}{d} = \dfrac{1000 \, V}{10^{-2} \, m} = 1 \times 10^5 \, V \cdot m^{-1};$

or

$\qquad E = \dfrac{\sigma}{\epsilon} = \dfrac{Q}{A\epsilon} = \dfrac{53.1 \times 10^{-8} \, C}{(2 \times 10^{-1} \, m^2)(26.6 \times 10^{-12} \, C^2 \cdot N^{-1} \cdot m^{-2})}$

$\qquad\qquad = 1 \times 10^5 \, V \cdot m^{-1};$

or
$$E = \frac{\sigma - \sigma_i}{\epsilon_0} = \frac{Q - Q_i}{A\epsilon_0}$$
$$= \frac{(53.1 - 35.4) \times 10^{-8} \text{ C}}{(2 \times 10^{-1} \text{ m}^2)(8.85 \times 10^{-12} \text{ C}^2 \cdot \text{N}^{-1} \cdot \text{m}^{-2})}$$
$$= 1 \times 10^5 \text{ V} \cdot \text{m}^{-1};$$

or, from Eq. (27-12),
$$E = \frac{E_0}{K} = \frac{3 \times 10^5 \text{ V} \cdot \text{m}^{-1}}{3} = 1 \times 10^5 \text{ V} \cdot \text{m}^{-1}.$$

Any dielectric material, when subjected to a sufficiently strong electric field, becomes a conductor, a phenomenon known as *dielectric breakdown*. The onset of conduction, associated with cumulative ionization of molecules of the material, is often quite sudden, and may be characterized by spark or arc discharges. When a capacitor is subjected to excessive voltage, an arc may be formed through a layer of dielectric, burning or melting a hole in it, permitting the two metal foils to come in contact, creating a short circuit, and rendering the device permanently useless as a capacitor.

The maximum electric field a material can withstand without the occurrence of breakdown is called the *dielectric strength*. Because dielectric breakdown is affected significantly by impurities in the material, small irregularities in the metal electrodes, and other factors difficult to control, only approximate figures can be given for dielectric strengths. The dielectric strength of dry air is about 0.8×10^6 V·m^{-1}. Typical values for plastic and ceramic materials commonly used to insulate capacitors and current-carrying wires are of the order of 10^7 V·m^{-1}. For example, a layer of such a material, 10^{-4} m in thickness, could withstand a maximum voltage of 1000 V, for 1000 V/10^{-4} m = 10^7 V·m^{-1}.

*27-6 Molecular theory of induced charges

It remains to be discussed how a dielectric, which supposedly has no electric charges that are free to move within the material, can nevertheless acquire a surface charge distribution as described in the preceding section. When a *conductor* is placed in an electric field, the free charges within it are displaced by the forces exerted on them by the field. In the final steady state, the conductor has an induced charge on its surface, distributed in such a way that the field of this induced charge neutralizes the original field at all internal points, and the net electric field within the conductor is reduced to zero. A *dielectric,* however, contains no free charges. How can induced charge appear on the surfaces of a dielectric when it is inserted in the electric field between the plates of a charged capacitor?

The explanation involves the redistribution of electric charge in the material at the molecular level. First, the molecules of a dielectric may be classified as either *polar* or *nonpolar*. A nonpolar molecule is one in which the "centers of gravity" of the positive nuclei and the electrons normally coincide, while a polar molecule is one in which they do not.

Symmetrical molecules such as H_2, N_2, and O_2 are nonpolar. In the molecules N_2O and H_2O, on the other hand, both nitrogen atoms or both hyrogen atoms lie on the same side of the oxygen atom. These molecules are *polar,* and each is a tiny electric dipole.

Under the influence of an electric field, the charges of a *nonpolar* molecule become displaced relative to each other, as indicated schematically in Fig. 27-8. The molecules are said to become *polarized* by the field and are called *induced dipoles*. When a nonpolar molecule becomes polarized, restoring forces come into play on the displaced charges, pulling them together much as if they were connected by a spring. Under the influence of a given external field, the charges separate until the restoring force is equal and opposite to the force exerted on the charges by the field. Naturally, the restoring forces vary in magnitude from one kind of molecule to another, with corresponding differences in the displacement produced by a given field.

When a dielectric consists of polar molecules or *permanent dipoles,* these dipoles are oriented at random when no electric field is present, as in Fig. 27-9a. When an electric field is present, as in Fig. 27-9b, the forces on a dipole give rise to a torque whose effect is to tend to orient the dipole in the same direction as the field. The stronger the field, the greater is the aligning effect.

Whether the molecules of a dielectric are polar or nonpolar, the net effect of an external field is substantially the same, as shown in Fig. 27-10. Within the two extremely thin surface layers indicated by light color lines there is an excess charge, negative in one layer and positive in the other. It is these layers of charge that give rise to the induced charge on the surface of a dielectric. The charges are not free, but each is *bound* to a molecule lying in or near the surface. Within the remainder of the dielectric the net charge per unit volume remains zero.

The four parts of Fig. 27-11 illustrate the behavior of a sheet of dielectric when inserted in the field between a pair of oppositely charged plane parallel plates. Part (a) shows the original field. Part (b) is the situation after the dielectric has been inserted but before any rearrangement of charges has occurred. Part (c) shows by dotted lines the field set up in the dielectric by its induced surface charges. This field is opposite to the original field but, since the charges in the dielectric are not free to move indefinitely, their displacement does not proceed to such an extent that the induced field is equal in magnitude to the original field. The field in the dielectric is therefore *weakened* but not reduced to zero, as it would be in the interior of a conductor.

The resultant field is shown in Fig. 27-11d. Some of the field lines leaving the positive plate penetrate the dielectric; others terminate on the induced charges on the faces of the dielectric.

The charges induced on the surface of a dielectric in an external field afford an explanation of the attraction of an *uncharged* pith ball or bit of paper by a charged rod of rubber or glass. Figure 27-12 shows an uncharged dielectric sphere B in the radial field of a positive charge A. The induced positive charges on B experience a force toward the right, while the force on the negative charges is toward the left. Since the negative charges are closer to A and therefore in a stronger field than are the positive, the force toward the left exceeds that toward the right, and B, although its net charge is zero, experiences a resultant force toward

27-8 Behavior of nonpolar molecules (a) in the absence and (b) in the presence of an electric field.

27-9 Behavior of polar molecules (a) in the absence and (b) in the presence of an electric field.

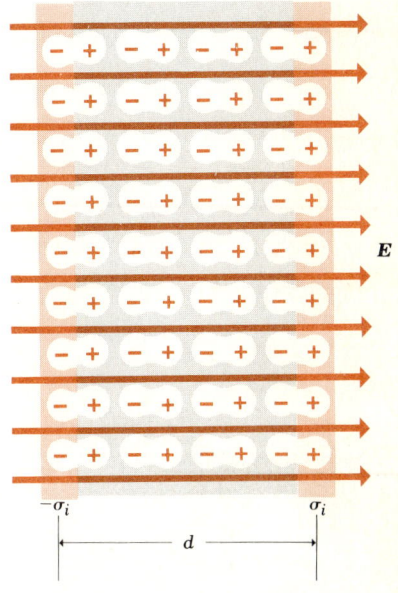

27-10 Polarization of a dielectric in an electric field gives rise to thin layers of bound charges on the surfaces.

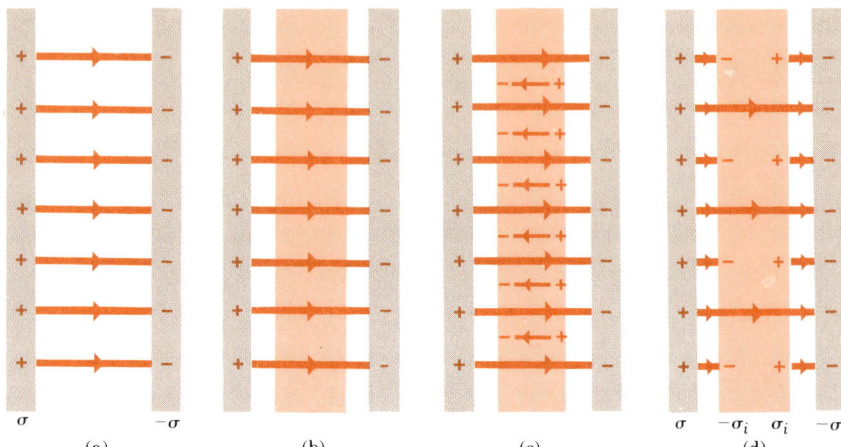

27-11 (a) Electric field between two charged plates. (b) Introduction of a dielectric. (c) Induced surface charges and their field. (d) Resultant field when a dielectric is between charged plates.

A. The sign of A's charge does not affect the conclusion, as may readily be seen. Furthermore, the effect is not limited to dielectrics; a conducting sphere would be similarly attracted.

More general arguments based on energy considerations show that a dielectric body in a nonuniform field *always* experiences a force urging it from a region where the field is weak toward a region where it is stronger, provided the dielectric constant of the body is greater than that of the medium in which it is immersed. If the dielectric constant is less, the reverse is true.

*27–7 Polarization and displacement

The extent to which the molecules of a dielectric become polarized by an electric field, or oriented in the direction of the field, is described by a vector quantity called the *polarization* $\boldsymbol{P}$. If $\boldsymbol{p}$ is the component of the vector dipole moment (defined in Sec. 25–1, Example 5) of each molecule in the direction of the applied field, and there are n molecules per unit volume, the polarization is defined as

$$\boldsymbol{P} = n\boldsymbol{p}. \tag{27–18}$$

Polarization is therefore *dipole moment per unit volume*. The polarization vector has the same direction as the molecular dipole moments, from left to right in Fig. 27–10. For the special case in Fig. 27–10, the magnitude of $\boldsymbol{P}$ is the same at all points of the dielectric. In other cases it can vary from point to point and the quantities n and $\boldsymbol{p}$ then refer to a small volume including the point. The SI unit of $\boldsymbol{P}$ is one *coulomb meter per cubic meter,* or one *coulomb per square meter* ($1\ \mathrm{C \cdot m^{-2}}$).

The polarized dielectric in Fig. 27–10 can be considered a single large dipole, consisting of the induced bound charges Q_i at the opposite faces, separated by the thickness d of the dielectric. The dipole moment of the dielectric is then $Q_i d$, and since the volume of the dielectric is the product of its cross-sectional area A and thickness d, the dipole moment per unit volume, or the polarization P, is

$$P = \frac{Q_i d}{A d} = \frac{Q_i}{A} = \sigma_i, \tag{27–19}$$

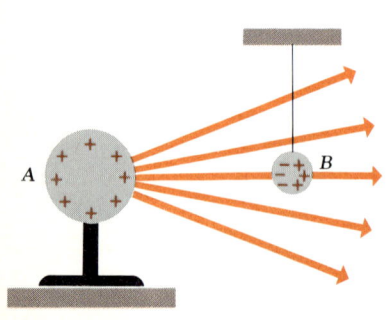

27-12 An uncharged dielectric sphere B in the radial field of a positive charge A.

where σ_i is the surface density of bound charge. In this special case, the polarization is numerically equal to the surface density of bound charge. More generally, the surface density of bound charge equals the normal component of P at the surface.

Figure 27-13 shows in cross section a sheet of polarized dielectric between two oppositely charged conducting plates. The thickness is exaggerated; we assume the sheet is very thin compared with its other dimensions, so that the electric field is uniform throughout the dielectric. The broken line is the outline of a Gaussian surface in the form of a cylinder with its ends parallel to the sheet and having cross-sectional area A. We consider the surface integral $\int P \cdot dA$ of the polarization vector P over this closed surface. P is zero at the left face, which is inside the left conducting plate. On the sides of the cylinder, P is parallel to the surface and makes no contribution to the integral. Only the right face, perpendicular to P, contributes, and gives simply PA. But, from Eq. (27-19), this is equal to Q_i, the total induced bound charge enclosed by the surface.

27-13

In the general case, we have

$$\oint P \cdot dA = -Q_i. \qquad (27\text{-}20)$$

The minus sign must be included because, from the diagram, the flux of P is *outward* (and hence positive) when the enclosed bound charge is negative. Equation (27-20) is Gauss's law for the polarization vector P: *the surface integral of P over any closed surface* (the flux of P) *equals the negative of the total bound charge within the surface.*

The resultant electric field E at any point, when bound charges are present, is due to *both* the free and bound charges. The general form of Gauss's law for E is therefore

$$\oint E \cdot dA = \frac{1}{\epsilon_0}(Q + Q_i). \qquad (27\text{-}21)$$

We may use Eq. (27-20) to eliminate the induced charge Q_i from Eq. (27-21); the result is

$$\oint E \cdot dA = \frac{1}{\epsilon_0}\left(Q - \oint P \cdot dA\right) \qquad (27\text{-}22)$$

or

$$\oint (\epsilon_0 E + P) \cdot dA = Q. \qquad (27\text{-}23)$$

Let us define a new quantity D called the *displacement* as the vector sum

$$D = \epsilon_0 E + P. \qquad (27\text{-}24)$$

Equation (27-23) then takes the simple form

$$\oint D \cdot dA = Q, \qquad (27\text{-}25)$$

which is Gauss's law for the displacement vector: *the surface integral of D over any closed surface* (the flux of D) *is equal to the free charge* **only** *within the surface.*

In summary, for any closed surface, the flux of E equals the *total* enclosed charge (divided by ϵ_0), the flux of P equals the (negative of the) *bound* charge, and the flux of D equals the *free* charge.

Applying Eq. (27-25) to the Gaussian surface of Fig. 27-13, we see that the magnitude of D is given simply by $D = \sigma$, where σ is the surface density of free charge on the conducting plate. Similarly, applying Gauss's law for E to this same surface gives $E = (\sigma - \sigma_i)/\epsilon_0$, where σ_i is the magnitude of the induced surface charge. Thus we find

$$\frac{D}{E} = \frac{\epsilon_0 \sigma}{\sigma - \sigma_i}.$$

Comparing this with Eq. (27-12), we obtain the simple relation

$$D = K\epsilon_0 E = \epsilon E. \qquad (27\text{-}26)$$

More generally, in any dielectric where the polarization is proportional to the electric field (so σ_i is proportional to σ), the vector quantities D and E are related by

$$D = \epsilon E.$$

This relation will be useful in our study of electromagnetic waves in later chapters.

Questions

27-1 Two identical capacitors are connected in series. Is the resulting capacitance greater or less than that of each individual capacitor? What if they are connected in parallel?

27-2 Could one define a capacitance for a single conductor? What would be a reasonable definition?

27-3 Suppose the two plates of a capacitor have different areas. When the capacitor is charged by connecting it to a battery, do the charges on the two plates have equal magnitude, or may they be different?

27-4 Can you think of a situation in which the two plates of a capacitor *do not* have equal magnitudes of charge?

27-5 A capacitor is charged by connecting it to a battery, and is then disconnected from the charging agency. The plates are then pulled apart a little. How does the electric field change? The potential difference? The total energy?

27-6 According to the text, the energy in a charged capacitor can be considered to be located in the field between the plates. But suppose there is vacuum between the plates; can there be energy in vacuum?

27-7 The charged plates of a capacitor attract each other. To pull the plates farther apart therefore requires work by some external force. What becomes of the energy added by this work?

27-8 A solid slab of metal is placed between the plates of a capacitor without touching either plate. Does the capacitance increase, decrease, or remain the same?

27-9 The two plates of a capacitor are given charges $\pm Q$, and then they are immersed in a tank of oil. Does the electric field between them increase, decrease, or remain the same? How might this field be measured?

27-10 Is dielectric strength the same thing as dielectric constant?

27-11 Liquid dielectrics having polar molecules (such as water) always have dielectric constants that decrease with increasing temperature. Why?

27-12 A capacitor made of aluminum foil strips separated by Mylar film was subjected to excessive voltage, and the resulting dielectric breakdown melted holes in the Mylar. After this the capacitance was found to be about the same as before, but the breakdown voltage was much less. Why?

27-13 Two capacitors have equal capacitance, but one has a higher maximum voltage rating than the other. Which one is likely to be bulkier? Why?

27-14 A capacitor is made by rolling a sandwich of aluminum foil and Mylar, as described in Sec. 27-5. A student claimed that the capacitance when the sandwich is rolled up is twice the value when it is flat. Discuss his allegation.

Problems

27-1 An air capacitor is made from two flat parallel plates 0.5 mm apart. The magnitude of charge on each plate is 0.01 μC when the potential difference is 200 V.

a) What is the capacitance?
b) What is the area of each plate?
c) What maximum voltage can be applied without dielectric breakdown?
d) When the charge is 0.01 μC, what total energy is stored?

27-2 A parallel-plate air capacitor has a capacitance of 0.001 μF.

a) What potential difference is required for a charge of 0.5 μC on each plate?
b) In (a), what is the total stored energy?
c) If the plates are 1.0 mm apart, what is the area of each plate?
d) What potential difference is required for dielectric breakdown?

27-3 An air capacitor consisting of two closely spaced parallel plates has a capacitance of 1000 pF. The charge on each plate is 1 μC.

a) What is the potential diference between the plates?
b) If the charge is kept constant, what will be the potential difference between the plates if the separation is doubled?
c) How much work is required to double the separation?

27-4 A capacitor has a capacitance of 8.5 μF. How much charge must be removed to lower the potential difference of its plates by 50 V?

27-5 The capacitance of a variable capacitor can be changed from 50 pF to 950 pF by turning the dial from 0° to 180°. With the dial set at 180°, the capacitor is connected to a 400-V battery. After charging, the capacitor is disconnected from the battery and the dial is turned to 0°.

a) What is the charge on the capacitor?
b) What is the potential difference across the capacitor when the dial reads 0°?
c) What is the energy of the capacitor in this position?
d) How much work is required to turn the dial, if friction is neglected?

27-6 A 20-μF capacitor is charged to a potential difference of 1000 V. The terminals of the charged capacitor are then connected to those of an uncharged 5-μF capacitor. Compute

a) the original charge of the system,
b) the final potential difference across each capacitor,
c) the final energy of the system, and
d) the decrease in energy when the capacitors are connected.

27-7 In Fig. 27-14, each capacitance C_3 is 3 μF and each capacitance C_2 is 2 μF.

a) Compute the equivalent capacitance of the network between points a and b.
b) Compute the charge on each of the capacitors nearest a and b when $V_{ab} = 900$ V.
c) With 900 V across a and b, compute V_{cd}.

Figure 27-14

27-8 A number of 0.5-μF capacitors are available. The voltage across each is not to exceed 400 V. A capacitor of capacitance 0.5 μF is required to be connected across a potential difference of 600 V.

a) Show in a diagram how an equivalent capacitor having the desired properties can be obtained.
b) No dielectric is a perfect insulator, of infinite resistance. Suppose that the dielectric in one of the capacitors in your diagram is a moderately good conductor. What will happen?

27-9 A 1-μF capacitor and a 2-μF capacitor are connected in series across a 1200-V supply line.

a) Find the charge on each capacitor and the voltage across each.
b) The charged capacitors are disconnected from the line and from each other, and reconnected with terminals of like sign together. Find the final charge on each and the voltage across each.

27-10 A 1-μF capacitor and a 2-μF capacitor are connected in parallel across a 1200-V supply line.

a) Find the charge on each capacitor and the voltage across each.
b) The charged capacitors are then disconnected from the line and from each other, and reconnected with terminals of unlike sign together. Find the final charge on each and the voltage across each.

27-11 In Fig. 27-4a, let $C_1 = 6\,\mu F$, $C_2 = 3\,\mu F$, and $V_{ab} = 18$V. Suppose that the charged capacitors are disconnected from the source and from each other, and reconnected with plates of *opposite* sign connected together. By how much does the energy of the system decrease?

27-12 Three capacitors having capacitances of 8, 8, and 4 μF are connected in series across a 12-V line.

a) What is the charge on the 4-μF capacitor?
b) What is the total energy of all three capacitors?
c) The capacitors are disconnected from the line and reconnected in parallel with the positively charged plates connected together. What is the voltage across the parallel combination?
d) What is the energy of the combination?

27-13 A 500-μF capacitor is charged to 120 V. How many calories are produced on discharging the capacitor if all of the energy goes into heating the wire?

27-14 The capacitors in Fig. 27-15 are initially uncharged, and are connected as in the diagram with switch S open.

a) What is the potential difference V_{ab}?
b) What is the potential of point b after switch S is closed?
c) How much charge flowed through the switch when it was closed?

Figure 27-15

27-15 The plates of a parallel-plate capacitor in vacuum have charges $+Q$ and $-Q$ and the distance between the plates is x. The plates are disconnected from the charging voltage and pulled apart a short distance dx.

a) What is the change dC in the capacitance of the capacitor?
b) What is the change dW in its energy?
c) Equate the work $F\,dx$ to the increase in energy dW and find the force of attraction F between the plates.
d) Explain why F is not equal to QE, where E is the electric field between the plates.

27-16 A parallel-plate capacitor is to be constructed using, as a dielectric, rubber having a dielectric constant of 3 and a dielectric strength of 2×10^5 V·cm^{-1}. The capacitor is to have a capacitance of 0.15 μF and must be able to withstand a maximum potential difference of 6000 V. What is the minimum area the plates of the capacitor may have?

27-17 The paper dielectric in a paper and foil capacitor is 0.005 cm thick. Its dielectric constant is 2.5 and its dielectric strength is 50×10^6 V·m^{-1}.

a) What area of paper, and of tinfoil, is required for a 0.1-μF capacitor?
b) If the electric intensity in the paper is not to exceed one-half the dielectric strength, what is the maximum potential difference that can be applied across the capacitor?

27-18 A Mylar–aluminum foil capacitor is to be designed for a capacitance of 0.1 μF and a voltage rating of 500 V. It is to be made from strips of aluminum foil and Mylar about 4 cm wide, and the foil has a thickness of 0.02 mm.

a) What thickness Mylar film is required?
b) What total area of Mylar and aluminum is required?
c) What is the diameter of the rolled-up capacitor?

27-19 A parallel-plate air capacitor is made using two plates 0.2 m square, spaced 1 cm apart. It is connected to a 50-V battery.

a) What is the capacitance?
b) What is the charge on each plate?
c) What is the electric field between the plates?
d) What is the energy stored in the capacitor?
e) If the battery is disconnected and then the plates are pulled apart to a separation of 2 cm, what are the answers to parts (a), (b), (c), and (d)?

27-20 In Problem 27-19, suppose the battery remains connected while the plates are pulled apart. What are the answers to parts (a), (b), (c), and (d)?

27-21 A parallel-plate capacitor with plate area A and separation x is charged to a charge of magnitude q on each plate.

a) What is the total energy stored in the capacitor?
b) The plates are now pulled apart an additional distance dx; now what is the total energy?
c) If F is the force with which the plates attract each other, then the difference in the two energies above must equal the work $dW = F\,dx$ done in pulling the plates apart. Hence show that $F = q^2/2\epsilon_0 A$.

27-22 A parallel-plate capacitor has the space between the plates filled with a slab of dielectric with constant K_1 and one with constant K_2, each of thickness $d/2$, where d is the plate separation. Show that the capacitance is

$$C = \frac{2\epsilon_0 A}{d}\left(\frac{K_1 K_2}{K_1 + K_2}\right).$$

27-23 Three square metal plates A, B, and C, each 10 cm on a side and 3 mm thick, are arranged as in Fig. 27-16. The plates are separated by sheets of paper 0.5 mm thick and of

Figure 27-16

dielectric constant 5. The outer plates are connected together and connected to point b. The inner plate is connected to point a.

a) Copy the diagram, and show by + and − signs the charge distribution on the plates when point a is maintained at a positive potential relative to point b.

b) What is the capacitance between points a and b?

27-24 Two parallel plates have equal and opposite charges. When the space between the plates is evacuated, the electric field is 2×10^5 V·m^{-1}. When the space is filled with dielectric, the electric field is 1.2×10^5 V·m^{-1}.

a) What is the charge density on the surface of the dielectric?

b) What is its dielectric constant?

27-25 Two oppositely charged conducting plates, having numerically equal quantities of charge per unit area, are separated by a dielectric 5 mm thick, of dielectric constant 3. The resultant electric field in the dielectric is 10^6 V·m^{-1}. Compute

a) the charge per unit area on the conducting plate and

b) the charge per unit area on the surfaces of the dielectric.

27-26 A capacitor consists of two parallel plates of area 25 cm^2 separated by a distance of 0.2 cm. The material between the plates has a dielectric constant of 5. The plates of the capacitor are connected to a 300-V battery.

a) What is the capacitance of the capacitor?

b) What is the charge on either plate?

c) What is the energy in the charged capacitor?

d) What is the energy density in the dielectric?

27-27 Two parallel plates of 100 cm^2 area are given equal and opposite charges of 10^{-7} C. The space between the plates is filled with a dielectric material, and the electric field within the dielectric is 3.3×10^5 V·m^{-1}.

a) What is the dielectric constant of the dielectric?

b) What is the total induced charge on either face of the dielectric?

27-28 An air capacitor is made using two flat plates of area A separated by a distance d. Then a metal slab having thickness a (less than d) and the same shape and size as the plates is inserted between them, parallel to the plates and not touching either plate.

a) What is the capacitance of this arrangement?

b) Express the capacitance as a multiple of the capacitance when the metal slab is not present.

27-29 A spherical capacitor consists of an inner metal sphere of radius r_a supported on an insulating stand at the center of a hollow metal sphere of inner radius r_b; there is a charge $+Q$ on the inner sphere and a charge $-Q$ on the outer. (See Problem 26-24.)

a) What is the potential difference V_{ab} between the spheres?

b) Prove that the capacitance is

$$C = 4\pi\epsilon_0 \cdot \frac{r_b r_a}{r_b - r_a}.$$

27-30 A coaxial cable consists of an inner solid cylindrical conductor of radius r_a supported by insulating disks on the axis of a conducting tube of inner radius r_b. The two cylinders are oppositely charged with a charge λ per unit length. (See Problem 26-25.)

a) What is the potential difference between the two cylinders?

b) Prove that the capacitance of a length L of the cable is

$$C = \frac{2\pi\epsilon_0 L}{\ln(r_b/r_a)}.$$

Neglect any effect of the supporting disks.

CURRENT, RESISTANCE, AND ELECTROMOTIVE FORCE

28

28-1 Current

When there is a net flow of charge perpendicular to any area (passing through it from one side to the other), we say there is a *current* across the area. If an isolated conductor is placed in an electrostatic field, the charges in the conductor rearrange themselves so as to make the interior of the conductor a field-free region throughout which the potential is constant. This motion of the charges constitutes a *transient* current, of short duration only, and the current ceases when the field in the conductor becomes zero. To maintain a *continuous* current, we must in some way maintain a force on the mobile charges in a conductor. The force may result from an electrostatic field or from other causes that will be described later. For the present, we assume that within a conductor an electric field E is maintained such that a charged particle in the conductor is acted on by a force $F = qE$. We shall refer to this force as the *driving force* on the particle.

The motion of a free charged particle in a conductor is very different from that of a particle in empty space. After a momentary acceleration, the particle makes an inelastic collision with one of the fixed particles in the conductor and makes a fresh start. Thus *on the average* it moves in the direction of the driving force with an average velocity called its *drift velocity*. The inelastic collisions with the fixed particles result in a transfer of energy to them. This increases their energy of vibration and causes a rise in temperature if the conductor is thermally insulated, or results in a flow of heat from the conductor to its surroundings if it is not.

The current across an area is defined quantitatively as *the net charge flowing across the area per unit time*. Thus if a net charge ΔQ flows across a certain area in a time interval Δt, the average current I_{av} across this area is

$$I_{av} = \frac{\Delta Q}{\Delta t}.$$

The rate of flow of charge may not be constant, in which case we generalize the definition of current in a natural way, using the derivative. The instantaneous current I is defined as

$$I = \frac{dQ}{dt}. \qquad (28\text{-}1)$$

Current is a *scalar* quantity.

The SI unit of current, *one coulomb per second*, is called *one ampere* (1 A), in honor of the French scientist André Marie Ampère (1775–1836). Small currents are more conveniently expressed in *milliamperes* (1 mA = 10^{-3} A) or in *microamperes* (1 μA = 10^{-6} A).

The current across an area can be expressed in terms of the drift velocity of the moving charges as follows. Consider a portion of a conductor of cross-sectional area A within which there is a resultant electric field E from left to right. We suppose first that the conductor contains free *positively* charged particles; these move in the same direction as the field. A few positive particles are shown in Fig. 28–1. Suppose there are n such particles per unit volume, all moving with a drift velocity v. In a time Δt each advances a distance $v\,\Delta t$. Hence, all of the particles within the shaded cylinder of length $v\,\Delta t$, and only those particles, will flow across the end of the cylinder in time Δt. The volume of the cylinder is $Av\,\Delta t$, the number of particles within it is $nAv\,\Delta t$, and if each has a charge q, the charge ΔQ flowing across the end of the cylinder in time Δt is

$$\Delta Q = nqvA\,\Delta t,$$

The current carried by the positively charged particles is therefore

$$I = \frac{\Delta Q}{\Delta t} = nqvA. \qquad (28\text{-}2)$$

If the moving charges are negative rather than positive, the electric-field force is opposite to E, and so the drift velocity is right to left, opposite to the direction shown in Fig. 28–1. Positive particles crossing from left to right *increase* the *positive* charge at the right of the section, while negative particles crossing from right to left *decrease* the *negative* charge at the right of the section. But a *decrease* of *negative* charge is equivalent to an *increase* of *positive* charge, so the motion of *both* kinds of charge has the same effect, namely, to increase the positive charge at the right of the section. In either case, particles flowing out through an end of the cylindrical section are continuously replaced by particles flowing *in* through the oposite end.

In general, a conductor may contain a number of different kinds of charged particles having charges q_i, densities n_i, and drift velocities v_i; the total current is then

$$I = A \sum n_i q_i v_i. \qquad (28\text{-}3)$$

In metals the moving charges are always (negative) electrons, while an ionized gas has moving electrons and positively charged ions. In a semiconductor material such as germanium or silicon, conduction is partly by electrons and partly by motion of vacancies, also known as *holes*; these are sites of missing electrons and act like positive charges.

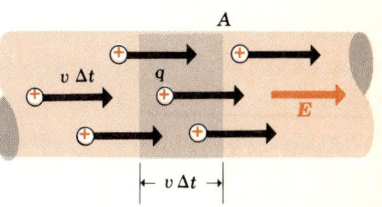

28–1 All of the particles, and only those particles, within the shaded cylinder will cross its base in time Δt.

The *current per unit cross-section area* is called the *current density J*:

$$J = \frac{I}{A} = \sum nqv. \qquad (28\text{-}4)$$

The *vector current density* **J** is defined by the equation

$$\mathbf{J} = \sum n_i q_i \mathbf{v}_i. \qquad (28\text{-}5)$$

Current, by definition, is a scalar quantity and thus it is not correct to speak of the "direction of a current." This expression is often used, however, for brevity, meaning thereby the direction of the *vector* current density **J**.

The direction of the drift velocity **v** of a positive charge is the same as that of the electric field **E**, and the direction of the velocity of a negative charge is opposite to **E**. But since the charge q is negative, each of the vectors $nq\mathbf{v}$ is in the same direction as **E**, and hence the *vector current density* **J** *always has the same direction as the field* **E**. Thus even in a metallic conductor, where the moving charges are negative electrons only and move in the *opposite* direction to **E**, the *vector* current density **J** is in the *same* direction as **E**.

The above discussion has shown that the effect of a current is the same, whether it consists of positive charges moving in the direction of **E**, of negative charges moving in the opposite direction, or a combination of the two. In describing circuit behavior it is customary to describe currents as though they consisted entirely of positive charge flow, even in cases where the actual current is known to be due to electrons. This convention will be followed consistently in the following sections. In Chapter 31, in studying the effect of a *magnetic* field on a moving charge, we shall consider a phenomenon, the Hall effect, in which the sign of the moving charges *is* important.

When there is a steady current in a closed loop (a "complete circuit"), the total charge in every portion of a conductor remains constant. Hence, if we consider a portion between two fixed cross sections, the rate of flow of charge *out* of the portion at one end equals the rate of flow of charge *into* the portion at the other end. In other words, the current is the same at any two cross sections and hence is *the same at all cross sections*. Current is *not* something that squirts out of the positive terminal of a battery and gets all used up by the time it reaches the negative terminal! These statements are direct consequences of the principle of conservation of charge, introduced in Sec. 24-2.

Example A copper conductor of square cross section 1 mm on a side carries a constant current of 20 A. The density of free electrons is 8×10^{28} electrons per cubic meter. Find the current density and the drift velocity.

Solution The current density in the wire is

$$J = \frac{I}{A} = 20 \times 10^6 \text{ A} \cdot \text{m}^{-2}.$$

From Eq. (28-4),

$$v = \frac{J}{nq} = \frac{(20 \times 10^6 \text{ A} \cdot \text{m}^{-2})}{(8 \times 10^{28} \text{ m}^{-3})(1.6 \times 10^{-19} \text{ C})}$$
$$= 1.6 \times 10^{-3} \text{ m} \cdot \text{s}^{-1},$$

or about 1.6 mm·s⁻¹. At this speed, an electron would require 625 s or about 10 min to travel the length of a wire 1 m long. Thus the drift velocity is *very* small compared with the velocity of propagation of a current pulse along a wire, about 3×10^8 m·s⁻¹.

28-2 Resistivity

The current density J in a conductor depends on the electric field E, and on the nature of the conductor. In general the dependence of J on E can be quite complex but, for some materials, especially the metals, it can be represented quite well by a direct proportionality. For such materials the ratio of E to J is constant; we define the *resistivity* ρ of a particular material as the ratio of electric field to current density:

$$\rho = \frac{E}{J}. \tag{28-6}$$

That is, the resistivity is the *electric field per unit current density*. The greater the resistivity, the greater the field needed to establish a given current density, or the smaller the current density for a given field. Representative values are given in Table 28-1. The unit $\Omega \cdot$m (ohm·meter) will be explained in the following section. A "perfect" conductor would have zero resistivity and a "perfect" insulator an infinite resistivity. Metals and alloys have the lowest resistivities and are the best conductors. The resistivities of insulators exceed those of the metals by a factor of the order of 10^{22}.

Table 28-1 Resistivities at room temperature

Substance		ρ, $\Omega \cdot$m	Substance		ρ, $\Omega \cdot$m
Conductors			Semiconductors		
Metals	Silver	1.47×10^{-8}	Pure	Carbon	3.5×10^{-5}
	Copper	1.72×10^{-8}		Germanium	0.60
	Gold	2.44×10^{-8}		Silicon	2300
	Aluminum	2.63×10^{-8}	Insulators		
	Tungsten	5.51×10^{-8}		Amber	5×10^{14}
	Steel	20×10^{-8}		Glass	10^{10}–10^{14}
	Lead	22×10^{-8}		Lucite	$> 10^{13}$
	Mercury	95×10^{-8}		Mica	10^{11}–10^{15}
Alloys	Manganin	44×10^{-8}		Quartz (fused)	75×10^{16}
	Constantan	49×10^{-8}		Sulfur	10^{15}
	Nichrome	100×10^{-8}		Teflon	$> 10^{13}$
				Wood	10^8–10^{11}

Comparison with Table 16-1 shows that *thermal* insulators have thermal resistivities (the reciprocals of their thermal conductivities) that differ from those of good thermal conductors by factors of only about 10^3. By the use of electrical insulators, electric currents can be confined to well-defined paths in good electrical conductors, while it is impossible to confine heat currents to a comparable extent. It is also interesting to note that the metals, as a class, are also the best *thermal* conductors. The free electrons in a metal that carry charge in electrical conduction also play an important role in the conduction of heat; hence, a correlation can be expected between electrical and thermal conductivity. It is a familiar fact that good electrical conductors, such as the metals, are also good conductors of heat, while poor electrical conductors, such as ceramic and plastic materials, are also poor thermal conductors.

The *semiconductors* form a class intermediate between the metals and the insulators. They are of importance not primarily because of their resistivities, but because of the way in which these are affected by temperature and by small amounts of impurities.

The discovery that ρ is a constant for a metallic conductor at constant temperatures was made by G. S. Ohm (1789-1854) and is called *Ohm's law*. A material obeying Ohm's law is called an *ohmic* conductor or a *linear* conductor. If Ohm's law is *not* obeyed, the conductor is called *nonlinear*. Thus Ohm's law, like the ideal gas equation, Hooke's law, and many other relations describing the properties of materials, is an *idealized model* that describes the behavior of certain materials reasonably well but is by no means a general property of all matter.

The resistivity of all *metallic* conductors increases with increasing temperature, as shown in Fig. 28-2a. Over a temperature range that is not too great, the resistivity of a metal can be represented approximately by the equation

$$\rho_T = \rho_0[1 + \alpha(T - T_0)], \tag{28-7}$$

where ρ_0 is the resistivity at a reference temperature T_0 (often taken as 0°C or 20°C) and ρ_T the resistivity at temperature T°C. The factor α is called the *temperature coefficient of resistivity*. Some representative values are given in Table 28-2. The resistivity of carbon (a nonmetal) *decreases* with increasing temperature and its temperature coefficient of resistivity is negative. The resistivity of the alloy manganin is practically independent of temperature.

A number of materials have been found to exhibit the property of *superconductivity*. As the temperature is decreased, the resistivity at first decreases regularly, like that of any metal. At the so-called *critical temperature*, usually in the range 0.1 K to 20 K, a phase transition occurs, and the resistivity suddenly drops to zero, as shown in Fig. 28-2b. A current once established in a superconducting ring will continue of itself, apparently indefinitely, without the presence of any driving field.

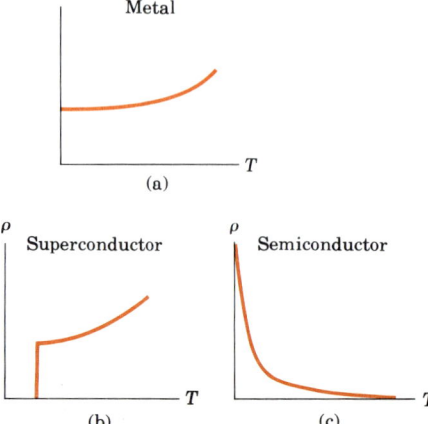

28-2 Variation of resistivity with temperature for three conductors: (a) an ordinary metal, (b) a superconducting metal, alloy, or compound, and (c) a semiconductor.

Table 28-2 Temperature coefficients of resistivity (approximate values near room temperature)

Material	α, °C^{-1}	Material	α, °C^{-1}
Aluminum	0.0039	Lead	0.0043
Brass	0.0020	Manganin (Cu 84,	0.000000
Carbon	−0.0005	Mn 12, Ni 4)	
Constantan (Cu 60, Ni 40)	+0.000002	Mercury	0.00088
Copper (Commercial	0.00393	Nichrome	0.0004
annealed)		Silver	0.0038
Iron	0.0050	Tungsten	0.0045

The resistivity of a *semiconductor* decreases rapidly with increasing temperature, as shown in Fig. 28-2c. A tiny bead of semiconducting material, called a *thermistor,* serves as a sensitive thermometer.

28-3 Resistance

The current density J, at a point within a conductor where the electric field is E, is given by Eq. (28-6):

$$E = \rho J.$$

It is often difficult to measure E and J directly, and it is useful to put this relation in a form involving readily measured quantities such as total current and potential difference. To do this we consider a conductor with uniform cross-section area A and length l, as shown in Fig. 28-3. Assuming a constant current density over a cross section, and a uniform electric field along the length of the conductor, the total current I is given by

$$I = JA,$$

and the potential difference V between the ends is

$$V = El. \tag{28-8}$$

Solving these equations for J and E, respectively, and substituting the results in Eq. (28-6), we obtain

$$\frac{V}{l} = \frac{\rho I}{A}. \tag{28-9}$$

Thus the total current is proportional to the potential difference.

The quantity $\rho l/A$ for a particular specimen of material is called its *resistance R*:

$$R = \frac{\rho l}{A}. \tag{28-10}$$

Equation (28-9) then becomes

$$V = IR. \tag{28-11}$$

This relation is often referred to as *Ohm's law;* in this form it refers to a specific piece of material, not to a general property of the material as with Eq. (28-6).

Equation (28-10) shows that the resistance of a wire or other conductor of uniform cross section is directly proportional to its length and

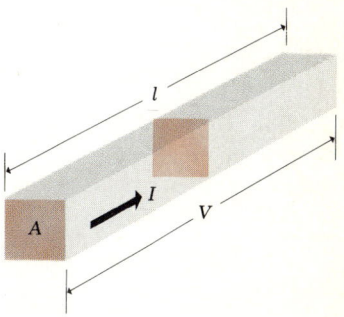

28-3 A conductor of uniform cross section. The current density is uniform over any cross section, and the electric field is constant along the length.

inversely proportional to its cross-section area. It is of course also proportional to the resistivity of the material of which the conductor is made.

The SI unit of resistance is one *volt per ampere* ($1 \text{ V} \cdot \text{A}^{-1}$). A resistance of $1 \text{ V} \cdot \text{A}^{-1}$ is called *1 ohm* ($1 \text{ }\Omega$). The unit of resistivity is therefore one *ohm·meter* ($1 \text{ }\Omega \cdot \text{m}$).

Large resistances are conveniently expressed in *kilohms* ($1 \text{ k}\Omega = 10^3 \text{ }\Omega$) or *megohms* ($1 \text{ M}\Omega = 10^6 \text{ }\Omega$), and small resistances in microhms ($1 \text{ }\mu\Omega = 10^{-6} \text{ }\Omega$). Resistivities are also expressed in a variety of hybrid units, most common of which is the *ohm·centimeter* ($1 \text{ }\Omega \cdot \text{cm} = 10^{-2} \text{ }\Omega \cdot \text{m}$).

Because the resistance of any specimen of material is proportional to its resistivity, which varies with temperature, resistance also varies with temperature. For temperature ranges that are not too great, this variation may be represented approximately as a linear relation analogous to Eq. (28-7):

$$R_T = R_0[1 + \alpha(T - T_0)]. \tag{28-12}$$

Here R_T is the resistance at temperature T, and R_0 is the resistance at the temperature T_0, often taken to be 20°C or 0°C. Within the limits of validity of Eq. (28-12), the *change* in resistance resulting from a temperature change $T - T_0$ is given by $\alpha(T - T_0)$, where α is the temperature coefficient of resistivity, given for several common materials in Table 28-2.

Example 1 For the example at the end of Sec. 28-1, find the electric field, and the potential difference between two points 100 m apart.

Solution From Eq. (28-6), the electric field is given by

$$E = \rho J = (1.72 \times 10^{-8} \text{ }\Omega \cdot \text{m})(20 \times 10^6 \text{ A} \cdot \text{m}^{-2})$$
$$= 0.344 \text{ V} \cdot \text{m}^{-1}.$$

The potential difference is given by

$$V = El = (0.344 \text{ V} \cdot \text{m}^{-1})(100 \text{ m}) = 34.4 \text{ V}.$$

Thus the resistance of a piece of this wire 100 m in length is

$$R = \frac{V}{I} = \frac{34.4 \text{ V}}{20 \text{ A}} = 1.72 \text{ }\Omega.$$

This result can also be obtained directly from Eq. (28-10):

$$R = \frac{\rho l}{A} = \frac{(1.72 \times 10^{-8} \text{ }\Omega \cdot \text{m})(100 \text{ m})}{(1 \times 10^{-3} \text{ m})^2} \times 1.72 \text{ }\Omega.$$

Example 2 In the previous example, suppose the resistance is $1.72 \text{ }\Omega$ at a temperature of 20°C. Find the resistance at 0°C and at 100°C.

Solution We use Eq. (28-12). In this instance $T_0 = 20°\text{C}$ and $R_0 = 1.72 \text{ }\Omega$. From Table 28-2, the temperature coefficient of resistivity of copper is $\alpha = 0.00393(°\text{C})^{-1}$. Thus at $T = 0°\text{C}$,

$$R = 1.72 \text{ }\Omega[1 + (0.00393°\text{C}^{-1})(0°\text{C} - 20°\text{C})]$$
$$= 1.58 \text{ }\Omega,$$

and at $T = 100°C$,

$$R = 1.72\,\Omega[1 + (0.00393°C^{-1})(100°C - 20°C)]$$
$$= 2.26\,\Omega.$$

Example 3 The space between two metallic coaxial cylinders of radii r_a and r_b is filled with a material of resistivity ρ. What is the resistance between the cylinders?

Solution Equation (28–10) cannot be used directly because the cross section through which the charge travels varies from $2\pi r_a l$ at the inner cylinder to $2\pi r_b l$ at the outer cylinder. Instead, we consider a cylindrical shell of inner radius r and thickness dr. The area A is then $2\pi rl$, and the length of the current path through the shell is dr. Thus, the resistance dR of the shell is

$$dR = \frac{\rho\,dr}{2\pi rl}$$

and the total resistance between the cylinders is

$$R = \frac{\rho}{2\pi l}\int_{r_a}^{r_b}\frac{dr}{r} = \frac{\rho}{2\pi l}\ln\frac{r_b}{r_a}.$$

28-4 Electromotive Force

In order for a steady current to exist in a conducting path, that path must form a closed loop or *complete circuit*. Otherwise charge would accumulate at the ends of the conductor, the resulting electric field would change with time, and the current could not be constant.

However, such a path cannot consist solely of resistance. Current in a resistor requires an electric field and an associated potential. The field always does *positive* work on the charge, which moves always in the direction of *decreasing* potential. But after a complete trip around the loop, a charge returns to its starting point, and the potential then must be the same as when it left that point. This cannot be so if its travel around the loop involves only *decreases* in potential.

Thus there must be some portion of the loop where a charge travels from lower to higher potential, despite the electrostatic force trying to push it from higher to lower potential. The influence that makes charge move from lower to higher potential is called *electromotive force*. Every complete circuit in which there is a steady current must have some device which provides electromotive force.

Batteries, generators, photovoltaic cells, and thermocouples are examples of such devices, which are called *seats of electromotive force*. Any such device can transfer energy into a circuit in which it is connected; thus such a device is sometimes called a *source,* although the term *energy converter* is preferable. Electromotive force is usually abbreviated emf, pronounced "ee-em-eff."

Figure 28–4 is a schematic representation of a seat of emf, such as a battery or generator. Such a device has the property that it can maintain a potential difference between conductors a and b, called the *terminals* of the device. In Fig. 28–4, there is no conducting path *outside* the device connecting a and b, and the device is said to be on *open circuit*.

28–4 Schematic diagram showing the general directions of the electrostatic field $\mathbf{E}_e$ and the non-electrostatic field $\mathbf{E}_n$ within a source on open circuit. In this case, $\mathbf{E}_n = -\mathbf{E}_e$ and $V_{ab} = \mathcal{E}$.

Terminal a, marked $+$, is maintained by the source at a *higher* potential than terminal b, marked $-$. Associated with this potential difference is an electrostatic field $\mathbf{E}_e$, at all points between and around the terminals, both inside and outside the source. The electrostatic field $\mathbf{E}_e$ inside the device is directed from a toward b, as shown. The source is itself a conductor, however, and if the *only* force on the free charges within it were that exerted by the electrostatic field, the positive charges would move from a toward b (or the negative charges from b toward a). The excess charges on the terminals would decrease, and the potential difference between them would decrease and eventually become zero.

But this is not the way batteries and generators actually act; in fact, they maintain a potential difference even when there is a steady current. From this we conclude that there must be some additional force on the charges within the source, tending to push them from a lower-potential point to a higher, opposite to the tendency of the electrostatic force. The origin of the non-electrostatic force depends on the nature of the source. In a generator it results from the action of a *magnetic* field on moving charges. In a battery it is associated with varying electrolyte concentrations arising from chemical reactions. In an electrostatic machine such as a Van de Graaff or Wimshurst generator, it is a mechanical force applied by a moving belt or wheel.

Whatever the origin of the non-electrostatic force, which we may call $\mathbf{F}_n$, its effect is the same as though there were an additional electric field $\mathbf{E}_n$, of non-electrostatic origin, related to the force by $\mathbf{F}_n = q\mathbf{E}_n$. That is, the non-electrostatic force is the same *as if* there were a non-electrostatic field $\mathbf{E}_n$ in addition to the purely electrostatic field $\mathbf{E}_e$.

When the source is on open circuit, as in Fig. 28–4, the charges are in equilibrium, and the *resultant* field $\mathbf{E}$, the vector sum of $\mathbf{E}_e$ and $\mathbf{E}_n$, must be zero at every point:

$$\mathbf{E}_e + \mathbf{E}_n = 0.$$

Now the electrostatic potential difference V_{ab} is defined as the work per unit charge performed by the *electrostatic* field $\mathbf{E}_e$ on a charge moving from a to b. Similarly, we may consider the work done by the *non-electrostatic* field $\mathbf{E}_n$. It is customary to speak of the (positive) work of this field during a displacement from b to a, rather than the reverse. Specifically, the work performed by $\mathbf{E}_n$, per unit charge, when a charge moves from b to a, is called the *electromotive force* $\mathcal{E}$ *of the source.*

When $\mathbf{E}_e = -\mathbf{E}_n$, we have $V_{ab} = \mathcal{E}$. Hence *for a source on open circuit,* the potential difference V_{ab}, or the *open-circuit terminal voltage,* is equal to the electromotive force:

$$V_{ab} = \mathcal{E} \quad \text{(source on open circuit)}. \tag{28–13}$$

The term *electromotive force,* although widely used, is somewhat unfortunate, in that the concept to which it refers is not a *force* but a *work per unit charge.* The term *electromotance* is sometimes used, but the concept is usually referred to simply as *emf,* pronounced "ee-em-eff."

The SI unit of $\mathbf{E}_n$ is the same as that of $\mathbf{E}_e$, namely, one volt per meter, so the unit of emf is the same as that of potential or potential difference, namely, 1 V. However, an electromotive force is not the *same thing* as a potential difference, since the latter is the work of an *electrostatic* field and the former is the work of a *non-electrostatic* field.

As we shall see later, the *electrostatic* field within a source, and hence the potential difference between its terminals, depend on the current in the source. The *non*-electrostatic field, and hence the emf of the source, is in many cases a *constant* independent of the current, and hence the emf represents a definite property of a source. Unless stated otherwise, we shall assume, in what follows, that the emf of a source is constant.

Now suppose that the terminals of a source are connected by a wire, as shown schematically in Fig. 28-5, forming a *complete circuit*. The driving force on the free charges *in the wire* is due solely to the *electrostatic* field E_e set up by the charged terminals a and b of the source. This field sets up a current *in the wire* from a toward b. The charges on the terminals decrease slightly and the electrostatic fields, both within the wire and within the source, decrease also. As a result, the *electrostatic* field within the source becomes smaller than the (constant) *non-electrostatic* field. Hence positive charges within the source are driven toward the positive terminal, and there is a current within the source from b toward a. The circuit settles down to a steady state in which the current is the same at all cross sections.

If current could travel through the source without impediment (that is, if the source had no *internal* resistance), charge entering the external circuit through terminal a would be replaced immediately by charge flow through the source. In this case the internal electrostatic field in the source would not change under complete-circuit conditions, and the terminal potential difference V_{ab} would still be equal to $\mathcal{E}$. Since V_{ab} is also related to the current and resistance in the external circuit by Eq. (28-11), we would then have

$$\mathcal{E} = IR, \quad \text{no internal resistance} \qquad (28\text{-}14)$$

where R is the resistance of the external circuit. This relation determines the current in the circuit, once $\mathcal{E}$ and R are specified.

We say "if" in the above paragraph because every real source has some *internal* resistance, which we may denote as r. Under closed-circuit conditions the total electric field $E_e + E_n$ inside the source cannot then be precisely zero because some net field is required in order to push the charge through the internal resistance. Thus E_e must have somewhat smaller magnitude than E_n, and correspondingly V_{ab} is less than $\mathcal{E}$; the difference is equal to the work per unit charge done by the total field, which is simply Ir. Thus the terminal potential difference under closed-circuit conditions is given by

$$V_{ab} = \mathcal{E} - Ir, \qquad (28\text{-}15)$$

where r is the *internal resistance* of the source. The equation determining the current in the complete circuit is then

$$\mathcal{E} - Ir = IR,$$

or

$$I = \frac{\mathcal{E}}{R + r}. \qquad (28\text{-}16)$$

That is, the current equals the source emf divided by the *total* circuit resistance, external plus internal.

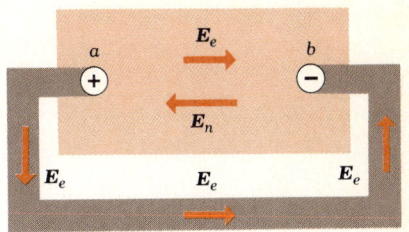

28-5 Schematic diagram of a source with a complete circuit. The vectors E_n and E_e represent the directions of the corresponding fields. The current is everywhere the same and is in the direction of the *resultant* field, that is, from a to b in the external circuit and from b to a within the source. $V_{ab} = IR = \mathcal{E} - Ir$.

If the terminals of a source are connected by a conductor of zero (or negligible) resistance, the source is said to be *short circuited*. (This would be an *extremely* dangerous procedure to carry out with the storage battery of your car, or with the terminals of the power line!) Then $R = 0$, and from the circuit equation the *short-circuited current* I_s is

$$I_s = \frac{\mathcal{E}}{r}. \qquad (28\text{-}17)$$

The terminal voltage is then zero:

$$V_{ab} = \mathcal{E} - \left(\frac{\mathcal{E}}{r}\right)r = 0. \qquad (28\text{-}18)$$

The *electrostatic* field within the source is zero, and the driving force on the charges within it is due to the *non*-electrostatic field only.

A source is completely described by its emf $\mathcal{E}$ and its internal resistance r. These properties may be found (at least in principle) from measurements of the open-circuit terminal voltage, which equals $\mathcal{E}$, and the short-circuit current, which enables r to be calculated from Eq. (28-17).

We must consider one more special case. If a source is connected to an external circuit containing other sources, it is possible that the electrostatic field within the given source will be *greater* than the non-electrostatic field, as in Fig. 28-6. When this is the case, the current *within* the source is from terminal a toward terminal b. This is the case when the storage battery of an automobile is being "charged" by the alternator. Equation (28-15), then becomes:

$$V_{ab} = \mathcal{E} + Ir. \qquad (28\text{-}19)$$

The terminal voltage is then *greater* than the emf $\mathcal{E}$.

28-6

The diagrams in the preceding sections have been more or less representational, so as to show the electric fields within sources and conductors. Every conductor (except a superconductor) has resistance, and is therefore a *resistor*, also. In fact, the terms "conductor" and "resistor" may be used interchangeably. Resistance units constructed to introduce into a circuit lumped resistances that are large compared with those of leads and contacts are called *resistors*. A resistor is represented by the symbol

Conductors having negligible resistance are shown by straight lines.

A variable resistor is called a *rheostat* or, particularly in electronics, a *potentiometer*. A common type consists of a resistor with a sliding contact that can be moved along its length and is represented by the symbol

Connections are made to either end of the resistor and to the sliding contact. The symbol

is also used for a variable resistor.

A source is represented by the symbol

The longer vertical line always corresponds to the + terminal. We shall modify this in the following examples to

or

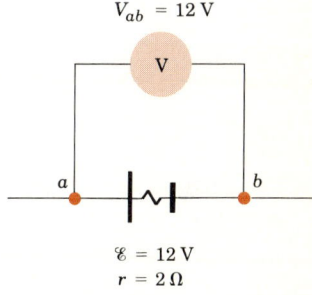

so as to show explicitly that a source has an internal resistance.

Example 1 Consider a source whose emf $\mathcal{E}$ is constant and equal to 12 V, and whose internal resistance r is 2 Ω. (The internal resistance of a commercial 12-V lead storage battery is only a few thousandths of an ohm.) Figure 28-7 represents the source with a voltmeter V connected between its terminals a and b. A voltmeter reads the potential difference between its terminals. If it is of the conventional type, the voltmeter provides a conducting path between the terminals and so there is a current in the source (and through the voltmeter). We shall assume, however, that the resistance of the voltmeter is so large (essentially infinite) that it draws no appreciable current. The source is then on *open circuit*, corresponding to the source in Fig. 28-4, and the voltmeter reading V_{ab} equals the emf $\mathcal{E}$ of the source, or 12 V.

28-7 A source on open circuit.

Example 2 In Fig. 28-8, an ammeter A and a resistor of resistance $R = 4\,\Omega$ have been connected to the terminals of the source to form a complete circuit. The total resistance of the circuit is the sum of the resistance R, the internal resistance r, and the resistance of the ammeter. The ammeter resistance, however, can be made very small, and we shall assume it so small (essentially zero) that it can be neglected. The ammeter (whatever its resistance) reads the current I through it. The circuit corresponds to that in Fig. 28-5.

The wires connecting the resistor to the source and the ammeter, shown by the straight lines, have zero resistance and hence there is no potential difference between their ends. Thus, points a and a' are at the

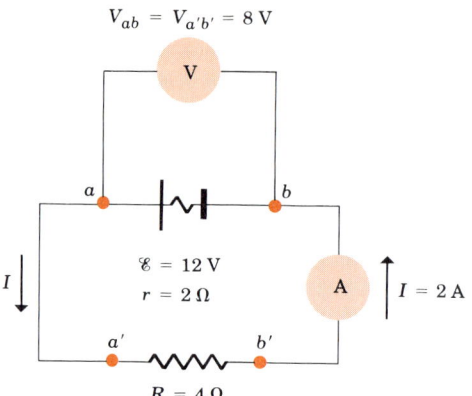

28-8 A source in a complete circuit.

same potential and are electrically equivalent, as are points b and b'. The potential differences V_{ab} and $V_{a'b'}$ are therefore equal. In the future, we shall use the same symbol to represent all points in a circuit that are connected by resistanceless conductors and are at the same potential.

The current I in the resistor (and hence at all points of the circuit) could be found from the relation $I = V_{ab}/R$, if the potential difference V_{ab} were known. However, V_{ab} is the terminal voltage of the source, equal to $\mathcal{E} - Ir$, and since this depends on I it is unknown at the start. We can, however, calculate the current from the circuit equation:

$$I = \frac{\mathcal{E}}{R + r} = \frac{12 \text{ V}}{4\,\Omega + 2\,\Omega} = 2 \text{ A}.$$

The potential difference V_{ab} can now be found by considering a and b either as the terminals of the resistor or as those of the source. If we consider them as the terminals of the resistor,

$$V_{a'b'} = IR = (2 \text{ A})(4\,\Omega) = 8 \text{ V}.$$

If we consider them as the terminals of the source,

$$V_{ab} = \mathcal{E} - Ir = 12 \text{ V} - (2 \text{ A})(2\,\Omega) = 8 \text{ V}.$$

The voltmeter therefore reads 8 V and the ammeter reads 2 A.

Example 3 In Fig. 28–9, the source is short circuited. The current is

$$I = \frac{\mathcal{E}}{r} = \frac{12 \text{ V}}{2\,\Omega} = 6 \text{ A}.$$

The terminal voltage is

$$V_{ab} = \mathcal{E} - Ir = 12 \text{ V} - (6 \text{ A})(2\,\Omega) = 0.$$

The ammeter reads 6 A and the voltmeter reads zero.

The rule represented by Eq. (28–16), for finding the current in a circuit, can be recast in a form which is particularly useful in more complex circuits having several seats of emf or several branches. The electrostatic field is a *conservative* force field. Suppose we go around a loop, measuring potential differences across successive circuit elements. When we return to the starting point we must find that the *algebraic sum* of

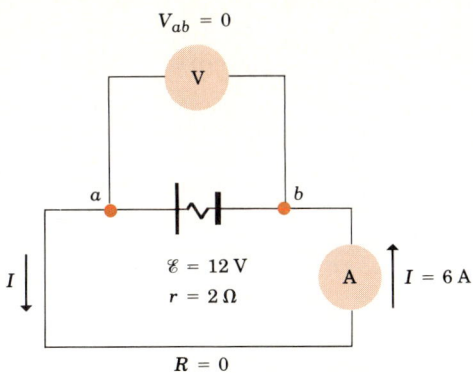

28-9 A source on short circuit.

these differences is zero; otherwise we could not say that the potential at this point has a definite value.

Thus, in Fig. 28-8, if we start at point b and travel counterclockwise (the same direction as I) we find a *rise* in potential (a *positive* change) due to the battery emf, a *drop* (a *negative* change) due to the battery's internal resistance, and an additional drop due to the 4-Ω resistor. The algebraic sum of these must be zero. Thus

$$12 \text{ V} - I(2\,\Omega) - I(4\,\Omega) = 0, \qquad I = 2 \text{ A},$$

in agreement with the result obtained in Example 2.

To allow for the possibility of several seats of emf in the loop, we generalize this procedure as follows: *The algebraic sum of the emf's around a series circuit, minus the sum of the IR products for the resistive elements in the circuit, must equal zero.* Formally,

$$\sum \mathcal{E} - \sum IR = 0. \qquad (28\text{-}20)$$

This is called *Kirchhoff's loop rule.* In applying it, we need some sign conventions. We first assume a direction for the current, and mark it on the diagram. Then, starting at any point in the circuit, we go around the circuit in the direction of the assumed current, adding emf's and IR products as we come to them. When a source is traversed in the direction from $(-)$ to $(+)$ (the direction of $\mathbf{E}_n$ in that source), the emf is considered *positive*, and when from $(+)$ to $(-)$, negative. The IR terms are all negative because the direction of current is always that of *decreasing* potential. Of course, we could also traverse the loop in the direction *opposite* to that of the assumed current. In that case all the emf's have opposite sign, and all the IR terms are positive because in going in the opposite direction to the current we are going "up-hill" from lower to higher potential.

For the circuit in Fig. 28-8, if we start at point b and go clockwise, the resulting equation is

$$I(4\,\Omega) + I(2\,\Omega) - 12 \text{ V} = 0,$$

which is the same as the previous equation except for an overall factor of (-1) which does not change the value of I. But if we assume that I is clockwise, then starting at b and going counterclockwise around the circuit yields the equation

$$12 \text{ V} + I(2\,\Omega) + I(4\,\Omega) = 0, \qquad I = -2 \text{ A}.$$

Fig. 28-10

In this case the negative sign on the result shows that our initial assumption about the current direction was wrong; the actual direction is counterclockwise.

This same bookkeeping system may be used to determine the potential difference between any two points a and b in a circuit. To find $V_{ab} = V_a - V_b$ (the potential at a with respect to b), we start at b and add the potential changes encountered in going from b to a. Again an emf is considered positive when we go from $(-)$ to $(+)$, negative otherwise. An IR term is positive when we go "up-hill," against the current direction, negative when in the same direction as the current.

Example 4 The circuit shown in Fig. 28-10 contains two batteries, each having an emf and an internal resistance, and two resistors. Find the current in the circuit and the potential difference V_{ab}.

Solution We assume a direction for the current, as shown; then, starting at a and going counterclockwise, we add potential increases and decreases to compute $\Sigma \mathcal{E} - \Sigma IR$. The resulting equation is

$$-I(4\,\Omega) - 4\text{ V} - I(7\,\Omega) + 12\text{ V} - I(2\,\Omega) - I(3\,\Omega) = 0.$$

Collecting terms containing I and solving for I, we find

$$8\text{ V} = I(16\,\Omega) \qquad \text{and} \qquad I = 0.5\text{ A}.$$

The result for I is positive, showing that our assumed current direction is correct. The reader may verify that if one assumes the opposite direction the result is $I = -0.5$ A, indicating that the actual current is opposite to this assumption.

To find V_{ab}, the potential at a with respect to b, we start at b and go toward a, adding potential changes. There are two possible paths from b to a; taking the lower one first, we find

$$V_{ab} = (0.5\text{ A})(7\,\Omega) + 4\text{ V} + (0.5\text{ A})(4\,\Omega) = 9.5\text{ V}.$$

Point a is at 9.5 V higher potential than b. All the terms in this sum are positive because each represents an *increase* in potential as we go from b toward a. If instead we use the upper path, the resulting equation is

$$V_{ab} = 12\text{ V} - (0.5\text{ A})(2\,\Omega) - (0.5\text{ A})(3\,\Omega) = 9.5\text{ V}.$$

Here the IR terms are negative because our path goes in the direction of the current, with potential decreases through the resistors. The result is the same as for the other path, as it must be in order for the total potential change around the complete loop to be zero. In each case potential rises are taken as positive, and drops as negative.

In the foregoing analysis, emf's have been treated on the same basis as potential differences in circuit elements. This procedure might well be questioned; an emf is, strictly speaking, *not* a potential difference, inasmuch as it represents work done (per unit charge) by a force of *non*-electrostatic origin. But as we have seen the (electrostatic) potential difference between terminals of a source can always be expressed in terms of its emf and the drop across its internal resistance. Thus the emf terms in Eq. (28-20) really *do* represent true potential differences.

28-5 Current-voltage relations

The current through a device such as a resistor depends on the potential difference between its terminals. For a device obeying Ohm's law, the current is directly proportional to voltage, as shown in Eq. (28-11). But as mentioned in Sec. 28-2, there are many devices for which this simple model is not an adequate description. Current may depend on voltage in a more complicated way, and the current resulting from a given potential difference may depend on the polarity of the potential difference. This is the case with *diodes,* devices constructed deliberately to conduct much better in one direction than the other.

It is convenient to represent the current-voltage relation as a graph, and Fig. 28-11 shows several examples. Part (a) shows the behavior of a resistor that obeys Ohm's law, for which the graph is a straight line. Part (b) shows the relation for a vacuum diode. For positive potentials of anode with respect to cathode, I is approximately proportional to $V^{3/2}$, while for negative potentials the current is several orders of magnitude smaller and for most purposes may be assumed to be zero. Germanium diode behavior (c) is somewhat different but still strongly asymmetric, acting as a one-way valve in a circuit. Diodes are used to convert alternating current to direct and to perform a wide variety of logic functions in computer circuitry. The microscopic basis of diode behavior will be explored in some detail in later chapters.

An additional consideration is that for nearly all materials the current-voltage relation is temperature-dependent. Thus at low temperatures the curve in Fig. 28-11c rises more steeply for positive V than at higher temperatures, and at successively higher temperatures the asymmetry in the curve becomes less and less pronounced.

The current-voltage relation for a source may also be represented graphically. For a source represented by Eq. (28-15), that is,

$$V = \mathcal{E} - Ir,$$

the graph appears as in Fig. 28-12a. The intercept on the V-axis, corresponding to the open-circuit condition ($I = 0$), is at $V = \mathcal{E}$, and the intercept on the I-axis, corresponding to a short-circuit situation ($V = 0$), is at $I = \mathcal{E}/r$.

This line may be used to find the current in a circuit containing a nonlinear device, as in Fig. 28-12b. Its current-voltage relation is shown in Fig. 28-12c, and Eq. (28-15) is also plotted on this graph. Each curve represents a current-voltage relation that must be satisfied, so the intersection represents the only possible values of V and I. This amounts to a graphical solution of two simultaneous equations for V and I, one of which is nonlinear.

When a device has a nonlinear voltage-current relation, the quantity V/I is not constant. This ratio may still be called resistance, but now it varies with current; it is constant only for a device obeying Ohm's law. Often a more useful quantity is dV/dI, which expresses the relation between a small change in current and the resulting voltage change. This is called the *dynamic* or *incremental* resistance, and in general it too depends on current. For the voltage-current relation shown in Fig. 28-13, the dynamic resistance at current I_0 is the slope of the tangent line to the curve at I_0, while the ordinary resistance V/I is the slope of the line from the origin to this point.

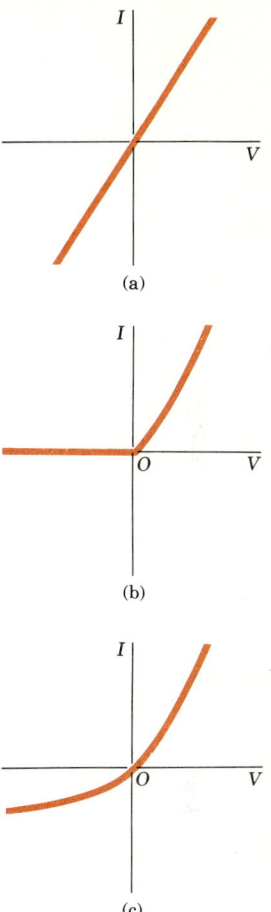

28-11 Current-voltage relations for (a) a resistor obeying Ohm's law; (b) a vacuum diode; (c) a semiconductor diode.

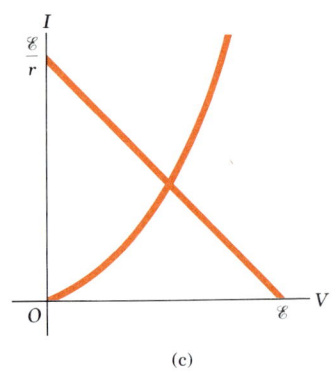

28-12 (a) Current–voltage relation for a source with emf $\mathcal{E}$ and internal resistance r; (b) a circuit containing a source and a nonlinear element; (c) simultaneous solution of I–V equations for this circuit.

Finally, we remark that Eq. (28–15) is not always an adequate representation of the behavior of a source. What we have described as an internal resistance may actually be a more complex voltage–current relation, and the emf may vary somewhat. Nevertheless, the concept of internal resistance frequently provides an adequate description of batteries, generators, and other energy converters. The difference between a fresh flashlight battery and an old one is not in the emf, which decreases only slightly with use, but principally in the internal resistance, which may increase from a few ohms when fresh to as much as 1000 Ω or more after long use. Similarly, the current a car battery can deliver to the starter motor on a cold morning is less than when the battery is warm, partly because the emf is somewhat less but principally because the internal resistance is temperature-dependent, decreasing with increasing temperature.

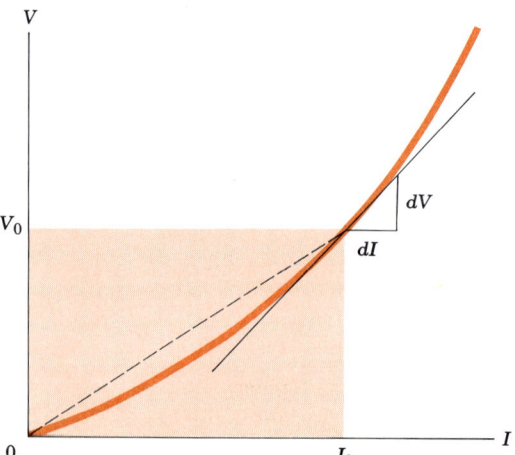

28-13 Dynamic resistance is the slope of the voltage–current curve.

28-6 Work and power in electrical circuits

We are now ready to consider in detail the energy relations in electric circuits. The rectangle in Fig. 28–14 represents a portion of a circuit

28–14 The power input P to the portion of the circuit between a and b is $P = V_{ab}I$.

having current I and potential difference $V_a - V_b = V_{ab}$ between the two conductors leading to and from this point of the circuit. The detailed nature of this circuit element need not be specified at present. As charge passes through, the electric field does work on it. In a time interval Δt, an amount of charge $\Delta Q = I\,\Delta t$ passes through, and the work ΔW done by the electric field is given by the product of the potential difference (work per unit charge) and the quantity of charge:

$$\Delta W = V_{ab}\,\Delta Q = V_{ab}I\,\Delta t.$$

By means of this work the electric field transfers energy into this portion of the circuit.

Rate of transfer of energy is *power*, denoted by P; dividing the above relation by Δt, we obtain the rate at which energy enters this part of the circuit:

$$P = \frac{\Delta W}{\Delta t} = V_{ab}I. \tag{28-21}$$

It may happen that the potential at b is higher than that at a; in this case V_{ab} is negative. The charge then *gains* potential energy (at the expense of some other form of energy), and there is a corresponding transfer of electrical energy *out of* this portion of the circuit.

Equation (28–21) is the general expression for the magnitude of the electrical power input to (or the power output from) any portion of an electrical circuit. The unit of V_{ab} is one volt, or one joule per coulomb, and the unit of I is one ampere or one coulomb per second. The SI unit of power is therefore

$$(1\,\text{J}\cdot\text{C}^{-1})(1\,\text{C}\cdot\text{s}^{-1}) = 1\,\text{J}\cdot\text{s}^{-1} = 1\,\text{W} = 1\,\text{watt}.$$

We now consider some special cases.

1. Pure resistance. If the portion of the circuit in Fig. 28–14 is a pure resistance, the potential difference is given by $V_{ab} = IR$ and

$$P = V_{ab}I = I^2R = \frac{V^2}{R}. \tag{28-22}$$

The potential at a is necessarily higher than at b and there is a power *input* to the resistor. The circulating charges give up energy to the atoms of the resistor when they collide with them, and the temperature of the resistor increases unless there is a flow of heat out of it. We say that energy is *dissipated* in the resistor at a rate I^2R.

Because of this heat, every resistor has a maximum power rating, the maximum power that can be dissipated without overheating the device. When this rating is exceeded the resistance may change unpredictably; in more extreme cases the resistor may melt or even explode. In practical applications, the power rating of a resistor is just as important a characteristic as its resistance value.

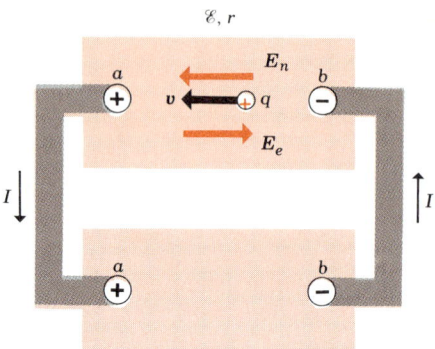

28-15 The rate of conversion of non-electrical to electrical energy in the source equals $\mathcal{E}I$. The rate of energy dissipation in the source is I^2r. The difference $\mathcal{E}I - I^2r$ is the power output of the source.

2. Power output of a source. The upper rectangle in Fig. 28-15 represents a source having emf $\mathcal{E}$ and internal resistance r, connected by ideal (resistanceless) conductors to an external circuit represented by the lower rectangle; the precise nature of the external circuit does not matter. We assume only that there is a current I in the circuit in the direction shown, from a to b in the external circuit and from b to a within the source. The letters a and b can be considered to represent either the terminals of the source or those of the external circuit, and $V_a > V_b$. The external circuit then corresponds to the rectangle in Fig. 28-14, and the power input to it is

$$P = V_{ab}I.$$

If a and b are considered as the terminals of the source, then as we have shown,

$$V_{ab} = \mathcal{E} - Ir$$

and hence,

$$P = V_{ab}I = \mathcal{E}I - I^2r. \tag{28-23}$$

The terms $\mathcal{E}I$ and I^2r have direct significance. The emf $\mathcal{E}$ has been defined as the work per unit charge performed on the charges by the non-electrostatic field $\boldsymbol{E}_n$ as the charges move from b to a in the source. Hence, if a charge ΔQ flows in time Δt, this field does work $\Delta W = \mathcal{E}\,\Delta Q$, and its *rate* of doing work, or power, is

$$P = \frac{\Delta W}{\Delta t} = \mathcal{E}\frac{\Delta Q}{\Delta t} = \mathcal{E}I. \tag{28-24}$$

Hence, the product $\mathcal{E}I$ is the rate at which work is done on the circulating charges by the agency that maintains the non-electrostatic field. The term I^2r is the rate at which energy is *dissipated* in the internal resistance of the source, and the difference $\mathcal{E}I - I^2r$ is the rate at which energy is delivered by the source to the remainder of the circuit. In other words, the power P in Eq. (28-23) represents the *power output* of the source, or the *power input* to the remainder of the circuit.

3. Power input to a source. Suppose that the lower rectangle in Fig. 28-15 is itself a source of emf larger than that of the upper source and with its emf opposite to that of the upper source. The current I in the circuit is then *opposite* to that shown in Fig. 28-15. The lower output of

the lower source is
$$P = V_{ab}I.$$

Considering a and b as the terminals of the upper source, we have
$$V_{ab} = \mathcal{E} + Ir,$$
and
$$P = V_{ab}I = \mathcal{E}I + I^2 r. \tag{28-25}$$

The charges in the upper source now move from left to right, in a direction opposite the field $\boldsymbol{E}_n$, and the work done by the non-electrostatic force $\boldsymbol{F}_n$ is *negative*. That is, work is done *on* the agent maintaining the non-electrostatic field. The product $\mathcal{E}I$ equals the rate at which work is done on this agent, and the term $I^2 r$ again equals the rate of dissipation of energy in the internal resistance of the source. The sum $\mathcal{E}I + I^2 r$ is therefore the total *power input* to the upper source. The source converts electrical energy to non-electrical energy.

To illustrate these relations, let us apply them to the circuits in Figs. 28-8 through 28-10.

Example 1 The rate of energy conversion in the source in Fig. 28-8 is
$$\mathcal{E}I = (12 \text{ V})(2 \text{ A}) = 24 \text{ W}.$$

The rate of dissipation of energy in the source is
$$I^2 r = (2 \text{ A})^2 (2 \text{ }\Omega) = 8 \text{ W}.$$

The power *output* of the source is the difference between these, or 16 W.

The power output is also given by
$$V_{ab}I = (8 \text{ V})(2 \text{ A}) = 16 \text{ W}.$$

The power input to the resistor is
$$V_{a'b'}I = (2 \text{ A})(8 \text{ V}) = 16 \text{ W}.$$

This equals the rate of dissipation of energy in the resistor:
$$I^2 R = (2 \text{ A})^2 (4 \text{ }\Omega) = 16 \text{ W}.$$

Example 2 The rate of energy conversion in the source in Fig. 28-9 is
$$\mathcal{E}I = (12 \text{ V})(6 \text{ A}) = 72 \text{ W}.$$

The rate of dissipation of energy in the source is
$$I^2 r = (6 \text{ A})^2 (2 \text{ }\Omega) = 72 \text{ W}.$$

The power *output* of the source (also given by $V_{ab}I$) equals zero. *All of the energy converted is dissipated within the source.*

Example 3 The rate of energy conversion in the upper source in Fig. 28-10 is
$$\mathcal{E}I = (12 \text{ V})(0.5 \text{ A}) = 6 \text{ W}.$$

The rate of dissipation of energy in the source is
$$I^2 r = (0.5 \text{ A})^2 (2 \text{ }\Omega) = 0.5 \text{ W}.$$

The power output of the source is $6\text{ W} - 0.5\text{ W} = 5.5\text{ W}$.

The rates of dissipation of energy in the 3-Ω and 7-Ω resistors are, respectively,

$$(0.5\text{ A})^2(3\text{ }\Omega) = 0.75\text{ W},$$
$$(0.5\text{ A})^2(7\text{ }\Omega) = 1.75\text{ W}.$$

The rate of energy conversion in the lower source is

$$\mathcal{E}I = (4\text{ V})(0.5\text{ A}) = 2\text{ W},$$

and the rate of energy dissipation in this source is

$$I^2 r = (0.5\text{ A})^2(4\text{ }\Omega) = 1\text{ W}.$$

The power *input* to this source is $2\text{ W} + 1\text{ W} = 3\text{ W}$. Thus, of the 5.5-W output of the upper source, 2.5 W is dissipated in the two resistors, and the remaining 3 W is partly converted, partly dissipated in the lower source. Conversion of energy in the sources is often reversible, as in a storage battery where electrical energy is converted to chemical for later retrieval as electrical energy. Energy dissipated in resistors is converted irreversibly to heat.

28-7 Physiological effects of currents

Electrical potential differences and currents play a vital role in the nervous systems of animals. Conduction of nerve impulses is basically an electrical process, although the mechanism of conduction is much more complex than in simple material such as metals. A nerve fiber or *axon*, along which an electrical impulse can travel, includes a cylindrical membrane with one conducting fluid (electrolyte) inside and another outside. By mechanisms similar to those in batteries, a potential difference of the order of 0.1 V is maintained between these fluids.

When a pulse is initiated, the membrane temporarily becomes more permeable to the ions in the fluids, leading to a local drop in potential. As the pulse passes, with a typical speed of the order of $30\text{ m}\cdot\text{s}^{-1}$, the membrane recovers and the potential returns to its initial value. There are several aspects of this process that are not yet well understood.

The basically electrical nature of nerve-impulse conduction is responsible for the great sensitivity of the body to externally supplied electrical currents. Currents through the body as small as 0.1 A, much too small to produce significant heating, are fatal because they interfere with nerve processes essential for vital functions such as heartbeat. The *resistance* of the human body is highly variable, principally because of the fairly low conductivity of the skin, but the resistance between two electrodes grasped by dry hands is of the order 5 kΩ to 10 kΩ. For $R = 10\text{ k}\Omega$, a current of 0.1 A requires a potential difference $V = IR = (0.1\text{ A})(10\text{ k}\Omega) = 1000\text{ V}$.

Even much smaller currents can be very dangerous. A current of 0.01 A causes strong, convulsive muscle action and considerable pain, and with 0.02 A the person typically is unable to release a conductor inflicting the shock. Currents of this magnitude and even as small as 0.001 A can cause ventricular fibrillation, a disorganized twitching of heart muscles that occurs instead of regular beating, and that unfortunately pumps very little blood. Surprisingly, very large currents (over

0.1 A) are somewhat *less* likely to cause fatal fibrillation because the heart muscle is "clamped" in one position and is more likely to resume normal beating when the current is removed. Severe burns are, of course, more likely with large currents.

The moral of this rather morbid story, if there is one, is that under certain conditions voltages as small as 10 V can be dangerous, and should not be regarded with anything but respect and caution.

On the positive side, rapidly alternating currents can have beneficial effects. Alternating currents with frequencies the order of 10^6 Hz do not interfere appreciably with nerve processes, and can be used for therapeutic heating for arthritic conditions, sinusitis, and a variety of other disorders. If one electrode is made very small, the resulting concentrated heating can be used for local destruction of tissue, such as tumors, or even for cutting tissue in certain surgical procedures.

Study of particular nerve impulses is also an important *diagnostic* tool in medicine. The most familiar examples are electrocardiography (EKG) and electroencephalography (EEG). Electrocardiograms, obtained by attaching electrodes to the chest and back and recording the regularly varying potential differences, are used to study heart function. Similarly, electrodes attached to the scalp permit study of potentials in the brain, and the resulting patterns can be helpful in diagnosing epilepsy, brain tumors, and other disorders.

*28-8 The electric field of the earth

If the molecules of the earth's atmosphere were electrically neutral, the atmosphere would be a nonconductor or insulator. Actually, because of the bombardment of the earth by cosmic rays (e.g., high-speed protons from outer space), a few ions, both positive and negative, are always present. With increasing elevation, the ionization produced by cosmic rays increases; and since the density of the atmosphere decreases, ions can move more freely and the conductivity increases. Above an elevation of about 50 km, the atmosphere is a relatively good conductor; and the earth itself is also a relatively good conductor. An extremely simplified model of the earth and its lower atmosphere consists of a conducting sphere surrounded by a conducting spherical shell, the two being separated by a poorly conducting layer about 50 km thick.

Near the earth's surface, there is found to be a radial electrostatic field or potential gradient of about 100 V·m^{-1}. The field becomes weaker with increasing elevation, and the total potential difference between the earth's surface and the outer conducting layer is about 400,000 V. The direction of the field is downward, so the earth has a negative surface charge and the outer conducting layer has a positive charge. The surface charge density and the electric field are related by Eq. (25-22),

$$\sigma = \epsilon_0 E;$$

and hence, if $E = 100$ V·m^{-1}, the surface density is, roughly, 10^{-9} C·m^{-2}.

Approximating the earth by a sphere of radius 5000 km, we find its surface area is about 3×10^{14} m^2, so the total surface charge is about 3×10^5 C.

Because the atmosphere is not a perfect insulator, there is a current in it, directed downward in the same direction as the electrostatic field. (Positive ions drift downward; negative ions drift upward.) The total current over the entire earth's surface is fairly constant and is about 1800 A or 1800 C·s^{-1}. The time required to neutralize the entire surface charge of 3×10^5 C, if it were not replenished in some way, would therefore be only about two or three minutes.

The charge, however, remains constant, so there must be some mechanism that pumps positive charges upward and negative charges downward, opposite to the directions of the forces exerted on them by the electrostatic field. The nature of this mechanism is not fully understood, but it is believed to be the processes that go on in the development of thunderstorms. In some way, positive and negative ions are developed in a thunderstorm. The positive ions are carried upward and the negative ions downward by the air currents in the storm. These currents, in turn, are driven by pressure differences resulting from nonuniform temperatures in the atmosphere, so that the ultimate source of the energy input is the radiant energy reaching the earth from the sun.

The earth's atmosphere can be compared to an enormous Van de Graaff generator immersed in a conducting medium. Instead of a moving belt, the motion of the atmosphere transports charge carriers in a direction opposite to the electrostatic force acting on them. The charges then flow back by conduction through the lower atmosphere. The terminal voltage of the generator is 400,000 V and the current is 1800 A, so the power is about 700 MW. By way of comparison, a modern power plant may develop on the order of 1000 MW, and others to be built are in the GW range.

*28-9 Theory of metallic conduction

Additional insight into the phenomenon of conduction can be gained by examining the microscopic mechanisms of conductivity. Here we consider only a crude and primitive model that treats the electrons as classical particles and ignores their inherently quantum-mechanical behavior in solids. Thus this model is not correct conceptually; yet it is useful in helping to develop an intuitive idea of the microscopic basis of conduction.

In the simplest microscopic model of metallic conduction, each atom in the crystal lattice is assumed to give up one or more of its outer electrons. These electrons are then free to move through the crystal lattice, colliding at intervals with the stationary positive ions. Their motion is like that of the molecules of gas in a container and they are often referred to as an "electron gas." In the absence of an electric field, the electrons move in straight lines between collisions, but if there is an electric field, the paths are slightly curved, as in Fig. 28–16, which represents schematically a few free paths of an electron in an electric field directed from right to left. At each collision, the electron is assumed to lose any energy it may have acquired from the field and to make a fresh start. The energy given up in these collisions increases the thermal energy of vibration of the positive ions.

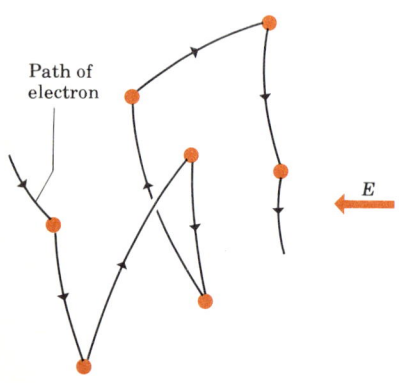

28–16 Random motion of an electron, with superimposed drift under action of electric field.

A force $F = eE$ is exerted on each electron by the field, and produces an acceleration a in the direction of the force given by

$$a = \frac{F}{m} = \frac{eE}{m},$$

where m is the electron mass. Let u represent the average *random* speed of an electron, and λ the mean free path. The average time t between collisions, called the *mean free time,* is

$$t = \frac{\lambda}{u}.$$

In this time, the electron acquires a final velocity component v_f in the direction of the force, given by

$$v_f = at = \frac{eE}{m}\frac{\lambda}{u}.$$

Its *average* velocity v in the direction of the force, which is superposed on its random velocity and which we interpret as the *drift* velocity, is one-half the final velocity, so

$$v = \frac{1}{2}v_f = \frac{1}{2}\frac{e\lambda}{mu}E.$$

The drift velocity is therefore proportional to the electric field E.

The current density is

$$J = nev = \frac{ne^2\lambda}{2mu}E,$$

and the resistivity is

$$\rho = \frac{E}{J} = \frac{2mu}{ne^2\lambda}. \qquad (28\text{-}26)$$

This theoretical expression for resistivity is in *qualitative* agreement with experiment. At a given temperature, the quantities m, u, n, e, and λ are constant. The resistivity is then constant, and Ohm's law is obeyed. When the temperature is increased, the random speed u increases and the theory predicts that the resistivity of a metal increases with increasing temperature.

In a semiconductor, the number of charge carriers per unit volume, n, increases rapidly with increasing temperature. The increase in n far outweighs any increase in u, and the resistivity *decreases*. At low temperatures, n is very small and the resistivity becomes so large that the material can be considered an insulator.

The modern theory of superconductivity predicts that, in effect, at temperatures below the critical temperature the electrons move freely throughout the lattice. The mean free path λ then becomes very large and the resistivity very small.

Questions

28-1 A rule of thumb used to determine the internal resistance of a source is that it is the open-circuit voltage divided by the short-circuit current. Is this correct?

28-2 The energy that can be extracted from a storage battery is always less than the energy that goes into it while it is being charged. Why?

28-3 In circuit analysis one often assumes that a wire connecting two circuit elements has no potential difference between its ends; yet there must be an electric field within the wire to make the charges move, and so there must be a potential difference. How do you resolve this discrepancy?

28-4 Long-distance electric-power transmission lines always operate at very high voltage, sometimes as much as 750 kV. What are the advantages of such high voltages? The disadvantages?

28-5 Ordinary household electric lines usually operate at 110 V. Why is this a desirable voltage, rather than a value considerably larger or smaller? What about cars, which usually have 12-V electrical systems?

28-6 What is the difference between an emf and a potential difference?

28-7 Electric power for household and commercial use always uses *alternating current,* which reverses direction 120 times each second. A student claimed that the power conveyed by such a current would have to average out to zero, since it is going one way half the time and the other way the other half. What is your response?

28-8 As discussed in the text, the drift velocity of electrons in a good conductor is very slow. Why then does the light come on so quickly when the switch is turned on?

28-9 A fuse is a device designed to break a circuit, usually by melting, when the current exceeds a certain value. What characteristics should the material of the fuse have?

28-10 What considerations determine the maximum current-carrying capacity of household wiring?

28-11 The text states that good thermal conductors are also good electrical conductors. If so, why don't the cords used to connect toasters, irons, and similar heat-producing appliances get hot by conduction of heat from the heating element?

28-12 Eight flashlight batteries in series have an emf of about 12 V, about the same as that of a car battery. Could they be used to start a car with a dead battery?

28-13 High-voltage power supplies are sometimes designed intentionally to have rather large internal resistance, as a safety precaution. Why is such a power supply with a large internal resistance safer than one with the same voltage but lower internal resistance?

28-14 How would you expect the resistivity of a good insulator such as glass or polystyrene to vary with temperature? Why?

Problems

28-1 A silver wire 1 mm in diameter transfers a charge of 90 C in 1 hr and 15 min. Silver contains 5.8×10^{28} free electrons per m^3.

a) What is the current in the wire?

b) What is the drift velocity of the electrons in the wire?

28-2 When a sufficiently high potential difference is applied between two electrodes in a gas, the gas ionizes; electrons move toward the positive electrode and positive ions toward the negative electrode.

a) What is the current in a hydrogen discharge if, in each second, 4×10^{18} electrons and 1.5×10^{18} protons move in opposite directions past a cross section of the tube?

b) What is the direction of the current?

28-3 A vacuum diode can be approximated by a plane cathode and a plane anode, parallel to each other and 5 mm apart. The area of both cathode and anode is 2 cm^2. In the region between cathode and anode the current is carried solely by electrons. If the electron current is 50 mA, and the electrons strike the anode surface with a speed of 1.2×10^7 m·s^{-1}, find the number of electrons per cubic millimeter in the space just outside the surface of the anode.

28-4 In the Bohr model of the hydrogen atom the electron makes about 6×10^{15} rev·s^{-1} around the nucleus. What is the average current at a point on the orbit of the electron?

28-5 A copper wire has a square cross section, 2.0 mm on a side. It is 4 m long and carries a current of 10 A. The density of free electrons is 8×10^{28} m^{-3}.

a) What is the current density in the wire?

b) What is the electric field?

c) How much time is required for an electron to travel the length of the wire?

28-6 A wire 100 m long and 2 mm in diameter has a resistivity of 4.8×10^{-8} Ω·m.

a) What is the resistance of the wire?

b) A second wire of the same material has the same mass as the 100-m length, but twice its diameter. What is its resistance?

28-7 A certain electrical conductor has a square cross section, 2.0 mm on a side, and is 12 m long. The resistance between its ends is 0.072 Ω.

a) What is the resistivity of the material?

b) If the electric field magnitude in the conductor is 0.12 V·m^{-1}, what is the total current?

c) If the material has 8.0×10^{28} free electrons per cubic meter, find the average drift velocity under conditions of part (b).

28-8 The following measurements of current and potential difference were made on a resistor constructed of Nichrome wire:

I, A	V_{ab}, V
0.5	2.18
1.0	4.36
2.0	8.72
4.0	17.44

a) Make a graph of V_{ab} as a function of I.
b) Does the Nichrome obey Ohm's law?
c) What is the resistance of the resistor, in ohms?

28-9 The following measurements were made on a Thyrite resistor:

I, A	V_{ab}, V
0.5	4.76
1.0	5.81
2.0	7.05
4.0	8.56

a) Make a graph of V_{ab} as a function of I. Does Thyrite have constant resistance?
b) Construct a graph of the resistance R as a function of I.
c) Find the dynamic resistance at 2.0 A, and compare with the resistance values found in part (b).

28-10 The current in a wire varies with time according to the relation

$$i = 4 \text{ A} + (2 \text{ A·s}^{-2})t^2.$$

a) How many coulombs pass a cross section of the wire in the time interval between $t = 5$ s and $t = 10$ s?
b) What constant current would transport the same charge in the same time interval?

28-11 The current in a wire varies with time according to the relation

$$i = (20 \text{ A}) \sin (377 \text{ s}^{-1}t).$$

a) How many coulombs pass a cross section of the wire in the time interval between $t = 0$ and $t = 1/120$ s?
b) In the interval between $t = 0$ and $t = 1/60$ s?
c) What constant current would transport the same charge in each of the intervals above?

28-12 An aluminum bar 2.5 m long has a rectangular cross section 1 cm by 5 cm.

a) What is its resistance?
b) What would be the length of an iron wire 15 mm in diameter having the same resistance?

28-13 A solid cube of silver has a mass of 84.0 g. What is its resistance between opposite faces?

28-14 The two parallel plates of a capacitor have equal and opposite charges Q. The dielectric has a dielectric constant K and a resistivity ρ. Show that the "leakage" current carried by the dielectric is given by the relationship $i = Q/K\epsilon_0\rho$.

28-15

a) What is the resistance of a Nichrome wire at 0°C, if its resistance is 100.00 Ω at 12°C?
b) What is the resistance of a carbon rod at 30°C, if its resistance is 0.0150 Ω at 0°C?

28-16 A copper wire is initially at 0°C. How much must its temperature rise for the resistance to increase by 10 percent?

28-17 Refer to the third example in Sec. 28-3. Let the resistivity of the material between the cylinders be 10 Ω·m and let $r_a = 10$ cm, $r_b = 20$ cm, and $l = 5$ cm.

a) Find the resistance between the cylinders.
b) Find the current between the cylinders if $V_{ab} = 8$ V.

28-18 The region between two concentric conducting spheres of radii r_a and r_b is filled with a conducting material of resistivity ρ.

a) Show that the resistance between the spheres is given by

$$R = \frac{\rho}{4\pi}\left(\frac{1}{r_a} - \frac{1}{r_b}\right).$$

b) Derive an expression for the current density as a function of radius, if the potential difference between the spheres is V_{ab}.

28-19 A carbon resistor is to be used as a thermometer. On a winter day when the temperature is 0°C, its resistance is 217.3 Ω. What is the temperature on a hot summer day when the resistance is 214.2 Ω?

28-20 The resistance of a coil of copper wire is 200 Ω at 20°C. What is its resistance at 50°C?

28-21 A toaster using a Nichrome heating element operates on 120 volts. When it is switched on at 0°C, it carries an initial current of 1.5 A. A few seconds later the current reaches the steady value of 1.33 A. What is the final temperature of the element? The average value of the temperature coefficient of Nichrome over the temperature range is 0.00045 (C°)$^{-1}$.

28-22 A certain resistor has a resistance of 150.4 Ω at 20°C and a resistance of 162.4 Ω at 40°C. What is its temperature coefficient of resistivity?

28-23 A resistance thermometer using a platinum wire is used to measure the temperature of a liquid. The resistance is 2.42 Ω at 0°C, and when immersed in the liquid it is 2.98 Ω. The temperature coefficient of resistivity of platinum is 0.0038 (C°)$^{-1}$. What is the temperature of the liquid?

28-24 A piece of wire has a resistance R. It is cut into three pieces of equal length, and the pieces are twisted together in parallel. What is the resistance of the resulting wire?

28-25 What diameter must an aluminum wire have if it is to have the same resistance as an equal length of copper wire of diameter 2.0 mm?

28-26 When switch S is open, the voltmeter V, connected across the terminals of the dry cell in Fig. 28-17, reads 1.52 V. When the switch is closed, the voltmeter reading drops to 1.37 V and the ammeter A reads 1.5 A. Find the emf and internal resistance of the cell. Neglect meter corrections.

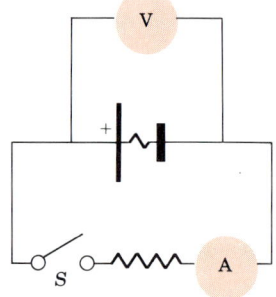

Figure 28-17

28-27 The potential difference across the terminals of a battery is 8.5 V when there is a current of 3 A in the battery from the negative to the positive terminal. When the current is 2 A in the reverse direction, the potential difference becomes 11 V.
a) What is the internal resistance of the battery?
b) What is the emf of the battery?

28-28
a) What is the potential difference V_{ad} in the circuit of Fig. 28-18?

Figure 28-18

b) What is the terminal voltage of the 4-V battery?
c) A battery of emf 17 V and internal resistance 1 Ω is inserted in the circuit at d, with its positive terminal connected to the positive terminal of the 8-V battery. What is now the difference of potential V_{bc} between the terminals of the 4-V battery?

28-29 A closed circuit consists of a 12-V battery, a 3.7-Ω resistor, and a switch. The internal resistance of the battery is 0.3 Ω. The switch is opened. What would a high-resistance voltmeter read when placed
a) across the terminals of the battery,
b) across the resistor,
c) across the switch?
Repeat (a), (b), and (c) for the case when the switch is closed.

28-30 The internal resistance of a dry cell increases gradually with age, even though the cell is not used. The emf, however, remains fairly constant at about 1.5 V. Dry cells may be tested for age at the time of purchase by connecting an ammeter directly across the terminals of the cell and reading the current. The resistance of the ammeter is so small that the cell is practically short circuited.
a) The short-circuit current of a fresh No. 6 dry cell (1.5-V emf) is about 30 A. Approximately what is the internal resistance?
b) What is the internal resistance if the short-circuit current is only 10 A?
c) The short-circuit current of a 6-V storage battery may be as great as 1000 A. What is its internal resistance?

28-31 The open-circuit terminal voltage of a source is 10 V and its short-circuit current is 4.0 A.
a) What will be the current when the source is connected to a linear resistor of resistance 2 Ω?
b) What will be the current in the Thyrite resistor of Problem 28-9 when connected across the terminals of this source?
c) What is the terminal voltage at this current?

28-32 A "660-W" electric heater is designed to operate from 120-V lines.
a) What is its resistance?
b) What current does it draw?
c) What is the rate of dissipation of energy, in calories per second?
d) If the line voltage drops to 110 V, what power does the heater take, in watts? (Assume the resistance constant. Actually, it will change because of the change in temperature.)

28-33 A resistor develops heat at the rate of 360 W when the potential difference across its ends is 180 V. What is its resistance?

28-34 A motor operating on 120 V draws a current of 2 A. If heat is developed in the motor at the rate of 9 cal·s⁻¹, what is its efficiency?

28-35

a) Express the rate of dissipation of energy in a resistor in terms of (1) potential difference and current, (2) resistance and current, (3) potential difference and resistance.

b) Energy is dissipated in a resistor at the rate of 40 W when the potential difference between its terminals is 60 V. What is its resistance?

28-36 A storage battery whose emf is 12 V and whose internal resistance is 0.1 Ω is to be charged from a 112-V dc supply. [*Caution:* Don't try to charge a battery directly from a power line, either dc or the more usual ac in household wiring. Real 12-V car batteries have internal resistance of only a few milliohms, and a serious explosion is possible.]

a) Should the + or the − terminal of the battery be connected to the + side of the line?

b) What will be the charging current if the battery is connected directly across the line?

c) Compute the resistance of the series resistor required to limit the current to 10 A.

With this resistor in the circuit, compute

d) the potential difference between the terminals of the battery,

e) the power taken from the line,

f) the power dissipated in the series resistor, and

g) the *useful* power input to the battery.

h) If electrical energy costs 3 cents per kWh, what is the cost of operating the circuit for 2 hr when the current is 10 A?

28-37 In the circuit in Fig. 28-19, find

a) the rate of conversion of internal energy to electrical energy within the battery,

b) the rate of dissipation of energy in the battery,

c) the rate of dissipation of energy in the external resistor.

Figure 28-19

28-38 A source whose emf is $\mathcal{E}$ and whose internal resistance is r is connected to an external circuit.

a) Show that the power output of the source is maximum when the current in the circuit is one-half the short-circuit current of the source.

b) If the external circuit consists of a resistance R, show that the power output is maximum when $R = r$, and that the maximum power is $\mathcal{E}^2/4r$.

DIRECT-CURRENT CIRCUITS AND INSTRUMENTS

29-1 Four different ways of connecting three resistors.

29-1 Resistors in series and in parallel

Most electrical circuits consist not merely of a single source and a single external resistor, but comprise a number of sources, resistors, or other elements such as capacitors, motors, etc., interconnected in a more or less complicated manner. The general term applied to such a circuit is a *network*. We shall next consider a few of the simpler types of network.

Figure 29-1 illustrates four different ways in which three resistors having resistances R_1, R_2, and R_3 might be connected between points a and b. In part (a), the resistors provide only a single path between the points, and are said to be connected in *series* between these points. Any number of circuit elements such as resistors, cells, motors, etc., are similarly said to be in series with one another between two points if connected as in (a) so as to provide only a single path between the points. The *current* is the same in each element.

The resistors in Fig. 29-1b are said to be in *parallel* between points a and b. Each resistor provides an alternative path between the points, and any number of circuit elements similarly connected are in parallel with one another. The *potential difference* is the same across each element.

In Fig. 29-1c, resistors R_2 and R_3 are in parallel with each other, and this combination is in series with the resistor R_1. In Fig. 29-1d, R_2 and R_3 are in series, and this combination is in parallel with R_1.

It is always possible to find a single resistor that could replace a combination of resistors in any given circuit and leave unaltered the potential difference between the terminals of the combination and the current in the rest of the circuit. The resistance of this single resistor is called the *equivalent* resistance of the combination. If any one of the networks in Fig. 29-1 were replaced by its equivalent resistance R, we could write

$$V_{ab} = IR \quad \text{or} \quad R = \frac{V_{ab}}{I},$$

where V_{ab} is the potential difference between the terminals a and b of the network and I is the current at point a or b. Hence the method of computing an equivalent resistance is to assume a potential difference V_{ab} across the actual network, compute the corresponding current I, and take the ratio V_{ab}/I. The simple series and parallel connections of resistors are sufficiently common so that it is worthwhile to develop formulas for these two special cases.

If the resistors are in *series*, as in Fig. 29-1a, the current in each must be the same and equal to the line current I. Hence

$$V_{ax} = IR_1, \quad V_{xy} = IR_2, \quad V_{yb} = IR_3,$$

and

$$V_{ab} = V_{ax} + V_{xy} + V_{yb} = I(R_1 + R_2 + R_3),$$
$$\frac{V_{ab}}{I} = R_1 + R_2 + R_3.$$

But V_{ab}/I is, by definition, the equivalent resistance R. Therefore

$$R = R_1 + R_2 + R_3. \tag{29-1}$$

The equivalent resistance of any number of resistors in series equals the *sum of their individual resistances*.

If the resistors are in *parallel*, as in Fig. 29-1b, the potential difference between the terminals of each must be the same and equal to V_{ab}. If the currents in each are denoted by I_1, I_2, and I_3, respectively,

$$I_1 = \frac{V_{ab}}{R_1}, \quad I_2 = \frac{V_{ab}}{R_2}, \quad I_3 = \frac{V_{ab}}{R_3}.$$

Charge is delivered to point a by the line current I, and removed from a by the currents I_1, I_2, and I_3. Since charge is not accumulating at a, it follows that

$$I = I_1 + I_2 + I_3 = V_{ab}\left(\frac{1}{R_1} + \frac{1}{R_2} + \frac{1}{R_3}\right),$$

or

$$\frac{I}{V_{ab}} = \frac{1}{R_1} + \frac{1}{R_2} + \frac{1}{R_3}.$$

But

$$\frac{I}{V_{ab}} = \frac{1}{R},$$

so

$$\frac{1}{R} = \frac{1}{R_1} + \frac{1}{R_2} + \frac{1}{R_3}. \tag{29-2}$$

For any number of resistors in parallel, the *reciprocal* of the equivalent resistance equals the *sum of the reciprocals* of their individual resistances.

For the special case of *two* resistors in parallel,

$$\frac{1}{R} = \frac{1}{R_1} + \frac{1}{R_2} = \frac{R_2 + R_1}{R_1 R_2}$$

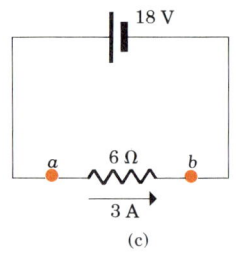

29-2 Reduction of a network of resistors to a single equivalent resistor.

and
$$R = \frac{R_1 R_2}{R_1 + R_2}.$$

Also, since $V_{ab} = I_1 R_1 = I_2 R_2$,

$$\frac{I_1}{I_2} = \frac{R_2}{R_1}, \tag{29-3}$$

and the currents carried by two resistors in parallel are *inversely proportional* to their resistances.

The equivalent resistances of the networks in Figs. 29-1c and 29-1d could be found by the same general method, but it is simpler to consider them as combinations of series and parallel arrangements. Thus, in (c) the combination of R_2 and R_3 in parallel is first replaced by its equivalent resistance, which then forms a simple series combination with R_1. In (d), the combination of R_2 and R_3 in series forms a simple parallel combination with R_1. Not all networks, however, can be reduced to simple series-parallel combinations, and special methods must be used for handling such networks.

Example Compute the equivalent resistance of the network in Fig. 29-2, and find the current in each resistor.

Solution Successive stages in the reduction to a single equivalent resistance are shown in parts (b) and (c). The 6-Ω and the 3-Ω resistors in part (a) are equivalent to the single 2-Ω resistor in part (b), and the series combination of this with the 4-Ω resistor results in the single equivalent 6-Ω resistor in part (c).

In the simple series circuit of part (c), the current is 3 A, and hence the current in the 4-Ω and 2-Ω resistors in part (b) is 3 A also. The potential difference V_{cb} is therefore 6 V, and since it must be 6 V in part (a) as well, the currents in the 6-Ω and 3-Ω resistors in part (a) are 1 A and 2 A, respectively.

29-2 Kirchhoff's rules

Not all networks can be reduced to simple series–parallel combinations. An example is a resistance network with a cross connection, as in Fig. 29-3a. A circuit like that in Fig. 29-3b, which contains sources in parallel paths, is another example. No new *principles* are required to compute the currents in these networks, but there are a number of techniques that enable such problems to be handled systematically. We shall describe only one of these, first developed by Gustav Robert Kirchhoff (1824–1887).

We first define two terms. A *branch point* in a network is a point where three or more conductors are joined. A *loop* is any closed conducting path. In Fig. 29-3a, for example, points a, d, e, and b are branch points but c and f are not. In Fig. 29-3b there are only two branch points, a and b.

Some possible loops in Fig. 29-3a are the closed paths *aceda*, *defbd*, *hadbgh*, and *hadefbgh*.

Kirchhoff's rules consist of the following two statements:

Point rule *The algebraic sum of the currents **toward** any branch point is zero:*

$$\sum I = 0. \qquad (29\text{-}4)$$

Loop rule *The algebraic sum of the potential differences in any loop, including those associated with emfs and those of resistive elements, must equal zero:*

$$\sum \mathcal{E} - \sum IR = 0. \qquad (29\text{-}5)$$

The point rule is an application of the principle of conservation of electric charge. Since no charge can accumulate at a branch point, the total current entering the point must equal the total leaving, or (considering those entering as positive and those leaving as negative) the algebraic sum of currents into a point must be zero.

The loop rule, which we have already encountered in Sec. 28-4, is an expression of conservation of *energy*; as a charge traverses a loop and returns to its starting point, the sum of the *rises* in potential associated with emf's in the loop must equal the sum of *drops* in potential associated with resistors (or, in more general problems, other circuit elements).

These basic rules are sufficient for the solution of a wide variety of network problems. Usually in such problems some of the emf's, currents, and resistances are known and others are unknown. The number of equations obtained from Kirchhoff's rules must always be equal to the number of unknowns, to permit simultaneous solution of the equations. The principal difficulty is not in understanding the basic ideas but in keeping track of algebraic signs! The following procedures should be followed carefully.

First, all quantities, known and unknown, should be labeled carefully, including an assumed sense of direction for each unknown current and emf. Often one does not know in advance the *actual* direction of an unknown current or emf, but this does not matter. The solution is carried out using the assumed directions, and if the actual direction of a particular quantity is opposite to the assumed direction, the value of the quantity will emerge from the analysis with a negative sign. Hence Kirchhoff's rules, correctly used, give the directions as well as the magnitudes of unknown currents and emf's. This point will be illustrated in the examples to follow.

Usually in labeling currents it is advantageous to use the point rule immediately to express the currents in terms of as few quantities as possible. For example, Fig. 29-4a shows a circuit correctly labeled, and Fig. 29-4b shows the same circuit, relabeled by applying the point rule to point a to eliminate I_3.

The following guidelines will help with the problem of signs:

1. Choose any closed loop in the network, and designate a direction (clockwise or counterclockwise) to traverse the loop in applying the loop rule.

29-3 Two networks that cannot be reduced to simple series-parallel combinations of resistors.

29-4 Application of the point rule to point a reduces the number of unknown currents from three to two.

29-5

2. Go around the loop in the designated direction, adding emf's and potential differences. An emf is counted as positive when it is traversed from $(-)$ to $(+)$ (in the direction of the non-electrostatic field E_n in the source) and negative when from $(+)$ to $(-)$. An IR term is counted as negative if the resistor is traversed in the *same* direction as the assumed current, positive if in the opposite direction.

3. Equate the sum of step (2) to zero.

4. If necessary, choose another loop to obtain a different relation between the unknowns, and continue until there are as many equations as unknowns, or until every circuit element has been included in at least one of the chosen loops.

Example 1 In the circuit shown in Fig. 29-5, find the unknown current I, resistance R, and emf $\mathcal{E}$.

Solution Application of the point rule to point a yields the relation

$$I + 1\text{ A} - 6\text{ A} = 0; \qquad I = 5\text{ A}.$$

To determine R we apply the loop rule to the loop labeled (1), obtaining

$$18\text{ V} - (5\text{ A})R + (1\text{ A})(2\text{ }\Omega) = 0,$$
$$R = 4\text{ }\Omega.$$

The term for resistance R is negative because our loop traverses that element in the same direction as the current and hence finds a potential *drop*, while the term for the 2-Ω resistor is positive because in traversing it in the direction opposite to the current we find a potential *rise*. If we had chosen to traverse loop (1) in the opposite direction, every term would have had the opposite sign, and the result for R would have been the same.

To determine $\mathcal{E}$, we apply the loop rule to loop (2);

$$\mathcal{E} + (6\text{ A})(2\text{ }\Omega) + (1\text{ A})(2\text{ }\Omega) = 0,$$
$$\mathcal{E} = -14\text{ V}.$$

This shows that the actual polarity of this emf is opposite to that assumed, and that the positive terminal of this source is really on the left side. Alternatively, one could use loop (3), obtaining the equation

$$\mathcal{E} + (6\text{ A})(2\text{ }\Omega) + (5\text{ A})(4\text{ }\Omega) - 18\text{ V} = 0,$$

from which again $\mathcal{E} = -14\text{ V}$.

Example 2 In Fig. 29-6, find the current in each resistor, and find the equivalent resistance.

Solution As pointed out at the beginning of this section, it is not possible to represent this network in terms of series and parallel combinations. There are five different currents to determine, but by applying the point rule to junctions a and b we represent them in terms of three unknown currents, as indicated in the figure. The current in the battery is of course $(I_1 + I_2)$.

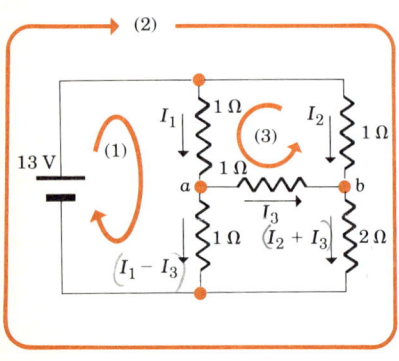

29-6

We now apply the loop rule to the three loops shown, obtaining the following three equations:

$$13\text{ V} - I_1(1\,\Omega) - (I_1 - I_3)(1\,\Omega) = 0; \tag{1}$$

$$-I_2(1\,\Omega) - (I_2 + I_3)(2\,\Omega) + 13\text{ V} = 0; \tag{2}$$

$$-I_1(1\,\Omega) - I_3(1\,\Omega) + I_2(1\,\Omega) = 0. \tag{3}$$

This is a set of three simultaneous equations for the three unknown currents. They may be solved by various methods; one straightforward procedure is to solve the third for I_2, obtaining $I_2 = I_1 + I_3$, and then substitute this expression into the first two equations to eliminate I_2. When this is done one obtains the two equations

$$13\text{ V} = I_1(2\,\Omega) - I_3(1\,\Omega), \tag{1'}$$

$$13\text{ V} = I_1(3\,\Omega) + I_3(5\,\Omega). \tag{2'}$$

Now I_3 may be eliminated by multiplying the first of these by 5 and adding the two to obtain

$$78\text{ V} = I_1(13\,\Omega),$$

from which $I_1 = 6$ A. This result may be substituted back into (1′) to obtain $I_3 = -1$ A, and finally from (3) we find $I_2 = 5$ A. We note that the direction of I_3 is opposite that of the initial assumption.

The total current through the network is $I_1 + I_2 = 11$ A, and the potential drop across it is equal to the battery emf, namely, 13 V. Thus the equivalent resistance of the network is

$$R = \frac{13\text{ V}}{11\text{ A}} = 1.18\,\Omega.$$

29-3 Ammeters and voltmeters

The most common instruments for measuring potential or current use a device called a *d'Arsonval galvanometer*. A pivoted coil of fine wire is placed in the magnetic field of a permanent magnet, as shown in Fig. 29-7. When there is a current in the coil, the magnetic field exerts on the coil a *torque* that is proportional to the current. (This magnetic interaction is discussed in detail in Chapter 31.) This torque is opposed by a spring, similar to the hairspring on the balance wheel of a watch, which exerts a restoring torque proportional to the angular displacement.

Thus the angular deflection of the indicator needle attached to the pivoted coil is directly proportional to the coil current, and the device can be calibrated to measure current. The maximum deflection for which the meter is designed, typically 90° to 120°, is called *full-scale deflection*. The current required to produce full-scale deflection (typically of the order of 10 μA to 10 mA) and the resistance of the coil (typically of the order of 10 to 1000 Ω) are the essential electrical characteristics of the meter.

The meter deflection is proportional to the *current* in the coil, but if the coil obeys Ohm's law, the current is proportional to the *potential difference* between the terminals of the coil. Thus the deflection is also proportional to this potential difference. For example, consider a meter whose coil has a resistance of 20 Ω, and which deflects full scale with a

29-7 A d'Arsonval meter movement, showing pivoted coil with attached pointer, permanent magnet supplying uniform magnetic field, and spring to provide restoring torque, which opposes magnetic-field torque.

29-8 (a) Internal connections of an ammeter. (b) Internal connections of a voltmeter.

current of 1 mA in its coil. The corresponding potential difference is

$$V_{ab} = IR = (10^{-3} \text{ A})(20 \text{ }\Omega) = 0.020 \text{ V} = 20 \text{ mV}.$$

Let us consider further the use of the d'Arsonval meter as a current-measuring instrument, often called an *ammeter.* To measure the current in a circuit, an ammeter must be inserted in *series* in the circuit so that the current to be measured actually passes *through* the meter. If the galvanometer above is inserted in this way, it will measure any current from zero to 1 mA. However, the resistance of the *coil* adds to the total resistance of the circuit, with the result that the current *after* the galvanometer is inserted, although it is correctly indicated by the instrument, may be less than it was *before* insertion of the instrument. It is evidently desirable that the resistance of the instrument should be much *smaller* than that of the remainder of the circuit, so that when the instrument is inserted it does not change the very thing we wish to measure. An *ideal* ammeter would have *zero* resistance.

Furthermore, the *range* of the galvanometer, if it is used without modification, is limited to a maximum current of 1 mA. The range can be extended, and at the same time the equivalent resistance can be reduced, by connecting a low resistance R_{sh} in parallel with the moving coil, as in Fig. 29-8a. The parallel resistor is called a *shunt*; its effect is to permit some of the circuit current I to bypass the meter and pass through the shunt instead. Usually the meter and the shunt are mounted inside a case, with binding posts or plug connectors for external connections at a and b. Occasionally, when several interchangeable shunts are used, they are mounted outside the case.

Suppose we wish to modify the meter described above for use as an ammeter with a range of 0 to 10 A. That is, the coil is to deflect full-scale when the current I in the *circuit* in which the ammeter is inserted equals 10 A. The *coil* current I_c must then be 1 mA, so the current I_{sh} in the

shunt is 9.999 A. The potential difference V_{ab} is
$$V_{ab} = I_c R_c = I_{sh} R_{sh}.$$
Hence
$$R_{sh} = R_c \left(\frac{I_c}{I_{sh}}\right) = 20\,\Omega \left(\frac{0.001}{9.999}\right) = 0.00200\,\Omega.$$

The equivalent resistance R of the instrument is
$$\frac{1}{R} = \frac{1}{R_c} + \frac{1}{R_{sh}},$$
and
$$R = 0.00200\,\Omega.$$

Thus we have a low-resistance instrument with the desired range of 0 to 10 A. Of course, if the current I is *less* than 10 A, the coil current (and the deflection) is correspondingly less, also.

Now let us consider the construction of a *voltmeter*. A voltmeter measures the potential difference beween two points, and its terminals must be connected to these points. A moving-coil meter cannot be used to measure the potential differences between, say, two charged spheres. When its terminals are connected to the spheres, the coil provides a conducting path from one sphere to the other. There will be a *momentary* current in the coil, but the charges on the sphere will change until the entire system is at the *same* potential. Only if the resistance of the instrument is so great that a very long time is required to reach equilibrium can a voltmeter be used for this purpose. Thus an *ideal* voltmeter has an *infinite* resistance, but a moving-coil galvanometer can be deflected only by a *current* in its coil; its resistance must be finite.

A moving-coil galvanometer *can* be used to measure the potential difference between the terminals of a *source*, or between *two points of a circuit containing a source*, because the source *maintains* a difference of potential between the points. However, the *range* of the galvanometer in our example, if used without modification, is limited to a maximum value of 20 mV. The range can be extended, and at the same time the equivalent resistance can be increased, by connecting a resistance R_s in *series* with the moving coil, as in Fig. 29–8b.

Suppose we wish to modify the galvanometer for use as a voltmeter with a range of 0 to 10 V. The coil is to deflect full-scale when the potential difference between the terminals of the instrument is 10 V. In other words, the current in the instrument is to be 1 mA when the potential difference between its terminals is 10 V.

The terminal potential difference is
$$V_{ab} = I(R_c + R_s),$$
and the necessary series resistance is
$$R_s = \frac{V_{ab}}{I} - R_c = \frac{10\,\text{V}}{0.001\,\text{A}} - 20\,\Omega = 9980\,\Omega.$$

The equivalent resistance R is
$$R = R_c + R_s = 10{,}000\,\Omega.$$

Thus we have a high-resistance instrument with the desired range of 0 to 10 V.

A voltmeter, like an ammeter, can disturb the circuit to which it is connected. For example, when a source is on *open* circuit, the potential difference between its terminals equals its emf. It would seem, therefore, that to measure the emf we need only measure this potential difference. But when the terminals of a voltmeter are connected to those of a source, the meter and source form a *complete* circuit in which there is a current. The potential difference *after* the meter is connected, although it is correctly indicated by the instrument, is not equal to $\mathcal{E}$, but to $\mathcal{E} - Ir$, and is *less* than it was *before* the instrument was connected. Again, the measuring instrument alters the quantity it is intended to measure. It is evidently desirable that the resistance of a voltmeter, even though not infinite, should be as *large* as possible.

A voltmeter and an ammeter can be used together to measure *resistance* and *power*. The resistance of a resistor equals the potential difference V_{ab} between its terminals, divided by the current I:

$$R = \frac{V_{ab}}{I},$$

and the power input to any portion of a circuit equals the product of the potential difference across this portion and the current:

$$P = V_{ab}I.$$

The most straightforward method of measuring R or P is therefore to measure V_{ab} and I simultaneously.

In Fig. 29-9a, ammeter A reads correctly the current I in the resistor R. Voltmeter V, however, reads the *sum* of the potential difference V_{ab} across the resistor and the potential difference V_{bc} across the ammeter.

If we transfer the voltmeter terminal from c to b, as in Fig. 29-9b, the voltmeter reads correctly the potential difference V_{ab} but the ammeter now reads the *sum* of the current I in the resistor and the current I_V in the voltmeter. Thus, whichever connection is used, we must correct the reading of one instrument or the other to obtain the true values of V_{ab} or I (unless, of course, the corrections are small enough to be neglected).

Example 1 The circuit of Fig. 29-9a is to be used to measure an unknown resistor R. The meter resistances are $R_V = 10{,}000\,\Omega$ and $R_A = 2.0\,\Omega$. If the voltmeter reads 12.0 V and the ammeter reads 0.10 A, what is the true resistance?

Solution If the meters were ideal (i.e., $R_V = \infty$ and $R_A = 0$), the resistance would be simply $R = V/I = (12.0\text{ V})/(0.10\text{ A}) = 120\,\Omega$. But the voltmeter reading includes the potential V_{bc} across the ammeter as well as that (V_{ab}) across the resistor. We have $V_{bc} = IR_A = (0.10\text{ A})(2.0\,\Omega) = 0.2\text{ V}$, so the actual potential drop V_{ab} across the resistor is $12.0\text{ V} - 0.2\text{ V} = 11.8\text{ V}$, and the resistance is

$$R = \frac{V_{ab}}{I} = \frac{11.8\text{ V}}{0.10\text{ A}} = 118\,\Omega.$$

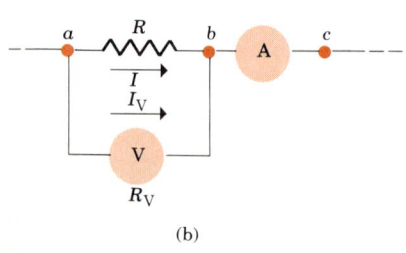

29-9 Ammeter–voltmeter method for measuring resistance or power.

Example 2 Suppose the same meters are connected instead as shown in Fig. 29-9b, and the above readings are obtained. What is the true resistance?

Solution In this case the voltmeter measures the potential across the resistor correctly; the difficulty is that the ammeter measures the voltmeter current I_V as well as the current I in the resistor. We have $I_V = V/R_V = (12.0 \text{ V})/(10,000 \text{ }\Omega) = 1.2$ mA; so the actual current I in the resistor is $I = 0.10 \text{ A} - 0.0012 \text{ A} = 0.0988 \text{ A}$. Thus the resistance is

$$R = \frac{V_{ab}}{I} = \frac{12.0 \text{ V}}{0.0988 \text{ A}} = 121.5 \text{ }\Omega.$$

29-10

29-4 The ohmmeter

Although not a precision instrument, the ohmmeter is a useful device for rapid measurement of resistance. It consists of a meter, a resistor, and a source (often a flashlight cell) connected in series, as in Fig. 29-10. The resistance R to be measured is connected between terminals x and y.

The series resistance R_s is chosen so that, when terminals x and y are short-circuited (that is, when $R = 0$), the galvanometer deflects full scale. When the circuit between x and y is open (that is, when $R = \infty$), the galvanometer shows no deflection. For a value of R between zero and infinity, the galvanometer deflects to some intermediate point depending on the value of R, and hence the galvanometer scale can be calibrated to read the resistance R.

*29-5 The R-C series circuit

In the circuits considered thus far, we have assumed that the emf's and resistances are constant, so that all potentials and currents are constant, independent of time. Figure 29-11 shows a simple example of a circuit in which the current and voltages are *not* constant. When the double-pole, double-throw (dpdt) switch is in the upper position, points a and b are connected to the battery, which charges the capacitor through resistor R.

If the capacitor is initially uncharged, then the initial potential difference across the capacitor is zero, and the entire battery voltage appears across the resistor, causing an initial current $I = V/R$. As the capacitor charges, its voltage increases, and the potential difference across the resistor decreases, corresponding to a decrease in current. After a long time the capacitor has become fully charged, the entire battery voltage V appears across the capacitor, there is no potential difference across the resistor, and the current becomes zero.

Let q represent the charge on the capacitor and i the charging current at some instant after the switch is thrown in the "up" position. (As is customary in circuit analysis, we use small letters for quantities that vary with time, and capital letters for constant quantities.) The instantaneous potential differences v_{ac} and v_{cb} are

$$v_{ac} = iR, \qquad v_{cb} = \frac{q}{C}. \tag{29-6}$$

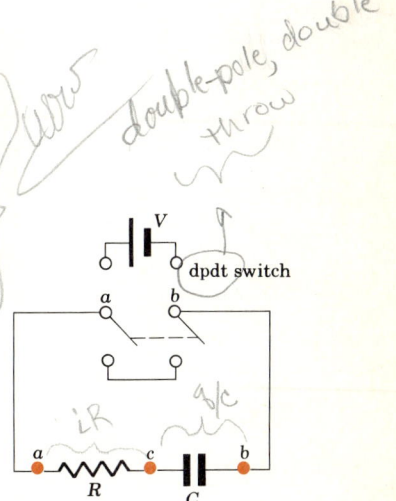

29-11

Therefore,

$$V_{ab} = V = v_{ac} + v_{cb} = iR + \frac{q}{C}, \qquad (29\text{-}7)$$

where V = constant. The current i is then

$$i = \frac{V}{R} - \frac{q}{RC}. \qquad (29\text{-}8)$$

At the instant connections are made, $q = 0$ and the *initial current* $I_0 = V/R$, which would be the steady current if the capacitor were not present.

As the charge q increases, the term q/RC becomes larger, and the current decreases and eventually becomes zero. When $i = 0$,

$$\frac{V}{R} = \frac{q}{RC}, \qquad q = CV = Q_f,$$

where Q_f is the final charge on the capacitor.

To obtain a quantitative description of the variation with time of the charge and current in this circuit, we replace i in Eq. (29-8) with dq/dt, obtaining

$$\frac{dq}{dt} = \frac{V}{R} - \frac{q}{RC}, \qquad (29\text{-}9)$$

which may be rearranged as

$$\frac{dq}{VC - q} = \frac{dt}{RC}.$$

Integrating both sides, we obtain

$$-\ln(VC - q) = t/RC + \text{constant}.$$

To evaluate the constant, we note that at $t = 0$, $q = 0$, so

$$-\ln(VC - 0) = 0 + \text{constant}.$$

Rearranging again, we obtain

$$\ln(VC - q) - \ln VC = \ln\frac{VC - q}{VC} = -\frac{t}{RC},$$

$$1 - \frac{q}{VC} = e^{-t/RC},$$

$$q = VC(1 - e^{-t/RC}) = Q_f(1 - e^{-t/RC}). \qquad (29\text{-}10)$$

The time derivative of this expression is the current:

$$i = \frac{dq}{dt} = \frac{V}{R}e^{-t/RC} = I_0 e^{-t/RC} \qquad (29\text{-}11)$$

The charge and current are therefore both *exponential* functions of time. Figure 29-12a is a graph of Eq. (29-11), and Fig. 29-12b is a graph of Eq. (29-10). At a time $t = RC$, the current has decreased to $1/e$ (about 0.368) of its initial value and the charge has increased to *within* $1/e$ of its final value. The product RC is called the *time constant*, or the *relaxation*

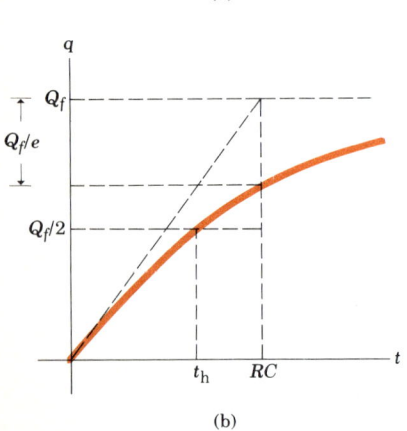

29-12

time, of the circuit, denoted by τ:

$$\tau = RC. \tag{29-12}$$

The *half-life* of the circuit, t_h, is the time for the current to decrease to half its initial value, or for the capacitor to acquire half its final charge. Setting $i = I_0/2$ in Eq. (29-11), we find

$$t_h = RC \ln 2 = 0.693\, RC.$$

The half-life depends only on the time constant RC, and not on the initial current. Thus if the current decreases from I_0 to $I_0/2$ in a time t_h, as shown in Fig. 29-12a, it decreases to half this value, or to $I_0/4$, in another half-life, and so on.

Example A resistor of resistance $R = 10$ MΩ is connected in series with a capacitor of capacitance $1\,\mu$F. The time constant is

$$\tau = RC = (10 \times 10^6\,\Omega)(10^{-6}\,\text{F}) = 10\,\text{s},$$

and the half-life is

$$t_h = (10\,\text{s})(\ln 2) = 6.9\,\text{s}.$$

On the other hand, if $R = 10\,\Omega$, the time constant is only 10×10^{-6} s, or $10\,\mu$s.

Suppose next that the capacitor has acquired a charge Q_0 and that the switch is thrown to the "down" position. The capacitor then *discharges* through the resistor and its charge eventually decreases to zero. (We designate the charge by Q_0 because this is the *initial* charge in the discharge process. It is not necessarily equal to the charge Q_f defined above.)

Again let i and q represent the time-varying current and charge at some instant after the switch is thrown. Since V_{ab} is now zero, we have, from Eq. (29-7),

$$0 = v_{ac} + v_{cb}.$$

The direction of the current in the resistor is now from c to a, so $v_{ca} = -v_{ac} = iR$ and

$$i = \frac{q}{RC}. \tag{29-13}$$

When $t = 0$, $q = Q_0$ and the initial current I_0 is

$$I_0 = \frac{Q_0}{RC} = \frac{V_0}{R},$$

where V_0 is the initial potential difference across the capacitor. As the capacitor discharges, both q and i decrease.

The same procedures as above can be followed to obtain $i(t)$ and $q(t)$. If we replace i in Eq. (29-13) by $-dq/dt$ (the charge q is now *decreasing*), we get

$$\frac{dq}{dt} = -\frac{q}{RC}. \tag{29-14}$$

Integration of this equation gives $q(t)$, and by differentiation we find $i(t)$.

Alternatively, differentiation of Eq. (29-13) gives

$$\frac{di}{dt} = -\frac{i}{RC}, \tag{29-15}$$

from which we can get $i(t)$ and, by a second integration, get $q(t)$. It is left as a problem to show that

$$i = I_0 e^{-t/RC}, \tag{29-16}$$

$$q = Q_0 e^{-t/RC}. \tag{29-17}$$

Both the current and the charge decrease exponentially with time.

*29-6 Displacement current

For a *conducting* circuit, in the steady state, the total current *into* any given portion must be equal to the current *out of* that portion. This statement forms the basis of Kirchhoff's point rule, discussed in Sec. 29-2. However, this rule is *not* obeyed for a capacitor that is being charged. In Fig. 29-13, there is a conduction current *into* the left plate but no conduction current *out of* this plate; similarly, there is conduction current out of the right plate, but none into it.

James Clerk Maxwell (1831–1879) showed that it is possible to generalize the definition of current so that one can still say that the current out of each plate is equal to the current into it. As the capacitor charges, the conduction current increases the charge on each plate, and this in turn increases the electric field between the plates. The *rate* of increase of field is proportional to the conduction current; Maxwell's scheme was to associate an equivalent current density with this rate of increase of field.

Assuming, for simplicity, that the capacitor consists of two parallel plates with uniform charge density σ, the field E between the plates is given by

$$E = \frac{\sigma}{\epsilon_0} = \frac{Q}{\epsilon_0 A}. \tag{29-18}$$

In a time interval dt, the charge Q increases by $dQ = I\,dt$; the corresponding change in E is

$$dE = \frac{dQ}{\epsilon_0 A} = \frac{I\,dt}{\epsilon_0 A},$$

and the *rate* of change of E is

$$\frac{dE}{dt} = \frac{I}{\epsilon_0 A}. \tag{29-19}$$

We now define an *equivalent current density* J_D between the plates as

$$J_D = \epsilon_0 \frac{dE}{dt}. \tag{29-21}$$

Then the total equivalent current between the plates, which we may call I_D is

$$I_D = J_D A = \epsilon_0 \left(\frac{dE}{dt}\right) A = I.$$

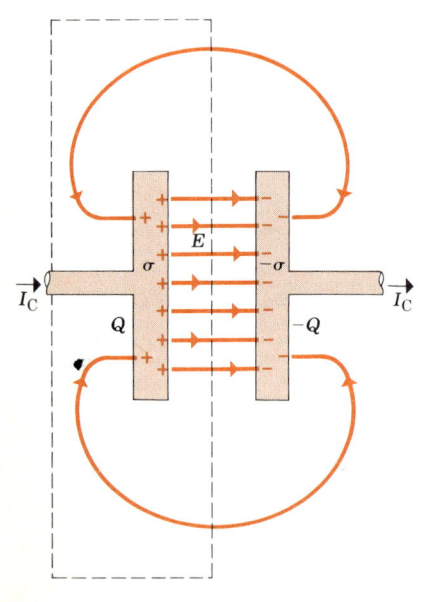

29-13 The conduction current into the left plate of the capacitor equals the displacement current between the plates.

Thus, if we include the "effective current" as well as the conduction current, the current *into* the region bounded by the broken line in Fig. 29-13 equals the current *out of* this region. The subscript D is chosen because Maxwell originally called this effective current the *displacement current;* this term is still used, although the reasons for its use are dubious.

Displacement current may also be expressed in terms of *electric flux,* defined in Sec. 25-4, Eqs. (25-16) and (25-17). For a uniform field E perpendicular to an area A, $\Psi = EA$, and Eq. (29-21) may be rewritten as

$$I_D = \epsilon_0 \frac{d\Psi}{dt}. \tag{29-22}$$

If the space between capacitor plates contains a dielectric, Eq. (29-18) must be replaced by the more general relation derived in Sec. 27-5:

$$E = \frac{\sigma}{\epsilon} = \frac{Q}{\epsilon A} = \frac{Q}{k\epsilon_0 A}. \tag{29-23}$$

The corresponding modification of the definition of displacement current density is

$$J_D = \epsilon \left(\frac{dE}{dt}\right) = K\epsilon_0 \left(\frac{dE}{dt}\right). \tag{29-24}$$

Maxwell's generalized view of current and current density may appear to be merely an artifice introduced to preserve Kirchhoff's current rule even in cases where charge is accumulating in a certain region of space, such as a capacitor plate. However, it is much more than this. In a later chapter we shall study the role of current as a source of *magnetic* field, and we shall see that a *displacement* current sets up a magnetic field in exactly the same way as an ordinary *conduction* current. Thus displacement current, far from being an artifice, is a fundamental fact of nature, and Maxwell's discovery of it was the bold step of an extraordinary genius. As we shall see, the concept of displacement current provided the necessary basis for the understanding of electromagnetic waves in the last third of the nineteenth century.

Questions

29-1 Can the potential difference between terminals of a battery ever be opposite in direction to the emf?

29-2 Why do the lights on a car become dimmer when the starter is operated?

29-3 What determines the maximum current that can be carried safely by household wiring? (Typical limits are 15 A for 14-gauge wire, 20 A for 12-gauge, and so on.)

29-4 Lights in a house often dim momentarily when a motor, such as a washing machine or a power saw, is turned on. Why does this happen?

29-5 Compare the formulas for resistors in series and parallel with those for capacitors in series and parallel. What similarities and differences do you see? Sometimes in circuit analysis one uses the quantity *conductance,* denoted as g and defined as the reciprocal of resistance: $g = 1/R$. What is the corresponding comparison for conductance and capacitance?

29-6 Is it possible to connect resistors together in a way that cannot be reduced to some combination of series and parallel combinations? If so, give examples; if not, state why not.

29-7 In a two-cell flashlight, the batteries are usually connected in series. Why not connect them in parallel?

29-8 Some Christmas-tree lights have the property that, when one bulb burns out, all the lights go out, while with others only the burned-out bulb goes out. Discuss this difference in terms of series and parallel circuits.

29-9 What possible advantage could there be in connecting several identical batteries in parallel?

29-10 Two 110-V lightbulbs, one 25-W and one 200-W, were connected in series across a 220-V line. It seemed like a good idea at the time, but one bulb burned out almost instantaneously. Which one burned out, and why?

29-11 When the direction of current in a battery reverses, does the direction of its emf also reverse?

29-12 Under what conditions would the terminal voltage of a battery be zero?

29-13 What sort of meter should be used to test the condition of a dry cell (such as a flashlight battery) having constant emf but internal resistance that increases with age and use?

29-14 For very large resistances, it is easy to construct RC circuits having time constants of several seconds or minutes. How might this fact be used to measure very large resistances, too large to measure by more conventional means?

Problems

29-1 Prove that, when two resistors are connected in parallel, the equivalent resistance of the combination is always smaller than that of either resistor.

29-2

a) A resistance R_2 is connected in parallel with a resistance R_1. What resistance R_3 must be connected in series with the combination of R_1 and R_2 so that the equivalent resistance is equal to the resistance R_1? Draw a diagram.

b) A resistance R_2 is connected in series with a resistance R_1. What resistance R_3 must be connected in parallel with the combination of R_1 and R_2 so that the equivalent resistance is equal to R_1? Draw a diagram.

29-3

a) Calculate the equivalent resistance of the circuit of Fig. 29-14 between x and y.

b) What is the potential difference between x and a if the current in the 8-Ω resistor is 0.5 A?

Figure 29-14

29-4 Each of three resistors in Fig. 29-15 has a resistance of 2 Ω and can dissipate a maximum of 18 W without becoming excessively heated. What is the maximum power the circuit can dissipate?

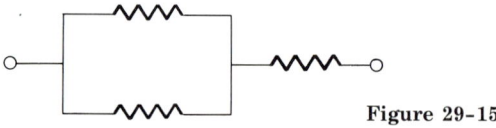

Figure 29-15

29-5 Two lamps, marked "60 W, 120 V" and "40 W, 120 V," are connected in series across a 120-V line. What power is consumed in each lamp? Assume that the resistance of the filaments does not vary with current.

29-6 Three equal resistors are connected in series. When a certain potential difference is applied across the combination, the total power dissipated is 10 W. What power would be dissipated if the three resistors were connected in parallel across the same potential difference?

29-7

a) The power rating of a 10,000-Ω resistor is 2 W. (The power rating is the maximum power the resistor can safely dissipate without too great a rise in temperature.) What is the maximum allowable potential difference across the terminals of the resistor?

b) A 20,000-Ω resistor is to be connected across a potential difference of 300 V. What power rating is required?

c) It is desired to connect an equivalent resistance of 1000 Ω across a potential difference of 200 V. A number of 10-W, 1000-Ω resistors are available. How should they be connected?

29-8 A 60-Ω resistor and a 90-Ω resistor are connected in parallel and the combination is connected across a 120-V dc line.

a) What is the resistance of the parallel combination?

b) What is the total current through the parallel combination?

c) What is the current through each resistor?

29-9 A 1000-Ω 2-W resistor is needed, but only several 1000-Ω 1-W resistors are available.

a) How can the required resistance and power rating be obtained by a combination of the available units?

b) What power is then dissipated in each resistor?

29-10 A 25-W, 120-V lightbulb and a 100-W, 120-V lightbulb are connected in series across a 240-V line. Assume that the resistance of each bulb does not vary with current.

a) Find the current through the bulbs.
b) Find the power dissipated in each bulb.
c) One bulb burns out very quickly. Which one? Why?

29-11 An electric heating element was designed for a power rating of 550 W when connected to a 110-V line. If the line voltage is 120 V, what power is consumed?

29-12 Prove that the resistance of the infinite network shown in Fig. 29-16 is equal to $(1 + \sqrt{3})r$.

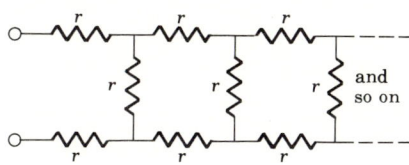

Figure 29-16

29-13

Note: Figure 29-17 employs a convention often used in circuit diagrams. The battery (or other power supply) is not shown explicitly. It is understood that the point at the top, labeled "36 V," is connected to the positive terminal of a 36-V battery having negligible internal resistance, and the "ground" symbol at the bottom to its negative terminal. The circuit is completed through the battery, even though it is not shown on the diagram.

a) In Fig. 29-17a, what is the potential difference V_{ab} when switch S is open?
b) What is the current through switch S, when it is closed?
c) In Fig. 29-17b, what is the potential difference V_{ab} when switch S is open?
d) What is the current through switch S when it is closed?

What is the equivalent resistance in Fig. 29-17b,
e) when switch S is open? f) when switch S is closed?

Figure 29-17

29-14 (See note with Problem 29-13.)
a) What is the potential difference between points a and b in Fig. 29-18 when switch S is open?
b) Which point, a or b, is at the higher potential?

Figure 29-18

c) What is the final potential of point b when switch S is closed?
d) How much charge flows through switch S when it is closed?

29-15 (See note with Problem 29-13.)
a) What is the potential difference between points a and b in Fig. 29-19 when switch S is open?
b) Which point, a or b, is at the higher potential?
c) What is the final potential of point b when switch S is closed?
d) How much does the charge on each capacitor change when S is closed?

Figure 29-19

29-16 Calculate the three currents indicated in the circuit diagram of Fig. 29-20.

Figure 29-20

29-17 Find the emf's $\mathcal{E}_1$ and $\mathcal{E}_2$ in the circuit of Fig. 29-21, and the potential difference between points a and b.

Figure 29-21

29-18

a) Find the potential difference between points a and b in Fig. 29-22.

b) If a and b are connected, find the current in the 12-V cell.

Figure 29-22

29-19 Find the current in each branch of the circuit shown in Fig. 29-23.

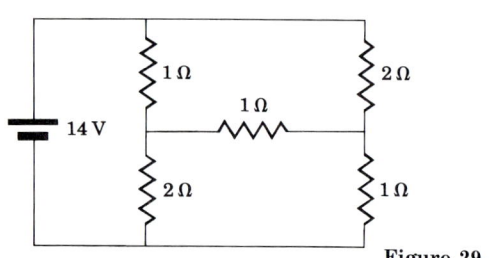

Figure 29-23

29-20 In the circuit shown in Fig. 29-24, find

a) the current in resistor R;
b) the resistance R;
c) the unknown emf $\mathcal{E}$.
d) If the circuit is broken at point x, what is the current in the 28-V battery?

Figure 29-24

29-21 In the circuit shown in Fig. 29-25, find

a) the current in each branch;
b) the potential difference V_{ab}.

Figure 29-25

29-22 Suppose a resistor R lies along each edge of a cube (12 resistors in all) with connections at the corners. Find the equivalent resistance between two diagonally opposite corners of the cube.

29-23 A 600-Ω resistor and a 400-Ω resistor are connected in series across a 90-V line. A voltmeter across the 600-Ω resistor reads 45 V.

a) Find the voltmeter resistance.
b) Find the reading of the same voltmeter if connected across the 400-Ω resistor.

29-24 Point a in Fig. 29-26 is maintained at a constant potential of 300 V above ground. (See note with Problem 29-13.)

a) What is the reading of a voltmeter of the proper range, and of resistance 3×10^4 Ω, when connected between point b and ground?
b) What would be the reading of a voltmeter of resistance 3×10^6 Ω?
c) Of a voltmeter of infinite resistance?

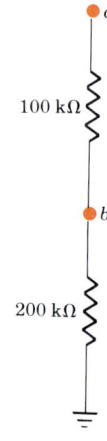

Figure 29-26

29-25 Two 150-V voltmeters, one of resistance 15,000 Ω and the other of resistance 150,000 Ω, are connected in series across a 120-V dc line. Find the reading of each voltmeter.

29-26 A 150-V voltmeter has a resistance of 20,000 Ω. When connected in series with a large resistance R across a 110-V line, the meter reads 5 V. Find the resistance R. (This problem illustrates one method of measuring large resistances.)

29-27 A 100-V battery has an internal resistance of 5 Ω.

a) What is the reading of a voltmeter having a resistance of 500 Ω when placed across the terminals of the battery?

b) What maximum value may the ratio r/R_V have if the error in the reading of the emf of a battery is not to exceed 5 percent?

29-28 The resistance of a galvanometer coil is 50 Ω and the current required for full-scale deflection is 500 μA.

a) Show in a diagram how to convert the galvanometer to an ammeter reading 5 A full-scale, and compute the shunt resistance.

b) Show how to convert the galvanometer to a voltmeter reading 150 V full-scale, and compute the series resistance.

29-29 The resistance of the coil of a pivoted-coil galvanometer is 10 Ω, and a current of 0.02 A causes it to deflect full-scale. It is desired to convert this galvanometer to an ammeter reading 10 A full-scale. The only shunt available has a resistance of 0.03 Ω. What resistance R must be connected in series with the coil? (See Fig. 29–27.)

Figure 29–27

29-30 The resistance of the moving coil of the galvanometer G in Fig. 29–28 is 25 Ω; the meter deflects full-scale with a current of 0.010 A. Find the magnitudes of the resistances R_1, R_2, and R_3, required to convert the galvanometer to a multirange ammeter deflecting full-scale with currents of 10 A, 1 A, and 0.1 A.

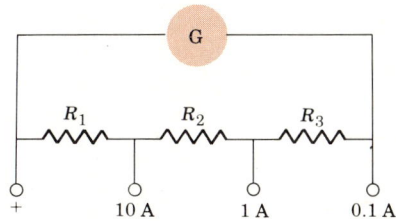

Figure 29–28

29-31 Figure 29–29 shows the internal wiring of a "three-scale" voltmeter whose binding posts are marked +, 3 V,

Figure 29–29

15 V, 150 V. The resistance of the moving coil, R_G, is 15 Ω, and a current of 1 mA in the coil causes it to deflect full-scale. Find the resistances R_1, R_2, R_3, and the overall resistance of the meter on each of its ranges.

29-32 A certain dc voltmeter is said to have a resistance of "one thousand ohms per volt of full-scale deflection." What current, in milliamperes, is required for full-scale deflection?

29-33 Let V and I represent the readings of the voltmeter and ammeter, respectively, shown in Fig. 29–9, and R_V and R_A their equivalent resistances.

a) When the circuit is connected as in Fig. 29–9a, show that
$$R = \frac{V}{I} - R_A.$$

b) When the connections are as in Fig. 29–9b, show that
$$R = \frac{V}{I - (V/R_V)}.$$

c) Show that the power delivered to the resistor in part (a) is $IV - I^2 R_A$, and in part (b) is $IV - (V^2/R_V)$.

29-34 A certain galvanometer has a resistance of 200 Ω and deflects full-scale with a current of 1 mA in its coil. It is desired to replace this with a second galvanometer whose resistance is 50 Ω and which deflects full-scale with a current of 50 μA in its coil. Devise a circuit incorporating the second galvanometer such that

a) the equivalent resistance of the circuit equals the resistance of the first galvanometer, and

b) the second galvanometer will deflect full-scale when the current into and out of the circuit equals the full-scale current of the first galvanometer.

29-35 Suppose the galvanometer of the ohmmeter in Fig. 29–10 has a resistance of 50 Ω and deflects full-scale with a current of 1 mA in its coil. The emf $\mathcal{E} = 1.5$ V.

a) What should be the value of the series resistance R_s?

b) What values of R correspond to galvanometer deflections of $\frac{1}{4}$, $\frac{1}{2}$, and $\frac{3}{4}$ full-scale?

c) Does the ohmmeter have a linear scale?

29-36 In the ohmmeter in Fig. 29–30, M is a 1-mA meter having a resistance of 100 Ω. The battery B has an emf of 3 V and negligible internal resistance. R is so chosen that, when the terminals a and b are shorted ($R_x = 0$), the meter reads full scale. When a and b are open ($R_x = \infty$), the meter reads zero.

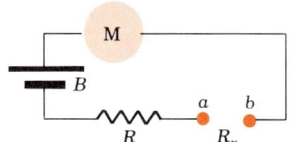

Figure 29-30

a) What should be the value of the resistor R?
b) What current would indicate a resistance R_x of 600 Ω?
c) What resistances correspond to meter deflections of $\frac{1}{4}$, $\frac{1}{2}$, and $\frac{3}{4}$ full-scale?

29-37 In Fig. 29-31, a resistor of resistance 75 Ω is connected between points a and b. The resistance of the galvanometer G is 90 Ω. What should be the resistance between b and the sliding contact c, if the galvanometer current I_G is to be $\frac{1}{3}$ of the current I?

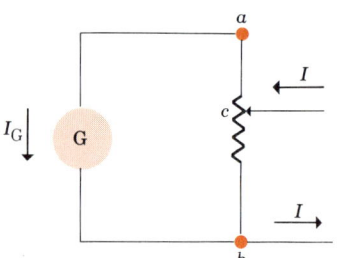

Figure 29-31

29-38 The circuit shown in Fig. 29-32 is called a *Wheatstone bridge*; it is used to determine the value of an unknown resistor X by comparison with three precisely known resistors M, N, and P. With switches K_1 and K_2 closed, these resistors are varied until the current in the galvanometer G is zero; the bridge is then said to be *balanced*. Show that under this condition the unknown resistance is given by $X = MP/N$. (This method permits very high precision in comparing resistors.)

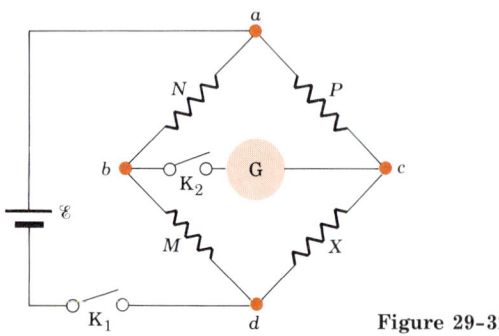

Figure 29-32

29-39 In Fig. 29-32, the galvanometer G shows zero deflection when $M = 1000$ Ω, $N = 10.00$ Ω, and $P = 27.49$ Ω. What is the unknown resistance X?

29-40 The circuit shown in Fig. 29-33 is called a *potentiometer*. It permits measurements of potential difference without drawing current from the circuit being measured, and hence acts as an infinite-resistance voltmeter. The resistor between a and b is a uniform wire of length l, with a sliding contact c at a distance x from b. An unknown potential difference V_x is measured by sliding the contact until the galvanometer G reads zero.

a) Show that under this condition the unknown potential difference is given by $V_x = (x/l)\mathcal{E}$.
b) Why is the internal resistance of the galvanometer not important?

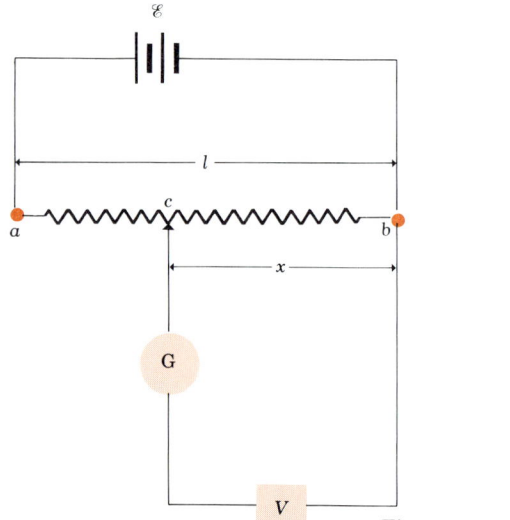

Figure 29-33

29-41 In Fig. 29-33, suppose $\mathcal{E} = 12.00$ V and $l = 1.000$ m. The galvanometer G reads zero when $x = 0.793$ m.

a) What is the potential difference V?
b) Suppose V is the emf of a battery; can its internal resistance be determined by this method?

29-42 A 0.05-μF capacitor is charged to a potential of 200 V and is then permitted to discharge through a 10-MΩ resistor. How much time is required for the charge to decrease to

a) $1/e$,   b) $1/e^2$

of its initial value?

29-43 A capacitor is charged to a potential of 10 V and is then connected to a voltmeter having an internal resistance of 1.0 MΩ. After a time of 5 s, the voltmeter reads 5 V. What is the capacitance?

29-44 A 10-μF capacitor is connected through a 1-MΩ resistor to a constant potential difference of 100 V.

a) Compute the charge on the capacitor at the following times after the connections are made: 0, 5 s, 10 s, 20 s, 100 s.
b) Compute the charging current at the same instants.

c) How long a time would be required for the capacitor to acquire its final charge if the charging current remained constant at its initial value?

d) Find the time required for the charge to increase from zero to 5×10^{-4} C.

e) Construct graphs of the results of parts (a) and (b) for a time interval of 20 s.

29–45 Two capacitors are charged in series by a 12-V battery (Fig. 29–34).

a) What is the time constant of the charging circuit?

b) After being closed for the length of time determined in (a), the switch S is opened. What is the voltage across the 6-μF capacitor?

Figure 29–34

29–46 A capacitor of capacitance C is charged by connecting it through a resistance R to the terminals of a battery of emf $\mathcal{E}$ and of negligible internal resistance.

a) How much energy is supplied by the battery in the charging process?

b) What fraction of this energy appears as heat in the resistor?

29–47 The current in a discharging capacitor is given by Eq. (29–16).

a) Using Eq. (29–16), derive an expression for the instantaneous power $P = i^2 R$ dissipated in the resistor.

b) Integrate the expression for P to find the total energy dissipated in the resistor, and show that this is equal to the total energy initially stored in the capacitor.

29–48 The current in a charging capacitor is given by Eq. (29–11).

a) The instantaneous power supplied by the battery is Vi. Integrate this to find the total energy supplied by the battery.

b) The instantaneous power dissipated in the resistor is $i^2 R$. Integrate this to find the total energy dissipated in the resistor.

c) Find the final energy stored in the capacitor, and show that this equals the total energy supplied by the battery, less the energy dissipated in the resistor, as obtained in parts (a) and (b).

d) What fraction of the energy supplied by the battery is stored in the capacitor? How does this fraction depend on R?

29–49

a) The differential equation for the instantaneous charge q of a capacitor a moment after its terminals have been disconnected from a source and connected to a resistance R is given by Eq. (29–14), namely,

$$\frac{dq}{dt} = -\frac{q}{RC}.$$

Show that

$$q = Q_0 e^{-t/RC}.$$

b) The current in the circuit of part (a) is given by Eq. (29–15), namely,

$$\frac{di}{dt} = -\frac{i}{RC}.$$

Show that

$$i = I_0 e^{-t/RC}.$$

29–50 Suppose that the parallel plates in Fig. 29–13 have an area of 2 m^2 and are separated by a sheet of dielectric 1 mm thick, of dielectric constant 3. (Neglect edge effects.) At a certain instant, the potential difference between the plates is 100 V and the current I_C equals 2 mA.

a) What is the charge Q on each plate?

b) What is the rate of change of charge on the plates?

c) What is the displacement current in the dielectric?

29–51 In a certain copper conductor ($\rho = 2 \times 10^{-8}\,\Omega\cdot$m) carrying a current, the electric field varies sinusoidally with time according to $E = E_0 \sin \omega t$, where $E_0 = 0.1$ V m^{-1} and $\omega = (2\pi)(60$ Hz$)$.

a) Find the magnitude of the maximum conduction current density in the wire.

b) Assuming $\epsilon = \epsilon_0$, find the maximum displacement current density in the wire, and compare with the result of part (a).

29–52

a) Repeat the calculations of Problem 29–51 for a rod of pure silicon having $\rho = 2000\,\Omega\cdot$m.

b) At what frequency would the maximum conduction and displacement densities become equal, if $\epsilon = \epsilon_0$ (which is not actually the case)?

c) At the frequency determined in part (b), what is the relative *phase* of the conduction and displacement currents?

29–53 A capacitor is made from two cylindrical rods of radius R, placed end to end with a gap of width d between them, where d is much smaller than R. There is a current i in each rod, and the charge q on each end surface changes at a rate given by $dq/dt = i$.

a) Find the conduction current density in each rod.

b) Find the electric field between the surfaces, in terms of q.
c) Find the displacement current density in the gap, and show that it is equal to the conduction current density in the rods.

29-54 A parallel-plate capacitor is made from two plates, each having area A, spaced a distance d apart. The space between plates is filled with a material having dielectric constant K. This material is not a perfect insulator but has resistivity ρ. The capacitor is initially charged with charge of magnitude Q_0 on each plate, which gradually discharges by conduction through the dielectric. Show that at any instant the displacement current density in the dielectric is equal in magnitude to the conduction current density but opposite in direction, so that the *total* current density is zero at every instant.

THE MAGNETIC FIELD 30

30-1 Magnetism

The first magnetic phenomena observed were those associated with naturally occurring magnets, fragments of iron ore found near the ancient city of Magnesia (whence the term "magnet"). These natural magnets attract unmagnetized iron; the effect is most pronounced at certain regions of the magnet known as its *poles*. It was known to the Chinese as early as 121 A.D. that an iron rod, after being brought near a natural magnet, would acquire and retain this property of the natural magnet, and that such a rod, when freely suspended about a vertical axis, would set itself approximately in the north–south direction. The use of magnets as aids to navigation can be traced back at least to the eleventh century.

The study of magnetic phenomena was confined for many years to magnets made in this way. Not until 1819 was any connection between electrical and magnetic phenomena shown. In that year, the Danish scientist Hans Christian Oersted (1777–1851) observed that a pivoted magnet (a compass needle) was deflected when in the neighborhood of a wire carrying a current. Twelve years later, Michael Faraday (1791–1867), an English physicist, found that a momentary current existed in a circuit while the current in a nearby circuit was being *started or stopped*. Shortly afterward it was discovered that the *motion* of a magnet toward or away from the circuit would produce the same effect. Joseph Henry (1797–1878), an American scientist who later became the first director of the Smithsonian Institution, discovered some of the same phenomena independently. The work of Oersted thus demonstrated that *magnetic* effects could be produced by moving *electric* charges, and that of Faraday and Henry showed that currents could be produced by moving magnets.

It is now known that all magnetic phenomena result from forces between electric charges in motion. That is, charges in motion relative to an observer set up a *magnetic* field as well as an *electric* field, and this magnetic field exerts a force on a second charge in motion relative to the observer. The electrons in atoms are in motion about the atomic nuclei, and each electron has additional motion ("spin") that can be visualized roughly as rotation about an axis passing through it. Thus all atoms can be expected to exhibit magnetic effects, and in fact they do. The possibility that the magnetic properties of matter are the result of tiny atomic currents was first suggested by Ampère in 1820. Not until recent years has the verification of these ideas been possible.

The material through which the charges are moving may have a significant effect on the observed magnetic forces between them. In the present chapter we shall assume that the charges or conductors are in otherwise empty space. For all practical purposes, the results will apply equally well to other charges and conductors in air.

30-2 The magnetic field

In describing the interaction of two charges *at rest,* we found it useful to introduce the concept of *electric field,* and to describe the interaction in two stages:

1. One charge sets up or creates an electric field E in the space surrounding it;
2. The electric field E exerts a force $F = qE$ on a charge q placed in the field.

We shall follow the same pattern in describing the interactions of charges in motion:

1. A *moving* charge or a current sets up or creates a *magnetic field* in the space surrounding it;
2. The magnetic field exerts a *force* on a *moving* charge or a current in the field.

Like electric field, magnetic field is a *vector field,* that is, a vector quantity associated with each point in space. We shall use the symbol B for magnetic field.

In this chapter and the next we consider the *second* aspect of the interaction: Given the presence of a magnetic field, what force does it exert on a moving charge or a current? Then in Chapter 32 we return to the problem of how magnetic fields are *created* by moving charges and currents.

Some aspects of the magnetic force on a moving charge are analogous to corresponding properties of the electric-field force. Both have magnitude proportional to the charge. If a 1-μC charge and a 2-μC charge move through a given magnetic field with the same velocity, the force on the 2-μC charge is twice as great as that on the 1-μC charge. Also, both are proportional to the magnitude or "strength" of the magnetic field; if a given charge moves with the same velocity in two magnetic fields, one having twice the magnitude (and the same direction) as the other, it experiences twice as great a force in the larger field as in the smaller.

The dependence of the magnetic force on the particle's *velocity* is quite different from the electric-field case. The *electric* force on a charge *does not* depend on velocity; it is the same whether the charge is moving or not. The *magnetic* force, conversely, is found to have a magnitude that increases with speed. Furthermore, the *direction* of the force depends on the directions of the magnetic field B and the velocity v in an interesting way. The force **F** *does not* have the same direction as **B**, but instead of always *perpendicular* to both **B** and **v**. The magnitude F of the force is found to be proportional to the component of **v** perpendicular to the field; when that component is zero, that is, when **v** and **B** are parallel or antiparallel, there is *no* force!

These characteristics of the magnetic force can be summarized, with reference to Fig. 30–1, as follows: The direction of **F** is perpendicular to the plane containing **v** and **B**; its magnitude is given by

$$F = qv_\perp B = qvB \sin \phi, \qquad (30\text{–}1)$$

where q is the charge and ϕ is the angle between the vectors **v** and **B**, as shown in the figure.

This prescription does not specify the direction of **F** completely; there are always two opposite directions both perpendicular to the plane of **v** and **B**. To complete the description we use the same righthand-thread rule used to define the vector product in Sec. 1–8 and for angular mechanical quantities in Sec. 9–14. We imagine turning **v** into **B**, through the smaller of the two possible angles; the direction of **F** is the direction in which a righthand-thread screw would advance if turned the same way. Alternatively, one can wrap the fingers of the right hand around the line perpendicular to the plane of **v** and **B** so that they curl around with this sense of rotation; the thumb then points in the direction of **F**.

Thus the force on a charge q moving with velocity **v** in a magnetic field **B** is given, both in magnitude and in direction, by

$$F = qv \times B. \qquad (30\text{–}2)$$

The vector product is extremely useful in formulating relationships involving magnetic fields; this is our first example of its application in this area.

Spatial relations for magnetic forces can be troublesome; the following review of the definition of the vector product may help. First, always draw the vectors **v** and **B** from a common point, that is, with their tails together. Second, visualize and if possible draw the *plane* in which these two vectors lie. The force vector always lies along a line perpendicular to this plane. To determine its direction along this line, imagine rotating **v** into **B**, as in Fig. 30–1. The direction of **F** is the direction of advance of a righthand-thread screw rotated in this direction, or the direction of the extended thumb of the right hand if the fingers are curled around the perpendicular line with this direction of rotation.

Finally, we note that Eq. (30–1) can be interpreted in a different but equivalent way. Recalling that ϕ is the angle between the direction of vectors **v** and **B**, we may interpret $(B \sin \phi)$ as the component of **B** perpendicular to **v**, that is, $B_\perp$. With this notation, the force expression becomes

$$F = qvB_\perp. \qquad (30\text{–}3)$$

Although equivalent to Eq. (30–1), this is sometimes more convenient, especially in problems involving currents rather than individual particles. Forces on currents in conductors are discussed in Chapter 31.

The *units* of B can be deduced from Eq. (30–1); they must be the same as the units of F/qv. Therefore the SI unit of B is one newton second per coulomb, or $1 \text{ N} \cdot \text{s} \cdot \text{C}^{-1} \cdot \text{m}^{-1}$, or, since one ampere is one

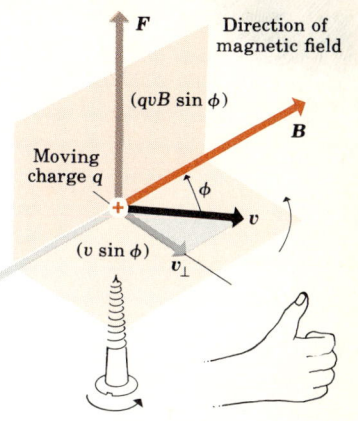

30–1 The magnetic force **F** acting on a charge q moving with velocity **v** is perpendicular to both the magnetic field **B** and to **v**.

coulomb per second, $1 \text{ N} \cdot \text{A}^{-1} \cdot \text{m}^{-1}$. This unit is called *one tesla* (1 T). The cgs unit of B, the *gauss* ($1 \text{ G} = 10^{-4}$ T) is also in common use. To summarize,

$$1 \text{ T} = 1 \text{ N} \cdot \text{A}^{-1} \cdot \text{m}^{-1} = 10^4 \text{ G}.$$

In this discussion we have assumed that q is a *positive* charge. If q is negative, the direction of $\mathbf{F}$ is opposite to that shown in Fig. 30–1 and given by the righthand rule. Thus if two charges of equal magnitude and opposite sign move in the same $\mathbf{B}$ field with the same velocity, the forces on the two charges have equal magnitude and opposite direction.

An unknown *electric* field can be "explored" by measuring the magnitude and direction of the force on a test charge. To "explore" an unknown *magnetic* field, we must measure the magnitude and direction of the force on a *moving* test charge. The cathode-ray tube, discussed in Sec. 26–8, is a convenient device for studying the behavior of moving charges in a magnetic field. At one end of this tube an electron gun shoots out a narrow electron beam at a speed that can be controlled and calculated. At the other end is a fluorescent screen that emits light at the small spot where the electron beam strikes it. In the absence of any deflecting force, the beam strikes the center of the screen. If the spot does not move when the tube is moved or its orientation changed, we may conclude that there is no detectable magnetic field. But if, as we move the tube, the spot of light changes position because the electron beam is deflected, we conclude that a magnetic field is present.

At a given point in a magnetic field, the electron beam will, in general, be deflected. By rotating the cathode-ray tube, however, one finds that there are always two directions for which *no* deflection takes place. *One of the directions of motion of a charge on which a magnetic field exerts no force is the direction of the $\mathbf{B}$-vector.* Thus, in Fig. 30–2, the electron beam is undeflected when its direction is parallel to the z-axis and hence the $\mathbf{B}$-vector must point either up or down.

Let us now place the cathode-ray tube so that the electron beam moves in a plane perpendicular to this direction. Experiment shows that a deflection always takes place in such a manner as to indicate a force acting in this plane, but at right angles to the velocity of the electron beam. Thus, in Fig. 30–2, when the electron beam lies in the xy-plane and its direction is initially along the positive x-axis, the beam is deflected in the direction of the positive y-axis. That is, *when the velocity of the moving charge is perpendicular to the magnetic field, the force is perpendicular to both the magnetic field and the velocity.* Thus this observation corroborates our previous discussion. Additional experiments in which $\mathbf{B}$ and $\mathbf{v}$ are at various angles can be performed, and the results are always consistent with Eq. (30–2) and the accompanying discussion.

Although this experiment uses electrons, which have negative charge, similar experiments can be carried out with beams of positively charged particles, typically ionized atoms, and again the results are consistent with our general statements about magnetic forces.

Finally, when a charged particle moves through a region of space where *both* electric and magnetic fields are present, both fields exert forces on the particle, and the total force is the vector sum of the electric-field and magnetic-field forces:

$$\mathbf{F} = q(\mathbf{E} + \mathbf{v} \times \mathbf{B}). \qquad (30\text{–}4)$$

30–2 The electron beam of the cathode-ray tube is undeflected when the beam is parallel to the z-axis. The $\mathbf{B}$-vector then points either up or down. When the tube axis is parallel to the x-axis, the beam is deflected in the positive y-direction. Then the $\mathbf{B}$-vector points upward, and the force $\mathbf{F}$ on the electrons points along the positive y-axis, opposite to $\mathbf{v} \times \mathbf{B}$ because q is negative.

Example A proton beam moves through a region of space where there is a uniform magnetic field of magnitude 2.0 T, with direction along the positive z-axis, as in Fig. 30–3. The protons have velocity of magnitude 3×10^5 m·s^{-1}, in the xz-plane, at an angle of 30° to the positive z-axis. Find the force on a proton. ($q = 1.6 \times 10^{-19}$ C.)

Solution The righthand rule shows that the direction of the force is along the negative y-axis. The magnitude of the force, from Eq. (30–1), is

$$F = (1.6 \times 10^{-19} \text{ C})(3 \times 10^5 \text{ m·s}^{-1})(2.0 \text{ T})(\sin 30°)$$
$$= 4.8 \times 10^{-14} \text{ N}.$$

Alternatively, in vector language, with Eq. (30–2),

$$v = (3 \times 10^5 \text{ m·s}^{-1})(\sin 30°)i + (3 \times 10^5 \text{ m·s}^{-1})(\cos 30°)k,$$
$$B = (2.0 \text{ T})k,$$
$$F = qv \times B = (1.6 \times 10^{-19} \text{ C})(3 \times 10^5 \text{ m·s}^{-1})(2.0 \text{ T})$$
$$(\sin 30° \, i + \cos 30° \, k) \times k$$
$$= (-4.8 \times 10^{-14} \text{ N})j,$$

since $i \times k = -j$ and $k \times k = 0$.

If the beam consists of *electrons* rather than protons, the charge is negative ($q = -1.6 \times 10^{-19}$ C), and the direction of the force is reversed. The solution proceeds just as before; the force is now directed along the *positive* y-axis: $F = +(4.8 \times 10^{-14} \text{ N})j$.

30–3

30–3 Magnetic field lines; magnetic flux

A magnetic field can be represented by lines, such that the direction of a line through a given point is the same as that of the magnetic field vector **B** at that point. We shall call these *magnetic field lines;* they are sometimes called magnetic lines of force, but this term is unfortunate because, unlike electric field lines, they *do not* point in the direction of the force on a charge.

In a uniform magnetic field, where the **B** vector has the same magnitude and direction at every point in a region, the field lines are straight and parallel. If the poles of an electromagnet are large, flat, and close together, there is a region between them where the magnetic field is approximately uniform.

The *magnetic flux* across a surface is defined in analogy to the electric field flux used with Gauss's law (Sec. 25–4). Any surface can be divided into elements of area dA, as shown in Fig. 30–4. For each element we obtain the components of **B** normal and tangent to the surface at the position of that element, as shown. In general these components will vary from point to point on the surface. From the figure, $B_\perp = B \cos \theta$. The magnetic flux $d\Phi$ through this area is defined as

$$d\Phi = B_\perp dA = B \cos \theta \, dA = \mathbf{B} \cdot d\mathbf{A}. \quad (30\text{–}5)$$

The total magnetic flux through the surface is the sum of the contributions from the individual area elements, given by

$$\Phi = \int B_\perp dA = \int \mathbf{B} \cdot d\mathbf{A}. \quad (30\text{–}6)$$

In the special case where **B** is uniform over a plane surface with total

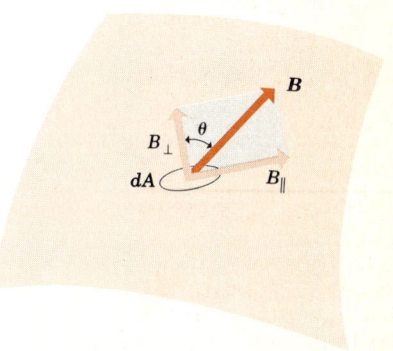

30–4 The magnetic flux through an area element dA is defined to be $\Phi = B_\perp dA$.

area A,

$$\Phi = B_\perp A = BA \cos\theta. \tag{30-7}$$

If $\boldsymbol{B}$ happens to be perpendicular to the surface, $\cos\theta = 1$, and this expression reduces to $\Phi = BA$. The chief usefulness of the concept of magnetic flux is in the study of electromagnetic induction, to be considered in Chapter 33.

The above definitions of magnetic flux have an ambiguity of sign associated with the direction of the vector area element $d\boldsymbol{A}$. In Gauss's law we always took $d\boldsymbol{A}$ as pointing *out of* a closed surface; some applications of magnetic flux involve *open* surfaces with an edge, and then we shall have to take care to define which is the positive side of the surface.

The SI unit of magnetic field B is one tesla = one newton per ampere meter; hence the unit of magnetic flux is one newton meter per ampere (1 N·m·A^{-1}). In honor of Wilhelm Weber (1804–1890), 1 N·m·A^{-1} is called one *weber* (1 Wb).

If the element of area $d\boldsymbol{A}$ in Eq. (30-5) is at right angles to the field lines, $B_\perp = B$, and hence

$$B = \frac{d\Phi}{dA}. \tag{30-8}$$

That is, the magnetic field equals the *flux per unit area* across an area at right angles to the magnetic field. Since the unit of flux is 1 weber, the unit of field, 1 tesla, is equal to one *weber per square meter* (1 Wb·m^{-2}). The magnetic field $\boldsymbol{B}$ is sometimes called *flux density*.

The total flux across a surface can then be pictured as proportional to the number of field lines crossing the surface, and the field (the flux density) as the number of lines *per unit area*.

In the cgs system, the unit of magnetic flux is one *maxwell* and the corresponding unit of flux density, one maxwell per square centimeter, is called one *gauss* (1 G). (Instruments for measuring flux density are often referred to as *gaussmeters*.)

The largest values of steady magnetic field that have been achieved in the laboratory are of the order of 30 T = 300,000 G; some pulsed-current electromagnets can produce fields of the order of 120 T = 1.2×10^6 G for short time intervals of the order of a millisecond. By comparison, the magnetic field of the earth is of the order of 10^{-4} T or 1 G.

30-4 Motion of charged particles in magnetic fields

Let a positively charged particle at point O in a uniform magnetic field $\boldsymbol{B}$ be given a velocity $\boldsymbol{v}$ in a direction at right angles to the field (Fig. 30-5). An upward force $\boldsymbol{F}$, equal in magnitude to qvB, is exerted on the particle at this point. As explained in Sec. 5-4, when the force acting on a particle (and thus also its acceleration) has a direction perpendicular to the particle's velocity, the velocity changes only in direction, not in magnitude. Thus the particle moves with constant speed. At points such as P and Q the directions of force and velocity will have changed as shown; the magnitude of the force is constant, since the magnitudes of q, v, and $\boldsymbol{B}$ are constant. The particle therefore moves under the influence of a force whose *magnitude* is constant but whose *direction* is always at right angles to the velocity of the particle. The orbit of the particle is therefore a

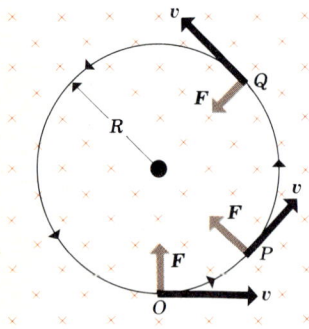

30-5 The orbit of a charged particle in a uniform magnetic field is a circle when the initial velocity is perpendicular to the field. The crosses represent a uniform magnetic field directed *away from* the reader.

circle described with constant tangential speed v. Since the centripetal acceleration is v^2/R, as shown in Sec. 5-6, we have, from Newton's second law,

$$F = qvB = m\left(\frac{v^2}{R}\right),$$

where m is the mass of the particle. The radius of the circular orbit is

$$R = \frac{mv_\perp}{Bq}. \qquad (30\text{-}9)$$

If the direction of the initial velocity is *not* perpendicular to the field, the velocity component parallel to the field remains constant and the particle moves in a helix. In this case, v in Eq. (30-9) is the component of velocity perpendicular to the field.

Note that the radius is proportional to the *momentum* of the particle, mv. Note also that the *work* of the magnetic force acting on a charged particle is always *zero*, because the force is always at *right angles* to the motion. The only effect of a magnetic force is to change the *direction* of motion, never to increase or decrease the *magnitude* of the velocity. Thus, *motion of a charged particle under the action of a magnetic field alone is always motion with constant speed.*

Figure 30-6 is a photograph of the circular tracks made in a cloud chamber by charged particles moving in a magnetic field perpendicular to the plane of the paper. The photograph shows three pairs of tracks originating at common points but curving in opposite directions. A study of the density of droplets along the paths shows that both particles ionize like electrons, but since the tracks curve in opposite directions, the charges must be of opposite sign. These tracks are in fact made by *electron–positron pairs*, created at the points from which the tracks originate by the annihilation of a high-energy gamma ray in the process known as *pair production*. Photographs such as this provided, in the early 1930s, the first experimental evidence of the existence of positrons.

The other tracks at the top of the photograph are portions of the circular paths of photoelectrons ejected from the lead sheet in the chamber. One complete circular track appears in the lower part of the photograph.

The spiral in Fig. 30-7 is the track made by a 4-MeV electron as it slows down in a liquid hydrogen bubble chamber in which there is a magnetic field. Many more ions are produced per unit length in a liquid than in a gas, so that, although an electron may pass completely through a cloud chamber without appreciable energy loss, as in Fig. 30-6, the electron in Fig. 30-7 loses all its energy and comes to rest at the end of the spiral. The shape of the path shows in a striking way how the radius of curvature decreases as the velocity decreases.

30-5 Thomson's measurement of *e/m*

The charge-to-mass ratio of an electron, e/m, was first measured by Sir J. J. Thomson in 1897 at the Cavendish Laboratory in Cambridge, England. The discovery that this ratio is constant provided the best experimental evidence available at that time of the *existence* of electrons, particles with definite mass and charge. Thomson's term for these particles was "cathode corpuscles."

30-6 An early photograph showing cloud-chamber tracks of three electron–positron pairs in a magnetic field. The gamma-ray photons entering at the top materialize into pairs within a lead sheet. The coiled tracks are due to low-energy photoelectrons ejected from the lead. (Courtesy of Lawrence Berkeley Laboratory, University of California.)

30-7 A 4-MeV electron slowing down in a liquid-hydrogen bubble chamber traversed by a magnetic field. The shape of the path shows how the radius of curvature decreases with velocity. (Courtesy of Lawrence Berkeley Laboratory, University of California.)

Thomson's apparatus (Fig. 30–8) was very similar in principle to the cathode-ray tube discussed in Sec. 26–8. It consisted of a highly evacuated glass tube into which several metal electrodes were sealed. Electrode C is the hot cathode from which the electrons emerged. Electrode A is the anode, which was maintained at a high positive potential. Most of the electrons accelerated toward A by this potential hit electrode A, but there was a small hole in A through which some of them passed. They were further restricted by an electrode A′ in which there was another hole. Thus a narrow *beam* of electrons, all having the same speed, passed into the region between the two plates P and P′. After passing between the plates, the electrons struck the end of the tube, where they caused fluorescent material at S to glow. The speed of the electrons is related to the accelerating potential V; we simply equate the kinetic energy $\frac{1}{2}mv^2$ to the loss of potential energy eV (where e is the magnitude of the electron charge):

$$\tfrac{1}{2}mv^2 = eV \quad \text{or} \quad v = \sqrt{\frac{2eV}{m}}. \tag{30-10}$$

If now a potential difference is established between the two deflecting plates P and P′, as shown, the resulting downward electric field deflects the negatively charged electrons upward. Alternatively, we may impose a *magnetic* field directed into the plane of the figure, as shown by the x's; this field results in a downward deflection of the beam; the reader should verify this direction. Finally, if *both* E and B fields are applied simultaneously, it is possible to adjust their relative magnitudes so that the two forces cancel and the beam is undeflected. The condition that must be satisfied, obtained by equating the two force magnitudes, is

$$eE = evB, \quad \text{or} \quad v = \frac{E}{B}. \tag{30-11}$$

Finally, we may combine this with Eq. (30–10) to eliminate v and obtain an expression for the charge-to-mass ratio e/m in terms of the other quantities:

$$\frac{E}{B} = \sqrt{\frac{2eV}{m}},$$
$$\frac{e}{m} = \frac{E^2}{2B^2V}. \tag{30-12}$$

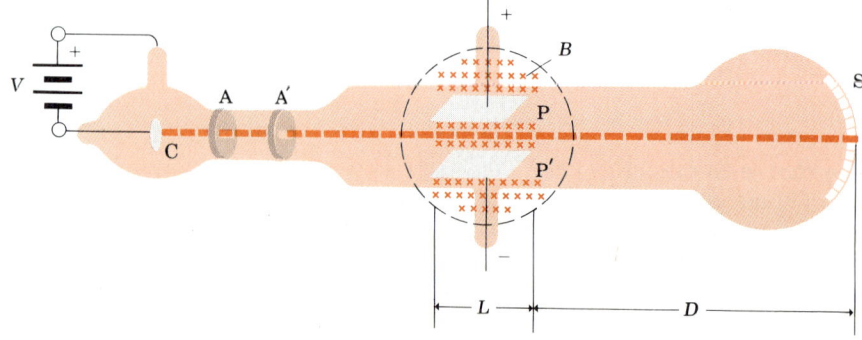

30-8 Thomson's apparatus for measuring the ratio e/m for cathode rays.

All the quantities on the right side can be measured, so e/m can be determined. We note that it is *not* possible to measure e or m separately by this method, but only their ratio.

Thomson measured e/m for his "cathode corpuscles" and found a unique value for this quantity that was independent of the cathode material and the residual gas in the tube. This independence indicated that cathode corpuscles are a common constituent of all matter. The modern accepted value of e/m is $(1.758803 \pm 0.000003) \times 10^{11}$ C·kg^{-1}. Thus Thomson is credited with discovery of the first subatomic particle, the electron. He also found that the speed of the electrons in the beam was about one-tenth the speed of light, much larger than any previously measured material particle speed.

Fifteen years after Thomson's experiments, Millikan succeeded in measuring the charge of the electron with his famous oil-drop experiment, described in Sec. 26-6. The magnitude of the electron charge e is 1.602×10^{-19} C. Thus the *mass* of the electron can be obtained:

$$m = \frac{1.602 \times 10^{-19} \text{ C}}{1.759 \times 10^{11} \text{ C} \cdot \text{kg}^{-1}} = 9.109 \times 10^{-31} \text{ kg}.$$

*30-6 Isotopes and mass spectroscopy

Thomson devised a method similar to the above e/m measurement, for measuring the charge–mass ratio for positive ions. An added difficulty was that in Thomson's day it was difficult to produce a beam of ions all having the same speed. Because the e/m electron experiment depends on the particles having a common speed (in order for the electric and magnetic field forces to balance), this method is not directly applicable for a beam of particles having various velocities. Thomson's method was to make the electric and magnetic fields *parallel,* so that the deflections due to these fields are in perpendicular directions. The net deflection can then never be zero, but it turns out that the relation between the x- and y-deflections for a beam permits determination of the charge–mass ratio of the particles.

Thomson assumed that each positive ion had a charge equal in magnitude to that of the electron because each was an atom that had lost one electron. He could then identify particular values of q/m with particular ions. Positive particles move more slowly than electrons and have lower values of q/m because they are much more massive. The *largest* q/m for positive particles is that for the *lightest* element, hydrogen. From the value of q/m it was found that the mass of the *hydrogen ion* or *proton* is 1836.13 ± 0.01 times the mass of an electron. Electrons contribute only a small amount of the mass of material objects.

The most striking result of these experiments was that certain chemically pure gases have *more than one* value of q/m. Most notable was the case of neon, which has atomic mass 20.2 g·mol^{-1}. Thomson obtained *two* values of q/m, corresponding to 20 and 22 g·mol^{-1}, and after trying and discarding various explanations he concluded that there must be two kinds of neon atoms with different masses.

Soon afterward Aston, a student of Thomson, succeeded in actually separating these two atomic species. Aston permitted the gas to diffuse repeatedly through a porous plug between two containers. At a given temperature T, each atom has an average kinetic energy $3kT/2$, as discussed in Sec. 20-4. Thus the more massive atoms have, on the average,

30-9 Bainbridge's mass spectrograph, utilizing a velocity selector.

somewhat smaller speeds; because of this speed difference, the gas emerging from the plug had a slightly greater concentration of the less massive atoms than the gas entering the plug. Thus Aston demonstrated directly the existence of two species of neon atoms with different masses.

Since these experiments in the early years of the twentieth century, many other elements have been shown to have several kinds of atoms, identical in their chemical behavior but differing in mass. Such forms of an element are called *isotopes*. We shall see in a later chapter that the mass differences are due to differing numbers of neutrons in the *nuclei* of the atoms.

A detailed search for the isotopes of all the elements required precise experimental technique. Aston built the first of many instruments called *mass spectrometers* in 1919. His instrument had a precision of one part in 10,000, and he found that many elements have isotopes. We shall describe here a variation built by Bainbridge. The Bainbridge mass spectrometer (Fig. 30–9) has a source of ions (not shown) situated above S_1. The ions under study pass through slits S_1 and S_2 and move down into the electric field between the two plates P and P'. In the region of the electric field there is also a magnetic field B, perpendicular to the paper. Thus the ions enter a region of crossed electric and magnetic fields like those used by Thomson to measure the velocity of electrons in his determination of e/m. Only those ions whose speed is equal to E/B, as in Eq. (30–11), pass undeviated through this region; ions with other speeds are stopped by the slit S_3. Thus all ions that emerge from S_3 have the same velocity. The region of crossed fields is called a *velocity selector*.

Below S_3 the ions enter a region where there is another magnetic field B', perpendicular to the page, but no electric field. Here the ions move in circular paths of radius R. From Eq. (30–9), we find that

$$m = \frac{qB'R}{v}. \quad (30\text{-}13)$$

Assuming equal charges on each ion, the mass of each ion is proportional to the radius of its path. Ions of different isotopes converge at different points on the photographic plate. The relative abundance of the isotopes is measured from the densities of the photographic images they produce.

Figure 30–10 shows the mass spectrum of germanium. The numbers shown beside the isotope images are not the atomic masses of the isotopes but the integers nearest the atomic masses. These integers are called *mass numbers*; isotopes are written with the mass number as a superscript to the chemical symbol. Thus the isotopes shown would be written Ge^{70}, Ge^{72}, etc. The mass number is represented by the letter A. As will be discussed in a later chapter, A is equal to the total number of protons and neutrons in the nucleus of the atom; different isotopes of an element have the same number of protons but differing numbers of neutrons in their nuclei.

Masses of atoms are often expressed in *atomic mass units*. By definition, one atomic mass unit (1 u) is $\frac{1}{12}$ the mass of one atom of the most abundant isotope of carbon, C^{12}. Since the mass of an atom in grams is equal to its atomic mass divided by Avogadro's number, it follows that

$$1\text{ u} = \frac{(1/12)(12\text{ g} \cdot \text{mol}^{-1})}{6.022 \times 10^{23} \text{ mol}^{-1}}$$

$$= 1.661 \times 10^{-24}\text{ g} = 1.661 \times 10^{-27}\text{ kg}.$$

30-10 The mass spectrum of germanium, showing the isotopes of mass numbers 70, 72, 73, 74, 76.

Questions

30-1 The direction of the E field is defined to be the direction of the force on a positive charge; why can't we define the direction of B similarly?

30-2 Does the earth's magnetic field have a significant effect on the electron beam in a TV picture tube?

30-3 If an electron beam in a cathode-ray tube travels in a straight line, can you be sure there is no magnetic field present?

30-4 A charge moving in a magnetic field experiences a force. But suppose the phenomenon is viewed in a frame of reference moving with the charge, in which the charge is therefore at rest. Then presumably *no* magnetic force would be observed. How is this apparent inconsistency to be reconciled?

30-5 If the magnetic-field force does no work on a charged particle, how can it have any effect on the particle's motion? Are there other examples of forces that do no work but have a significant effect on a particle's motion?

30-6 A permanent magnet can be used to pick up a string of nails, tacks, or paper clips, even though these are not magnets by themselves. How can this be?

30-7 There are regions on earth where the direction of the magnetic field is nearly perpendicular to the earth's surface. What difficulties would this cause in the use of a compass? Where would you expect to find such regions?

30-8 Can one build a compass that does not depend on magnetism for its operation?

30-9 How could a compass be used for a *quantitative* determination of magnitude and direction of magnetic field at a point?

30-10 The direction in which a compass points (magnetic north) is not in general exactly the same as the direction toward the north pole (true north). The difference is called *magnetic declination;* it varies from point to point on the earth and also varies with time. What are some possible explanations for magnetic declination?

30-11 Can a charged particle move through a magnetic field without experiencing any force? How?

30-12 Could the electron beam in an oscilloscope tube (cathode-ray tube) be used as a compass? How? What advantages and disadvantages would it have, compared with a conventional compass?

30-13 Does a magnetic field exert forces on charges moving within conductors? What evidence is there for your answer?

30-14 Does a magnetic field exert forces on the electrons within atoms? What observable effect might such interaction have on the behavior of the atom?

Problems

30-1 Each of the lettered circles at the corners of the cube in Fig. 30-11 represents a positive charge q moving with a velocity of magnitude v in the direction indicated. The region in the figure is a uniform magnetic field B, parallel to the x-axis and directed toward the right. Copy the figure, find the magnitude and direction of the force on each charge, and show the force in your diagram.

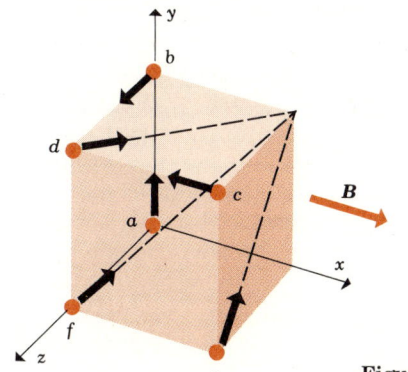

Figure 30-11

30-2 The magnetic field B in a certain region is 2 T and its direction is that of the positive x-axis in Fig. 30-12.

a) What is the magnetic flux across the surface $abcd$ in the figure?
b) What is the magnetic flux across the surface $befc$?
c) What is the magnetic flux across the surface $aefd$?

Figure 30-12

30-3 A particle having a mass of 0.5 g carries a charge of 2.5×10^{-8} C. The particle is given an initial horizontal velocity of 6×10^4 m·s^{-1}. What is the magnitude and direction of the minimum magnetic field that will keep the particle moving in a horizontal direction?

30-4 A deuteron, an isotope of hydrogen whose mass is very nearly 2 u, travels in a circular path of radius 40 cm in a magnetic field of magnitude 1.5 T.

a) Find the speed of the deuteron.
b) Find the time required for it to make one-half a revolution.
c) Through what potential difference would the deuteron have to be accelerated to acquire this velocity?

30-5 An electron at point A in Fig. 30–13 has a speed v_0 of 10^7 m·s^{-1}. Find

a) the magnitude and direction of the magnetic field that will cause the electron to follow the semicircular path from A to B, and
b) the time required for the electron to move from A to B.

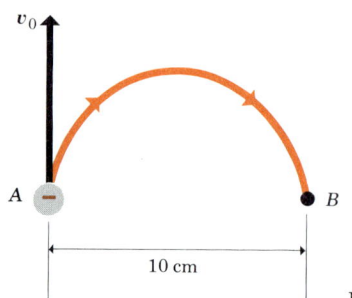

Figure 30–13

30-6 In Problem 30–5, suppose the particle is a proton rather than an electron. Answer the same questions as in that problem.

30-7 In a magnetic field directed vertically upward, a particle initially moving north is deflected toward the west. What is the sign of the charge on the particle?

30-8 In a TV picture tube, an electron in the beam is accelerated by a potential difference of 20,000 V. Then it passes through a region of transverse magnetic field where it moves in a circular arc with radius 12 cm. What is the magnitude of the field?

30-9 Estimate the effect of the earth's magnetic field on the electron beam in a TV picture tube. Suppose the accelerating voltage is 20,000 V; calculate the approximate deflection of the beam over a distance of 0.4 m from the electron gun to the screen, under the action of a transverse field of magnitude 0.5×10^{-4} T (comparable to the magnitude of the earth's field), assuming there are no other deflecting fields. Is this deflection significant?

30-10 A particle carries a charge of 4×10^{-9} C. When it moves with a velocity v_1 of 3×10^4 m·s^{-1} at 45° above the x-axis in the xy-plane, a uniform magnetic field exerts a force F_1 along the negative z-axis. When the particle moves with a velocity v_2 of 2×10^4 m·s^{-1} along the z-axis, there is a force F_2 of 4×10^{-5} N exerted on it along the x-axis. What are the magnitude and direction of the magnetic field? (See Fig. 30–14.)

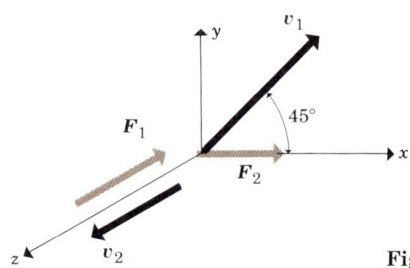

Figure 30–14

30-11 An electron moves in a circular path of radius 1.2 cm perpendicular to a uniform magnetic field. The speed of the electron is 10^6 m·s^{-1}. What is the total magnetic flux encircled by the orbit?

30-12 An electron and an alpha particle (a doubly ionized helium atom) both move in circular paths in a magnetic field with the same tangential speed. Compare the number of revolutions they make per second. The mass of the alpha particle is 6.68×10^{-27} kg.

30-13 In the Bainbridge mass spectrometer (Fig. 30–9), suppose the magnetic field B in the velocity selector is 1.0 T, and ions having a speed of 0.4×10^7 m·s^{-1} pass through undeflected.

a) What should be the electric field between the plates P and P′?
b) If the separation of the plates is 0.5 cm, what is the potential difference between plates?

30-14

a) What is the velocity of a beam of electrons when the simultaneous influence of an electric field of 34×10^4 V·m^{-1} and a magnetic field of 2×10^{-2} T, both fields being normal to the beam and to each other, produces no deflection of the electrons?
b) Show in a diagram the relative orientation of the vectors v, E, and B.
c) What is the radius of the electron orbit when the electric field is removed?

30-15 A singly charged Li^7 ion has a mass of 1.16×10^{-23} g. It is accelerated through a potential difference of 500 V and then enters a magnetic field of 0.4 T, moving perpendicular to the field. What is the radius of its path in the magnetic field?

30-16 Suppose the electric field between the plates P and P′ in Fig. 30–9 is 150 V·cm^{-1}, and the magnetic field is 0.5 T. If the source contains the three isotopes of magnesium, $_{12}$Mg^{24}, $_{12}$Mg^{25}, and $_{12}$Mg^{26}, and the ions are singly charged, find the distance between the lines formed by the three isotopes on the photographic plate. Assume the atomic masses of the isotopes equal to their mass numbers.

30-17 The electric field between the plates of the velocity selector in a Bainbridge mass spectrograph is 1200 V·cm^{-1}, and the magnetic field in both regions is 0.6 T. A stream of singly charged neon moves in a circular path of 7.28-cm radius in the magnetic field. Determine the mass number of the neon isotope.

30-18 A particle of mass m and charge $+q$ starts from rest at the origin in Fig. 30-15. There is a uniform electric field E in the positive y-direction and a uniform magnetic field B directed toward the reader. It is shown in more advanced books that the path is a *cycloid* whose radius of curvature at the top points is twice the y-coordinate at that level.

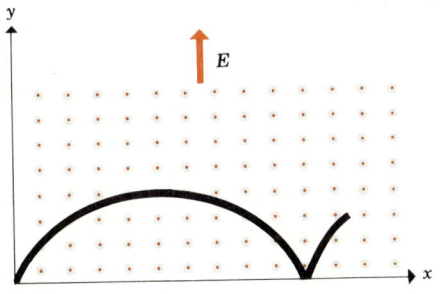

Figure 30-15

a) Explain why the path has this general shape and why it is repetitive.
b) Prove that the speed at any point is equal to $\sqrt{2qEy/m}$. [*Hint.* Use energy conservation.]
c) Applying Newton's second law at the top point, prove that the speed at this point is $2E/B$.

30-19 Two positive ions having the same charge q but different masses, m_1 and m_2, are accelerated horizontally from rest through a potential difference V. They then enter a region where there is uniform magnetic field B normal to the plane of the trajectory.

a) Show that if the beam entered the magnetic field along the x-axis, the value of the y-coordinate for each ion at any time t is approximately

$$y = Bx^2 \left(\frac{q}{8mV}\right)^{1/2},$$

provided x remains much smaller than y.

b) Can this arrangement be used for isotope separation?

MAGNETIC FORCES ON CURRENT-CARRYING CONDUCTORS

31-1 Force on a conductor

When a current-carrying conductor lies in a magnetic field, magnetic forces are exerted on the moving charges within the conductor. These forces are transmitted to the material of the conductor, and the conductor as a whole experiences a force distributed along its length. The electric motor and the moving-coil galvanometer both depend for their operation on the magnetic forces on conductors carrying currents.

Figure 31-1 represents a portion of a conducting wire of length l and cross-sectional area A, in which the current density $\boldsymbol{J}$ is from left to right. The wire is in a magnetic field $\boldsymbol{B}$, perpendicular to the plane of the diagram, and directed *into* the plane. A positive charge q_1 within the wire, moving with drift velocity $\boldsymbol{v}_1$, is acted on by a force $\boldsymbol{F}_1$ given by Eq. (30-2), $\boldsymbol{F}_1 = q_1(\boldsymbol{v}_1 \times \boldsymbol{B})$. As the figure shows, the direction of this force is upward, and since in this case $\boldsymbol{v}_1$ and $\boldsymbol{B}$ are perpendicular, the magnitude of the force is $F_1 = q_1 v_1 B$. Similarly, a negative charge q_2, with drift velocity $\boldsymbol{v}_2$ in a direction opposite to that of $\boldsymbol{J}$, experiences a force $\boldsymbol{F}_2 = q_2 \boldsymbol{v}_2 \times \boldsymbol{B}$. Because q_1 and q_2 have opposite signs and $\boldsymbol{v}_1$ and $\boldsymbol{v}_2$ opposite directions, $\boldsymbol{F}_2$ has the *same* direction as $\boldsymbol{F}_1$, as shown.

The *total* force on all the moving charges in a length l of conductor can be expressed in terms of the current, using the considerations of Sec. 28-1, Eqs. (28-3), (28-4), and (28-5). Let n_1 and n_2 represent the numbers of positive and negative charge carriers, respectively, per unit volume. The numbers of carriers in the portion are then $n_1 A l$ and $n_2 A l$; the

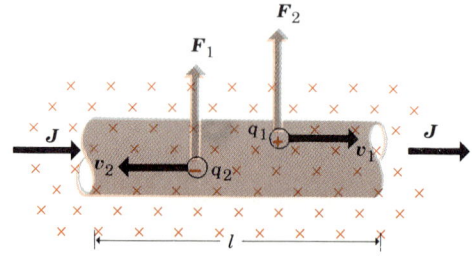

31-1 Forces on the moving charges in a current-carrying conductor. The forces on both positive and negative charges are in the same direction.

total force **F** on all carriers (and hence the total force on the wire) has magnitude

$$F = (n_1 A l)(q_1 v_1 B) + (n_2 A l)(q_2 v_2 B)$$
$$= (n_1 q_1 v_1 + n_2 q_2 v_2) A l B.$$

But $n_1 q_1 v_1 + n_2 q_2 v_2$ (or more generally, $\Sigma \, nqv$) equals the current density J, and the product JA equals the current I, so finally

$$F = IlB. \qquad B \perp \text{wire} \qquad (31\text{-}1)$$

If the **B** field is not perpendicular to the wire but makes an angle ϕ with it, the situation is just like that discussed in Sec. 30-2 for a single charge. The component of **B** parallel to the wire (and thus to the drift velocities of the charges) exerts no force; the component perpendicular to the wire is given by $B_\perp = B \sin \phi$ so, in general,

$$F = IlB_\perp = IlB \sin \phi. \qquad (31\text{-}2)$$

To find the direction of the force on a current-carrying conductor placed in a magnetic field, we may use the same righthand-screw rule that was used in the case of a moving positive charge. Rotate a righthand screw from the direction of I toward **B**; the direction of advance will be the direction of **F**. The situation is the same as that shown in Fig. 30-1, but with the direction of **v** replaced by the direction of I.

Thus this force, like the force on a single moving charge, may be expressed as a vector product. We represent the section of wire by a vector **l** along the wire and in the direction of the current. Then the force on this section is given by

$$F = I\mathbf{l} \times \mathbf{B}. \qquad (31\text{-}3)$$

Example A straight horizontal wire carries a current of 50 A from west to east, in a region where the magnetic field is toward the northeast (i.e., 45° north of east) with magnitude 1.2 T. Find the magnitude and direction of the force on a 1-m section of wire.

Solution The angle ϕ between the directions of current and field is 45°. From Eq. (31-2) we obtain

$$F = (50 \text{ A})(1 \text{ m})(1.2 \text{ T})(\sin 45°) = 42.4 \text{ N}.$$

Consistency of units may be checked by observing, as noted in Sec. 30-2, that $1 \text{ T} = 1 \text{ N} \cdot \text{A}^{-1} \cdot \text{m}^{-1}$. The direction of the force is perpendicular to the plane of the current and the field, both of which lie in the horizontal plane. Thus the force must be along the vertical; the righthand rule shows that it is vertically upward.

31-2 The Hall effect

The reality of the forces on the moving charges in a conductor in a magnetic field is strikingly demonstrated by the *Hall effect*. The conductor in Fig. 31-2 is in the form of a flat strip. The current carriers within it are driven toward the upper edge of the strip by the magnetic force $q\mathbf{v} \times \mathbf{B}$.

This force is an example of the *non-electrostatic* forces F_n discussed in Chapter 28, and in this case the equivalent *non-electrostatic field* E_n

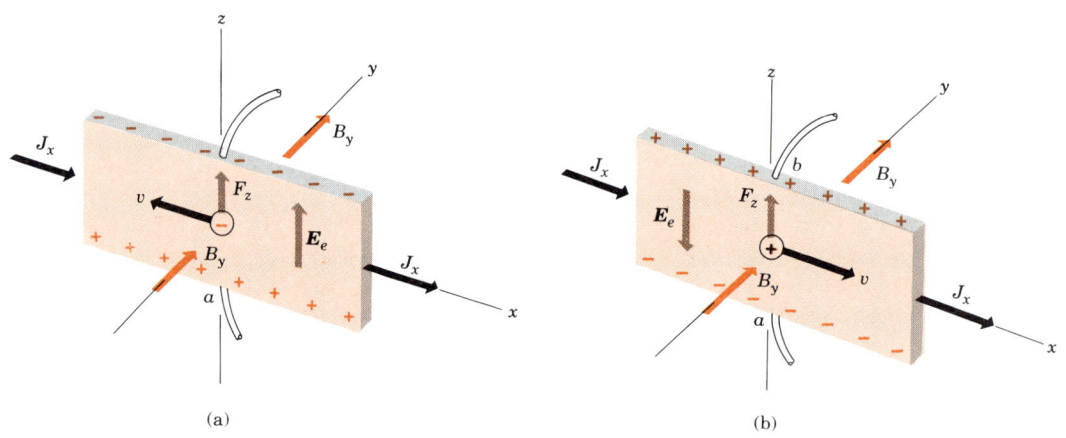

31-2 Forces on charge carriers in a conductor in a magnetic field. (a) Negative current carriers (electrons) are pushed toward top of slab, leading to charge distribution as shown. Point a is at higher potential than point b. (b) Positive current carriers; polarity of potential difference is opposite to that of (a).

(the force per unit charge) has magnitude

$$E_n = vB. \tag{31-4}$$

If the charge carriers are electrons, as in Fig. 31–2a, an excess negative charge accumulates at the upper edge of the strip, leaving an excess positive charge at its lower edge, until the transverse electrostatic field $\boldsymbol{E}_e$ within the conductor is equal and opposite to the *non-electrostatic* field $\boldsymbol{E}_n$. Because the final transverse current is zero, the conductor is on "open circuit" in the transverse direction, and the potential difference between the edges of the strip, which can be measured with a potentiometer, is equal to the *Hall emf* in the strip. The study of this Hall emf has provided much information regarding the conduction process. It is found that, for the metals, the upper edge of the strip in Fig. 31–2a *does* become charged negatively relative to the lower, which justifies our belief that the charge carriers in a metal are negative electrons.

Suppose, however, that the charge carriers are *positive*, as in Fig. 31–2b. Then *positive* charge accumulates at the upper edge, and the potential difference is *opposite* to that resulting from the deflection of negative charges. Soon after the discovery of the Hall effect, in 1879, it was observed that many materials, notably the *semiconductors*, exhibited a Hall emf opposite to that of the metals, *as if* their charge carriers were *positively* charged. We now know that these metals conduct by a process known as *hole conduction*. There are sites within the material that would normally be occupied by an electron but are actually empty, and a *missing negative* charge is equivalent to a *positive* charge. When an electron moves in one direction to fill a hole, it leaves another hole behind it, and the result is that the *hole* (equivalent to a positive charge) migrates in the direction *opposite* to that of the electron.

In terms of the set of coordinate axes in Fig. 31–2a, the electrostatic field E_e is in the z-direction, and we write it as E_z. The magnetic field is in the y-direction, and we write it as B_y. The non-electrostatic field E_n (in the negative z-direction) equals vB_y. The current density, J_x, is in the x-direction.

In the final steady state, when the fields E_e and E_n are equal,

$$E_z = vB_y.$$

The current density J_x is

$$J_x = nqv.$$

When v is eliminated, we have

$$nq = \frac{J_x B_y}{E_z}. \tag{31-5}$$

Thus, from measurements of J_x, B_y, and E_z, one can compute the product nq. In both metals and semiconductors, q is equal in magnitude to the electron charge, so the Hall effect permits a direct measurement of n, the density of current-carrying charges in the material.

31–3 Force and torque on a complete circuit

The *total* force and torque on any conductor in a magnetic field may be calculated by dividing it into segments and applying Eq. (31–2) or (31–3) to each segment. In general, this requires an integration, but when the magnetic field is uniform the calculation is much simpler. As we shall see, the total *force* on a complete circuit in a uniform field is always zero, but the *torque* in general is not. Three simple examples will be analyzed in detail, a rectangular loop, a circular loop, and a cylindrical coil or solenoid.

1. *Rectangular loop.* Figure 31–3 shows a rectangular loop of wire with sides of lengths a and b. The normal to the plane of the loop (a line perpendicular to the plane) makes an angle α with the direction of a uniform magnetic field, and the loop carries a current I. (Provision must be made for leading the current into and out of the loop, or for inserting a seat of emf. This is omitted from the diagram, for simplicity.)

The force F on the right side, of length a, is in the direction of the x-axis, toward the right, as shown. In this side B is perpendicular to the current direction, and the total force on this side (the sum of the forces

31–3 Forces on the sides of a current-carrying loop in a magnetic field. The resultant of the set of forces is a couple of moment $\Gamma = IAB \sin \alpha$.

distributed along the side) has magnitude
$$F = IaB.$$
A force of the same magnitude but in the opposite direction acts on the opposite side, as shown in the figure.

The forces on the sides of length b, represented by the vectors $\mathbf{F'}$ and $-\mathbf{F'}$, have magnitude
$$IbB \sin(90° - \alpha), \quad \text{or} \quad IbB \cos \alpha.$$
The lines of action of both lie along the y-axis.

The total force on the loop is zero, since the forces on opposite sides cancel out in pairs. The two forces $\mathbf{F'}$ and $-\mathbf{F'}$ lie along the same line and so have no torque with respect to any axis, but the two forces $\mathbf{F}$ and $-\mathbf{F}$ constitute a *couple*, as defined in Sec. 8-4. The torque of a couple has the same value with respect to any point, and is given by the magnitude of either force multiplied by the distance between lines of action of the two forces. From Fig. 31-3, this distance is $b \sin \alpha$, so the torque is
$$\Gamma = (IBa)(b \sin \alpha). \tag{31-6}$$
The torque is greatest when $\alpha = 90°$ (i.e., when the plane of the coil is parallel to the field), and it is zero when α is zero or $180°$ and the plane of the coil is perpendicular to the field. One of the latter positions is a position of *stable* equilibrium, the other of *unstable* equilibrium; which is which?

Since ab is the area A of the coil, Eq. (31-6) may also be written
$$\Gamma = IBA \sin \alpha. \tag{31-7}$$
The product IA is called the *magnetic moment m* of the loop. It is analogous to the dipole moment of an electric dipole, introduced in Sec. 25-1. Thus the torque on a magnetic moment corresponding to a current loop can be expressed simply as
$$\Gamma = mB \sin \alpha. \tag{31-8}$$
Because of the directional relations indicated, the effect of the torque Γ is to tend to rotate the loop in the direction of *decreasing* α, toward its equilibrium position, in which it lies in the xy-plane, perpendicular to the direction of the field $\mathbf{B}$.

The interaction of a conducting loop with a magnetic field may be expressed more compactly in terms of vector torque, introduced in Sec. 8-5. We first define a vector area $\mathbf{A}$, as in Sec. 25-4, defined as having magnitude A and direction perpendicular to its plane; the sense is determined by the righthand rule applied to the direction of circulation of current around the loop, as shown in Fig. 31-3. Then, as Eq. (31-7) and the figure show, the vector torque Γ is given by
$$\mathbf{\Gamma} = I\mathbf{A} \times \mathbf{B}. \tag{31-9}$$
We may also define a vector magnetic moment
$$\mathbf{m} = I\mathbf{A}; \tag{31-10}$$
in terms of this the torque is expressed simply as
$$\mathbf{\Gamma} = \mathbf{m} \times \mathbf{B}. \tag{31-11}$$

Formulas apply regardless of shape

The effect of the torque Γ is to tend to rotate the loop toward its equilibrium position, in which it lies with its vector magnetic moment $\mathbf{m}$ in the same direction as the field $\mathbf{B}$. The torque is greatest when $\mathbf{m}$ and $\mathbf{B}$ are perpendicular, and zero when they are parallel or antiparallel. Equation (31-11) is the analog of Eq. (25-3) in Sec. 25-1, for the torque on an *electric* dipole in an electric field.

When a magnetic dipole changes its orientation in a magnetic field, the field does work on it, given by $\int \Gamma \, d\theta$, and there is a corresponding potential energy. As the above discussion suggests, the potential energy is least when $\mathbf{m}$ and $\mathbf{B}$ are parallel and greatest when they are antiparallel. If the potential energy U is taken as zero when $\alpha = \pi/2$, then the potential energy $U(\alpha)$ at any other position is given by

$$U(\alpha) = \int_\alpha^{\pi/2} \Gamma \, d\theta.$$

The torque Γ is expressed in terms of the variable angle θ by use of Eq. (31-8):

$$\Gamma = -mB \sin \theta.$$

The extra minus sign is included here because the torque is in the direction of *decreasing* θ. Combining the above relations and evaluating the integral, we find

$$U(\alpha) = -mB \int_\alpha^{\pi/2} \sin \theta \, d\theta$$
$$= -mB \cos \theta,$$

or

$$U = -\mathbf{m} \cdot \mathbf{B}. \tag{31-12}$$

2. *Circular loop.* A circular loop may be approximated by a very large number of small rectangular loops; then the sum of the areas of the rectangular loops may be made to approach the area of the circular one as closely as we please. Furthermore, the boundary of the rectangular loops will approximate the circular loop with any desired accuracy. Currents in the same sense in all the rectangular loops will give rise to forces that will cancel at all points except on the boundary. It can therefore be proved quite rigorously that, *not only for a circular loop, but for a plane loop of any shape whatever,* carrying a current I in a magnetic field of magnitude B, the torque is given by Eqs. (31-7) through (31-11). In particular, for a circular loop of radius R,

$$\Gamma = \pi I B R^2 \sin \alpha \tag{31-13}$$

and

$$\Gamma = \pi I R^2 \mathbf{n} \times \mathbf{B}, \tag{31-14}$$

where $\mathbf{n}$ is a unit vector perpendicular to the plane of the loop as determined by the sense of circulation of current.

3. *Solenoid.* A helical winding of wire, such as that obtained by winding wire around the surface of a circular cylinder, is called a *solenoid.* If the windings are closely spaced, the solenoid can be approximated by a number of circular loops lying in planes at right angles to its long axis. The total torque on a solenoid in a magnetic field is simply the sum of

the torques on the individual turns. Hence, for a solenoid of N turns in a uniform field B,

$$\Gamma = NIAB \sin \alpha, \qquad (31\text{-}15)$$

where α is the angle between the axis of the solenoid and the direction of the field. The torque is maximum when the magnetic field is parallel to the planes of the individual turns or perpendicular to the long axis of the solenoid. The effect of this torque is to tend to rotate the solenoid into a position in which each turn is perpendicular to the field and the axis of the solenoid is parallel to the field.

Although little has been said thus far regarding permanent magnets, the behavior of the solenoid as described above closely resembles that of a bar magnet or compass needle, in that both the solenoid and the magnet will, if free to turn, set themselves with their axes parallel to a magnetic field. The behavior of a bar magnet or compass is sometimes "explained" by ascribing the torque on it to magnetic forces exerted on "poles" at its ends. We see, however, that no such interpretation is demanded in the case of the solenoid. The similarity suggests, in fact, that the orbiting and spinning electrons in a bar of magnetized iron are equivalent to the current in the windings of a solenoid, and that the observed torque arises from the same cause in both instances. We shall return to this matter in Chapter 35.

Example 1   A circular coil of wire 0.05 m in radius, having 30 turns, lies in a horizontal plane, as shown in Fig. 31–4. It carries a current of 5 A, in a counterclockwise sense when viewed from above. The coil is in a magnetic field directed toward the right, with magnitude 1.2 T. Find the magnetic moment and the torque on the coil.

31–4

Solution   The area of the coil is

$$A = \pi r^2 = \pi (0.05 \text{ m})^2 = 7.85 \times 10^{-3} \text{ m}^2.$$

The magnetic moment of each turn of the coil is

$$m = IA = (5 \text{ A})(7.85 \times 10^{-3} \text{ m}^2) = 3.93 \times 10^{-2} \text{ A} \cdot \text{m}^2,$$

and the total magnetic moment of all 30 turns is

$$m = (30)(3.93 \times 10^{-2} \text{ A} \cdot \text{m}^2) = 1.18 \text{ A} \cdot \text{m}^2.$$

The angle α between the direction of B and the normal to the plane of the coil is 90°. From Eq. (31–7) the torque on each turn of the coil is

$$\Gamma = IBA \sin \alpha = (5 \text{ A})(1.2 \text{ T})(7.85 \times 10^{-3} \text{ m}^2)(\sin 90°) = 0.047 \text{ N} \cdot \text{m},$$

and the total torque on the coil is
$$\Gamma = (30)(0.047 \text{ N}\cdot\text{m}) = 1.41 \text{ N}\cdot\text{m}.$$
Alternatively, from Eq. (31–8),
$$\Gamma = mB \sin \alpha = (1.18 \text{ A}\cdot\text{m}^2)(1.2 \text{ T})(\sin 90°) = 1.41 \text{ N}\cdot\text{m}.$$

The direction of the torque is such as to tend to push the right side of the coil down and the left side up and to rotate it into a position where the normal to its plane is parallel to **B**.

Example 2 If the coil in Example 1 rotates from its initial position to a position where its magnetic moment is parallel to **B**, what is the change in potential energy?

Solution From Eq. (31–12), the initial potential energy U_1 is
$$U_1 = -(1.18 \text{ A}\cdot\text{m}^2)(1.2 \text{ T})(\cos 90°) = 0,$$
and the final potential energy U_2 is
$$U_2 = -(1.18 \text{ A}\cdot\text{m}^2)(1.2 \text{ T})(\cos 0°) = -1.41 \text{ J}.$$
Thus the change in potential energy is -1.41 J.

Example 3 What vertical forces applied to the left and right edges of the coil of Fig. 31–4 would be required to hold it in equilibrium in its initial position?

Solution An upward force of magnitude F at the right side and a downward force of equal magnitude on the left side would have a total torque (with respect to the coil's center) of
$$\Gamma = (2)(0.05 \text{ m})F.$$
This must be equal to the magnitude of the magnetic-field torque of 1.41 N·m, and we find that the required forces have magnitude 14.1 N.

31–4 The pivoted-coil galvanometer

The pivoted-coil galvanometer, a modified version of an instrument invented by d'Arsonval, has been described in Sec. 29–3. It is widely used in a variety of electrical measurements. A moving coil is pivoted between two bearings, which in high-quality instruments are jeweled bearings. The field of the permanent magnet is shaped by the configuration of its poles and the iron slug in the center of the coil, so that the coil rotates in a field that is everywhere *radial*, as shown in Fig. 29–7. Thus the side thrusts on the coil are always perpendicular to the plane of the coil, and the magnetic-field torque is directly proportional to the current, no matter what the position of the coil.

A restoring torque proportional to the angular displacement of the coil is provided by two hairsprings, which also serve as current leads to the coil. When current is supplied to the coil, it rotates, along with its attached pointer, until the restoring spring torque just balances the magnetic-field torque; hence the deflection is proportional to the current. The frictional torque of the bearings, while small, is not entirely negligible, and it is one of the limitations on the sensitivity of such in-

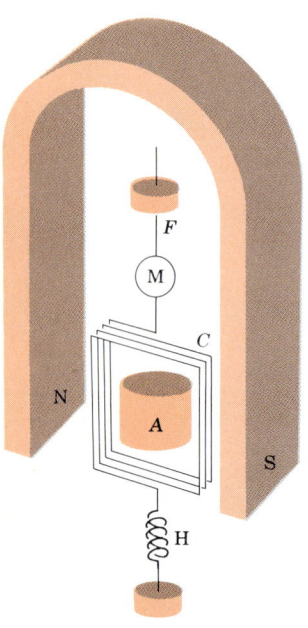

31-5 Principle of the d'Arsonval galvanometer.

struments; the smallest currents that can be measured with these instruments is of the order of 0.1 μA (10^{-7} A).

When greater current sensitivity is needed, an instrument closer to d'Arsonval's original conception can be used, as shown schematically in Fig. 31–5. Instead of being pivoted, the coil is suspended by a fine conducting wire or thin flat strip F, which provides a restoring torque when the coil is deflected from its normal position, and which also serves as one current lead to the coil. The other terminal of the coil is connected to the loosely wound helix H which serves as the second lead, but which exerts a negligible torque on the coil.

The angle of deflection is observed with the aid of a beam of light reflected from a small mirror M cemented to the upper suspension. The light beam serves as a weightless pointer, just as with the Cavendish balance, described in Sec. 4–4.

Because of the nature of its suspension, this type of galvanometer can be made much more sensitive than the pivoted-coil type; currents as small as 10^{-10} A can be detected. However, it is fragile and not readily portable; in fact, such instruments are usually mounted permanently on rigid walls. For measurements involving great sensitivity or great precision, galvanometers have been largely supplanted by electronic instruments using solid-state amplifying devices and often direct digital display for the read-out.

*31-5 The direct-current motor

The direct-current motor is illustrated schematically in Fig. 31–6. The armature or rotor, A, is a cylinder of soft steel mounted on a shaft so that it can rotate about its axis. Embedded in longitudinal slots in the surface of the rotor are a number of copper conductors C. Current is led into and out of these conductors through graphite brushes making contact with a cylinder on the shaft called the *commutator* (not shown in Fig. 31–6). The commutator is an automatic switching arrangement, which maintains the currents in the conductors in the directions shown in the figure, whatever the position of the rotor. The current in the field coils F and F' sets up a magnetic field in the motor frame M and in the gap between the pole pieces P and P' and the rotor. Some of the magnetic field lines are shown as broken lines. With the directions of field and rotor currents shown, the side thrust on each conductor in the rotor is such as to produce a counterclockwise torque on the rotor.

If the rotor and the field windings are connected in series, we have a *series* motor; if they are connected in parallel, we have a *shunt* motor. In some motors the field windings are in two parts, one in series with the rotor and the other in parallel with it; the motor is then *compound*.

A motor converts electrical energy to mechanical energy or work, and thus requires electrical energy input. If the potential difference between terminals is V_{ab} and the current is I, then the power input is $P = V_{ab}I$. Even if the motor coils have no resistance, there must be a potential difference if P is to be different from zero. This potential difference results principally from magnetic forces exerted on the charges in the conductors of the rotor as they rotate through the magnetic field. These forces are an example of non-electrostatic forces, and give rise to a corresponding equivalent non-electrostatic field (concepts introduced in Sec. 28–4). In the present context the resulting electromotive force $\mathcal{E}$ is

31-6 Schematic diagram of a dc motor.

called an *induced* emf or sometimes a *back* emf, referring to the fact that its sense is opposite to that of the current. Induced emf's resulting from motion of conductors in magnetic fields will be considered in greater detail in Chapter 33.

The situation is analogous to that of a battery being charged, as discussed in Sec. 28-4. In that case, electrical energy is converted to chemical rather than mechanical energy. But in either case, if the device has internal resistance r, then V_{ab} is greater than $\mathcal{E}$, the difference being the drop Ir across the internal resistance of the device. Thus, for either a motor or a battery being charged,

$$V_{ab} = \mathcal{E} + Ir. \tag{31-16}$$

For a battery, $\mathcal{E}$ is often constant or nearly so; for a motor, $\mathcal{E}$ is *not* constant but depends on the speed of rotation of the rotor.

Example A dc motor with its rotor and field coils connected in series has an internal resistance of 2.0 Ω. When running at full load on a 120-V line, it draws a current of 4.0 A.

a) What is the emf in the rotor?

$$V_{ab} = \mathcal{E} + Ir,$$
$$120 \text{ V} = \mathcal{E} + (2.0 \text{ Ω})(4.0 \text{ A}),$$
$$\mathcal{E} = 112 \text{ V}.$$

b) What is the power delivered to the motor?

$$P = IV_{ab} = (4.0 \text{ A})(120 \text{ V}) = 480 \text{ W}.$$

c) What is the rate of dissipation of energy in the motor?

$$P = I^2 r = (4.0 \text{ A})^2 (2.0 \text{ Ω}) = 32 \text{ W}.$$

d) What is the mechanical power developed?

Mechanical power = total power − rate of dissipation of energy
$$= 480 \text{ W} - 32 \text{ W} = 448 \text{ W}.$$

The mechanical power may also be calculated from the relation

Mechanical power $= \mathcal{E}I = (112 \text{ V})(4.0 \text{ A}) = 448 \text{ W}.$

*31-6 The electromagnetic pump

Magnetic forces acting on conducting fluids provide a convenient means of pumping these fluids. An example is found in nuclear-reactor design; in some types of reactors, heat is transferred from the reactor core to the point where it is to be utilized, by a circulating flow of liquid metal. (Sodium, lithium, bismuth, and sodium-potassium alloys have been used.) The flow is maintained by an *electromagnetic pump*, one form of which is shown schematically in Fig. 31-7.

A current is sent transversely through the liquid metal, in a direction at right angles to a transverse magnetic field. The resulting side thrust on the current-carrying metal drives it along the pipe in which it is contained. The system is completely sealed and the only moving part is the metal itself.

31-7 An electromagnetic pump.

Magnetic pumps are also finding a variety of applications in medical technology, particularly for pumping blood. Ordinary mechanical pumps with moving parts can damage blood cells. Blood contains ions, making it an electrical conductor, and magnetic pumping is possible. The pump is again completely sealed, reducing danger of contamination, and there are no moving parts except the blood itself. Magnetic pumps are now in use in some heart–lung machines and artificial kidney machines.

Questions

31-1 How might a loop of wire carrying a current be used as a compass? Could such a compass distinguish between north and south?

31-2 How could the direction of a magnetic field be determined by making only *qualitative* observations of the magnetic force on a straight wire carrying a current?

31-3 Do the currents in the electrical system of a car have a significant effect on a compass placed in the car?

31-4 A student claimed that, if lightning strikes a metal flagpole, the force exerted by the earth's magnetic field on the current in the pole can be large enough to bend it. Typical lightning currents are of the order of 10^4 to 10^5 A; is the student's opinion justified?

31-5 Is it possible that, in a Hall-effect experiment, *no* transverse potential difference will be observed? Under what circumstances might this happen?

31-6 Can a circular current loop in a uniform magnetic field ever experience a torque tending to rotate it about an axis perpendicular to the plane of the loop?

31-7 A student tried to make an electromagnetic compass by suspending a coil of wire from a thread (with the plane of the coil vertical) and passing a current through it. He expected the coil to align itself perpendicular to the horizontal component of the earth's magnetic field; but instead, the coil went into what appeared to be angular simple harmonic motion (cf. Sec. 11-7), turning back and forth past the expected direction. What was happening? Was the motion truly simple harmonic?

31-8 In the pivoted-coil galvanometer shown in Fig. 29-7, is the magnetic field uniform? Is Eq. (31-7) directly applicable to this situation? If not, how *is* the torque on the coil determined?

31-9 Hall-effect voltages are much *larger* for relatively poor conductors such as germanium than for good conductors such as copper, for comparable currents, fields, and dimensions. Why?

31-10 When the polarity of the voltage applied to a dc motor is reversed, the direction of rotation *does not* reverse. Why not? How *could* the direction of rotation be reversed?

31-11 If an emf is produced in a dc motor, would it be possible to use the motor somehow as a *generator* or *source*, taking power out of it instead of putting power into it? How might this be done?

31-12 In an electromagnetic pump used to pump liquid sodium, as in Fig. 31-7, it was found that there was a difference in pressure between the top and bottom of the pipe passing through the pump. Gravity is one reason for this difference, but there is also an electromagnetic effect. What is it?

31-13 Sodium chloride (common salt) is not a good conductor, yet melted sodium chloride can be pumped by an electromagnetic pump. How is this possible?

31-14 It was noted in Chapter 30 that a magnetic force on a moving charge does no work on the charge. Yet work is required to push a fluid through a pipe, and so a magnetic pump must do work on the pumped fluid. How can this apparent inconsistency be resolved?

Problems

31-1 The cube in Fig. 31-8, 0.5 m on a side, is in a uniform magnetic field of 0.6 T, parallel to the x-axis. The wire *abcdef* carries a current of 4 A in the direction indicated. Determine the magnitude and direction of the force acting on the segments *ab*, *bc*, *cd*, *de*, and *ef*.

31-2 Figure 31-9 shows a portion of a silver ribbon with $z_1 = 2$ cm and $y_1 = 1$ mm, carrying a current of 200 A in the positive x-direction. The ribbon lies in a uniform magnetic field, in the y-direction, of magnitude 1.5 T. If there are 7.4×10^{28} free electrons per m^3, find

a) the drift velocity of the electrons in the x-direction,
b) the magnitude and direction of the electric field in the z-direction due to the Hall effect, and
c) the Hall emf.

Figure 31-8

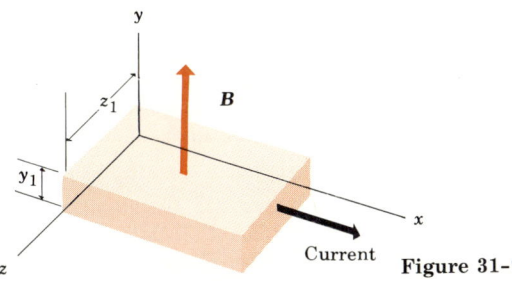

Figure 31-9

31-3 Let Fig. 31-9 represent a strip of copper of the same dimensions as those of the silver ribbon of the preceding problem. When the magnetic field is 5 T and the current is 100 A, the Hall emf is found to be 45.4 μV. What is the density of free electrons in the copper?

31-4 The plane of a rectangular loop of wire 5 cm × 8 cm is parallel to a magnetic field of magnitude 0.15 T.

a) If the loop carries a current of 10 A, what torque acts on it?

b) What is the magnetic moment of the loop?

c) What is the maximum torque that can be obtained with the same total length of wire carrying the same current in this magnetic field?

31-5 An electromagnet produces a magnetic field of 1.2 T in a cylindrical region of radius 5 cm between its poles. A wire carrying a current of 20 A passes through this region, intersecting the axis of the cylinder and perpendicular to it. What force is exerted on the wire?

31-6 A wire 0.5 m long lies along the y-axis and carries a current of 10 A in the $+y$-direction. The magnetic field is uniform and has components $B_x = 0.3$ T, $B_y = -1.2$ T, and $B_z = 0.5$ T.

a) Find the components of force on the wire.

b) What is the magnitude of the total force on the wire?

31-7 A horizontal rod 0.2 m long is mounted on a balance and carries a current. In the vicinity of the rod is a uniform horizontal magnetic field of magnitude 0.05 T, perpendicular to the rod. The force on the rod is measured by the balance and is found to be 0.24 N. What is the current?

31-8 In Problem 31-7, suppose the magnetic field is horizontal but makes an angle of 30° with the rod. What is the current in the rod?

31-9 The rectangular loop in Fig. 31-10 is pivoted about the y-axis and carries a current of 10 A in the direction indicated.

a) If the loop is in a uniform magnetic field of magnitude 0.2 T, parallel to the x-axis, find the force on each side of the loop and the torque required to hold the loop in the position shown.

b) Same as (a) except that the field is parallel to the z-axis.

c) What torque would be required if the loop were pivoted about an axis through its center, parallel to the y-axis?

Figure 31-10

31-10 The rectangular loop of wire in Fig. 31-11 has a mass of 0.1 g per centimeter of length, and is pivoted about

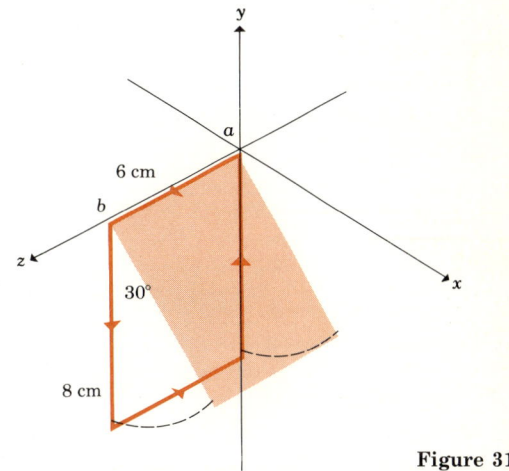

Figure 31-11

side ab as a frictionless axis. The current in the wire is 10 A in the direction shown.

a) Find the magnitude and sense of the magnetic field, parallel to the y-axis, that will cause the loop to swing up until its plane makes an angle of 30° with the yz-plane.

b) Discuss the case where the field is parallel to the x-axis.

31–11 What is the maximum torque on a coil 5×12 cm, of 600 turns, when carrying a current of 10^{-5} A in a uniform field of magnitude 0.1 T?

31–12 The coil of a pivoted-coil galvanometer has 50 turns and encloses an area of 6 cm². The magnetic field in the region in which the coil swings is 0.01 T and is radial. The torsional constant of the hairsprings is

$$k' = 1.0 \times 10^{-8} \text{ N} \cdot \text{m} \cdot (\text{degree})^{-1}.$$

Find the angular deflection of the coil for a current of 1 mA.

31–13 A circular coil of wire 8 cm in diameter has 12 turns and carries a current of 5 A. The coil is in a region where the magnetic field is 0.6 T.

a) What is the maximum torque on the coil?

b) In what position would the torque be one-half as great as in (a)?

31–14 In a shunt-wound dc motor (Fig. 31–12), the resistance R_f of the field coils is 150 Ω and the resistance R_f of the rotor is 2 Ω. When a difference of potential of 120 V is applied to the brushes, and the motor is running at full speed delivering mechanical power, the current supplied to it is 4.5 A.

a) What is the current in the field coils?

b) What is the current in the rotor?

c) What is the emf developed by the motor?

d) How much mechanical power is developed by this motor?

31–15 Figure 31–12 is a diagram of a shunt-wound dc motor, operating from a 120-V dc power line. The resistance of the field windings, R_f, is 240 Ω. The resistance of the rotor, R_r, is 3 Ω. When the motor is running the rotor develops an emf $\mathcal{E}$. The motor draws a current of 4.5 A from the line. Compute

a) the field current,

b) the rotor current,

c) the emf $\mathcal{E}$,

d) the rate of development of heat in the field windings,

e) the rate of development of heat in the rotor,

f) the power input to the motor, and

g) the efficiency of the motor,

if friction and windage losses amount to 50 W.

31–16 A horizontal tube of rectangular cross section (height h, width w) is placed at right angles to a uniform magnetic field of induction B, so that a length l is in the field. (See Fig. 31–7.) The tube is filled with liquid sodium and an electric current of density J is maintained in the third mutually perpendicular direction.

a) Show that the difference of pressure between a point in the liquid on a vertical plane through ab (Fig. 31–7) and a point in the liquid on another vertical plane through cd, under conditions in which the liquid is prevented from flowing, is

$$\Delta p = JlB.$$

b) What current density would be needed to provide a pressure difference of 1 atm between these two points if $B = 1$ T and $l = 0.1$ m?

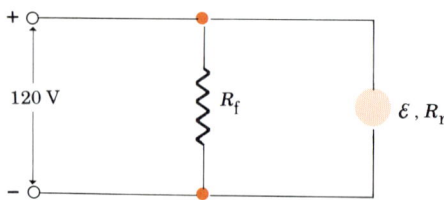

Figure 31–12

MAGNETIC FIELD OF A CURRENT 32

32-1 Sources of magnetic field

In Chapters 30 and 31 we discussed one aspect of the magnetic interaction of moving charges; we took the existence of the magnetic field as a given fact and studied the *forces* exerted by the magnetic field on moving charges and on currents in conductors. We now return to the other aspect of this interaction, the principles governing the *creation* of magnetic fields. As noted in the beginning of Chapter 30, the fundamental source of magnetic field is electric charge in motion; in many cases this takes the form of electric current in a conductor.

The first recorded observations of magnetic fields set up by currents were those of Oersted, who discovered that a pivoted compass needle, beneath a wire in which there was a current, set itself with its long axis perpendicular to the wire. Later experiments by Biot and Savart, and by Ampère, led to a relation which can be used to compute the magnetic field at any point in space around a circuit in which there is a current.

32-2 Magnetic field of a moving charge

We consider first the **B** field produced by a single moving point charge, as in Fig. 32-1. Experiment shows that this field is proportional to the charge q and to its speed v, and inversely proportional to the square of the distance r from the charge. The direction of the magnetic field at a point P is *not* along the line from the charge to P, but rather is perpendicular to the plane containing this line and the velocity vector v. Furthermore, the field magnitude is proportional to the sine of the angle θ between these two directions. Thus the magnitude of the magnetic field at point P is given by

$$B = k' \frac{qv \sin \theta}{r^2}, \tag{32-1}$$

where k' is a proportionality constant with a numerical value that depends on the system of units being used.

32-1 Magnetic field vectors due to a moving positive point charge q.

The magnitude and direction of **B** can be incorporated into a single vector equation by use of the vector product. We introduce a *unit vector* $\hat{r} = \mathbf{r}/r$, pointing in the direction from charge q to point P. Then the above description is summarized by the equation

$$\mathbf{B} = k'\frac{q\mathbf{v} \times \hat{r}}{r^2}. \tag{32-2}$$

Figure 32-1 shows the magnetic field at several points in the vicinity of the charge. The field is zero at all points along a line through the charge in the direction of the velocity $\mathbf{v}$, since at all such points $\sin\theta = 0$. At a given distance r from the charge, **B** has maximum magnitude at points lying in the plane through the charge perpendicular to $\mathbf{v}$, since at all such points $\theta = 90°$ and $\sin\theta = 1$. The charge also produces an *electric* field in its vicinity, just as when it is at rest; the electric-field vectors are not shown in the figure.

The field lines describing the *electric* field of a point charge radiate outward from the charge, but the *magnetic* field lines are completely different in character; they are *circles* in planes perpendicular to $\mathbf{v}$, with centers along the line of $\mathbf{v}$.

As discussed in Sec. 30-2, the unit of B is

$$1\,\text{T} = 1\,\text{N}\cdot\text{s}\cdot\text{C}^{-1}\cdot\text{m}^{-1} = 1\,\text{N}\cdot\text{A}^{-1}\cdot\text{m}^{-1}.$$

Using this with Eq. (32-1) or (32-2) and solving for k', we find that the unit of the magnetic constant k' is

$$1\,\text{N}\cdot\text{s}^2\cdot\text{C}^{-2} = 1\,\text{N}\cdot\text{A}^{-2} = 1\,\text{Wb}\cdot\text{A}^{-1}\cdot\text{m}^{-1}$$
$$= 1\,\text{T}\cdot\text{A}^{-1}\cdot\text{m}.$$

In SI units, the numerical value of k' is arbitrarily assigned to be *exactly* 10^{-7}. Thus

$$k' = 10^{-7}\,\text{N}\cdot\text{s}^2\cdot\text{C}^{-2} = 10^{-7}\,\text{N}\cdot\text{A}^{-2}$$
$$= 10^{-7}\,\text{Wb}\cdot\text{A}^{-1}\cdot\text{m}^{-1} = 10^{-7}\,\text{T}\cdot\text{A}^{-1}\cdot\text{m} \quad\text{(exactly)}.$$

The reason for this choice of the numerical value of k' will be explained later in the chapter.

It will be recalled that the electrical constant k has the value 8.98755×10^9 N·m²·C⁻². The ratio k/k' is therefore

$$\frac{k}{k'} = \frac{8.98755 \times 10^9 \text{ N·m}^2\text{·C}^{-2}}{10^{-7} \text{ N·s}^2\text{·C}^{-2}}$$

$$= 8.98755 \times 10^{16} \text{ m}^2\text{·s}^{-2},$$

which is equal to the square of the speed of light, c. This suggests that there may be a close relation between electricity, magnetism, and light. We shall return to this question in Chapter 37.

In electrostatics we found it convenient to express electric-field relations not in terms of the constant k in Eq. (24–1) but in terms of ϵ_0, where $k = 1/4\pi\epsilon_0$. Similarly, in magnetic-field relations it is convenient to introduce a constant μ_0, defined by the relation

$$k' = \frac{\mu_0}{4\pi}.$$

[margin note: $\mu_0 = 4\pi k'$]

Thus

$$\mu_0 = 4\pi \times 10^{-7} \text{ Wb·A}^{-1}\text{·m}^{-1} = 4\pi \times 10^{-7} \text{ T·A}^{-1}\text{·m}.$$

In terms of μ_0, Eqs. (32–1) and (32–2) become

$$B = \frac{\mu_0}{4\pi} \frac{qv \sin\theta}{r^2}, \tag{32-3}$$

$$\mathbf{B} = \frac{\mu_0}{4\pi} \frac{q\mathbf{v} \times \hat{\mathbf{r}}}{r^2}. \tag{32-4}$$

We can now write the expression for the magnetic force between two point charges, both of which are in motion relative to an observer. Thus a charge q', moving with velocity $\mathbf{v}'$ in a magnetic field $\mathbf{B}$, experiences a force

$$\mathbf{F} = q'\mathbf{v}' \times \mathbf{B},$$

and if the field $\mathbf{B}$ is set up by a charge q moving with velocity $\mathbf{v}$,

[margin note: two moving charges]

$$\mathbf{F} = \frac{\mu_0}{4\pi} \frac{qq'\mathbf{v}' \times (\mathbf{v} \times \hat{\mathbf{r}})}{r^2}, \tag{32-5}$$

which corresponds to Coulomb's law for the electrical force between the charges.

As noted above, the constants k and k' are related to the speed of light c. The corresponding relation for ϵ_0 and μ_0, as the reader may verify, is

$$\epsilon_0 \mu_0 = \frac{1}{c^2}. \tag{32-6}$$

The *principle of superposition,* which has been used to calculate the total electric field from several charges, is found to be valid also for magnetic-field calculations. *The total magnetic field caused by several moving charges is the vector sum of the fields caused by the individual charges*. This fact makes it possible to calculate the field caused by a current in a circuit in terms of fields caused by individual segments of the conductors. This useful relationship is developed in the next section.

32-3 Magnetic field of a current element. The Biot law

The magnetic field set up at any point by the current in a conductor is the resultant (vector sum) of the fields due to all of the moving charges in the conductor. The conductor is to be divided, in imagination, into short segments of length dl, one of which is shown in Fig. 32-2. The volume of a segment equals $A\,dl$, where A is its cross-section area. If there are n charges q per unit volume, the total moving charge dQ in the element is

$$dQ = nqA\,dl.$$

The moving charges are therefore equivalent to a single charge dQ, traveling with a velocity equal to the *drift* velocity $\boldsymbol{v}$. (Fields due to the *random* velocities of the carriers will, on the average, cancel out at every point.) From Eq. (32-3), the magnitude of dB at any point is

$$dB = \frac{\mu_0}{4\pi} \frac{dQ\,v\sin\theta}{r^2} = \frac{\mu_0}{4\pi} \frac{nqvA\,dl\sin\theta}{r^2}.$$

But ($nqvA$) equals the current I in the element, so

$$dB = \frac{\mu_0}{4\pi} \frac{I\,dl\sin\theta}{r^2}, \tag{32-7}$$

or in vector form,

$$d\boldsymbol{B} = \frac{\mu_0}{4\pi} \frac{I\,d\boldsymbol{l} \times \hat{\boldsymbol{r}}}{r^2}, \tag{32-8}$$

where $d\boldsymbol{l}$ is a vector of length dl.

As shown in Fig. 32-2, the field vectors $d\boldsymbol{B}$ are exactly like those set up by a finite positive charge Q, moving in the direction of the drift velocity $\boldsymbol{v}$.

The resultant $\boldsymbol{B}$ at any point in space, due to the current in a complete circuit, is the *vector integral* of the values of $d\boldsymbol{B}$ due to all elements of the circuit. Thus

$$\boldsymbol{B} = \frac{\mu_0}{4\pi} \int \frac{I\,d\boldsymbol{l} \times \hat{\boldsymbol{r}}}{r^2}. \tag{32-9}$$

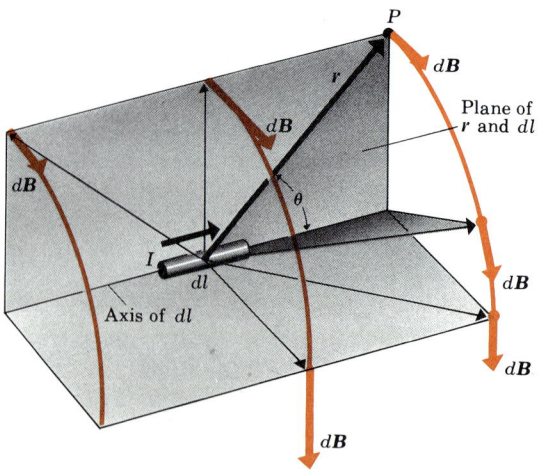

32-2 Magnetic field vectors due to a current element.

It should be pointed out that it is impossible to verify Eq. (32-8) experimentally, since one can never obtain an isolated element of a current-carrying circuit. Only the *resultant* field **B**, given by the integral in Eq. (32-9), can be measured experimentally. Equation (32-8) was *deduced* originally by the French physicist Jean Biot, in 1820, from experimental studies of the magnetic fields around circuits of various shapes. It is known as the *Biot law* (pronounced "bee-OH").

If there is matter in the space around a circuit, the field at a point will be due not entirely to currents in conductors, but in part to the *magnetization* of this matter. The magnetic properties of matter are discussed more fully in Chapter 35. However, unless iron or some other ferromagnetic material is present, this effect is so small that, while strictly speaking Eq. (32-9) holds only for conductors in vacuum, as a practical matter it can be used without correction for conductors in air, or in the vicinity of any nonferromagnetic material.

32-4 Magnetic field of a long straight conductor

Let us use the Biot law to compute the magnetic field **B** at point P in Fig. 32-3 due to a long straight conductor carrying a current I. We consider a conductor of total length $2L$, with its midpoint at the origin. The various factors in Eq. (32-9) are:

$$d\mathbf{l} = dy\,\mathbf{j},$$
$$\hat{\mathbf{r}} = \frac{\mathbf{r}}{r} = \frac{-y\mathbf{j} + z\mathbf{k}}{\sqrt{y^2 + z^2}},$$
$$r^2 = y^2 + z^2.$$

Thus for our problem Eq. (32-9) takes the form

$$\mathbf{B} = \frac{\mu_0}{4\pi} \int_{-L}^{L} \frac{I(dy\,\mathbf{j}) \times (-y\mathbf{j} + z\mathbf{k})}{(y^2 + z^2)^{3/2}}$$
$$= \frac{\mu_0 I}{4\pi} \int_{-L}^{L} \frac{z\,dy}{(y^2 + z^2)^{3/2}} \mathbf{i}. \tag{32-10}$$

32-3

Integrating by trigonometric substitution or use of a table of integrals, we find

$$\mathbf{B} = \frac{\mu_0 I}{4\pi} \frac{2L}{z\sqrt{L^2 + z^2}} \mathbf{i}. \tag{32-11}$$

Or Eq. (32-7) may be used, with $dl = dy$, $\sin\theta = z/\sqrt{y^2 + z^2}$, and $r^2 = y^2 + z^2$, to obtain

$$\mathbf{B} = \frac{\mu_0 I}{4\pi} \int_{-L}^{L} \frac{z\, dy}{(y^2 + z^2)^{3/2}},$$

which is the same integral as Eq. (32-10).

When the length of the conductor is very large compared to its distance from point P, it may be considered to be infinitely long. In Eq. (32-11) the quantity $\sqrt{L^2 + z^2}$ becomes approximately equal to L, and in the limit as $L \to \infty$, Eq. (32-11) becomes

$$\mathbf{B} = \frac{\mu_0 I}{2\pi z} \mathbf{i} \quad \text{(long straight wire)}. \tag{32-12}$$

Because the physical situation has axial symmetry about the y-axis, B has the same *magnitude* at all points on a circle centered on the conductor and lying in a plane perpendicular to the conductor, and its *direction* is everywhere tangent to such a circle, as shown in Fig. 32-3. Thus, at all points on a circle of radius r around the conductor (shown as a broken line in the figure), the magnitude B is given by

$$B = \frac{\mu_0 I}{2\pi r} \quad \text{(long straight wire)}. \tag{32-13}$$

This relation was deduced from experimental observations by Biot and Savart before the differential form, Eq. (32-8), had been discovered. It is called the *Biot–Savart law*, although this name is sometimes applied also to Eqs. (32-8) and (32-9).

Example A long straight conductor carries a current of 100 A. At what distance from the conductor is the magnetic field caused by the current equal in magnitude to the earth's magnetic field in Pittsburgh (about 0.5×10^{-4} T)?

Solution We use Eq. (32-13); everything except r is known, so we solve for r and insert the appropriate numbers:

$$r = \frac{\mu_0 I}{2\pi B} = \frac{(4\pi \times 10^{-7}\,\text{T·m·A}^{-1})(100\,\text{A})}{(2\pi)(0.5 \times 10^{-4}\,\text{T})} = 0.4\,\text{m}.$$

At smaller distances the field becomes stronger; when $r = 0.2$ m, $B = 1.0 \times 10^{-4}$ T, and so on.

A portion of the magnetic field around a long straight conductor is shown in Fig. 32-4. The shape of the magnetic-field lines in this situation is completely different from that of the electric-field lines in the analogous electrical situation. Electric-field lines radiate outward from the charges that are their sources (or inward for negative charges). By contrast, magnetic-field lines *encircle* the current that acts as their source.

32-4 Magnetic field around a long straight conductor. The field lines are circles, with directions determined by the righthand rule.

Electric-field lines begin and end at charges, while magnetic-field lines *never* have endpoints, irrespective of the shape of the conductor which sets up the field.

If an imaginary closed surface is constructed in a magnetic field, no field line can start or end inside this surface. Thus the number of lines emerging from the surface must equal the number entering it. We have shown that the number of lines crossing a surface is proportional to the flux Φ across the surface. Hence in a magnetic field, the flux across any *closed* surface is zero, or

$$\oint B_\perp \, dA = \oint \mathbf{B} \cdot d\mathbf{A} = 0. \tag{32-14}$$

This should be compared with Gauss's law for electrostatic fields, in which the surface integral of $\mathbf{E}$ over a closed surface equals $1/\epsilon_0$ times the enclosed charge. Thus Eq. (32-14) can be interpreted as saying that there is no such thing as "magnetic charge" to act as a source of $\mathbf{B}$. The sources of $\mathbf{B}$ are moving electric charges, as outlined above.

Finally, we note that the direction of the $\mathbf{B}$ lines around a straight conductor is given by a righthand rule: Grasp the conductor, with the thumb extended in the direction of the current; the fingers then curl around the conductor in the direction of the $\mathbf{B}$ lines. This rule is illustrated in Fig. 32-4.

32-5 Force between parallel conductors; the ampere

Figure 32-5 shows portions of two long straight parallel conductors separated by a distance a and carrying currents I and I', respectively, in the same direction. Since each conductor lies in the magnetic field set up by the other, each experiences a force. The diagram shows some of the field lines set up by the current in the lower conductor.

From Eq. (32-13), the magnitude of the $\mathbf{B}$-vector at the upper conductor is

$$B = \frac{\mu_0 I}{2\pi r}.$$

32-5 Parallel conductors carrying currents in the same direction attract each other.

From Eq. (31-1), the force on a length l of the upper conductor is

$$F = IlB = \frac{\mu_0 II'l}{2\pi r},$$

and the force *per unit length* is therefore

$$\frac{F}{l} = \frac{\mu_0 II'}{2\pi r}. \qquad (32\text{-}15)$$

The righthand rule shows that the direction of the force on the upper conductor is downward. There is an equal and opposite force per unit length on the lower conductor, as may be seen by considering the field set up by the upper conductor. Hence, the conductors *attract* one another.

If the direction of either current is reversed, the forces reverse also. Parallel conductors carrying currents in *opposite* directions *repel* one another.

The fact that two straight parallel conductors exert forces of attraction or repulsion on one another is made the basis of the SI definition of the ampere. The ampere is defined as follows:

> *One ampere is that unvarying current which, if present in each of two parallel conductors of infinite length and one meter apart in empty space, causes each conductor to experience a force of exactly 2×10^{-7} newtons per meter of length.*

It follows from this and the preceding equation that, by definition, the constant μ_0 is *exactly* $4\pi \times 10^{-7}$ N·A^{-2}, as stated in Sec. 32-2.

From the definition above, the ampere can be established, in principle, with the help of a meter stick and a spring balance. For the practical standardization of the ampere, coils of wire are used instead of straight wires, and their separation is made only a few centimeters. The complete instrument, which is capable of measuring currents with a high degree of precision, is called a *current balance*.

Having defined the ampere, we can now define the coulomb as *the quantity of charge that in one second crosses a section of a circuit in which there is a constant current of one ampere.*

Mutual forces of attraction exist not only between *wires* carrying currents in the same direction, but between each of the longitudinal elements into which a single current-carrying conductor may be subdivided. If the conductor is a liquid or an ionized gas (a plasma), these forces result in a constriction of the conductor as if its surface were acted on by an external, inward, pressure force. The constriction of the conductor is called the *pinch effect,* and attempts are being made to utilize the high temperature produced by the pinch effect in a plasma to bring about nuclear fusion.

32-6 Magnetic field of a circular loop

In many devices in which a current is used to establish a magnetic field, as in an electromagnet or a transformer, the wire carrying the current is wound into a *coil,* often consisting of many circular loops. We therefore consider next the magnetic field set up by a single circular loop of wire carrying a current. Figure 32-6 shows a circular loop of wire of radius a,

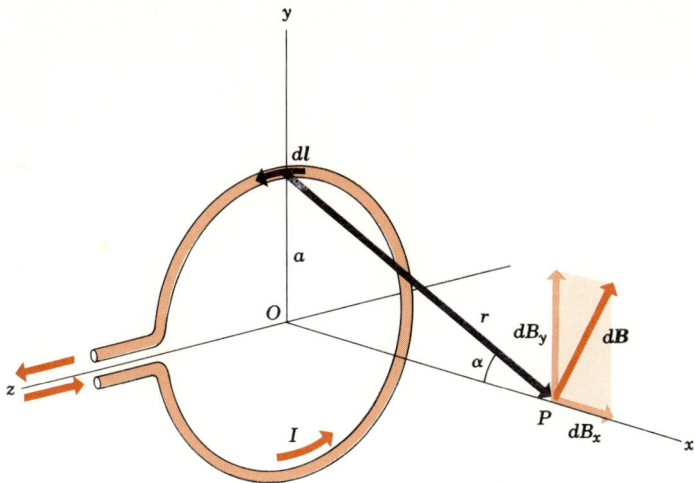

32-6 Magnetic field of a circular loop. The segment dl causes the field $d\mathbf{B}$, lying in the xy-plane. Other dl's have different components perpendicular to the x-axis; these add to zero, while the x-components combine to give the total $\mathbf{B}$ field at point P.

carrying a current I that is led into and out of the loop through two long straight wires side by side. The currents in the straight wires are in opposite directions and annul each other's magnetic effects.

We may use the law of Biot, Eq. (32–9), to calculate the magnetic field at a point P along a line perpendicular to the plane of the loop, a distance x from its center. As the figure shows, dl and r are perpendicular, and the direction of the field $d\mathbf{B}$ caused by element dl lies in the xy-plane. Also, $r = (x^2 + a^2)^{1/2}$. Thus for the element dl we find

$$dB = \frac{\mu_0 I}{4\pi} \frac{dl}{x^2 + a^2}. \tag{32-16}$$

In terms of components,

$$dB_x = dB \sin \alpha = \frac{\mu_0 I}{4\pi} \frac{dl}{x^2 + a^2} \cdot \frac{a}{(x^2 + a^2)^{1/2}}, \tag{32-17}$$

$$dB_y = dB \cos \alpha = \frac{\mu_0 I}{4\pi} \frac{dl}{x^2 + a^2} \cdot \frac{x}{(x^2 + a^2)^{1/2}}. \tag{32-18}$$

Because of the symmetry about the x-axis, the components of $\mathbf{B}$ perpendicular to this axis must add to zero. In fact, an element dl on the opposite side of the loop, having direction opposite to that of the dl shown, gives a contribution to the field having the same x-component as given by Eq. (32–17) but an *opposite* y-component. Similarly, the components of $\mathbf{B}$ perpendicular to the x-axis set up by pairs of diametrically opposite elements cancel, leaving only the x-components. The total x-component is obtained by integrating Eq. (32–17); everything is constant in this integration except dl, and the integral of dl for the entire loop is simply the circumference of the circle, $2\pi a$. Thus we obtain

$$B_x = \int \frac{\mu_0 I}{4\pi} \cdot \frac{a\,dl}{(x^2 + a^2)^{3/2}} = \frac{\mu_0 I a}{4\pi (x^2 + a^2)^{3/2}} \int dl,$$

or

$$B_x = \frac{\mu_0 I a^2}{2(x^2 + a^2)^{3/2}} \quad \text{(circular loop)}. \tag{32-19}$$

Equations (32–17) and (32–18) may be obtained somewhat more compactly by use of unit vectors. We write $dl = dl\,\mathbf{k}$, $\mathbf{r} = x\mathbf{i} - a\mathbf{j}$, and

$$\frac{d\mathbf{l} \times \hat{\mathbf{r}}}{r^2} = \frac{d\mathbf{l} \times \mathbf{r}}{r^3} = \frac{(dl\,\mathbf{k}) \times (x\mathbf{i} - a\mathbf{j})}{(x^2 + a^2)^{3/2}}$$

$$= \frac{x\,dl\,\mathbf{j} + a\,dl\,\mathbf{i}}{(x^2 + a^2)^{3/2}}.$$

As before, the components perpendicular to the x-axis add to zero, and only the term containing $\mathbf{i}$ survives. Using the above expression in Eq. (32–9) and again using the fact that $\int dl = 2\pi a$, we obtain

$$\mathbf{B} = \int \frac{\mu_0 I}{4\pi} \frac{a\,dl\,\mathbf{i}}{(x^2 + a^2)^{3/2}} = \frac{\mu_0 I a^2 \mathbf{i}}{2(x^2 + a^2)^{3/2}}, \quad (32\text{–}20)$$

which is equivalent to Eq. (32–19).

At the center of the loop, $x = 0$, and Eq. 32–19 reduces to

$$B_x = \frac{\mu_0 I}{2a} \quad \text{(center of circular loop)}. \quad (32\text{–}21)$$

If, instead of a single loop as in Fig. 32–6, one has a coil of N closely spaced loops, all having the same radius, each loop contributes equally to the field, and Eq. (32–21) becomes

$$B_x = \frac{\mu_0 N I}{2a} \quad (N \text{ circular loops}). \quad (32\text{–}22)$$

Similarly, Eqs. (32–19) and (32–20) may be adapted to the case of N loops by the insertion of a factor N.

Some of the magnetic-field lines surrounding a circular loop and lying in planes through the axis are shown in Fig. 32–7. Again we see that the field lines encircle the conductor, and that their directions are given by the righthand rule, as for the long straight conductor. The field lines for the circular loop are *not* circles, but they are closed curves that link the conductor.

It was pointed out in Chapter 31 that a current-carrying conducting loop of *any* shape is acted on by a *torque* when placed in a magnetic field

32–7 Field lines surrounding a circular loop.

caused by sources other than the loop. The position of stable equilibrium is one in which the plane of the loop is perpendicular to the externally caused field. Now we see that this position is such that *within the area enclosed by the loop, the **B** field produced by the loop itself is in the same direction as that of the external field*. In other words, a loop, if free to turn, will set itself in such a plane that the flux passing through the area enclosed by it has its *maximum* possible value. This is found to be true in all instances and is a useful general principle. For example, if a current is sent through an irregular loop of flexible wire in an external magnetic field, the loop assumes a circular form with its plane perpendicular to the field and with its own flux adding to that of the field. The same conclusion can, of course, be drawn by analyzing the side thrusts on the elements of the conductor.

32-7 Ampère's law

Ampère's law is a useful relation that is analogous to Gauss's law. The latter, it will be recalled, is a relation between the integral of the normal component of electric field over a closed surface and the net electric charge enclosed by the surface. Ampère's law is a relation between the *tangential* component of *magnetic* field at points on a closed curve and the net current through the *area* bounded by the curve.

Ampère's law is formulated in terms of the *line integral* of **B** around a closed path, denoted by

$$\oint \boldsymbol{B} \cdot d\boldsymbol{l}.$$

This is the same sort of integral used to define work in Chapter 6 and electric potential in Chapter 26. We divide the path into infinitesimal segments $d\boldsymbol{l}$ and for each one calculate the scalar product of **B** and $d\boldsymbol{l}$. In general **B** varies from point to point, and the **B** at the location of each $d\boldsymbol{l}$ must be used. An alternative notation is $\int B_\parallel dl$, where $B_\parallel$ is the component of **B** parallel to $d\boldsymbol{l}$ at each point. The circle on the integral sign indicates that this integral will always be computed for a *closed* path whose beginning and end points are the same. As with Gauss's law, the path need not be an actual physical object; usually it is an imaginary curve constructed to permit the application of Ampère's law to a specific situation.

We consider first a long, straight conductor carrying a current I, passing through the center of a circle of radius r in a plane perpendicular to the conductor. In Fig. 32–3, such a circle is shown as a broken line. From Eq. (32–13), the field has magnitude $\mu_0 I/2\pi r$ at every point on the circle, and it is tangent to the circle at each point. Thus for every point on the circle $B_\parallel = B = \mu_0 I/2\pi r$. The line integral of **B** around the circle is

$$\oint \boldsymbol{B} \cdot d\boldsymbol{l} = \oint \frac{\mu_0 I}{2\pi r} dl = \frac{\mu_0 I}{2\pi r} \oint dl,$$

or, since $\oint dl$ is just the circumference ($2\pi r$) of the circle,

$$\oint \boldsymbol{B} \cdot d\boldsymbol{l} = \mu_0 I. \qquad (32\text{–}23)$$

That is, the line integral of **B** is equal to μ_0 multiplied by the current passing through the area bounded by the circle.

This result may also be derived for a more general integration path, such as the one in Fig. 32–8. At the position of the line element dl, the angle between dl and **B** is θ, and

$$\mathbf{B} \cdot d\mathbf{l} = B\, dl \cos\theta.$$

From the figure, $dl \cos\theta = r\, d\phi$, where $d\phi$ is the angle subtended by dl at the position of the conductor. Thus

$$\oint \mathbf{B} \cdot d\mathbf{l} = \oint \left(\frac{\mu_0 I}{2\pi r}\right)(r\, d\phi) = \frac{\mu_0 I}{2\pi} \oint d\phi.$$

But $\oint d\phi$ is just the total angle swept out by the radial line from the conductor to dl during a complete trip around the path, that is, 2π. Thus

$$\oint \mathbf{B} \cdot d\mathbf{l} = \mu_0 I. \quad (32\text{–}24)$$

The line integral does not depend on the shape of the path, or on the position of the wire within it. If the current in the wire were opposite to that shown, the integral would have the opposite sign. Hence if any number of long straight conductors pass through the surface bounded by the path, the line integral equals μ_0 times the *algebraic sum* of the currents. The positive direction for current through the surface is determined by the direction in which the path is traversed, in accordance with the usual righthand rule. Thus in Fig. 32–8, a current directed into the plane of the figure is considered positive, and one out of the figure is negative.

If a closed path does not encircle the wire (or if a wire lies outside the path), the line integral of the **B** field of that wire is zero, because the angle ϕ then has the same value at the start and finish of any round trip. Hence if there are other conductors present that do not pass through a given path, they may contribute to the value of **B** at every point, but the *line integrals* of their fields are zero.

It follows that if we interpret I in Eq. (32–24) to mean the *algebraic sum* of the currents across the area bounded by a closed path, this equation implies all of the statements above and is the analytic form of Ampère's law:

*The line integral of the magnetic field **B** around any closed path is equal to μ_0 times the net current across the area bounded by the path.*

Although derived only for the special case of the field of a number of long straight parallel conductors, the law is true for conductors and paths of any shape. The general derivation is no different in principle from that above, but it is complicated geometrically and will not be given.

Finally, we emphasize that the statement $\oint \mathbf{B} \cdot d\mathbf{l} = 0$ does not necessarily mean that $B = 0$ everywhere along the path, but only that no current is linked by the path.

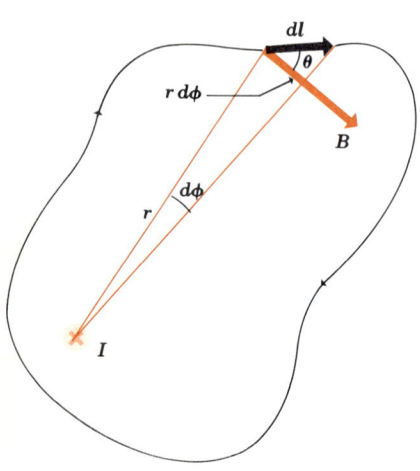

32–8 Line integral of **B** around a closed path linking a long straight conductor, carrying current I *into* the plane of the figure.

32-8 Applications of Ampère's law

In some cases of practical importance, symmetry considerations make it possible to use Ampère's law to compute the magnetic field caused by a certain current-carrying conductor. A few guiding principles, analogous to those stated for Gauss's law at the end of Sec. 25–4, are the following:

1. If **B** is everywhere tangent to the integration path and has the same magnitude B at every point on the path, then its line integral is equal to B multiplied by the circumference of the path.
2. If **B** is everywhere perpendicular to the path, for all or some portion of the path, that portion of the path makes no contribution to the line integral.
3. In the integral $\oint \mathbf{B} \cdot d\mathbf{l}$, **B** is always the *total* magnetic field at each point on the path. In general this field is caused partly by currents linked by the path and partly by currents outside. Even when *no* current is linked by the path, the field at points on the path need not be zero. In that case, however, $\oint \mathbf{B} \cdot d\mathbf{l}$ is always zero.
4. Some judgment is required in choosing an integration path. Two useful guiding principles are that the point or points at which the field is to be determined must lie on the path, and that the path must have enough symmetry so that the integral can be evaluated.

1. Field of a solenoid. A solenoid is constructed by winding wire in a helix around the surface of a cylindrical form, usually of circular cross section. The turns of the winding are ordinarily closely spaced and may consist of one or more layers. For simplicity, we have represented a solenoid in Fig. 32–9 by a relatively small number of circular turns, each carrying a current I. The resultant field at any point is the vector sum of the **B**-vectors due to the individual turns. The diagram shows the field lines in the xy- and yz-planes. Exact calculations show that for a long, closely wound solenoid, half of the lines passing through a cross section at the center emerge from the ends and half "leak out" through the windings between center and end.

32-9 Magnetic-field lines surrounding a solenoid. The dashed rectangle $abcd$ is used to compute the flux density **B** in the solenoid from Ampere's law.

If the length of the solenoid is large compared with its cross-sectional diameter, the *internal* field near its center is very nearly uniform and parallel to the axis, and the *external* field near the center is very small. The internal field at or near the center can then be found by use of Ampère's law.

We select as a closed path the dashed rectangle *abcd* in Fig. 32-9. Side *ab*, of length *l*, is parallel to the axis of the solenoid. Sides *bc* and *da* are to be taken very long so that side *cd* is far from the solenoid and the field at this side is negligibly small.

By symmetry, the **B** field along side *ab* is parallel to this side and is constant, so that for this side $B_\parallel = B$ and

$$\int \mathbf{B} \cdot d\mathbf{l} = Bl.$$

Along sides *bc* and *da*, $B_\parallel = 0$ since *B* is perpendicular to these sides; and along side *cd*, $B_\parallel = 0$ also since $B = 0$. The sum around the entire closed path therefore reduces to *Bl*.

Let *n* be the number of turns *per unit length* in the windings. The number of turns in length *l* is then *nl*. Each of these turns passes once through the rectangle *abcd* and carries a current *I*, where *I* is the current in the windings. The total current through the rectangle is then *nlI* and, from Ampère's law,

$$Bl = \mu_0 nlI,$$
$$B = \mu_0 n I \quad \text{(solenoid)}. \tag{32-25}$$

Since side *ab* need not lie on the axis of the solenoid, the field is uniform over the entire cross section.

2. *Field of a toroidal solenoid.* Figure 32-10a represents a *toroid*, wound with wire carrying a current *I*. The black lines in Fig. 32-10b are paths to which we wish to apply Ampère's law. Consider first path 1. By symmetry, if there is any field at all in this region, it will be *tangent* to the path at all points, and $\oint \mathbf{B} \cdot d\mathbf{l}$ will equal the product of *B* and the circumference $l = 2\pi r$ of the path. The current through the path, however, is zero, and hence, from Ampère's law (since the circumference is *not* zero), the field **B** must be zero.

Similarly, if there is any field at path 3, it will also be tangent to the path at all points. Each turn of the winding passes *twice* through the area bounded by this path, carrying equal currents in opposite directions. The *net* current through the area is therefore zero, and hence $B = 0$ at all points of the path. *The field of the toroidal solenoid is therefore confined wholly to the space enclosed by the windings*! The toroid may be thought of as a solenoid that has been bent into a circle.

Finally, we consider path 2, a circle of radius *r*. Again by symmetry, the **B** field is tangent to the path, and $\int \mathbf{B} \cdot d\mathbf{l}$ equals $2\pi rB$. Each turn of the winding passes *once* through the area bounded by path 2, and the total current through the area is *NI*, where *N* is the *total* number of turns in the winding. Then, from Ampère's law,

$$2\pi r B = \mu_0 NI,$$

(a)

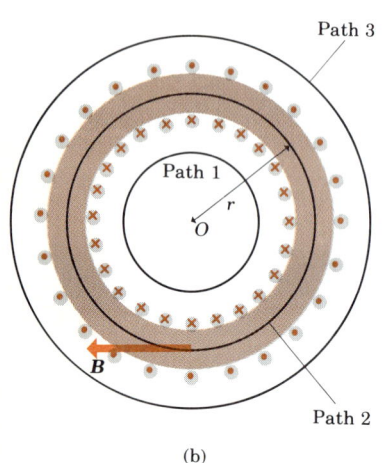

(b)

32-10 (a) A toroidal winding. (b) Closed paths (black circles) used to compute the magnetic field **B** set up by a current in a toroidal winding. The field is very nearly zero at all points except those within the space enclosed by the windings.

and

$$B = \frac{\mu_0 NI}{2\pi r} \quad \text{(toroidal solenoid)}. \tag{32-26}$$

The magnetic field is *not* uniform over a cross section of the core, because the path length l is larger at the outer side of the section than at the inner side. However, if the radial thickness of the core is small compared with the toroid radius r, the field varies only slightly across a section. In that case, considering that $2\pi r$ is the circumferential length of the toroid and that $N/2\pi r$ is the number of turns per unit length n, the field may be written

$$B = \mu_0 nI,$$

just as at the center of a long *straight* solenoid.

The equations derived above for the field in a closely wound straight or toroidal solenoid are strictly correct only for windings in *vacuum*. For most practical purposes, however, they can be used for windings in air, or on a core of any nonferromagnetic material. We shall show in Chapter 35 how they are modified if the core is of iron.

32-9 Magnetic fields and displacement currents

When the Biot and Savart law was discussed in Sec. 32-3, it was assumed that the current I in an element $d\mathbf{l}$ was a *conduction* current. Experiment shows, however, that a *displacement* current I_D (Sec. 29-6) contributes to a magnetic field in exactly the same way as a conduction current, I_C, so the general form of the Biot law is

$$d\mathbf{B} = \frac{\mu_0}{4\pi} \frac{(I_C + I_D) \, d\mathbf{l} \times \hat{\mathbf{r}}}{r^2}, \tag{32-27}$$

and the general form of Ampère's law is

$$\oint \mathbf{B} \cdot d\mathbf{s} = \mu_0 (I_C + I_D). \tag{32-28}$$

Note that in the Biot law, the symbols I_C and I_D refer to the currents in an element $d\mathbf{l}$, while in Ampère's law they refer to the *total* currents across an area bounded by the path of integration.

As an example, Fig. 32-11 is a sectional view of two circular conducting plates of radius R separated by a narrow gap in vacuum (i.e., a parallel-plate capacitor). If the gap is small compared to R, the $\mathbf{E}$ field caused by the charge distributions on the plates is confined almost entirely to the region between plates and is nearly uniform, as shown. Suppose we wish to calculate the $\mathbf{B}$ field at some point in the midplane, shown in the figure as a broken line. In principle, this could be calculated from the Biot law. It would be necessary to subdivide the whole of space (including the wires and the plates) into elementary filaments $d\mathbf{l}$, find the conduction and displacement currents in each filament, and perform a vector integration over all space. Such a calculation would be extremely complicated, if not impossible. The result can be obtained quite easily,

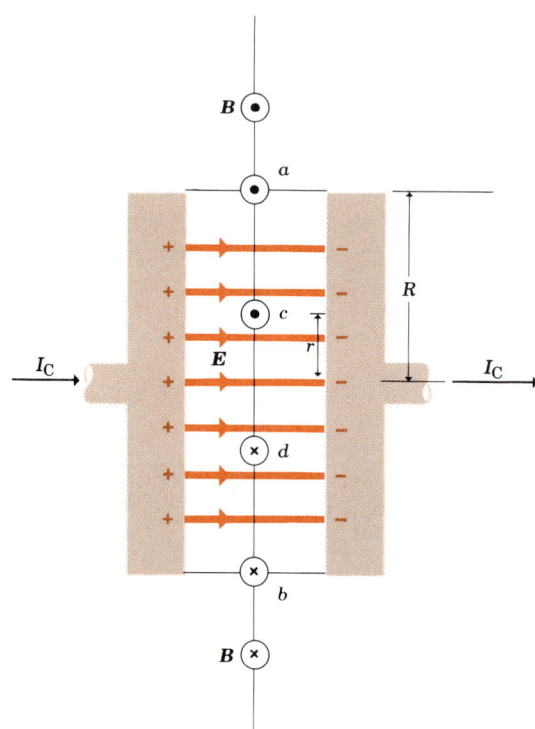

32-11 A capacitor being charged by a current I_C has a displacement current I_D between its plates equal to I_C, with displacement current density $J_D = \epsilon_0\, dE/dt$.

however, if we use Ampère's law and take advantage of symmetry considerations.

As is evident from the diagram, there is no *conduction* current across the midplane; the current across this plane is a *displacement* current only. Also, since every line crosses the plane at some point, the *total* displacement current across the plane is the *displacement* current *out of* the left plate; from the definition of displacement current in Sec. 29-6, this must equal the *conduction* current *into* the plate. Furthermore, by symmetry, the ***B*** field lines are circles with centers on the axis. Thus, at any point on a circle perpendicular to the axis and passing through points such as a and b, the ***B***-vector has the same magnitude and is tangent to the circle. At point a, the ***B***-vector points toward the reader and at point b it points away from the reader, as indicated by the symbols ⊙ and ⊗.

The displacement current density J_D is given by Eq. (29-20):

$$J_D = \epsilon_0 \frac{dE}{dt}. \qquad (32\text{-}29)$$

At any instant E is uniform over the region between plates, so J_D is also uniform; the *total* displacement current I_D between plates is just J_D times the area πR^2 of the plates. As remarked above, this must also equal the total *conduction* current I_C. Thus

$$I_D = I_C = \pi R^2 J_D$$
$$J_D = \frac{I_C}{\pi R^2}. \qquad (32\text{-}30)$$

Now we may find the magnetic field at a point in the region between plates, at a distance r from the axis. We apply Ampère's law to a circle of radius r passing through the point; such a circle passes through points c and d in Fig. 32–11. The total current enclosed by the circle is J_D times its area, or $(I_C/\pi R^2)(\pi r^2)$. The integral $\oint \mathbf{B} \cdot d\mathbf{l}$ in Ampère's law is just B times the circumference $2\pi r$ of the circle, and Ampère's law becomes

$$\oint \mathbf{B} \cdot d\mathbf{l} = 2\pi r B = \mu_0 \left(\frac{r^2}{R^2}\right) I_C,$$

or

$$B = \frac{\mu_0}{2\pi} \left(\frac{r}{R^2}\right) I_C. \tag{32–31}$$

This result shows that in the region between the plates $\mathbf{B}$ is zero at the axis, increasing linearly with distance from the axis. A similar calculation shows that *outside* the region between the plates, $\mathbf{B}$ is the same as though the wire were continuous and the plates not present at all.

The role of displacement current as a source of magnetic field can be expressed in a different way using the concept of *electric flux* Ψ, introduced in Secs. 24–4 and 29–6. As stated in Eq. (29–22), the displacement current may be expressed as

$$I_D = \epsilon_0 \frac{d\Psi}{dt}.$$

Thus in empty space, where there is no conduction current ($I_C = 0$), Ampère's law, Eq. (32–28), may be rewritten

$$\oint \mathbf{B} \cdot d\mathbf{l} = \mu_0 \epsilon_0 \frac{d\Psi}{dt}. \tag{32–31}$$

A significant feature of this equation is its resemblance to another relation, *Faraday's law*, to be introduced in Chapter 33. Faraday's law describes the *electric* field induced by a changing *magnetic* field, and one form of the relation is

$$\oint \mathbf{E} \cdot d\mathbf{l} = -\frac{d\Phi}{dt}, \tag{32–32}$$

where Φ is *magnetic* flux. Apart from constants, Eqs. (32–31) and (32–32) have the same form, but the roles of electric and magnetic fields are reversed. Thus the concept of displacement current introduces a kind of symmetry into electric- and magnetic-field relations.

The fact that *displacement* current acts as a source of magnetic field plays an essential role in the understanding of electromagnetic *waves*. A changing *electric* field in a region of space induces a *magnetic* field in neighboring regions, even when no conduction current and no matter are present. This relationship, first proposed by Maxwell in 1865, provides the key to a theoretical understanding of electromagnetic radiation and of light as a particular example of this radiation. We return to this topic in Chapter 37.

Questions

32-1 Streams of charged particles emitted from the sun during unusual sunspot activity create a disturbance in the earth's magnetic field. How does this happen?

32-2 A topic of current interest in physics research is the search (thus far unsuccessful) for an isolated magnetic pole or magnetic *monopole*. If such an entity were found, how could it be recognized? What would its properties be?

32-3 What are the relative advantages and disadvantages of Ampère's Law and the Biot–Savart Law for practical calculations of magnetic fields?

32-4 A student proposed to obtain an isolated magnetic pole by taking a bar magnet (N pole at one end, S at the other) and breaking it in half. Will this work?

32-5 Pairs of conductors carrying current into and out of the power-supply components of electronic equipment are sometimes twisted together to reduce magnetic-field effects. Why does this help?

32-6 The text discusses the magnetic field of an infinitely long straight conductor carrying a current. Of course, there's no such thing as an infinitely long *anything*. How do you decide whether a particular wire is long enough to be considered infinite?

32-7 Suppose one has three long parallel wires, arranged so that in cross section they are at the corners of an equilateral triangle. Is there any way to arrange the currents so that all three wires attract each other? So that all three wires repel each other?

32-8 Two parallel conductors carrying current in the same direction attract each other. If they are permitted to move toward each other, the forces of attraction do work. Where does the energy come from? Does this contradict the assertion in Chapter 30 that magnetic forces on moving charges do no work?

32-9 Considering the magnetic field of a circular loop of wire, would you expect the field to be greatest at the center, or is it greater at some points in the plane of the loop but off-center?

32-10 Two concentric circular loops of wire, of different diameter, carry currents in the same direction. Describe the nature of the forces exerted on the inner loop.

32-11 A current was sent through a helical coil spring. The spring appeared to contract, as though it had been compressed. Why?

32-12 Using the fact that magnetic-field lines never have a beginning or end, explain why it is reasonable for the field of a toroidal solenoid to be confined entirely to its interior, while a straight solenoid *must* have some field outside it.

32-13 In the discussion of the **B** field of a solenoid (Sec. 32–8), suppose the cross section of the solenoid is not circular but some other shape, such as elliptical or square. Is Eq. (32-25) still valid? Why or why not?

32-14 Can one have a displacement current as well as a conduction current within a conductor?

32-15 A character in a popular comic strip has at various times proposed the possibility of "harnessing the earth's magnetic field" as a nearly inexhaustible source of energy. Comment on this concept.

Problems

$$1\,\text{T} = 1\,\text{Wb}\cdot\text{m}^{-2} = 1\,\text{N}\cdot\text{A}^{-1}\cdot\text{m}^{-1},$$
$$\mu_0 = 4\pi \times 10^{-7}\,\text{N}\cdot\text{A}^{-2} = 4\pi \times 10^{-7}\,\text{T}\cdot\text{A}^{-1}\cdot\text{m}$$
$$= 4\pi \times 10^{-7}\,\text{Wb}\cdot\text{A}^{-1}\cdot\text{m}^{-1}.$$

32-1 An overhead transmission line 5.0 m above the ground carries a current of 400 A in a direction from south to north. Find the <u>magnitude</u> and <u>direction</u> of the magnetic field at a point on the ground directly under the conductor. Would this field cause appreciable error in a magnetic compass located at this point?

32-2 A long straight conductor passes vertically through the center of a laboratory; the direction of the current is upward. The magnetic field at a point 0.20 m from the wire is found to have magnitude 5.0×10^{-4} T. What is the current in the conductor?

32-3 A long straight wire, carrying a current of 200 A, runs through a cubical wooden box, entering and leaving through holes in the centers of opposite faces, as in Fig. 32-12. The length of each side of the box is 20 cm. Consider

Figure 32-12

an element of the wire 1 cm long at the center of the box. Compute the magnitude of the magnetic field ΔB produced by this element at the points lettered a, b, c, d, and e in Fig. 32-12. Points a, c, and d are at the centers of the faces of the cube, point b is at the midpoint of one edge, and point e

is at a corner. Copy the figure and show by vectors the directions and relative magnitudes of the field vectors.

32-4 A long straight wire carries a current of 10 A directed along the negative y-axis, as shown in Fig. 32–13. A uniform magnetic field $\boldsymbol{B}_0$ of magnitude 10^{-6} T is directed parallel to the x-axis. What is the resultant magnetic field at the following points?

a) $x = 0$, $z = 2$ m;
b) $x = 2$ m, $z = 0$;
c) $x = 0$, $z = -0.5$ m.

Figure 32–14

Figure 32–13

32-5 A magnetic field of magnitude 5.0×10^{-4} T is to be produced at a distance of 5 cm from a long straight wire.

a) What current is required to produce this field?
b) With the current found in (a), what is the magnitude of the field at a distance of 10 cm from the wire? At 20 cm?

32-6 A long straight telephone cable contains six wires, each carrying a current of 0.5 A. The distances between wires can be neglected.

a) If the currents in all six wires are in the same direction, what is the magnitude of the magnetic field 10 cm from the cable?
b) If four wires carry currents in one direction and the other two in the opposite direction, what is the field magnitude 10 cm from the cable?

32-7 Figure 32–14 is an end view of two long parallel wires perpendicular to the xy-plane, each carrying a current I, but in opposite directions.

a) Copy the diagram, and show by vectors the $\boldsymbol{B}$ field of each wire, and the resultant $\boldsymbol{B}$ field at point P.
b) Derive the expression for the magnitude of $\boldsymbol{B}$ at any point on the x-axis, in terms of the coordinate x of the point.
c) Construct a graph of the magnitude of $\boldsymbol{B}$ at any point on the x-axis.
d) At what value of x is B a maximum?

32-8 Same as Problem 32–7, except that the current in both wires is *away* from the reader.

32-9 Two long, straight, horizontal parallel wires, one above the other, are separated by a distance $2a$. If the wires carry equal currents in opposite directions, what is the field magnitude in the plane of the wires at a point

a) midway between them, and
b) at a distance a above the upper wire?

If the wires carry equal currents in the same direction, what is the field magnitude in the plane of the wires at a point,

c) midway between them, and
d) at a distance a above the upper wire?

32-10 Two long, straight, parallel wires are 1.0 m apart, as in Fig. 32–15. The upper wire carries a current I_1 of 6 A into the plane of the paper.

a) What must be the magnitude and direction of the current I_2 for the resultant field at point P to be zero?

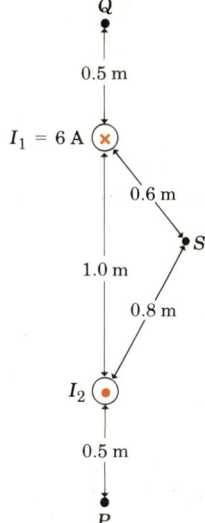

Figure 32–15

b) What is then the resultant field at Q?

c) At S?

32-11 In Fig. 32-14 suppose a third long straight wire, parallel to the other two, passes through point P, and that each wire carries a current $I = 20$ A. Let $a = 30$ cm and $x = 40$ cm. Find the magnitude and direction of the force per unit length on the third wire

a) if the current in it is away from the reader,

b) if the current is *toward* the reader.

32-12 A long straight wire carries a current of 1.5 A. An electron travels with a speed of 5×10^6 cm·s^{-1} parallel to the wire, 10 cm from it, and in the same direction as the current. What force does the magnetic field of the current exert on the moving electron?

32-13 A long horizontal wire AB rests on the surface of a table. (See Fig. 32-16). Another wire CD vertically above the first is 100 cm long and is free to slide up and down on the two vertical metal guides C and D. The two wires are connected through the sliding contacts and carry a current of 50 A. The mass per unit length of the wire CD is 0.05 g·cm^{-1}. To what equilibrium height h will the wire CD rise, assuming the magnetic force on it to be due wholly to the current in the wire AB?

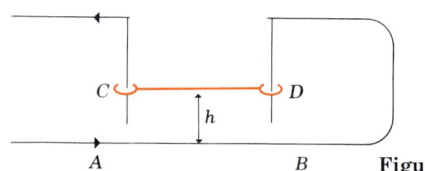

Figure 32-16

32-14 Two long parallel wires are hung by cords of 4 cm length from a common axis, as shown in Fig. 32-17. The wires have a mass per unit length of 50 g·m^{-1} and carry the same current in opposite directions. What is the current if the cords hang at an angle of 6° with the vertical?

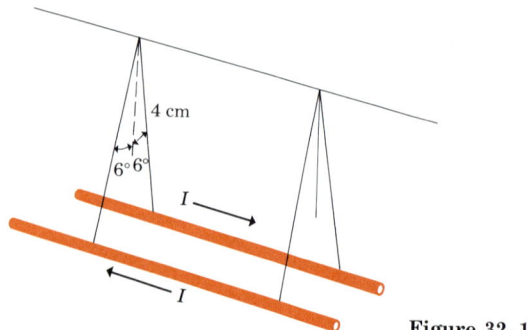

Figure 32-17

32-15 The long straight wire AB in Fig. 32-18 carries a current of 20 A. The rectangular loop whose long edges are parallel to the wire carries a current of 10 A. Find the magnitude and direction of the resultant force exerted on the loop by the magnetic field of the wire.

Figure 32-18

32-16 The long straight wire AB in Fig. 32-19 carries a constant current I.

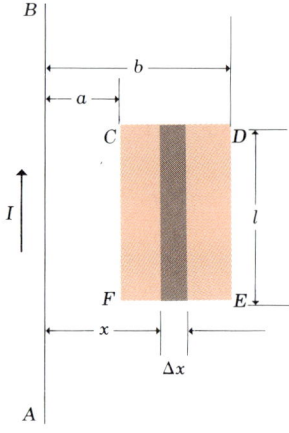

Figure 32-19

a) What is the magnetic field at the shaded area at a perpendicular distance x from the wire?

b) What is the magnetic flux $\Delta \Phi$ through the shaded area?

c) What is the total flux through the rectangle $CDEF$?

32-17 Refer to Fig. 32-7. Sketch a graph of the magnitude of the B field on the axis of the loop, from $x = -3R$ to $x = +3R$.

32-18 Figure 32-20 is a sectional view of two circular coils of radius a, each wound with N turns of wire carrying a current I, circulating in the same direction in both coils. The coils are separated by a distance a equal to their radii.

a) Derive the expression for the magnetic field at point P, midway between the coils.

b) Calculate the magnitude of B if $N = 100$ turns, $I = 5$ A, and $a = 30$ cm?

32-19 Considering the magnetic field along the axis of a circular loop of radius R, at what distance from the center of the loop is the field $\frac{1}{10}$ of its value at the center?

32-20 A circular coil of radius 5 cm has 200 turns and carries a current of 0.2 A. What is the magnetic field

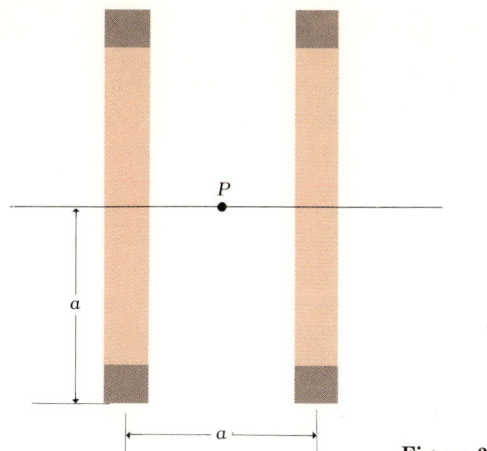

Figure 32-20

a) at the center of the coil?

b) At a point on the axis of the coil, 10 cm from its center?

32-21 A closely wound coil has a diameter of 40 cm and carries a current of 2.5 A. How many turns does it have if the magnetic field at the center of the coil is 1.26×10^{-4} T?

32-22 A thin disk of dielectric material, having a total charge $+Q$ distributed uniformly over its surface, and of radius a, rotates n times per second about an axis perpendicular to the surface of the disk and passing through its center. Find the magnetic field at the center of the disk.

32-23 A solenoid is 30 cm long and is wound with two layers of wire. The inner layer consists of 300 turns, the outer layer of 250 turns. The current is 3 A in the same direction in both layers. What is the magnetic field at a point near the center of the solenoid?

32-24 A wire of circular cross section and radius R carries a current I, uniformly distributed over its cross-sectional area.

a) In terms of I, R, and r_1, what is the current through a circular area of radius r_1, inside the wire?

b) Use Ampère's law to find B inside the wire, at a distance r_1 from the axis.

c) What is B outside the wire, at a distance r_2 from the axis?

d) What would the field be at this distance if the current were concentrated in a very fine wire along the axis?

e) Sketch a graph of the magnitude of $\boldsymbol{B}$ as a function of r, from $r = 0$ to $r = 2R$.

32-25 A coaxial cable consists of a solid conductor of radius R_1, supported by insulating disks on the axis of a tube of inner radius R_2 and outer radius R_3. If the central conductor and the tube carry equal currents in opposite directions, find the magnetic field

a) at points outside the axial conductor but inside the tube, and

b) at points outside the tube.

32-26 A solenoid of length 20 cm and radius 2 cm is closely wound with 200 turns of wire. The current in the winding is 5 A. Compute the magnetic field at a point near the center of the solenoid.

32-27 A wooden ring whose mean diameter is 10 cm is wound with a closely spaced toroidal winding of 500 turns. Compute the field at a point on the mean circumference of the ring when the current in the windings is 0.3 A.

32-28 A conductor is made in the form of a hollow cylinder with inner and outer radii a and b, respectively. It carries a current I, uniformly distributed over the cross section. Derive expressions for the magnetic field in the regions $r < a$, $a < r < b$, and $r > b$.

32-29 A solenoid is to be designed to produce a magnetic field of 0.1 T at its center. The radius is to be 5 cm and the length 50 cm, and the available wire can carry a maximum current of 10 A.

a) How many turns per unit length should the solenoid have?

b) What total length of wire is required?

32-30 A copper wire with circular cross section of area 4 mm^2 carries a current of 50 A. The resistivity of the material is 2.0×10^{-8} $\Omega \cdot$m.

a) What is the electric field in the material?

b) What is the magnetic field 5 cm from the wire?

c) If the current is changing at the rate of 5000 A·s^{-1}, at what rate is the electric field in the material changing?

d) What is the displacement current density in the material in (c)?

e) What is the additional magnetic field 5 cm from the wire, due to the displacement current? Is it significant compared to that due to the conduction current?

32-31 A capacitor has two parallel plates of area A separated by a distance d. The capacitor is given an initial charge Q but, because the damp air between the plates is slightly conductive, the charge slowly leaks through the air. The charge initially changes at a rate dQ/dt.

a) In terms of dQ/dt, what is the initial rate of change of electric field between the plates?

b) Show that the displacement current density has the same magnitude as the conduction current density, but the opposite direction. Hence show that the magnetic field in the material is exactly zero at all times.

33 INDUCED ELECTROMOTIVE FORCE

33-1 Motional electromotive force

The present large-scale production, distribution, and use of electrical energy would not be economically feasible if the only seats of emf available were those of *chemical* nature, such as dry cells. The development of electrical engineering, as we now know it, began with Faraday and Henry who, independently and at nearly the same time, discovered the principles of *magnetically* induced emf's and the methods by which mechanical energy can be converted directly to electrical energy.

Figure 33–1 represents a conductor of length l in a uniform magnetic field perpendicular to the plane of the diagram and directed away from the reader. If the conductor is set in motion toward the right with a velocity v, perpendicular both to its own length and to the magnetic field, a charged particle q within it experiences a force F equal to $qv \times B$ directed along the length of the conductor. The direction of the force on a positive charge is from b toward a in Fig. 33–1, while the force on a negative charge is from a toward b. Because this force is of non-electrostatic origin, we denote it by F_n:

$$F_n = qv \times B. \qquad (33\text{-}1)$$

The state of affairs within the conductor is the same as though it had been inserted in an *electric* field equal to $v \times B$, directed from b toward a. Paralleling the discussion of Sec. 28–4, we define the equivalent non-electrostatic field E_n as the non-electrostatic force per unit charge:

$$E_n = v \times B. \qquad (33\text{-}2)$$

The free charges in the conductor move in the direction of the force acting on them until the accumulation of excess charges at the ends of the conductor establishes an electrostatic field E_e such that the *resultant* force on every charge within the conductor is zero. The charges are then in equilibrium. The upper end of the wire acquires an excess positive charge, and the lower end an excess negative charge.

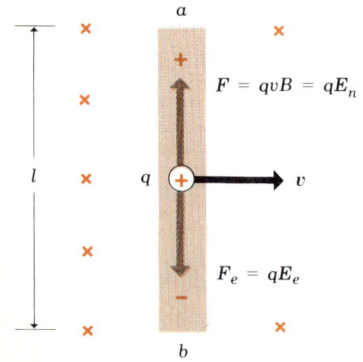

33–1 Conducting rod in uniform magnetic field.

33-2 Current produced by the motion of a conductor in a magnetic field.

Suppose now that the moving conductor slides along a stationary U-shaped conductor, as in Fig. 33-2. There is no magnetic force on the charges within the stationary conductor but, since it lies in the electrostatic field surrounding the moving conductor, a current will be established within it; the direction of this current (defined as usual as the direction of positive-charge motion) is counterclockwise, or from b toward a. As a result of this current, the excess charges at the ends of the moving conductor are reduced, the electrostatic field within the moving conductor is weakened, and the magnetic forces cause a further displacement of the free electrons within the conductor from a toward b. As long as the motion of the conductor is maintained there is, therefore, a continual current in a counterclockwise direction. The moving conductor corresponds to a seat of electromotive force, and is said to have *induced* within it a *motional electromotive force*.

The magnitude of this emf can be found as follows. When a charge q moves from b to a through a distance l, the work of the force F is

$$W = Fl = vBql.$$

The emf $\mathcal{E}$ is the work per unit charge, so

$$\mathcal{E} = \frac{W}{q} = vBl. \qquad (33\text{-}3)$$

If the velocity of the conductor makes an angle ϕ with the field, we must replace v by $v \sin \phi$, and the induced emf becomes

$$\mathcal{E} = v \sin \phi \, Bl. \qquad (33\text{-}4)$$

More generally, the electromotive force $\mathcal{E}$ may be defined as in Sec. 28-4, that is, as the line integral of the non-electrostatic field:

$$\mathcal{E} = \int_b^a \mathbf{E}_n \cdot d\mathbf{l} = \int_b^a \mathbf{v} \times \mathbf{B} \cdot d\mathbf{l}. \qquad (33\text{-}5)$$

In Fig. 33-2, $\mathbf{v}$ and $\mathbf{B}$ are perpendicular, and their vector product is parallel to $d\mathbf{l}$; thus in this case the right side of Eq. (33-5) reduces to vBl, in agreement with Eq. (33-3) or (33-4).

The emf associated with the moving conductor in Fig. 33-2 is analogous to that of a battery with its positive terminal at a and negative at b. In each the direction of the non-electrostatic field is from b and a, and when the device is connected to an external conductor, the direction of current is from a to b in the external circuit. Electromotive force is not a vector quantity, but the quantity defined by Eq. (33-5) may be positive or negative, depending on the direction of E_n.

If v is expressed in m·s^{-1}, B in T, and l in meters, $\mathcal{E}$ is in joules per coulomb or volts, as the reader should verify.

Example 1 Let the length l in Fig. 33-2 be 0.1 m, the velocity v be 0.1 m·s^{-1}, the resistance of the loop be 0.01 Ω, and let $B = 1$ T.

Solution The emf $\mathcal{E}$ is

$$\mathcal{E} = vBl = 0.01 \text{ V}.$$

The current in the loop is

$$I = \frac{\mathcal{E}}{R} = 1 \text{ A}.$$

Because of this current, there is a force F on the loop, in the opposite direction to its motion, and equal to

$$F = IBl = (1 \text{ A})(1 \text{ T})(0.1 \text{ m})$$
$$= 0.1 \text{ N}.$$

The power necessary to move the loop against this force is

$$P = Fv = (0.1 \text{ N})(0.1 \text{ m·s}^{-1}) = 0.01 \text{ W}.$$

The product $\mathcal{E}I$ is

$$\mathcal{E}I = (0.01 \text{ V})(1 \text{ A}) = 0.01 \text{ W}.$$

Thus, the rate of energy conversion, $\mathcal{E}I$, equals the mechanical power input, Fv, to the system.

Example 2 The rectangular loop in Fig. 33-3, of length a and width b, is rotating with uniform angular velocity ω about the y-axis. The entire loop lies in a uniform, constant B field, parallel to the z-axis. We wish to calculate the induced emf in the loop, from Eq. (33-3).

Solution The velocity v of either of the sides of the loop of length a has magnitude

$$v = \omega\left(\frac{b}{2}\right).$$

The motional emf in each of the two sides of length a is

$$\mathcal{E} = vB \sin \theta \, a = \tfrac{1}{2}\omega B \, ab \sin \theta.$$

These two emf's add, so the total emf due to these two sides is

$$\mathcal{E} = \omega B \, ab \sin \theta.$$

The magnetic forces on the other two sides of the loop are transverse to these sides and so contribute nothing to the emf. As an alternative approach, the emf may be calculated from Eq. (33-5). The direction of the motional field $E_n = v \times B$ in each of the sides of length a is shown in the diagram. Its magnitude is

$$E_n = vB \sin \theta = \omega \frac{b}{2} B \sin \theta.$$

33-3 Rectangular loop rotating with constant angular velocity in a uniform magnetic field.

The motional fields in the other two sides of the loop are transverse to these sides and contribute nothing to the emf. The line integral of $\mathbf{E}_n$ around the loop reduces to $2E_n a$, so

$$\mathcal{E} = \oint \mathbf{E}_n \cdot d\mathbf{l} = 2E_n a = 2\left(\omega \frac{b}{2} B \sin\theta\right) a$$

$$= \omega B a b \sin\theta.$$

The product ab equals the area A of the loop, and if the loop lies in the xy-plane at $t = 0$, then $\theta = \omega t$. Hence

$$\mathcal{E} = \omega A B \sin \omega t. \qquad (33\text{-}6)$$

The emf therefore varies sinusoidally with time. The *maximum* emf $\mathcal{E}_m$, which occurs when $\sin \omega t = 1$, is

$$\mathcal{E}_m = \omega A B,$$

so we can write Eq. (33-6) as

$$\mathcal{E} = \mathcal{E}_m \sin \omega t. \qquad (33\text{-}7)$$

The rotating loop is the prototype of one type of *alternating current generator*, or *alternator*; we say it develops a *sinusoidal alternating emf*.

The emf is maximum (in absolute value) when $\theta = 90°$ or $270°$ and the long sides are moving at right angles to the field. The emf is zero when $\theta = 0$ or $180°$ and the long sides are moving parallel or antiparallel to the field. The emf depends only on the *area A* of the loop and not on its shape. This is most easily verified by use of Faraday's law, to be introduced in the next section.

33-4 (a) Schematic diagram of an alternator, using a conducting loop rotating in a magnetic field. Connections to the external circuit are made by means of the slip rings S. (b) The resulting emf at terminals ab.

The rotating loop in Fig. 33-3 can be utilized as the source in an external circuit by making connections to *slip rings* S, S, which rotate with the loop, as shown in Fig. 33-4a. Stationary brushes bearing against the rings are connected to the output terminals a and b. The instantaneous terminal voltage v_{ab}, on open circuit, equals the instantaneous emf. Figure 33-4b is a graph of v_{ab} as a function of time.

A similar scheme may be used to obtain an emf which, though varying with time, always has the same sign. The loop is connected to a split ring or commutator, as in Fig. 33-5a. At angular positions where the emf reverses, the connections to the external circuit are interchanged. The resulting emf is shown in Fig. 33-5b.

This device is the prototype of a dc generator. Commercial dc generators have a large number of coils and commutator segments; their terminal voltage is not only unidirectional but also practically constant.

Example 3 A disk of radius R, shown in Fig. 33-6, lies in the xy-plane and rotates with uniform angular velocity ω about the z-axis. The disk is in a uniform, constant $\mathbf{B}$ field parallel to the z-axis. Consider a short portion of a narrow radial segment of the disk, of length dr. Its velocity is $v = \omega r$, and since $\mathbf{v}$ is at right angles to $\mathbf{B}$, the motional field $\mathbf{E}_n$ in the segment is

$$E_n = vB = \omega r B.$$

The direction of $\mathbf{E}_n$ is radially outward.

33-5 (a) Schematic diagram of a dc generator, using a split-ring commutator. (b) The resulting induced emf at terminals ab.

The motional emf due to this segment is

$$d\mathcal{E} = \mathbf{E}_n \cdot d\mathbf{l} = E_n\, dr = \omega r B\, dr.$$

The total emf between center and rim is

$$\mathcal{E} = \int_0^R \mathbf{E}_n \cdot d\mathbf{l} = \omega B \int_0^R r\, dr = \tfrac{1}{2}\omega B R^2. \tag{33-8}$$

All of the radial segments of the disk are in *parallel*, so the emf between center and rim equals that in any radial segment. The entire disk can therefore be considered a *source* for which the emf between center and rim equals $\omega B R^2 / 2$. The source can be included in a closed circuit by completing the circuit through sliding contacts of brushes b, b.

The emf in such a disk was studied by Faraday, and the device is called a *Faraday disk dynamo*.

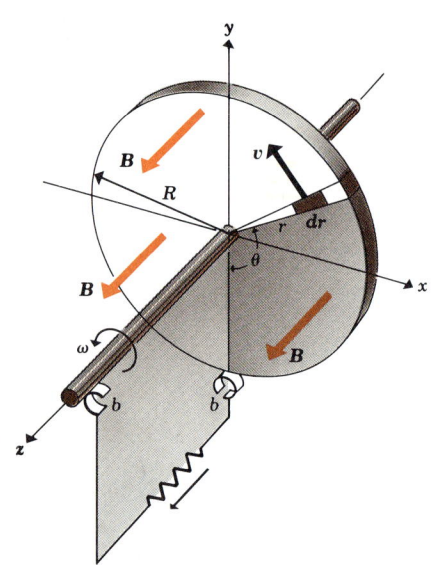

33-6 A Faraday disk dynamo. The emf is induced along radial lines of the rotating disk and is connected to an external circuit through the sliding contacts bb.

It was pointed out in Sec. 30-4 that the magnetic-field force on a moving charged particle never does any *work* on the particle because $\mathbf{F}$ and $\mathbf{v}$ are always perpendicular. Yet in defining motional emf we have spoken about the work done by the equivalent non-electrostatic field $\mathbf{E}_n = \mathbf{v} \times \mathbf{B}$. How is this apparent paradox resolved?

The resolution is somewhat subtle and cannot be discussed in detail here. Very briefly, the vertical motion of the charges in the moving conductor in Fig. 33-2 causes a transverse (horizontal) magnetic-field force and thus a transverse displacement of charge, corresponding to the Hall effect discussed in Sec. 31-2. Thus there is a transverse *electrostatic* field in the conductor, from left to right in the example of Fig. 33-2. As the conductor moves to the right, it is this electrostatic force which actually does work on the moving charges. Detailed analysis shows that this work is the same *as though* it had been done by the vertical non-electrostatic or motional force $q\mathbf{v} \times \mathbf{B}$ during the vertical displacement of the moving charges.

33-2 Faraday's law

The induced emf in the circuit of Fig. 33-2 may be considered from another viewpoint. While the conductor moves toward the right a distance ds, the *area* enclosed by the circuit $abcd$ increases by

$$dA = l\, ds,$$

and the change in magnetic *flux* through the circuit is

$$d\Phi = B\, dA = Bl\, ds.$$

The *rate of change* of flux is therefore

$$\frac{d\Phi}{dt} = \left(\frac{ds}{dt}\right) Bl = vBl. \tag{33-9}$$

But the product vBl equals the induced emf $\mathcal{E}$, so this equation states that *the induced emf in the circuit is numerically equal to the rate of change of the magnetic flux through it.* That is,

$$\mathcal{E} = \frac{d\Phi}{dt}. \tag{33-10}$$

This equation does not tell us the *sense* of the emf, that is, which end is at higher potential or the direction of the non-electrostatic force. The sense of the emf can be determined by other considerations, to be discussed later.

Equation (33–10) is known as *Faraday's law*. It may appear to be merely an alternative form of Eq. (33–3) for the emf in a moving conductor. It turns out, however, that the relation has a much broader significance than might be expected from its derivation. That is, it is found to apply to all circuits through which the flux is caused to vary, even when there is no *motion* of any part of the circuit and hence no emf directly attributable to a force on a moving charge.

Suppose, for example, that two loops of wire are located as in Fig. 33–7. A current in circuit 1 sets up a magnetic field whose magnitude at all points is proportional to this current. A part of this flux passes through circuit 2, and if the current in circuit 1 is increased or decreased, the flux through circuit 2 will also vary. Circuit 2 is not moving in a magnetic field and hence no "motional" emf is induced in it, but there is a change in the flux through it, and it is found experimentally that an emf appears in circuit 2 of magnitude $\mathcal{E} = d\Phi/dt$. In such a situation, no one portion of circuit 2 can be considered the seat of emf; the *entire circuit* constitutes the seat.

33–7 As the current in circuit 1 is varied, the magnetic flux through circuit 2 changes.

Here is another example. Suppose we set up a magnetic field within the toroidal winding of Fig. 33–8, link the toroid with a conducting ring, and vary the current in the winding of the toroid. We have shown that the flux lines set up by a steady current in a toroidal winding are wholly confined to the space enclosed by the winding; not only is the ring not *moving* in a magnetic field, but if the current were steady it would not even be *in* a magnetic field! However, field lines do pass through the area bounded by the ring, and the flux changes as the current in the windings changes. Equation (33–10) predicts an induced emf in the ring, and we find by experiment that the emf actually exists. In case the reader has not identified the apparatus in Fig. 33–8, it may be pointed out that it is merely a *transformer* with a one-turn secondary, so that the phenomenon we are now discussing is the basis of the operation of every transformer.

To sum up, then, an emf is induced in a circuit whenever the magnetic flux through the circuit varies with time. The flux may be caused to change in various ways: For example, (1) a conductor may move through a stationary magnetic field, as in Figs. 33–2 and 33–3, or (2) the magnetic field through a stationary conducting loop may change with time, as in Figs. 33–7 and 33–8. For case (1), the emf may be computed *either* from

$$\mathcal{E} = vBl$$

or from

$$\mathcal{E} = \frac{d\Phi}{dt}.$$

For case (2), the emf may be computed *only* by

$$\mathcal{E} = \frac{d\Phi}{dt}.$$

Conducting ring

33–8 An emf is induced in the ring when the flux in the toroid varies.

If we have a coil of N turns and the flux varies at the same rate through each, the induced emf's in the turns are in *series*, and the total emf is

$$\mathcal{E} = N\frac{d\Phi}{dt}. \qquad (33\text{-}11)$$

Example 1 A certain coil of wire consists of 500 circular loops of radius 4 cm. It is placed between the poles of a large electromagnet, where the magnetic field is uniform, perpendicular to the plane of the coil, and increasing at a rate of $0.2\ \text{T}\cdot\text{s}^{-1}$. What is the magnitude of the resulting induced emf?

Solution The flux Φ at any time is given by $\Phi = BA$, and the rate of change of flux by $d\Phi/dt = (dB/dt)A$. In our problem, $A = \pi \times (0.04\ \text{m})^2 = 0.00503\ \text{m}^2$ and

$$\frac{d\Phi}{dt} = (0.2\ \text{T}\cdot\text{s}^{-1})(0.00503\ \text{m}^2)$$
$$= 0.00101\ \text{T}\cdot\text{m}^2\cdot\text{s}^{-1} = 0.00101\ \text{Wb}\cdot\text{s}^{-1}.$$

From Eq. (33-11), the magnitude of the induced emf is

$$\mathcal{E} = N\frac{d\Phi}{dt} = (500)(0.00101\ \text{Wb}\cdot\text{s}^{-1}) = 0.503\ \text{V}.$$

If the coil is tilted so that a line perpendicular to its plane makes an angle of $30°$ with **B,** then only the component $B\cos 30°$ contributes to the flux through the coil. In that case the induced emf has magnitude $\mathcal{E} = (0.503\ \text{V})(\cos 30°) = 0.435\ \text{V}$.

Example 2 We consider again the rotating rectangular loop in Fig. 33-3, discussed in Example 2 of Section 33-1; this time we compute the emf from Eq. (33-10). The flux through the loop equals its area A multiplied by the component of B perpendicular to the area, that is, $B\cos\theta$. This is also equal to the magnitude B multiplied by the projected area shaded in Fig. 33-3 on a plane (the xy-plane) perpendicular to **B**.

$$\Phi = \mathbf{B}\cdot\mathbf{A} = BA\cos\theta = BA\cos\omega t,$$

Then

$$\mathcal{E} = \frac{d\Phi}{dt} = -\omega BA\sin\omega t.$$

The negative sign is associated with the direction or sense of $\mathcal{E}$, to be discussed later. Apart from this sign, the result agrees with Eq. (33-6).

Note that the maximum value of $\mathcal{E}$ occurs when $\theta = 90°$ and the flux through the loop is zero, and that $\mathcal{E} = 0$ when $\theta = 0$ and the flux is maximum. That is, the emf depends not on the flux through the loop, but on its *rate of change*.

Example 3 We consider again the rotating disk in Fig. 33-6, discussed in Example 3 of Section 33-1. To compute the emf from Faraday's law, Eq. (33-10), we consider the circuit to be the periphery of the shaded areas in Fig. 33-6. The rectangular portion in the yz-plane is fixed. The area of the shaded section in the xy-plane is $\tfrac{1}{2}R^2\theta$, and the flux through it is

$$\Phi = \tfrac{1}{2}BR^2\theta.$$

As the disk rotates, the shaded area increases. In a time dt, the angle θ increases by $d\theta = \omega\,dt$. The flux increases by

$$d\Phi = \tfrac{1}{2}Br^2\,d\theta = \tfrac{1}{2}BR^2\omega\,dt,$$

and the induced emf is

$$\mathcal{E} = \frac{d\Phi}{dt} = \tfrac{1}{2}BR^2\omega,$$

in agreement with the previous result.

Example 4 *The search coil.* A useful experimental method of measuring magnetic field strength uses a small, closely wound coil of N turns called a *search coil* or a *snatch coil*. Assume first, for simplicity, that the search coil is placed with its plane perpendicular to a magnetic field $\mathbf{B}$. If the area enclosed by the coil is A, the flux Φ through it is $\Phi = BA$. Now if the coil is quickly given a quarter-turn about one of its diameters so that its plane becomes parallel to the field, or if it is quickly snatched from its position to another where the field is known to be zero, the flux through it decreases rapidly from BA to zero. During the time that the flux is decreasing, an emf of short duration is induced in the coil and there is a momentary induced current in the external circuit to which the coil is connected. As we shall see, the total flux change is proportional to the total *charge* which flows around the circuit.

The current at any instant is

$$i = \frac{\mathcal{E}}{R},$$

where R is the combined resistance of external circuit and search coil, $\mathcal{E}$ is the instantaneous induced emf, and i the instantaneous current.

From Faraday's law,

$$\mathcal{E} = N\frac{d\Phi}{dt}, \qquad i = \frac{N}{R}\frac{d\Phi}{dt}.$$

The total charge Q which flows through the circuit is given by

$$Q = \int_0^t i\,dt = \frac{N}{R}\int d\Phi = \frac{N\Phi}{R}.$$

Thus Φ and B are given by

$$\Phi = \frac{RQ}{N} \quad \text{and} \quad B = \frac{\Phi}{A} = \frac{RQ}{NA}. \tag{33-12}$$

It is not difficult to construct an instrument that measures the total charge which passes through it. Thus if the external circuit contains such an instrument, Q may be measured directly. Equation (33-12) may then be used to compute B.

Strictly speaking, while this method gives correctly the total flux through the coil, it is only the *average* field over the area of the coil which is measured. However, if the area is sufficiently small, this approximates closely the field at, say, the center of the coil.

This discussion has assumed the plane of the coil to be initially perpendicular to the direction of the field. If one is "exploring" a field whose direction is not known in advance, the same apparatus may be used to find the direction by performing a series of experiments in which the coil is placed at a given point in the field in various orientations, and snatched out of the field from each orientation. The charge flow Q will be maximum for the particular orientation in which the plane of the coil was perpendicular to the field. Thus both the magnitude and direction of an unknown field can be found by this method.

33-3 Induced electric fields

The examples of induced emf considered thus far have all involved conductors moving in a magnetic field, but we have also pointed out, in introducing Faraday's law, that induced emf can also occur with *stationary* conductors. An example is shown in Fig. 33-9, a solenoid encircled by a conducting loop of arbitrary shape. A current I in the windings of the solenoid sets up a magnetic field B along the solenoid axis and a magnetic flux $\Phi = BA$ passes through any surface bounded by the loop. Suppose that a small galvanometer G is inserted in the loop. It is found that if the current I is changed (and hence if the flux Φ is changed), the galvanometer indicates an emf $\mathcal{E}$ in the wire *during the time that the flux is changing*, and that this emf is equal to the *time rate of change of flux* through the surface bounded by the wire. That is,

$$\mathcal{E} = \frac{d\Phi}{dt}. \qquad (33\text{-}13)$$

Equation (33-13) is Faraday's law; the essential difference between this and preceding examples is that here it is applied not to a moving conductor but to a stationary one in which there is a changing flux because the magnetic field in the area bounded by the conductor is changing.

Example Suppose the long solenoid in Fig. 33-9 is wound with 1000 turns per meter, and the current in its windings is increasing at the rate of $100\ \text{A}\cdot\text{s}^{-1}$. The cross-sectional area of the solenoid is $4\ \text{cm}^2 = 4 \times 10^{-4}\ \text{m}^2$.

The magnetic field magnitude inside the solenoid, at points not too near its ends, is given by Eq. (32-25). The flux Φ in the solenoid is

$$\Phi = BA = \mu_0 nIA.$$

The rate of change of flux is

$$\begin{aligned}
\frac{d\Phi}{dt} &= \mu_0 nA \frac{dI}{dt} \\
&= (4\pi \times 10^{-7}\ \text{Wb}\cdot\text{A}^{-1}\cdot\text{m}^{-1})(1000\ \text{turns}\cdot\text{m}^{-1}) \times \\
&\quad (4 \times 10^{-4}\ \text{m}^2)(100\ \text{A}\cdot\text{s}^{-1}) \\
&= 16\pi \times 10^{-6}\ \text{Wb}\cdot\text{s}^{-1}.
\end{aligned}$$

The magnitude of the induced emf is

$$|\mathcal{E}| = \frac{d\Phi}{dt} = 16\pi \times 10^{-6}\ \text{V} = 16\pi\ \mu\text{V}.$$

We have previously associated electromotive force with the line integral of an electric field E_n of nonelectrostatic origin, and the same point

33-9 The windings of a long solenoid carry a current I that is increasing at a rate dI/dt. The magnetic flux in the solenoid is increasing at a rate $d\Phi/dt$, and this changing flux passes through a wire loop of arbitrary size and shape. An emf $\mathcal{E}$ is induced in the loop, given by $\mathcal{E} = d\Phi/dt$.

of view must be taken here. The emf is the line integral of E_n around the loop:

$$\mathcal{E} = \oint E_n \cdot dl$$

Hence the Faraday law states that

$$\oint E_n \cdot dl = \frac{d\Phi}{dt}. \tag{33-14}$$

As an example, suppose the loop in Fig. 33-9 is a circle of radius r. Because of the axial symmetry, the non-electrostatic field E_n has the same magnitude at all points on the circle and is everywhere tangent to this circle. The line integral in Eq. (33-14) becomes simply the magnitude E_n times the circumference $2\pi r$ of the circle. Thus, the induced electric field at a distance r from the axis is given by

$$E_n = \frac{1}{2\pi r} \frac{d\Phi}{dt}, \tag{33-15}$$

where Φ is the flux through a circle of radius r.

If it should happen that at a particular instant the B field is uniform across this circle (as is *not* the case with the solenoid above), then

$$\Phi = \pi r^2 B,$$
$$\frac{d\Phi}{dt} = \pi r^2 \frac{dB}{dt},$$

and

$$E_n = \frac{1}{2\pi r} \pi r^2 \frac{dB}{dt} = \frac{r}{2} \frac{dB}{dt}. \tag{33-16}$$

In summary, Eq. (33-10) is valid for two rather different situations. In the first, an emf is induced by the magnetic forces on charges in a conductor moving through a magnetic field, while in the second a time-varying magnetic field induces an electric field of non-electrostatic nature in a stationary conductor and hence induces an emf. The E_n field in the latter case differs from an electro*static* field in two significant ways. First, its line integral around a closed path is *not* zero, so it is not a *conservative* field. In contrast, an electro*static* field is *always* conservative, as discussed in Section 26-1. Second, the non-electrostatic field is not produced by static charges; it can be shown that the surface integral of E_n over a closed surface (the same integral that appears in Gauss's law) always has the value zero, whether charges are enclosed or not.

Thus, a changing magnetic field acts as a source of electric field, but one that differs from an electrostatic field in two ways: It is nonconservative, and its integral over a closed surface is zero. The reader may also note that this situation is analogous to that of the displacement current discussed in Section 29-6, in which a changing *electric* field acts as a source of *magnetic* field. These relations thus exhibit a kind of symmetry in the behavior of the two fields. We shall return to this relationship in Chapter 37, in connection with the analysis of electromagnetic waves.

33-4 Lenz's law

We return now to the question of determining the sign or direction of an induced emf or current or the direction of the associated non-electrostatic field. A very useful principle is *Lenz's law*. H. F. E. Lenz (1804–1864) was a German scientist who, without knowledge of the work of Faraday and Henry, duplicated many of their discoveries nearly simultaneously. The law states:

> *The direction of an induced current is such as to oppose the cause producing it.*

The "cause" of the current may be the motion of a conductor in a magnetic field, or it may be the change of flux through a stationary circuit. In the first case, the direction of the induced current in the moving conductor is such that the direction of the side-thrust exerted on the conductor by the magnetic field is opposite in direction to its motion. The motion of the conductor is therefore "opposed."

In the second case, the induced current sets up a magnetic field of its own which, within the area bounded by the circuit, is (a) *opposite* to the original field if this is *increasing*, but (b) is in the *same* direction as the original field if the latter is *decreasing*. Thus it is the *change in flux* through the circuit (not the flux itself) which is "opposed" by the induced current.

In order to have an induced current, we must have a closed circuit. If a conductor does not form a closed circuit, then we mentally complete the circuit between the ends of the conductor and use Lenz's law to determine the direction of the current. The polarity of the ends of the open-circuit conductor may then be deduced.

Example 1 In Fig. 33–2, when the conductor moves to the right, a counterclockwise current is induced in the loop. The force exerted by the field on the moving conductor as a result of this current is to the left, *opposing* the conductor's motion.

Example 2 In Fig. 33–3, the direction of the induced current in the loop is the same as that of the non-electrostatic field E_n. The magnetic field exerts forces on the loop as a result of this current; the force on the right side of the loop (of length a) is in the $+x$-direction, and that on the left side is in the $-x$-direction. The resulting torque is opposite in direction to ω and thus opposes the motion.

Example 3 In Fig. 33–9, when the solenoid current is increasing, the induced current in the loop is counterclockwise. The additional field caused by this current, at points inside the loop, is opposite in direction to that of the solenoid; hence the induced current tends to oppose the increase in flux through the loop by causing flux in the opposite sense.

Lenz's law is also directly related to energy conservation. For instance, in Example 1 above, the induced current in the loop dissipates energy at the rate I^2R, and this energy must be supplied by the force that makes the conductor move despite the magnetic force opposing its motion. The work done by this applied force, in fact, must equal the energy dissipated in the circuit resistance. If the induced current were to have the opposite direction, the resulting force on the moving conductor would make it move faster and faster, violating energy conservation.

The content of Lenz's law can be incorporated into the formal statement of Faraday's law by means of a set of sign rules. We first designate one side of the surface bounded by the conducting circuit to be the positive; which side is chosen is immaterial. The positive side may be identified in diagrams by drawing a unit vector n away from the surface on the positive side.

A magnetic flux through this surface is positive if the B field has a component normal to the surface in the same direction as n, and negative if opposite. This defines a sign for $d\Phi/dt$. For example, if B has the same direction as n and is increasing in magnitude, or if the area is increasing, $d\Phi/dt$ is positive. If it has the same direction as n but is decreasing, or if the area is decreasing, $d\Phi/dt$ is negative. If the direction of B is opposite to n but its magnitude is decreasing, $d\Phi/dt$ is positive because Φ is becoming less negative. The reader may check other possibilities.

Finally, we specify the direction used to evaluate $\oint E_n \cdot dl$; we choose the direction around the boundary circuit defined by applying the right-hand rule to n. Thus if the circuit lies in the plane of this page and n points toward the reader, the circuit is to be traversed in the counterclockwise direction. With these conventions, the correct statement of Faraday's law is

$$\mathcal{E} = -\frac{d\Phi}{dt}. \qquad (33\text{--}17)$$

This differs from Eqs. (33–10) and (33–13) in the negative sign.

Example 4 In Fig. 33–2 let us take n as parallel to B, away from the reader. Then Φ and $d\Phi/dt$ are both positive. The integral $\oint E_n \cdot dl$ is to be taken clockwise around the loop. E_n is different from zero only along the line ab, where its direction is upward. Thus

$$\oint E_n \cdot dl = -E_n l.$$

This quantity is negative because E_n and dl have opposite directions, and it is consistent with Eq. (33–17), in which both sides are positive. Alternatively, we could have taken n as toward the reader. In that case Φ and $d\Phi/dt$ are both negative, the loop is traversed in the counterclockwise sense, and $\oint E_n \cdot dl$ is positive. Either way, consistency with Eq. (33–17) requires that the direction of E_n must be from b to a, and this is the direction of current in the moving conductor.

Example 5 In Fig. 33–9, suppose the current has the direction shown and is decreasing. What is the direction of the induced current in the loop?

Solution Take the normal to the loop's surface as pointing to the right, parallel to B. Then the flux is positive but decreasing, and $d\Phi/dt$ is negative. Hence, according to Eq. (33–17), when we calculate $\mathcal{E} = \oint E_n \cdot dl$ by traversing the loop clockwise (looking in the direction of B) the result must be positive. Hence E_n points clockwise around the

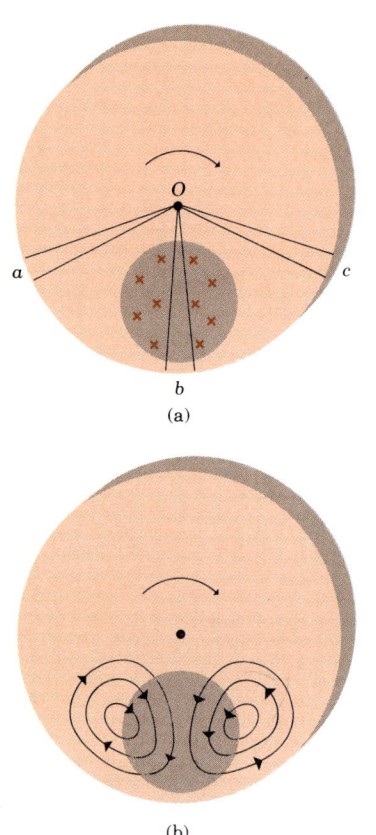

33-10 Eddy currents in a rotating disk.

33-5 Eddy currents

Thus far we have considered only instances in which the currents resulting from induced emf's were confined to well-defined paths provided by the wires and apparatus of the external circuit. In many pieces of electrical equipment, however, one finds masses of metal moving in a magnetic field or located in a changing magnetic field, with the result that induced currents circulate throughout the volume of the metal. Because of their general circulatory nature, these are referred to as *eddy currents*.

Consider a disk rotating in a magnetic field perpendicular to the plane of the disk but confined to a limited portion of its area, as in Fig. 33-10a. Element Ob is moving across the field and has an emf induced in it. Elements Oa and Oc are not in the field but, in common with all other elements located outside the field, provide return conducting paths along which charges displaced along Ob can return from b to O. A general eddy-current circulation is therefore set up in the disk, somewhat as sketched in Fig. 33-10b.

The currents in the neighborhood of radius Ob experience a side thrust that *opposes* the motion of the disk, while the return currents, since they lie outside the field, do not experience such a thrust. The interaction between the eddy currents and the field therefore results in a braking action on the disk. The apparatus finds some technical applications and is known as an "eddy current brake."

As a second example of eddy currents, consider the core of an alternating current transformer, shown in Fig. 33-11. The alternating current in the primary winding P sets up an alternating flux within the core, and an induced emf develops in the secondary winding S because of the continual change in flux through it. The iron core, however, is also a conductor, and any section such as that at AA can be thought of as a number of closed conducting circuits, one within the other. The flux through each of these circuits is continually changing, so that there is an eddy-current circulation in the entire volume of the core, the lines of flow lying in planes perpendicular to the flux. These eddy currents are very undesirable both because of the energy they dissipate and because of the flux they themselves set up.

33-11 Reduction of eddy currents by use of a laminated core.

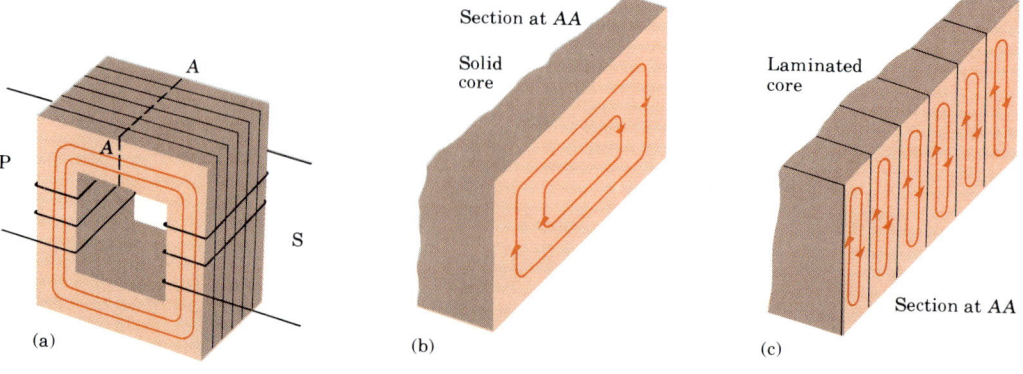

In all actual transformers, the eddy currents are greatly reduced by the use of a *laminated* core, that is, one built up of thin sheets or laminae. The electrical resistance between the surfaces of the laminations (due either to a natural coating of oxide or to an insulating varnish) effectively confines the eddy currents to individual laminae. The resulting length of path is greatly increased, with consequent increase in resistance. Hence, although the induced emf is not altered, the currents and their heating effects are minimized.

In small transformers where eddy-current losses must be kept to an absolute minimum, the cores are sometimes made of *ferrites*, which are complex oxides of iron and other metals. These materials are ferromagnetic but have relatively high resistivity.

Questions

33-1 In most parts of the northern hemisphere the earth's magnetic field has a vertical component directed *into* the earth. An airplane flying east generates an emf between its wingtips. Which wingtip acquires an excess of electrons and which a deficiency?

33-2 A sheet of copper is placed between the poles of an electromagnet, so the magnetic field is perpendicular to the sheet. When it is pulled out, a considerble force is required, and the force required increases with speed. What's happening?

33-3 In Fig. 33-4, if the angular velocity ω of the loop is doubled, the frequency with which the induced current changes direction doubles, and the maximum emf also doubles. Why? Does the torque required to turn the loop change?

33-4 Comparing the conventional dc generator (Fig. 33-5) and the Faraday disk dynamo (Fig. 33-6), what are some advantages and disadvantages of each?

33-5 Some alternating-current generators operate on a principle similar to that shown in Fig. 33-4 but using a rotating permanent magnet and stationary coils. What advantages does this scheme have? What disadvantages?

33-6 When a conductor moves through a magnetic field, the magnetic forces on the charges in the conductor cause an emf. But if this phenomenon is viewed in a frame of reference moving with the conductor, there is no motion, but there is still an emf. How is this paradox resolved?

33-7 Two circular loops lie adjacent to each other. One is connected to a source that supplies an increasing current; the other is a simple closed ring. Is the induced current in the ring in the same direction as that in the ring connected to the source, or opposite? What if the current in the first ring is decreasing?

33-8 A farmer claimed that the high-voltage transmission lines running parallel to his fence induced dangerously large voltages on the fence. Is this within the realm of possibility?

33-9 Small one-cylinder gasoline engines sometimes use a device called a *magneto* to supply current to the spark plug. A permanent magnet is attached to the flywheel, and there is a stationary coil mounted adjacent to it. What happens when the magnet passes the coil?

33-10 A current-carrying conductor passes through the center of a metal ring, perpendicular to its plane. If the current in the conductor increases, is a current induced in the ring?

33-11 A student asserted that, if a permanent magnet is dropped down a vertical copper pipe, it eventually reaches a terminal velocity, even if there is no air resistance. Why should this be? Or should it?

Problems

33-1 In Fig. 33-2, a rod of length $l = 0.40$ m moves in a magnetic field of magnitude $B = 1.2$ T. The emf induced in the moving rod is found to be 2.40 V.

a) What is the speed of the rod?
b) If the total circuit resistance is 1.2 Ω, what is the induced current?
c) What force (magnitude and direction) does the field exert on the rod as a result of this current?

33-2 In Fig. 33-1 a rod of length $l = 0.25$ m moves with constant speed of 6.0 m·s^{-1} in the direction shown. The induced emf is found to be 3.0 V.

a) What is the magnitude of the magnetic field?
b) Which point is at higher potential, a or b?
c) What is the direction of the electrostatic field in the rod? The non-electrostatic field?
d) What is the magnitude of the non-electrostatic field?

33-3 In Fig. 33-1, let $l = 1.5$ m, $B = 0.5$ T, $v = 4$ m·s^{-1}.

a) What is the motional emf in the rod?

b) What is the potential difference between its terminals?

c) Which end is at the higher potential?

33-4 The cube in Fig. 33-12, 1 m on a side, is in a uniform magnetic field of 0.2 T, directed along the positive y-axis. Wires A, C, and D move in the directions indicated, each with a speed of 0.5 m·s^{-1}.

a) What is the motional emf in each wire?

b) What is the potential difference between the terminals of each?

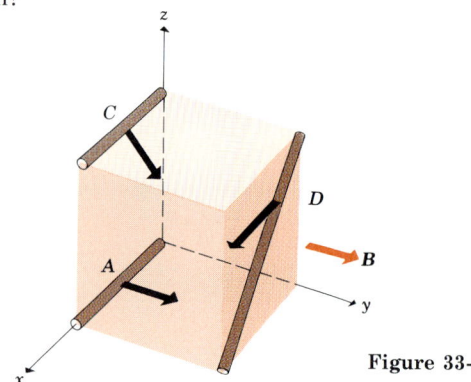

Figure 33-12

33-5 A conducting rod AB in Fig. 33-13 makes contact with the metal rails CA and DB. The apparatus is in a uniform magnetic field 0.5 T, perpendicular to the plane of the diagram.

a) Find the magnitude and direction of the emf induced in the rod when it is moving toward the right with a speed 4 m·s^{-1}.

b) If the resistance of the circuit $ABCD$ is 0.2 Ω (assumed constant), find the force required to maintain the rod in motion. Neglect friction.

c) Compare the rate at which mechanical work is done by the force (Fv) with the rate of development of heat in the circuit (i^2R).

Figure 33-13

33-6 A closely wound rectangular coil of 50 turns has dimensions of 12 cm × 25 cm. The plane of the coil is rotated from a position where it makes an angle of 45° with a magnetic field 2 T to a position perpendicular to the field in time $t = 0.1$ s. What is the average emf induced in the coil?

33-7 A coil of 1000 turns enclosing an area of 20 cm^2 is rotated from a position where its plane is perpendicular to the earth's magnetic field to one where its plane is parallel to the field, in 0.02 s. What average emf is induced if the earth's magnetic field is 6×10^{-5} T?

33-8 A square loop of wire with resistance R is moved at constant velocity v across a uniform magnetic field confined to a square region whose sides are twice the length of those of the square loop. (See Fig. 33-14.)

a) Sketch a graph of the external force F needed to move the loop at constant velocity, as a function of the distance x, from $x = -2l$ to $x = +2l$.

b) Sketch a graph of the induced current in the loop as a function of x, plotting clockwise currents upward and counterclockwise currents downward.

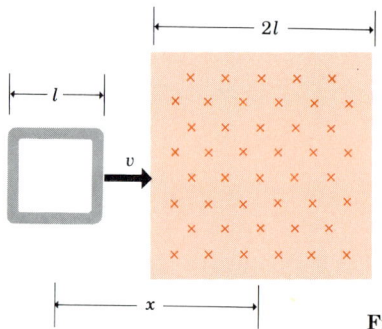

Figure 33-14

33-9 The long rectangular loop in Fig. 33-15, of width l, mass m, and resistance R, starts from rest in the position shown, and is acted on by a constant force $\mathbf{F}$. At all points in the colored area there is a uniform magnetic field $\mathbf{B}$, perpendicular to the plane of the diagram.

a) Sketch a graph of the velocity of the loop as a function of time.

b) Find the terminal velocity.

Figure 33-15

33-10 A slender rod 1 m long rotates about an axis through one end and perpendicular to the rod, with an angular velocity of 2 rev·s^{-1}. The plane of rotation of the rod is perpendicular to a uniform magnetic field 0.5 T.

a) What is the induced emf in the rod?

b) What is the potential difference between its terminals?

33-11 A flat square coil of 10 turns has sides of length 12 cm. The coil rotates in a magnetic field 0.025 T.

a) What is the angular velocity of the coil if the maximum emf produced is 20 mV?

b) What is the average emf at this velocity?

33-12 A rectangular coil of wire having 10 turns with dimensions of 20 cm × 30 cm rotates at a constant speed of 600 rpm in a magnetic field 0.10 T. The axis of rotation is perpendicular to the field. Find the maximum emf produced.

33-13 The rectangular loop in Fig. 33-16, of area A and resistance R, rotates at uniform angular velocity ω about the y-axis. The loop lies in a uniform magnetic field $\mathbf{B}$ in the direction of the x-axis. Sketch the following graphs:

a) the flux Φ through the loop as a function of time (let $t = 0$ in the position shown in Fig. 33-16);

b) the rate of change of flux $d\Phi/dt$;

c) the induced emf in the loop;

d) the torque Γ needed to keep the loop rotating at constant angular velocity;

e) the induced emf if the angular velocity is doubled. (Neglect the self-inductance of the loop.)

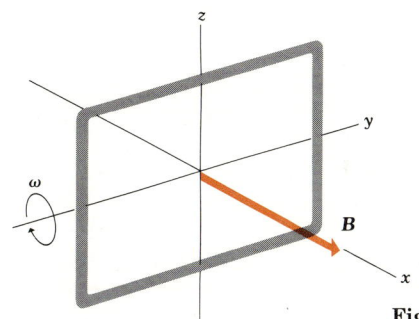

Figure 33-16

33-14 In Problem 33-13 and Fig. 33-16, let $A = 400$ cm², $R = 2\,\Omega$, $\omega = 10$ rad·s⁻¹, $B = 0.5$ T. Find

a) the maximum flux through the loop,

b) the maximum induced emf,

c) the maximum torque.

d) Show that the work of the external torque in one revolution is equal to the energy dissipated in the loop.

33-15 Suppose the loop in Fig. 33-16 is

a) rotated about the z-axis;

b) rotated about the x-axis;

c) rotated about an edge parallel to the y-axis.

What is the maximum induced emf in each case, if the angular velocity is the same as in Problem 33-14?

33-16 A flexible circular loop 10 cm in diameter lies in a magnetic field 1.2 T, directed into the plane of the diagram in Fig. 33-17. The loop is pulled at the points indicated by the arrows, forming a loop of zero area in 0.2 s.

a) Find the average induced emf in the circuit.

b) What is the direction of the current in R?

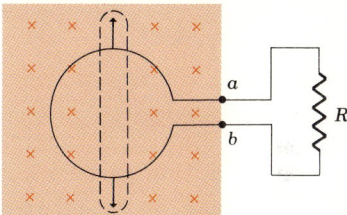

Figure 33-17

33-17 The magnetic field within a long, straight solenoid of circular cross section and radius R is increasing at a rate dB/dt.

a) What is the rate of change of flux through a circle of radius r_1 inside the solenoid, normal to the axis of the solenoid, and with center on the solenoid axis?

b) Find the induced electric field $\mathbf{E}_n$ inside the solenoid, at a distance r_1 from its axis. Show the direction of this field in a diagram.

c) What is the induced electric field *outside* the solenoid, at a distance r_2 from the axis?

d) Sketch a graph of the magnitude of $\mathbf{E}_n$ as a function of the distance r from the axis, from $r = 0$ to $r = 2R$. Compare with part (e) of Problem 32-24.

e) What is the induced emf in a circular turn of radius $R/2$?

f) Of radius R?

g) Of radius $2R$?

33-18 A long, straight solenoid of cross-sectional area 6 cm² is wound with 10 turns of wire per centimeter, and the windings carry a current of 0.25 A. A secondary winding of 2 turns encircles the solenoid. When the primary circuit is opened, the magnetic field of the solenoid becomes zero in 0.05 s. What is the average induced emf in the secondary?

33-19 The magnetic field B at all points within the colored circle of Fig. 33-18 is 0.5 T. It is directed into the plane of the diagram and is decreasing at the rate of 0.1 T·s⁻¹.

a) What is the shape of the field lines of the induced E_n-field in Fig. 33-18, within the colored circle?

b) What are the magnitude and direction of this field at any point of the circular conducting ring of radius 10 cm, and what is the emf in the ring?

c) What is the current in the ring, if its resistance is $2\,\Omega$?

d) What is the potential difference between points a and b of the ring?

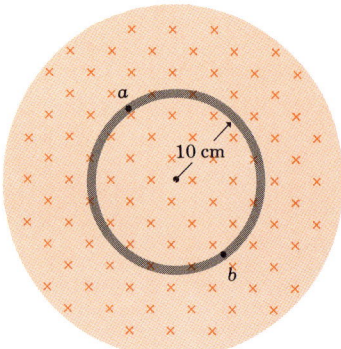

Figure 33-18

e) How do you reconcile your answers to (c) and (d)?

f) If the ring is cut at some point and the ends separated slightly, what will be the potential difference between the ends?

33-20 A square conducting loop, 20 cm on a side, is placed in the same magnetic field as in Problem 33-19. (See Fig. 33-19.)

a) Copy Fig. 33-19, and show by vectors the directions and relative magnitudes of the induced electric field E_n at points a, b, and c.

b) Prove that the component of E_n along the loop has the same value at every point of the loop and is equal to that at the ring of Fig. 33-18.

c) What is the current in the loop, if its resistance is 2 Ω?

d) What is the potential difference between points a and b?

Figure 33-19

33-21 A square conducting loop, 20 cm on a side, is placed in the same magnetic field as in Problem 33-19 with side ac along a diameter and with point b at the center of the field. (See Fig. 33-20.)

a) Copy Fig. 33-20, and show by vectors the directions and relative magnitudes of the induced electric field E_n at the lettered points.

b) What is the induced emf in side ac?

c) What is the induced emf in the loop?

d) What is the current in the loop, if its resistance is 2 Ω?

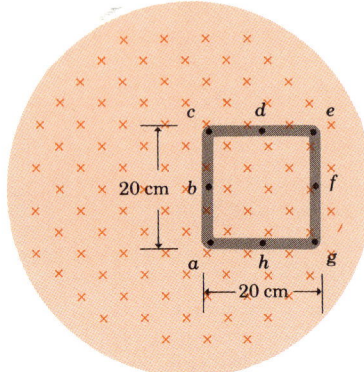

Figure 33-20

e) What is the potential difference between points a and c? Which is the higher potential?

33-22

a) Find the induced field E_n in each side of the square loop in Problem 33-21 and Fig. 33-20.

b) Find the induced emf in each side.

c) Find the *electrostatic* field E_e in each side.

d) Find the potential differences V_{ac}, V_{ce}, V_{eg}, and V_{ga}. What should be the sum of these potential differences?

33-23

a) What is the direction of the drift velocity of an electron at point b in the wire loop of Fig. 33-20?

b) What would be the direction of the drift velocity if the loop were at the left side of the center line instead of at the right side?

c) What would be the direction of the force on an electron if it were at rest at the center of the magnetic field, with no wire loop present?

d) How do you reconcile the answers to (a), (b), and (c)?

33-24 The magnetic flux in a toroid of small cross-sectional area and radius R is increasing at a constant rate $d\Phi/dt$.

a) What are the magnitude and direction of the induced E_n-field at a point on the axis of the toroid at a distance x from its center? (See Sec. 32-6 for the corresponding expression for the B-field at a point on the axis of a circular loop of wire carrying a constant current.)

b) Sketch a graph of the magnitude of E_n as a function of x.

c) Evaluate $\int_{-\infty}^{+\infty} E_n \, dx$ to find the induced emf in a wire that extends along the axis of the toroid from $x = -\infty$ to $x = +\infty$.

d) If the ends of the wire are joined by a conductor very far from the toroid, what is the induced emf in this circuit?

e) What is the induced emf in a ring that links the toroid closely?

Figure 33-21

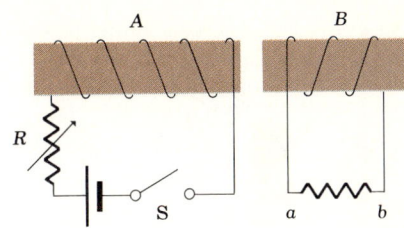

Figure 33-23

33-25 The current in the wire AB of Fig. 33-21 is upward and is increasing steadily at a rate dI/dt.

a) At an instant when the current is i, what are the magnitude and direction of the field B at a distance r from the wire?
b) What is the flux $d\Phi$ through the narrow shaded strip?
c) What is the total flux through the loop?
d) What is the induced emf in the loop?
e) Evaluate the numerical value of the induced emf if $a = 10$ cm, $b = 30$ cm, $l = 20$ cm, and $dI/dt = 2$ A·s^{-1}.

33-26 A cardboard tube is wound with two windings of insulated wire, as in Fig. 33-22. Terminals a and b of winding A may be connected to a battery through a reversing switch. State whether the induced current in the resistor R is from left to right, or from right to left, in the following circumstances:

a) the current in winding A is from a to b and is increasing;
b) the current is from b to a and is decreasing;
c) the current is from b to a and is increasing.

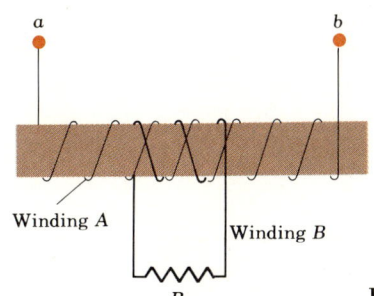

Figure 33-22

33-27 Using Lenz's law, determine the direction of the current in resistor ab of Fig. 33-23 when

a) switch S is opened,
b) coil B is brought closer to coil A,
c) the resistance of R is decreased.

33-28 The cross-section area of a closely wound search coil having 20 turns is 1.5 cm^2 and its resistance is 4 Ω. The coil is connected through leads of negligible resistance to a charge-measuring instrument having internal resistance 16 Ω. Find the quantity of charge displaced when the coil is pulled quickly out of a region where $B = 1.8$ T to a point where the magnetic field is zero. The plane of the coil, when in the field, makes an angle of 90° with the magnetic field.

33-29 A solenoid 50 cm long and 8 cm in diameter is wound with 500 turns. A closely wound coil of 20 turns of insulated wire surrounds the solenoid at its midpoint, and the terminals of the coil are connected to a charge-measuring instrument. The total circuit resistance is 25 Ω.

a) Find the quantity of charge displaced through the instrument when the current in the solenoid is quickly decreased from 3 A to 1 A.
b) Draw a sketch of the apparatus, showing clearly the directions of windings of the solenoid and coil, and of the current in the solenoid. What is the direction of the current in the coil when the solenoid current is decreased?

33-30 A closely wound search coil has an area of 4 cm^2, 160 turns, and a resistance of 50 Ω. It is connected to a charge-measuring instrument whose resistance is 30 Ω. When the coil is rotated quickly from a position parallel to a uniform magnetic field to one perpendicular to the field, the instrument indicates a charge of 4×10^{-5} C. What is the magnitude of the field?

33-31 A Faraday disk dynamo is to be used to supply current to a large electromagnet that requires 20,000 A at 1.0 V. The disk is to be 0.6 m in radius, and it turns in a magnetic field of 1.2 T, supplied by a smaller electromagnet.

a) How many revolutions per second must the disk turn?
b) What torque is required to turn the disk, assuming that all the mechanical energy is dissipated as heat in the electromagnet?

33-32 A search coil used to measure magnetic fields is to be made with a radius of 2 cm. It is to be designed so that flipping it 180° in a field of 0.1 T causes a total charge of 10^{-4} C to flow in a charge-measuring instrument when the total circuit resistance is 50 Ω. How many turns should the coil have?

33-33 A coil 4 cm in radius, containing 500 turns, turns with constant angular velocity about an axis along a diameter perpendicular to the earth's magnetic field, which may be taken as 0.5×10^{-4} T. What angular velocity must it have for the induced emf to have a maximum value of 1.0×10^{-3} V?

33-34 The coil described in Problem 33-33 is placed in a magnetic field which varies with time according to $B = 0.01\,t + (2 \times 10^{-4})t^3$, where B is in teslas and t in seconds. It is connected to a 500-Ω resistor.

a) Find the induced emf in the coil as a function of time.

b) What is the current in the resistor at time $t = 10$ s?

INDUCTANCE 34

34-1 Mutual inductance

An emf is induced in a stationary circuit whenever the magnetic flux through the circuit varies with time. If the variation in flux is brought about by a varying current in a second circuit, it is convenient to express the induced emf in terms of the varying *current*, rather than in terms of the varying *flux*. We shall use the symbol i to represent the instantaneous value of a varying current.

Figure 34-1 is a sectional view of two closely wound coils of wire. A current i_1 in coil 1 sets up a magnetic field, as indicated by the color lines, and some of these lines pass through coil 2. Let the resulting flux through coil 2 be Φ_2. The magnetic field is proportional to i_1, so Φ_2 is also proportional to i_1. When i_1 changes, Φ_2 changes; this changing flux induces an emf $\mathcal{E}_2$ in coil 2, given by

$$\mathcal{E}_2 = N_2 \frac{d\Phi_2}{dt}. \tag{34-1}$$

The proportionality of Φ_2 and i_1 could be represented in the form $\Phi_2 = (\text{constant})\, i_1$, but it is more convenient to include the number of turns N_2 in the relation. Introducing a proportionality constant M, we write

$$N_2 \Phi_2 = M i_1, \tag{34-2}$$

From this,

$$N_2 \frac{d\Phi_2}{dt} = M \frac{di_1}{dt},$$

and Eq. (34-1) may be rewritten

$$\mathcal{E}_2 = M \frac{di_1}{dt}. \tag{34-3}$$

The constant M, which depends only on the geometry of the two coils,

34-1 A portion of the flux set up by a current in coil 1 links with coil 2.

is called their *mutual inductance*. It is defined by Eq. (34–2), which may be rewritten

$$M = \frac{N_2 \Phi_2}{i_1}. \qquad (34\text{–}4)$$

The entire discussion can be repeated for the case where a changing current i_2 in coil 2 causes a changing flux Φ_1 and hence an emf $\mathcal{E}_1$ in coil 1. It might be expected that the constant M would be different in this case, since the two coils are not, in general, symmetric. It turns out, however, that M is always the same in this case as in the case considered above, so a single value of the mutual inductance characterizes completely the induced-emf interaction of two coils.

The SI unit of mutual inductance, from Eq. (34–4), is *one weber per ampere*. An equivalent unit, obtained by reference to Eq. (34–3), is *one volt-second per ampere*. These two equivalent units are called *one henry* (1 H), in honor of Joseph Henry (1797–1878), a pioneer in the development of electromagnetic theory and, incidentally, the first director of the Smithsonian Institution. Thus the unit of mutual inductance is

$$1\text{ H} = 1\text{ Wb}\cdot\text{A}^{-1} = 1\text{ V}\cdot\text{s}\cdot\text{A}^{-1}.$$

Example A long solenoid of length l and cross-sectional area A is closely wound with N_1 turns of wire. A small coil of N_2 turns surrounds it at its center, as in Fig. 34–2. A current i_1 in the solenoid sets up a **B** field at its center, of magnitude

$$B = \mu_0 n i_1 = \frac{\mu_0 N_1 i_1}{l}.$$

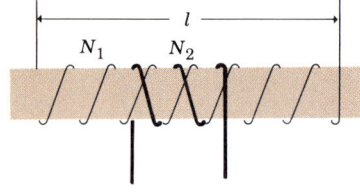

34–2

The flux through the central section is equal to BA, and since all of this flux links with the small coil, the mutual inductance is

$$M = \frac{N_2 \Phi_2}{i_1} = \frac{N_2}{i_1}\left(\frac{\mu_0 N_1 i_1}{l}\right)A = \frac{\mu_0 A N_1 N_2}{l}.$$

If $l = 0.50$ m, $A = 10$ cm$^2 = 10^{-3}$m^2, $N_1 = 1000$ turns, $N_2 = 10$ turns,

$$M = \frac{(4\pi \times 10^{-7}\text{ Wb}\cdot\text{A}^{-1}\cdot\text{m}^{-1})(10^{-3}\text{m}^2)(1000)(10)}{0.5\text{ m}}$$

$$= 25.1 \times 10^{-6}\text{ Wb}\cdot\text{A}^{-1} = 25.1 \times 10^{-6}\text{ H} = 25.1\ \mu\text{H}.$$

34–2 Self-inductance

In the preceding section, the circuit acting as the source of the magnetic field linking a circuit was assumed to be independent of the circuit in which the induced emf appeared. But whenever there is a current in any circuit, this current sets up a magnetic field that links with *the same* circuit and varies when the current varies. Hence any circuit in which there is a varying current has induced in it an emf due to the variation in *its own* magnetic field. Such an emf is called a *self-induced electromotive force*.

Suppose that a circuit has N turns of wire and that a flux Φ passes through each turn, as in Fig. 34–3. In analogy to Eq. (34–4), we define

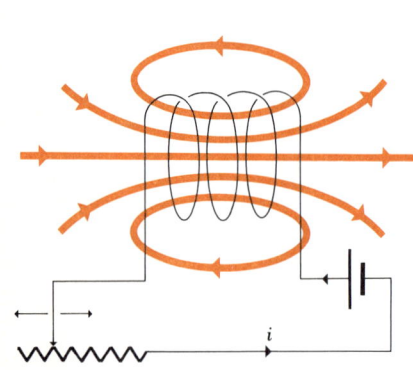

34–3 A flux Φ linking a coil of N turns. When the current in the circuit changes, the flux changes also, and a self-induced emf appears in the circuit.

the *self-inductance* L of the circuit:

$$L = \frac{N\Phi}{i}. \tag{34-5}$$

This can be written as

$$N\Phi = Li.$$

If Φ and i change with time, then

$$N\frac{d\Phi}{dt} = L\frac{di}{dt},$$

and, since the self-induced emf $\mathcal{E}$ is

$$\mathcal{E} = N\frac{d\Phi}{dt},$$

it follows that

$$\mathcal{E} = L\frac{di}{dt}. \tag{34-6}$$

The self-inductance of a circuit is therefore *the self-induced emf per unit rate of change of current*. The SI unit of self-inductance is 1 henry.

A circuit, or part of a circuit, that has inductance is called an *inductor*. An inductor is represented by the symbol

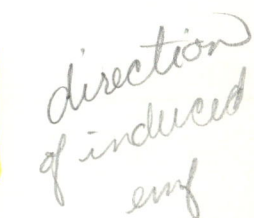

The *direction* of a self-induced (non-electrostatic) field can be found from Lenz's law. We consider the "cause" of this field, and hence of the emf associated with it, to be the *changing current* in the conductor. If the current is *increasing*, the direction of the induced field is *opposite* to that of the current. If the current is *decreasing*, the induced field is in the *same* direction as the current. Thus it is the *change* in current, not the current itself, that is "opposed" by the induced field.

Example 1 An air-core toroidal solenoid of cross-section area A and mean radius r is closely wound with N turns of wire. We neglect the variation of B across the cross section, assuming its average value to be very nearly equal to the value at the center of the cross section, as given by Eq. (32-26). Then the flux in the toroid is

$$\Phi = BA = \frac{\mu_0 NiA}{2\pi r}.$$

Since all of the flux links with each turn, the self-inductance is

$$L = \frac{N\Phi}{i} = \frac{\mu_0 N^2 A}{2\pi r}.$$

Thus, if $N = 100$ turns, $A = 10 \text{ cm}^2 = 10^{-3} \text{m}^2$, $r = 0.10$ m,

$$L = \frac{(4\pi \times 10^{-7} \text{ Wb}\cdot\text{A}^{-1}\cdot\text{m}^{-1})(100)^2(10^{-3}\text{m}^2)}{2\pi(0.10 \text{ m})}$$
$$= 20 \times 10^{-6} \text{ H} = 20\mu\text{H}.$$

Example 2 If the current in the coil above increases uniformly from zero to 1 A in 0.1 s, find the magnitude and direction of the self-induced emf.

Solution

$$\mathcal{E} = L\frac{di}{dt} = (20 \times 10^{-6} \text{ H})\frac{1 \text{ A}}{0.1 \text{ s}}$$
$$= 2.0 \times 10^{-4} \text{ V}.$$

Since the current is increasing, the direction of this emf is opposite to that of the current. For example, if the inductor terminals are a and b and an increasing current is in the direction a to b in the inductor, then E_n and the emf in the inductor are in the direction b to a, tending to cause a current *in the external circuit* in the direction a to b.

When Kirchhoff's loop rule (Sec. 28-4) is used with circuits containing inductors, this emf is treated as though it were a potential difference, with a at higher potential than b. That is, if the defined positive direction of the current i is from a to b in the inductor, then

$$V_{ab} = L\frac{di}{dt}. \qquad (34\text{-}7)$$

The self-inductance of a circuit depends on its size, shape, number of turns, etc. It also depends on the magnetic properties of the material enclosed by the circuit. For example, the self-inductance of a solenoid of given dimensions is much greater if it has an iron core than if it is in vacuum, and when ferromagnetic materials are present, the self-inductance varies in a complicated way as the current varies. For simplicity, we shall consider here only circuits of constant self-inductance.

34-3 Energy in an inductor

Because a changing current in an inductor causes an emf, the source supplying the current must maintain a potential difference between its terminals and hence must supply energy to the inductor. When an inductor carries an instantaneous current i which is changing at the rate di/dt, the induced emf is equal to $L\,di/dt$, and the power P supplied to the inductor is

$$P = \mathcal{E}i = Li\frac{di}{dt}.$$

The energy dW supplied in time dt is $P\,dt$, or

$$dW = Li\,di,$$

and the total energy supplied while the current increases from zero to a final value I is

$$W = L\int_0^I i\,di = \tfrac{1}{2}LI^2. \qquad (34\text{-}8)$$

After the current has reached its final steady value, $di/dt = 0$, and the power input is zero. The energy that has been supplied to the inductor is used to establish the magnetic field in and around the inductor, where it is stored as a form of potential energy as long as the current is maintained. When the current is reduced to zero, this energy is returned to the circuit which supplied it. If the current is suddenly interrupted by opening a switch, the energy may be dissipated in an arc across the switch contacts.

The energy can be considered as associated with the magnetic field itself, and a relationship can be developed which is analogous to that obtained for electric-field energy in Sec. 27–4, Eq. (27–9). As in that discussion, we consider only one simple case, the toroidal solenoid. As in the preceding example, we assume that the cross-section area A is small enough so the magnetic field may be considered constant over the area, and so the volume in the toroid is approximately equal to the circumferential length $l = 2\pi r$ multiplied by the area A. From the preceding example, the self-inductance of the toroidal solenoid is

$$L = \frac{\mu_0 N^2 A}{l},$$

and the energy stored in the toroid when the current in the windings is I is

$$W = \frac{1}{2}LI^2 = \frac{1}{2}\left(\frac{\mu_0 N^2 A}{l}\right)I^2.$$

We can think of this energy as localized in the volume enclosed by the windings, equal to lA. The energy *per unit volume* u is then

$$u = \frac{W}{lA} = \frac{1}{2}\mu_0\left(\frac{N^2 I^2}{l^2}\right).$$

But $N^2 I^2 / l^2 = B^2/\mu_0^2$, so

$$u = \frac{B^2}{2\mu_0}, \tag{34-9}$$

which is the analog of the expression for the energy per unit volume in the electric field of an air capacitor, $\frac{1}{2}\epsilon_0 E^2$, discussed in Sec. 27–4.

34–4 The *R-L* circuit

Every inductor necessarily has some resistance (unless its windings are superconducting). To distinguish between the effects of the resistance R and the self-inductance L, we represent the inductor as in Fig. 34–4, replacing it with an ideal resistanceless inductor in series with a noninductive resistor. The same diagram can also represent a resistor in series with an inductor, in which case R is the *total* resistance of the combination. By means of the switch, the *R-L* circuit may be connected to a source having constant terminal voltage V, or it may be shorted by the conductor across the lower switch terminals.

Suppose the switch in the diagram is suddenly closed in the "up" position. Because of the self-induced emf, the current does not immediately rise to its final value at the instant the circuit is closed, but grows at a rate that depends on the inductance and resistance of the circuit.

At some instant after the switch is closed, let i represent the current in the circuit and di/dt its rate of increase. The potential difference

34–4

across the inductor is

$$v_{cb} = L\frac{di}{dt},$$

and that across the resistor is

$$v_{ac} = iR.$$

Since $V = v_{ac} + v_{cb}$, it follows that

$$V = L\frac{di}{dt} + iR. \tag{34-10}$$

The rate of increase of current is therefore

$$\frac{di}{dt} = \frac{V - iR}{L} = \frac{V}{L} - \frac{R}{L}i. \tag{34-11}$$

At the instant the circuit is first closed, $i = 0$ and the current starts to grow at the rate

$$\left(\frac{di}{dt}\right)_{\text{initial}} = \frac{V}{L}.$$

The greater the self-inductance L, the more slowly does the current start to increase.

As the current increases, the term Ri/L increases also, and hence the *rate* of increase of current becomes smaller and smaller, as Eq. (34-11) shows. When the current reaches its final *steady-state* value I, its rate of increase is zero. Then

$$0 = \frac{V}{L} - \left(\frac{R}{L}\right)I$$

and

$$I = \frac{V}{R}.$$

That is, the *final* current does not depend on the self-inductance and is the same as it would be in a pure resistance R connected to a source having emf V.

To obtain an expression for the current as a function of time, we proceed just as we did for the problem of the charging capacitor, discussed in Sec. 29-5. We first rearrange Eq. (34-11):

$$\frac{di}{(V/R) - i} = \frac{R}{L}dt.$$

Integrating both sides, we find

$$-\ln\left(\frac{V}{R} - i\right) = \left(\frac{R}{L}\right)t + \text{constant}.$$

The integration constant is evaluated by noting that the initial current is zero, so $i = 0$ at time $t = 0$.

$$\text{constant} = -\ln\frac{V}{R}.$$

Rearranging again, we obtain

$$\ln\left(\frac{V}{R} - i\right) - \ln\frac{V}{R} = \ln\left(1 - \frac{Ri}{V}\right) = -\left(\frac{R}{L}\right)t,$$

$$i = \frac{V}{R}(1 - e^{-(R/L)t}). \qquad (34\text{-}12)$$

From this,

$$\frac{di}{dt} = \frac{V}{L}e^{-(R/L)t}. \qquad (34\text{-}13)$$

We note that at time $t = 0$, $i = 0$ and $di/dt = V/L$, and that as $t \to \infty$, $i \to V/R$ and $di/dt \to 0$, as predicted above.

Figure 34-5a is a graph of Eq. (34-12) and shows the variation of current with time. The instantaneous current i first rises rapidly, then increases more slowly and approaches asymptotically the final value $I = V/R$. At a time τ equal to L/R the current has risen to $(1 - 1/e)$ or about 0.63 of its final value. This time is called the *time constant* or the *decay constant* for the circuit.

$$\tau = \frac{L}{R}. \qquad (34\text{-}14)$$

The time required for the current to reach *half* its final value is obtained by setting

$$e^{-(R/L)t} = \tfrac{1}{2},$$

from which $T_{1/2} = (L/R)\ln 2$. For a given value of R, these times increase when the inductance L increases. Thus, although the graph of i vs. t has the same general shape whatever the inductance, the current rises rapidly to its final value if L is small, and slowly if L is large. For example, if $R = 100\ \Omega$ and $L = 10$ H,

$$\frac{L}{R} = \frac{10\text{ H}}{100\ \Omega} = 0.1\text{ s},$$

and the current increases to about 63% of its final value in 0.1 s. On the other hand, if $L = 0.01$ H,

$$\frac{L}{R} = \frac{0.01\text{ H}}{100\ \Omega} = 10^{-4}\text{ s},$$

and only 10^{-4} s is required for the current to increase to 63% of its final value.

If there is a steady current I in the circuit of Fig. 34-4 and the switch is then quickly thrown to the "down" position, the current decays, as shown in Fig. 34-5(b). The equation of the decaying current is

$$i = Ie^{-Rt/L}, \qquad (34\text{-}15)$$

and the time constant, L/R, is the time for the current to decrease to $1/e$ or about 37% of its original value. Derivation of Eq. (34-15) is left as a problem. The energy necessary to maintain the current during this decay is provided by the energy stored in the magnetic field of the inductor.

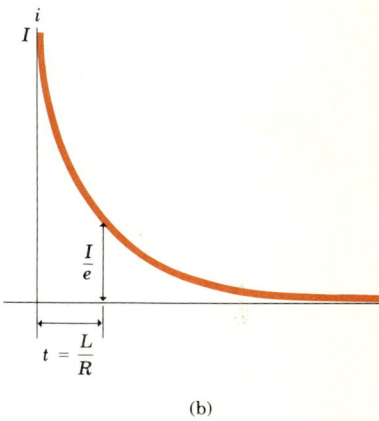

34-5 (a) Growth of current in a circuit containing inductance and resistance. (b) Decay of current in a circuit containing inductance and resistance.

34-5 The *L-C* circuit

The behavior of an *R-C* circuit was discussed in Sec. 29-5, and that of an *R-L* circuit in Sec. 34-4. We now consider the *L-C* circuit shown in Fig. 34-6, a resistanceless inductor connected between the terminals of a charged capacitor. At the instant connections are made, in Fig. 34-6a, the capacitor starts to discharge through the inductor. At a later instant, represented in Fig. 34-6b, the capacitor has completely discharged and the potential difference between its terminals (and those of the inductor) has decreased to zero. The current in the inductor has meanwhile established a magnetic field in the space around it. This magnetic field now decreases, inducing an emf in the inductor in the same direction as the current. The current therefore persists, although with diminishing magnitude, until the magnetic field has disappeared and the capacitor has been charged in the opposite sense to its initial polarity, as in Fig. 34-6c. The process now repeats itself in the reversed direction, and, in the absence of energy losses, the charges on the capacitor surge back and forth indefinitely. This process is called an *electrical oscillation*.

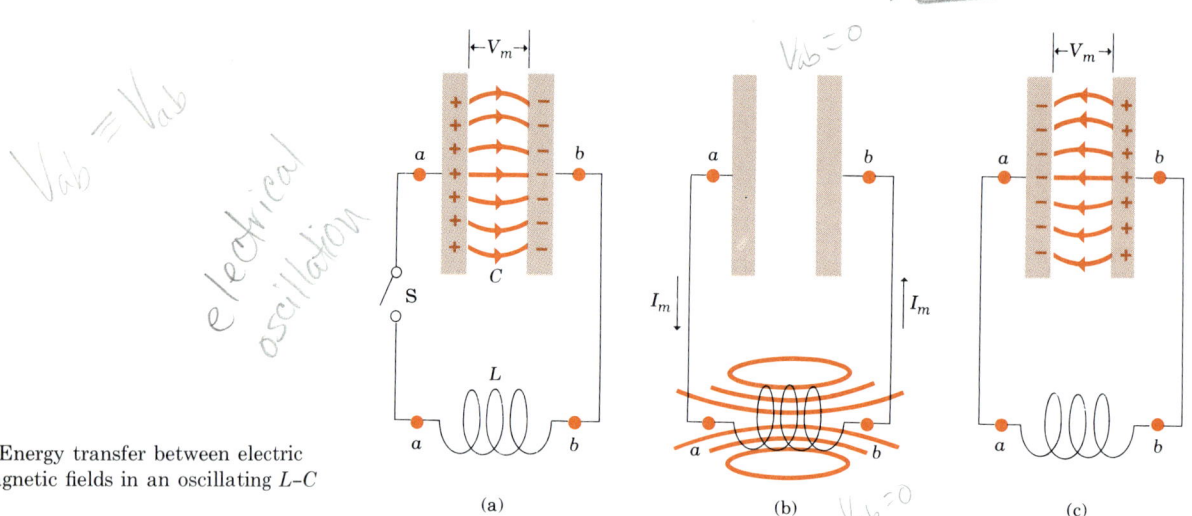

34-6 Energy transfer between electric and magnetic fields in an oscillating *L-C* circuit.

From the energy standpoint, the oscillations of an electrical circuit consist of a transfer of energy back and forth from the electric field of the capacitor to the magnetic field of the inductor, the *total* energy associated with the circuit remaining constant. This is analogous to the transfer of energy in an oscillating mechanical system from kinetic to potential, and vice versa.

The frequency of the electrical oscillations of a circuit containing inductance and capacitance only (a so-called *L-C* circuit) may be calculated in exactly the same way as the frequency of oscillation of a body suspended from a spring, discussed in Chapter 11. In the mechanical problem, a body of mass m is attached to a spring of force constant k. Let the body be displaced a distance A from its equilibrium positions and released from rest at time $t = 0$. Then, as shown in the left column of Table 34-1, the kinetic energy of the system at any later time is $\frac{1}{2}mv^2$, and its elastic potential energy is $\frac{1}{2}kx^2$. Because the system is conservative, the sum of these equals the initial energy of the system, $\frac{1}{2}kA^2$. The

Table 34-1 Oscillation of a mass on a spring compared with the electrical oscillation in an L-C circuit

Mass on a spring	Circuit containing inductance and capacitance
Kinetic energy = $\frac{1}{2}mv^2$	Magnetic energy = $\frac{1}{2}Li^2$
Potential energy = $\frac{1}{2}kx^2$	Electrical energy $\frac{1}{2}\frac{1}{C}q^2$
$\frac{1}{2}mv^2 + \frac{1}{2}kx^2 = \frac{1}{2}kA^2$	$\frac{1}{2}Li^2 + \frac{1}{2}\frac{1}{C}q^2 = \frac{1}{2}\frac{1}{C}Q^2$
$v = \pm\sqrt{k/m}\sqrt{A^2 - x^2}$	$i = \pm\sqrt{1/LC}\sqrt{Q^2 - q^2}$
$v = \frac{dx}{dt}$	$i = \frac{dq}{dt}$
$x = A\cos\sqrt{k/m}\,t = A\cos\omega t$	$q = Q\cos\sqrt{1/LC}\,t = Q\cos\omega t$
$v = -\omega A\sin\omega t = -v_{max}\sin\omega t$	$i = -\omega Q\sin\omega t = -I\sin\omega t$

velocity v at any coordinate x is therefore

$$v = \pm\sqrt{\frac{k}{m}}\sqrt{A^2 - x^2}. \qquad (34\text{-}16)$$

The velocity equals dx/dt, and the coordinate x as a function of t has been shown (Chapter 11) to be

$$x = A\cos\left(\sqrt{\frac{k}{m}}\right)t = A\cos\omega t, \qquad (34\text{-}17)$$

where the angular frequency ω is

$$\omega = \sqrt{\frac{k}{m}}.$$

We recall also that the ordinary frequency f, the number of cycles per unit time, is given by $f = \omega/2\pi$.

In the electrical problem, also a conservative system, a capacitor of capacitance C is given an initial charge Q and, at time $t = 0$, is connected to the terminals of an inductor of self-inductance L. The magnetic energy of the inductor at any later time corresponds to the kinetic energy of the vibrating body and is given by $\frac{1}{2}Li^2$. The electrical energy of the capacitor corresponds to the elastic potential energy of the spring and is given by $q^2/2C$, where q is the charge on the capacitor. The sum of these equals the initial energy of the system, $Q^2/2C$. That is,

$$\frac{1}{2}Li^2 + \frac{q^2}{2C} = \frac{Q^2}{2C}.$$

Solving for i, we find that when the charge on the capacitor is q, the current i is

$$i = \sqrt{\frac{1}{LC}}\sqrt{Q^2 - q^2}. \qquad (34\text{-}18)$$

Comparing this with Eq. (34-16), we see that the current $i = dq/dt$ varies with time in the same way as the velocity $v = dx/dt$ in the mechanical problem. Continuing the analogy, we conclude that q is given as a

function of time by

$$q = Q \cos\left(\sqrt{\frac{1}{LC}}\right)t = Q \cos \omega t. \tag{34-19}$$

The angular frequency ω of the electrical oscillations is therefore

$$\omega = \sqrt{\frac{1}{LC}}. \tag{34-20}$$

This is called the *natural frequency* of the L-C circuit.

These results may also be derived directly from an analysis of the L-C circuit. At each instant the capacitor voltage $v_{ab} = q/C$ must equal that of the inductor, $v_{ab} = L\,di/dt$. Also, because of the choice of positive direction for i, we have $i = -dq/dt$. Combining these relations, we find

$$\frac{d^2q}{dt^2} + \frac{1}{LC}q = 0. \tag{34-21}$$

The solutions of this differential equation, functions whose second derivative is equal to $-1/LC$ times the original function, are

$$q = Q \cos \omega t, \tag{34-22a}$$
$$q = Q \sin \omega t, \tag{34-22b}$$
$$q = Q \cos(\omega t + \phi), \tag{34-22c}$$

where $\omega = 1/(LC)^{1/2}$, and Q and ϕ are constants. Just as with the harmonic oscillator, the choice of one of these functions is determined by the initial conditions. If at time $t = 0$, the capacitor has maximum charge and $i = 0$, as in the above discussion, then Eq. (34-22a) is to be used. If at $t = 0$, $q = 0$ but i is different from zero, we use Eq. (34-22b), and if both q and i are different from zero at time $t = 0$, the more general form, Eq. (34-22c), must be used.

34-6 The R-L-C Circuit

In the above discussion we have assumed that the L-C circuit contains no *resistance*. This is an idealization, of course; for every real inductor there is resistance associated with the windings, and there may be resistance in the connecting wires as well. The effect of resistance is to dissipate the electromagnetic energy and convert it to heat; thus, resistance in an electric circuit plays a role analogous to that of friction in a mechanical system.

Suppose an inductor of self-inductance L and a resistor of resistance R are connected in series across the terminals of a capacitor. If the capacitor is initially charged, it starts to discharge at the instant the connections are made but, because of i^2R losses in the resistor, the energy of the inductor when the capacitor is completely discharged is *less* than the original energy of the capacitor. In the same way, the energy of the capacitor when the magnetic field has collapsed is still smaller, and so on.

If the resistance R is relatively small, the circuit oscillates, but with *damped harmonic motion*, as illustrated in Fig. 34-7a. As R is increased, the oscillations die out more rapidly. At a sufficiently large value of R,

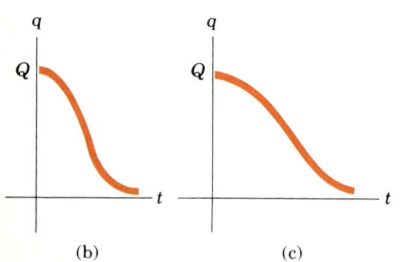

34-7 Graphs of q versus t in an R-L-C circuit. (a) Small damping. (b) Critically damped. (c) Overdamped.

34-8 An R-L-C series circuit.

the circuit no longer oscillates and is said to be *critically damped*, as in Fig. 34-7b. For still larger resistances it is *overdamped*, as in Fig. 34-7c.

With the proper electronic circuitry, energy can be fed *into* an R-L-C circuit at the same rate as that at which it is dissipated by i^2R or radiation losses. In effect, a *negative resistance* is inserted in the circuit so that its total resistance is zero. The circuit then oscillates with *sustained* oscillations, as does the idealized circuit with no resistance, discussed in Sec. 34-5.

The circuit of Fig. 34-8 may be analyzed in detail to find the charge on the capacitor and the current in the circuit as functions of time; that analysis is sketched below. Suppose the switch is closed in the upward position for a long time, so the capacitor acquires a charge $Q = CV$. Then at time $t = 0$ the switch is flipped to the downward position. Application of Kirchhoff's loop rule to the circuit yields the equation

$$iR + L\frac{di}{dt} + \frac{q}{C} = 0.$$

We replace i with dq/dt and rearrange to obtain

$$\frac{d^2q}{dt^2} + \frac{R}{L}\frac{dq}{dt} + \frac{1}{LC}q = 0. \qquad (34\text{-}23)$$

We note that when $R = 0$ this reduces to Eq. (34-21).

Solutions of Eq. (34-23) can be obtained by general methods of differential equations. The form of the solution depends on whether R is large or small. When R is less than $2(L/C)^{1/2}$, the solution has the form

$$q = VCe^{-(R/2L)t} \cos \sqrt{\frac{1}{LC} - \left(\frac{R}{2L}\right)^2} \, t. \qquad (34\text{-}24) \quad \text{underdamped}$$

When R is greater than $2(L/C)^{1/2}$, the solution is

$$q = e^{-(R/2L)t}\left[Ae^{\sqrt{\left(\frac{R}{2L}\right)^2 - \frac{1}{LC}}\,t} + Be^{-\sqrt{\left(\frac{R}{2L}\right)^2 - \frac{1}{LC}}\,t}\right], \qquad (34\text{-}25) \quad \text{overdamped}$$

where A and B are constants determined by V, R, L, and C.

Equation (34-24) corresponds to the *underdamped* behavior shown in Fig. 34-7a; the function represents a sinusoidal oscillation with an exponentially decaying amplitude. We note that the angular frequency of the oscillation is no longer $1/(LC)^{1/2}$ but is *less* than this because of the term containing R. The frequency of the damped oscillations is thus given by

$$\omega = \sqrt{\frac{1}{LC} - \frac{R^2}{4L^2}}. \qquad (34\text{-}26)$$

As R increases, ω becomes smaller and smaller; when $R^2 = 2L/C$ the quantity under the radical becomes zero and the case of *critical damping* has been reached (Fig. 34-7b). For still larger values of R the behavior is no longer oscillatory but is described as the sum of two exponential functions, as in Fig. 34-7c; the circuit is then *overdamped*.

Additional interesting aspects of the behavior of this circuit emerge when a sinusoidally varying emf is included in the circuit. This leads us into the area of alternating-current (ac) circuit analysis, a topic discussed in detail in Chapter 36.

Questions

34-1 A resistor is to be made by winding a wire around a cylindrical form. In order to make the inductance as small as possible, it is proposed that we wind half the wire in one direction and the other half in the opposite direction. Would this achieve the desired result? Why or why not?

34-2 In Fig. 34-1, if coil 2 is turned 90° so that its axis is vertical, does the mutual inductance increase or decrease?

34-3 The toroidal solenoid is one of the few configurations for which it is easy to calculate self-inductance. What features of the toroidal solenoid give it this simplicity?

34-4 Two identical closely wound circular coils, each having self-inductance L, are placed side by side, close together. If they are connected in series, what is the self-inductance of the combination? What if they are connected in parallel? Can they be connected so that the total inductance is zero?

34-5 If two inductors are separated enough so that practically no flux from either links the coils of the other, show that the equivalent inductance of two inductors in series or parallel is obtained by the same rules used for combining resistance.

34-6 Two closely wound circular coils have the same number of turns, but one has twice the radius of the other. How are the self-inductances of the two coils related?

34-7 One of the great problems in the field of energy resources and utilization is the difficulty of storing electrical energy in large quantities economically. Discuss the possibility of storing large amounts of energy by means of currents in large inductors.

34-8 In what regions in a toroidal solenoid is the energy density greatest? Least?

34-9 Suppose there is a steady current in an inductor. If one attempts to reduce the current to zero instantaneously by opening a switch, a big fat arc appears at the switch contacts. Why? What happens to the induced emf in this situation? Is it physically possible to stop the current instantaneously?

34-10 In the R-L circuit of Fig. 34-4, is the current in the resistor always the same as that in the inductor? How do you know?

34-11 In the R-L circuit of Fig. 34-4, when the switch is in the up position and is then thrown to the down position, the potential V_{ab} changes suddenly and discontinuously, but the current does not. Why can the voltage change suddenly but not the current?

34-12 In the R-L-C circuit, what criteria could be used to decide whether the system is overdamped or underdamped? For example, could one compare the maximum energy stored during one cycle to the energy dissipated during one cycle?

Problems

34-1 A solenoid of length 10 cm and radius 2 cm is wound uniformly with 1000 turns. A second coil of 50 turns is wound around the solenoid at its center. What is the mutual inductance of the two coils?

34-2 A toroidal solenoid (cf. Sec. 32-8, Example 2) has a radius of 10 cm and a cross-sectional area of 5 cm², and is wound uniformly with 1000 turns. A second coil with 500 turns is wound uniformly on top of the first. What is the mutual inductance?

34-3 Find the self-inductance of the toroidal solenoid in Problem 34-2 if only the 1000-turn coil is used. How would your answer change if the two coils were connected in series?

34-4 A toroidal solenoid has two coils with n_1 and n_2 turns, repectively; it has radius r and cross-sectional area A.

a) Derive an expression for the self-inductance L_1 when only the first coil is used, and that for L_2 when only the second soil is used.

b) Derive an expression for the mutual inductance of the two coils.

c) Show that $M^2 = L_1 L_2$. This result is valid whenever all the flux linked by one coil is also linked by the other.

34-5 Two coils have mutual inductance $M = 0.01$ H. The current i_1 in the first coil increases at a uniform rate of 0.05 A·s^{-1}.

a) What is the induced emf in the second coil? Is it constant?

b) Suppose that the current described is that in the second coil rather than the first; what is the induced emf in the first coil?

34-6 An inductor of inductance 5 H carries a current that decreases at a uniform rate, $di/dt = -0.02$ A·s^{-1}. Find the self-induced emf; what is its polarity?

34-7 An inductor used in a dc power supply has an inductance of 20 H and a resistance of 200 Ω, and carries a current of 0.1 A.

a) What is the energy stored in the magnetic field?

b) At what rate is energy dissipated in the resistor?

34-8

a) Show that the two expressions for self-inductance, namely
$$\frac{N\Phi}{i} \quad \text{and} \quad \frac{\mathcal{E}}{di/dt},$$
have the same units.

b) Show that L/R and RC both have the units of time.

c) Show that 1 Wb·s^{-1} equals 1 V.

34-9 An inductor with $L = 40$ H carries a current i that varies with time according to $i = (0.1 \text{ A}) \sin(120\pi \text{ s}^{-1})t$. Find an expression for the induced emf. What is the phase of $\mathcal{E}$ relative to i?

34-10 A coaxial cable consists of a small solid conductor of radius r_a supported by insulating disks on the axis of a thin-walled tube of inner radius r_b. Show that the self-inductance of a length l of the cable is

$$L = l \frac{\mu_0}{2\pi} \ln \frac{r_b}{r_a}.$$

Assume the inner and outer conductors to carry equal currents in opposite directions. [*Hint:* Use Ampère's law to find the magnetic field at any point in the space between the conductors. Write the expression for the flux $d\Phi$ through a narrow strip of length l parallel to the axis, of width dr, at a distance r from the axis of the cable and lying in a plane containing the axis. Integrate to find the total flux linking a current i in the central conductor.]

34-11 A certain toroidal solenoid has a rectangular cross section, as shown in Fig. 34-9. It has N uniformly spaced turns, with air inside. The magnetic field at a point inside the toroid is given by Eq. (32-26). *Do not* assume the field to be uniform over the cross section.

a) Show that the magnetic flux through a cross section of the toroid is

$$\Phi = \frac{\mu_0 N I h}{2\pi} \ln \frac{b}{a}.$$

b) Show that the self-inductance of the toroidal solenoid is given by

$$L = \frac{\mu_0 N^2 h}{2\pi} \ln \frac{b}{a}.$$

c) The fraction b/a may be written as

$$\frac{b}{a} = \frac{a+b-a}{a} = 1 + \frac{b-a}{a}.$$

The power series expansion for $\ln(1 + x)$ is $\ln(1 + x) = x + x^2/2 + \cdots$. Hence show that when $b - a$ is much less

Figure 34-9

than a, the self-inductance is approximately equal to

$$L = \frac{\mu_0 N^2 h(b-a)}{2\pi a}.$$

Compare this result with the result obtained in Sec. 34–1, Example 1.

34–12 For the toroidal solenoid described in Prob. 34–11, calculate the total energy when the current is I. Do not assume the field to be uniform over the cross section. Obtain the total energy in two ways:

a) Use Eq. (34–8), with the expression for L obtained in Prob. 34–11;

b) Integrate Eq. (34–9) through the volume of the toroid.

Compare the results of parts (a) and (b).

34–13 The current in a resistanceless inductor is caused to vary with time as in the graph of Fig. 34–10.

a) Sketch the pattern that would be observed on the screen of an oscilloscope connected to the terminals of the inductor. (The oscilloscope spot sweeps horizontally across the screen at constant speed, and its vertical deflection is proportional to the potential difference between the inductor terminals.)

b) Explain why the inductor can be described as a "differentiating circuit."

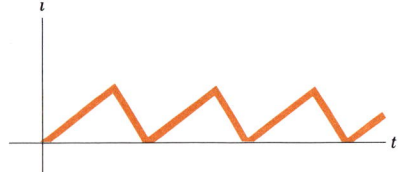

Figure 34–10

34–14 The 1000-turn toroidal solenoid described in Problem 34–2 carries a current of 5.0 A.

a) What is the energy density in the magnetic field.

b) What is the total magnetic-field energy? Find using Eq. (34–8) and also by multiplying the energy density from (a) by the volume of the toroid, which is $2\pi rA$; compare the two results.

34–15 A toroidal solenoid has a mean radius of 0.12 m and a cross-sectional area of 20×10^{-4} m²? It is found that when the current is 20 A, the energy stored is 0.1 J. How many turns does the winding have?

34–16 An inductor of inductance 3 H and resistance 6 Ω is connected to the terminals of a battery of emf 12 V and of negligible internal resistance. Find

a) the initial rate of increase of current in the circuit,

b) the rate of increase of current at the instant when the current is 1 A,

c) the current 0.2 s after the circuit is closed,

d) the final steady-state current.

34–17 The resistance of a 10-H inductor is 200 Ω. The inductor is suddenly connected across a potential difference of 10 V.

a) What is the final steady current in the inductor?

b) What is the initial rate of increase of current?

c) At what rate is the current increasing when its value is one-half the final current?

d) At what time after the circuit is closed does the current equal 99% of its final value?

e) Compute the current at the following times after the circuit is closed: 0, 0.025 s, 0.05 s, 0.075 s, 0.10 s. Show the results in a graph.

34–18 An inductor of resistance R and self-inductance L is connected in series with a noninductive resistor of resistance R_0 to a constant potential difference V (Fig. 34–11).

a) Find the expression for the potential difference v_{cb} across the inductor at any time t after switch S_1 is closed.

b) Let $V = 20$ V, $R_0 = 50$ Ω, $R = 150$ Ω, $L = 5$ H. Compute a few points, and construct graphs of v_{ac} and v_{cb} over a time interval from zero to twice the time constant of the circuit.

Figure 34–11

34–19 After the current in the circuit of Fig. 34–11 has reached its final steady value the switch S_2 is closed, thus short-circuiting the inductor. What will be the magnitude and direction of the current in S_2, 0.01 s after S_2 is closed?

34–20 Refer to Problem 34–16.

a) What is the power input to the inductor at the instant when the current in it is 0.5 A?

b) What is the rate of dissipation of energy at this instant?

c) What is the rate at which the energy of the magnetic field is increasing?

d) How much energy is stored in the magnetic field when the current has reached its final steady value?

34-21 An inductor having inductance L and resistance R carries a current I. Show that the time constant is equal to *twice the ratio* of the energy stored in the magnetic field to the rate of dissipation of energy in the resistance.

34-22 Show that the quantity $(L/C)^{1/2}$ has units of resistance (ohms).

34-23 The maximum capacitance of a variable air capacitor is 35 pF.

a) What should be the self-inductance of a coil to be connected to this capacitor if the natural frequency of the L-C circuit is to be 550×10^3 Hz, corresponding to one end of the broadcast band?

b) The frequency at the other end of the broadcast band is 1550×10^3 Hz. What must be the minimum capacitance of the capacitor if the natural frequency is to be adjustable over the range of the broadcast band?

34-24 Write an equation corresponding to Eq. (34–11) for the current in Fig. 34–4 just after the switch is thrown to the down position, if the initial current is I. Solve the resulting differential equation and verify Eq. (34–15).

34-25 In Fig. 34–6, equate the voltage across the inductor to that across the capacitor, to show that

$$L\frac{d^2q}{dt^2} = -\frac{q}{C}.$$

Show that this differential equation is satisfied by the function $q = Q \cos \omega t$, with ω given by $1/(LC)^{1/2}$.

34-26 An inductor having $L = 40$ mH is to be combined with a capacitor to make an L-C circuit with natural frequency 2×10^6 Hz. What value of capacitance should be used?

34-27 An inductor is made with two coils wound close together on a form, so all the flux linking one coil also links the other. The number of turns is the same in each. If the inductance of one coil is L, what is the inductance when the two coils are connected

a) in series;

b) in parallel?

c) If an L-C circuit using this inductor has natural frequency ω using one coil, what is the natural frequency when the two coils are used in series?

34-28 An L-C circuit in an AM radio tuner uses an inductor with inductance 0.1 mH and a variable capacitor. If the natural frequency of the circuit is to be adjustable over the range 0.5 to 1.5 MHz, corresponding to the AM broadcast band, what range of capacitance must the variable capacitor cover?

34-29 The equation preceding Eq. (34–23) may be converted into an energy relation. Multiply both sides of this equation by $i = dq/dt$. Show that the first term on the right is i^2R, the second can be written $d(\frac{1}{2}Li^2)/dt$, and the third can be written $d(q^2/2C)/dt$. Hence, show that the rate at which the battery delivers energy, $V_{ab}i$, is equal to the rate of energy dissipation in the resistor plus the sum of the rates of change of energy stored in the inductor and the capacitor.

MAGNETIC PROPERTIES OF MATTER

35-1 Magnetic materials

In Chapter 32 we discussed the magnetic fields set up by moving charges or by currents in conductors, when the charges or conductors are in air (or, strictly speaking, in vacuum). However, pieces of technical equipment, such as transformers, motors, generators, or electromagnets, always incorporate iron or an iron alloy in their structures, both to increase the magnetic flux and to confine it to a desired region. The magnetic field of a galvanometer or loudspeaker is set up by a *permanent* magnet, which produces this field without any apparent circulation of charge. The coating on magnetic tape makes a record of information supplied to it by the extent to which it becomes permanently magnetized.

We therefore turn next to a consideration of the magnetic properties that make iron and a few other ferromagnetic materials so useful. We shall find that magnetic properties are not confined to ferromagnetic materials but are exhibited (to a much smaller extent, to be sure) by *all* substances. A study of the magnetic properties of materials affords another means of gaining an insight into the nature of matter in general.

Magnetic properties of a substance can be demonstrated by supporting a small spherically shaped specimen by a fine thread near the poles of a powerful electromagnet. If the specimen is of iron or one of the *ferromagnetic* substances, it is attracted into the strong part of the magnetic field. Not so familiar is the fact that *every* substance is influenced by the field, although to an extent that is extremely small compared with a substance like iron. Some substances, like iron, are forced into the strong part of the field, while others are pushed toward the weak part of the field. The first type is called *paramagnetic*; the second, *diamagnetic*. All substances, including liquids and gases, fall into one or the other of these classes.

In our discussion of induced emf's in Sec. 33–3, it was assumed that the core of the solenoid in Fig. 33–9 was empty. Suppose, however, that the solenoid is wound on a solid rod. It is found that, for a given change of current in the solenoid windings, the induced emf is not the same as when the core is empty, and hence the change in *flux* is not the same. If the rod is *ferromagnetic*, the change in flux for a given change in current is very much *larger*; if paramagnetic, it will be *slightly* larger; and if diamagnetic, slightly *smaller* than if the core were empty.

35-2 Magnetic permeability

In Chapter 27, our formulation of the properties of a dielectric substance was based on a specimen in the form of a flat slab, inserted in the field between oppositely charged parallel plates. The electric field is almost entirely confined to the region between the plates if their separation is small, and therefore a flat slab between the plates will completely occupy all points of space at which an electric field exists. A specimen of this shape is not as well suited for a study of magnetic effects, however, since magnetic field lines are always *closed* lines and there is no way of producing a magnetic field that is confined to the region between two closely spaced surfaces. The magnetic field within a closely spaced toroidal winding, however, *is* wholly confined to the space enclosed by the winding. We shall accordingly use such a field as a basis for our discussion of magnetic properties of materials, using a specimen in the form of a ring on whose surface the wire is wound. Such a specimen is often called a *Rowland ring*, after J. H. Rowland, who made much use of it in his experimental and theoretical work on electricity and magnetism. The winding of wire around the specimen is called the *magnetizing winding*, and the current in the winding, the *magnetizing current*.

The magnetic field (flux density) within the space enclosed by a toroidal winding *in vacuum* is, from Sec. 32-8,

$$B_0 = \frac{\mu_0 N i}{2\pi r}, \quad (35\text{-}1)$$

where $2\pi r$ is the mean circumference.

Suppose now that the same coil is wound on a Rowland ring, and that a second winding is placed on the ring, as in Fig. 35-1, with its terminals connected to a charge-measuring instrument. The **B** field within the ring may be measured by a procedure essentially the same as that of the search coil, discussed in Sec. 33-2. When switch S is opened and the magnetizing current suddenly drops to zero, the deflection of the charge-measuring instrument is proportional to the total flux change, and the initial flux Φ and magnetic field B can be determined.

The results are found *not* to agree with that computed from Eq. (35-1). If the core is made of a ferromagnetic material, the measured flux density is very much *larger*; if made of a paramagnetic material, very slightly larger; and if made of a diamagnetic material, it is very slightly *smaller* than the calculated value.

Let B represent the magnetic field in a material ring and B_0 the field in a "ring of vacuum." The ratio of B to B_0 is called the *relative permeability* of the material and is represented by K_m:

$$K_m = \frac{B}{B_0}, \quad (35\text{-}2)$$

35-1 Magnetic specimen in the form of a ring with a toroidal winding.

where

K_m is $\begin{cases} \text{equal to 1 for vacuum,} \\ \text{slightly larger than 1 for paramagnetic materials,} \\ \text{slightly smaller than 1 for diamagnetic materials,} \\ \text{often much larger than 1 for ferromagnetic materials.} \end{cases}$

Substituting for B_0 its value $\mu_0 Ni/2\pi r$, we get

$$B = K_m \mu_0 \left(\frac{Ni}{2\pi r}\right).$$

The product $K_m \mu_0$ is called the *permeability* of the material and is denoted by μ:

$$\mu = K_m \mu_0, \qquad (35\text{-}3)$$

where

μ is $\begin{cases} \text{equal to } \mu_0 \text{ for vacuum,} \\ \text{slightly greater than } \mu_0 \text{ for paramagnetic materials,} \\ \text{slightly smaller than } \mu_0 \text{ for diamagnetic materials,} \\ \text{often much larger than } \mu_0 \text{ for ferromagnetic materials.} \end{cases}$

The unit in which μ is expressed is the same as that for μ_0, namely, webers per ampere·meter or tesla·meters per ampere.

The expression for the magnetic field B in a material substance in the form of a Rowland ring may now be written

$$B = \frac{\mu Ni}{2\pi r}, \qquad (35\text{-}4)$$

The presence of a magnetic material in a toroidal coil increases the flux for a given current by a factor of K_m, the relative permeability of the material. The *self-inductance* of such a coil, defined by Eq. (34-5) in terms of the ratio of flux to current, also increases by this factor. Thus the derivation of the expression for self-inductance of a toroidal coil, Example 1 of Sec. 34-2, must include an additional factor K_m. Since $\mu = K_m \mu_0$, the inductance of a toroidal coil with core of permeability μ is given by

$$L = \frac{\mu N^2 A}{2\pi r}. \qquad (35\text{-}5)$$

The self-inductance of an iron-core coil is therefore very much *larger* than that of the same winding in vacuum.

The expression for the energy density in a magnetic field is modified in the same way. It was shown in Sec. 34-3 that the energy density in a magnetic field in vacuum equals $B^2/2\mu_0$. In a magnetic material of permeability μ, the energy density is

$$u = \frac{1}{2}\frac{B^2}{\mu}. \qquad (35\text{-}6)$$

35-3 Molecular theory of magnetism

When it was discovered that magnetic effects can be produced by currents as well as by permanent magnets, Ampère proposed the theory that the magnetic properties of a material arise from a large number of tiny closed current loops within the material. The total magnetic field in a material would then be the sum of the field caused by the external current and the additional field caused by these microscopic currents. In diamagnetic materials the direction of the additional field would be *opposite* to that of the field of the external current, *decreasing* the total field, while in paramagnetic materials the two fields would *add* to produce a total field *greater than* that due to the external current alone.

It is now known that this picture is essentially correct. The microscopic current loops are associated with the motion of electrons within atoms. Each current loop has an associated magnetic moment, just as with macroscopic current loops as discussed in Sec. 31-3. In a *paramagnetic* material each atom or molecule has a certain net current distribution and magnetic moment resulting from the combined effect of all its electron motions. In the absence of any externally caused field, these currents and magnetic moments are randomly oriented and cause no net magnetic field; but in the presence of a magnetic field caused by an external current, the magnetic moments tend to become *aligned* with this field, and the additional field caused by the associated currents is in the same direction as the field of the external current, just as for the macroscopic current loop discussed in Sec. 32-6. Thus the total field in the material is *greater* than that of the external current alone.

In a *diamagnetic* material with no external field, the electron currents add to zero for each molecule, and there is no net magnetic moment. But when a field is applied by an external current the electron motion is altered by an effect that is essentially the same as induction of a macroscopic current in a conducting loop by a changing magnetic flux through the loop, as discussed in Sec. 33-2. Lenz's law (Sec. 33-4) requires that the additional magnetic field caused by such a current loop always *oppose* the flux change that caused it. Thus the change in the electron motion in a diamagnetic material in a magnetic field always causes an additional field in the *opposite* direction to that of the applied field. This effect is always very small, typically one part in 10^5, and it acts to decrease the total field in the material.

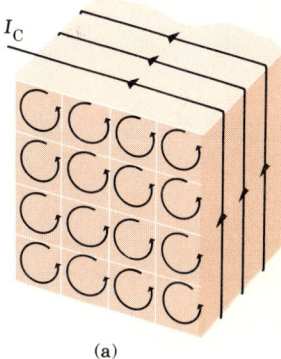

(a)

Diamagnetic effects are in fact present in *all* materials. However, paramagnetic effects, when they occur, are always much *larger* than diamagnetic effects, so a material having diamagnetism and paramagnetism always appears paramagnetic, not diamagnetic.

The alignment of electron currents in a paramagnetic material is shown schematically in Fig. 35-2a. At interior points the currents in adjacent loops are in opposite directions and cancel, so there is no net current in the interior of the material. The parts of the loops adjacent to the outer surface are not canceled however, and the entire collection of loops is equivalent to a current I_S circulating around the outside of the body, as shown in Fig. 35-2b. This equivalent current has the same shape as an additional magnetizing winding around the body, and since its direction is the same as that of the external magnetizing current I_C, the effect is to increase the field within the material. A similar figure could be drawn for a diamagnetic material; in that case, the direction of the current loops is reversed, and the direction of I_S is opposite to that of I_C, leading to a decreased field.

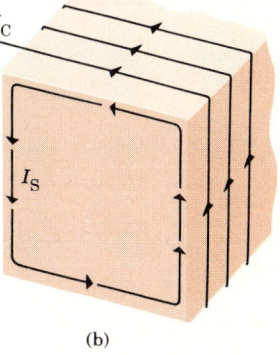

(b)

35-2 (a) Alignment of atomic current loops in a paramagnetic substance.
(b) Surface current equivalent to part (a).

*35-4 Magnetization and magnetic intensity

The discussion of the preceding section has shown that the magnetic field in a material is due to the combined effect of the current in external conductors and the microscopic currents in the material, with their associated magnetic moments. For a more complete understanding of magnetic materials, it is useful to develop relationships governing these two separate contributions to the total field. For this purpose we first introduce a quantity called *magnetization*, which is the *density* of magnetic moment in a material.

We consider a small volume V; let a typical molecular magnetic moment in this volume be $\mathbf{m}$. Then the *total* magnetic moment within V is the vector sum $\Sigma \mathbf{m}$. The magnetization, denoted by $\mathbf{M}$, is defined as

$$M = \frac{\Sigma m}{V}. \tag{35-7}$$

Magnetization is therefore *magnetic moment per unit volume*, or *density of magnetic moment*. If the molecular magnetic moments are randomly oriented, or if there are none, as in a diamagnetic material in the absence of an external field, then the vector sum is zero and $\mathbf{M} = \mathbf{0}$.

Let us now consider a long rod magnetized by a current I_C in a solenoidal conductor, as shown in Fig. 35-3. At points not too near the ends, the $\mathbf{B}$ field is uniform through the material. As shown in the figure, this field is caused jointly by the conduction current I_C and the equivalent surface current I_S due to the microscopic currents. The case shown corresponds to a paramagnetic material, in which I_S has the same direction as I_C.

The magnetization $\mathbf{M}$ is directly related to the surface current, and we now need to develop that relationship in detail. The magnetic moment of a current loop is defined as the product of the current and the area of the loop. Let I_S represent the total surface current around the periphery of a portion of the rod of length l. (The surface current is distributed *continuously* along the surface of the rod.) The magnetic moment of this portion is then $I_S A$, where A is the cross-sectional area

35-3 Magnetic field in a paramagnetic material is caused jointly by the conduction current I_C in the winding and the surface current I_S due to microscopic currents. For a *diamagnetic* material I_C and I_S have opposite directions, and $\mathbf{B}$ and $\mathbf{M}$ are opposite.

of the rod, and since the volume of the portion is Al, the magnetic moment per unit volume, or the magnetization M, has magnitude

$$M = \frac{I_S A}{Al} = \frac{I_S}{l} = j_S, \tag{35-8}$$

where j_S is the *surface current per unit length*. In this special case, the magnetization is numerically *equal* to the surface current per unit length. More generally, the surface current per unit length equals the tangential component of M at the surface.

The broken line in Fig. 35-3 (compare Fig. 32-9) is a closed path of length l. In analogy to Ampère's law, we may consider the integral

$$\oint M \cdot dl$$

around the broken line. Now the magnetization M is uniform within the rod and is zero outside; the sides are perpendicular to M and do not contribute to the integral, which is therefore equal simply to the product Ml. But from Eq. (35-8), $Ml = I_S$, so the integral is also equal to the total surface current I_S through the area bounded by this path. Thus,

$$\oint M \cdot dl = I_S. \tag{35-9}$$

Although we have obtained this relation only for a special case, it is found to be true in general; it may be called Ampère's law for the magnetization vector M.

The total B field in the rod is due both to the conduction current in the windings of the solenoid, and to the surface currents. Hence if I_C is the total *conduction current* through the rectangle, Ampère's law for B states that

$$\oint B \cdot dl = \mu_0 (I_C + I_S). \tag{35-10}$$

When I_S is eliminated between the preceding equations, we have

$$\oint B \cdot dl = \mu_0 \left(I_C + \oint M \cdot dl \right)$$

or

$$\oint \left(\frac{B}{\mu_0} - M \right) \cdot dl = I_C.$$

Let us define a new quantity called the *magnetic intensity* H (analogous to the displacement D) as the vector difference

$$H = \frac{B}{\mu_0} - M. \tag{35-11}$$

Equation (35-10) then takes the simple form,

$$\oint H \cdot dl = I_C. \tag{35-12}$$

Thus the magnetic intensity H obeys a relation similar in form to Ampère's law but containing *only* the conduction current I_C, *not* the microscopic currents. In other words, the sources of H are only conduction currents, not magnetization currents. This property is very useful in the analysis of magnetic behavior.

In summary, for any closed path, the line integral of **B** equals the product of μ_0 and the *total* current across any surface bounded by the path, the line integral of **M** equals the *surface* current, and the line integral of **H** equals the *conduction* current. (If displacement currents are also present, they must be added to the conduction current.)

Like the magnetic field **B**, the magnetic intensity **H** can be represented by lines, called in this case lines of magnetic intensity. In free space where there is no magnetization, $M = 0$, $B = \mu_0 H$, and the lines of **H** have the same shape as those of **B**. In matter, however, the two can be quite different.

Although the symbols **H** and **B** are used nearly universally, these fields are given various names by various authors. In some books **H** is called the *magnetic field* and **B** the *magnetic induction* or the *magnetic flux density*. Other authors avoid confusion by referring simply to "the **H** field" and "the **B** field." In this book we usually call **B** the *magnetic field* and **H** the *magnetic intensity*. In the following discussion of the relation of these fields, we shall sometimes call **B** the *magnetic flux density* to emphasize that the **B** field is always used in computing magnetic flux for applications of Faraday's law.

As Eq. (35–12) shows, the units of **H** are amperes per meter ($A \cdot m^{-1}$). Also, from Eq. (35–11), the units of **H** must be the same as those of **M**. This may be verified by noting that **M** is magnetic moment per unit volume, with corresponding units $(A \cdot m^2)/(m^3) = A \cdot m^{-1}$.

In a diamagnetic or paramagnetic material the three vector fields **H**, **M**, and **B** are all proportional, and the proportionality can be represented in various ways. The relation of **M** to **H** for any particular material is represented by the *magnetic susceptibility* χ of the material:

$$M = \chi H. \tag{35-13}$$

Since **H** and **M** have the same units, χ is a dimensionless quantity.

Magnetic susceptibilities for a few materials are shown in Table 35–1. Paramagnetic susceptibilities decrease with increasing temperature because of the disorienting effect of thermal motion; diamagnetic susceptibilities are nearly independent of temperature.

In terms of χ, **B** is

$$B = \mu_0(H + M) = \mu_0(1 + \chi)H. \tag{35-14}$$

The relative permeability K_m, defined in Sec. 35–2, is

$$K_m = 1 + \chi, \tag{35-15}$$

so

$$B = \mu_0 K_m H.$$

The product $\mu_0 K_m$ is called the *permeability* μ:

$$\mu = \mu_0 K_m. \tag{35-16}$$

and hence

$$B = \mu H.$$

In vacuum, $K_m = 1$ and $\mu = \mu_0$. For this reason, the magnetic constant μ_0 is often referred to as "the permeability of vacuum" or "the

Table 35-1 Magnetic susceptibilities of paramagnetic and diamagnetic materials

Materials	Temperature, °C	$\chi_m = K_m - 1$
Paramagnetic		
Iron ammonium alum	−269	4830×10^{-5}
Iron ammonium alum	−183	213
Oxygen, liquid	−183	152
Iron ammonium alum	+20	66
Uranium	20	40
Platinum	20	26
Aluminum	20	2.2
Sodium	20	0.72
Oxygen gas	20	0.19
Diamagnetic		
Bismuth	20	-16.6×10^{-5}
Mercury	20	−2.9
Silver	20	−2.6
Carbon (diamond)	20	−2.1
Lead	20	−1.8
Rock salt	20	−1.4
Copper	20	−1.0

permeability of free space." The "relative permeability of vacuum," where $\chi = 0$, is equal to 1. The relative permeability of a paramagnetic substance is slightly greater than 1, and that of a diamagnetic substance is slightly less than 1.

*35-5 Ferromagnetism

The behavior of ferromagnetic materials differs from that of diamagnetics and paramagnetics in three important ways. First, the magnetic field in such a material may be several hundred or thousand times as large as that due to the external field alone. Second, the relation between magnetization and magnetic intensity (or magnetic field) is no longer a direct proportion; thus the magnetic susceptibility, defined as the ratio of **M** to **H,** is very large but no longer constant. Third, in some cases ferromagnetic materials exhibit large magnetization even when there is no externally applied field at all; that is, they can become permanently magnetized.

Any substance that exhibits these properties is called *ferromagnetic*. Iron, nickel, cobalt, and gadolinium are the only ferromagnetic *elements* at room temperature, but a number of elements at low temperatures and several alloys whose components are not ferromagnetic also show these effects.

Because of the complicated relation between the flux density B and the magnetic intensity H in a ferromagnetic material, it is not possible to express B as a simple function of H. Instead, the relation between these quantities is either given in tabular form or represented by a graph of B versus H, called the *magnetization curve* of the material.

The initial magnetization curve of a specimen of annealed iron is shown in Fig. 35-4 in the curve labeled B versus H. The permeability μ, equal to the ratio of B to H, can be found at any point of the curve by dividing the flux density B, at the point, by the corresponding magnetic

35-4 Magnetization curve and permeability curve of annealed iron.

intensity H. For example, when $H = 150$ A·m^{-1}, $B = 1.01$ T = 1.01 Wb·m^{-2}, and

$$\mu = \frac{B}{H} = \frac{1.01 \text{ T}}{150 \text{ A·m}^{-1}} = 67{,}300 \times 10^{-7} \text{ Wb·A}^{-1}\text{·m}^{-1}.$$

The relative permeability $K_m = \mu/\mu_0$ at this point is

$$K_m = \frac{67{,}300 \times 10^{-7} \text{ Wb·A}^{-1}\text{·m}^{-1}}{4\pi \times 10^{-7} \text{ Wb·A}^{-1}\text{·m}^{-1}} = 5356.$$

For some purposes a more useful quantity is the *differential* permeability, defined as dB/dH and represented by the *slope* of the curve of B versus H in Fig. 35-4. For the point just mentioned, the differential permeability for this material is about $25{,}000 \times 10^{-7}$ Wb·A^{-1}·m^{-1}, and the relative differential permeability is about 2000. Like the permeability, the differential permeability of a particular material varies with B and H.

Figure 35-5 covers a wider range of values for the same specimen, and also shows how the two quantities $\mu_0 H$ and $\mu_0 M$ contribute to B. When H is relatively small, practically all of the flux density B is due to the magnetization M (or to the equivalent surface currents). Beyond the point where H is of the order of 100,000 A·m^{-1} there is very little further increase in the magnetization M, and the iron is said to become *saturated*. Further increases in B are due almost wholly to increases in H resulting from increased current in the windings.

This behavior can be understood qualitatively on the basis of *magnetic domains*, which will be discussed in the next section.

*35-6 Magnetic domains

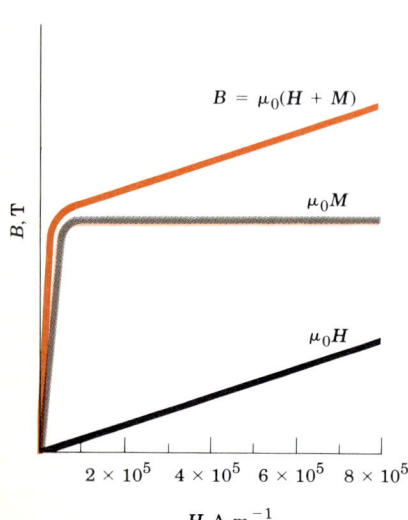

35-5 Graph of B, $\mu_0 H$, and $\mu_0 M$ as functions of H, for annealed iron.

In most metals, the electrons in the metallic ions that form the crystal lattice structure are paired off, with half spinning one way and half the other way. The ion is thus "magnetically neutral." The ions of the ferromagnetic elements, iron, nickel, cobalt, etc., are exceptions. In particular, that of iron has an excess of two uncompensated electrons; the magnetization of iron is due almost entirely to the alignment of the magnetic moments of these electrons. When a sample of iron is magnetically *saturated*, all of the unpaired electrons are spinning with their axes in the direction of the magnetizing field.

In *small* fields, ferromagnetic materials typically have much larger susceptibility, and therefore larger permeability, than paramagnetic materials. Ferromagnetism results because of a spontaneous, self-aligning, *cooperative* interaction among relatively large numbers of iron atoms in regions called *domains*. As a result of molecular interactions, the molecular magnetic moments in each domain are all aligned parallel to one another. In other words, each domain is *spontaneously magnetized to saturation* even in the absence of any external field. The directions of magnetization in different domains are not necessarily parallel to one another, so that in an unmagnetized specimen the *resultant* magnetization is zero. When the specimen is placed in a magnetic field the resultant magnetization may increase in two different ways, either by an increase in the volume of those domains that are favorably oriented with respect to the field at the expense of unfavorably oriented domains, or by *rotation* of the direction of magnetization toward the direction of the field.

We can now understand the reason for the shape of the graph of $\mu_0 M$ versus H in Fig. 35-5. As H is increased from zero, more and more of the spinning electrons become aligned with the H field and M increases steadily. When H is about 100,000 $A \cdot m^{-1}$, practically *all* of the electrons have lined up with the H field. No further increase in M can occur, and the iron is *saturated*.

Saturation can also occur with paramagnetic materials, but in the absence of the cooperative spontaneous alignment that occurs in ferromagnetics, the field required to cause saturation is larger than can be attained with present laboratory equipment. At present, saturation of paramagnetics can be observed only at extremely low temperatures, where the disorienting effect of thermal motion is greatly reduced.

Every ferromagnetic material has a critical temperature called the *Curie temperature*, above which the material is paramagnetic but not ferromagnetic. As the Curie temperature is reached, the energy associated with thermal motion in the material becomes large enough (compared to the interaction energies responsible for the spontaneous alignment of magnetic moments in a domain) so that this alignment disappears. Above the Curie treatment there is no spontaneous magnetization. The transition from ferromagnetic to paramagnetic behavior is a *phase change* analogous to the transitions between solid, liquid, and gaseous phases of matter, discussed in Chapter 15.

*35-7 Hysteresis

A magnetization curve such as that in Fig. 35-4 expresses the relation between the magnetic field B in a ferromagnetic material and the corresponding magnetic intensity H, *provided the sample is initially unmagnetized and the magnetic intensity is steadily increased from zero.* Thus, in Fig. 35-6, if the magnetizing current in the windings of an unmagnetized ring sample, as in Sec. 35-2, is steadily increased, the relation of B to H follows the curve Oab. If the current is now *decreased* until point c is

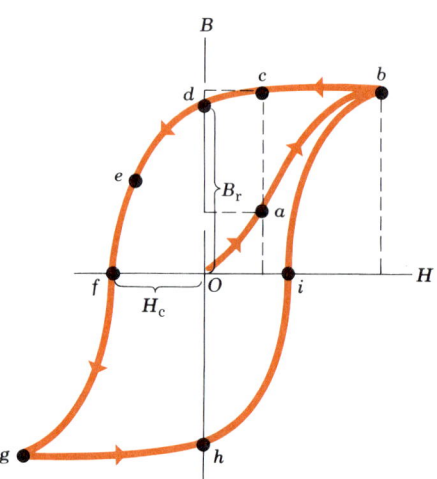

35-6 Hysteresis loop.

reached, B is much larger than at a, even though H is the same. When the current decreases to zero, H is zero, and we reach point d. But at this point B is *not* zero; the material is magnetized even in the absence of a magnetizing current, and thus has become a permanent magnet. In fact, B does not decrease to zero until H has reversed direction and reached point f. As H becomes larger in the reverse direction, point g is reached and the material approaches saturation magnetization in the reverse direction. As H decreases to zero and then increases in the original direction, the path $ghib$ is followed.

The magnetic field B that remains after the material has been magnetized to saturation and then H has been reduced to zero is called the *remanance*; it is denoted by B_r in Fig. 35-6. Because at d, H is zero and so $B = \mu_0 M$, from Eq. (35-14), B_r/μ_0 is also called the *remanent magnetization*. The reverse H needed to reduce B to zero, denoted in the figure by H_c, is called the *coercive field* or *coercivity*.

A significant consequence of hysteresis is the dissipation of *energy* within a ferromagnetic material as it traverses its hysteresis loop. It can be shown that the energy dissipated per unit volume of material in one cycle around the loop is represented by the *area* of the loop.

Different magnetic properties are important for different applications. A permanent magnet material should have large remanent magnetization and, to prevent demagnetization in external fields, large coercivity. Properties of a few permanent-magnet materials are shown in Table 35-2.

For iron cores in inductors, transformers, motors, and other devices, it is usually desirable to have as little hysteresis as possible because of the attendant energy loss and heating when the field undergoes repeated reversals in the presence of an alternating current. In such cases the remanent magnetization and coercivity should be as small as possible. For such materials the curves $bcdefg$ and $bihg$ in Fig. 35-6 are very close together.

A material having small remanence and coercivity is said to be magnetically *soft*, while one having large values is said to be magnetically *hard*. It is interesting to note that soft iron is soft mechanically as well as magnetically, while some (but not all) steels are hard magnetically.

Table 35-2 **Remanence and coercivity of permanent magnet materials**

Material	Composition percent	B_r, T	H_c A·m^{-1}
Carbon steel	98 Fe, 0.86 C, 0.9 Mn	0.95	3.6×10^3
Cobalt steel	52 Fe, 36 Co, 7 W, 3.5 Cr, 0.5 Mn, 0.7 C	0.95	18×10^3
Alnico 2	55 Fe, 10 Al, 17 Ni, 12 Co, 6 Cu	0.76	42×10^3
Alnico 5	51 Fe, 8 Al, 14 Ni, 24 Co, 3 Cu	1.25	44×10^3

*35-8 The magnetic field of the earth

The magnetic field of the earth, at distances up to about five earth radii, is approximately the same as that outside a uniformly magnetized sphere. Figure 35–7a represents a section through the earth. The broken line is its axis of rotation, and the *geographic* north and south poles are lettered N_g and S_g. The direction of the (presumed) internal magnetization makes an angle of about 15° with the earth's axis. The dashed line indicates the plane of the magnetic equator, and the letters N_m and S_m represent the so-called *magnetic* north and south poles. Note carefully that B-field lines emerge from the earth's surface over the entire southern magnetic hemisphere and enter its surface over the entire northern magnetic hemisphere. Hence, if we wish to attribute the earth's field to magnetic poles, we must assume that magnetic N-poles are distributed over the entire *southern* magnetic hemisphere, and magnetic S-poles over the entire *northern* magnetic hemisphere!

It is interesting to note that the same field at *external* points would result if the earth's magnetism were due to a short bar magnet near its center, as in Fig. 35–7b, with the S-pole of the magnet pointing toward the north magnetic pole. The field within the earth is different in the two cases, but for obvious reasons experimental verification of either hypothesis is impossible.

Except at the magnetic equator, the earth's field is not horizontal. The angle that the field makes with the horizontal is called the *angle of dip* or the *inclination*. At Pittsburgh, Pennsylvania (about 40° N latitude), the magnitude of the earth's field is about 6.1×10^{-5} T and the angle of dip about 72°. Hence, the horizontal component at Pittsburgh is about 1.9×10^{-5} T and the vertical component about 5.8×10^{-5} T. In northern magnetic latitudes, the vertical component is directed downward; in southern magnetic latitudes it is upward. The angle of dip is, of course, 90° at the magnetic poles.

Out to a distance from the earth of about five earth radii, the magnetic field is governed almost entirely by the earth. At larger distances, the motions of ionized particles play an important role. These motions are strongly influenced by the *solar wind*, a thin hot gas expelled by the

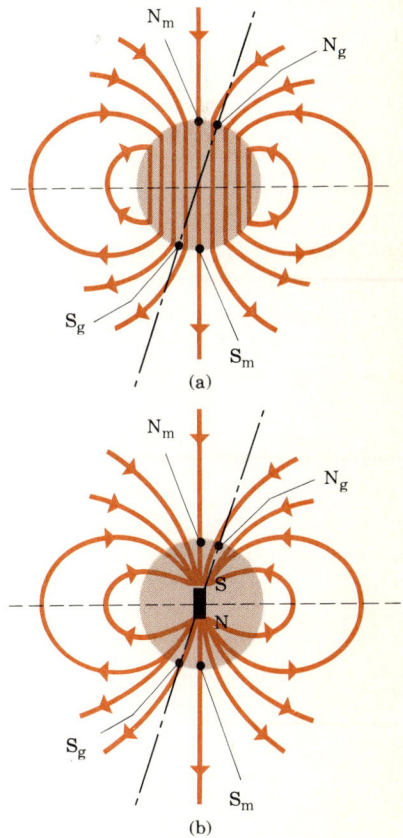

35-7 Simplified diagram of the magnetic field of the earth. The broken lines represent the axis of rotation.

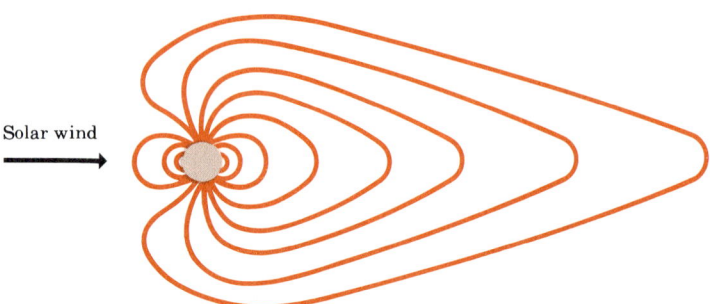

35-8 Probable distortion of the geomagnetic field by the solar wind.

sun. Recent studies using satellites and space probes suggest that at large distances the field is distorted, as suggested in Fig. 35-8.

The angle between the horizontal component of **B** and the true north–south direction is called the *variation* or *declination*. At Pittsburgh, the declination at present is about 5° W, that is, a compass needle points about 5° to the west of true north.

Questions

35-1 Why should the permeability of a paramagnetic material be expected to decrease with increasing temperature?

35-2 In the discussion of magnetic forces on current loops in Sec. 31-3, it was found that no net force is exerted on a complete loop in a uniform magnetic field, only a torque. Yet magnetized materials, which contain atomic current loops, certainly *do* experience net forces in magnetic fields. How is this discrepancy resolved?

35-3 What features of atomic structure determine whether an element is diamagnetic or paramagnetic?

35-4 The magnetic susceptibility of paramagnetic materials is quite strongly temperature-dependent, while that of diamagnetic materials is nearly independent of temperature. Why the difference?

35-5 When a piece of magnetic material is attracted to a permanent magnet, it is moving and hence has kinetic energy just before it strikes the magnet and sticks to it. Where does the energy come from? Where does it go?

35-6 Would a paramagnetic substance have a curved magnetization curve similar to Fig. 35-4 in a sufficiently strong field, or is this behavior limited to ferromagnetic materials?

35-7 You are given two iron bars and told that one is a permanent magnet, the other is not. They look identical. How can you tell which is which without using any equipment other than the bars themselves?

35-8 The magnetic declination at a fixed point on the earth usually varies with time, often by a few degrees per hundred years. What does this tell you about the interior of the earth?

35-9 Are there any points on earth where the magnetic declination is 180°, that is, where a compass points exactly backward?

35-10 You are lost in the woods with only a map, a compass, a fishhook, and a piece of string. Which is more important for you to know, the angle of dip or the declination?

35-11 A student proposed to obtain a single magnetic pole by taking a bar magnet with an N-pole at one end and an S-pole at the other end, and breaking off the N-end. Does this work?

35-12 In a laboratory electromagnet designed to produce very strong fields, what property of the iron core is most important, its permeability, remanence, or coercivity?

Problems

$$\mu_0 = 4\pi \times 10^{-7} \text{ Wb}\cdot\text{A}^{-1}\cdot\text{m}^{-1}.$$

35-1 Experimental measurements of the magnetic susceptibility of iron ammonium alum are given in Table 35-3. Make a graph of $1/\chi$ against Kelvin temperature; what conclusion can you draw?

Table 35-3

T, °C	χ
-258.15	129×10^{-4}
-173	19.4×10^{-4}
-73	9.7×10^{-4}
27	6.5×10^{-4}

35-2 A toroid having 500 turns of wire and a mean circumferential length of 50 cm carries a current of 0.3 A. The relative permeability of the core is 600.

a) What is the magnetic field in the core?
b) What is the magnetic intensity?
c) What part of the magnetic field is due to surface currents?

35-3 The current in the windings on a toroid is 2.0 A. There are 400 turns and the mean circumferential length is 40 cm. With the aid of a search coil and charge-measuring instrument, the magnetic field is found to be 1.0 T. Calculate

a) the magnetic intensity,
b) the magnetization,
c) the magnetic susceptibility,
d) the equivalent surface current, and
e) the relative permeability.

35-4 Each of the two preceding coils (Problems 35-2 and 35-3) has a cross-sectional area of 8 cm². Calculate the self-inductance L of each coil.

35-5 In 1911, Kamerlingh-Onnes discovered that at low temperatures some metals lose their electrical resistance and become superconductors. Thirty years later, Meissner showed that the magnetic induction inside a superconductor is zero. If the current in the windings of a toroidal superconductor is increased, a critical value H_c may be reached at which the metal suddenly becomes normal with practically zero magnetization.

a) Plot a graph of B/μ_0 against H from $H = 0$ to $H = 2H_c$.
b) Plot a graph of M against H in the same range.
c) Is a superconductor paramagnetic, diamagnetic, or ferromagnetic?
d) If the current in the windings is 10 A, how large are the surface currents on the superconducting ring and in what direction?
e) What happens when you try to place a small permanent magnet on a superconducting plate?

35-6 Table 35-4 lists corresponding values of H and B for a specimen of commercial hot-rolled silicon steel, a material widely used in transformer cores.

a) Construct graphs of B and μ as functions of H, in the range from $H = 0$ to $H = 1000$ A·m⁻¹.
b) What is the maximum permeability?

Table 35-4 Magnetic properties of silicon steel

Magnetic intensity H, A·m⁻¹	Flux density B, T
0	0
10	0.050
20	0.15
40	0.43
50	0.54
60	0.62
80	0.74
100	0.83
150	0.98
200	1.07
500	1.27
1,000	1.34
10,000	1.65
100,000	2.02
800,000	2.92

c) What is the initial permeability ($H = 0$)?
d) What is the permeability when $H = 800,000$ A·m⁻¹?

35-7 Construct graphs of $\mu_0 H$, $\mu_0 M$, and K_m against H, like those in Fig. 35-5 for the specimen of silicon steel in Table 35-4.

35-8 Suppose the ordinate of the point b in Fig. 35-6 corresponds to a flux density of 1.6 T, and the abscissa to a magnetic intensity H of 1000 A·m⁻¹. Approximately what is the relative permeability at points a, b, c, d, e, and f.

35-9 A bar magnet has a coercivity of 4×10^3 A·m⁻¹. It is desired to demagnetize it by inserting it inside a solenoid 12 cm long and having 60 turns. What current should be carried by the solenoid?

35-10 The horizontal component of the earth's magnetic field at Cambridge, Mass., is 1.7×10^{-5} Wb·m⁻². What is the horizontal component of the magnetic intensity?

35-11 A Rowland ring has a cross section of 2 cm², a mean length of 30 cm, and is wound with 400 turns. Find the current in the winding that is required to set up a flux density of 0.1 T in the ring,

a) if the ring is of annealed iron (Fig. 35-4);
b) if the ring is of silicon steel (Table 35-4).
c) Repeat the computations above if a flux density of 1.2 T is desired.

36 ALTERNATING CURRENTS

36-1 Introduction

In Chapters 28 and 29 we have studied the behavior of *direct-current* (dc) circuits, in which all the currents, voltages, and emfs are *constant*, that is, not varying with time. However, many electric circuits of practical importance, including nearly all large-scale electric-power distribution systems and much electronic equipment, use *alternating current* (ac), in which the voltages and currents vary with time, often in a *sinusoidal* manner. In this chapter we study the properties of circuits in which there are sinusoidally varying voltages and currents. Many of the same principles used in Chapters 28 and 29 are applicable here as well, and there are some new ideas related to the circuit behavior of inductors and capacitors.

We have already seen several sources of alternating emf or voltage. A coil of wire, rotating with constant angular velocity in a magnetic field, develops a sinusoidal alternating emf, as explained in Sec. 33–1. This simple device is the prototype of the commercial alternating current generator, or *alternator*. An *L–C* circuit, as discussed in Sec. 34–5, oscillates sinusoidally, and with the proper circuitry provides an alternating potential difference between its terminals having a frequency, depending on the purpose for which it is designed, that may range from a few hertz to many millions of hertz.

We consider first a number of circuits connected to an alternator or oscillator that maintains between its terminals a sinusoidal alternating potential difference

$$v = V \cos \omega t. \tag{36-1}$$

In this expression V is the maximum potential difference or the *voltage amplitude*, v is the *instantaneous* potential difference, and ω the *angular frequency*, equal to 2π times the frequency f. For brevity, the alternator or oscillator will be referred to as an *ac source*. The circuit-diagram symbol for an ac source is ⊙.

Analysis of alternating-current circuits is facilitated by use of vector diagrams similar to those used in the study of harmonic motion (Sec.

11-5). In such diagrams, the instantaneous value of a quantity that varies sinusoidally with time is represented by the *projection* onto a horizontal axis of a vector of length corresponding to the amplitude of the quantity and rotating counterclockwise with angular velocity ω. In the context of ac-circuit analysis, these rotating vectors are often called *phasors*, and diagrams containing them are called *phasor diagrams*.

Alternating currents are of utmost importance in technology and industry. Transmission of power over long distance is very much easier and more economical with alternating than with direct currents. Circuits used in modern communication equipment, including radio and television, make extensive use of alternating currents. Many life processes involve alternating voltages and currents. The beating of the heart induces alternating currents in the surrounding tissues; the detection and study of these currents, called electrocardiography, provide valuable information concerning the health or pathology of the heart. Electroencephalograms, recordings of alternating currents in the brain, provide analogous information regarding brain function. Both electrocardiograms and electroencephalograms are invaluable diagnostic tools in modern medicine.

36-2 Circuits containing resistance, inductance, or capacitance

Let a resistor of resistance R be connected between the terminals of an ac source, as in Fig. 36-1a. The instantaneous potential of point a with respect to point b is $v_{ab} = V \cos \omega t$, and the instantaneous current in the resistor is

$$i = \frac{v_{ab}}{R} = \frac{V}{R} \cos \omega t.$$

The maximum current I, or the *current amplitude*, is evidently

$$I = \frac{V}{R}, \quad (36\text{-}2)$$

and we can therefore write

$$i = I \cos \omega t. \quad (36\text{-}3)$$

The current and voltage are both proportional to $\cos \omega t$, so the current is *in phase* with the voltage. The current and voltage amplitudes, from Eq. (36-2), are related in the same way as in a dc circuit.

Figure 36-1b shows graphs of i and v as functions of time. The fact that the curve representing the current has the greater amplitude in the diagram is of no significance, because the choice of vertical scales for i and v is arbitrary. The corresponding phasor diagram is given in Fig. 36-1c. Because i and v are *in phase* and have the same frequency, the current and voltage phasors rotate together.

Next, suppose that a capacitor of capacitance C is connected across the source, as in Fig. 36-2a. The instantaneous charge q on the capacitor is

$$q = Cv_{ab} = CV \cos \omega t. \quad (36\text{-}4)$$

In this case the instantaneous current is equal to the *rate of change* of

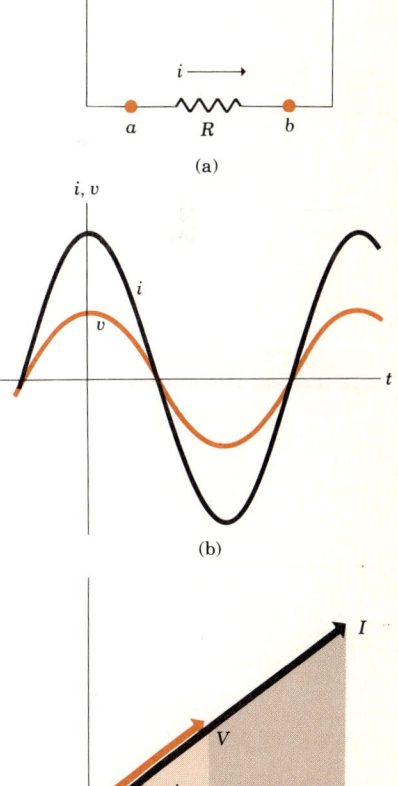

36-1 (a) Resistance R connected across an ac source. (b) Graphs of instantaneous voltage and current. (c) Phasor diagram; current and voltage in phase.

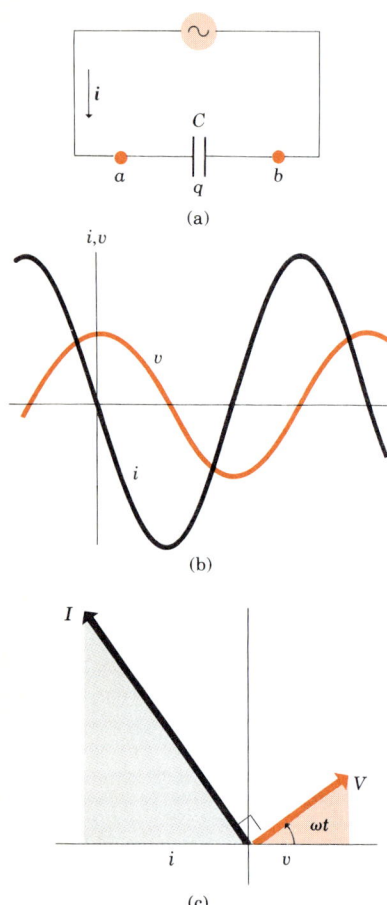

36-2 (a) Capacitor C connected across an ac source. (b) Graphs of instantaneous voltage and current. (c) Phasor diagram; current *leads* voltage by 90°.

the capacitor charge and is thus proportional to the rate of change of voltage. Taking the derivative of Eq. (36-4), we find

$$i = \frac{dq}{dt} = -\omega CV \sin \omega t. \qquad (36\text{-}5)$$

Thus if the voltage is represented by a *cosine* function, the current is represented by a *negative sine* function, as shown in Fig. 36-2b. The current is *not* in phase with the voltage; the current is greatest at times when the v curve is rising or falling most steeply, and is zero at times when the v curve "levels off," that is, when it reaches its maximum and minimum values.

The voltage and current are "out of step" or *out of phase* by a quarter-cycle, with the current a quarter-cycle ahead. The peaks of current occur a quarter-cycle *before* the corresponding voltage peaks. This result is also represented by the vector diagram of Fig. 36-2c, which shows the current vector ahead of the voltage vector by a quarter-cycle, or 90°. This *phase difference* between current and voltage can also be obtained by rewriting Eq. (36-5), using the trigonometric identity $\cos (A + 90°) = -\sin A$:

$$i = \omega CV \cos (\omega t + 90°). \qquad (36\text{-}6)$$

This shows that the expression for i can be viewed as a cosine function with a "head start" of 90° compared with that for the voltage. We say that the current *leads* the voltage by 90° or that the voltage *lags* the current by 90°.

Equations (36-5) and (36-6) show that the maximum current I is given by

$$I = \omega CV. \qquad (36\text{-}7)$$

This expression can be put in the same *form* as that for the maximum current in a resistor ($I = V/R$) if we write Eq. (36-7) as

$$I = \frac{V}{1/\omega C}.$$

and define a quantity X_C, called the *capacitive reactance* of the capacitor, as

$$X_C = \frac{1}{\omega C}. \qquad (36\text{-}8)$$

Then

$$I = \frac{V}{X_C}. \qquad (36\text{-}9)$$

From Eq. (36-9), the unit of capacitive reactance is one *volt per ampere* (1 V·A^{-1}), or one *ohm* (1 Ω).

The reactance of a capacitor is inversely proportional both to the capacitance C and to the angular frequency ω; the greater the capacitance, and the higher the frequency, the *smaller* is the reactance X_C.

Example 1 At an angular frequency of 1000 rad·s^{-1}, the reactance of a 1-μF capacitor is

$$X_C = \frac{1}{\omega C} = \frac{1}{(10^3 \text{ rad·s}^{-1})(10^{-6} \text{ F})} = 1000 \text{ Ω}.$$

At frequency of 10,000 rad·s⁻¹, the reactance of the same capacitor is only 100 Ω, and at a frequency of 100 rad·s⁻¹ it is 10,000 Ω.

Finally, suppose a pure inductor having a self-inductance L and zero resistance is connected to an ac source as in Fig. 36-3. Since the potential difference between the terminals of an inductor equals $L\,di/dt$, we have

$$L\frac{di}{dt} = V\cos\omega t \qquad (36\text{-}10)$$

and

$$di = \frac{V}{L}\cos\omega t\,dt.$$

Integration of both sides gives

$$i = \frac{V}{\omega L}\sin\omega t + C.$$

If $i = 0$ at time $t = 0$, then $C = 0$ and

$$i = \frac{V}{\omega L}\sin\omega t. \qquad (36\text{-}11)$$

Again, the voltage and current are a quarter-cycle out of phase, but this time the current *lags* the voltage (or the voltage *leads* the current) by 90°. This may also be seen by rewriting Eq. (36-11) using the trigonometric identity $\cos(A - 90°) = \sin A$:

$$i = \frac{V}{\omega L}\cos(\omega t - 90°). \qquad (36\text{-}12)$$

This result and the vector diagram of Fig. 36-3c show that the current can be viewed as a cosine function with a "late start" of 90°.

In Fig. 36-3b, the points of maximum voltage correspond on the graphs to points of maximum steepness of the current curve, and the points of zero voltage correspond to the points where the current curve levels off at its maximum and minimum values.

The maximum current is

$$I = \frac{V}{\omega L}, \qquad (36\text{-}13)$$

and

$$i = I\sin\omega t. \qquad (36\text{-}14)$$

The *inductive reactance* X_L of an inductor is defined as

$$X_L = \omega L, \qquad (36\text{-}15)$$

and Eq. (36-13) can also be written in the same form as that for a resistor:

$$I = \frac{V}{X_L}. \qquad (36\text{-}16)$$

The unit of inductive reactance is also 1 *ohm*.

The reactance of an inductor is directly proportional both to its inductance L and to the angular frequency ω; the greater the inductance, and the higher the frequency, the *larger* is the reactance.

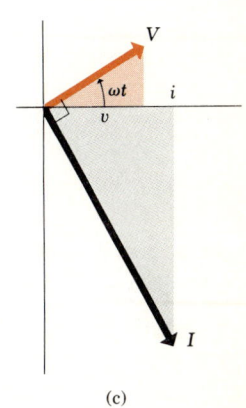

36-3 (a) Inductance L connected across an ac source. (b) Graphs of instantaneous voltage and current. (c) Phasor diagram; current lags voltage by 90°.

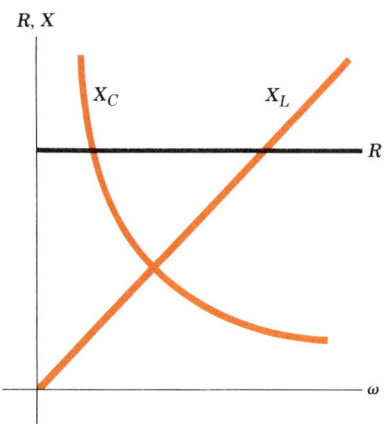

36-4 Graphs of R, X_L, and X_C as functions of frequency.

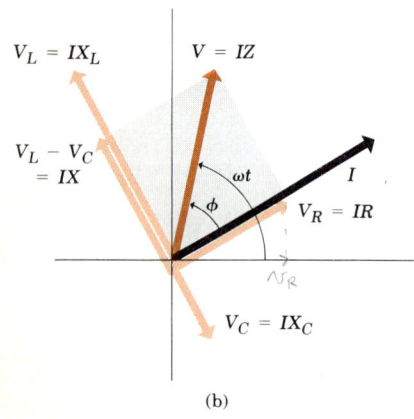

36-5 (a) A series R-L-C circuit. (b) Phasor diagram.

Example 2 At an angular frequency of 1000 rad·s⁻¹, the reactance of a 1-H inductor is

$$X_L = \omega L = (10^3 \text{ rad·s}^{-1})(1 \text{ H}) = 1000 \text{ }\Omega.$$

At a frequency of 10,000 rad·s⁻¹ the reactance of the same inductor is 10,000 Ω, while at a frequency of 100 rad·s⁻¹ it is only 100 Ω.

The graphs in Fig. 36-4 summarize the variations with frequency of the resistance of a resistor, and of the reactances of an inductor and of a capacitor. As the frequency increases, the reactance of the inductor approaches infinity, and that of the capacitor approaches zero. As the frequency decreases, the inductive reactance approaches zero and the capacitive reactance approaches infinity. The limiting case of zero frequency corresponds to a dc circuit.

36-3 The R-L-C series circuit

In many instances, ac circuits include resistance, inductive reactance, and capacitive reactance. A simple series circuit is shown in Fig. 36-5a. Analysis of this and similar circuits is facilitated by use of a phasor diagram, which includes the voltage and current vectors for the various components. In this instance the instantaneous *total* voltage across all three components is equal to the source voltage at that instant, and its phasor is the *vector sum* of the phasors for the individual voltages. The complete phasor diagram for this circuit is shown in Fig. 36-5b. This may appear complex, so we shall explain it step by step.

The instantaneous current i has the same value at all points of the circuit. Thus, a *single phasor I*, of length proportional to the current amplitude, suffices to represent the current in each circuit element.

Let us use the symbols v_R, v_L, and v_C for the instantaneous voltages across R, L, and C, and V_R, V_L, and V_C for their maximum values. The instantaneous and maximum voltages across the source will be represented by v and V. Then $v = v_{ab}$, $v_R = v_{ac}$, $v_L = v_{cd}$, and $v_C = v_{db}$.

We have shown that the potential difference between the terminals of a resistor is *in phase* with the current in the resistor, and that its maximum value V_R is

$$V_R = IR.$$

Thus the phasor V_R in Fig. 36-5b, in phase with the current vector, represents the voltage across the resistor. Its projection on the horizontal axis, at any instant, gives the instantaneous potential difference v_R.

The voltage across an inductor *leads* the current by 90°. The voltage amplitude is

$$V_L = IX_L.$$

The phasor V_L in Fig. 36-5b represents the voltage across the inductor, and its projection at any instant onto the horizontal axis equals v_L.

The voltage in a capacitor *lags* the current by 90°. The voltage amplitude is

$$V_C = IX_C.$$

The phasor V_C in Fig. 36-5b represents the voltage across the capacitor, and its projection at any instant onto the horizontal axis equals v_C.

The instantaneous potential difference v between terminals a and b equals at every instant the (algebraic) sum of the potential differences v_R, v_L, and v_C. That is, it equals the sum of the projections of the phasors V_R, V_L, and V_C. But the *projection* of the *vector sum* of these phasors is equal to the *sum* of their *projections*, so this vector sum V must be the phasor that represents the source voltage. To form the vector sum, we first subtract the phasor V_C from the phasor V_L (since these always lie in the same straight line), giving the phasor $V_L - V_C$. Since this is at right angles to the phasor V_R, the magnitude of the phasor V is

$$V = \sqrt{V_R^2 + (V_L - V_C)^2} = \sqrt{(IR)^2 + (IX_L - IX_C)^2}$$
$$= I\sqrt{R^2 + (X_L - X_C)^2}.$$

The quantity $X_L - X_C$ is called the *reactance of the circuit*, denoted by X:

$$X = X_L - X_C. \tag{36-17}$$

Finally, we define the *impedance* Z of the circuit as

$$Z = \sqrt{R^2 + (X_L - X_C)^2} = \sqrt{R^2 + X^2}, \tag{36-18}$$

so we can write

$$V = IZ \quad \text{or} \quad I = \frac{V}{Z}. \tag{36-19}$$

Thus, the *form* of the equation relating current and voltage *amplitudes* is the same as that for a dc circuit, the impedance Z playing the same role as the resistance R of the dc circuit. Note, however, that the impedance is actually a function of R, L, and C, as well as of the frequency ω. The complete expression for Z, for a series circuit, is

$$Z = \sqrt{R^2 + X^2}$$
$$= \sqrt{R^2 + (X_L - X_C)^2}$$
$$= \sqrt{R^2 + [\omega L - (1/\omega C)]^2}. \tag{36-20}$$

The unit of impedance, from Eq. (36-19) or (36-20), is one *volt per ampere* ($1\ \text{V}\cdot\text{A}^{-1}$) or one *ohm*.

The expressions for the impedance Z of (a) an R–L series circuit, (b) an R–C series circuit, and (c) an L–C series circuit can be obtained from Eq. (36-20) by letting (a) $X_C = 0$, (b) $X_L = 0$, and (c) $R = 0$.

Equation (36-20) gives the impedance Z only for a *series* R–L–C circuit. But whatever the nature of R–L–C network, its impedance can be *defined* by Eq. (36-19) as the ratio of the voltage amplitude to the current amplitude.

The angle ϕ, in Fig. 36-5b, is the phase angle of the source voltage V with respect to the current I. From the diagram,

$$\tan \phi = \frac{V_L - V_C}{V_R} = \frac{I(X_L - X_C)}{IR} = \frac{X}{R}. \tag{36-21}$$

Hence, if the source voltage is represented by a cosine function,

$$v = V \cos \omega t,$$

the current *lags* by an angle ϕ between 0 and 90°, and its equation is

$$i = I \cos (\omega t - \phi).$$

Figure 36–5b has been constructed for a circuit in which $X_L > X_C$. If $X_L < X_C$, vector V lies on the opposite side of the current vector I and the current *leads* the voltage. In this case, $X = X_L - X_C$ is a *negative* quantity, tan ϕ is negative, and ϕ is a negative angle between 0 and $-90°$.

To summarize, we can say that the *instantaneous* potential differences in an ac series circuit add *algebraically*, just as in a dc circuit, while the voltage *amplitudes* add *vectorially*.

Example In the series circuit of Fig. 36–5, let $R = 300\,\Omega$, $L = 0.9$ H, $C = 2.0\,\mu$F, and $\omega = 1000$ rad·s^{-1}. Then

$$X_L = \omega L = 900\,\Omega, \qquad X_C = \frac{1}{\omega C} = 500\,\Omega.$$

The reactance X of the circuit is

$$X = X_L - X_C = 400\,\Omega,$$

and the impedance Z is

$$Z = \sqrt{R^2 + X^2} = 500\,\Omega.$$

If the circuit is connected across an ac source of voltage amplitude 50 V, the current amplitude is

$$I = \frac{V}{Z} = 0.10 \text{ A}.$$

The lag angle ϕ is

$$\phi = \tan^{-1}\frac{X}{R} = 53°.$$

The voltage amplitude across the resistor is

$$V_R = IR = 30 \text{ V}.$$

The voltage amplitudes across the inductor and capacitor are, respectively,

$$V_L = IX_L = 90 \text{ V}, \qquad V_C = IX_C = 50 \text{ V}.$$

The above analysis describes the *steady-state* condition of a circuit, the condition that prevails after the circuit has been connected to the source for a long time. When the source is first connected, there may be additional voltages and currents, called *transients*, whose nature depends on the time in the cycle when the circuit is initially completed. A detailed analysis of transients is beyond our scope. In any event, they always die out after a sufficiently long time, and therefore do not affect the steady-state behavior of the circuit.

36-4 Average and root-mean-square values; ac instruments

The *instantaneous* potential difference between two points of an ac circuit can be measured by connecting a calibrated oscilloscope between the points, and the instantaneous current by connecting an oscilloscope across a resistor in the circuit. The usual moving-coil galvanometer, however, has too large a moment of inertia to follow the instantaneous values of an alternating current. It averages out the fluctuating torque on its coil, and its deflection is proportional to the *average* current.

The *average* value f_{av} of any quantity $f(t)$ that varies with time, over a time interval from t_1 to t_2, is defined as

$$f_{av} = \frac{1}{t_2 - t_1} \int_{t_1}^{t_2} f(t)\, dt. \quad (36\text{–}22)$$

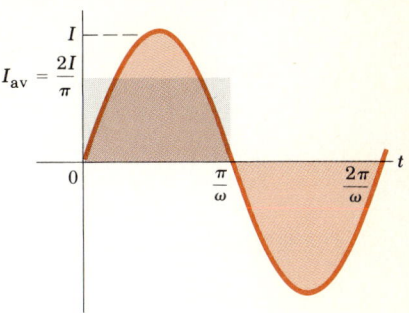

36–6 The average value of a sinusoidal current over a half-cycle is $2I/\pi$. The average over a complete cycle is zero.

This may be interpreted graphically. The integral represents the *area* bounded by the curve of $f(t)$ and by the two vertical lines t_1 and t_2. The product $f_{av}(t_2 - t_1)$ is the area of a rectangle of height f_{av} and base $(t_2 - t_1)$. Thus f_{av} is the height of a rectangle having the same area as the area under the curve between the same two values of t.

Let us apply this definition to a sinusoidally varying quantity, for example a current given by

$$i = I \sin \omega t.$$

The period τ of this sinusoidal current is given by $\tau = 1/f = 2\pi/\omega$. Let us find the average value of i for a half-period or half-cycle, from $t = 0$ to $t = \pi/\omega$. From Eq. (36–22), this is

$$I_{av} = \frac{\omega}{\pi} \int_0^{\pi/\omega} I \sin \omega t\, dt = \frac{2}{\pi} I. \quad (36\text{–}23)$$

That is, the average current is $2/\pi$ (about $\frac{2}{3}$) times the maximum current, and the area under the rectangle in Fig. 36–6 equals the area under one loop of the sine curve.

The average current for a complete cycle (or any number of complete cycles) is

$$I_{av} = \frac{\omega}{2\pi} \int_0^{2\pi/\omega} I \sin \omega t\, dt = 0,$$

as would be expected, since the *positive* area of the loop between 0 and π/ω is equal to the *negative* area of the loop between π/ω and $2\pi/\omega$. Hence if a sinusoidal current is sent through a moving-coil galvanometer, the meter reads zero!

Such a meter can be used in an ac circuit, however, if it is connected in the *full-wave rectifier circuit* shown in Fig. 36–7a. As explained in Sec. 28–5, an ideal rectifier offers a constant finite resistance to current in the forward direction and an infinite resistance to current in the opposite direction. When the line current is in the direction shown, two of the rectifiers are conducting and two are nonconducting, and the current in the galvanometer is upward. When the line current is in the opposite direction, the rectifiers that are nonconducting in Fig. 36–7a carry the current, and the current in the galvanometer is still upward. Thus, if the line current alternates sinusoidally, the current in the galvanometer has the waveform shown as a solid curve in Fig. 36–7b. Although pulsating, it is always in the same direction and its average value is *not* zero. Then, when provided with the necessary series resistance or shunt, as for a dc voltmeter or ammeter, the galvanometer can serve as an ac voltmeter or ammeter.

The average value of the rectified current, in any number of complete cycles, is the same as the average current in the first half-cycle in

36–7 (a) A full-wave rectifier. (b) Graph of a full-wave rectified current and its average value.

Fig. 36–6, or $2/\pi$ times the maximum current I. Hence, if the meter deflects full-scale with a steady current I_0 through it, it also deflects full-scale when the *average value* of the rectified current, $2I/\pi$, is equal to I_0. The current amplitude I, when the meter deflects full-scale, is then

$$I = \frac{\pi I_0}{2}.$$

For example, if $I_0 = 1$ A, $I = 1.57$ A.

Most ac meters are calibrated to read not the maximum value of the current or voltage, but the *root-mean-square* value, that is, the *square root* of the *average* value of the *square* of the current or voltage. This special kind of average is usually called simply the rms value. Because i^2 is always positive, the average of i^2 is never zero, even when the average of i itself is zero. Also, rms values of voltage and current are useful in relationships for *power* in ac circuits; we return to this point in Sec. 36–5.

Figure 36–8 shows graphs of a sinusoidally varying current and of its square. If $i = I \sin \omega t$, then

$$i^2 = I^2 \sin^2 \omega t = I^2 \left[\frac{1}{2}(1 - \cos 2\omega t) \right]$$

$$= \frac{1}{2}I^2 - \frac{1}{2}I^2 \cos 2\omega t.$$

The average value of i^2, or the *mean square current*, is equal to the constant term $\frac{1}{2}I^2$, since the average value of $\cos 2\omega t$, over any number of complete cycles, is zero:

$$(I^2)_{av} = \frac{I^2}{2}.$$

The root-mean-square current is the square root of this, or

$$I_{rms} = \sqrt{(I^2)_{av}} = \frac{I}{\sqrt{2}}. \tag{36–24}$$

In the same way, the root-mean-square value of a sinusoidal voltage is

$$V_{rms} = \frac{V}{\sqrt{2}}. \tag{36–25}$$

Voltages and currents in power distribution systems are always referred to in terms of their rms values. Thus, when we speak of our household power supply as "115-volt ac," this means that the rms voltage is 115 V. The voltage amplitude is

$$V = \sqrt{2} V_{rms} = 163 \text{ V}.$$

All the voltage-current relations developed in preceding sections have been stated in terms of amplitudes (maximum values) of voltage and current, but they remain valid when rms values are used; in each case the difference is simply a multiplicative factor of $1/\sqrt{2}$.

36–5 Power in ac circuits

The *instantaneous* power input p to an ac circuit is

$$p = vi,$$

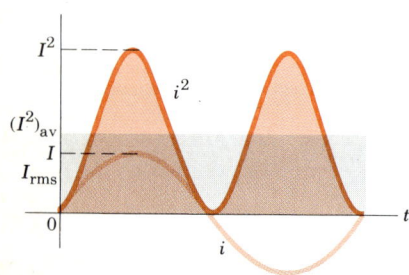

36–8 The average value of the square of a sinusoidally varying current, over any number of half-cycles, is $I^2/2$. The root-mean-square value is $I/\sqrt{2}$.

where v is the instantaneous source potential difference and i is the instantaneous current. We consider first some special cases.

If the circuit consists of a pure resistance R, as in Fig. 36–1, i and v are *in phase*. The graph representing p is obtained by multiplying together at every instant the ordinates of the graphs of v and i in Fig. 36–1b, and it is shown by the full curve in Fig. 36–9a. (The product vi is positive when v and i are both positive or both negative.) That is, energy is supplied *to* the resistor at all instants, whatever the direction of the current, although the *rate* at which it is supplied is not constant.

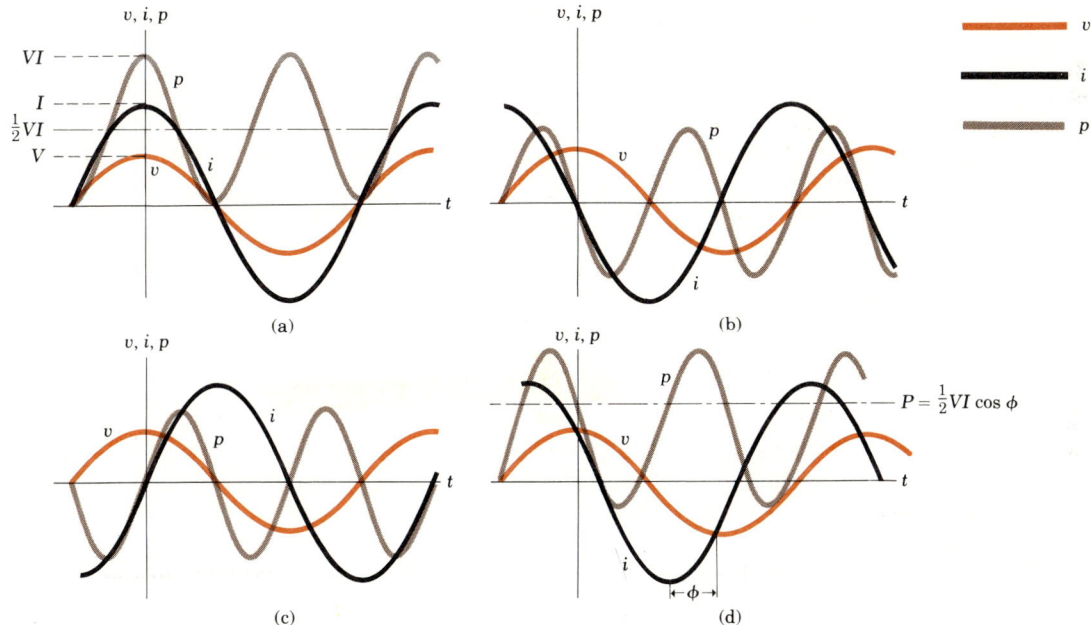

36–9 Graphs of voltage, current, and power as functions of time, for various circuits. (a) Instantaneous power input to a resistor. The average power is $\frac{1}{2}VI$. (b) Instantaneous power input to a capacitor. The average power is zero. (c) Instantaneous power input to a pure inductor. The average power is zero. (d) Instantaneous power input to an arbitrary ac circuit. The average power is $\frac{1}{2}VI \cos\phi = V_{rms}I_{rms}\cos\phi$.

The power curve is symmetrical about a value equal to one-half its maximum ordinate VI, so the *average* power P is

$$P = \frac{1}{2}VI. \qquad (36\text{–}26)$$

The average power can also be written

$$P = \frac{VI}{\sqrt{2}}\left(\frac{I}{\sqrt{2}}\right) = V_{rms}I_{rms}. \qquad (36\text{–}27)$$

Furthermore, since $V_{rms} = I_{rms}R$, we have

$$P = I_{rms}^2 R. \qquad (36\text{–}28)$$

Note that Eqs. (36–27) and (36–28) have the same *form* as those for a dc circuit.

Suppose next that the circuit consists of a capacitor, as in Fig. 36–2. The current and voltage are then 90° out of phase. When the curves of v and i are multiplied together (the product vi is *negative* when v and i have *opposite* signs), we get the power curve in Fig. 36–9b, which is symmetrical about the horizontal axis. The average power is therefore zero.

To see why this is so, we recall that positive power means that energy is supplied *to* a device and that negative power means that energy is supplied *by* a device. The process we are considering is merely the charge and discharge of a capacitor. During the intervals when p is positive, energy is supplied to charge the capacitor, and when p is negative the capacitor is discharging and returning energy to the source.

Figure 36–9c is the power curve for a pure inductor. As with a capacitor, the current and voltage are out of phase by 90°, and the average power is zero. Energy is supplied to establish a magnetic field around the inductor and is returned to the source when the field collapses.

In the most general case, current and voltage differ in phase by an angle ϕ and

$$p = [V \sin \omega t][I \sin(\omega t - \phi)]. \qquad (36\text{--}29)$$

The instantaneous power curve has the form shown in Fig. 36–9d. The area under the positive loops is greater than that under the negative loops and the net average power is positive.

The preceding analyses have shown that when v and i are *in phase*, the average power equals $\tfrac{1}{2}VI$; and when v and i are 90° *out of phase*, the average power is zero. Hence, in the general case, when v and i differ by an angle ϕ, the average power equals $\tfrac{1}{2}V$, multiplied by $I \cos \phi$, the component of I that is *in phase* with V. That is,

$$P = \frac{1}{2} VI \cos \phi = V_{\text{rms}} I_{\text{rms}} \cos \phi. \qquad (36\text{--}30)$$

This is the general expression for the power input to *any* ac circuit. The factor $\cos \phi$ is called the *power factor* of the circuit. For a pure resistance, $\phi = 0$, $\cos \phi = 1$, and $P = V_{\text{rms}} I_{\text{rms}}$. For a capacitor or inductor, $\phi = 90°$, $\cos \phi = 0$, and $P = 0$.

A low power factor (large angle of lag or lead) is usually undesirable in power circuits because, for a given potential difference, a large current is needed to supply a given amount of power with correspondingly large heat losses in the transmission lines. Since many types of ac machinery draw a lagging current, this situation is likely to arise. It can be corrected by connecting a capacitor in parallel with the load. The leading current drawn by the capacitor compensates for the lagging current in the other branch of the circuit. The capacitor itself takes no net power from the line.

36–6 Series resonance

The impedance of an R–L–C series circuit depends on the frequency, since the inductive reactance is directly proportional to frequency, and the capacitive reactance inversely proportional. This dependence is illustrated in Fig. 36–10a, where a logarithmic frequency scale has been used because of the wide range of frequencies covered. Note that there is one particular frequency at which X_L and X_C are numerically equal. At this frequency, $X = X_L - X_C$ is zero. Hence the impedance Z, equal to $\sqrt{R^2 + X^2}$, is *minimum* at this frequency and is equal to the resistance R.

If an ac source of constant voltage amplitude but variable frequency is connected across the circuit, the current amplitude I varies with frequency, as shown in Fig. 36-10b, and is *maximum* at the frequency at which the impedance Z is *minimum*. The same diagram also shows the phase difference ϕ of the current with respect to the voltage, as a function of frequency. At low frequencies, where capacitive reactance X_C predominates, the current *leads* the voltage. At the frequency where $X = 0$, the current and voltage are *in phase*, and at high frequencies, where X_L predominates, the current *lags* the voltage.

The behavior of the current in an ac series circuit, as the source frequency is varied, is exactly analogous to the response of a spring-mass system having a viscous damping force as the frequency of the driving force is varied. The frequency ω_0 at which the current is maximum is called the *resonant frequency* and is easily computed from the fact that, at this frequency, $X_L = X_C$:

$$X_L = X_C, \qquad \omega_0 L = \frac{1}{\omega_0 C}, \qquad \omega_0 = \frac{1}{\sqrt{LC}}. \qquad (36\text{-}31)$$

Note that this is equal to the natural frequency of oscillation of an L-C circuit, as derived in Sec. 34-5.

If the inductance L or the capacitance C of a circuit can be varied, the resonant frequency can be varied also. This is the procedure by which a radio or television receiving set may be "tuned" to receive the signal from a desired station.

Example The series circuit in Fig. 36-11 is connected to the terminals of an ac source whose frequency is variable but whose rms terminal voltage is constant and equal to 100 V. The resonant frequency is

$$\omega_0 = \frac{1}{\sqrt{LC}} = \frac{1}{\sqrt{(2 \text{ H})(0.5 \times 10^{-6} \text{ F})}} = 1000 \text{ rad} \cdot \text{s}^{-1}.$$

At this frequency,

$$X_L = (1000 \text{ rad} \cdot \text{s}^{-1})(2 \text{ H}) = 2000 \text{ }\Omega,$$

$$X_C = \frac{1}{(1000 \text{ rad} \cdot \text{s}^{-1})(0.5 \times 10^{-6} \text{ F})} = 2000 \text{ }\Omega,$$

$$X = X_L - X_C = 0.$$

The impedance Z is then equal to the resistance R, and the rms current is

$$I = \frac{V}{Z} = \frac{V}{R} = 0.20 \text{ A}.$$

The rms potential difference across the resistor is

$$V_R = IR = 100 \text{ V}.$$

The rms potential differences across the inductor and capacitor are, respectively,

$$V_L = IX_L = (0.20 \text{ A})(2000 \text{ }\Omega) = 400 \text{ V},$$
$$V_C = IX_C = (0.20 \text{ A})(2000 \text{ }\Omega) = 400 \text{ V}.$$

The rms potential difference across the inductor–capacitor combination (V_{cb}) is

$$V = IX = I(X_L - X_C) = 0.$$

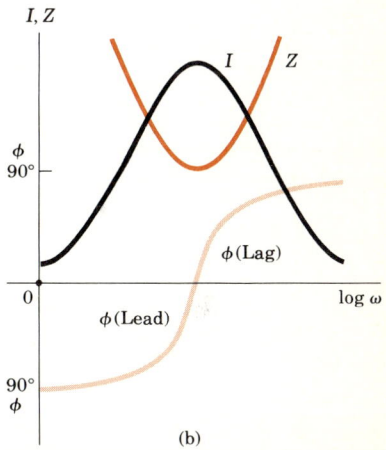

36-10 (a) Reactance, resistance, and impedance as functions of frequency (logarithmic frequency scale). (b) Impedance, current, and phase angle as functions of frequency (logarithmic frequency scale).

36-11 Series resonant circuit.

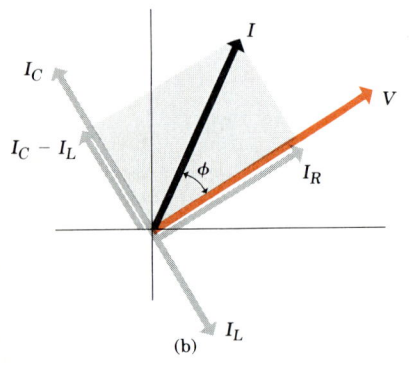

36-12 (a) A parallel R-L-C circuit. (b) Phasor diagram showing current phasors for the three branches. The single voltage phasor V represents the voltage across all three branches.

This is true because the instantaneous potential differences across the inductor and the capacitor have equal amplitudes but are 180° out of phase. Although the rms value of each is substantial, their *resultant* at each instant is zero.

36-7 Parallel circuits

Circuit elements connected in parallel across an ac source can be analyzed by the same procedure as for elements in series. The phasor diagram for the circuit in Fig. 36–12a is given in Fig. 36–12b. In this case, the instantaneous potential difference across each element is the same, and a *single* phasor V represents the common terminal voltage. The phasor I_R, of amplitude V/R and in phase with V, represents the current in the resistor. Phasor I_L, of amplitude V/X_L and lagging V by 90°, represents the current in the inductor, and phasor I_C, of amplitude V/X_C and leading V by 90°, represents the current in the capacitor.

The instantaneous current i, by Kirchhoff's point rule, equals the (algebraic) sum of the instantaneous currents i_R, i_L, and i_C, and is represented by the phasor I, the vector sum of phasors I_R, I_L, and I_C. Angle ϕ is the phase angle of current with respect to source voltage.

From Fig. 36–12,

$$I = \sqrt{I_R^2 + (I_C - I_L)^2} = \sqrt{\left(\frac{V}{R}\right)^2 + \left(\omega C V - \frac{V}{\omega L}\right)^2}$$
$$= V\sqrt{\frac{1}{R^2} + \left(\omega C - \frac{1}{\omega L}\right)^2} \qquad (36\text{-}32)$$

The maximum current I is frequency-dependent, as expected. It is *minimum* when the second factor in the radical is zero; this occurs when the two reactances have equal magnitudes, at the resonant frequency ω_0 given by Eq. (36–31).

Thus at resonance the total current in the parallel R-L-C circuit is *minimum*, in contrast to the R-L-C *series* circuit, which has *maximum* current at resonance. This can be understood by noting that, in the parallel circuit, the currents in L and C are *always* exactly a half-cycle out of phase; when they also have equal *magnitudes*, they cancel each other completely, and the total current is simply that through R. Indeed, when $\omega C = 1/\omega L$, Eq. (36–32) becomes simply $I = V/R$. This does *not* mean that there is *no* current in L or C at resonance, but only that the two currents cancel. If R is large, the equivalent impedance of the circuit near resonance is much *larger* than the individual reactances X_L and X_C.

*36-8 The transformer

For reasons of efficiency, it is desirable to transmit electrical power at high voltages and small currents, with consequent reduction of I^2R heating in the transmission line. On the other hand, considerations of safety and of insulation of moving parts require relatively low voltages in generating equipment and in motors and household appliances. One of the most useful features of ac circuits is the ease and efficiency with which voltages (and currents) may be changed from one value to another by *transformers*.

In principle, the transformer consists of two coils electrically insulated from each other and wound on the same iron core. An alternating current in one winding sets up an alternating flux in the core, and the induced electric field produced by this varying flux (see Sec. 33–2) induces an emf in the other winding. Energy is thus transferred from one winding to another via the core flux and its associated induced electric field. The winding to which power is supplied is called the *primary;* that from which power is delivered is called the *secondary*. The circuit symbol for an iron core transformer is

The power output of a transformer is necessarily less than the power input because of unavoidable losses. These losses consist of I^2R losses in the primary and secondary windings, and hysteresis and eddy-current losses in the core. Hysteresis losses are minimized by the use of iron having a narrow hysteresis loop; and eddy currents are minimized by laminating the core. In spite of these losses, transformer efficiencies are usually well over 90%, and in large installations may reach 99%.

For simplicity, we shall consider only an idealized transformer in which there are *no* losses and in which all of the flux is confined to the iron core, so that the same flux links both primary and secondary. The transformer is shown schematically in Fig. 36–13. A primary winding of N_1 turns and a secondary winding of N_2 turns both encircle the core in the same sense. An ac source with voltage amplitude V_1 is connected to the primary and, to begin with, we assume the secondary to be open, so that there is no secondary current. The primary winding then functions merely as an inductor.

The primary current, which is small, lags the primary voltage by 90° and is called the *magnetizing* current. The power input to the transformer is zero. The core flux is in phase with the primary current. Since the same flux links both primary and secondary, the induced emf *per turn* is the same in each. The ratio of primary to secondary induced emf is therefore equal to the ratio of primary to secondary turns, or

$$\frac{\mathcal{E}_2}{\mathcal{E}_1} = \frac{N_2}{N_1}.$$

Since the windings are assumed to have zero resistance, the induced emf's $\mathcal{E}_1$ and $\mathcal{E}_2$ are numerically equal to the corresponding terminal voltages V_1 and V_2, and

$$\frac{V_2}{V_1} = \frac{N_2}{N_1}. \tag{36-33}$$

Hence by properly choosing the turn ratio N_2/N_1, we may obtain any desired secondary voltage from a given primary voltage. If $V_2 > V_1$, we have a *step-up* transformer; if $V_2 < V_1$, a *step-down* transformer.

Consider next the effect of closing the secondary circuit. The secondary current I_2 and its phase angle ϕ_2 will, of course, depend on the nature of the secondary circuit. As soon as the latter is closed, some power must be delivered by the secondary (except when $\phi_2 = 90°$); and, from energy

36-13 Schematic diagram of a transformer with secondary open.

considerations, an equal amount of power must be supplied to the primary. The process by which the transformer is enabled to draw the requisite amount of power is as follows. When the secondary circuit is open, the core flux is produced by the primary current only. But when the secondary circuit is closed, *both* primary and secondary currents set up a flux in the core. The secondary current, by Lenz's law, tends to weaken the core flux and therefore to decrease the back-emf in the primary. But (in the absence of losses) the back-emf in the primary must equal the primary terminal voltage, which is assumed to be fixed. The primary current therefore increases until the core flux is restored to its original no-load magnitude.

If the secondary circuit is completed by a resistance R, $I_2 = V_2/R$. From energy considerations, the power delivered to the primary equals that taken out of the secondary (neglecting losses), so

$$V_1 I_1 = V_2 I_2. \tag{36-34}$$

Combining this with the above expression for I_2 and Eq. (36-33) to eliminate V_2 and I_2, we find

$$I_1 = \frac{V_1}{(N_1/N_2)^2 R}. \tag{36-35}$$

Thus, when the secondary circuit is completed through a resistance R, the result is the same as if the *source* had been connected directly to a resistance equal to R multiplied by the reciprocal of the *square* of the turns ratio. In other words, the transformer "transforms" not only voltages and currents, but resistances (more generally, impedances) as well. It can be shown that maximum power is supplied by a source to a resistor when its resistance equals the internal resistance of the source. The same principle applies in ac circuits, with resistance replaced by impedance. When a high-impedance ac source must be connected to a low-impedance circuit, as when an audio amplifier is connected to a loudspeaker, the impedance of the source can be *matched* to that of the circuit by the insertion of a transformer having the correct turns ratio.

Questions

36-1 Some electric-power systems formerly used 25-Hz alternating current instead of the 60-Hz that is now standard. The lights flickered noticeably. Why is this not a problem with 60-Hz?

36-2 Power-distribution systems in airplanes sometimes use 400-Hz ac. What advantages and disadvantages does this have compared to the standard 60-Hz?

36-3 Fluorescent lights often use an inductor, called a "ballast," to limit the current through the tubes. Why is it better to use an inductor than a resistor for this purpose?

36-4 At high frequencies a capacitor becomes a short circuit. Discuss.

36-5 At high frequencies an inductor becomes an open circuit. Discuss.

36-6 Household electric power in most of western Europe is 220 volts, rather than the 110-V which is standard in the United States and Canada. What advantages and disadvantages does each system have?

36-7 The current in an ac power line changes direction 120 times per second, and its average value is zero. So how is it possible for power to be transmitted in such a system?

36-8 Electric-power connecting cords, such as lamp cords, always have two conductors which carry equal currents in opposite directions. How might one determine, using measurements only at the midpoints along the lengths of the wires, the direction of power transmission in the cord?

36-9 Are the equations for the average and rms values of current, Eqs. (36-23) and (36-24), correct when the variation with time is not sinusoidal? Explain.

36-10 Electric power companies like to have their power factors (cf. Sec. 36–5) as close to unity as possible. Why?

36-11 Some electrical appliances operate equally well on ac or dc, while others work only on ac or only on dc. Give examples of each, and explain the differences.

36-12 When a series-resonant circuit is connected across a 110-V ac line, the voltage rating of the capacitor may be exceeded, even if it is rated at 200 or 400 V. How can this be?

36-13 In a parallel-resonant circuit connected across a 110-V line, it is possible for the maximum current rating in the inductor to be exceeded, even if the total current through the circuit is very small. How can this be?

36-14 Can a transformer be used with dc? What happens if a transformer designed for 110-V ac is connected to a 110-V dc line?

36-15 During the last quarter of the nineteenth century there was a great and acrimonious controversy concerning whether ac or dc should be used for power transmission. Edison favored dc, George Westinghouse ac. What arguments might each proponent have used to promote his scheme?

Problems

Angular frequency = 2π (frequency),
$$\omega(\text{rad}\cdot\text{s}^{-1}) = 2\pi f(\text{Hz}).$$

36-1
a) At what frequency would a 5-H inductor have a reactance of 4000 Ω?
b) At what frequency would a 5-μF capacitor have the same reactance?

36-2 What is the reactance of a 0.015-μF capacitor at (a) 1 Hz? (b) 5 kHz? (c) 2 MHz?

36-3
a) What is the reactance of a 1-H inductor at a frequency of 60 Hz?
b) What is the inductance of an inductor whose reactance is 1 Ω at 60 Hz?
c) What is the reactance of a 1-μF capacitor at a frequency of 60 Hz?
d) What is the capacitance of a capacitor whose reactance is 1 ohm at 60 Hz?

36-4
a) Compute the reactance of a 10-H inductor at frequencies of 60 Hz and 600 Hz.
b) Compute the reactance of a 10-μF capacitor at the same frequencies.
c) At what frequency is the reactance of a 10-H inductor equal to that of a 10-μF capacitor?

36-5 A 1-μF capacitor is connected across an ac source whose voltage amplitude is kept constant at 50 V, but whose frequency can be varied. Find the current amplitude when the angular frequency is
a) 100 rad$\cdot$s^{-1},
b) 1000 rad$\cdot$s^{-1},
c) 10,000 rad$\cdot$s^{-1}.
d) Construct a log-log plot of current amplitude vs. frequency.

36-6 The voltage amplitude of an ac source is 50 V and its angular frequency is 1000 rad$\cdot$s^{-1}. Find the current amplitude if the capacitance of a capacitor connected across the source is
a) 0.01 μF,
b) 1.0 μF,
c) 100 μF.
d) Construct a log-log plot of current amplitude vs. capacitance.

36-7 An inductor of self-inductance 10 H and of negligible resistance is connected across the source in Problem 36–5. Find the current amplitude when the angular frequency is
a) 100 rad$\cdot$s^{-1},
b) 1000 rad$\cdot$s^{-1},
c) 10,000 rad$\cdot$s^{-1}.
d) Construct a log-log plot of current amplitude vs. frequency.

36-8 Find the current amplitude if the self-inductance of a resistanceless inductor connected across the source of Problem 36–6 is
a) 0.01 H,
b) 1.0 H,
c) 100 H.
d) Construct a log-log plot of current amplitude versus self-inductance.

36-9 The expression for the impedance Z of an R–L series circuit can be obtained from Eq. (36–20) by setting $X_C = 0$, which corresponds to $C = \infty$. Explain.

36-10 In an R–L–C series circuit, the source has a constant voltage amplitude of 50 V and a frequency of 1000 rad$\cdot$s^{-1}. $R = 300\,\Omega$, $L = 0.9$ H, and $C = 2.0\,\mu$F. Suppose a series circuit contains only the resistor and the inductor in series.
a) What is the impedance of the circuit?
b) What is the current amplitude?

c) What are the voltage amplitudes across the resistor and across the inductor?

d) What is the phase angle ϕ? Does the current lag or lead?

e) Construct the phasor diagram.

36-11 Same as Problem 36-10, except that the circuit consists of the resistor and the capacitor in series.

36-12 Same as Problem 36-10, except that the circuit consists of the inductor and capacitor in series.

36-13

a) Compute the impedance of an R-L-C series circuit at an angular frequency of 500 rad·s^{-1}, using the numbers in Problem 36-10.

b) Describe how the current amplitude varies as the frequency of the source is slowly reduced from 1000 rad·s^{-1} to 500 rad·s^{-1}.

c) What is the phase angle when $\omega = 500$ rad·s^{-1}? Construct the phasor diagram when $\omega = 500$ rad·s^{-1}.

36-14 Five infinite impedance voltmeters, calibrated to read rms values, are connected as shown in Fig. 36-14. What does each voltmeter read? Use the numbers given in Problem 36-10.

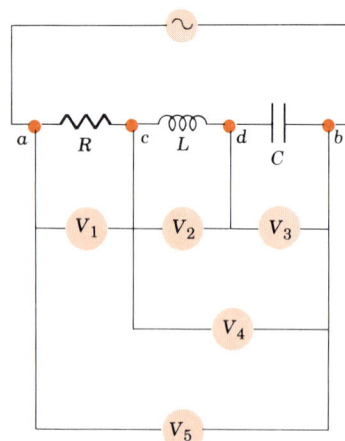

Figure 36-14

36-15 What is the reading of each voltmeter in Fig. 36-14 if the angular frequency $\omega = 500$ rad·s^{-1}? Use the values of V, R, L, and C in Problem 36-10, and assume that the voltmeters measure rms values.

36-16 A 400-Ω resistor is in series with a 0.1-H inductor and a 0.5-μF capacitor. Compute the impedance of the circuit and draw the vector impedance diagram (a) at a frequency of 500 Hz, and (b) at a frequency of 1000 Hz. Compute, in each case, the phase angle between line current and line voltage, and state whether the current lags or leads.

36-17

a) Construct a graph of the current amplitude in the circuit of Fig. 36-5 as the angular frequency of the source is increased from 500 rad·s^{-1} to 1000 rad·s^{-1}. Use the values of V, R, L, and C in Problem 36-10.

b) At what frequency is the circuit in resonance?

c) What is the power factor at resonance?

d) What is the reading of each voltmeter in Fig. 36-14 when the frequency equals the resonant frequency?

e) What would be the resonant frequency if the resistance were reduced to 100 Ω?

f) What would then be the rms current at resonance?

36-18

a) At what frequency is the voltage amplitude across the inductor in an R-L-C series circuit a maximum?

b) At what frequency is the voltage amplitude across the capacitor a maximum?

36-19 In an R-L-C series circuit, $R = 250\ \Omega$, $L = 0.5$ H, and $C = 0.02\ \mu$F.

a) What is the resonant frequency of the circuit?

b) The capacitor can withstand a peak voltage of 350 V. What maximum effective terminal voltage may the generator have at resonant frequency?

36-20 A resistor of 500 Ω and a capacitor of 2 μF are connected in parallel to an ac generator supplying a constant voltage amplitude of 282 V and having an angular frequency of 377 rad·s^{-1}. Find

a) the current amplitude in the resistor,

b) the current amplitude in the capacitor,

c) the phase angle, and

d) the line current amplitude.

36-21 A coil has a resistance of 20 Ω. At a frequency of 100 Hz, the voltage across the coil leads the current in it by 30°. Determine the inductance of the coil.

36-22 An inductor having a reactance of 25 Ω gives off heat at the rate of 2.39 cal·s^{-1} when it carries a current of 0.5 A (rms). What is the impedance of the inductor?

36-23 The circuit of Problem 36-16 carries an rms current of 0.25 A with frequency 100 Hz.

a) What power is consumed in the circuit?

b) In the resistor?

c) In the capacitor?

d) In the inductor?

e) What is the power factor of the circuit?

36-24 A series circuit has a resistance of 75 Ω and an impedance of 150 Ω. What power is consumed in the circuit when a voltage of 120 V (rms) is impressed across it?

36-25 A circuit draws 330 W from a 110-V, 60-Hz ac line. The power factor is 0.6 and the current lags the voltage.

a) Find the capacitance of the series capacitor that will result in a power factor of unity.

b) What power will then be drawn from the supply line?

36–26 A series circuit has an impedance of 50 Ω and a power factor of 0.6 at 60 Hz, the voltage lagging the current.

a) Should an inductor or a capacitor be placed in series with the circuit to raise its power factor?

b) What size element will raise the power factor to unity?

36–27 The internal resistance of an ac source is 10,000 Ω.

a) What should be the turns ratio of a transformer to match the source to a load of resistance 10 Ω?

b) If the voltage amplitude of the source is 100 V, what is the voltage amplitude of the secondary on open circuit?

36–28 A 100-Ω resistor, a 0.1-μF capacitor, and a 0.1-H inductor are connected in parallel to a voltage source with amplitude 100 V.

a) What is the resonant frequency? The resonant angular frequency?

b) What is the maximum total current through the parallel combination at the resonant frequency?

c) What is the maximum current in the resistor at resonance?

d) What is the maximum current in the inductor at resonance?

e) What is the maximum energy stored in the inductor at resonance? In the capacitor?

36–29 The same three components as in Problem 36–28 are connected in *series* to a voltage source with amplitude 100 V.

a) What is the resonant frequency? The resonant angular frequency?

b) What is the maximum current in the resistor at resonance?

c) What is the maximum voltage across the capacitor at resonance?

d) What is the maximum energy stored in the capacitor at resonance?

36–30 A transformer connected to a 120-V ac line is to supply 12 V to a low-voltage lighting system for a model-railroad village. The total equivalent resistance of the system is 2 Ω.

a) What should be the turns ratio of the transformer?

b) What current must the secondary supply?

c) What power is delivered to the load?

d) What resistance connected directly across the 120-V line would draw the same power as the transformer? Show that this is equal to 2 Ω times the square of the turns ratio.

36–31 A step-up transformer connected to a 120-V ac line is to supply 18,000 V for a neon sign. To reduce shock hazard, a fuse is to be inserted in the primary circuit; the fuse is to blow when the current in the secondary circuit exceeds 10 mA.

a) What is the turns ratio of the transformer?

b) What power must be supplied to the transformer when the secondary current is 10 mA?

c) What current rating should the fuse in the primary circuit have?

36–32 The current in a certain circuit varies with time as shown in Fig. 36–15. Find the average current and the rms current, in terms of I_0.

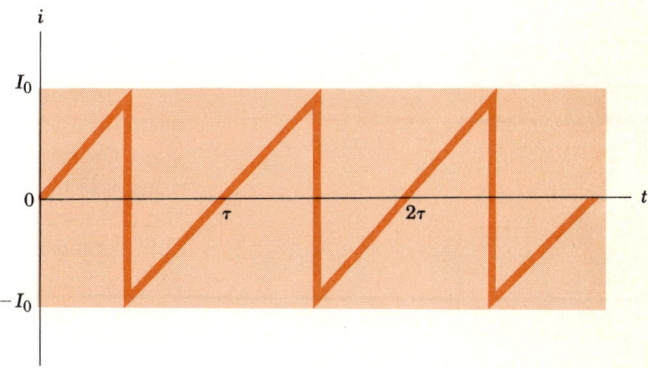

Figure 36–15

ELECTROMAGNETIC WAVES
37

37-1 Introduction

Our discussion of electric and magnetic fields in the last several chapters can be classified in two general categories. The first includes fields that do not vary with time. The electrostatic field of a distribution of charges at rest and the magnetic field of a steady current in a conductor are examples of fields which, while they may vary from point to point in space, do not vary with time at any individual point. For such situations we found it possible to treat the electric and magnetic fields independently, without worrying unduly about interactions between the two fields.

The second category includes situations in which the fields *do* vary with time, and in all such cases we have found that it is *not* possible to treat the fields independently. Faraday's law tells us that a time-varying magnetic field acts as a source of electric field. This field is manifested in the induced emf's in inductances and transformers. Similarly, in developing the general formulation of Ampère's law (Sec. 32–9), which is valid for charging capacitors and similar situations as well as for ordinary conductors, we found it necessary to regard a changing electric field as a source of magnetic field.

Thus, when *either* field is changing with time, a field of the other kind is induced in adjacent regions of space. We are thus led naturally to consider the possibility of an electromagnetic disturbance, consisting of time-varying electric and magnetic fields, which can propagate through space from one region to another, even when there is no matter in the intervening region. Such a disturbance, if it exists, will have the properties of a *wave,* and an appropriate descriptive term is *electromagnetic wave.* Such waves do exist, of course, and it is a familiar fact that radio and television transmission, light, x-rays, and many other phenomena are examples of electromagnetic radiation. In the following pages we shall show how the existence of such waves is related to the principles of electromagnetism studied thus far, and shall examine their properties.

As so often happens in the development of science, the theoretical understanding of electromagnetic waves originally took a considerably more devious path than the one outlined above. In the early days of electromagnetic theory (the early nineteenth century), two different units of electric charge were used, one for electrostatics, the other for magnetic phenomena involving currents. The particular system of units in common use at the time had the property that these two units of charge had different physical dimensions; their ratio turned out to have units of velocity. This in itself is not so astounding, but experimental measurements revealed that the ratio had a numerical value precisely equal to the speed of light, 3.00×10^8 m·s^{-1}. At the time, physicists regarded this as an extraordinary coincidence, and had no idea how to explain it. It was the search for an explanation of this result that led Maxwell, in 1864, to prove by theoretical reasoning that an electrical disturbance should propagate in free space with a speed equal to that of light, and hence to postulate that light waves were *electromagnetic* waves.

During the course of his analysis, Maxwell also discovered that all the basic principles of electromagnetism can be formulated in terms of four fundamental equations, now called *Maxwell's equations*. We have now studied all four of these principles; they are (1) Gauss's law, (2) the absence of magnetic charge, (3) Ampère's law, including displacement current, and (4) Faraday's law. To summarize, Maxwell's equations, in the integral form used in this text, are:

$$\int \boldsymbol{E} \cdot d\boldsymbol{A} = \frac{Q}{\epsilon_0}, \qquad \text{(Gauss's law)} \qquad (25\text{--}20)$$

$$\int \boldsymbol{B} \cdot d\boldsymbol{A} = 0, \qquad \text{(No magnetic charge)} \qquad (32\text{--}14)$$

$$\oint \boldsymbol{B} \cdot d\boldsymbol{l} = \mu_0 \left(I + \epsilon_0 \frac{d\Psi}{dt} \right), \qquad \text{(Ampère's law)} \qquad (32\text{--}28)$$

$$\oint \boldsymbol{E} \cdot d\boldsymbol{l} = -\frac{d\Phi}{dt}. \qquad \text{(Faraday's law)} \qquad (33\text{--}14)$$

In this form, these equations apply to electric and magnetic fields *in vacuum;* they may also be generalized to include fields in matter. Maxwell's synthesis of electromagnetism in these four equations is one of the great milestones of theoretical physics, comparable in importance to Newton's laws of motion in mechanics.

It remained for Heinrich Hertz, in 1887, actually to *produce* electromagnetic waves, with the aid of oscillating circuits, and to receive and detect these waves with other circuits tuned to the same frequency. Hertz then produced standing electromagnetic waves and measured the distance between adjacent nodes, to measure the wavelength. Knowing the frequency of his resonators, he then found the velocity of the waves from the fundamental wave equation $c = f\lambda$, and verified Maxwell's theoretical value directly.

The possible use of electromagnetic waves for purposes of long-distance communication does not seem to have occurred to Hertz. It remained for the enthusiasm and energy of Marconi and others to make "wireless telegraphy" a familiar household phenomenon.

The unit of frequency, one cycle per second, is named one *hertz* (1 Hz) in honor of Hertz.

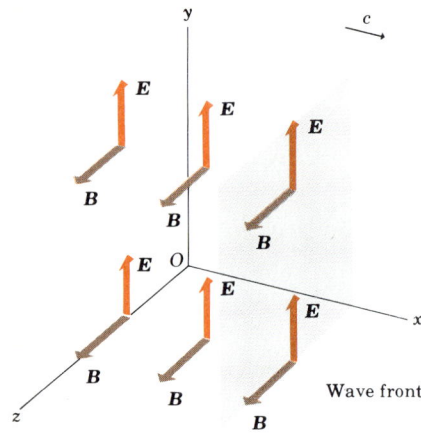

37-1 An electromagnetic wave front. The E and B fields are uniform over the region to the left of the plane, but are zero everywhere to the right of it. The plane representing the wave front moves to the right with speed c.

*37-2 Speed of an electromagnetic wave

In developing the relation of electromagnetic wave propagation to familiar electromagnetic principles, we begin with a particularly simple and somewhat artificial example of an electromagnetic wave. We shall postulate a particular configuration of fields, and then test whether it is consistent with the principles mentioned above, Faraday's law and Ampère's law including displacement current.

Using an x-y-z-coordinate system, as shown in Fig. 37–1, we imagine that all space is divided into two regions by a plane perpendicular to the x-axis (parallel to the yz-plane). At all points to the right of this plane there are no electric or magnetic fields; at all points to the left there is a uniform electric field E in the $+y$-direction and a uniform magnetic field B in the $+z$-direction, as shown. Furthermore, we suppose that the boundary surface, which may also be called the *wave front*, moves to the right with a constant speed c, as yet unknown. Thus the E and B fields travel to the right into previously field-free regions, with a definite speed. The situation, in short, does describe a rudimentary electromagnetic wave, provided it is consistent with the laws of electromagnetism.

We shall not worry about how such a field configuration can be produced. It can be shown that, in principle, an infinitely large sheet of charge in the yz-plane, which is initially at rest and suddenly starts moving with constant velocity in the $-y$-direction, gives such fields, but of course there is no practical way to realize an infinitely large charge sheet. At any rate, we now apply Faraday's law to a rectangle in the xy-plane, as in Fig. 37–2, located so that at some instant the wave front has progressed partway through the rectangle, as shown. In a time dt, the boundary surface moves a distance $c\,dt$ to the right, sweeping out an area $ac\,dt$ of the rectangle, and in this time the magnetic flux increases by $B(ac\,dt)$.

Thus the rate of change of magnetic flux is given by

$$\frac{d\Phi}{dt} = Bac. \qquad (37\text{-}1)$$

According to Faraday's law, this must equal the line integral $\oint \mathbf{E} \cdot d\mathbf{l}$ around the boundary of the area. The field at the right end is zero, and the top and bottom sides do not contribute because there the component of E along dl is zero. Thus the integral becomes simply Ea, and Faraday's law gives

$$E = cB. \qquad (37\text{-}2)$$

Thus the postulated wave is consistent with Faraday's law only if E, B, and c are related as in Eq. (37-2).

Next, we consider a rectangle in the yz-plane, as shown in Fig. 37–3. We wish to apply Ampère's law to this rectangle; the general form of this law, including displacement current, was stated in Chapter 32, Eq. (32-27):

$$\oint \mathbf{B} \cdot d\mathbf{l} = \mu_0 (I_C + I_D). \qquad (37\text{-}3)$$

In the present situation there is no conduction current, so $I_C = 0$. Previously we have defined the displacement current I_D in terms of a displace-

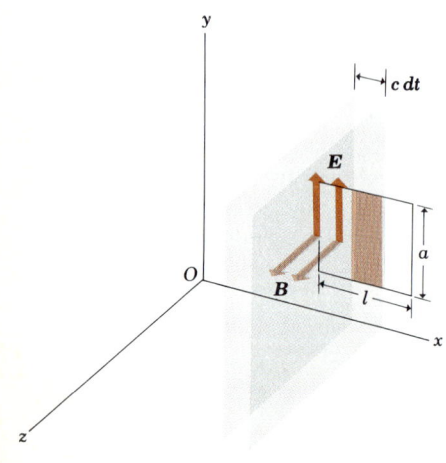

37-2 In time dt the wave front moves to the right a distance $c\,dt$. The magnetic flux through the rectangle in the xy-plane increases by an amount $d\Phi$ equal to the flux through the shaded rectangle of area $ac\,dt$, that is, $d\Phi = Bac\,dt$.

ment current *density*, as in Eq. (29-20):

$$J_D = \epsilon_0 \frac{dE}{dt}. \tag{37-4}$$

Then if the **E** field lines pass through an area A normal to the direction of **E**, we have

$$I_D = J_D A = \epsilon_0 \frac{dE}{dt} A. \tag{37-5}$$

Displacement current may also be expressed in terms of electric flux, as shown in Secs. 29-6 and 32-9. If an area A lies perpendicular to an electric field **E**, the electric flux through the area, denoted by Ψ, is defined as

$$\Psi = EA.$$

In terms of electric flux, Eq. (37-5) can be rewritten as

$$I_D = \epsilon_0 \frac{d\Psi}{dt}, \tag{37-6}$$

and Ampère's law with displacement current but no conduction current is

$$\oint \mathbf{B} \cdot d\mathbf{l} = \mu_0 \epsilon_0 \frac{d\Psi}{dt}. \tag{37-7}$$

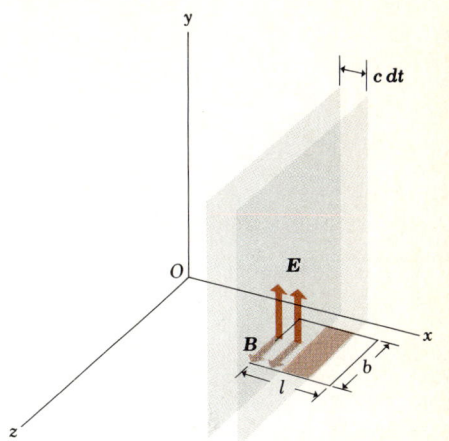

37-3 In time dt the electric flux through the rectangle in the xz-plane increases by an amount equal to E times the area $(bc\,dt)$ of the shaded rectangle; that is, $d\Psi = Ebc\,dt$. Thus $d\Psi/dt = Ebc$.

This is the form we shall apply to the rectangle in Fig. 37-3. The change in electric flux $d\Psi$ in time dt is the area $bc\,dt$ swept out by the wave front, multiplied by E. In evaluating the integral $\oint \mathbf{B} \cdot d\mathbf{l}$, we note that B is zero on the right end and $B_\parallel = 0$ on the front and back sides. Thus only the left end contributes, and we find $\oint \mathbf{B} \cdot d\mathbf{l} = Bb$. Combining these results with Eq. (37-7) and dividing through by b, we obtain

$$B = \mu_0 \epsilon_0 cE. \tag{37-8}$$

Thus Ampère's law is obeyed only if **B**, c, and **E** are related as in Eq. (37-8).

Since *both* Ampère's law and Faraday's law must be obeyed simultaneously, Eqs. (37-2) and (37-8) must both be satisfied. This can occur only when $\mu_0 \epsilon_0 c = 1/c$, or

$$c = \frac{1}{\sqrt{\epsilon_0 \mu_0}}. \tag{37-9}$$

Inserting the numerical values of these quantities, we find

$$c = \frac{1}{\sqrt{(8.85 \times 10^{-12})(4\pi \times 10^{-7})}} = 3.00 \times 10^8 \text{ m} \cdot \text{s}^{-1}.$$

The postulated field configuration *is* consistent with the laws of electrodynamics, provided the wave front moves with the speed given above, which of course is recognized as the speed of light.

We have chosen a particularly simple wave for study in order to avoid undue mathematical complexity, but this special case nevertheless illustrates several important features of *all* electromagnetic waves:

1. The wave is *transverse;* both **E** and **B** are perpendicular to the direction of propagation of the wave, and to each other.
2. There is a definite ratio between **E** and **B**.
3. The wave travels in vacuum with a definite and unchanging speed.

It is not difficult to generalize the above discussion to a more realistic situation. Suppose we have several wave fronts in the form of parallel planes perpendicular to the axis and all moving to the right with speed c. Suppose that, within a single region between two planes, the **E** and **B** fields are the same at all points in the region, but that they differ from region to region. An extension of the above development shows that such a situation is also consistent with Ampère's and Faraday's laws, provided the wave fronts all move with the speed c given by Eq. (37-9). From this picture it is only a short additional step to a wave picture in which the **E** and **B** fields at any instant vary smoothly, rather than in steps, as we move along the x-axis and the entire field pattern moves to the right with speed c. In Sec. 37-5 we consider waves in which the dependence of **E** and **B** on position and time is *sinusoidal;* first, however, we consider the *energy* associated with an electromagnetic wave.

37-3 Energy in electromagnetic waves

Analysis of energy needed to charge a capacitor and to establish a current in an inductor has led us (Sections 27-4 and 34-3) to associate an energy density (energy per unit volume) with electric and magnetic fields. Specifically, Eqs. (27-9) and (34-9) show that the total energy density u in a region of space where **E** and **B** fields are present is given by

$$u = \frac{1}{2}\epsilon_0 E^2 + \frac{1}{2\mu_0}B^2. \qquad (37\text{-}10)$$

This and subsequent expressions become somewhat more symmetrical when expressed in terms of the field $\mathbf{H} = \mathbf{B}/\mu_0$ introduced in Chapter 35. In terms of **E** and **H**, the energy density is

$$u = \tfrac{1}{2}\epsilon_0 E^2 + \tfrac{1}{2}\mu_0 H^2. \qquad (37\text{-}11)$$

Now we have found above that **E** and **B** in an electromagnetic wave are not independent but are related by

$$B = \frac{E}{c} = \sqrt{\epsilon_0 \mu_0}\, E \quad \text{or} \quad H = \sqrt{\frac{\epsilon_0}{\mu_0}}\, E. \qquad (37\text{-}12)$$

Thus the energy density may also be expressed

$$u = \frac{1}{2}\epsilon_0 E^2 + \frac{1}{2}\mu_0 \left(\sqrt{\frac{\epsilon_0}{\mu_0}}E\right)^2 = \frac{1}{2}\epsilon_0 E^2 + \frac{1}{2}\epsilon_0 E^2 = \epsilon_0 E^2, \quad (37\text{-}13)$$

which shows that, in a wave, the energy density associated with the **E** field is equal to that of the **H** field.

Equation (37-12) may also be written

$$\frac{E}{H} = \sqrt{\frac{\mu_0}{\epsilon_0}}. \qquad (37\text{-}14)$$

The unit of E is 1 V·m^{-1} and that of H is 1 A·m^{-1}, so the unit of the ratio E/H is $1\text{ V}\cdot\text{A}^{-1}$ or $1\text{ }\Omega$. When numerical values of ϵ_0 and μ_0 are inserted, we find

$$\frac{E}{H} = 377\text{ }\Omega. \tag{37-15}$$

For reasons that we cannot discuss in detail, this ratio is called the *impedance of free space*.

Because the $\boldsymbol{E}$ and $\boldsymbol{H}$ fields in the simple wave considered above advance with time into regions where originally there were no fields, it is clear that the wave transports energy from one region to another. This energy transfer is conveniently characterized by considering the energy transferred *per unit time, per unit cross-sectional area* for an area perpendicular to the direction of wave travel. This quantity will be denoted by S. This is analogous to the concept of current density, the charge per unit time transferred across unit area perpendicular to the direction of flow.

To see how the energy flow is related to the fields, we consider a stationary plane perpendicular to the x-axis, which at a certain time coincides with the wave front. In a time dt after this time, the wave front moves a distance $c\,dt$ to the right. Considering an area A on the stationary plane, we note that the energy in the space to the right of this area must have passed through it to reach its new location. The volume dV of the relevant region is the base area A times the length $c\,dt$, and the total energy dU in this region is the energy density u times this volume:

$$dU = \epsilon_0 E^2 A c\,dt. \tag{37-16}$$

Since this much energy passed through area A in time dt, the energy flow S per unit time, per unit area is

$$S = \frac{dU}{A\,dt} = \epsilon_0 c E^2.$$

Using Eqs. (37-9) and (37-12), we obtain the alternative forms

$$S = \frac{\epsilon_0}{\sqrt{\epsilon_0 \mu_0}} E^2 = \sqrt{\frac{\epsilon_0}{\mu_0}} E^2 = EH. \tag{37-17}$$

The unit of S is energy per unit time, per unit area. The SI unit of S is $1\text{ J}\cdot\text{s}^{-1}\cdot\text{m}^{-2}$ or $1\text{ W}\cdot\text{m}^{-2}$.

We can also define a *vector* quantity that describes both the magnitude and direction of the energy flow rate. We define

$$\boldsymbol{S} = \boldsymbol{E} \times \boldsymbol{H}. \tag{37-18}$$

$\boldsymbol{S}$ is called the *Poynting vector;* its magnitude is given by Eq. (37-17), and its direction is the direction of propagation of the wave. The magnitude EH gives the flow of energy across a cross section perpendicular to the propagation direction, per unit area and per unit time.

The electric and magnetic fields at any point in a wave vary with time, so the Poynting vector at any point is also a function of time. The *average* value of the magnitude of the Poynting vector at a point is called the *intensity* of the radiation at that point. As mentioned above, the SI unit of intensity is the watt per square meter ($\text{W}\cdot\text{m}^{-2}$). In Sec. 37-5 we consider the relation of the intensity of a sinusoidal wave to the amplitudes of the sinusoidally varying electric and magnetic fields.

Example In the wave discussed above, suppose
$$E = 100 \text{ V} \cdot \text{m}^{-1} = 100 \text{ N} \cdot \text{C}^{-1}.$$
Find the values of B and H, the maximum energy density, and the maximum energy flow S.

Solution From Eq. (37–2),
$$B = \frac{E}{c} = \frac{100 \text{ V} \cdot \text{m}^{-1}}{3.0 \times 10^8 \text{ m} \cdot \text{s}^{-1}} = 3.33 \times 10^{-7} \text{ T};$$

$$H = \frac{B}{\mu_0} = \frac{3.33 \times 10^{-7} \text{ T}}{4\pi \times 10^{-7} \text{ Wb} \cdot \text{A}^{-1} \cdot \text{m}^{-1}} = 0.265 \text{ A} \cdot \text{m}^{-1}.$$

From Eq. (37–13),
$$u = \epsilon_0 E^2 = (8.85 \times 10^{-12} \text{ C}^2 \cdot \text{N}^{-1} \cdot \text{m}^{-2})(100 \text{ N} \cdot \text{C}^{-1})^2$$
$$= 8.85 \times 10^{-8} \text{ N} \cdot \text{m}^{-2} = 8.85 \times 10^{-8} \text{ J} \cdot \text{m}^{-3};$$

$$S = EH = (100 \text{ V} \cdot \text{m}^{-1})(0.265 \text{ A} \cdot \text{m}^{-1})$$
$$= 26.5 \text{ W} \cdot \text{m}^{-2}.$$

The idea that energy can travel through empty space without the aid of any matter in motion may seem strange, yet this is the very mechanism by which energy reaches us from the sun. The conclusion that electromagnetic waves transport energy is as inescapable as the conclusion that energy is required to establish electric and magnetic fields. It is also possible to show that electromagnetic waves carry *momentum*, with a corresponding momentum density of magnitude

$$\frac{EH}{c^2} = \frac{S}{c^2}. \qquad (37\text{–}19)$$

This momentum is a property of the field alone and is not associated with moving mass. There is also a corresponding momentum flow rate; just as the energy density u corresponds to S, the rate of energy flow per unit area, the momentum density given by Eq. 37–19 corresponds to the momentum flow rate

$$\frac{1}{A}\frac{dp}{dt} = \frac{S}{c} = \frac{EH}{c}, \qquad (37\text{–}20)$$

which represents the momentum transferred per unit surface area, per unit time.

This momentum is responsible for the phenomenon of *radiation pressure*; when an electromagnetic wave is absorbed by a surface perpendicular to the propagation direction, the time rate of change of momentum is equal to the force on the surface. Thus the force per unit area, or presssure, is equal to S/c. If the wave is totally reflected, the momentum change is twice as great, and the pressure is $2S/c$. For example, the value of S for direct sunlight is about $1.4 \text{ kW} \cdot \text{m}^{-2}$, and the corresponding pressure on a completely absorbing surface is

$$p = \frac{1.4 \times 10^3 \text{ W} \cdot \text{m}^{-2}}{3.0 \times 10^8 \text{ m} \cdot \text{s}^{-1}} = 4.7 \times 10^{-6} \text{ Pa}$$
$$= 4.7 \times 10^{-6} \text{ N} \cdot \text{m}^{-2}.$$

Radiation pressure is important in the structure of stars. Gravitational attractions tend to shrink a star, but this tendency is balanced by radiation pressure in maintaining the size of the star through most stages of its evolution.

*37-4 Electromagnetic waves in matter

The above analysis can be extended to include electromagnetic waves in dielectrics. The wave speed is not the same as in vacuum, and we denote it by w instead of c. Faraday's law is unaltered, but Eq. (37-2) is replaced by $E = wB$. In Ampère's law, the displacement current density is given not by $\epsilon_0\, dE/dt$ but by $\epsilon\, dE/dt = K\epsilon_0\, dE/dt$. In addition, the constant μ_0 in Ampère's law must be replaced by $\mu = K_m \mu_0$. Thus, Eq. (37-8) is replaced by

$$B = \mu \epsilon w E,$$

and the wave speed is given by

$$w = \frac{1}{\sqrt{\epsilon\mu}} = \frac{1}{\sqrt{KK_m}} \frac{1}{\sqrt{\epsilon_0\mu_0}}. \qquad (37\text{-}21)$$

For many dielectrics the relative permeability K_m is very nearly equal to unity; in such cases we can say

$$w \simeq \frac{1}{\sqrt{K}} \frac{1}{\sqrt{\epsilon_0\mu_0}} = \frac{c}{\sqrt{K}}.$$

Because K is always greater than unity, the speed of electromagnetic waves in a dielectric is always *less* than the speed in vacuum, by a factor of $1/\sqrt{K}$. The ratio of the speed in vacuum to the speed in a material is known in optics as the *index of refraction n* of the material; we see that this is given by

$$\frac{c}{w} = n = \sqrt{KK_m} \simeq \sqrt{K}. \qquad (37\text{-}22)$$

Corresponding modifications are required in the expressions for the energy density and the Poynting vector; these are developed in detail in more advanced texts on electromagnetic theory.

The waves described above cannot propagate in a *conducting* material because the **E** field leads to currents that provide a mechanism for dissipating the energy of the wave. For an ideal conductor with zero resistivity, **E** must be zero everywhere inside the material. When an electromagnetic wave strikes such a material, it is totally reflected. Real conductors with finite resistivity permit some penetration of the wave into the material, with partial reflection. A polished metal surface is usually a good *reflector* of electromagnetic waves, but metals are not *transparent* to radiation.

37-5 Sinusoidal waves

Sinusoidal electromagnetic waves are closely analogous to sinusoidal transverse mechanical waves on a stretched string, as discussed in Chapter 21. At any point in space, the **E** and **H** fields are sinusoidal functions of time and, at any instant of time, the spatial variation of the fields is also sinusoidal.

The simplest sinusoidal electromagnetic waves share with the wave of Sec. 37-2 the property that at any instant the fields are uniform over any plane perpendicular to the direction of propagation. Such a wave is called a *plane wave*. The entire pattern travels to the right with speed c. Since E and H are at right angles to the direction of propagation, the wave is *transverse*.

The frequency f, the wavelength λ, and the speed of propagation c are related by the equation applicable to any sort of periodic wave motion, namely, $c = f\lambda$. If the frequency f is the power-line frequency of 60 Hz, the wavelength is

$$\lambda = \frac{c}{f} = \frac{3 \times 10^8 \text{ m} \cdot \text{s}^{-1}}{60 \text{ Hz}} = 5 \times 10^6 \text{ m} = 5000 \text{ km},$$

which is of the order of the earth's radius! Hence at this frequency even a distance of many miles includes only a small fraction of a wavelength. On the other hand, if the frequency is 10^8 Hz (100 MHz), typical of commercial FM radio stations, the wavelength is

$$\lambda = \frac{3 \times 10^8 \text{ m} \cdot \text{s}^{-1}}{10^8 \text{ Hz}} = 3 \text{ m},$$

and a moderate distance can include a number of complete waves.

Figure 37-4 represents a sinusoidal electromagnetic wave traveling in the $+x$-direction. The E and H vectors are shown only for a few points on the x-axis; if a plane is constructed perpendicular to the x-axis at a particular point, the fields have the same values at all points in that plane. The values are of course different on different planes. In those planes in which the E vector is in the $+y$-direction, H is in the $+z$-direction; where E is in the $-y$-direction, H is in the $-z$-direction. In both cases the direction of the Poynting vector, given by Eq. (37-18), is along the $+x$-direction.

We recall from Sec. 21-3, Eq. (21-10), that one form of the equation of a transverse wave traveling to the right along a stretched string is $y = A \sin(\omega t - kx)$, where y is the transverse displacement from its equilibrium position, at the time t, of a point of the string whose coordinate is x. The quantity A is the maximum displacement or the *amplitude* of the

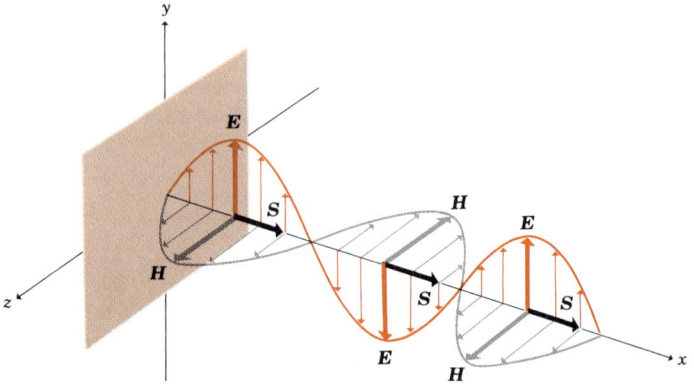

37-4 E-, H-, and S-vectors in a sinusoidal electromagnetic wave traveling in the x-direction. The position of the wave at time $t = 0$ is shown.

wave, ω is its *angular frequency*, equal to 2π times the frequency f, and k is the *propagation constant*, equal to $2\pi/\lambda$, where λ is the wavelength.

Let E and H represent the instantaneous values, and E_{max} and H_{max} the maximum values, or *amplitudes*, of the electric and magnetic fields in Fig. 37-4. The equations of the traveling electromagnetic wave are then

$$E = E_{max} \sin(\omega t - kx), \quad H = H_{max} \sin(\omega t - kx). \quad (37\text{-}23)$$

The sine curves in Fig. 37-4 represent instantaneous values of E and H, as functions of x at the time $t = 0$. The wave travels to the right with speed c.

The instantaneous value S of the Poynting vector is

$$S = EH = E_{max}H_{max} \sin^2(\omega t - kx)$$
$$= \tfrac{1}{2}E_{max}H_{max}[1 - \cos 2(\omega t - kx)].$$

The time average value of $\cos 2(\omega t - kx)$ is zero, so the average value S_{av} of the Poynting vector is

$$S_{av} = \tfrac{1}{2}E_{max}H_{max}. \quad (37\text{-}24)$$

This is the *average power* transmitted per unit area, and, as noted in Sec. 37-3, it is called the *intensity* of the radiation.

Figure 37-5 represents schematically the electric and magnetic fields of a wave traveling in the *negative x*-direction. At points where **E** is in the positive *y*-direction, **H** is in the *negative z*-direction, and where **E** is in the negative *y*-direction, **H** is in the *positive z*-direction. The Poynting vector is in the negative *x*-direction at all points. (Compare with Fig. 37-4, which represents a wave traveling in the *positive x*-direction.) The equations of the wave are

$$E = -E_{max} \sin(\omega t + kx), \quad H = H_{max} \sin(\omega t + kx). \quad (37\text{-}25)$$

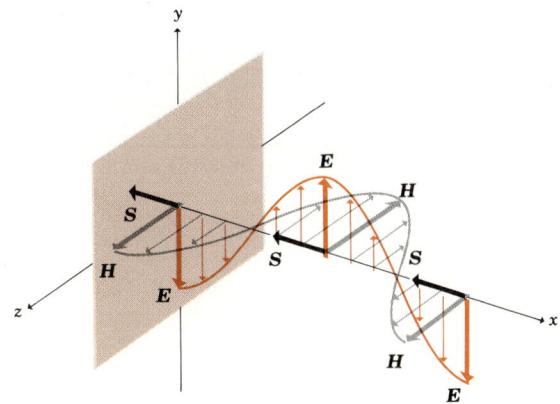

37-5 Electric and magnetic fields of a sinusoidal wave traveling in the negative *x*-direction. The position of the wave at time $t = 0$ is shown.

37-6 Standing waves

Electromagnetic waves can be reflected, and the superposition of an incident wave and a reflected wave can form a *standing wave* analogous to standing waves on a stretched string, discussed in Sec. 22-2. A conducting surface can serve as a reflector for electromagnetic waves.

Suppose a sheet of an ideal conductor, having zero resistivity, is placed in the yz-plane of Fig. 37–5, and that the wave shown, traveling in the negative x-direction, is incident on it. The essential characteristic of an ideal conductor is that there can never be any electric field in it; any attempt to establish a field is immediately canceled by rearrangement of the mobile charges in the conductor. Thus ***E*** must always be zero everywhere in this plane, and the ***E*** field of the incident wave induces sinusoidal currents in the conductor so that ***E*** is zero inside it.

These induced currents then produce a reflected wave, traveling out from the plane, to the right. From the superposition principle, the total ***E*** field at any point to the right of the plane is the vector sum of the ***E*** fields of the incident and reflected waves; the same is true for the total ***H*** field.

Suppose the incident wave is described by Eqs. (37–25) and the reflected wave by Eqs. (37–23). Then the total fields at any point are given by

$$E = E_{max}[-\sin(\omega t + kx) + \sin(\omega t - kx)],$$
$$H = H_{max}[\sin(\omega t + kx) + \sin(\omega t - kx)].$$

These expressions may be expanded and simplified, using the identity

$$\sin(A \pm B) = \sin A \cos B \pm \cos A \sin B.$$

The results are

$$E = -2E_{max} \cos \omega t \sin kx, \qquad (37\text{–}26a)$$
$$H = 2H_{max} \sin \omega t \cos kx. \qquad (37\text{–}26b)$$

The first of these is analogous to Eq. (22–1) for the stretched string. We see that at $x = 0$, E is *always* zero; this is required by the nature of the ideal conductor, which plays the same role as a fixed point at the end of the string. Furthermore, E is zero at all times in those planes for which $\sin kx = 0$, that is, $kx = 0, \pi, 2\pi, \ldots$, or

$$x = 0, \frac{\lambda}{2}, \lambda, \frac{3\lambda}{2}, \ldots.$$

These are called *nodal planes* of the ***E*** field.

The magnetic field is zero at all times in those planes for which $\cos kx = 0$, or at which

$$x = \frac{\lambda}{4}, 3\frac{\lambda}{4}, 5\frac{\lambda}{4}, \text{ etc.}$$

These are the nodal planes of the ***H*** field. The magnetic field is *not* zero at the conducting surface ($x = 0$), and there is no reason it should be. The nodal planes of one field are midway between those of the other, and the nodal planes of either field are separated by one-half wavelength. Figure 37–6 shows a standing-wave pattern at one instant of time.

The electric field is a *cosine* function of t and the magnetic field is a *sine* function of t. The fields are therefore 90° out of phase. At times when $\cos \omega t = 0$, the electric field is zero *everywhere* and the magnetic field is maximum. When $\sin \omega t = 0$, the magnetic field is zero *everywhere* and the electric field is maximum.

37–6 E- and H-vectors in a standing wave. The pattern does not move along the x-axis, but the E- and H-vectors grow and diminish with time at each point. At each point E is maximum when H is minimum, and conversely. The position of the wave at time $t = 0$ is shown.

Pursuing the stretched-string analogy, we may now insert a second conducting plane, parallel to the first and a distance L from it, along the $+x$-axis. This is analogous to the stretched string held at the points $x = 0$ and $x = L$. A standing wave can exist only when L is an integer multiple of $\lambda/2$. Thus the possible wavelengths are

$$\lambda_n = \frac{2L}{n}, \qquad n = 1, 2, 3, \ldots, \tag{37-27}$$

and the corresponding frequencies are

$$f_n = \frac{c}{\lambda_n} = n\frac{c}{2L}, \qquad n = 1, 2, 3, \ldots. \tag{37-28}$$

Thus there is a set of *normal modes,* each with a characteristic frequency, wave shape, and node pattern. Measurement of the node positions makes it possible to measure the wavelength. If the frequency is known, the wave speed can be determined. This technique was in fact used by Hertz in his pioneering investigations of electromagnetic waves.

Conducting surfaces are not the only reflectors of electromagnetic waves; reflections also occur at an interface between two insulating materials having different dielectric or magnetic properties. The mechanical analog is a junction of two strings with equal tension but different linear mass density. In general, a wave incident on such a boundary surface is partly transmitted into the second material and partly reflected back into the first. The partial transmission and reflection of light at a glass surface is a familiar phenomenon.

*37–7 Radiation from an antenna

The waves discussed above are called *plane waves,* referring to the fact that at any instant the fields are uniform over a plane perpendicular to the direction of propagation of the wave. Although these waves are the simplest to describe and analyze, they are by no means the simplest to produce experimentally. Any charge or current distribution that oscillates sinusoidally with time produces sinusoidal electromagnetic waves. The simplest example of such a source is an oscillating dipole, a pair of electric charges of equal magnitude and opposite sign, with the charge magnitude varying sinusoidally with time. Such an oscillating dipole can

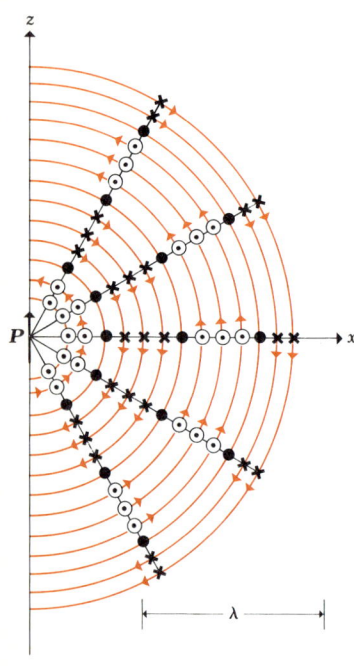

be constructed in a number of ways, the details of which need not concern us.

The radiation from an oscillating dipole is *not* a plane wave, but travels out in all directions from the source. At points far from the source the **E** and **H** fields are perpendicular to the direction from the source and to each other; in this sense the wave is still transverse. The value of S drops off as the square of the distance from the source and also depends on the direction from the source; the radiation is most intense in directions perpendicular to the dipole axis, and $S = 0$ in directions parallel to the axis. The radiation pattern from a dipole source is shown schematically in Fig. 37-7.

37-7 Cross section in the xz-plane of radiation from an oscillating electric dipole, showing electric-field vectors at one instant of time. At the points with circles the **B**-field comes out of the plane of the figure; at the points with crosses, it is into the plane.

Questions

37-1 In Ampère's law, is it possible to have both a conduction current and a displacement current at the same time? Is it possible for the effects of the two kinds of current to cancel each other exactly, so that there is *no* magnetic field produced?

37-2 By measuring the electric and magnetic fields at a point in space where there is an electromagnetic wave, can one determine the direction from which the wave came?

37-3 Sometimes neon signs located near a powerful radio station are seen to glow faintly at night, even though they aren't turned on. What is happening?

37-4 Light can be *polarized;* is this a property of all electromagnetic waves, or is it unique to light? What about polarization of sound waves? What fundamental distinction in wave properties is involved?

37-5 How does a microwave oven work? Why does it heat materials that are conductors of electricity, including most foods, but not insulators, such as glass or ceramic dishes?

37-6 Electromagnetic waves can travel through vacuum, where there is no matter. We usually think of vacuum as empty space, but is it *really* empty if electric and magnetic fields are present? What *is* vacuum, anyway?

37-7 Give several examples of electromagnetic waves that are encountered in everyday life. How are they all alike? How do they differ?

37-8 We are surrounded by electromagnetic waves emitted by many radio and television stations. How is a radio or television receiver able to select a single station among all this mishmash of waves? What happens inside a radio receiver when the dial is turned to change stations?

37-9 The metal conducting rods on a television antenna are always in a horizontal plane. Would they work as well if they were vertical?

37-10 If a light beam carries momentum, should a person holding a flashlight feel a recoil analogous to the recoil of a rifle when it is fired? Why is this recoil not actually observed?

37-11 The nineteenth-century inventor Nikolai Tesla proposed to transmit large quantities of electrical energy across space using electromagnetic waves instead of conventional transmission lines. What advantages would this scheme have? What disadvantages?

37-12 Does an electromagnetic *standing wave* have energy? Momentum? What distinction can be drawn between a standing wave and a propagating wave on this basis?

37-13 If light is an electromagnetic wave, what is its frequency? Is this a proper question to ask?

37-14 When an electromagnetic wave is reflected from a moving reflector, the frequency of the reflected wave is different from that of the initial wave. Explain physically how this can happen. (Some radar systems used for highway speed control operate on this principle.)

Problems

37-1 A certain radio station broadcasts at a frequency of 1020 kHz. At a point some distance from the transmitter, the maximum magnetic field of the electromagnetic wave it emits is found to be 1.6×10^{-11} T.
a) What is the speed of propagation of the wave?
b) What is the wavelength?
c) What is the maximum electric field?

37-2 A certain plane electromagnetic wave emitted by a microwave antenna has a wavelength of 3.0 cm and a maximum magnitude of electric field of 2.0×10^{-4} V·cm^{-1}.
a) What is the frequency of the wave?
b) What is the maximum magnetic field?
c) What is the intensity (power per unit area) if the *average* power is half the power computed from the *maximum* values of E and B?

37-3 An electromagnetic wave propagates in a ferrite material having $K = 10$ and $K_M = 1000$. Find
a) the speed of propagation;
b) the wavelength

of a wave having a frequency of 100 MHz.

37-4 The maximum electric field in the vicinity of a certain radio transmitter is 1.0×10^{-3} V·m^{-1}. What is the maximum magnitude of the B field? How does this compare in magnitude with the earth's field?

37-5 The energy flow to the earth associated with sunlight is about 1.4 kW·m^{-2}.
a) Find the maximum values of E and B for a wave of this intensity.
b) The distance from the earth to the sun is about 1.5×10^{11} m. Find the total power radiated by the sun.

37-6 For a 50,000-W radio station, find the maximum magnitudes of E and B at a distance of 100 km from the antenna, assuming that the antenna radiates equally in all directions (which is probably not actually the case).

37-7 The nineteenth-century inventor Nikolai Tesla proposed to transmit electric power via electromagnetic waves. Suppose power is to be transmitted in a beam of cross-sectional area 100 m²; what E and B strengths are required to transmit an amount of power comparable to that handled by modern transmission lines (of the order of 500 kV and 1000 A)?

37-8 A bar magnet is mounted on an insulating support, as in the side and end views of Fig. 37-8(a) and (b), and is given a positive electric charge. Copy the diagram, sketch the lines of the E and H fields around the magnet, and draw the Poynting vector at a number of points.
a) What is the general nature of the energy flow in the electromagnetic field?

Figure 37-8

b) What can you say about the "hidden momentum" in this case?

37-9 A cylindrical conductor of circular cross section has a radius a and a resistivity ρ and carries a constant current I.
a) What are the magnitude and direction of the E-vector at a point inside the wire, at a distance r from the axis?
b) What are the magnitude and direction of the H-vector at the same point?
c) What are the magnitude and direction of the Poynting vector S at the same point?
d) Compare your answer to (c) with the rate of dissipation of energy within a volume of the conductor of length l and radius r.

37-10 A wire of radius 1 mm whose resistance per unit length is 3×10^{-3} Ω·m^{-1} carries a current of 25.1 A. At a point very near the surface of the wire, calculate
a) the magnitude of H,
b) the component of E parallel to the wire,
c) the component of S perpendicular to the wire.

37-11 Assume that 10% of the power input to a 100-W lamp is radiated uniformly as light of wavelength 500 nm (1 nm = 10^{-9} m). At a distance of 2 m from the source, the electric and magnetic intensities vary sinusoidally according to the equations $E = E_{max} \sin(\omega t + \phi)$ and $H = H_{max} \sin(\omega t + \phi)$. Calculate E_{max} and H_{max}.

37-12 A capacitor consists of two circular plates of radius r separated by a distance l. Neglecting fringing, show that while the capacitor is being charged, the rate at which energy flows into the space between the plates is equal to the rate at which the electrostatic energy increases.

37-13 A very long solenoid of n turns per unit length and radius a carries an increasing current i.
a) Calculate the induced electric field at a point inside the solenoid at a distance r from the solenoid axis.
b) Compute the magnitude and direction of the Poynting vector at this point.

37-14 An FM radio station antenna radiates a power of 10 kW at a wavelength of 3 m. Assume for simplicity that the radiated power is confined to, and is uniform over, a hemisphere with the antenna at its center. What are the amplitudes of E_{max} and H_{max} in the radiation field at a distance of 10 km from the antenna?

37-15 In a TV picture, ghost images are formed when the signal from the transmitter travels directly to the receiver and also indirectly after reflection from a building or other large metallic mass. In a 25-inch set the ghost is about 1 cm to the right of the principal image if the reflected signal arrives 1 μs after the principal signal. In this case, what is the difference in path length for the two signals?

37-16 For a sinusoidal electromagnetic wave in vacuum, such as that described by Eqs. (37–23), show that the average density of energy in the electric field is the same as that in the magnetic field.

37-17 A plane electromagnetic wave has a wavelength of 3.0 cm and an E-field amplitude of 30 V·m^{-1}.

a) What is the frequency?
b) What is the B-field amplitude?
c) What is the intensity?
d) What average force does the radiation pressure exert on a totally absorbing surface of area 0.5 m^2 perpendicular to the direction of propagation?

37-18 If the intensity of direct sunlight is 1.4 kW·m^{-2}, find the radiation pressure (in pascals) on

a) a totally absorbing surface;
b) a totally reflecting surface.

Also express your results in atmospheres.

THE NATURE AND PROPAGATION OF LIGHT 38

38-1 Nature of light

Until the middle of the seventeenth century, light was generally thought to consist of a stream of some sort of particles or *corpuscles* emanating from light sources. Newton and many other scientists of the day supported the corpuscular theory of light. But about this time the idea was proposed by Huygens and others that light might be a *wave* phenomenon. Indeed, diffraction effects that are now known to be associated with the wave nature of light were observed by Grimaldi as early as 1665, but the significance of his observations was not understood at this time.

Early in the nineteenth century, evidence for the wave theory of light grew more persuasive. The experiments of Fresnel and Thomas Young on interference and diffraction showed conclusively that there are many optical phenomena that can be understood on the basis of the wave theory but for which the corpuscular theory is inadequate. These phenomena will be discussed further in Chapter 41. Young's experiments enabled him to measure the wavelength of the waves, and Fresnel showed that the rectilinear propagation of light, as well as the diffraction effects observed by Grimaldi and others, could be accounted for by the behavior of waves of short wavelength.

The next great forward step in the theory of light was the work of Maxwell, discussed in the preceding chapter. In 1873, Maxwell showed that an oscillating electrical circuit should radiate electromagnetic waves. The speed of propagation of the waves could be computed from purely electrical and magnetic measurements. It turned out to be equal, within the limits of experimental error, to the measured speed of propagation of light. The evidence seemed inescapable that light consisted of electromagnetic waves of extremely short wavelength. In 1887 Heinrich Hertz, using an oscillating circuit of small dimensions, succeeded in producing short-wavelength electromagnetic waves and showed that they possessed all the properties of light waves. They could be reflected, refracted, focused by a lens, polarized, and so on, just as could waves of light. Maxwell's electromagnetic theory of light and its experimental justification by Hertz constituted one of the triumphs of physical science.

By the end of the nineteenth century researchers believed that little, if anything, would be added in the future to our knowledge of the nature of light. How wrong they were!

The classical electromagnetic theory failed to account for several phenomena associated with *emission* and *absorption* of light. One example is the phenomenon of photoelectric emission, that is, the ejection of electrons from a conductor by light incident on its surface. In 1905, Einstein extended an idea proposed five years earlier by Planck, and postulated that the energy in a light beam was concentrated in packets or *photons*. The wave picture was retained, in that a photon was still considered to have a frequency and the energy of a photon was proportional to its frequency. Experiments by Millikan confirmed Einstein's predictions. These experiments will be discussed in more detail in Chapter 44.

Another striking confirmation of the photon nature of light is the Compton effect. A. H. Compton, in 1921 succeeded in determining the motion of a photon and a single electron, both before and after a "collision" between them, and found that they behaved like material bodies having kinetic energy and momentum, both of which were conserved in the collision. The photoelectric effect and the Compton effect, then, both seem to demand a return to a corpuscular theory of light.

The reconciliation of these apparently contradictory experiments has come only since 1930, with the development of quantum electrodynamics, a comprehensive theory that includes *both* wave and particle properties. The phenomena of light *propagation* may best be described by the electromagnetic *wave* theory, while the interaction of light with matter, in the processes of emission and absorption, is a *corpuscular* phenomenon.

38-2 Sources of light

All bodies emit electromagnetic radiation as a result of thermal motion of their molecules; this radiation, called *thermal radiation,* is a mixture of different wavelengths. At a temperature of 300°C the most intense of these waves has a wavelength of 5000×10^{-9} m or 5000 nm, which is the *infrared* region. At a temperature of 800°C a body emits enough visible radiant energy to be self-luminous and appears "red hot." By far the larger part of the energy emitted, however, is still carried by infrared waves. At 3000°C, which is about the temperature of an incandescent lamp filament, the radiant energy contains enough of the "visible" wavelengths, between 400 nm and 700 nm, that the body appears "white hot." In modern incandescent lamps, the filament is a coil of fine tungsten wire. An inert gas such as argon is introduced to reduce evaporation of the filament. Incandescent lamps vary in size from one no larger than a grain of wheat to one with a power input of 5000 W, used for illuminating airfields.

One of the brightest sources of light is the *carbon arc*. Two carbon rods, typically 10 to 20 cm long and 1 cm in diameter, are connected to a 110-V or 220-V dc source. They are touched together momentarily and then are pulled apart a few millimeters. Intense electron bombardment of the positive rod causes an extremely hot crater to form at its end; this crater, whose temperature is typically 4000°C, is the source of light. Carbon-arc lights are used in most theater motion-picture projectors and in large searchlights and lighthouses.

Some light sources use an arc discharge in a metal vapor, such as mercury or sodium, in a sealed bulb containing two electrodes, which are connected to a power source. Some argon is sometimes added to permit a glow discharge that helps vaporize and ionize the metal. The bluish light of mercury-arc lamps and the bright orange-yellow of sodium-vapor lamps are familiar in highway and other outdoor lighting.

An important variation of the mercury-arc lamp is the *fluorescent* lamp. This consists of a glass tube containing argon and a droplet of mercury. The electrodes are of tungsten. When an electric discharge takes place in the mercury–argon mixture, only a small amount of visible light is emitted by the mercury and argon atoms. There is, however, considerable *ultraviolet* light (light of wavelength shorter than that of visible violet). This ultraviolet light is absorbed in a thin layer of material, called a *phosphor,* which is the white coating on the interior walls of the glass tube. The phosphor has the property of *fluorescence,* which means that it emits visible light when illuminated by light of shorter wavelength. Lamps may be obtained that will fluoresce with any desired color, depending on the nature of the phosphor.

A special light source that has attained prominence in the last twenty years is the *laser,* which can produce a very narrow beam of enormously intense radiation. High-intensity lasers have been used to cut through steel, fuse high-melting-point materials, and bring about many other effects that are important in physics, chemistry, biology, and engineering. An equally significant characteristic of laser light is that it is much more nearly *monochromatic,* or single-frequency, than any other light source. The operation of one type of laser will be discussed in Chapter 44.

38-3 The speed of light

The speed of light in free space is one of the fundamental constants of nature. Its magnitude is so great (about $186{,}000 \text{ mi} \cdot \text{s}^{-1}$ or $3 \times 10^8 \text{ m} \cdot \text{s}^{-1}$) that it evaded experimental measurement until 1676. Up to that time it was generally believed that light traveled with an infinite speed.

The first recorded attempt to measure the speed of light involved a method proposed by Galileo. Two experimenters were stationed on the tops of two hills about a mile apart. Each was provided with a lantern, the experiment being performed at night. One man was first to uncover his lantern and, observing the light from this lantern, the second was to uncover his. The velocity of light could then be computed from the known distance between the lanterns and the time elapsing between the instant when the first observer uncovered his lantern and when he observed the light from the second. While the experiment was correct in principle, we know now that the speed is much too great for the time interval to be measured in this way.

In 1676, the Danish astronomer Olaf Roemer, from astronomical observations made on one of the satellites of the planet Jupiter, obtained the first definite evidence that light travels with finite speed. Jupiter has twelve small satellites or moons, four of which are sufficiently bright to be seen with a moderately good telescope or a pair of field glasses. The satellites appear as tiny bright points at one side or the other of the planet. These satellites revolve about Jupiter just as does our moon about the earth and, since the plane of their orbits is nearly the same as that in which the earth and Jupiter revolve, each is *eclipsed* by the planet during a part of every revolution.

Roemer measured the time of revolution of one of the satellites by taking the time interval between consecutive eclipses (about 42 hr). He found, by a comparison of results over a long period of time, that while the earth was *receding* from Jupiter the periodic times were all somewhat *longer* than the average, and that while it was *approaching* Jupiter the times were all somewhat *shorter*. He concluded rightly that the cause of these variations was the varying distance between Jupiter and the earth.

Roemer concluded from his observations that a time of about 22 min was required for light to travel a distance equal to the diameter of the earth's orbit. The best figure for this distance, in Roemer's time, was about 172,000,000 mi. Although there is no record that Roemer actually made the computation, if he *had* used the data above he would have found a speed of about 130,000 mi·s^{-1} or 2.1×10^8 m·s^{-1}.

The first successful determination of the speed of light from purely *terrestrial* measurements was made by the French scientist Fizeau in 1849. A schematic diagram of his apparatus is given in Fig. 38–1. Lens L_1 forms an image of the light source S at a point near the rim of a toothed wheel T, which can be set into rapid rotation. G is an inclined plate of clear glass. Suppose first that the wheel is stationary and the light passes through one of the openings between the teeth. Lenses L_2 and L_3, which are separated by about 8.6 km, form a second image on the mirror M. The light is reflected from M, retraces its path, and is in part reflected from the glass plate G through the lens L_4 into the eye of an observer at E.

If the wheel T is set in rotation, the light from S is "chopped up" into a succession of wave trains of limited length. If the speed of rotation is such that by the time the front of one wave train has traveled to the mirror and returned, an opaque segment of the wheel has moved into the position formerly occupied by an open portion, *no* reflected light will reach the observer E. At twice this angular velocity, the light transmitted through any one opening will return through the next and an image

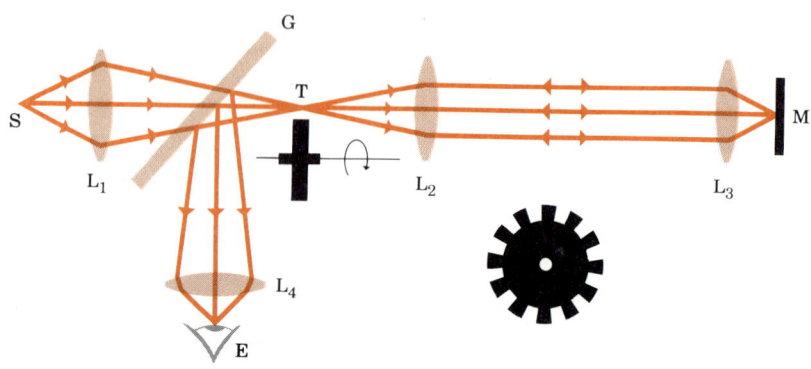

38–1 Fizeau's toothed-wheel method for measuring the velocity of light. S is a light source, L_1, L_2, L_3, and L_4 are lenses. T is the toothed wheel, M is a mirror, and G is a glass plate.

of S will again be observed. From a knowledge of the angular velocity and radius of the wheel, the distance between openings, and the distance from wheel to mirror, the speed of light may be computed. Fizeau's measurements were not of high precision. He obtained a value of

$$3.15 \times 10^8 \text{ m·s}^{-1}.$$

Fizeau's apparatus was modified by Foucault, who replaced the toothed wheel with a rotating mirror. The most precise measurements by the Foucault method were made by the American physicist Albert A. Michelson (1852-1931). His first experiments were performed in 1878 while he was on the staff of the Naval Academy at Annapolis. The last, under way at the time of his death, were completed in 1935 by Pease and Pearson.

From analysis of all measurements up to 1976, the most probable value for the speed of light is

$$c = 2.9979246 \times 10^8 \text{ m·s}^{-1},$$

which is believed to be correct within ± 2 m·s^{-1}.

As shown in Chapter 37, the speed of any electromagnetic wave in free space is given by

$$c = \sqrt{\frac{1}{\epsilon_0 \mu_0}},$$

and since, in SI units, μ_0 is assigned a value of *exactly* $4\pi \times 10^{-7}$ N·s^2·C^{-2}, the preceding equation provides the most precise means for finding the value of the electrical constant ϵ_0:

$$\epsilon_0 = 1/\mu_0 c^2,$$

since the speed of light can be measured with much higher precision than the force between two charged bodies.

38-4 The electromagnetic spectrum

As indicated in preceding sections, it is now well established that light consists of electromagnetic waves. In this respect it is a small part of a very broad class of electromagnetic radiations, all having the general characteristics described in Chapter 37 (including the common speed of propagation $c = 3.00 \times 10^8$ m·s^{-1}) but differing in frequency (f) and wavelength (λ). The general wave relation $c = \lambda f$ holds for each.

The wavelengths of visible light (i.e., of electromagnetic waves that are perceived by the sense of sight) can be measured by methods to be discussed in Chapter 41, and are found to lie in the range 4 to 7×10^{-7} m. The corresponding range of frequencies is about 7.5 to 4.3×10^{14} Hz. The position of visible light in relation to the entire electromagnetic spectrum is shown in Fig. 38-2.

Because of the very small magnitudes of light wavelengths, it is convenient to measure them in small units of length. Three such units are commonly used: the micrometer (1 μm), the nanometer (1 nm) (both accented on the *first* syllable), and the angstrom (1 Å):

$$1 \, \mu\text{m} = 10^{-6} \text{ m} = 10^{-4} \text{ cm},$$
$$1 \text{ nm} = 10^{-9} \text{ m} = 10^{-7} \text{ cm},$$
$$1 \text{ Å} = 10^{-10} \text{ m} = 10^{-8} \text{ cm} = 0.1 \text{ nm}.$$

38-2 A chart of the electromagnetic spectrum.

In older literature the micrometer is sometimes called the *micron,* and the nanometer is sometimes called the *millimicron;* these terms are now obsolete. Most workers in the fields of optical instrument design, color, and physiological optics express wavelengths in *nanometers*. For example, the wavelength of the yellow light from a sodium-vapor lamp is 589 nm, but many spectroscopists would identify this same wavelength as 5890 Å.

Different parts of the visible spectrum evoke the sensations of different colors. Wavelengths for colors in the visible spectrum are (very approximately) as follows:

400 to 450 nm	Violet
450 to 500 nm	Blue
500 to 550 nm	Green
550 to 600 nm	Yellow
600 to 650 nm	Orange
650 to 700 nm	Red

By the use of special sources or special filters, it is possible to limit the wavelength spread to a small band, say from 1 to 10 nm. Such light is called roughly *monochromatic light,* meaning light of a single color. Light consisting of only one wavelength is an idealization that is useful in theoretical calculations but represents an experimental impossibility. When the expression "monochromatic light of wavelength 550 nm" is used in theoretical discussions, it refers to one wavelength, but in descriptions of laboratory experiments it means a small band of wavelengths *around* 550 nm. One distinguishing characteristic of light from a *laser* is that it is much more nearly monochromatic than light obtainable in any other way.

38-5 Waves, wave fronts, and rays

It is convenient to represent a wave of any sort by means of *wave fronts*. *A wave front is defined as the locus of all points at which the phase of vibration of a physical quantity is the same.* Thus, in the case of sound waves spreading out in all directions from a point source, any spherical surface concentric with the source is a possible wave front. Some of these spherical surfaces are the loci of points at which the pressure is a *maximum,* others where it is a *minimum,* and so on; but the *phase* of the pressure variations is the same over any spherical surface. It is customary to draw only a few wave fronts, usually those that pass through the maxima and minima of the disturbance. Such wave fronts are separated from one another by one-half wavelength.

If the wave is a light wave, the quantity corresponding to the pressure in a sound wave is the electric or magnetic field. It is usually unnecessary to indicate in a diagram either the magnitude or direction of the field, but simply to show the *shape* of the wave by drawing the wave fronts or their intersections with some reference plane. For example, the electromagnetic waves radiated by a small light source may be represented by *spherical* surfaces concentric with the source or, as in Fig. 38–3a, by the intersections of these surfaces with the plane of the diagram. At a sufficiently great distance from the source, where the radii of the spheres have become very large, the spherical surfaces can be considered planes and we have a *plane* wave, as in Fig. 38–3b.

For some purposes, especially in that branch of optics called *geometrical* optics, it is convenient to represent a light wave by *rays* rather than by wave fronts. Indeed, rays were used to describe light long before its wave nature was firmly established; in a corpuscular theory of light, rays are merely the paths of the corpuscles. From the wave viewpoint, *a ray is an imaginary line drawn in the direction in which the wave is traveling*. Thus, in Fig. 38–3a the rays are the radii of the spherical wave fronts, and in Fig. 38–3b they are straight lines perpendicular to the wave fronts. In fact, in every case where waves travel in a homogeneous isotropic medium, the rays are straight lines normal to the wave fronts. At a boundary surface between two media, such as the surface between a glass plate and the air outside it, the direction of a ray may change suddenly, but it is a straight line both in the air and in the glass. If the medium is *not* homogeneous, for instance, if one is considering the passage of light through the earth's atmosphere, where the density and hence the velocity vary with elevation, the rays are curved but are still normal to the wave fronts.

Wave fronts
(a)

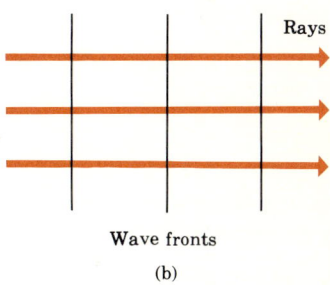

Wave fronts
(b)

38–3 Wave fronts and rays.

38-6 Reflection and refraction

Many familiar optical phenomena involve the behavior of a wave that strikes an interface between two optical materials, such as air and glass or water and glass. When the interface is smooth, i.e., when its irregulari-

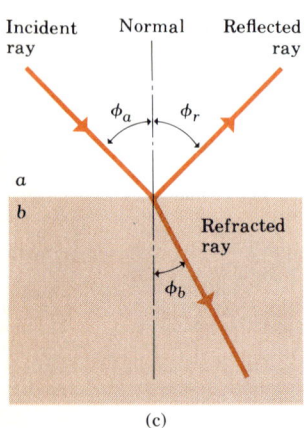

38-4 (a) A plane wave is in part reflected and in part refracted at the boundary between two media. (b) The waves in (a) are represented by rays. (c) For simplicity, only one example each of incident, reflected, and refracted rays is drawn.

ties are small compared with the wavelength, the wave is in general partly reflected and partly transmitted into the second medium, as shown in Fig. 38-4a. As an example of this behavior, when one looks into a store window from the street, one sees light coming from inside by transmission through the glass and also a reflection of the street scene caused by the air–glass interface.

The segments of plane waves shown in Fig. 38-4a can be represented by bundles of rays forming *beams* of light, as in Fig. 38-4b; and for simplicity in discussing the various angles we often consider only one ray in each beam, as in Fig. 38-4c. Representing these waves in terms of rays is the basis of that branch of optics called *geometrical optics*. This chapter and the two following chapters are devoted entirely to geometrical optics and are concerned with optical phenomena that can be understood on the basis of rays, without having to use the wave nature of light explicitly. It must be understood that geometrical optics, like any idealized model, has its limitations, and that there are optical phenomena that require a more detailed model embodying the *wave* properties of light for their understanding. We return to such phenomena in Chapters 41 and 42.

The directions of the incident, reflected, and refracted beams of light are specified in terms of the angles they make with the *normal* to the surface at the point of incidence. For this purpose it is sufficient to indicate one ray, as in Fig. 38-4c, although a single ray of light is a geometrical abstraction. A careful experimental study of the directions of the incident, reflected, and refracted beams leads to the following results:

1. *The incident, reflected, and refracted rays, and the normal to the surface, all lie in the same plane.* Thus, if the incident ray is in the plane of the diagram, and the surface of separation is perpendicular to this plane, the reflected and refracted rays are in the plane of the diagram.

2. *The angle of reflection ϕ_r is equal to the angle of incidence ϕ_a for all colors and any pair of substances.* Thus

$$\phi_r = \phi_a. \tag{38-1}$$

The experimental result that $\phi_r = \phi_a$, and that the incident and reflected rays and the normal all lie in the same plane, is known as the *law of reflection.*

3. For monochromatic light and for a given pair of substances, a and b, on opposite sides of the surface of separation, *the ratio of the sine of the angle ϕ_a* (between the ray in substance a and the normal) *and the sine of the angle ϕ_b* (between the ray in substance b and the normal) *is a constant.* Thus

$$\frac{\sin \phi_a}{\sin \phi_b} = \text{constant}. \tag{38-2}$$

This experimental result, together with the fact that the incident and refracted rays and the normal to the surface all lie in the same plane, is known as the *law of refraction.* The discovery of this law is usually credited to Willebrord Snell (1591–1626), although there is some doubt that it was actually original with him. It is called Snell's law.

In the above discussion, the angles of the various rays are by convention always measured with respect to the *normal* to the surface, not the surface itself. It is essential to keep this in mind when applying Eq. (38–2).

The laws of reflection and refraction relate only to the *directions* of the corresponding rays but say nothing about an equally important question, namely, the *intensities* of the reflected and refracted rays. These depend on the angle of incidence; for the present we simply state that the fraction reflected is smallest at *normal* incidence, where it is a few percent, and that it increases with increasing angle of incidence to almost 100% at grazing incidence or when $\phi_a = 90°$.

When a ray of light is directed from *below* the surface in Fig. 38–4, there are again reflected and refracted rays; these, together with the incident ray and the normal, all lie in the same plane. The same law of reflection applies as when the ray is originally traveling in air, and the same law of refraction, Eq. (38–2). *The passage of a ray of light in going from one medium to another is reversible*. It follows the same path in going from b to a as when going from a to b.

Let us now consider a beam of monochromatic light traveling *in vacuum*, making an angle of incidence ϕ_0 with the normal to the surface of a substance a, and let ϕ_a be the angle of refraction in the substance. The constant in Snell's law is then called the *index of refraction* of substance a and is designated by n_a:

$$\frac{\sin \phi_0}{\sin \phi_a} = n_a. \quad (38-3)$$

The index of refraction (also called *refractive index*) is always greater than unity; it depends not only on the substance but on the wavelength of the light. If no wavelength is stated, the index is usually assumed to be that corresponding to the yellow light from a sodium lamp, of wavelength 589 nm. This wavelength is near the middle of the visible spectrum.

The index of refraction of most of the common glasses used in optical instruments lies between 1.46 and 1.96. There are very few substances having indices larger than 1.96, diamond being one, with an index of 2.42, and rutile (crystalline titanium dioxide) another, with an index of 2.7. The values for a number of solids and liquids are given in Table 38–1.

The index of refraction of *air* at standard conditions is about 1.0003; for most purposes the *index of refraction of air can be assumed to be unity*. The index of refraction of a gas increases uniformly as the density of the gas increases.

It follows from Eq. (38–3) that the angle of refraction ϕ_a is always *less than* the angle of incidence ϕ_0 for a ray passing from a vacuum into one of the materials listed in Table 38–1. In such a case, the ray is bent *toward* the normal. If the light is traveling in the opposite direction, the reverse is true and the ray is bent *away from* the normal.

To relate the index of refraction to the constant in Eq. (38–2), we now consider two parallel-sided plates of substances a and b placed parallel to each other with an arbitrary space between them, as shown in Fig. 38–5a. Let the medium surrounding both plates be vacuum, although the behavior of light would be practically the same if the plates

Table 38–1 Index of refraction for yellow sodium light ($\lambda = 589$ nm)

Substance	Index of refraction
Solids	
Ice (H$_2$O)	1.309
Fluorite (CaF$_2$)	1.434
Rock salt (NaCl)	1.544
Quartz (SiO$_2$)	1.544
Zircon (ZrO$_2$ · SiO$_2$)	1.923
Diamond (C)	2.417
Fabulite (SrTiO$_3$)	2.409
Glasses (typical values)	
Crown	1.52
Light flint	1.58
Medium flint	1.62
Dense flint	1.66
Lanthanum flint	1.80
Liquids at 20°C	
Methyl alcohol (CH$_3$OH)	1.329
Water (H$_2$O)	1.333
Ethyl alcohol (C$_2$H$_5$OH)	1.36
Carbon tetrachloride (CCl$_4$)	1.460
Turpentine	1.472
Glycerine	1.473
Benzene	1.501
Carbon disulfide (CS$_2$)	1.628

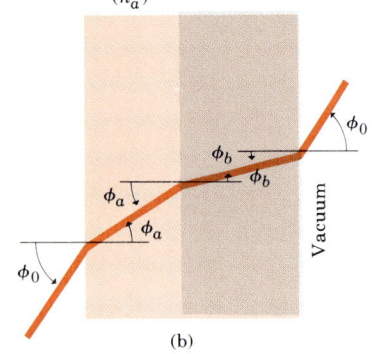

38-5 The transmission of light through parallel plates of different substances. The incident and emerging rays are parallel, regardless of direction and regardless of the thickness of the space between adjacent slabs.

were surrounded by air. If a ray of monochromatic light starts at the lower left with an angle of incidence ϕ_0, the angle between ray and normal in substance a is ϕ_a, and the light emerges from substance a at an angle ϕ_0 equal to its incident angle. The light ray therefore enters plate b with an angle of incidence ϕ_0, makes an angle ϕ_b in substance b, and emerges again at an angle ϕ_0. Exactly the same path would be traversed if the same light ray were to start at the upper right and enter substance b at an angle ϕ_0. Moreover, *the angles are independent of the thickness of the space between the two plates*, and are the same when the space shrinks to nothing, as in Fig. 38-5b.

Applying Snell's law to the refractions at the surface between vacuum and substance a, and at the surface between vacuum and substance b, we have

$$\frac{\sin \phi_0}{\sin \phi_a} = n_a, \qquad \frac{\sin \phi_0}{\sin \phi_b} = n_b.$$

Dividing the second equation by the first, we obtain

$$\frac{\sin \phi_a}{\sin \phi_b} = \frac{n_b}{n_a}, \qquad (38\text{-}4)$$

which shows that *the constant in Snell's law for the refraction between substances a and b is the ratio of the indices of refraction*. From Eq. (38-4) we see that the simplest and most symmetrical way of writing Snell's law for any two substances a and b, and for any direction, is

$$n_a \sin \phi_a = n_b \sin \phi_b. \qquad (38\text{-}5)$$

Example In Fig. 38-4, let material a be water and material b a glass with index of refraction 1.50. If the incident ray makes an angle of 60° with the normal, find the direction of the reflected and refracted rays.

Solution The reflected-ray angle with the normal is the same as that of the incident ray, according to Eq. (38-1). Hence $\phi_r = \phi_a = 60°$. To find the direction of the refracted ray we use Eq. (38-5), with $n_a = 1.33$, $n_b = 1.50$, and $\phi_a = 60°$. We find

$$(1.33)(\sin 60°) = (1.50)(\sin \phi_b),$$
$$\phi_b = 50.2°.$$

The second material has larger refractive index than the first, and the refracted ray is bent toward the normal.

Two final comments about reflection and refraction need to be added. First, reflection occurs at a highly polished surface of an *opaque* material such as a metal. There is then no refracted ray, but the reflected ray still behaves according to Eq. (38-1). Second, if the reflecting surface of either a transparent or an opaque material is rough, with irregularities on a scale comparable to or larger than the wavelength of light, reflection occurs not in a single direction but in all directions; such reflection is called *diffuse* reflection. Conversely, reflections in a single direction from smooth surfaces are called *regular* reflections or *specular* reflections.

38-7 Total internal reflection

Figure 38-6 shows a number of rays diverging from a point source P in medium a of index n_a and striking the surface of a second medium b of index n_b, where $n_a > n_b$. From Snell's law,

$$\sin \phi_b = \frac{n_a}{n_b} \sin \phi_a.$$

Since n_a/n_b is greater than unity, $\sin \phi_b$ is larger than $\sin \phi_a$. Thus there must be some value of ϕ_a less than 90° for which $\sin \phi_b = 1$ and $\phi_b = 90°$. This is illustrated by ray 3 in the diagram, which emerges just grazing the surface at an angle of refraction of 90°. The angle of incidence for which the refracted ray emerges tangent to the surface is called the *critical angle* and is designated by ϕ_{crit} in the diagram. If the angle of incidence is greater than the critical angle, the sine of the angle of refraction, as computed by Snell's law, would have to be greater than unity. This may be interpreted to mean that beyond the critical angle the ray does not pass into the upper medium but is *totally internally reflected* at the boundary surface. Total internal reflection can occur only when a ray is incident on the surface of a medium whose index is *smaller* than that of the medium in which the ray is traveling.

The critical angle for two given substances may be found by setting $\phi_b = 90°$ or $\sin \phi_b = 1$ in Snell's law. We then have

$$\sin \phi_{\text{crit}} = \frac{n_b}{n_a}. \tag{38-6}$$

The critical angle of a glass–air surface, taking 1.50 as a typical index of refraction of glass, is

$$\sin \phi_{\text{crit}} = \frac{1}{1.50} = 0.67, \qquad \phi_{\text{crit}} = 42°.$$

The fact that this angle is slightly less than 45° makes it possible to use a prism of angles 45°–45°–90° as a totally reflecting surface. The advantages of totally reflecting prisms over metallic surfaces as reflec-

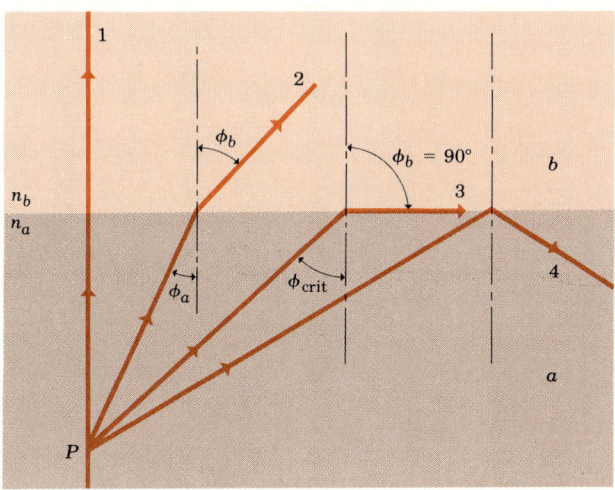

38-6 Total internal reflection. The angle of incidence ϕ_a for which the angle of refraction is 90°, is called the critical angle.

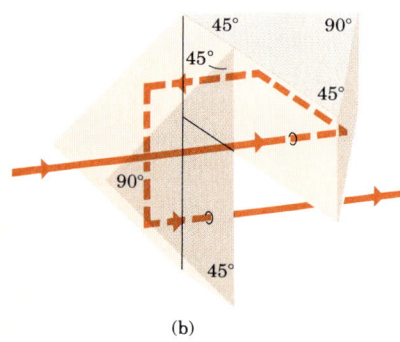

38-7 (a) A Porro prism. (b) A combination of two Porro prisms.

tors are, first, that the light is *totally* reflected, while no metallic surface reflects 100% of the light incident on it, and second, the reflecting properties are permanent and not affected by tarnishing. Offsetting these is the fact that there is some loss of light by reflection at the surfaces where light enters and leaves the prism, although coating the surfaces with so-called "nonreflecting" films can reduce this loss considerably.

A 45°–45°–90° prism, used as in Fig. 38–7a, is called a *Porro prism*. Light enters and leaves at right angles to the hypotenuse and is totally reflected at each of the shorter faces. The deviation is 180°. Two Porro prisms are sometimes combined, as in Fig. 38–7b, an arrangement often found in binoculars.

If a beam of light enters one end of a transparent rod, as in Fig. 38–8, the light is totally reflected internally and is "trapped" within the rod even if it is curved, provided the curvature is not too great. The rod is sometimes referred to as a *light pipe*. A bundle of fine fibers will behave in the same way and has the advantage of being flexible. The study of the properties of such a bundle is an active field of research known as *fiber optics*.

A bundle may consist of thousands of individual fibers, each the order of 0.002 to 0.01 mm in diameter. If the fibers can be assembled in the bundle so that the relative positions of the ends are the same at both ends, the bundle can transmit an image, as shown in Fig. 38–9. Bundles several feet in length have been made. Devices using fiber optics are finding a wide range of applications in medical science. The interior of lungs and other passages in the human body can be viewed by insertion of a fiber bundle. A bundle can be enclosed in a hypodermic needle for the study of tissues and blood vessels far beneath the skin. Despite the enormous technical difficulties of manufacturing fiber-optic components, such systems promise to become an extremely important class of optical systems.

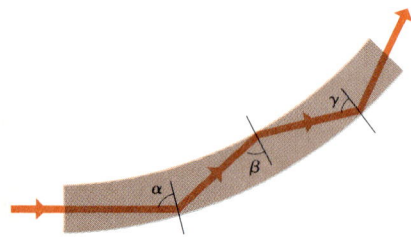

38-8 A light ray "trapped" by internal reflections.

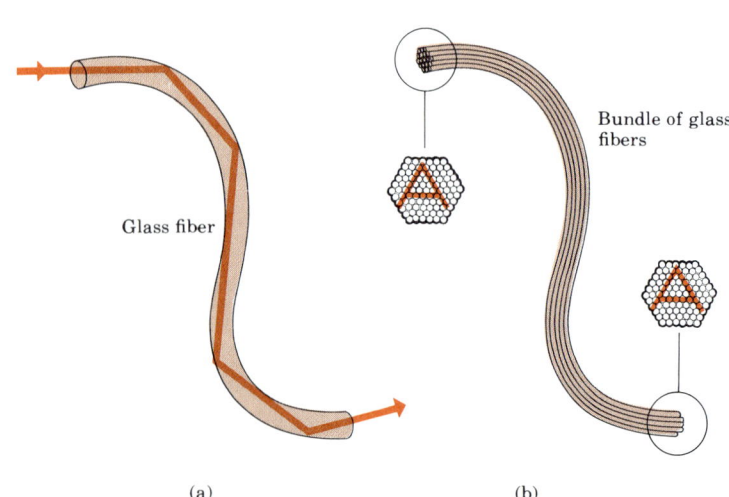

38-9 (a) Total internal reflection in a single fiber. (b) Image transmission by a bundle of fibers.

38-8 Huygens' principle

The principles governing reflection and refraction of light rays, discussed in the preceding section, were discovered experimentally long before the wave nature of light was firmly established. These principles may however be derived from *wave* considerations and thus shown to be consistent with the wave nature of light. To establish this connection we use a principle called *Huygens' principle*; as stated originally by Christian Huygens in 1678, this principle is a geometrical method for finding, from the known shape of a wave front at some instant, the shape of the wave front at some later time. Huygens assumed that *every point of a wave front may be considered the source of secondary wavelets, which spread out in all directions with a speed equal to the speed of propagation of the waves*. The new wave front is then found by constructing a surface *tangent* to the secondary wavelets or, as it is called, the *envelope* of the wavelets.

38-10 Huygens' principle.

Huygens' principle is illustrated in Fig. 38-10. The original wave front, AA', is traveling as indicated by the small arrows. We wish to find the shape of the wave front after a time interval t. Let v represent the speed of propagation. We construct a number of circles (traces of spherical wavelets) of radius $r = vt$, with centers along AA'. The trace of the envelope of these wavelets, which is the new wave front, is the curve BB'. The speed v has been assumed the same at all points in all directions.

To derive the law of reflection from Huygens' principle, we consider a plane wave approaching a plane reflecting surface. In Fig. 38-11a, the lines AA', BB', and CC' represent successive positions of a wave front approaching the surface MM'. The actual planes are perpendicular to the plane of the figure. Point A on the wave front AA' has just arrived at the reflecting surface. The position of the wave front after a time interval t may be found by applying Huygens' principle. With points on AA' as centers, draw a number of secondary wavelets of radius vt, where v is the speed of propagation of the wave. Those wavelets originating near the upper end of AA' spread out unhindered, and their envelope gives that portion of the new wave surface OB'. The wavelets originating near the lower end of AA', however, strike the reflecting surface. If the latter had not been there, they would have occupied the positions shown by the dotted circular arcs. The effect of the reflecting surface is to *reverse the direction* of travel of those wavelets that strike it, so that that part of a wavelet that would have penetrated the surface actually lies to the left of it, as shown by the full lines. The envelope of these reflected wavelets is then that portion of the wave front OB. The trace of the entire wave front at this instant is the broken line BOB'. A similar construction gives the line CPC' for the wave front after another interval t.

The angle ϕ between the incident *wave front* and the *surface* is the same as that between the incident *ray* and the *normal* to the surface, and is therefore the angle of incidence. Similarly, ϕ_r is the angle of reflection. To find the relation between these angles, we consider Fig. 38-11b. From O, draw $OP = vt$, perpendicular to AA'. Now OB, by construction, is tangent to a circle of radius vt with center at A. Hence, if AQ is drawn from A to the point of tangency, the triangles APO and OQA are equal (right triangles with the side AO in common and with $AQ = OP$). The angle ϕ therefore equals the angle ϕ_r, and we have the law of reflection.

The law of refraction is obtained by a similar procedure. In Fig. 38-12a, we consider a wave front, represented by line AA', for which point A has just arrived at the boundary surface MM' between two transparent media a and b, of indices of refraction n_a and n_b. (The *reflected* waves are not shown in the figure; they proceed exactly as in Fig.

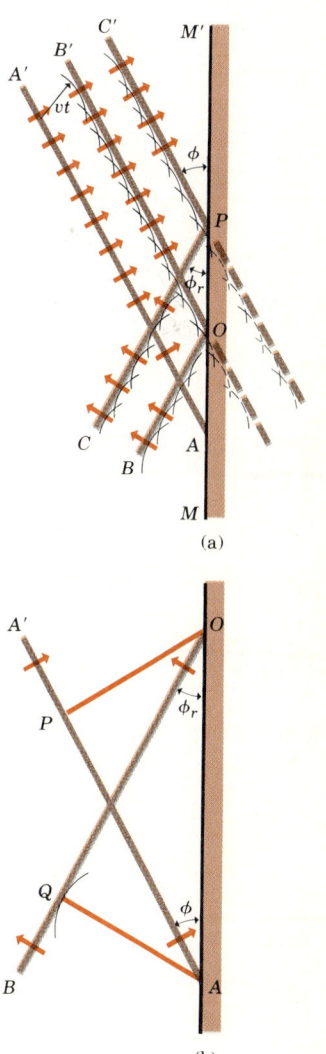

38-11 (a) Successive positions of a plane wave AA' as it is reflected from a plane surface. (b) A portion of (a).

38-11.) Let us apply Huygens' principle to find the position of the refracted wave front after a time t.

With points on AA' as centers, we draw a number of secondary wavelets. Those originating near the upper end of AA' travel with speed v_a and, after a time interval t, are spherical surfaces of radius $v_a t$. The wavelet originating at point A, however, is traveling in the second medium b with speed v_b and at time t is a spherical surface of radius $v_b t$. The envelope of the wavelets from the original wave front is the plane whose trace is the broken line BOB'. A similar construction leads to the trace CPC' after a second interval t.

The angles ϕ_a and ϕ_b between the surface and the incident and refracted wave fronts are, respectively, the angle of incidence and the angle of refraction. To find the relation between these angles, refer to Fig. 38-12b. Draw $OQ = v_a t$, perpendicular to AQ, and draw $AB = v_b t$, perpendicular to BO. From the right triangle AOQ,

$$\sin \phi_a = \frac{v_a t}{AO},$$

and from the right triangle AOB,

$$\sin \phi_b = \frac{v_b t}{AO}.$$

Hence

$$\frac{\sin \phi_a}{\sin \phi_b} = \frac{v_a}{v_b} \tag{38-7}$$

Since v_a/v_b is a constant, Eq. (38-7) expresses Snell's law, and we have derived Snell's law from a wave theory.

The most general form of Snell's law is given by Eq. (38-4), namely

$$\frac{\sin \phi_a}{\sin \phi_b} = \frac{n_b}{n_a}.$$

Comparing this with Eq. (38-7), we see that

$$\frac{v_a}{v_b} = \frac{n_b}{n_a},$$

and

$$n_a v_a = n_b v_b.$$

When either medium is vacuum, $n = 1$ and the speed is c. Hence

$$n_a = \frac{c}{v_a}, \qquad n_b = \frac{c}{v_b}, \tag{38-8}$$

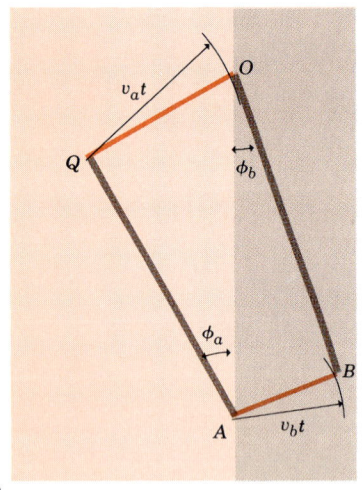

38-12 (a) Successive positions of a plane wave front AA' as it is refracted by a plane surface. (b) A portion of (a). The case $v_b < v_a$ is shown.

showing that *the index of refraction of any medium is the ratio of the speed of light in vacuum to the speed in the medium*. For any material, n is always greater than unity; hence the speed of light in a material is always *less* than in vacuum.

In Fig. 38-12, if t is chosen to be the period τ of the wave, the spacing is $v\tau$, which is the wavelength λ. The figure shows that when v_b is less than v_a, the wavelength in the second medium is smaller than that in the first. When a light wave proceeds from one medium to another, where the speed is different, the wavelength changes *but not the frequency*. Since

$$v_a = f\lambda_a \quad \text{and} \quad v_b = f\lambda_b,$$
$$\frac{\lambda_a}{v_a} = \frac{\lambda_b}{v_b} \quad \text{and} \quad \lambda_a \frac{c}{v_a} = \lambda_b \frac{c}{v_b}.$$

Therefore,

$$\lambda_a n_a = \lambda_b n_b.$$

If either medium is vacuum, the index is 1 and the wavelength in vacuum is represented by λ_0. Hence

$$\lambda_a = \frac{\lambda_0}{n_a}, \quad \lambda_b = \frac{\lambda_0}{n_b}, \quad (38\text{-}9)$$

showing that *the wavelength in any medium is the wavelength in vacuum divided by the index refraction of the medium.*

38-13 Variation of index with wavelength.

38-9 Dispersion

Most light beams are a mixture of waves whose wavelengths extend throughout the visible spectrum. While the speed of light waves in a vacuum is the same for all wavelengths, the speed in a material substance is different for different wavelengths. Hence the index of refraction of a substance is a function of wavelength. The dependence of wave speed on wavelength was mentioned in connection with water waves, in Sec. 21-7. A substance in which the speed of a wave varies with wavelength is said to exhibit *dispersion*. Figure 38-13 is a diagram showing the variation of index of refraction with wavelength for a number of the more common optical materials.

Consider a ray of white light, a mixture of all visible wavelengths, incident on a prism, as in Fig. 38-14. Since the deviation produced by the prism increases with increasing index of refraction, violet light is deviated most and red least, with other colors occupying intermediate positions. On emerging from the prism, the light is spread out into a fan-shaped beam, as shown. The light is said to be *dispersed* into a spectrum. A simple measure of the dispersion is provided by the angular separation of the red and violet rays. Dispersion depends on the *difference* between the refractive index for violet light and that for red light. From Fig. 38-13 it can be seen that, for a substance such as fluorite, whose index for yellow light is small, the difference between the indices for red and violet is also small. On the other hand, in the case of silicate flint glass, both the index for yellow light and the difference between extreme indices are large. In other words, for most transparent materials the greater the deviation, the greater the dispersion.

The brilliance of diamond is due in part to its large dispersion. In recent years synthetic crystals of titanium dioxide and of strontium titanate, with about eight times the dispersion of diamond, have been produced.

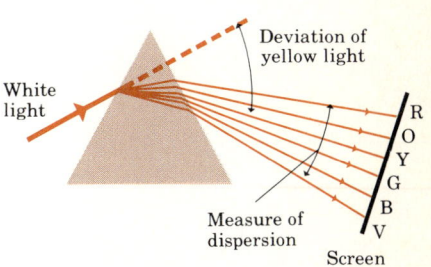

38-14 Dispersion by a prism. The band of colors on the screen is called a spectrum.

*38-10 Absorption

No material is perfectly transparent; as light passes through any optical medium (except vacuum) its energy is partially absorbed, increasing the internal energy in the material, and the intensity (power per unit area) is correspondingly attenuated.

When a beam of light passes through a thin sheet of material, of thickness dx, the decrease dI in its intensity I is found to be proportional to the initial intensity I and to the thickness dx. Thus

$$dI = -\alpha I \, dx. \tag{38-10}$$

The proportionality constant α, which depends on the material, is called the *absorption coefficient*. The intensity after passage through a slab of finite thickness x can be obtained by integrating Eq. (38-10):

$$I = I_0 e^{-\alpha x}, \tag{38-11}$$

where I_0 is the intensity at $x = 0$. Equation (38-11) is called *Lambert's law*.

Example A certain glass has $\alpha = 10 \text{ m}^{-1}$. Light passing through a flat slab of this glass at normal incidence is decreased in intensity by 5 percent. What is the thickness of the slab?

Solution We divide both sides of Eq. (38-11) by I_0 and then take natural logs of both sides to obtain

$$\ln \frac{I}{I_0} = -\alpha x, \quad \text{or} \quad x = -\frac{1}{\alpha} \ln \frac{I}{I_0}.$$

In our example, $I/I_0 = 0.95$, and we find

$$x = -\frac{1}{10 \text{ m}^{-1}} \ln 0.95 = 0.00513 \text{ m}$$

$$= 0.513 \text{ cm}.$$

The absorption coefficient is often strongly wavelength-dependent. A clear optical glass for which $\alpha = 4 \text{ m}^{-1}$ (a typical value in the middle of the visible spectrum) may have $\alpha = 1000 \text{ m}^{-1}$ at $\lambda = 250$ nm (the near ultraviolet) and hence be essentially opaque to ultraviolet radiation. If α varies substantially *within* the visible spectrum, white light incident on the material appears colored when it emerges. For example, if α is larger at 500 nm than at 600 nm, proportionally more green and blue light is absorbed, and the emergent light appears red.

In some materials, absorption depends on the *polarization* of the incident light, that is, the plane in which the electric-field vector oscillates. A familiar example is the Polaroid filter, discussed in Sec. 42-5, used for sunglasses and many other applications. This material is transparent for light with one plane of polarization but absorbent for light in the perpendicular plane; the two absorption coefficients may differ by as much as a factor of 100.

A beam of light passing through an optical medium may also be attenuated by *scattering*. In contrast to absorption, in which the energy is ordinarily converted to internal energy, scattering simply redirects some of the radiation into directions other than that of the beam. The visibility from the side of a searchlight beam at night results from scattering of light out of the beam by dust particles and water droplets in the air and, to a small extent, by the air molecules themselves.

Scattering of visible light is usually wavelength-dependent, usually increasing with increasing frequency (decreasing wavelength). The sky is blue because the shorter-wavelength (blue) visible light from the sun is scattered more strongly in the earth's atmosphere than is the longer-wavelength (red) light. Similarly, at sunset the sun appears red because proportionally more of the blue light has been scattered out of the beam.

*38-11 Illumination

We have defined the *intensity* of light and other electromagnetic radiation as power per unit area, measured in watts per square meter. Similarly, the *total* rate of radiation of energy from any of the sources of light discussed in Sec. 38-2 is called the *radiant power* or *radiant flux*, measured in watts. These quantities are not adequate to measure the visual sensation of *brightness*, however, for two reasons: First, not all the radiation from a source lies in the visible spectrum; an ordinary incandescent light bulb radiates more energy in the infrared than in the visible spectrum. Second, the eye is not equally sensitive to all wavelengths; a bulb emitting 1 watt of yellow light appears brighter than one emitting one watt of blue light.

The quantity analogous to radiant power, but compensated to include the above effects, is called *luminous flux*, denoted by F. The unit of luminous flux is the *lumen*, abbreviated lm, defined as that quantity of light emitted by $\frac{1}{60}$ cm^2 surface area of pure platinum at its melting temperature (about 1770°C), within a solid angle of 1 steradian (1 sr). As an example, the total light output (luminous flux) of a 40-watt incandescent light bulb is about 500 lm, while that of a 40-watt fluorescent tube is about 2300 lm.

When luminous flux strikes a surface, the surface is said to be *illuminated*. The intensity of illumination, analogous to the intensity of electromagnetic radiation (which is power per unit area) is the *luminous flux per unit area*, called the *illuminance*, denoted by E. The unit of illuminance is the lumen per square meter, also called the *lux*:

$$1 \text{ lux} = 1 \text{ lm} \cdot \text{m}^{-2}.$$

An older unit, the lumen per square foot, or foot-candle, has become obsolete. If luminous flux F falls at normal incidence on an area A, the illuminance E is given by

$$E = \frac{F}{A}. \tag{38-12}$$

Most light sources do not radiate equally in all directions; it is useful to have a quantity that describes the intensity of a source in a specific direction, without using any specific distance from the source. We place the source at the center of an imaginary sphere of radius R. A small area A of the sphere subtends a solid angle Ω given by $\Omega = A/R^2$. If the luminous flux passing through this area is F, we define the *luminous intensity* I in the direction of the area as

$$I = \frac{F}{\Omega}. \tag{38-13}$$

The unit of luminous intensity is one lumen per steradian, also called one *candela,* abbreviated cd:

$$1 \text{ cd} = 1 \text{ lm} \cdot \text{sr}^{-1}.$$

The term "luminous intensity" is somewhat misleading. The usual usage of *intensity* connotes power per unit area, and the intensity of radiation from a point source decreases as the square of the distance. Luminous intensity, however, is flux per unit *solid angle,* not per unit *area,* and the luminous intensity of a source in a particular direction *does not* decrease with increasing distance.

Example A certain 100-watt lightbulb emits a total luminous flux of 1200 lm, distributed uniformly over a hemisphere. Find the illuminance and the luminous intensity at a distance of 1 m, and at 5 m.

Solution The area of a half-sphere of radius 1 m is

$$(2\pi)(1 \text{ m})^2 = 6.28 \text{ m}^2.$$

The illuminance at 1 m is

$$E = \frac{1200 \text{ lm}}{6.28 \text{ m}^2} = 191 \text{ lm} \cdot \text{m}^{-2} = 191 \text{ lux}.$$

Similarly, the area of a half-sphere of radius 5 m is

$$(2\pi)(5 \text{ m})^2 = 157 \text{ m}^2,$$

and the illuminance at 5 m is

$$E = \frac{1200 \text{ lm}}{157 \text{ m}^2} = 7.64 \text{ lm} \cdot \text{m}^{-2} = 7.64 \text{ lux}.$$

This is smaller by a factor of 5^2 than the illuminance at 1 m, and illustrates the inverse-square law for illuminance from a point source.

The solid angle subtended by a hemisphere is 2π sr. The luminous intensity is

$$I = \frac{1200 \text{ lm}}{2\pi \text{ sr}} = 191 \text{ lm} \cdot \text{sr}^{-1} = 191 \text{ cd}.$$

The luminous intensity does not depend on distance.

Questions

38-1 During a thunderstorm one always sees the flash of lightning before hearing the accompanying thunder. Discuss this in terms of the various wave speeds. Can this phenomenon be used to determine how far away the storm is?

38-2 When hot air rises around a radiator or from a heating duct, objects behind it appear to shimmer or waver. What is happening?

38-3 Light requires about 8 minutes to travel from the sun to the earth. Is it delayed appreciably by the earth's atmosphere?

38-4 Sometimes when looking at a window one sees two reflected images, slightly displaced from each other. What causes this?

38-5 An object submerged in water appears to be closer to the surface than it actually is. Why? Swimming pools are always deeper than they look; is this the same phenomenon?

38-6 Can radio or television waves be reflected? Refracted? Give examples of situations where this might occur.

38-7 A ray of light in air strikes a glass surface. Is there a range of angles for which total reflection occurs?

38-8 As shown in Table 38-1, diamond has a much larger refractive index than glass. Is there a larger or smaller range of angles for which total internal reflection occurs for diamond, than for glass? Does this have anything to do with the fact that a real diamond has more sparkle than a glass imitation?

38-9 Light is usually observed to travel in straight lines, while radio waves seem to be able to bend around obstacles. If both are electromagnetic waves, why the difference?

38-10 Sunlight or starlight passing through the earth's atmosphere is always bent toward the vertical. Why? Does this mean that a star isn't really where it appears to be?

38-11 The sun or moon usually appears flattened just before it sets. Is this related to refraction in the earth's atmosphere, mentioned in Question 38-10?

38-12 A student claimed that, because of atmospheric refraction (cf. Question 38-10), the sun can be seen after it has set, and that the day is therefore longer than it would be if the earth had no atmosphere. First, what does he mean by saying the sun can be seen after it has set? Second, comment on the validity of his conclusion.

38-13 Can sound waves be reflected? Refracted? Give examples. Does Huygens' principle apply to sound waves?

38-14 Why should the wavelength of light change, but not its frequency, in passing from one material to another.

38-15 When light is incident on an interface between two materials, the angle of the refracted ray depends on the wavelength, but the angle of the reflected ray does not. Why should this be?

38-16 A room is completely lined with mirrors. A light source inside the room is turned on and then off. Does the light continue to reflect forever? If not, what happens to it?

38-17 When light slows down as it enters glass from vacuum, and speeds up again as it emerges, does it gain and lose energy in the process? Momentum?

Problems

38-1 What is the wavelength in meters, microns, nanometers, and angstrom units of (a) soft x-rays of frequency 2×10^{17} Hz? (b) green light of frequency 5.6×10^{14} Hz?

38-2 The visible spectrum includes a wavelength range from about 400 nm to about 700 nm. Express these wavelengths in inches.

38-3 Assuming the radius of the earth's orbit to be 92,900,000 mi, and taking the best value of the speed of light, compute the time required for light to travel a distance equal to the diameter of the earth's orbit. Compare with Roemer's value of 22 min.

38-4 Fizeau's measurements of the speed of light were continued by Cornu, using Fizeau's apparatus but with the distance between mirrors increased to 22.9 km. One of the toothed wheels used was 40 mm in diameter and had 180 teeth. Find the angular velocity at which it should rotate so that light transmitted through one opening will return through the next.

38-5 A ray of light traveling with speed c leaves point 1 of Fig. 38-15 and is reflected to point 2. Show that the time required for the light to travel from 1 to 2 is $(y_1 \sec\theta_1 + y_2 \sec\theta_2)/c$.

38-6 Prove that a ray of light reflected from a plane mirror rotates through an angle 2θ when the mirror rotates through an angle θ about an axis perpendicular to the plane of incidence.

38-7 A parallel beam of light is incident on a prism, as shown in Fig. 38-16. Part of the light is reflected from one face and part from another. Show that the angle θ between the two reflected beams is twice the angle A between the two reflecting surfaces.

Figure 38-16

38-8 A parallel beam of light makes an angle of 30° with the surface of a glass plate having a refractive index of 1.50.
a) What is the angle between the refracted beam and the surface of the glass?
b) What should be the angle of incidence ϕ with this plate for the angle of refraction to be $\phi/2$?

Figure 38-15

38-9 Light strikes a glass plate at an angle of incidence of 60°, part of the beam being reflected and part refracted. It is observed that the reflected and refracted portions make an angle of 90° with each other. What is the index of refraction of the glass?

38-10 A ray of light is incident on a plane surface separating two transparent substances of indices 1.60 and 1.40. The angle of incidence is 30° and the ray originates in the medium of higher index. Compute the angle of refraction.

38-11 A parallel-sided plate of glass having a refractive index of 1.60 is held on the surface of water in a tank. A ray coming from above makes an angle of incidence of 45° with the top surface of the glass.

a) What angle does the ray make with the normal in the water?

b) How does this angle vary with the refractive index of the glass?

38-12 An inside corner of a cube is lined with mirrors. A ray of light is reflected successively from each of three mutually perpendicular mirrors; show that its final direction is always exactly opposite to its initial direction. This principle is used in tail-light lenses and reflecting highway signs.

38-13

a) What is the speed of light of wavelength 500 nm (in vacuum) in glass whose index at this wavelength is 1.50?

b) What is the wavelength of these waves in the glass?

38-14 A glass plate 3 mm thick, of index 1.50, is placed between a point source of light of wavelength 600 nm (in vacuum) and a screen. The distance from source to screen is 3 cm. How many waves are there between source and screen?

38-15 The speed of light of wavelength 656 nm in heavy flint glass is 1.60×10^8 m·s^{-1}. What is the index of refraction of this glass?

38-16 Light of a certain frequency has a wavelength in water of 442 nm. What is the wavelength of this light when it passes into carbon disulfide?

38-17 A ray of light goes from point A in a medium where the velocity of light is v_1 to point B in a medium where the velocity is v_2, as in Fig. 38–17.

a) Show that the time required for the light to go from A to B is

$$t = \frac{h_1 \sec \theta_1}{v_1} + \frac{h_2 \sec \theta_2}{v_2}.$$

b) Take the derivative of t with respect to θ_1, noting that θ_1 and θ_2 are functionally related by the fact that $l = h_1 \tan \theta_1 + h_2 \tan \theta_2$. Set this derivative equal to zero to show that this time reaches its *minimum* value when $n_1 \sin \theta_1 = n_2 \sin \theta_2$. This is an example of Fermat's *principle of least time*, which states that

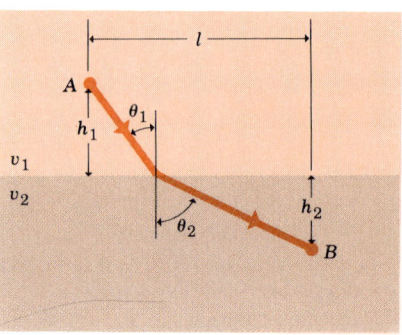

Figure 38–17

among all possible paths between two points, the one actually taken by a ray of light is that for which the time of travel is a *minimum*. (In fact, there are some cases where the time is a maximum rather than a minimum.)

38-18 A point light source is 5 cm below a water–air surface. Compute the angles of refraction of rays from the source making angles with the normal of 10°, 20°, 30°, and 40°, and show these rays in a carefully drawn full-size diagram.

38-19 A glass cube in air has a refractive index of 1.50. Parallel rays of light enter the top obliquely and then strike a side of the cube. Is it possible for the rays to emerge from this side?

38-20 A point source of light is 20 cm below the surface of a body of water. Find the diameter of the largest circle at the surface through which light can emerge from the water.

38-21 The index of refraction of the prism shown in Fig. 38–18 is 1.56. A ray of light enters the prism at point a and follows in the prism the path ab, which is parallel to the line cd.

a) Sketch carefully the path of the ray from a point outside the prism at the left, through the glass, and out some distance into the air again.

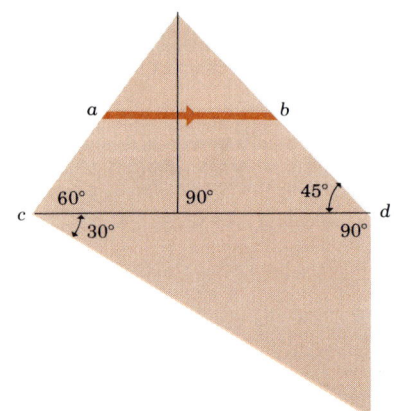

Figure 38–18

b) Compute the angle between the original and final directions in air.

(Black lines are construction lines only.)

38-22 Light is incident normally on the short face of a 30°-60°-90° prism, as in Fig. 38-19. A drop of liquid is placed on the hypotenuse of the prism. If the index of the prism is 1.50, find the maximum index the liquid may have if the light is to be totally reflected.

Figure 38-19

38-23 A 45°-45°-90° prism is immersed in water. What is the minimum index of refraction the prism may have if it is to totally reflect a ray incident normally on one of its shorter faces?

38-24 The velocity of a sound wave is $330 \text{ m} \cdot \text{s}^{-1}$ in air and $1320 \text{ m} \cdot \text{s}^{-1}$ in water.

a) What is the critical angle for a sound wave incident on the surface between air and water?

b) Which medium has the higher "index of refraction" for sound?

38-25 Light is incident at an angle ϕ_1 (as in Fig. 38-20) on the upper surface of a transparent plate, the surfaces of the plate being plane and parallel to each other.

a) Prove that $\phi_1 = \phi_2'$.

b) Show that this is true for any number of different parallel plates.

c) Prove that the lateral displacement d of the emergent beam is given by the relation

$$d = t \frac{\sin(\phi_1 - \phi_1')}{\cos \phi_1'},$$

where t is the thickness of the plate.

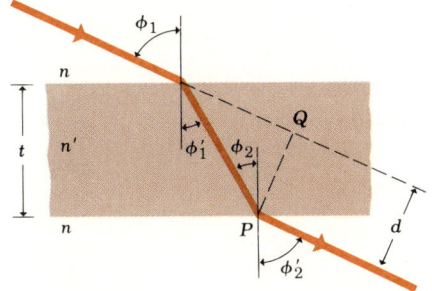

Figure 38-20

38-26 A parallel beam of light containing wavelengths A and B is incident on the face of a triangular glass prism having a refracting angle of 60°. The indices of refraction are $n_A = 1.40$ and $n_B = 1.60$. If beam A goes through the prism at minimum deviation, find (a) the angle of emergence of each beam and (b) the angle of deviation of each.

38-27 A ray of light is incident at an angle of 60° on one surface of a glass plate 2 cm thick, of index 1.50. The medium on either side of the plate is air. Find the transverse displacement between the incident and emergent rays.

38-28 The prism of Fig. 38-21 has a refractive index of 1.414, and the angles A are 30°. Two light rays m and n are parallel as they enter the prism. What is the angle between them after they emerge?

Figure 38-21

38-29 Light passes symmetrically through a prism having apex angle A, as shown in Fig. 38-22. Show that the angle of deviation δ (the angle between the initial and final directions of the ray) is given by

$$\sin \frac{A + \delta}{2} = n \sin \frac{A}{2}.$$

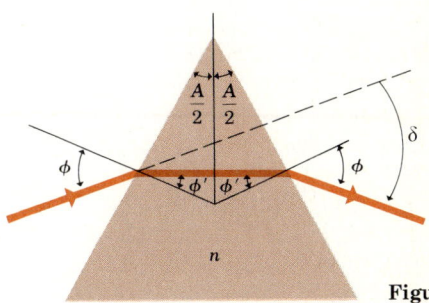

Figure 38-22

38-30 Use the result of Problem 38-29 to find the angle of deviation for a ray of light passing symmetrically through a prism having three equal angles ($A = 60°$) and $n = 1.50$.

38-31 A certain glass has a refractive index of 1.50 for red light (700 nm) and 1.52 for violet light (400 nm). If both colors pass through symmetrically, as described in Problem 38-30, and if $A = 60°$, find the difference between the angles of deviation for the two colors, using the result of Problem 38-29.

(a) (b)

Figure 38-23

38-32 The glass vessel shown in Fig. 38-23a contains a large number of small, irregular pieces of glass and a liquid. The dispersion curves of the glass and of the liquid are shown in Fig. 38-23b. Explain the behavior of a parallel beam of white light as it traverses the vessel. (This is known as a *Christiansen filter*.)

38-33 The density of the earth's atmosphere increases as the surface of the earth is approached.

a) Draw a diagram showing how the light from a star or planet is bent as it goes through the atmosphere.
b) Indicate the apparent position of the light source.
c) Explain how one can see the sun after it has set.
d) Explain why the setting sun appears flattened.

38-34 A lens for sunglasses is to be made from a gray-tinted glass having an absorption coefficient of 500 m^{-1}. If the intensity of light passing through the lens is to be decreased by a factor of $\frac{1}{4}$, what should be the thickness of the lens?

38-35 If a swimming pool is 5 m deep and the water has an absorption coefficient of 2 m^{-1}, by what factor is the intensity of light attenuated as it travels from the surface to the bottom of the pool?

38-36 If the absorption coefficient of sea water is 2 m^{-1}, and if the eye can perceive light of intensity smaller than that of sunlight by a factor of 10^{-18}, what is the greatest depth below the ocean surface at which light can be seen?

38-37 The illuminance of direct sunlight is about 10^5 lux. If a photoflash lamp has an intensity in a certain direction of 5×10^6 cd, at what distance from a surface should it be placed to produce illuminance equal to that of sunlight?

38-38 The *luminous efficiency* of a lamp is defined as the ratio of luminous flux to electric power input. A certain lamp mounted 3 m above a desk top has a luminous efficiency of 20 lm·W^{-1}. What is the power input to the lamp if the illuminance on the desk is equal to that of sunlight, about 10^5 lux? Assume that the lamp radiates uniformly over its lower half-sphere.

38-39 A certain baseball field in the shape of a square 140 m on a side is to be illuminated for night games by six towers supporting banks of 1000-watt incandescent lamps with luminous efficiency (see Problem 38-38) of 30 lm·W^{-1}. The illuminance required on the playing field is 200 lux. Assume that 50% of the luminous flux from the lamps reaches the field.

a) How many lamps are required in each tower?
b) What is the electric power input to each tower?
c) If power for all six towers is supplied by a generator driven by a gasoline engine, what must be the power capacity of the engine?

IMAGES FORMED BY A SINGLE SURFACE 39

39-1 Introduction

In the preceding chapter we discussed reflection and refraction of a ray of light, or of a bundle of parallel rays, at a surface separating two materials having different refractive indices. We now consider the analysis of the paths of several rays that diverge from a common point and strike a reflecting or refracting surface. The point may be self-luminous, or it may be part of a rough surface that reflects light in various directions.

Such a situation is shown in Fig. 39–1. Rays diverge from point P and are reflected or refracted (or both) at the interface. The direction of each *reflected* ray is given by the law of reflection, and that of each transmitted or *refracted* ray by Snell's law. Here and in the following discussion we denote the two refractive indexes as n and n', without using the more elaborate subscript notation of Chapter 38.

A central concept in this discussion is that of *image*. As we shall see, the rays after reflection or refraction emerge with directions characteristic of having passed through some common point, which we call the *image point*. In some cases the emerging rays really *do* meet at a common point and then diverge again after passing it; such a point is called a *real* image point. In other cases the rays diverge *as though* they had passed through such a point, which is then called a *virtual* image point. In many situations the image point exists only in an approximate sense, that is, when certain approximations are used in the calculations. This chapter and the next are devoted primarily to a study of the formation and properties of images.

39-2 Reflection at a plane surface

We begin our analysis by considering reflected rays. The reflection can occur at an interface between two transparent materials or at a highly polished surface of an opaque material such as a metal, in which case the surface is usually called a *mirror*. We shall often use the term mirror to include all these possibilities.

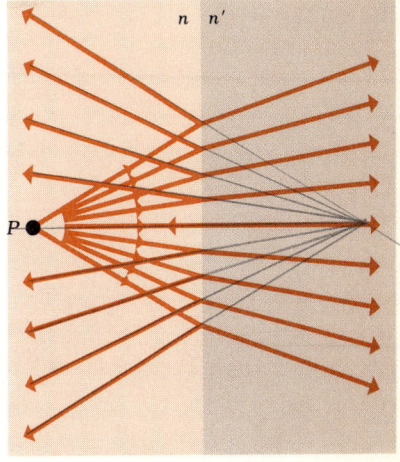

39–1 Reflection and refraction of rays at a plane interface between two transparent materials.

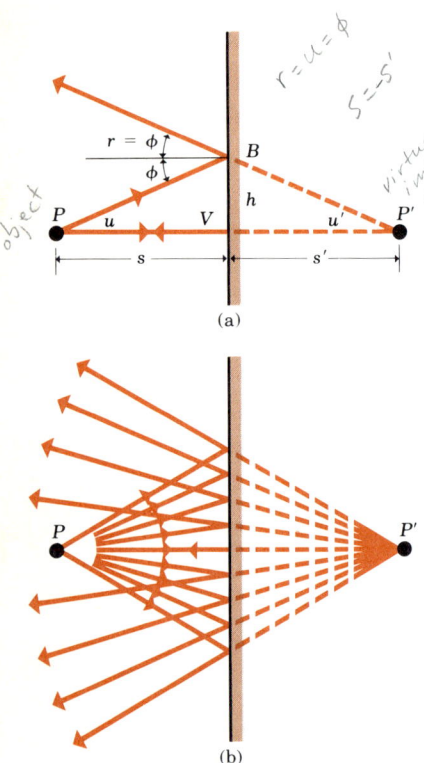

Figure 39–2a shows two rays diverging from a point P at a distance s to the left of a plane mirror. The ray PV, incident normally on the mirror, returns along its original path. The ray PB, making an angle u with PV, strikes the mirror at an angle of incidence $\phi = u$, and is reflected at an angle $r = \phi = u$. When extended backward, the reflected ray intersects the normal to the surface at point P'. The angle u' is equal to r and hence to u.

Figure 39–2b shows several rays diverging from P. The construction of Fig. 39–2a can be repeated for each of them, and we see that the directions of the outgoing rays are as though they had originated at point P', which is therefore the *image* of P. The rays do not, of course, actually pass through this point; in fact, if the mirror is opaque there is no light at all on the right side. Thus P' is a *virtual* image. Nevertheless P' is a very real point in the sense that it describes the final directions of all the rays that originally diverged from P.

From the symmetry of the figure, we see that P' lies on a line perpendicular to the mirror passing through P, and that P and P' are equidistant from the mirror, on opposite sides. Thus *for a plane mirror the image of an object point lies on the extension of the normal from the object point to the mirror, and the object and image points are equidistant from the mirror.*

Before proceeding further, we pause here to introduce some conventions that anticipate later situations in which the object and image may be on either side of a reflecting or refracting surface. We adopt the following:

1. *When the object is on the same side of the reflecting surface as the incoming light, the object distance s is positive; otherwise it is negative.*

2. *When the image is on the same side of the reflecting surface as the outgoing light, the image distance s' is positive; otherwise it is negative.*

For a mirror the incoming and outgoing sides are always the same; in Fig. 39–2 they are both the left side. The above rules have been stated in this form so they may be applied also to *refracting* surfaces, for which light comes in one side and goes out the opposite side.

In Fig. 39–2 the object distance s is *positive* because the object point P is on the incoming side, that is, the left side, of the reflecting surface. The image distance s' is *negative* because the image point P' is *not* on the outgoing side of the surface. Thus object and image distances are related simply by

$$s = -s' \qquad (39\text{–}1)$$

Next we consider an object of finite size, parallel to the mirror, represented by the arrow PQ in Fig. 39–3. Point P', the image of P, is found as in Fig. 39–2. Two of the rays from Q are shown, and *all* rays from Q diverge from its image Q' after reflection. Other points of the object PQ have image points between P' and Q'.

Let y and y' represent the lengths of object and image, respectively. The ratio y'/y is called the *lateral magnification m:*

$$m = \frac{y'}{y}.$$

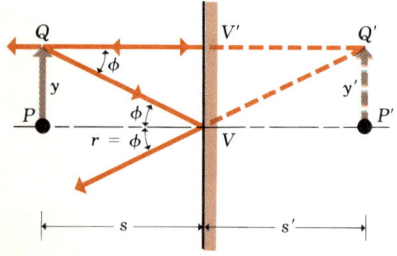

39–2 After reflection at a plane surface, all rays originally diverging from the object point P now diverge from the point P', although they do not *originate* at P'. Point P' is called the *virtual image* of point P.

39–3 Construction for determining the height of an image formed by reflection at a plane surface.

From the triangles PQV and $P'Q'V$,

$$\tan\phi = \frac{y}{s} = \frac{y'}{-s'}$$

and since $s' = -s$,

$$m = \frac{y'}{y} = -\frac{s'}{s} = +1. \qquad (39\text{-}2)$$

39-4 The image formed by a plane mirror is virtual, erect, and perverted, and is the same size as the object.

The lateral magnification for a plane mirror is unity.

In general, if a transverse object is represented by an arrow, its image may point in the same direction as the object, or in the opposite direction. If the directions are the same, the image is called *erect*; if they are opposite, the image is *inverted*. When the lateral magnification m is positive, the object and image arrows point in the same direction and the image is erect.

The three-dimensional virtual image of a three-dimensional object, formed by a plane mirror, is shown in Fig. 39–4. The image of every object point lies on the normal from that point to the mirror, and the distances from object and image to the mirror are equal. The images $P'Q'$ and $P'S'$ are parallel to their objects, but $P'R'$ is reversed relative to PR. We can think of the arrow PR as generated by the displacement of an object point P from P to R, through a distance Δs. Since

$$s' = -s,$$

it follows that the corresponding displacement $\Delta s'$ of the image, from P' to R', is equal to $-\Delta s$. The ratio $\Delta s'/\Delta s$ (or more generally ds'/ds) is called the *longitudinal magnification* m'. In this case,

$$m' = \frac{\Delta s'}{\Delta s} = -1. \qquad (39\text{-}3)$$

The negative sign means that the arrows PR and $P'R'$ point in opposite directions. Hence the image of a three-dimensional object formed by a plane mirror is the same size as the object in both its lateral and transverse dimensions. However, the image and object are not identical in all respects but are related in the same way as are a right hand and a left hand. To verify this, point your thumbs along PR and $P'R'$, your forefingers along PQ and $P'Q'$, and your middle fingers along PS and $P'S'$. When an object and its image are related in this way the image is said to be *perverted*. When the transverse dimensions of object and image are in the same direction, the image is erect. Thus a plane mirror forms an erect but perverted image.

39-3 Reflection at a spherical mirror

Next we consider the formation of an image by a *spherical* mirror. Figure 39–5a shows a spherical mirror of radius of curvature R, with its concave side facing the incident light. The center of curvature of the surface is at C. Point P is an object point; for the moment we assume that P is farther from V than is the center of curvature. The ray PV, passing through C, strikes the mirror normally and is reflected back on itself. Point V is called the *vertex* of the mirror, and the line PCV the *optic axis*.

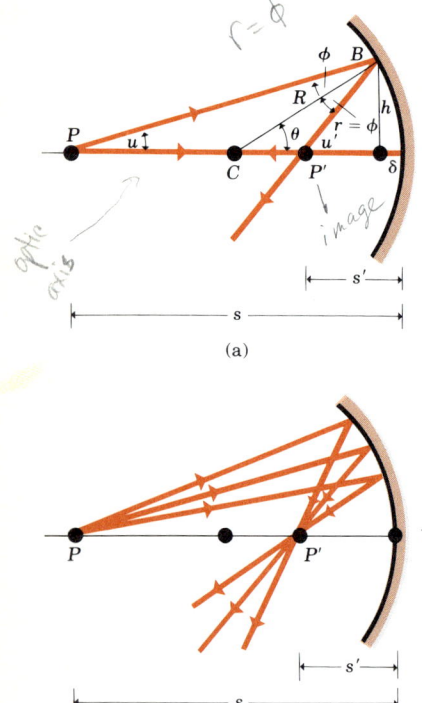

39-5 (a) Construction for finding the position of the image P' of a point object P, formed by a concave spherical mirror. (b) If the angle u is small, *all* rays from P intersect at P'.

Ray PB, at an angle u with the axis, strikes the mirror at B, where the angle of incidence is ϕ and the angle of reflection is $r = \phi$. The reflected ray intersects the axis at point P'. We shall show shortly that if the angle u is small, *all* rays from P intersect the axis at the *same* point P', as in Fig. 39-5b, no matter what u is. Point P' is therefore the *image* of the object point P. The object distance, measured from the vertex V, is s, and the image distance is s'.

The object point P is on the same side as the incident light, so the object distance s is positive. The image point P' is on the same side as the reflected light, so the image distance s' is also positive.

Unlike the reflected rays in Fig. 39-2, the reflected rays in Fig. 39-5b actually intersect at point P', and then diverge from P' *as if* they had originated at this point. The image P' is called *real*, and a real image corresponds to a *positive* image distance.

Making use of the fact that an exterior angle of a triangle equals the sum of the two opposite interior angles, and considering the triangles PBC and $P'BC$ in Fig. 39-5a, we have

$$\theta = u + \phi, \qquad u' = \theta + \phi.$$

Eliminating ϕ between these equations gives

$$u + u' = 2\theta. \tag{39-4}$$

We may now introduce a sign convention for radii of curvature.

When the center of curvature C is on the same side as the outgoing (reflected) light, the radius of curvature is positive; otherwise, it is negative.

In Fig. 39-5, R is positive, because the center of curvature C is on the same side of the mirror as the reflected light. This is always the case for reflection from the concave side of a surface; for a convex surface, the center of curvature is on the opposite side from the reflected light, and R is negative.

We may now compute the image position s'. Let h represent the height of point B above the axis, and δ the short distance from V to the foot of this vertical line. Now write expressions for the tangents of u, u', and θ, remembering that s, s', and R are all positive quantities:

$$\tan u = \frac{h}{s - \delta}, \qquad \tan u' = \frac{h}{s' - \delta}, \qquad \tan \theta = \frac{h}{R - \delta}.$$

These *trigonometric* equations cannot be solved as simply as the corresponding *algebraic* equations for a plane mirror. However, *if the angle u is small*, the angles u' and θ will be small also. Since the tangent of a small angle is nearly equal to the angle itself (in radians), we can replace $\tan u'$ by u', etc., in the equations above. Also if u is small, the distance δ can be neglected compared with s', s, and R. Hence, approximately, for small angles,

$$u = \frac{h}{s}, \qquad u' = \frac{h}{s'}, \qquad \theta = \frac{h}{R}.$$

Substituting in Eq. (39-4) and canceling h, we obtain

$$\frac{1}{s} + \frac{1}{s'} = \frac{2}{R} \tag{39-5}$$

as a general relation among the three quantities s, s', and R. Since the equation above does not contain the angle u, *all* rays from P making sufficiently small angles with the axis will, after reflection, intersect at P'. Such rays, nearly parallel to the axis, are called *paraxial* rays.

It must be understood that Eq. (39-5), as well as many similar relations to be derived later, is the result of a calculation containing approximations and is valid only for paraxial rays. (The term *paraxial approximation* is also commonly used.) As the angle increases, the point P' moves closer to the vertex; a spherical mirror, unlike a plane mirror, does not form precisely a point image of a point object. This property of a spherical mirror is called *spherical aberration*.

If $R = \infty$, the mirror becomes *plane* and Eq. (39-5) reduces to Eq. (39-1), previously derived for this special case.

Now suppose we have an object of finite size, represented by the arrow PQ in Fig. 39-6, perpendicular to the axis PV. The image of P formed by paraxial rays is at P'. Since the object distance for point Q is very nearly equal to that for the point P, the image $P'Q'$ is nearly straight and is perpendicular to the axis. Herein lies another approximation; if the height PQ is not sufficiently small compared to the object distance s, the image $P'Q'$ is not a straight line but is curved; this is another aberration of spherical surfaces, called *curvature of field*. We shall assume height PQ is small enough that this effect is negligible.

We now compute the lateral magnification m. In Fig. 39-6 the ray QCQ' passes through the center of curvature; hence it strikes the spherical reflecting surface at normal incidence and is reflected back along the same radial line on which it approached the mirror. The ray QV makes an angle of incidence ϕ and an angle of reflection $r = \phi$. From the triangles PQV and $P'Q'V$,

$$\tan \phi = \frac{y}{s} = \frac{-y'}{s'}.$$

The negative sign arises because object and image are on opposite sides of the optic axis; if y is positive, y' is negative. Hence,

$$m = \frac{y'}{y} = -\frac{s'}{s}. \tag{39-6}$$

A negative value of m indicates that the image is *inverted* relative to the object, as the figure shows. In cases to be considered later, where m may

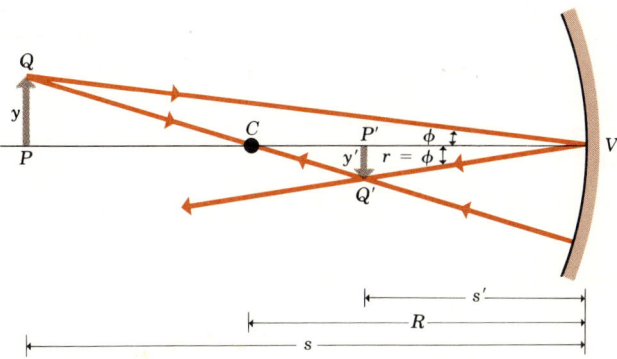

39-6 Construction for determining the height of an image formed by a concave spherical mirror.

be either positive or negative, a positive value always corresponds to an erect image, a negative value to an inverted one. For a plane mirror, $s = -s'$ and hence $y' = y$, as we have already shown.

The longitudinal magnification m' is defined as before, as ds'/ds. Taking the derivative of Eq. (39–5) with respect to s, we find

$$-\frac{1}{s^2} - \frac{1}{s'^2}\frac{ds'}{ds} = 0,$$

$$m' = \frac{ds'}{ds} = -\frac{s'^2}{s^2} = -m^2, \tag{39-7}$$

since the lateral magnification is $m = -s'/s$. Thus the longitudinal magnification equals the *square* of the lateral magnification and is always *negative*, regardless of the sign of m, since m^2 is always positive. The negative sign means that when the object is displaced slightly along the axis, image and object always move in *opposite* directions.

Although the ratio of image size to object size is referred to as the *magnification,* the image formed by a mirror or lens may be *smaller* than the object. The magnification is then a small fraction and might more appropriately be called the *reduction*. The image formed by an astronomical telescope mirror, or by a camera lens, is much smaller than the object. Since the longitudinal magnification is the square of the lateral magnification, the two are not equal except in the special case of *unit* magnification when $m = 1$. In particular, if m is a small fraction, m^2 is very small and the three-dimensional image of a three-dimensional object is reduced *longitudinally* much more than it is reduced *transversely*. Figure 39–7 illustrates this effect. Also, the image formed by a spherical mirror, like that of a plane mirror, is always perverted.

39-7 Schematic diagram of an object and its real, inverted, reduced image formed by a concave mirror.

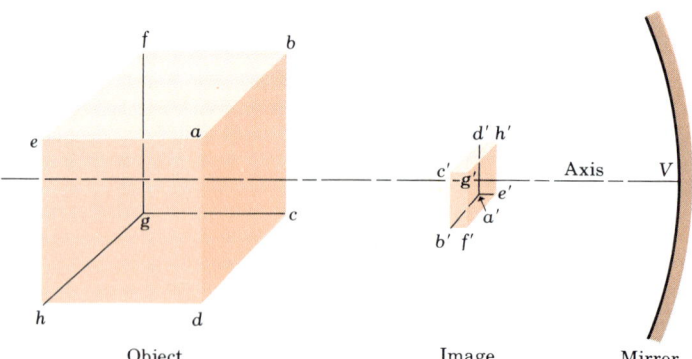

Example 1 (a) What type of mirror is required to form an image, on a wall 3 m from the mirror, of the filament of a headlight lamp 10 cm in front of the mirror? (b) What is the height of the image if the height of the object is 5 mm?

Solution

a)
$$s = 10 \text{ cm}, \quad s' = 300 \text{ cm},$$

$$\frac{1}{10 \text{ cm}} + \frac{1}{300 \text{ cm}} = \frac{2}{R},$$

$$R = 19.4 \text{ cm}.$$

Since the radius is positive, a concave mirror is required.

b) $$m = \frac{y'}{y} = -\frac{s'}{s} = -\frac{300 \text{ cm}}{10 \text{ cm}} = -30.$$

The image is therefore inverted (m is negative) and is 30 times the height of the object, or $(30)(5 \text{ mm}) = 150 \text{ mm}$.

Example 2 A wire frame in the form of a small cube 3 cm on a side is placed with its center on the axis of a concave mirror of radius of curvature 30 cm. The sides of the cube are parallel or perpendicular to the axis. The face toward the mirror is 60 cm to the left of the vertex. Find (a) the position of the image of the cube, (b) the lateral magnification, (c) the longitudinal magnification.

Solution
a) Let us calculate the position of the image of a point at the center of the right face of the cube. From the given data, $s = +60$ cm, $R = +30$ cm:

$$\frac{1}{s} + \frac{1}{s'} = \frac{2}{R}, \qquad \frac{1}{60} + \frac{1}{s'} = \frac{2}{30}, \qquad s' = +20 \text{ cm}.$$

The image of this point is therefore 20 cm to the left of the vertex (s' is positive) and is real.

b) $$m = \frac{y'}{y} = -\frac{s'}{s} = -\frac{1}{3}, \qquad y' = -\frac{1}{3}y.$$

The image of the right face is therefore inverted (m is negative) and is a square 1 cm on a side.

c) The object distance for the left face of the cube is 63 cm. To find the position and magnification of the left face of the cube, we could repeat the calculations above with $s = 63$ cm, but since the dimensions of object and image are small compared with the object distance, let us instead find the longitudinal magnification.

$$m' = \frac{\Delta s'}{\Delta s} = -m^2 = -\frac{1}{9}, \qquad \Delta s' = -\frac{1}{9}\Delta s.$$

The longitudinal magnification is negative, so the image of the left face is nearer the mirror than the image of the right face, and the images of the cube sides parallel to the axis are $\frac{1}{3}$ cm in length.

Figure 39-7 (not to scale) represents schematically the object and image, with corresponding points designated by unprimed and primed letters. It has been assumed that the lateral magnification is the same for both transverse faces.

In Fig. 39-8a, the *convex* side of a spherical mirror faces the incident light so that R is negative. Ray PB is reflected with the angle of reflection r equal to the angle of incidence ϕ, and the reflected ray, projected backward, intersects the axis at P'. As in the case of a concave mirror, *all* rays from P will, after reflection, diverge from the same point P', provided that the angle u is small; so P' is the image of P. The object distance s is positive, the image distance s' is negative, and the radius of curvature R is negative.

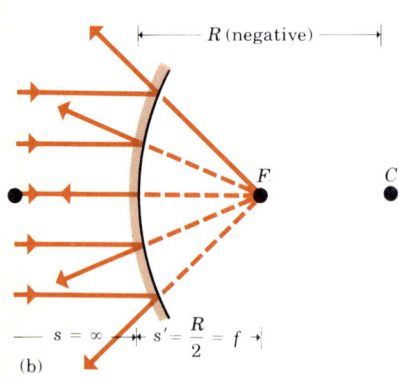

39-8 Construction for finding (a) the position and (b) the magnification of the image formed by a convex mirror.

Figure 39-8b shows two rays diverging from the head of the arrow PQ, and the virtual image $P'Q'$ of this arrow. It is left to the reader to show, by the same procedure as that used for a concave mirror, that:

$$\frac{1}{s} + \frac{1}{s'} = \frac{2}{R}.$$

The lateral magnification is

$$m = \frac{y'}{y} = -\frac{s'}{s};$$

and the longitudinal magnification again equals the negative square of the lateral magnification. These expressions are exactly the same as those for a concave mirror, as they should be when a consistent sign convention is used.

39-4 Focal point and focal length

When an object point is a very large distance from a mirror, all rays from that point that strike the mirror are parallel to one another. The object distance is $s = \infty$ and, from Eq. (39-5),

$$\frac{1}{\infty} + \frac{1}{s'} = \frac{2}{R}, \qquad s' = \frac{R}{2}.$$

The image distance s' then equals one-half the radius of curvature and has the same sign as R. This means that, if R is positive, as in Fig. 39-9a, the image point F lies to the left of the mirror and is real, while if R is negative, as in Fig. 39-9b, the image point F lies to the right of the mirror and is virtual.

Conversely, if the image distance s' is very large, the object distance s is

$$\frac{1}{s} + \frac{1}{\infty} = \frac{2}{R}, \qquad s = \frac{R}{2}.$$

The object distance then equals half the radius of curvature. If R is positive, as in Fig. 39-10a, s is positive. If R is negative, as in Fig. 39-10b, s is negative and the *object is behind* the mirror. That is, rays previously made converging by some other surface are converging toward F.

The point F in Figs. 39-9 and 39-10 is called the *focal point* of the mirror, or simply the *focus*. It may be considered either as the image point of an infinitely distant object point on the mirror axis, or as the

39-9 Incident rays parallel to the axis (a) converge to the focal point F of a concave mirror, (b) diverge as though coming from the focal point F of a convex mirror.

object point of an infinitely distant image point. Thus the mirror of an astronomical telescope forms at its focal point an image of a star on the axis of the mirror.

The distance between the vertex of a mirror and the focal point is called the *focal length* of the mirror and is represented by *f*. As seen from the preceding discussion, the magnitude of the focal length equals one-half the radius of curvature:

$$f = \frac{R}{2}. \tag{39-8}$$

The relation between object and image distances, Eq. (39-5), for a mirror may now be written as

$$\frac{1}{s} + \frac{1}{s'} = \frac{1}{f}. \tag{39-9}$$

39-5 Graphical methods

The position and size of the image formed by a mirror may be found by a simple graphical method. This method consists of finding the point of intersection, after reflection from the mirror, of a few particular rays diverging from some point of the object *not* on the mirror axis, such as point Q in Fig. 39-11. Then (neglecting aberrations) *all* rays from this point that strike the mirror will intersect at the same point. Four rays whose paths may readily be traced are shown in Fig. 39-11. These are often called *principal rays*.

1. *A ray parallel to the axis*, after reflection, passes through the focal point of a concave mirror or appears to come from the focal point of a convex mirror.

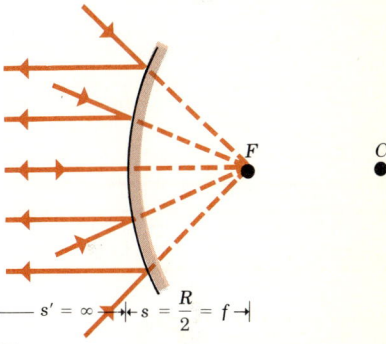

39-10 Rays from a point object at the focal point of a spherical mirror are parallel to the axis after reflection. The object in part (b) is virtual.

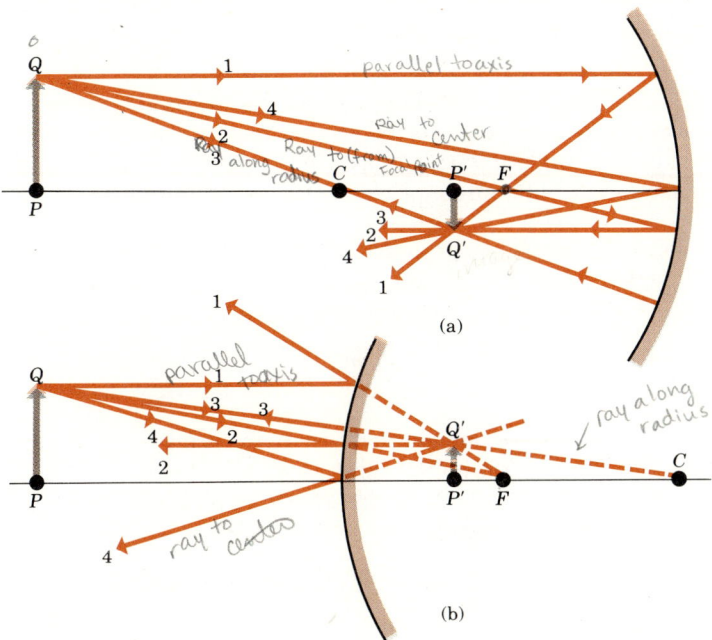

39-11 Rays used in the graphical method of locating an image.

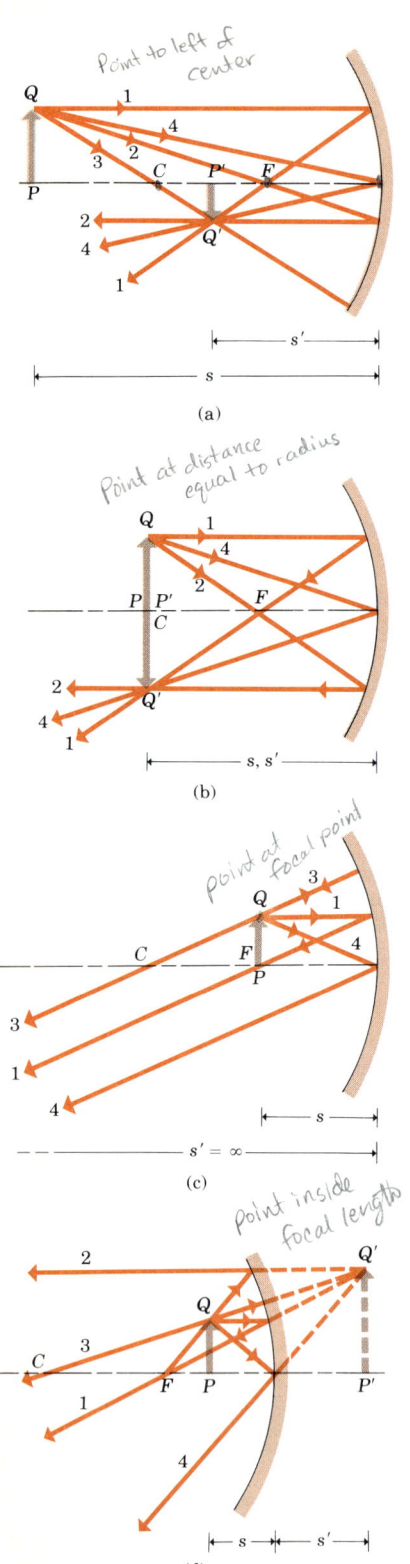

39-12 Image of an object at various distances from a concave mirror, showing principal-ray construction.

2. *A ray from* (or *proceeding toward*) *the focal point* is reflected parallel to the axis.
3. *A ray along the radius* (extended if necessary) intersects the surface normally and is reflected back along its original path.
4. *A ray to the center* is reflected with equal angles with the optic axis.

Once the position of the image point has been found by means of the intersection of any two of these principal rays (1, 2, 3, 4) the paths of all other rays from the same object point may be drawn.

Example A concave mirror has a radius of curvature of magnitude 20 in. Find graphically the image of an object in the form of an arrow perpendicular to the axis of the mirror at each of the following object distances: 30 in., 20 in., 10 in., and 5 in. Check the construction by computing the size and magnification of the image.

The graphical construction is indicated in the four parts of Fig. 39-12. (Note that in (b) and (c) only three of the four rays can be used.) The calculations are left as a problem.

In every problem involving formation of an image by a reflecting surface or a lens, the student should *always* draw a principal-ray diagram. This is useful not only as a graphical check on numerical calculations but also as an aid to understanding the basic concepts, especially the concept of *image*.

39-6 Refraction at a plane surface

The method of finding the image of a point object formed by rays *refracted* at a plane or spherical surface is essentially the same as for reflection; the only difference is that Snell's law replaces the law of reflection. We let n represent the refractive index of the medium on the "incoming" side of the surface and n' that of the medium on the "outgoing" side. The same convention of signs is used as in reflection.

Consider first a plane surface, shown in Fig. 39-13, and assume $n' > n$. A ray from the object point P toward the vertex V is incident normally and passes into the second medium without deviation. A ray making an angle u with the axis is incident at B with an angle of incidence $\phi = u$. The angle of refraction, ϕ', is found from Snell's law,

$$n \sin \phi = n' \sin \phi'.$$

The two rays both appear to come from the image point P' after refraction. From the triangles PVB and $P'VB$,

$$\tan \phi = \frac{h}{s}, \qquad \tan \phi' = \frac{h}{-s'}. \qquad (39\text{-}10)$$

We must write $-s'$, since the image point is on the side *opposite* to that of the refracted (outgoing) light.

If the angle u is small, the angles ϕ, u', and ϕ' are small also, and therefore, approximately,

$$\tan \phi = \sin \phi, \qquad \tan \phi' = \sin \phi'.$$

Then Snell's law can be written as

$$n \tan \phi = n' \tan \phi',$$

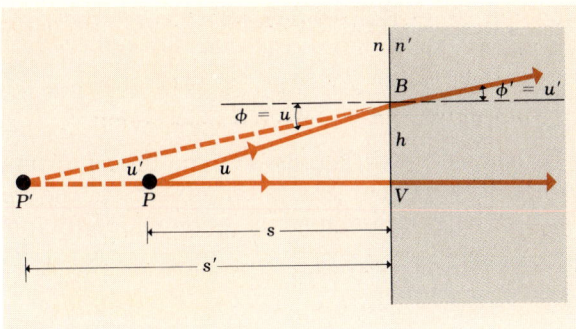

39-13 Construction for finding the position of the image P' of a point object P, formed by refraction at a plane surface.

and from Eq. (39-10), after cancelling h, we have

$$\frac{n}{s} = \frac{-n'}{s'},$$

or

$$\frac{s'}{s} = -\frac{n'}{n}. \qquad (39\text{-}11)$$

This is an *approximate* relation, *valid for paraxial rays only*. That is, a plane refracting surface does *not* image all rays from a point object at the same image point.

Consider next the image of a finite object, as in Fig. 39-14. The two rays diverging from point Q appear to diverge from its image Q' after refraction. From the triangles PQV and $P'Q'V$,

$$\tan\phi = \frac{y}{s}, \qquad \tan\phi' = \frac{y'}{-s'}.$$

Combining with Snell's law and using the small-angle approximation,

$$\sin\phi = \tan\phi, \qquad \sin\phi' = \tan\phi',$$

we obtain

$$\frac{ny}{s} = -\frac{n'y'}{s'},$$

and

$$m = \frac{y'}{y} = -\frac{ns'}{n's}. \qquad (39\text{-}12)$$

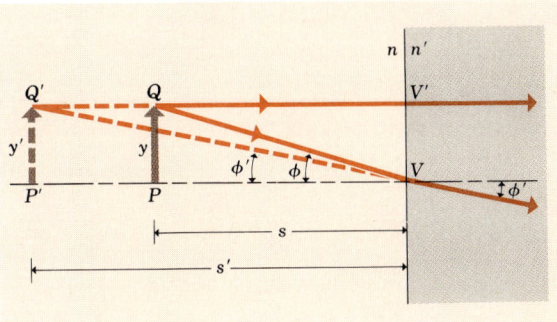

39-14 Construction for determining the height of an image formed by refraction at a plane surface.

39-15 Arrow $P'Q'$ is the image of the underwater object PQ.

However, from Eq. (39-11), $ns' = -n's$, so

$$m = \frac{y'}{y} = 1, \quad (39\text{-}13)$$

in agreement with Fig. 39-14. The image distance is greater than the object distance, but image and object are the same height.

The longitudinal magnification m' is easily found from Eq. (39-11):

$$m' = \frac{ds'}{ds} = -\frac{n'}{n}. \quad (39\text{-}14)$$

A common example of refraction at a plane surface is afforded by looking vertically downward into the quiet water of a pond or a swimming pool; the apparent depth is less than the actual depth. Figure 39-15 illustrates this situation. Two rays are shown diverging from a point Q at a distance s below the surface. Here, n' (air) is less than n (water) and the ray through V is deviated *away from* the normal. The rays after refraction appear to diverge from Q', and the arrow PQ, to an observer looking vertically downward, appears lifted to the position $P'Q'$. From Eq. (39-11),

$$s' = -\frac{n'}{n}s = -\frac{1.00}{1.33}s = -0.75\,s.$$

The apparent depth s' is therefore only three-fourths of the actual depth s. The same phenomenon accounts for the apparent sharp bend in an oar when a portion of it extends below a water surface. The submerged portion appears lifted above its actual position.

39-7 Refraction at a spherical surface

Finally, we consider refraction at a spherical surface. In Fig. 39-16, P is an object point at a distance s to the left of a spherical surface of radius R, with center of curvature C. The refractive indexes at the left and right of the surface are n and n', respectively. Ray PV, incident normally, passes into the second medium without deviation. Ray PB, making an angle u with the axis, is incident at an angle ϕ with the normal and is refracted at an angle ϕ'. These rays intersect at P' at a distance s' to the right of the vertex.

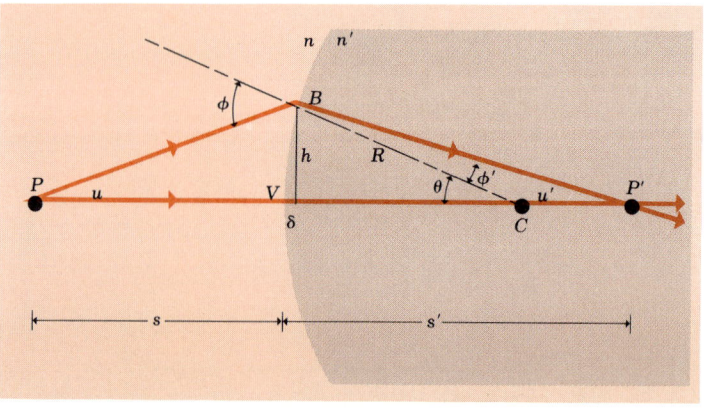

39-16 Construction for finding the position of the image P' of a point object P, formed by refraction at a spherical surface.

We shall show that if the angle u is small, all rays from P intersect at the same point P', so P' is the *real image* of P. The object and image distances are both positive. The radius of curvature is positive also, because the center of curvature is on the side of the outgoing or refracted light.

When considering rays from P that are *reflected* at the surface, as in Fig. 39–8, the radius of curvature is negative. Thus the radius of curvature of a given surface has one sign for reflected light and the opposite sign for refracted light. This apparent inconsistency is resolved by noting that in both cases R is positive when the center of curvature is on the "outgoing" side of the reflecting or refracting surface and negative when it is on the "incoming" side.

From the triangles PBC and $P'BC$, we have

$$\phi = \theta + u, \qquad \theta = u' + \phi'. \qquad (39\text{-}15)$$

From Snell's law,

$$n \sin \phi = n' \sin \phi'.$$

Also, the tangents of u, u', and θ are

$$\tan u = \frac{h}{s+\delta}, \qquad \tan u' = \frac{h}{s'-\delta}, \qquad \tan \theta = \frac{h}{R-\delta}.$$

For paraxial rays we may approximate both the sine and tangent of an angle by the angle itself, and neglect the small distance δ. Snell's law then becomes

$$n\phi = n'\phi',$$

and, combining with the first of Eqs. (39-15), we get

$$\phi' = \frac{n}{n'}(u + \theta).$$

Substituting this in the second of Eqs. (39-15) gives

$$nu + n'u' = (n' - n)\theta.$$

Using the small-angle approximations and cancelling h, we obtain

$$\frac{n}{s} + \frac{n'}{s'} = \frac{n' - n}{R}. \qquad (39\text{-}16)$$

This equation does not contain the angle u, so the image distance is the same for all paraxial rays from P.

If the surface is plane, $R = \infty$ and this equation reduces to Eq. (39-11), already derived for the special case of a plane surface.

The magnification is found from the construction in Fig. 39-17. From point Q draw two rays, one through the center of curvature C and the other incident at the vertex V. From the triangles PQV and $P'Q'V$,

$$\tan \phi = \frac{y}{s}, \qquad \tan \phi' = \frac{-y'}{s'},$$

and from Snell's law,

$$n \sin \phi = n' \sin \phi'.$$

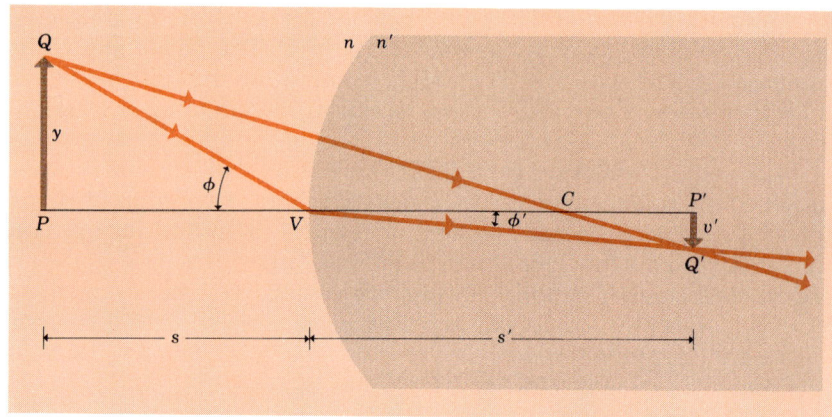

39-17 Construction for determining the height of an image formed by refraction at a spherical surface.

For small angles,
$$\tan \phi = \sin \phi, \quad \tan \phi' = \sin \phi',$$
and hence
$$\frac{ny}{s} = -\frac{n'y'}{s'},$$
or
$$m = \frac{y'}{y} = -\frac{ns'}{n's}, \qquad (39\text{-}17)$$

which agrees with the relation previously derived for a plane surface.

The longitudinal magnification is
$$m' = -\frac{n'}{n}m^2. \qquad (39\text{-}18)$$

Example 1 One end of a cylindrical glass rod (Fig. 39–18) is ground to a hemispherical surface of radius $R = 20$ mm. Find the image distance of a point object on the axis of the rod, 80 mm to the left of the vertex. The rod is in air.

Solution

$$n = 1, \quad n' = 1.5,$$
$$R = +20 \text{ mm}, \quad s = +80 \text{ mm}.$$
$$\frac{1}{80 \text{ mm}} + \frac{1.5}{s'} = \frac{1.5 - 1}{+20 \text{ mm}},$$
$$s' = +120 \text{ mm}.$$

The image is therefore formed at the right of the vertex (s' is positive) and at a distance of 120 mm from it. Suppose that the object is an arrow 1 mm high, perpendicular to the axis. Then

$$m = -\frac{ns'}{n's} = -\frac{(1)(120 \text{ mm})}{(1.5)(80 \text{ mm})} = -1.$$

That is, the image is the same height as the object, but is inverted.

Example 2 Let the same rod be immersed in water of index 1.33, the other quantities having the same values as before. Find the image distance (Fig. 39–19).

39-19

Solution

$$\frac{1.33}{80 \text{ mm}} + \frac{1.5}{s'} = \frac{1.5 - 1.33}{+20 \text{ mm}}, \qquad s' = -180 \text{ mm}.$$

The fact that s' is negative means that the rays, after refraction by the surface, are not converging but *appear* to diverge from a point 180 mm to the *left* of the vertex. We have met a similar case before in the refraction of spherical waves by a plane surface, and have called the point a *virtual image*. In this example, then, the surface forms a virtual image 180 mm to the left of the vertex.

Equations (39-16), (39-17), and (39-18) can be applied to both convex and concave refracting surfaces when a consistent sign convention is used, and they apply whether n' is greater or less than n. The reader should construct diagrams like Figs. 39-16 and 39-17, when R is negative and $n' < n$, and use them to derive Eqs. (39-16) and (39-17).

The concepts of focal point and focal length can also be applied to a refracting surface. Such a surface is found to have *two* focal points. The first is the *object* point when the image is at infinity, the second is the *image* point of an infinitely distant object. These points lie on opposite sides of the surface and at different distances from it, so that a single refracting surface has two focal lengths. The positions of the focal points can readily be found from Eq. (39-16) by setting s or s' equal to infinity.

39-8 Summary

The results of this chapter are summarized in Table 39-1. Note that if we let $R = \infty$, the equation for a plane surface follows immediately from the appropriate equation for a curved surface.

Table 39-1

	Plane mirror	Spherical mirror	Plane refracting surface	Spherical refracting surface
Object and image distances	$\frac{1}{s} + \frac{1}{s'} = 0$	$\frac{1}{s} + \frac{1}{s'} = \frac{2}{R} = \frac{1}{f}$	$\frac{n}{s} + \frac{n'}{s'} = 0$	$\frac{n}{s} + \frac{n'}{s'} = \frac{n' - n}{R}$
Lateral magnification	$m = -\frac{s'}{s} = 1$	$m = -\frac{s'}{s}$	$m = -\frac{ns'}{n's} = 1$	$m = -\frac{ns'}{n's}$
Longitudinal magnification	$m' = \frac{\Delta s'}{\Delta s} = -1$	$m' = -\frac{s'^2}{s^2} \; (= -m^2)$	$m' = -\frac{n'}{n}$	$m' = -\frac{ns'^2}{n's^2} \; (= -m^2)$

Questions

39-1 Can a person see a real image by looking backward along the direction from which the rays come? A virtual image? Can you tell by looking whether an image is real or virtual? How *can* the two be distinguished?

39-2 Why does a plane mirror reverse left and right but not top and bottom?

39-3 For a spherical mirror, if $s = f$, then $s' = \infty$, and the lateral magnification m is infinite. Does this make sense? If so, what does it mean?

39-4 According to the discussion of the preceding chapter, light rays are reversible. Are the formulas in Table 39–1 still valid if object and image are interchanged? What does reversibility imply with respect to the *forms* of the various formulas?

39-5 If a spherical mirror is immersed in water, does its focal length change?

39-6 For what range of object positions does a concave spherical mirror form a real image. What about a convex spherical mirror?

39-7 If a piece of photographic film is placed at the location of a real image, the film will record the image. Can this be done with a virtual image? How might one record a virtual image?

39-8 When a room has mirrors on two opposing walls, an infinite series of reflections can be seen. Discuss this phenomenon in terms of images. Why do the distant images appear darker?

39-9 When observing fish in an aquarium filled with water, one can see clearly only when looking nearly perpendicularly to the glass wall; objects viewed at an oblique angle always appear blurred. Why? Do the fish have the same problem when looking at you?

39-10 Can an image formed by one reflecting or refracting surface serve as an object for a second reflection or refraction? Does it matter whether the first image is real or virtual?

39-11 A concave mirror (sometimes surrounded by lights) is often used as an aid for applying cosmetics to the face. Why is such a mirror always concave rather than convex? What considerations determine its radius of curvature?

39-12 A student claimed that one can start a fire on a sunny day by use of the sun's rays and a concave mirrror. How is this done? Is the concept of image relevant? Could one do the same thing with a convex mirror?

39-13 A person looks at his reflection in the concave side of a shiny spoon. Is it right side up or inverted? What if he looks in the convex side?

Problems

39-1 What is the size of the smallest vertical plane mirror in which an observer standing erect can see his full-length image?

39-2 The image of a tree just covers the length of a 5-cm plane mirror when the mirror is held 30 cm from the eye. The tree is 100 m from the mirror. What is its height?

39-3 An object is placed between two mirrors arranged at right angles to each other.

  a) Locate all of the images of the object.
  b) Draw the paths of rays from the object to the eye of an observer.

39-4 An object 1 cm high is 20 cm from the vertex of a concave spherical mirror whose radius of curvature is 50 cm. Compute the position and size of the image. Is it real or virtual? Erect or inverted?

39-5 A concave mirror is to form an image of the filament of a headlight lamp on a screen 4 m from the mirror. The filament is 5 mm high, and the image is to be 40 cm high.

  a) What should be the radius of curvature of the mirror?
  b) How far in front of the vertex of the mirror should the filament be placed?

39-6 The diameter of the moon is 3480 km and its distance from the earth is 386,000 km. Find the diameter of the image of the moon formed by a spherical concave telescope mirror of focal length 4 m.

39-7 A spherical concave shaving mirror has a radius of curvature of 30 cm.

  a) What is the magnification when the face is 10 cm from the vertex of the mirror?
  b) Where is the image?

39-8 A concave spherical mirror has a radius of curvature of 10 cm. Make a diagram of the mirror to scale, and show rays incident on it parallel to the axis and at distances of 1, 2, 3, 4, and 5 cm from the axis. Using a protractor, construct the reflected rays and indicate the points at which they cross the axis.

39-9 An object is 16 cm from the center of a silvered spherical glass Christmas tree ornament 8 cm in diameter. What are the position and magnification of its image?

39-10 A concave mirror of radius 5 cm has a radius of curvature of 20 cm.

  a) What is its focal length?
  b) If the mirror is immersed in water (refractive index 1.33), what is its focal length?

39-11 An object 2 cm high is placed 5 cm away from a concave spherical mirror having radius of curvature of 20 cm.

a) Draw a principal-ray diagram showing formation of the image.

b) Determine the position, size, orientation, and nature of the image.

39-12 Prove that the image formed of a real object by a convex mirror is always virtual, no matter what the object position.

39-13 If light striking a convex mirror does not diverge from an object point but instead is converging toward a point at a (negative) distance s to the right of the mirror, this point is called a *virtual object*.

a) For a convex mirror having radius of curvature 10 cm, for what range of virtual-object positions is a real image formed?

b) What is the orientation of the image?

c) Draw a principal-ray diagram showing formation of such an image.

39-14 A tank whose bottom is a mirror is filled with water to a depth of 20 cm. A small object hangs motionless 8 cm under the surface of the water. What is the apparent depth of its image when viewed at normal incidence?

39-15 A ray of light in air makes an angle of incidence of 45° at the surface of a sheet of ice. The ray is refracted within the ice at an angle of 30°.

a) What is the critical angle for the ice?

b) A speck of dirt is embedded 2 cm below the surface of the ice. What is its apparent depth when viewed at normal incidence?

39-16 A microscope is focused on the upper surface of a glass plate. A second plate is then placed over the first. In order to focus on the bottom surface of the second plate, the microscope must be raised 1 mm. In order to focus on the upper surface it must be raised 2 mm *farther*. Find the index of refraction of the second plate. (This problem illustrates one method of measuring index of refraction.)

39-17 A layer of ether ($n = 1.36$) 2 cm deep floats on water ($n = 1.33$) 4 cm deep. What is the apparent distance from the ether surface to the bottom of the water layer, when viewed at normal incidence?

39-18 The end of a long glass rod 8 cm in diameter has a hemispherical surface 4 cm in radius. The refractive index of the glass is 1.50. Determine each position of the image if an object is placed on the axis of the rod at the following distances from its end:

a) infinitely far,   b) 16 cm,   c) 4 cm.

39-19 The rod of Problem 39-18 is immersed in a liquid. An object 60 cm from the end of the rod and on its axis is imaged at a point 100 cm inside the rod. What is the refractive index of the liquid?

39-20 What should be the index of refraction of a transparent sphere in order that paraxial rays from an infinitely distant object will be brought to a focus at the vertex of the surface opposite the point of incidence?

39-21 The left end of a long glass rod 10 cm in diameter, of index 1.50, is ground and polished to a convex hemispherical surface of radius 5 cm. An object in the form of an arrow 1 mm long, at right angles to the axis of the rod, is located on the axis 20 cm to the left of the vertex of the convex surface. Find the position and magnification of the image of the arrow formed by paraxial rays incident on the convex surface.

39-22 A transparent rod 40 cm long is cut flat at one end and rounded to a hemispherical surface of 12 cm radius at the other end. A small object is embedded within the rod along its axis and halfway between its ends. When viewed from the flat end of the rod the apparent depth of the object is 12.5 cm. What is its apparent depth when viewed from the curved end?

39-23 A solid glass hemisphere having a radius of 10 cm and a refractive index of 1.50 is placed with its flat face downward on a table. A parallel beam of light of circular cross section 1 cm in diameter travels directly downward and enters the hemisphere along its diameter. What is the diameter of the circle of light formed on the table?

39-24 A small tropical fish is at the center of a spherical fish bowl 30 cm in diameter. Find its apparent position and magnification to an observer outside the bowl. The effect of the thin walls of the bowl may be neglected.

39-25 In Fig. 39-20a the first focal length f is seen to be the value of s corresponding to $s' = \infty$; in (b) the second focal length f' is the value of s' when $s = \infty$.

a) Prove that $n/n' = f/f'$.

b) Prove that the general relation between object and image distances is

$$\frac{f}{s} + \frac{f'}{s'} = 1.$$

(a)

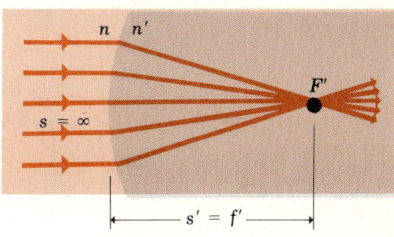

(b)

Figure 39-20

LENSES AND OPTICAL INSTRUMENTS

40-1 Images as objects

Most optical systems include more than one reflecting or refracting surface. The image formed by the first surface serves as the object for the second; the image formed by the second surface serves as the object for the third; etc. Figure 40-1 illustrates the various situations that may arise, and the following discussion of them should be studied carefully.

In Fig. 40-1, the arrow at point O represents a small object at right angles to the axis. A narrow cone of rays diverging from the head of the arrow is traced through the system. Surface 1 forms a real image of the arrow at point P. Distance OV_1 is the object distance for the first surface and distance V_1P is the image distance. Both of these are positive.

The image at P, formed by surface 1, serves as the object for surface 2. The object distance is PV_2 and is positive, since the direction from P to V_2 is the same as that of the incident light. The second surface forms a virtual image at point Q. The image distance is V_2Q and is negative because the direction from V_2 to Q is opposite to that of the refracted light.

The image at Q, formed by surface 2, serves as the object for surface 3. The object distance is QV_3 and is positive. The image at Q, although virtual, constitutes a *real object* so far as surface 3 is concerned. The rays incident on surface 3 are rendered converging and, except for the interposition of surface 4, would converge to a real image at point R. Even though this image is never formed, distance V_3R is the image distance for surface 3 and is positive.

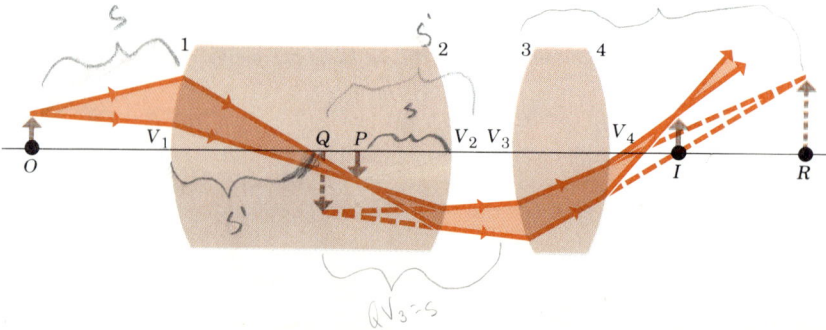

40-1 The object for each surface, after the first, is the image formed by the preceding surface.

The rays incident on surfaces 1, 2, and 3 have all been diverging, and the object distance has been the distance from the surface to the point from which the rays were actually or apparently diverging. The rays incident on surface 4, however, are *converging* and there is no point at the left of the vertex from which they diverge or appear to diverge. The *image at R, toward which the rays are converging*, is the object for surface 4, and since the direction from R to V_4 is opposite to that of the incident light, the object distance RV_4 is negative. The image at R is called a *virtual object* for surface 4. In general, whenever a converging cone of rays is incident on a surface, the point toward which the rays are converging serves as the object, the object distance is negative, and the point is called a virtual object.

Finally, surface 4 forms a real image at I, the image distance being $V_4 I$ and positive.

The meanings of virtual object and of virtual image are further exemplified by Fig. 40–2.

40–2 The thin lens

A lens is an optical system including two refracting surfaces. The general problem of refraction by a lens is solved by applying the methods of Sec. 39–7 to each surface in turn, the object for the second surface being the image formed by the first. Figure 40–3 shows a pencil of rays diverging from point Q of an object PQ. The first surface of lens L forms a virtual image of Q at Q'. This virtual image serves as a real object for the second surface of the lens, which forms a real image of Q' at Q''. Distance s_1 is the object distance for the first surface; s_1' is the corresponding image distance. The object distance for the second surface is s_2, equal to the sum of s_1' and the lens thickness t, and s_2' is the image distance for the second surface.

If, as is often the case, the lens is so thin that its thickness t is negligible in comparison with the distances s_1, s_1', s_2, s_2', we may assume that s_1' equals $-s_2$, and measure object and image distances from either vertex of the lens. We shall also assume the medium on both sides of the lens to be air, with index of refraction 1.00. For the first refraction, Eq. (39–16) becomes

$$\frac{1}{s_1} + \frac{n}{s_1'} = \frac{n-1}{R_1}.$$

(a)

(b)

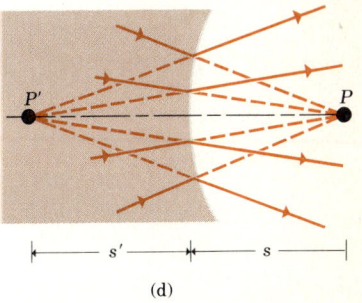

(c)

(d)

40–2 (a) A real image of a real object. (b) A virtual image of a real object. (c) A real image of a virtual object. (d) A virtual image of a virtual object.

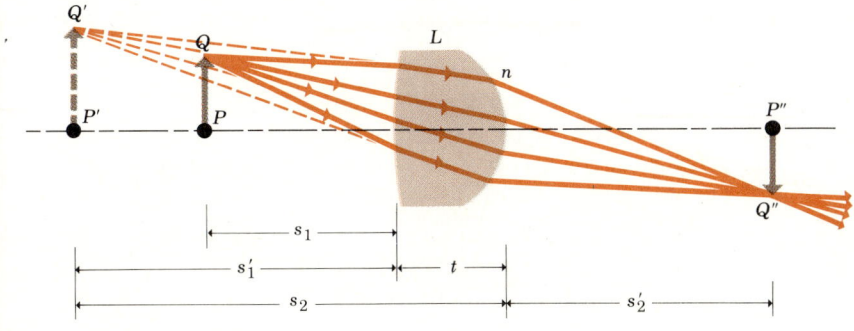

40–3 The image formed by the first surface of a lens serves as the object for the second surface.

Refraction at the second surface yields the equation

$$\frac{n}{s_2} + \frac{1}{s_2'} = \frac{1-n}{R_2}.$$

Adding these two equations, and remembering that the lens is so thin that $s_2 = -s_1'$, we get

$$\frac{1}{s_1} + \frac{1}{s_2'} = (n-1)\left(\frac{1}{R_1} - \frac{1}{R_2}\right).$$

Since s_1 is the object distance for the thin lens and s_2' is the image distance, the subscripts may be omitted, and we get finally

$$\frac{1}{s} + \frac{1}{s'} = (n-1)\left(\frac{1}{R_1} - \frac{1}{R_2}\right). \tag{40-1}$$

The usual sign conventions apply to this equation. Thus, in Fig. 40-4, s, s', and R_1 are positive quantities, but R_2 is negative.

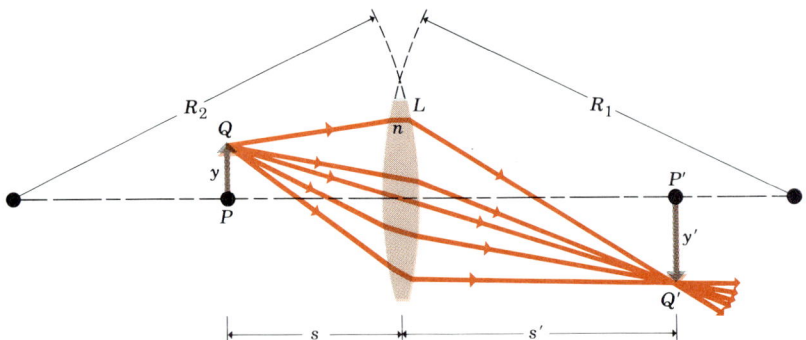

40-4 A thin lens.

The *focal length f* of a thin lens may be defined either as (a) the object distance of a point object on the lens axis whose image is at infinity, or (b) the image distance of a point object on the lens axis at an infinite distance from the lens. When we set either s or s' equal to infinity in Eq. (40-1) we find, for the focal length,

$$\frac{1}{f} = (n-1)\left(\frac{1}{R_1} - \frac{1}{R_2}\right), \tag{40-2}$$

which is known as the *lensmaker's equation.*

Example In Fig. 40-4, let the absolute magnitudes of the radii of curvature of the lens surfaces be respectively 20 cm and 5 cm. Since the center of curvature of the first surface is on the side of the outgoing light, $R_1 = +20$ cm, and since that of the second surface is not, $R_2 = -5$ cm. Let $n = 1.50$. Then

$$\frac{1}{f} = (1.50 - 1)\left(\frac{1}{20 \text{ cm}} - \frac{1}{-5 \text{ cm}}\right),$$
$$f = +8 \text{ cm}.$$

Substituting Eq. (40-2) in Eq. (40-1), we obtain the thin-lens equation

$$\frac{1}{s} + \frac{1}{s'} = \frac{1}{f}. \qquad (40\text{-}3)$$

This is known as the *gaussian* form of the thin-lens equation, after Karl F. Gauss, the same mathematician responsible for the law in electrostatics bearing his name. Note that Eq. (40-3) has exactly the same form as the equation for a spherical mirror.

The object point for which the image is at infinity is called the *first focal point* or *focus* of the lens and is lettered F in Fig. 40-5a. The image point for an infinitely distant object is called the *second focal point* or *focus* and is lettered F' in Fig. 40-5b. The focal points (foci) of a thin lens lie on opposite sides of the lens at distances from it equal to its focal length f.

Figure 40-6 corresponds to Fig. 40-5, except that it is drawn for object and image points not on the axis of the lens. Planes through the first and second focal points of a lens, perpendicular to the axis, are called the first and second *focal planes*. Paraxial rays from point Q in Fig. 40-6a, in the first focal plane of the lens, are parallel to one another after refraction. In other words, they converge to an infinitely distant image of point Q. In Fig. 40-6b a bundle of parallel rays from an infinitely distant point, not on the lens axis, converges to an image Q' lying in the second focal plane of the lens.

The lateral magnification produced by a lens may be obtained by inspection of Fig. 40-4; the object PQ, the image $P'Q'$, and the lines PP' and QQ' form two similar triangles. Hence

$$\frac{-y'}{y} = \frac{s'}{s},$$

and, since m, as usual, is y'/y, we obtain

$$m = -\frac{s'}{s}. \qquad (40\text{-}4)$$

Taking the derivative of Eq. (40-3) with respect to s, we find that the *longitudinal* magnification is

$$m' = \frac{ds'}{ds} = -\left(\frac{s'}{s}\right)^2 = -m^2. \qquad (40\text{-}5)$$

Although Eqs. (40-3) and (40-4) were derived for the special case of rays making small angles with the axis and, in general, do not apply to rays making large angles, they may be used for any lens that has been corrected so that all rays are imaged at the same point. These are therefore two of the most important equations in geometrical optics.

The three-dimensional image of a three-dimensional object, formed by a lens, is shown in Fig. 40-7. Since point R is nearer the lens than point P, its image, from Eq. (40-3), is farther from the lens than is point P', and the image $P'R'$ points in the same direction as the object PR. Arrows $P'S'$ and $P'Q'$ are reversed in space, relative to PS and PQ. Although we speak of the image as "inverted," only its transverse dimensions are reversed.

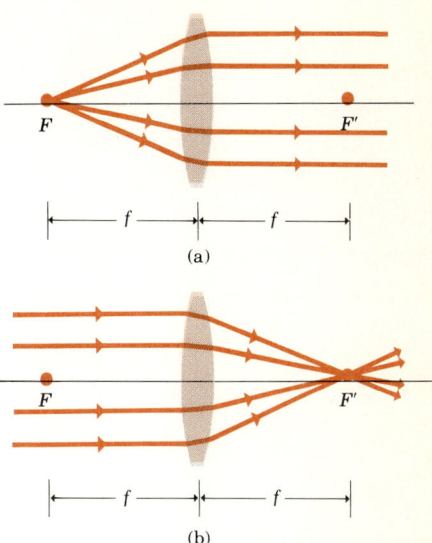

40-5 First and second focal points of a thin lens.

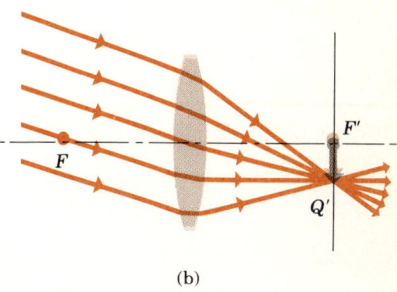

40-6 Planes through the focal points of a lens are called focal planes.

40-7 A lens forms a three-dimensional image of a three-dimensional object.

Figure 40-7 should be compared with Fig. 39-4, showing the image formed by a plane mirror. Note that the image formed by a lens, although it is inverted, is *not* perverted. That is, if the object is a left hand, its image is a left hand also. This may be verified by pointing the left thumb along PR, the left forefinger along PQ, and the left middle finger along PS. A rotation of 180° about the thumb as an axis then brings the fingers into coincidence with $P'Q'$ and $P'S'$. In other words, *inversion* of an image is equivalent to a rotation of 180° about the lens axis.

40-3 Diverging lenses

A bundle of parallel rays incident on the lens shown in Figs. 40-5 and 40-6 converges to a real image after passing through the lens. The lens is called a *converging lens*. Its focal length, as computed from Eq. (40-2), is a positive quantity and therefore the lens is also called a *positive lens*.

A bundle of parallel rays incident on the lens in Fig. 40-8 becomes diverging after refraction, and the lens is called a *diverging lens*. Its focal length, computed by Eq. (40-2), is a negative quantity and therefore the lens is also called a *negative lens*. The focal points of a negative lens are reversed, relative to those of a positive lens. The second focal point, F', of a negative lens is the point from which rays, originally parallel to the axis, appear to diverge after refraction, as in Fig. 40-8a. Incident rays converging toward the first focal point F, as in Fig. 40-8b, emerge from the lens parallel to its axis. That is, just as for a positive lens, the second focal point is the (virtual) image of an infinitely distant object on the axis of the lens, while the first focal point is the object point (a virtual

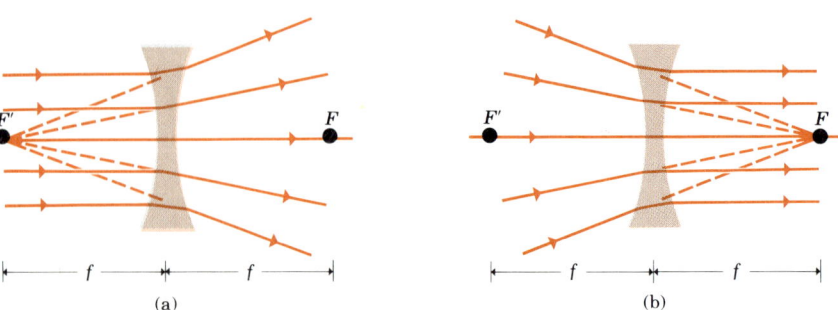

40-8 Focal points of a diverging lens.

object if the lens is diverging) for which an image is formed at infinity. Equations (40-2) through (40-5) apply both to negative and to positive lenses. Various types of lenses, both converging and diverging, are illustrated in Fig. 40-9.

40-9 (a) Meniscus, plano-convex, and double-convex converging lenses. (b) Meniscus, plano-concave, and double concave diverging lenses.

40-4 Graphical methods

The position and size of the image of an object formed by a thin lens may be found by a simple graphical method. This method consists of finding the point of intersection, after passing through the lens, of a few rays (called *principal rays*) diverging from some chosen point of the object *not* on the lens axis, such as point Q in Fig. 40–10. Then (neglecting lens aberrations) all rays from this point that pass through the lens will intersect at the same point. In using the graphical method, the entire deviation of any ray is assumed to take place at a plane through the center of the lens. Three principal rays whose paths may readily be traced are shown in Fig. 40–10.

1. *A ray parallel to the axis,* after refraction by the lens, passes through the second focal point of a converging lens, or appears to come from the second focal point of a diverging lens.

2. *A ray through the center of the lens* is not appreciably deviated, since the two lens surfaces through which the central ray passes are very nearly parallel if the lens is thin. We have seen that a ray passing through a plate with parallel faces is not deviated, but only dis-

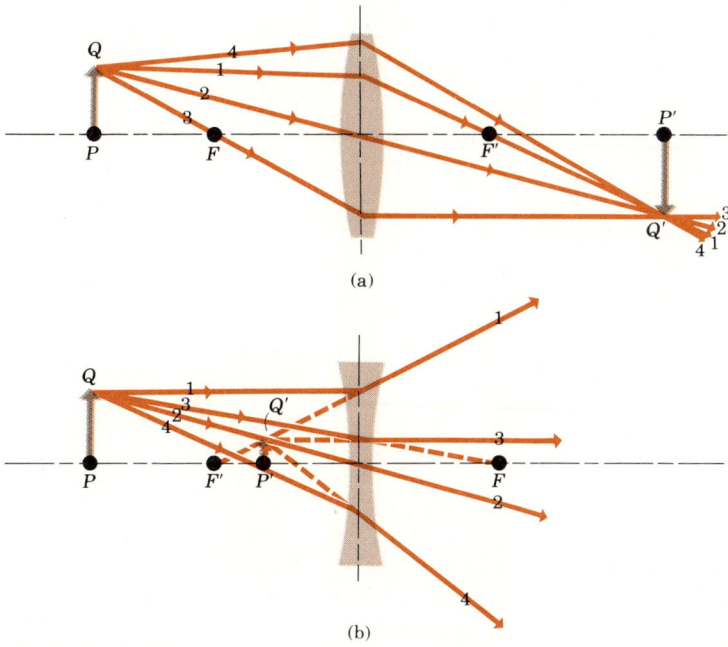

40-10 Principal-ray diagram, showing graphical method of locating an image.

placed. For a thin lens, the displacement may be neglected.

3. A *ray through (or proceeding toward) the first focal point* emerges parallel to the axis.

Once the position of the image point has been found by means of the intersection of any two rays 1, 2, and 3, the paths of all other rays from the same point, such as ray 4 in Fig. 40–10, may be drawn. A few examples of this procedure are given in Fig. 40–11. Not all the rays in Fig. 40–11 are principal rays.

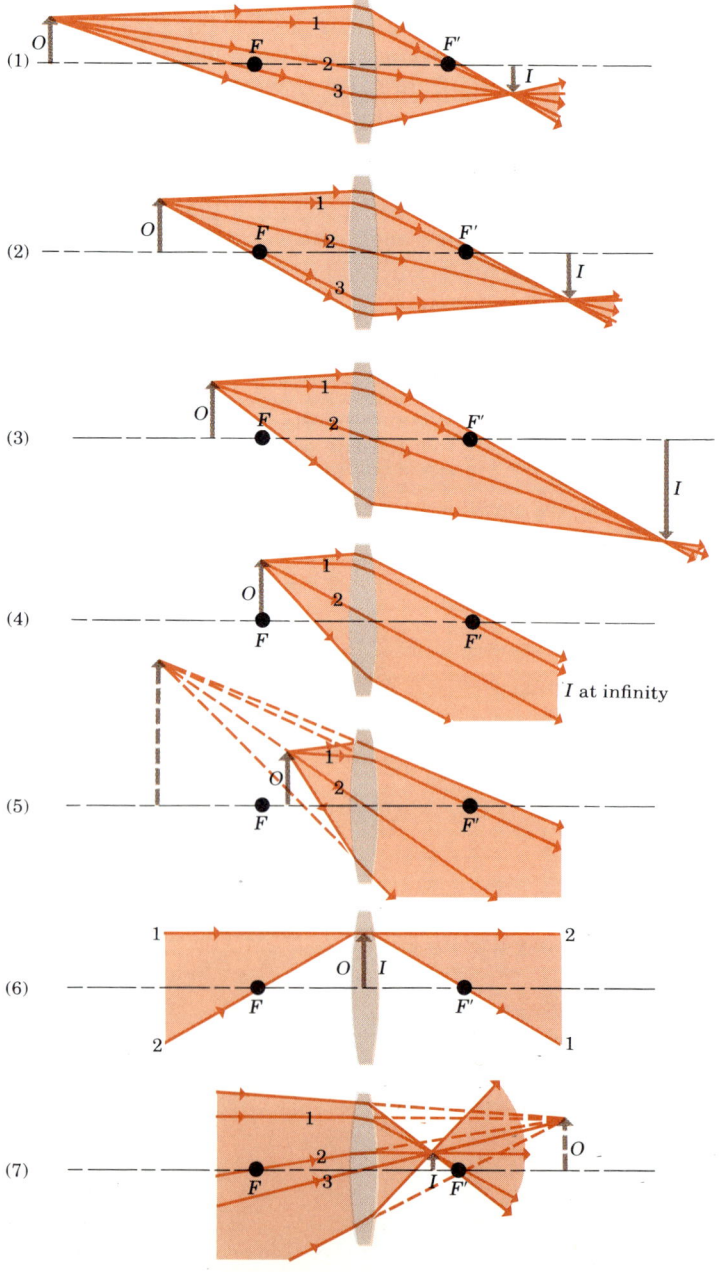

40–11 Formation of an image by a thin lens. The principal rays are labeled.

In every problem involving formation of an image by a lens, the student should *always* draw a principal-ray diagram; this serves not only as a graphical check on numerical calculations but also as an aid to the understanding of basic concepts, especially the concept of *image*.

40-5 Images as objects for lenses

It was shown in Sec. 40-1 and Fig. 40-1 that the image formed by any one *surface* in an optical system serves as the object for the next surface. This principle was used to derive the thin-lens relations, Eqs. (40-1) through (40-3), with the assumption that the distance between the two surfaces is negligible. A similar analysis can be carried out for a thick lens, where this distance (which we may call d) is *not* negligible compared to the other distances. In this case the object distance s_2 for the second surface is not equal to the negative of the image distance s_1' for the first, but instead we have

$$s_2 = -(s_1' - d), \tag{40-6}$$

and the resulting formulas are not so simple. Examples of thick-lens analysis are found in the problems for this chapter.

The same principle can be used in systems containing several *lenses*. Many optical systems, such as camera lenses, microscopes, and telescopes, use more than one lens. In each case the image formed by any one lens serves as the object for the next lens. Figure 40-12 shows various possibilities. Lens 1 forms a real image at P of a real object at O. This real image serves as a real object for lens 2. The virtual image at Q formed by lens 2 is a real object for lens 3. If lens 4 were not present, lens 3 would form a real image at R. Although this image is never formed, it serves as a *virtual object* for lens 4, which forms a final real image at I.

40-12 The object for each lens, after the first, is the image formed by the preceding lens.

*40-6 Lens aberrations

The relatively simple equations we have derived relating object and image distances, focal lengths, radii of curvature, etc., were based upon the *paraxial approximation;* all rays were assumed to be paraxial, that is, to make small angles with the axis. In general, however, a lens must image not only points on its axis, but points that lie off the axis as well. Furthermore, because of the finite size of the lens, the cone of rays that forms the image of any point is of finite size. Nonparaxial rays proceeding from a given object point *do not*, in general, all intersect at the same point after refraction by a lens. Consequently, the image formed by these rays is not a sharp one. Furthermore, the focal length of a lens depends upon its index of refraction, which varies with wavelength. Therefore, if the light proceeding from an object is not monochromatic, a lens forms a number of colored images, which lie in different positions and are of different sizes, even if formed by paraxial rays.

The departures of an actual image from the predictions of simple theory are called *aberrations*. Those caused by the variation of index with wavelength are the *chromatic aberrations*. The others, which would arise even if the light were monochromatic, are the *monochromatic aberrations*. Lens aberrations are not caused by faulty construction of the lens, such as the failure of its surfaces to conform to a truly spherical shape, but are simply consequences of the laws of refraction at spherical surfaces.

The monochromatic aberrations are all related to the failure of the paraxial-ray approximation for lenses of finite aperture, but it is customary to distinguish various aspects of this difficulty, each with its characteristic effect. *Spherical aberration* is the failure of rays from a point object on the optic axis to converge to a point image; instead, the rays converge to a circle of minimum radius, called *the circle of least confusion,* and then diverge again, as shown in Fig. 40–13. The corresponding effect for points off the axis produces images that are comet-shaped figures rather than circles; this is called *coma*.

Astigmatism is the imaging of a point off the axis as a *line;* in this aberration the rays from a point object converge at some distance from the lens to a line in the place defined by the optic axis and the object point and, at a somewhat different distance from the lens, to a line *perpendicular* to this plane. The circle of least confusion appears between these two positions, at a location that depends on the object point's distance from the axis as well as its distance from the lens. As a result, object points lying in a plane are, in general, imaged not in a plane but in some curved surface; this effect is called *curvature of field*. Finally, the image of a straight line that does not pass through the axis may be curved; as a result the image of a square with the axis through its center may resemble a barrel (sides bent outward) or a pincushion (sides bent inward). This effect, called *distortion,* is not related to lack of sharpness of the image but results from a change in lateral magnification with distance from the axis.

Chromatic aberrations result directly from the variation of index of refraction with wavelength. Even in the absence of all monochromatic aberration, different wavelengths are imaged at different points, and when an object is illuminated with white light containing a mixture of wavelengths, there is no single point at which a point object is imaged. The magnification of a lens also varies with wavelength; this effect is responsible for the rainbow-fringed images seen with inexpensive binoculars or telescopes.

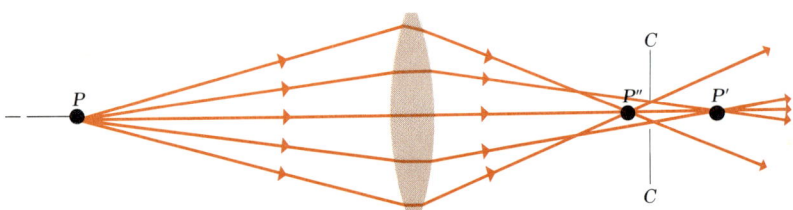

40–13 Spherical aberration. The circle of least confusion is shown by *C–C*.

It is impossible to eliminate these aberrations from a single lens, but in a compound lens of several elements, the aberrations of one element may partially cancel those of another element. Design of such lenses is an extremely complex problem, which has been aided greatly in recent years by the use of high-speed computers. It is still impossible to eliminate all aberrations, but it *is* possible to decide which ones are most troublesome for a particular application and to design accordingly.

40-7 The eye

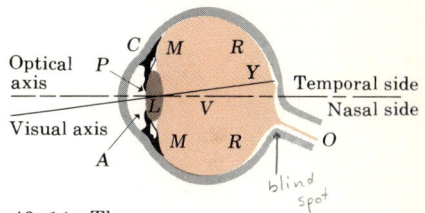

40-14 The eye.

Since the purpose of most optical instruments is to enable us to see better, the logical place to begin a discussion of such instruments is with the eye. The essential parts of the eye, considered as an optical system, are shown in Fig. 40-14.

The eye is very nearly spherical in shape, and about an inch in diameter. The front portion is somewhat more sharply curved, and is covered by a tough, transparent membrane C, called the *cornea*. The region behind the cornea contains a liquid A called the *aqueous humor*. Next comes the *crystalline lens*, L, a capsule containing a fibrous jelly, hard at the center and progressively softer at the outer portions. The crystalline lens is held in place by ligaments that attach it to the ciliary muscle M. Behind the lens, the eye is filled with a thin jelly V consisting largely of water, called the *vitreous humor*. The indices of refraction of both the aqueous humor and the vitreous humor are nearly equal to that of water, about 1.336. The crystalline lens, while not homogeneous, has an "average" index of 1.437. This is not very different from the indices of the aqueous and vitreous humors, so that most of the refraction of light entering the eye occurs at the cornea.

A large part of the inner surface of the eye is covered with a delicate film of nerve fibers, R, called the *retina*. A cross section of the retina is shown in Fig. 40-15a. Nerve fibers branching out from the *optic nerve* O terminate in minute structures called *rods and cones*. The rods and cones, together with a bluish liquid called the *visual purple*, which circulates among them, receive the optical image and transmit it along the optic nerve to the brain. There is a slight depression in the retina at Y called the *yellow spot* or *macula*. At its center is a minute region, about 0.25 mm in diameter, called the *fovea centralis*, which contains cones exclusively. Vision is much more acute at the fovea than at other portions of the retina, and the muscles controlling the eye always rotate the eyeball until the image of the object toward which attention is directed

40-15 (a) Section of the human retina (500X). Light is incident from the left. (b) Figure for demonstrating the blind spot.

(a)

(b)

falls on the fovea. The outer portion of the retina merely serves to give a general picture of the field of view. The fovea is so small that motion of the eye is necessary to focus distinctly two points as close together as the dots in a colon (:).

There are no rods or cones at the point where the optic nerve enters the eye, and an image formed at this point cannot be seen. This region is called the *blind spot*. The existence of the blind spot can be demonstrated by closing the left eye and looking with the right eye at the cross in Fig. 40–15b. When the diagram is about 25 cm from the eye, the image of the square falls on the blind spot and the square disappears. At a smaller distance, the square reappears while the circle disappears. At a still smaller distance, the circle again appears.

In front of the crystalline lens is the iris, at the center of which is an opening P called the *pupil*, which regulates the quantity of light entering the eye, dilating if the brightness of the field is low, and contracting if the brightness is increased. This process is known as *adaptation*. However, the range of pupillary diameter is only about fourfold (hence, the range in area is about sixteenfold) over a range of brightness that is 100,000-fold. The receptive mechanism of the retina can also adapt itself to large differences in quantity of light.

To see an object distinctly, a sharp image of it must be formed on the retina. If all the elements of the eye were rigidly fixed in position, there would be but one object distance for which a sharp retinal image would be formed, while in fact the normal eye can focus sharply on an object at any distance from infinity up to about 25 cm in front of the eye. This is made possible by the action of the crystalline lens and the ciliary muscle to which it is attached. When relaxed, the normal eye is focused on objects at infinity, i.e., the second focal point is at the retina. When it is desired to view an object nearer than infinity, the ciliary muscle tenses and the crystalline lens assumes a more nearly spherical shape. This process is called *accommodation*.

The extremes of the range over which distinct vision is possible are known as the *far point* and the *near point* of the eye. The far point of a normal eye is at infinity. The position of the near point evidently depends on the extent to which the curvature of the crystalline lens may be increased in accommodation. The range of accommodation gradually diminishes with age as the crystalline lens loses its flexibility. For this reason the near point gradually recedes as one grows older. This recession of the near point with age is called *presbyopia*, and should not be considered a defect of vision, since it proceeds at about the same rate in all normal eyes. The following is a table of the approximate average position of the near point at various ages:

Age, years	Near point, cm
10	7
20	10
30	14
40	22
50	40
60	200

*40-8 Defects of vision

Several common defects of vision result from an incorrect relation between the parts of the optical system of the eye. A normal eye forms an image on the retina of an object at infinity when the eye is relaxed, as in Fig. 40-16a. In the *myopic* (nearsighted) eye, the eyeball is too long from front to back in comparison with the radius of curvature of the cornea, and rays from an object at infinity are focused in front of the retina. The most distant object for which an image can be formed on the retina is then nearer than infinity. In the *hyperopic* (farsighted) eye, the eyeball is too short and the image of an infinitely distant object would be formed behind the retina. By accommodation, these parallel rays may be made to converge on the retina, but evidently, if the range of accommodation is normal, the near point will be more distant than that of a normal eye. The myopic eye produces too much convergence in a parallel bundle of rays for an image to be formed on the retina; the hyperopic eye, not enough convergence.

Astigmatism refers to a defect in which the surface of the cornea is not spherical, but is more sharply curved in one plane than another. (It should not be confused with the lens aberration of the same name, which applies to the behavior, after passing through a lens having spherical surfaces, of rays making a large angle with the axis.) Astigmatism makes it impossible, for example, to focus clearly on the horizontal and vertical bars of a window at the same time.

These defects can be corrected by the use of corrective lenses ("glasses"). The near point of either a presbyopic or a hyperopic eye is farther from the eye than normal. To see clearly an object at normal reading distance (usually assumed to be 25 cm) we must place in front of the eye a lens of such focal length that it forms an image of the object at or beyond the near point. Thus the function of the lens is not to make the object appear larger, but in effect to move the object farther away from the eye to a point where a sharp retinal image can be formed.

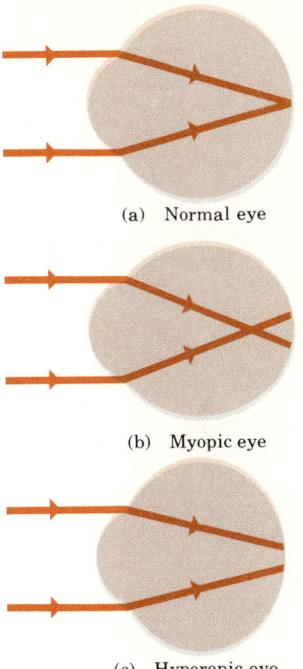

(a) Normal eye

(b) Myopic eye

(c) Hyperopic eye

40-16

Example 1 The near point of a certain eye is 100 cm in front of the eye. What lens should be used to see clearly an object 25 cm in front of the eye?

Solution We have

$$s = +25 \text{ cm}, \quad s' = -100 \text{ cm},$$

$$\frac{1}{f} = \frac{1}{s} + \frac{1}{s'} = \frac{1}{+25 \text{ cm}} + \frac{1}{-100 \text{ cm}},$$

$$f = +33 \text{ cm}.$$

That is, a converging lens of focal length 33 cm is required.

The far point of a *myopic* eye is nearer than infinity. To see clearly objects beyond the far point, a lens must be used that will form an image of such objects, not farther from the eye than the far point.

Example 2 The far point of a certain eye is 1 m in front of the eye. What lens should be used to see clearly an object at infinity?

Solution Assume the image to be formed at the far point. Then

$$s = \infty, \quad s' = -100 \text{ cm},$$
$$\frac{1}{f} = \frac{1}{s} + \frac{1}{s'} = \frac{1}{\infty} + \frac{1}{-100 \text{ cm}},$$
$$f = -100 \text{ cm}.$$

A *diverging* lens of focal length 100 cm is required.

Astigmatism is corrected by means of a *cylindrical* lens. The curvature of the cornea in a horizontal plane may have the proper value such that rays from infinity are focused on the retina. In the vertical plane, however, the curvature may not be sufficient to form a sharp retinal image. When a cylindrical lens with axis horizontal is placed before the eye, the rays in a horizontal plane are unaffected, while the additional convergence of the rays in a vertical plane now causes these to be sharply imaged on the retina.

The optometrist describes the converging or diverging effect of lenses in terms, not of the focal length, but of its *reciprocal*. The reciprocal of the focal length of a lens is called its *power*, and if the focal length is in meters the power is in *diopters*. Thus the power of a positive lens whose focal length is 1 m is 1 diopter; if the focal length is 2 m the power is 0.5 diopter, and so on. If the focal length is negative, the power is negative also. Thus in the two examples above, the required powers are $+3.0$ diopter and -1.0 diopter, respectively.

40-9 The magnifier

The apparent size of an object is determined by the size of its retinal image; if the eye is unaided, this depends upon the *angle* subtended by the object at the eye. When one wishes to examine a small object in detail one brings it close to the eye, in order that the angle subtended and the retinal image may be as large as possible. Since the eye cannot focus sharply on objects closer than the near point, a given object subtends the maximum possible angle at an unaided eye when placed at this point. (We shall assume hereafter that the near point is 25 cm from the eye.) By placing a converging lens in front of the eye, the accommodation may, in effect, be increased. The object may then be brought closer to the eye than the near point and will subtend a correspondingly larger angle. A lens used for this purpose is called a *magnifying glass*, a *simple microscope*, or a *magnifier*. The magnifier forms a virtual image of the object and the eye "looks at" this virtual image. Since a (normal) eye can focus sharply on an object anywhere between the near point and infinity, the image can be seen equally clearly if it is formed anywhere within this range. We shall assume that the image is formed at infinity.

The magnifier is illustrated in Fig. 40–17. In (a), the object is at the near point, where it subtends an angle u at the eye. In (b), a magnifier in front of the eye forms an image at infinity, and the angle subtended at the magnifier is u'. The *angular magnification M* (not to be confused with the *lateral magnification m*) is defined as the ratio of the angle u' to the angle u. The value of M may be found as follows.

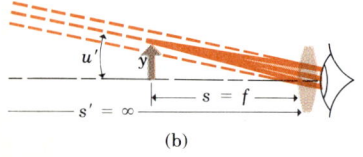

40–17 A simple magnifier.

From Fig. 40–17, u and u' are given (in radians) by

$$u = \frac{y}{25 \text{ cm}} \quad \text{(approximately)},$$

$$u' = \frac{y}{f} \quad \text{(approximately)}.$$

Hence,

$$M = \frac{u'}{u} = \frac{y/f}{y/25} = \frac{25}{f} \quad (f \text{ in centimeters}). \tag{40–7}$$

While it appears at first that the angular magnification may be made as large as desired by decreasing the focal length f, the aberrations of a simple double convex lens set a limit to M of about 2X or 3X. If these aberrations are corrected, the magnification may be carried as high as 20X.

40–10 The camera

The essential elements of a camera are a lens equipped with a shutter, a light-tight enclosure, and a light-sensitive film to record an image. An example is shown in Fig. 40–18. The lens forms a real image in the plane of the film of the object being photographed. The lens may be moved closer to or farther from the film to provide proper image distances for various object distances. All but the most inexpensive lenses have several elements, to permit partial correction of various aberrations. Many modern camera lenses are variations of the Zeiss "Tessar" design, shown in Fig. 40–19.

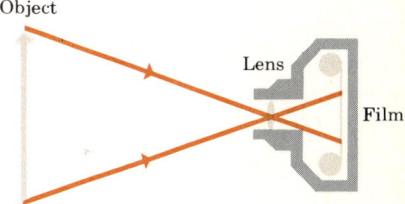

40–18 Essential elements of a camera.

In order for the image to be recorded properly on the film, the total light energy per unit area reaching the film (the "exposure") must fall within certain limits; this is controlled by the shutter and the lens aperture. The shutter controls the time during which light enters the lens, typically adjustable in steps corresponding to factors of about two, from 1 s to $\frac{1}{1000}$ s or thereabouts. The light-gathering capacity of the lens is proportional to its effective area; this may be varied by means of an adjustable aperture or *diaphragm*, which is a nearly circular hole of variable diameter. The aperture size is usually described in terms of its "f-number," which is the focal length of the lens divided by the diameter of the aperture. Thus a lens having $f = 50$ mm and an aperture diameter of 25 mm would be said to have an aperture of $f/2$.

Because the light-gathering capacity of the lens is proportional to its area and thus to the *square* of its diameter, changing the diameter by a factor of $\sqrt{2}$ corresponds to a factor of two in exposure. Thus, adjustable apertures usually have scales labeled with successive numbers related by factors of $\sqrt{2}$, such as

$$f/2, \quad f/2.8, \quad f/4, \quad f/5.6, \quad f/8, \quad f/11, \quad f/16,$$

and so on, with the larger numbers representing smaller apertures and exposures.

The choice of focal length for a camera lens depends on the film size and the desired angle of view, or *field*. For the popular 35-mm cameras,

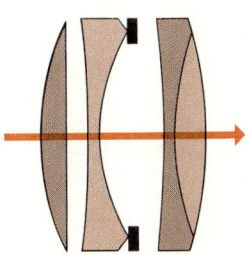

40–19 Zeiss "Tessar" lens design.

with image size of 24 × 36 mm, the normal lens is usually about 50 mm in focal length and permits an angle of about 45°. A longer focal-length lens, used with the same film size, provides a smaller angle of view and a larger image of part of the object, compared with a normal lens; this gives the impression that the camera is closer than it really is, and such a lens is called a *telephoto* lens. At the other extreme, a lens of shorter focal length, such as 35 mm or 28 mm, permits a wider angle of view and is called a *wide-angle* lens.

For a given position of the photographic film only those points lying in the corresponding object plane are sharply focused; objects at greater or lesser distance appear somewhat blurred. However, because of lens aberrations, a point of a given object will be imaged as a small circle, called the *circle of confusion,* even with the best focusing. The circles of confusion of points at other distances will be larger. If extremely sharp definition of the image is not essential, there is evidently a certain range of object distances, called the *depth of field,* such that all objects within this range are simultaneously "in focus" on the plate. That is, the circles of confusion of points within this range are not so large that the image is unsatisfactory. The diameter of the circle of confusion of a particular point is proportional to the aperture diameter; thus, depth of field is increased by "stopping down" the lens to a smaller aperture and larger f-number.

40-11 The projector

The optical system of a projector for slides or motion pictures is shown in Fig. 40-20. The arrow at the left represents the light source, for example, the filament of a projection lamp. For simplicity, the slide to be projected is represented as opaque except for a single transparent aperture.

The diagram traces the course of three pencils of rays originating at the ends and at the midpoint of the source. The function of the condensing lens is to deviate the light from the source inward, so that it can pass through the projecting lens. If the condensing lens were omitted, light passing through the outer portions of the slide would not strike the projecting lens and only a small portion of the slide near its center would be imaged on the screen.

A study of the figure shows that, for the three selected points of the source, only those rays within the shaded pencils can pass through the aperture; all others striking the condensing lens are intercepted by the opaque portions of the slide. Similar pencils of rays could be drawn from all other points of the source.

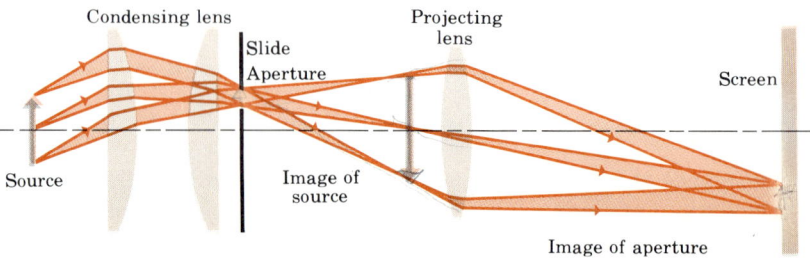

40-20 Essential elements of a projector for slides or motion pictures.

Each of these pencils converges, after passing through the aperture, to form an image of its point of origin just to the left of the projecting lens. In practice, this image would be formed *at* the projecting lens, but for clarity in the diagram the image and the lens have been displaced slightly. The focal length of the condensing lens should be such that the image of the source just fills the projecting lens. If the image of the source is larger than the projecting lens, some of the light passing through the slide is wasted. If it is smaller, the area of the projecting lens is not being fully utilized. Thus, in the diagram, the outer portions of the projecting lens serve no useful purpose.

Three rays tangent to the upper edge of the aperture have been emphasized in the figure. These rays originate at *different* points of the source. Hence, although they intersect at the edge of the aperture, this point of intersection does not constitute an image of any point of the source; but these three rays diverge from a common point of the *slide*, and therefore this point of the slide is imaged, as shown on the screen. Similarly, rays tangent to any point of the edge of the aperture are imaged at a corresponding point on the screen. Thus, if the aperture is circular, a circular spot of light appears on the screen. Note that light from *all* points of the source illuminates *every* point of the image of the aperture, and would do the same were the aperture at any other point of the slide.

The preceding discussion has explained the conditions that determine the *focal length* of the condensing lens and the *diameter* of the projecting lens (the image of the source formed by the condensing lens should just fill the projecting lens). The *diameter* of the condensing lens must evidently be at least as great as the diagonal of the largest slide to be projected, while the *focal length* of the projecting lens is determined by the magnification desired and the distance of the projector from the screen.

40-12 The compound microscope

When an angular magnification larger than that attainable with a simple magnifier is desired, it is necessary to use a *compound microscope*, usually called merely a *microscope*. The essential elements of a microscope are illustrated in Fig. 40-21. The object O to be examined is placed just beyond the first focal point F_1 of the *objective* lens, which forms a real and enlarged image I. This image lies just within the first focal point F_2 of the *eyepiece*, which forms a virtual image of I at I'. As was stated earlier, the position of I' may be anywhere between the near and far points of the eye. Although both the objective and eyepiece of an actual microscope are highly corrected compound lenses, they are shown as simple thin lenses for simplicity.

Since the objective lens merely forms an enlarged real image that is viewed by the eyepiece, the overall angular magnification M of the compound microscope is the product of the lateral magnification m_1 of the objective and the angular magnification M_2 of the eyepiece. The former is given by

$$m_1 = -\frac{s_1'}{s_1},$$

40-21 The microscope.

where s_1 and s_1' are the object and image distances for the objective lens. Ordinarily the object is very close to the focus, resulting in an image whose distance from the objective is much larger than its focal length f_1. Thus s_1 is approximately equal to f_1, and $m_1 = -s_1'/f_1$, approximately. The angular magnification of the eyepiece, from Eq. (40-7), is $M_2 = (25 \text{ cm})/f_2$, where f_2 is the focal length of the eyepiece, considered as a simple lens. Hence the overall magnification M of the compound microscope is, apart from a negative sign, which is customarily ignored,

$$M = m_1 M_2 = \frac{(25 \text{ cm})s_1'}{f_1 f_2}, \qquad (40\text{-}8)$$

where s_1', f_1, and f_2 are measured in centimeters. Microscope manufacturers customarily specify the values of m_1 and M_2 for microscope components rather than the focal lengths of the objective and eyepiece.

40-13 Telescopes

The optical system of a refracting telescope is essentially the same as that of a compound microscope. In both instruments, the image formed by an objective is viewed through an eyepiece. The difference is that the telescope is used to examine large objects at large distances and the microscope to examine small objects close at hand.

The *astronomical* telescope is illustrated in Fig. 40-22. The objective lens forms a real, reduced image I of the object, and a virtual image of I is formed by the eyepiece. As with the microscope, the image I' may be formed anywhere between the near and far points of the eye. In practice, the objects examined by a telescope are at such large distances from the instrument that the image I is formed very nearly at the second focal point of the objective. Furthermore, if the image I' is at infinity, the image I is at the first focal point of the eyepiece. The distance between objective and eyepiece, or the length of the telescope, is therefore the *sum* of the focal lengths of objective and eyepiece, $f_1 + f_2$.

The angular magnification M of a telescope is defined as the ratio of the angle subtended at the eye by the final image I', to the angle subtended at the (unaided) eye by the object. This ratio may be expressed in terms of the focal lengths of objective and eyepiece as follows. In Fig. 40-22, the ray passing through F_1, the first focal point of the objective, and through F_2', the second focal point of the eyepiece, has been emphasized. The object (not shown) subtends an angle u at the objective and would subtend essentially the same angle at the unaided eye. Also, since the observer's eye is placed just to the right of the focal point F_2', the

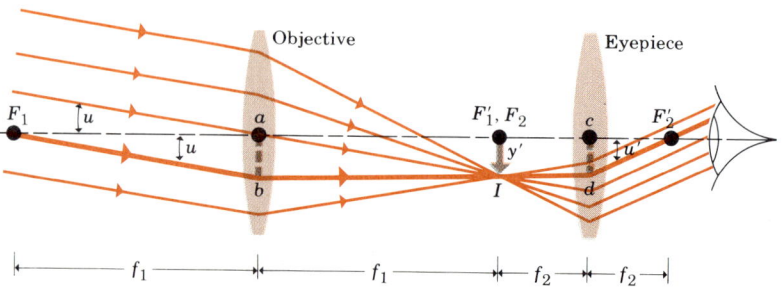

40-22 Telescope; final image at infinity.

angle subtended at the eye by the final image is very nearly equal to the angle u'. The distances ab and cd are evidently equal to each other and to the height y' of the image I. Since u and u' are small, they may be approximated by their tangents. From the right triangle F_1ab and $F_2'cd$,

$$u = \frac{-y'}{f_1}, \qquad u' = \frac{y'}{f_2},$$

Hence,

$$M = \frac{u'}{u} = -\frac{y'/f_2}{y'/f_1} = -\frac{f_1}{f_2}. \qquad (40\text{-}9)$$

The angular magnification M of a telescope is therefore equal to the ratio of the focal length of the objective to that of the eyepiece. The negative sign denotes an inverted image.

An inverted image is not a disadvantage if the instrument is to be used for astronomical observations, but it is desirable that a terrestrial telescope form an erect image. This is accomplished in the *prism binocular*, of which Fig. 40–23 is a cutaway view, by a pair of 45°–45°–90° totally reflecting prisms inserted between objective and eyepiece. The image is inverted by the four reflections from the inclined faces of the prisms. It is customary to stamp on a flat metal surface of a binocular two numbers separated by a multiplication sign, thus, 7 × 50. The first number is the magnification and the second is the diameter of the objective lenses in millimeters.

40–23 The prism binocular. (Courtesy of Bausch & Lomb Optical Co.)

40–24 The reflecting telescope.

In the *reflecting telescope* the objective lens is replaced by a concave mirror, as shown in Fig. 40–24. In large telescopes this scheme has many advantages, both theoretical and practical. The mirror is intrinsically free of chromatic aberrations, and spherical aberrations are much easier to correct than with a lens. The material need not be transparent, and the reflector can be made more rigid than a lens, which has to be supported only at its edges. The largest reflecting telescope in the world has a mirror over 5 m in diameter.

Because the image is formed in a region traversed by incoming rays, this image can be observed directly with an ocular only by blocking off part of the incoming beam; this is practical only for the very largest telescopes. Alternative schemes use a mirror to reflect the image out the side or through a hole in the mirror, as shown in Figs. 40–24b and 40–24c.

Questions

40-1 Sometimes a wine glass filled with white wine forms an image of an overhead light on a white tablecloth. Would the same image be formed with an empty glass? With a glass of gin? Gasoline?

40-2 How could one very quickly make an approximate measurement of the focal length of a converging lens? Could the same method be applied if we wished to use a diverging lens?

40-3 If you look closely at a shiny Christmas-tree ball, you can see nearly the entire room. Does the room appear right-side-up or upside-down? Discuss your observations in terms of images.

40-4 A student asserted that any lens with spherical surfaces has a positive focal length if it is thicker at the center than at the edge, and negative if it is thicker at the edge. Do you agree?

40-5 The focal length of a simple lens depends on the color (wavelength) of light passing through it. Why? Is it possible for a lens to have a positive focal length for some colors and negative for others?

40-6 The human eye is often compared to a camera. In what ways is it similar to a camera? In what ways does it differ?

40-7 How could one make a lens for sound waves?

40-8 A student proposed to use a plastic bag full of air, immersed in water, as an underwater lens. Is this possible? If the lens is to be a converging lens, what shape should the air pocket have?

40-9 When a converging lens is immersed in water, does its focal length increase or decrease, compared with the value in air?

40-10 You are marooned on a desert island and want to use your eyeglasses to start a fire. Can this be done if you are nearsighted? If you are farsighted?

40-11 While lost in the mountains a person who was nearsighted in one eye and farsighted in the other made a crude emergency telescope from the two lenses of his eyeglasses. How did he do this?

40-12 In using a magnifying glass, is the magnification greater when the glass is close to the object or when it is close to the eye?

40-13 When a slide projector is turned on without a slide in it, and the focus adjustment is moved far enough in one direction, a gigantic image of the lightbulb filament can be seen on the screen. Explain how this happens.

40-14 A spherical air bubble in water can function as a lens. Is it a converging or diverging lens? How is its focal length related to its radius?

40-15 As discussed in the text, some binoculars use prisms to invert the final image. Why are prisms better than ordinary mirrors for this purpose?

40-16 There have been reports of round fishbowls starting fires by focusing the sun's rays coming in a window. Is this possible?

40-17 How does a person judge distance? Can a person with vision in only one eye judge distance? What is meant by "binocular vision?"

40-18 Zoom lenses are widely used in television cameras and conventional photography. Such a lens has, effectively, a variable focal length; changes in focal length are accomplished by moving some lens elements relative to others. Try to devise a scheme to accomplish this effect.

Problems

40-1 A thin-walled glass sphere of radius R is filled with water. An object is placed a distance $3R$ from the surface of the sphere. Determine the position of the final image. The effect of the glass wall may be neglected.

40-2 A transparent rod 40 cm long is cut flat at one end and rounded to a hemispherical surface of 12 cm radius at the other end. An object is placed on the axis of the rod, 10 cm from the hemispherical end.

a) What is the position of the final image?

b) What is its magnification? Assume the refractive index to be 1.50.

40-3 Both ends of a glass rod 10 cm in diameter, of index 1.50, are ground and polished to convex hemispherical surfaces of radius 5 cm at the left end and radius 10 cm at the right end. The length of the rod between vertices is 60 cm.

An arrow 1 mm long, at right angles to the axis and 20 cm to the left of the first vertex, constitutes the object for the first surface.

a) What constitutes the object for the second surface?
b) What is the object distance for the second surface?
c) Is the object real or virtual?
d) What is the position of the image formed by the second surface?
e) What is the height of the final image?

40-4 The same rod as in Problem 40-3 is now shortened to a distance of 10 cm between its vertices, the curvatures of its ends remaining the same.

a) What is the object distance for the second surface?
b) Is the object real or virtual?
c) What is the position of the image formed by the second surface?
d) Is the image real or virtual, erect or inverted, with respect to the original object?
e) What is the height of the final image?

40-5 A glass rod of refractive index 1.50 is ground and polished at both ends to hemispherical surfaces of 5 cm radius. When an object is placed on the axis of the rod, 20 cm from one end, the final image is formed 40 cm from the opposite end. What is the length of the rod?

40-6 A solid glass sphere of radius R and index 1.50 is silvered over one hemisphere, as in Fig. 40-25. A small object is located on the axis of the sphere at a distance $2R$ from the pole of the unsilvered hemisphere. Find the position of the final image after all refractions and reflections have taken place.

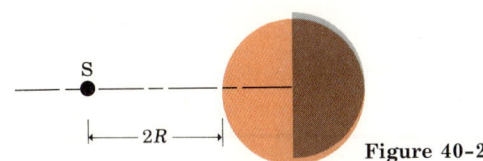

Figure 40-25

40-7 A narrow beam of parallel rays enters a solid glass sphere in a radial direction. At what point outside the sphere are these rays brought to a focus? The radius of the sphere is 3 cm and its index is 1.50.

40-8 A glass plate 2 cm thick, of index 1.50, having plane parallel faces, is held with its faces horizontal and its lower face 8 cm above a printed page. Find the position of the image of the page, formed by rays making a small angle with the normal to the plate.

40-9

a) Show that the equation

$$\frac{1}{s} + \frac{1}{s'} = \frac{1}{f}$$

is that of an equilateral hyperbola having as asymptotes the lines $x = f$ and $y = f$.

b) Construct a graph with object distance s as abscissa, and image distance s' as ordinate for a lens of focal length f, and for object distances from 0 to ∞.
c) On the same set of axes, construct a graph of magnification (ordinate) vs. object distance.

40-10 A converging lens has a focal length of 10 cm. For object distances of 30 cm, 20 cm, 15 cm, and 5 cm determine

a) image position,
b) magnification,
c) whether the image is real or virtual,
d) whether the image is erect or inverted.

40-11 Sketch the various possible thin lenses obtainable by combining two surfaces whose radii of curvature are, in absolute magnitude, 10 cm and 20 cm. Which are converging and which are diverging? Find the focal length of each lens if made of glass of index 1.50.

40-12 The radii of curvature of the surfaces of a thin lens are +10 cm and +30 cm. The index is 1.50.

a) Compute the position and size of the image of an object in the form of an arrow 1 cm high, perpendicular to the lens axis, 40 cm to the left of the lens.
b) A second similar lens is placed 160 cm to the right of the first. Find the position of the final image.
c) Same as (b), except the second lens is 40 cm to the right of the first.
d) Same as (c), except the second lens is diverging, of focal length −40 cm.

40-13 An object is placed 18 cm from a screen.

a) At what points between object and screen may a lens of 4 cm focal length be placed to obtain an image on the screen?
b) What is the magnification of the image for these positions of the lens?

40-14 An object is imaged by a lens on a screen placed 12 cm from the lens. When the lens is moved 2 cm farther from the object, the screen must be moved 2 cm closer to the object to refocus it. What is the focal length of the lens?

40-15 Three thin lenses, each of focal length 20 cm, are aligned on a common axis and adjacent lenses are separated by 30 cm. Find the position of the image of a small object on the axis, 60 cm to the left of the first lens.

40-16 An equiconvex thin lens made of glass of index 1.50 has a focal length in air of 30 cm. The lens is sealed into an opening in one end of a tank filled with water (index = 1.33). At the end of the tank opposite the lens is a plane mirror, 80 cm distant from the lens. Find the position of the image formed by the lens–water–mirror system, of a small object outside the tank on the lens axis and 90 cm to the left of the lens. Is the image real or virtual, erect or inverted?

40-17 A diverging meniscus lens of 1.48 refractive index has concave spherical surfaces whose radii are 2.5 and 4 cm. What would be the position of the image if an object were placed 15 cm in front of the lens?

40-18 Figure 40-26 represents a compound lens consisting of two thin lenses L_1 and L_2, each of focal length 6 cm, separated by a distance of 3 cm.

a) Find the position of the image formed by lens L_1 of a point on the axis at an infinite distance to the left of the lens.

b) Find the position of the image of this image formed by lens L_2. This locates the second focal point F' of the compound lens and, by symmetry, the first focal point F lies at the same distance to the left of L_1.

c) Find the position of the image formed by L_1 of a point on the axis at a distance $x = 2$ cm to the left of the first focal point F.

d) Find the distance x' from the second focal point F' to the image of this image formed by lens L_2.

e) What is the focal length f of the compound lens?

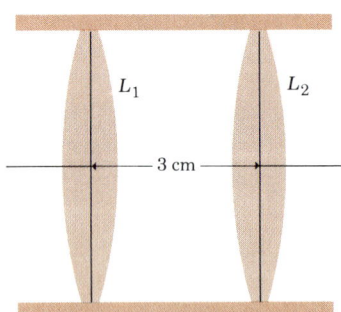

Figure 40-26

40-19 A lens having one convex and one plane surface is 2 cm thick along its axis. The refractive index is 1.50 and the radius of curvature of the convex surface is 10 cm. The convex surface faces toward the left.

a) Find the distance from the first focal point to the vertex of the convex surface.

b) Find the distance from the vertex on the plane surface to the second focal point.

40-20 When an object is placed at the proper distance in front of a converging lens, the image falls on a screen 20 cm from the lens. A diverging lens is now placed halfway between the converging lens and the screen, and it is found that the screen must be moved 20 cm farther away from the lens to obtain a sharp image. What is the focal length of the diverging lens?

40-21

a) Prove that when two thin lenses of focal lengths f_1 and f_2 are placed *in contact,* the focal length f of the combination is given by the relation

$$\frac{1}{f} = \frac{1}{f_1} + \frac{1}{f_2}.$$

b) A converging meniscus lens has an index of refraction of 1.50, and the radii of its surfaces are 5 and 10 cm. The concave surface is placed upward and filled with water. What is the focal length of the water–glass combination?

40-22 Two thin lenses, both of 10 cm focal length, the first converging, the second diverging, are placed 5 cm apart. An object is placed 20 cm in front of the first (converging) lens.

a) How far from this lens will the image be formed?

b) Is the image real or virtual?

40-23 Rays from a lens are converging toward a point image P, as in Fig. 40-27. What thickness t of glass of index 1.50 must be interposed, as in the figure, in order that the image shall be formed at P'?

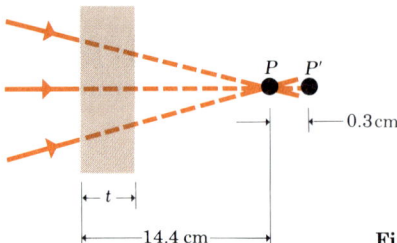

Figure 40-27

40-24 An object 2.4 m in front of a camera lens is sharply imaged on a photographic film 12 cm behind the lens. A glass plate 1 cm thick, of index 1.50, having plane parallel faces, is interposed between lens and plate, as shown in Fig. 40-28.

a) Find the new position of the image.

b) At what distance in front of the lens will an object be in sharp focus on the film with the plate in place, the distance from lens to film remaining 12 cm? Consider the lens as a simple thin lens.

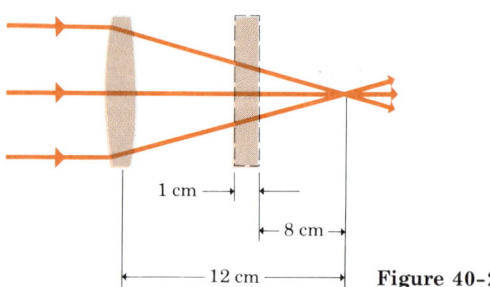

Figure 40-28

40-25 The picture size on ordinary 35-mm camera film is 24 × 36 mm. Focal lengths of lenses available for 35-mm cameras typically include 28 mm, 35 mm, 50 mm (the "standard" lens), 85 mm, 100 mm, 135 mm, 200 mm, and 300 mm, among others. Which of these lenses should be used to photograph the following objects, assuming the object is to fill most of the picture area?

a) A cathedral 100 m high and 150 m long, at a distance of 150 m?

b) An eagle with a wingspan 2.0 m, at a distance of 15 m?

40-26 During a lunar eclipse, a picture of the moon (diameter 3.48×10^6 m, distance from earth 3.8×10^8 m) is taken with a camera whose lens has focal length 50 mm. What is the diameter of the image on the film?

40-27 The *resolution* of a camera lens can be defined as the maximum number of lines per millimeter in the image that can barely be distinguished as separate lines. A certain lens has a focal length of 50 mm and resolution of 100 lines mm^{-1}. What is the minimum separation of two lines in an object 100 m away if they are to be visible in the image as separate lines?

40-28 Show that when two thin lenses are placed in contact, the *power* of the combination in diopters, as defined in Sec. 40-8, is the sum of the powers of the separate lenses. Is this relation valid even when one lens has positive power and the other negative?

40-29 An eyepiece consists of two similar positive thin lenses having focal lengths of 6 cm, separated by a distance of 3 cm. Where are the focal points of the eyepiece?

40-30 When two thin lenses are closely spaced, the power of the combination is the sum of the powers of the individual lenses. Two thin lenses of 25 cm and 40 cm focal lengths are in contact. What is the power of the combination?

40-31 What is the power of the spectacles required (a) by a hyperopic eye whose near point is at 125 cm? (b) by a myopic eye whose far point is at 50 cm?

40-32

a) What spectacles are required for reading purposes by a person whose near point is at 200 cm?

b) The far point of a myopic eye is at 30 cm. What spectacles are required for distant vision?

40-33

a) Where is the near point of an eye for which a spectacle lens of power $+2$ diopters is prescribed?

b) Where is the far point of an eye for which a spectacle lens of power -0.5 diopter is prescribed for distant vision?

40-34 A thin lens of focal length 10 cm is used as a simple magnifier.

a) What angular magnification is obtainable with the lens?

b) When an object is examined through the lens, how close may it be brought to the eye?

40-35 The focal length of a simple magnifier is 10 cm.

a) How far in front of the magnifier should an object to be examined be placed if the image is formed at the observer's near point, 25 cm in front of his eye?

b) If the object is 1 mm high, what is the height of its image formed by the magnifier? Assume the magnifier to be a thin lens.

40-36 A camera lens is focused on a distant point source of light, the image forming on a screen at a (Fig. 40-29). When the screen is moved backward a distance of 2 cm to b, the circle of light on the screen has a diameter of 4 mm. What is the f/number of the lens?

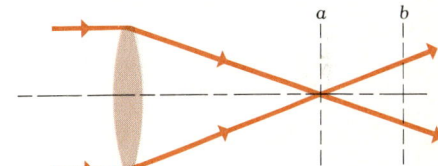

Figure 40-29

40-37 Camera A, having an $f/8$ lens 2.5 cm in diameter, photographs an object using the correct exposure of $\frac{1}{100}$ s. What exposure should camera B use in photographing the same object if it has an $f/4$ lens 5 cm in diameter?

40-38 The focal length of an $f/2.8$ camera lens is 8 cm.

a) What is the diameter of the lens?

b) If the correct exposure of a certain scene is $\frac{1}{200}$ s at $f/2.8$, what would be the correct exposure at $f/5.6$?

40-39 The dimensions of the picture on a 35-mm color slide are 24 mm $\times$ 36 mm. It is desired to project an image of the slide, enlarged to 2 m $\times$ 3 m, on a screen 10 m from the projection lens.

a) What should be the focal length of the projection lens?

b) Where should the slide be placed?

40-40 The image formed by a microscope objective of focal length 4 mm is 180 mm from its second focal point. The eyepiece has a focal length of 31.25 mm.

a) What is the magnification of the microscope?

b) The unaided eye can distinguish two points as separate if they are about 0.1 mm apart. What is the minimum separation, using this microscope?

40-41 A certain microscope is provided with objectives of focal lengths 16 mm, 4 mm, and 1.9 mm, and with eyepieces of angular magnification 5X and 10X. What is (a) the largest, and (b) the least overall magnification obtainable? Each objective forms an image 160 mm beyond its second focal point.

40-42 The focal length of the eyepiece of a certain microscope is 2.5 cm. The focal length of the objective is 16 mm. The distance between objective and eyepiece is 22.1 cm. The final image formed by the eyepiece is at infinity. Treat all lenses as thin.

a) What should be the distance from the objective to the object viewed?

b) What is the linear magnification produced by the objective?

c) What is the overall magnification of the microscope?

40-43 A microscope with an objective of focal length 9 mm and an eyepiece of focal length 5 cm is used to project an image on a screen 1 m from the eyepiece. What is the lateral magnification of the image? Let the image distance of the objective be 18 cm.

40-44 The moon subtends an angle at the earth of approximately $\frac{1}{2}°$. What is the diameter of the image of the moon produced by the objective of the Lick Observatory telescope, a refractor having a focal length of 18 m?

40-45 The eyepiece of a telescope has a focal length of 10 cm. The distance between objective and eyepiece is 2.1 m. What is the angular magnification of the telescope?

40-46 A crude telescope is constructed of two spectacle lenses of focal lengths 100 cm and 20 cm, respectively.

a) Find its angular magnification.

b) Find the height of the image formed by the objective of a building 80 m high and distant 2 km.

40-47 Figure 40-30 is a diagram of a *Galilean telescope,* or *opera glass,* with both the object and its final image at infinity. The image I serves as a virtual object for the eyepiece. The final image is virtual and erect. Prove that the angular magnification $M = -f_1/f_2$.

40-48 A Galilean telescope is to be constructed, using the same objective as in Problem 40-46.

a) What type lens should be used as an eyepiece and what focal length should it have, if the telescopes are to have the same magnification?

b) Compare the lengths of the telescopes.

40-49 A reflecting telescope is made using a mirror of radius of curvature 0.50 m and an eyepiece of focal length 1.0 cm. What is the angular magnification? What should be the position of the eyepiece if both the object and the final image are at infinity?

40-50 A certain reflecting telescope has a mirror 10 cm in diameter, with radius of curvature 1.0 m, and an eyepiece of focal length 1.0 cm. If the angular magnification is 48 and the object is at infinity, find the position of the lens and the position and nature of the final image.

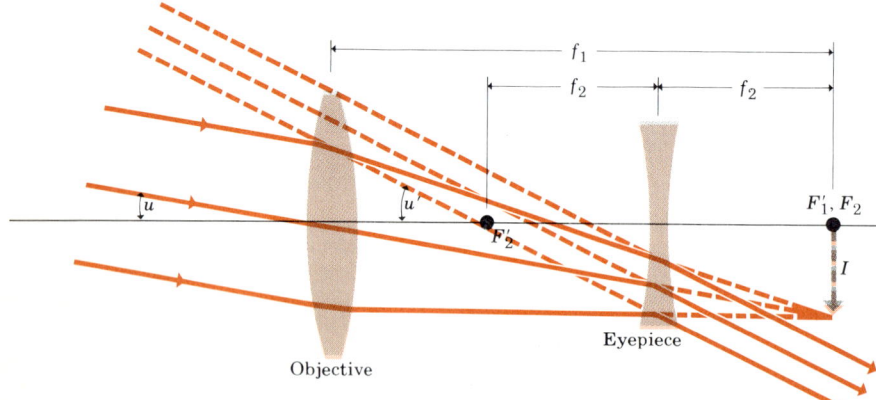

Figure 40-30

INTERFERENCE AND DIFFRACTION 41

41-1 Interference and coherent sources

In analyzing the formation of images by lenses and mirrors, we have represented light as *rays* which travel in straight lines in a homogeneous medium and which are deviated in accordance with simple laws at a reflecting surface or an interface between two optical media. This simple model forms the basis of *geometrical optics,* which, as we have seen, is adequate for understanding a wide variety of phenomena involving lenses and mirrors.

In this chapter we shall discuss the phenomena of *interference* and *diffraction,* for whose understanding the principles of geometrical optics *do not* suffice. Instead, we must return to the more fundamental point of view that light is a *wave motion,* and that the total effect of a number of waves arriving at one point depends on the *phases* of the waves as well as upon their amplitudes. This part of the subject is called *physical optics*.

In our discussions of mechanical waves in Chapter 21 and electromagnetic waves in Chapter 37, we have often considered *sinusoidal* waves having a single frequency and a single wavelength. Such a wave is called a *monochromatic* (single-color) wave. Common sources of light, such as an incandescent lightbulb or a flame, *do not* emit monochromatic light but rather a continuous distribution of wavelengths. Indeed, a strictly monochromatic light wave is an unattainable idealization, like a frictionless plane, a point mass, an inductor with zero resistance, and many other idealizations in physics. Nevertheless, it remains a useful concept for the analysis of phenomena of light interference.

Monochromatic light can be *approximated* in the laboratory. Continuous-spectrum light can be passed through a filter that blocks all but a narrow range of wavelengths. Gas-discharge lamps, such as the mercury-arc lamp, emit line spectra in which the light consists of a discrete set of colors, each having a narrow band of wavelengths called a *spectrum line*. For example, the bright green line in the mercury spectrum has an average wavelength of 546.1 nm, with a spread of wavelength of the order of ± 0.001 nm, depending on the pressure and temperature of the mercury vapor in the lamp. By far the most nearly monochromatic source available at present is the *laser,* to be discussed in Chapter 44. The familiar helium–neon laser, inexpensive and readily available, emits visible light at 632.8 nm with a line width (wavelength range) of the order of ±0.000001 nm, or about one part in 10^9. Laser light also has much greater *coherence* (to be discussed later) than ordinary light.

The term *interference* refers to any situation in which two or more waves overlap in space. This term was introduced in Sec. 22-2 in connection with standing waves on a stretched string, formed by the superposition of two sinusoidal waves traveling in opposite directions. In such cases, the total displacement at any point at any instant of time is governed by the *principle of linear superposition,* introduced in Sec. 22-2. This principle, the most important in all of physical optics, states that *when two or more waves overlap, the resultant displacement at any point and at any instant may be found by adding the instantaneous displacements that would be produced at the point by the individual waves if each were present alone*. The term "displacement," as used here, is a general one. If one is considering surface ripples on a liquid, the displacement means the actual displacement of the surface above or below its normal level. If the waves are sound waves, the term refers to the excess or deficiency of pressure. If the waves are electromagnetic, the displacement means the magnitude of the electric or magnetic field. When light of extremely high intensity passes through matter, the principle of linear superposition is *not* precisely obeyed, and the resulting phenomena are classified under the heading *nonlinear optics*. (These effects are beyond the scope of this book.)

To introduce the essential ideas of interference, we consider first the problem of two identical sources of monochromatic waves, S_1 and S_2, separated in space by a certain distance. The two sources are permanently *in phase,* so that at every point in space there is a definite and unchanging phase relation between waves from the two sources. They might be, for example, two loudspeakers driven by the same amplifier, or two radio antennas powered by the same transmitter, or two small apertures in an opaque screen, illuminated by the same monochromatic light source.

We locate the sources S_1 and S_2 along the y-axis and equidistant from the origin, as shown in Fig. 41-1. Let P_0 be any point on the x-axis. From symmetry, the two distances S_1P_0 and S_2P_0 are equal; waves from the two sources thus require equal times to travel to P_0, and having left S_1 and S_2 in phase, they arrive at P_0 in phase. The total amplitude at P_0 is thus twice the amplitude of each individual wave. Next, we consider a point P_1, located so that its distance from S_2 is exactly one wavelength greater than its distance from S_1. That is,

$$S_2P_1 - S_1P_1 = \lambda.$$

Then any given wave crest from S_1 arrives at P_1 exactly one cycle earlier than the crest emitted at the same time from S_2, and again the two waves arrive in phase. Similarly, waves arrive in phase at all points P_2 for which the path difference is two wavelengths ($S_2P_2 - S_1P_2 = 2\lambda$), or indeed for *any* positive or negative integer number of wavelengths.

The addition of amplitudes that results when waves from two or more sources arrive at a point in a phase is often called *constructive interference* or *reinforcement,* and the above discussion shows that con-

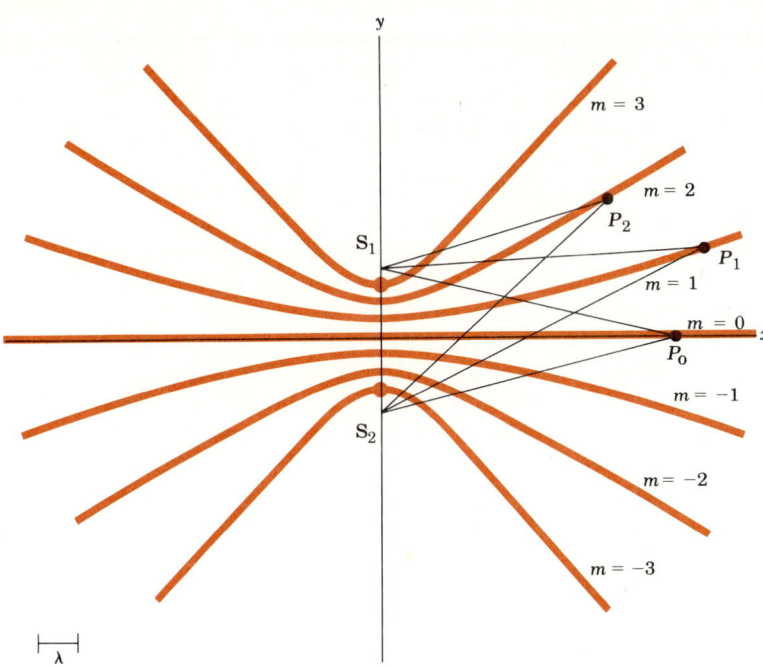

41-1 Curves of maximum intensity in the interference pattern of two monochromatic point sources. In this example the distance between sources is four times the wavelength.

structive interference occurs whenever the path difference for the two sources is an integer multiple of the wavelength:

$$S_2P - S_1P = m\lambda \qquad (m = 0, \pm 1, \pm 2, \pm 3, \ldots). \qquad (41\text{-}1)$$

In our example, the locus of points satisfying this condition is a set of hyperbolas, as shown in Fig. 41-1. Intermediate between these lines is a set of other lines for which the path difference for the two sources is a half-integer number of wavelengths. Waves from the two sources arrive at a point on one of these lines exactly a half-cycle out of phase, and the resultant amplitude is the *difference* of the two individual amplitudes. If the amplitudes are equal, which is approximately the case when the distance from either source to P is much greater than the distance between sources, then the total amplitude at such a point is zero! This condition is called *destructive interference* or *cancellation.* In our example, the condition for destructive interference is

$$S_2P - S_1P = (m + \tfrac{1}{2})\lambda \qquad (m = 0, \pm 1, \pm 2, \pm 3, \ldots). \qquad (41\text{-}2)$$

Thus far we have considered only points in the xy-plane, but it should be clear that the discussion can be extended to include other points in the space surrounding the two sources. If the lines in Fig. 41-1 are rotated about the y-axis, they trace out surfaces called *hyperboloids of revolution,* and every point on such a surface is a point of maximum intensity of radiation. Similarly, the intermediate lines of zero intensity generate surfaces on which the intensity is zero everywhere.

In the above discussion the constant *phase* relationship between the sources is an essential requirement. If the relative phase of the sources changes, the positions of the maxima and minima in the resulting interference pattern also change. When the radiation is light, it is possible for the two sources to have a definite and constant phase relation *only when they both emit light coming from a single primary source;* it is *not* possible with two separate sources. The reason is a fundamental one associated with the mechanism of light emission.

In ordinary light sources, atoms of the material of the source are given excess energy by thermal agitation or impact with accelerated electrons. An atom thus "excited" begins to radiate and continues until it has lost all the energy it can, typically in a time of the order of 10^{-8} s. A source ordinarily contains a very large number of atoms, which radiate in an unsynchronized and random phase relationship. Thus emission from two such sources has a rapidly varying phase relation; the result is a constantly changing interference pattern which, with ordinary observations, does not reveal a visible interference pattern at all.

However, if the light from a single source is split so that parts of it emerge from two or more regions of space, forming two or more *secondary sources*, any random phase change in the source affects these secondary sources equally and does not change their *relative* phase. Two such sources derived from a single source and having a definite phase relation are said to be *coherent*.

The distinguishing feature of light from a *laser* is that the emission of light from many atoms is *synchronized* in frequency and phase, by mechanisms to be discussed in Chapter 44. As a result, the random phase changes mentioned above occur *much* less frequently. Definite phase relations are preserved over correspondingly much greater lengths in the beam. Accordingly, laser light is said to be much more *coherent* than ordinary light.

41-2 Young's experiment and Pohl's experiment

One of the earliest demonstrations of the fact that light can produce interference effects was performed in 1800 by the English scientist Thomas Young. The experiment was a crucial one at the time, since it added further evidence to the growing belief in the wave nature of light. A corpuscular theory was quite inadequate to account for the effects observed.

Young's apparatus is shown in Fig. 41-2a. Monochromatic light issuing from a narrow slit S_0 is divided into two parts by falling upon a screen in which are cut two other narrow slits S_1 and S_2, very close together. The dimensions in this figure are distorted for clarity. The distance from the source slit S_0 to the screen containing S_1 and S_2 is 20 cm to 100 cm. The distance from the double-slit screen to the final screen is usually from 1 m to 5 m. The slits are 0.1 mm to 0.2 mm wide

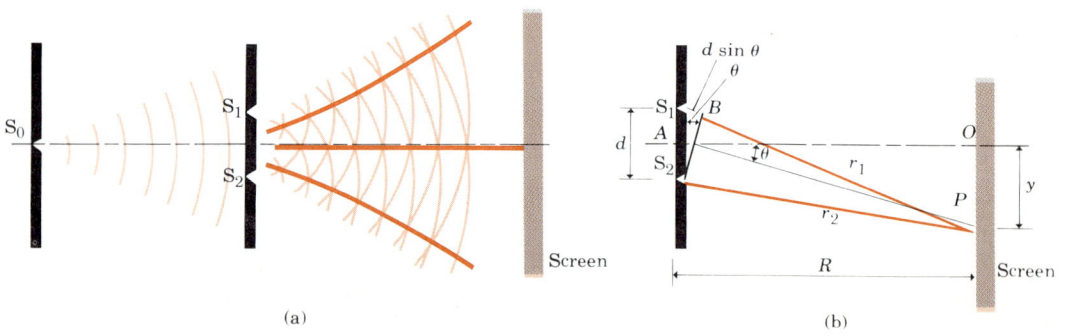

41-2 (a) Interference of light waves passing through two slits. (b) Young's experiment.

and the separation of the slits S_1 and S_2 is less than 1 mm. In short, all slit widths and slit separations are fractions of millimeters; all other distances are hundreds or thousands of millimeters.

According to Huygens' principle (Sec. 38-8), cylindrical wavelets spread out from slit S_0 and reach slits S_1 and S_2 in phase, because they travel equal distances from S_0. A succession of Huygens wavelets diverges from each slit; and the two sets of wavelets leave in phase, and therefore act as *coherent* sources. But they do not necessarily arrive at point P in phase because of the path difference $(r_1 - r_2)$ for the two waves.

When the distance R is very large compared to the distance d between slits, as is usually the case, the path difference can be expressed simply in terms of the angle θ. With P as a center and PS_2 as radius, we draw an arc which intersects line PS_1 at point B. The path difference is then the distance S_1B. When R is much larger than d, the arc S_2B is very nearly a straight line, and it is perpendicular to PS_2, PA, and PS_1. Then triangle BS_1S_2 is a right triangle, similar to POA, and the distance S_1B is equal to $d \sin \theta$. Thus the path difference is given by

$$r_1 - r_2 = d \sin \theta. \tag{41-3}$$

According to the principles discussed in the preceding section, complete reinforcement occurs at point P (that is, P lies at the center of a bright fringe) only when the path difference $d \sin \theta$ is some integral number of wavelengths, say $m\lambda$ ($m = 0, 1, 2, 3$, etc.). Thus,

$$d \sin \theta = m\lambda \quad \text{or} \quad \sin \theta = \frac{m\lambda}{d}. \tag{41-4}$$

Now λ is of the order of 5×10^{-5} cm, while d cannot be made much smaller than about 10^{-2} cm. As a rule, only the first five to ten fringes are bright enough to be seen, so that m is at most, say, 10. Therefore the very largest value of $\sin \theta$ is

$$\sin \theta \text{ (maximum)} = \frac{(10)(5 \times 10^{-5} \text{ cm})}{10^{-2} \text{ cm}} = 0.05,$$

which corresponds to an angle of only 3°. Figure 41-3 shows a typical pattern.

The central bright fringe at point O, or zeroth fringe ($m = 0$), corresponds to zero path difference, or $\sin \theta = 0$. If point P is at the center of the mth fringe, the distance y_m from the zeroth to the mth fringe is, from Fig. 41-2b,

$$y_m = R \tan \theta_m.$$

Zeroth fringe

41-3 Interference fringes produced by Young's double-slit interferometer.

If, as is often the case in optical interference experiments, y is much smaller than R, then the angle θ_m for all values of m is extremely small. In that case, $\tan\theta_m \approx \sin\theta_m$ and

$$y_m = R\sin\theta_m.$$

Therefore,

$$y_m = R\frac{m\lambda}{d},$$

and

$$\lambda = \frac{y_m d}{mR}. \tag{41-5}$$

Hence, by measuring the distance d between the slits, the distance R to the screen, and the distance y_m from the center of the zeroth fringe to the center of the mth fringe on either side, one may compute the wavelength of the light producing the interference pattern. Such an experiment, first performed by Thomas Young in 1800, provided the first direct measurement of wavelengths of light, and of course also gave very strong confirmation of the wave nature of light.

Example With two slits spaced 0.2 mm apart, and a screen at a distance of 1 m, the third bright fringe is found to be displaced 7.5 mm from the central fringe. Find the wavelength of the light used.

Solution Let λ be the unknown wavelength. Then

$$\lambda = \frac{y_m d}{mR} = \frac{(0.75 \text{ cm})(0.02 \text{ cm})}{(3)(100 \text{ cm})} = 5 \times 10^{-5} \text{ cm}$$
$$= 500 \times 10^{-9} \text{ m} = 500 \text{ nm}.$$

Circular interference fringes may be produced very easily with the aid of a simple apparatus suggested by Robert Pohl and shown in Fig. 41-4. A small arc lamp S_0 is placed a few centimeters away from a sheet of mica of thickness about 0.05 mm. Some light is reflected from the first surface, as though it were issuing from the virtual image S_1. An approximately equal amount of light is reflected from the back surface, as though it were coming from the virtual image S_2. The circular interference fringes formed by the light issuing from these two mutually coherent sources may be shown on the entire wall of a room, as shown in Fig. 41-4b.

*41-3 Intensity distribution in interference fringes

In Sec. 41-2 we computed the positions of the maxima (bright fringes) in the two-slit interference pattern; we may also compute the intensity at *any* point in the pattern. In Fig. 41-2b, each source produces a sinusoidally varying disturbance at point P; if the sources are in phase, the waves arriving at P differ in phase by an amount proportional to the path difference $(r_1 - r_2)$. If the path difference is one wavelength, the phase difference is $\delta = 2\pi$ (or 360°). If $(r_1 - r_2)/\lambda = \frac{1}{2}$, the phase difference is π, and so on. The general relation is that a path difference $r_1 - r_2$

41-4 (a) Pohl's mica-sheet interferometer. (b) Circular interference fringes produced by Pohl's interferometer. The dark rectangle is the shadow of the mercury arc housing.

yields a phase difference δ given by

$$\delta = \frac{2\pi}{\lambda}(r_1 - r_2) = k(r_1 - r_2), \qquad (41\text{-}6)$$

where $k = 2\pi/\lambda$ is the *wave number* introduced in Sec. 21-3.

If the medium in the space between the sources and P is other than vacuum, the wavelength *in the medium* must be used in Eq. (41-5). If the medium has refractive index n, then

$$\lambda = \frac{\lambda_0}{n} \quad \text{and} \quad k = nk_0, \qquad (41\text{-}7)$$

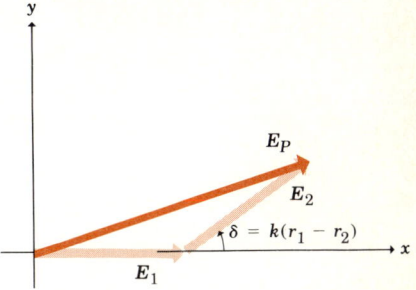

41-5 Phasor diagram of the variations of the electric vector at a point P where two waves meet.

where λ_0 and k_0 are the values of λ and k, respectively, in vacuum.

The total wave arriving at P is the superposition of two waves from slits S_1 and S_2, having amplitudes E_1 and E_2, respectively, and with a phase difference δ resulting from their path difference. To find the amplitude of the resulting wave, we represent each wave at P as a rotating vector or *phasor*, just as in Chapters 11 and 36. The appropriate phasor diagram is shown in Fig. 41-5, where the amplitude of the resultant is labeled E_P. Applying the law of cosines, we find

$$E_P{}^2 = E_1{}^2 + E_2{}^2 + 2E_1 E_2 \cos \delta.$$

When the two coherent sources S_1 and S_2 are equally intense, $E_1 = E_2 = E$, and

$$E_P{}^2 = 2E^2 + 2E^2 \cos \delta = 2E^2(1 + \cos \delta)$$
$$= 4E^2 \cos^2 \frac{\delta}{2}. \qquad (41\text{-}8)$$

When the two waves are in phase, $\delta = 0$ and $E_P = 2E$. When they are exactly a half-cycle out of phase, $\delta = \pi$ (180°), $\delta/2 = \pi/2$, $\cos^2(\delta/2) = 0$, and $E = 0$. Thus the superposition of two sinusoidal waves with the same amplitude and frequency but a phase difference yields a sinusoidal wave with amplitude between zero and twice the individual amplitudes, depending on the phase difference.

Now the analysis of electromagnetic waves in Chapter 37, particularly Sec. 37-3, has shown that the *intensity* of an electromagnetic wave is proportional to the *square* of the amplitude of the wave. Thus the intensity is proportional to the final expression in Eq. (41-8). This shows, incidentally, that at points of maximum constructive interference the intensity is *four times* (not twice) as great as it would be from either source alone.

The phase angle δ may be expressed in terms of geometrical quantities in Fig. 41-2 by use of Eqs. (41-3) and (41-6). The path difference is given by

$$r_1 - r_2 = d \sin \theta,$$

provided R is much larger than d. From Eq. (41-6), the phase difference is

$$\delta = k(r_1 - r_2) = kd \sin \theta = \frac{2\pi d}{\lambda} \sin \theta.$$

When this is substituted into Eq. (41-8), we obtain

$$E_P{}^2 = 4E^2 \cos^2(\tfrac{1}{2}kd \sin \theta) = 4E^2 \cos^2\left(\frac{\pi d}{\lambda} \sin \theta\right).$$

Thus the ratio of intensity I at any angle θ to the intensity I_0 at $\theta = 0$ (where $\delta = 0$) is given by

$$\frac{I}{I_0} = \frac{4E^2 \cos^2(\tfrac{1}{2}kd \sin \theta)}{4E^2 \cos^2(\tfrac{1}{2}kd \sin 0)} = \cos^2(\tfrac{1}{2}kd \sin \theta),$$

and finally

$$I = I_0 \cos^2(\tfrac{1}{2}kd \sin \theta) = I_0 \cos^2\left(\frac{\pi d}{\lambda}\sin \theta\right). \qquad (41\text{-}9)$$

In cases where the further approximation $y/R = \sin \theta$ is justified, i.e., when $y \ll R$ and θ is small, we obtain the simpler expressions

$$I = I_0 \cos^2\left(\frac{kdy}{2R}\right) = I_0 \cos^2\left(\frac{\pi dy}{\lambda R}\right), \qquad (41\text{-}10)$$

which enable us to calculate the intensity of light at any point and to reproduce the pattern of Fig. 41-3.

Example In Fig. 41-1, suppose the two sources are identical radio antennas 10 m apart, radiating waves in all directions with a frequency of $f = 30$ MHz. If the intensity in the $+x$-direction (corresponding to $\theta = 0$ in Fig. 41-2) is $I_0 = 0.02$ W·m^{-2}, what is the intensity in the direction $\theta = 45°$? In what direction is the intensity zero?

Solution We want to use Eq. (41-9); the approximate relation of Eq. (41-10) cannot be used in this case because θ is not small. First we must find the wavelength, using the relation $c = \lambda f$. We find

$$\lambda = \frac{c}{f} = \frac{3.0 \times 10^8 \text{ m·s}^{-1}}{30 \times 10^6 \text{ s}^{-1}} = 10 \text{ m}.$$

The spacing between sources is $d = 10$ m, and Eq. (41-9) becomes

$$I = (0.02 \text{ W·m}^{-2}) \cos^2\left[\frac{\pi (10 \text{ m})}{(10 \text{ m})}\sin \theta\right]$$
$$= (0.02 \text{ W·m}^{-2}) \cos^2(\pi \sin \theta).$$

When $\theta = 45°$,

$$I = (0.02 \text{ W·m}^{-2}) \cos^2(\pi \sin 45°) = 0.0073 \text{ W·m}^{-2}.$$

This is about 37% of the intensity at $\theta = 0$.

The intensity is zero when $\cos(\pi \sin \theta) = 0$; this occurs when $\pi \sin \theta = \pi/2$, $\sin \theta = \tfrac{1}{2}$, and $\theta = 30°$. By symmetry, the intensity is also zero when $\theta = 150°, 210°,$ and $330°$ (or $-150°, -30°$).

41-4 Interference in thin films. Newton's rings

The brilliant colors that are often seen when light is reflected from a soap bubble or from a thin layer of oil floating on water are produced by interference effects between the two light waves reflected at opposite surfaces of the thin films of soap solution or of oil. In Fig. 41-6, the line ab is one ray in a beam of monochromatic light incident on the upper surface of a thin film. A part of the incident light is reflected at the first surface, as indicated by ray bc, and a part, represented by bd, is transmitted. At the second surface a part is again reflected, and, of this, a part

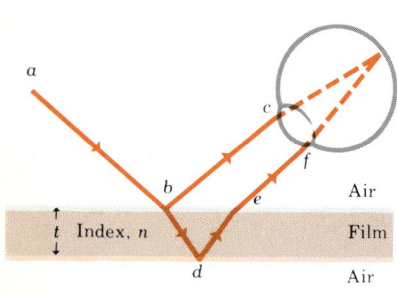

41-6 Interference between rays reflected from the upper and lower surface of a thin film.

emerges as represented by ray *ef*. The rays *bc* and *ef* come together at a point on the retina of the eye. Depending on the phase relationship, they may interfere constructively or destructively. Because different colors have different wavelengths, the interference may be constructive for some colors and destructive for others; hence the appearance of colored rings or fringes.

To discuss these phenomena in as simple a context as possible, let us consider interference of *monochromatic* light reflected from two nearly parallel surfaces. Figure 41–7 shows two plates of glass separated by a wedge of air; we want to consider interference between the two light waves reflected from the surfaces adjacent to the air wedge, as shown. If the observer is at a great distance compared to the other dimensions of the experiment and the rays are nearly perpendicular to the reflecting surfaces, then the path difference between the two waves (corresponding to $r_1 - r_2$ in the preceding section) is just twice the thickness d of the air wedge at each point. At points for which this path difference is an integer number of wavelengths, we expect to see constructive interference and a bright area, and where it is a half-integer number of wavelengths, destructive interference and a dark area. Along the line where the plates are in contact there is *no* path difference and we expect a bright area.

41–7 Interference between two light waves reflected from the two sides of an air wedge separating two glass plates. The path difference is $2d$.

It is easy enough to carry out this experiment; the bright and dark fringes appear as expected, but they are interchanged! Along the line of contact a *dark* fringe, not a bright one, is found. What happened? The inescapable conclusion is that one of the waves has undergone a half-cycle phase shift during its reflection, so that the two reflected waves are a half-cycle out of phase even though they have the same path length.

This phase change can be demonstrated directly with an arrangement called Lloyd's mirror, shown in Fig. 41–8. In this arrangement, the two coherent sources are the actual source slit S_0 and its virtual image S_1. The fringes formed by interference between the light waves from these coherent sources may be viewed on a ground-glass screen placed anywhere beyond the mirror. If, instead, an eyepiece of high magnification is used to view the fringes that form in space in a plane passing through the edge B, the fringe nearest this edge is seen to be black. This is the fringe corresponding to zero path difference, and if the two waves giving rise to this fringe had both traveled in air or had both been reflected from glass, this fringe would have been bright. The zeroth fringe, however, was formed by two waves, of which one had undergone reflection from the glass and one had proceeded directly from S_1. The fact that the zeroth fringe is black indicates that *the waves reflected from glass have undergone a phase shift of* 180°. In other words, the wave has gained (or lost) half a cycle in phase during the process of reflection.

Further experiments show that a half-cycle phase change occurs whenever the material in which the wave is initially traveling before reflection has a smaller refractive index than the second material forming the interface, such as the glass in Fig. 41–8. But when the first material has *greater* refractive index than the second, such as a wave in glass

41–8 Lloyd's mirror. When the eyepiece is focused on the edge B of the glass block, the fringe nearest the edge is black.

reflected internally at a glass–air interface, there is found to be *no* phase change. Thus, in Fig. 41-6 the waves reflected at points *b* and *e* undergo the half-cycle phase shift, while that reflected at *d* does not.

Example Suppose the two glass plates in Fig. 41-7 are two microscope slides 10 cm long; at one end they are in contact, while at the other end they are separated by a thin piece of tissue paper 0.02 mm thick. What is the spacing of the resulting interference fringes? Is the fringe adjacent to the line of contact bright or dark? Assume monochromatic light with $\lambda = 500$ nm.

Solution To answer the second question first, the fringe at the line of contact is dark because the wave reflected from the lower surface of the air wedge has undergone a half-cycle phase shift, while that from the upper surface has not. For this reason, the condition for *destructive* interference (a dark fringe) is that the path difference ($2d$) be an integer number of wavelengths:

$$2d = m\lambda \quad (m = 0, 1, 2, 3, \ldots). \tag{41-11}$$

From similar triangles, d is proportional to the distance x from the line of contact:

$$\frac{d}{x} = \frac{h}{l}.$$

Combining this with Eq. (41-11), we find

$$\frac{2xh}{l} = m\lambda,$$

or

$$x = m\frac{l\lambda}{2h} = m\frac{(0.1 \text{ m})(500 \times 10^{-9} \text{ m})}{(2)(0.02 \times 10^{-3} \text{ m})} = m(1.25 \text{ mm}).$$

Thus successive dark fringes, corresponding to successive integer values of m, are spaced 1.25 mm apart.

In this example, if the space between plates contains water ($n = 1.33$) instead of air, the phase changes are the same but the wavelength is $\lambda = \lambda_0/n = 376$ nm, and the fringe spacing is 0.94 mm. But now suppose the top plate is a glass with $n = 1.4$, the wedge is filled with a silicone grease having $n = 1.5$, and the bottom plate has $n = 1.6$. In this case there are half-cycle phase shifts at *both* surfaces bounding the wedge, and the line of contact corresponds to a *bright* fringe, not a dark one. The fringe spacing is again obtained using the wavelength in the wedge, $\lambda = 500$ nm$/1.5 = 333$ nm, and is found to be 0.83 nm.

If the arrangement in Fig. 41-7 is illuminated first by blue, then by red light, the spacing of the red fringes is greater than that of the blue, as is to be expected from the greater wavelength of the red light. The fringes produced by intermediate wavelengths occupy intermediate positions. If it is illuminated by white light, the color at any point is that due to the mixture of those colors that may be reflected at that point, while the colors for which the thickness is such as to result in destructive interference are absent. Just those colors that are absent in the reflected light, however, are found to predominate in the transmitted light. At any point, the color of the wedge by reflected light is *complementary* to its color by transmitted light!

If the convex surface of a lens is placed in contact with a plane glass plate, as in Fig. 41-9, a thin film of air is formed between the two surfaces. The thickness of this film is very small at the point of contact, gradually increasing as one proceeds outward. The loci of points of equal thickness are circles concentric with the point of contact. Such a film is found to exhibit interference colors, produced in the same way as the colors in a thin soap film. The interference bands are circular, concentric with the point of contact. When viewed by reflected light, the center of the pattern is black, as is a thin soap film. Note that in this case there is no phase reversal of the light reflected from the upper surface of the film (which here is of smaller index than that of the medium in which the light is traveling before reflection), but the phase of the wave reflected from the lower surface is reversed. When viewed by transmitted light, the center of the pattern is bright. If white light is used, the color of the light reflected from the film at any point is complementary to the color transmitted.

These interference fringes were studied by Newton, and are called *Newton's rings*. Figure 41-10 is a photograph of Newton's rings formed by the air film between a convex and a plane surface.

The surface of an optical part that is being ground to some desired curvature may be compared with that of another surface, known to be correct, by bringing the two in contact and observing the interference fringes. Figure 41-11 is a photograph made at one stage of the process of manufacturing a telescope objective. The lower, larger-diameter, thicker disk is the master. The smaller upper disk is the objective under test. The "contour lines" are Newton's interference fringes, and each one indicates an additional departure of the specimen from the master of $\frac{1}{2}$ wavelength of light. That is, at 10 lines from the center spot the space between the specimen and master is 5 wavelengths, or about 0.0001 inch. This specimen is very poor; high-quality lenses are routinely ground with a precision of less than a wavelength.

41-5 Thin coatings on glass

The phenomenon of interference is utilized in nonreflective coatings for glass. A thin layer or film of hard transparent material with an index of refraction smaller than that of the glass is deposited on the surface of the glass, as in Fig. 41-12. If the coating has the proper index of refraction, equal quantities of light will be reflected from its outer surface and from the boundary surface between it and the glass. Furthermore, since in both reflections the light is reflected from a medium of greater index than that in which it is traveling, the same phase change occurs in each reflection. It follows that if the film thickness is $\frac{1}{4}$ wavelength *in the film* (normal incidence is assumed), the light reflected from the first surface will be 180° out of phase with that reflected from the second, and complete destructive interference will result.

The thickness can, of course, be $\frac{1}{4}$ wavelength for only one particular wavelength. This is usually chosen in the yellow-green portion of the spectrum (about 550 mm) where the eye is most sensitive. Some reflec-

41-9 Air film between a convex and a plane surface.

41-10 Newton's rings formed by interference in the air film between a convex and a plane surface. (Courtesy of Bausch & Lomb Optical Co.)

41-11 The surface of a telescope objective under inspection during manufacture. (Courtesy of Bausch & Lomb Optical Co.)

41-12 Destructive interference results when the film thickness is one-quarter of the wavelength in the film.

tion then takes place at both longer and shorter wavelengths, and the reflected light has a purple hue. The overall reflection from a lens or prism surface can be reduced in this way from 4 or 5% to a fraction of 1%. The treatment is extremely effective in eliminating stray reflected light and increasing the contrast in an image formed by highly corrected lenses having a large number of air-glass surfaces. A commonly used material is magnesium fluoride, MgF_2, with an index of 1.38. With this coating, the wavelength of green light in the coating is

$$\lambda = \frac{\lambda_0}{n} = \frac{550 \times 10^{-9} \text{ m}}{1.38} = 4 \times 10^{-5} \text{ cm},$$

and the thickness of a "nonreflecting" film of MgF_2 is 10^{-5} cm.

If a material whose index of refraction is *greater* than that of glass is deposited on glass to a thickness of $\frac{1}{4}$ wavelength, then the reflectivity is *increased*. For example, a coating of index 2.5 will allow 38% of the incident energy to be reflected, instead of the usual 4% when there is no coating. With the aid of multiple coatings it is possible to achieve reflectivity for a particular wavelength of almost 100%.

*41-6 The Michelson interferometer

Young's double slit and Lloyd's mirror are examples of optical interferometers in which a wavefront from a very narrow source slit is subdivided into two wavefronts by reflecting or transmitting regions of the interferometer. Some interferometers can be used in conjunction with a large extended source. Of these, the Michelson interferometer has been most important in the past and is still of some significance.

Figure 41-13 is a diagram of the principal features of the Michelson interferometer. The figure shows the path of one ray from a point A of an extended source. Light from the source strikes a glass plate C, the right side of which has a thin coating of silver. Part of the light is reflected from the silvered surface at point P to the mirror M_2 and back through C to the observer's eye. The remainder of the light passes through the silvered surface and the compensator plate D, and is re-

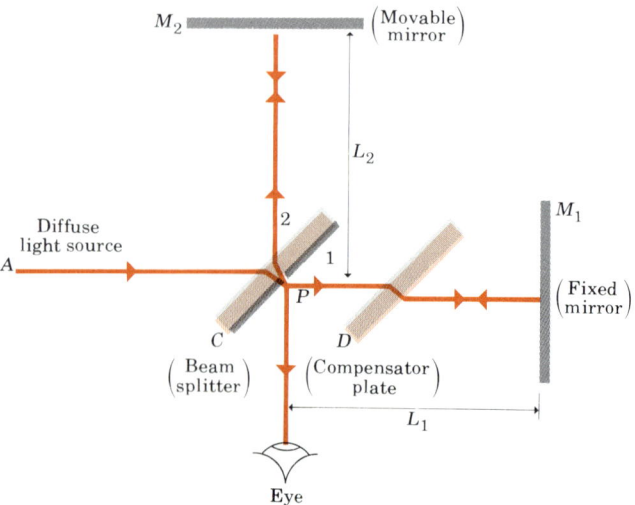

41-13 The Michelson interferometer.

flected from mirror M_1. It then returns through D and is reflected from the silvered surface of C to the observer. The compensator plate D is cut from the same piece of glass as plate C, so that its thickness will not differ from that of C by more than a fraction of a wavelength. Its purpose is to ensure that rays 1 and 2 pass through the same thickness of glass. Plate C is called a *beam splitter*.

The whole apparatus is mounted on a heavy rigid frame and a fine, very accurate screw thread is used to move the mirror M_2. A common commercial model of the interferometer is shown in Fig. 41-14. The source is placed to the left, and the observer is directly in front of the handle that turns the screw.

If the distances L_1 and L_2 in Fig. 41-13 are exactly equal, and the mirrors M_1 and M_2 are exactly at right angles, the virtual image of M_1 formed by reflection at the silvered surface of plate C will coincide with mirror M_2. If L_1 and L_2 are not exactly equal, the image of M_1 will be displaced slightly from M_2; and if the angle between the mirrors is not exactly a right angle, the image of M_1 will make a slight angle with M_2. Then the mirror M_2, and the virtual image of M_1, play the same roles as the two surfaces of a thin film, discussed in Sec. 41-4, and the same sort of interference fringes result from the light that is reflected from these surfaces.

Suppose that the extended source in Fig. 41-13 is monochromatic with wavelength λ, and that the angle between mirror M_2 and the virtual image of M_1 is such that five or six vertical fringes are present in the field of view. If the mirror M_2 is now moved slowly either backward or forward by a distance $\lambda/2$, the effective film thickness will change by λ and each of the fringes will move either to the right or to the left through a distance equal to the spacing of the fringes. If the fringes are observed through a telescope whose eyepiece is equipped with a crosshair, and m fringes cross the crosshair when the mirror is moved a distance x, then

$$x = m\frac{\lambda}{2} \quad \text{or} \quad \lambda = \frac{2x}{m}. \tag{41-12}$$

41-14 A common type of Michelson interferometer.

If m is as large as several thousand, the distance x is sufficiently great so that it can be measured with good precision and hence a precise value of the wavelength λ can be obtained.

It was stated in Chapter 1 that the meter is defined as a length equal to a specified number of wavelengths of the orange-red light of krypton-86. Before this standard could be established, it was necessary to measure as accurately as possible the number of these wavelengths in the *former* standard meter, defined as the distance between two scratches on a bar of platinum-iridium. The measurement was made with a modified Michelson interferometer, many times and under very carefully controlled conditions. The number of wavelengths in a distance equal to the old standard meter was found to be 1,650,763.73 wavelengths. The meter is now defined as *exactly* this number of wavelengths.

Another application of the Michelson interferometer of considerable historical interest is the Michelson–Morley experiment. To understand the purpose of this experiment, we must recall that, before the electromagnetic theory of light and Einstein's special theory of relativity became established, physicists believed that light waves were propagated in a medium called *the ether*.

Although the ether was assumed to be very rigid, in order to propagate waves with the enormous speed of light, it was assumed also to be very tenuous, so the planets could move freely through it. A light wave was considered to travel with speed c relative to the ether, just as sound waves in a medium travel with the speed of sound relative to the medium. If a medium is in motion relative to the earth, the velocity of a *sound* wave relative to the earth, v_{WE}, is the vector sum of its velocity relative to the medium, v_{WM}, and the velocity of the medium relative to the earth, v_{ME}:

$$v_{\text{WE}} = v_{\text{WM}} + v_{\text{ME}}. \tag{41-13}$$

This point of view has been substantiated so many times with sound waves and water waves that it seemed obvious that the same results should hold for light waves.

In 1887, Michelson and Morley utilized the Michelson interferometer in an attempt to detect the relative motion of the earth and the ether. Suppose the interferometer in Fig. 41-13 is moving from left to right relative to the ether. According to nineteenth-century theory, this would lead to changes in the speed of light in the horizontal portions of the path, and corresponding fringe shifts relative to the positions the fringes would have if the instrument were at rest in the ether. Then, when the entire instrument was rotated 90°, the vertical paths would be similarly affected, giving a fringe shift in the opposite direction.

A reasonable guess as to the velocity of the earth relative to the ether is that it is of the same order as the orbital velocity of the earth around the sun, about 3×10^4 m·s^{-1}. Michelson and Morley calculated that, for green light, this velocity should cause a fringe shift of about four-tenths of a fringe when the instrument was rotated. The shift actually observed was less than a hundredth of a fringe, and within the limits of experimental uncertainty appeared in fact to be zero! Despite its orbital velocity the earth appeared to be at rest relative to the ether. This negative result baffled physicists of the time, and to this day the Michelson–Morley experiment is the most significant "negative-result" experiment ever performed.

Understanding of this result had to wait for Einstein's special theory of relativity, published in 1905. Einstein realized that the classical equation for combining relative velocities, Eq. (41–13), is only the limiting case of a more general equation, and that this general equation leads to the result that the velocity of a light wave has the same magnitude c relative to *all* reference frames, whatever their velocity may be relative to other frames. Even if there is an ether breeze past an interferometer, this breeze has no effect on the velocity of light relative to the interferometer. The result is that rays 1 and 2 in Fig. 41–13 travel, relative to the interferometer, with the same speed c, regardless of the orientation of the interferometer. There is no path difference, and no shift when the interferometer is rotated. The presumed ether then plays no role, and the very concept of an ether has been given up. The underlying principle of Einstein's theory of special relativity may be stated as follows: *There is no preferred frame of reference for the measurement of the speed of light; the speed is the same for every observer, without regard to the magnitude or direction of his velocity relative to other observers.* This theory, a well-established cornerstone of modern physics, is discussed in detail in Chapter 43.

41–7 Fresnel diffraction

According to *geometrical* optics, if an opaque object is placed between a point light source and a screen, as in Fig. 41–15, the edges of the object cast a sharp shadow on the screen. No light reaches the screen at points within the geometrical shadow, while outside the shadow the screen is uniformly illuminated. Geometrical optics is, however, an idealized model of the behavior of light. There are situations in which the representation in terms of straight-line or ray propagation is inadequate; we now proceed to examine some of these situations. An important class of phenomena in which the ray model of geometrical optics is *not* adequate is grouped under the heading *diffraction*.

The photograph in Fig. 41–16 was made by placing a razor blade halfway between a pinhole illuminated by monochromatic light and a photographic film, so that the film made a record of the shadow cast by the blade. Figure 41–17 is an enlargement of a region near the shadow of an edge of the blade. The boundary of the *geometrical* shadow is indicated by the short arrows. Note that a small amount of light has "bent" around the edge, into the geometrical shadow, which is bordered by alternating bright and dark bands. Note also that in the first bright band, just outside the geometrical shadow, the illumination is actually *greater* than in the region of uniform illumination to the extreme left. This simple experimental setup serves to give some idea of the true complexity of what is often considered the most elementary of optical phenomena, the shadow cast by a small source of light. The distribution of light and dark on any screen after a beam of light has been partially blocked by a perforated diaphragm is called a *diffraction pattern*.

Diffraction patterns such as that in Fig. 41–16 are not commonly observed in everyday life because most ordinary light sources are not

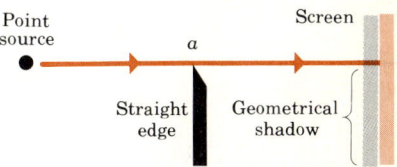

41–15 Geometrical shadow of a straight edge.

41-16 Shadow of a razor blade.

point sources of monochromatic light. If the shadow of a razor blade is cast by a frosted-bulb incandescent lamp, for example, the light from every point of the surface of the lamp forms its own diffraction pattern, but these overlap to such an extent that no individual pattern can be observed.

The term *diffraction* is applied to problems in which one is concerned with *the resultant effect produced by a limited portion of a wave front.* Since in most diffraction problems some light is found within the region of geometrical shadow, diffraction is sometimes defined as "the bending of light around an obstacle." It should be emphasized, however, that the process by which diffraction effects are produced is going on continuously in the propagation of *every* wave. Only if a part of the wave is cut off by some obstacle are the effects commonly called "diffraction effects" observed. But since every optical instrument does, in fact, make use of only a limited portion of a wave (a telescope, for example, utilizes only that portion of a wave admitted by the objective lens), it is evident that a clear comprehension of the nature of diffraction is essential for a complete understanding of practically all optical phenomena.

41-17 Shadow of a straight edge.

41-18 Fresnel diffraction pattern of a small circular obstacle. A bright spot is at the center of the shadow.

A circular *opening* in an opaque screen transmits only a small circular patch of a wave surface; the remainder of the wave is obscured. An interesting effect is observed if we reverse this procedure and insert a small circular *obstacle* in the light from a distant point source. A small circular patch of the wave is then obscured, while the remainder is allowed to proceed. Figure 41–18 is a photograph of the shadow of a small steel ball, supported from the tip of a magnetized sewing needle. Constructive interference of the wavelets from the unobstructed portion of the incident wave results in the small bright spot at the center of the geometrical shadow.

The essential features observed in diffraction effects can be predicted with the help of Huygens' principle, according to which every point of a wave surface can be considered the source of a secondary wavelet that spreads out in all directions. However, instead of finding the new wave surface by the simple process of constructing the envelope of all the secondary wavelets, we must combine these wavelets according to the principles of interference. That is, at every point we must combine the displacements that would be produced by the secondary wavelets, taking into account their amplitudes and relative phases. The mathematical operations are often quite complicated.

In Fig. 41–15 both the point source and the screen are at large but finite distances (say several meters) from the obstacle forming the diffraction pattern, and no lenses are used. This situation is described as *Fresnel diffraction* (after Augustin Jean Fresnel, 1788–1827), and the resulting pattern on the screen is called a *Fresnel diffraction pattern*. If the source is far enough away so that the diffraction pattern appears on a screen in the second focal plane of the lens, the phenomenon is called *Fraunhofer diffraction* (after Joseph von Fraunhofer, 1787–1826.) The latter situation is simpler to treat in detail, and we shall consider it first.

41-8 Fraunhofer diffraction from a single slit

Suppose a beam of parallel monochromatic light is incident from the left on an opaque plate having a narrow vertical slit. According to geometri-

41-19 (a) Geometrical "shadow" of a slit. (b) Diffraction pattern of a slit. The slit width has been greatly exaggerated.

cal optics, the transmitted beam should have been the same cross section as the slit, and a screen in the path of the beam would be illuminated uniformly over an area of the same size and shape as the slit. What is actually observed is the pattern shown in Fig. 41–19b. The beam spreads out horizontally after passing through the slit, and the diffraction pattern consists of a central bright band, which may be much wider than the slit width, bordered by alternating dark bands and bright bands of decreasing intensity. A diffraction pattern of this nature can readily be observed by looking at a point source such as a distant street light through a narrow slit formed between two fingers in front of the eye. The retina of the eye then corresponds to the screen.

Let us now apply Huygens' principle to compute the distribution of light on the screen. We consider a plane wave front at the moment it reaches the space between the edges of a slit such as that shown in section in Fig. 41–20a. Small elements of area are obtained by subdividing the wave front into narrow strips, parallel to the long edges of the slit, perpendicular to the page. From each of these strips, secondary Huygens wavelets spread out in all directions, as shown.

In Fig. 41–20b, a screen is placed at the right of the slit and P is one point on a line in the screen, the line being parallel to the long edges of the slit and perpendicular to the plane of the diagram. The light reaching a point on the line is calculated by applying the principle of superposition to all the wavelets arriving at the point, from all the elementary strips of the original wave surface. Because of the varying distances to the point, and the varying angles with the original direction of the light, the amplitudes and phases of the wavelets at the point will be different.

The problem is greatly simplified when the screen is sufficiently distant, or the slit sufficiently narrow, so that all rays from the slit to a point on the screen can be considered parallel, as in Fig. 41–20c. The former case, where the screen is relatively close to the slit (or the slit is relatively wide) is Fresnel diffraction, whereas the latter is Fraunhofer

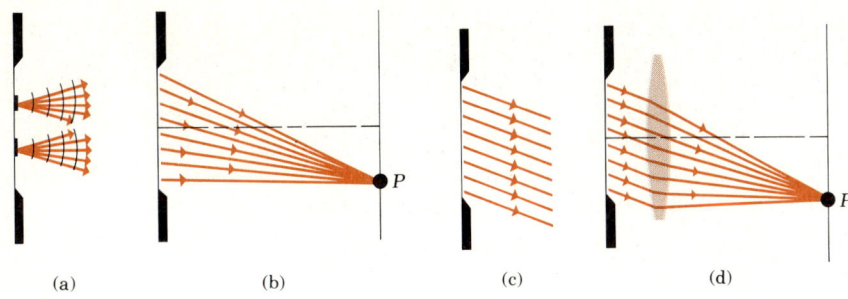

41-20 Diffraction by a single slit. Relation of Fresnel diffraction to Fraunhofer diffraction by a single slit.

diffraction. There is, of course, no difference in the *nature* of the diffraction process in the two cases, and Fresnel diffraction merges gradually into Fraunhofer diffraction as the screen is moved away from the slit, or as the slit width is decreased.

Fraunhofer diffraction occurs also if a lens is placed just beyond the slit, as in Fig. 41-20d, since the lens brings to a focus, in its second focal plane, all light traveling in a specified direction. That is, the lens forms in its focal plane a reduced *image* of the pattern that would appear on an infinitely distant screen in the absence of the lens.

When a lens is used in this manner to observe a diffraction pattern, the question arises whether the lens may introduce additional phase shifts that are different for different parts of the wave front. If it does, the interference phenomena that result may be quite different from those we have described. But it can be shown from very general considerations that *no* such additional phase shifts occur. For example, when a beam of parallel rays is brought to a focus by a converging lens or a concave mirror, the wavelets that are in phase across a plane perpendicular to the beam direction are still in phase when they arrive at the focal point after reflection or refraction. Alternatively, one may say that all rays require the same *time* to travel from a given cross-section plane perpendicular to the beam to the focal point. Thus the lenses in Figs. 41-20, 41-21, and 41-24 do not introduce additional phase shifts.

Some aspects of Fraunhofer diffraction from a single slit can be deduced easily. We first consider two narrow strips, one just below the top edge of the slit and one at its center. The difference in path length to point P in Fig. 41-21 is $(a/2) \sin \theta$, where a is the slit width. Suppose this

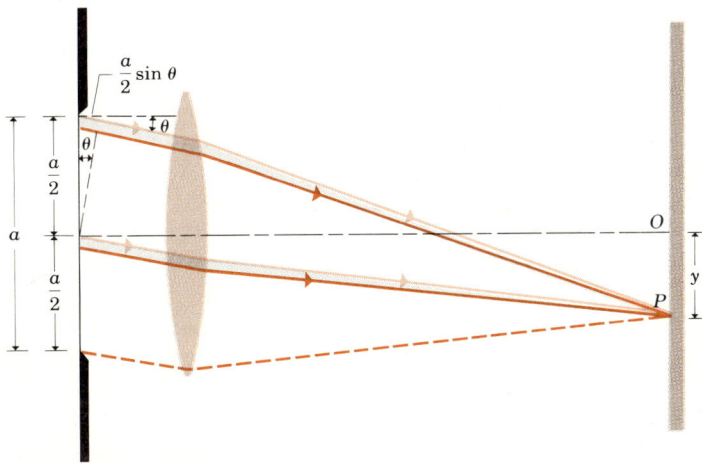

41-21 The wavefront is divided into a large number of narrow strips.

path difference happens to be equal to $\lambda/2$; then light from these two strips arrives at point P with a half-cycle phase difference, and cancellation occurs. Similarly, light from two strips just below these two will also arrive a half-cycle out of phase; and, in fact, light from *every* strip in the top half cancels out that from a corresponding strip in the bottom half, resulting in complete cancellation, and giving a dark fringe in the interference pattern. Thus, a dark fringe occurs whenever

$$\left(\frac{a}{2}\right)\sin\theta = \frac{\lambda}{2} \quad \text{or} \quad \sin\theta = \frac{\lambda}{a}. \tag{41-14}$$

We may also divide the screen into quarters, sixths, and so on, and use the above argument to show that a dark fringe occurs whenever $\sin\theta = 2\lambda/a$, $3\lambda/a$, and so on. Thus the condition for a *dark* fringe is

$$\sin\theta = \frac{n\lambda}{a}, \quad \text{where } n = 1, 2, 3, \ldots \tag{41-15}$$

For example, if the slit width is equal to ten wavelengths, dark fringes occur at $\sin\theta = \frac{1}{10}, \frac{2}{10}, \frac{3}{10}, \ldots$ Midway between the dark fringes are bright fringes. We also note that $\sin\theta = 0$ is a *bright* band, since then light from the entire slit arrives at P in phase. Thus the central bright fringe is twice as wide as the others, as Fig. 41–19 shows.

The complete intensity distribution for the single-slit pattern may be calculated by the same method used to obtain Eq. (41–9) for the two-slit pattern. We again imagine a plane wavefront at the slit subdivided into a large number of strips each of which sends out rays in *all* directions toward the lens, shown in Fig. 41–21. If we choose an arbitrary point P on a screen in the focal plane of the lens, only those rays making an angle θ with the axis of the lens will arrive at P. Point O on the screen is the special point at which all rays making the angle $\theta = 0$ arrive. Figure 41–22a is a phasor diagram, showing that when the slit is subdivided into 14 sections and each section emits a Huygens wavelet at the angle $\theta = 0$, all wavelets arrive in phase. The resultant amplitude at O is denoted by S.

With the same subdivision of the wavefront into 14 strips, the wavelets that make the angle θ and arrive at P have a slight phase difference between succeeding wavelets, and the corresponding phasor diagram is shown in Fig. 41–22b. The sum S is now the perimeter of a portion of a many-sided polygon and E_P, the amplitude of the resultant electric intensity at P, is the *chord*. The angle δ is the total phase difference between the wave from the bottom strip of Fig. 41–21 and the wave from the top strip.

In the limit, as the number of strips into which the slit is subdivided is increased, the phasor diagram becomes an *arc of a circle,* as shown in Fig. 41–22c, with the length of arc S equal to a constant. By constructing perpendiculars at A and B, the center of the circular arc C is found. The radius of the circle of which S is an arc is S/δ, and the resultant amplitude E_P (distance AB) is $2(S/\delta)\sin\delta/2$. We have then

$$E_P = S\frac{\sin\delta/2}{\delta/2},$$

where δ is the phase difference between the two rays at the extreme edges of the slit.

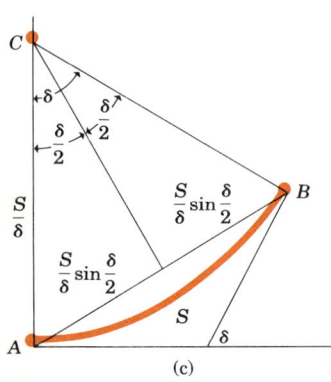

41-22 (a) Phasor diagram when all elementary electric intensities are in phase ($\theta = 0$, $\delta = 0$). (b) Phasor diagram when each elementary electric intensity differs in phase slightly from the preceding one. (c) Limit reached by the phasor diagram when the slit is subdivided infinitely.

41-23 (a) Intensity distribution. (b) Photograph of the Fraunhofer diffraction pattern of a single slit. (c) Fraunhofer diffraction pattern of a double slit.

From Eq. (41-6), the phase difference is $2\pi/\lambda$ times the path difference. From Fig. 41-21, the path difference between the ray from the top of the slit and that from the bottom is $a \sin \theta$. Therefore

$$\delta = \frac{2\pi}{\lambda} a \sin \theta, \qquad (41\text{--}16)$$

and

$$E_P = S \frac{\sin[\pi a (\sin \theta)/\lambda]}{\pi a (\sin \theta)/\lambda}.$$

Since the intensity I is proportional to the *square* of the amplitude,

$$I = I_0 \left\{ \frac{\sin[\pi a (\sin \theta)/\lambda]}{\pi a (\sin \theta)/\lambda} \right\}^2, \qquad (41\text{--}17)$$

where I_0 is the intensity at O in Fig. 41-21 where $\theta = 0$.

Equation (41-17) is plotted in Fig. 41-23a, and a photograph of the diffraction pattern is shown directly underneath. Note that most of the light exists in the region close to the point where $\theta = 0$, or the geometrical focus. From Eq. (41-17), the smallest value of $\pi a (\sin \theta)/\lambda$ at which the intensity becomes zero is the value π. This corresponds to a value of θ equal to θ_1 and given by

$$\sin \theta_1 = \frac{\lambda}{a}.$$

Since λ is of the order of 5×10^{-5} cm, and a typical slit width is 10^{-2} cm, $\sin \theta_1$ is ordinarily so small that $\sin \theta_1 = \theta_1$, and

$$\theta_1 = \frac{\lambda}{a}.$$

When a is several centimeters, θ_1 is so small that one can consider practically all the light to be concentrated at the geometrical focus.

The photograph in Fig. 41–23c shows the Fraunhofer diffraction pattern of *two* slits each of the same width a as the one above, but separated by a distance $d = 4a$. Note that the interference fringes due to cooperation between the slits have intensities that follow the diffraction pattern of each separate slit. The "cosine-squared" interference fringes are modulated by the shape of the curve shown in part (a) of the figure, because of the finite width of the slits S_1 and S_2.

41–9 The diffraction grating

Suppose that, instead of a single slit, or two slits side by side as in Young's experiment, we have a very large number of parallel slits, all of the same width and spaced at regular intervals. Such an arrangement, known as a *diffraction grating,* was first constructed by Fraunhofer. The earliest gratings were of fine wires, 0.04 mm to 0.6 mm in diameter. Gratings are now made by using a diamond point to rule a large number of equidistant grooves on a glass or metal surface.

Let GG, in Fig. 41–24, represent the grating, the slits of which are perpendicular to the plane of the paper. While only five slits are shown in the diagram, an actual grating contains several thousand, with a grating spacing d of the order of 0.002 mm. Let a train of plane waves be incident normally on the grating from the left. The problem of finding the intensity of the light transmitted by the grating then combines the principles of interference and diffraction. That is, each slit gives rise to a diffracted beam whose nature, as we have seen, depends on the slit width. These diffracted beams then interfere with one another to produce the final pattern.

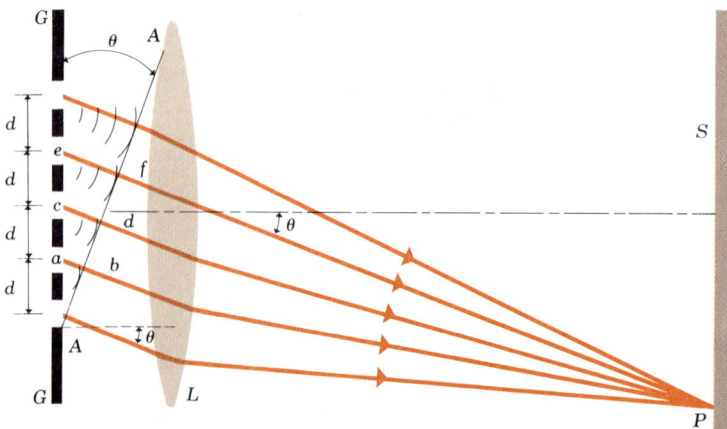

41–24 The plane diffraction grating.

Let us assume that the slits are so narrow that the diffracted beam from each spreads out over a sufficiently wide angle for it to interfere with all the other diffracted beams. Consider first the light proceeding from elements of infinitesimal width at the lower edges of each opening, and traveling in a direction making an angle θ with that of the incident beam, as in Fig. 41-24. A lens at the right of the grating forms in its focal plane a diffraction pattern similar to that which would appear on a screen at infinity.

Suppose the angle θ in Fig. 41-24 is taken so that the distance ab equals λ, the wavelength of the incident light. Then $cd = 2\lambda$, $ef = 3\lambda$, etc. The waves from all of these elements, since they are in phase at the plane of the grating, are also in phase along the plane AA and therefore reach the point P in phase. The same holds true for any set of elements in corresponding positions in the various slits.

If the angle θ is increased slightly, the disturbances from the grating elements no longer arrive at AA in phase with one another, and even an extremely small change in angle results in almost complete destructive interference among them, provided there is a large number of slits in the grating. Hence, the maximum at the angle θ is an extremely sharp one, differing from the rather broad maxima and minima that result from interference or diffraction effects with a small number of openings.

As the angle θ is increased still further, a position is eventually reached in which the distance ab in Fig. 41-24 becomes equal to 2λ. Then cd equals 4λ, cf equals 6λ, and so on. The disturbances at AA are again all in phase, the path difference between them now being 2λ, and another maximum results. Evidently still others will appear when $ab = 3\lambda, 4\lambda, \ldots$ Maxima will also be observed at corresponding angles on the opposite side of the grating normal, as well as along the normal itself, since in the latter position the phase difference between disturbances reaching AA is zero.

The angles of deviation for which the maxima occur may readily be found from Fig. 41-25. Consider the right angle Aba. Let d be the distance between successive grating elements, called the *grating spacing*. The necessary condition for a maximum is that $ab = m\lambda$, where $m = 0, 1, 2, 3$, etc. It follows that

$$\sin \theta = m\frac{\lambda}{d} \quad (m = 0, 1, 2, \ldots). \tag{41-18}$$

is the necessary condition for a maximum. The angle θ is also the angle by which the rays corresponding to the maxima have been *deviated* from the direction of the incident light.

In practice, the parallel beam incident on the grating is usually produced by a lens having a narrow illuminated slit at its first focal point. Each of the maxima is then a sharp image of the slit, of the same color as that of the light illuminating the slit, assumed thus far to be monochromatic. If the slit is illuminated by light consisting of a mixture of wavelengths, the lens will form a number of images of the slit in different positions; each wavelength in the original light gives rise to a set of slit images deviated by the appropriate angles. If the slit is illuminated with white light, a continuous group of images is formed side by side or, in other words, the white light is dispersed into continuous spectra. In contrast with the single spectrum produced by a prism, a grating forms a

41-25 First-order maximum when $ab = \lambda$, second-order maximum when $ab = 2\lambda$.

number of spectra on either side of the normal. Those that correspond to $m = 1$ in Eq. (41–18) are called *first-order,* those that correspond to $m = 2$ are called *second-order,* and so on. Since for $m = 0$ the deviation is zero, all colors combine to produce a white image of the slit in the direction of the incident beam.

In order that an appreciable deviation of the light may be produced, it is necessary that the grating spacing be of the same order of magnitude as the wavelength of light. Gratings for use in or near the visible spectrum are ruled with from about 500 to 1500 lines per millimeter.

The diffraction grating is widely used in spectrometry, instead of a prism, as a means of dispersing a light beam into spectra. If the grating spacing is known, then from a measurement of the angle of deviation of any wavelength, the value of this wavelength may be computed. In the case of a prism this is not so; the angles of deviation are not related in any simple way to the wavelengths but depend on the characteristics of the material of which the prism is constructed. Since the index of refraction of optical glass varies more rapidly at the violet than at the red end of the spectrum, the spectrum formed by a prism is always spread out more at the violet end than it is at the red. Also, while a prism deviates red light the least and violet the most, the reverse is true of a grating, since in the latter case the deviation increases with increasing wavelength.

Example 1 The wavelengths of the visible spectrum are approximately 400 nm to 700 nm. Find the angular breadth of the first-order visible spectrum produced by a plane grating having 6,000 lines per centimeter, when light is incident normally on the grating.

Solution The grating spacing d is

$$d = \frac{1}{6000 \text{ lines} \cdot \text{cm}^{-1}} = 1.67 \times 10^{-6} \text{ m}.$$

The angular deviation of the violet is

$$\sin \theta = \frac{400 \times 10^{-9} \text{ m}}{1.67 \times 10^{-6} \text{ m}} = 0.240,$$
$$\theta = 13.9°.$$

The angular deviation of the red is

$$\sin \theta = \frac{700 \times 10^{-9} \text{ m}}{1.67 \times 10^{-6} \text{ m}} = 0.420,$$
$$\theta = 24.8°.$$

Hence, the first-order visible spectrum includes an angle of

$$24.8° - 13.9° = 10.9°.$$

Example 2 Show that the violet of the third-order spectrum overlaps the red of the second-order spectrum.

Solution The angular deviation of the third-order violet is

$$\sin \theta = \frac{(3)(400 \times 10^{-9} \text{ m})}{d}$$

and of the second-order red it is

$$\sin \theta = \frac{(2)(700 \times 10^{-9} \text{ m})}{d}.$$

Since the first angle is smaller than the second, whatever the grating spacing, the third order will *always* overlap the second.

*41-10 Diffraction of x-rays by a crystal

Although x-rays were discovered by Roentgen in 1895, it was not until 1913 that x-ray wavelengths were measured with any degree of precision. Experiments had indicated that these wavelengths might be of the order of 10^{-8} cm, which is about the same as the interatomic spacing in a solid. It occurred to Laue in 1913 that if the atoms in a crystal were arranged in a regular way, a crystal might serve as a three-dimensional diffraction grating for x-rays. The experiment was performed by Friederich and Knipping and it succeeded, thus verifying in a single stroke both the hypothesis that x-rays *are* waves (or, at any rate, wavelike in some of their properties) and that the atoms in a crystal *are* arranged in a regular manner. Since that time, the phenomenon of x-ray diffraction by a crystal has proved an invaluable tool of the physicist, both as a way to measure x-ray wavelengths and as a method of studying the structure of crystals.

Figure 41-26 is a diagram of a simple type of crystal, that of sodium chloride (NaCl). The black circles represent the sodium, and the color circles the chlorine ions. Figure 41-27 is a diagram of a section through the crystal. Planes such as those parallel to *aa, bb, cc,* etc., can be constructed through the crystal in such a way that they pass through relatively large numbers of atoms.

Figure 41-28 is a photograph made by directing a narrow beam of x-rays at a thin section of a crystal of quartz and allowing the diffracted beams to strike a photographic plate. Each atom in the crystal scatters some of the incident radiation, and interference occurs between the waves scattered by the various atoms. Just as with the diffraction grating, nearly complete cancellation occurs for all but certain very specific directions for which constructive interference is possible; hence, the spots in Fig. 41-28. The interference pattern is often described in terms of *reflections* from various planes of atoms as in Fig. 41-27, but the basic phenomenon is one of *interference*.

41-26 Model of arrangement of ions in a crystal of NaCl. Black circles, Na; color circles, Cl.

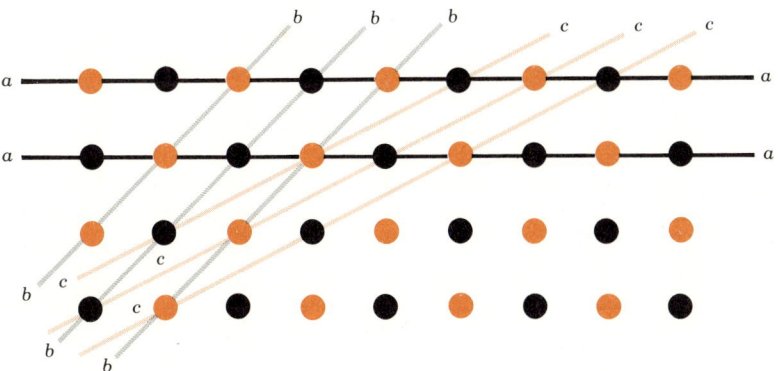

41-27 Crystal planes such as *aa, bb,* and *cc* serve as a three-dimensional diffraction grating for x-rays.

41-28 Laue diffraction pattern formed by directing a beam of x-rays at a thin section of quartz crystal. (Courtesy of Dr. B. E. Warren.)

*41-11 The resolving power of optical instruments

The expressions for the magnification of a telescope or a microscope involved (except for certain numerical factors) only the focal lengths of the lenses making up the optical system of the instrument. It appears, at first glance, as though any desired magnification might be attained by a proper choice of these focal lengths. Beyond a certain point, however, while the image formed by the instrument becomes larger (or subtends a larger angle) it *does not gain in detail*, even though all lens aberrations have been corrected. This limit to the useful magnification is set by the fact that light is a wave motion and the laws of geometrical optics do not hold strictly for a wave surface of limited extent. Physically, the image of a point source is not the intersection of *rays* from the source, but the diffraction pattern of those *waves* from the source that pass through the lens system.

It is an important experimental fact that the light from a point source, diffracted by a circular opening, is focused by a lens not as a geometrical point, but as a disk of finite radius surrounded by dark and bright rings. This diffraction pattern is, in fact, identical to the acoustic radiation pattern from a circular piston, which was discussed in Sec. 23-8. The larger the wave surface admitted (i.e., the larger the lenses or diaphragms in an optical system), the smaller the diffraction pattern of a point source and the closer together may two point sources be before their diffraction disks overlap and become indistinguishable. An optical system is said to be able to *resolve* two point sources if the corresponding diffraction patterns are sufficiently small or sufficiently separated to be distinguished. The numerical measure of the ability of the system to resolve two such points is called its *resolving power* or its *resolution*.

Figure 41-29a is a photograph of four point sources made with the camera lens "stopped down" to an extremely small aperture. The nature of the diffraction patterns is clearly evident, and it is obvious that further magnification of the picture would not aid in resolving the sources.

What is necessary is not to make the image *larger*, but to make the diffraction patterns *smaller*. Figures 41–29b and 41–29c show how the resolving power of the lens is increased by increasing its aperture. In (b) the diffraction patterns are sufficiently small for all four sources to be distinguished. In (c) the full aperture of the lens was utilized.

An arbitrary criterion proposed by Lord Rayleigh is that two point sources are just resolvable if the central maximum of the diffraction pattern of one source just coincides with the first *minimum* of the other. Let P_1 and P_2 in Fig. 41–30 be the central rays in the two parallel beams coming from two very distant point sources. Let P_1' and P_2' be the cen-

41-29 Diffraction patterns of four "point" sources, with a circular opening in front of the lens. In (a), the opening is so small that the patterns at the right are just resolved, by Rayleigh's criterion. Increasing the aperture decreases the size of the diffraction patterns, as in (b) and (c).

(a)

(b)

(c)

41-30 Two point sources are just resolved if the maximum of the diffraction pattern of one source just coincides with the first minimum of the other (Rayleigh's criterion).

ters of their diffraction patterns, formed by some optical instrument such as a microscope or telescope, which for simplicity is represented in the diagram as a single lens. If the images are just resolved, that is, if according to Rayleigh's criterion the first minimum of one pattern coincides with the center of the other, the separation of the centers of the patterns equals the radius of the central bright disk. From a knowledge of the focal lengths and separations of the lenses in any particular instrument, one can compute the corresponding distance between the two point objects. This distance, the minimum separation of two points that can just be resolved, is called the *limit of resolution* of the instrument. The smaller this distance, the greater is said to be the *resolving power*. The resolving power increases with the solid angle of the cone of rays intercepted by the instrument, and is inversely proportional to the wavelength of the light used. It is as if the light waves were the "tools" with which our optical system is provided, and the smaller the tools the finer the work the system can do.

*41-12 Holography

Holography is a technique for recording and reproducing an image of an object without the use of lenses or mirrors. Unlike the two-dimensional images recorded by an ordinary photograph or television system, a holographic image is truly three-dimensional. Such an image can be viewed from different directions to reveal different sides, and from various distances to reveal changing perspective.

The basic procedure for making a hologram is very simple in principle. A possible arrangement is shown in Fig. 41–31a. The object to be holographed is illuminated by monochromatic light, and a photographic film is located so that it is struck by scattered light from the object and also by direct light from the source. In practice the source must be a laser, for reasons to be discussed later. Interference between the direct and scattered light leads to the formation and recording of a complex interference pattern on the film.

To form the images, one simply projects laser light through the developed film, as shown in Fig. 41–31b. Two images are formed, a virtual image on the side of the film nearer the source, and a real image on the opposite side.

A complete analysis of holography is beyond our scope, but we can gain some insight into the process by examining how a single point is holographed and imaged. We consider the interference pattern formed

41-31 (a) The hologram is the record on film of the interference pattern formed with light directly from the source and light scattered from the object. (b) Images are formed when light is projected through the hologram.

on a screen by the superposition of an incident plane wave and a spherical wave, as shown in Fig. 41-32a. The spherical wave originates at a point source P a distance d_0 from the screen; P may in fact be a small object that scatters part of the incident plane wave. In any event, we assume that the two waves are monochromatic and coherent, and that the phase relation is such that constructive interference occurs at point O on the diagram. Then constructive interference will *also* occur at any point Q on the screen that is farther from P than O is, by an integer number of wavelengths. That is, if $d_n - d_0 = n\lambda$, where n is an integer, constructive interference occurs. The points where this condition is satisfied form circles centered at O, with radii r_n given by

$$d_n - d_0 = \sqrt{d_0^2 + r_n^2} - d_0 = n\lambda, \quad (n = 1, 2, 3, \ldots). \quad (41\text{-}19)$$

Solving this for r_n^2, we find

$$r_n^2 = \lambda\,(2nd_0 + n^2\lambda).$$

41-32 (a) Constructive interference of the plane and spherical waves occurs in the plane of the film at every point Q for which the distance d_n from P is greater than the distance d_0 from P to O by an integer number of wavelengths $n\lambda$. For the point shown, $n = 2$. (b) When a plane wave strikes the developed film, the diffracted wave consists of a wave converging to P' and then diverging again, and a diverging wave that appears to originate at P. These waves form the real and virtual images, respectively.

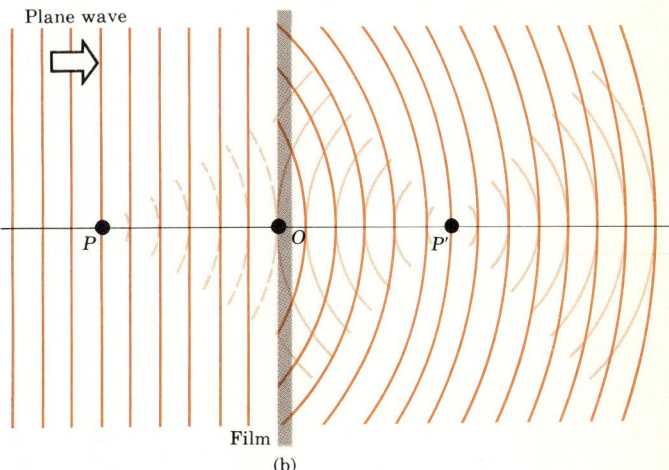

Ordinarily d_0 is very much larger than λ, so we neglect the second term in parentheses, obtaining

$$r_n = \sqrt{2n\lambda d_0}. \tag{41-20}$$

Since n may be any integer, the interference pattern consists of a series of concentric bright circular fringes, with the radii of the brightest regions given by Eq. (41-20). Between these bright fringes will be darker fringes. This pattern may be recorded on film by simply placing the film in the position of the screen.

Now the film is developed and a transparent positive print is made, so the bright-fringe areas have the greatest transparency on the film. It is then illuminated with monochromatic plane-wave light of the same wavelength as that used initially. In Fig. 41-32b, consider a point P' at a distance d_0 along the axis from the film; the centers of successive bright fringes differ in their distances from P' by an integer number of wavelengths, and therefore a strong *maximum* in the diffracted wave occurs at P'. That is, light converges to P' and then diverges from it on the opposite side, and P' is therefore a *real image* of point P.

This is not the entire diffracted wave, however; there is also a diverging spherical wave, which would represent a continuation of the wave originally emanating from P if the film had not been present. Thus the *total* diffracted wave is a superposition of a converging spherical wave forming a real image at P', and a diverging spherical wave shaped as though it had originated at P, forming a virtual image at P.

Because of the principle of linear superposition, what is true for the imaging of a single point is also true for the imaging of any number of points. The film records the superposed interference pattern from the various points, and when light is projected through the film the various image points are reproduced simultaneously. Thus the images of an extended object can be recorded and reproduced just as for a single point object.

Several practical problems must be overcome. First, the light used must be *coherent* over the distances large compared to the dimensions of the object and its distance from the film. Ordinary light sources *do not* satisfy this requirement, for reasons discussed in Sec. 41-1, and laser light is essential. Second, extreme mechanical stability is needed. If there is any relative motion of source, object, or film during exposure, even by as much as a wavelength, the interference pattern on the film is blurred enough to prevent satisfactory image formation. These obstacles are not insurmountable, however, and holography promises to become increasingly important in research, entertainment, and a wide variety of technological applications.

Questions

41-1 Could an experiment similar to Young's two-slit experiment be performed with sound? How might this be carried out? Does it matter that sound waves are longitudinal and electromagnetic waves transverse?

41-2 At points of constructive interference between waves of equal amplitude, the intensity is four times that of either individual wave. Does this violate energy conservation? If not, why not?

41-3 In using the superposition principle to calculate intensities in interference and diffraction patterns, could one add the intensities of the waves instead of their amplitudes? What's the difference?

41-4 A two-slit interference experiment is set up and the fringes displayed on a screen. Then the whole apparatus is immersed in the nearest swimming pool. How does the fringe pattern change?

41-5 Would the headlights of a distant car form a two-source interference pattern? If so, how might it be observed? If not, why not?

41-6 A student asserted that it is impossible to observe interference fringes in a two-source experiment if the distance between sources is less than half the wavelength of the wave. Do you agree? Explain.

41-7 An amateur scientist proposed to record a two-source interference pattern using only one source, placing it first in position S_1 in Fig. 41-1 and turning it on for a certain time, then placing it at S_2 and turning it on for an equal time. Does this work?

41-8 When a thin oil film spreads out on a puddle of water, the thinnest part of the film looks lightest in the resulting interference pattern. What does this tell you about the relative magnitudes of the refractive indexes of oil and water?

41-9 A glass windowpane with a thin film of water on it reflects less than when it is perfectly dry. Why?

41-10 In high-quality camera lenses, the resolution in the image is determined by diffraction effects. Is the resolution best when the lens is "wide open" or when it is "stopped down" to a smaller aperture? How does this behavior compare with the effect of aperture size on the depth of focus, that is, on the limit of resolution due to imprecise focusing?

41-11 If a two-slit interference experiment were done with white light, what would be seen?

41-12 Why is a diffraction grating better than a two-slit setup for measuring wavelengths of light?

41-13 Would the interference and diffraction effects described in this chapter still be seen if light were a longitudinal wave instead of transverse?

41-14 One sometimes sees rows of evenly spaced radio antenna towers. A student remarked that these act like diffraction gratings. What did he mean? Why would one *want* them to act like a diffraction grating?

41-15 Could x-ray diffraction effects with crystals be observed using visible light instead of x-rays? Why or why not?

41-16 Does a microscope have better resolution with red light or blue light?

41-17 How could an interference experiment, such as one using a Michelson interferometer or fringes caused by a thin air space between glass plates, be used to measure the refractive index of air?

Problems

41-1 Two slits are spaced 0.3 mm apart and are placed 50 cm from a screen. What is the distance between the second and third dark lines of the interference pattern when the slits are illuminated with light of 600-nm wavelength?

41-2 In Problem 41-1, suppose the entire apparatus is immersed in water. Then what is the distance between the second and third dark lines?

41-3 Young's experiment is performed with sodium light ($\lambda = 589$ nm). Fringes are measured carefully on a screen 100 cm away from the double slit, and the center of the twentieth fringe is found to be 11.78 mm from the center of the zeroth fringe. What is the separation of the two slits?

41-4 An FM radio station has a frequency of 100 MHz and uses two identical antennas mounted at the same elevation, 12 m apart. The resulting radiation pattern has maximum intensity along a horizontal line perpendicular to the line joining the antennas. At what other angles (measured from the line of maximum intensity) is the intensity maximum? At what angles is it zero?

41-5 An AM radio station has a frequency of 1000 kHz; it uses two identical antennas at the same elevation, 150 m apart.

a) In what directions is the intensity maximum, considering points in a horizontal plane?

b) Calling the maximum intensity in (a) I_0, determine in terms of I_0 the intensity in directions making angles of 30°, 45°, 60°, and 90° to the direction of maximum intensity.

41-6 Light from a mercury-arc lamp is passed through a filter that blocks everything except for one spectrum line in the green region of the spectrum. It then falls on two slits separated by 0.6 mm. In the resulting interference pattern on a screen 2.5 mm away, adjacent bright fringes are separated by 2.27 mm. What is the wavelength?

41-7 Light of wavelength 500 nm is incident perpendicularly from air on a film 10^{-4} cm thick and of refractive index 1.375. Part of the light is reflected from the first surface of the film, and part enters the film and is reflected back at the second face.

a) How many waves are contained along the path of this light in the film?

b) What is the phase difference between these waves as they leave the film?

41-8 A glass plate, 0.40 μm thick, is illuminated by a beam of white light normal to the plate. The index of refraction of the glass is 1.50. What wavelengths within the limits of the visible spectrum ($\lambda = 400$ nm to $\lambda = 700$ nm) will be intensified in the reflected beam?

41-9 Figure 41-33 shows an interferometer known as *Fresnel's biprism*. The magnitude of the prism angle A is extremely small

a) If S_0 is a very narrow source slit, show that the separation of the two virtual coherent sources S_1 and S_2 is given by $d = 2aA(n-1)$, where n is the index of refraction of the material of the prism.

b) Calculate the spacing of the fringes of green light of wavelength 500 nm on a screen 2 m from the biprism. Take $a = 20$ cm, $A = 0.005$ rad, and $n = 1.5$.

41-10 In Lloyd's mirror, the source slit S_0 and its virtual image S_1 lie in a plane 20 cm behind the left edge of the mirror. (See Fig. 41-8.) The mirror is 30 cm long and a ground-glass screen is placed at the right edge. Calculate the distance from this edge to the first light maximum, if the perpendicular distance from S_0 to the mirror is 2 mm and $\lambda = 7.2 \times 10^{-5}$ cm.

41-11 How far must the mirror M_2 of the Michelson interferometer be moved in order that 3000 fringes of krypton-86 light ($\lambda = 606$ nm) move across a line in the field of view?

41-12 Two rectangular pieces of plane glass are laid one upon the other on a table. A thin strip of paper is placed between them at one edge so that a very thin wedge of air is formed. The plates are illuminated by a beam of sodium light at normal incidence ($\lambda = 589$ nm). Interference fringes are formed, there being ten per centimeter length of wedge measured normal to the edges in contact. Find the angle of the wedge.

41-13 A sheet of glass 10 cm long is placed in contact with a second sheet, and is held at a small angle with it by a metal strip 0.1 mm thick placed under one end. The glass is illuminated from above with light of 546-nm wavelength. How many interference fringes are observed per cm in the reflected light?

41-14 The radius of curvature of the convex surface of a plano-convex lens is 120 cm. The lens is placed convex side down on a plane glass plate, and illuminated from above with red light of wavelength 650 nm. Find the diameter of the third bright ring in the interference pattern.

41-15 What is the thinnest film of 1.40 refractive index in which destructive interference of the violet component (400 nm) of an incident white beam in air can take place by reflection? What is, then, the residual color of the beam?

41-16 The surfaces of a prism of index 1.52 are to be made "nonreflecting" by coating them with a thin layer of transparent material of index 1.30. The thickness of the layer is such that, at a wavelength of 550 nm (in vacuum), light reflected from the first surface is $\frac{1}{2}$ wavelength out of phase with that reflected from the second surface. Find the thickness of the layer.

41-17

a) Is a thin film of quartz suitable as a nonreflecting coating for fabulite? (See Table 38-1.)

b) If so, how thick should the film be?

41-18 Parallel rays of green mercury light of wavelength 546 nm pass through a slit of width 0.437 mm covering a lens of focal length 40 cm. In the focal plane of the lens, what is the distance from the central maximum to the first minimum?

41-19 Light of wavelength 589 nm from a distant source is incident on a slit 1.0 mm wide, and the resulting diffraction pattern is observed on a screen 2.0 m away. What is the distance between the two dark fringes on either side of the central bright fringe?

41-20 In Problem 41-19, suppose the entire apparatus is immersed in water. Then what is the distance between the two dark fringes?

41-21 Plane monochromatic waves of wavelength 600 nm are incident normally on a plane transmission grating having 500 lines·mm^{-1}. Find the angles of deviation in the first, second, and third orders.

41-22 A plane transmission grating is ruled with 4000 lines·cm^{-1}. Compute the angular separation in degrees, in

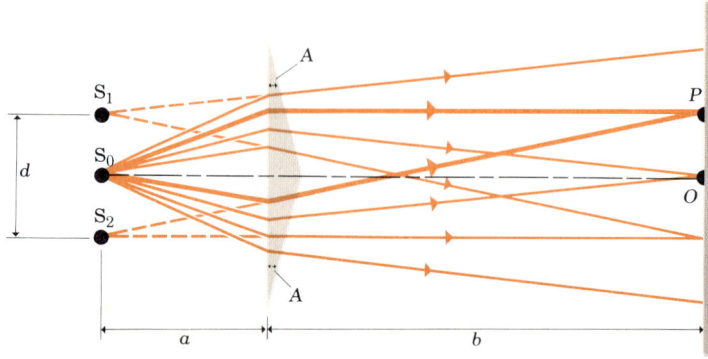

41-33 The Fresnel biprism.

the second-order spectrum, between the α and δ lines of atomic hydrogen, whose wavelengths are, respectively, 656 nm and 410 nm. Assume normal incidence.

41-23

a) What is the wavelength of light that is deviated in the first order through an angle of 20° by a transmission grating having 6000 lines·cm^{-1}?

b) What is the second-order deviation of this wavelength? Assume normal incidence.

41-24 A slit of width a was placed in front of a lens of focal length 80 cm. The slit was illuminated by parallel light of wavelength 600 nm and the diffraction pattern of Fig. 41-23b was formed on a screen in the second focal plane of the lens. If the photograph of Fig. 41-23b represents an enlargement to twice the actual size, what was the slit width?

41-25 The intensity of light in the Fraunhofer diffraction pattern of a single slit is

$$I = I_0(\sin \beta/\beta)^2,$$

where

$$\beta = (\pi a \sin \theta)/\lambda.$$

Show that the equation for the values of β at which I is a maximum is $\tan \beta = \beta$. How can you solve such an equation graphically?

41-26 What is the longest wavelength that can be observed in the fourth order for a transmission grating having 5000 lines · cm^{-1}? Assume normal incidence.

41-27 In Fig. 41-34, two point sources of light, a and b, at a distance of 50 m from lens L and 6 mm apart, produce images at c that are just resolved by Rayleigh's criterion. The focal length of the lens is 20 cm. What is the diameter of the diffraction circles at c?

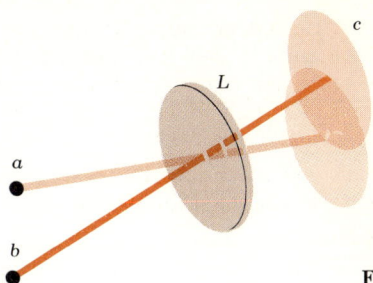

Figure 41-34

41-28 A telescope is used to observe two distant point sources 1 m apart. ($\lambda = 500$ nm.) The objective of the telescope is covered with a slit of width 1 mm. What is the maximum distance in meters at which the two sources may be distinguished?

41-29 A typical lens in a 35-mm camera has a focal length of 50 mm and a diameter of 25 mm ($f/2$). The resolution of such lenses is expressed as the number of lines per millimeter in the image that can be resolved. If the resolution of this lens is limited by diffraction effects, approximately what is its resolution in lines·mm^{-1}?

41-30 If a hologram is made using 600-nm light and then viewed using 500-nm light, how will the images look compared to those observed with 500-nm light?

41-31 A hologram is made using 600-nm light and is then viewed using continuous-spectrum white light from an incandescent bulb. What will be seen?

41-32 Ordinary photographic film reverses black and white, in the sense that the most brightly illuminated areas become blackest upon development (hence the term *negative*). Suppose a hologram negative is viewed directly, without making a positive transparency. How will the resulting images differ from those obtained with the positive hologram?

42 POLARIZATION

(a)

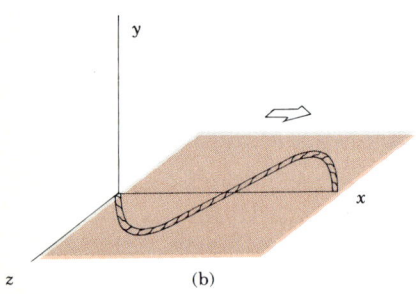

(b)

42-1 Polarization

Interference and diffraction effects occur with all kinds of waves, including sound waves and surface waves on a liquid as well as electromagnetic waves. These effects are, in general, independent of whether the waves are longitudinal or transverse. In this chapter we consider *polarization* phenomena, a group of wave phenomena that occur only with *transverse* waves. Our principal area of interest will be with electromagnetic waves, but we begin with a mechanical example.

To introduce basic ideas in a familiar context, we recall the discussion of mechanical waves on a string in Chapter 21. Displacements of points on the string are perpendicular to the length of the string, and to the direction of propagation of the wave, so the wave is transverse. For a string whose equilibrium position is along the x-axis, we considered principally displacements along the y-direction, in which case the string always lies entirely in the xy-plane; an example of such a wave is shown in Fig. 42–1a. But the displacement might instead be along the z-axis, so that the string lies always in the xz-plane, as in Fig. 42–1b. Furthermore, in view of the principle of linear superposition, any superposition of waves having displacements along the y- and z-directions is also possible; conversely, *any* transverse wave can be represented as a superposi-

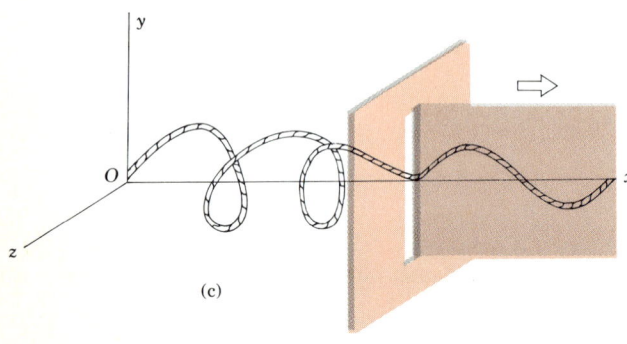

(c)

42–1 (a) Transverse wave on a string, polarized in the y-direction. (b) Wave polarized in the z-direction. (c) Barrier with a frictionless vertical slot passes components polarized in the y-direction but blocks those polarized in the z-direction, acting as a polarizing filter.

tion of two component waves, one having displacements only along the y-direction, the other only along the z-direction.

The wave having only y-displacements in the above discussion is said to be *linearly polarized* in the y-direction, and the one with only z-displacements is linearly polarized in the z-direction. The qualifying words *linear* and *linearly* refer to the fact that each particle of the string moves back and forth along a straight line. These words are often omitted when it is clear from the context that linear polarization is meant.

Any transverse wave on a string may be regarded as a superposition of waves polarized in the y- and z-directions, respectively. It is also easy in principle to construct a mechanical *filter* that permits only waves with a certain polarization direction to pass. An example is shown in Fig. 42-1c; the string can slide vertically in the slot without friction, but no horizontal motion is possible. Thus this filter passes waves polarized in the y-direction but blocks those polarized in the z-direction.

This same language can be applied to light and other electromagnetic waves, which also exhibit polarization. An electromagnetic wave consists of fluctuating electric and magnetic fields, perpendicular to each other and to the direction of propagation, as described in detail in Chapter 37. By convention, the direction of polarization is taken to be that of the *electric*-field vector, not the magnetic field, partly because most mechanisms for detecting electromagnetic waves employ principally the electric-field forces on electrons in materials. The most common manifestations of electromagnetic radiation are due chiefly to the electric-field force, not the magnetic-field force.

Polarizing filters can be made for electromagnetic waves; the details of construction depend on the wavelength. For microwaves having a wavelength of a few centimeters, a grid of closely spaced, parallel conducting wires insulated from each other will pass waves whose E-fields are perpendicular to the wires but not those with E-fields parallel to the wires. For light, the most common polarizing filter is a material known by the trade name Polaroid, widely used for sunglasses and polarizing filters for camera lenses. This material works on a principle of preferential absorption, passing waves polarized parallel to a certain axis in the material (called the *polarizing axis*) with 80 percent or more transmission, but offering only one percent or less transmission to waves with polarization perpendicular to this axis. The action of such a polarizing filter is shown schematically in Fig. 42-2. Other polarizing filters are discussed in Sec. 42-5.

Waves emitted by a radio transmitter are usually linearly polarized; a vertical rod antenna of the type widely used for CB radios emits waves which, in a horizontal plane around the antenna, are polarized in the vertical direction (parallel to the antenna). Light from ordinary sources is *not* polarized, for a slightly subtle reason. The "antennas" that radiate light waves are the molecules of which light sources are composed. The electrically charged particles in the molecules acquire energy in some way, and radiate this energy as electromagnetic waves of short wavelength. The waves from any one molecule may be linearly polarized, like those from a radio antenna; but since any actual light source contains a tremendous number of molecules, oriented at random, the light emitted is a random mixture of waves linearly polarized in all possible transverse directions.

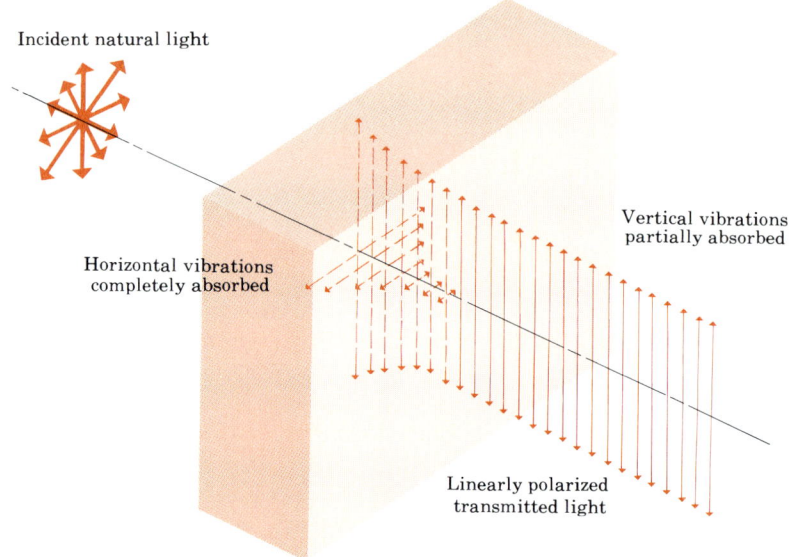

42-2 Linearly polarized light transmitted by a polarizing filter.

42-2 Malus' law and percent polarization

To clarify basic concepts, we introduce in this section the idea of an ideal polarizing filter or *polarizer* having the property that it passes 100 percent of incident light polarized in the direction of the filter's polarizing axis, but blocks completely all light polarized perpendicular to this axis. We recognize that such a polarizer is an unattainable idealization, like the frictionless plane and the zero-resistance wire, but it can be approximated closely by real devices and is helpful in clarifying the discussion.

For quantitative experiments, the intensity of transmitted light can be measured by means of a photocell connected to an amplifier and a current-measuring device. In Fig. 42–3, unpolarized light is incident on a polarizer whose axis is represented by the broken line. As the polarizer is rotated about an axis parallel to the incident ray, the intensity does not change. The polarizer transmits the components of the incident waves in which the E-vector is parallel to the transmission direction of the polarizer, and by symmetry the components are equal for all azimuths.

The transmitted intensity is found to be exactly one-half that of the incident light. This may be understood as follows: The incident light can always be resolved into components polarized parallel to the polarizer

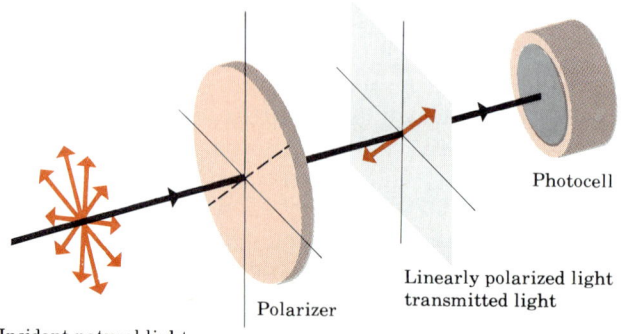

42-3 The intensity of the transmitted linearly polarized light is the same at all azimuths of the polarizer.

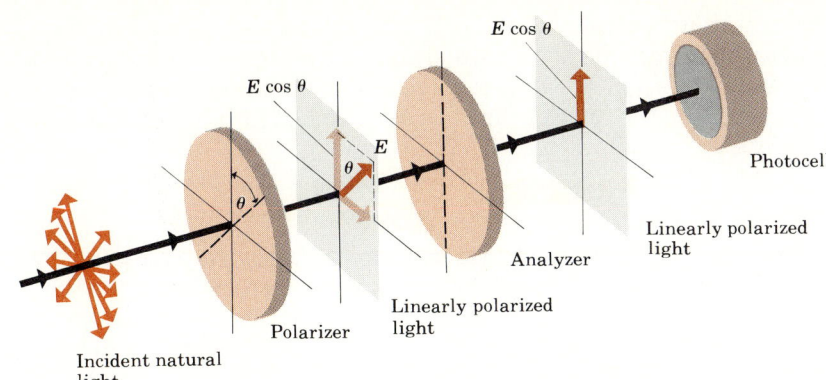

42-4 The analyzer transmits only the component parallel to its transmission direction or polarizing axis.

axis and components polarized perpendicular to it. Because the incident light is a random mixture of all states of polarization, these two components are, on the average, equal. Thus (in the ideal polarizer) exactly half of the incident intensity, that corresponding to the component parallel to the polarizer axis, is transmitted.

Suppose now that a second polarizer is inserted between the first polarizer and the photocell, as in Fig. 42-4. Let the transmission direction of the second polarizer, or *analyzer,* be vertical, and let that of the first polarizer make an angle θ with the vertical. The linearly polarized light transmitted by the polarizer may be resolved into two components as shown, one parallel and the other perpendicular to the transmission direction of the analyzer. Evidently only the parallel component, of amplitude $E \cos \theta$, will be transmitted by the analyzer. The transmitted intensity is maximum when $\theta = 0$, and is zero when $\theta = 90°$, or when polarizer and analyzer are *crossed*. At intermediate angles, since the quantity of energy is proportional to the *square* of the amplitude, we have

$$I = I_{\max} \cos^2 \theta, \qquad (42\text{-}1)$$

where $I_{\max}$ is the maximum amount of light transmitted and I is the amount transmitted at the angle θ. This relation, discovered experimentally by Etienne Louis Malus in 1809, is called *Malus' Law.*

The angle θ is the angle between the transmission directions of polarizer and analyzer. If either the analyzer or the polarizer is rotated, the amplitude of the transmitted beam varies with the angle between them according to Eq. (42-1).

Example In Fig. 42-4, let the incident unpolarized light have intensity I_0. Find the intensity transmitted by the first polarizer and by the second, if the angle θ is 30°.

Solution As explained above, the intensity after the first filter is $(I_0/2)$. According to Eq. (42-1), the second filter reduces the intensity by a factor of $\cos^2 30° = \frac{3}{4}$. Thus the intensity transmitted by the second polarizer is $(I_0/2)(\frac{3}{4}) = 3I_0/8$.

If, in Fig. 42-3, the incident light is completely linearly polarized rather than unpolarized, then of course the transmitted intensity varies from zero to some maximum value as the polarizer axis is turned from perpendicular to parallel to the polarization direction of the light. Or the incident light may be a mixture of linearly polarized and unpolarized, with intensities I_{pol} and I_{un}, respectively. The *percent polarization* is then defined as the ratio of the polarized intensity to the total intensity, times 100%:

$$\% \text{ polarization} = \frac{I_{\text{pol}}}{I_{\text{pol}} + I_{\text{un}}} \times 100\%. \tag{42-2}$$

In this case, in Fig. 42–3 the transmitted intensity varies from a minimum (but not zero) when the polarizer axis is perpendicular to the polarized component of incident light, to a maximum when it is parallel. The percent polarization can be expressed in terms of these two intensities, which may be denoted as $I_{\min}$ and $I_{\max}$. Recalling that half the unpolarized intensity is transmitted by the polarizer, we may write

$$I_{\min} = \tfrac{1}{2} I_{\text{un}}, \qquad I_{\max} = \tfrac{1}{2} I_{\text{un}} + I_{\text{pol}}. \tag{42-3}$$

These equations can be solved for I_{pol} and I_{un} and the results substituted into Eq. (42–2). The details of this calculation are left as an exercise; the result is

$$\% \text{ polarization} = \frac{I_{\max} - I_{\min}}{I_{\max} + I_{\min}} \times 100\%. \tag{42-4}$$

42–3 Polarization by reflection

There are various processes by which an initially unpolarized beam of light can yield a partially polarized beam. One of these is the familiar process of *reflection*. When natural light strikes a reflecting surface between two optical media, there is found to be a preferential reflection for those waves in which the electric-field vector is perpendicular to the plane of incidence (the plane containing the incident ray and the normal to the surface). An exception is that at *normal* incidence all directions of polarization are reflected equally. At one particular angle of incidence, known as the *polarizing angle* ϕ_p, no light whatever is reflected except that in which the electric vector is perpendicular to the plane of incidence. This case is illustrated in Fig. 42–5.

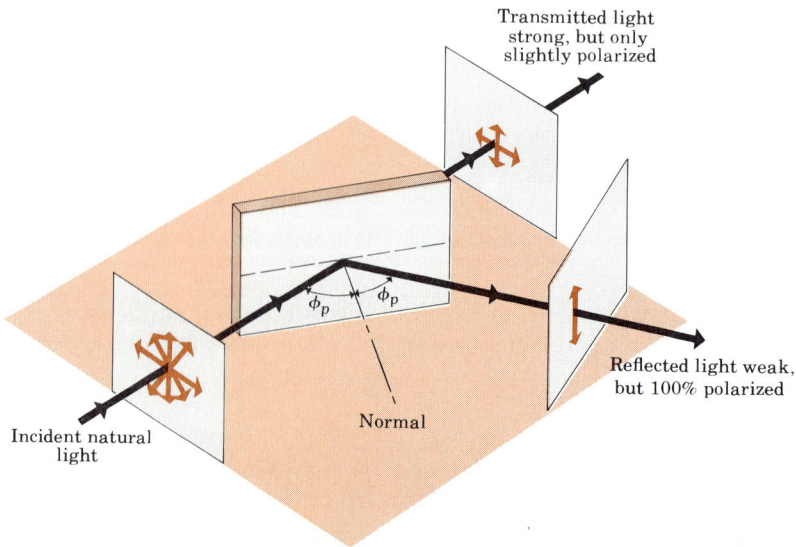

42–5 When light is incident at the polarizing angle, the reflected light is linearly polarized.

When light is incident at the polarizing angle, *none* of the component parallel to the plane of incidence is reflected; this component is 100% transmitted in the *refracted* beam. Of the component perpendicular to the plane of incidence, about 15% is reflected if the reflecting surface is glass. The fraction reflected depends on the index of the reflecting material. Hence the *reflected* light is weak and *completely* linearly-polarized. The *refracted* light is a mixture of the parallel component, all of which is refracted, and the remaining 85% of the perpendicular component. It is therefore strong, but only *partially* polarized.

At angles of incidence other than the polarizing angle, some of the component parallel to the plane of incidence is reflected, so that, except at the polarizing angle, the reflected light is not completely linearly-polarized.

In 1812, Sir David Brewster noticed that when the angle of incidence is equal to ϕ_p, the reflected ray and the refracted ray are *perpendicular* to each other, as shown in Fig. 42–6. When this is the case, the angle of refraction ϕ' becomes the complement of ϕ_p, so that $\sin \phi' = \cos \phi_p$. Since

$$n \sin \phi_p = n' \sin \phi',$$

we find $n \sin \phi_p = n' \cos \phi_p$, and

$$\tan \phi_p = \frac{n'}{n}, \qquad (42\text{–}5)$$

a relation known as *Brewster's law*.

42-6 At the polarizing angle the reflected and transmitted rays are perpendicular to each other.

*42-4 Double refraction

The progress of a wave train through a homogeneous isotropic medium, such as glass, may be determined graphically by Huygens' construction. The secondary wavelets in such a medium are spherical surfaces. There exist, however, many transparent crystalline substances which, while homogeneous, are *anisotropic*. That is, the velocity of a light wave in them is not the same in all directions. Crystals having this property are said to be *doubly refracting,* or *birefringent*. *Two* sets of Huygens wavelets propagate from every wave surface in such a crystal, one set being spherical and the other ellipsoidal.

A consequence of this property of anisotropic crystals is that a ray of light striking such a crystal at normal incidence is broken up into two rays as it enters the crystal. The ray that corresponds to wave surfaces tangent to the spherical wavelets is undeviated and is called the *ordinary ray*. The ray corresponding to the wave surfaces tangent to the ellipsoids is deviated even though the incident ray is normal to the surface, and is called the *extraordinary ray*. If the crystal is rotated about the incident ray as an axis, the ordinary ray remains fixed but the extraordinary ray revolves around it, as shown in Fig. 42–7. Furthermore, for angles of incidence other than 0°, Snell's law (that is, $\sin \phi/\sin \phi' =$ constant) holds for the ordinary but *not* for the extraordinary ray, since the velocity of the latter is different in different directions.

The index of refraction for the extraordinary ray is therefore a function of direction. There is always one direction in the crystal for which there is no distinction between the *O*- and *E*-rays; this direction is called the *optic axis*. It is customary to state the index for the direction at right

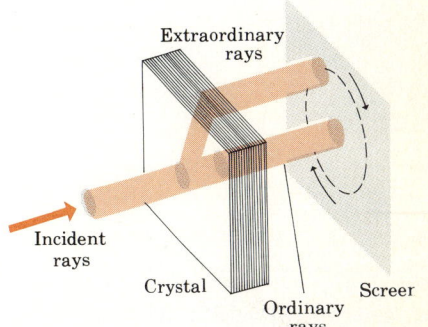

42-7 A narrow beam of natural light can be split into two beams by a doubly refracting crystal.

Table 42-1 Indexes of refraction of doubly refracting crystals (for light wavelength 589 nm)		
Material	n_O	n_E
Calcite	1.658	1.486
Quartz	1.544	1.553
Tourmaline	1.64	1.62
Ice	1.306	1.307

angles to the optic axis, in which the velocity is a maximum or a minimum. Some values of n_O and n_E, the indices for the ordinary and extraordinary rays, are listed in Table 42-1.

42-5 Polarizers

Experiment shows that the ordinary and extraordinary waves in a doubly refracting crystal are linearly polarized in mutually perpendicular directions. Consequently, if some means can be found to separate one wave from the other, a doubly refracting crystal may be used to obtain plane-polarized light from natural light. There are several ways in which this separation may be accomplished:

1. One of the rays may be made to undergo internal reflection and be deflected to one side, allowing the other to proceed undeflected.

2. Both rays may be separated slightly so that, at sufficient distance from the separating prism, only one ray is intercepted.

3. One of the rays may be absorbed while the other is unaffected.

Many ingenious composite prisms of doubly refracting crystals have been developed as polarizers in the last 150 years. They are known by the names of the discoverers: Nicol, Rochon, Wollaston, Glan, Foucault, Ahrens, and many others. The best known is the Nicol prism, invented in 1828 by the Scottish physicist William Nicol. It is complicated and difficult to construct, uses a large amount of optically clear calcite, produces a lateral displacement of the transmitted beam, gives rise to a distorted image, and produces less than 100% linearly polarized light.

A prism originally developed by Glan and Thompson and later modified by Ammann and Massey (*Journal of the Optical Society of America,* November 1968) is constructed on a similar principle but avoids the defects of the Nicol prism. The modified Glan–Thompson prism consists of a glass prism with index of refraction 1.655 joined to a calcite prism, with a cement that has the same index of refraction, as shown in Fig. 42-8. The natural unpolarized light incident at the left is equivalent to two equal linearly polarized beams. The direction of polarization of one (represented by dots) is perpendicular to the page and that of the other (represented by short vertical lines) is parallel to the page. Both beams travel the same path with equal speed in the glass. In calcite, however, with the optic axis vertical as shown, perpendicular vibrations constitute

42-8 Glan–Thompson polarizing prism, as modified by Ammann and Massey. The thickness of the layer of cement is exaggerated.

the ordinary ray for which calcite has an index of refraction 1.6583. This value is so close to 1.655 that the perpendicular vibrations proceed without deflection from glass to cement to calcite and emerge into the air as 100% linearly polarized light.

The parallel vibrations, if they existed in the calcite crystal, would constitute the extraordinary ray with an index of refraction 1.4864. In traveling *in the cement* toward the calcite, this ray proceeds toward a medium in which its speed is greater, and therefore a critical angle exists. With the dimensions of the prism as shown, the angle of incidence of the ray with parallel polarization exceeds the critical angle, and this ray is therefore internally reflected and deflected from its original direction. This polarizing prism functions properly for all incident rays that make an angle of not more than 10° above or below the horizontal.

Certain doubly refracting crystals exhibit *dichroism;* that is, one of the polarized components is *absorbed* much more strongly than the other. Hence, if the crystal is cut of the proper thickness, one of the components is practically extinguished by absorption, while the other is transmitted in appreciable amount, as indicated in Fig. 42-2. Tourmaline is one example of such a dichroic crystal.

An early form of Polaroid, invented by Edwin H. Land in 1928, consists of a thin layer of tiny needlelike dichroic crystals of herapathite (iodoquinine sulfate), in parallel orientation, embedded in a plastic matrix and enclosed for protection between two transparent plates. A more recent modification, developed by Land in 1938 and known as an H-sheet, is a *molecular* polarizer. It consists of long polymeric molecules of polyvinyl alcohol (PVA) that have been given a preferred direction by stretching, and have been stained with an ink containing iodine that causes the sheet to exhibit dichroism. The PVA sheet is laminated to a support sheet of cellulose acetate butyrate.

Polaroid disks do not polarize all wavelengths equally. When two such disks are crossed, small amounts of red and of violet (the two ends of the visible spectrum) are transmitted. When white light passes through one Polaroid sheet, the transmitted light is slightly colored. The large area of such plates, however, and their moderate cost more than compensate for these small deficiencies. The existence of Polaroid sheets has stimulated the development and applications of polarized light to an extent that was out of the question when reflecting surfaces, Nicol prisms, and other costly devices had to be used.

Polaroid sheet is widely used in sunglasses where, from the standpoint of its polarizing properties, it plays the role of the analyzer in Fig. 42-4. We have seen that, when unpolarized light is reflected, there is a preferential reflection for light polarized perpendicular to the plane of incidence. When sunlight is reflected from a horizontal surface, the plane of incidence is vertical. Hence, in the reflected light there is a preponderance of light polarized in the horizontal direction, the proportion being greater the nearer the angle of incidence is to the polarizing angle. When such reflection occurs at smooth asphalt road surfaces, the surface of a lake, or a similar situation, it causes unwanted "glare," and vision is improved by eliminating it. The transmission direction of the Polaroid sheet in the sunglasses is vertical, so none of the horizontally polarized light is transmitted to the eyes.

Apart from this polarizing feature, these glasses serve the same purpose as any dark glasses, absorbing 50% of the incident light; in an unpolarized beam, half the light can be considered as polarized horizontally and half vertically, and only the vertically polarized light is transmitted. The sensitivity of the eye is independent of the state of polarization of the light.

*42-6 The scattering of light

The sky is blue. Sunsets are red. Skylight is partially polarized, as can readily be verified by looking at the sky directly overhead through a polarizing filter. It turns out that one phenomenon is responsible for all three of these effects.

In Fig. 42-9, sunlight (unpolarized) comes from the left along the z-axis and passes over an observer looking vertically upward along the y-axis. Molecules of the earth's atmosphere are located at point O. The electric field in the beam of sunlight sets the electric charges in the molecules in vibration. Since light is a transverse wave, the direction of the electric field in any component of the sunlight lies in the xy-plane and the motion of the charges takes place in this plane. There is no field, and hence no vibration, in the direction of the z-axis.

An arbitrary component of the incident light, vibrating at an angle θ with the x-axis, sets the electric charges in the molecules vibrating in the same direction, as indicated by the heavy line through point O. In the usual way, we can resolve this vibration into two components, one along the x-axis and the other along the y-axis. The result, then, is that each component in the incident light produces the equivalent of two molecular "antennas," oscillating with the frequency of the incident light, and lying along the x- and y-axis.

It has been explained in Sec. 37-7 that an antenna does not radiate in the direction of its own length. Hence, the antenna along the y-axis

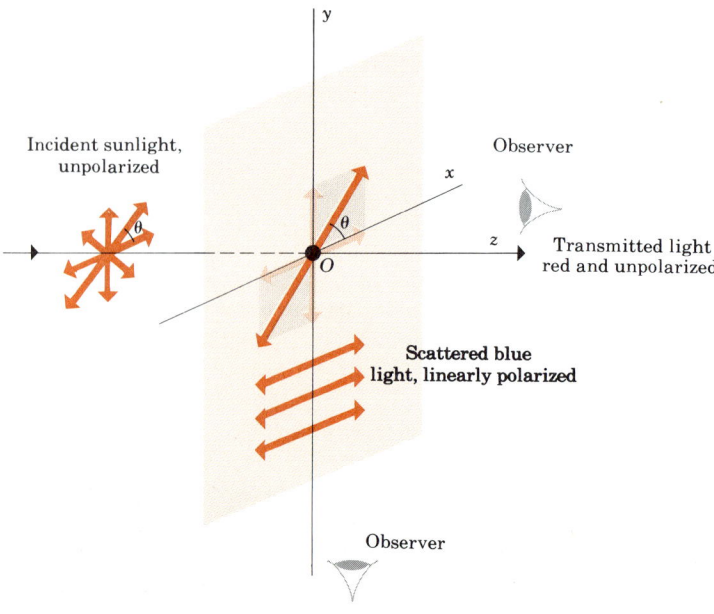

42-9 Scattered light is linearly polarized.

does not send any light to the observer directly below it. It does, of course, send out light in other directions. The only light reaching the observer comes from the component of vibration along the x-axis, and, as is the case with the waves from any antenna, this light is linearly polarized, with the electric field parallel to the antenna. The vectors on the y-axis below point O show the direction of polarization of the light reaching the observer.

The process described above is called *scattering*. The energy of the scattered light is removed from the original beam, which becomes weakened in the process. Analysis of the details of the scattering process shows that, for a given intensity of incident light, the intensity of the scattered light increases with increasing frequency. In other words, blue light is scattered more than red, with the result that the hue of the scattered light is blue.

Polarizers are commonly used in photography. Because skylight is partially polarized, the sky can be darkened in a photograph by appropriate orientation of the polarizer axis. The effect of atmosphere haze can be reduced in exactly the same way, and unwanted reflections can be controlled just as with polarizing sunglasses, discussed in Sec. 42–5.

Toward evening, when sunlight has to travel a large distance through the earth's atmosphere to reach a point over or nearly over an observer, a large proportion of the blue light in sunlight is removed from it by scattering. White light minus blue light is yellow or red in hue. Thus when sunlight, with the blue component removed, is incident on a cloud, the light reflected from the cloud to the observer has the yellow or red hue so commonly seen at sunset.

From the discussion above, it follows that if the earth had no atmosphere we would receive *no* skylight at the earth's surface, and the sky would appear as black in the daytime as it does at night. Thus, to an astronaut in a spaceship or on the moon, the sky appears black, not blue.

42–7 Circular and elliptical polarization

Up to this point we have discussed polarization phenomena in terms of *linearly polarized* light. This is the simplest kind of polarization, but not the only kind. Light (and all other electromagnetic radiation) may also have *circular* or *elliptical* polarization. To introduce these new concepts, we return again to mechanical waves on a stretched string.

In Fig. 42–1, suppose the two linearly polarized waves in parts (a) and (b) have equal amplitude and are *superimposed*; then each point in the string has simultaneously y- and z-displacements of equal magnitude, and a little thought shows that the resultant wave lies in a plane oriented at 45° to the xy-plane. The amplitude of the resultant wave is larger by a factor of $\sqrt{2}$ than that of either component wave, and the resultant wave is again linearly polarized. The motion of each point on the string is a superposition of two simple harmonic motions at right angles and in phase, which is a simple harmonic motion at 45° to the component motions.

But now suppose one component wave differs in phase by a quarter-cycle from the other. Then the resultant motion of each point corresponds to a superposition of two simple harmonic motions at right angles with a quarter-cycle phase difference. As a specific example, suppose that the y- and z-components of motion of a certain point are given by

$$y = A \sin \omega t,$$
$$z = A \sin \left(\omega t + \frac{\pi}{2}\right) = A \cos \omega t. \qquad (42\text{-}6)$$

Then z is maximum at times when y is zero, and conversely, and the resultant motion of the point is no longer back and forth along a line perpendicular to the direction of propagation of the wave. In fact, the point moves in a *circle*; to show this, we square the two equations and add, obtaining

$$y^2 + z^2 = A^2 \sin^2 \omega t + A^2 \cos^2 \omega t = A^2. \qquad (42\text{-}7)$$

Now $y^2 + z^2 = A^2$ is the equation of a circle of radius A in the yz-plane, and the circle is traversed with constant angular velocity equal to the angular frequency ω. Similarly, every other point in the string undergoes circular motion, but with phase differences depending on position relative to the end of the string.

The overall motion of the string then has the appearance of a rotating helix. This particular superposition of two linearly polarized waves is called *circular polarization*. By convention, the wave is said to be *right circularly polarized* when, as in the present instance, the sense of motion of a particle, to an observer looking *backwards* along the direction of propagation, is *clockwise,* and *left circularly polarized* when it appears counterclockwise to that observer. The latter would be the result if the phase difference between y- and z-components were opposite to that in our example.

More generally, the phase difference need not be $\pi/2$. Figure 42–10 shows the result of combining a horizontal and a vertical simple harmonic motion with equal amplitudes, but with the horizontal having various phase differences with respect to the vertical. We note the following results:

1. When the phase difference is 0, 2π, or any even multiple of π, the result is a linear vibration at 45° to both original vibrations.

2. When the phase difference is π, 3π, or any *odd* multiple of π, the result is also a linear vibration, but at right angles to that corresponding to even multiples of π.

3. When the phase difference is $\pi/2$, $3\pi/2$, or any odd multiple of $\pi/2$, the resulting vibration is *circular*.

4. At all other phase differences, the resulting vibration is *elliptical*.

For electromagnetic waves of radio frequencies, circular or elliptical polarization can be produced by using two antennas at right angles, fed from the same transmitter but with a phase-shifting circuit that intro-

42–10 Vibrations that result from the combination of a horizontal and a vertical simple harmonic motion of the same frequency and the same amplitude, for various values of the phase difference.

0	$\frac{\pi}{4}$	$\frac{\pi}{2}$	$\frac{3\pi}{4}$	π	$\frac{5\pi}{4}$	$\frac{3\pi}{2}$	$\frac{7\pi}{4}$	2π

duces the desired phase difference between the two waves. For light, the necessary phase shift can be introduced by use of a birefringent material. In Fig. 42–7, when the crystal faces are parallel to the optic axis of the crystal, the ordinary and extraordinary rays are not spatially separated, but they travel through the crystal with different *speeds*, corresponding to the two different indexes of refraction. Thus if the O-wave and E-wave are in phase as they enter the crystal, they are in general *not* in phase when they emerge.

Let us consider the optical apparatus shown in Fig. 42–11. After unpolarized light traverses the polarizer, it is linearly polarized, with the polarization direction parallel to the broken line on the polarizer. This linearly polarized light then enters a crystal plate, cut so that the light travels perpendicular to the optic axis and oriented so that the optic axis makes an angle of 45° with the direction of vibration of the linearly polarized light incident upon it. Since the E-vibration is, in this case, parallel to the optic axis, and the O-vibration is perpendicular to it, it follows that the amplitudes of the E- and O-beams are identical. The E- and O-beams travel through the crystal along the same path but with different speeds and, as they are about to emerge from the second crystal face, they combine to form one of the vibrations shown in Fig. 42–10, depending on the phase difference.

The phase difference between the E- and O-vibrations at the second face of the crystal depends on the following: (1) the frequency of the light, (2) the indexes of refraction of the crystal for E- and O-light, and (3) the thickness of the crystal.

If a given crystal has such a thickness as to give rise to a phase difference of $\pi/2$ for a given frequency, then, according to Fig. 42–10, a *circular* vibration results, and the light emerging from this crystal is said to be *circularly polarized light*. The crystal itself is called a *quarter-wave plate*. If the crystal plate shown in Fig. 42–11 is a quarter-wave plate, the intensity of light transmitted by the analyzer will remain constant as the analyzer is rotated. In other words, if an analyzer alone is used to analyze circularly polarized light, it will give the same result as when used to analyze unpolarized light.

A quarter-wave plate for, say, green light is not a quarter-wave plate for any other color. Other colors have different frequencies and different E- and O-indices. Hence the phase difference would not be $\pi/2$. (See Problem 42–19.)

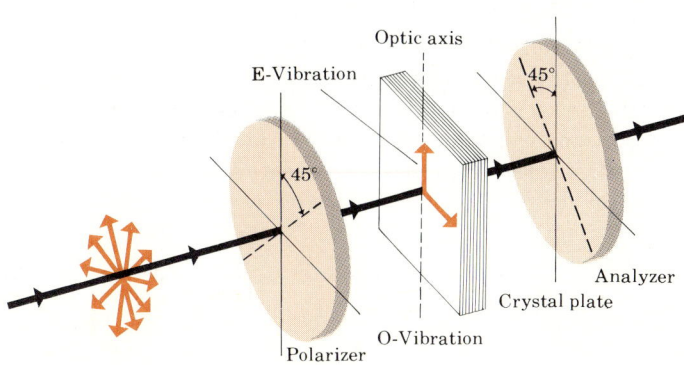

42–11 Crystal plate between crossed Polaroids. Since the optic axis makes an angle of 45° to the vibration direction of the linearly polarized light transmitted by the polarizer, the amplitudes of the O- and E-vibrations in the crystal are equal.

If the crystal has such a thickness as to give rise to a phase difference of π for a given frequency, then, according to Fig. 42–10, a linear vibration *perpendicular* to the incident vibration direction results. The light that emerges from this crystal is linearly polarized; and by rotating the analyzer, we may find a position of the analyzer at which this light will be completely stopped. A crystal plate of this sort is called a *half-wave plate* for the given frequency of light. For any other frequency it would not be a half-wave plate.

If the phase difference produced by a crystal plate is such as to produce an elliptical vibration, the emerging light is said to be *elliptically polarized*.

42-8 Optical stress analysis

When a polarizer and an analyzer are mounted in the "crossed" position, i.e., with their transmission directions at right angles to each other, no light is transmitted through the combination. But if a doubly refracting crystal is inserted between polarizer and analyzer, the light after passing through the crystal is, in general, elliptically polarized, and some light will be transmitted by the analyzer. Thus the field of view, dark in the absence of the crystal, becomes light when the crystal is inserted.

Some substances, such as glass and various plastics, while not normally doubly refracting, become so when subjected to mechanical stress. From a study of the specimen between crossed polarizers, much information regarding the stresses can be obtained. Improperly annealed glass, for example, may be internally stressed to an extent that might cause it later to develop cracks. It is evidently important that optical glass should be free from such a condition before it is subjected to expensive grinding and polishing. Hence, such glass is always examined between crossed polarizers before grinding operations are begun.

The double refraction produced by stress is the basis of the science of *photoelasticity*. The stresses in opaque engineering materials, such as girders, boiler plates, gear teeth, etc., can be analyzed by constructing a transparent model of the object, usually of a plastic, and examining it between a polarizer and an analyzer in the crossed position. Very complicated stress distributions, such as those around a hole or a gear tooth, which it would be practically impossible to analyze mathematically, may thus be studied by optical methods. Figure 42–12 is a photograph of a photoelastic model under stress.

42-12 Photoelastic stress analysis. (Courtesy of Dr. W. M. Murray, Massachusetts Institute of Technology.)

Liquids are not normally doubly refracting, but some become so when an electric field is established with them. This phenomenon is known as the *Kerr effect*. The existence of the Kerr effect makes it possible to construct an electrically controlled "light valve." A cell with transparent walls contains the liquid between a pair of parallel plates. The cell is inserted between crossed Polaroid disks. Light is transmitted when an electric field is set up between the plates and is cut off when the field is removed.

42-9 Optical activity

When a beam of linearly polarized light is sent through certain types of crystals and certain liquids, the direction of polarization of the emerging linearly polarized light is found to be different from the original direction. This phenomenon is called *rotation of the direction of polarization,* and substances that exhibit the effect are called *optically active.* Those that rotate the direction of polarization to the right, looking along the advancing beam, are called *dextrorotatory* or righthanded; those that rotate it to the left, *levorotatory* or lefthanded.

Optical activity may be due to an asymmetry of the molecules of a substance, or it may be a property of a crystal as a whole. For example, solutions of cane sugar are dextrorotatory, indicating that the optical activity is a property of the sugar molecule. The molecules of the sugars dextrose and levulose are mirror images, and their optical activities are opposite. The rotation of the plane of polarization by a sugar solution is used commercially as a method of determining the proportion of cane sugar in a given sample. Crystalline quartz is also optically active, some natural crystals being righthanded and others lefthanded. Here the optical activity is a consequence of the crystalline structure, since it disappears when the quartz is melted and allowed to resolidify into a glassy, noncrystalline state called fused quartz.

Questions

42-1 It has been proposed that automobile windshields and headlights should have polarizing filters, to reduce the glare of oncoming lights during night driving. Would this work? How should the polarizing axes be arranged? What advantages would this scheme have. What disadvantages?

42-2 A salesperson at a bargain counter claims that a certain pair of sunglasses has Polaroid filters; you suspect they are just tinted plastic. How could you find out for sure?

42-3 The fact that light consists of transverse rather than longitudinal waves was known even before the electromagnetic nature of light was firmly established. How was this determined?

42-4 When unpolarized light is incident on two crossed polarizers, no light is transmitted. A student asserted that if a third polarizer is inserted between the other two, some transmission may occur. Does this make sense? How can adding a third filter *increase* transmission?

42-5 How could you determine the direction of the polarizing axis of a single polarizer?

42-6 When unpolarized light is incident on a polarizer, only half the intensity is transmitted. What becomes of the energy of the untransmitted part? Is the answer different for the various polarizers described in Sec. 42-5?

42-7 In three-dimensional movies, two images are projected on the screen, and the viewers wear special glasses to sort them out. How does this work?

42-8 Suppose a two-slit interference experiment, as described in Chapter 41, is carried out with a polarizer in front of each slit. Would interference effects be observed when the two polarizing axes are parallel? When they are perpendicular? When they are at 45°?

42-9 The phenomenon of polarization is usually cited as evidence for the transverse nature of electromagnetic radiation. But how can we be sure it does not have *both* longitudinal and transverse components? How could one test for this possibility?

42-10 In Fig. 42-9, since the light scattered out of the incident beam is polarized, why is the transmitted beam not also partially polarized?

42-11 Light from the blue sky is strongly polarized because of the nature of the scattering process described in Sec. 42-6. But light from white clouds is usually *not* polarized. Why not?

42-12 When a sheet of plastic food wrap is placed between two crossed polarizers, no light is transmitted. When the sheet is stretched in one direction, some light passes through. What is happening?

42-13 Television transmission usually uses plane-polarized waves. It has been proposed to use circularly polarized waves to improve reception. Why? [*Hint:* See Problem 42-23.]

42-14 Given two polarizers and a quarter-wave plate, describe how one might determine completely the composition of a given light beam, with respect to its state of polarization, such as unpolarized, partially circularly polarized, completely elliptically polarized, and so on.

Problems

42-1 Unpolarized light of intensity I_0 is incident on a polarizing filter, and the emerging light strikes a second polarizing filter with its axis at 45° to that of the first. Determine

a) the intensity of the emerging beam, and

b) its state of polarization.

42-2 Three polarizing filters are stacked, with the polarizing axes of the second and third at 45° and 90°, respectively, with that of the first.

a) If unpolarized light of intensity I_0 is incident on the stack, find the intensity and state of polarization of light emerging from each filter.

b) If the second filter is removed, how does the situation change?

42-3 Three polarizing filters are stacked, with the polarizing axes of the second and third at angles θ and 90°, respectively, with that of the first. Unpolarized light of intensity I_0 is incident on the stack.

a) Derive an expression for the intensity of light transmitted through the stack, as a function of I_0 and θ.

b) Take the derivative of this expression with respect to θ to show that maximum transmission occurs when $\theta = 45°$.

42-4 Complete the derivation of Eq. (42-4) in the text.

42-5 It is desired to rotate the direction of polarization of linearly polarized light 90°, using two Polaroid filters. Explain how this can be done, and find the final intensity in terms of the incident intensity.

42-6 A polarizer and an analyzer are oriented so that the maximum amount of light is transmitted. To what fraction of its maximum value is the intensity of the transmitted light reduced when the analyzer is rotated through (a) 30°, (b) 45°, (c) 60°?

42-7 A beam of light is incident on a liquid of 1.40 refractive index. The reflected rays are completely polarized. What is the angle of refraction of the beam?

42-8 The critical angle of light in a certain substance is 45°. What is the polarizing angle?

42-9

a) At what angle above the horizontal must the sun be in order that sunlight reflected from the surface of a calm body of water shall be completely polarized?

b) What is the plane of the **E**-vector in the reflected light?

42-10 A parallel beam of "natural" light is incident at an angle of 58° on a plane glass surface. The reflected beam is completely linearly polarized.

a) What is the angle of refraction of the transmitted beam?

b) What is the refractive index of the glass?

42-11 The refractive index of a certain flint glass is 1.65. For what incident angle is light reflected from the surface of this glass completely polarized if the glass is immersed in (a) air? (b) water?

42-12 The Glan–Thompson prism shown in Fig. 42-8 has a horizontal length of 10 cm.

a) What is the critical angle for a ray in the cement approaching the calcite?

b) If a horizontal ray approaching the calcite is to make an angle of incidence 10° larger than the critical angle, what should be the vertical height of the prism?

42-13 A certain birefringent material has a refractive index of 1.71 for the ordinary ray and 1.74 for the extraordinary ray, for 600-nm light. What thickness of material is needed for a quarter-wave plate?

42-14 A parallel beam of linearly polarized light of wavelength 589 nm (in vacuum) is incident on a calcite crystal. Find the wavelengths of the ordinary and extraordinary waves in the crystal.

42-15 A beam of linearly polarized light strikes a calcite crystal, the direction of the electric vector making an angle of 60° with the optic axis.

a) What is the ratio of the amplitudes of the two refracted beams?

b) What is the ratio of their intensities?

42-16 Figure 42-13 represents a Wollaston prism made of two prisms of quartz cemented together. The optic axis of the righthand prism is perpendicular to the page, whereas that of the lefthand prism is parallel. The incident light is normal to the surface and gives rise to O- and E-beams that travel along the same path in the lefthand prism but with different speeds. Copy Fig. 42-13 and show on your diagram how the O- and E-beams are bent in going into the righthand prism and thence into the air.

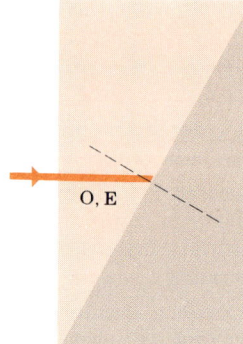

Figure 42-13

42-17 A beam of light, after passing through the Polaroid disk P_1 in Fig. 42-14, traverses a cell containing a scatter-

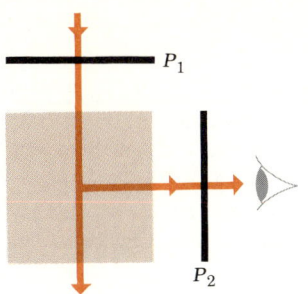

Figure 42-14

ing medium. The cell is observed at right angles through another Polaroid disk P_2. Originally, the disks are oriented until the brightness of the field as seen by the observer is a maximum.

a) Disk P_2 is now rotated through 90°. Is extinction produced?
b) Disk P_1 is now rotated through 90°. Is the field bright or dark?
c) Disk P_2 is then restored to its original position. Is the field bright or dark?

42-18 In Fig. 42-15, A and C are Polaroid sheets whose transmission directions are as indicated. B is a sheet of doubly refractive material whose optic axis is vertical. All three sheets are parallel. Unpolarized light enters from the left. Discuss the state of polarization of the light at points 2, 3, and 4.

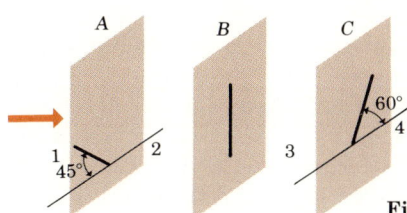

Figure 42-15

42-19 The phase difference δ between the E- and O-rays emerging from a birefringent crystal as in Fig. 42-11 is equal to 2π times the difference in the number of wavelengths in the path for the two rays. The wavelength for the O-ray is $\lambda_O = \lambda/n_O$, where λ is the wavelength in vacuum and n_O is the refractive index for the ordinary ray. Hence the number of wavelengths in a crystal of thickness t is tn_O/λ. A similar expression holds for the extraordinary ray.

a) Show that the phase difference is given by

$$\delta = \frac{2\pi}{\lambda} t(n_O - n_E).$$

b) Show that the minimum thickness of a quarter-wave plate is given by $t = \lambda/4(n_O - n_E)$.
c) What is this minimum thickness for a quarter-wave calcite plate and light of 400-nm wavelength?

42-20 What is the state of polarization of the light transmitted by a quarter-wave plate when the electric vector of the incident linearly polarized light makes an angle of 30° with the optic axis?

42-21 Assume the values of n_O and n_E for quartz to be independent of wavelength. A certain quartz crystal is a quarter-wave plate for light of wavelength 800 nm (in vacuum). What is the state of polarization of the transmitted light when linearly polarized light of wavelength 400 nm (in vacuum) is incident on the crystal, the direction of polarization making an angle of 45° with the optic axis?

42-22 Consider two vibrations, one along the y-axis,

$$y = a \sin(\omega t - \alpha),$$

and the other along the z-axis, of equal amplitude and frequency, but differing in phase,

$$z = a \sin(\omega t - \beta).$$

Let us write them as follows:

$$\frac{y}{a} = \sin \omega t \cos \alpha - \cos \omega t \sin \alpha, \quad (1)$$

$$\frac{z}{a} = \sin \omega t \cos \beta - \cos \omega t \sin \beta. \quad (2)$$

a) Multiply Eq. (1) by $\sin \beta$ and Eq. (2) by $\sin \alpha$ and then subtract the resulting equations.
b) Multiply Eq. (1) by $\cos \beta$ and Eq. (2) by $\cos \alpha$ and then subtract the resulting equations.
c) Square and add the results of (a) and (b).
d) Derive the equation $y^2 + z^2 - 2yz \cos \delta = a^2 \sin^2 \delta$, where $\delta = \alpha - \beta$.
e) Justify the diagrams in Fig. 42-10.

42-23 A beam of right circularly polarized light is reflected at normal incidence from a reflecting surface. Is the reflected beam right or left circularly polarized? Explain.

43 RELATIVISTIC MECHANICS

43-1 Invariance of physical laws

In preceding chapters we have stressed the importance of inertial frames of reference. Newton's laws of motion are valid only in inertial frames, but they are valid in *all* inertial frames. Any frame moving with constant velocity with respect to an inertial frame is itself an inertial frame, and all such frames are equivalent with respect to expressing the basic principles of mechanics. *The laws of mechanics are the same in every inertial frame of reference.*

Einstein proposed in 1905 that this principle should be extended to include *all* the basic laws of physics. This innocent-sounding proposition has far-reaching and startling consequences, a few of which have already been mentioned. If the principle of conservation of momentum is to be valid in all inertial systems, for example, the definition of momentum, for particles moving at speeds comparable to the speed of light, must be changed from mv to $mv/(1 - v^2/c^2)^{1/2}$, as discussed in Sec. 7-9. Even more fundamental are the modifications needed in the *kinematic* aspects of motion, as we shall see. Nevertheless, Einstein's *principle of relativity*, as it has come to be called, is now accepted as an essential requirement for a physical theory. The principle of relativity states that *the laws of physics are the same in every inertial frame of reference.*

A familiar example of this principle in electromagnetism is the induced electromotive force in a coil of wire resulting from motion of a nearby permanent magnet. In a frame of reference where the *coil* is stationary, the moving magnet causes a change of magnetic flux through the coil and hence an induced emf. In a frame where the *magnet* is stationary, the coil is moving through a magnetic field, and this motion causes magnetic-field forces on the mobile charges in the conductor, inducing an emf. According to the principle of relativity, both of these points of view have equal validity and both must predict the same result for the induced emf. As we have seen in Chapter 33, Faraday's law of electromagnetic induction does indeed satisfy this requirement.

Of equal significance is the prediction of the speed of light and other electromagnetic radiation, emerging from the development in Chapter

37. The principle of relativity requires that the speed of propagation of this radiation must be independent of the frame of reference and must be the same for all inertial frames.

Indeed, the speed of light plays a special role in the theory of relativity. Light travels in vacuum with speed c that is independent of the motion of the source. Its numerical value is known very precisely; to six significant figures it is

$$c = 2.99792 \times 10^8 \text{ m} \cdot \text{s}^{-1}.$$

The approximate value $c = 3.00 \times 10^8$ m·s^{-1} is in error by less than one part in 1000 and is often used when greater precision is not required.

At one time it was thought that light traveled through a hypothetical medium called *the ether,* just as sound waves travel through air. During the late nineteenth and early twentieth centuries, intensive efforts were made to find experimental evidence for the existence of the ether. The Michelson–Morley experiment, described in Sec. 41–6, was an effort to detect motion of the earth relative to the ether. This and all similar experiments yielded consistently negative results, and it is now known that *there is no ether.* With this result in mind, let us suppose the speed of light is measured by two observers, one at rest with respect to the source, the other moving away from it. Both are in inertial frames of reference, and according to Einstein's principle of relativity, the laws of physics, and in particular the speed of light, must be the same in both frames.

For example, suppose a light source is located in a spaceship moving with respect to earth. An observer moving with the spaceship measures the speed of the light and obtains the value c. But after the light has left the source its motion cannot be influenced by the motion of the source, so an observer on earth measuring the speed of this same light must also obtain the value c, despite the fact that there is relative motion between the two observers. This conclusion may appear not to be consistent with common sense; but it is important to recognize that "common sense" is intuition based on everyday experience, and this does not usually include measurements of the speed of light. We must be prepared to accept results that seem to be in conflict with common sense when they involve realms far removed from everyday observation.

Thus the speed of light (in vacuum) is independent of the motion of the source and is the same in all frames of reference. To explore the consequences of this statement, we consider the newtonian relationship between two inertial frames, labeled S and S' in Fig. 43–1. Let the x-axes of the two frames lie along the same line, but let the origin O' of S' move relative to the origin O of S with constant velocity u along the common x-axis. If the two origins coincide at time $t = 0$, then their separation at a later time t is ut.

A point P may be described by the coordinates (x, y, z) in S or by coordinates (x', y', z') in S'. Reference to the figure shows that these are related by

$$x = x' + ut, \quad y = y', \quad z = z'. \tag{43–1}$$

These equations are called the *galilean coordinate transformation.*

If the point P moves in the x-direction, its velocity v relative to S is given by $v = \Delta x/\Delta t$, and its velocity v' relative to S' is $v' = \Delta x'/\Delta t$. Intu-

43–1 The position of point P can be described by the coordinates x and y in frame of reference S, or by x' and y' in S'. S' moves relative to S with constant velocity u along the common x–x' axis. The two origins O and O' coincide at time $t = t' = 0$.

itively it is clear that these are related by

$$v = v' + u. \tag{43-2}$$

This relation may also be obtained formally from Eqs. (43–1). Suppose the particle is at a point described by coordinate x_1 or x_1' at time t_1, and at x_2 or x_2' at time t_2. Then $\Delta t = t_2 - t_1$, and, from Eq. (43–1),

$$\Delta x = x_2 - x_1 = (x_2' - x_1') + u(t_2 - t_1) = \Delta x' + u\,\Delta t,$$
$$\frac{\Delta x}{\Delta t} = \frac{\Delta x'}{\Delta t} + u,$$

and, in the limit as $\Delta t \to 0$,

$$v = v' + u,$$

in agreement with Eq. (43–2).

A fundamental problem now appears. Applied to the speed of light, Eq. (43–2) says $c = c' + u$. Einstein's principle of relativity, supported by experimental observations, says $c = c'$. This is a genuine inconsistency, not an illusion, and it demands resolution. If we accept the principle of relativity, we are forced to conclude that Eqs. (43–1) and (43–2), intuitively appealing as they are, cannot be correct, but need to be modified to bring them into harmony with this principle.

The resolution involves modifications of our basic kinematic concepts. The first of these involves an assumption so fundamental that it might seem unnecessary, namely the assumption that the same *time scale* is used in frames S and S'. This may be stated formally by adding to Eqs. (43–1) a fourth equation $t = t'$. Obvious though this assumption may seem, it turns out not to be correct when the relative speed u of the two frames of reference is comparable to the speed of light. The difficulty lies in the concept of *simultaneity*, which we examine next.

43–2 Relative nature of simultaneity

Measuring times and time intervals involves the concept of *simultaneity*. When a person says he awoke at seven o'clock he means that two *events*, his awakening and the arrival of the hour hand of his clock at the number seven, occurred *simultaneously*. The fundamental problem in measuring time intervals is that, in general, two events that appear simultaneous in one frame of reference *do not* appear simultaneous in a second frame that is moving relative to the first, even if both are inertial frames.

The following thought experiment, devised by Einstein, illustrates this point. Consider a long train moving with uniform velocity, as shown in Fig. 43–2a. Two lightning bolts strike the train, one at each end. Each bolt leaves a mark on the train and one on the ground at the same instant. The points on the ground are labeled A and B in the figure, and the corresponding points on the train A' and B'. An observer on the ground is located at O, midway between A and B; another observer is at O', moving with the train and midway between A' and B'. Both observers use the light signals from the lightning to observe the events.

Suppose the two light signals reach the observer at O simultaneously; he concludes that the two events took place at A and B simultaneously. But the observer at O' is moving with the train, and the light pulse from B' reaches him before the light pulse from A' does; he con-

43–2 (a) To the stationary observer at point O, two lightning bolts appear to strike simultaneously. (b) The moving observer at point O' sees the light from the front of the train first and thinks that the bolt at the front struck first. (c) The two light pulses arrive at O simultaneously.

cludes that the event at the front of the train happened *earlier* than that at the rear. Thus the two events appear simultaneous to one observer, but not to the other. *Whether or not two events at different space points are simultaneous depends on the state of motion of the observer.* It follows that *the time interval between two events at different space points is in general different for two observers in relative motion.*

It might be argued that, in this example, the lightning bolts really *are* simultaneous, and that if the observer at O' could communicate with the distant points without time delay, he would realize this. But the finite speed of information transmission is not the problem. If O' is really midway between A' and B', then, in his frame of reference, the time for a signal to travel from A' to O' is the same as from B' to O'. Two signals arrive simultaneously at O' only if they were emitted simultaneously at A' and B'; in this example they *do not* arrive simultaneously at O', and so O' must conclude that the events at A' and B' were *not* simultaneous.

Furthermore, there is no basis for saying either that O is right and O' is wrong, or the reverse since, according to the principle of relativity, no inertial frame of reference is preferred over any other in the formulation of physical laws. Each observer is correct *in his own frame of reference,* but simultaneity is not an absolute concept. Whether or not two events are simultaneous depends on the frame of reference, and the time interval between two events depends on the frame of reference.

43-3 Relativity of time

To derive a quantitative relation between time intervals in different coordinate systems, we consider another thought experiment. As before, a frame of reference S' moves with velocity u relative to a frame S. An observer in S' directs a source of light at a mirror a distance d away, as shown in Fig. 43-3, and measures the time interval $\Delta t'$ for light to make the "round trip" to the mirror and back. The total distance is $2d$, so the time interval is

$$\Delta t' = \frac{2d}{c}. \tag{43-3}$$

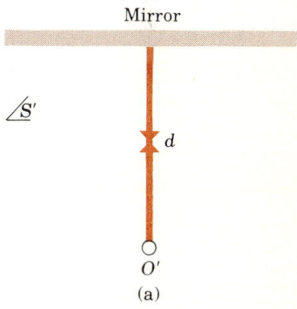

As measured in frame S, the time for the round trip is a different interval Δt. During this time, the source moves relative to S a distance $u\,\Delta t$, and the total round-trip distance is not just $2d$ but is $2l$, where

$$l = \sqrt{d^2 + \left(\frac{u\,\Delta t}{2}\right)^2}.$$

The speed of light is the same for both observers, so the relation in S analogous to Eq. (43-3) is

$$\Delta t = \frac{2l}{c} = \frac{2}{c}\sqrt{d^2 + \left(\frac{u\,\Delta t}{2}\right)^2}. \tag{43-4}$$

To obtain a relation between Δt and $\Delta t'$ that does not contain d, we solve Eq. (43-3) for d and substitute the result into Eq. (43-4), obtaining

$$\Delta t = \frac{2}{c}\sqrt{\left(\frac{c\,\Delta t'}{2}\right)^2 + \left(\frac{u\,\Delta t}{2}\right)^2}.$$

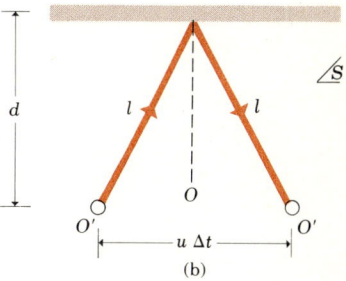

43-3 (a) Light pulse emitted from source at O' and reflected back along the same line, as observed in S'. (b) Path of the same light pulse, as observed in S. The positions of O' at the times of departure and return of the pulse are shown. The speed of the pulse is the same in S as in S', but the path is longer in S.

This may now be squared and solved for Δt; the result is

$$\Delta t = \frac{\Delta t'}{\sqrt{1 - u^2/c^2}}. \quad (43\text{-}5)$$

We may generalize this important result: If a time interval $\Delta t'$ separates two events occurring at the same space point in a frame of reference S' (in this case, the departure and arrival of the light signal at O'), then the time interval Δt between these two events as observed in S is *larger* than $\Delta t'$, and the two are related by Eq. (43-5). Thus when the rate of a clock at rest in S' is measured by an observer in S, the rate measured in S is *slower* than the rate observed in S'. This effect is called *time dilation*.

It is important to note that the observer in S measuring the time interval Δt cannot do so with a single clock. In Fig. 43-3 the points of departure and return of the light pulse are different space points in S, although they are the same point in S'. If S tries to use a single clock, the finite time of communication between two points will becloud the issue To avoid this, S may use two clocks at the two relevant points. There is no difficulty in synchronizing two clocks in the same frame of reference; one procedure is to send a light pulse simultaneously to two clocks from a point midway between them, with the two operators setting their clocks to a predetermined time when the pulses arrive. In thought experiments it is often helpful to imagine a large number of synchronized clocks distributed conveniently in a single frame of reference. Only when a clock is moving relative to a given frame of reference do ambiguities of synchronization or simultaneity arise.

Example A spaceship flies past earth with a speed of $0.99c$ (about 2.97×10^8 m·s^{-1}). A high-intensity signal light (perhaps a pulsed laser) blinks on and off, each pulse lasting 2×10^{-6} s. At a certain instant the ship appears to an earthling observer to be directly overhead at an altitude of 1000 km, and to be traveling perpendicular to the line of sight. What is the duration of each light pulse, as measured by this observer, and how far does the ship travel relative to earth during each pulse?

Solution The observer does not see the pulse at the instant it is emitted, because the light signal requires a time equal to $(1000 \times 10^3 \text{ m})/(3 \times 10^8 \text{ m·s}^{-1})$, or $(1/300)$ s, to travel from the ship to earth. But if the distance from the spaceship to observer is essentially constant during the emission of a pulse, the time delays at beginning and end of the pulse are equal and the time *interval* is not affected.

Let S be the earth's frame of reference, S' that of the spaceship. Then, in the notation of Eq. (43-5), $\Delta t' = 2 \times 10^{-6}$ s. This interval refers to two events occurring at the same point relative to S', namely, the starting and stopping of the pulse. The corresponding interval in S is given by Eq. (43-5):

$$\Delta t = \frac{\Delta t'}{\sqrt{1 - u^2/c^2}} = \frac{2 \times 10^{-6} \text{ s}}{\sqrt{1 - (0.99)^2}} = 14.2 \times 10^{-6} \text{ s}.$$

Thus the time dilation in S is about a factor of seven. The distance D traveled in S during this interval is

$$D = u \, \Delta t = (0.99)(3 \times 10^8 \text{ m·s}^{-1})(14.2 \times 10^{-6} \text{ s})$$
$$= 4220 \text{ m} = 4.22 \text{ km}.$$

If the spaceship is traveling directly *toward* the observer, the time interval cannot be measured directly by a single observer because the time delay is not the same at the beginning and end of the pulse. One possible scheme, at least in principle, is to use *two* observers at rest in S, with synchronized clocks, one at the position of the ship when the pulse starts, the other at its position at the end of the pulse. These observers will again measure a time interval in S of 14.2×10^{-6} s.

Time-dilation effects are not observed in everyday life because the speeds of all practical modes of transportation are much less than the speed of light. For example, for a jet airplane flying at 600 mi·hr^{-1} (about 270 m·s^{-1}),

$$\frac{v^2}{c^2} = \left(\frac{270 \text{ m}\cdot\text{s}^{-1}}{3 \times 10^8 \text{ m}\cdot\text{s}^{-1}}\right)^2 = 8.1 \times 10^{-13},$$

and the time-dilation factor in Eq. (43-5) is approximately $1 + (4 \times 10^{-13})$. Thus to observe time dilation in this situation requires a clock with a precision of the order of one part in 10^{13}. But, as noted in Sec. 1-2, atomic clocks capable of this precision have recently been developed, and in the past few years experiments with such clocks in jet airplanes have verified Eq. (43-5) directly.

From the derivation of Eq. (43-5) and the spaceship example, it can be seen that a time interval between two events occurring *at the same point* in a given frame of reference is a more fundamental quantity than an interval between events at different points. The term *proper time* is used to denote an interval between two events occurring at the same space point. Thus Eq. (43-5) may be used *only* when $\Delta t'$ is a proper time interval in S', in which case Δt is *not* a proper time interval in S. If, instead, Δt is proper in S, then Δt and $\Delta t'$ must be interchanged in Eq. (43-5).

When the relative velocity u of S and S' is very small, the factor $(1 - u^2/c^2)$ is very nearly equal to unity, and Eq. (43-5) approaches the newtonian relation $\Delta t = \Delta t'$ (i.e., the same time scale for all frames of reference). This assumption, therefore, retains its validity in the limit of small relative velocities.

43-4 Relativity of length

Just as the time interval between two events depends on the frame of reference, the *distance* between two points also depends on the frame of reference. To measure a distance one must, in principle, observe the positions of two points, such as the two ends of a ruler, simultaneously; but what is simultaneous in one frame is not in another.

To develop a relation between lengths in various coordinate systems, we consider another thought experiment. We attach a source of light pulses to one end of a ruler and a mirror to the other, as shown in Fig. 43-4. Let the ruler be at rest in S' and its length in this frame be l'. Then the time $\Delta t'$ required for a light pulse to make the round trip from source to mirror and back is given by

$$\Delta t' = \frac{2l'}{c}. \tag{43-6}$$

43-4 (a) A light pulse is emitted from a source at one end of a ruler, reflected from a mirror at the opposite end, and returned to the source position. (b) Motion of the light pulse as seen by an observer in S. The distance traveled from source to mirror is greater than the length l measured in S, by the amount $u \Delta t_1$, as shown.

This is a proper time interval, since departure and return occur at the same point in S'.

In S the ruler is displaced during this travel of the light pulse. Let the length of the ruler in S be l, and let the time of travel from source to mirror, as measured in S, be Δt_1. During this interval the mirror moves a distance $u\,\Delta t_1$, and the total length of path d from source to mirror is not l but

$$d = l + u\,\Delta t_1. \tag{43-7}$$

But since the pulse travels with speed c, it is also true that

$$d = c\,\Delta t_1. \tag{43-8}$$

Combining Eqs. (43-7) and (43-8) to eliminate d,

$$c\,\Delta t_1 = l + u\,\Delta t_1,$$

or

$$\Delta t_1 = \frac{l}{c - u}. \tag{43-9}$$

In the same way it can be shown that the time Δt_2 for the return trip from mirror to source is

$$\Delta t_2 = \frac{l}{c + u}. \tag{43-10}$$

The *total* time $\Delta t = \Delta t_1 + \Delta t_2$ for the round trip, as measured in S, is

$$\Delta t = \frac{l}{c - u} + \frac{l}{c + u} = \frac{2l}{c(1 - u^2/c^2)}. \tag{43-11}$$

It is also known that Δt and $\Delta t'$ are related by Eq. (43-5), since $\Delta t'$ is proper in S'. Thus Eq. (43-6) becomes

$$\Delta t\sqrt{1 - \frac{u^2}{c^2}} = \frac{2l'}{c}. \tag{43-12}$$

Finally, combining this with Eq. (43-11) to eliminate Δt, and simplifying we obtain

$$l = l'\sqrt{1 - \frac{u^2}{c^2}}. \tag{43-13}$$

Thus the length measured in S, in which the ruler is moving, is *shorter* than in S', where it is at rest. A length measured in the rest frame of the body is called a proper length; thus, l' above is a proper length in S', and the length measured in any other frame is less than l'. This effect is called *contraction of length*.

Example In the spaceship example of Sec. 43-3, what distance does the spaceship travel during emission of a pulse, as measured in its rest frame?

Solution The question is somewhat ambiguous since, of course, in its own frame of reference the ship is at rest. But suppose it leaves markers in space, such as small smoke bombs, at the instants when the pulse starts and stops, and measures the distance between these markers, with the aid of observers behind the ship but moving with it, each with a clock synchronized with that of the ship. The distance d between the markers is a proper length in the earth's frame S. In the spaceship's frame S', the distance d' is contracted by the factor given in Eq. (43-13):

$$d' = d\sqrt{1 - \frac{u^2}{c^2}} = (4220\text{ m})\sqrt{1 - (0.99)^2}$$
$$= 595\text{ m}.$$

(Note that because d, not d', is a proper length, we must reverse the roles of l and l'.) An observer in the spaceship can calculate his speed relative to earth from this data:

$$u = \frac{d'}{\Delta t'} = \frac{595\text{ m}}{2 \times 10^{-6}\text{ s}} = 2.97 \times 10^8\text{ m}\cdot\text{s}^{-1},$$

which agrees with the initial data.

When u is very small compared to c, the contraction factor in Eq. (43-13) approaches unity, and in the limit of small speeds we recover the newtonian relation $l = l'$. This and the corresponding result for time dilation show that Eqs. (43-1) retain their validity in the limit of small speeds; only at speeds comparable to c are modifications needed.

Lengths measured perpendicular to the direction of motion are *not* contracted; this may be verified by constructing a thought experiment for measuring in S and S' the length of a ruler oriented perpendicular to the direction of relative motion. The details of this discussion are not essential for our further work and will not be given here.

*43-5 The Lorentz transformation

The galilean coordinate transformation given by Eqs. (43-1) is valid only in the limit when u is much smaller than c, but we are now in position to derive a more general transformation not subject to this limitation. The more general relations are called the *Lorentz transformation*. When u is small they reduce to the galilean transformation, but they may also be used when u is comparable to c.

The basic problem is this: When an event occurs at point (x, y, z) at time t, as observed in a frame of reference S, what are the coordinates (x', y', z') and time t' of the event as observed in a second frame S' moving relative to S with constant velocity u along the x-direction?

To derive the transformation equations, we return to Fig. 43-1. As before, we assume that the origins coincide at the initial time $t = t' = 0$. Then in S the distance from O to O' at time t is still ut. The coordinate x' is a proper length in S', so in S it appears contracted by the factor given in Eq. (43-13). Thus the distance x from O to P in S is given not simply by $x = ut + x'$ as in the galilean transformation, but by

$$x = ut + x'\sqrt{1 - \frac{u^2}{c^2}}. \qquad (43\text{-}14)$$

Solving this equation for x', we obtain

$$x' = \frac{x - ut}{\sqrt{1 - u^2/c^2}}. \qquad (43\text{-}15)$$

This is half of the Lorentz transformation; the other half is the equation giving t' in terms of x and t. To obtain this note that the principle of relativity requires that the *form* of the transformation from S to S' must be identical to that from S' to S, the only difference being a change in the sign of the relative velocity u. Thus, from Eq. (43–14), it must be true that

$$x' = -ut' + x\sqrt{1 - \frac{u^2}{c^2}}. \tag{43-16}$$

We may now equate Eqs. (43–15) and (43–16) to eliminate x' from this expression, obtaining the desired relation for t' in terms of x and t. The algebraic details will be omitted; the result is

$$t' = \frac{t - ux/c^2}{\sqrt{1 - u^2/c^2}}. \tag{43-17}$$

As remarked previously, lengths perpendicular to the direction of relative motion are not affected by the motion, so $y' = y$ and $z' = z$.

Collecting all the transformation equations, we have

$$\begin{aligned} x' &= \frac{x - ut}{\sqrt{1 - u^2/c^2}}, \\ y' &= y, \\ z' &= z, \\ t' &= \frac{t - ux/c^2}{\sqrt{1 - u^2/c^2}}. \end{aligned} \tag{43-18}$$

These are the *Lorentz tranformation equations*, the relativistic generalization of the galilean transformation, Eqs. (43–1). When u is much smaller than c, the two transformations become identical.

Next we consider the relativistic generalization of the velocity transformation relation, Eq. (43–2), which, as peviously noted, is valid only when u is very small. The relativistic expression can easily be obtained from the Lorentz transformation. Suppose that a body observed in S' is at point x'_1 at time t'_1 and point x'_2 at time t'_2. Then its speed v' in S' is given by

$$v' = \frac{x'_2 - x'_1}{t'_2 - t'_1} = \frac{\Delta x'}{\Delta t'}. \tag{43-19}$$

To obtain the speed in S we use Eqs. (43–18) to translate this expression into terms of the corresponding positions x_1 and x_2 and times t_1 and t_2 observed in S. We find

$$x'_2 - x'_1 = \frac{x_2 - x_1 - u(t_2 - t_1)}{\sqrt{1 - u^2/c^2}} = \frac{\Delta x - u\,\Delta t}{\sqrt{1 - u^2/c^2}},$$

$$t'_2 - t'_1 = \frac{t_2 - t_1 - u(x_2 - x_1)/c^2}{\sqrt{1 - u^2/c^2}} = \frac{\Delta t - u\,\Delta x/c^2}{\sqrt{1 - u^2/c^2}}.$$

Using these results in Eq. (43–19), we find

$$v' = \frac{\Delta x - u\,\Delta t}{\Delta t - u\,\Delta x/c^2} = \frac{(\Delta x/\Delta t) - u}{1 - (u/c^2)(\Delta x/\Delta t)}.$$

Now $\Delta x/\Delta t$ is just the velocity v measured in S, so we finally obtain

$$v' = \frac{v - u}{1 - uv/c^2}. \qquad (43\text{-}20)$$

We note that when u and v are much smaller than c, the denominator becomes equal to unity, and we obtain the nonrelativistic result $v' = v - u$. The opposite extreme is the case $v = c$; then we find

$$v' = \frac{c - u}{1 - uc/c^2} = c.$$

That is, anything moving with speed c relative to S also has speed c relative to S', despite the relative motion of the two frames. This result demonstrates the consistency of Eq. (43-20) with our initial assumption that the speed of light is the same in all frames of reference.

Equation (43-20) may also be rearranged to give v in terms of v'. The algebraic details are left as a problem; the result is

$$v = \frac{v' + u}{1 + uv'/c^2}. \qquad (43\text{-}21)$$

Example A spaceship moving away from earth with a speed $0.9c$ fires a missile in the same direction as its motion, with a speed $0.9c$ relative to the spaceship. What is the missile's speed relative to earth?

Solution Let the earth's frame of reference be S, the spaceship's S'. Then $v' = 0.9c$ and $u = 0.9c$. The nonrelativistic velocity addition formula would give a velocity relative to earth of $1.8\ c$. The correct relativistic result, obtained from Eq. (43-21), is

$$v = \frac{0.9c + 0.9c}{1 + (0.9c)(0.9c)/c^2} = 0.994c.$$

When u is less than c, a body moving with a speed less than c in one frame of reference also has a speed less than c in *every other* frame of reference. This is one reason for thinking that no material body may travel with a speed greater than that of light, relative to any frame of reference. The relativistic generalizations of energy and momentum, to be considered next, give further support to this hypothesis.

43-6 Momentum

We have discussed the fact that Newton's laws of motion are *invariant* under the galilean coordinate transformation, but that to satisfy the principle of relativity this transformation must be replaced by the more general Lorentz transformation. This requires corresponding generalizations in the laws of motion and the definitions of momentum and energy.

The principle of conservation of momentum states that *when two bodies collide, the total momentum is constant,* provided there is no interaction except that of the two bodies with each other. However, when one considers a collision in one coordinate system S, in which momentum is conserved, and then uses the Lorentz transformation to obtain the velocities in a second system S', it is found that if the newtonian definition of momentum ($\boldsymbol{p} = m\boldsymbol{v}$) is used, momentum is *not* conserved in the second system. Thus if the Lorentz transformation is correct, and if we believe in the principle of relativity (i.e., momentum conservation must hold in *all* systems), the only choice remaining is to modify the *definition* of momentum.

Deriving the correct relativistic generalization is beyond our scope, and we simply quote the result, which has already been mentioned in Sec. 7-9:

$$p = \frac{mv}{\sqrt{1 - v^2/c^2}}. \qquad (43\text{-}22)$$

We note that, as usual when the particle's speed v is much less than c, this reduces to the newtonian expression $p = mv$, but that in general the momentum is greater in magnitude than mv.

In newtonian mechanics the second law of motion can be stated in the form

$$F = \frac{dp}{dt}. \qquad (43\text{-}23)$$

That is, force equals time rate of change of momentum. Experiment shows that this result is still valid in relativistic mechanics, provided we use the relativistic momentum given by Eq. (43-22). That is, the relativistically correct generalization of Newton's second law is found to be

$$F = \frac{d}{dt}\frac{mv}{\sqrt{1 - v^2/c^2}}. \qquad (43\text{-}24)$$

This relation has the consequence that constant force no longer causes constant acceleration. For example, when force and velocity are both along a single line, causing straight-line motion, it is not difficult to show that Eq. (43-24) becomes

$$F = \frac{m}{(1 - v^2/c^2)^{3/2}} \frac{dv}{dt},$$

or

$$\frac{dv}{dt} = a = \frac{F}{m}(1 - v^2/c^2)^{3/2}. \qquad (43\text{-}25)$$

The derivation of Eq. (43-25) is left as a problem; the equation shows that as a particle's speed increases, the acceleration for a given force continuously *decreases*. As the speed approaches c, the acceleration approaches zero, no matter how great the force. Thus it is impossible to accelerate a particle from a state of rest to a speed equal to or greater than c, and the speed of light is sometimes referred to as "the ultimate speed."

43-7 Work and energy

The work–energy relation developed in Chapter 6 made use of Newton's laws of motion. Since these must be generalized to bring them into accord with the principle of relativity, it is not surprising that the work–energy relation also requires generalization.

Because a constant force on a body no longer causes a constant acceleration (except at very small velocities), even the simplest dynamics problems require the use of the calculus. We may, however, follow the same pattern as that of Sec. 6–3 in deriving the relativistic generalization of the work-energy principle. We begin with the definition of work: $W = \int F\,dx$. According to Eq. (43–23), $F = dp/dt$. By repeated application of the chain rule for derivatives, we obtain

$$F = \frac{dp}{dt} = \frac{dp}{dv}\frac{dv}{dx}\frac{dx}{dt} = \frac{dp}{dv}\frac{dv}{dx}v. \qquad (43\text{--}26)$$

Thus if the particle has speed v_1 at point x_1 and v_2 at x_2, the work done by F during the motion from x_1 to x_2 may be expressed as

$$W = \int_{x_1}^{x_2} F\,dx = \int_{x_1}^{x_2} \frac{dp}{dv}\frac{dv}{dx}v\,dx.$$

This may now be converted into an integral on v, with $dv = (dv/dx)\,dx$:

$$W = \int_{v_1}^{v_2} \frac{dp}{dv} v\,dv. \qquad (43\text{--}27)$$

An expression for dp/dv may be obtained by taking the derivative of Eq. (43–22). The reader may verify that the result is

$$\frac{dp}{dv} = \frac{m}{(1 - v^2/c^2)^{3/2}}. \qquad (43\text{--}28)$$

Substituting this result in Eq. (43–27) and integrating, we find

$$\begin{aligned} W &= \int_{v_1}^{v_2} \frac{mv\,dv}{(1 - v^2/c^2)^{3/2}} \\ &= \frac{mc^2}{\sqrt{1 - v_2^2/c^2}} - \frac{mc^2}{\sqrt{1 - v_1^2/c^2}}, \end{aligned} \qquad (43\text{--}29)$$

where v_1 and v_2 are the initial and final velocities of the particle.

This result suggests defining kinetic energy as

$$\frac{mc^2}{\sqrt{1 - v^2/c^2}}. \qquad (43\text{--}30)$$

But this expression is not zero when $v = 0$; instead it becomes equal to mc^2. Thus the correct relativistic generalization of kinetic energy K is:

$$K = \frac{mc^2}{\sqrt{1 - v^2/c^2}} - mc^2. \qquad (43\text{--}31)$$

This expression, if correct, must reduce to the newtonian expression $K = \tfrac{1}{2}mv^2$ when v is much smaller than c. It is not obvious that this is the case; to demonstrate that it is so, we can expand the radical using the binomial theorem:

$$\left(1 - \frac{v^2}{c^2}\right)^{-1/2} = 1 + \frac{1}{2}\frac{v^2}{c^2} + \frac{3}{8}\frac{v^4}{c^4} + \frac{5}{16}\frac{v^6}{c^6} + \cdots.$$

Combining this with Eq. (43–31), we find

$$\begin{aligned} K &= mc^2\left(1 + \frac{1}{2}\frac{v^2}{c^2} + \frac{3}{8}\frac{v^4}{c^4} + \cdots\right) - mc^2 \\ &= \frac{1}{2}mv^2 + \frac{3}{8}m\frac{v^4}{c^2} + \cdots. \end{aligned} \qquad (43\text{--}32)$$

In each expression the dots stand for omitted terms. When v is much smaller than c, all the terms in the series except the first are negligibly small, and we obtain the classical $\frac{1}{2}mv^2$.

But what is the significance of the term mc^2 that had to be subtracted in Eq. (43–31)? Although Eq. (43–30) does not give the *kinetic energy* of the particle, perhaps it represents some kind of *total* energy, including both the kinetic energy and an additional energy mc^2, which the particle possesses even when it is not moving. Calling this total energy E and using Eq. (43–31), we find

$$E = K + mc^2 = \frac{mc^2}{\sqrt{1 - v^2/c^2}}. \qquad (43\text{–}33)$$

The hypothetical energy mc^2 associated with mass rather than motion may be called the *rest energy* of the particle. This speculation does not prove that the concept of rest energy is meaningful, but it points the way toward further investigation.

There is in fact direct experimental evidence of the existence of rest energy. The simplest example is the decay of the π^0 meson, an unstable particle that "decays"; in the decay process, the particle disappears and electromagnetic radiation appears. When the particle is at rest (and therefore with no kinetic energy) before its decay, the total energy of the radiation produced is found to be exactly equal to mc^2. There are many other examples of fundamental particle transformations in which the total mass of the system changes, and in every case there is a corresponding energy change consistent with the assumption of a rest energy mc^2 associated with a mass m.

Although the principles of conservation of mass and of energy originally developed quite independently, the theory of relativity shows that they are but two special cases of a single broader conservation principle, the *principle of conservation of mass and energy*. There are physical phenomena where neither mass nor energy is separately conserved, but where the changes in these quantities are governed by the more general relation that a change m in the mass of the system must be accompanied by an opposite change mc^2 in its energy.

The term *mass* as used here always means the rest mass of a particle, the inertial mass m measured through Eq. (43–22). For a given particle, m is a constant, independent of the state of motion of the particle. It is sometimes useful to introduce the concept of a variable, velocity-dependent relativistic mass. This concept is not needed in the present analysis, and it is not used in this discussion.

The possibility of conversion of mass into energy is the fundamental principle involved in the generation of power through nuclear reactions, a subject to be discussed in later chapters. When a uranium nucleus undergoes fission in a nuclear reactor, the total mass of the resulting fragments is less than that of the parent nucleus, and the total kinetic energy of the fragments is equal to this mass deficit times c^2. This kinetic energy can be used to produce steam to operate turbines for electric power generators or in a variety of other ways.

The total energy E (kinetic plus rest) of a particle is related simply to its momentum, as shown by combining Eqs. (43–22) and (43–33) to eliminate the particle's velocity. This is most easily accomplished by rewriting these equations in the following forms:

$$\left(\frac{E}{mc^2}\right)^2 = \frac{1}{1 - v^2/c^2}; \qquad \left(\frac{p}{mc}\right)^2 = \frac{v^2/c^2}{1 - v^2/c^2}.$$

Subtracting the second of these from the first and rearranging, we find

$$E^2 = (mc^2)^2 + (pc)^2. \tag{43-34}$$

Again we see that for a particle at rest ($p = 0$), $E = mc^2$. Equation (43-34) also suggests that a particle may have energy and momentum even when it has no rest mass. In such a case, $m = 0$ and

$$E = pc. \tag{43-35}$$

Massless particles, including photons, the quanta of electromagnetic radiation, and others, were mentioned in Sec. 7-9. The existence of such particles is well established, and they will be discussed in greater detail in later chapters. These particles always travel with the speed of light; they are emitted and absorbed during changes of state of atomic systems, accompanied by corresponding changes in the energy and momentum of these systems.

43-8 Relativity and newtonian mechanics

The sweeping changes required by the principle of relativity go to the very roots of newtonian mechanics, including the concepts of length and time, the equations of motion, and the conservation principles. Thus it may appear that foundations on which Newton's mechanics are built have been destroyed. While this is true in one sense, it is essential to keep in mind that the newtonian formulation still retains its validity whenever speeds are small compared with the speed of light. In such cases time dilation, length contraction, and the modifications of the laws of motion are so small that they are unobservable. In fact, every one of the principles of newtonian mechanics survives as a special case of the relativistic formulation.

Relativity does not *contradict* the older mechanics but *generalizes* it. After all, Newton's laws rest on a very solid base of experimental evidence, and it would be very strange indeed to advance a new theory inconsistent with this evidence. So it always is with the development of physical theory. Whenever a new theory is in partial conflict with an older, established theory, it nevertheless must yield the same predictions as the old in areas where the old theory is supported by experimental evidence. Every new physical theory must pass this test, called the *correspondence principle*, which has come to be regarded as a fundamental procedural rule in all physical theory. There are many problems for which newtonian mechanics is clearly inadequate, including all situations where particle speeds approach that of light or there is direct conversion of mass to energy. But there is still a large area, including nearly all the behavior of macroscopic bodies in mechanical systems, in which newtonian mechanics is still perfectly adequate.

At this point it is legitimate to ask whether the relativistic mechanics just discussed is the final word on this subject or whether *further* generalizations are possible or necessary. For example, inertial frames of reference have occupied a privileged position in all our discussion thus far. Should the principle of relativity be extended to non-inertial frames as well?

Here is an example to illustrate some implications of this question. A man decides to go over Niagara Falls while enclosed in a large wooden box. During his free fall over the falls he can, in principle, perform experiments inside the box. An object released inside the box does not fall to the floor because both the box and the object are in free fall with a downward acceleration of $9.8 \text{ m} \cdot \text{s}^{-2}$. But an alternative interpretation, from this man's point of view, is that the force of gravity has suddenly been turned off. Provided he remains in the box and it remains in free fall, he cannot tell whether he is indeed in free fall or whether the force of gravity has vanished. A similar problem appears in a space station in orbit around the earth. Objects in the spaceship appear weightless, but without going outside the ship there is no way to determine whether gravity has disappeared or the spaceship is in an accelerated (i.e., non-inertial) frame of reference.

These considerations form the basis of Einstein's *general theory of relativity*. If one cannot distinguish experimentally between a gravitational field and an accelerated reference system, then there can be no real distinction between the two. Pursuing this concept, we may try to represent *any* gravitational field in terms of special characteristics of the coordinate system. This turns out to require even more sweeping revisions of our space–time concepts than did the special theory of relativity, and we find that, in general, the geometric properties of the space are non-euclidean.

The basic ideas of the general theory of relativity are now well established, but some of the details remain speculative in nature. Its chief application is in cosmological investigations of the structure of the universe, the formation and evolution of stars, and related matters. It is not believed to have any relevance for atomic or nuclear phenomena or macroscopic mechanical problems of less than astronomical dimensions.

Questions

43-1 What do you think would be different in everyday life if the speed of light were $10 \text{ m} \cdot \text{s}^{-1}$ instead of its actual value?

43-2 The average life span in the United States is about 70 years. Does this mean that it is impossible for an average person to travel a distance greater than 70 light-years away from the earth? (A light-year is the distance light travels in a year.)

43-3 What are the fundamental distinctions between an inertial frame of reference and a noninertial frame?

43-4 A physicist claimed that it is impossible to define what is meant by a rigid body in a relativistically correct way. Why?

43-5 Two events occur at the same space point in a particular frame of reference and appear simultaneous in that frame. Is it possible that they may not appear simultaneous in another frame?

43-6 If simultaneity is not an absolute concept, does this also destroy the concept of *causality*? If event A is to *cause* event B, A must occur first. Is it possible that in some frames A may appear to cause B, and in others B appears to cause A?

43-7 A social scientist who has done distinguished work in fields far removed from physics has written a book purporting to refute the special theory of relativity. He begins with a premise that might be paraphrased as follows: "Either two events occur at the same time, or they don't; that's just common sense." How would you respond to this in the light of our discussion of the relative nature of simultaneity?

43-8 When an object travels across an observer's field of view at a relativistic speed, it appears not only foreshortened but also slightly rotated, with the side toward the observer shifted in the direction of motion relative to the side away from him. How does this come about?

43-9 According to the famous "twin paradox," if one twin stays on earth while the other takes off in a spaceship at relativistic speed and then returns, one will be older than the other. Which is older? Does this really happen? Can you think of a practical experiment, perhaps using two very precise atomic clocks, that would test this conclusion?

43-10 When a monochromatic light source moves toward an observer, its wavelength appears to be shorter than the value measured when the source is at rest. Does this contradict the hypothesis that the speed of light is the same for all observers? What about the apparent frequency of light from a moving source?

43-11 A student asserted that a massive particle must always have a speed less than that of light, while a massless particle must always travel at exactly the speed of light. Is he correct? If so, how do massless particles such as photons and neutrinos acquire this speed? Can't they start from rest and accelerate?

43-12 The theory of relativity sets an upper limit on the speed a particle can have. Are there also limits on its energy and momentum?

43-13 In principle, does a hot gas have more mass than the same gas when it is cold? Explain. In practice would this be a measurable effect?

43-14 Why do you think the development of newtonian mechanics preceded the more refined relativistic mechanics by so many years?

Problems

43-1 The π^+ meson, an unstable particle, lives, on the average, about 2.6×10^{-8} s (measured in its own frame of reference) before decaying.
 a) If such a particle is moving with respect to the laboratory with a speed of $0.8\,c$, what lifetime is measured in the laboratory?
 b) What distance, measured in the laboratory, does the particle move before decaying?

43-2 The μ^+ meson (or positive muon) is an unstable particle with a lifetime of about 2.3×10^{-6} s (measured in the rest frame of the muon).
 a) If a muon travels with a speed of $0.99\,c$ relative to a laboratory, what is the lifetime as measured in the laboratory?
 b) What distance, measured in the laboratory, does the particle travel during its lifetime?

43-3 For the two trains discussed in Sec. 43-2, suppose the two lightning bolts appear simultaneous to an observer on the train. Show that they *do not* appear simultaneous to an observer on the ground. Which appears to come first?

43-4 Solve Eqs. (43-18) to obtain x and t in terms of x' and t', and show that the resulting transformation has the same form as the original one except for a change of sign for u.

43-5 A light pulse is emitted at the origin of a frame of reference S' at time $t' = 0$. Its distance x' from the origin after a time t' is given by $x'^2 = c^2 t'^2$. Use the Lorentz transformation to transform this equation to an equation in x and t, and show that the result is $x^2 = c^2 t^2$; that is, the motion appears exactly the same in the frame of reference S of x and t, as it does in S'.

43-6 Two events observed in a frame of reference S have positions and times given by (x_1, t_1) and (x_2, t_2), respectively. Show that in a frame S' moving just fast enough so the two events occur at the same point in S', the time interval $\Delta t'$ between the two events is given by

$$\Delta t' = \sqrt{(\Delta t)^2 - \left(\frac{\Delta x}{c}\right)^2},$$

where $\Delta x = x_2 - x_1$, and $\Delta t = t_2 - t_1$. Hence show that, if $\Delta x \geq c\,\Delta t$, there is *no* frame S' in which the two events occur at the same point. The interval $\Delta t'$ is sometimes called the *proper time interval* for the events; is this term appropriate?

43-7 For the two events in Problem 43-6, show that if $\Delta x > c\,\Delta t$ there is a frame of reference S' in which the two events occur *simultaneously*. Find the distance between the two events in S'. This distance is sometimes called a *proper length*; is this term appropriate?

43-8 Two events are observed in a frame of reference S to occur at the same space point, the second occurring 2 s after the first. In a second frame S' moving relative to S, the second event is observed to occur 3 s after the first. What is the distance between the positions of the two events as measured in S'?

43-9 Two events are observed in a frame of reference S to occur simultaneously, at points separated by a distance of 1 m. In a second frame S' moving relative to S along the line joining the two points in S, the two events appear to be separated by 2 m. What is the time interval between the events, as measured in S'?

43-10 A particle is said to be in the *extreme relativistic range* when its kinetic energy is much larger than its rest energy.

a) What is the speed of a particle (expressed as a fraction of c) such that the total energy is ten times the rest energy?

b) For such a particle, what percent error in the energy–momentum relation of Eq. (43-34) results if the term $(mc^2)^2$ is neglected?

43-11 A photon of energy E is emitted by an atom of mass m, which recoils in the opposite direction. Assuming that the atom can be treated nonrelativistically, compute the recoil velocity of the atom. Hence, show that the recoil velocity is much smaller than c whenever E is much smaller than the rest energy mc^2 of the atom.

43-12 Two particles emerge from a high-energy accelerator in opposite directions, each with a speed $0.6\,c$. What is the relative velocity of the particles?

43-13 At what speed is the momentum of a particle twice as great as the result obtained from the nonrelativistic expression mv?

43-14

a) At what speed does the momentum of a particle differ from the value obtained using the nonrelativistic expression mv by 1 percent?

b) Is the correct relativistic value greater or less than that obtained from the nonrelativistic expression?

43-15 The mass of an electron is 9.11×10^{-31} kg. Comparing the classical definition of momentum with its relativistic generalization, by how much is the classical expression in error if (a) $v = 0.01\,c$; (b) $v = 0.5\,c$; (c) $v = 0.9\,c$?

43-16 A radioactive isotope of cobalt, ^{60}Co, emits an electromagnetic photon (γ ray) of wavelength 0.932×10^{-12} m. The cobalt nucleus contains 27 protons and 33 neutrons, each with a mass of about 1.66×10^{-27} kg.

a) If the nucleus is at rest before emission, what is the speed afterward?

b) Is it necessary to use the relativistic generalization of momentum?

43-17 In Problem 43-16, suppose the cobalt atom is in a metallic crystal containing 0.01 mol of cobalt (about 6.02×10^{21} atoms) and that the entire crystal recoils as a unit, rather than just the single nucleus. Find the recoil velocity. (This recoil of the entire crystal rather than a single nucleus is called the *Mössbauer effect*, in honor of its discoverer, who first observed it in 1958.)

43-18 What is the speed of a particle whose kinetic energy is equal to its rest energy?

43-19 At what speed is the kinetic energy of a particle equal to $10\,mc^2$?

43-20 How much work must be done to accelerate a particle from rest to a speed $0.1\,c$? From a speed $0.9\,c$ to a speed $0.99\,c$?

43-21 In *positron annihilation*, an electron and a positron (a positively charged electron) collide and disappear, producing electromagnetic radiation. If each particle has a mass of 9.1×10^{-31} kg and they are at rest just before the annihilation, find the total energy of the radiation.

43-22 The total consumption of electrical energy per year in the United States is of the order of 10^{19} joules. If matter could be converted completely into energy, how many kilograms of matter would have to be converted to produce this much energy?

43-23 Compute the kinetic energy of an electron (mass 9.11×10^{-31} kg) using both the nonrelativistic and relativistic expressions, and compare the two results, for speeds of (a) 1.0×10^8 m·s^{-1}; (b) 2.0×10^8 m·s^{-1}.

43-24 In a hypothetical nuclear-fusion reactor, two deuterium nuclei combine or "fuse" to form one helium nucleus. The mass of a deuterium nucleus, expressed in atomic mass units (u), is 2.0147 u; that of a helium nucleus is 4.0039 u. (1 u = 1.66×10^{-27} kg.)

a) How much energy is released when 1 kg of deuterium undergoes fusion?

b) The annual consumption of electrical energy in the United States is of the order of 10^{19} J. How much deuterium must react to produce this much energy?

43-25 A nuclear bomb containing 20 kg of plutonium explodes. The rest mass of the products of the explosion is less than the original rest mass by one part in 10^4.

a) How much energy is released in the explosion?

b) If the explosion takes place in 1 μs, what is the average power developed by the bomb?

c) How much water could the released energy lift to a height of 1 km?

43-26 Construct a right triangle in which one of the angles is α, where $\sin \alpha = v/c$. (v is the speed of a particle, c the speed of light.) If the base of the triangle (the side adjacent to α) is the rest energy mc^2, show that

a) the hypotenuse is the total energy and

b) the side opposite α is c times the relativistic momentum.

c) Describe a simple graphical procedure for finding the kinetic energy K.

43-27 An electron in a certain x-ray tube is accelerated from rest through a potential difference of 180,000 V in going from the cathode to the anode. When it arrives at the anode, what is its

a) kinetic energy in eV,

b) its total energy,

c) its velocity?

d) What is the velocity of the electron, calculated classically?

43-28 The Cambridge electron accelerator accelerates electrons through a potential difference of 6.5×10^9 V, so that their kinetic energy is 6.5×10^9 eV.

a) What is the ratio of the speed v of an electron having this energy to the speed of light, c?

b) What would the speed be if computed from the principles of classical mechanics?

43-29 A particle moves along a straight line under the action of a force lying along the same line. Show that the acceleration $a = dv/dt$ of the particle is given by Eq. (43-25).

43-30 A particle moves along the x-axis with speed v. A force in the $+y$-direction, having magnitude F, is applied. Show that the magnitude of the acceleration initially is

$$a = \frac{F}{m}(1 - v^2/c^2)^{1/2}.$$

This result and that of Problem 43-29 were sometimes interpreted in the early days of relativity as meaning that a particle has a "longitudinal mass" given by $m/(1 - v^2/c^2)^{3/2}$ and a "transverse mass" given by $m/(1 - v^2/c^2)^{1/2}$. These terms are no longer in common use.

43-31 Calculate, relativistically, the amount of work in MeV that must be done

a) to bring an electron from rest to a velocity of $0.4c$, and

b) to increase its velocity from $0.4c$ to $0.8c$.

c) What is the ratio of the kinetic energy of the electron at the velocity of $0.8c$ to that of $0.4c$ when computed (1) from relativistic values and (2) from classical values?

44 PHOTONS, ELECTRONS, AND ATOMS

44-1 Emission and absorption of light

The past several chapters have been concerned with understanding various phenomena associated with the propagation of light, on the basis of an electromagnetic wave theory. The work of Hertz established the existence of electromagnetic waves and the fact that light is an electromagnetic wave. The phenomena of interference, diffraction, and polarization are easily understood on the basis of a wave model; and, when interference effects can be neglected, the further simplification of ray optics permits analysis of the behavior of lenses and mirrors. These phenomena are collectively referred to as *classical optics*, and insofar as the phenomena of classical optics are concerned, the electromagnetic wave theory is *complete*.

There are, however, many phenomena that are *not* so readily understood on this basis. An example is the emission of light from matter. The electromagnetic waves of Hertz, with frequencies of the order of 10^8 Hz, were produced by oscillations in a resonant L–C circuit similar to those studied in Chapter 34. Frequencies of visible light are much larger, of the order of 10^{15} Hz, far higher than the highest frequencies attainable with conventional electronic equipment.

In the mid-nineteenth century it was speculated that visible light might be produced by motion of electric charge within individual atoms rather than in macroscopic circuits. In fact, in 1862 Faraday placed a light source in a strong magnetic field in an attempt to determine whether the emitted radiation was changed by the field. He was not able to detect any change, but when his experiments were repeated thirty years later by Zeeman with greatly improved equipment, changes *were* observed.

Particularly puzzling was the existence of *line spectra*. Light emitted from atoms heated in a flame, or excited electrically in a glow tube such as the familiar neon sign or mercury-vapor light, does not contain a continuous spread of wavelengths, but only certain well-defined wavelengths. In a spectrometer using a narrow slit in conjunction with a prism or a diffraction grating, the spectrum pattern appears as a series of bright lines, and hence has come to be known as a *line spectrum*. It was learned early in the nineteenth century that each element emits a *characteristic spectrum*, suggesting that there is a direct relation between the characteristics and internal structure of an atom and its spectrum. Attempts to understand this relation on the basis of newtonian mechanics and classical electricity and magnetism were not successful, however.

There were other mysteries. The *photoelectric effect*, discovered by Hertz in 1887 during his investigations of electromagnetic wave propagation, is the liberation of electrons from the surface of a conductor when light strikes the surface. This phenomenon can be understood qualitatively on the basis that when light is absorbed by the surface it transfers energy to electrons near the surface and that some of the electrons acquire enough energy to surmount the potential-energy barrier at the surface and escape from the material into space. More detailed investigation revealed some puzzling features that could *not* be understood on the basis of classical optics.

Still another area of unsolved problems centered around the production and scattering of *x-rays*, electromagnetic radiation with wavelengths shorter than those of visible light by a factor of the order of 10^4 and with correspondingly greater frequencies. These rays were produced in high-voltage glow discharge tubes, but the details of this process eluded understanding. Even worse, when these rays collided with matter, the scattered ray sometimes had a longer wavelength than the original ray. This is like directing a beam of blue light at a mirror and having it reflect red!

All these phenomena, and several others, pointed forcefully to the conclusion that classical optics, successful though it was in explaining ray optics, interference, and polarization, nevertheless had its limitations. Understanding the phenomena cited above would require at least some generalization of the classical theory. In fact, it has required something much more radical than that. All these phenomena are concerned with the *quantum* theory of radiation, which includes the assumption that despite the *wave* nature of electromagnetic radiation, it nevertheless has some properties akin to those of *particles*. In particular, the *energy* conveyed by an electromagnetic wave is always carried in units whose magnitude is proportional to the frequency of the wave. These units of energy are called *photons* or *quanta*.

Thus, electromagnetic radiation emerges as an entity with a dual nature, having both wave and particle aspects. The remainder of this chapter will be devoted to the applications of this duality to some of the phenomena mentioned above, and to a study of this seemingly (but not actually) inconsistent nature of electromagnetic radiation.

44-2 Thermionic emission

As a prelude to the discussion of the photoelectric effect, we consider briefly a related phenomenon, *thermionic emission*, discovered by Thomas Edison in 1883 during his experiments on electric lightbulbs. A glassblower had sealed into the bulb of an ordinary filament lamp an extra metal electrode or plate, shown in Fig. 44-1. The glass bulb was then evacuated and the filament heated as usual. When the plate was

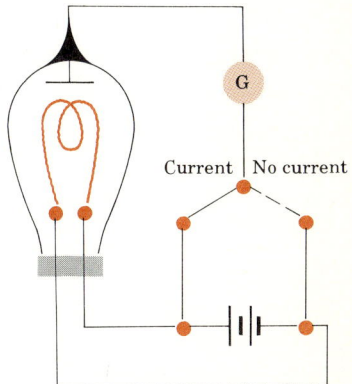

44-1 Edison's original thermionic-emission experiment.

connected through a galvanometer to the *positive* terminal of the 110-V dc source, a galvanometer deflection indicated the existence of a current, despite the fact that there was no conducting path from the plate to the other terminal of the battery. When the plate was connected to the *negative* terminal there was no current. Edison was very much interested in this phenomenon at the time but could not explain it.

The effect is caused by the escape of electrons from the hot filament. Ordinarily electrons in a conductor are prevented from escaping from the surface by a potential-energy barrier. When an electron starts to move away from the surface it induces a corresponding positive charge in the material, which tends to pull it back into the surface. To escape, the electron must somehow acquire enough energy to surmount this energy barrier; the minimum energy needed to escape is called the *work function*, and varies from one material to another. Typical work functions are of the order of a few electronvolts. The electronvolt (eV) is defined in Sec. 26–7, and it would do no harm for the reader to review that section in preparation for this chapter.

At ordinary temperatures almost none of the electrons can acquire enough energy to escape, but when the filament is very hot the electron energies are greatly increased by thermal motion, and at sufficiently high temperatures considerable numbers are able to escape. Once out of the material, the electrons are attracted to the plate if it is positively charged, but repelled if it is negatively charged. The liberation of electrons from a hot surface is called *thermionic emission*.

The electrons that have escaped from the hot conductor form a cloud of negative charge near it, called a *space charge*. If a second conductor (the plate) is maintained by a battery at a higher potential than the first, the electrons in the cloud are attracted to it, and so long as the potential difference between the conductors is maintained, there will be a steady drift of electrons from the emitter or *cathode* to the other body, which is called the plate or *anode*.

In the common thermionic tube the cathode and anode (and often other electrodes as well) are enclosed within an evacuated glass or metal container, and leads to the various electrodes are brought out through the base or walls of the tube. The simplest thermionic tube, in which the only electrodes are the cathode and anode, is called a *diode*.

The diode is shown schematically in Fig. 44–2. The cathode and plate are represented by K and P. The cathode is often in the form of a hollow cylinder, which is heated by a fine resistance wire H within it. Electrons emitted from the outer surface of the cathode are attracted to the plate, which is a larger cylinder surrounding the cathode and coaxial with it. The electron current to the plate is read on the milliammeter MA. The potential difference between plate and cathode can be controlled by the slide wire and read on voltmeter V.

If the potential difference between cathode and anode is small (a few volts), only a few of the emitted electrons reach the plate, the majority penetrating a short distance into the cloud of space charge and then returning to the cathode. As the plate potential is increased, more and more electrons are drawn to it, and with sufficiently high potentials (of the order of 100 V), *all* of the emitted electrons arrive at the plate. Further increase of plate potential does not increase the plate current, which is then said to become *saturated*.

44–2 Circuit for measuring plate current and plate voltage in a vacuum diode.

A graph of plate current, I_p, versus plate potential, V_p, is shown in Fig. 44–3a. Note that I_p is not zero even when V_p is zero. This is because the electrons leave the cathode with an initial speed and the more rapidly moving ones may penetrate the cloud of space charge and reach the plate even with no accelerating field. In fact, a *retarding* field is necessary to prevent their reaching the plate, an effect which may be used to measure their speeds of emission.

The saturation current I_s, in Fig. 44–3a, is equal to the current from the cathode, and for a given tube its magnitude increases markedly with cathode temperatue. Figure 44–3b shows three plate current curves at three different temperatures, where $T_3 > T_2 > T_1$.

The work function ϕ of a surface may be considerably reduced by the presence of impurities. A small amount of thorium, for example, reduces the work function of pure tungsten by about 50%. Since the smaller the work function the larger the current density at a given temperature (or the lower the temperature at which a given emission can be attained), most vacuum tubes now use cathodes having composite surfaces.

Because thermionic emission occurs only at the heated cathode, not at the anode, electron flow occurs in only one direction, from cathode to anode. Most circuit applications of the diode make use of this one-directional characteristic; the simplest example is a *rectifier* for converting alternating current to direct current.

Lee de Forest discovered in 1907 that the electron flow can be modified by inserting a third electrode called a *grid* between cathode and anode. The grid is usually an open structure, either a screen or a coil of wire. When the grid is negative with respect to the cathode the resulting field enhances the space charge near the cathode and decreases the electron flow through the grid to the plate. Because of the grid's proximity to the cathode, the current is much more sensitive to changes in grid voltage than to changes in anode voltage, and thus the device can function as an *amplifier*. In the form just described it is called a *triode*. This basic device and various elaborations of it are very widely used in electronic equipment, although in the past two decades they have been supplanted in many applications by transistors, which will be discussed in Chapter 46.

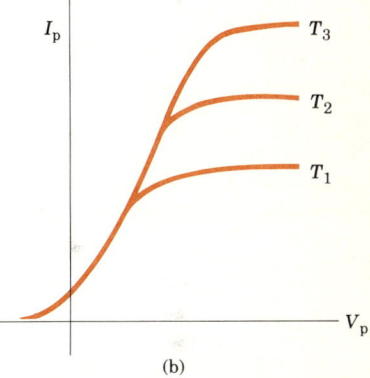

44-3 (a) Plate current-plate voltage characteristic of a diode. (b) Plate current curves at three different cathode temperatures, $T_3 > T_2 > T_1$.

44-3 The photoelectric effect

In the thermionic emission of electrons from metals, the energy needed by an electron to escape from the metal surface is furnished by the energy of thermal agitation. Electrons may also acquire enough energy to escape from a metal, even at low temperatures, if the metal is illuminated by light of sufficiently short wavelength. This phenomenon is called the *photoelectric effect*. It was first observed by Heinrich Hertz in 1887, who noticed that a spark would jump more readily between two spheres when their surfaces were illuminated by the light from another spark. The effect was investigated in detail in the following years by Hallwachs and Lenard.

A modern phototube is shown schematically in Fig. 44–4. A beam of light, indicated by the arrows, falls on a photosensitive surface K called the *cathode*. The battery or other source of potential difference creates an electric field in the direction from A (called the *collector* or *anode*)

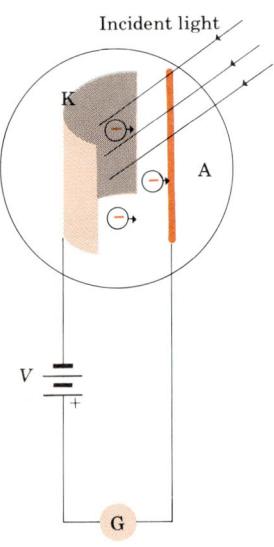

44-4 Schematic diagram of a photocell circuit.

toward K, and electrons emitted from K are pushed by this field to the anode A. Anode and cathode are enclosed in an evacuated container; the photoelectric current is measured by the galvanometer G.

It is found that with a given material as emitter, no photoelectrons at all are emitted unless the wavelength of the light is *shorter* than some critical value. The corresponding *minimum* frequency is called the *threshold frequency* of the particular surface. The threshold frequency for most metals is in the ultraviolet (critical wavelength 200 to 300 nm), but for potassium and cesium oxide it lies in the visible spectrum (400 to 700 nm).

Just as in the case of thermionic emission, the photoelectrons form a cloud of space charge around the emitter. Some of the electrons are emitted with an initial speed; this is shown by the fact that, even with *no* emf in the external circuit, a few electrons penetrate the cloud of space charge and reach the collector, causing a small current in the external circuit. Indeed, even when the polarity of the potential difference V is reversed and the associated electric-field force on the electrons is back toward the cathode, some electrons still reach the anode. Only when the reversed potential V is made large enough so that the potential energy eV is greater than the maximum kinetic energy $\frac{1}{2}mv_{\max}^2$ with which the electrons leave the cathode, does the electron flow stop completely. This critical reversed potential is called the *stopping potential*, denoted by V_0, and it provides a direct measurement of the maximum kinetic energy with which electrons leave the cathode, through the relation

$$\tfrac{1}{2}mv_{\max}^2 = eV_0. \tag{44-1}$$

Surprisingly, it turns out that the maximum kinetic energy of a photoelectron does not depend on the intensity of the incident light, but does depend on its wavelength. When the light intensity increases, the photoelectric current also increases, but only because the *number* of emitted electrons increases, not the energy of an individual electron.

The correct explanation of the photoelectric effect was given by Einstein in 1905. Extending a proposal made two years earlier by Planck, Einstein postulated that a beam of light consisted of small bundles of energy that are now called *light quanta* or *photons*. The energy E of a photon is proportional to its frequency f, or is equal to its frequency multiplied by a constant. That is,

$$E = hf, \tag{44-2}$$

where h is a universal constant, called *Planck's constant*, whose value is 6.626×10^{-34} J·s. When a photon collides with an electron at or just within the surface of a metal, it may transfer its energy to the electron. This transfer is an "all-or-none" process, the electron getting all the photon's energy or none at all. The photon then simply drops out of existence. The energy acquired by the electron may enable it to escape from the surface of the metal if it is moving in the right direction.

In leaving the surface of the metal, the electron loses an amount of energy ϕ (the work function of the surface). Some electrons may lose more than this if they start at some distance below the metal surface, but the *maximum* energy with which an electron can emerge is the energy gained from a photon minus the work function. Hence the maximum kinetic energy of the photoelectrons ejected by light of frequency f is

$$\tfrac{1}{2}mv_{\max}^2 = hf - \phi. \tag{44-3}$$

Combining this with Eq. (44-1) leads to the relation

$$eV_0 = hf - \phi. \tag{44-4}$$

Thus by measuring the stopping potential V_0 required for each of several values of frequency f for a given cathode material, one can determine both the work function ϕ for the material and the value of the quantity h/e. Thus this experiment provides a direct measurement of the value of Planck's constant in addition to a direct confirmation of Einstein's interpretation of photoelectric emission.

Example For a certain cathode material used in a photoelectric-effect experiment, a stopping potential of 3.0 V was required for light of wavelength 300 nm, 2.0 V for 400 nm, and 1.0 V for 600 nm. Determine the work function for this material, and the value of Planck's constant.

Solution According to Eq. (44-4), a graph of V_0 as a function of f should be a straight line. We rewrite this equation as

$$V_0 = \frac{h}{e}f - \frac{\phi}{e}.$$

In this form we see that the *slope* of the line is h/e and the *intercept* on the vertical axis (corresponding to $f = 0$) is at $-\phi/e$. The frequencies, obtained from $f = c/\lambda$ and $c = 3.00 \times 10^8 \text{ m}\cdot\text{s}^{-1}$, are 0.5, 0.75, and $1.0 \times 10^{15} \text{ s}^{-1}$, respectively. The graph is shown in Fig. 44-5. From it we find

$$-\frac{\phi}{e} = -1.0 \text{ V}, \qquad \phi = 1.0 \text{ eV},$$

and

$$\frac{h}{e} = \frac{1.0 \text{ V}}{0.25 \times 10^{15} \text{ s}^{-1}} = 4 \times 10^{-15} \text{ J}\cdot\text{C}^{-1}\cdot\text{s},$$

$$h = (4.0 \times 10^{-15} \text{ J}\cdot\text{C}^{-1}\cdot\text{s})(1.6 \times 10^{-19} \text{ C})$$
$$= 6.4 \times 10^{-34} \text{ J}\cdot\text{s}.$$

The particle-like nature of electromagnetic radiation, used in the above analysis of the photoelectric effect, has been established beyond any reasonable doubt. A photon of electromagnetic radiation of frequency f and corresponding wavelength $\lambda = c/f$ has energy E given by

$$E = hf = \frac{hc}{\lambda}. \tag{44-5}$$

Furthermore, according to relativity theory, every particle having energy must also have momentum, even if it has no rest mass. Photons have zero rest mass; according to Eq. (43-35), the momentum $\boldsymbol{p}$ of a photon of energy E has magnitude p given by

$$E = pc. \tag{44-6}$$

44-5 Stopping potential as a function of frequency. For a different cathode material having a different work function, the line would be displaced up or down but would have the same slope.

Thus the wavelength of a photon and its momentum are related simply by

$$p = \frac{h}{\lambda}. \tag{44-7}$$

These relations will be used frequently in the remainder of this chapter.

44-4 Line spectra

The quantum hypothesis, used in the preceding section for the analysis of the photoelectric effect, also plays an important role in the understanding of atomic spectra. We have seen how a prism or grating spectrograph functions to disperse a beam of light into a *spectrum*. If the light source is an incandescent solid or liquid, the spectrum is *continuous*; that is, light of all wavelengths is present. If, however, the source is a gas through which an electrical discharge is passing, or a flame into which a volatile salt has been introduced, the spectrum is of an entirely different character. Instead of a continuous band of color, only a few colors appear, in the form of isolated parallel lines. (Each "line" is an image of the spectrograph slit, deviated through an angle dependent on the frequency of the light forming the image.) A spectrum of this sort is termed a *line spectrum*. The wavelengths of the lines are characteristic of the element emitting the light. That is, hydrogen always gives a set of lines in the same position, sodium another set, iron still another, and so on. The line structure of the spectrum extends into both the ultraviolet and infrared regions, where photographic or other means are required for its detection.

It might be expected that the frequencies of the light emitted by a particular element would be arranged in some regular way. For instance, a radiating atom might be analogous to a vibrating string, emitting a fundamental frequency and its harmonics. At first sight there does not seem to be any semblance of order or regularity in the lines of a typical spectrum; for many years unsuccessful attempts were made to correlate the observed frequencies with those of a fundamental and its overtones. Finally, in 1885, Johann Jakob Balmer (1825–1898) found a simple formula which gave the frequencies of a group of lines emitted by atomic hydrogen. Since the spectrum of this element is relatively simple, and fairly typical of a number of others, we shall consider it in more detail.

Under the proper conditions of excitation, atomic hydrogen may be made to emit the sequence of lines illustrated in Fig. 44–6. This sequence is called a *series*. There is evidently a certain order in this spectrum, the

44-6 The Balmer series of atomic hydrogen. (Reproduced by permission from *Atomic Spectra and Atomic Structures* by Gerhard Herzberg. Copyright 1937 by Prentice-Hall, Inc.)

lines becoming crowded more and more closely together as the limit of the series is approached. The line of longest wavelength or lowest frequency, in the red, is known as H_α, the next, in the blue-green, as H_β, the third as H_γ, and so on. Balmer found that the wavelengths of these lines were given accurately by the simple formula

$$\frac{1}{\lambda} = R\left(\frac{1}{2^2} - \frac{1}{n^2}\right), \qquad (44\text{-}8)$$

where λ is the wavelength, R is a constant called the Rydberg constant, and n may have the integral values 3, 4, 5, etc. If λ is in meters,

$$R = 1.097 \times 10^7 \text{ m}^{-1}.$$

Letting $n = 3$ in Eq. (44-5), one obtains the wavelength of the H_α-line:

$$\frac{1}{\lambda} = 1.097 \times 10^7 \text{ m}^{-1}\left(\frac{1}{4} - \frac{1}{9}\right)$$
$$= 1.524 \times 10^6 \text{ m}^{-1},$$

whence

$$\lambda = 656.3 \text{ nm}.$$

For $n = 4$, one obtains the wavelength of the H_β-line, etc. For $n = \infty$, one obtains the limit of the series, at $\lambda = 364.6$ nm. This is the *shortest* wavelength in the series.

Other series spectra for hydrogen have since been discovered. These are known, after their discoverers, as the Lyman, Paschen, Brackett, and Pfund series. The formulas for these are

Lyman series:

$$\frac{1}{\lambda} = R\left(\frac{1}{1^2} - \frac{1}{n^2}\right), \qquad n = 2, 3, \ldots,$$

Paschen series:

$$\frac{1}{\lambda}\ R\left(\frac{1}{3^2} - \frac{1}{n^2}\right), \qquad n = 4, 5, \ldots,$$

Brackett series:

$$\frac{1}{\lambda} = R\left(\frac{1}{4^2} - \frac{1}{n^2}\right), \qquad n = 5, 6, \ldots,$$

Pfund series:

$$\frac{1}{\lambda} = R\left(\frac{1}{5^2} - \frac{1}{n^2}\right), \qquad n = 6, 7, \ldots.$$

The Lyman series is in the ultraviolet, and the Paschen, Brackett, and Pfund series are in the infrared. The Balmer series evidently fits into the scheme between the Lyman and the Paschen series.

The Balmer formula, Eq. (44-8), may also be written in terms of the frequency of the light, by use of the relation

$$c = f\lambda \qquad \text{or} \qquad \frac{1}{\lambda} = \frac{f}{c}.$$

Thus, Eq. (44-8) becomes

$$f = Rc\left(\frac{1}{2^2} - \frac{1}{n^2}\right) \quad (44\text{-}9)$$

or

$$f = \frac{Rc}{2^2} - \frac{Rc}{n^2}. \quad (44\text{-}10)$$

Each of the fractions on the right side of Eq. (44-10) is called a *term*, and the frequency of every line in the series is given by the difference between two terms.

There are only a few elements (hydrogen, singly ionized helium, doubly ionized lithium) whose spectra can be represented by a simple formula of the Balmer type. Nevertheless, it is possible to separate the more complicated spectra of other elements into series, and to express the frequency of each line in the series as the difference of two *terms*. The first term is constant for any one series, while the various values of the second term can be labeled by values of an integer index n analogous to the n appearing in Eq. (44-10). In a few simple cases the numerical values of the terms can be *calculated* from theoretical considerations, as we shall see in Chapter 45. For complex atoms, however, the term values must be determined experimentally by analysis of spectra, often an extremely complex problem.

44-5 Energy levels

Every element has a characteristic line spectrum, which must somehow result from the characteristics and structure of the *atoms* of that element. Part of the key to understanding the relation of atomic structure to atomic spectra was supplied in 1913 by the Danish physicist Niels Bohr, who applied to spectra the same concept of light quanta or *photons* that Einstein had used earlier in analysis of the photoelectric effect.

Bohr's hypothesis was as follows: Each atom, as a result of its internal structure (and presumably internal motion), can have a variable amount of *internal energy*. But the energy of an atom cannot change by any arbitrary amount; rather, each atom has a series of discrete *energy levels,* such that an atom can have an amount of internal energy corresponding to any one of these levels, but it cannot have an energy *intermediate* between two levels. All atoms of a given element have the same set of energy levels, but atoms of different elements have different sets. While an atom is in one of these states corresponding to a definite energy, it does not radiate, but an atom can make a transition from one energy level to a lower level by emitting a photon, whose energy is equal to the energy difference between the initial and final states. If E_i is the initial energy of the atom, before such a transition, and E_f its final energy, after the transition, then, since the photon's energy is hf, we have

$$hf = E_i - E_f. \quad (44\text{-}11)$$

For example, a photon of orange light of wavelength 600 nm has a frequency f given by

$$f = \frac{c}{\lambda} = \frac{3.00 \times 10^8 \text{ m} \cdot \text{s}^{-1}}{600 \times 10^{-9} \text{ m}} = 5.00 \times 10^{14} \text{ Hz}$$
$$= 5.00 \times 10^{14} \text{ s}^{-1}.$$

The corresponding photon energy is

$$E = hf = (6.63 \times 10^{-34} \text{ J} \cdot \text{s})(500 \times 10^{14} \text{ s}^{-1})$$
$$= 3.31 \times 10^{-19} \text{ J} = 2.07 \text{ eV}.$$

Thus, this photon must be emitted in a transition between two states of the atom differing in energy by 2.07 eV.

The Bohr hypothesis, if correct, would shed new light on the analysis of spectra on the basis of *terms,* as described in Sec. 44-4. For example, Eq. (44-10) gives the frequencies of the Balmer series in the hydrogen spectrum. Multiplied by Planck's constant h, this becomes

$$hf = \frac{Rch}{2^2} - \frac{Rch}{n^2}. \tag{44-12}$$

If we now compare Eqs. (44-11) and (44-12), identifying $-Rch/n^2$ with the initial energy of the atom E_i and $-Rch/2^2$ with its final energy, E_f, before and after a transition in which a photon of energy $hf = E_i - E_f$ is emitted, then Eq. (44-12) takes on the same form as Eq. (44-11). More generally, if we assume that the possible energy levels for the hydrogen atom are given by

$$E_n = -\frac{Rch}{n^2}, \qquad n = 1, 2, 3, \ldots, \tag{44-13}$$

then *all* the series spectra of hydrogen can be understood on the basis of transitions from one energy level to another. For the Lyman series the final state is always $n = 1$, for the Paschen series it is $n = 3$, and so on. Similarly, complex spectra of other elements, represented by *terms,* are understood on the basis that each term corresponds to an energy level; and a frequency, represented as a difference of two terms, corresponds to a transition between the two corresponding energy levels.

In 1914 a series of experiments by Franck and Hertz provided more direct experimental evidence for energy levels in atoms. In studying the motion of electrons through mercury vapor, under the action of an electric field, they found that a spectrum line at 254 nm was emitted by the vapor when the electron kinetic energy was greater than 4.9 eV but not when it was less. This strongly suggests the existence of an energy level 4.9 eV above the ground state; a mercury atom is excited to this level by collision with an electron, and subsequently decays to the ground state by emitting a photon. The energy of the photon, according to Eq. (44-2), should be

$$E = hf = \frac{hc}{\lambda} = \frac{(6.63 \times 10^{-34} \text{ J} \cdot \text{s})(3.00 \times 10^{8} \text{ m} \cdot \text{s}^{-1})}{254 \times 10^{-9} \text{ m}}$$
$$= 7.82 \times 10^{-19} \text{ J} = \frac{7.82 \times 10^{-19} \text{ J}}{1.60 \times 10^{-19} \text{ J} \cdot \text{eV}^{-1}}$$
$$= 4.9 \text{ eV},$$

in excellent agreement with the measured electron energy.

Although the Bohr hypothesis permits partial understanding of line spectra on the basis of energy levels of atoms, it is not yet complete because it provides no basis for *predicting* what the energy levels for any particular kind of atom should be. We shall return to this problem in

Chapter 45, where the new mechanical principles required for the understanding of the structure and energy levels of atoms will be developed. These principles constitute the subject of *quantum mechanics*.

44-6 Atomic spectra

As we have seen, the key to the understanding of atomic spectra is the concept of atomic *energy levels*. Every spectrum line corresponds to a specific transition between two energy levels of an atom, and the corresponding frequency is given in each case by Eq. (44-11).

Thus the fundamental problem of the spectroscopist is to determine the energy levels of an atom from the measured values of the wavelengths of the spectral lines emitted when the atom proceeds from one energy level to another. In the case of complicated spectra emitted by the heavier atoms, this is a task requiring tremendous ingenuity. Nevertheless, almost all atomic spectra have been analyzed, and the resulting energy levels have been tabulated or plotted with the aid of diagrams similar to the one shown for sodium in Fig. 44-7.

As we shall see, every atom has a lowest energy level, representing the *minimum* energy the atom can have. This lowest energy level is called the *ground state,* and all higher levels are called *excited states*. As we have seen, a spectral line is emitted when an atom proceeds from an excited state to a lower state. The only means discussed so far for raising the atom from the normal state to an excited state has been with the aid of an electric discharge. Let us consider now another method, involving absorption of radiant energy.

From Fig. 44-7 it may be seen that a sodium atom emits the characteristic yellow light of wavelengths 589.0 and 589.6 nm (the D_1- and D_2-lines) when it undergoes the transitions from the two levels marked *resonance levels* to the ground state. Suppose a sodium atom in the ground state were to *absorb* a quantum of radiant energy of wavelength 589.0 or 589.6 nm. It would then undergo a transition in the opposite direction and be raised to one of the resonance levels. After a short time, the average value of which is called the *lifetime* of the excited state, the atom returns to the ground state and emits this quantum. For the resonance levels of the sodium atom, the lifetime is about 1.6×10^{-8} s.

This emission process is called *resonance radiation* and may be easily demonstrated as follows. A strong beam of the yellow light from a sodium arc is concentrated on a glass bulb that has been highly evacuated and into which a small amount of pure metallic sodium has been distilled. If the bulb is warmed to increase the sodium vapor pressure, resonance radiation will take place throughout the whole bulb, which glows with the yellow light characteristic of sodium.

A sodium atom in the ground state may absorb radiant energy of wavelengths other than the yellow resonance lines. All wavelengths corresponding to spectral lines *emitted* when the sodium atom returns to its normal state may also be absorbed. Thus, from Fig. 44-7, wavelengths 330.2 nm, 285.3 nm, etc., may be *absorbed* by a normal sodium atom. If, therefore, the continuous-spectrum light from a carbon arc is sent through an absorption tube containing sodium vapor, and then examined with a spectroscope, there will be a series of dark lines corresponding to the wavelengths absorbed, as shown in Fig. 44-8. This is known as an *absorption spectrum.*

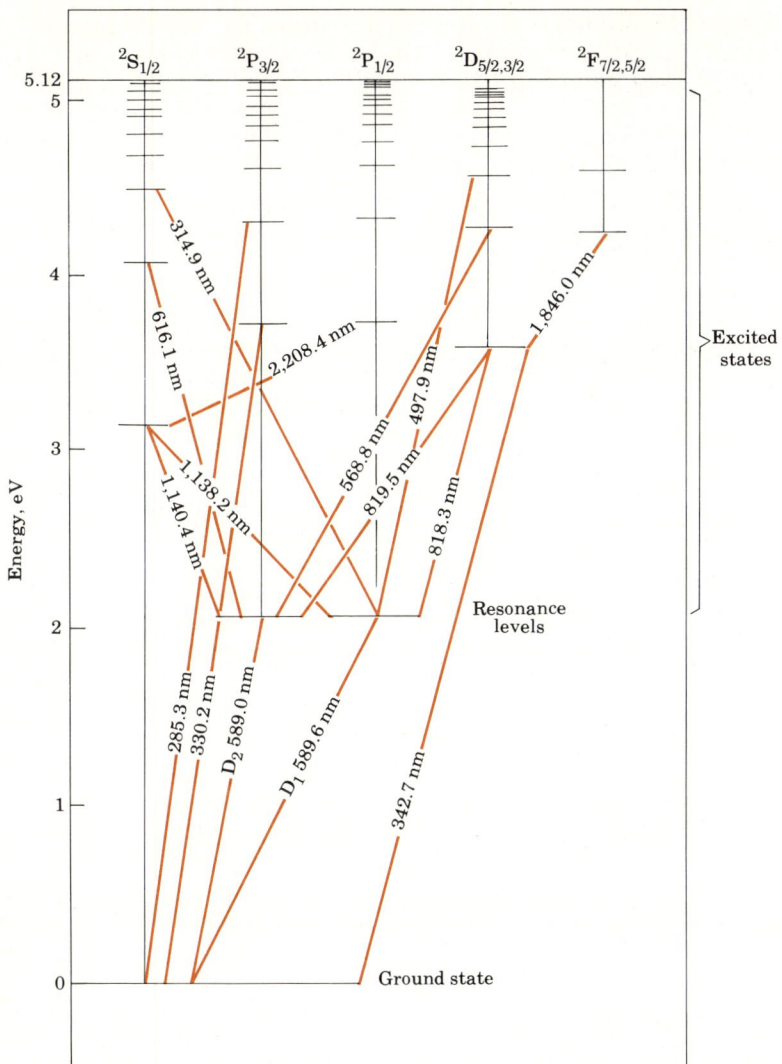

44-7 Energy levels of the sodium atom. Numbers on the lines between levels are wavelengths.

The sun's spectrum is an absorption spectrum. The main body of the sun emits a continuous spectrum, whereas the cooler vapors in the sun's atmosphere emit line spectra corresponding to all the elements present. When the intense light from the main body of the sun passes through the cooler vapors, the lines of these elements are absorbed. The light *emitted* by the cooler vapors is so small compared with the unabsorbed continuous spectrum that the continuous spectrum appears to be crossed by many faint *dark* lines. These were first observed by Fraunhofer and are therefore called *Fraunhofer lines*. They may be observed with any student spectroscope pointed toward any part of the sky.

44-8 Absorption spectrum of sodium.

44-7 The laser

If the energy difference between the normal and the first excited state of an atom is E, the atom is capable of absorbing a photon whose frequency f is given by the Planck equation $E = hf$. The *absorption* of a photon by a normal atom A is depicted schematically in Fig. 44–9a. After absorbing the photon, the atom becomes an excited atom A^*. A short time later, *spontaneous emission* takes place and the excited atom becomes normal again by emitting a photon of the same frequency as that which was originally absorbed but in a random direction and with a random phase, as shown in Fig. 44–9b. There is also a third process, first proposed by Einstein, called *stimulated emission*, shown schematically in Fig. 44–9c. *Stimulated emission takes place when a photon encounters an excited atom and forces it to emit another photon of the same frequency, in the same direction, and in the same phase.* The two photons go off together as *coherent* radiation.

Consider an absorption cell containing a large number of atoms of the type depicted in Fig. 44–9. In the absence of an external beam of radiation, most of the atoms are in the ground state; there are only a few excited atoms present in the cell. The ratio of the number n_E of excited atoms to the number n_0 of normal atoms is extremely small.

Now suppose a beam of radiation is sent through the cell with frequency f corresponding to the energy difference E. The ratio of the numbers n_E and n_0, that is, the ratio of the *populations* of the energy levels, is increased. Since the population of the normal state was so much larger than that of the excited state, an enormously intense beam of light would be required to increase the population of the excited state to a value comparable to or greater than that of the normal state. Therefore, the rate at which energy is extracted from the beam by absorption of normal atoms far outweighs the rate at which energy is added to the beam by stimulated emission of excited atoms.

If a condition can be created in which n_E is substantially increased compared to the normal equilibrium value, creating a condition known as *population inversion,* the rate of energy radiation by stimulated emission may *exceed* the rate of absorption. The system then acts as a *source* of radiation with photon energy E. Furthermore, since the photons are

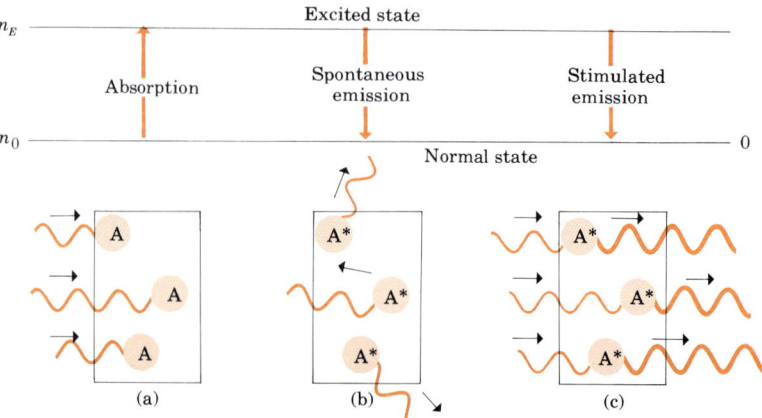

44–9 Three interaction processes between an atom and radiation.

the result of stimulated emission, they all have the same frequency, phase, polarization, and direction. The resulting radiation is therefore very much more *coherent* than light from ordinary sources, in which the emissions of individual atoms are *not* coordinated.

The necessary population inversion can be achieved in a variety of ways. As an example we consider the helium–neon laser, a simple, inexpensive laser available in many undergraduate laboratories. A mixture of helium and neon, each typically at a pressure of the order of 10^2 Pa (or 10^{-3} atm), is sealed in a glass enclosure provided with two electrodes. When a sufficiently high voltage is applied, a glow discharge occurs. Collisions between ionized atoms and electrons carrying the discharge current excite atoms to various energy states.

Figure 44–10 shows an energy-level diagram for the system. The notation used to label the various energy levels, such as $1s2s$ or $5s$, will be discussed in Sec. 46–1, and need not concern us here. Helium atoms excited to the $1s2s$ state cannot return to the ground state by emitting a 20.61-eV photon, as might be expected, because both the states have zero total angular momentum, while a photon must carry away at least one unit ($h/2\pi$) of angular momentum. Such a state, in which radiative decay is impossible, is called a *metastable state*.

The helium atoms *can,* however, lose energy by energy-exchange collisions with neon atoms initially in the ground state. A $1s2s$ helium atom, with its internal energy of 20.61 eV and a little additional kinetic

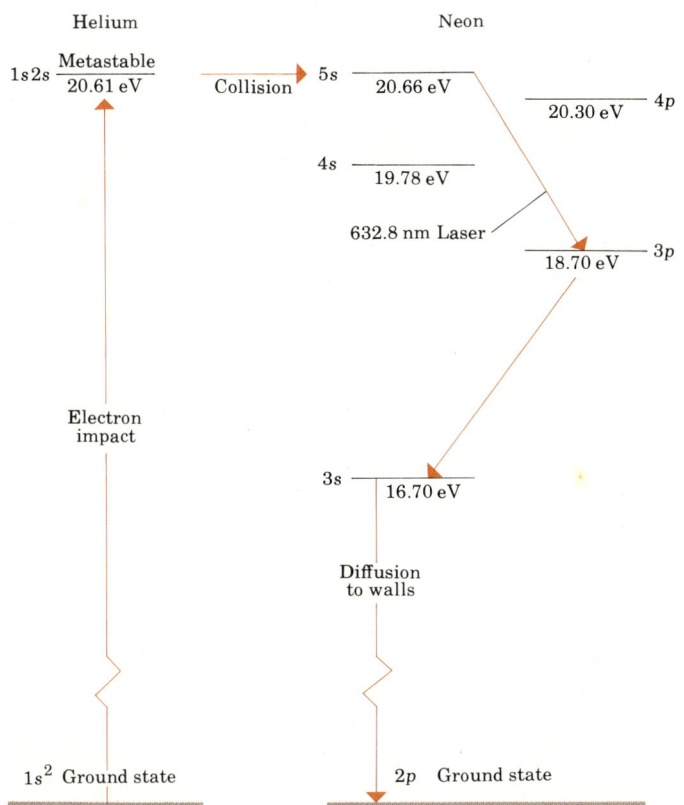

44–10 Energy-level diagram for helium–neon laser.

energy, can collide with a neon atom in the ground state, exciting it to the 5s excited state at 20.66 eV and leaving the helium atom in the $1s^2$ ground state. Thus, we have the necessary mechanism for a population inversion in neon, with the population in the 5s state substantially enhanced. Stimulated emission from this state then results in the emission of highly coherent light at 632.8 nm, as shown on the diagram. In practice the beam is sent back and forth through the gas many times by a pair of parallel mirrors, so as to stimulate emission from as many excited atoms as possible. One of the mirrors is partially transparent, so a portion of the beam emerges as an external beam.

The net effect of all the processes taking place in a laser tube is a beam of radiation that is (1) very intense, (2) almost perfectly parallel, (3) almost monochromatic, and (4) spatially *coherent* at all points within a given cross section. To understand this fourth characteristic, we recall the simple double-slit interference experiment. A mercury arc placed directly behind the double slit would not give rise to interference fringes because the light issuing from the two slits would come from different points of the arc and would not retain a constant phase relationship. In the use of the usual laboratory arc-lamp sources, it is necessary to use the light from a very small portion of the source to illuminate the double slit. The slightly diverging beam from a laser, however, may be allowed to fall directly on a double slit (or other interferometer) because the light rays from any two points of a cross section are in phase, and are said to exhibit "spatial coherence."

In recent years lasers have found a wide variety of practical applications. The high intensity of a laser beam makes it a convenient drill. A very small hole can be drilled in a diamond for use as a die in drawing very small-diameter wire. The ability of a laser beam to travel long distances without appreciable spreading makes it a very useful tool for surveyors, especially in situations where great precision is required over long distances, as in the case of a long tunnel drilled from both ends.

Lasers are finding increasing application in medical science. A laser can produce a very *narrow* beam with extremely *high intensity*, high enough to vaporize anything in its path. This property is used in the treatment of a detached retina; a short burst of radiation damages a small area of the retina, and the resulting scar tissue "welds" the retina back to the choroid from which it has become detached. Laser beams are also used in surgery; blood vessels cut by the beam tend to seal themselves off, making it easier to control bleeding. The use of laser radiation in the treatment of skin cancer is an active area of research.

44-8 X-ray production and scattering

X-rays are produced when rapidly moving electrons that have been accelerated through potential differences of the order of 10^3 to 10^6 V are allowed to strike a metal target. They were first observed by Wilhelm K. Röntgen (1845–1923) in 1895, and were originally called *Röntgen rays*.

X-rays are of the same nature as light or any other electromagnetic wave and, like light waves, they are governed by quantum relations in their interaction with matter. Hence, one may speak of x-ray photons or quanta, the energy of such a photon being given by the relation $E = hf$. Wavelengths of x-rays range from approximately 0.001 to 1 nm (10^{-12} to

44-11 Coolidge-type x-ray tube.

10^{-9} m). X-ray wavelengths can be measured quite precisely by crystal diffraction techniques, described in Sec. 41–10.

A common x-ray tube is the Coolidge type, invented by W. D. Coolidge of the General Electric laboratories in 1913. A Coolidge tube is shown in Fig. 44–11. A thermionic cathode and an anode are enclosed in a glass tube that has been pumped down to an extremely low pressure. Electrons emitted from the cathode can then travel directly to the anode with only a small probability of a collision on the way, and they reach the anode with a speed corresponding to the full potential difference across the tube. X-radiation is emitted from the anode surface as a consequence of its bombardment by the electron stream.

Two distinct processes are involved in x-ray emission. Some of the electrons are stopped by the target and their kinetic energy is converted directly to x-radiation. Others transfer their energy in whole or in part to the atoms of the target, which retain it temporarily as "energy of excitation" but very shortly emit it as x-radiation. The latter is characteristic of the material of the target, while the former is not.

The atomic energy levels associated with x-ray excitation are rather different in character from those associated with visible spectra. To understand them we need some understanding of the arrangement of electrons in complex atoms, a topic to be discussed in greater detail in Chapter 46. For the present we state simply that in a many-electron atom the electrons are always arranged in concentric *shells* at increasing distances from the nucleus. These shells are labeled *K, L, M, N,* etc., the *K* shell being closest to the nucleus, the *L* shell next, and so on. For any given atom in the ground state there is a definite number of electrons in each shell.

For reasons to be discussed later, each shell has a maximum number of electrons it can accommodate, and we may speak of *filled shells* and *partially filled shells*. The *K* shell can contain at most two electrons. The next, the *L* shell, can contain eight. The third, the *M* shell, has a capacity for 18 electrons, while the *N* shell may hold 32. The sodium atom, for example, which contains 11 electrons, has two in the *K* shell, eight in the *L* shell, and a single electron in the *M* shell. Molybdenum, with 42 electrons, has two in the *K* shell, eight in the *L* shell, 18 in the *M* shell, 13 in the *N* shell, and one in the *O* shell.

The *outer* electrons of an atom are the ones responsible for the optical spectra of the elements. Relatively small amounts of energy suffice to remove these to excited states, and on their return to their normal states, wavelengths in or near the visible region are emitted. The inner electrons, being closer to the nucleus, are more tightly bound, and much more energy is required to displace them from their normal levels. As a result, we would expect a photon of much larger energy, and hence much higher frequency, to be emitted when the atom returns to its normal state after the displacement of an inner electron. This is, in fact, the case, and it is the displacement of the inner electrons that gives rise to the emission of x-rays.

On colliding with the atoms of the anode, some of the electrons accelerated in an x-ray tube, provided they have acquired sufficient energy, will dislodge one of the inner electrons of a target atom, say one of the K electrons. This leaves a vacant space in the K shell, which is immediately filled by an electron from either the L, M, or N shell. The readjustment of the electrons is accompanied by a decrease in the energy of the atom, and an x-ray photon is emitted with energy just equal to this decrease. Since the energy change is perfectly definite for atoms of a given element, the emitted x-rays should have definite frequencies. In other words, the x-ray spectrum should be a *line spectrum*. We can predict further that there should be just three lines in the series, corresponding to the three possibilities that the vacant space may have been filled by an L, M, or N electron.

This is precisely what is observed. Figure 44–12 illustrates the so-called K series of the elements tungsten, molybdenum, and copper. Each series consists of three lines, known as the K_α-, K_β-, and K_γ-lines. The K_α-line is produced by the transition of an L electron to the vacated space in the K shell, the K_β-line by an M electron, and the K_γ-line by an N electron.

In addition to the K series, there are other series known as the L, M, and N series, produced by the ejection of electrons from the L, M, and N shells rather than the K shell. As would be expected, the electrons in these outer shells, being farther away from the nucleus, are not held as firmly as those in the K shell. Consequently, the other series may be excited by more slowly moving electrons, and the photons emitted are of lower energy and longer wavelength.

In addition to the x-ray *line* spectrum there is a background of *continuous* x-radiation from the target of an x-ray tube. This is due to the sudden deceleration of those "cathode rays" (bombarding electrons) that do not happen to eject an atomic electron. The remarkable feature of the continuous spectrum is that while it extends indefinitely toward the *long* wavelength end, it is cut off very sharply at the *short* wavelength end. The quantum theory furnishes a simple explanation of the short-wave limit of the continuous x-ray spectrum.

A bombarding electron may be brought to rest in a single process if the electron happens to collide head on with an atom of the target; or

44–12 Wavelengths of the K_α, K_β, and K_γ lines of copper, molybdenum, and tungsten.

it may make a number of collisions before coming to rest, giving up part of its energy each time. If we assume that the energy lost at each collision is radiated as an x-ray photon, these photons may have any energy up to a certain maximum, namely, that of an electron that gives up all of its energy in a single collision. Hence there is a short-wave limit to the spectrum. The frequency of this limit is found by setting the energy of the electron equal to the energy of the x-ray photon:

$$hf = \tfrac{1}{2}mv^2. \tag{44-14}$$

This is precisely the same equation as that for the photoelectric effect except for the work-function term, which is negligible here since the energies of the x-ray photons are so large. In fact, the emission of x-rays may be described as an *inverse photoelectric effect*. In photoelectric emission the energy of a photon is transformed into kinetic energy of an electron; here, the kinetic energy of an electron is transformed into that of a photon.

Example Compute the potential difference through which an electron must be accelerated in order that the short-wave limit of the continuous x-ray spectrum shall be exactly 0.1 nm.

Solution The frequency corresponding to 0.1 nm (10^{-10} m) is given by

$$f = \frac{c}{\lambda} = \frac{3 \times 10^8 \text{ m} \cdot \text{s}^{-1}}{10^{-10} \text{ m}} = 3 \times 10^{18} \text{ s}^{-1} = 3 \times 10^{18} \text{ Hz}.$$

The energy of the photon is

$$hf = (6.63 \times 10^{-34} \text{ J} \cdot \text{s})(3 \times 10^{18} \text{ s}^{-1}) = 19.9 \times 10^{-16} \text{ J}.$$

This must equal the kinetic energy of the electron, $\tfrac{1}{2}mv^2$, which is also equal to the product of the electronic charge and the accelerating voltage, V:

$$\tfrac{1}{2}mv^2 = eV = 19.9 \times 10^{-16} \text{ J}.$$

Since

$$e = 1.60 \times 10^{-19} \text{ C},$$
$$V = \frac{19.9 \times 10^{-16} \text{ J}}{1.60 \times 10^{-19} \text{ C}} = 12{,}400 \text{ V}.$$

A phenomenon called *Compton scattering,* first observed in 1924 by A. H. Compton, provides additional direct confirmation of the quantum nature of electromagnetic radiation. When x-rays impinge on matter, some of the radiation is *scattered,* just as visible light falling on a rough surface undergoes diffuse reflection. Observation shows that some of the scattered radiation has smaller frequency and longer wavelength than the incident radiation, and that the change in wavelength depends on the angle through which the radiation is scattered. Specifically, if the scattered radiation emerges at an angle ϕ with respect to the incident direction, and if λ and λ' are the wavelengths of the incident and scattered radiation, respectively, it is found that

$$\lambda' - \lambda = \frac{h}{mc}(1 - \cos \phi) \tag{44-15}$$

where m is the electron mass.

Compton scattering cannot be understood on the basis of classical electromagnetic theory. On the basis of classical principles, the scattering mechanism is induced motion of electrons in the material, caused by the incident radiation. This motion must have the same frequency as that of the incident wave, and so the scattered wave radiated by the oscillating charges should have the same frequency. There is no way the frequency can be *shifted* by this mechanism.

The quantum theory, by contrast, provides a beautifully simple explanation. We imagine the scattering process as a collision of two *particles,* the incident photon and an electron initially at rest, as in Fig. 44-13. The photon gives up some of its energy and momentum to the electron, which recoils as a result of this impact; and the final photon has less energy, smaller frequency, and longer wavelength than the initial one.

Equation (44-15) can be derived from the principles of conservation of energy and of momentum. We sketch the derivation below; the details of the calculation are left as a problem. Because relativistic energies may be involved, we use relativistic energy and momentum relations for the electron; its initial energy is mc^2, and its final energy E is given by $E^2 = (mc^2)^2 + (Pc)^2$. The energies of the incident and scattered photons are respectively pc and $p'c$; Thus energy conservation yields the relation

$$pc + mc^2 = p'c + E,$$

or

$$(pc - p'c - mc^2)^2 = E^2 = (mc^2)^2 + (Pc)^2. \qquad (44\text{-}16)$$

We may eliminate the electron momentum P from this equation by using momentum conservation:

$$\boldsymbol{p} = \boldsymbol{p}' + \boldsymbol{P},$$

or

$$\boldsymbol{p} - \boldsymbol{p}' = \boldsymbol{P}. \qquad (44\text{-}17)$$

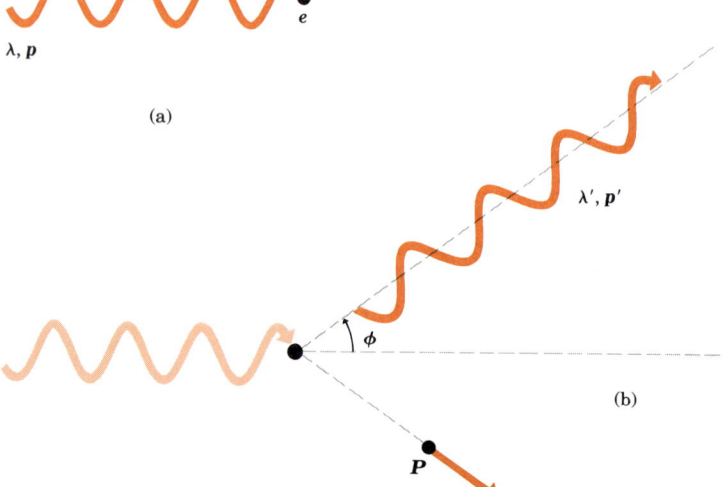

44-13 Schematic diagram of Compton scattering showing (a) electron initially at rest with incident photon of wavelength λ and momentum $\boldsymbol{p}$; (b) scattered photon with longer wavelength λ' and momentum $\boldsymbol{p}'$ and recoiling electron with momentum $\boldsymbol{P}$. The direction of the scattered photon makes an angle ϕ with that of the incident photon, and the angle between $\boldsymbol{p}$ and $\boldsymbol{p}'$ is also ϕ.

We take the scalar product of this quantity with itself, noting that $\mathbf{p} \cdot \mathbf{p}' = pp' \cos \phi$; we obtain

$$P^2 = p^2 + p'^2 - 2pp' \cos \phi. \tag{44-18}$$

This expression for P^2 may now be substituted into Eq. (44-16) and the left side multiplied out. A common factor c^2 is divided out; several terms cancel, and when the resulting equation is divided through by (pp'), the result is

$$\frac{mc}{p'} - \frac{mc}{p} = 1 - \cos \phi. \tag{44-19}$$

Finally, we substitute $p = h/\lambda$ and $p' = h/\lambda'$ and rearrange again to obtain Eq. (44-15).

Questions

44-1 In analyzing the photoelectric effect, how can we be sure that each electron absorbs only *one* proton?

44-2 In what ways do photons resemble other particles such as electrons? In what ways do they differ? Do they have mass? Electric charge? Can they be accelerated? What mechanical properties do they have?

44-3 Considering a two-slit interference experiment, if the photons are not synchronized with each other (i.e., are not coherent) and if half go through each slit, how can they possibly interfere with each other? Is there any way out of this paradox?

44-4 Can you devise an experiment to measure the work function of a material?

44-5 How might the energy levels of an atom be measured directly, i.e., without recourse to analysis of spectra?

44-6 Would you expect quantum effects to be generally more important at the low-frequency end of the electromagnetic spectrum (radio waves) or at the high-frequency end (x-rays and gamma rays)? Why?

44-7 Most black-and-white photographic film (with the exception of some special-purpose films) is less sensitive at the far red end of the visible spectrum than at the blue end, and has almost no sensitivity to infrared. How can these properties be understood on the basis of photons?

44-8 Human skin is relatively insensitive to visible light, but ultraviolet radiation can be quite destructive. Does this have anything to do with photon energies?

44-9 Does the concept of photon energy shed any light (no pun intended) on the question of why x-rays are so much more penetrating than visible light?

44-10 The phosphorescent materials that coat the inside of a fluorescent lamp tube convert ultraviolet radiation (from the mercury-vapor discharge inside the tube) to visible light. Could one also make a phosphor that converts visible light to ultraviolet?

44-11 As a body is heated to very high temperature and becomes self-luminous, the apparent color of the emitted radiation shifts from red to yellow and finally to blue as the temperature increases. Why the color shift?

44-12 Elements in the gaseous state emit line spectra with well-defined wavelengths; but hot solid bodies usually emit a continuous spectrum, that is, a continuous smear of wavelengths. Can you account for this difference?

44-13 Could Compton scattering occur with protons as well as electrons? Suppose, for example, one directed a beam of x-rays at a liquid-hydrogen target. What similarities and differences in behavior would be expected?

Problems

$e = 1.602 \times 10^{-19}$ C
$m = 9.109 \times 10^{-31}$ kg
$h = 6.626 \times 10^{-34}$ J·s
$N_0 = 6.022 \times 10^{23}$ atoms·mol^{-1}
Energy equivalent of 1 u = 931.5 MeV

$\frac{e}{m} = 1.758 \times 10^{11}$ C·kg^{-1}
$k = 1.381 \times 10^{-23}$ J·K^{-1}
1 eV $= 1.602 \times 10^{-19}$ J
1 u $= 1.661 \times 10^{-27}$ kg
$\varepsilon_0 = 8.854 \times 10^{-12}$ C^2·N^{-1}·m^{-2}

44-1 In the photoelectric effect, what is the relation between the threshold frequency f_0 and the work function ϕ?

44-2 The photoelectric threshold wavelength of tungsten is 2.73×10^{-5} cm. Calculate the maximum kinetic energy of the electrons ejected from a tungsten surface by ultraviolet radiation of wavelength 1.80×10^{-5} cm. (Express the answer in electron volts.)

44-3 A photoelectric surface has a work function of 4.00 eV. What is the maximum velocity of the photoelectrons emitted by light of frequency 3×10^{15} Hz?

44-4 When ultraviolet light of wavelength 2.54×10^{-5} cm from a mercury arc falls upon a clean copper surface, the retarding potential necessary to stop the emission of photoelectrons is 0.59 V. What is the photoelectric threshold wavelength for copper?

44-5 When a certain photoelectric surface is illuminated with light of different wavelengths, the stopping potentials in the table below are observed:

Wavelength, nm	Stopping potential, V
366	1.48
405	1.15
436	0.93
492	0.62
546	0.36
579	0.24

Plot the stopping potential as ordinate against the frequency of the light as abscissa. Determine

a) the threshold frequency,
b) the threshold wavelength,
c) the photoelectric work function of the material, and
d) the value of Planck's constant h (the value of e being known).

44-6 The photoelectric work function of potassium is 2.0 eV. If light having a wavelength of 360 nm falls on potassium, find

a) the stopping potential,
b) the kinetic energy in electron volts of the most energetic electrons ejected, and
c) the velocities of these electrons.

44-7 What will be the change in the stopping potential for photoelectrons emitted from a surface if the wavelength of the incident light is reduced from 400 nm to 360 nm?

44-8 The photoelectric work functions for particular samples of certain metals are as follows: cesium, 2.00 eV; copper, 4.00 eV; potassium, 2.25 eV; and zinc, 3.60 eV.

a) What is the threshold wavelength for each metal?
b) Which of these metals could not emit photoelectrons when irradiated with visible light?

44-9 The light-sensitive compound on most photographic films is silver bromide, AgBr. A film is "exposed" when the light energy absorbed dissociates this molecule into its atoms. (The actual process is more complex, but the quantitative result does not differ greatly.) The energy of or heat of dissociation of AgBr is 23.9 kcal·mol^{-1}. Find

a) the energy in electron volts,
b) the wavelength, and
c) the frequency of the photon that is just able to dissociate a molecule of silver bromide.
d) What is the energy in electron volts of a quantum of radiation having a frequency of 100 MHz?
e) Explain the fact that light from a firefly can expose a photographic film, whereas the radiation from a TV station transmitting 50,000 W at 100 MHz cannot.
f) Will photographic films stored in a light-tight container be ruined (exposed) by the radio waves constantly passing through them? Explain.

44-10

a) Show that the energy E (in electron volts) of a photon of wavelength λ (in nanometers) is given by E (eV) $= (1240/\lambda)$(nm).
b) What is the energy in electron volts of a photon having a wavelength of 91.2 nm?

44-11 If 5% of the energy supplied to an incandescent lightbulb is radiated as visible light, how many visible quanta are emitted per second by a 100-W bulb? Assume the wavelength of all the visible light to be 560 nm.

44-12 The directions of emission of photons from a source of radiation are random. According to the wave theory, the intensity of radiation from a point source varies inversely as the square of the distance from the source. Show that the number of photons from a point source passing out through a unit area is also given by an inverse-square law.

44-13 Calculate (a) the frequency, and (b) the wavelength of the H_β-line of the Balmer series for hydrogen. This line is emitted in the transition from $n = 4$ to $l = 2$.

44-14

a) What is the least amount of energy in electron volts that must be given to a hydrogen atom so that it can emit the H_β-line (see Problem 44-13) in the Balmer series?
b) How many different possibilities of spectral line emission are there for this atom when the electron goes from $n = 4$ to the ground state?

44-15 In Fig. 44-10, compute the energy difference for the $5s$-$3p$ transition; express your result in electron volts and in joules. Compute the wavelength of a photon having this energy, and compare your result with the observed wavelength of the laser light.

44-16 In the helium–neon laser, what wavelength corresponds to the 3p–3s transition in neon? Why is this not observed in the beam with the same intensity as the 632.8-nm laser line?

44-17
a) What is the minimum potential difference between the filament and the target of an x-ray tube if the tube is to produce x-rays of wavelength 0.05 nm?
b) What is the shortest wavelength produced in an x-ray tube operated at 2×10^6 V?

44-18 An electron in a certain x-ray tube is accelerated from rest through a potential difference of 180,000 V in going from the cathode to the anode. When it arrives at the anode, what is
a) its kinetic energy in electron volts,
b) its relativistic velocity?
c) What is the velocity of the electron, calculated classically?

44-19 An x-ray tube is operating at 150,000 V and 10 mA.
a) If only 1% of the electric power supplied is converted into x-rays, at what rate is the target being heated in calories per second?
b) If the target has a mass of 300 g and a specific heat of 0.035 cal·g^{-1}·°C^{-1}, at what average rate would its temperature rise if there were no thermal losses?
c) What must be the physical properties of a practical target material? What would be some suitable target elements?

44-20 If electrons in a metal had the same energy distribution as molecules in a gas at the same temperature (which is not actually the case), at what temperature would the average electron kinetic energy equal 1 eV, typical of work functions of metals?

44-21 If hydrogen were monatomic, at what temperature would the average translational kinetic energy be equal to the energy required to raise a hydrogen atom from the ground state to the $n = 2$ excited state?

44-22 Complete the derivation of the Compton-scattering formula, Eq. (44-15), following the outline given in Eqs. (44-16) through (44-19).

44-23 Calculate the maximum increase in x-ray wavelength that can occur during Compton scattering.

44-24 X-rays with initial wavelength 0.5×10^{-10} m undergo Compton scattering. For what scattering angle is the wavelength of the scattered x-rays greater than that of the incident x-rays by one percent?

44-25 X-rays are produced in a tube operating at 50 kV. After emerging from the tube, some x-rays strike a target and are Compton-scattered through an angle of 20°.
a) What is the original x-ray wavelength?
b) What is the wavelength of the scattered x-rays?
c) What is the energy of the scattered x-rays (in electron-volts)?

44-26 What is the energy (in electronvolts) of the smallest-energy x-ray photon for which Compton scattering could result in doubling the original wavelength?

45 QUANTUM MECHANICS

45-1 The Bohr atom

We have seen in the preceding chapter that some aspects of atomic spectra can be understood on the basis of the quantum or photon concept of electromagnetic radiation, together with the concept of discrete energy levels of atoms. Of course, any theory of atomic spectra that aspires to completeness should also offer some means of *predicting,* on theoretical grounds, the values of these energy levels for any given atom. At the same time that Bohr advanced his hypothesis about the relation of spectrum-line frequencies to energy levels, he also proposed a mechanical model of the simplest atom, hydrogen; he was able to calculate the energy levels of hydrogen and obtain agreement with values determined from spectra.

Bohr's was not by any means the first attempt to understand the internal structure of atoms. Starting in 1906, Rutherford and his co-workers had performed experiments on the scattering of alpha particles (helium nuclei emitted from radioactive elements) by thin metallic foils. These experiments, which will be discussed in Chapter 47, showed that each atom contains a massive nucleus whose size (of the order of 10^{-15} m) is very much *smaller* than the overall size of the atom (of the order of 10^{-10} m). The nucleus is surrounded by a swarm of electrons.

To account for the fact that the electrons remain at relatively large distances from the positively charged nucleus despite the electrostatic attraction the nucleus exerts on the electrons, Rutherford postulated that the electrons *revolve* about the nucleus in orbits, more or less as the planets in the solar system revolve around the sun, but with the electrical attraction providing the necessary centripetal force.

This assumption, however, has an unfortunate consequence. A body moving in a circle is continuously accelerated toward the center of the circle and, according to classical electromagnetic theory, an accelerated electron radiates energy. The total energy of the electrons would therefore gradually decrease, their orbits would become smaller and smaller, and eventually they would spiral into the nucleus and come to rest. Furthermore, according to classical theory, the *frequency* of the electromagnetic waves emitted by a revolving electron is equal to the frequency of revolution. As the electrons radiated energy, their angular velocities would change continuously and they would emit a *continuous* spectrum (a mixture of all frequencies), in contradiction to the *line* spectrum actually observed.

Faced with the dilemma that electromagnetic theory predicted an unstable atom emitting radiant energy of all frequencies, while observation showed stable atoms emitting only a few frequencies, Bohr concluded that, in spite of the success of electromagnetic theory in explaining large-scale phenomena, it could not be applied to processes on an atomic scale. He therefore postulated that an electron in an atom can revolve in certain stable orbits, each having a definite associated energy, *without* emitting radiation, contrary to the predictions of classical electromagnetic theory. According to Bohr, an atom radiates only when it makes a transition from one of these special orbits to another, emitting (or absorbing) a photon of appropriate energy, given by Eq. (44–8), at the same time.

To determine the radii of the "permitted" orbits, Bohr introduced what must be regarded in hindsight as a brilliant intuitive guess. He noted that the *units* of Planck's constant h, usually written as J·s, are the same as the units of angular momentum, usually written as kg·m^2·s^{-1}, and he postulated that only those orbits are permitted for which the angular momentum is an integer multiple of $h/(2\pi)$. We recall from Chapter 9 that the angular momentum of a particle of mass m, moving with tangential speed v in a circle of radius r, is mvr. Hence the above condition may be stated as

$$mvr = n\frac{h}{2\pi}, \qquad (45\text{–}1)$$

where $n = 1, 2, 3,$ etc.

We now incorporate this condition into the analysis of the hydrogen atom. This atom consists of a single electron of charge $-e$, revolving about a single proton of charge $+e$. The proton, being nearly 2000 times as massive as the electron, will be assumed stationary in this discussion. The electrostatic force of attraction between the charges,

$$F = \frac{1}{4\pi\varepsilon_0}\frac{e^2}{r^2},$$

provides the centripetal force and, from Newton's second law,

$$\frac{1}{4\pi\varepsilon_0}\frac{e^2}{r^2} = \frac{mv^2}{r}. \qquad (45\text{–}2)$$

When Eqs. (45–1) and (45–2) are solved simultaneously for r and v, we obtain

$$r = \varepsilon_0 \frac{n^2 h^2}{\pi m e^2}, \qquad (45\text{–}3)$$

$$v = \frac{1}{\varepsilon_0}\frac{e^2}{2nh}. \qquad (45\text{–}4)$$

Let

$$\varepsilon_0 \frac{h^2}{\pi m e^2} = r_0. \qquad (45\text{–}5)$$

Then Eq. (45-3) becomes
$$r = n^2 r_0,$$
and the permitted, nonradiating orbits have radii r_0, $4r_0$, $9r_0$, etc. The appropriate value of n is called the *quantum number* of the orbit.

The numerical values of the quantities on the left side of Eq. (45-5) are

$$\varepsilon_0 = 8.854 \times 10^{-12} \, \text{C}^2 \cdot \text{N}^{-1} \cdot \text{m}^{-2},$$
$$h = 6.626 \times 10^{-34} \, \text{J} \cdot \text{s},$$
$$m = 9.109 \times 10^{-31} \, \text{kg},$$
$$e = 1.602 \times 10^{-19} \, \text{C}.$$

Hence r_0, the radius of the first Bohr orbit, is

$$r_0 = \frac{(8.854 \times 10^{-12} \, \text{C}^2 \cdot \text{N}^{-1} \cdot \text{m}^{-2})(6.626 \times 10^{-34} \, \text{J} \cdot \text{s})^2}{(3.14)(9.109 \times 10^{-31} \, \text{kg})(1.602 \times 10^{-19} \, \text{C})^2}$$
$$= 0.53 \times 10^{-10} \, \text{m} = 0.53 \times 10^{-8} \, \text{cm}.$$

This is in good agreement with atomic diameters as estimated by other methods, namely, about 10^{-8} cm.

The kinetic energy of the electron in any orbit is

$$K = \tfrac{1}{2} m v^2 = \frac{1}{\varepsilon_0^2} \frac{me^4}{8n^2 h^2},$$

and the potential energy is

$$U = -\frac{1}{4\pi \varepsilon_0} \frac{e^2}{r} = -\frac{1}{\varepsilon_0^2} \frac{me^4}{4n^2 h^2}.$$

The total energy, E, is therefore

$$E = K + U = -\frac{1}{\varepsilon_0^2} \frac{me^4}{8n^2 h^2}. \tag{45-6}$$

The total energy has a negative sign because the reference level of potential energy is taken with the electron at an infinite distance from the nucleus. Since we are interested only in energy *differences*, this is not of importance.

The energy of the atom is least when its electron is revolving in the orbit for which $n = 1$, for then E has its largest negative value. For $n = 2, 3, \ldots$, the absolute value of E is smaller; hence the energy is progressively larger in the outer orbits. The *normal* state of the atom, called the *ground state,* is that of lowest energy, with the electron revolving in the orbit of smallest radius, r_0. As a result of collisions with rapidly moving electrons in an electrical discharge, or for other causes, the atom may temporarily acquire sufficient energy to raise the electron to some outer orbit. The atom is then said to be in an *excited* state. This state is an unstable one, and the electron soon falls back to a state of lower energy, emitting a photon in the process.

Let n be the quantum number of some excited state, and l the quantum number of the lower state to which the electron returns after the emission process. Then E_i, the initial energy, is

$$E_i = -\frac{1}{\varepsilon_0^2} \frac{me^4}{8n^2 h^2},$$

and E_f, the final energy, is

$$E_f = -\frac{1}{\varepsilon_0^2}\frac{me^4}{8l^2h^2}.$$

The decrease in energy, $E_i - E_f$, which we place equal to the energy hf of the emitted photon, is

$$E_i - E_f = hf = -\frac{1}{\varepsilon_0^2}\frac{me^4}{8n^2h^2} + \frac{1}{\varepsilon_0^2}\frac{me^4}{8l^2h^2},$$

or

$$f = \frac{1}{\varepsilon_0^2}\frac{me^4}{8h^3}\left(\frac{1}{l^2} - \frac{1}{n^2}\right). \tag{45-7}$$

This equation is of precisely the same form as the Balmer formula (Eq. 44-6) for the frequencies in the hydrogen spectrum if we place

$$\frac{1}{\varepsilon_0^2}\frac{me^4}{8h^3} = Rc, \tag{45-8}$$

and let $l = 1$ for the Lyman series, $l = 2$ for the Balmer series, etc. The Lyman series is therefore the group of lines emitted by electrons returning from some excited state to the ground state. The Balmer series is the group emitted by electrons returning from some higher state, but stopping in the *second orbit* instead of falling at once to that of lowest energy. That is, an electron returning from the third orbit ($n = 3$) to the second orbit ($l = 2$) emits the H_α-line. One returning from the fourth orbit ($n = 4$) to the second ($l = 2$) emits the H_β-line, etc. These transitions are shown in Fig. 45-1.

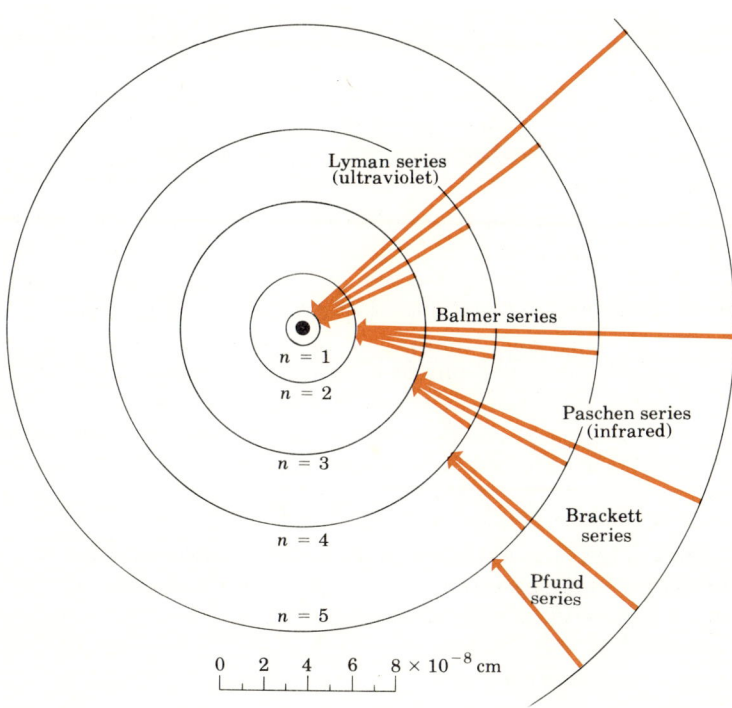

45-1 "Permitted" orbits of an electron in the Bohr model of the hydrogen atom. The transitions responsible for some of the lines of the various series are indicated by arrows.

Every quantity in Eq. (45-8) may be determined quite independently of the Bohr theory, and apart from this theory we have no reason to expect these quantities to be related in this particular way. The quantities m and e, for instance, are found from experiments on free electrons, h may be found from the photoelectric effect, and R by measurements of wavelengths, while c is the speed of light. However, if we substitute in Eq. (45-8) the values of these quantities, obtained by such diverse means, we find that it *does* hold exactly, within the limits of experimental error, providing direct confirmation of Bohr's theory.

The interaction of atoms with a magnetic field can be analyzed with the Bohr model. It was suggested first by Faraday that shifts in spectrum wavelengths might occur when atoms are placed in a magnetic field. Faraday's spectroscopic techniques were not refined enough to observe such shifts, but Zeeman, using improved instruments, *was* able to detect shifts, an effect now called the *Zeeman effect*.

The interaction is most easily treated using the concept of magnetic moment, introduced in Sec. 31-3. Let us consider an electron in the first Bohr orbit ($n = 1$). The orbiting charge is equivalent to a current loop of radius r and area πr^2. The average charge per unit time passing a point of the orbit is the average current I, and this is given by e/τ, where τ is the time for one revolution: $\tau = 2\pi r/v$. The magnetic moment, which we denote by μ to avoid confusion with the electron mass m, is given by

$$\mu = IA = \tfrac{1}{2}evr.$$

But according to Bohr's theory, the angular momentum mvr is equal to $h/2\pi$, or

$$vr = \frac{h}{2\pi m}.$$

Therefore

$$\mu = \frac{h}{4\pi} \cdot \frac{e}{m}$$

$$= \frac{(6.626 \times 10^{-34}\,\text{J}\cdot\text{s})(1.758 \times 10^{11}\,\text{C}\cdot\text{kg}^{-1})}{12.57}$$

$$= 9.27 \times 10^{-24}\,\text{A}\cdot\text{m}^2.$$

Example Find the interaction potential energy when the hydrogen atom described above is placed in a magnetic field of 2 T.

Solution According to Eq. (31-12), the interaction energy U when $\alpha = 0$ is

$$U = -\mu B = -(9.27 \times 10^{-24}\,\text{A}\cdot\text{m}^2)(2\,\text{T})$$
$$= -1.85 \times 10^{-23}\,\text{J}$$
$$= -1.16 \times 10^{-4}\,\text{eV}.$$

When μ and B are antiparallel, the energy is $+1.16 \times 10^{-4}$ eV. We note that these energies are much *smaller* than the energy of the electron.

Although the Bohr model was successful in predicting the energy levels of the hydrogen atom, it raised as many questions as it answered. It combined elements of classical physics with new postulates inconsistent with classical ideas. It provided no insight into what happens during a transition from one orbit to another, and the stability of certain orbits was bought at the expense of discarding the only picture available at the time of the mechanism of radiation of energy. There was no clear justification (except that it led to the right answer) for restricting the angular momentum to multiples of $h/2\pi$. Furthermore, attempts to extend the model to atoms with two or more electrons were not successful. We shall see in the next section that an even more radical departure from classical concepts was required before the understanding of atomic structure could progress further.

45-2 Wave nature of particles

The next advance in atom building came in 1923, about 10 years after the Bohr theory. This was a suggestion by de Broglie that, since light is dualistic in nature, behaving in some aspects like waves and in others like particles, the same might be true of matter. That is, electrons and protons, which until that time had been thought to be purely corpuscular, might in some circumstances behave like *waves*. Specifically, de Broglie postulated that a free electron of mass m, moving with speed v, should have a wavelength λ given by

$$\lambda = \frac{h}{mv}, \tag{45-9}$$

where h is the same Planck's constant that appears in the frequency-energy relation for photons.

This wave hypothesis, unorthodox though it seemed at the time, almost immediately received very direct experimental confirmation. We have described in Sec. 41-10 how the layers of atoms in a crystal serve as a diffraction grating for x-rays. An x-ray beam is strongly reflected when it strikes a crystal at such an angle that the waves scattered from the atomic layers combine to reinforce one another. The essential point here is that the existence of these strong reflections is evidence of the *wave* nature of x-rays.

In 1927, Davisson and Germer, working in the Bell Telephone Laboratories, were studying the nature of the surface of a crystal of nickel by directing a beam of *electrons* at the surface and observing the electrons reflected at various angles. It might be expected that even the smoothest surface attainable would still look rough to an electron, and that the electron beam would therefore be diffusely reflected. But Davisson and Germer found that the electrons were reflected in almost the same way that x-rays would be reflected from the same crystal. The wavelengths of the electrons in the beam were computed from their known speed, with the help of Eq. (45-9); and the angles at which strong reflection took place were found to be the same as those at which x-rays of the same wavelength would be reflected. This result gave strong support to de Broglie's hypothesis.

This wave hypothesis clearly requires sweeping revisions of our fundamental concepts regarding the description of matter. What we are accustomed to calling a *particle* actually looks like a particle only if we do not look too closely. In general, a particle has to be regarded as a spread-out entity that is not entirely localized in space; and at least in some cases this spreading out appears as a periodic pattern suggesting wavelike properties. The wave and particle aspects are not inconsistent, but the particle model is an *approximation* of a more general wave picture. We are reminded of the ray picture of geometrical optics, a special case of the more general wave picture of physical optics; indeed, there is a very close analogy between optics and the description of particles.

Within a few years, after 1923, the wave hypothesis of de Broglie was developed by Heisenberg, Schrödinger, and many others, into a complete theory called *wave mechanics* or *quantum mechanics*. In the following sections we shall attempt to outline the main lines of thought in a nonmathematical way, describe some of the experimental evidence of the wave nature of material particles, and show how the quantum numbers that were introduced in such an artificial way by Bohr now enter naturally into the theory of atomic structure.

One of the essential features of quantum mechanics is that material particles are no longer regarded as geometrical points, localized in space, but are intrinsically spread-out entities. The spatial distribution of a *free* electron may have a recurring pattern characteristic of a wave that propagates through space. Electrons *in atoms* are visualized as diffuse clouds surrounding the nucleus. The idea that the electrons in an atom move in definite orbits such as those in Fig. 45–1 has been abandoned. The orbits themselves, however, were never an essential part of Bohr's theory, since the quantities that determine the frequencies of the emitted photons are the *energies* corresponding to the orbits. The new theory still assigns definite energy states to an atom. In the hydrogen atom the energies are the same as those given by Bohr's theory; in more complicated atoms where the Bohr theory did not work, the quantum mechanical picture is in excellent agreement with observation.

We shall illustrate how quantization arises in atomic structure by an analogy with the classical mechanical problem of a vibrating string fixed at its ends. When the string vibrates, the ends must be nodes, but nodes may occur at other points also, and the general requirement is that the length of the string shall equal some *integer* number of half-wavelengths. The point of interest is that the solution of the problem of the vibrating string leads to the appearance of *integral numbers*.

In a similar way, the principles of quantum mechanics lead to a wave equation (Schrödinger's equation) that must be satisfied by an electron in an atom, subject also to certain boundary conditions. Let us think of an electron as a wave extending in a circle around the nucleus. In order that the wave may "come out even," the circumference of this circle must include some *integer number* of wavelengths. The wavelength of a particle of mass m, moving with speed v, is given, according to wave mechanics, by Eq. (45–9), $\lambda = h/mv$. Then if r is the radius and $2\pi r$ the circumference of the circle occupied by the wave, we must have $2\pi r = n\lambda$, where $n = 1, 2, 3$, etc. Since $\lambda = h/mv$, this equation becomes

$$2\pi r = n\frac{h}{mv}, \qquad mvr = n\frac{h}{2\pi}. \qquad (45\text{--}10)$$

But mvr is the angular momentum of the electron, so we see that the wave-mechanical picture leads naturally to Bohr's postulate that the angular momentum equals some integral multiple of $h/2\pi$.

To be sure, the idea of wrapping a wave around in a circular orbit is a rather vague notion. But the agreement of Eq. (45–10) with Bohr's hypothesis is much too remarkable to be a coincidence; and it strongly suggests that the wave properties of electrons do indeed have something to do with atomic structure.

45-3 The electron microscope

The shorter the wavelength, the smaller the limit of resolution of a microscope. The wavelengths of electron waves can easily be made very much *shorter* than the wavelengths of visible light. Hence, the limit of resolution of a microscope may be extended to a value several hundred times smaller than that obtainable with an optical instrument, by using electrons, rather than light waves, to form an image of the object being examined.

A beam of electrons can be focused by either a magnetic or an electric field of the proper configuration, and both types are used in electron microscopes. Figure 45-2 illustrates an electrostatic lens. Two hollow cylinders are maintained at different potentials. A few of the equipotentials are indicated, and the trajectories of a beam of electrons traveling from left to right are shown by the colored lines. The optical analog of this electrostatic lens is shown in Fig. 45-3. It will be evident, without going into further details, that by the proper design of such lenses the elements of an optical microscope, such as its condensing lens, objective, and eyepiece, can all be duplicated electrically.

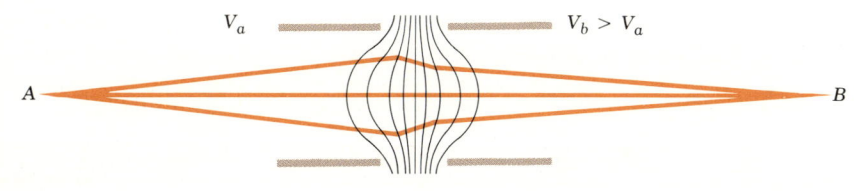

45-2 An electrostatic electron lens. The cylinders are at different potentials V_a and V_b. A beam of electrons diverging from point A is focused at point B.

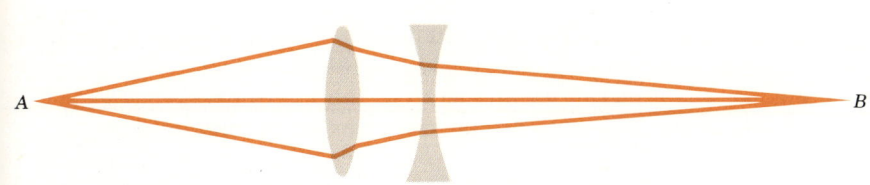

45-3 Optical analog of the electron lens in Fig. 45-2.

The source of electrons in an electron microscope is a heated filament. Electrons emitted by the filament are accelerated by an electron gun and strike the object to be examined. This must necessarily be a thin section so that some of the electrons can pass through it. The thicker portions of the section absorb more of the electron stream than do the thinner portions, just as would a slide in a slide projector. The entire apparatus must, of course, be evacuated.

The final image may be formed on a photographic plate, or on a fluorescent screen that can be examined visually or photographed with a still further gain in magnification. Commercial electron microscopes give satisfactory definition at an overall magnification (electronic, followed by photographic) as great as 50,000 times. Figure 45-4 is an electron micrograph of aluminum oxide, magnified 53,500 times.

45-4 Electron micrograph of aluminum oxide, magnified 53,500 times. (Courtesy of Radio Corporation of America.)

It should be pointed out that the ability of the electron microscope to form an image does *not* depend on the wave properties of the electrons; their trajectories can be computed by treating them as charged particles, deflected by the electric or magnetic fields through which they move. It is only when considerations of *resolving power* arise that the electron wavelengths come into the picture. The situation is analogous to that in the optical microscope. The paths of light rays through an optical microscope can be computed by the principles of geometrical optics, but the resolving power of the microscope is determined by the wavelength of the light used.

Example What is the speed of an electron having a wavelength of 10^{-10} m (the order of magnitude of atomic spacing in crystals)? Through what potential difference must an electron be accelerated to acquire this speed?

Solution From the de Broglie relation, Eq. (45–9), we have

$$v = \frac{h}{m\lambda} = \frac{6.626 \times 10^{-34} \text{ J} \cdot \text{s}}{(9.109 \times 10^{-31} \text{ kg})(10^{-10} \text{ m})} = 7.27 \times 10^6 \text{ m} \cdot \text{s}^{-1}.$$

To find the required potential difference V, we equate the gain in kinetic energy, $\frac{1}{2}mv^2$, to the loss in potential energy, eV:

$$V = \frac{mv^2}{2e} = \frac{(9.109 \times 10^{-31} \text{ kg})(7.27 \times 10^6 \text{ m} \cdot \text{s}^{-1})^2}{(2)(1.602 \times 10^{-19} \text{ C})} = 150 \text{ V}.$$

*45–4 Probability and uncertainty

As we have seen, the entities that we are accustomed to calling *particles* can, in some experiments, behave like *waves*. Figure 45–5 shows a *wave packet* or *wave pulse* that has both wave and particle properties. The regular spacing λ_{av} between successive maxima is characteristic of a wave, but there is also a particle-like localization in space. To be sure, the wave pulse is not localized at a single point, but in any experiment that detects only dimensions much larger than Δx, the wave pulse appears to be a localized particle. Although it would be simplistic to say that Fig. 45–5 is a picture of an electron, the figure does suggest that wave and particle properties are not necessarily incompatible.

It must be emphasized that it would *not* be correct to regard the wave of Fig. 45–5 as having only a single wavelength. A sinusoidal wave with a definite wavelength has no beginning and no end; to make a wave *pulse* it is necessary to superpose many sinusoidal waves having various wavelengths. Thus Fig. 45–5 is a wave having such a distribution of wavelengths, of which λ_{av} is an average value. This distribution has additional important implications, which will be explored at the end of this section.

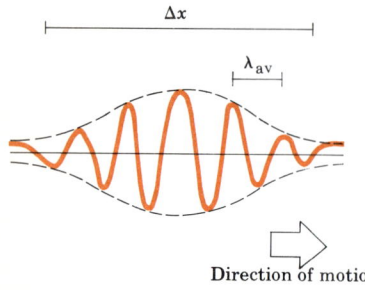

45–5 A wave pulse or packet. There is an average wavelength λ_{av}, the distance between adjacent peaks, but the wave is localized at any instant in a region with length of the order of Δx. The broken lines are called the envelope of the pulse; all the peaks lie on the envelope curves, which approach zero at both ends of the pulse.

The discovery of the dual wave–particle nature of matter has forced a drastic revision of the language used to describe the behavior of a particle. In classical newtonian mechanics we think of a particle as an idealized geometrical point, which, at any instant of time, has a perfectly definite location in space and is moving with a perfectly definite velocity. As we shall see, such a specific description is, in general, not possible; on a sufficiently small scale, there are fundamental limitations on the precision with which the position and velocity of a particle can be described;

and some aspects of a particle's behavior can be stated only in terms of *probabilities*.

To illustrate the nature of the problem, let us consider again the single-slit diffraction experiment described in Sec. 41-8. As Fig. 41-23 suggests, most of the intensity in the diffraction pattern is concentrated in the central maximum; the angular size of this maximum is determined by the positions of the first intensity minimum on either side of the central maximum. Using Eq. (41-15), with $n = 1$, we find that the angle θ in Fig. 45-6 between the central peak and the minimum on either side is given by $\sin \theta = \lambda/a$, where a is the slit width. If λ is much smaller than a, then θ is very small, $\sin \theta$ is very nearly equal to θ, and this may be simplified further to

$$\theta = \frac{\lambda}{a}. \tag{45-11}$$

Now we perform the same experiment again, but using a beam of *electrons* instead of a beam of monochromatic light. The apparatus must be evacuated to avoid collisions of electrons with air molecules; and there are other experimental details that need not concern us. The electron beam can be produced with a setup similar in principle to the electron gun in a cathode-ray oscilloscope, described in Sec. 26-8, which produces a narrow beam of electrons all having the same direction and speed, and therefore also the same wavelength. Such an experiment is shown schematically in Fig. 45-6.

The result of this experiment, as again recorded on photographic film or by means of more sophisticated detectors, is a diffraction pattern identical to that shown in Fig. 41-23, providing additional direct evidence of the wave nature of electrons. Most of the electrons strike the film in the vicinity of the central maximum, but a few strike farther from the center, near the edges of that maximum and also in the subsidiary maxima on both sides. Thus the wave behavior in this experiment presents no surprises.

Interpreted in terms of *particles,* however, this experiment poses very serious problems. First, although the electrons all have the same initial state of motion, they do not all follow the same path; and in fact the trajectory of an individual electron cannot be predicted from knowledge of its initial state. The best we can do is to say that most of the electrons go to a certain region, fewer to other regions, and so on; alter-

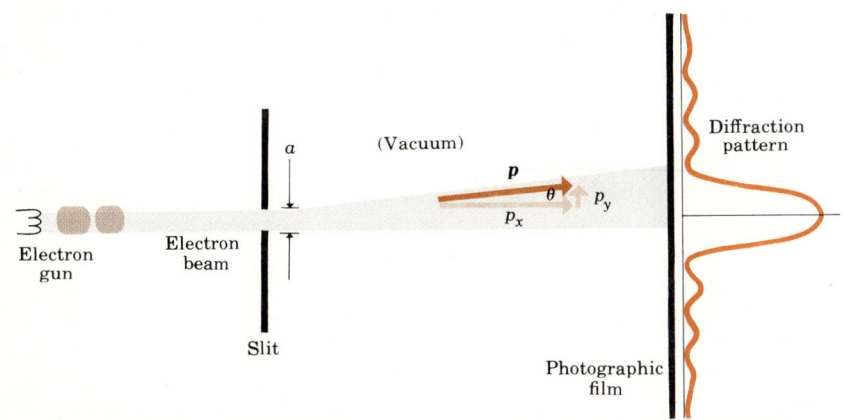

45-6 An electron diffraction experiment. The graph at the right shows the degree of blackening of the film, which in any region is proportional to the number of electrons striking that region. The components of momentum of an electron striking the outer fringe of the central maximum are shown.

natively, we can describe the *probability* for an individual electron to strike each of various areas on the film. This fundamental indeterminacy has no counterpart in newtonian mechanics, where the motion of a particle or a system is always completely predictable if the initial position and motion are known with sufficient precision.

Second, there are fundamental *uncertainties* in both position and momentum of an individual particle, and these two uncertainties are related inseparably. To illustrate this point, we note that, in Fig. 45–6, an electron striking the film at the outer edge of the central maximum, at angle θ, must have a component of momentum p_y in the y-direction, as well as a component p_x in the x-direction, despite the fact that initially the beam was directed along the x-axis. From the geometry of the situation, the two components are related by $p_y/p_x = \tan\theta$; and if θ is small we may approximate $\tan\theta = \theta$, obtaining

$$p_y = p_x \theta. \tag{45-12}$$

Neglecting any electrons striking the film outside the central maximum (that is, at angles greater than λ/a), we see that the y-component of momentum may be as large as

$$p_y = p_x\left(\frac{\lambda}{a}\right). \tag{45-13}$$

Hence the *uncertainty* Δp_y in the y-component of momentum is of the order of $p_x \lambda/a$. Thus the narrower the slit width a, the broader is the diffraction pattern and the greater the uncertainty in the y-component of momentum.

Now the electron wavelength λ is related to the momentum $p_x = mv_x$ by the de Broglie relation, Eq. (45–9), which may be rewritten as $\lambda = h/p_x$. Using this result in Eq. (45–13) and simplifying, we find

$$\Delta p_y = p_x\left(\frac{h}{ap_x}\right) = \frac{h}{a},$$

or

$$\Delta p_y\, a = h. \tag{45-14}$$

To interpret this result, we note that the slit width a represents the uncertainty in *position* of an electron as it passes through the slit; we do not know through which particular part of the slit each particle passes. Thus the y-components of *both* position and momentum have uncertainties, and the two uncertainties are related by Eq. (45–14). We can reduce the *momentum* uncertainty only by increasing the slit width, which increases the *position* uncertainty; and conversely, when we decrease the position uncertainty by narrowing the slit, the diffraction pattern broadens and the corresponding momentum uncertainty increases.

All of this may be bitter medicine for a reader steeped in the tradition of nearly three centuries of newtonian mechanics; but the weight of experimental evidence leaves us no alternatives. To those who protest that the lack of a definite position and momentum is contrary to common sense, we reply that what we call common sense is based on familiarity gained through experience, and that our usual experience includes very little contact with the microscopic behavior of particles. Thus we must sometimes be prepared to accept conclusions that seem not to make sense when we are dealing with areas far removed from everyday experience.

Equation (45-14) is one form of the *Heisenberg uncertainty principle*; it states that, in general, neither the momentum nor the position of a particle can be known with arbitrarily great precision, but that the two play complementary roles as described above. It might be protested that greater precision could be attained by using more sophisticated particle detectors in various areas of the slit or by other means, but this turns out not to be possible. To detect a particle the detector must *interact* with it, and this interaction unavoidably changes the state of motion of the particle. A more detailed analysis of such hypothetical experiments shows that the uncertainty is a fundamental and intrinsic one that cannot be circumvented even in principle by any experimental technique, no matter how sophisticated.

Additional insight into the uncertainty principle is provided by the wave packet shown in Fig. 45-5. Such a packet can be constructed by superposing several sinusoidal waves (each of which has no beginning or end but extends indefinitely in both directions) with various wavelengths, choosing the wavelengths and the amplitude of each component wave so that constructive interference occurs only in a small region of width Δx, as shown in the figure, and the interference is destructive everywhere else. It turns out that a rather broad wave packet can be obtained by superposing waves with a relatively small range of wavelengths; but to make a narrow packet requires a wider range of wavelengths. But a range of wavelengths also means a corresponding range of values of momentum, because of the de Broglie relation. So again we see that a small uncertainty in position must be accompanied by a large uncertainty in momentum, and conversely.

There is also an uncertainty principle involving *energy*. It turns out that the energy of a system as well as its position and momentum always has uncertainty. The uncertainty ΔE is found to depend on the time interval Δt during which the system remains in the given state. The relation is

$$\Delta E \, \Delta t = \frac{h}{2\pi}. \tag{45-15}$$

Thus a system that remains in a certain state for a very long time can have a very well-defined energy, but if it remains in that state for only a short time, the uncertainty in energy must be correspondingly greater.

Example A sodium atom in one of the "resonance levels" shown in Fig. 44-7 remains in that state for an average time of 1.6×10^{-8} s before making a transition to the ground state by emitting a photon of wavelength 589 nm and energy 2.109 eV. What is the uncertainty in energy of the resonance level?

Solution From Eq. (45-15),

$$\Delta E = \frac{h}{2\pi \, \Delta t} = \frac{(6.626 \times 10^{-34} \text{ J} \cdot \text{s})}{(2\pi)(1.6 \times 10^{-8} \text{ s})}$$
$$= 6.59 \times 10^{-27} \text{ J} = 4.11 \times 10^{-8} \text{ eV}.$$

The atom remains an indefinitely long time in the ground state, so there is *no* uncertainty there; the uncertainty of the resonance level energy and of the photon energy amounts to about two parts in 10^8. This irreducible uncertainty is called the *natural line width* of this particular spectrum line. Ordinarily the natural line width is much smaller than line broadening from other causes such as collisions among atoms.

*45-5 Wave functions

As we have seen, the dual wave-particle nature of electrons and other fundamental particles requires generalization of the kinematic language used to describe the position and motion of a particle. The classical notion of a particle as a point having at each instant a definite position in space (described by three coordinates) and a definite velocity (described by three components) must be replaced by a more general language.

In making the needed generalizations we are guided by the language of classical wave motion. In Chapter 21, in studying transverse waves on a string, we described the motion of the string by specifying the position of each point on the string at each instant of time. This was done by means of a *wave function,* introduced in Sec. 21–3. If y represents the displacement from equilibrium at time t of a point on the string whose equilibrium position is a distance x from the origin, then the function $y = f(x, t)$ or, more briefly, $y(x, t)$ represents the displacement of point x at time t. If we know the wave function for a given motion, we know everything there is to know about the motion; from the function the shape of the string at any time, the slope at each point, the velocity and acceleration of each point, and any other needed information can be obtained.

Similarly, in Chapter 23 we discussed a sound wave propagating in the x-direction. Letting p represent the variation in air pressure from its equilibrium value at any point, we write $p(x, t)$ as the pressure variation at any point x at any time t; again this is a *wave function*. If the wave is three-dimensional, we can describe p at a space point with coordinates (x, y, z) at any time t by means of a wave function $p(x, y, z, t)$, which contains all the space coordinates and time. The same pattern reappears once more in the description of *electromagnetic* waves in Sec. 37–5, where we use two wave functions to describe the electric and magnetic fields at any point in space, at any time.

Thus it is natural to use a wave function as the central element of our generalized language for describing particles. The symbol usually used for this wave function is Ψ, and it is, in general, a function of all the space coordinates and time. Just as the wave function $y(x, t)$ for mechanical waves on a string provides a complete description of the motion, the wave function $\Psi(x, y, z, t)$ for a particle contains all the information that can be known about the particle.

Two questions immediately arise. First, what is the *meaning* of the wave function Ψ for a particle? Second, how is Ψ determined for any given physical situation? Our answers to both these questions must be qualitative and incomplete. The wave function describes the distribution of the particle in space. It is related to the *probability* of finding the particle in each of various regions; the particle is most likely to be found in regions where Ψ is large, and so on. If the particle has charge, the wave function can be used to find the *charge density* at any point in space. In addition, from Ψ one can calculate the *average* position of the particle, its average velocity, and dynamic quantities such as momentum, energy, and angular momentum. The required techniques are far beyond the scope of this discussion, but they are well established and no longer subject to any reasonable doubt.

The answer to the second question is that the wave function must be one of a set of solutions of a certain differential equation called the *Schrödinger equation,* developed by Erwin Schrödinger in 1925. One can set up a Schrödinger equation for any given physical situation, such as the electron in a hydrogen atom; the functions that are solutions of this equation represent various possible physical states of the system. Furthermore, it turns out for some systems that acceptable solutions exist only when some physical quantity such as the energy of the system has certain special values. Thus the solutions of the Schrödinger equation are also associated with *energy levels.* This discovery is of the utmost importance; before the discovery of the Schrödinger equation there was no way to predict energy levels from any fundamental theory, except for the very limited success of the Bohr model for hydrogen.

Soon after its discovery the Schrödinger equation was applied to the problem of the hydrogen atom. The predicted energy levels turned out to be identical to those from the Bohr model, Eq. (45-6), and thus to agree with experimental values from spectrum analysis. The energy levels are labeled with the quantum number n. In addition, the solutions have *quantized* values of angular momentum; that is, only certain discrete values of the magnitude of angular momentum and its components are possible. We recall that angular momentum quantization was put into the Bohr model as an *ad hoc* assumption with no fundamental justification; with the Schrödinger equation it comes out automatically!

Specifically, it is found that the magnitude L of the angular momentum of an electron in the hydrogen atom in a state with energy E_n and principal quantum number n must be given by

$$L = \sqrt{l(l+1)}\left(\frac{h}{2\pi}\right), \tag{45-16}$$

where l is zero or a positive integer no larger than $n - 1$. The *component* of L in a given direction, say the z-component L_z, can have only the set of values

$$L_z = mh/2\pi, \tag{45-17}$$

where m can be zero or a positive or negative integer up to but no larger than l.

The quantity $h/2\pi$ appears so often in quantum mechanics that it is given a special symbol, $\hbar$. That is,

$$\hbar = \frac{h}{2\pi} = 1.054 \times 10^{-34} \text{ J} \cdot \text{s}.$$

In terms of $\hbar$, the two preceding equations become

$$L = \sqrt{l(l+1)}\,\hbar \qquad (l = 0, 1, 2, \ldots, n-1), \tag{45-18}$$

and

$$L_z = m\hbar \qquad (m = 0, \pm 1, \pm 2, \ldots, \pm l). \tag{45-19}$$

We note that the component L_z can never be quite as large as L. For example, when $l = 4$ and $m = 4$, we find

$$L = \sqrt{4(4+1)}\,\hbar = 4.47\hbar,$$
$$L_z = 4\hbar.$$

This inequality arises from the uncertainty principle, which makes it impossible to know the *direction* of the angular momentum vector with complete certainty. Thus the component of L in a given direction can never be quite as large as the magnitude L, except when $l = 0$ and both L and L_z are zero. Unlike the Bohr model, the Schrödinger equation gives values for the magnitude L of angular momentum which are *not* integer multiples of $\hbar$.

Another interesting feature of Eqs. (45–18) and (45–19) is that there are states for which the angular momentum is *zero*. This is a result that has no classical analog; in the Bohr model the electron always moved in an orbit and thus had nonzero angular momentum; but in the new mechanics we find states having zero angular momentum.

The possible wave functions for the hydrogen atom can be labeled according to the values of the three integers n, l, and m, called, respectively, the *principal* quantum number, the *angular momentum* quantum number, and the *magnetic* quantum number. For each energy level E_n there are several distinct states with the same energy but different values of l and m, the only exception being the ground state $n = 1$, for which only $l = 0$, $m = 0$ is possible. Examination of the spatial extent of the wave functions shows that, in each case, the wave function is confined primarily to a region of space around the nucleus having a radius of the same order of magnitude as the corresponding Bohr radius. For increasing values of n the electron is, on the average, farther away from the nucleus, and hence has less negative potential energy and becomes less tightly bound.

The new mechanics described above is called *quantum mechanics*, and it is clearly much more complex, both conceptually and mathematically, than newtonian mechanics. To a reader who yearns for the comparative simplicity of newtonian mechanics, we can only reply that quantum mechanics enables us to understand physical phenomena and to analyze physical problems for which classical mechanics is completely powerless. Added complexity is the price we pay for this greatly expanded understanding.

45–6 Electron spin

One additional concept, that of *electron spin,* is needed for the analysis of more complex atoms, and even for certain details of the spectrum of hydrogen. To illustrate this concept, we consider the motion of the earth around the sun. The earth travels in a nearly circular orbit and at the same time *rotates* on its axis. There is angular momentum associated with each motion, called the *orbital* and *spin* angular momentum, respectively, and the total angular momentum of the system is the sum of the two. If we were to model the earth as a single point, no spin angular momentum would be possible; but a more refined model including an earth of finite size includes the possibility of spin angular momentum.

This discussion can be translated to the language of the Bohr model. Suppose the electron is not a point charge moving in an orbit but a small spinning sphere in orbit. Then again there is additional angular momentum associated with the spin motion. Because the sphere carries an electric charge, the spinning motion leads to current loops and to a magnetic moment, as discussed in Sec. 31–3. If this magnetic moment really exists, then when the atom is placed in a magnetic field there is an interaction with the field. Associated with the interaction is a potential energy that causes shifts in the energy levels of the atom and thus in the wavelengths of the spectrum lines.

Such shifts *are* indeed observed in precise spectroscopic analysis; this and other experimental evidence have shown conclusively that the electron *does* have angular momentum and magnetic moment that are not related to the orbital motion but are intrinsic to the particle itself. Like orbital angular momentum, spin angular momentum (usually denoted by S) is found to be *quantized*. Denoting the z-component of S by S_z, we find that the only possible values are

$$S_z = \pm \tfrac{1}{2}\hbar. \tag{45-20}$$

This relation is reminiscent of Eq. (45–19) for the z-component of orbital angular momentum, but the component is one-half of $\hbar$ instead of an *integer* multiple.

In quantum mechanics, where the Bohr orbits are superseded by wave functions, it is not so easy to picture electron spin. If the wave functions are visualized as clouds surrounding the nucleus, then one can imagine many tiny arrows distributed through the cloud, all pointing in the same direction, either all $+z$ or all $-z$. Of course, this picture should not be taken too seriously; there is no hope of actually seeing an atom's structure because it is thousands of times smaller than wavelengths of light and because interactions with light photons would seriously disturb the very structure one is trying to observe.

In any event, the concept of electron spin is well established by a variety of experimental evidence. To label completely the state of the electron in a hydrogen atom, we now need a fourth quantum number s, which specifies the electron spin orientation. If s can take the values $+1$ or -1, then the z-component of spin angular momentum is given by

$$S_z = \tfrac{1}{2}s\hbar. \tag{45-21}$$

The corresponding component of magnetic moment, which we may denote by μ_z, turns out to be related to S_z, in a simple way:

$$\mu_z = \frac{e}{m_e} S_z, \tag{45-22}$$

where e and m_e are the charge and mass of the electron, respectively. The ratio of μ to S, in this case (e/m_e), is called the *gyromagnetic ratio*.

In this connection it should be mentioned that there is also magnetic moment associated with the *orbital* motion of the electron; in this case the gyromagnetic ratio is found to be just *half* as great as for electron spin, $e/2m_e$. Thus the z-component of magnetic moment associated with the orbital angular momentum component L_z is given by

$$\mu_z = \frac{e}{2m_e} L_z = m \frac{e\hbar}{2m_e}. \tag{45-23}$$

(Care must be taken to distinguish between the electron mass m_e and the magnetic quantum number m.) This magnetic moment leads to additional interactions with a magnetic field and to further contributions to shifts and splitting of spectrum lines.

Questions

45-1 In analyzing the absorption spectrum of hydrogen at room temperature, one finds absorption lines corresponding to wavelengths in the Lyman series but not to those in the Balmer series. Why not?

45-2 A singly ionized helium atom has one of its two electrons removed, and the energy levels of the remaining electron are closely related to those of the hydrogen atom. The nuclear charge for helium is $+2e$ instead of just $+e$; exactly how are the energy levels related to those of hydrogen? How is the size of the ion in the ground state related to that of the hydrogen atom?

45-3 Consider the line spectrum emitted from a gas discharge tube such as a neon sign or a sodium-vapor or mercury-vapor lamp. It is found that when the pressure of the vapor is increased the spectrum lines spread out, i.e., are less sharp and less monochromatic. Why?

45-4 Suppose a two-slit interference experiment is carried out using an electron beam. Would the same interference pattern result if one slit at a time is uncovered instead of both at once? If not, why not? Doesn't each electron go through one slit or the other? Or does every electron go through both slits? Does the latter possibility make sense?

45-5 Is the wave nature of electrons significant in the function of a television picture tube? For example, do diffraction effects limit the sharpness of the picture?

45-6 A proton and an electron have the same speed. Which has longer wavelength?

45-7 A proton and an electron have the same kinetic energy. Which has longer wavelength?

45-8 Does the uncertainty principle have anything to do with marksmanship? That is, is the accuracy with which a bullet can be aimed at a target limited by the uncertainty principle?

45-9 Is the Bohr model of the hydrogen atom consistent with the uncertainty principle?

45-10 If the energy of a system can have uncertainty, as stated by Eq. (45-15), does this mean that the principle of conservation of energy is no longer valid?

45-11 If quantum mechanics replaces the language of newtonian mechanics, why do we not have to use wave functions to describe the motion of macroscopic objects such as baseballs and cars?

45-12 Why is the analysis of the helium atom much more complex than that of the hydrogen atom, either in a Bohr-type model or using the Schrödinger equation?

45-13 Do gravitational forces play a significant role in atomic structure?

Problems

$e = 1.602 \times 10^{-19}$ C
$m = 9.109 \times 10^{-31}$ kg
$h = 6.626 \times 10^{-34}$ J·s
$N_0 = 6.022 \times 10^{23}$ atoms·mol^{-1}
Energy equivalent of 1 u = 931.5 MeV
$\frac{e}{m} = 1.758 \times 10^{11}$ C·kg^{-1}
$k = 1.381 \times 10^{-23}$ J·K^{-1}
1 eV = 1.602×10^{-19} J
1 u = 1.661×10^{-27} kg
$\varepsilon_0 = 8.854 \times 10^{-12}$ C^2·N^{-1}·m^{-2}

45-1 For a hydrogen atom in the ground state, determine, in electron volts:

a) the kinetic energy;
b) the potential energy;
c) the total energy;
d) the energy required to remove the electron completely.

45-2 Referring to Problem 45-1, what are the energies if the atom is a singly ionized helium atom (i.e., a helium atom with one electron removed)?

45-3 A singly ionized helium ion (a helium atom with one electron removed) behaves very much like a hydrogen atom, except that the nuclear charge is twice as great.

a) How do the energy levels differ in magnitude from those of the hydrogen atom?
b) Which spectral series for He$^+$ have lines in the visible spectrum?

45-4 Refer to the example in Sec. 45-1. Suppose a hydrogen atom makes a transition from the $n = 3$ state to the $n = 2$ state (the Balmer Hα line at 656.3 nm) while in a magnetic field of magnitude 2 T. If the magnetic moment of the atom is parallel to the field in both the initial and final states,

a) by how much is each enery level shifted from the zero-field value;
b) by how much is the wavelength of the spectrum line shifted?

45-5 The negative μ-meson (or muon) has a charge equal to that of an electron but a mass about 207 times as great. Consider a hydrogenlike atom consisting of a proton and a muon.

a) What is the ground-state energy?
b) What is the radius of the $n = 1$ Bohr orbit?
c) What is the wavelength of the radiation emitted in the transition from the $n = 2$ state to the $n = 1$ state?

45-6 According to Bohr, the Rydberg constant R is equal to $me^4/8\varepsilon_0^2 h^3 c$. Calculate R in m^{-1} and compare with the experimental value.

45-7

a) Show that the frequency of revolution of an electron in its circular orbit in the Bohr model of the hydrogen atom is $f = me^4 \varepsilon_0^2 n^3 h^3$.
b) Show that when n is very large, the frequency of revolution equals the radiated frequency calculated from Eq. (45-7) for a transition from

$$n = n' + 1 \quad \text{to} \quad l = n'.$$

(This problem illustrates Bohr's *correspondence principle*, which is often used as a check on quantum calculations. When n is small, quantum physics gives results that are very different from those of classical physics. When n is large, the differences are not significant and the two methods then "correspond.")

45-8 A 10-kg satellite circles the earth once every 2 hr in an orbit having a radius of 8000 km.

a) Assuming that Bohr's angular-momentum postulate applies to satellites just as it does to an electron in the hydrogen atom, find the quantum number of the orbit of the satellite.
b) Show from Bohr's first postulate and Newton's law of gravitation that the radius of an earth-satellite orbit is directly proportional to the square of the quantum number, $r = kn^2$, where k is the constant of proportionality.
c) Using the result from part (b), find the distance between the orbit of the satellite in this problem and its next "allowed" orbit.
d) Comment on the possibility of observing the separation of the two adjacent orbits.
e) Do quantized and classical orbits correspond for this satellite? Which is the "correct" method for calculating the orbits?

45-9 A hydrogen atom initially in the ground state absorbs a photon, which excites it to the $n = 4$ state. Determine the wavelength and frequency of the photon.

45-10 In Problem 45-3,

a) how much energy would be required to remove the remaining electron completely from a helium ion?

b) What wavelength would a photon of this energy have?
c) In what region of the electromagnetic spectrum would it lie?

45-11 Calculate the energy of the longest-wavelength line in the Balmer series for hydrogen, and of the shortest-wavelength line.

45-12 Approximately what range of photon energies (in electron volts) corresponds to the visible spectrum? Approximately what range of wavelengths would electrons in this energy range have?

45-13 For crystal diffraction experiments, wavelengths of the order of 0.1 nm are often appropriate. Find the energy, in electron volts, for a particle with this wavelength if the particle is

a) a photon;
b) an electron.

45-14

a) An electron moves with a velocity of 3×10^8 cm·s^{-1}. What is its de Broglie wavelength?
b) A proton moves with the same velocity. Determine its de Broglie wavelength.

45-15 The average kinetic energy of a thermal neutron is $(3/2)kT$. What is the de Broglie wavelength associated with the neutrons in thermal equilibrium with matter at 300 K? (The mass of a neutron is approximately 1 u.)

45-16

a) What is the de Broglie wavelength of an electron that has been accelerated through a potential difference of 200 V?
b) Would this electron exhibit particlelike or wavelike characteristics on meeting an obstacle or opening 1 mm in diameter?

45-17 What is the de Broglie wavelength of an electron accelerated through 20,000 V, a typical voltage for color-television picture tubes?

45-18 Suppose the uncertainty of position of an electron is equal to the radius of the $n = 1$ Bohr orbit, about 0.5×10^{-10} m. Find the uncertainty in its momentum, and compare this with the magnitude of the momentum of the electron in the $n = 1$ Bohr orbit.

45-19 Suppose that the uncertainty in position of a particle is equal to its de Broglie wavelength; show that in this case the uncertainty in its momentum is equal to its momentum.

45-20 In a certain television picture tube the accelerating voltage is 20,000 V and the electron beam passes through an aperture 0.5 mm in diameter to a screen 0.3 m away.

a) What is the uncertainty in position of the point where the electrons strike the screen?
b) Does this uncertainty affect the clarity of the picture significantly?

45-21 A certain atom has an energy level 2.0 eV above the ground state. When excited to this state it remains on the average 2.0×10^{-6} s before emitting a photon and returning to the ground state.

a) What is the energy of the photon? Its wavelength?

b) What is the uncertainty in energy of the photon? Of its wavelength?

45-22 The π^0 meson is an unstable particle produced in high-energy particle collisions. Its mass is about 264 times that of the electron, and it exists for an average lifetime of 0.8×10^{-16} s before decaying into two gamma-ray photons. Assuming that the mass and energy of the particle are related by the Einstein relation $E = mc^2$, find the uncertainty in the mass of the particle, and express it as a fraction of the mass.

45-23 Suppose an unstable particle produced in a high-energy collision has a mass three times that of the proton and an uncertainty in mass that is 1% of the particle's mass. Assuming mass and energy are related by $E = mc^2$, find the lifetime of the particle.

45-24 A hydrogen atom is placed in a magnetic field of magnitude 1.2 T. Find the interaction energy with the field due to the electron spin.

45-25 Make a chart showing all the possible sets of quantum numbers for the states of the electron in the hydrogen atom when $n = 3$.

45-26 Show that the total number of hydrogen-atom states for a given value of the principal quantum number n is $2n^2$.

ATOMS, MOLECULES, AND SOLIDS 46

46-1 The exclusion principle

The hydrogen atom, discussed in Chapter 45, is the simplest of all atoms, containing one electron and one proton. Analysis of atoms with more than one electron increases in complexity very rapidly; each electron interacts not only with the positively charged nucleus but also with all the other electrons. In principle the motion of the electrons is governed by the Schrödinger equation, mentioned in Sec. 45–5, but the mathematical problem of finding appropriate solutions of this equation is so complex that it has not been accomplished exactly even for the helium atom, with two electrons.

Various approximation schemes can be used; the simplest (and most drastic) is to ignore completely the interactions between electrons, and regard each electron as influenced only by the electric field of the nucleus, considered to be a point charge. A less drastic and more useful approximation is to think of all the electrons together as making up a charge cloud that is, on the average, *spherically symmetric,* and to think of each individual electron as moving in the total electric field due to the nucleus and this averaged-out electron cloud. This is called the *central-field* approximation; it provides a useful starting point for the understanding of atomic structure.

An additional principle is also needed, the *exclusion principle*. To understand the need for this principle, we consider the lowest-energy state or *ground state* of a many-electron atom. The central-field model suggests that each electron has a lowest energy state (roughly corresponding to the $n = 1$ state for the hydrogen atom). It might be expected that, in the ground state of a complex atom, all the electrons should be in this lowest state. If this is the case, then when we examine the behavior of atoms with increasing numbers of electrons, we should find gradual changes in physical and chemical properties of elements as the number of electrons in the atoms increases.

A variety of evidence shows conclusively that this is *not* what happens at all. For example, the elements of fluorine, neon, and sodium have, respectively, 9, 10, and 11 electrons per atom. Fluorine is a halogen and tends strongly to form compounds in which each atom acquires an extra electron. Sodium, an alkali metal, forms compounds in which it loses an electron, and neon is an inert gas, forming no compounds at all. This and many other observations show that, in the ground state of a complex atom, the electrons *cannot* all be in the lowest energy states.

The key to this puzzle, discovered by the Swiss physicist Wolfgang Pauli in 1925, is called the *Pauli exclusion principle*. Briefly, it states that *no two electrons can occupy the same quantum-mechanical state*. Since different states correspond to different spatial distributions, including different distances from the nucleus, this means that in a complex atom there is not room for all the electrons in states near the nucleus; some are forced into states farther away, having higher energies.

To apply the Pauli principle to atomic structure, we first review some results quoted in Sec. 45–5. The quantum-mechanical state of the electron in the hydrogen atom is identified by the four quantum numbers n, l, m, and s, which determine the energy, angular momentum, and components of orbital and spin angular momentum in a particular direction. It turns out that this scheme can also be used when the electron moves not in the electric field of a point charge, as in the hydrogen atom, but in the electric field of *any spherically symmetric charge distribution*, as in the central-field approximation. One important difference is that the energy of a state is no longer given by Eq. (45–6); in general the energy corresponding to a given set of quantum numbers depends on *both n and l*, usually increasing with increasing l for a given value of n.

We can now make a list of the possible sets of quantum numbers and thus of the possible states of electrons in an atom. Such a list is given in Table 46–1, which also indicates two alternative notations. It is customary to designate the value of l by a letter, according to this scheme:

$$l = 0; \quad s \text{ state}$$
$$l = 1; \quad p \text{ state}$$
$$l = 2; \quad d \text{ state}$$
$$l = 3; \quad f \text{ state}$$
$$l = 4; \quad g \text{ state}$$

The origins of these letters are rooted in the early days of spectroscopy, and need not concern us. A state for which $n = 2$ and $l = 0$ is called a $2s$ state, and so on, as shown in Table 46–1. This table also shows the relation between values of n and the x-ray levels ($K, L, M, \ldots$) described in Sec. 44–8. The $n = 1$ levels are designated as K, $n = 2$ as L, and so on. Because the average electron distance from the nucleus increases with n, each value of n corresponds roughly to a region of space around the nucleus in the form of a spherical shell. Hence, one speaks of the *L shell* as that region occupied by the electrons in the $n = 2$ states, and so on. States with the same n but different l are said to form *subshells,* such as the $3p$ subshell.

We are now ready for a more precise statement of the exclusion principle: *In any atom, only one electron can occupy any given quantum state.* That is, no two electrons in an atom can have all four quantum numbers the same. Since each quantum state corresponds to a certain distribution of the electron "cloud" in space, the principle says in effect: "Not more than two electrons (with opposite values of the spin quantum

Table 46-1

n	l	m	Spectroscopic notation	Maximum number of electrons		Shell
1	0	0	$1s$	2		K
2	0	0	$2s$	2		
2	1	−1			8	L
2	1	0	$2p$	6		
2	1	1				
3	0	0	$3s$	2		
3	1	−1				
3	1	0	$3p$	6		
3	1	1				
3	2	−2			18	M
3	2	−1				
3	2	0	$3d$	10		
3	2	1				
3	2	2				
4	0	0	$4s$	2		
4	1	−1				
4	1	0	$4p$	6	32	N
4	1	1				
etc.						

number s) can occupy the same region of space." This statement must be interpreted rather broadly, since the clouds and associated wave functions describing electron distributions do not have definite, sharp boundaries; but the exclusion principle limits the degree of overlap of electron wave functions that is permitted. The maximum numbers of electrons in each shell and subshell are shown in Table 46-1.

The exclusion principle plays an essential role in the understanding of the structure of complex atoms. In the next section we shall see how the periodic table of the elements can be understood on the basis of this principle.

46-2 Atomic structure

The number of electrons in an atom in its normal state is called the *atomic number,* denoted by Z. The nucleus contains Z protons and some number of neutrons. The proton and electron charges have the same magnitude but opposite sign, so in the normal atom the net electric charge is zero. Because the electrons are attracted to the nucleus, we expect the quantum states corresponding to regions near the nucleus to have lowest energy. We may imagine starting with a bare nucleus with Z protons, and adding electrons one by one until the normal complement of Z electrons for a neutral atom is reached. We expect the lowest-energy states, ordinarily those with the smallest values of n and l, to fill first, and we use successively higher states until all electrons are accommodated.

The chemical properties of an atom are determined principally by interactions involving the outermost electrons, so it is of particular in-

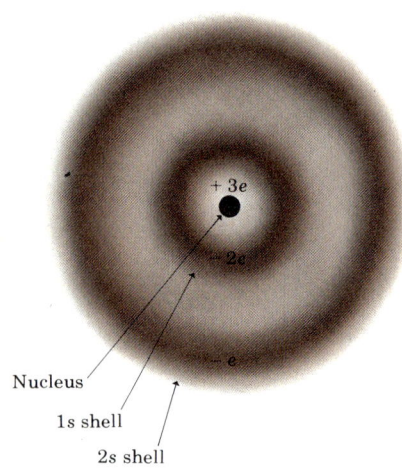

46-1 Schematic representation of charge distribution in lithium atom. The nucleus has a charge $3e$; the two $1s$ electrons are closer to the nucleus than the $2s$ electron, which moves in a field approximately equal to that of a point charge $3e - 2e$, or simply e.

terest to find out how these are arranged. For example, when an atom has one electron considerably farther from the nucleus (on the average) than the others, this electron will be rather loosely bound; the atom will tend to lose this electron and form what chemists call an *electrovalent* or *ionic* bond, with valence $+1$. This behavior is characteristic of the alkali metals lithium, sodium, potassium, and so on.

We now proceed to describe ground-state electron configurations for the first few atoms (in order of increasing Z). For hydrogen the ground state is $1s$; the single electron is in the state $n = 1$, $l = 0$, $m = 0$, and $s = \pm 1$. In the helium atom ($Z = 2$) *both* electrons are in $1s$ states, with opposite spins; this state is denoted as $1s^2$. For helium, the K shell is completely filled and all others are empty.

Lithium ($Z = 3$) has three electrons; in the ground state two are in the $1s$ state, and one in a $2s$ state. We denote this state as $1s^2 2s$. On the average, the $2s$ electron is considerably farther from the nucleus than the $1s$ electrons, as shown schematically in Fig. 46–1. Thus, according to Gauss's law, the *net* charge influencing the $2s$ electron is $+e$, rather than $+3e$ as it would be without the $1s$ electrons present. Thus the $2s$ electron is loosely bound, as the chemical behavior of lithium suggests. An alkali metal, it forms ionic compounds with a valence of $+1$, in which each atom loses an electron.

Next is beryllium ($Z = 4$); its ground-state configuration is $1s^2 2s^2$, with two electrons in the L shell. Beryllium is the first of the *alkaline-earth* elements, forming ionic compounds with a valence of $+2$.

Table 46-2 shows the ground-state electron configurations of the first twenty elements. The L shell can hold a total of eight electrons, as the reader should verify from the rules in Sec. 46-1; at $Z = 10$ the K and L shells are filled and there are no electrons in the M shell. This is

Table 46-2 Ground-state electron configurations

Element	Symbol	Atomic number (Z)	Electron configuration
Hydrogen	H	1	$1s$
Helium	He	2	$1s^2$
Lithium	Li	3	$1s^2 2s$
Beryllium	Be	4	$1s^2 2s^2$
Boron	B	5	$1s^2 2s^2 2p$
Carbon	C	6	$1s^2 2s^2 2p^2$
Nitrogen	N	7	$1s^2 2s^2 2p^3$
Oxygen	O	8	$1s^2 2s^2 2p^4$
Fluorine	F	9	$1s^2 2s^2 2p^5$
Neon	Ne	10	$1s^2 2s^2 2p^6$
Sodium	Na	11	$1s^2 2s^2 2p^6 3s$
Magnesium	Mg	12	$1s^2 2s^2 2p^6 3s^2$
Aluminum	Al	13	$1s^2 2s^2 2p^6 3s^2 3p$
Silicon	Si	14	$1s^2 2s^2 2p^6 3s^2 3p^2$
Phosphorus	P	15	$1s^2 2s^2 2p^6 3s^2 3p^3$
Sulfur	S	16	$1s^2 2s^2 2p^6 3s^2 3p^4$
Chlorine	Cl	17	$1s^2 2s^2 2p^6 3s^2 3p^5$
Argon	Ar	18	$1s^2 2s^2 2p^6 3s^2 3p^6$
Potassium	K	19	$1s^2 2s^2 2p^6 3s^2 3p^6 4s$
Calcium	Ca	20	$1s^2 2s^2 2p^6 3s^2 3p^6 4s^2$

expected to be a particularly stable configuration, with little tendency to gain or lose electrons, and in fact this element is neon, an inert gas with no known compounds. The next element after neon is sodium ($Z = 11$), with filled K and L shells and one electron in the M shell. Thus its "filled-shell-plus-one-electron" structure resembles that of lithium; both are alkali metals. The element *before* neon is fluorine, with $Z = 9$. It has a vacancy in the L shell and might be expected to have an affinity for an electron, forming ionic compounds with a valence of -1. This behavior is characteristic of the *halogens* (fluorine, chlorine, bromine, iodine, astatine), all of which have "filled-shell-minus-one" configurations.

By similar analysis, one can understand all the regularities in chemical behavior exhibited by the periodic table of the elements, on the basis of electron configurations. With the M and N shells there is a slight complication because the $3d$ and $4s$ subshells ($n = 3$, $l = 2$, and $n = 4$, $l = 0$, respectively) overlap in energy. Thus argon ($Z = 18$) has all the $1s$, $2s$, $2p$, $3s$, and $3p$ states filled, but in potassium ($Z = 19$) the additional electron goes into a $4s$ rather than $3d$ level. The next several elements have one or two electrons in the $4s$ states and increasing numbers in the $3d$ states. These elements are all metals with rather similar properties, and form the first *transition series,* starting with scandium ($Z = 21$) and ending with zinc ($Z = 30$), for which the $3d$ levels are filled.

Thus, the similarity of elements in each *group* of the periodic table reflects corresponding similarity in electron configuration. All the inert gases (helium, neon, argon, krypton, xenon, and radon) have filled-shell configurations. All the alkali metals (lithium, sodium, potassium, rubidium, cesium, and francium) have "filled-shell-plus-one" configurations. All the alkaline-earth metals (beryllium, magnesium, calcium, strontium, barium, and radium) have "filled-shell-plus-two" configurations, and all the halogens (fluorine, chlorine, bromine, iodine, and astatine) have "filled-shell-minus-one" structures. And so it goes.

This whole theory can, of course, be refined to account for differences between elements in a group and to account for various aspects of chemical behavior. Indeed, some physicists like to claim that all of chemistry is contained in the Schrödinger equation! This statement may be a bit extreme, but we can see that even this qualitative discussion of atomic structure takes us a considerable distance toward understanding the atomic basis of many chemical phenomena. Next we shall examine in somewhat more detail the nature of the chemical bond.

46-3 Diatomic molecules

As indicated in Sec. 46-2, the study of electron configurations in atoms provides valuable insight into the nature of the chemical bond, that is, the interaction that holds atoms together to form stable structures such as molecules and crystalline solids. There are a number of types of chemical bonds; the simplest to understand is the *ionic* bond, also called the *electrovalent* or *heteropolar* bond. The most familiar example is sodium chloride (NaCl), in which the sodium atom gives its $3s$ electron to the chlorine atom, filling the vacancy in the $3p$ subshell of chlorine. Energy is required to make this transfer if the atoms are far apart, but the result is two ions, one positively charged and one negatively, which attract each other. As they come together, their potential energy decreases, so that the final bound state of Na⁺Cl⁻ has lower total energy than the state in which the two atoms are separated and neutral.

Removing the $3s$ electron from the sodium requires 5.1 eV of energy; this is called the *ionization energy* or *ionization potential*. Chlorine actually has an electron *affinity* of 3.8 eV. That is, the neutral chlorine atom can attract an extra electron which, once it takes its place in the $3p$ level, requires 3.8 eV for its removal. Thus, creating the separate Na⁺ and Cl⁻ ions requires a net expenditure of only $5.1 - 3.8 = 1.3$ eV. The potential energy associated with the mutual attraction of the ions is about -5 eV, more than enough to repay the initial energy investment of creating the ions. This potential energy is determined by the closeness with which the ions can approach each other, and this in turn is detemined by the exclusion principle, which forbids extensive overlap of the electron clouds of the two atoms.

Ionic bonds can involve more than one electron per atom. The alkaline-earth elements form ionic compounds in which each atom loses *two* electrons; an example is $Mg^{++}Cl_2^-$. Loss of more than two electrons is relatively rare, and instead a *different kind of bond* comes into operation.

The *covalent* or *homopolar* bond is characterized by a more nearly symmetric participation of the two atoms, as contrasted with the complete asymmetry involved in the electron-transfer process of the ionic bond. The simplest example of the covalent bond is the hydrogen molecule, a structure containing two protons and two electrons. As a preliminary to understanding this bond we consider first the interaction of two electric dipoles, as in Fig. 46–2. In (a) the dipoles are far apart; in (b) the like charges are farther part than the unlike charges, and there is a net *attractive* force. In (c) and (d) the interaction is repulsive.

In a molecule the charges are, of course, not at rest, but Fig. 46–2b at least makes it seem plausible that if the electrons are in the region between the protons, the attractive force the electrons exert on each proton may more than counteract the repulsive interactions of the protons on each other and the electrons on each other. That is, in the covalent bond in the hydrogen molecules, the attractive interaction is supplied by a pair of electrons, one contributed by each atom, whose charge clouds are concentrated primarily in the region between the two atoms. Thus this bond may also be thought of as a shared-electron or electron-pair bond.

According to the exclusion principle, two electrons can occupy the same region of space only when they have opposite spin orientations. When the spins are parallel, the state that would be most favorable from energy considerations is forbidden by the exclusion principle, and the lowest-energy state permitted is one in which the electron clouds are concentrated *outside* the central region between atoms. The nuclei then repel each other and the interaction is repulsive rather than attractive. Thus, opposite spins are an essential requirement for an electron-pair bond, and no more than two electrons can participate in such a bond.

This is not to say, however, that an atom cannot have several electron-pair bonds. On the contrary, an atom having several electrons in its outermost shell can form covalent bonds with several other atoms. The bonding of carbon and hydrogen atoms, of central importance in organic

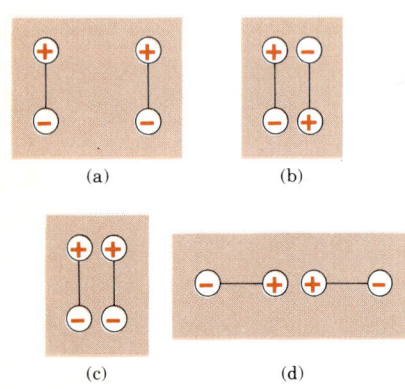

46–2 When two dipoles (a) are brought together, the interaction may be (b) attractive, or (c and d) repulsive.

chemistry, is an example. In the *methane* molecule (CH_4) the carbon atom is at the center of a regular tetrahedron, with a hydrogen atom at each corner. The carbon atom has four electrons in its L shell, and one of these electrons forms a covalent bond with each of the four hydrogen atoms, as shown in Fig. 46–3. Similar patterns occur in more complex organic molecules.

Both the ionic and covalent bonds are important in the structure of solids. In many respects a crystalline solid is really a very large molecule held together by chemical bonds. A third type of bonding, the *metallic* bond, is also important in the structure of some solids; we return to this subject in Sec. 46–5.

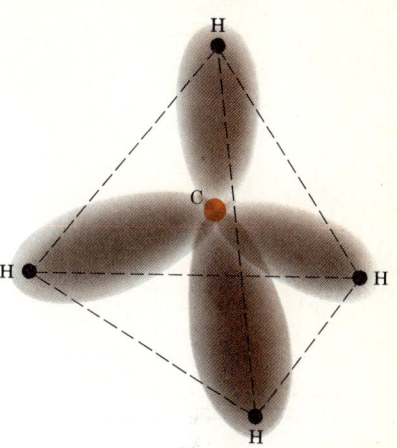

46–3 Methane (CH_4) molecule, showing four covalent bonds. The electron cloud between the central carbon atom and each of the four hydrogen nuclei represents the two electrons of a covalent bond. The hydrogen nuclei are at the corners of a regular tetrahedron.

46–4 Molecular spectra

The energy levels of atoms are associated with the kinetic and potential energies of electrons with respect to the nucleus. Energy levels of molecules have additional features resulting from relative motion of the nuclei of the atoms, and there are characteristic spectra associated with these levels.

For the sake of simplicity we consider only diatomic molecules. Viewed as a rigid dumbbell, a diatomic molecule can *rotate* about an axis through its center of mass. According to classical mechanics, the kinetic energy of a rotating body may be expressed as $\frac{1}{2}I\omega^2$, as given by Eq. (9–13). This energy may also be expressed in terms of the magnitude L of angular momentum, which is given by $L = I\omega$. Combining this with the preceding expression, we find that the energy E is given by

$$E = \frac{L^2}{2I}. \qquad (46\text{--}1)$$

Now in a quantum-mechanical discussion of this motion, it is reasonable to assume that the angular momentum is quantized in the same way as for electrons in an atom, as given by Eq. (45–18):

$$L^2 = l(l+1)\hbar^2 \qquad (l = 0, 1, 2, 3, \ldots). \qquad (46\text{--}2)$$

Combining Eqs. (46–1) and (46–2), we obtain the result

$$E = l(l+1)\left(\frac{\hbar^2}{2I}\right) \qquad (l = 0, 1, 2, 3, \ldots). \qquad (46\text{--}3)$$

A more detailed analysis, using the Schrödinger equation, confirms that the angular momentum is indeed given by Eq. (46–2) and the energy levels by Eq. (46–3).

Considering the magnitudes involved, we note that, for an oxygen molecule, $I = 5 \times 10^{-46}$ kg·m². Thus, the constant $\hbar^2/2I$ in Eq. (46–3) is approximately equal to

$$\frac{\hbar^2}{2I} = \frac{(1.05 \times 10^{-34}\,\text{J·s})^2}{2(5 \times 10^{-46}\,\text{kg·m}^2)} = 1.11 \times 10^{-23}\,\text{J} = 0.69 \times 10^{-4}\,\text{eV}.$$

We note that this energy is much *smaller* than typical atomic energy levels associated with optical spectra. The energies of photons emitted and absorbed in transitions among rotational levels are correspondingly small, and they fall in the far infrared region of the spectrum.

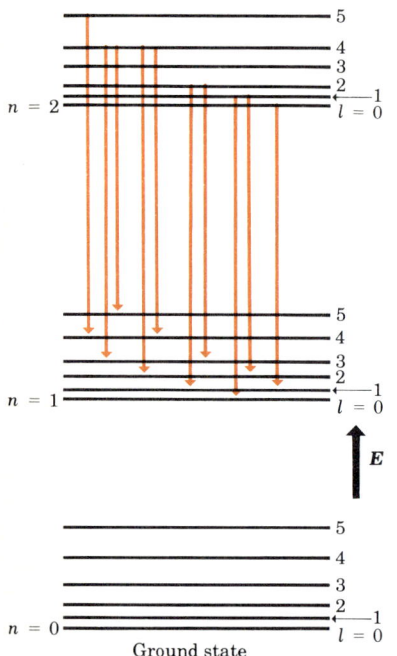

46–4 Energy-level diagram for vibrational and rotational energy levels of a diatomic molecule. For each vibrational level (n) there is a series of more closely spaced rotational levels (l). Several transitions corresponding to a single band in a band spectrum are shown.

The rigid-dumbbell model of a diatomic molecule suggests that the distance between the two nuclei is fixed; in fact, a more realistic model would represent the connection as a *spring* rather than a rigid rod. The atoms can undergo *vibrational* motion relative to the center of mass, and there is additional kinetic and potential energy associated with this motion. Application of the Schrödinger equation shows that the corresponding energy levels are given by

$$E = \left(n + \frac{1}{2}\right)hf \qquad (n = 0, 1, 2, 3, \ldots), \tag{46-4}$$

where f is the frequency of vibration. For typical diatomic molecules this turns out to be of the order of 10^{13} Hz; thus the constant hf in Eq. (46–4) is of the order of

$$hf = (6.6 \times 10^{-34} \text{ J} \cdot \text{s})(10^{13} \text{ s}^{-1}) = 6.6 \times 10^{-21} \text{ J}$$
$$= 0.041 \text{ eV}.$$

Thus the vibrational energies, while still much smaller than those of the optical spectra, are typically considerably *larger* than the rotational energies.

An energy-level diagram for a diatomic molecule has the general appearance of Fig. 46–4. For each value of n, there are many values of l, leading to series of closely spaced levels. Transition between different pairs of n values gives different series of spectrum lines, and the resulting spectrum has the appearance of a series of *bands*; each band corresponds to a particular vibrational transition, and each individual line in a band, to a particular rotational transition. A typical band spectrum is shown in Fig. 46–5.

The same considerations can be applied to more complex molecules. A molecule with three or more atoms has several different kinds or *modes* of vibratory motion, each with its own set of energy levels related to the frequency by Eq. (46–4). The resulting energy-level scheme and associated spectra can be quite complex, but the general considerations discussed above still apply. In nearly all cases the associated radiation lies in the infrared region of the electromagnetic spectrum. Analysis of molecular spectra has proved to be an extremely valuable analytical tool, providing a great deal of information about the strength and rigidity of molecular bonds and the structure of complex molecules.

46–5 Typical band spectrum. (Courtesy of R. C. Herman.)

46–5 Structure of solids

At ordinary temperatures and pressures most materials are in the solid state, a condensed state of matter characterized by interatomic interactions of sufficient strength to give the material a definite volume and shape, which change relatively little with applied stress. The separation of adjacent atoms in a solid is of the same order of magnitude as the diameter of the electron cloud around each atom.

A solid may be *amorphous* or *crystalline*. Crystalline solids have been discussed briefly in Sec. 20-7, and it would do no harm for the reader to review that discussion. A crystalline solid is characterized by *long-range order,* a recurring pattern in the arrangement of atoms; this pattern is called the *crystal structure* or the *lattice structure*. Amorphous solids have no long-range order but only short-range order, a state of affairs in which each atom has other atoms arranged around it in a more or less regular pattern, but without the recurring pattern characteristic of crystals. Amorphous solids have more in common with liquids than with crystalline solids; liquids also have short-range order but not long-range order.

46-6 Symbolic representation of a sodium chloride crystal, with exaggerated distances between ions.

The forces responsible for the regular arrangement of atoms in a crystal are, in some cases, the same as those involved in molecular binding. Corresponding to the classes of chemical bonds, there are ionic and covalent crystals. The alkali halides, of which ordinary salt (NaCl) is the most common variety, are the most familiar ionic crystals. The positive sodium ions and the negative chlorine ions occupy alternate positions in a cubic crystal lattice, as in Fig. 46-6. The forces are the familiar Coulomb's-law forces between charged particles; these forces are not directional, and the particular arrangement in which the material crystallizes is determined by the relative size of the two ions.

The simplest example of a *covalent* crystal is the *diamond structure,* a structure found in the diamond form of carbon and also in silicon, germanium, and tin, all elements in Group IV of the periodic table, with four electrons in the outermost shell. Each atom in this structure is situated at the center of a regular tetrahedron, with four nearest-neighbor atoms at the corners, and it forms a covalent bond with each of these atoms. These bonds are strongly directional because of the asymmetrical electron distribution, leading to the tetrahedral structure.

A third crystal type, which is less directly related to the chemical bond than ionic or covalent crystals, is the *metallic crystal*. In this structure the outermost electrons are not localized at individual lattice sites but are detached from their parent atoms and free to move through the crystal. The corresponding charge clouds extend over many atoms. Thus a metallic crystal can be visualized roughly as an array of positive ions from which one or more electrons have been removed, immersed in a sea of electrons whose attraction for the positive ions holds the crystal together. This sea has many of the properties of a gas, and indeed one speaks of the *electron-gas model* of metallic solids.

In a metallic crystal the situation is as though the atoms would like to form shared-electron bonds but there are not enough valence electrons. Instead, electrons are shared among *many* atoms. This bonding is not strongly directional in nature, and the shape of the crystal lattice is determined primarily by considerations of *close packing,* i.e., the maximum number of atoms that can fit into a given volume. The two most common metallic crystal lattices, the face-centered cubic and the hexagonal close-packed, are shown in Fig. 46-7. In each of these, each atom has 12 nearest neighbors.

The discussion in this section has centered around *perfect crystals,* crystals in which the crystal lattice extends uninterrupted through the

46-7 Close-packed crystal lattice structures. (a) Face-centered cubic; (b) hexagonal close-packed.

(a) (b)

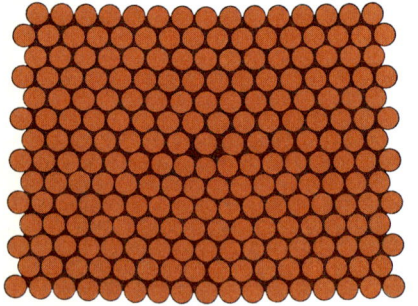

46-8 A dislocation. The concept is seen most easily by viewing the figure from various directions at a grazing angle with the page.

entire material. Real crystals exhibit a variety of departures from this idealized structure. Materials are often *polycrystalline,* composed of many small single crystals bonded together at *grain boundaries.* Within a single crystal there may be *interstitial* atoms in places where they do not belong, and *vacancies,* lattice sites that should be occupied by an atom but are not. An imperfection of particular interest in semiconductors, to be discussed in Sec. 46-7, is the *impurity atom,* a foreign atom (e.g., arsenic in a silicon crystal) occupying a regular lattice site.

A more complex kind of imperfection is the *dislocation,* illustrated schematically in Fig. 46-8, in which one plane of atoms slips relative to another. The mechanical properties of metallic crystals are influenced strongly by the presence of dislocations; the ductility and malleability of some metals depends on the presence of dislocations that move through the lattice during plastic deformations.

46-6 Properties of solids

Many *macroscopic* properties of solids, including mechanical, thermal, electrical, magnetic, and optical properties, can be understood by considering their relation to the *microscopic* structure of the material, and various aspects of the relation of structure to properties are part of the vigorous program of research in the physics of solids being carried out throughout the world. Although these topics cannot be discussed in detail in a book such as this one, a few examples will indicate the kinds of insights to be gained through study of the microscopic structure of solids.

We have already discussed, in Sec. 20-8, the subject of specific heat capacities of crystals, using the same principle of equipartition of energy as in the kinetic theory of gases. A simple analysis permits understanding of the empirical rule of Dulong and Petit on the basis of a microscopic model. This analysis has its limitations, to be sure; it does not include the energy of electron motion, which in metals makes a small additional contribution to specific heat, nor does it predict the temperature dependence of specific heats resulting from the quantization of the lattice-vibration energy discussed in Sec. 20–8. But these additional refinements *can* be included in the model to permit more detailed comparison of observed macroscopic properties with theoretical predictions.

Electrical conductivity is understood on the basis of the mobility or lack of mobility of electrons in the material. In a metallic crystal the valence electrons are not bound to individual lattice sites but are free to move through the crystal. Thus we expect metals to be good conductors, and they usually are. In a covalent crystal the valence electrons are involved in the bonds responsible for the crystal structure and are therefore *not* free to move. Thus there are no mobile charges available for conduction, and such materials are expected to be insulators. Similarly, an ionic crystal such as NaCl has no charges that are free to move, and solid NaCl is an insulator. However, when salt is melted, the ions are no longer locked to their individual lattice sites but are free to move, and *molten* NaCl is a good conductor.

There are, of course, no perfect conductors or insulators, although the resistivity of good insulators is greater than that of good conductors by an enormous factor, the order of at least 10^{15}. This great difference is one of the factors that make extremely precise electrical measurements possible. In addition, the resistivity of all materials depends on temperature; in general, the small conductivity of insulators *increases* with temperature, but that of good conductors usually *decreases* at increased temperatures.

Two competing effects are responsible for this difference. In metals the *number* of electrons available for conduction is nearly independent of temperature, and the resistivity is determined by the frequency of collisions of electrons with the lattice. Roughly speaking, lattice vibrations increase with increased temperature, and the ion cores present a larger target area for collisions with electrons. In insulators, what little conduction does take place is due to electrons that have gained enough energy from thermal motion of the lattice to break away from their "home" atoms and wander through the lattice. The number of electrons able to acquire the needed energy is very strongly temperature-dependent; a twofold increase for a 10° temperature rise is typical. There is also increased scattering at higher temperatures, as with metals, but the increased number of carriers is a far larger effect; thus insulators invariably become better conductors at higher temperatures.

A similar analysis can be made for *thermal* conductivity, which involves transport of microscopic mechanical energy rather than electric charge. The wave motion associated with lattice vibrations is one mechanism for energy transfer, and in metals the mobile electrons also carry kinetic energy from one region to another. This effect turns out to be much larger than that of the lattice vibrations, with the result that metals are usually much better thermal conductors than are electrical insulators, which have at most very few free electrons available to transport energy.

Optical properties are also related directly to microscopic structure. Good electrical conductors cannot be transparent to electromagnetic waves, for the electric fields of the waves induce currents in the material, and these dissipate the wave energy into heat as the electrons collide with the atoms in the lattice. All transparent materials are very good insulators. Metals are, however, good *reflectors* of radiation, and again the reason is the presence of free electrons at the surface of the material, which can move in response to the incident wave and generate a reflected wave. Reflection from the polished surface of an insulator is a somewhat more subtle phenomenon, dependent on polarization of the material, but again it is possible to relate macroscopic properties to microscopic structure.

46-7 Semiconductors

As the name implies, the electrical resistivity of a semiconductor is intermediate between that of good conductors and of good insulators. This is, however, only one aspect of the behavior of this important class of materials, so vital to present-day electronics. This discussion will be confined, for simplicity, to the elements germanium and silicon, the simplest semiconductors, but these examples serve to illustrate the principal concepts.

Silicon and germanium, having four electrons in the outermost subshell, crystallize in the diamond structure described in Sec. 46-5; each atom lies at the center of a regular tetrahedron, with four nearest neighbors at the corners and a covalent bond with each. Thus all the electrons are involved in the bonding and the materials should be insulators. However, an unusually small amount of energy is needed to break one of the bonds and set an electron free to roam around the lattice, 1.1 eV for silicon and only 0.7 eV for germanium. Thus even at room temperature a substantial number of electrons are dissociated from their parent atoms, and this number increases rapidly with temperature.

Furthermore, when an electron is removed from a covalent bond, it leaves a positively charged vacancy where there would ordinarily be an electron. An electron in a neighboring atom can drop into this vacancy, leaving the neighbor with the vacancy. In this way the vacancy, usually called a *hole,* can travel through the lattice and serve as an additional current-carrier. In a pure semiconductor, holes and electrons are always present in equal numbers; the resulting conductivity is called *intrinsic conductivity* to distinguish it from conductivity due to impurities, to be discussed later.

An analogy is often used to clarify the mechanism of conduction in a semiconductor. A crystal with no bonds broken is like a completely filled floor of a parking garage. No cars (electrons) can move because there is nowhere for them to go. But if one car is removed to the empty floor above, it can move around freely, and the vacancy it leaves also permits cars to move on the nearly filled floor; this motion is most easily described in terms of *motion of the vacant space* from which the car has been removed. The analogy can be drawn even more closely by considering the quantum states available to electrons in the solid, using the concept of energy bands, groups of closely spaced energy levels separated by "forbidden bands."

Now suppose we mix into melted germanium a small amount of arsenic, a Group V element having five electrons in its outermost subshell. When one of these electrons is removed, the remaining electron structure is essentially that of germanium; the only difference is that it is scaled down in size by the insignificant factor $\frac{32}{33}$ because the arsenic nucleus contains a charge $+33e$ rather than $+32e$. Thus an arsenic atom can take the place of a germanium atom in the lattice. Four of its five valence electrons form the necessary four covalent bonds with the nearest neighbors; the fifth is very loosely bound (energy only about 0.01 eV) and even at ordinary temperature can very easily escape and wander about the lattice. The corresponding positive charge is associated with the nuclear charge and is *not* free to move, in contrast to the situation with electrons and holes in pure germanium.

Because at ordinary temperatures only a very small fraction of the valence electrons are able to escape their sites and participate in the conduction process, a concentration of arsenic atoms as small as one part in 10^{10} can increase the conductivity so drastically that conduction due to impurities becomes by far the dominant mechanism. In such a case the conductivity is due almost entirely to *negative* charge motion; the material is called an *n-type* semiconductor or is said to have *n-type* impurities.

Adding atoms of an element in Group III, with only three electrons in its outermost subshell, has an analogous effect. An example is gallium; placed in the lattice, the gallium atom would like to form four covalent bonds, but it has only three outer electrons. It can, however, steal an electron from a neighboring germanium atom to complete the bonding. This leaves the neighbor with a *hole* or missing electron, and this hole can then move through the lattice just as with intrinsic conductivity. In this case the corresponding negative charge is associated with the deficiency of positive charge of the gallium nucleus ($+31e$ instead of $+32e$), so it is not free to move. This state of affairs is characteristic of *p-type* semiconductors, materials with *p-type* impurities. The two types of impurities, n and p, are also called *donors* and *acceptors*, respectively.

The assertion that in n and p semiconductors the current *is* actually carried by electrons and holes, respectively, can be verified using the Hall effect, discussed in Sec. 31-2. The direction of the Hall emf is opposite in the two cases, and measurements of the Hall effect in various semiconductor materials confirm the above analysis of the conduction mechanisms.

46-8 Semiconductor devices

The tremendous importance of semiconductor devices in practical electronic equipment results directly from the fact that the conductivity can be controlled within wide limits and changed from one region of a device to another, by varying impurity concentrations. The simplest example is the *p-n junction,* a crystal of germanium or silicon with *p*-type impurities in one region and *n*-type in the other, with the two separated by a boundary region called the *junction*. The details of how such a crystal is produced need not concern us; one way, not necessarily the most economical, is to grow a crystal by pulling a seed crystal very slowly away from the surface of a melted semiconductor. By varying the concentration of impurities in the melt while the crystal is grown, one can make a crystal with two or more regions of varying conductivity.

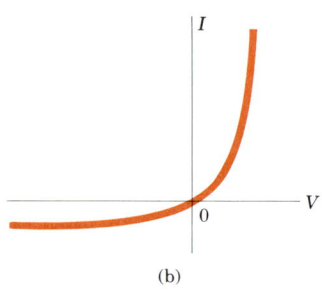

46-9 (a) A semiconductor *p-n* junction in a circuit; (b) graph showing the asymmetric voltage–current relationship given by Eq. (46-5).

When a *p-n* junction is connected in an external circuit as shown in Fig. 46-9a and the potential V across the device is varied, the behavior of the current is as shown in Fig. 46-9b. The device conducts much more readily in the direction $p \rightarrow n$ than in the reverse, in striking contrast to the behavior of materials that obey Ohm's law. In the language of electronics, such a one-way device is called a *diode*; a different kind of diode was discussed in Sec. 44-2 in connection with thermionic emission.

The behavior of a *p-n* junction diode can be understood at least roughly on the basis of the conductivity mechanisms in the two regions. When the *p* region is at higher potential than the *n*, holes in the *p* region flow into the *n* region and electrons in the *n* region into the *p* region, so both contribute substantially to current. When the polarity is reversed, the resulting electric fields tend to push electrons from *p* to *n* and holes from *n* to *p*. But there are very few electrons in the *p* region, only those associated with intrinsic conductivity and some that diffuse over from the *n* region. A similar condition prevails in the *n* region, and the current is much smaller than with the opposite polarity.

A more detailed analysis of this process, taking into account the effects of drift under the applied field and the diffusion that would take place even in the absence of a field, shows that the voltage–current relationship is given by

$$I = I_0(e^{eV/kT} - 1), \qquad (46\text{-}5)$$

where I_0 is a constant characteristic of the device, e is the electron charge, k is Boltzmann's constant, and T is absolute temperature.

The *transistor* includes two *p-n* junctions in a "sandwich" configuration which may be either *p-n-p* or *n-p-n*. One of the former type is shown in Fig. 46-10. The three regions are usually called the emitter, base, and collector, as shown. In the absence of current in the left loop of the circuit, there is only a very small current through the resistor R because the voltage across the base–collector junction is in the "reverse" direction, i.e., the direction of small current flow, as in a simple *p-n* junction. But when a voltage is applied between emitter and base, as shown, the holes traveling from emitter to base can travel *through* the base to the second junction, where they come under the influence of the collector-to-base potential difference and thus flow on around through the resistor.

Thus the current in the collector circuit is *controlled* by the current in the emitter circuit. Furthermore, since V_c may be considerably larger than V_e, the power dissipated in R may be much larger than that supplied to the emitter circuit by the battery V_e. Thus the device functions as a power amplifier. If the potential drop across R is greater than V_e, it may also be a voltage amplifier.

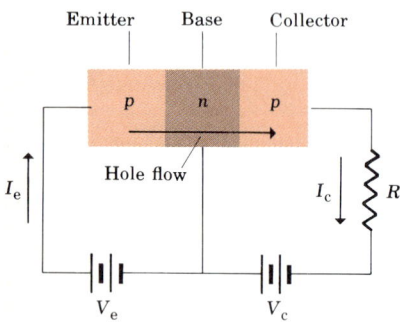

46-10 Schematic diagram of a *p-n-p* transistor and circuit. When $V_e = 0$ the current in the collector circuit is very small. When a potential V_e is applied between emitter and base, holes travel from emitter to base, as shown; when V_c is sufficiently large, most of them continue into the collector. The collector current I_c is controlled by the emitter current I_e.

In this configuration the *base* is the common element between the "input" and "output" sides of the circuit. Another widely used arrangement is the *common-emitter* circuit, shown in Fig. 46-11. In this circuit the current in the collector side of the circuit is much larger than in the base side, and the result is current amplification.

Transistors can perform many of the functions for which vacuum tubes have been used ever since the invention of the triode by de Forest in 1907; and where applicable they offer many advantages over vacuum tubes. These include mechanical ruggedness (since no vacuum container or fragile electrodes are involved), small size, long life (because they operate at relatively low temperatures and deteriorate very little with age), and efficiency (because no power is wasted heating a cathode for thermionic emission). Since their invention in 1948, semiconductor devices have completely revolutionized the electronics industry, including applications to communications, computer systems, control systems, and many other areas.

A further refinement in semiconductor technology is the *integrated circuit*. By successive depositing of layers of material and etching patterns to define current paths, one can combine the functions of several transistors, capacitors, and resistors on a single square of semiconductor material that may be only a few millimeters on a side. A further elaboration of this idea leads to the *large-scale integrated circuit*. On a base consisting of a silicon chip, various layers are built up, including evaporated metal layers for conducting paths and silicon-dioxide layers for insulators and for dielectric layers in capacitors. Etching appropriate patterns into each layer by use of photosensitive etch-resistant materials and optically reduced patterns makes it possible to build up a circuit containing the functional equivalent of thousands of transistors, diodes, resistors, and capacitors on a single chip a few millimeters on a side. These so-called MOS (metal-oxide-semiconductor) chips are the heart of pocket calculators and microprocessors, as well as of many larger-scale computers.

46-11 A common-emitter circuit. When $V_b = 0$, I_c is very small and most of the voltage V_c appears across the base–collector junction. As V_b increases, the base–collector potential decreases and more holes can diffuse into the collector; thus I_c increases. Ordinarily I_c is much larger than I_b.

Questions

46-1 In the ground state of the helium atom, the electrons must have opposite spins. Why?

46-2 The central-field approximation is more accurate for alkali metals than for transition metals (Group IV of the periodic table). Why?

46-3 The outermost electron in the potassium atom is in a 4s state. What does this tell you about the relative positions of the 3d and 4s states for this atom?

46-4 A student asserted that any filled shell (i.e., all the levels for a given n occupied by electrons) must have zero total angular momentum and hence must be spherically symmetric. Do you believe this? What about a filled *subshell* (all values of m for given values of n and l)?

46-5 What factors determine whether a material is a conductor of electricity or an insulator?

46-6 The nucleus of a gold atom contains 79 protons. How would you expect the energy required to remove a 1s electron completely from a gold atom to compare with the energy required to remove the electron from a hydrogen atom? In what region of the electromagnetic spectrum would a photon of the appropriate energy lie?

46-7 Elements can be identified by their visible spectra. Could analogous techniques be used to identify compounds from their molecular spectra? In what region of the electromagnetic spectrum would the appropriate radiation lie?

46-8 The ionization energies of the alkali metals (i.e., the energy required to remove an outer electron) are in the range from 4 to 6 eV, while those of the inert gases are in the range from 15 to 25 eV. Why the difference?

46-9 The energy required to remove the 3s electron from a sodium atom in its ground state is about 5 eV. Would you expect the energy required to remove an additional electron to be about the same, or more or less? Why?

46-10 Electrical conductivities of most metals decrease gradually with increasing temperature, but the intrinsic conductivity of semiconductors always *increases* rapidly with increasing temperature. Why the difference?

46-11 What are some advantages of transistors compared to vacuum tubes, for electronic devices such as amplifiers? What are some disadvantages? Are there any situations in which vacuum tubes *cannot* be replaced by solid-state devices?

Problems

46-1 Make a list of the four quantum numbers for each of the six electrons in the ground state of the carbon atom, and for each of the 14 electrons in the silicon atom.

46-2 Cobalt ($Z = 27$) has two electrons in $4s$ states. How many does it have in $3d$ states?

46-3 For germanium ($Z = 32$) make a list of the number of electrons in each state ($1s$, $2s$, $2p$, etc.).

46-4 Calculate the magnitude of the potential energy ($e^2/4\pi\varepsilon_0 r$) of interaction of a charge $+e$ with a charge $-e$ (where e is the electron charge magnitude) at a distance of 2×10^{-10} m, roughly the magnitude of the lattice spacing in the NaCl crystal.

46-5 For the sodium chloride molecule discussed at the beginning of Sec. 46-3, what is the maximum separation of the ions for stability if they may be regarded as point charges? (The potential energy of interaction must be at least -1.3 eV.)

46-6 For magnesium, the first ionization potential is 7.6 eV; the second (additional energy required to remove a second electron) is almost exactly twice this, 15 eV, and the third ionization potential is much larger, about 80 eV. How can these numbers be understood?

46-7 The ionization potentials of the alkali metals are:

Li	5.4 eV	Rb	4.2 eV
Na	5.1	Cs	3.9
K	4.3		

Which of these would you expect to be most active chemically?

46-8 The rotation spectrum of HCl contains the following wavelengths:

60.4 μm

69.0

80.4

96.4

120.4

Find the moment of inertia of the HCl molecule.

46-9 Show that the frequencies in a pure rotation spectrum (disregarding vibrational levels) are all integer multiples of the quantity $h/2\pi I$.

46-10 If the distance between atoms in a diatomic oxygen molecule is 2×10^{-10} m, calculate the moment of inertia about an axis through the center of mass perpendicular to the line joining the atoms.

46-11 The dissociation energy of the hydrogen molecule (i.e., the energy required to separate the two atoms) is 4.72 eV. At what temperature is the average kinetic energy of a molecule equal to this energy?

46-12 The vibrational frequency of the hydrogen molecule is 1.29×10^{14} Hz.

a) What is the spacing of adjacent vibrational energy levels, in electron volts?

b) What is the wavelength of radiation emitted in the transition from the $n = 2$ to $n = 1$ vibrational state of the hydrogen molecule?

c) From what initial values of n do transitions to the ground state of vibrational motion yield radiation in the visible spectrum?

46-13 The distance between atoms in a hydrogen molecule is 0.074 nm.

a) Calculate the moment of inertia of a hydrogen molecule about an axis perpendicular to the line joining the nuclei, at its center.

b) Find the energies of the $l = 0$, $l = 1$, and $l = 2$ rotational states.

c) Find the wavelength and frequency of the photon emitted in the transition from $l = 2$ to $l = 0$.

46-14 The spacing of adjacent atoms in a sodium-chloride crystal is 0.282 nm. Calculate the density of sodium chloride.

46-15 In the hexagonal close-packed crystal structure, regarding the atoms as rigid spheres, show that an atom can accommodate twelve nearest neighbors (all touching it) if six are placed around it in a plane hexagonal array, with three additional atoms above and three below the plane.

46-16 Suppose a piece of very pure germanium is to be used as a light detector by observing the increase in conductivity resulting from generation of electron–hole pairs by absorption of photons. If each pair requires 0.7 eV of energy, what is the maximum wavelength that can be detected? In what portion of the spectrum does it lie?

NUCLEAR PHYSICS 47

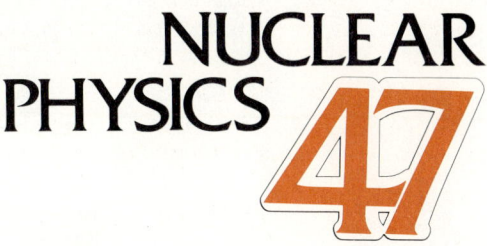

47-1 The nuclear atom

In preceding chapters we have frequently made use of the fact that every atom contains a massive, positively charged nucleus, much smaller than the overall dimensions of the atom but nevertheless containing most of the total mass of the atom. It is instructive to review the earliest experimental evidence for the existence of the nucleus, the *Rutherford scattering experiments*. The experiments were carried out in 1910–1911 by Sir Ernest Rutherford and two of his students, Hans Geiger and Ernest Marsden, at Cambridge, England.

The electron had been "discovered" in 1897 by Sir J. J. Thomson, and by 1910 its mass and charge were quite accurately known. It had also been well established that, with the sole exception of hydrogen, all atoms contain more than one electron. Thomson had proposed a model of the atom consisting of a relatively large sphere of positive charge (about 2 or 3×10^{-8} cm in diameter) within which were embedded, like plums in a pudding, the electrons.

What Rutherford and his coworkers did was to project other particles at the atoms under investigation; and, from observations of the way in which the projected particles were deflected or *scattered*, they drew conclusions about the distribution of charge within the "target" atoms.

At this time the high-energy particle accelerators now in common use for nuclear physics research had not yet been developed, and Rutherford had to use as projectiles the particles produced in natural radioactivity, to be discussed later in this chapter. Some radioactive disintegrations result in the emission of *alpha particles;* these particles are now known to be identical with the nuclei of helium atoms, each consisting of two protons and two neutrons bound together, but without the two electrons normally present in a neutral helium atom. Alpha particles are ejected from unstable nuclei with speeds of the order of 10^7 m·s^{-1}, and they can travel several centimeters through air, or on the order of 0.1 mm through solid matter, before they are brought to rest by collisions.

47-1 The scattering of alpha particles by a thin metal foil.

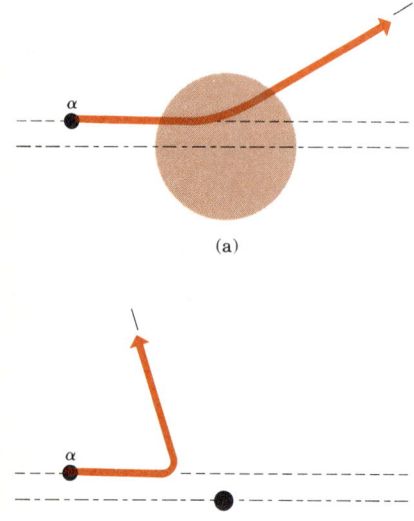

47-2 (a) Alpha particle scattered through a small angle by the Thomson atom. (b) Alpha particle scattered through a large angle by the Rutherford nuclear atom.

The experimental setup is shown schematically in Fig. 47–1. A radioactive source at the left emits alpha particles. Thick lead screens stop all particles except those in a narrow *beam* defined by small holes. The beam then passes through a thin metal foil (gold, silver, and copper were used) and strikes a plate coated with zinc sulfide. A momentary flash or scintillation can be observed visually on the screen whenever it is struck by an alpha particle, and the number of particles that have been deflected through any angle from their original direction can therefore be determined.

According to the Thomson model, the atoms of a solid are packed together like marbles in a box. The experimental fact that an alpha particle can pass right through a sheet of metal foil forces one to conclude, if this model is correct, that the alpha particle is capable of actually penetrating the spheres of positive charge. Granted that this is possible, we can compute the deflection it would undergo. The Thomson atom is electrically neutral, so outside the atom no force would be exerted on the alpha particle. Within the atom, the electrical force would be due in part to the electrons and in part to the sphere of positive charge. However, the mass of an alpha particle is about 7400 times that of an electron, and from momentum considerations it follows that the alpha particle can suffer only a negligible scattering as a consequence of forces between it and the much less massive electrons. It is only interactions with the *positive* charge, which makes up most of the atomic mass, that can deviate the alpha particle.

When electric charge is distributed uniformly inside a spherical volume, the electric field at points inside the sphere is proportional to the distance from the center of the sphere; this can be proved using Gauss's law, as in Problem 25–25. Thus the positively charged alpha particle inside the Thomson atom would be *repelled* from the center of the sphere with a force proportional to its distance from the center, and its trajectory can be computed for any initial direction of approach such as that in Fig. 47–2a. On the basis of such calculations, Rutherford predicted the number of alpha particles that should be scattered at any angle with respect to the original direction.

The experimental results did not agree with the calculations based on the Thomson atom. In particular, many more particles were scattered through large angles than were predicted. To account for the observed large-angle scattering, Rutherford concluded that the positive charge, instead of being spread through a sphere of atomic dimensions (2 or 3×10^{-8} cm) was concentrated in a much *smaller* volume, which he called a *nucleus*. When an alpha particle approaches the nucleus, the entire nuclear charge exerts a repelling effect on it down to extremely small separations, with the consequence that much larger deviations can be produced. Figure 47-2b shows the trajectory of an alpha particle deflected by a Rutherford nuclear atom, for the same original path as that in part (a) of the figure.

Rutherford again computed the expected number of particles scattered through any angle, assuming an inverse-square law of force between the alpha particle and the nucleus of the scattering atom. Within the limits of experimental accuracy, the computed and observed results were in agreement down to distances of approach of about 10^{-12} cm. These experiments thus indicate that the size of the nucleus is no larger than the order of 10^{-12} cm.

47-2 Properties of nuclei

As indicated in Sec. 47-1, the most obvious feature of the atomic nucleus is its size, of the order of 20,000 to 200,000 times smaller than the atom itself. Since Rutherford's initial experiments, many additional scattering experiments have been performed, using high-energy protons, electrons, and neutrons as well as alpha particles. Although the "surface" of a nucleus is not a sharp boundary, these experiments determine an approximate radius for each nucleus. The radius is found to depend on the mass, which in turn depends on the total number A of neutrons and protons, usually called the *mass number*. The radii of most nuclei are represented fairly well by the empirical equation

$$r = r_0 A^{1/3} \qquad (47\text{-}1)$$

where r_0 is an empirical constant equal to 1.2×10^{-15} m, the same for all nuclei.

Since the volume of a sphere is proportional to r^3, Eq. (47-1) shows that the *volume* of a nucleus is proportional to A (i.e., to the total mass) and therefore that the mass per unit volume (proportional to A/r^3) is the same for all nuclei. That is, *all nuclei have approximately the same density*. This fact is of crucial importance in understanding nuclear structure.

Two additional important properties of nuclei are angular momentum and magnetic moment. The particles in the nucleus are in motion, just as the electrons in an atom are in motion; there is angular momentum associated with this motion, and, because circulating charge constitutes a current, there is also magnetic moment. Experimental evidence for the existence of nuclear angular momentum (often called *nuclear spin*) and magnetic moment came originally from spectroscopy. Some spectrum lines are found to be split into series of very closely spaced lines, called *hyperfine structure,* which can be understood on the basis of interactions between the electrons in an atom and the magnetic field produced by a nuclear magnetic moment. Detailed analysis of this phenomenon indicates that the nuclear angular momentum is *quantized,* just as it is for electrons and for molecular rotation; the component of angular momentum in a specified axis direction is a multiple of $h/2\pi$, but some nuclei seem to require *integer* multiples (as with orbital angular momentum of electrons) and some *half-integer* multiples, as with electron spin. As we shall see later, the nuclear spin can be understood on the basis of the particles making up the nucleus.

Although it was once believed that nuclei were made of protons and electrons, the discovery of the neutron (discussed in Sec. 47-10) and many other experiments have established that the basic building blocks of the nucleus are the proton and the neutron. The total number of protons, equal in a neutral atom to the number of electrons, is the *atomic number Z*. The total number of *nucleons* (protons and neutrons) is called the mass number A. The number of neutrons, denoted by N, is called the *neutron number*. For any nucleus these are related by

$$A = Z + N \qquad (47\text{-}2)$$

Table 47-1 lists values of A, Z, and N for several nuclei. As the table shows, some nuclei have the same Z but different N. Since the electron structure, which determines the chemical properties, is determined by the charge of the nucleus, these are nuclei of the same element, but they have different masses and can be distinguished in precise experiments. The experimental investigation of *isotopes* through mass spectroscopy is discussed in Sec. 30-6, and the reader would do well to review this section. Nuclei of a given element having different mass numbers are called *isotopes* of the element, and a single nuclear species (single values of both Z and N) is called a *nuclide*.

Table 47-1 also shows the notation usually used to denote individual nuclides; the symbol of the element is used, with a pre-subscript equal to

Table 47-1 Nuclear particles

Nucleus	Mass number (total number of nuclear particles), A	Atomic number (number of protons), Z	Neutron number, $N = A - Z$
$_1\text{H}^1$	1	1	0
$_1\text{D}^2$	2	1	1
$_2\text{He}^4$	4	2	2
$_3\text{Li}^6$	6	3	3
$_3\text{Li}^7$	7	3	4
$_4\text{Be}^9$	9	4	5
$_5\text{B}^{10}$	10	5	5
$_5\text{B}^{11}$	11	5	6
$_6\text{C}^{12}$	12	6	6
$_6\text{C}^{13}$	13	6	7
$_7\text{N}^{14}$	14	7	7
$_8\text{O}^{16}$	16	8	8
$_{11}\text{Na}^{23}$	23	11	12
$_{29}\text{Cu}^{65}$	65	29	36
$_{80}\text{Hg}^{200}$	200	80	120
$_{92}\text{U}^{235}$	235	92	143
$_{92}\text{U}^{238}$	238	92	146

the atomic number Z and a post-superscript equal to the mass number A. This is of course redundant, since the element is determined by the atomic number, but the notation is a useful aid to memory.

The total mass of a nucleus is always *less* than the total mass of its constituent parts because of the mass equivalent ($E = mc^2$) of the negative potential energy associated with the attractive forces that hold the nucleus together. This mass difference is sometimes called the *mass defect*. In fact, the best way to determine the total potential energy, or *binding energy*, is to compare the mass of a nucleus with the masses of its constituents. The proton and neutron masses are

$$m_p = 1.673 \times 10^{-27} \text{ kg},$$
$$m_n = 1.675 \times 10^{-27} \text{ kg}.$$

Since these are nearly equal, it is not surprising that many nuclear masses are approximately integer multiples of the proton or neutron mass.

This observation suggests defining a new mass unit equal to the proton or neutron mass. Instead, it has been found more convenient to define the new unit, called the *atomic mass unit* (u), as $\frac{1}{12}$ the mass of the neutral carbon atom having mass number $A = 12$. It is found that

$$1 \text{ u} = 1.660566 \times 10^{-27} \text{ kg}.$$

In atomic units, the masses of the proton, neutron, and electron are found to be:

$$m_p = 1.007276 \text{ u},$$
$$m_n = 1.008665 \text{ u},$$
$$m_e = 0.000549 \text{ u}.$$

The masses of some common atoms, including their electrons, are shown in Table 47-2. The masses of the bare nuclei are obtained by subtracting Z times the electron mass. The energy equivalent of 1 u is found from the relation $E = mc^2$:

$$E = (1.660566 \times 10^{-27} \text{ kg})(2.998 \times 10^8 \text{ m} \cdot \text{s}^{-1})^2$$
$$= 1.492 \times 10^{-10} \text{ J} = 931.5 \text{ MeV}.$$

Example Find the mass defect, the total binding energy, and the binding energy per nucleon for carbon (12).

Solution The mass of the neutral carbon atom, including the nucleus and the six electrons, is, according to Table 47-2, 12.00000 u. The mass of the bare nucleus is obtained by subtracting the mass of the six electrons:

$$m = 12.00000 \text{ u} - (6)(0.000549 \text{ u}) = 11.996706 \text{ u}.$$

The total mass of the six protons and six neutrons in the nucleus is

$$(6)(1.007276 \text{ u}) + (6)(1.008665 \text{ u}) = 12.095646 \text{ u}.$$

The mass defect is therefore

$$12.095646 \text{ u} - 11.996706 \text{ u} = 0.09894 \text{ u}.$$

The energy equivalent of this mass is

$$(0.09894 \text{ u})(931.5 \text{ MeV} \cdot \text{u}^{-1}) = 92.16 \text{ MeV}.$$

Table 47-2 Atomic data*

Element		Atomic number, Z	Neutron number N	Atomic mass, u	Mass number, A
Hydrogen	H^1	1	0	1.00783	1
Deuterium	H^2	1	1	2.01410	2
Helium	He^3	2	1	3.01603	3
Helium	He^4	2	2	4.00260	4
Lithium	Li^6	3	3	6.01513	6
Lithium	Li^7	3	4	7.01601	7
Beryllium	Be^8	4	4	8.00508	8
Beryllium	Be^9	4	5	9.01219	9
Boron	B^{10}	5	5	10.01294	10
Boron	B^{11}	5	6	11.00931	11
Carbon	C^{12}	6	6	12.00000	12
Carbon	C^{13}	6	7	13.00335	13
Nitrogen	N^{14}	7	7	14.00307	14
Nitrogen	N^{15}	7	8	15.00011	15
Oxygen	O^{16}	8	8	15.99491	16
Oxygen	O^{17}	8	9	16.99913	17
Oxygen	O^{18}	8	10	17.99916	18

* American Institute of Physics Handbook, 1963

Thus the total binding energy for the twelve nucleons is 92.16 MeV. To pull the carbon nucleus completely apart into twelve separate nucleons would require a minimum of 92.16 MeV. The binding energy *per nucleon* is $\frac{1}{12}$ of this, or 7.68 MeV per nucleon. Nearly all stable nuclei, from the lightest to the most massive, have binding energies in the range of 6 to 8 MeV per nucleon.

Because the nucleus is a composite structure, it can have internal motion, with corresponding energy levels. Thus each nucleus has a set of allowed energy levels, corresponding to a *ground state* (state of lowest energy) and several *excited states*. Because of the strength of nuclear interactions, excitation energies of nuclei are typically of the order of 1 MeV, compared with a few eV for atomic energy levels. In ordinary physical and chemical transformations the nucleus always remains in its ground state. When a nucleus is placed in an excited state, either by bombardment by high-energy particles or by a radioactive transformation, it can decay to the ground state by emission of one or more photons, called in this case *gamma rays* or *gamma photons*.

The forces that hold protons and neutrons together in the nucleus despite the electrical repulsion of the protons are not of any familiar sort but are unique to the nucleus. Some aspects of the nuclear force are still incompletely understood, but several qualitative features can be described. First, it does not depend on charge, since neutrons as well as protons must be bound. Second, it must be of short range; otherwise the nucleus would pull in additional protons and neutrons. But within the range it must be stronger than electrical forces; otherwise the nucleus could never be stable. Third, the nearly constant density of nuclear matter and the nearly constant binding energy per nucleon indicate that a given nucleon cannot interact simultaneously with all the other nucleons but only with those few in its immediate vicinity. This is again in contrast to the behavior of electrical forces, in which *every* proton in the nucleus repels every other one. This limitation on the maximum number of nucleons with which a nucleon can interact is called *saturation;* it is analogous in some respects to covalent bonding in solids. Finally, the nuclear forces appear to favor binding of pairs of particles (e.g., two protons with opposite spin, or two neutrons with opposite spin) and of *pairs of pairs,* a pair of protons and a pair of neutrons, with total spin zero. Thus the alpha particle is an exceptionally stable nuclear structure.

These qualitative features of the nuclear force are helpful in understanding the various kinds of nuclear instability, to be discussed in the following sections.

47-3 Natural radioactivity

In studying the fluorescence and phosphorescence of compounds irradiated with visible light, Becquerel, in 1896, performed a crucial experiment that led to a deeper understanding of the properties of the nucleus of an atom. After illuminating some pieces of uranium–potassium sulfate with visible light, Becquerel wrapped them in black paper and separated the package from a photographic plate by a piece of silver. After several hours' exposure, the photographic plate was developed and showed a blackening due to something that must have been emitted from the compound and was able to penetrate both the black paper and the silver.

Rutherford showed later that the emanations given off by uranium sulfate were capable of ionizing the air in the space between two oppositely charged metallic plates (an ionization chamber). The current registered by a galvanometer in series with the circuit was taken to be a measure of the "activity" of the compound.

A systematic study of the activity of various elements and compounds led Mme. Curie to the conclusion that this activity was an atomic phenomenon; and by the methods of chemical analysis, she and her husband, Pierre Curie, found that "ionizing ability" or "activity" was associated not only with uranium but with two other elements that they discovered, radium and polonium. The activity of radium was found to be more than a million times that of uranium. Since the pioneer days of the Curies, many more radioactive substances have been discovered.

The activity of radioactive material may be easily shown to be the result of three different kinds of emanations. An early experiment is shown in Fig. 47-3. A small piece of radioactive material is placed at the bottom of a long groove in a lead block. At some distance above the lead block a photographic plate is placed, and the whole apparatus is highly evacuated. A strong magnetic field is applied at right angles to the plane of the diagram. When the plate is developed, three distinct spots are found, one in the direct line of the groove in the lead block, one deflected to one side, and one to the other side. From a knowledge of the direction of the magnetic field, it is concluded that one of the emanations is positively charged (alpha particles), one is negatively charged (beta particles), and one is neutral (gamma rays).

47-3 The three emanations from a radioactive material and their paths in a magnetic field perpendicular to the plane of the diagram.

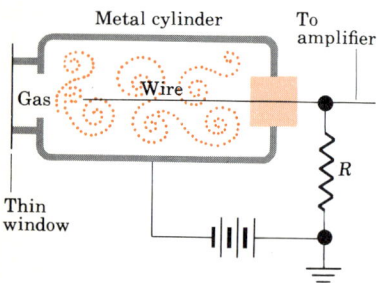

47-4 Schematic diagram of a Geiger counter.

Futher investigation showed that not all three emanations are emitted simultaneously by all radioactive substances. Some elements emit alpha particles, others emit beta particles, while gamma rays sometimes accompany one and sometimes the other. Furthermore, no simple macroscopic physical or chemical process, such as raising or lowering the temperature, chemical combination with other nonradioactive substances, etc., could change or affect in any way the activity of a given sample. As a result, it was suspected from the beginning that radioactivity is a *nuclear* process and that the emission of a charged particle from the nucleus of an atom results in leaving behind a different atom, occupying a different place in the periodic table. In other words, radioactivity involves the *transmutation of elements*.

The first measurement of the charge of the alpha particle used a device called a *Geiger counter*, still an important instrument of modern physics.

As shown in Fig. 47-4, a Geiger counter consists of a metal cylinder and a wire along its axis. The cylinder contains a gas such as air or argon, at a pressure of from 50 to 100 mm of mercury. A difference of potential slightly less than that necessary to produce a discharge is maintained between the wire and the cylinder wall. Alpha particles (or, for that matter, any particles to be studied) can enter through a thin glass or mica window. The particle entering the counter produces ionization of the gas molecules. These ions are accelerated by the electric field and produce more ions by collisions, causing the ionization current to build up rapidly. The current, however, decays rapidly since the circuit has a small time constant. There is therefore a momentary surge of current or a momentary potential surge across R, which may be amplified and made to advance an electronic counter, or to activate a counting-rate meter.

Placing a known mass of radium at a known distance from the window of a Geiger counter, Rutherford and Geiger counted the number of alpha particles emitted in a known time interval. They found that 3.57×10^{10} alpha particles were emitted per second per gram of radium. They then allowed the alpha particles from the same source to fall upon a plate and measured its rate of increase of charge. Dividing the rate of increase of charge by the number emitted per second, Rutherford and Geiger determined the charge on an alpha particle to be 3.19×10^{-19} C, or practically twice the charge on an electron, but opposite in sign.

The next problem was to determine the mass of an alpha particle. This was accomplished by measuring the ratio of charge to mass by the electric and magnetic deflection method described in Chapter 30. The ratio was found by Rutherford and Robinson to be 4.82×10^7 C·kg^{-1}. Combining this result with the charge of an alpha particle, the mass was found to be 6.62×10^{-27} kg, almost exactly four times the mass of a hydrogen atom.

Since a helium atom has a mass four times that of a hydrogen atom and, stripped of its two electrons (as a bare nucleus), has a charge equal in magnitude and opposite in sign to two electrons, it seemed certain that alpha particles were helium nuclei. To make the identification certain, however, Rutherford and Royds collected the alpha particles in a glass discharge tube over a period of about six days and then established an electric discharge in the tube. Examining the spectrum of the emitted light, they identified the characteristic helium spectrum and established without doubt that alpha particles *are* helium nuclei.

The speed of an alpha particle emitted from a given radioactive source such as radium may be measured by observing the radius of the circle traversed by the particle in a magnetic field perpendicular to the motion. Such experiments show that alpha particles are emitted with very high speeds. For example, the alpha particles emitted by radium, $_{88}\text{Ra}^{226}$, have speeds of about $1.5 \times 10^7 \text{ m} \cdot \text{s}^{-1}$. The corresponding kinetic energy is

$$K = \tfrac{1}{2}(6.62 \times 10^{-27} \text{ kg})(1.5 \times 10^7 \text{ m} \cdot \text{s}^{-1})^2$$
$$= 7.4 \times 10^{-13} \text{ J} = 4.7 \times 10^6 \text{ eV}$$
$$= 4.7 \text{ MeV}.$$

We note that the speed, although large, is only five percent of the speed of light, so the nonrelativistic kinetic-energy expression may be used. We also note that the energy is larger than typical energies of atomic electrons by a factor of the order of a million. Because of these large energies, alpha particles are capable of traveling several centimeters in air, or a few tenths or hundredths of a millimeter through solids, before they are brought to rest by collisions.

Beta particles are *negatively* charged and are therefore deflected in an electric or magnetic field. Deflection experiments similar to those described in Chapter 30 prove conclusively that beta particles have the same charge and mass as electrons. They are emitted with tremendous speeds, some reaching a value of 0.9995 that of light. Thus relativistic relations must be used in analyzing their motion. Unlike alpha particles, which are emitted from a given nucleus with one speed or a few definite speeds, beta particles are emitted with a *continuous* range of speeds, from zero up to a maximum that depends on the nature of the emitting nucleus. If the principles of conservation of energy and of momentum are to hold in nuclear processes, it is necessary to assume that the emission of a beta particle is accompanied by the emission of another particle with no charge. This particle, called a *neutrino,* has zero rest mass and zero charge and therefore, even in traversing the densest matter, produces very little measurable effect. In spite of this, Reines and Cowan, in 1953 and again in 1956, succeeded in detecting its existence in a series of extraordinary experiments. Subsequent investigation has shown that in fact there are at least three varieties of neutrinos, the one associated with beta decay and others associated with the decay of unstable particles.

Since gamma rays are not deflected by a magnetic field, they cannot consist of charged particles. They are, however, diffracted at the surface of a crystal in a manner similar to that of x-rays but with extremely small angles of diffraction. Experiments of this sort lead to the conclusion that gamma rays are actually electromagnetic waves of extremely short wavelength, about $\frac{1}{100}$ that of x-rays.

The gamma-ray spectrum of any one element is a line spectrum, suggesting that a gamma-ray photon is emitted when a nucleus proceeds from a state of higher to a state of lower energy. This view is substantiated in the case of radium by the following facts. When alpha particles are emitted from radium they are found to consist of two groups, those with a kinetic energy of 4.879 MeV and those with an energy of 4.695 MeV. When a radium atom emits an alpha particle of the smaller energy, the resulting nucleus (which corresponds to the element *radon*) has a *greater* amount of energy than if the higher-speed alpha particle had been emitted. This represents an excited state of the radon nucleus. If now the radon nucleus undergoes a transition from this excited state to the lower energy state, a gamma-ray photon of energy

$$(48.79 - 46.95) \times 10^5 = 1.84 \times 10^5 \text{ eV}$$

should be emitted. The *measured* energy of the gamma-ray photon emitted by radium is $1.89 = 10^5$ eV, in excellent agreement.

Thus, by correlating alpha-particles energies and gamma-ray energies, it is possible in some cases to construct *nuclear energy-level diagrams* similar to atomic energy-level diagrams.

When a radioactive nucleus decays by alpha or beta emission, the resulting nucleus may also be unstable, and there may be a chain of successive decays until a stable configuration is reached. The most abundant radioactive nucleus found in nature is that of uranium $_{92}U^{238}$, which undergoes a series of 14 decays, including eight alpha emissions and six beta emissions, terminating at the stable isotope of lead, $_{82}Pb^{206}$. Decay series are discussed further in Sec. 47-5.

In alpha decay the neutron number N and the charge number Z each decrease by two, and the mass number A decreases by four, corresponding to the values $N = 2, Z = 2, A = 4$ for the alpha particle. The situation in beta decay is less obvious; one might well question how a nucleus composed of protons and neutrons can emit an *electron*. The answer is that, in beta decay, a neutron in the nucleus is transformed into a proton, an electron, and a neutrino. Such transformations of fundamental particles are discussed further in Sec. 47-10. The effect is to *increase* the charge number Z by one, decrease the neutron number N by one, and leave the mass number A unchanged. Finally, gamma emission leaves all three numbers unchanged. Both alpha and beta emission are often accompanied by gamma emission.

47-4 Nuclear stability

Of about 1500 different nuclides now known, only about one-fifth are stable. The others are radioactive, with lifetimes ranging from a small fraction of a second to many years. The stable nuclei are indicated by dots on the graph in Fig. 47-5, where the neutron number N is plotted against the charge number Z. Such a graph is called a *Segre chart,* after its inventor.

Since the mass number A is the sum $N + Z$, a curve of constant A is a straight line perpendicular to the line $N = Z$. In general, lines of constant A pass through only one or two stable nuclei (see $A = 20, A = 40, A = 60$, etc.), but there are four cases when such lines pass through three stable isotopes, namely at $A = 96, 124, 130,$ and 136. Only four stable nuclei have both odd Z and odd N:

$$_1H^2, \quad _3Li^6, \quad _5B^{10}, \quad _7N^{14};$$

these are called *odd-odd nuclei*. Also, there is *no* stable nucleus with $A = 5$ or $A = 8$.

47-5 Segre chart, showing neutron number and proton number for stable nuclei.

The points representing stable isotopes define a rather narrow stability region. For low mass numbers $N/Z = 1$. This ratio increases and becomes about 1.6 at large mass numbers. Points to the right of the stability region represent nuclei that have an excess of protons or a deficiency of neutrons; to the left of the stability region are the points representing nuclei with an excess of neutrons or a deficiency of protons. The graph also shows that there is a maximum value of A; no nucleus with A greater than 209 is stable.

The stability of nuclei can be understood qualitatively on the basis of the nature of the nuclear force and the competition between the attractive nuclear force and the repulsive electrical force. As pointed out in Sec. 47-2, the nuclear force favors pairs of nucleons, and pairs of pairs. In the absence of electrical interactions, the most stable nuclei would be those having equal numbers of neutrons and protons, $N = Z$. The electrical repulsion shifts the balance to favor greater numbers of neutrons, but a nucleus with *too many* neutrons is unstable because not enough of them are paired with protons. A nucleus with too many *protons* has too much repulsive electrical interaction, compared with the attractive nuclear interaction, to be stable.

Furthermore, as the number of nucleons increases, the total energy of electrical interaction increases faster than that of the nuclear interaction. To understand this, we recall the discussion of electrostatic energy in Sec. 27-4. The energy of a capacitor with a charge Q is proportional to Q^2. It can be shown that to place a total charge Q on the surface of a sphere of radius a requires a total energy $Q^2/8\pi\epsilon_0 a$.

Thus the (positive) electric potential energy in the nucleus increases approximately as Z^2, while the (negative) nuclear potential energy increases approximately as A, with corrections for pairing effects. Thus the competition of electric and nuclear forces accounts for the fact that the neutron–proton ratio in stable nuclei increases with Z, and also for the fact that there is a maximum A (and a maximum Z) for stability. At large A the electric energy *per nucleon* grows faster than the nuclear energy per nucleon, until the point is reached where stability is impossible. Unstable nuclei respond to these conditions in various ways; the next several sections discuss various types of decay of unstable nuclei.

47-5 Radioactive transformations

The number of radioactive nuclei in any sample of radioactive material decreases continuously as some of the nuclei disintegrate. The *rate* at which the number decreases, however, varies widely for different nuclei. Let N represent the number of radioactive nuclei in a sample at time t, and dN the number that undergo transformations in a short time interval dt. Since every transformation results in a *decrease* in the number N, the corresponding change in N is $-dN$ and the rate of change of N is $-dN/dt$. The larger the number of nuclei in the sample, the larger the number that will undergo transformations, so that the rate of change of N is proportional to N, or is equal to a constant λ multiplied by N. Therefore

$$\frac{dN}{dt} = -\lambda N. \tag{47-3}$$

If we denote by N_0 the number of nuclei at time $t = 0$, then the solution of this differential equation is an exponential function:

$$N = N_0 e^{-\lambda t}. \tag{47-4}$$

A graph of this function is shown in Fig. 47-6.

The *half-life* $T_{1/2}$ of a radioactive sample is defined as the time at which the number of radioactive nuclei has decreased to one-half the number at $t = 0$. At this time,

$$e^{-\lambda T_{1/2}} = \tfrac{1}{2}.$$

47-6 Decay curve for the radioactive element polonium. Polonium has a half-life of 140 days.

Taking natural logarithms of both sides and solving for $T_{1/2}$, we find

$$\lambda T_{1/2} = \ln 2,$$
$$T_{1/2} = \frac{\ln 2}{\lambda} = \frac{0.693}{\lambda}. \qquad (47\text{-}5)$$

Half of the original nuclei in a radioactive sample decay in a time interval $T_{1/2}$, half of those remaining at this time decay in a second interval $T_{1/2}$, and so on; thus the fraction remaining after successive intervals of $T_{1/2}$ is $\frac{1}{2}, \frac{1}{4}, \frac{1}{8}$, and so on. We also note that when $t = 1/\lambda$, N has decreased to $1/e$ of its initial value.

The *activity* of a sample is defined to be the number of disintegrations per unit time. A commonly used unit is the *curie,* defined to be 3.70×10^{10} decays per second. This is approximately equal to the activity of one gram of radium. Since the number of disintegrations is proportional to the number of radioactive nuclei in the sample, the activity decreases exponentially with time in the same way as the number N. Figure 47-6 is a graph of the activity of polonium, $_{84}\text{Po}^{210}$, which has a half-life of 140 days.

In studying radioactivity the following questions are relevant:

1. What is the parent nucleus?
2. What particle is emitted from this nucleus?
3. What is the half-life of the parent nucleus?
4. What is the resulting nucleus (often called the *daughter nucleus*)?
5. Is the daughter nucleus radioactive and if so, what are the answers to questions 2, 3, and 4 for this nucleus?

Very extensive investigations have been carried on in the last 75 years, and these questions have been answered for many nuclides. The results are most conveniently presented on a Segre chart such as that shown in Fig. 47-7. The neutron number N is plotted along the vertical axis and the atomic number (or charge number) Z on the horizontal axis. Unit increase of Z with unit decrease of N indicates beta emission; decrease of two in both N and Z indicates alpha emission. The half-lives are given either in years (y), days (d), hours (h), minutes (m), or seconds (s).

Figure 47-7 shows the uranium decay series, which begins with the common uranium isotope U^{238}. Each arrow represents a decay in which an alpha or beta particle is emitted. The decays can also be represented in equation form; the first two decays in the series are written as

$$U^{238} \longrightarrow Th^{234} + \alpha,$$
$$Th^{234} \longrightarrow Pa^{234} + \beta,$$

or, even more briefly, as

$$U^{238} \xrightarrow{\alpha} Th^{234},$$
$$Th^{235} \xrightarrow{\beta} Pa^{234}.$$

In the second decay a gamma photon is emitted following the beta; the reason for this is that the beta decay leaves the daughter nucleus not in

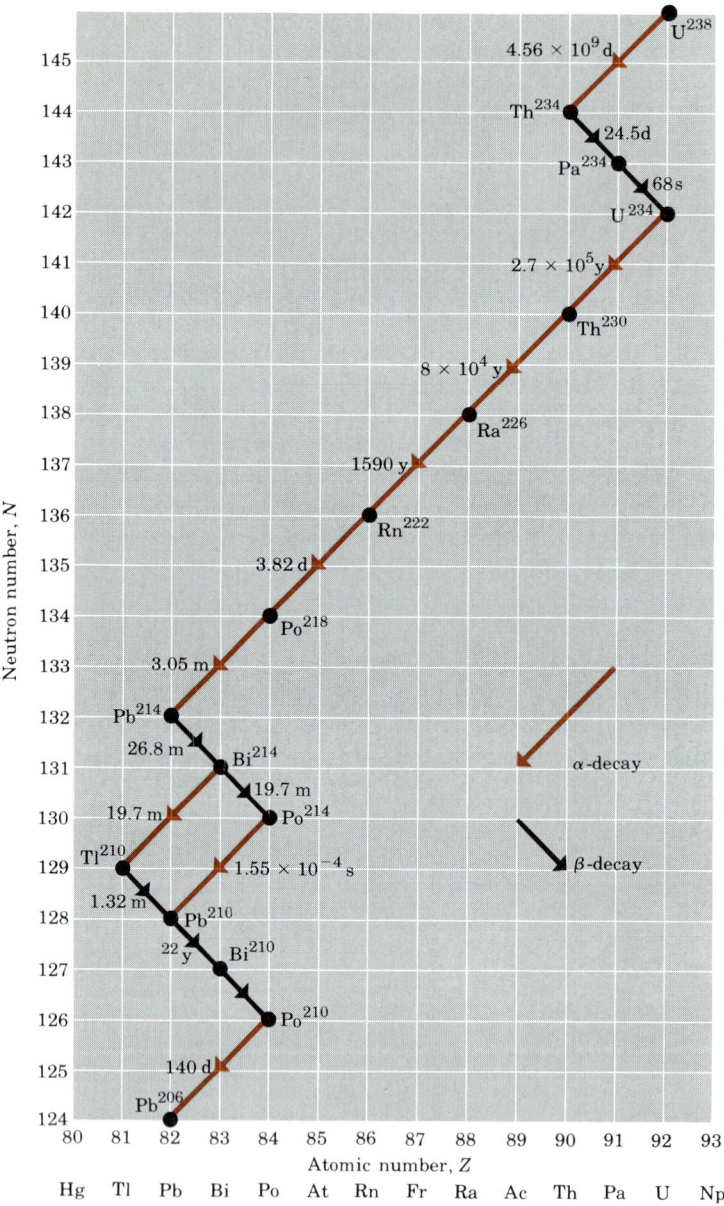

47-7 Segre chart showing the uranium U^{238} decay series, terminating with the stable nuclide $_{82}Pb^{206}$.

its lowest energy state but in an excited state, from which it decays to the ground state by emitting a photon.

An interesting feature of the U^{238} decay series is the branching that occurs at Bi^{214}; this nuclide decays to Pb^{210} by emission of an alpha and a beta, which can occur in either order. We also note that the series includes unstable isotopes of several elements that also have stable isotopes, including Tl, Pb, and Bi. The unstable isotopes of these elements that occur in the U^{238} series all have too many neutrons to be stable, as discussed in Sec. 47-4.

Three other decay series are known; two of these occur in nature, one starting with the uncommon isotope U^{235}, the other with thorium

(Th^{232}). The last series starts with neptunium (Np^{237}), an element not found in nature but produced in nuclear reactors. In each case the series continues until a stable nucleus is reached; for these series the final members are Pb^{206}, Pb^{207}, Pb^{208}, and Bi^{209}, respectively.

An interesting application of radioactivity is the dating of archeological and geological specimens by measurement of the concentration of radioactive isotopes. The most familiar example is carbon dating; C^{14}, an unstable isotope, is produced by nuclear reactions in the atmosphere caused by cosmic-ray bombardment, and there is a small portion of C^{14} in the CO_2 in the atmosphere. Plants that obtain their carbon from this source contain the same proportion of C^{14} as the atmosphere, but when a plant dies it stops taking in carbon, and the C^{14} it has already taken in decays, with a half-life of 5,568 years. Thus by measuring the proportion of C^{14} in the remains, one can determine how long ago the organism died. Similar techniques are used with other isotopes for dating of geologic specimens. A difficulty with carbon dating is that the C^{14} concentration in the atmosphere changes with time, over long intervals.

47-6 Nuclear reactions

The nuclear disintegrations described up to this point have consisted exclusively of a natural, uncontrolled emission of either an alpha or beta particle. Nothing was done to initiate the emission, and nothing could be done to stop it. It occurred to Rutherford in 1919 that it ought to be possible to penetrate a nucleus with a massive high-speed particle such as an alpha particle and thereby either produce a nucleus with greater atomic number and mass number or induce an artificial nuclear disintegration. Rutherford was successful in bombarding nitrogen with alpha particles and obtaining, as a result, an oxygen nucleus and a proton, according to the reaction

$$_2He^4 + {}_7N^{14} \rightarrow {}_8O^{17} + {}_1H^1. \qquad (47\text{-}6)$$

Note that the sum of the initial atomic numbers is equal to the sum of the final atomic numbers, a condition imposed by conservation of charge. The sum of the initial mass numbers is also equal to the sum of the final mass numbers, but the initial rest mass is *not* equal to the final rest mass.

The difference between the rest masses corresponds to the *nuclear reaction energy*, according to the mass–energy relation $E = mc^2$. If the sum of the final rest masses *exceeds* the sum of the initial rest masses, energy is *absorbed* in the reaction. Conversely, if the final sum is less than the initial sum, energy is released in the form of kinetic energy of the final particles. (1 u = 931 MeV.)

For example, in the nuclear reaction represented by Eq. (47–6), the rest masses of the various particles, in u, are found from Table 47–2 to be

$$\begin{array}{ll} {}_2He^4 = 4.00260 \text{ u} & {}_8O^{17} = 16.99913 \text{ u} \\ {}_7N^{14} = 14.00307 \text{ u} & {}_1H^1 = 1.00783 \text{ u} \\ \hline \phantom{{}_7N^{14} = }18.00567 \text{ u} & \phantom{{}_1H^1 = }18.00696 \text{ u} \end{array}$$

(These values include nine electron masses in each case.) The total rest mass of the final products exceeds that of the initial particles by

0.00129 u, which is equivalent to 1.20 MeV. This amount of energy is *absorbed* in the reaction. If the initial particles do not have this much kinetic energy, the reaction cannot take place.

On the other hand, in the proton bombardment of lithium and consequent formation of two alpha particles,

$$_1H^1 + {_3Li^7} \rightarrow {_2He^4} + {_2He^4}, \qquad (47\text{-}7)$$

the sum of the final masses is *smaller* than the sum of the initial values, as shown by the following data:

$$\begin{array}{ll} _1H^1 = 1.00783 \text{ u} & _2He^4 = 4.00260 \text{ u} \\ _3Li^7 = 7.01601 \text{ u} & _2He^4 = 4.00260 \text{ u} \\ \hline 8.02384 \text{ u} & 8.00520 \text{ u} \end{array}$$

(Four electron masses on each side are included.) Since the decrease in mass is 0.01864 u, 17.4 MeV of energy is liberated and appears as kinetic energy of the two separating alpha particles. This computation may be verified by observing the distance the alpha particles travel in air at atmospheric pressure before being brought to rest by collisions with molecules. The distance is found to be 8.31 cm. A series of independent experiments is then performed in which the range of alpha particles of known energy is measured. These experiments show that, in order to travel 8.31 cm, an alpha particle must have an initial kinetic energy of 8.64 MeV. The energy of the two alphas together is therefore $2 \times 8.64 = 17.28$ MeV, in excellent agreement with the value 17.4 MeV obtained from the mass decrease.

Alpha particles and protons are not the only particles used to initiate artificial nuclear disintegration. The nucleus of a deuterium atom, known as the *deuteron* and represented by the symbol $_1H^2$, may be accelerated to high energy in one of a variety of particle accelerators. In order that positively charged particles such as the alpha particle, the proton, and the deuteron can be used to penetrate the nuclei of other atoms, they must travel with very high speeds to avoid being repelled or deflected by the positive charge of the nucleus they are approaching.

Absorption of neutrons by nuclei is an important class of nuclear reactions. Heavy nuclei bombarded by neutrons in a nuclear reactor may undergo a series of neutron absorptions alternating with beta decays, in which the mass number A increases by as much as 25. The *transuranic* elements, having Z larger than 92, are produced in this way; they do not occur in nature. Thirteen transuranic elements, having Z up to 105 and A up to about 260, have been identified.

The analytical technique of *neutron activation analysis* uses similar reactions. When stable nuclei are bombarded by neutrons, some absorb neutrons and then undergo beta decay. The energies of the beta and gamma emissions depend on the parent nucleus and hence provide a means of identifying it. The presence of quantities of elements far too small for conventional chemical analysis can be detected in this way.

47-7 Particle accelerators

Many important experiments in nuclear and high-energy physics during the last 60 years or so have made use of beams of charged particles, such as protons or electrons, which have been accelerated to high speeds. Any device that uses electric and magnetic fields to guide and accelerate a beam of charged particles to high speed is called a *particle accelerator*. In a sense, the cathode-ray tubes of Thomson and his contemporaries were the first accelerators. In more recent times, accelerators have grown enormously in size, complexity, and energy range.

The *cyclotron*, developed in 1931 by Lawrence and Livingston at the University of California, is important historically because it was the first accelerator to use a magnetic field to guide particles in a nearly circular path so that they could be accelerated repeatedly by an electric field in a cyclic process. Lawrence was awarded the Nobel prize for physics in 1939 for his invention of the cyclotron.

The heart of the cyclotron is a pair of hollow metal chambers, labeled D_1 and D_2 in Fig. 47–8, resembling the halves of a hollow cylinder cut parallel to its axis. These are called *dees,* referring to their shape. A source of ions, often simply protons (ionized hydrogen atoms), is located near the midpoint of the gap between the dees, and the dees are connected to the terminals of a circuit generating an alternating voltage. The potential between the dees thus alternates rapidly, some millions of times per second; so the electric field in the gap between the dees is directed first toward one and then toward the other. But, because of the electrical shielding effect of the dees, the space *within* each is a region of nearly zero electric field.

47–8 Schematic diagram of a cyclotron.

The two dees are enclosed within, but insulated from, a somewhat larger cylindrical metal container from which the air is exhausted, and the whole apparatus is placed between the poles of a powerful electromagnet, which provides a uniform magnetic field perpendicular to the ends of the cylindrical container.

Consider an ion of charge $+q$ and mass m, emitted from the ion source S at an instant when D_1 in Fig. 47–8 is positive. The ion is accelerated by the electric field in the gap between the dees and then enters the region within D_2, where $E = 0$, with a speed v_1. Since its motion is at right angles to the magnetic field, the ion travels in a circular arc of radius

$$r_1 = \frac{mv_1}{Bq}.$$

If, now, in the time required for the ion to complete a half-circle, the *electric* field has reversed so that its direction is toward D_1, the ion is again accelerated as it crosses the gap between the dees and enters D_1 with a greater speed v_2. It therefore moves in a half circle of larger radius within D_1 to emerge again into the gap.

The *angular* velocity ω of the ion is

$$\omega = \frac{v}{r} = B\left(\frac{q}{m}\right). \tag{47-8}$$

Hence the angular velocity is *independent of the speed of the ion and of the radius of the circle* in which it travels, depending only on the magnetic field and the charge-to-mass ratio (q/m) of the ion. If, therefore, the electric field reverses at regular intervals, each equal to the time required for the ion to make a half-revolution, the field in the gap will always be in the proper direction to accelerate an ion each time the gap is crossed. It is this feature of the motion, that the time of rotation is independent of the radius, that makes the cyclotron feasible, since the regularly timed reversals are accomplished automatically by the oscillator circuit to which the dees are connected.

The path of an ion is a sort of spiral, composed of semicircular arcs of progressively larger radius connected by short segments along which the radius is increasing. If R represents the outside radius of the dees, and $v_{\max}$ the speed of the ion when traveling in a path of this radius, then

$$v_{\max} = BR\left(\frac{q}{m}\right),$$

and the corresponding kinetic energy of the ion is

$$K = \frac{1}{2}mv_{\max}^2 = \frac{1}{2}m\left(\frac{q}{m}\right)^2 B^2 R^2. \tag{47-9}$$

If the ions are protons,

$$\frac{q}{m} = \frac{1.60 \times 10^{-19} \text{ C}}{1.67 \times 10^{-27} \text{ kg}} = 9.58 \times 10^7 \text{ C} \cdot \text{kg}^{-1}.$$

Taking values typical of early-day cyclotrons, we assume $B = 1.5$ T and $R = 0.5$ m. Then the maximum kinetic energy is, from Eq. (47–9),

$$K = \tfrac{1}{2}(1.67 \times 10^{-27} \text{ kg})(9.58 \times 10^7 \text{ C} \cdot \text{kg}^{-1})^2 (1.5 \text{ T})^2 (0.5 \text{ m})^2$$
$$= 4.31 \times 10^{-12} \text{ J} = 2.69 \times 10^7 \text{ eV} = 26.7 \text{ MeV}.$$

This energy, considerably larger than the average binding energy per nucleon, is sufficient to cause a variety of interesting nuclear reactions.

The energy attainable with the cyclotron is limited by relativistic effects. For Eq. (47–8) to be relativistically correct, m should be replaced by $m/(1 - v^2/c^2)^{1/2}$. As the particles speed up, the angular velocity *decreases;* if the decrease is appreciable, the particle motion is no longer in the correct phase relative to the alternating dee voltage. In a variation called the *synchrocyclotron,* the particles are accelerated in bursts, and for each burst the frequency of the alternating voltage is decreased at just the right rate to maintain the correct phase relation with the particles' motion. A practical limitation of the cyclotron is the expense of building very large electromagnets. The largest synchrocyclotron ever built has a vacuum chamber about 8 m in diameter and accelerates protons to energies of about 600 MeV.

To attain higher energies, another type of machine, called the *synchrotron*, is more practical. In a synchrotron, the vacuum chamber in which the particles move is in the form of a thin doughnut, called the *accelerating ring*. The particles are forced to move within this chamber by a series of magnets placed around it. As the particles speed up, the magnetic field is increased so that the particles retrace the same trajectory over and over. The synchrotron located at the Fermi National Accelerator Laboratory (Fermilab) in Batavia, Illinois, can accelerate protons to an energy of 500 GeV (500×10^9 eV), and modifications are underway which will permit a maximum energy of 1000 GeV. It uses an accelerating ring 2 km in diameter, costs about $400 million to build, and has an operating budget of over $50 million per year. In each machine cycle, of a few seconds' duration, it accelerates approximately 10^{13} protons.

As higher and higher energies are sought in the attempt to investigate new phenomena in particle interactions, a new problem emerges. In an experiment where a beam of high-energy particles collides with a stationary target (such as protons onto a liquid-hydrogen target), not all the kinetic energy of the incident particles is available to form new particle states. Because momentum must be conserved, the particles emerging from the collision must have some motion and thus some kinetic energy. The energy E_a available for creating new particles or particle configurations is the *difference* between initial and final kinetic energies.

In the extreme-relativistic range, where the kinetic energies of the particles are large compared to their rest energies, this is a severe limitation. When bombarding and target particles have equal mass, as with protons onto a hydrogen target, it can be shown from relativistic mechanics that the available energy E_a is related to the total energy E of the bombarding particle and to its mass m by

$$E_a = \sqrt{2mc^2 E}. \qquad (47\text{-}10)$$

For example, for the proton, $mc^2 = 931$ MeV $= 0.931$ GeV. If $E = 500$ GeV, as for the Fermilab accelerator, then

$$E_a = \sqrt{2(0.931 \text{ GeV})(500 \text{ GeV})} = 30.5 \text{ GeV},$$

and if $E = 1000$ GeV, $E_a = 43.1$ GeV. Thus, increasing the beam energy by 500 GeV increases the available energy by only 12.6 GeV.

This limitation may be circumvented in part by *colliding-beam* experiments, in which there is no stationary target but in which beams of particles circulate in opposite directions in arrangements called *storage rings*. In regions where the rings intersect, the beams are focused sharply onto one another, and collisions can occur. Because the total momentum in such a two-particle collision is zero, the available energy E_a is the total kinetic energy of the two particles.

An example is the storage ring facility at the Stanford Linear Accelerator Center (SLAC), where electron and positron beams collide with total available energy of up to 36 GeV. Some other storage-ring facilities are located at DESY in Hamburg, West Germany (E_a up to 38 GeV in e^+e^- collisions), and at the Cornell Electron Storage Ring Facility (CESR), where the maximum energy *per beam* is 8 GeV. At Brookhaven

National Laboratory there is under construction a machine called ISABELLE that is designed to have 400 GeV per beam, in a proton–proton storage-ring facility with high intensity in each beam.

47-8 Nuclear fission

Up to this point, all nuclear reactions considered have involved the ejection of relatively light particles, such as alpha particles, beta particles, protons, or neutrons. That this is not always the case was discovered by Hahn and Strassman in Germany in 1939. These scientists bombarded uranium ($Z = 92$) with neutrons, and after a careful chemical analysis discovered barium ($Z = 56$) and krypton ($Z = 36$) among the products. Cloud-chamber photographs showed the two heavy particles traveling in opposite directions with tremendous speed. The uranium nucleus is said to undergo *fission*. Measurement showed that an enormous amount of energy, 200 MeV, is released when uranium splits up in this way. The rest mass of a uranium atom exceeds the sum of the rest masses of the fission products. Thus it follows, from the Einstein mass–energy relation, that the extra energy released during fission is transformed into kinetic energy of the fission fragments. Uranium fission may be accomplished by either fast or slow neutrons. Of the two most abundant isotopes of uranium, $_{92}U^{238}$ and $_{92}U^{235}$, both may be split by a fast neutron, whereas only $_{92}U^{235}$ is split by a slow neutron.

When uranium undergoes fission, barium and krypton are not the only products. Over 100 different isotopes of more than 20 different elements have been detected among fission products. All of these atoms are, however, in the middle of the periodic table, with atomic numbers ranging from 34 to 58. Because the neutron–proton ratio needed for stability in this range is much *smaller* than that of the original uranium nucleus, the *fission fragments,* as the residual nuclei are called, always have too many neutrons for stability. A few free neutrons are liberated during fission, and the fission fragments undergo a series of beta decays (each of which increases Z by one and decreases N by one) until a stable nucleus is reached. During decay of the fission fragments, an average of 15 MeV of additional energy is liberated.

Discovery of the facts that 200 MeV of energy is released when uranium undergoes fission, and that other neutrons are liberated from the uranium nucleus during fission, suggested the possibility of a *chain reaction,* that is, a self-sustaining series of events which, once started, will continue until much of the uranium in a given sample is used up (provided the sample stays together). In the case of a uranium chain reaction, a neutron causes one uranium atom to undergo fission, during which a large amount of energy and several neutrons are emitted. These neutrons then cause fission in neighboring uranium nuclei, which also give out energy and more neutrons. The chain reaction may be made to proceed slowly and in a controlled manner, and the device for accomplishing this is called a *nuclear reactor*. If the chain reaction is fast and uncontrolled, the device is a bomb.

In a nuclear reactor, the fissionable element is contained in *fuel elements,* whose configuration must be designed so that sufficient neutrons are slowed down in the surrounding material (often water) and cause further fissions, rather than escaping from the fuel region. On the average each fission produces about 2.5 free neutrons, so 40% of the neutrons are needed to sustain a chain reaction. The *rate* of the reaction is controlled by inserting or withdrawing *control rods* made of elements (often cadmium) whose nuclei *absorb* neutrons without undergoing any additional chain reaction.

The most familiar application of nuclear reactors is for the generation of electric power. To illustrate some of the numbers involved, we consider a hypothetical nuclear power plant with a generating capacity of 1000 MW; this figure is typical of large plants currently being built. As noted above, the fission energy appears as kinetic energy of the fission fragments, and its immediate result is to heat the fuel elements and the surrounding water. This heat generates steam to drive turbines, which in turn drive the electrical generators. The turbines, being heat engines, are subject to the efficiency limitations imposed by the second law of thermodynamics, as discussed in Chapter 19. In modern nuclear plants the overall efficiency is about one-third, so 3000 MW of thermal power from the fission reaction are needed for 1000 MW of electrical power.

It is easy to calculate how much uranium must undergo fission per unit time to provide 3000 MW of thermal power. Each second we need 3000 MJ or 3000×10^6 J. Each fission provides 200 MeV, which is

$$200 \text{ MeV} = (200 \text{ MeV})(1.6 \times 10^{-13} \text{ J} \cdot \text{MeV}^{-1}) = 3.2 \times 10^{-11} \text{ J}.$$

Thus the number of fissions needed per second is

$$\frac{3000 \times 10^6 \text{ J}}{3.2 \times 10^{-11} \text{ J}} = 0.94 \times 10^{20}.$$

Each uranium atom has a mass of about $(235)(1.67 \times 10^{-27} \text{ kg}) = 3.9 \times 10^{-25}$ kg, so the mass of uranium needed per second is

$$(0.94 \times 10^{20})(3.9 \times 10^{-25} \text{ kg}) = 3.7 \times 10^{-5} \text{ kg} = 37 \text{ mg}.$$

In one day (86,400 seconds) the consumption of uranium is

$$(3.7 \times 10^{-5} \text{ kg} \cdot \text{s}^{-1})(86,400 \text{ s} \cdot \text{d}^{-1}) = 3.2 \text{ kg} \cdot \text{d}^{-1}.$$

For comparison, we note that the 1000-MW coal-fired power plant discussed in Sec. 19–11 burns 10,600 tons of coal per day!

Nuclear fission reactors have several other practical uses; among these are the production of artificial radioactive isotopes for medical and other research; production of high-intensity neutron beams for research in nuclear structure, and production of fissionable transuranic elements such as plutonium from the common isotope U^{238}. The last is the function of so-called *breeder reactors*.

As noted above, about 15 MeV of the energy liberated as a result of fission of a U^{235} nucleus comes from the subsequent beta decay of the fission fragments rather than the kinetic energy of the fragments themselves. This fact poses a serious problem with respect to control and safety of reactors. Even after the chain fission reaction has been completely stopped by insertion of control rods into the core, heat continues to be evolved by the beta decays, which cannot be stopped. For the 1000-MW reactor described above, this heat power amounts to about 200 MW, which, in the event of total loss of cooling water, is more than enough to cause a catastrophic "melt-down" of the reactor core and possible penetration of the containment vessel. The difficulty in achieving a "cold shut-down" following the accident at Three Mile Island in March 1979 was due to the continued evolution of heat due to beta decays.

Fission appears to set an upper limit on the production of transuranic nuclei, discussed in Sec. 47-6. When a nucleus with $Z = 105$ is bombarded with neutrons, fission occurs essentially instantaneously; no $Z = 106$ nucleus is formed even for a short time. There are theoretical reasons to expect that nuclei in the vicinity of $Z = 114$, $N = 184$, might be stable with respect to spontaneous fission. These numbers correspond to *filled shells* in the nuclear energy-level structure, analogous to the filled shells of electrons in the inert gases, as discussed in Sec. 46-2. Such nuclei, called *superheavy nuclei,* would still be unstable with respect to alpha emission, but they might live long enough to be identified. Attempts to produce superheavy nuclei in the laboratory have not been successful; whether or not they exist in nature is still an open question.

47-9 Nuclear fusion

There are two types of nuclear reactions in which large amounts of energy may be liberated. In both types, the rest mass of the products is less than the original rest mass. The fission of uranium, already described, is an example of one type. The other involves the combination of two light nuclei to form a nucleus that is more complex but whose rest mass is less than the sum of the rest masses of the original nuclei. Examples of such energy-liberating reactions are as follows:

$$_1H^1 + {_1H^1} \rightarrow {_1H^2} + {_1e^0},$$

$$_1H^2 + {_1H^1} \rightarrow {_2He^3} + \gamma\text{-radiation},$$

$$_2He^3 + {_2He^3} \rightarrow {_2He^4} + {_1H^1} + {_1H^1}.$$

In the first, two protons combine to form a deuteron and a positron (a positively charged electron, to be discussed in Sec. 47–10). In the second, a proton and a deuteron unite to form the light isotope of helium. For the third reaction to occur, the first two reactions must occur twice, in which case two nuclei of light helium unite to form ordinary helium. These reactions, known as the *proton–proton chain,* are believed to take place in the interior of the sun and also in many other stars that are known to be composed mainly of hydrogen.

The positrons produced during the first step of the proton–proton chain collide with electrons; annihilation takes place, and their energy is converted into gamma radiation. The net effect of the chain, therefore, is the combination of four hydrogen nuclei into a helium nucleus and gamma radiation. The net amount of energy released may be calculated from the mass balance as follows:

Mass of four hydrogen atoms (including electrons)	$= 4.03132$ u
Mass of one helium plus two additional electrons	$= 4.00370$ u
Difference in mass	$= 0.02762$ u
	$= 25.7$ MeV

In the case of the sun, 1 g of its mass contains about 2×10^{23} protons. Hence, if all of these protons were fused into helium, the energy released would be about 57,000 kWh. If the sun were to continue to radiate at its present rate, it would take about 30 billion years to exhaust its supply of protons.

For fusion to occur, the two nuclei must come together to within the range of the nuclear force, typically of the order of 2×10^{-15} m. To do this they must overcome the electrical repulsion of their positive charges; for two protons at this distance the corresponding potential energy is of the order of 1.1×10^{-13} J or 0.7 MeV, which thus represents the initial *kinetic* energy the fusing nuclei must have.

Such energies are available at extremely high temperatures. According to Sec. 20-4, the average translational kinetic energy of a gas molecule at temperature T is $3kT/2$, where k is Boltzmann's constant. For this to be equal to 1.1×10^{-13} J, the temperature must be of the order of 5×10^9 K. Of course, not all the nuclei have to have this energy, but this calculation shows that the temperature must be of the order of millions of kelvins if any appreciable fraction of the nuclei are to have enough kinetic energy to surmount the electrical repulsion and achieve fusion.

Such temperatures occur in stars as a result of gravitational contraction and its associated liberation of gravitational potential energy. When the temperature gets high enough, the reactions occur, more energy is liberated, and the pressure of the resulting radiation prevents further contraction. Only after most of the hydrogen has been converted into helium will further contraction and an accompanying increase of temperature result. Conditions are then suitable for the formation of heavier elements.

Temperatures and pressures similar to those in the interior of stars may be achieved on earth at the moment of explosion of a uranium or plutonium fission bomb. If the fission bomb is surrounded by proper proportions of the hydrogen isotopes, these may be caused to combine into helium and liberate still more energy. This combination of uranium and hydrogen is called a "hydrogen bomb."

Intensive efforts are underway in many laboratories to achieve controlled fusion reactions, which potentially represent an enormous new energy resource. In one kind of experiment, a plasma is heated to extremely high temperature by an electrical discharge, while being contained by appropriately shaped magnetic fields. In another, pellets of the material to be fused are heated by a high-intensity laser beam. Reactions being studied include the following:

$$_1\text{H}^2 + {_1\text{H}^2} \rightarrow {_1\text{H}^3} + {_1\text{H}^1} + 4 \text{ MeV}, \tag{1}$$

$$_1\text{H}^3 + {_1\text{H}^2} \rightarrow {_2\text{He}^4} + {_0\text{n}^1} + 17.6 \text{ MeV}, \tag{2}$$

$$_1\text{H}^2 + {_1\text{H}^2} \rightarrow {_2\text{He}^3} + {_0\text{n}^1} + 3.3 \text{ MeV}, \tag{3}$$

$$_2\text{He}^3 + {_1\text{H}^2} \rightarrow {_2\text{He}^4} + {_1\text{H}^1} + 18.3 \text{ MeV}. \tag{4}$$

In the first, two deuterons combine to form tritium and a proton. In the second, the tritium nucleus combines with another deuteron to form helium and a neutron. The result of both of these reactions together is the conversion of three deuterons into a helium-4 nucleus, a proton, and a neutron, with the liberation of 21.6 MeV of energy. Reactions (3) and (4) together achieve the same conversion. In a plasma containing deuterium, the two pairs of reactions occur with roughly equal probability. As yet no one has succeeded in producing these reactions under controlled conditions in such a way as to yield a net surplus of usable energy, but the practical problems do not appear to be insurmountable.

47-10 Fundamental particles

The physics of fundamental particles has been a recognized field of research only in the past 40 years. The electron and the proton were known by the turn of the century, but the existence of the neutron was not established definitely until 1930; its discovery is an interesting story and a useful illustration of nuclear reactions.

In 1930, Bothe and Becker in Germany observed that when beryllium, boron, or lithium was bombarded by fast alpha particles, the bombarded material emitted something, either particles or electromagnetic waves, of much greater penetrating power than the original alpha particles. Further experiments in 1932 by Curie and Joliot in Paris confirmed these results, but all attempts to explain them in terms of gamma rays were unsuccessful. Chadwick in England repeated the experiments and found that they could be satisfactorily interpreted on the assumption that *uncharged* particles of mass approximately equal to that of the proton were emitted from the nuclei of the bombarded material. He called the particles *neutrons*. The emission of a neutron from a beryllium nucleus takes place according to the reaction

$$_2\text{He}^4 + {_4\text{Be}^9} \rightarrow {_6\text{C}^{12}} + {_0\text{n}^1}, \tag{47-10}$$

where $_0\text{n}^1$ is the symbol for a neutron.

Since neutrons have no charge, they produce no ionization in their passage through gases. They are not deflected by the electric field around a nucleus and can be stopped only by colliding with a nucleus in a direct hit, in which case they may either undergo an elastic impact or penetrate the nucleus. It was shown in Chapter 7 that, if an elastic body strikes a motionless elastic body of the same mass, the first is stopped and the second moves off with the same speed as the first. Since the proton and neutron masses are almost the same, fast neutrons are slowed down most effectively by collisions with the hydrogen atoms in hydrogenous materials such as water or paraffin. The usual laboratory method of obtaining slow neutrons is to surround the fast neutron source with water or blocks of paraffin.

Once the neutrons are moving slowly, they may be detected by means of the alpha particles they eject from the nucleus of a boron atom, according to the reaction

$$_0\text{n}^1 + {_5\text{B}^{10}} \rightarrow {_3\text{Li}^7} + {_2\text{He}^4}. \tag{47-11}$$

The ejected alpha particle then produces ionization that may be detected in a Geiger counter or an ionization chamber.

The discovery of the neutron gave the first real clue to the structure of the nucleus. Before 1930 it had been thought that the total mass of a nucleus was due to protons only. We now know that a nucleus consists of both protons and neutrons (except hydrogen, whose nucleus consists only of one proton) and that (1) the mass number A equals the total number of nuclear particles and (2) the atomic number Z equals the number of protons.

The study of *cosmic rays* has been a very fertile field of particle physics. It has been known since the early years of this century that air and other gases are slightly ionized, and therefore slightly conductive, at all times, even in the absence of obvious causes of ionization such as x-rays, ultraviolet light, or radioactivity. For example, if a charged electroscope is left standing, it will eventually lose its charge no matter how well it is insulated. Ionization of air inside a vessel is decreased slightly if the vessel is lowered into a lake, but increases considerably if the vessel is transported in a balloon high into the stratosphere. Hess suggested that the ionization is due to some kind of penetrating waves or particles from outer space, and called them *cosmic rays*.

It is fairly certain that cosmic rays consist largely of high-speed protons with energies of the order of billions of electron volts. A collision between such a proton and the nucleus of a nitrogen or oxygen atom in the upper atmosphere gives rise to so many interesting secondary phenomena that the study of cosmic rays has become one of the richest sources of knowledge of the behavior and properties of fundamental particles.

An instrument called the *bubble chamber* has been used extensively not only to study cosmic rays but also to render visible the paths of particles involved in nuclear and fundamental-particle reactions; this instrument was described in Sec. 17–9. Bubble chambers are sometimes made several feet in diameter; particle tracks are photographed with two or three cameras to obtain stereoscopic pictures. Normally the chamber is placed in a magnetic field so that a charged particle will be deflected; by measuring the radius of curvature of the path we may determine the particle's momentum. The *cloud chamber,* also described in Sec. 17–9, was an important historical predecessor of the bubble chamber.

The positive electron or *positron* was first observed during the course of an investigation of cosmic rays by Dr. Carl D. Anderson in 1932, in a historic cloud-chamber photograph reproduced in Fig. 47–9. The photograph was made with the cloud chamber in a magnetic field perpendicular to the plane of the paper. A lead plate crosses the chamber and evidently the particle has passed through it. Since the curvature of the track is greater above the plate than below it, the velocity is less above than below; the inference is that the particle was moving upward, since it could not have *gained* energy going through the lead.

The *density* of droplets along the path is the same as would be expected if the particle were an electron. But the direction of the magnetic field and the direction of motion are consistent only with a particle of *positive* charge. Hence Anderson concluded that the track had been made by a positive electron or *positron*. Since the time of this discovery, the positron's existence has been definitely established. Its mass equals that of a negative electron and its charge is equal in magnitude but of opposite sign to that of the electron.

Positrons have only a transitory existence and do not form a part of ordinary matter. They are produced in high-energy collisions of charged particles or gamma photons with matter in a process called *pair production,* in which an ordinary electron and a positron are produced simultaneously. Electric charge is conserved in this process, but enough energy must be available to account for the energy equivalent of the rest masses

47-9 Track of a positive electron traversing a lead plate 6 mm thick. (Photograph by C. D. Anderson.)

of the two particles, about 0.5 MeV each. The inverse process, e^+e^- *annihilation,* occurs when a positron collides with an ordinary electron. Both particles disappear and two or three gamma photons appear, with total energy $2mc^2$, where m is the electron rest mass. Decay into a *single* gamma photon is impossible because such a process cannot possibly conserve both energy and momentum.

Positrons also occur in the decay of some unstable nuclei. We recall that nuclei having too many neutrons for stability often emit a beta particle (electron), decreasing N by one and increasing Z by one. Similarly, a nucleus having *too few neutrons* for stability may respond by emitting a positron, increasing N by one and decreasing Z by one. Such nuclides do not occur in nature, but they can be produced artificially by neutron bombardment of stable nuclides in nuclear reactors. An example is the unstable nuclide $_{11}\text{Na}^{22}$, which has one less neutron than the stable $_{11}\text{Na}^{23}$. It emits a positron, leaving the stable nuclide $_{10}\text{Ne}^{22}$.

In 1935, the Japanese physicist Hideki Yukawa inferred, from theoretical considerations, the existence of a particle of mass intermediate between that of the electron and the proton. A particle of intermediate mass, but *not* identical with that predicted by Yukawa, was discovered one year later by Anderson and Neddermeyer as a component of cosmic radiation. This particle is now known as a μ *meson* (or *muon*). The μ^- has charge equal to that of the electron, and its antiparticle the μ^+ has a positive charge of equal magnitude. The two particles have equal mass, about 207 times the electron mass. The muons are unstable; each decays into an electron of the same sign, plus two neutrinos, with a half-life of about 2.3×10^{-6} s.

Yukawa first proposed the mesons as a basis for nuclear forces; he suggested that nucleons could interact by emitting and absorbing unstable particles, just as two basketball players interact by tossing the ball back and forth, or by snatching it away from each other. It was established soon after the discovery of the muons that they could not be Yukawa's particles because their interactions with nuclei were far too weak. But in 1947 another family of mesons was discovered; called π *mesons* or *pions,* these can be positive, negative, and neutral. The charged pions have masses of about 273 times the electron mass and decay into muons with the same sign, plus a neutrino, with a half-life of about 2.6×10^{-8} s. The neutral pion has a smaller mass, about 264 electron masses, and decays, with a very short half-life of about 0.8×10^{-16} s, into two gamma-ray photons.

In the years since 1947, *high-energy physics* has emerged as a distinct branch of physics. These years have witnessed the attainment of higher and higher energies in particle accelerators, the discovery of a whole array of new particles, and intensive efforts to understand the properties of these new particles and their interactions.

*47-11 High-energy physics

It was recognized, even in the early years of high-energy physics, that the fundamental particles are not *permanent* entities but can be created and destroyed in interactions with other particles. The earliest such interaction to be observed was that of creation and destruction of electron–positron pairs. Such pairs are *created* in collisions of high-energy cosmic-ray particles with stationary targets; when an electron and a positron collide, both *disappear* and two or three gamma-ray photons are created to carry away the energy. This transitory nature of the fundamental particles may seem disturbing, but in one sense it is a welcome development. We have seen that photons and electrons (and indeed all particles) share the dual wave-particle nature discussed in Sec. 45-2, and photons are known to be created and destroyed (or emitted and absorbed) in atomic transitions. Thus it seems natural that other particles can also be created and destroyed.

As an example, it was speculated as early as 1932 that there might be an *antiproton,* bearing the same relation to the ordinary proton as the positron does to the electron, that is, a particle with the same mass as the proton but negatively charged. Finally in 1955 proton–antiproton pairs were created by impact on a stationary target of a beam of protons with kinetic energy 6 GeV (6×10^9 eV) from the Bevatron at the University of California at Berkeley.

In the years after 1960, as higher-energy accelerators and more sophisticated detectors were developed, a veritable blizzard of new unstable particles occurred. Initially they were classified according to *mass.* The electrons, muons, and their associated neutrinos were called *leptons;* particles with masses between those of muons and nucleons were called *mesons,* and the nucleons and more massive particles were called *baryons.* Particles were further classified according to electric charge, spin, and two additional quantum numbers, *isospin* (the number determining the number of different charges for a particular type of particle) and *strangeness* (a number needed to account for the production and decay modes of certain particles). A partial list of some of the particles known around 1965 is shown in Table 47–3.

During this period it became clear that particles could also be classified in terms of the types of *interactions* in which they participate, and in terms of the conservation laws associated with these interactions. There appear to be four classes of interactions; in order of decreasing strength, they are

1. Strong interactions;
2. Electromagnetic interactions;
3. Weak interactions;
4. Gravitational interactions.

Particles which experience strong interactions are called *hadrons;* these include all the mesons and baryons in Table 47–3. The strong interactions are responsible for the emission and absorption of pions and heavy mesons by nucleons, and hence for the nuclear force. They are also responsible for the *creation* of pions, heavy mesons, and hyperons in high-energy collisions. Electrons, muons, and neutrinos, collectively called *leptons,* have *no* strong interactions.

The *electromagnetic* interactions are those associated directly with electric charge; as noted previously, the electromagnetic interaction between two protons is weaker at distances of the order of nuclear dimensions than the strong interaction, but it has longer range. Neutral particles have no electromagnetic interactions, with the exception of effects due to the magnetic moments of neutral baryons; these magnetic mo-

Table 47-3 Some particles known in 1965

	Particle	Mass (MeV)	Charge	Spin	Isospin	Strange-ness	Lifetime (s)	Typical decay modes	Quark content
leptons	e^-	0.511	−1	$\frac{1}{2}$	—	0	stable	—	—
	ν_e	0	0	$\frac{1}{2}$	—	0	stable	—	—
	μ^-	105.7	−1	$\frac{1}{2}$	—	0	2.2×10^{-6}	$e^- \bar{\nu}_e \nu_\mu$	—
	ν_μ	0	0	$\frac{1}{2}$	—	0	stable	—	—
	τ^-	1784	−1	$\frac{1}{2}$	—	0	$<2.3 \times 10^{-12}$	$\mu^- \bar{\nu}_\mu \nu_\tau$	—
	ν_τ	0	0	$\frac{1}{2}$	—	0	stable	—	—
mesons	π^0	135.0	0	0	1	0	0.83×10^{-16}	$\gamma\gamma$	$(u\bar{u} - d\bar{d})/\sqrt{2}$
	π^+	139.6	+1	0	1	0	2.6×10^{-8}	$\mu^+ \nu_\mu$	$u\bar{d}$
	π^-	139.6	−1	0	1	0	2.6×10^{-8}	$\mu^- \bar{\nu}_\mu$	$\bar{u}d$
	K^+	493.7	+1	0	$\frac{1}{2}$	+1	1.24×10^{-8}	$\mu^+ \nu_\mu$	$u\bar{s}$
	K^-	493.67	−1	0	$\frac{1}{2}$	−1	1.24×10^{-8}	$\mu^- \bar{\nu}_\mu$	$\bar{u}s$
	η^0	548.8	0	0	0	0	$\sim 10^{-18}$	$\gamma\gamma$	$(u\bar{u} + d\bar{d} - 2s\bar{s})/\sqrt{6}$
baryons	p	938.3	+1	$\frac{1}{2}$	$\frac{1}{2}$	0	stable	—	uud
	n	939.6	0	$\frac{1}{2}$	$\frac{1}{2}$	0	917	$pe^- \bar{\nu}_e$	udd
	Λ	1115	0	$\frac{1}{2}$	0	−1	2.63×10^{-10}	$p\pi^-$ or $n\pi^0$	uds
	Σ^+	1189	+1	$\frac{1}{2}$	1	−1	0.80×10^{-10}	$p\pi^0$ or $n\pi^+$	uus
	Δ^{++}	1232	+2	$\frac{3}{2}$	$\frac{3}{2}$	0	$\sim 10^{-23}$	$p\pi^+$	uuu
	Ξ^-	1321	−1	$\frac{1}{2}$	$\frac{1}{2}$	−2	1.64×10^{-10}	$\Lambda \pi^-$	dss
	Ω^-	1672	−1	$\frac{3}{2}$	0	−3	0.82×10^{-10}	ΛK^-	sss
	Λ_c^+	2273	1	$\frac{1}{2}$	0	0	$\sim 7 \times 10^{-13}$	$\Lambda \pi \pi \pi$	udc

ments are believed to be associated with the emission and absorption of charged pions and heavy mesons.

The *weak* interaction is responsible for beta decay such as the conversion of a neutron into a proton, an electron, and a neutrino. It is also responsible for the *decay* of many unstable particles (pions into muons, muons into electrons, Λ particles into protons, and so on). The *gravitational* interaction, although of central importance for the large-scale structure of celestial bodies, is not believed to be of significance in the analysis of fundamental-particle interactions. For example, the gravitational attraction of two electrons is smaller than their electrical repulsion by a factor of about 2.4×10^{-43}.

Several conservation laws are believed to be obeyed by *all* of the above interactions. These include the laws growing out of classical physics: energy, momentum, angular momentum, and electric charge. In addition, several new quantities having no classical analog have been introduced to help characterize the properties of particles. These include *baryon number* (the number of baryons minus the number of antibaryons), *isospin* (used also to describe the charge-independence of nuclear forces), *parity* (the comparative behavior of two systems that are mirror images of each other), and *strangeness* (a quantum number used to classify particle production and decay reactions). Baryon number is conserved in *all* interactions; isospin is conserved in strong but not in electromagnetic or weak interactions. Parity and strangeness are conserved in strong and electromagnetic but not in weak interactions. Thus, the new conservation laws are not absolute but, instead, serve as a means for *classifying* interactions.

The large number of supposedly fundamental particles discovered in the last twenty years (well over a hundred) suggests strongly that these particles *do not* represent the most fundamental level of the structure of matter, but that there is at at least one additional level of structure. There is now fairly general agreement among physicists concerning the nature of this level; the theory is based on a proposal made initially in 1964 by Gell-Mann and his collaborators. We cannot discuss this theory in detail, but the following is a very brief sketch of some of its features.

Leptons are indeed fundamental particles. In addition to the electrons and muons and their associated neutrinos, there is a third kind of lepton called the τ, having spin 1/2 and mass of about 1780 MeV. Its associated neutrino is the τ-neutrino, ν_τ. These six leptons are definitely established; there may be others, yet undiscovered.

Hadrons are *not* fundamental particles but are composite structures whose constituents are truly fundamental particles called *quarks*. In the original quark picture there were three quarks labeled **u**, **d**, and **s** ("up," "down," and "strange"), with three corresponding antiquarks $\bar{u}$, $\bar{d}$, and $\bar{s}$. Some properties of these quarks, and of three others to be discussed later, are shown in Table 47–4. We note that the charge and baryon number of each quark are fractions, unlike any particle observed in nature.

Each meson is a quark–antiquark pair; for example, the π^+ is $u\bar{d}$, and the K^+ is $u\bar{s}$. Each baryon is a combination of three quarks; the proton is **uud**, the neutron is **udd**, the Σ^+ is **uus**, and so on. The quark structure of each of the hadrons listed in Table 47–3 is shown in the last column of that table. Leptons, on the other hand, are *not* made of quarks, and no quark structure is shown for them.

Quarks are expected to be very massive, although their masses are not known at present. They are expected to have a very strong affinity for each other, so individual quarks should be seen at most rarely. As noted at the end of Sec. 26–6, there has been at least one report of observation of a fractionally charged particle. On the other hand, many theo-

Table 47-4 Properties of quarks

Quark	Charge	Baryon number	Strangeness	Isospin	Charm
u	$\frac{2}{3}$	$\frac{1}{3}$	0	$\frac{1}{2}$	0
d	$-\frac{1}{3}$	$\frac{1}{3}$	0	$\frac{1}{2}$	0
s	$-\frac{1}{3}$	$\frac{1}{3}$	-1	0	0
c	$\frac{2}{3}$	$\frac{1}{3}$	0	0	1
b	$-\frac{1}{3}$	$\frac{1}{3}$	0	0	0
t	$\frac{2}{3}$	$\frac{1}{3}$	0	0	0

For each antiquark, the values of charge, baryon number, strangeness, and charm have the same magnitude but opposite sign, compared to the corresponding quark. States involving the *t* quark have not yet been observed.

rists believe that phenomena associated with creation of quark–antiquark pairs make it impossible *ever* to observe an individual quark. Thus the direct observation of a single quark is still an open question.

In the quark model the strong interaction between quarks is mediated by particles called *gluons,* which play a role in the strong interaction analogous to that of *photons* for the electromagnetic interaction. The weak and gravitational interactions also have corresponding mediating particles, called the *weak boson* and the *graviton,* respectively.

Although three quarks sufficed to describe the particles known in 1970, more recent discoveries require the existence of additional quarks. It is now believed that there are six quarks, each existing in three varieties or *colors.* The three additional quarks are called the *charmed* quark c and the *top* t and *bottom* b (or *truth* and *beauty*) quarks. The existence of c was discovered in 1974 in experiments at Brookhaven and at SLAC, where the Ψ particle (3095 MeV) and some of its excited states were observed. These are believed to be states of the system $c\bar{c}$. Subsequently, charmed mesons such as D^+ (a $c\bar{d}$ state) have been observed through their decay products. More recently, evidence has been found at Fermilab for the existence of $b\bar{b}$ states. States involving t have not yet been observed.

The fact that all known particles can be represented as either leptons or combinations of these few quarks is indeed reason to believe that they are truly fundamental particles. Experimental and theoretical work is proceeding in many different directions. In a theoretical scheme called the Weinberg–Salam model, the weak and electromagnetic interactions are represented as two aspects of a single fundamental interaction. Some theorists are searching for a "grand unification scheme" that would unify *all* of the four basic interactions. Although such schemes are still speculative in nature, the entire area is a very active field of present-day theoretical and experimental research.

47-12 Radiation and the life sciences

The interaction of radiation with living organisms is a topic that grows daily in interest and usefulness. In this context we construe *radiation* to include radiation emitted as a result of nuclear instability (alpha, beta, gamma, and neutrons) as well as electromagnetic radiation such as microwaves and x-rays. Space permits only brief mention of a few examples; the two general classes of phenomena to be considered here are: (1) the use of radioactive isotopes as an analytical tool, and (2) the beneficial and harmful effects of radiation in changing living tissue.

Radioactive isotopes of an element have the same electron configuration as the stable isotopes and therefore exhibit the same chemical behavior. The location and concentration of radioactive isotopes can, however, be detected easily, even at a distance, by measuring the radiation emitted. For example, an unstable isotope of iodine, I^{131}, can be used to study thyroid function. It is known that nearly all the iodine in food that is not eliminated eventually reaches the thyroid; by feeding the patient minute, measured quantities of I^{131} and subsequently measuring the radiation from the thyroid, one can measure the activity of this organ.

More subtle applications occur in complex chemical reactions. By use of radioactive tracers one can "tag" specific parts of molecules and "follow" the radioactive atoms through complex reactions.

A different analytical technique, neutron activation analysis, makes use of the fact that when stable nuclei are bombarded with neutrons, some nuclei absorb neutrons, becoming radioactive beta-emitters. Each element has characteristic energies for the gamma radiation that usually follows beta emission, and measurement of these energies permits determination of the original elements present, even if the quantities are extremely minute.

The interactions of radiation with living tissue are extremely complex. It has been known for many years that excessive exposure to radiation, including sunlight, x-rays, and all the nuclear radiations, can cause destruction of tissues. In mild cases this destruction is manifested as a burn, as with common sunburn; greater exposure can cause very severe illness or death by a variety of mechanisms, one of which is the destruction of the components in bone marrow that produce red blood cells.

On the other hand, radiation is present everywhere from sunlight, cosmic rays, and natural radioactivity, and some exposure to radiation is unavoidable. Exactly what constitutes a *safe* level of radiation exposure is open to considerable question, but the available evidence has been interpreted to show that exposure to the extent of 10 to 100 times that from natural sources is very rarely harmful.

There has been a great deal of hysteria concerning alleged radiation hazards from nuclear power plants. It is certainly true that the radiation level from these plants is *not* zero. But to make a meaningful evaluation of hazards one must compare these levels with the alternatives, such as coal-powered plants. The health hazards of coal smoke are serious and well documented, and even the radioactivity in the smoke from a coal-fired power plant is believed to be greater than that from a properly operating nuclear plant of similar power capacity. It is clearly impossible to eliminate *all* hazards to health, and the next best alternative is an intelligent approach to the problem of *minimizing* hazards.

Radiation also has many *beneficial* effects; the great usefulness of x-rays in medical diagnosis is well known, and there is no doubt that, when x-rays are used properly, the benefits from this diagnostic usefulness almost always outweigh the small radiation hazard. Higher-energy radiation is used for intentional selective destruction of tissue, such as cancerous tumors. The hazards are considerable, but if the disease would be fatal without treatment, considerable hazard may be tolerable.

For sources of radiation for the treatment of cancers and related diseases, artificially produced isotopes are often used. One of the most commonly used is an isotope of cobalt, Co^{60}. This is prepared by bombarding the stable isotope Co^{59} with neutrons in a nuclear reactor. Neutron absorption leads to the unstable Co^{60}; with $Z = 27$ and $N = 33$, this is an "odd–odd" nucleus, which decays to Ni^{60} by beta and gamma emission, with a half-life of about 5 years. Such artificial sources have several advantages over naturally radioactive sources. Having shorter half-lives, they are more *intense* sources. They do not emit alpha particles, which are usually not wanted, and the electrons emitted are easily stopped by thin metal sheets without appreciably attenuating the intensity of the desired gamma radiation.

Questions

47-1 How can one be sure that nuclei are not made of protons and electrons, rather than protons and neutrons?

47-2 In calculations of nuclear binding energies such as those in the examples of Sec. 47-2 and 47-6, should the binding energies of the *electrons* in the atoms be included?

47-3 If different isotopes of the same element have the same chemical behavior, how can they be separated?

47-4 In beta decay a neutron becomes a proton, an electron, and a neutrino. This decay also occurs with free neutrons, with a half-life of about 15 minutes. Could a free *proton* undergo a similar decay?

47-5 Since lead is a stable element, why doesn't the U^{238} decay series shown in Fig. 47-7 stop at lead, Pb^{214}?

47-6 In the U^{238} decay chain shown in Fig. 47-7, some nuclides in the chain are found much more abundantly in nature than others, despite the fact that every U^{238} nucleus goes through every step in the chain before finally becoming Pb^{206}. Why are the abundances of the intermediate nuclides not all the same?

47-7 Radium has a half-life of about 1600 years. If the universe was formed five billion or more years ago, why is there any radium left now?

47-8 Why is the decay of an unstable nucleus unaffected by the *chemical* situation of the atom, such as the nature of the molecule in which it is bound, and so on?

47-9 Fusion reactions, which liberate energy, occur only with light nuclei, and fission reactions only with heavy nuclei. A student asserted that this shows that the binding energy per nucleon increases with A at small A but decreases at large A and hence must have a maximum somewhere in between. Do you agree?

47-10 Nuclear power plants use nuclear fission reactions to generate steam to run steam-turbine generators. How does the nuclear reaction produce heat?

47-11 There are cases where a nucleus having too few neutrons for stability can capture one of the electrons in the K shell of the atom. What is the effect of this process on N, A, and Z? Is this the same effect as that of β^+ emission? Might there be situations where K-capture is energetically possible while β^+ emission is not? Explain.

47-12 Is it possible that some parts of the universe contain antimatter whose atoms have nuclei made of antiprotons and antineutrons, surrounded by positrons? How could we detect this condition without actually going there? What problems might arise if we actually *did* go there?

47-13 When x-rays are used to diagnose stomach disorders such as ulcers, the patient first drinks a thick mixture of (insoluble) barium sulfate and water. What does this do? What is the significance of the choice of barium for this purpose?

47-14 Why are there so many health hazards associated with fission fragments that are produced during fission of heavy nuclei?

Problems

47-1 Find the potential energy of an alpha particle 10^{-14} m away from a gold nucleus. Express your results in joules and in electronvolts.

47-2 A 4.7-MeV alpha particle from a radium Ra^{226} decay makes a head-on collision with a gold nucleus. What is the distance of closest approach of the alpha particle to the center of the nucleus?

47-3 The radius of the uranium U^{238} nucleus is about 7.4×10^{-15} m. What energy is required to push an alpha particle to the "surface" of this nucleus?

47-4 Compute the approximate density of nuclear matter, and compare your result with typical densities of ordinary matter.

47-5 How much energy would be required to add a proton to a nucleus with $Z = 91$ and $A = 234$? Express your results in joules, and in MeV.

47-6 Using the data in Table 47-2, calculate the binding energy of deuterium.

47-7 Calculate the mass defect, the binding energy, and the binding energy per nucleon, for the common isotope of oxygen, O^{16}.

47-8 Calculate the energy in MeV of an electron whose wavelength is of the order of nuclear dimensions, say 2×10^{-15} m. This calculation is one argument used to show that there cannot be any electrons in the nucleus.

47-9 Tritium is an unstable isotope of hydrogen, $_1H^3$; its mass, including one electron, is 3.01647 u.

a) Show that it must be unstable with respect to beta decay because $_2He^3$ plus an emitted electron has less total mass.

b) Determine the total kinetic energy of the decay products, taking care to account for the electron masses correctly.

47-10 A carbon specimen found in a cave believed to have been inhabited by cavemen contained $\frac{1}{8}$ as much C^{14} as an equal amount of carbon in living matter. Find the approximate age of the specimen.

47-11 Radium (Ra^{226}) undergoes alpha emission, leading to radon (Rn^{222}). The masses, including all electrons in each atom, are 226.0254 u and 222.0163 u, respectively. Find the maximum kinetic energy that the emitted alpha particle can have.

47-12 The common isotope of uranium, U^{238}, has a half-life of 4.50×10^9 years, decaying by alpha emission.

a) What is the decay constant?

b) What mass of uranium would be required for an activity of one curie?

c) How many alpha particles are emitted per second by 1 g of uranium?

47-13 The unstable isotope K^{40} is used for dating of rock samples. Its half-life is 2.4×10^8 years.

a) How many decays occur per second in a sample containing 2×10^{-6} g of K^{40}?

b) What is the activity of the sample, in curies?

47-14 An unstable isotope of cobalt, Co^{60}, has one more neutron in its nucleus than the stable Co^{59}, and is a beta emitter with a half-life of 5.3 years. This isotope is widely used in medicine. A certain radiation source in a hospital contains 0.01 g of Co^{60}.

a) What is the decay constant for this isotope?

b) How many atoms are in the source?

c) How many decays occur per second?

d) What is the activity of the source, in curies? How does this compare with the activity of an equal mass of radium?

47-15 A free neutron decays into a proton, an electron, and a neutrino, with a half-life of about 15 minutes. Calculate the total kinetic energy of the decay products.

47-16 The magnetic field in a cyclotron that is accelerating protons is 1.5 T.

a) How many times per second should the potential across the dees reverse?

b) The maximum radius of the cyclotron is 0.35 m. What is the maximum velocity of the proton?

c) Through what potential difference would the proton have to be accelerated to give it the maximum cyclotron velocity?

47-17 Deuterons in a cyclotron describe a circle of radius 32.0 cm just before emerging from the dees. The frequency of the applied alternating voltage is 10 MHz. Find (a) the magnetic field, and (b) the energy and speed of the deuterons upon emergence.

47-18 In the fission of U^{238}, 200 MeV of energy is released. Express this energy in joules per mole, and compare with typical heats of combustion.

47-19 Consider the fusion reaction.

$$H^2 + H^2 \rightarrow He^4 + \text{energy}.$$

Compute the energy liberated in this reaction, in MeV and in joules. Compute the energy *per mole* of deuterium, remembering that the gas is diatomic, and compare with the heat of combustion of hydrogen, about 2.9×10^5 J·mol^{-1}.

47-20 Why do elements with mass numbers of 210 and above decay by emission of alpha particles rather than single protons or neutrons.

47-21 If two gamma-ray photons are produced in positron annihilation, find the energy, frequency, and wavelength of each photon.

47-22 A neutral pion at rest decays into two gamma-ray photons. Find the energy, frequency, and wavelength of each photon.

APPENDIXES

APPENDIX A THE INTERNATIONAL SYSTEM OF UNITS

The Système International d'Unités, abbreviated SI, is the system developed by the General Conference on Weights and Measures and adopted by nearly all the industrial nations of the world. It is based on the mksa (meter-kilogram-second-ampere) system. The following material is reproduced from Publication SP–7012 of the Scientific and Technical Information Office of the National Aeronautics and Space Administration.

Quantity	Name of unit	Symbol	
SI Base Units			
length	meter	m	
mass	kilogram	kg	
time	second	s	
electric current	ampere	A	
thermodynamic temperature	kelvin	K	
luminous intensity	candela	cd	
amount of substance	mole	mol	
SI Derived Units			
area	square meter	m^2	
volume	cubic meter	m^3	
frequency	hertz	Hz	s^{-1}
mass density (density)	kilogram per cubic meter	$kg \cdot m^{-3}$	
speed, velocity	meter per second	$m \cdot s^{-1}$	
angular velocity	radian per second	$rad \cdot s^{-1}$	
acceleration	meter per second squared	$m \cdot s^{-2}$	
angular acceleration	radian per second squared	$rad \cdot s^{-2}$	
force	newton	N	$kg \cdot m \cdot s^{-2}$
pressure (mechanical stress)	pascal	Pa	$N \cdot m^{-2}$
kinematic viscosity	square meter per second	$m^2 \cdot s^{-1}$	
dynamic viscosity	newton-second per square meter	$N \cdot s \cdot m^{-2}$	
work, energy, quantity of heat	joule	J	$N \cdot m$
power	watt	W	$J \cdot s^{-1}$
quantity of electricity	coulomb	C	$A \cdot s$
potential difference, electromotive force	volt	V	$W \cdot A^{-1}, J \cdot C^{-1}$
electric field strength	volt per meter	$V \cdot m^{-1}$	$N \cdot C^{-1}$
electric resistance	ohm	Ω	$V \cdot A^{-1}$
capacitance	farad	F	$A \cdot s \cdot V^{-1}$
magnetic flux	weber	Wb	$V \cdot s$
inductance	henry	H	$V \cdot s \cdot A^{-1}$
magnetic flux density	tesla	T	$Wb \cdot m^{-2}$
magnetic field strength	ampere per meter	$A \cdot m^{-1}$	
magnetomotive force	ampere	A	
luminous flux	lumen	lm	$cd \cdot sr$
luminance	candela per square meter	$cd \cdot m^2$	
illuminance	lux	lx	$lm \cdot m^{-2}$
wave number	1 per meter	m^{-1}	
entropy	joule per kelvin	$J \cdot K^{-1}$	

(continued)

Quantity	Name of unit	Symbol
specific heat capacity	joule per kilogram kelvin	J·kg^{-1}·K^{-1}
thermal conductivity	watt per meter kelvin	W·m^{-1}·K^{-1}
radiant intensity	watt per steradian	W·sr^{-1}
activity (of a radioactive source)	1 per second	s^{-1}
SI Supplementary Units		
plane angle	radian	rad
solid angle	steradian	sr

Definitions of SI Units

meter (m) The *meter* is the length equal to 1,650,763.73 wavelengths in vacuum of the radiation corresponding to the transition between the levels $2p_{10}$ and $5d_5$ of the krypton-86 atom.

kilogram (kg) The *kilogram* is the unit of mass; it is equal to the mass of the international prototype of the kilogram. (The international prototype of the kilogram is a particular cylinder of platinum-iridium alloy which is preserved in a vault at Sèvres, France, by the International Bureau of Weights and Measures.)

second (s) The *second* is the duration of 9,192,631,770 periods of the radiation corresponding to the transition between the two hyperfine levels of the ground state of the cesium-133 atom.

ampere (A) The *ampere* is that constant current which, if maintained in two straight parallel conductors of infinite length, of negligible circular cross section, and placed 1 meter apart in vacuum, would produce between these conductors a force equal to 2×10^{-7} newton per meter of length.

kelvin (K) The *kelvin*, unit of thermodynamic temperature, is the fraction 1/273.16 of the thermodynamic temperature of the triple point of water.

candela (cd) The *candela* is the luminous intensity, in the perpendicular direction, of a surface of 1/600,000 square meter of a blackbody at the temperature of freezing platinum under a pressure of 101,325 newtons per square meter.

mole (mol) The *mole* is the amount of substance of a system which contains as many elementary entities as there are carbon atoms in 0.012 kg of carbon 12. The elementary entities must be specified and may be atoms, molecules, ions, electrons, other particles, or specified groups of such particles.

newton (N) The *newton* is that force which gives to a mass of 1 kilogram an acceleration of 1 meter per second per second.

joule (J) The *joule* is the work done when the point of application of 1 newton is displaced a distance of 1 meter in the direction of the force.

watt (W) The *watt* is the power which gives rise to the production of energy at the rate of 1 joule per second.

volt (V) The *volt* is the difference of electric potential between two points of a conducting wire carrying a constant current of 1 ampere, when the power dissipated between these points is equal to 1 watt.

ohm (Ω) The *ohm* is the electric resistance between two points of a conductor when a constant difference of potential of 1 volt, applied between these two points, produces in this conductor a current of 1 ampere, this conductor not being the source of any electromotive force.

coulomb (C) The *coulomb* is the quantity of electricity transported in 1 second by a current of 1 ampere.

farad (F) The *farad* is the capacitance of a capacitor between the plates of which there appears a difference of potential of 1 volt when it is charged by a quantity of electricity equal to 1 coulomb.

henry (H) The *henry* is the inductance of a closed circuit in which an electromotive force of 1 volt is produced when the electric current in the circuit varies uniformly at a rate of 1 ampere per second.

weber (Wb) The *weber* is the magnetic flux which, linking a circuit of one turn, produces in it an electromotive force of 1 volt as it is reduced to zero at a uniform rate in 1 second.

lumen (lm) The *lumen* is the luminous flux emitted in a solid angle of 1 steradian by a uniform point source having an intensity of 1 candela.

radian (rad) The *radian* is the plane angle between two radii of a circle which cut off on the circumference an arc equal in length to the radius.

steradian (sr) The *steradian* is the solid angle which, having its vertex in the center of a sphere, cuts off an area of the surface of the sphere equal to that of a square with sides of length equal to the radius of the sphere.

SI Prefixes The names of multiples and submultiples of SI units may be formed by application of the prefixes listed in Table 1–1, page 3.

APPENDIX B USEFUL MATHEMATICAL RELATIONS

Algebra

$$a^{-x} = \frac{1}{a^x} \qquad a^{(x+y)} = a^x a^y \qquad a^{(x-y)} = \frac{a^x}{a^y}$$

Logarithims:

$\log a + \log b = \log(ab) \qquad \log a - \log b = \log(a/b)$
$\log(a^n) = n \log a$

Quadratic formula:

If $ax^2 + bx + c = 0$,
$$x = \frac{-b \pm \sqrt{b^2 - 4ac}}{2a}$$

Binomial theorem

$$(a+b)^n = a^n + na^{n-1}b + \frac{n(n-1)a^{n-2}b^2}{2!} + \frac{n(n-1)(n-2)a^{n-3}b^3}{3!} + \cdots$$

Trigonometry

In the right triangle ABC, $x^2 + y^2 = r^2$.

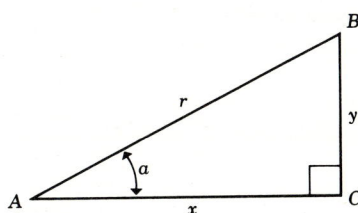

Definitions of the trigonometric functions:

$\sin a = y/r \qquad \cos a = x/r \qquad \tan a = y/x$

Identities:

$\sin^2 a + \cos^2 a = 1 \qquad\qquad \tan a = \dfrac{\sin a}{\cos a}$

$\sin 2a = 2 \sin a \cos a$

$\cos 2a = \cos^2 a - \sin^2 a = 2 \cos^2 a - 1$

$\sin \tfrac{1}{2}a = \sqrt{\dfrac{1 - \cos a}{2}} \qquad\qquad \cos \tfrac{1}{2}a = \sqrt{\dfrac{1 + \cos a}{2}}$

$\sin(a \pm b) = \sin a \cos b \pm \cos a \sin b$

$\cos(a \pm b) = \cos a \cos b \mp \sin a \sin b$

$\sin a + \sin b = 2 \sin \tfrac{1}{2}(a+b) \cos \tfrac{1}{2}(a-b)$

$\cos a + \cos b = 2 \cos \tfrac{1}{2}(a+b) \cos \tfrac{1}{2}(a-b)$

$\sin(-a) = -\sin a \qquad\qquad \cos(-a) = \cos a$

$\sin(a \pm \pi/2) = \pm\cos a \qquad\qquad \cos(a \pm \pi/2) = \mp\sin a$

T-4

Geometry

Circumference of circle of radius r: $\quad C = 2\pi r$
Area of circle of radius r: $\quad A = \pi r^2$
Volume of sphere of radius r: $\quad V = 4\pi r^3/3$
Surface area of sphere of radius r: $\quad A = 4\pi r^2$

Calculus

Derivatives:

$$\frac{d}{dx} x^n = n x^{n-1}$$

$$\frac{d}{dx} \sin ax = a \cos ax$$

$$\frac{d}{dx} \cos ax = -a \sin ax$$

$$\frac{d}{dx} e^{ax} = a e^{ax}$$

$$\frac{d}{dx} \ln ax = \frac{1}{x}$$

Integrals:

$$\int x^n \, dx = \frac{x^{n+1}}{n+1}$$

$$\int \frac{dx}{x} = \ln x$$

$$\int \sin ax \, dx = -\frac{1}{a} \cos ax$$

$$\int \cos ax \, dx = \frac{1}{a} \sin ax$$

$$\int e^{ax} \, dx = \frac{1}{a} e^{ax}$$

$$\int \frac{dx}{\sqrt{a^2 - x^2}} = \arcsin \frac{x}{a}$$

$$\int \frac{dx}{\sqrt{x^2 + a^2}} = \ln\left(x + \sqrt{x^2 + a^2}\right)$$

$$\int \frac{dx}{x^2 + a^2} = \frac{1}{a} \arctan \frac{x}{a}$$

$$\int \frac{dx}{(x^2 + a^2)^{3/2}} = \frac{1}{a^2} \frac{x}{\sqrt{x^2 + a^2}}$$

$$\int \frac{x \, dx}{(x^2 + a^2)^{3/2}} = -\frac{1}{\sqrt{x^2 + a^2}}$$

Power series (convergent for range of x shown):

$$\sin x = x - \frac{x^3}{3!} + \frac{x^5}{5!} - \frac{x^7}{7!} + \cdots \quad \text{(all } x\text{)}$$

$$\cos x = 1 - \frac{x^2}{2!} + \frac{x^4}{4!} - \frac{x^6}{6!} + \cdots \quad \text{(all } x\text{)}$$

$$\tan x = x + \frac{x^3}{3} + \frac{2x^5}{15} + \frac{17x^7}{315} + \cdots \quad (|x| < \pi/2)$$

$$e^x = 1 + x + \frac{x^2}{2!} + \frac{x^3}{3!} + \cdots \quad \text{(all } x\text{)}$$

$$\ln(1 + x) = x - \frac{x^2}{2} + \frac{x^3}{3} - \frac{x^4}{4} + \cdots \quad (|x| < 1)$$

APPENDIX C THE GREEK ALPHABET

Name	Capital	Lowercase	Name	Capital	Lowercase
Alpha	A	α	Nu	N	ν
Beta	B	β	Xi	Ξ	ξ
Gamma	Γ	γ	Omicron	O	o
Delta	Δ	δ	Pi	Π	π
Epsilon	E	ϵ	Rho	P	ρ
Zeta	Z	ζ	Sigma	Σ	σ
Eta	H	η	Tau	T	τ
Theta	Θ	θ	Upsilon	Υ	υ
Iota	I	ι	Phi	Φ	ϕ
Kappa	K	κ	Chi	X	χ
Lambda	Λ	λ	Psi	Ψ	ψ
Mu	M	μ	Omega	Ω	ω

APPENDIX D NATURAL TRIGONOMETRIC FUNCTIONS

Angle					Angle				
Degree	Radian	Sine	Cosine	Tangent	Degree	Radian	Sine	Cosine	Tangent
0°	.000	0.000	1.000	0.000					
1°	.017	.018	1.000	.018	46°	0.803	0.719	0.695	1.036
2°	.035	.035	0.999	.035	47°	.820	.731	.682	1.072
3°	.052	.052	.999	.052	48°	.838	.743	.669	1.111
4°	.070	.070	.998	.070	49°	.855	.755	.656	1.150
5°	.087	.087	.996	.088	50°	.873	.766	.643	1.192
6°	.105	.105	.995	.105	51°	.890	.777	.629	1.235
7°	.122	.122	.993	.123	52°	.908	.788	.616	1.280
8°	.140	.139	.990	.141	53°	.925	.799	.602	1.327
9°	.157	.156	.988	.158	54°	.942	.809	.588	1.376
10°	.175	.174	.985	.176	55°	.960	.819	.574	1.428
11°	.192	.191	.982	.194	56°	.977	.829	.559	1.483
12°	.209	.208	.978	.213	57°	.995	.839	.545	1.540
13°	.227	.225	.974	.231	58°	1.012	.848	.530	1.600
14°	.244	.242	.970	.249	59°	1.030	.857	.515	1.664
15°	.262	.259	.966	.268	60°	1.047	.866	.500	1.732
16°	.279	.276	.961	.287	61°	1.065	.875	.485	1.804
17°	.297	.292	.956	.306	62°	1.082	.883	.470	1.881
18°	.314	.309	.951	.325	63°	1.100	.891	.454	1.963
19°	.332	.326	.946	.344	64°	1.117	.899	.438	2.050
20°	.349	.342	.940	.364	65°	1.134	.906	.423	2.145
21°	.367	.358	.934	.384	66°	1.152	.914	.407	2.246
22°	.384	.375	.927	.404	67°	1.169	.921	.391	2.356
23°	.401	.391	.921	.425	68°	1.187	.927	.375	2.475
24°	.419	.407	.914	.445	69°	1.204	.934	.358	2.605
25°	.436	.423	.906	.466	70°	1.222	.940	.342	2.747
26°	.454	.438	.899	.488	71°	1.239	.946	.326	2.904
27°	.471	.454	.891	.510	72°	1.257	.951	.309	3.078
28°	.489	.470	.883	.532	73°	1.274	.956	.292	3.271
29°	.506	.485	.875	.554	74°	1.292	.961	.276	3.487
30°	.524	.500	.866	.577	75°	1.309	.966	.259	3.732
31°	.541	.515	.857	.601	76°	1.326	.970	.242	4.011
32°	.559	.530	.848	.625	77°	1.344	.974	.225	4.331
33°	.576	.545	.839	.649	78°	1.361	.978	.208	4.705
34°	.593	.559	.829	.675	79°	1.379	.982	.191	5.145
35°	.611	.574	.819	.700	80°	1.396	.985	.174	5.671
36°	.628	.588	.809	.727	81°	1.414	.988	.156	6.314
37°	.646	.602	.799	.754	82°	1.431	.990	.139	7.115
38°	.663	.616	.788	.781	83°	1.449	.993	.122	8.144
39°	.681	.629	.777	.810	84°	1.466	.995	.105	9.514
40°	.698	.643	.766	.839	85°	1.484	.996	.087	11.43
41°	.716	.658	.755	.869	86°	1.501	.998	.070	14.30
42°	.733	.669	.743	.900	87°	1.518	.999	.052	19.08
43°	.751	.682	.731	.933	88°	1.536	.999	.035	28.64
44°	.768	.695	.719	.966	89°	1.553	1.000	.018	57.29
45°	.785	.707	.707	1.000	90°	1.571	1.000	.000	

APPENDIX E PERIODIC TABLE OF THE ELEMENTS

Period	IA	IIA	IIIB	IVB	VB	VIB	VIIB	VIIIB	VIIIB	VIIIB	IB	IIB	IIIA	IVA	VA	VIA	VIIA	Noble gases
1	1 H 1.008																	2 He 4.003
2	3 Li 6.939	4 Be 9.012											5 B 10.811	6 C 12.011	7 N 14.007	8 O 15.999	9 F 18.998	10 Ne 20.183
3	11 Na 22.990	12 Mg 24.312											13 Al 26.982	14 Si 28.086	15 P 30.974	16 S 32.064	17 Cl 35.453	18 Ar 39.948
4	19 K 39.102	20 Ca 40.08	21 Sc 44.956	22 Ti 47.90	23 V 50.942	24 Cr 51.996	25 Mn 54.938	26 Fe 55.847	27 Co 58.933	28 Ni 58.71	29 Cu 63.54	30 Zn 65.37	31 Ga 69.72	32 Ge 72.59	33 As 74.922	34 Se 78.96	35 Br 79.909	36 Kr 83.80
5	37 Rb 85.47	38 Sr 87.62	39 Y 88.905	40 Zr 91.22	41 Nb 92.906	42 Mo 95.94	43 Tc (99)	44 Ru 101.07	45 Rh 102.91	46 Pd 106.4	47 Ag 107.87	48 Cd 112.40	49 In 114.82	50 Sn 118.69	51 Sb 121.75	52 Te 127.60	53 I 126.90	54 Xe 131.30
6	55 Cs 132.91	56 Ba 137.34	57 La 138.91	72 Hf 178.49	73 Ta 180.95	74 W 183.85	75 Re 186.2	76 Os 190.2	77 Ir 192.2	78 Pt 195.09	79 Au 196.97	80 Hg 200.59	81 Tl 204.37	82 Pb 207.19	83 Bi 208.98	84 Po (210)	85 At (210)	86 Rn (222)
7	87 Fr (223)	88 Ra (226)	89 Ac (227)	104 Rf(?) (259)	105 Ha(?) (260)													

58 Ce 140.12	59 Pr 140.91	60 Nd 144.24	61 Pm (145)	62 Sm 150.35	63 Eu 151.96	64 Gd 157.25	65 Tb 158.92	66 Dy 162.50	67 Ho 164.93	68 Er 167.26	69 Tm 168.93	70 Yb 173.04	71 Lu 174.97
90 Th 232.04	91 Pa (231)	92 U 238.03	93 Np (237)	94 Pu (242)	95 Am (243)	96 Cm (247)	97 Bk (249)	98 Cf (251)	99 Es (254)	100 Fm (253)	101 Md (256)	102 No (253)	103 Lr (257)

APPENDIX F UNIT CONVERSION FACTORS

Length

$1 \text{ m} = 100 \text{ cm} = 1000 \text{ mm} = 10^6 \text{ }\mu\text{m} = 10^9 \text{ nm}$
$1 \text{ km} = 1000 \text{ m} = 0.6214 \text{ mi}$
$1 \text{ m} = 3.281 \text{ ft} = 39.37 \text{ in.}$
$1 \text{ cm} = 0.3937 \text{ in.}$
$1 \text{ ft} = 30.48 \text{ cm}$
$1 \text{ in.} = 2.540 \text{ cm}$
$1 \text{ mi} = 5280 \text{ ft} = 1.609 \text{ km}$
$1 \text{ Å} = 10^{-10} \text{ m} = 10^{-8} \text{ cm} = 10^{-1} \text{ nm}$

Area

$1 \text{ cm}^2 = 0.155 \text{ in}^2$
$1 \text{ m}^2 = 10^4 \text{ cm}^2 = 10.76 \text{ ft}^2$
$1 \text{ in}^2 = 6.452 \text{ cm}^2$
$1 \text{ ft}^2 = 144 \text{ in}^2 = 0.0929 \text{ m}^2$

Volume

$1 \text{ liter} = 1000 \text{ cm}^3 = 10^{-3} \text{ m}^3 = 0.0351 \text{ ft}^3 = 61.02 \text{ in.}^3$
$1 \text{ ft}^3 = 0.02832 \text{ m}^3 = 28.32 \text{ liters} = 7.477 \text{ gallons}$

Time

$1 \text{ min} = 60 \text{ s}$
$1 \text{ hr} = 3600 \text{ s}$
$1 \text{ da} = 86{,}400 \text{ s}$
$1 \text{ yr} = 3.156 \times 10^7 \text{ s}$

Velocity

$1 \text{ cm} \cdot \text{s}^{-1} = 0.03281 \text{ ft} \cdot \text{s}^{-1}$
$1 \text{ ft} \cdot \text{s}^{-1} = 30.48 \text{ cm} \cdot \text{s}^{-1}$
$1 \text{ mi} \cdot \text{min}^{-1} = 60 \text{ mi} \cdot \text{hr}^{-1} = 88 \text{ ft} \cdot \text{s}^{-1}$
$1 \text{ km} \cdot \text{hr}^{-1} = 0.2778 \text{ m} \cdot \text{s}^{-1}$
$1 \text{ mi} \cdot \text{hr}^{-1} = 0.4470 \text{ m} \cdot \text{s}^{-1}$

Acceleration

$1 \text{ m} \cdot \text{s}^{-2} = 100 \text{ cm} \cdot \text{s}^{-2} = 3.281 \text{ ft} \cdot \text{s}^{-2}$
$1 \text{ cm} \cdot \text{s}^{-2} = 0.01 \text{ m} \cdot \text{s}^{-2} = 0.03281 \text{ ft} \cdot \text{s}^{-2}$
$1 \text{ ft} \cdot \text{s}^{-2} = 0.3048 \text{ m} \cdot \text{s}^{-2} = 30.48 \text{ cm} \cdot \text{s}^{-2}$
$1 \text{ mi} \cdot \text{hr}^{-1} \cdot \text{s}^{-1} = 1.467 \text{ ft} \cdot \text{s}^{-2}$

Mass

$1 \text{ kg} = 10^3 \text{ g} = 0.0685 \text{ slug}$
$1 \text{ g} = 6.85 \times 10^{-5} \text{ slug}$
$1 \text{ slug} = 14.59 \text{ kg}$
$1 \text{ u} = 1.661 \times 10^{-27} \text{ kg}$

Force

$1 \text{ N} = 10^5 \text{ dyn} = 0.2247 \text{ lb}$
$1 \text{ lb} = 4.45 \text{ N} = 4.45 \times 10^5 \text{ dyn}$

Pressure

$1 \text{ Pa} = 1 \text{ N} \cdot \text{m}^{-2} = 1.451 \times 10^{-4} \text{ lb} \cdot \text{in}^{-2} = 0.209 \text{ lb} \cdot \text{ft}^{-2}$
$1 \text{ lb} \cdot \text{in}^{-2} = 6891 \text{ Pa}$
$1 \text{ lb} \cdot \text{ft}^{-2} = 47.85 \text{ Pa}$
$1 \text{ atm} = 1.013 \times 10^5 \text{ Pa} = 14.7 \text{ lb} \cdot \text{in}^{-2} = 2117 \text{ lb} \cdot \text{ft}^{-2}$

Energy

$1 \text{ J} = 10^7 \text{ ergs} = 0.239 \text{ cal}$
$1 \text{ cal} = 4.186 \text{ J}$ (based on 15° calorie)
$1 \text{ ft} \cdot \text{lb} = 1.356 \text{ J}$
$1 \text{ Btu} = 1055 \text{ J} = 252 \text{ cal}$
$1 \text{ eV} = 1.602 \times 10^{-19} \text{ J}$
$1 \text{ kWh} = 3.600 \times 10^6 \text{ J}$

Mass-Energy Equivalence

$1 \text{ kg} \leftrightarrow 8.988 \times 10^{16} \text{ J}$
$1 \text{ u} \leftrightarrow 931.5 \text{ MeV}$
$1 \text{ eV} \leftrightarrow 1.073 \times 10^{-9} \text{ u}$

Power

$1 \text{ W} = 1 \text{ J} \cdot \text{s}^{-1}$
$1 \text{ hp} = 746 \text{ W} = 550 \text{ ft} \cdot \text{lb} \cdot \text{s}^{-1}$
$1 \text{ Btu} \cdot \text{hr}^{-1} = 0.293 \text{ W}$

APPENDIX G NUMERICAL CONSTANTS

Fundamental physical constants

Name	Symbol	Value
Speed of light	c	2.9979×10^8 m·s^{-1}
Charge of electron	e	1.602×10^{-19} C
Gravitational constant	G	6.673×10^{-11} N·m^2·kg^{-2}
Planck's constant	h	6.626×10^{-34} J·s
Boltzmann's constant	k	1.381×10^{-23} J·K^{-1}
Avogadro's number	N_0	6.022×10^{23} molecules·mol^{-1}
Gas constant	R	8.314 J·mol^{-1}·K^{-1}
Mass of electron	m_e	9.110×10^{-31} kg
Mass of neutron	m_n	1.675×10^{-27} kg
Mass of proton	m_p	1.673×10^{-27} kg
Permittivity of free space	ϵ_0	8.854×10^{-12} C^2·N^{-1}·m^{-2}
	$\frac{1}{4\pi\epsilon_0}$	8.987×10^9 N·m^2·C^{-2}
Permeability of free space	μ_0	$4\pi \times 10^{-7}$ Wb·A^{-1}·m^{-1}

Other useful constants

Name	Symbol	Value
Mechanical equivalent of heat		4.185 J·cal^{-1}
Standard atmospheric pressure	1 atm	1.013×10^5 Pa
Absolute zero	0 K	-273.15 °C
Electronvolt	1 eV	1.602×10^{-19} J
Atomic mass unit	1 u	1.661×10^{-27} kg
Electron rest energy	mc^2	0.511 MeV
Energy equivalent of 1 u	Mc^2	931.5 MeV
Volume of ideal gas (0°C and 1 atm)	V	22.4 liter·mol^{-1}
Acceleration due to gravity (sea level, at equator)	g	9.78049 m·s^{-2}

Astronomical data

Body	Mass, kg	Radius, m	Orbit radius, m	Orbit period
Sun	1.99×10^{30}	6.95×10^8	—	—
Moon	7.36×10^{22}	1.74×10^6	0.38×10^9	27.3 d
Mercury	3.28×10^{23}	2.57×10^6	5.8×10^{10}	88.0 d
Venus	4.82×10^{24}	6.31×10^6	1.08×10^{11}	224.7 d
Earth	5.98×10^{24}	6.38×10^6	1.49×10^{11}	365.3 d
Mars	6.34×10^{23}	3.43×10^6	2.28×10^{11}	687.0 d
Jupiter	1.88×10^{27}	7.18×10^7	7.78×10^{11}	11.86 yr
Saturn	5.63×10^{26}	6.03×10^7	1.43×10^{12}	29.46 yr
Uranus	8.61×10^{25}	2.67×10^7	2.87×10^{12}	84.02 y
Neptune	9.99×10^{25}	2.48×10^7	4.49×10^{12}	164.8 y
Pluto	5×10^{23}	4×10^5	5.90×10^{12}	247.7 y

ANSWERS TO ODD-NUMBERED PROBLEMS

Part I Mechanics, Heat, and Sound

Chapter 1

1-1 1 mi = 1.6093 km
1-3 a) 88.0 ft·s^{-1} b) 27.8 m·s^{-1}
1-5 86,400; 3.15×10^7
1-7 0.39%
1-9 5.50×10^3 kg·m^{-3}
1-11 1 mi = 1.609344 km (exactly)
1-13 $[(x_2 - x_1)^2 + (y_2 - y_1)^2]^{1/2}$, $\tan^{-1}\dfrac{y_2 - y_1}{x_2 - x_1}$
1-15 a) 5 cm, −53.1°   b) 13 m, −112°   c) 3.61 km, 124°
1-17 6 in., at 45° to initial direction of line
1-19 230 N, 55° above negative x-axis
1-21 15.0 N, 53.3° above $+x$-axis; 25.9 N, 27.7° below $-x$-axis
1-25 5 mi, 53.1° south of east
1-27 a) −1 cm, +1 cm   b) 1.41 cm, 135°
1-29 When the angle between A and B is 60°; when it is 120°.
1-31 a) $\sqrt{29}$, $\sqrt{14}$ b) $3i + j + 7k$ c) $\sqrt{59}$
   d) $i + 5j + k$ e) $\sqrt{27}$; yes
1-33 $i \times i = 0$ $j \times i = -k$ $k \times i = j$
   $i \times j = k$ $j \times j = 0$ $k \times j = -i$
   $i \times k = -j$ $j \times k = i$ $k \times k = 0$
1-35 8
1-37 $\pm(17i - 2j - 7k)/\sqrt{342}$
1-39 a) No   b) Vector product is not associative.

Chapter 2

2-1 25.7 N, 30.6 N
2-3 a) 18.5 N   b) 9.2 N
2-5 308 N, 25° above $+x$-axis
2-7 a) 7.01 N, 2.93 N   b) 7.60 N   c) 11.0 N
2-9 46.6 N in $-y$-direction
2-11 a) The earth   b) 4 N, the book   c) No
   d) 4 N, the earth, the book, upward
   e) 4 N, the hand, the book, downward   f) First
   g) Third   h) No   i) No   j) Yes   k) Yes   l) One
   m) No   n) Nothing
2-13 a) 10 N   b) 20 N
2-15 a) $T_1 = 50$ N, $T_2 = 28.9$ N, $T_3 = 57.7$ N
   b) $T_1 = 50$ N, $T_2 = 50$ N, $T_3 = 70.7$ N
2-17 828 N, 214 N
2-19 765 N, 750 N
2-21 14.1 N
2-23 a) 20 N   b) 30 N
2-25 a) $w/(2 \sin \theta)$ at angle θ below horizontal
   b) $w/(2 \tan \theta)$
2-29 22.3 N
2-31 a) 76.0 N   b) 24.0 N   c) 15.4 to 84.6 N
2-33 a) Held back   b) 480 N
2-35 a) 3.00 N   b) 4.00 N   c) 5.00 N
2-37 b) 10.0 N   c) 30.0 N

Chapter 3

3-1 a) 15 mi·hr^{-1} b) 22.0 ft·s^{-1} c) 6.71 m·s^{-1}
3-3 61.0 cm·s^{-1}, 60.1 cm·s^{-1}, 60.01 cm·s^{-1}; 60.0 cm·s^{-1}
3-5 a) 0, 6.25 m·s^{-2}, −11.2 m·s^{-2}
   b) 100 m, 230 m, 320 m
3-9 a) 3.95 ft·s^{-2} b) 620 ft
3-11 a) 2.67 km·hr^{-1}·s^{-1} b) 7.50 s   c) 83.3 m, 104.2 m
3-13 a) 12.5 cm·s^{-2} b) 7840 cm
3-15 a) 100 m   b) 20 m·s^{-1}
3-17 a) 24 ft·s^{-1}, 29 ft·s^{-1}, 34 ft·s^{-1} b) 2.5 ft·s^{-2}
   c) 21.5 ft·s^{-1} d) 8.58 s   e) 22.7 ft   f) 1.05 s
   g) 24.1 ft·s^{-1}
3-19 a) 32 m·s^{-1} b) 2.83 s   c) 16 m   d) 22.6 m·s^{-1}
   e) 16 m·s^{-1}

A-2 ANSWERS TO ODD-NUMBERED PROBLEMS

3-21 a) 8.66 s b) 37.5 m c) 26.0 m·s^{-1}, 17.3 m·s^{-1}
3-23 a) 29.6 m·s^{-1} b) 39.6 m c) 17.2 m·s^{-1}
  d) 50.0 m·s^{-2} e) 2.01 s f) 29.7 m·s^{-1}
3-25 10.7 m
3-27 a) 48.0 ft·s^{-1} b) 36.0 ft c) zero d) 32.0 ft·s^{-2}, downward e) 80 ft·s^{-1}
3-29 a) 10.2 m b) 4.34 m·s^{-1}, -9.8 m·s^{-2}
  c) -15.3 m·s^{-1}, -9.8 m·s^{-2} d) 11.7 m·s^{-1}
3-31 a) 11.3 m b) 0.51 s c) 9.98 m·s^{-1}, -9.8 m·s^{-2}
3-33 a) 4 m·s^{-2} b) 6 m·s^{-1} c) 4.5 m
3-35 a) 46.8 s b) 102 mi
3-37 a) 6000 m·s^{-1} b) 18.4 hr c) 0.99
3-39 $\frac{1}{6}$ m, $\frac{2}{3}$ m·s^{-1}, 2 m·s^{-2}
3-41 a) $\omega A \cos \omega t$ b) $-\omega^2 A \sin \omega t$
  c) $\pm \omega \sqrt{A^2 - x^2}$; $x = 0$; $x = \pm A$
  d) $-\omega^2 x$; $x = \pm A$, $x = 0$ e) $\pm A$ f) $\pm \omega A$
  g) $\pm \omega^2 A$
3-43 a) A b) A/B
3-45 a) 50 s b) 150 s
3-47 a) 17.3 s b) 421 m c) 28.7 m·s^{-2}

Chapter 4

4-1 a) 0.102 kg b) 0.00102 g
4-3 The two forces in an action-reaction pair never act on the same body.
4-5 a) 5 kg b) 200 m c) 0.5 slug, 500 ft
4-7 a) 2 m·s^{-2} b) 100 m c) 20 m·s^{-1}
4-9 a) 19.6 N b) 2.55 s
4-11 a) 1.62×10^{-15} N b) 3.33×10^{-9} s
  c) 1.8×10^{15} m·s^{-2}
4-13 49.8 N, 2.54%
4-15 6.18×10^{24} kg
4-17 1.94 m·s^{-2}
4-19 21,600 N
4-21 4500 lb
4-23 a) 59.0 N b) Constant velocity c) 6.26 m·s^{-1}
4-25 a) 40 N b) 2.45 m·s^{-2}, downward c) Zero
4-27 242 ft
4-29 $W = 2wa/(g + a)$
4-31 3.13 s, 9.8 m
4-33 a) 2000 N, 1000 N b) 5000 N
4-35 a) 36.9° b) 1.95 m·s^{-2} c) 2.5 s
4-37 a) 36.8 N b) 2.45 m·s^{-2}
4-39 a) 156 N b) 46.8 N c) 1.12 m·s^{-2} d) 57.9 N
4-41 a) 319.7 N b) 95.9 N c) 3.30 m·s^{-2} d) 208 N
4-43 a) 10.54 N b) 10.0 N
4-45 a) 1.96 m·s^{-2} b) 3.14 N
4-47 a) To the left b) 0.658 m·s^{-2} c) 424 N
4-49 $2m_2 g/(4m_1 + m_2)$, $m_2 g/(4m_1 + m_2)$
4-51 a) 2.70 m·s^{-2} b) 112.5 N c) 87.5 N
4-53 2.43 m
4-55 g/μ
4-57 a) $1.6\text{ N} - (2.4\text{ N·s}^{-2})t^2$ b) 1.6 N, $t = 0$, 6 m
4-59 c) $v_T = (mg/k)^{1/2}$
  d) $v = v_T(e^{2t/\tau} - 1)/(e^{2t/\tau} + 1)$, $\tau = (m/kg)^{1/2}$
4-61 a) 800 N b) 200 N c) 200 N

Chapter 5

5-1 a) 0.452 s b) 3.32 m·s^{-1}
  c) 5.53 m·s^{-1}, 53.1° below horizontal
5-3 a) 20.2 s b) 2020 m c) 100 m·s^{-1}, 198 m·s^{-1}
5-5 a) 30.6 m b) 62.5 m c) 25 m·s^{-1}, 24.5 m·s^{-1}; 35.0 m·s^{-1}, 44.4° below horizontal
5-7 a) 0.444 ft b) 1.78 ft c) 4.00 ft d) 16.0 ft
5-9 a) 122 ft·s^{-1} b) 57.7 ft c) 3.80 s
5-11 a) 1.01 m b) -3.91 m c) Yes
5-13 17.0 ft·s^{-1}
5-15 a) 21.8 m·s^{-1}
  b) 28.3 m·s^{-1}, 58.8° below horizontal
5-17 1.79 s or 170.8 s
5-21 Straight line, 27.0° to vertical
5-23 a) $(8\text{ N})\mathbf{i} + (12\text{ N·s}^{-1})t\mathbf{j}$
  b) 25.3 N, 71.6° to $+x$-axis
  c) ± 0.816 s d) always
5-25 a) $(2\text{ m·s}^{-4})t^2\mathbf{i} + (\frac{4}{3}\text{ m·s}^{-3})t\mathbf{j}$;
  $(\frac{2}{3}\text{ m·s}^{-4})t^3\mathbf{i} + (\frac{2}{3}\text{ m·s}^{-3})t\mathbf{j}$;
  $(\frac{1}{6}\text{ m·s}^{-4})t^4\mathbf{i} + (\frac{2}{9}\text{ m·s}^{-3})t^2\mathbf{j}$
  c) 19.0 m·s^{-1}, 18.4° to $+x$-axis
5-27 a) 2.96×10^4 m·s^{-1} b) 0.0594 m·s^{-2}
5-29 6.00°
5-31 22.4 m·s^{-1}
5-33 a) 0.127 b) 2.19 in.
5-35 a) 6.73 s b) No
5-37 a) 51.6° b) No
  c) The bead remains at the bottom.
5-39 a) 28.1° b) 1910 m c) $0.534w$
5-41 a) 1537 ft b) 1440 lb
5-43 a) $0.403w$ b) 4.49 s c) $2w$ d) He would strike the ground 10.0 m from the center of the wheel.
5-47 1.98 hr
5-49 7560 m·s^{-1}
5-51 1.41 hr
5-53 6.55 knots
5-55 a) 100 km·hr^{-1}, 53.1° S of W b) 30° N of W
5-57 a) 3.6 m·s^{-1}, 42° N of E b) 667 m c) 333 s
5-59 11.6 mi·hr^{-1}, 59° N of E

Chapter 6

6-1 6.00×10^7 J
6-3 985 J
6-5 a) 6 N b) 48 N c) 22.5 J
6-7 a) 1.85×10^4 J b) 4 times
6-9 4.5×10^{-17} J
6-11 14.1 m·s^{-1}
6-13 4.0 m
6-15 9.8 J
6-17 a) 10 N, 20 N, 40 N b) 0.5 J, 2 J, 8 J
6-19 a) 33 J b) 13 J
6-21 a) 200 J b) 8.12 m·s^{-1}, 200 J
6-23 a) 2400 J b) 515 J c) 1415 J d) 470 J
  c) $b + c + d = a$
6-25 a) 240 N b) 238 J

6-27 a) 11.2 m·s^{-1} b) 6.71 m·s^{-1}
6-29 0.10 m
6-31 a) 0.272 b) 3.60 J
6-35 1.81 m·s^{-1}
6-37 8.02 m·s^{-1}
6-39 a) 5.51 m·s^{-1} b) 0.0091 m
6-41 a) 4.43 m·s^{-1} b) 14.7 N
6-43 $1.12 \times 10^4 \text{ m·s}^{-1}$
6-45 a) $38 \times 10^6 \text{ m}$ from center of moon
   b) $1.10 \times 10^4 \text{ m·s}^{-1}$ c) $2.24 \times 10^3 \text{ m·s}^{-1}$
6-47 5.34 m·s^{-1}; yes, barely
6-49 0.63 hp, 470 W, 0.470 kW
6-51 148 hp
6-53 $\$1.07$
6-55 a) 1070 N b) 5 hp, 80 hp
6-57 a) 8232 J b) 8232 J c) 0.552 hp
6-59 a) $27,600 \text{ N}$ b) 10.9 m·s^{-1} c) $1.44 \times 10^5 \text{ J}$
   d) $3.53 \times 10^5 \text{ J}$ e) 296 hp
6-61 a) 1070 N b) 41.8 hp c) 15.6 hp d) 9.2%
6-63 227 km^2
6-65 a) $1.80 \times 10^{14} \text{ J}$ b) $1.80 \times 10^{20} \text{ W}$
   c) $1.83 \times 10^{10} \text{ kg}$
6-67 a) $4.56 \times 10^{-17} \text{ J}$ b) $1.82 \times 10^{-14} \text{ J}, 2.80 \times 10^{-14} \text{ J}$

Chapter 7

7-1 a) $2.0 \times 10^5 \text{ kg·m·s}^{-1}$ b) 40 m·s^{-1} c) 28.3 m·s^{-1}
7-3 a) $8.0 \times 10^5 \text{ m·s}^{-2}$ b) $4.0 \times 10^4 \text{ N}$
   c) $5.0 \times 10^{-4} \text{ s}$ d) 20 N·s
7-5 No; no; transferred to earth
7-7 180 N
7-9 $4600 \text{ N}, 1500 \text{ N}$
7-11 a) 0.667 m·s^{-1} b) $1.33 \times 10^4 \text{ J}$ c) 1.0 m·s^{-1}
7-13 237 m·s^{-1}
7-15 a) 0.163 b) 240 J c) 0.320 J
7-17 595 m·s^{-1}
7-19 24.0 km·h^{-1}, $33.7°$ S of E
7-21 a) 10.0 cm·s^{-1} b) 0.135 J
   c) -70 cm·s^{-1}, 80 cm·s^{-1}
7-23 -25 cm·s^{-1}, 5 cm·s^{-1}
7-25 a) 50 cm·s^{-1}, $53.1°$ below x-axis b) 0.037 J
7-27 d) $\rho A(v_J - v_B)$ e) $\rho A(v_J - v_B)^2$ f) $\rho A v_B(v_J - v_B)^2$
7-29 a) $v_A = 22.0 \text{ m·s}^{-1}$, $v_B = 15.5 \text{ m·s}^{-1}$ b) No; 0.196
7-33 a) $21,208 \text{ m}$ b) $4.0 \times 10^5 \text{ J}$
7-35 a) 7.20 ft·s^{-1} b) 375
7-37 0.91 m·s^{-1}
7-39 a) $150°$ from direction of electron
   b) $1.07 \times 10^{-20} \text{ kg·m·s}^{-1}$ c) $1.45 \times 10^{-16} \text{ J}$
7-41 16 N
7-43 a) 20 m ahead of 1000-kg car b) $50,000 \text{ kg·m·s}^{-1}$
   c) 16.7 m·s^{-1} d) $50,000 \text{ kg·m·s}^{-1}$
7-45 On bisector of angle, 0.707 m from vertex
7-47 2.18 m
7-49 15.9
7-51 a) $1.4 \, v_r$ b) $1.2 \, v_r$ c) $2.6 \, v_r$ d) 3 km·s^{-1}
7-55 a) 0.005% b) 13.4% c) 56.3%
7-57 $1.19 \times 10^{-18} \text{ m·s}^{-1}$

Chapter 8

8-1 1000 N, 0.8 m from 600-N end
8-3 $350 \text{ N}, 750 \text{ N}$
8-5 $3\tfrac{1}{3}$ ft from other end
8-7 b) 2.5 m c) 2.89 m
8-9 125 N, 100 N to the right, 25 N up
8-11 a) 2.31 m b) 213 N
8-13 0.3 m from the 1-kg end
8-15 $0.0667 \text{ m}, 0.0667 \text{ m}$; yes
8-17 $0.646 \times 10^{-10} \text{ m}$ from carbon atom
8-19 $833 \text{ kg}, 1367 \text{ kg}$
8-21 a) $-30 \text{ N}, 50 \text{ N}$ b) $5/3$ c) 58.3 N
   d) 1.0 m from right end
8-23 a) 270 N b) 120 N
8-25 a) $326 \text{ N}, 374 \text{ N}$ b) 224 N c) 236 N d) 864 N
8-27 a) $16.7°$ b) 24 cm c) 0.53
8-29 a) $66.7 \text{ N}, 100 \text{ N}$ b) $120.2 \text{ N}, 33.7°$ to vertical
8-31 b) Yes; it tips at $26.6°$ c) Tip before sliding; tip and slide at same angle
8-33 a) $100 \text{ N}, 700 \text{ N}$ b) 2.0 m

Chapter 9

9-1 a) 1.5 rad b) $1.57 \text{ rad}, 90°$ c) $120 \text{ cm}, 120 \text{ ft}$
9-3 a) 19.6 ft·s^{-1} b) 229 rev·min^{-1}
9-5 5 rad·s^{-2}, 1000 rad
9-7 20 rad·s^{-2}
9-9 32.5 in.·s^{-1}, 6.98 in.·s^{-1};
   33.2 in.·s^{-1}, $12.1°$ from radial line
9-11 a) At the bottom
   b) 35.6 m·s^{-2} at $3°$ from vertical
9-13 a) $135°$ b) 0.942 m·s^{-2}, 4.44 m·s^{-2}
9-15 $164,000 \text{ rev·min}^{-1}$
9-17 a) $a_\perp = 2\alpha\theta r$ b) 0.289 rad
9-19 a) 0.18 kg·m^2 b) 0.32 kg·m^2
   c) 0.50 kg·m^2 ($=0.18 \text{ kg·m}^2 + 0.32 \text{ kg·m}^2$)
9-21 a) 2.67 kg·m^2 b) 10.7 kg·m^2 c) $1.6 \times 10^{-3} \text{ kg·m}^2$
9-23 $11ml^2/16$
9-25 0.748 kg·m^2
9-27 $2(g/3k)^{1/2}$
9-29 a) 247 rev·min^{-1} b) 1.07 hp
9-31 a) 19.6 N·m b) 3084 J
9-33 a) 283 N·m b) $1.33 \times 10^5 \text{ J}$
9-35 a) 49.0 N b) 40.0 N c) 8.16 s
9-37 a) 6.25 rad·s^{-2} b) 250 J
   c) 4.74 rad·s^{-2}; tension is less than weight.
9-39 a) 65.3 N b) 16.2 m·s^{-1} c) 2.47 s
9-41 a) 3.27 m·s^{-2}, 0 rad·s^{-2}, $26.1 \text{ N}, 26.1 \text{ N}$
   b) 0.754 m·s^{-2}, 7.54 rad·s^{-2}, $21.1 \text{ N}, 36.2 \text{ N}$
9-43 a) 1.97 m·s^{-2} b) 9.85 N
9-45 $(2/3)g \sin\theta$, $(1/3)mg \sin\theta$, $(1/3)\tan\theta$ (3.27 m·s^{-2}, $mg/6$, 0.192)
9-47 $F/2m$, $F/2$
9-49 $2Ma^2/3$
9-51 a) 2.4 m·s^{-1}, 3.2 m·s^{-1} b) $5.0 \times 10^{-2} \text{ kg·m}^2$
   c) 16 m·s^{-2} d) 7.84 m·s^{-2} e) 0.506 N, tension
9-53 0.60 rad·s^{-1}

9-55 a) $5.61 \text{ rad} \cdot \text{s}^{-1}$ b) 0.204 m c) $918 \text{ m} \cdot \text{s}^{-1}$
9-57 1.44 J
9-59 a) $12.0 \text{ rad} \cdot \text{s}^{-1}$ b) 0.027 J
9-61 $(mg \pm \mu Mg)/M\omega^2$
9-63 $0.0789 \text{ rev} \cdot \text{s}^{-1}$
9-65 a) $-0.05 \text{ rad} \cdot \text{s}^{-1}$ b) $32.7°$ c) $36.0°$
9-67 a) $4.0 \text{ kg} \cdot \text{m}^2$ b) 5260 J
9-69 a) 1.76 N b) $4290 \text{ rev} \cdot \text{min}^{-1}$
9-71 a) $26{,}180 \text{ rad} \cdot \text{s}^{-2}$ b) $1.01 \text{ N} \cdot \text{m}$

Chapter 10

10-1 $4.8 \times 10^{11} \text{ Pa}$
10-3 $4.11 \times 10^8 \text{ Pa}, 0.054 \text{ mm}$
10-5 b) $1.02 \times 10^{11} \text{ Pa}$ c) $7.64 \times 10^7 \text{ Pa}$
10-7 a) 62.8 N b) 5.71 mm c) $1.83 \times 10^{-5} \text{ cm}$
10-9 1.66 mm
10-11 a) 1.82 m b) $1.5 \times 10^8 \text{ Pa}, 3.0 \times 10^8 \text{ Pa}$
   c) $1.36 \times 10^{-3}, 1.50 \times 10^{-3}$
10-13 $QP/2 = P^2Y/2 = Q^2/2Y$
10-15 $1.0 \times 10^8 \text{ Pa}, 0.5 \times 10^{-3}, 2.5 \text{ mm}, 9.5 \times 10^{-5}$
10-17 $4.0 \times 10^9 \text{ Pa}, 0.25 \times 10^{-9} \text{ Pa}^{-1}$
10-19 $4.71 \times 10^4 \text{ N}$
10-21 $0.633 \times 10^{-6} \text{ atm}^{-1}, 49.6 \times 10^{-6} \text{ atm}^{-1}$; water is 78 times more compressible.
10-23 0.17 m
10-29 1.66 cm

Chapter 11

11-1 $B = A \cos \theta_0, C = A \sin \theta_0$
11-3 a) $94.8 \text{ m} \cdot \text{s}^{-2}, 3.77 \text{ m} \cdot \text{s}^{-1}$
   b) $56.9 \text{ m} \cdot \text{s}^{-2}, 3.02 \text{ m} \cdot \text{s}^{-1}$ c) 0.0369 s
11-5 a) $7106 \text{ m} \cdot \text{s}^{-2}$ b) 3553 N c) $67.9 \text{ km} \cdot \text{hr}^{-1}$
11-7 22.5 kg
11-9 a) 7.88 Hz b) $9.8 \times 10^{10} \text{ Pa}$
11-11 a) 3.13 kg b) 4.33 cm below equilibrium, moving upward c) 36 N
11-13 a) $31.4 \text{ cm} \cdot \text{s}^{-1}$ b) $49.3 \text{ cm} \cdot \text{s}^{-2}$ c) 0.333 s
   d) 99.3 cm
11-15 0.777 s
11-17 a) 17.8 N b) $57.0 \text{ N}, 39.2 \text{ N}, 21.4 \text{ N}$
   c) $0.373 \text{ J}, 0.0711 \text{ J}$
11-19 a) $k_1 + k_2$ b) $k_1 + k_2$ c) $k_1 k_2/(k_1 + k_2)$ d) $\sqrt{2}$
11-21 a) $1.41 \text{ s}, 3.53 \text{ cm}$ b) Yes; friction c) No
11-23 0.248 m
11-25 4.80 s
11-27 a) $1.80 \text{ g} \cdot \text{cm}^2$ or $1.80 \times 10^{-7} \text{ kg} \cdot \text{m}^2$ [UNITS WRONG]
   b) $284 \text{ dyn} \cdot \text{cm} \cdot \text{rad}^{-1}$ or $0.284 \times 10^{-4} \text{ N} \cdot \text{m} \cdot \text{rad}^{-1}$
11-29 a) $7ML^2/48$ b) $2\pi(7L/12g)^{1/2}$
11-31 a) 0.0402 m^2 b) $0.698 \text{ rad} \cdot \text{s}^{-1}$
11-33 24.8 m

Chapter 12

12-1 $2.08 \times 10^5 \text{ Pa}, 2.05 \text{ atm}$
12-3 a) $1.077 \times 10^5 \text{ Pa}$ b) $1.037 \times 10^5 \text{ Pa}$
   c) $1.037 \times 10^5 \text{ Pa}$ d) 5 cm Hg e) $68 \text{ cm H}_2\text{O}$
12-5 a) $128{,}000 \text{ lb} \cdot \text{ft}^{-2}$ or $889 \text{ lb} \cdot \text{in}^{-2}$ b) $25{,}132 \text{ lb}$
12-7 a) $1/4$ b) $3/4$
12-9 0.893
12-11 a) 98 N b) 1.95 N
12-13 2.41 m^2
12-15 $8333 \text{ m}^3, 9266 \text{ kg}$
12-17 a) 5 cm b) 490 Pa
12-19 a) $1667 \text{ kg} \cdot \text{m}^{-3}$ b) E reads 2.5 kg, D reads 7.5 kg
12-23 100.85 g
12-25 $5.68 \text{ lb} \cdot \text{ft}$
12-27 Air bubbles do cause errors.
12-29 $8.84 \times 10^4 \text{ N} \cdot \text{m}$
12-31 a) 9.26 m b) Yes c) No effect
12-33 a) Yes b) For rubber, not for soap c) No
12-35 1372 Pa
12-37 $72.8 \text{ Pa}, 1.46 \times 10^4 \text{ Pa}$
12-39 a) 70.7 cm Hg b) 71.3 cm Hg c) 10.8 cm
12-41 4.3 cm
12-43 $3.6 \times 10^{-4} \text{ mm}$

Chapter 13

13-1 a) $14 \text{ m} \cdot \text{s}^{-1}$ b) $4.4 \times 10^{-3} \text{ m}^3 \cdot \text{s}^{-1}$
13-3 10.0 cm
13-5 $11.8 \text{ m} \cdot \text{s}^{-1}$
13-7 a) $1.22 \times 10^{-3} \text{ m}^3 \cdot \text{s}^{-1}$ b) 0.90 m
13-9 a) $8 \text{ m} \cdot \text{s}^{-1}$ b) $0.0628 \text{ m}^3 \cdot \text{s}^{-1}$
13-11 $4.95 \times 10^5 \text{ Pa}$
13-13 a) $6.62 \times 10^4 \text{ Pa}$ b) $0.266 \text{ m}^3 \cdot \text{s}^{-1}$
13-15 $0.0268 \text{ m}^3 \cdot \text{min}^{-1}$
13-17 $0.438 \text{ m}^3 \cdot \text{min}^{-1}$
13-19 $0.816 \text{ m}^3 \cdot \text{s}^{-1}$
13-21 $38.8 \text{ m} \cdot \text{s}^{-1}$
13-23 $4.0 \times 10^4 \text{ N}$
13-25 8.04 Pa
13-29 a) $0.769 \text{ cm} \cdot \text{s}^{-1}$ b) $1.89 \text{ cm} \cdot \text{s}^{-1}$
13-31 $107 \text{ m} \cdot \text{s}^{-1}$
13-33 a) Turbulent b) $5.3 \text{ liter} \cdot \text{s}^{-1}$
13-35 a) Yes b) No

Chapter 14

14-1 600.4 K
14-3 a) $20°\text{C}$ b) $35°\text{C}$ c) $-15°\text{C}$
14-5 6.83 cm Hg
14-7 3.6×10^{-4}
14-9 0.2506 in.
14-11 $20 \text{ cm}, 10 \text{ cm}$
14-13 76.5 m
14-15 a) $1.16 \text{ C}°$ b) Lose
14-17 a) 1.50 cm^3 b) 1.47 cm^3
14-19 $1.44 \text{ cm}^3, -0.0039 \text{ g} \cdot \text{cm}^{-3}$
14-21 $1.0 \times 10^{-5}(\text{C}°)^{-1}$
14-25 $1.20 \times 10^6 \text{ N}$
14-27 $1.04 \times 10^8 \text{ Pa}$

14-29 a) 0.036 ft b) 1.74×10^4 lb·in^{-2}
14-31 a) 9.01×10^7 Pa b) 36.2°C
14-33 5.04×10^7 Pa
14-35 1.009×10^{-4}(C°)$^{-1}$

Chapter 15

15-1 a) Yes b) No
15-3 4480 cal
15-5 371 years
15-7 a) 80.6
15-9 a) 1:0.093:0.031 b) 1:0.83:0.35
15-11 0.1 Btu·lb^{-1}·(F°)$^{-1}$
15-13 0.091 cal·g^{-1}·(C°)$^{-1}$
15-15 a) 72.3°C b) Temperature of apparatus does not change.
15-17 a) 273 J b) 3.42 J·mol^{-1}·K^{-1} c) 10.9 J·mol^{-1}·K^{-1}
15-19 a) 0.806 b) 8.36×10^4 cal
15-21 116 W, 13
15-23 357 m·s^{-1}
15-25 b) 3.01 J·g^{-1}·(C°)$^{-1}$
15-27 107 g
15-29 0°C with 0.2 g of ice remaining
15-31 539 cal·g^{-1}
15-33 23.9°C
15-35 39.9°C
15-37 1.84 kg
15-39 a) 1602 ft^3 b) 310 ft^3
15-41 a) 12,000 Btu·hr^{-1} b) 3.51 kW
15-43 a) 8.05×10^6 J b) 805 N

Chapter 16

16-1 4.32×10^5 J
16-3 a) 7.70 W b) 19.6 cm
16-5 0.205 Btu·in·hr^{-1}·ft^{-2}·(F°)$^{-1}$
16-7 111°C
16-9 a) -4°C b) 8 W·m^{-2}
16-11 9.48°C, 82.2°C
16-13 a) 90.0°C b) 21.7 W
16-15 a) 40°C b) 2.4 cal·s^{-1} or 10.1 W
16-17 76.4 C°
16-19 a) 7.30×10^5 cal b) 7.83×10^5 cal
16-21 a) 459 W·m^{-2} b) 4.59×10^6 W·m^{-2}
16-23 1.78 W
16-25 20.027 K
16-27 6.89 W
16-29 d) 31.4 C°·cm^{-1} e) 29 cal·s^{-1} f) Zero g) 1.11×10^{-4} m^2·s^{-1} h) 10.9 C°·s^{-1} i) 3.33 s j) Decrease k) 5.4 C°·s^{-1}

Chapter 17

17-1 1.16 kg
17-3 3.0×10^4 Pa
17-5 a) 203 mol b) 0.812 kg
17-7 a) 7.60×10^4 Pa b) 1.95 g
17-9 $\rho = pM/RT$
17-11 3.21
17-13 a) 82.1 cm^3 b) 0.333 g
17-15 When the piston has moved 13.1 in.
17-17 2.24 mg
17-19 0.544×10^5 Pa
17-25 321 kg·m^{-3}, 0.321
17-27 a) 37.3% b) 868 Pa c) 6.41 g·m^{-3}
17-29 a) 0.616 kg b) 25.7×10^{-3} kg·m^{-3} c) 0.035 kg
17-31 10.8 kg·hr^{-1}

Chapter 18

18-1 a) Yes b) No c) Negative
18-3 a) No b) Yes c) Yes d) Work done on the resistor equals heat transferred to water.
18-5 171 m
18-7 10.8 ft^3
18-9 $U_1 = U_2$
18-11 a) 0.167×10^6 J b) 2.03×10^6 J
18-13 $nRT \ln \dfrac{V_2 - nb}{V_1 - nb} + an^2\left(\dfrac{1}{V_2} - \dfrac{1}{V_1}\right)$
18-15 a) 366 J b) Yes; 366 J out c) Zero
18-17 a) 2.45×10^{-4} m^3 b) -735 J c) 1.529×10^6 J d) 1.528×10^6 J
18-19 b) 831 J c) On the piston d) 2110 J e) 2940 J f) 831 J
18-23 a) 314 atm, 189 K b) 3.79 atm, 227 K
18-25 42.0 J, zero
18-27 254°C
18-29 a) When piston is 3.13 in. from bottom b) 477 K

Chapter 19

19-1 a) 1 atm, 2.46×10^{-3} m^3; 2 atm, 2.46×10^{-3} m^3; 1 atm, 3.73×10^{-3} m^3 b) 52.0 J
19-3 15.8%
19-5 a) 320 K b) 20%
19-7 a) 900 cal b) 1600 cal c) 400 cal
19-11 a) 610 J·K^{-1} b) -568 J·K^{-1} c) 41.6 J·K^{-1}
19-13 11.5 J·K^{-1}
19-15 a) 4.19×10^5 J b) 1306 J·K^{-1} c) 50°C d) 1306 J·K^{-1}, 1408 J·K^{-1}
19-17 0.0503
19-19 a) 1.19 km^2 b) 2.36 km^2

Chapter 20

20-1 a) 3.34×10^{-7} cm b) About 10 times as great
20-3 1.95×10^{-4} cm
20-5 10^8 Pa (1000 atm)
20-7 2.82×10^{-10} m
20-9 a) 51.5 K b) 100 K
20-11 1.00636
20-13 a) 517 m·s^{-1} b) 298 m·s^{-1}

ANSWERS TO ODD-NUMBERED PROBLEMS

20-15 a) 8650 m·s^{-1}, 1.25×10^{-19} J b) 61,800 m·s^{-1}; no
20-17 a) 1.1 cm·s^{-1} b) No
20-19 5.21×10^{-25} J; 0.0251 K
20-21 Bi: 0.99 cm; Bi$_2$: 1.4 cm
20-23 10.4 J·g^{-1}·K^{-1}, 2.48 times as great
20-25 24.9 J·mol^{-1}·K^{-1}; 0.124, 0.205, 0.039

Chapter 21

21-1 a) 10.75 m b) 282 Hz
21-3 4302 Hz
21-5 a) 2.00 cm b) 30.0 cm c) 100 Hz
  d) 3000 cm·s^{-1}
21-9 a) 0.50 Hz
  b) π s^{-1}
  c) $\pi/100$ cm^{-1}
  d) $y = 10 \sin[(\pi x/100) - \pi t]$ e) $y = -10 \sin \pi t$
  f) $y = 10 \sin[(3\pi/2) - \pi t]$ g) 31.4 cm·s^{-1}
  h) $y = 7.07$ cm, $v = -22.2$ cm·s^{-1}
21-11 316 m·s^{-1}
21-13 a) 200 m·s^{-1}, 20 m
  b) Increase by $2^{1/2}$
21-15 63.3 m·s^{-1}
21-17 4.82×10^{-5} atm or 4.76×10^{-10} Pa^{-1}
21-19 a) 323 m·s^{-1} b) 1316 m·s^{-1} c) 348 m·s^{-1}
21-21 a) 138 Hz b) 1380 Hz c) 13,800 Hz
21-23 10^{-2} Y
21-25 b) 291 m·s^{-1}
  c) Small λ, large thermal conductivity
21-27 a) 0.58 m·s^{-1}·(C°)$^{-1}$ b) No

Chapter 22

22-1 a) 200 Hz b) 49 (50th harmonic)
22-3 a) 36.0 m·s^{-1} b) 64.8 N
22-5 1.28
22-7 a) 4960 m·s^{-1} b) 342 m·s^{-1}
22-9 Diatomic
22-11 a) 1130, 2260, 3390, 4520, 5650 Hz
  b) 565, 1695, 2825, 3955, 5085 Hz c) 17, 35
22-13 a) 420 Hz b) 415 Hz
22-15 12.2%
22-17 a) 0.655 m b) 56°C

Chapter 23

23-1 a) 9 b) 4
23-3 a) 60 db b) 77 db
23-5 b) No; $\beta = 160 + 10 \log I$
23-7 10^{-6} W
23-9 a) 0.252 μm b) 0.290 Pa c) 50 m
23-11 a) π b) 3.98×10^{-6} W·m^{-2}, 11.9×10^{-6} W·m^{-2}
  c) 2.13×10^{-6} W·m^{-2}, 63.3 db
23-13 a) 0.029 m b) 0.21 m c) 475 Hz
23-15 1.5000, 1.4983; just fifth is 0.11% larger
23-17 a) 9.89 m·s^{-1}
23-19 a) 0.63 m b) 0.75 m c) 548 Hz d) 460 Hz
  e) 573 Hz f) 440 Hz
  g) 0.65 m, 0.73 m, 547, 458, 569, 439 Hz
23-21 a) 464.3 Hz b) 468.8 Hz c) 4.5 Hz
23-23 a) 0.345 m b) 0.315 m c) 1.48 wave
  d) 345 m·s^{-1} e) 0.305 m

Part II Electricity and Magnetism, Light, and Atomic Physics

Chapter 24

24-1 625
24-3 0.119 m
24-5 a) 8.34×10^{19} N b) 8.34×10^{5} N
24-7 a) 2.75×10^{26} b) 6.59×10^{15} c) 2.40×10^{-11}
24-9 2.16×10^{-7} N
24-11 a) $q^2/2\pi\epsilon_0 a^2$, vertically upward
  b) $q^2 a/2\pi\epsilon_0(a^2 + x^2)^{3/2}$
24-13 a) 4.04×10^6 b) 4.93×10^{-17}
24-15 a) 5.09×10^5 N b) 1.44×10^6 N
24-19 a) 5.40×10^{-5} N, -3.45×10^{-5} N;
  b) 6.41×10^{-5} N, 32.6° below $+x$-axis

Chapter 25

25-1 a) 4.0 N·C^{-1} upward b) 6.4×10^{-19} N downward
25-3 a) 1010 N·C^{-1} b) 2670 km·s^{-1}
25-5 a) 1.42 cm b) 9.86 cm
25-7 -24.9×10^{-9} C
25-9 a) 1.80×10^4 N·C^{-1}, $-x$-direction
  b) 7.99×10^3 N·C^{-1}, $+x$-direction
  c) 3.34×10^3 N·C^{-1}, 70° above $-x$-axis
  d) 6.36×10^3 N·C^{-1}, $-x$-direction
25-11 40.8°
25-13 5.58×10^{-11} N·C^{-1}
25-15 $\sqrt{2}\lambda/4\pi\epsilon_0 r$
25-17 8.85×10^{-10} C
25-19 a) 2400 N·m^2·C^{-1} b) Zero
  c) 101 N·C^{-1}, $-x$-direction
25-21 b) $q/4\pi\epsilon_0 r^2$
25-25 b) $q/4\pi\epsilon_0 r^2$ or $\rho R^3/3\epsilon_0 r^2$
25-27 a) $Q/4\pi\epsilon_0 r^2$, 0, $Q/4\pi\epsilon_0 r^2$ b) $-Q/4\pi a^2$

Chapter 26

26-1 30 V; positively charged plate
26-3 a) 180 V b) 180 V
26-5 a) -4.5×10^{-5} J b) 3.0×10^5 N·C^{-1}
  c) -1.5×10^4 V
26-7 b) 0.667 mm
26-9 a) 1798 V b) Zero c) 4.49×10^{-5} J
26-11 a) 3.00 m b) 0.200×10^{-6} C

26-13 b) $q/2\pi\epsilon_0 a$ e) $\pm a\sqrt{3}$
26-15 a) It remains at rest. b) It oscillates about the origin, along the y-axis. c) It accelerates away from the origin along the x-axis.
26-17 b) Zero
   c) $q/4\pi\epsilon_0[(x^2 + 2ax + 2a^2)^{-1/2} - (x^2 - 2ax + 2a^2)^{-1/2}]$
26-19 c) Zero
26-21 1.03×10^7 m·s^{-1}
26-23 a) $Q/4\pi\epsilon_0(x^2 + a^2)^{1/2}$ b) $Qx/4\pi\epsilon_0(x^2 + a^2)^{3/2}$
26-25 c) $(\lambda/2\pi\epsilon_0) \ln (r_b/r_a)$
26-27 a) Inside $(\rho/4\epsilon_0)(R^2 - r^2)$; outside $-(\lambda/2\pi\epsilon_0) \ln (r/R)$
26-29 a) -539 V, $+59.9$ V b) -2.40×10^{-6} J
26-31 a) 4.33×10^{-4} m·s^{-1}
   b) 2.74×10^{-4} m·s^{-1} upward
26-33 938 MeV
26-35 2.27×10^{-14} m
26-37 a) 5114 V b) $0.140c$ c) 9.38 MeV, $0.140c$
26-39 a) 0.704 cm b) 19.4° c) 4.92 cm

Chapter 27

27-1 a) 50 pF b) 2.82×10^{-3} m^2 c) 400 V
   d) 1.0×10^{-6} J
27-3 a) 1000 V b) 2000 V c) 0.5×10^{-3} J
27-5 a) 0.38×10^{-6} C b) 7600 V c) 1.43×10^{-3} J
   d) 1.37×10^{-3} J
27-7 a) 1 μF b) 900 μC c) 100 V
27-9 a) 800 μC, 800 V; 800 μC, 400 V
   b) 533 μC, 533 V; 1067 μC, 533 V
27-11 1.30×10^{-3} J
27-13 0.86 cal
27-15 a) $-(\epsilon_0 A/x^2) dx$ b) $(Q^2/2\epsilon_0 A) dx$
   c) $Q^2/2\epsilon_0 A = \frac{1}{2}QE$
27-17 a) 0.226 m^2 b) 1250 V
27-19 a) 35.4 pF b) 1.77 nC c) 5000 V·m^{-1}
   d) 4.42×10^{-8} J
   e) 17.7 pF, 1.77 nC, 5000 V·m^{-1}, 8.84×10^{-8} J
27-21 a) $Q^2 x/2\epsilon_0 A$ b) $Q^2(x + dx)/2\epsilon_0 A$
27-23 1770 pF
27-25 a) 26.6 μC·m^{-2} b) 17.7 μC·m^{-2}
27-27 a) 3.43 b) 0.708×10^{-5} C·m^{-2}
27-29 $(Q/4\pi\epsilon_0)(1/r_a - 1/r_b)$

Chapter 28

28-1 a) 20 mA b) 2.74×10^{-6} m·s^{-1}
28-3 1.30×10^5 mm^{-3}
28-5 a) 2.5×10^6 A·m^{-2} b) 0.0430 V·m^{-1}
   c) 5 hr 41 min (2.05×10^4 s)
28-7 a) 2.4×10^{-8} Ω·m b) 20 A c) 3.90×10^{-4} m·s^{-1}
28-9 a) No c) About 1.0 Ω; smaller
28-11 a) 0.106 C b) Zero c) 12.7 A, zero
28-13 0.735×10^{-6} Ω
28-15 a) 99.52 Ω b) 0.0148 Ω
28-17 a) 22.1 Ω b) 0.363 A
28-19 28.5°C
28-21 284°C
28-23 60.9°C
28-25 2.47 mm
28-27 a) 0.50 Ω b) 10 V
28-29 Open: a) 12 V b) 0 c) 12 V
   Closed: a) 11.1 V b) 11.1 V c) 0
28-31 a) 2.22 A b) 1.4 A c) 6.4 V
28-33 90 Ω
28-35 a) VI, $I^2 R$, V^2/R b) 90 Ω
28-37 a) 24 W b) 4 W c) 20 W

Chapter 29

29-3 a) 8 Ω b) 12 V
29-5 9.6 W in 60-W lamp, 14.4 W in 40-W lamp
29-7 a) 141 V b) 4.5 W c) Two in series, connected in parallel with two others in series
29-9 a) Two in series, connected in parallel with two others in series. b) 1/2 W, if total power is 2 W.
29-11 655 W
29-13 a) -12 V b) 3A c) -12 V
   d) 12/7 A from b to a e) 4.5 Ω f) 4.2 Ω
29-15 a) 18 V b) Point a c) 6 V d) 36 μC
29-17 18 V, 7 V, 13 V
29-19 6 A, 4 A, 2 A, 4 A, 6 A (top to bottom)
29-21 a) 0.8 A, 0.2 A, 0.6 A b) -3.2 V
29-23 a) 1200 Ω b) 30 V
29-25 10.9 V, 109.1 V
29-27 a) 99.0 V b) 0.0526
29-29 4.97 Ω
29-31 2985 Ω, 12,000 Ω; 135,000 Ω; 3000 Ω; 15,000 Ω, 150,000 Ω
29-35 a) 1450 Ω b) 4500, 1500, 500 Ω c) No
29-37 55 Ω
29-39 2749 Ω
29-41 a) 9.52 V b) No
29-43 7.21 μF
29-45 a) 12 μs b) 2.53 V
29-47 a) 5.31 μC b) 2.0 mA c) 2.0 mA
29-51 a) 5.0×10^6 A·m^{-2} b) 3.34×10^{-10} A·m^{-2}
29-53 a) $i/\pi R^2$ b) $q/\pi\epsilon_0 R^2$ c) $i/\pi R^2$

Chapter 30

30-1 a) qvB along $-z$-axis b) qvB along $+y$-axis
   c) Zero d) $qvB/\sqrt{2}$ parallel to $-y$-axis e) qvB in yz-plane at 45° to $-y$- and $-z$-axes
   f) $\sqrt{2}\,qvB/\sqrt{3}$ in yz-plane at 45° to $-y$- and $-z$-axes
30-3 3.27 T, perpendicular to direction of v
30-5 a) 1.14×10^{-3} T, into page b) 1.57×10^{-8} s
30-7 Negative
30-9 8 mm
30-11 2.15×10^{-7} Wb
30-13 a) 4.0×10^6 V·m^{-1} b) 20,000 V
30-15 2.13×10^{-2} m

A-8 ANSWERS TO ODD-NUMBERED PROBLEMS

30-17 21
30-19 b) Yes

Chapter 31

31-1 1.20 N, $-z$-direction; 1.20 N, $-y$-direction; 1.70 N, 45° up, parallel to yz-plane; 120 N, $-y$-direction; zero
31-3 6.88×10^{28} m^{-3}
31-5 2.4 N
31-7 24 A
31-9 a) 0.16 N, 0.06 N, 0.0083 N·m b) 0.16 N, 0.104 N, 0.0048 N·m c) Same torque
31-11 3.6×10^{-6} N·m
31-13 a) 0.181 N·m b) When normal to coil is at 30° to field.
31-15 a) 0.5 A b) 4 A c) 108 V d) 60 W e) 48 W f) 540 W g) 71%

Chapter 32

32-1 1.6×10^{-5} T; yes
32-3 a) 20.0 μT b) 7.07 μT c) 20.0 μT d) Zero e) 5.44 μT
32-5 a) 125 A b) 2.5×10^{-4} T, 1.25×10^{-4} T
32-7 b) $\mu_0 Ia/\pi(a^2 + x^2)$ d) Zero
32-9 a) $\mu_0 I/\pi a$ b) $\mu_0 I/3\pi a$ c) Zero d) $2\mu_0 I/3\pi a$
32-11 a) 1.92×10^{-4} N·m^{-1}, down b) 1.92×10^{-4} N·m^{-1}, up
32-13 1.02 cm
32-15 7.20×10^{-4} N to left
32-19 $1.91R$
32-21 16
32-23 6.91×10^{-3} T
32-25 a) $\mu_0 I/2\pi r$ b) Zero
32-27 6.0×10^{-4} T
32-29 a) 7958 m^{-1} b) 1250 m
32-31 a) $(1/\epsilon_0 A)(dQ/dt)$

Chapter 33

33-1 a) 5.0 m·s^{-1} b) 2.0 A c) 0.96 N, to left
33-3 a) 3 V b) 3 V c) Upper end
33-5 a) 1.0 V, from B to A b) 1.25 N c) 5.0 W
33-7 0.006 V
33-9 b) FR/B^2l^2
33-11 a) 5.56 rad·s^{-1} b) Zero
33-15 a) 0.2 V b) Zero c) 0.2 V
33-17 a) $\pi r_1^2 (dB/dt)$ b) $(r_1/2)(dB/dt)$ c) $(R^2/2r_2)(dB/dt)$ e) $(\pi R^2/4)(dB/dt)$ f) $(\pi R^2)(dB/dt)$ g) $(\pi R^2)(dB/dt)$
33-19 a) Circles, clockwise b) 0.005 V·m^{-1}, 3.14 mV c) 1.57 mA d) Zero f) 3.14 mV
33-21 b) Zero c) 4 mV d) 2 mA e) Zero
33-23 a) Down b) Up c) Zero
33-25 a) $\mu_0 I/2r$, into plane b) $\mu_0 Il\,dr/2r$
c) $(\mu_0 Il/2) \ln(b/a)$ d) $(\mu_0 l/2)(dI/dt) \ln(b/a)$ e) 1.38×10^{-6} V
33-27 a) a to b b) b to a c) b to a
33-29 a) 1.01×10^{-5} C
33-31 a) 0.737 rev·s^{-1} b) 4320 N·m
33-33 7.96 rad·s^{-1}

Chapter 34

34-1 0.790 mH
34-3 1.0 mH; 2.25 mH or 0.25 mH
34-5 a) 5.0×10^{-4} V; yes b) 5.0×10^{-4} V
34-7 a) 0.100 J b) 2.00 W
34-9 $(1508\text{ V})\cos(120\pi\text{s}^{-1}t)$; leads by 90°
34-15 387
34-17 a) 0.05 A b) 1.0 A·s^{-1} c) 0.5 A·s^{-1} d) 0.23 s e) 0.0197, 0.0317, 0.0389, 0.0433 A
34-19 0.326 A to right
34-23 a) 2.39 mH b) 4.41 pF
34-27 a) $4L$ b) L c) $\omega/2$

Chapter 35

35-1 χ_m varies inversely with T.
35-3 a) 2000 A·m^{-1} b) 7.94×10^5 A·m^{-1} c) 397 d) 3.18×10^5 A e) 398
35-5 c) Diamagnetic d) 10 A, opposite e) It is repelled.
35-9 8 A
35-11 a) 0.045 A b) 0.012 A c) 0.15 A, 0.30 A

Chapter 36

36-1 a) 127 Hz b) 7.96 Hz
36-3 a) 377 Ω b) 2.65 mH c) 2650 Ω d) 2650 μF
36-5 a) 5 mA b) 0.05 A c) 0.5 A
36-7 a) 0.05 A b) 5.0 mA c) 0.50 mA
36-11 a) 583 Ω b) 0.0858 A c) 25.7 V, 42.9 V d) 59°, lead
36-13 a) 626 Ω b) First larger, then smaller c) 61.4°, current leads
36-15 16.9, 25.4, 56.4, 31.0, 35.4 V (rms)
36-17 b) 745 rad·s^{-1} or 118 Hz c) 1 d) 35.4, 79.2, 79.2, 0, 35.4 V (rms) e) 745 s^{-1} f) 0.354 A
36-19 a) 1592 Hz b) 12.4 V (rms)
36-21 0.0184 H
36-23 a) 25 W b) 25 W c) Zero d) Zero e) 0.127
36-25 a) 151 μF b) 917 W
36-27 a) 31.6 b) 3.16 V
36-29 a) 1592 Hz, 10,000 s^{-1} b) 1.0 A c) 1000 V d) 0.05 J
36-31 a) 150 b) 180 W c) 1.5 A

Chapter 37

37-1 a) 3.0×10^8 m·s^{-1} b) 294 m
c) 4.8×10^{-3} V·m^{-1}
37-3 a) 3.0×10^6 m·s^{-1} b) 3 cm
37-5 a) 1027 V·m^{-1}, 3.42×10^{-6} T b) 3.96×10^{26} W
37-7 About 6×10^4 V·m^{-1}, 2×10^{-4} T
37-9 a) $\rho I/\pi a^2$, parallel to wire
b) $Ir/2\pi a^2$, perpendicular to wire c) $I^2 r\rho/2\pi^2 a^4$
d) $2\pi rlS$
37-11 24.5 V·m^{-1}, 0.0650 A·m^{-1}
37-13 a) $(\mu_0 nr/2)(dI/dt)$, tangent
b) $(\mu_0^2 n^2 r/2)(I\, dI/dt)$, radially inward
37-15 300 m
37-17 a) 10^{10} Hz b) 1.0×10^{-7} T c) 1.19 W·m^{-2}
d) 1.99×10^{-9} N

Chapter 38

38-1 a) 1.5×10^{-9} m, 1.5×10^{-3} μm, 1.50 nm, 15 Å
b) 5.36×10^{-7} m, 0.536 μm, 536 nm, 5360 Å
38-3 16.6 min
38-9 1.732
38-11 a) 32.1° b) Independent of n
38-13 a) 2×10^8 m·s^{-1} b) 333 nm
38-15 1.87
38-19 No
38-21 b) 90°
38-23 1.88
38-27 1.02 cm
38-31 1.75°
38-35 4.54×10^{-5}
38-37 7.07 m
38-39 a) 44 b) 44 kW c) 354 hp

Chapter 39

39-1 Half the observer's height
39-5 a) 9.88 cm b) 5 cm
39-7 a) 3 b) 30 cm behind mirror
39-9 1.71 cm, 0.143
39-11 b) 10 cm behind mirror, 4 cm, erect, virtual
39-13 a) $|s| < 5$ cm b) Erect
39-15 a) 45° b) 1.41 cm
39-17 4.48 cm
39-19 1.35
39-21 30 cm, -1
39-23 0.667 cm

Chapter 40

40-1 $4R$ from center of sphere
40-3 a) The first image b) 30 cm c) Real
d) At infinity
40-5 50 cm
40-7 1.5 cm

40-11

R_1	R_2	f
10 cm	20 cm	40 cm
10 cm	-20 cm	13.3 cm
-10 cm	20 cm	-13.3 cm
-10 cm	-20 cm	-40 cm

40-13 a) 6 cm or 12 cm from object b) $-2, -0.5$
40-15 60 cm to right of third lens
40-17 -7.21 cm
40-19 a) 20.0 cm b) 18.7 cm
40-21 b) 12 cm
40-23 0.9 cm
40-25 a) 35 mm b) 200 mm
40-27 2.0 cm
40-29 2 cm left of first lens, 2 cm right of second lens
40-31 a) $+3.2$ diopters b) -2 diopters
40-33 a) 50 cm b) 2 m
40-35 a) 7.14 cm b) 3.5 mm
40-37 1/400 s
40-39 a) 120 mm b) 121.4 mm from lens
40-41 a) 842× b) 50×
40-43 380×
40-45 20×
40-49 25×, 35 cm from mirror

Chapter 41

41-1 1.0 mm
41-3 1.0 mm
41-5 a) Perpendicular to line joining antenna
b) $0.5I_0$, $0.197I_0$, $0.044I_0$, 0
41-7 a) 5.5 b) zero
41-9 b) 1.1 mm
41-11 0.909 mm
41-13 36.7 fringes·cm^{-1}
41-15 143 nm
41-17 a) Yes b) 89 nm
41-19 2.36 mm
41-21 17.5°, 36.9°, 64.1°
41-23 a) 570 nm b) 43.2°
41-27 0.048 mm
41-29 800 lines·mm^{-1}
41-31 Rainbow-fringed images

Chapter 42

42-1 a) $I_0/4$ b) Parallel to second filter
42-3 a) $(I_0/2)\sin^2\theta \cos^2\theta$
42-5 First filter at 45° to polarization of beam, second at 90° to beam; $I_0/4$
42-7 35.5°
42-9 a) 37° b) Horizontal
42-11 a) 58.8° b) 51.1°
42-13 0.005 mm
42-15 a) 1.732 b) 3
42-17 a) Yes b) Dark c) Dark
42-19 c) 0.582 μm

42-21 Linear, rotated 90°
42-23 Left

Chapter 43

43-1 a) 4.3×10^{-8} s   b) 10.4 m
43-3 The bolt at A appears to come first.
43-7 $\Delta s' = [(\Delta s)^2 - c^2 (\Delta t)^2]^{1/2}$
43-9 5.77×10^{-9} s
43-11 $v/c = E/mc^2$
43-13 2.60×10^8 m·s^{-1}
43-15 a) 0.5×10^{-4} b) 15.5%   c) Factor of 2.29
43-17 1.18×10^{-18} m·s^{-1}
43-19 2.986×10^8 m·s^{-1}
43-21 1.64×10^{-13} J
43-23 a) 4.55×10^{-15} J, 4.92×10^{-15} J
   b) 18.2×10^{-15} J, 27.9×10^{-15} J
43-25 a) 1.8×10^{14} J
   b) 1.8×10^{20} W
   c) 1.84×10^{10} kg
43-27 a) 1.8×10^5 eV   b) 6.91×10^5 eV
   c) 2.02×10^8 m·s^{-1} d) 2.52×10^8 m·s^{-1}
43-31 a) 0.047 MeV   b) 0.294 MeV   c) 7.2, 4.0

Chapter 44

44-1 $hf_0 = \phi$
44-3 1.72×10^6 m·s^{-1}
44-5 a) 4.58×10^{14} Hz   b) 654 nm   c) 1.90 eV
   d) 6.62×10^{-34} J·s
44-7 0.34 V greater
44-9 a) 1.04 eV   b) 1190 nm
   c) 2.52×10^{14} Hz
   d) 4.14×10^{-7} eV   f) No
44-11 1.41×10^{19} s^{-1}
44-13 a) 6.16×10^{14} Hz   b) 486 nm
44-15 1.96 eV or 3.14×10^{-19} J
44-17 a) 2.48×10^4 V   b) 0.00062 nm or 0.62 pm
44-19 a) 355 cal·s^{-1} b) 33.8 C°·s^{-1}
   c) High melting temperature and good thermal conductivity
44-21 79,056 K
44-23 4.85×10^{-12} m
44-25 a) 2.481×10^{-11} m   b) 2.496×10^{-11} m
   c) 49.7 keV

Chapter 45

45-1 a) 13.6 eV   b) -26.2 eV   c) -13.6 eV   d) 13.6 eV
45-3 a) Larger by factor of 4   b) Series ending at $n = 4$, $n = 5$, and one line of series ending at $n = 3$.
45-5 a) -2.80 keV   b) 2.57×10^{-13} m or 257 fm
   c) 0.59 nm
45-9 97.2 nm, 3.09×10^{15} Hz
45-11 656.3 nm, 364.6 nm
45-13 a) 12,400 eV   b) 150 eV
45-15 0.145 nm
45-17 8.67 pm
45-21 a) 2 eV, 621 nm
   b) 3.29×10^{-10} eV, 1.02×10^{-7} nm
45-23 2.34×10^{-23} s
45-25
l	0	1	1	1	2	2	2	2	2
m	0	-1	0	1	-2	-1	0	1	2

Chapter 46

46-1 $(1s)^2(2s)^2(2p)^2$, $(1s)^2(2s)^2(2p)^6(3s)^2(3p)^2$
46-3 $(1s)^2(2s)^2(2p)^6(3s)^2(3p)^6(3d)^{10}(4s)^2(4p)^2$
46-5 1.11 nm
46-7 Cesium
46-11 36,480 K
46-13 a) 4.58×10^{-48} kg·m^2
   b) 0, 2.43×10^{-21} J, 7.28×10^{-21} J
   c) 27.3 µm, 1.10×10^{13} Hz

Chapter 47

47-1 22.7 MeV, 3.64×10^{-12} J
47-3 35.6 MeV, 5.72×10^{-12} J
47-5 17.6 MeV, 2.81×10^{-12} J
47-7 0.137 u, 127.6 MeV, 7.97 MeV
47-9 $\Delta m = 0.000440$ u, 0.410 MeV
47-11 6.05 MeV
47-13 2.76×10^6 s^{-1}, 7.46×10^{-5} curie
47-15 0.782 MeV
47-17 a) 1.32 T   b) 4.22 MeV, 2.01×10^7 m·s^{-1}
47-19 23.8 MeV, 3.82×10^{-12} J, 2.3×10^{12} J·mol^{-1}; larger by 10^7
47-21 0.511 MeV (8.19×10^{-14} J), 1.24×10^{20} Hz, 2.43×10^{-12} m

INDEX

Aberrations, 757-760
Absolute temperature, 376-378
Absolute zero, 377-378
Absorption, of light, 712, 726-727
Absorption spectrum, 851-852
Accelerated motion, 59-64
Acceleration, angular, 181-182
    and torque, 190-192
  angular and linear, 182-183
  average, 42-43, 85
  centripetal, 95-96
  components, 85-88
  due to gravity, 50-52, 66-68
    table, 106
    variation with location, 106
  instantaneous, 43-44, 85
  normal, 87, 95
  parallel, 87, 95
  tangential, 87, 95
Accelerator, particle, 912-916
Accelerometer, 74
Acceptor, 893
Accommodation, 760
Acoustic phenomena, applications, 451-452
Acoustics, 434-451
Action and reaction, 23-24
Adaptation, 760
Addition, of vectors, 6-11
Adiabatic nature of sound wave, 413-415
Adiabatic process, 352
  for ideal gas, 358-360
Adiabatic wall, 284
Air, composition of, 331
Air table, 31-32
Air track, 31-32
Airfoil, 270
Airplane wing, 270
Alkali metals, 884-885
Alkaline-earth metals, 884-885
Alpha particle, 459, 897, 903-905

Alternating current, 678-692
Alternator, 633, 678
Ammeter, 567-571
Amorphous solid, 397
Ampere, definition, 535, 616
Ampère's law, 619-623
  for magnetic intensity, 669
  for magnetization, 669
Amplifier, 843
Amplitude, 225
Angular acceleration, 181-182
  and torque, 190-192
Angular frequency, 228
Angular harmonic motion, 235-236
Angular impulse, 195-196
Angular magnification, 762
  compound microscope, 765-766
  telescope, 766-767
Angular momentum, 195-200
  conservation of, 196-200
  orbital, 875-878
  in quantum mechanics, 875-876
  spin, 876-878
  as vector, 200-202
Angular velocity, 179-180
Anisotropy, 398
Anode, 509, 842-843
Anomalous expansion, 292
Antiderivative, 46
Antinode, 423, 426
Antiproton, 923
Apparent weight, 73-74
Aqueous humor, 759
Arc lamp, 712
Archimedes' principle, 249-252
Aspirator pump, 269
Astigmatism, 761
Atmosphere, 248
Atmospheric pressure, variation with elevation, 332-333
Atomic mass unit, 592-593, 901

Atomic number, 459, 900
Atomic spectra, 846-848, 850-851
Atomic structure, 458-460, 883-885, 897-899
Atwood's machine, 207
Audibility, limits of, 436-439
Avogadro's number, 388-389

Back emf, 605
Ball, spinning, curve of, 271
Ballistic pendulum, 150-151
Balloon, 250
Balmer series, 847
Banked curve, 98-99
Bar, 247
Barometer, 247-248
Baryon, 923
Baryon number, 924
Base, 894
Beam splitter, 785
Beats, 442-443
Bel, 438
Bernoulli's equation 266-271
Beta particle, 905
Binding energy, nuclear, 901-902
Binocular, 767
Biot and Savart, law of, 612-615
Birefringence, 811-813
Blackbody radiator, 323
Blind spot, 760
Body temperature, 288
Bohr model, of atom, 459
  of hydrogen atom, 862-867
Boiling, 303-307
Boiling point, 340
Boundary conditions, 420, 426
Boundary layer, 277
Bourdon gauge, 248
Boyle's law, 330
Brackett series, 847
Branch point, 564

I-1

Breeder reactor, 917
Brewster's law, 810–811
British thermal unit, 299
British units, 63
Bubble, 256
Bubble chamber, 342–343, 589, 921
Bulk modulus, 217
   adiabatic, 413
   table, 216
Buoyancy, 249–252

Calorie, 299, 301
Calorimetry, 302
Camera, 763–764
Cancellation, 429–430, 775
Candela, 728
Capacitance, 516–520
Capacitor, 516, 520
   charge and discharge, 571–574
   energy in, 520–522
   in parallel, 519–520
   in series, 519
   variable, 518
Capillarity, 257–259
Carbon dating, 911
Carnot cycle, 371–375
Cathode, 509, 842–843
Cathode-ray oscilloscope, 509–511
Cavendish balance, 65
Celsius temperature, 288–289
Center of curvature, 736
Center of gravity, 170–172
Center of mass, 155–157
Centrifugal force, 98
Centripetal acceleration, 95–96
Centripetal force, 96–101
Change of phase, 303–307
Charge, 457–461
   conservation of, 460, 536
   of electron, measurement, 506–508
   induced, 524–528
   within conductor, 484–485
Charge density, linear, 476
   surface, 477
Chemical bond, 885–889
Chromatic aberration, 758
Circle of least confusion, 758
Circle of reference, 232–234
Circuit, 541–554, 562–574, 678–691
Circular motion, 95–101
Circular polarization, 815–818
Close packing, 397, 889–890
Cloud chamber, 343, 589, 921
Coating, on glass, 783–784
Coefficient of expansion, 290–292
   tables, 290, 291
Coefficient of friction, 30–31
   table, 31
Coercive force, 674
Coercivity, 674
Coherence, 854
Coherent light, 775–776, 802
Collector, 894
Collisions, 148
   elastic, 149, 151–154
   inelastic, 149–151

   semielastic, 153
   superelastic, 154
Coma, 758
Combination tones, 443
Combustion, heat of, 307
Commutative law, 8, 12–15
Commutator, 604
Compass, electromagnetic, 602
Compensator plate, 785
Complete circuit, 543
Compliance, 220
Components, of vectors, 9–12
Compound motor, 604
Compressibility, 217–218
   table, 218
Compton effect, 712
Compton scattering, 857–859
Condensation, 405
Condenser lens, 764–765
Conduction, electrical, theory of, 556–557
   of heat, 313–318
Conductivity, of solids, 891
   thermal, 313–318
   table, 315
Conductor, 891
   electrical, 460
Conic section, 102
Conical pendulum, 97–98
Conservation, of charge, 460, 536
   of energy, 124, 350
   of mass and energy, 834
   of momentum, 147–148
Conservative force, 131–132
Conservative force field, 494
Consonance, 440, 443
Constant acceleration, 45–52
Constants, fundamental, T-9
Constructive interference, 429–430, 774–775
Contact angle, 257–258
   table, 258
Contact force, 18
Continuity equation, 265–266
Contraction, of length, 827–829
Control rod, 917
Convection, 318–320
Convection coefficient, 320
   table, 320
Conversion factors, T-8
Corpuscles, 711
Correspondence principle, 835
Cosmic rays, 921
Coulomb, 463
Coulomb's law, 461–466
Couple, 172–173
Covalent bond, 886–887
Covalent crystal, 889
Critical angle, for total internal reflection, 721
Critical damping, 659–660
Critical point, 338–339
   table, 339
Cross product, 13–15
Crystal diffraction, 797
Crystal structure, 888–890
Crystals, 396–398
   heat capacity of, 398–399

Curie temperature, 673
Current, alternating, 678–692
   displacement, 574–575
   and magnetic field, 623–625
   electric, 534–537
   physiological effects, 554–555
Current amplitude, 679
Current density, 536
Current-voltage relations, 549–550
Curvature of field, 737, 758
Curve of spinning ball, 271
Cyclic process, 364–365
Cyclotron, 913–915
Cylindrical lens, 762

d'Arsonval galvanometer, 567–571, 603–604
Dalton, John, 388
Dam, force and torque, 252–253
Damped harmonic motion, in L-C circuit, 658–660
Daughter nucleus, 909
Davisson-Germer experiment, 867–869
de Forest, 843
Decay rate, 908
Decay series, radioactive, 909–911
Decibel, 438
Declination, 676
Defects of vision, 761–762
Deflecting plates, 509
Degree of freedom, 395
Density, 67, 244
   of nuclei, 899
   table, 244
Depth of field, 764
Derivative, and velocity, 41
Destructive interference, 429–430, 775
Deuteron, 912
Dew point, 342
Diamagnetism, 664–670
Diamond structure, 889
Diaphragm, 763
Diaphragm pump, 249
Diathermic wall, 284
Diatomic gas, 357, 394
Diatonic scale, 441–442
Dichroism, 813
Dielectric, 522–526
Dielectric breakdown, 523, 526
Dielectric constant, 523–526
   table, 523
Dielectric strength, 501
Diesel cycle, 367–368
Diesel engine, 359–360
Diffraction, 787–802
   of electrons, 867–869, 871–872
   grating, 793–797
   single slit, 789–793
   of sound, 449–450
   x-ray, 797
Diffuse reflection, 720
Diffusion pump, 249
Dilation of time, 825–827
Dimensional consistency, 4–5
Diode, 549, 842
   semiconductor, 894
Diopter, 762

Dip, angle of, 675
Dipole, electric, 472, 527
Dipole moment, electric, 472
Direct-current circuits, 541–554, 562–574
Dislocation, 890
Disorder, 380–382
Dispersion, 415–417
   of light, 725–726
Displacement, 7
   electric, 529–530
Displacement current, 574–575
   and magnetic field, 623–625
Dissipative force, 131–132
Dissonance, 443
Diverging lens, 754
Domain, magnetic, 672–673
Donor, 893
Doppler effect, light, 446–447
   sound, 443–446
Dot product, 12–13
Double refraction, 811–813
Drift velocity, 534
Dulong and Petit, law of, 302, 399
Dynamic resistance, 549
Dynamic similarity, 278
Dynamics, 59
Dyne, 18, 62–63

Earth, electric field of, 555–556
   magnetic field, 675–676
Eddy current, 642–643
Efficiency, luminous, 732
   Otto cycle, 366–367
   thermal, 366
Efflux, speed of, 268
Elastic limit, 215
Elastic modulus, 216–219
   table, 216
Elasticity, 211–220
   modulus of, 216, 219
      table, 216
   and plasticity, 214–216
Electric charge, 457–461
Electric dipole, 472
Electric field, 469–480
   of earth, 555–556
   induced, 638–639
   of long charged wire, 476–477
   of plane sheet of charge, 477–479
   of ring of charge, 475–476
Electric field lines, 479–480
Electric flux, and displacement current, 575
Electrical interactions, 466
Electrical oscillation, 656
Electrocardiography, 555
Electroencephalography, 555
Electrolytic capacitor, 522
Electromagnetic pump, 605–606
Electromagnetic spectrum, 715
Electromagnetic wave, 696–708
   energy in, 700–702
   in matter, 703
   momentum, 702
   sinusoidal, 703–705
   speed, 698–700

   standing, 705–706
Electromotance, 542
Electromotive force, 541–545
   induced, 605
   motional, 630–633
Electron, 458–460
   charge–mass ratio, 590–591
   charge of, 464
Electron affinity, 886
Electron charge, measurement, 506–508
Electron diffraction, 867–869, 871–872
Electron gun, 509
Electron microscope, 869–870
Electron spin, 876–878
Electron-gas model of solid, 889
Electron-pair bond, 886–887
Electronvolt, 508–509
Electrostatic unit, 463
Electrostatics, 457–466
Electrovalent bond, 885–886
Elevator, 70, 73
Elliptical polarization, 815–818
Emission, of light, 712
Emissivity, 322, 324
Emitter, 894
Energy, in charged capacitor, 520–522
   in collisions, 149–151
   conservation, 350
   elastic potential, 128–131
   in electric field, 521
   in electromagnetic waves, 700–702
   in fluid flow, 266–271
   of gas molecules, 391–395
   gravitational potential, 122–128
   in inductor, 652–653
   internal, 349–351
      of ideal gas, 355
   kinetic, 120–121
      of rotation, 183–186
   in magnetic field, 653
   of magnetic moment, 601
   and mass, 135–137
   potential, 122–131
      electrical, 494–497
      internal, 133
   relativistic, 135–137, 832–835
   in simple harmonic motion, 226–227
   in thermodynamics, 346
   and work, 120–122
Energy conversion, 364–366, 382–383
Energy density, in electric field, 521
   in magnetic field, 653
Energy levels, atomic, 848–852
   hydrogen atom, 864–865
   molecular, 887–888
   nucleus, 902
Engine, internal-combustion, 366–368
Entropy, 378–380
   and disorder, 380–382
   and second law, 380–382
Equal temperament, 441–442
Equation of state, 328–333
Equality, of vectors, 7
Equilibrium, phase, 336–338
   of point, 18–33
   of rigid body, 166–173

Equilibrium state, of thermodynamic system, 328
Equipartition of energy, 395
Equipotential surface, 503–504
Equivalent resistance, 562–564
Erg, 116
Ether, 786–787, 823
Excited state, 864
   of nucleus, 902
Exclusion principle, 881–883
Expansion, thermal, 289–292
Exploding universe, 447
External force, 147, 156
Extraordinary ray, 811
Eye, 759–762
Eyepiece lens, 765–767

Face-centered-cubic crystal, 397
Fahrenheit temperature, 288–289
Falling body, 50–52
Far point, 760
Farad, 516
Faraday disk dynamo, 634
Faraday's law, 625, 630–643
Farsightedness, 761
Fermat's principle, 730
Ferrite, 643
Ferromagnetism, 664–675
Fiber optics, 722
Field lines, electric, 479–480
   magnetic, 587–588
First law of thermodynamics, 350, 354–355
Fission, nuclear, 916–918
Fixed stars, 23
Fizeau, 714
Flow line, 264
Flow tube, 264
Fluid, 213–214, 244
Fluid dynamics, 264–278
Fluorescent lamp, 713
Fluorescent screen, 509
Flux, electric, 482
   and displacement current, 575
   magnetic, 587–588
   and induced emf, 634–639
Flux density, magnetic, 588, 670
Focal length, of lens, 752–753
   of mirror, 740–741
Focal plane, 753
Focal point, of lens, 752–753
   of mirror, 740–741
   virtual, 754
Focus, of lens, 752–753
   of mirror, 740–741
   virtual, 754
Foot-candle, 727
Foot-pound, 116
Force, 59–64
   centrifugal, 98
   centripetal, 96–101
   conservative, 131–132
   on a dam, 252–253
   dissipative, 131–132
   intermolecular, 386
   magnetic, between conductors, 615–616
      on current element, 596–597

Force (continued)
    on current loop, 599–602
    on moving charge, 584–587
Force constant, 219–220
Force diagram, 25
Forced oscillation, 430–431
Fovea centralis, 759
Franck-Hertz experiment, 849
Fraunhofer diffraction, 789–793
Fraunhofer lines, 851
Free fall, 50–52
Free-body diagram, 25, 69
Frequency, 225
    angular, 228
    of L-C circuit, 658
    of orbital motion, 103
    of wave, 404
Fresnel diffraction, 787–802
Friction, 29–33
    coefficient of, 30–31
        table, 31
Full-scale deflection, 567
Fundamental constants, T-9
Fundamental frequency, 425, 428
Fundamental particles, 920–926
Fusion, nuclear, 918–920

Galilean telescope, 772
Galilean transformation, 823
Galileo, 18, 50
Galvanometer, 567–571, 603–604
Gamma ray, 902, 905–906
Gas, kinetic theory, 389–395
Gas constant, 329
Gas thermometer, 286–287
Gauss, 586
Gauss's law, 480–489
    for displacement, 529–530
    for polarization, 529
General theory of relativity, 836
Generator, 633
Geometrical optics, 717–722, 773
Gradient, 505
Gram, 62–63
Grating, diffraction, 793–797
Gravitation, 64–68
Gravitational constant, 64
Gravity, acceleration due to, 50–52, 66–68
        effect of earth's rotation, 105–107
        table, 106
        variation with location, 106
    center of, 170–172
    specific, 245
Grand unification scheme, 926
Greek alphabet, T-5
Ground state, 864
    of nucleus, 902
Gyromagnetic ratio, 877
Gyroscope, 201–202

Hadron, 923
Half-life, 908
    for R-C circuit, 573
Half tone, 441
Hall effect, 597–599
    in semiconductors, 893

Halogens, 884–885
Harmonic motion, in L-C circuit, 658–660
Harmonic oscillator, 224–232
    angular, 235–236
Harmonics, 425, 428
Heat, 297–308
    in volume change, 348–349
Heat capacity, 299–303
    of gas, 394
    of ideal gas, 355–358
        table, 357
    of solid, 398–399, 890–891
Heat current, 314
Heat engine, 364–366
Heat of combustion, 307
Heat of fusion, 304
    table, 305
Heat of vaporization, 305
    table, 305
Heat reservoir, 365
Heat transfer, 297–298, 313–324
Heisenberg uncertainty principle, 872–874
Helium, 339–340
Henry, 650
Hertz, 225
Hertz, Heinrich, 697
Heteropolar bond, 885–886
High-energy physics, 922–926
Hole conduction, 598
Holography, 800–802
Homopolar bond, 886–887
Hooke's law, 215–216
Hovercraft, 31
Humidity, 340–342
Huygens' principle, 722–725
Hydraulic press, 246
Hydrodynamics, 264–278
Hydrogen atom, Bohr model, 862–867
Hydrometer, 250
Hydrotherapy, 252
Hyperfine structure, 899
Hyperopia, 761
Hypodermic syringe, 275
Hysteresis, 673–674
    elastic, 215

Ideal fluid, 264
Ideal gas, 329–334
    adiabatic process, 358–360
    heat capacities, 355–358
    internal energy of, 355
    kinetic theory of, 389–395
Ideal radiator, 323
Idealized model, 24–25
Illuminance, 727
Illumination, 727–728
Image, 733–747, 752–757
    hologram, 802
    as object, 750–751, 757
    real, 733
    virtual, 733
Impedance, 683, 688–690
    of free space, 701
    mechanical, 436

Impedance matching, 692
Impulse, 143–145
    angular, 195–196
Impulsive torque, 196
Incandescent lamp, 712
Inclination, magnetic, 675
Incremental resistance, 549
Indefinite integral, 46, 49
Index of refraction, 703, 719–720
    table, 719
Induced charge, in dielectric, 524–528
    molecular theory of, 526–528
Induced electric field, 638–639
Induced electromotive force, 605
Inductance, 649–653
Induction, charging by, 460–461
    electromagnetic, 630–643
Inert gases, 884–885
Inertia, 60
    moment of, 183–189
        table, 185
Inertial frame of reference, 822
Inertial reference system, 22–23
Initial velocity, 45
Insulator, 891
    electrical, 460
Integral, indefinite, 46, 49
    line, 119–120
Integrated circuits, 895
Integration, 48
Intensity, of electromagnetic wave, 701
    luminous, 728
    of sound, 436–439
    in two-slit interference, 778–780
Interactions, classes of, and conservation laws, 923
Interference, 423, 773–802
    and beats, 442–443
    of light, 773–802
    of sound waves, 429–430
    in thin films, 780–784
Interferometer, Michelson, 784–787
Internal energy, 349–351
    of ideal gas, 355
Internal force, 147, 156
Internal reflection, 721–722
Internal work, 132
Internal-combustion engine, 366–368
International System of Units, 2–5, T-1–T-2
Interstitial atom, 890
Intrinsic conductivity, 892
Invariance of physical laws, 822–824
Ionic bond, 885–886
Ionic crystal, 889
Ionization, 458
Ionization energy, 886
Irreversible process, 370–371
Isentropic process, 379
Isobar, 335
Isobaric process, 335, 353
Isochoric process, 352
Isochronous system, 236
Isospin, 924
Isothermal process, 335, 352–353
Isotopes, 591–592, 900

Jerk, 76
Joule, 116
Junction, *p-n*, 893–895
Just diatomic scale, 441

Kelvin, Lord, 1
Kelvin temperature, 287–289
Kelvin temperature scale, 376–378
Kepler's laws, 104–105
Kerr effect, 818
Kilogram, 62–63
    definition, 2
Kinematics, 39
Kinetic energy. *See* Energy, kinetic
Kinetic friction, 30
Kinetic theory of gases, 389–395
Kirchhoff's loop rule, 547
Kirchhoff's rules, 564–567

Lambert's law, 726–727
Laminar flow, 265, 272
Lamps, 712–713
Laser, 713, 852–854
Latent heat, 305
Lateral magnification, lens, 753
    plane mirror, 734–735
    plane refracting surface, 743
    spherical mirror, 737–738
    spherical refracting surface, 746
Law of reflection, 718
Law of refraction, 718
Leaning Tower of Pisa, 50
Length, contraction, 827–829
Lens, diverging, 754
    thin, 751–754
Lensmaker's equation, 752
Lenz's law, 640–642
Lepton, 923–925
Light, emission and absorption, 840–841
    monochromatic, 716–717
    nature of, 711–712
    sources of, 712–713
    speed, 713–715
Light pipe, 722
Line integral, 119–120, 497
Line of flow, 264
Line spectra, 840–841, 846–848, 850–851
Linear momentum. *See* Momentum
Lines of force, electric, 479
    magnetic, 587
Lloyd's mirror, 781
Longitudinal magnification, lens, 753
    plane mirror, 735
    plane refracting surface, 743
    spherical mirror, 738
    spherical refracting surface, 746
Loop, 564
Lorentz transformation, 829–830
Loschmidt's number, 389
Loudness, 437–439
Lumen, 727
Luminous efficiency, 732
Luminous intensity, 728
Lyman series, 847

Magnet, permanent, 602
Magnetic declination, 675
Magnetic domain, 672–673
Magnetic field, 584–587
    of circular loop, 616–619
    of current element, 612–613
    and displacement current, 623–625
    of earth, 675–676
    energy, 653
    of long straight conductor, 613–615
    of moving charge, 609–610
    of solenoid, 621–622
    sources, 609–625
    of toroidal solenoid, 622–623
Magnetic field lines, 587–588
Magnetic flux, 587–588
    and induced emf, 634–639
Magnetic flux density, 670
Magnetic force, between conductors, 615–616
    on current element, 596–597
    on current loop, 599–602
    on moving charge, 584–587
Magnetic induction, 670
Magnetic intensity, 668–671
Magnetic lines of force, 587
Magnetic moment, of current loop, 600
Magnetic permeability, 665–671
Magnetic properties of matter, 664–676
Magnetic susceptibility, 670
    table, 671
Magnetism, 583–584
    molecular theory, 666–668
Magnetization, 668–671
Magnetization curve, 671–672
Magneto, 643
Magnification, lateral. *See* Lateral magnification
Magnification, longitudinal. *See* Longitudinal magnification
Magnifier, 762–763
Magnitude, of vector quantity, 6–11
Malus' law, 808–810
Manometer, 247–248
Marconi, 697
Mass, 59–64
    center of, 155–157
    and energy, 135–137
    and weight, 66–68
Mass defect, 901
Mass spectrometer, 591–592
Mathematical relations, T-3–T-4
Maxwell, J. C., 574, 588
Maxwell-Boltzmann distribution, 396
Maxwell's equations, 697
Mechanical impedance, 436
Melting, 303–307
Mesons, 922–924
Metallic crystal, 889
Metastable state, 853
Meter, definition, 2
Methane, 887
Michelson, A. A., 715
Michelson interferometer, 784–787
Michelson-Morley experiment, 786–787
Micron, 716

Microscope, compound, 765–766
    electron, 869–870
Millikan oil-drop experiment, 506–508
Milliliter, 292
Millimicron, 716
Mirror, plane, 733–735
    spherical, 735–740
Mode, 425
Model, 24–25
Moderator, 152
Molar heat capactiy, 299–303
    table, 300
Molecular force, 386
Molecular mass, 300
Molecular speed, 391–396
Molecular theory of matter, 386–388
Molecular vibration, 888
Molecular weight, 300
Molecule, diatomic, 885–889
    structure of, 885–888
Moment, of force, 166–167, 173
    of momentum, 195
    vector, of force, 173
    of weight, 170–172
Moment of inertia, 183–189
    table, 185
Momentum, conservation of, 147–148
    in electromagnetic wave, 702
    relativistic, 159–160, 831–832
Monatomic gas, 357, 394
Monkey business, 82
Monochromatic aberrations, 758
Monochromatic light, 716–717, 773
Moons of Jupiter, 104
Mössbauer effect, 165
Motion, circular, 95–101
    constant-acceleration, 45–52
    in a plane, 83–109
    of planet, 101–105
    of projectile, 88–94
    rectilinear, 39–53
    satellite, 101–105
    straight-line, 39–53
Motion of hcarge particles, 588–592
Motional emf, 630–633
Motor, direct-current, 604–605
Moving axis, 192–194
Musical intervals, 440–442
Musical scales, 440–442
Muon, 922, 924
Mutual inductance, 649–650
Myopia, 761

Natural frequency, of L-C circuit, 658
Near point, 760
Nearsightedness, 761
Negative, of vector quantity, 7
Negative pressure, 259
Networks, electrical, 562–574
Neutron, 458–460, 900, 920
Neutron activation analysis, 912
Newton, 18, 62–63
Newtonian fluid, 273
Newtonian synthesis, 105
Newton's first law, 20–23
Newton's law of gravitation, 64–68

Newton's laws, 18
Newton's rings, 783
Newton's second law, 59–64
　applications, 68–75
Newton's third law, 23–24
Nodal plane, 706
Node, 423, 426
Noise levels, 438
Non-electrostatic field, 542–544
Non-electrostatic force, 597, 630–631, 638–639
Nonequilibrium process, 370–371
Nonlinear optics, 774
Nonreflective coatings, 783–784
Normal force, 29–31
Normal mode, 425, 707
Nuclear fission, 916–918
Nuclear force, 466
Nuclear fusion, 918–920
Nuclear reaction energy, 911
Nuclear reactions, 911–912
Nuclear reactor, 916–918
Nuclear spin, 899
Nucleon, 900
Nucleus, 897–911
Nuclide, 900

Objective lens, 765–767
Octane, 440
Ohm, 540
Ohmmeter, 571
Ohm's law, 538–539
Open circuit, 541–542
Operational definition, 1
Optic axis, 735
Optic nerve, 759
Optical activity, 818–819
Optical pyrometer, 286
Optical stress analysis, 818
Orbit, of planet or satellite, 101–105
Orbital angular momentum, 875–878
Order, in solids and liquids, 388
Ordinary ray, 811
Organ pipes, 427–428
Oscillation, in L-C circuit, 656–660
Oscillatory motion, 223
Oscilloscope, 509–511
Otto cycle, 366–368
Overdamping, 659–660
Overtone, 425, 428

Pair production, 589, 921–922
Pairing, and nuclear force, 903
Parabolic trajectory, 88–94
Parallel capacitors, 519–520
Parallel resistors, 562–564
Parallel-axis theorem, 194–195
Parallel-plate capacitor, 517–519
Paramagnetism, 664–670
Paraxial approximation, 737, 743, 745, 757–759
Paraxial ray, 737
Parent nucleus, 909
Parity, 924
Partial derivative, 407
Particle, 24–25

Particle accelerator, 912–916
Particles, wave nature, 867–869
Pascal, 247
Pascal's law, 246
Paschen series, 847
Pauli exclusion principle, 881–883
Pendulum, ballistic, 150–151
　conical, 97–98
　physical, 238–239
　simple, 236–237
Perfect gas. See Ideal gas
Performance coefficient, 368
Period, 225
　of orbital motion, 103
　of wave, 404
Periodic motion, 223–224
Periodic table of elements, 883–885, T–7
Periodic time, 225
Periodic wave, 404–408
Permanent magnet, 602, 664, 673–675
Permeability, 665–671
　relative, 665
Permittivity, 524–528
Pfund series, 847
Phase, in ac circuits, 680–683
Phase change, 303–307
　magnetic, 673
Phase diagram, 336–338
Phase equilibrium, 287, 336–339
Phase shift, in reflection, 781–782
Phasor, 232, 679, 779
Photoelasticity, 818
Photoelectric effect, 841, 843–846
Photon, 159–160, 841, 844, 855–859
Physical optics, 773
Physical pendulum, 238–239
Physical quantity, 1
Physiological effects of currents, 554–555
Picture tube, television, 509–511
Pitot tube, 270
Pion, 922, 924
Pitch, 439–440
Planck's constant, 159, 844, 863, 868
Plane, motion in, 83–109
Plane wave, 707
Planet, motion of, 101–105
Plasticity, 214–216
Pohl's experiment, 778
Poise, 273
Poiseuille's law, 273–275
Poisson's ratio, 216–217
　table, 216
Polar molecule, 526–527
Polarization, circular, 815–818
　of dielectric, 528–530
　elliptical, 815–818
　Gauss's law for, 529
　light, 806–819
　of molecule, 527
　percent, 808–810
　by reflection, 810–811
　in scattered light, 814–815
Polarizers, 812–814
Polarizing axis, 807
Polarizing filter, 807
Polaroid filter, 726, 813

Polyatomic gas, 394
Polycrystalline solid, 397
Population inversion, 852–853
Porro prism, 722
Position vector, 83
Positron, 589, 921–922
Potential, 494–506
Potential difference, 500–503
Potential energy. See Energy, potential
Potential gradient, 504–506
Potential well, 387
Pound, 63
　definition, 3–4
Pound-force, 18
Power, 134–135
　in ac circuit, 686–688
　in electric circuits, 550–554
　in rotational motion, 189–190
　and velocity, 135
Power factor, 687–688
Powers of ten, 6
Poynting vector, 701
Prandtl tube, 270
Precision, 5–6, 202
Prefixes, in SI units, 2–3
　for units, 3
Presbyopia, 760
Pressure, 213–214, 245
　absolute, 247
　gauge, 247
　inside bubble or drop, 256
　negative, 259
Pressure amplitude, 435
Pressure gauges, 247–248
Primary winding, 690
Principle of relativity, 822
Principal quantum number, 876
Principal-ray diagram, 741–742, 755–757
Prism, dispersion by, 725
Probability, in quantum mechanics, 870–876
Products of vectors, 12–15
Projectile motion, 88–94
Projector, 764–765
Propagation constant, 406
Proper length, 828, 837
Proper time, 827, 837
Properties of matter, 389
Proportional limit, 215
Proton, 458–460, 900
　charge of, 464
Pump, electromagnetic, 605–606
Pumps, 248–249
Pupil, 760
Pyrometer, 286

Quantum, 159–160, 841, 844, 855–859
Quantum number, 864, 875–878, 882–883
Quarks, 508, 925–926
Quarter-wave plate, 817

Radial heat flow, 317–318
Radian, 180
Radiant flux, 727
Radiant power, 727
Radiation, 321–324

Radiation (*continued*)
  acoustic, from piston, 447–451
  from antenna, 707–708
  medical applications, 926–927
Radiation hazards, 926–927
Radiation pressure, 702–703
Radioactive transformation, 908–911
Radioactivity, 903–911
Randomness, 380–382
Range, of projectile, 92
Rankine temperature, 288–289
Rappelling, 35
Rarefaction, 405
Rayleigh waves, 416
Rayleigh's criterion, 799–800
Rays, and waves, 717
Reactance, 680–684
  capacitive, 680
  inductive, 681
Recoil, 154–155
Rectifier, 686, 843
Rectilinear motion, 39–53
Red shift, 447
Reference circle, 232
Reference system, 22–23
Reflecting telescope, 767–768
Reflection, of light, 717–718
  of mechanical wave, 420–421, 426–427
  phase shift in, 781–782
  at a plane surface, 733–735
  polarization in, 810–811
  at a spherical surface, 735–740
  total internal, 721–722
Refraction, of light, 717–720
  at plane surface, 742–743
  at spherical surface, 743
Refractive index. *See* Index of refraction
Refrigerator, 368–369
Regular reflection, 720
Reinforcement, 429–430, 774–775
Relative humidity, 341
Relative nature of simultaneity, 824–825
Relative permeability, 665
Relative velocity, 52–53, 107–109
Relativity, 135–137, 159–160, 787, 822–836
  of distance, 827–829
  general theory, 836
  and newtonian mechanics, 835–836
  of time, 825–827
Relaxation time, of *R-C* circuit, 572–573
Remanance, 674
Remanant magnetization, 674
Reservoir, heat, 365
Resistance, 539–541
Resistance, dynamic, 549
Resistance thermometer, 285
Resistance-capacitance circuit, 571–574
Resistivity, 537–539
  table, 537
  temperature coefficient, 538
  table, 539
Resistors, in parallel, 562–564
  in series, 562–564
Resolution, of lens, 771 798–800
Resolving power, 798–800
Resonance, 430–431

  in ac circuits, 688–690
Resonance levels, 850
Resonant frequency, in ac circuit, 689
Rest energy, 137, 834
Rest mass, 834
Resultant, 8
  of forces, 20
Retina, 759
Reversible cycle, 375
Reynolds number, 276–278
Righthand rule, 14–15
  for magnetic field, 600, 610, 615
  for magnetic force, 585
Righthanded coordinate system, 15
Rigid body, equilibrium, 166–173
  rotation, 179–183
Ripple tank, 416
Rocket propulsion, 157–159
Roemer, 713–714
Rolling, 193
Röntgen rays, 854
Root-mean-square speed, 392
Root-mean-square values, 684–686
Rotation, 179–183
  kinetic energy, 183–186
  about moving axis, 192–194
Rowland ring, 665
Rutherford scattering, 897–899

Satellite, 101–105
Saturated liquid, 336
Saturated vapor, 336, 341
Saturation, 673
  of nuclear force, 903
Scalar product, 12–13
Scalar quantity, 6–7
Scattering, of light, 726–727, 814–815
  Rutherford, 897–899
  of x-rays, 857–859
Scientific notation, 6
Search coil, 637
Second, definition, 2
Second law of thermodynamics, 364–383
Secondary winding, 690
Sedimentation rate, 276
Segre chart, 906–907
Self-inductance, 650–653
Semiconductor, 538–539, 598, 892–895
Semiconductor devices, 893–895
Separation of variables, 227
Series capacitors, 519
Series motor, 604
Series resistors, 562–564
Shear modulus, 217
  table, 216
Shells, electrons in atoms, 855–857, 883–885
Shunt motor, 604
Sign conventions, for mirrors and lenses, 735–736
Significant figures, 5–6
Simple harmonic motion, 224–232
Simple pendulum, 236–237
Simultaneity, 824–825
Sinusoidal motion, 227
Sinusoidal wave, 406–408

  electromagnetic, 703–705
Sliding friction, 30
Slug, 63
Snell's law, 718
Solar wind, 675–676
Solenoid, 601–602
  toroidal, 622–623, 651–652
Solid, structure of, 888–890
Sone, 439
Sound intensity level, 437–439
Sound quality, 439–440
Sound waves, 413–415, 434–436
Source, of emf, 541–545
Sources of light, 712–713
Specific gravity, 245
Specific heat, 299–302
Specific heat capacity, 299–303
  table, 300
Specific volume, 335
Spectrum, absorption, 851–852
  band, 887–888
  electromagnetic, 715
  from grating, 795–797
  line, 846–848, 850–851
  molecular, 887–888
  visible, 716
Spectrum line, 773
Specular reflection, 720
Speed, 42
  of electromagnetic wave, 698–700
  of light, 713–715
  of longitudinal wave, 411–415
  molecular, 391–395
    distribution of, 395–396
  of wave, 404
  of wave on string, 408–410
Spherical aberration, 737, 758
Spherical mirror, 735–740
Sphygmomanometer, 247–248
Spin, nuclear, 899
Spin angular momentum, 876–878
Spinning ball, curve of, 271
Stability, of nucleus, 906–908
Standards, 1–4
Standing wave, 422–428
  electromagnetic, 705–707
State coordinate, 283
Statcoulomb, 463
Static friction, 30
Steady state, in ac circuit, 684
Steam engine, 368
Stefan-Boltzmann law, 321–323
Step-down transformer, 691
Step-up transformer, 691
Stiffness, 220
Stimulated emission, 852
Stokes' law, 276, 507
Stopping potential, 844
Storage rings, 915–916
Strain, 214
  bulk, 214, 217–218
  compressive, 214, 217–218
  shear, 214, 216
  tensile, 214, 216
  volume, 214, 217–218
Strangeness, 924

Streamline, 264
Stress, 211
    bulk, 213, 217
    compressive, 212–213, 217
    normal, 211
    shear, 212, 217
    thermal, 292–293
    tensile, 211, 216
    volume, 213, 217
Stress analysis, optical, 818
Stroboscopic photography, 52
Sublimation, 303–307
Subtraction, of vectors, 9
Superconductor, 538
Supercooling, 305–306, 343
Superfluid phase, 339–340
Superheating, 305–306, 342–343
Superheavy nuclei, 918
Superposition, for electrical forces, 463
    of magnetic fields, 611
Superposition principle, 422
Surface, pVT, 333–337
Surface current, 666–669
Surface integral, 482–483
Surface tension, 253–256
    table, 255
Susceptibility, magnetic, 670
    table, 671
Symmetry, and Gauss's law, 489
Synchrocyclotron, 914–915
Synchrotron, 915–916
Système International, 2–5, 62–63, 463
Systems of units, 62–64

Telephoto lens, 764
Telescope, 766–767
    Galilean, 772
    reflecting, 767–768
Television picture tube, 509–511
Temperature, 283–289
Temperature coefficient of resistivity, 538, 540
    table, 539
Temperature gradient, 314
Temperature scales, 287–289
Tensile stress, 211, 216
    in fluid, 259
Tension, 24
Terminal velocity, 276
Tesla, 586
Thermal conductivity, 314
    table, 315
Thermal efficiency, 366
    Carnot cycle, 371–373
    Otto cycle, 366–367
Thermal equilibrium, 284–285
Thermal expansion, 289–292
Thermal pollution, 376
Thermal stress, 292–293
Thermionic emission, 841–843
Thermistor, 539
Thermocouple, 285
Thermodynamic surface, 333–337
Thermodynamic system, 346
Thermodynamics, first law, 350, 354–355
    second law, 369–382

zeroth law, 285
Thermograph, 323
Thermometer, 285–289
Thermometric properties, 286
Thermonuclear reaction, 919
Thermos bottle, 323
Thin lens, 751–754
Third law of thermodynamics, 378
Thomson model of atom, 898
Thomson's e/m experiment, 590–591
Threshold, of audibility, 437–438
    of pain, 437–438
Throttling process, 353–354
Thyrite, 559
Timbre, 439–440
Time constant, for R-C circuit, 572–573
    of R-L circuit, 655
Time dilation, 825–827
Toroidal solenoid, inductance, 651–652
    magnetic field, 622–623
Torque, 166–167, 173
    and angular acceleration, 190–192
    on current loop, 599–602
    on a dam, 252–253
    external, 197
    impulsive, 196
    internal, 197
    as vector, 173, 200–202
    of weight, 170–172
Torque constant, 236
Torr, 247
Torricelli's theorem, 268
Torsion balance, 65
Total internal reflection, 721–722
Transducer, 451
Transformer, 643, 690
Transient, 684
Transient current, 534
Transistor, 894–895
Transverse waves, 403, 408–410, 704
    and polarization, 806–807
Triad, 440–442
Trigonometric functions, T-6
Triode, 843
Triple point, 287, 336, 338
    table, 338
Tube of flow, 264
Turbulence, 265, 276–278
Tyrolean traverse, 35–36

Ultrasonics, 451
Uncertainty principle, 870–874
Underdamping, 659
Uniform circular motion, 95
Unit conversion factors, T-8
Unit vectors, 11–12, 473
Units, consistency, 4–5
    conversion of, 4–5
    International System, T-1–T-2
    systems of, 1–4, 62–64
Universal gas constant, 329
Unpolarized light, 808–809

Vacancy, 890
Vapor, saturated, 341
Vapor pressure 340

Vector product, 13–15
Vector quantity, 6–9
Vector sum, 6–11
Velocity, angular, 179–180
    angular and linear, 182–183
    average, 39–40, 83–84
    instantaneous, 40–42, 83–84
    relative, 52–53, 107–109
    terminal, 276
Velocity addition, relativistic, 830–831
Velocity amplitude, 435
Velocity selector, 592
Vena contracta, 268
Ventricular fibrillation, 554–555
Venturi tube, 169
Vertex, of mirror, 735
Vibrating bodies, 420–429
Vibration, 223–225
    in molecule, 888
    of rods and plates, 428–429
Virtual object, 757
Viscosity, 271–276
    table, 273
Vision, defects, 761–762
Vitreous humor, 759
Volt, 497
Voltage amplitude, 678
Voltage–current relations, 549–550
Voltmeter, 498, 569–571
Volume, specific, 335
Vortices, 277

Water waves, 415–417
Watt, 134
Wave, electromagnetic, 696–708
    energy in, 700–702
    in matter, 703
    momentum, 702
    sinusoidal, 703–705
    speed, 698–700
    standing, 705–706
    longitudinal, 403, 411–415
    mechanical, 403–417
    periodic, 404–408
    and rays, 717
    sinusoidal, 406–408
    transverse, 403, 408–410
    water, 415–417
Wave fronts, 717
Wave function, 406
    in quantum mechanics, 874–876
Wave motion, 223–224
Wave number, 406
Wave packet, 870
Wave pulse, 870
Wavelength, 404
    of light in matter, 724
Weber, 588
Weighing the earth, 76
Weight, 18, 26
    apparent, 73–74, 105–107
    and mass, 66–68
Weightlessness, 73–74
Whole tone, 441
Wide-angle lens, 764
Work, 115–117

Work, (*continued*)
  in electric circuits, 550–554
  of electrical force, 495
  and energy, 120–122
  internal, 132
  in isothermal expansion, 347–348
  in rotational motion, 189–190
  in thermodynamics, 346
  of varying force, 118–120
  in volume change, 347–348
Work function, 844–845
Working substance, 364

X-ray diffraction, 797
X-rays, production, 854–857
  scattering, 857–859

Yield point, 215
Yo-yo, 192, 207
Young's experiment, 776–780
Young's modulus, 216–217
  table, 216

Zeeman effect, 840, 877
Zeroth law of thermodynamics, 284–285